KU-208-331

Soils: *Genesis and Geomorphology*

Soils: Genesis and Geomorphology is a comprehensive and accessible textbook on all aspects of soils.

The book's introductory chapters on soil morphology, physics, mineralogy and organisms prepare the reader for the more advanced and thorough treatment that follows. Unlike other books on soils, the authors devote considerable space to discussions of soil parent materials and soil mixing (pedoturbation), along with dating and paleoenvironmental reconstruction techniques. Theory and processes of soil genesis and geomorphology form the backbone of the book, rather than the emphasis on soil classification that permeates other soils textbooks. This refreshingly readable text takes a truly global perspective, with many examples from around the world sprinkled throughout.

Replete with hundreds of high-quality figures and a large glossary, this book will be invaluable for anyone studying soils, landforms and landscape change. *Soils: Genesis and Geomorphology* is an ideal textbook for mid- to upper-level undergraduate and graduate level courses in soils, pedology and geomorphology. It will also be an invaluable reference text for researchers.

RANDALL SCHAETZL is a Professor of Geography at Michigan State University, East Lansing. He has trained as a physical geographer at some of the top departments in the USA, and has established himself as a leading figure in soil genesis and geomorphology research. He has published in all the leading soils, geomorphology and geography journals. Schaetzl is an associate editor for the *Soil Science Society of America Journal.* His expertise on podzolization and pedoturbation has led him to publish papers that have advanced the theory behind both these widespread soil processes.

SHARON ANDERSON is an Associate Professor at California State University, Monterey Bay. Anderson has a broad educational background in geology, chemistry, plant–soil relations and soil chemistry, and has a publication record in all of these areas. Her research has spanned the areas of soil organic matter composition, soil mineralogy, pesticide fate in the environment and water quality. As Chair of the Earth Systems Science and Policy Program, her current work focusses on developing rigorous yet applied, interdisciplinary curricula.

Soils

Genesis and Geomorphology

Randall J Schaetzl
Michigan State University

and

Sharon Anderson
California State University

CAMBRIDGE UNIVERSITY PRESS
Cambridge, New York, Melbourne, Madrid, Cape Town, Singapore, São Paulo

CAMBRIDGE UNIVERSITY PRESS
The Edinburgh Building, Cambridge CB2 2RU, UK

Published in the United States of America by Cambridge University Press, New York

www.cambridge.org
Information on this title: www.cambridge.org/9780521812011

First published 2005

Printed in the United Kingdom at the University Press, Cambridge

A catalog record for this book is available from the British Library

Library of Congress Cataloging in Publication data

Schaetzl, Randall J., 1957–
Soils: genesis and geomorphology / Randall J. Schaetzl and Sharon Anderson.
p. cm.
Includes bibliographical references and index.
ISBN 0 521 81201 1
1. Soils. 2. Soil formation. 3. Soil Structure. 4. Geomorphology.
I. Anderson, Sharon, 1961– II. Title.
S591.S287 2005
631.4 – dc22 2004051845

ISBN-13 978-0-521-81201-6 hardback
ISBN-10 0-521-81201-1 hardback

We dedicate this volume to those who have inspired us to write it...
through their lifelong scholarship, insatiable curiosity about the world around them, and their willingness to share it with all who have an interest...
innovative thinkers who have made many, including us, stop and think about the world through different intellectual "filters"...

Francis Doan Hole (1913–2002)
and
Donald Lee Johnson (1934–)

Francis D. Hole in 1978. Image courtesy of the University of Wisconsin, Photo Media Center.

Donald L. Johnson in 1999. Image by RJS.

Soil is the hidden, secret friend ... the root domain of lively darkness and silence.

Francis D. Hole

Texture
Color
Types of soil horizons
Buried soils
Soil temperature

Contents

Preface — *page* xi
Acknowledgements — xii

Part I The building blocks of the soil — 1

1 Introduction — 3
Pioneers of soil science, soil survey and soil geography — 4
Things we hold self-evident... — 6
The framework for this book — 7

2 Basic concepts: soil morphology — 9
Texture — 9
Color — 14
Pores, voids and bulk density — 17
Structure — 18
Consistence — 20
Presentation of soil profile data — 22
Soil micromorphology — 22

3 Basic concepts: soil horizonation... the alphabet of soils — 32
Regolith, residuum and the weathering profile — 32
The soil profile, pedon, polypedon and map unit — 33
Soil horizons and the solum — 36
Types of soil horizons — 36
Buried soils — 52

4 Basic concepts: soil mineralogy — 54
Bonding and crystal structures — 54
Oxides — 55
Chlorides, carbonates, sulfates, sulfides, and phosphates — 60
Silicates — 61
Identification of phyllosilicates by X-ray diffraction — 73
Identification of iron and aluminum oxides — 80

5 Basic concepts: soil physics — 82
Soil water retention and energy — 82
Soil water movement — 85
Soil temperature — 87
Soil gas composition and transport — 91

6 Basic concepts: soil organisms 93
Primary producers 93
Soil fauna 96

7 Soil classification, mapping and maps 106
Soil geography, mapping and classification 106
The system of Soil Taxonomy 107
The Canadian system of soil classification 146
Soil mapping and soil maps 146
Soil landscape analysis 158

Part II Soil genesis: from parent material to soil 165

8 Soil parent materials 167
Effects of parent material on soils 167
The mutability of $time_{zero}$ 169
A classification of parent materials 170
Lithologic discontinuities in soil parent materials 215

9 Weathering 226
Physical weathering 227
Chemical and biotic weathering 231
Products of weathering 236
Controls on physical and chemical weathering 236
Assessing weathering intensity 238

10 Pedoturbation 239
Classifying pedoturbation: proisotropic vs. proanisotropic 239
Expressions of pedoturbation 244
Forms of pedoturbation 245
Lesser-studied forms of pedoturbation 293

11 Models and concepts of soil formation 295
Dokuchaev and Jenny: functional–factorial models 296
Simonson's process-systems model 320
Runge's energy model 323
Johnson's soil thickness model 324
Johnson and Watson-Stegner's soil evolution model 325
Phillips' deterministic chaos and uncertainty concepts 339
Other models 342
The geologic timescale and paleoclimates as applied to soils 342

12 Soil genesis and profile differentiation 347
Eluviation–illuviation 353
Process bundles 354

Surface additions and losses 456
Mass balance analysis, strain and self-weight collapse 460

Part III | Soil geomorphology 463

13 Soil geomorphology and hydrology 465
The geomorphic surface 466
Surface morphometry 468
The catena concept 469
Soil geomorphology case studies, models and paradigms 514

14 Soil development and surface exposure dating 547
Stratigraphic terminology, principles and geomorphic surfaces 547
Numerical dating 549
Relative dating 550
Principles of surface exposure dating (SED) 554
SED methods based on geomorphology, geology and biology 555
SED methods based on soil development 567
Chronosequences 587
Numerical dating techniques applicable to soils 596

15 Soils, paleosols and paleoenvironmental reconstruction 619
Paleosols and paleopedology 620
Environmental pedo-signatures 632

16 Conclusions and Perspectives 653

References 657
Glossary 741
Index 791

Preface

This book is about soil geography, which we think is a difficult and challenging area of study. Our purpose in writing this book is to assert that only through a study of the spatial interactions of soils *on landscapes* can soil and landscape evolution be truly resolved.

This book can be used in courses on soil geography, soil genesis, pedology and soil geomorphology. Our assumption is that the readers have had some background in the natural sciences, and are eager to learn more about soils. We do not assume, nor does the reader need, a substantial background in soils to read and comprehend this book. Difficult as the task may seem, our goal was to write a soils text that could serve both as an initial soils text and as a cutting-edge resource book of research grade. Only time will tell if we met that goal.

Our emphasis, beyond that of soil geography, is deliberately intended to be broad. Other books of similar ilk (Daniels and Hammer 1992, Birkeland 1999) focus on geomorphology and the initial geologic setting as a guiding framework for the understanding of soil landscape evolution. We emphasize these issues in later chapters. Buol *et al.* (1997) and Fanning and Fanning (1989) focus on soil genesis while at the same time emphasizing classification.

Our book relies heavily on concepts and imagery to convey ideas. We have compiled a suite of figures, images and graphics that, in and of themselves, convey messages that cannot be put into words. Throughout the text we include brief "outtakes" on soil landscapes from around the world. We call these excursions "Landscapes," and believe that they convey, with pictures and graphics, what would otherwise take many hundreds of words to tell.

We believe in the necessity of soil taxonomy and soil classification; we use its terminology in the book but do not focus on it. Taxonomy exists to serve those who study and communicate about soils; it is not an end in and of itself. Because we feel that one of the best ways to "learn" and use taxonomy is to examine it in the context of landscapes, we include taxonomic descriptions within many "Landscapes."

We are proud of the extensive literature listing that our book makes use of. We hope that we have cited all the major works, both the classic ones and the recent cutting-edge papers. If we have missed something, we urge our readers to call it to our attention; we will be receptive. Where possible, we have tried to cite mainly papers and studies that are readily accessible in most academic libraries. That is, we have steered clear of papers that are difficult to find or in the gray literature, as well as theses and dissertations, unless we felt that they were essential reading. The end result is a book that relies heavily on work published in national and international scholarly journals and books. If you wish to have a digital copy of our References Section entries, just email us and ask.

The glossary is rich in terms, many of which are only marginally touched upon in the text. Our philosophy with regard to the glossary was simple: if the reader needed to know a term to understand the book, include it in the glossary and define it clearly. The glossary adds length to the book but makes it more "readable."

We intend to continue to work at updating this book, without necessarily making it longer. We encourage you, the reader, to help us. For example, if you wish any topics added to the glossary or the body of the book, contact us with your request. More importantly, alert us to your papers, send reprints and citations, email or write to inform us of new findings or breakthroughs; we will include them as best we can. Contact us with your perceptions of the book, positive or otherwise. Help us make this book better and we promise to continue to work hard toward this goal.

Acknowledgements

We thank the many, many people who have made this book possible. We especially want to thank those who have inspired and taught us over the years. Too many to name, those of particular note are:

- My (RJS) family: my wife, Julie Brixie, has been a steadfast supporter of me, my work, my career, and this book, not to mention a solid proofreader and chapter/figure organizer. I could not have done it without her. My (RJS) children (Madeline, Annika and Heidi) have helped with many small tasks and have put up with their dad being at the office far too much; I will be home more now. My parents instilled within me, through example rather than spoken word, the importance and pay-offs of hard work.
- Don Johnson, a true academic free spirit and genuine thinker who is not afraid to look at the world through different glasses.
- Francis Hole, a one-of-a-kind scholar who will always hold a special place in my (RJS) heart and mind. I (RJS), like so many, would not have "found" the disciplines of soil science and soil geography were it not for Francis Hole.
- François Courchesne was a driving force behind the development of this book.
- Scott Isard, my (RJS) academic conscience and motivator, always willing to discuss academics and scholarship.
- Curt Sorenson, who taught me (RJS) to simply love soils and instilled within me a passion to excel.
- Leon Follmer, who taught me (RJS) to look closely at soils, and made me realize that soils and paleosols are truly remarkable things.
- Duke Winters, who put up with a young colleague (RJS) who was infatuated with soil science, while at the same time mentoring me to become a soil geographer first and foremost.

Those who assisted in the production, editing, or compilation of the book deserve special mention.

- Matt Mitroka, Ellen White, Beth Weisenborn, Peter Dimitriou and Beth Kaupa assisted Paul Delamater in the production of the figures, arguably the strength of the book. Paul was the consummate QA/QC person for graphics. His diligence and high standards permeate the book, and for that reason this is as much *his* book as it is ours. Thank you Paul!
- Ron Amundson, Dave Cremeens, Chris Evans, John Hunter, Don Johnson, Warren Lynn, Fritz Nelson, Jenny Olson, Paul Reich, John Tandarich, Charles Tarnocai, Pat Webber, Beth Weisenborn and Antoinette WinklerPrins provided images, graphs, charts and figures of soils and landscapes that have been reproduced within the book. Without these images, the book would have been much weaker.
- Bill Dollarhide, John Gagnon, Charles Gordon, Bill Johnson, Mike Risinger, Richard Schlepp, Bruce Thompson and Cleveland Watts (of the USDA–NRCS), and David Cremeens provided us with data and block diagrams from various NRCS soil surveys.
- Exceptional editing and reviews of individual chapters were provided by Bob Ahrens, Alan Arbogast, Linda Barrett, Janis Boettinger, Julie Brixie, Alan Busacca, François Courchesne, David Cremeens, Missy Eppes, Leon Follmer, Vance Holliday, Geoff Humphreys, Christina Hupy, Joe Hupy, Don Johnson, Christina Kulas, Johan Liebens, Jonathan Phillips, Greg Pope, Paul Rindfleisch, Mark Stolt, Julieann van Nest, Natasa Vidic, Beth Weisenborn, Gary Weissmann and Indrek Wichman.
- Patricia Brixie proofed the book in its entirety. It was then reviewed by Art Bettis, Donald Johnson, Vance Holliday and Dan Muhs.
- Judy Hibbler typed many of the tables, downloaded text from the web and was an invaluable "gal friday."
- Many others were quick to help or offer advice when a question or issue arose: Bob Ahrens, Alan Arbogast, Art Bettis, Steve Bozarth, George Brixie, David Cremeens, Bob Engel, Leon Follmer, Bill Frederick, Don Johnson, Bruce Knapp, Bruce Pigozzi, John Tandarich, Greg Thoen and Dan Yaalon.

Funding for the many costs associated with the development of a book of this type were provided by various agencies of Michigan State University: the Agricultural Experiment Station, the Office of the Vice President for Research and Graduate Studies, and the Department of Geography. Some of the data on Michigan soils and landforms was developed in conjunction with NSF grants made to RJS (NSF awards BCS-9819148 and SBR-9319967); any opinions, findings, and conclusions or recommendations expressed in this material are, however, those of the authors and do not necessarily reflect the views of the National Science Foundation or Michigan State University.

We thank the professional staff at Cambridge University Press (CUP) and their affiliates for their help on, and support of, this book project. Matt Lloyd, our editor, has been a strong supporter of the book from its inception. The staff at CUP, especially Jayne Aldhouse, Sally Thomas and Anna Hodson, made the typescript and figures into a book. We also thank the many "behind the scenes" people whose tireless work often goes unseen and unappreciated; we did it!

Last and most important, we acknowledge that we have approached this book from the perspective of St. John Vianney, the Curé of Ars, when he said, "I have been privileged to give great gifts from my empty hands."

Part I

The building blocks of the soil

Chapter 1

Introduction

Soils form and continually change, at different rates and along different pathways. They continually evolve and are never static for more than short periods of time. Along these lines, we embrace Daniels and Hammer's (1992) statement that soils are four-dimensional systems. They are not simply the two-dimensional profile, nor is the study of the spatial variation in soils (a three-dimensional effort) enough. Soils must be studied in space *and* time (the fourth dimension). We incorporate these ideas by synthesizing complex, overlapping topics and tying them into a cohesive message: soil landscapes – how they form and change through time. To do this, we necessarily take a process-based approach.

Soil genesis and geomorphology, the essence of this book, cannot be studied without a firm grasp on the processes that shape the *distributions* of soils. We will, however, never fully understand the complex patterns of the Earth's soils. Even if we do claim to understand it, we must be mindful that the pattern is ever-changing. Again we quote Daniels and Hammer (1992: xvi), "One cannot hope to interpret soil systems accurately without an understanding of how *the landscape and soils have coevolved* over time" (emphasis ours). Every percolation event translocates material within soils, while every runoff event moves material across their surfaces, changing them ever so slightly. The worms, termites and badgers that continually burrow, mix and churn soils make them more different tomorrow than they were yesterday. Biochemical reactions within soils weather minerals and enable microbes to decompose organic matter, perpetuating the cycle from living matter to humus to chemical elements and back again. Because this can all be quite complex, we provide information, tools, resources and background data to bring the reader closer to deciphering this most complicated of natural systems.

Whitehead (1925) wrote, "It takes a genius to undertake the analysis of the obvious." Soil is seemingly everywhere, yet, we would argue, comparatively few study it. Additionally, soils are usually hidden from view and require excavation to be revealed. Soils are not discrete like trees, insects, lakes or clouds, which have seemingly identifiable outer boundaries. Instead, they seem to grade continuously, one into another, until they end at the ocean, a sheer rock face or a lake. When broken into discrete entities, like a geologist might break apart a rock, soils appear to lose their identity. This soil science . . . it's not easy. But therein lies the challenge!

We argue that a geographic approach to the study of soils is absolutely necessary (Boulaine 1975). Soils are spatial things, varying systematically across space at all scales. To study them fully you must understand not only *what* they are, but also how they relate to their adjoining counterparts. Soil geography focusses upon the geographic distributions of soils with emphasis on their character and genesis, their interrelationships with the environment and humans, and their history and likely future changes. It is operationalized at many scales, from global to local. Soil geography *encompasses* soil genesis; it is not simply a part of it. One cannot explain soil patterns without knowing the genesis of the soils

Table 1.1 Some of the academic domains of soil geography

Distribution of soils and soil taxa across the landscape
Soil survey and mapping
Soil genesis, both within and among pedons
Interactions among soils and the natural and human environment
Paleopedology
Soil geomorphology
Soil-slope and soil catena studies
Soil landscape analysis and the study/explanation of soil pattern
Pedometrics
Cartographic representation of soils
Evolution of soils and landscapes

Not an exhaustive list. In no particular order. *Source:* Hole and Campbell (1985).

that comprise that pattern. Soil geography also incorporates geomorphology; one cannot fully explain soil patterns without knowledge of the evolution of the landforms and rocks of which they form the skin. Soil geography involves soil evolution; changing patterns of soils over time are a reflection of a multitude of interactions, processes and factors, replete with feedbacks, inertia and flows of energy and mass. Soil geography is manifested in soil survey (mapping) operations, which are extremely useful databases but are only as good as our understanding of the evolution of the soil pattern. This book, then, is about soil geography and all that it encompasses. Tandarich *et al.* (1988) used the term geopedology to refer to the interesction of the disciplines of geology, geography and soil science. We embrace that term and view it as a central component of this book.

Pioneers of soil science, soil survey and soil geography

Pedology is the science of soil genesis, classification and distribution; to many it is synonymous simply with *soil science*. Because soils have sustained human life since its inception, one may think that pedology has a long history. In fact, it was a late arrival among the natural sciences (Hole and Campbell 1985). Many attribute its founding to V. V. Dokuchaev (1846–1903), a Russian scholar and teacher. Others place emphasis on the work of Charles Darwin (1809–1882), perhaps the world's most underappreciated soil scientist. Regardless of who gets the credit for jump-starting this discipline, pedology is unquestionably little more than a century old! Our brief overview of the founders of soil science (below) should underscore that they were multifaceted thinkers who understood that the soil landscape was a complex system, requiring that it be studied using a geographic approach. More detailed accounts of the personalities involved in the development of the field are presented elsewhere (Kellogg 1974, Cline 1977, Tandarich and Sprecher 1994).

Vasili Vasilevich Dokuchaev is often called the father of soil science, although he acknowledged the influence of several others (particularly in the field of agricultural chemistry) in the development of his ideas (Tandarich and Sprecher 1994) (Fig. 1.1). Trained in Russia, he wrote his most reputed works on the soils of the Russian steppes, primarily Chernozems. He developed and used concepts on the nature and genesis of soil profiles, as well as soil landscapes, in his research. His geographic study of soils spanned local to regional scales. Dokuchaev and his students produced the first scientific classification of soils and developed soil mapping methods, laying the foundation for modern soil genesis and soil geography (Buol *et al.* 1997). He is known for developing the basic A–B–C horizon nomenclature, and

(a)

(b)

(c)

Fig. 1.1 Three influential scholars in the field of soil science. (a) Vasili V. Dokuchaev (1846–1903), Russian agriculturalist, geographer and pedologist. Image courtesy of John Tandarich. (b) Curtis F. Marbut (1863–1935), American agriculturalist, soil scientist and early developer of the US soil classification system. Image courtesy of John Tandarich. (c) Hans Jenny (1899–1992), Swiss pedologist and agricultural chemist; professor at the University of California. Image by R. Amundson.

a factorial model of soil development in which soils and soil patterns were seen as a function of independently varying state factors of the environment. Although not universal, this model remains, in various revised forms, the primary explanatory model for soils worldwide (see Chapter 11). Using this model, Dokuchaev's work allowed others to develop the concept of the *zonal soil* – one which characterized vast tracts of land and represented the epitome of soil development for that region. Zonal soil concepts, although conceptually flawed, essentially jump-started soil survey and mapping worldwide, and made the complex world of soils more understandable to the masses. Dokuchaev's teachings, carried across the Atlantic by E. W. Hilgard (1833–1903), were highly influential on many prominent soil scientists.

Unfortunately, by omitting the ideas of Charles Darwin from his writings, Dokuchaev would essentially bury them. Darwin's ideas focussed on local-scale biological origins of many soil properties, and on biomechanical processes in soils, such as mixing by worms (Darwin 1881). The lack of soil terminology in his works, coupled with the growing acceptance of Dokuchaev's factorial model for soil development, doomed *biomechanical soil processes* to the theoretical back seat, until resurrected years later.

In 1899, the United States started its soil survey program, under the direction of Professor Milton Whitney (1860–1927), primarily using geological concepts of soils, e.g., granite soils and alluvial soils (Shaler 1890). This practice continued for some time, e.g., Marbut *et al.* (1913). Shortly after this, Curtis Marbut (1863–1935), who earned his Ph.D. in geology at Harvard under the eminent geographer William Morris Davis (1850–1932), was appointed soil scientist in charge of the US Bureau of Soils (Tandarich *et al.* 1988) (Fig. 1.1). While at Harvard, Marbut had been influenced by the writings of Konstantin Glinka (1897–1927), a student of Dokuchaev, and the soils-related work of Nathaniel Shaler (1841–1906). He had translated Glinka's book *Die Typen der Bodenbildung* from German into English and applied many of the ideas within to the budding soil survey program (Cline 1977, Tandarich and Sprecher 1994). Marbut's impact on soil science in the USA proved to be strong and long-lasting. Indirectly but strongly influenced by the ideas of Dokuchaev, he changed the way soils were viewed, emphasizing that they should be

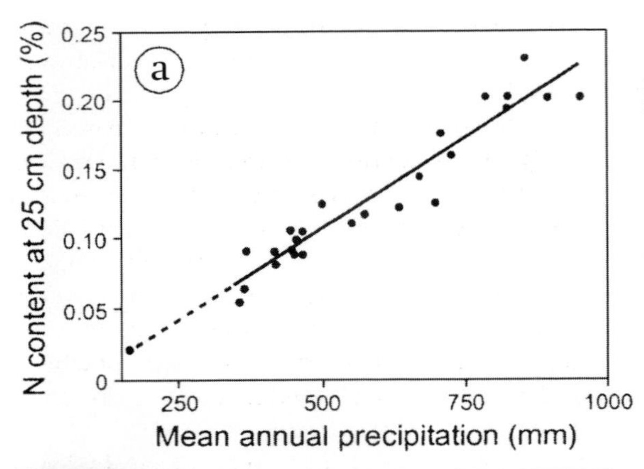

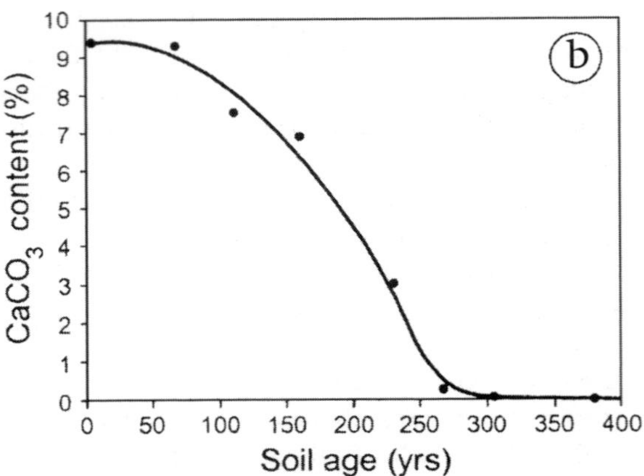

Fig. 1.2 Examples of two functional relationships that Hans Jenny produced for his 1941 book, *Factors of Soil Formation*.

classified and mapped based on horizon and profile characteristics, thereby reducing the influence of geology. Marbut eventually developed a multicategoric soil classification system (Marbut 1928, 1935; see Chapter 7). He thought about soils geographically, and his ideas translated into his classification system.

In 1941, Hans Jenny (1899–1992), at the University of California, published a landmark treatise entitled *Factors of Soil Formation*. Much of this book is devoted to his functional–factorial model of soil formation, in which soils are seen as the product of five interacting factors: climate, organisms, relief, parent material and time (see Chapter 11). Jenny developed many numerical soil functions in this book, each being an equation showing how soils change as four of the factors are held constant and one is allowed to vary (Figs. 1.1, 1.2). In this regard, Jenny (1941a: 262) noted that, "the goal of the soil geographer is the assemblage of soil knowledge in the form of a map. In contrast, the goal of the functionalist is the assemblage of soil knowledge in the form of a curve or an equation." He commented that soil maps display areal arrangement but give no insight into causal relationships, and that mathematical curves reveal dependency of soil properties on state factors but the conversion of such knowledge to the field is impossible without a soil map (Arnold 1994). Thus, Jenny proposed that the union of geographic and functional methods provided the most effective pedological research. Arnold (1994:105) restated this idea as follows – spatial soil patterns need to be understood through functional relationships of the soil-forming factors in *space and time*. Since Jenny's (1941a) model provided the theoretical framework for soil functional relationships, it stands today as perhaps one of the most geographic of the several soil models, because it is used subliminally or overtly by almost every soil mapper and geographer. More recent models, which refine and elaborate on Jenny's, as well as those that propose very different ways of looking at the soil landscape (Johnson and Hole 1994) are discussed in Chapter 11.

Things we hold self-evident...

Following the lead of Buol *et al.* (1997) and Hole and Campbell (1985), we provide below a listing of concepts or truisms in soil science and soil geography, slightly modified from their original sources.

- Complexity in soil genesis is more common than simplicity.
- Soils lie at the interface of the atmosphere, biosphere, hydrosphere and lithosphere. Therefore, a thorough understanding of soils requires some knowledge of meteorology, climatology, ecology, biology, hydrology, geomorphology, geology and many other earth sciences.
- Contemporary soils carry imprints of pedogenic processes that were active in the past, although in many cases these imprints are difficult to observe or quantify. Thus, knowledge

of paleoecology, paleogeography, glacial geology and paleoclimatology is important for the recognition and understanding of soil genesis and constitute a basis for predicting the future soil changes.
- Five major, external factors of soil formation (climate, organisms, relief, parent material and time), and several smaller, less identifiable ones, drive pedogenic processes and create soil patterns.
- Characteristics of soils and soil landscapes, e.g., the number, sizes, shapes and arrangements of soil bodies, each of which is characterized on the basis of horizons, degree of internal homogeneity, slope, landscape position, age and other properties and relationships, can be observed and measured.
- Distinctive bioclimatic regimes or combinations of pedogenic processes produce distinctive soils. Thus, distinctive, observable morphological features, e.g., illuvial clay accumulation in B horizons, are produced by certain combinations of pedogenic processes operative over varying periods of time.
- Pedogenic (soil-forming) processes act to both create and destroy order (anisotropy) within soils; these processes can proceed simultaneously. The resulting profile reflects the balance of these processes, present and past.
- The geological *Principle of Uniformitarianism* applies to soils, i.e., pedogenic processes active in soils today have been operating for long periods of time, back to the time of appearance of organisms on the land surface. These processes do, however, have varying degrees of expression and intensity over space and time.
- A succession of different soils may have developed, eroded and/or regressed at any particular site, as soil genetic factors and site factors, e.g., vegetation, sedimentation, geomorphology, change.
- There are very few old soils (in a geological sense) because they can be destroyed or buried by geological events, or modified by shifts in climate by virtue of their vulnerable position at the skin of the earth. Little of the soil continuum dates back beyond the Tertiary period and most soils and land surfaces are no older than the Pleistocene Epoch.
- Knowledge and understanding of the genesis of a soil is important in its classification and mapping.
- Soil classification systems cannot be based entirely on perceptions of genesis, however, because genetic processes are seldom observed and because pedogenic processes change over time.
- Knowledge of soil genesis is imperative and basic to soil use and management. Human influence on, or adjustment to, the factors and processes of soil formation can be best controlled and planned using knowledge about soil genesis.
- Soils are natural clay factories (clay includes both clay mineral structures and particles less than 2 μm in diameter). Shales worldwide are, to a considerable extent, simply soil clays that have been formed in the pedosphere and eroded and deposited in the ocean basins, to become lithified at a later date.

The framework for this book

In this book, we introduce the building blocks of soil in Part I, because we do not require that the reader be extremely well grounded in the fundamentals of soil; those with a strong background may choose to skim this section. We continue adding to the basic knowledge base in Part II (Chapters 8–12), but add a great deal more material on theory and soil genesis/processes. In Chapter 11, for example, we introduce a large dose of pedogenic and geomorphic *theory*, which in combination with the previous chapters allows us to discuss soil genesis and pedogenic *processes* at length in Chapter 12. Knowledge of soil genesis provides important information to scientists who classify them. Finally, we pay considerable attention in Part III (Chapters 13–15) to examining soil landscapes over time and how soils can be used as dating tools and as keys to past environments. This is how and when we really bring in the concept of change over time – the fourth dimension. Part III is the synthesis section, for within it we pull together concepts introduced previously and apply them to problems of dating landscapes and understanding their evolution. Lateral flows of materials and energy link soil bodies to adjoining ones on the landscape, helping to reinforce the all-important three-dimensional component – an emphasis of Part III. Thus, woven into the book are studies and examples of soil landscapes in three dimensions, often through the use of block diagrams. Hopefully, the reader will gain from such applications and discussions a *holistic* perspective on soils and begin to appreciate that they are integrated across and within landscapes, and that they have a history and a future. We also introduce, throughout the book, many classic studies and examples of how the evolution of soils has been effectively worked out, in order to tie certain concepts together and expose the reader to some of the classic literature. We also do our best to make this book truly global, by bringing in examples of soil studies and data from as many parts of the world as the literature allows. To be sure, our book has a North American focus – we live there, and it's the focus of a large proportion of the soil literature. However, we have gone to great lengths to serve the *global* soils community in this book. In sum, we think this book will be of use to "land lookers" worldwide (Hole 1980). We hope it is enjoyable, intellectually stimulating and, most importantly, useful to you, the reader.

Chapter 2

Basic concepts: soil morphology

Soil means diffferent things to different people. To a farmer or horticulturalist, it is a medium for plant growth. To an engineer, it is something to build on or remove before construction can occur, or it may actually be a type of engineering medium used for road building, house foundations or septic drain fields. To a hydrologist soil functions as a source of water purification and supply. To the pedologist or soil geographer, however, soil is a natural, three-dimensional body that has formed at the Earth's surface, through the interactions of at least five soil-forming factors (climate, biota, relief, parent materials and time). Its genesis involves past processes and it is likely to change in the future. It varies spatially in the horizontal and vertical dimensions. It is capable of being destroyed and yet it is resilient to perturbations.

Each soil also has a distinct morphology, defined as its structure or form. Soil morphology is all that can be seen and felt about a soil. It includes not only "what is there" but also how it is "put together" – its architecture. Soil's other defining characteristics, such as horizonation, chemistry and mineralogy, are discussed in later chapters.

Soils are composed of clastic particles (mineral matter), organic materials in various stages of decay, living organisms, water (or ice), and gases within pores of various sizes (Fig. 2.1). The absolute amounts of each, and their arrangement into a particular fabric, are the sum of soil morphology. We begin with the clastic materials that comprise the soil's *skeleton*.

Texture

Generally, the clastic mineral particles in a soil are divided into the fine earth fraction (<2 mm dia.) and a coarser fraction. Geologists commonly use the phi scale when referring to the sizes of individual particles, whereas pedologists usually refer to particle diameters in mm or μm (Fig. 2.2). Within the fine earth fraction, particles are divided, based on size, into sand, silt and clay (Soil Survey Division Staff 1993) (Fig. 2.2). Sand, silt and clay are each referred to as soil separates. Each of these three components imparts its own character to the soil and has distinct mineralogy (Table 2.1, Fig. 2.3). Sand and most of the silt fraction is composed of primary minerals, while many clay-sized particles are secondary minerals, formed from the weathering of primary minerals. This brings up an important point – *clay* is a size fraction irrespective of mineralogy, while the term *clay mineral* is thought of, by many, as a family of phyllosilicate minerals such as kaolinite, chlorite, smectite and vermiculite, along with oxide clays like hematite and goethite. Not all clay-size particles, however, are phyllosilicate minerals; many are quartz and/or amorphous materials.

Soil texture refers to the relative proportions of sand, silt and clay within the fine earth fraction (Fig. 2.4). It is commonly described as the "feel" of the sample. In the field, texture can be approximated by rubbing a sample between the thumb and forefinger. Clayey soils form a ribbon while sandy soils feel gritty. Silt imparts a smooth feel.

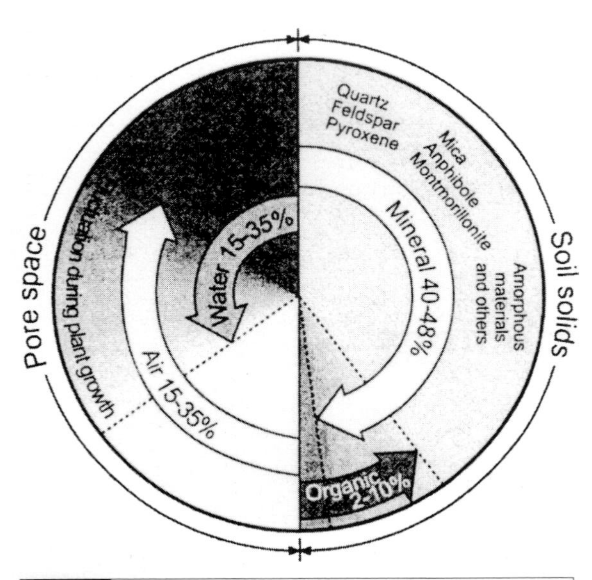

Fig. 2.1 Volumetric composition of a soil under normal conditions. The broken line between water and air indicates that these proportions fluctuate as the soil wets and dries. Similarly, organic matter contents of soils vary from zero to nearly 100%, although 2–8% is a common range for many mineral, i.e., non-organic soils.

After a bit of practice, most people can become quite proficient at determining soil texture by its feel alone.

Texture is quantitatively determined in the laboratory by first dispersing the sample so that the sand, silt and clay particles behave as an independent units, and then wet-sieving out the sand fractions. The silt and clay are both placed in a cylinder with water and mixed to disperse them evenly throughout the suspension. Then, the silt is allowed to settle and a sample is removed that is, in theory, clay and water only. The weight of the clay (and the sand, from the sieving) in the sample is, by use of an equation, used to determine the percentage of clay in the sample. Silt content is determined by subtraction.

Data on sand, silt and clay contents, when plotted on a type of ternary diagram called a *textural triangle*, place the sample within a specific texture class. The standard textural triangle (Fig. 2.4a) has been in use for decades. However, almost a half century ago, Elghamry and Elashkar (1962) realized that the textural class of any soil could be determined if one knows the percentages of only two of the three fractions. This led them to develop a textural triangle that looks quite different but functions similarly (Fig. 2.4b). It has the advantage of allowing textures to be determined by plotting data from only two variables, much like an *X–Y* plot in a traditional Cartesian coordinate system, making it adaptable to spreadsheet programs (Gerakis and Baer 1999).

Particle size classes that are totally dominated by one size fraction are simply named for that fraction, e.g., *sand*. Alternatively, *loamy* textures are not dominated by any one size fraction. Note that a sample with equal proportions of clay, silt

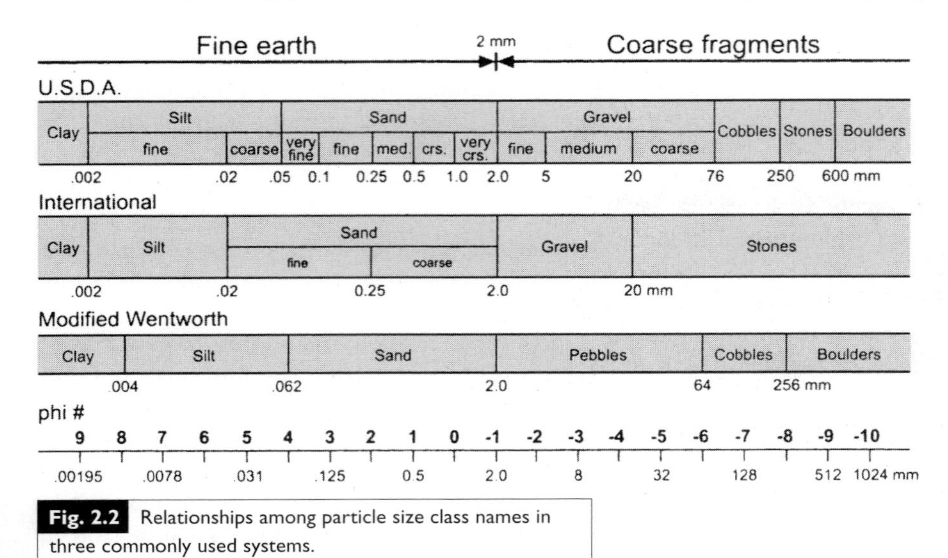

Fig. 2.2 Relationships among particle size class names in three commonly used systems.

Table 2.1 Some general properties of sand, silt and clay[a]

Property	Sand	Silt	Clay
Size range (mm)	2.0–0.05	0.05–0.002	<0.002
Means of observation	Naked eye	Light microscope	Electron microscope
Dominant mineral types	Primary	Primary and secondary	Mostly secondary
Attraction of particles for each other	Low	Medium	High
Attraction of particles for water	Low	Medium	High
Surface area	Very low	Low–medium	High–very high
Water-holding capacity	Low	Medium–high	High
Aeration	Good	Medium	Poor
Potential to be compacted	Low	Medium	High
Resistant to pH change	Low	Medium	High
Ability to retain chemicals and nutrients	Very low	Low	Medium–high
Susceptibility to wind erosion	Moderate (esp. fine sand)	High	Low
Susceptibility to water erosion	Low (unless fine sand)	High	Depends on degree of aggregation
Consistency when wet	Loose, gritty	Smooth	Sticky, malleable
Consistency when dry	Very loose, gritty	Powdery, some clods	Hard clods

[a] These are very generalized relationships and exceptions do occur.

Source: Brady and Weil (1999).

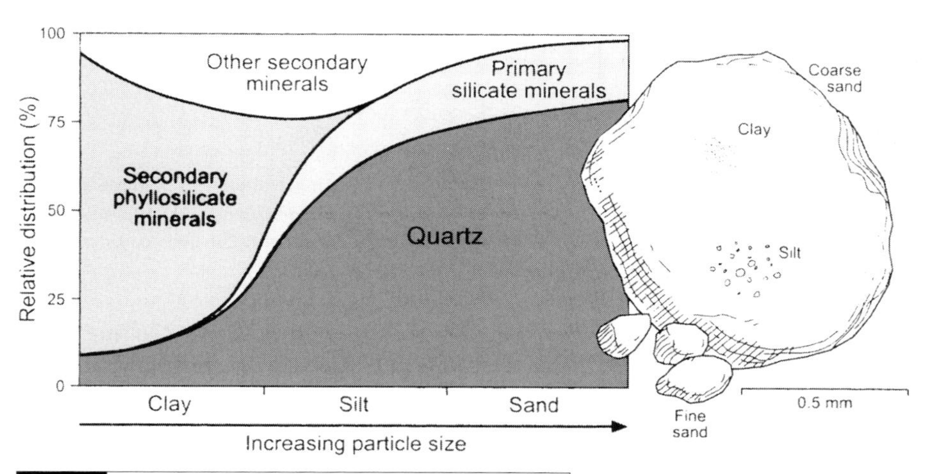

Fig. 2.3 General relationship between particle size and particle mineralogy.

Table 2.2 Terms for rock fragments, according to the USDA

Shape and size	Name	Adjectival term for soil texture class
Round (spherical, cube-like or equiaxial)		
2–75 mm diameter	Pebbles	Gravelly
2–5 mm diameter	Fine pebbles	Fine gravelly
5–20 mm diameter	Medium pebbles	Medium gravelly
20–75 mm diameter	Coarse pebbles	Coarse gravelly
75–250 mm diameter	Cobbles	Cobbly
250–600 mm diameter	Stones	Stony
>600 mm diameter	Boulders	Bouldery
Flat		
1–150 mm long	Channers	Channery
150–380 mm long	Flagstones, flags	Flaggy
380–600 mm long	Stones	Stony
>600 mm long	Boulders	Bouldery

Source: The Soil Survey Division Staff (1993).

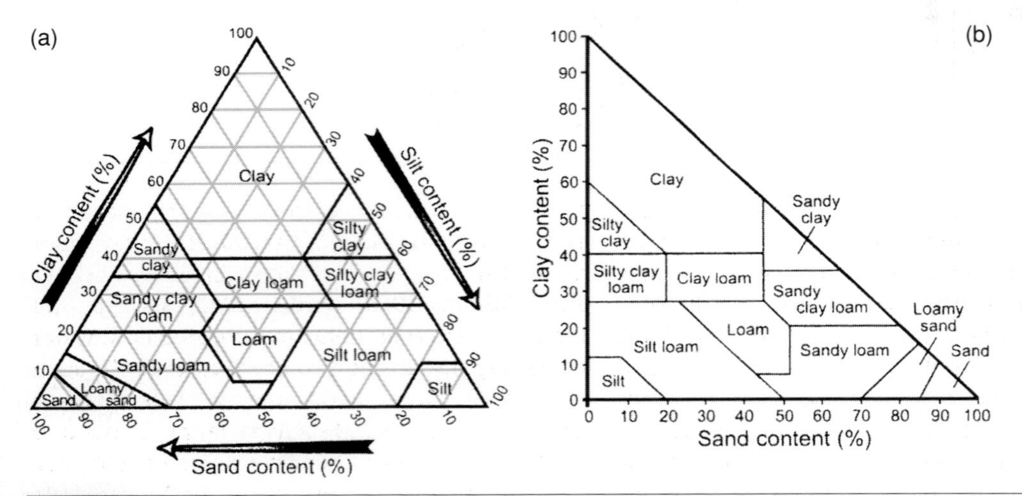

Fig. 2.4 Two variations on the soil textural triangle, based on the USDA, relating particle size distribution to texture classes. (a) The traditional textural triangle. (b) An alternative textural triangle, based on Elghamry and Elashkar (1962). This type of triangle has the advantage of allowing textures to be determined by plotting data from only two variables, much like an X–Y plot in a traditional Cartesian coordinate system.

and sand has a *clay loam* texture, because the larger surface area of the clay particles dominate the "feel" of the sample. Names of most texture classes have a modifier, such as loamy sand or silty clay. In these cases, the dominant feel of the sample is given by the last name, while the modifier implies that the its texture grades toward another texture class, loam or silt.

Recall that texture refers to only the fine earth fraction. The names given to coarse fragments vary between naming systems, depending on both size and shape and lithology (Alexander 1986, Poesen and Lavee 1994) (Table 2.2, Fig. 2.2). In any event, the fragments must be strongly cemented or more resistant to rupture to be considered coarse fragments. Aggregates of sand and smaller particles

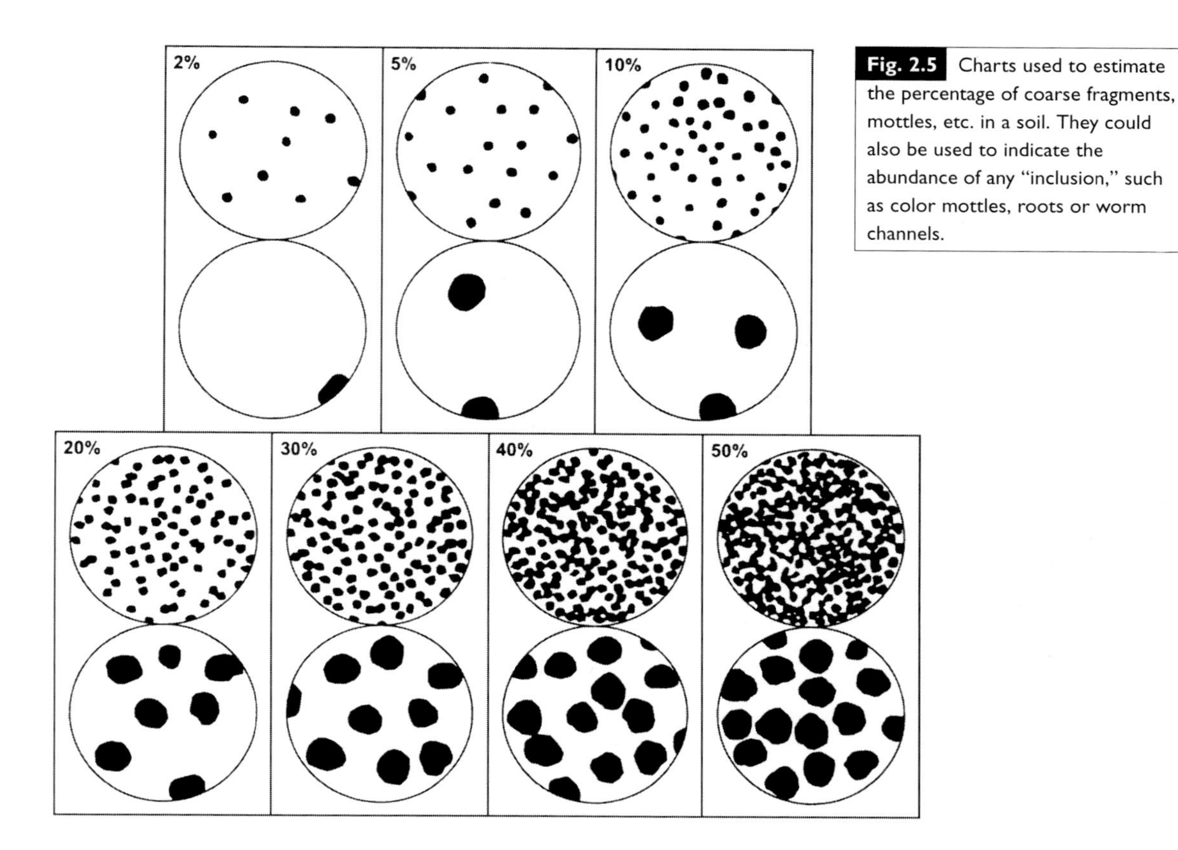

Fig. 2.5 Charts used to estimate the percentage of coarse fragments, mottles, etc. in a soil. They could also be used to indicate the abundance of any "inclusion," such as color mottles, roots or worm channels.

are not coarse fragments and should be disaggregated to determine their true size. Particles larger than 2 mm in diameter generally only figure into the textural class name, e.g., gravelly loamy sand, when they are present in large amounts. For example, a sandy loam with 17% gravel would be considered a gravelly sandy loam and a clay loam with 25% cobbles and 19% gravel would be a very cobbly, gravelly clay loam. The amount of the coarse fragments is a volume estimate, because it is too difficult to obtain data on coarse fragment content; very large samples are required for complete accuracy (Alexander 1982). Consult the Soil Survey Manual (Soil Survey Division Staff 1993) for the actual limits required for these coarse fragment modifiers (Fig. 2.5). Coarse fragments are very important in many other ways and probably should be given more consideration in soil characterization analyses (Corti *et al.* 1998).

Texture and coarse fragment content are important for a number of reasons (Jury and Bellantouni 1976, Poesen and Lavee 1994, Ugolini *et al.* 1996). Most importantly, they affect the way water moves through and is retained in the soil (Salter and Williams 1965, Harden 1988, Bennett and Entz 1989, Lin *et al.* 1999). In saturated flow, water moves rapidly through coarse-textured soils such as sands and those with large amounts of coarse fragments, because they have larger pores and little surface area to attract the water with matric (suction) forces (Brakensiek and Rawls 1994). In clays and fine-textured soils, pore space is small and usually not well interconnected, leading to low permeabilities. The high surface area of clayey soils, however, means that much water can be retained, although much is held so tightly that plants cannot extract it from the surfaces of the clays.

Texture and coarse fragment content greatly impact soil surface area. Surface area is important to soils because particle surfaces retain water, cations, anions and nutrients; it also acquires coatings which imparts color to the soil.

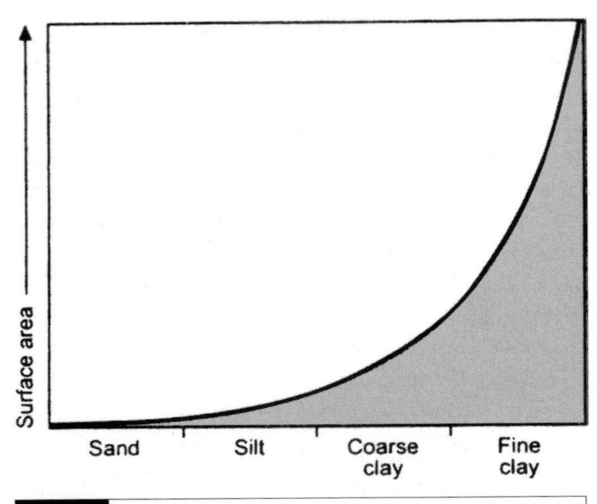

Fig. 2.6 General relationship between particle size and surface area in soils.

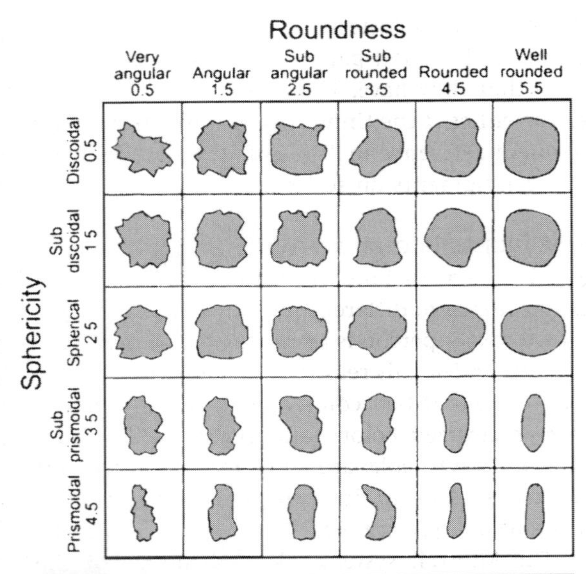

Fig. 2.7 Standard chart used to estimate the sphericity and roundness of clasts. After Schoeneberger *et al.* (1998).

Surface area increases as texture class gets finer, and is especially high in fine clays (Fig. 2.6), which makes them a chemically important particle size class. As a result, clayey soils tend to be the most reactive.

Another way in which coarse fragments impact soils is through potential void space. Rocks and other coarse fragments take up volume in soils. Thus, all the other soil processes are compressed into less space than if the same soil had no coarse fragments (Schaetzl 1991b). Rock fragments also help soils resist compaction and erosion and retain good structure (Poesen *et al.* 1990, van Wesemael *et al.* 1995). Indeed, soils with high amounts of coarse fragments tend to have lower bulk densities, probably because the fine earth fraction cannot pack as closely to the large particles as it can to itself (Stewart *et al.* 1970). Many coarse fragments are not impermeable and thus retain some soil water, thereby affecting soil water characteristics beyond just their impact on void space (Coile 1953, Hanson and Blevins 1979, Nichols *et al.* 1984, Ugolini *et al.* 1996).

It is often important to note not only the volume of the soil occupied by coarse fragments, but also their shape and lithology. Shape is captured by two variables: sphericity and roundness (Fig. 2.7). Sphericity relates to the overall shape of a feature irrespective of the sharpness of its edges; it is a measure of the degree to which it resembles a sphere. Roundness is a measure of how much a particle's corners and edges are rounded or smoothed. Coarse clasts often become more rounded and spherical as they get more weathered, and thus information of this kind may be a useful as a weathering or comminution index. Alternatively, fragments may be deposited in different stages of roundness or sphericity, and so this kind of data can provide discriminating information about parent materials' comminution. Information about lithology is important for can be used to discriminate among parent materials. It can also provide an estimate of the potential release of cations to the soil, as the coarse fragments weather.

Color

Color is perhaps the first thing that is noticed about soil. We see a soil's color long before we touch, smell or taste it.[1] The color we see is either the clean soil particles or the coatings (cutans) that they have. The late Francis D. Hole, a soil scientist at the University of Wisconsin, made frequent reference to "soil paint" when pointing out the pigments on soil particles.

The Munsell color system

Qualitative, verbal descriptions of color mean different things to different people. What is reddish brown to one person is brownish red or dusky red to another. For this reason, soil scientists have objectively quantified color by comparing samples to standardized color chips in Munsell© Color Charts. These books are standard equipment for anyone needing to describe accurately the color of a soil or rock sample.

The Munsell system takes advantage of the fact that color is composed of three elements: *hue*, *value* and *chroma*. Hue refers to the chromatic composition of the light, or wavelength of light, that emanates from the object; think of it as the actual spectral color, like red, yellow or brownish yellow. Each page of the Munsell charts contains several color chips, all with the same hue. *Hue* is most commonly represented by the abbreviations R for red, Y for yellow, and YR for yellow–red. Hue ranges from 2.5 to 10, e.g., going from red to yellow the hues are 10R, 2.5YR, 5YR, 7.5YR, 10YR, 2.5Y, 5Y, etc. The 10 point of each hue corresponds to the zero point of the next yellower hue. Most well-drained, midlatitude soils have 10YR or 7.5YR hues. The factor that most influences hue is mineralogy; red soils are hematite-rich, brown soils are goethite-rich and white soils can be rich in salts or carbonates (Table 2.3). *Value* describes the darkness or lightness of the color. Some refer to value as the intensity of color. It ranges from 0 to 10 and is displayed along the vertical axis of each Munsell page, in increments of two. As the color value changes, the amount of white or black pigment that is added to the color changes: 0 is black and 10 is pure white. Low color values imply dark colors, high values are very light, as if the chip has been faded by exposure to sunlight. Low values (dark colors) generally imply high amounts of organic carbon (humus) and/or wet soils (Fig. 2.8). For this reason, soil colors should always be measured at "field moist" wetness, unless a dry color is required (Soil Survey Staff 1999). *Chroma* refers to the purity, strength or grayness of the color; it is also ranked on a scale of 0 to 10, in increments of two. At a chroma of 0, all hues converge to a single scale of neutral grays, referred to as N0 (N for neutral). In essence, chroma is changed (reduced) as more and more gray is introduced into the color.

When referring to a soil color, the three components are listed as follows: hue value/chroma. For example, 5YR 5/3 is one of a few color chips that describes a reddish brown soil, while a 2.5Y 6/8 soil is olive yellow. Each color chip in the Munsell book has been assigned a color descriptor or adjective that can be used to facilitate communication of the color. Humans think in color names, not color chip numbers! Although these descriptors are provided to help us visualize a soil color, some of them are difficult to imagine without the book in front of one's nose, er . . . , eyes, e.g., light greenish gray.

Determining the color of a soil sample should be done in uniform, bright light, preferably sunlight. While it is impossible to match every soil sample exactly to a color chip, one should report the chip that is the closest match, or perhaps the two closest matches. Beware, however, that all Munsell books are not the same; the color chips can get soiled and faded with age, especially if they have been exposed to direct sunlight.

Origins of soil color

Most soil particles have some degree of soil "paint" on them. The color of a soil particle, soil

[1] Yes, some of us do smell and taste soil! Taste is a good way to differentiate silt from sand and clay. Perhaps "taste" is the wrong word, because most soil "tasters" do little more than gently pass the soil sample between their teeth, to test for the degree of grittiness. Taste is, nonetheless, a convenient way to distinguish (in the field) salic horizons from those that lack soluble salts.

And while you're at it, smell a newly exposed soil sometime. It smells great. Much of that "earthy" smell is the aroma of actinomycetes that are active in the soil, especially after a warm rain.

Table 2.3 Soil colors and the primary pigmenting agents that create them

Color of coating in soils[a]	Typical type of coating
Black or brown	Humus or magnetite
Black or bluish black	Reduced manganese (Mn^{2+})
White	Sodium salts, carbonates, silt-sized or smaller quartz grains
Amber yellow or brown (2.5YR)	Jarosite [$KFe_3(SO_4)_2(OH)_6$]
Light gray	Reduced iron (Fe^{2+}); salts
Brown and yellowish brown (7.5YR to 2.5Y hues; 10YR most common)[b]	Goethite [α-FeO(OH)]
Reds, browns and oranges[b]	Iron minerals or amorphous iron compounds deep red (5YR–2.5YR or redder)–hematite [α-Fe_2O_3] reddish brown (5YR [to 7.5YR?])–ferrihydrite [$Fe^{3+}{}_2O_3 \cdot 0.5(H_2O)$] reddish brown–maghemite [γ-$Fe_2^{3+}O_3$] orange (7.5YR 5/8, 6/8, 7/8 and other "bright orange" colors) –lepidocrocite [γ-FeO(OH)]

[a]White or light-colored soil horizons are often devoid of coatings, allowing the color of the quartz particles that dominate their mineralogy to show through.
[b]Based on data provided from various sources, including Davey *et al.* (1975), Hurst (1977), Soileau and McCracken (1967), Torrent *et al.* (1983) and Vepraskas (1999). Applies well to redox concentrations (see Chapter 13). These characteristics should be used as guidelines only; many exceptions can and do occur. Hues are not considered reliable indicators when color values and chromas are <3.5, due to masking by (usually) organic matter.

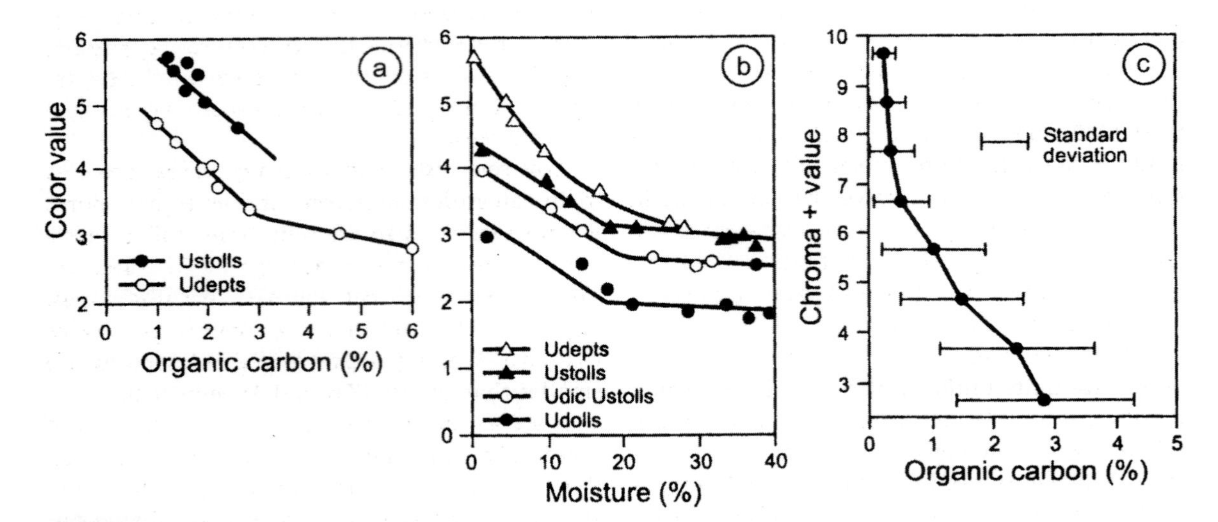

Fig. 2.8 Relationships between Munsell color and various soil properties. (a) Color value and organic carbon content, for some soils in Saskatchewan. After Shields *et al.* (1968). (b) Color value and soil moisture, for some soils in Saskatchewan. After Shields *et al.* (1968). (c) Color chroma + value and organic carbon content, for some soils in Ohio. After Calhoun *et al.* (2001).

sample or soil horizon is a function of the type and degree of its various coatings (Table 2.3). Dark colors usually imply organic matter, although manganese concretions are usually black. Red colors come from various iron-bearing minerals. Color provides a great deal of information about the particle and ped coatings in the soil, which immediately impart genetic clues. Only in E horizons are soil materials so clean that the color of the primary minerals shows through. Because most soils are dominated by quartz, at least in the sand fraction, and quartz is dominantly white or light pink in color, E horizons are white or nearly so. Dark colors suggest wetness and high organic matter content (Plice 1942). Bright (high chroma) reds and browns are generally associated with well-drained, upland soils in which the conditions are oxidizing (see Chapter 13). Prolonged anaerobic conditions, typical of wet soils rich in organic matter, develop muted gray colors and low chromas. This condition is referred to as *gleyed*.

As any painter knows, the less area one has to cover the more coats that can be applied and the deeper and richer the color. Thus, soils with low surface areas, i.e., the coarse-textured ones, are most easily and quickly colored. The color of sands can be changed rather quickly, pedogenically, which is why soil taxonomy (Soil Survey Staff 1999) does not allow cambic horizons to be sandy – the coloration associated with cambic-type development simply develops too quickly in sandy soils. Similarly, sandy horizons are more quickly and readily stripped of their soil "paint," explaining why E horizons form more quickly in sandy than in loamy soils.

While many soils are a pot-pourri of color, others are quite uniform. Variation in soil color, usually due to types of different coatings or variations in degree (completeness) of the coating, is worthy of description, for it can provide important clues about genesis. "Spots" of one color set amidst a matrix of another, e.g., mottles, might imply some sort of accumulation process, such as when sodium salts precipitate into a dark soil. For mottled soils, the colors of the matrix *and* the mottles should be described, and the abundance of the latter noted (Fig. 2.5).

Pores, voids and bulk density

About half of the volume of many soils is simply voids, filled with air or water or both (Fig. 2.1). A well-aerated soil may have up to 2/3 void space. If a soil is compacted, voids are reduced in size and number. The amount of void space in a soil or similar medium is termed its *porosity*. A soil with 45% porosity has 45% of its total volume composed of pores. Soils with few voids or pores are said to be densely packed. This characteristic should not be confused with sorting, which refers to the variety of grain sizes within a sample or soil. Well-sorted soils are dominated by a few grain sizes, while poorly sorted soils have a wide variety of grain sizes. In general, however, poorly sorted soils tend to be closely packed, as they have many small particles that can fill large pore spaces.

One of the main factors that maintains or creates high porosities in soils is soil biota, especially macroscopic soil fauna. Worms, termites and many other forms of fauna move through the soil and in so doing, leave behind preferred pathways as biopores (Dexter 1978). High amounts of aggregation also help to maintain high porosity, as large pores often exist between soil aggregates or peds. Pore space tends to decrease with depth, where soil biota and aggregation are increasingly less influential.

Soil scientists tend to group pores into two main categories: *macropores* and *micropores*; pores of intermediate size are sometimes called mesopores. There is no firmly agreed-on, predetermined size cut-off between the two pore types, but many assume that macropores are larger than 0.075 mm (75 μm) while micropores are smaller than 30 μm (Kay and Angers 2000). Gases and liquids move through large pores faster and more efficiently than through small pores, although connectivity and tortuosity of pores are also important properties that affect flow rates. Macropores are commonly formed by roots and biota, e.g., worms, and can readily accommodate most roots. They can occur between sand grains in a sandy soil, and between peds (aggregates) in medium- and fine-textured soils, where they are called *packing pores* and take on a crudely planar shape. When formed by burrowing biota or roots, the resulting *biopores* tend to be tubular in shape. Macropores conduct water and air rapidly, and drain quickly if filled with water after an infiltration event (Edwards *et al.* 1988). Micropores, on the other hand, retain water for long periods of time, due to surface tension and matric forces. They tend to be the main type of pore in clayey soils, with the exception of inter-ped pores.

The soil solution, consisting of both gases and liquids, or soil water, is stored within and passes through soil voids. Voids that are larger and well connected to each other allow for free passage of the soil solution. The property of *permeability* is meant to capture that attribute. Soils that are permeable allow for rapid interchange and movement of the soil solution through interconnected pores. Two factors that, therefore, contribute to high soil permeabilities are pore size and pore interconnectedness. Many clayey soils have high amounts of pore space but are very slowly permeable because the pores are too small to conduct water and gases at a rapid pace. As soil water is transmitted through soil pores, it impacts the walls of those pores in two possible ways: (1) microscale erosion of material from them, and/or (2) deposition of material, as cutans, that was either in suspension or solution onto the pore walls (see Chapter 3).

Permeability is measured as a length per unit time, e.g., cm of water per hour. The rate generally varies depending on the direction of flow. For example, vertical permeability may be faster than horizontal. It is also varies as a function of water/gas content, e.g., saturated permeability vs. unsaturated (see Chapter 5).

The opposite of porosity would be the volume occupied by clastic and other solid materials. The weight of these materials per unit volume of soil is referred to as *bulk density*. A soil with no voids or pores whatsoever would have a bulk density of about 2.65 g cm^{-3}, because the particle density of most soil-forming minerals is about 2.65–2.70 g cm^{-3}. Thus, 2.65 forms a theoretical upper limit of soil bulk density, although in reality only a few very dense, compacted soils achieve bulk densities in excess of 2.3 g cm^{-3}. Some organic soils have bulk densities below 1.0; in theory these soils would float on water, which has a bulk density of 1.0 g cm^{-3}. Most mineral soils, in a natural state, have bulk densities between 1.1 and 1.6 g cm^{-3} (Manrique and Jones 1991). Bulk density, by definition, is measured on oven-dry (100 °C) soil; the weights and volumes of any coarse fragments are excluded so that a large rock or rocks within the sample does not skew the data (Reinhart 1961, Curtis and Post 1964, Vincent and Chadwick 1994).

Any of several methods can be used to obtain bulk density information for soils (Shipp and Matelski 1965, Blake and Hartage 1986, Laundré 1989, Vincent and Chadwick 1994). In theory, a sample of known volume is removed from the soil, and after drying, the coarse fragments are removed. The weight of the remaining fine earth is divided by the initial sample volume (minus the volume of the coarse fragments), to arrive at the bulk density value in g cm^{-3}.

Factors that affect bulk density and porosity have been the subject of some study. Bulk densities tend to increase with depth in natural soils. Soils and horizons high in organic matter tend to have lower bulk densities, probably because many of these samples are from near-surface horizons where biological activity is high and because organic matter tends to attract soil fauna which create pores (Adams 1973, Alexander 1980) (Fig. 2.9). Somewhat surprisingly, clayey and silty soils tend to have low bulk densities while soils with sandy loam and sandy clay loam textures are more prone to compaction (Fig. 2.10). High bulk densities are often associated with loamy soils, where clay and silt can fill in the large voids between sand grains, optimizing the packing of the soil matrix. Clayey soils have many micropores, which helps to explain their low bulk densities (Heinonen 1960, Bernoux *et al.* 1998).

Structure

Structure refers to the arrangement of primary soil particles, e.g., sand, silt, clay, into natural aggregates called *peds*. Aggregates formed by tillage or other human-induced practices are called *clods*. Horizons have their own unique structures, depending on the amount of organic matter and biotic activity, texture, numbers of freeze–thaw

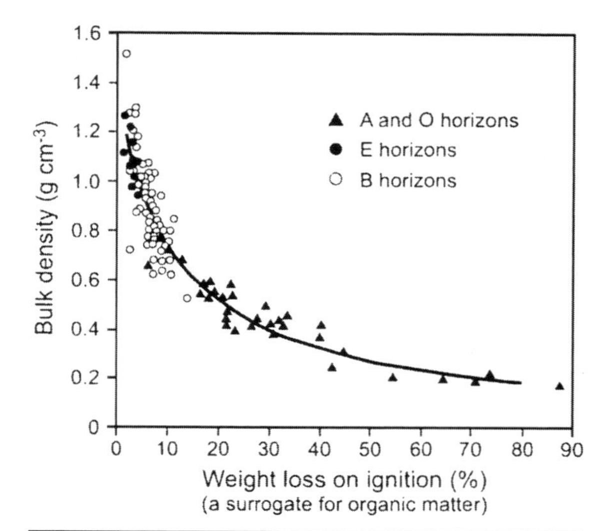

Fig. 2.9 Relationship between bulk density and loss on ignition in the O, A and B horizons of some soils of the Green Mountains, Vermont. After Curtis and Post (1964).

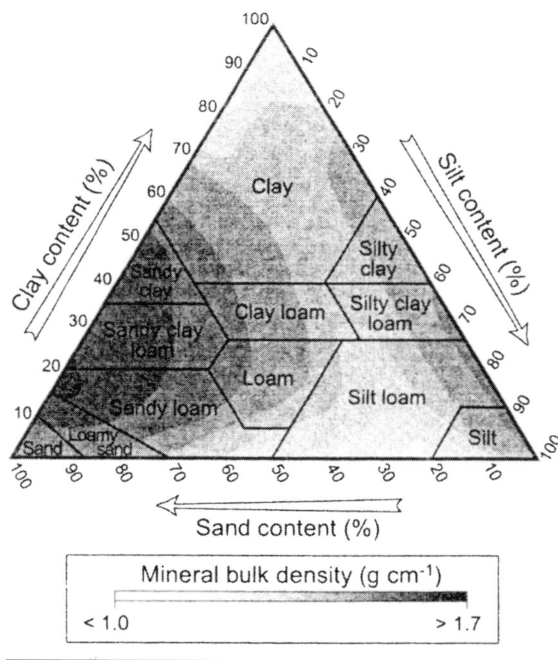

Fig. 2.10 Typical bulk densities for soils, arrayed by texture. After Rawls (1983).

and wet–dry cycles, etc. Soils that lack pedologic structure are termed massive (in fine-textured materials which break along no preferred planes) or single-grained (in sands where each grain behaves independently). How structure forms and is maintained, and what the particular structure types implies pedogenically, is our focus.

A distinguishing criterion of soil aggregates is that they have strong internal cohesion that prevents their being broken up by disruptive forces that continually occur in soils. That is, peds must be held together by some sort of binding agent or cohesive force. Humus (highly decomposed organic matter) is a common "glue" in A horizons, as are various fluids secreted by soil fauna, e.g., saliva and urine. Many peds are simply fecal pellets held together by body fluids and prior compaction, by some sort of microbial biomass (Sutton and Sheppard 1976, Zhang and Schrader 1993, Tisdall *et al.* 1997). In clayey soils, peds may be stable because of the attraction that clays have for each other, i.e., cohesion. Salts, iron, aluminum and other soluble substances may also act as binding agents (Giovannini and Sequi 1976).

Peds are essentially the most stable shape and size that the multitude of processes within a horizon can create. If those pieces (peds) are the most stable size and shape and if enough pedogenic glue is available, that type of ped will become the dominant one in the horizon. Opposing this natural aggregation are the forces that break peds apart, most of which decrease in intensity with depth. For that reason, ped sizes generally increase with depth, and at great depth soils can be structureless or massive, or at the very least they may have only inherited, geologic structure.

Wet–dry cycles, freeze–thaw cycles, root activity and fauna all play a role in the formation of soil structure (Materechera *et al.* 1992). Shrinking and swelling are common to certain types of clays when they undergo alternating desiccation and wetting. The primary measure of the degree to which a soil will expand upon wetting, and shrink upon drying, is the coefficient of linear extensibility (COLE). COLE is defined as the percent shrinkage in one dimension of a molded soil between two water contents, typically its plastic limit and its air-dry state; it is often used to represent a soil's shrink–swell potential. Finally, although any given structure type can form in a soil of any texture, structures formed in fine-textured soils tend to be more persistent and less likely to be destroyed by tillage, compaction, etc., because their cohesive forces are stronger.

When discussing structure, the two most frequently used adjectives refer to ped shape and size. The smallest peds are in the near-surface A horizons; here *granular* peds are typically about 1–5 mm in diameter. The granular (in old literature, *crumb*) structure common to A horizons owes much of its existence to soil biota, as many of these more-or-less spherical peds are worm casts or have been reworked many times by soil fauna (Blanchart 1992, Zhang and Schrader 1993). Buntley and Papendick (1960) described soils that are so intensively worked by worms that they have granular and small subangular blocky structure down to several dm below the surface. Dense networks of fine roots maintain the structure by proliferating on ped faces and only infrequently penetrating the ped itself (Fig. 2.11a). Their continued growth literally squeezes the peds and helps to maintain their integrity. High amounts of organic matter (humus) contribute to the strong intra-pedal cohesive forces. Granular peds tend to be porous, and because they do not fit together tightly, horizons of granular structure have high porosity and permeability; they can literally "fall apart" upon exposure (Buntley and Papendick 1960).

Platy structure consists of thin (<4 mm thick) plate-like peds that are aligned parallel to the soil surface (Fig. 2.11b). This type of structure is common in E horizons, and if well developed can impede infiltration. The mechanism of its formation is not well understood, but it may have something to do with wet–dry and/or freeze–thaw cycles, which are transmitted into the soil along planes that parallel the surface. High amounts of organic matter and the large numbers of soil fauna that go with it tend to disrupt platy structure, which may help explain why it is most common in humus-poor E horizons. One question must always be asked, when describing platy structure, "Is it pedogenic or geologic?" Many surficial sediments are initially stratified and if this stratification has not been pedogeneticized, it can appear as plates or laminae.

Continuing toward the lower part of the profile, peds get larger and more block-like. *Angular* and *subangular blocky* structure is common in B horizons. In angular blocky structure, the edges of the blocks are sharp and the rectangular facets between them are distinct, while the more common subangular blocky structure has more rounded edges. Many of these peds are held together by coatings (cutans) of material that has been translocated into this horizon from above. This material moves into the B horizon in suspension or solution, in the large pores between peds. Eventually, some of the infiltrating water soaks into the peds and in so doing any material being carried along is plastered on the ped face (see Chapter 12). As with granular structure, root proliferation between blocky peds helps to maintain them (Fig. 2.11c).

At still greater depths, the forces that act to break up the soil matrix into distinct peds not only become less frequent, but they tend to be more vertically oriented. Blocks that are taller than they are wide, prisms, will form under this scenario. *Prismatic structure* is, therefore, typical of the lower B horizon and C horizons (Fig. 2.11d). Prisms can reach mammoth proportions; widths of the flat prism tops can be up to 150 cm, and heights of 200 cm are not uncommon. A particular type of B horizon, a fragipan, consists of large, distinct prisms that many believe harken back to times when permafrost and freezing processes extended to several meters depth (see Chapter 12). In areas where soils have high amounts of exchangeable sodium, the tops of the prisms sometimes appear white from an abundance of clean quartz grains; this type of structure is called *columnar* (Fig. 2.11e). The whitish caps on the columns are said to resemble "biscuit tops." The tops of the prisms get rounded as soil water is forced to run between them, because permeabilities are slow within the peds. Thus, the edges are worn away and take on a rounded shape, but the bases may still be flat and angular, as in prismatic structure.

Consistence

Consistence refers to the way a soil "feels." The Soil Survey Division Staff (1993) provided a more comprehensive definition of consistence,

Fig. 2.11 Shapes of common soil structures. All photos by RJS unless otherwise indicated. (a) Granular soil structure, removed from an Alfisol in northern Michigan that had never been plowed. (b) Platy soil structure in an Ultisol in North Carolina. (c) Angular blocky structure. (d) Prismatic structure in an Entisol in Kansas. (e) Columnar structure in a Typic Natrustoll from Grant County, North Dakota. Photo by J. L. Arndt.

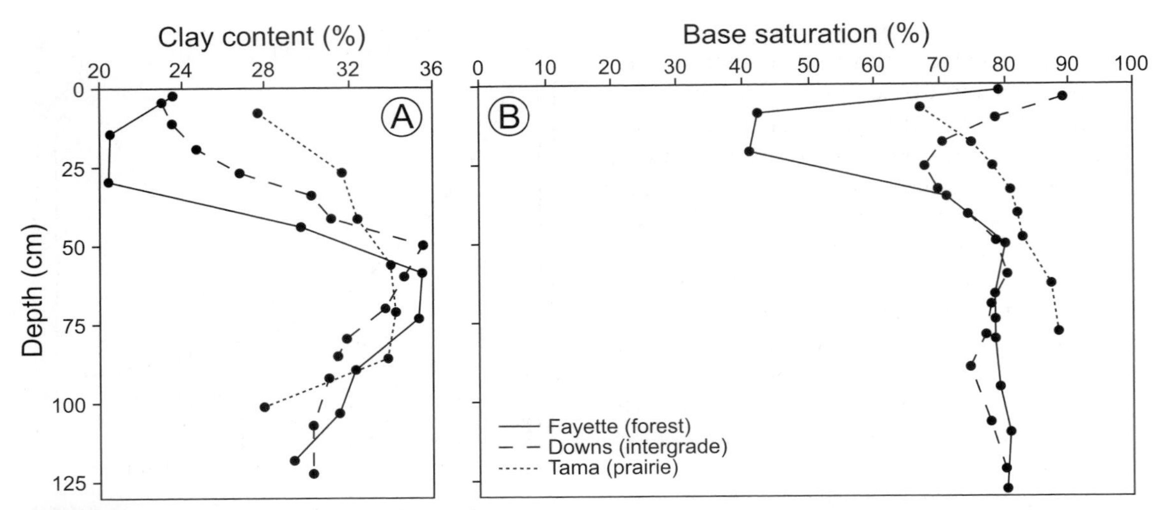

Fig. 2.12 Depth plots of clay contents and base saturation, for three loess-derived soils from mid-continent, USA. The soils represent a biosequence from a grassland soil to a forest soil, with one intergrade member. After White and Riecken (1955).

including: (1) resistance of soil material to rupture, (2) resistance to penetration, (3) plasticity, toughness and stickiness of puddled soil material, and (4) the manner in which the soil material behaves when subject to compression. Because consistence is highly dependent upon moisture content, it can be evaluated at wet, moist and dry conditions and the moisture content at the time of description must be noted. Examples of moist consistence are loose, very friable, friable, firm, very firm and extremely firm. *Friable* is a common type of moist consistence, referring to the ease with which a soil crumbles when pressure is applied; friable soils crumble easily between the fingers. Determining soil consistence is somewhat subjective.

Presentation of soil profile data

Soil data are usually generated, i.e., soils are usually sampled, on a per-horizon basis. In order to evaluate and compare these data, they are commonly presented either in a table or as a *depth plot* or function. In a traditional depth plot, the data for each horizon are shown as a single point at the depth-centroid of the horizon (Fig. 2.12). Depth plots are the standard within the field, providing a graphical way of visualizing changes in soil properties throughout the profile.

Soil micromorphology

Pedologists and soil geographers routinely use microscopic examinations of intact soil material to determine the arrangement of the various components of the soil matrix and to determine if a sediment sample is from a soil, i.e., does it have pedogenic fabric? The latter application is particularly useful in paleopedology – the study of buried soils (see Chapter 15). This aspect of soil science is called soil micromorphology – the study of the materials and fabric of the individual ped or parts thereof. In this section we provide a short introduction to the field; many fine reference works exist that provide further detail, e.g., Kubiëna (1970), Brewer (1976), Bullock and Murphy (1983), Douglas and Thompson (1985), Courty *et al.* (1989), Nahon (1991) and FitzPatrick (1993).

Why study the micromorphology of soils? Perhaps the most ingenious and straightforward answer to this question was provided by Kubiëna (1970), who imagined someone who wanted to learn about watches and how they work, yet had never seen one before. How would/should they begin to understand them? There would be any of several ways:

(1) Put each watch in a mortar, grind them to a powder and determine the chemical composition of the whole.[2] This method would provide information about bulk composition, but would not provide any insight into the working parts, such as chains, levers, gears, etc. The researcher would have learned only that watches are made of different metals and nonmetals. This analysis is similar to whole soil, elemental and oxide analyses done by soil scientists in the early and mid twentieth century.

(2) Take each watch apart, sort the myriad pieces into groups and determine the sizes and ratios of these groups. In this way, the researcher could classify the watches into logical groups, although this information will still not help answer questions about how they work. Nor might the researcher know even which watch parts interacted with others. This analysis is similar to fractionation procedures in soil science, where amounts of certain soil components, e.g., sand, clay, organic matter, feldspar, fecal pellets, are determined and possibly ratioed.

(3) Each watch could be carefully inspected, while intact, such that the placement of each part was determined, relative to each other and to the whole. Then the watches could carefully be dismantled and again, each part examined with respect to its interactions with other parts. This analysis is akin to traditional soil micromorphology, in which the parts of the soil are examined each in relation to the other.

(4) The last method mentioned by Kubiëna (1970) involves inspection of the watches as they operate, tick and chime, i.e., *while they work*. Stopping them for the sake of analysis (as was done in 1–3 above) deprives the investigator of important information about function. This fourth method of analysis is synonymous with the measurement of soil processes *in situ*. Examples include removal and analysis of soil solutions that are percolating through the soil, observing and measuring the mixing activities of ants or mobile tracers, or the continuous monitoring of soil water tables.

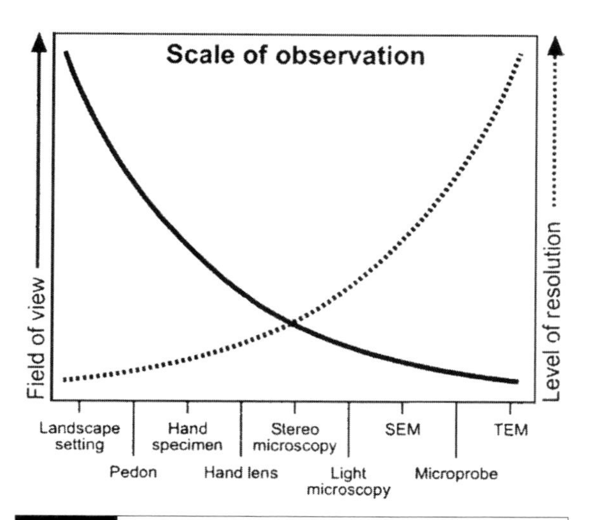

Fig. 2.13 Diagram illustrating the various scales at which soils and soil (or landscape) fabric can be examined, and the tools that are used at those scales. SEM, scanning electron microscopy; TEM, transmission electron microscopy. After Cady *et al.* (1986).

There exist several scales and types of imaging apparatus with which to view and study soils (Fig. 2.13). At the broadest scale, one might view the entire landscape. To examine soil fabric, however, one must get close enough to the soil to actually see individual components. This can be done by examining a hand sample with the naked eye or a hand lens. At increasingly finer scales of analysis, one might view (1) whole soil samples and natural aggregates under visible light in a dissecting microscope, (2) thin sections of soils under a plain light or polarized light microscope, (3) very small but otherwise intact fragments of soils with an electron microscope. In all cases, the goal is to observe various parts of the soil, in as natural a state as possible, and determine how they are related. For example, the micromorphologist will look for arrangement of sand and silt grains, coatings on voids and ped faces, and evidence of biotic materials, just as a start. The study of soil micromorphology is most useful when it is combined with data of other kinds and from various scales, such as a profile description or morphological and chemical data (Eswaran *et al.*, 1979) (Fig. 2.13).

[2] For example, see Sherman and Alexander (1959).

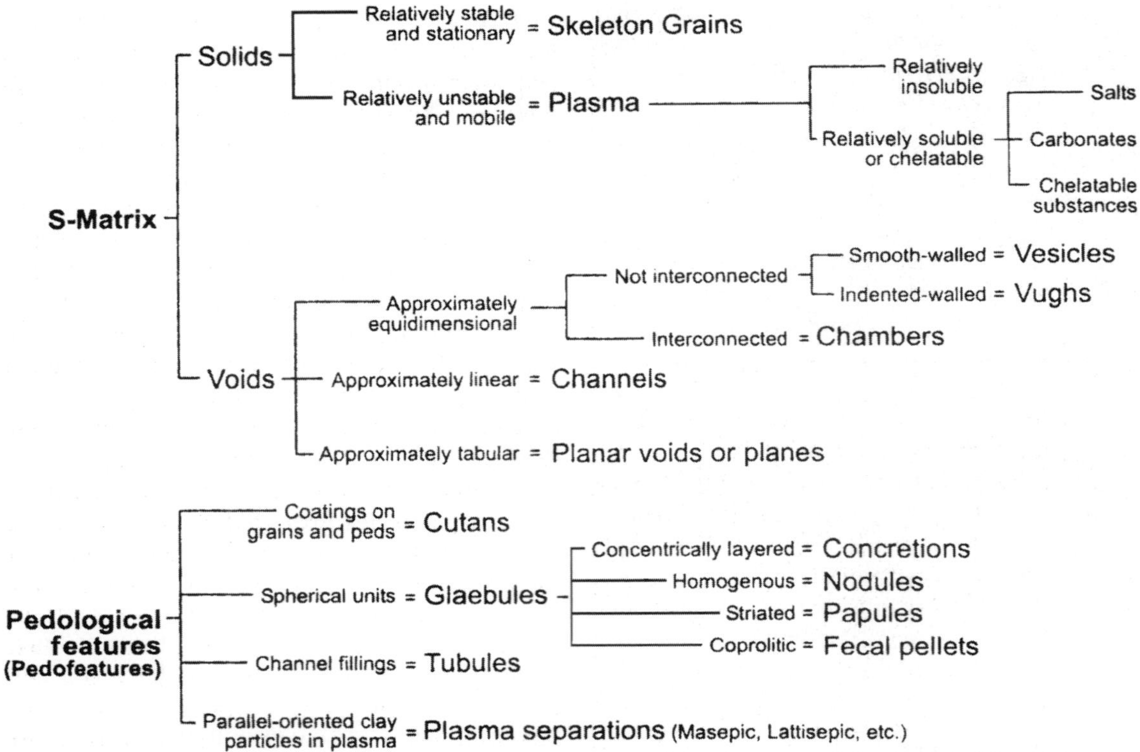

Fig. 2.14 A simplified classification flow chart of micromorphological terms and components. After Buol *et al.* (1997).

Terminology

The largest hurdle for many to get over, when it comes to soil micromorphology, is the voluminous and at times peculiar terminology that accompanies it. Most micromorphological terms originated with Brewer (1976) and Kubiëna (1938). To this end, we provide an outline of the more common terms used in the field of soil micromorphology (Fig. 2.14) (also see the Glossary).

Micromorphological techniques are used to examine the fabric of soils at various scales. Fabric includes the entire organization of the soil, as expressed by the spatial arrangement of its various constituents (solids, liquids, gases), including their shape, size and frequency (Bullock *et al.* 1985). The two major components of the soil fabric are *S-matrix* materials, non-pedological materials inherited from parent material, and *pedological features* or *pedofeatures* such as cutans (ped and void coatings), fecal pellets, glaebules (channel infillings), etc. The S-matrix is composed of voids and solids, and the solids can be divided into immobile skeletal grains, known as *skeleton*, and those solids that are also mobile, i.e., *plasma* (Fig. 2.15). The skeleton is that part of the S-matrix that is generally not mobile or readily translocated within the soil, except perhaps by large biota (see Chapter 10). Plasma materials include translocated clay and other materials, papules, glaebules, organic matter and other soluble materials that are subject to mobilization or have been mobilized in the past. Features that are distinctly pedological in origin are usually identified as such because of perceived origin, plasma concentrations, mineralogy or some type of rearrangement of constituents (Cady *et al.* 1986). The walls of voids (Table 2.4) are preferred locations for plasma materials to be pedogenically deposited as cutans. Generally, the cutan is

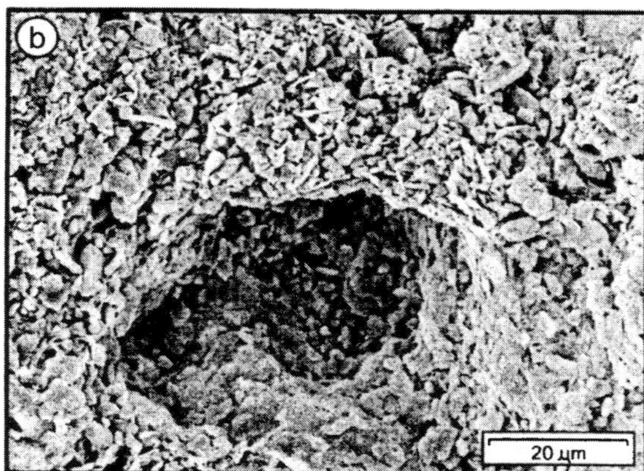

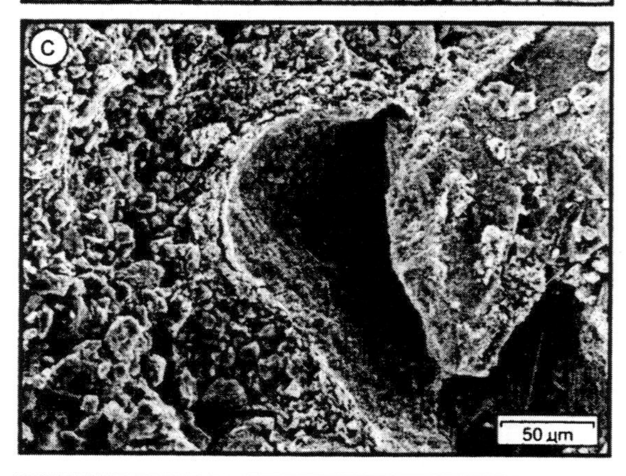

Fig. 2.15 Examples of various types of voids in soils (see Table 2.4 for terminology). SEM images by B. Weisenborn. (a) Packing voids between silt grains in the Bs horizon of a Alfic Oxyaquic Fragiorthod from northern Michigan. Every grain visible in the image is silt size or smaller; most of the individual clay particles are too small to be detected. (b) Vugh in the same horizon. Most of the particles are fine silt or coarse clay. (c) Channel void in the same horizon. The channel is coated with clay and an amorphous substance that is probably silica-rich.

named for its composition (e.g., calcan for calcite) (Table 3.4) and location (e.g., channel cutan).

Pedofeatures are parts of the soil fabric that, for various reasons, are interpreted to have a pedologic origin. Whether they date from contemporary or past processes does not matter, definitionally. Often, the interpretation is based on the presence of oriented clay minerals, organic matter, crystals of minerals interpreted to have been redeposited from solution, or internal fabric alone (Catt 1990). Determining what is and what is not pedofabric requires skill and experience; it is not something that can be readily gleaned from a book. Bullock *et al.* (1985) distinguished six main types of pedofeatures: (1) coatings on ped or clast surfaces, (2) areas of pedogenic depletion or eluviation/weathering, (3) features formed by crystallization of material from the soil solution, (4) amorphous features that do not appear to be part of the skeleton, (5) features whose fabric appears altered from that of the S-matrix and (6) obvious fecal pellets (Fig. 2.16). For each, he also described a number of terms related to size of the pedofeature, as well as various subtypes, e.g., coatings on grains, coatings on aggregates, quasi-coatings on voids, external hypo-coatings, etc. (Fig. 2.14). The rich and sometimes overwhelming terminology associated with describing soil fabric can also be a powerful qualitative descriptive tool when properly used.

Examples

Perhaps the least utilized but still very useful form of micromorphological research involves low-power magnification of intact soil aggregates and fragments, often using a binocular light microscope. Little pretreatment, other than drying (which is not even necessary), is required. At this scale, typically at magnifications of 10–100×, and under reflected light, it is possible to examine features that are too large for more traditional micromorphological techniques. This form of research and investigation provides information about the arrangement of skeletal grains and plasma, and has a great deal of merit, because (1) it requires little preparatory time and thus provides good return on (time) investment, and (2) it spans the scale between examination of soils in the field (macromorphology) and high

Table 2.4 Names traditionally given the various types of voids and pores in soils

Name	Characteristics
Packing void	Voids formed between larger particles that, because of their size and shape, do not adequately pack together
Vugh	Unconnected voids with irregular shape and walls, often most common in fine-textured soils
Vesicle	Unconnected voids with smooth walls and (usually) rounded shapes
Chambers and channels	Connecting passageways
Planes	Voids elongated along one plane or axis, commonly formed at ped faces

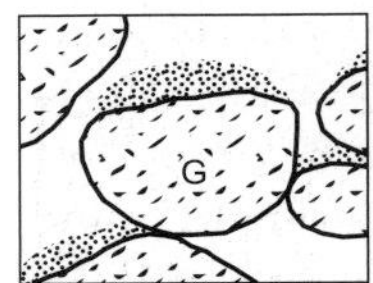

Textural pedofeatures
Fine silt cappings on grains (G)

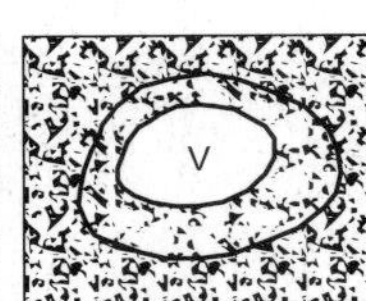

Depletion pedofeatures
Depletion within a zone around a void (V)

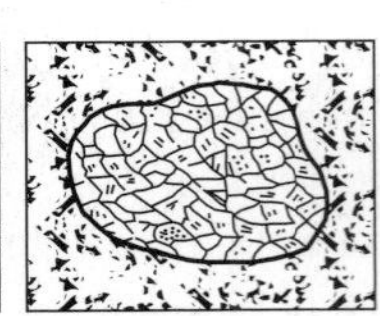

Crystalline pedofeatures
Typic crystalline pedofeature

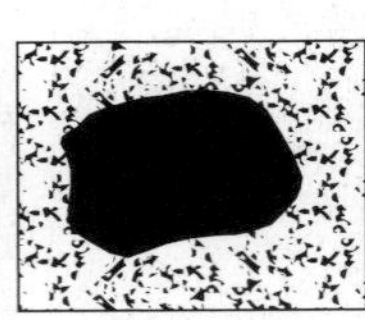

Amorphous and cryptocrystalline pedofeatures
Typic amorphous pedofeature

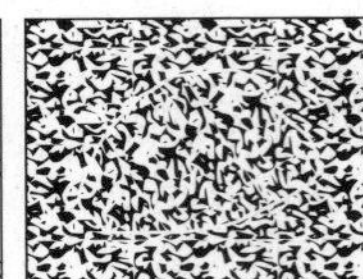

Fabric pedofeatures
Fabric pedofeature of low contrast

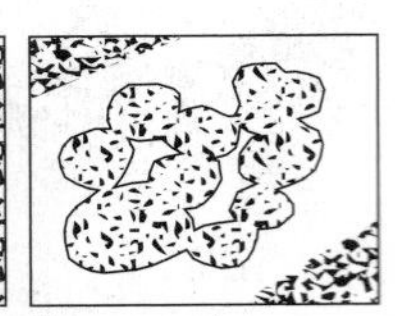

Excrement pedofeatures
Group of excrements (fecal matter)

Fig. 2.16 Six main types of pedofeatures. After Bullock *et al.* (1985).

magnification views provided by more traditional micromorphological methods.

Similar types of soil materials, i.e., peds and fragments of peds, can be examined at much higher magnifications with the scanning electron microscope (SEM). Electron microscopy uses a beam of electrons to form images of a sample's surface, unlike light microscopy which uses visible or polarized light. By using electrons, the method provides resolving power as much as a thousand times greater than a light microscope (Smart and Tovey 1982). In effect, the electron microscope represents about the same amount of improved resolution over the optical microscope as is attained by the optical microscope over the unaided eye. In addition, SEM has many advantages over optical methods, by providing a larger depth of field and greater resolution.

In SEM, a high-energy, focused electron beam is scanned across a sample in a raster (or pattern of parallel lines). The sample has usually been dried and gold- or carbon-coated within a large vacuum tube (the coating is used to make the sample surface more conductive) prior to imaging. When the electron beam hits the sample, it interacts in various ways, and secondary or backscattered electrons are emitted. The secondary electrons are captured by a detector and used to produce a detailed map of the sample surface. The intensity of emission of both secondary and backscattered electrons is very sensitive to the angle at which the electron beam strikes the surface, i.e. to topographical features on the specimen. Parts of the sample that are higher or stand above others are usually brighter because they give off more secondary electrons. Because secondary electrons that are produced

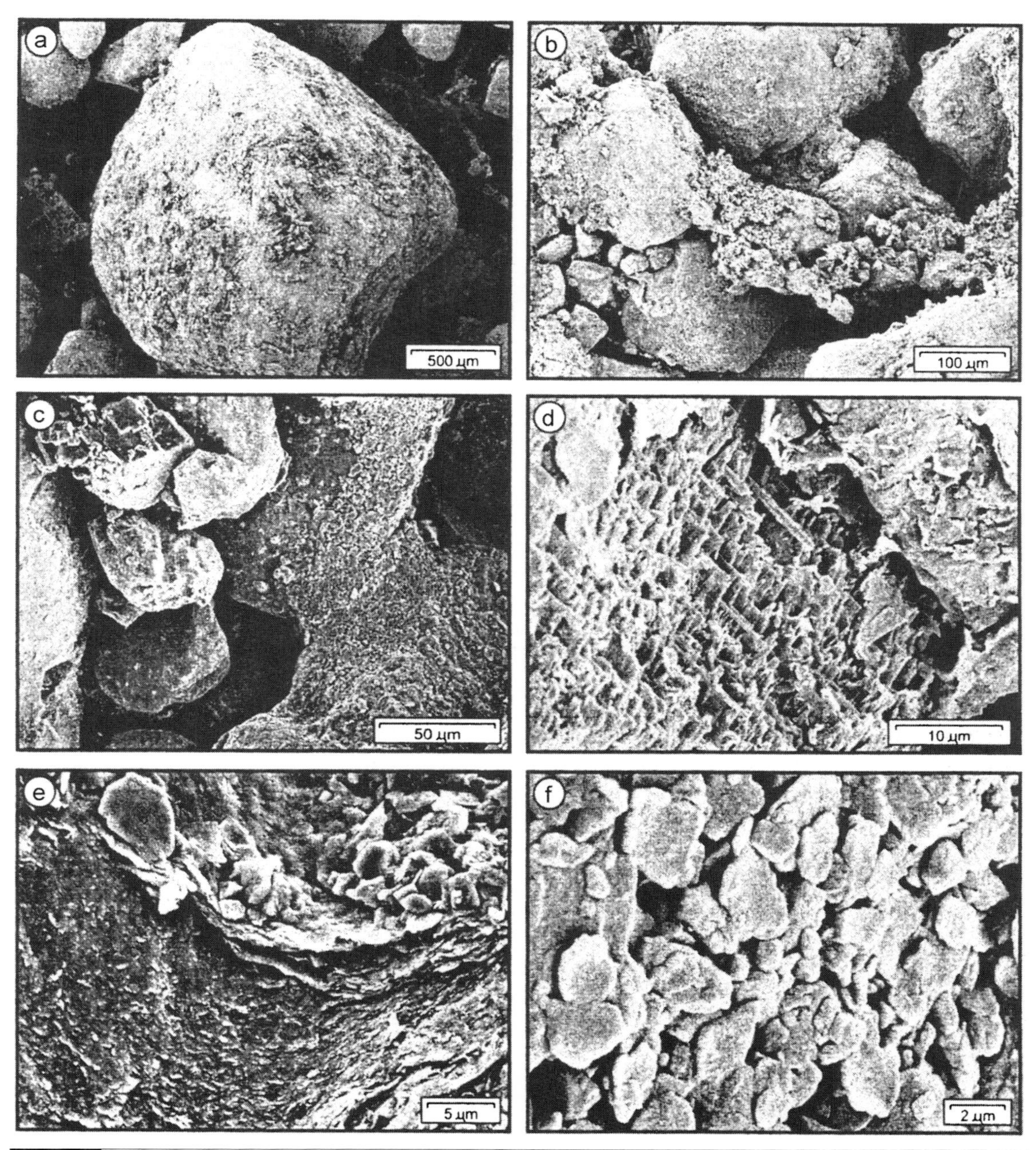

Fig. 2.17 SEM photomicrographs of Spodosols and Alfisols in Michigan, illustrating some of the various scales and resolutions that are possible in electron microscopy. SEM images by B. Weisenborn. (a) Coatings of amorphous and organic materials, including fecal pellets, on a very coarse sand grain. (b) Randomly packed fine sand grains with bridges of clay and fine silt. (c) Bridge of fine silt and clay, between two sand grains. The grain in the upper left is probably a detrital carbonate mineral. (d) Pore (probably a channel) lining of amorphous material filling in and binding clay platelets. (e) and (f) Close-up views of phyllosilicate clays in an argillan (oriented clay coating).

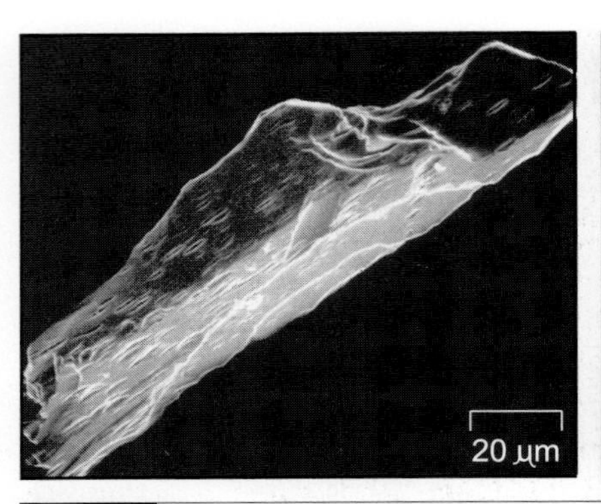

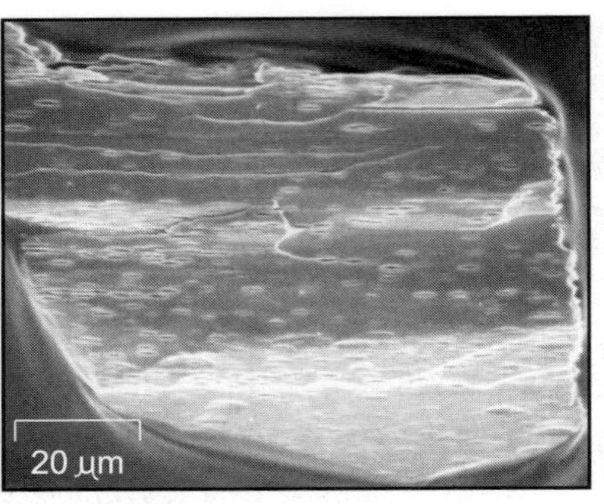

Fig. 2.18 Etch pits in some amphibole grains. The nature of this distinctive type of weathering is clearly seen using the SEM. Images by D. L. Cremeens.

deeper than about 5 nm into the sample are absorbed by it, the detectors capture primarily those electrons that interact with the very surface of the sample. Thus, the image created by the detector is essentially a relief map of the surface (Cremeens *et al.* 1988) (Figs. 2.17, 2.18). SEM can provide very high resolution images of particles down to fine clay size (<0.2 μm), but it alone cannot identify the minerals; this requires a petrographic microscope (see below). Therefore, although SEM images have excellent detail and resolution, they are limited in some respects.

One particular advantage of the SEM as a micromorphological analysis method, however, is that another very robust and useful tool can often be used in conjunction with it: energy dispersive spectroscopy (EDS) or energy dispersive X-ray analysis (EDXRA). SEM/EDS analysis is very useful in the identification of the locations and amounts of certain atomic elements within a sample (Hill and Sawhney 1971, Bisdom *et al.* 1983). This technique is performed within a modified electron microscope, taking advantage of the fact that the interaction of electron beam with

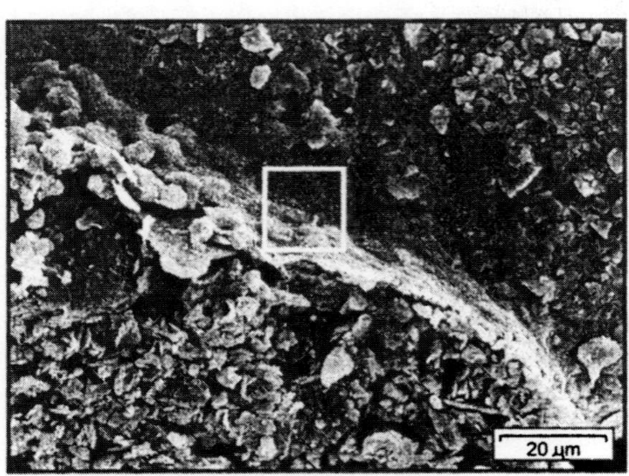

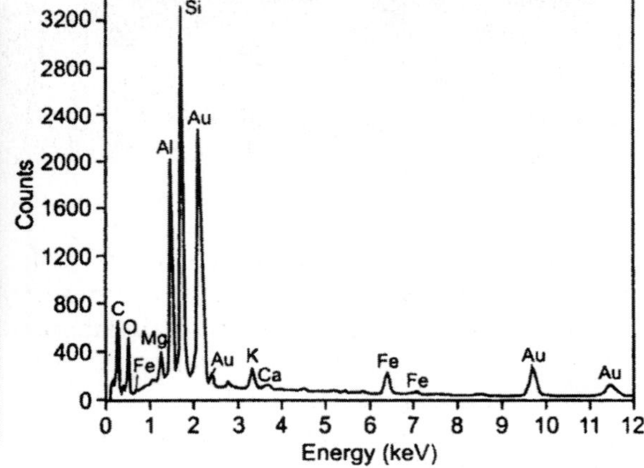

Fig. 2.19 Example of an energy dispersive spectroscopic (EDS) spectrum collected from a soil sample with an amorphous cutan overlying a matrix of fine silt and clay particles. SEM image and associated spectrum by B. Weisenborn.

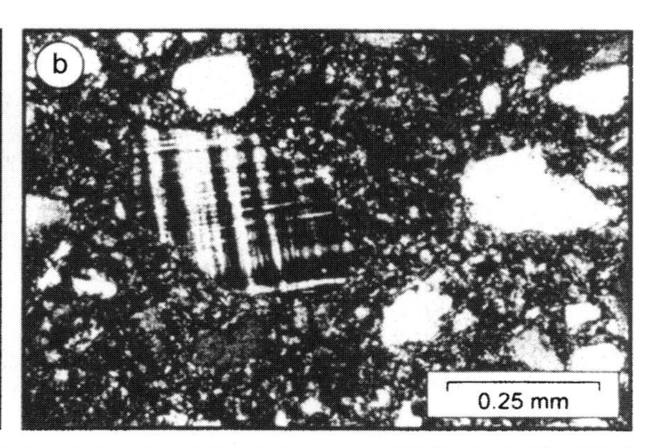

Fig. 2.20 Photomicrographs of twinned feldspar minerals in a soil thin section. The black–white stripes, or twinning pattern, alternate as the microscope stage is rotated. Cross polarized light. Images by D. L. Cremeens. (a) Plagioclase feldspar from the 3BCg horizon of a Typic Endoaqualf, Adams County, Illinois. Note the preferential weathering of one twin vs. the other. (b) Microcline feldspar from the Bt1 horizon of a Udollic Endoaqualf, Clinton County, Illinois. Note the apparent lack of weathering.

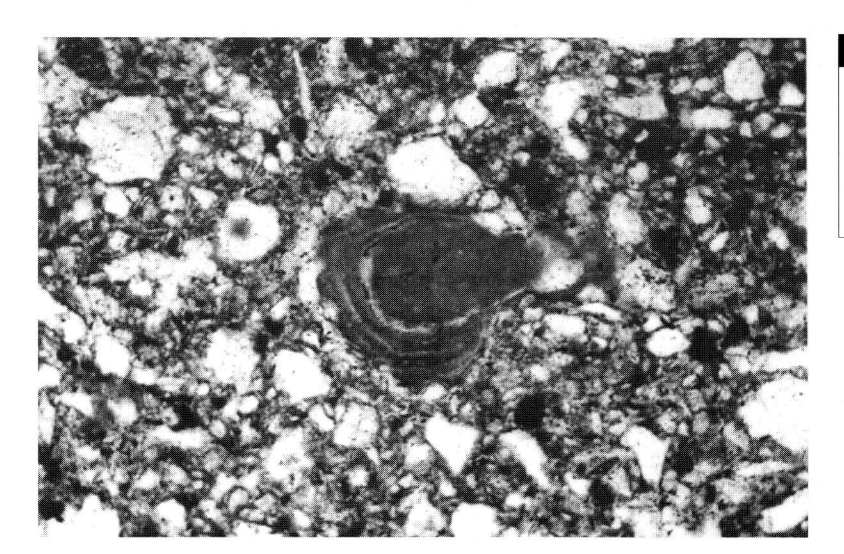

Fig. 2.21 Photomicrograph of a vugh or pore infilled with illuvial clay from the BC horizon of a Mollic Hapludalf, Clinton County, Pennsylvania. Plane polarized light. Width of the frame is 0.34 mm. Image by D. L. Cremeens.

the sample also generates X-rays characteristic of the elements present. It permits non-destructive, quantitative analysis of elements with atomic numbers >11 (Na) and semi-quantitative analysis of elements down to atomic number 4 (Be) (Sawhney 1986). These data can be displayed as element maps, showing where particular elements in the sample are more dominant than elsewhere (EDXRA), or they can be displayed as an EDS spectrum, illustrating the relative amounts of various elements within a given part of the sample (Norton *et al.* 1983) (Fig. 2.19). The energy or wavelength and intensity distribution of the X-ray signal emitted from the sample, and the precise elemental composition of materials (minerals, in this case) can be obtained with high spatial resolution. Thus, detection and mapping of the locations where various types of X-rays were produced on the sample provides a potential means whereby minerals and amorphous materials can be identified (Fig. 2.19). Commonly called microprobe analysis, EDS has many uses in soil micromorphological and paleoenvironmental analysis (Perry and Adams 1978, Sawhney 1986).

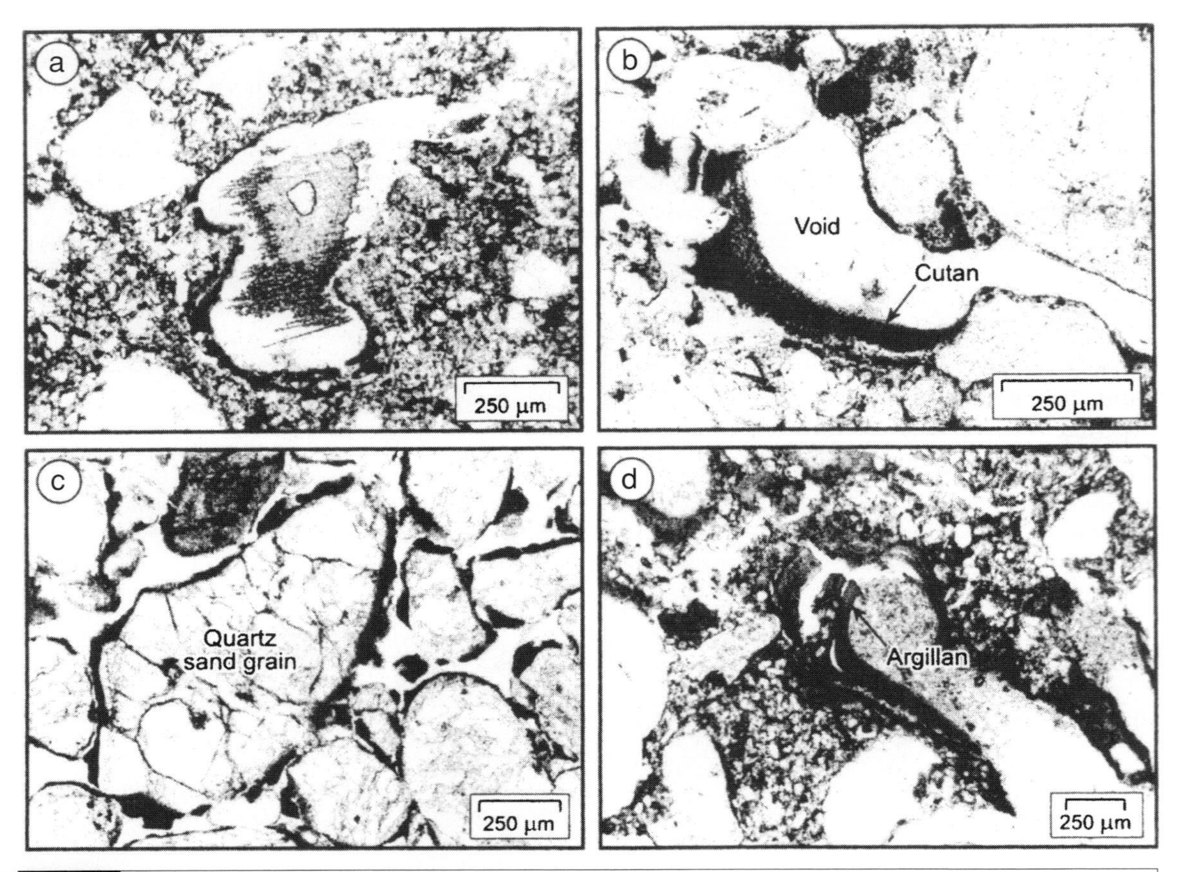

Fig. 2.22 Photomicrographs of some soil thin sections from Spodosols and Alfisols in Michigan, illustrating the various mineralogical and morphological features that can be viewed in this manner. Images by B. Weisenborn. (a) A weathered "ghost" of a mineral. Its previous outline remains, as indicated by a relict argillan. (b) A cutan rich in humus (organ or organan). (c) Cracked organo-metallic coatings on sand grains in a Spodosol. This type of accumulation is diagnostic for spodic horizons. (d) Thick, broken, laminated argillans set within a silt–sand matrix.

Another form of electron microscopy involves the transmission electron microscope (TEM). Very useful in studies of mineralogy, the TEM provides higher magnifications and greater detail than does the SEM (Chartres 1987). For that reason, its main application has been on weathering-related studies of minerals (Calvert *et al.* 1980b, Ahn and Peacor 1987, Dong *et al.* 1998, Vacca *et al.* 2003).

A third, and perhaps intermediate, type of analysis involves the examination of soil and rock thin sections under a petrographic microscope (FitzPatrick 1993). This method requires a significant amount of preparation, but the rewards can be great. Bulk samples a few centimeters in width and length are removed, intact and as little disturbed as possible, from the soil profile. They are dried and then impregnated with a clear polyester or epoxy resin (Innis and Pluth 1970, Middleton and Kraus 1980). Impregnation of the sample without significantly changing its morphology is always tricky. Usually it is accomplished by immersing the dry sample in the resin and then applying a vacuum, which slowly draws air out of the sample while the resin fills the pores. After the resin is allowed to harden, one or more thin sections of about 25–30 μm thickness are cut from it, polished, and then mounted on glass slides. Depending on the application, the thin sections can be oriented in any direction. For example, if most of the pores are oriented vertically and the investigator wants to examine pore linings, horizontally oriented thin sections would be useful. If, however, the S-matrix of ped interiors was the item of most interest, thin sections might be cut vertically. Thin sections can also be examined under the SEM.

When examined under the petrographic microscope, soil thin sections can provide a plethora of information (Fig. 2.20). Pores and pore linings are clearly visible, and minerals are readily discriminated in the polarized light of the microscope (Fig. 2.21). If desired, the thin section can be stained, bringing out the locations of certain minerals such as feldspars (Reeder and McAllister 1957, Norman 1974, Houghton 1980, Morris 1985, Ruzyla and Jezek 1987). Laminated coatings of clay (argillans) usually appear speckled with various colors as the stage is rotated. Edges of minerals provide information about weathering.

The disadvantage of this technique centers around the long preparatory times, the potential for failure during impregnation and the inability of the analyst to view the soil in three dimensions (Fig. 2.22). Advantages include the ability to see inside mineral grains and a much greater ability to identify minerals than either of the two previous methods provided.

Chapter 3

Basic concepts: soil horizonation . . . the alphabet of soils

Regolith, residuum and the weathering profile

Soil is different than sediment. Soil develops *from* sediment, *in* sediment. Knowing how to distinguish the two is critical to the application of soils knowledge (Mandel and Bettis 2001). Soils develop within weathered, unconsolidated (parent) materials at the Earth's surface, under the influence of biota and climate. Soil cannot form within solid rock, but can form within the weathered by-products of that rock.

Loose, unconsolidated material at the earth's surface is *regolith*. It can have one of two origins: (1) formed in place as bedrock weathers, or (2) transported to a site by gravity, water, wind, ice or another vector (Fig. 3.1) (see Chapter 8). The first material is referred to as residual regolith, or simply, *residuum*. It is most common on stable uplands in unglaciated parts of the world; in glaciated regions it is either deeply buried or long-since eroded. Transported regolith can take many forms, such as alluvium, glacial drift or eolian sand.

Regolith is exposed to the vagaries of climate and is acted upon by biota. In this process, it is eroded, transported, deposited, and most importantly (to soil formation) weathered, which encourages various components of it to be reorganized and translocated internally. Residual regolith is weathered out of rock, and therefore all parts of it can be considered weathered, but some types of transported regolith are little weathered, especially if buried deeply. Thus, immediately below the soil profile one might observe zones in transported regolith that are only slightly altered, primarily by leaching and/or oxidation. At still greater depth within transported regolith, certain parts may be essentially unmodified (unleached, unoxidized and not neoformed). This is particularly common in clay-rich glacial drift (Tandarich *et al.* 1994). Above this unweathered *D horizon* occur various zones of weathered regolith; together they constitute the *weathering profile* (Tandarich *et al.* 2002). The weathering profile was defined and developed by geologists and viewed as a part of their realm (Kay and Pearce 1920, Kay and Apfel 1928, Leighton 1958). Recently, Hallberg *et al.* (1978a) subdivided the weathering profile, based on work in Iowa. From the top down, they observed three zones: (1) oxidized and leached, (2) oxidized and unleached and (3) unoxidized and unleached (Bettis 1998), including the soil profile in only the uppermost of these three zones of the weathering profile. The lowermost of these is referred to as the D horizon (Fig. 3.2).

Weathering profiles grade downward into bedrock or unweathered sediment (Figs. 3.1, 3.2). The interface with the bedrock may be abrupt, as for example where glaciers have scoured it clean, or it may be intermittent or gradational as in sandstone where the weathering profile grades, with depth, to rock fragments of increasing size and, eventually, to solid rock. In residual settings, corestones of solid bedrock may occur within the lower parts of the residuum (Berry and Ruxton 1959) (Figs. 3.2, 3.3). In glacial landscapes, where

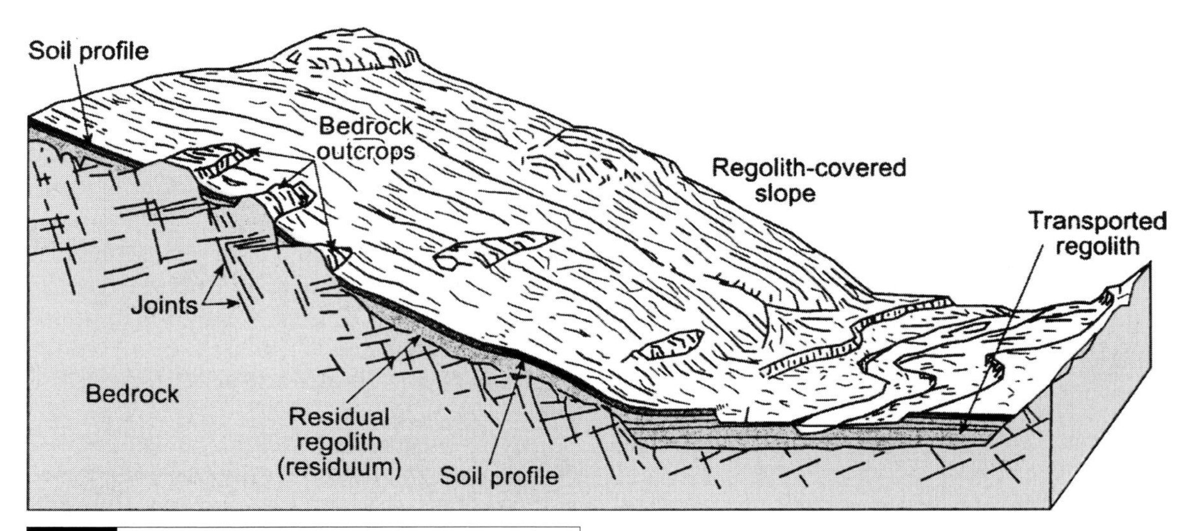

Fig. 3.1 Diagram of a regolith-covered slope, illustrating the difference between bedrock, residual regolith and transported regolith. After Strahler and Strahler (1992).

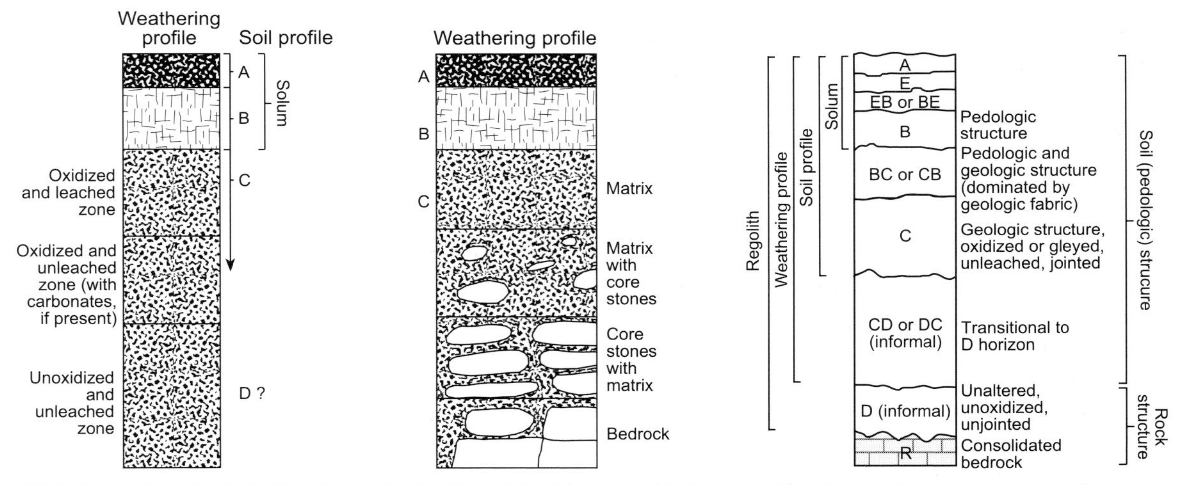

Fig. 3.2 Comparison of soil profile vs. weathering profile terminology. In two of the profiles, unaltered (D horizon) material exists at depth. After Ruhe (1975b) and Tandarich *et al.* (1994).

unaltered, unleached and unoxidized material can exist at depth, the weathering profile is more complex (Fig. 3.2). In general, within the weathering profile, zones closer to the surface are more weathered than are deeper zones (Phillips 2001b). Eventually, the sediments nearest the surface have been so modified by near-surface processes that they can be considered soil.

The soil profile, pedon, polypedon and map unit

Soil geographers and pedologists are interested in the soil profile and soil pedon (Anderson 1987). Originally called a soil section or vertical cut, the

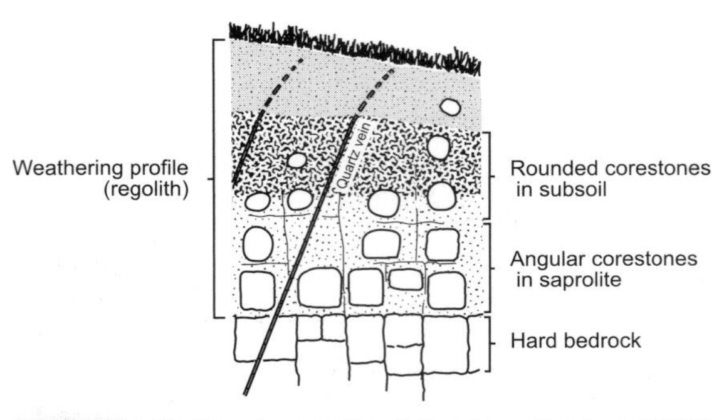

Fig. 3.3 Features of a weathering profile formed in residuum. After Ollier and Pain (1996).

soil profile is a side view of the soil from the surface down (Tandarich *et al.* 2002) (Fig. 3.4). The soil profile is exposed whenever a vertical cut is made through the uppermost few meters of the Earth's surface, as seen in house excavations, roadcuts and open graves. The profile is a two-dimensional entity. It can be viewed but not sampled, as it has no volume (Jones 1959). Within the soil profile one may be able to view various genetic layers (horizons, see below) and, if the profile extends deep enough, relatively unaltered material below.

The soil profile is the way one typically views soils, although the soil itself should be considered as something that has volume and variability in three dimensions. Such a fundamental soil body is called the *soil pedon*: the three-dimensional equivalent of the soil profile. It is the smallest soil body that still retains all of the major variability (nature and arrangement of the soil layers, or horizons) of the soil (Johnson 1963, Campbell and Edmonds 1984). The pedon extends downward, through all genetic horizons, and if the genetic horizons are thin, into the upper part of the underlying parent material. It includes the rooting zone of most plants. For all practical purposes, a lower limit of the pedon is bedrock or a depth of about 2 m, whichever is shallower. Many soil scientists now feel that the 2 m limit is much too shallow, as it does not capture all of the deep subsurface properties of the soil.

The pedon is a theoretical construct; it does not actually exist in nature. In theory, a pedon is roughly polygonal and ranges from 1 to 10 m^2 in area, depending on the nature of the soil variability. Where soil has great spatial variability, such as one with prominent horizon cyclicity, the pedon includes one-half of the cycle (1 to 3.5 m). In soils with less variability, the pedon has an area of approximately 1 m^2. If horizons are cyclic over an interval greater than 7 m, each cycle is considered to contain more than one pedon. The range in size, from 1 to 10 m^2, permits consistency of the pedon concept among different observers. The pedon and the biota that exist within and above it have been termed a *tessera* (Jenny 1958).

Pedons are too small to exhibit features associated with the landscape they are set within, such as slope and surface stoniness. To remedy this, the Soil Survey Staff (1975) created the *polypedon* concept (Johnson 1963). The polypedon is the basic unit of soil classification, large enough to exhibit all the soil characteristics considered in the description and classification of soils (see Chapter 7). Polypedons are taxonomically pure, i.e., they contain pedons all of the same soil series. It is a physical soil body bounded by either "not soil" or by dissimilar pedons (a different polypedon). It has a minimum area of 2 m^2 (two adjoining pedons). All of the pedons within a polypedon classify the same at all levels of Soil Taxonomy. In practice, the concept of polypedon is as fleeting and difficult to grasp as is the pedon. Thus, it is often ignored because of the extreme difficulty of finding the boundary of a polypedon on the ground. Soil scientists have therefore turned to classifying pedons, regardless of their limited size, by transferring any required, areally extensive properties of the surrounding landscape to the pedon.

The next logical step to take from profile to pedon to polypedon is the soil *map unit* (aka mapping unit). A soil map unit is an areal or cartographic representation of a polypedon (see Chapter 7). It is a mapping construct, defined and named in terms of the soil taxonomic units it is supposed to contain. Unlike a polypedon, which is a pure taxonomic unit and exists primarily as a concept, a map unit is a real entity. Therefore, map units almost always have inclusions of other soils (see Chapter 7).

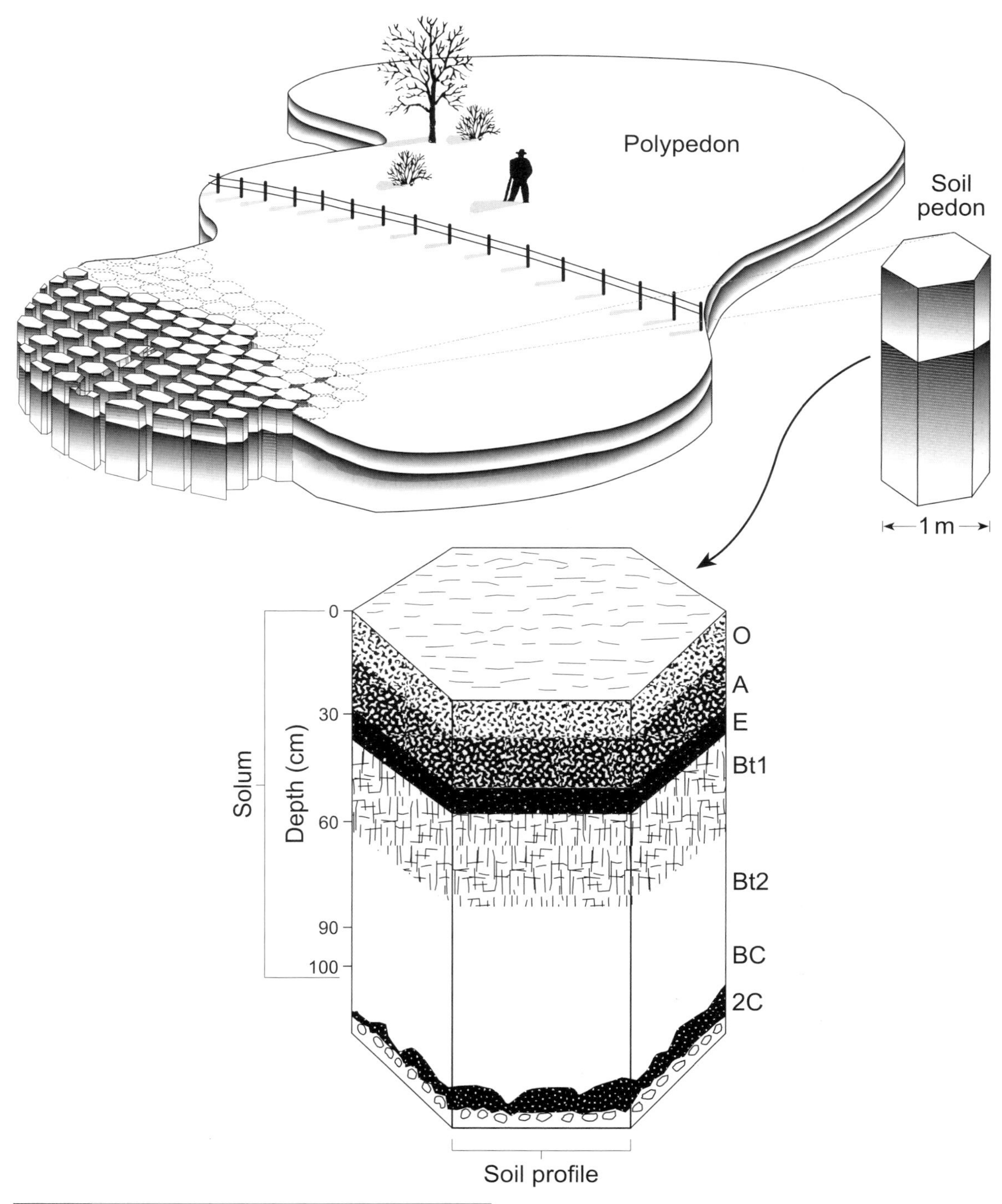

Fig. 3.4 Basic soil concepts: pedon, polypedon, profile, horizons and solum.

Soil horizons and the solum

Most soils exhibit some horizonation, or genetic layering; it is what distinguishes them from not-soil materials. Early soil scientists used terms like soil layer, vegetable mould, stratum and level to describe these genetic layers (Tandarich *et al.* 2002). Since about 1927 (Shaw 1927), a layer formed by pedogenic processes that is more or less parallel to the soil surface has been called a *soil horizon*. Moreover, because its origin is pedogenic, referring to it as a *genetic horizon* is appropriate. Anderson (1987: 56) described soil horizons as the "working aggregates of the whole (soil) system, and, like the organs of an organism, generally are adapted for the performance of specific functions." If a soil's evolutionary pathway dictates that a certain horizon performs certain functions, e.g., accumulate organic matter, for a long period of time that zone will acquire characteristics that set it apart from other zones. In short, this layer/zone will become uniquely different than the zones above and below, and in so doing, becomes a differentiable soil horizon.

Originally defined by Kellogg in 1936, the *solum* is essentially the O, A, E and B horizons (see below). In essence, the solum includes the part of the profile that has clearly been altered by pedogenic processes, i.e., all of the profile above the C horizon (Fig. 3.4).

Horizons may not have yet formed in very young sediments, in sediments that are essentially inert material lacking a vegetative cover or those in extremely dry (hyperarid) climates. On steep slopes, soil horizons are slow to form because of high erosion rates and surface instability. Thus, soil geomorphologists often use the presence of a soil profile with several intact horizons as an indication that the surface upon/within which it is forming has been relatively stable for a period of time (see Chapter 13).

Types of soil horizons

Soil horizons will generally form within unconsolidated materials on stable surfaces that have been subaerially exposed for a sufficient length of time, as material is *added* to or *removed* from parent material. They can also form as material is translocated (upward, downward or laterally) within the profile, or as it is transformed in place (Simonson 1959) (see Chapter 11). These sets of processes tend to form distinct layers within the upper mantle of unconsolidated materials.

Ever since Dokuchaev and Sibirtsev introduced them at the 1893 World's Columbian Exposition in Chicago, soil horizons have been divided into a few types of master horizons: O, A, E, B, C and R (Tandarich *et al.* 1988) (Table 3.1). We also recognize the D horizon but acknowledge that its use is informal. O horizons are organic horizons that form above a mineral soil or bedrock. A horizons (Ah in the Canadian classification system) have traditionally been assigned to "topsoil" layers rich in decomposed organic matter. B horizons have, likewise, been zones of accumulation or alteration in the subsoil, and C horizons have been seen as unaltered parent material. Recently, E (Ae in Canada) has been assigned to light-colored, eluvial horizons (Guthrie and Witty 1982). R horizons are hard bedrock.

Master horizons are often given descriptive suffixes that provide additional information about their characteristics, such as the presence of an illuviated ("washed in") substance, its degree of decomposition, or its density, among others (Table 3.2). Capital letters are used to denote master horizons, and lower-case letters are used as suffixes, e.g., Ap, Bx and Cd horizons. Horizons may have as many suffixes as are deemed appropriate, e.g., Btx, Bhsm and BCtg (Fig. 3.5).

Transitional horizons (between master horizons) can take two forms (Table 3.3). Gradational horizons occur where one horizon smoothly grades into, or is transitional to, another. The transitional horizon, designated with two capital letters, is characteristically more like the horizon designated first (e.g., AB, BC) (Figs. 3.2, 3.5). Another type of transitional horizon occurs where discrete, intermingled bodies (individual parts) of two horizons exist within one layer. When intermingled bodies of the two horizons are so "congested" that separation of each of the bodies into individual horizons is not justified, the capital

Table 3.1 Master soil horizons

Master horizon	Characteristics
O	Layers dominated by organic material (litter and humus) in various stages of decomposition.
A	Mineral horizons that formed at the surface or below an O horizon and (1) are characterized by an accumulation of humified organic matter intimately mixed with the mineral fraction, or (2) have properties resulting from cultivation, pasturing or similar kinds of disturbance.
E	Light-colored mineral horizons in which the main feature is loss of weatherable minerals, silicate clay, iron, aluminum, humus, or some combination, leaving a concentration of mostly uncoated quartz grains or other resistant materials.
B	Subsurface mineral horizons dominated by (1) illuvial accumulations of clay, iron, aluminum, humus, etc., (2) removal of primary carbonates, (3) residual concentrations of sesquioxides, (4) distinctive, non-geologic structure and/or (5) brittleness.
C	Mineral horizons, excluding hard bedrock, that have been little affected by pedogenic processes and lack properties of O, A, E or B horizons. Most C horizons are mineral soil layers and retain some rock structure (if developed in residuum) or sedimentary structure (if developed in transported regolith). Included as C horizons are deeply weathered, soft saprolite (see Chapter 8).
D	Deep horizons that show virtually no evidence of pedogenic alteration, such as leaching of carbonates or oxidation. D horizons retain geologic structure and are often dense and slowly permeable. Like C horizons, D horizons are formed in unconsolidated sediments.
R	Hard, continuous bedrock that is sufficiently coherent to make digging by hand impractical.

Source: Modified from Guthrie and Witty (1982).

letters of the two master horizons are separated by a slash, e.g., E/B.

Arabic numerals (1, 2, 3, etc.) are used both as suffixes to indicate vertical subdivisions within a horizon and to indicate changes in morphology or chemistry within a horizon (e.g., Bt1, Bt2 or Eg1, Eg2). Arabic numerals (Roman numerals in Canada) are also used as prefixes to indicate lithologic discontinuities, e.g., 2C, 3C. Lithologic discontinuities separate different parent materials based on criteria such as mineralogy or texture, e.g., when loess overlies glacial till (Fig. 3.5) (see Chapter 8). All horizons within the lower parent material must be preceded by the number 2, and if a third parent material should also be present, horizons formed in it would be preceeded with a 3. By convention, horizons formed in the uppermost parent material are not preceded by a 1 (I in Canada).

A prime (′) symbol is used to indicate the second occurrence of a horizon within the same profile. If the second occurrence of the horizon is immediately below the first, an Arabic numeral suffix is used (e.g., E1, E2, Bs, Bx), but if the two horizons are separated by different horizons, the prime is used to distinguish the two horizons (e.g., E, Bs, E′, Bx). The prime is used for horizons that have genetic links, or are forming in the same profile; if the lower E, for example, is from a buried profile from an earlier period of soil formation, the prime would not be used. Instead, it would be designated an Eb horizon (the b is for buried; see Table 3.2).

O horizons

O horizons are dominated by organic, not mineral material (Table 3.1). They include all organic materials and decomposing debris lying on the mineral soil surface and are usually located at the top of the profile, as the uppermost master horizon, if present at all. If the soil has been plowed,

Table 3.2 Suffixes for master soil horizons

Suffix	Used with which master horizons?	Characteristics/comments	Approximate Canadian equivalent
O horizon suffixes			
a	O	Highly decomposed organic material. "Sapric" material.	Oh in Organic soils, H in forest litter
e	O	Organic material of intermediate decomposition. "Hemic" material.	Om in Organic soils, F in forest litter
i	O	Slightly decomposed organic material. "Fibric" material.	Of in Organic soils, L in forest litter
A horizon suffixes			
p	A	Plowed, tilled or otherwise disturbed surface layer. Disturbed A, E, or B horizons are designated Ap.	p
v	A	Porous, vesicular horizon, common to desert regions. Informal designation, not recognized by the Natural Resources Conservation Service.	
B horizon suffixes associated with illuviation			
c	B, C	Presence of concretions or hard non-concretionary nodules, usually of Fe, Al, Mn or Ti.	cc
h	B	Dark, illuvial accumulations of organic matter and humus. The moist Munsell value and chroma of the horizon must be 3 or less.	h
j	B	Accumulation of jarosite, either as ped coatings or nodules.	
k	B, C	Accumulation of pedogenic carbonates, commonly $CaCO_3$, as ped coatings, filaments or nodules. From German *kalk* (lime).	ca
n	B, C	Accumulation of pedogenic, exchangeable sodium (Na), commonly as sodium salts. From German *natrium* (sodium).	n
q	B, C	Accumulation of pedogenic, secondary silica (quartz).	
s	B	Accumulation of sesquioxides of Fe and Al.	f
t	B, C	Accumulation of silicate clay, as evidenced by argillans on ped faces or lamellae (clay bands). From German *ton* (potter's clay).	t
y	B	Accumulation of pedogenic gypsum. From Spanish *yeso* (gypsum).	s or sa
z	B	Accumulation (pedogenic) of salt more soluble than gypsum.	
Other B horizon suffixes			
w	B	Development of color or structure in a horizon but with little or no apparent illuvial accumulation of materials.	m

(cont.)

Table 3.2 *(cont.)*

Suffix	Used with which master horizons?	Characteristics/comments	Approximate Canadian equivalent
x	E, B	Presence of fragipan characteristics (genetically developed firmness, brittleness and/or high bulk density). Generally, fragipans are root-restrictive.	x
Suffixes associated with cold soils			
f	Any except uppermost	Frozen subsoil horizon which contains permanent, continuous ice (permafrost). Not used for layers that are frozen only seasonally.	z
ff	Any except uppermost	Frozen subsoil horizon which contains "dry" permanent, continuous ice ("dry permafrost"). Not used for layers that are frozen only seasonally.	z
Suffixes indicating pedoturbation			
ss	A, B, C	Presence of slickensides.	v
jj	Any	Horizon showing evidence of cryoturbation.	y
Other suffixes			
g	E, B, C	Strong gleying in which Fe has been reduced and/or removed or in which Fe has been preserved in a reduced state because of saturation with stagnant water. Most gleyed layers have a moist Munsell chroma of 2 or less.	g
o	B, C	Residual, pedogenic accumulation of sesquioxides.	
v	B	Plinthite (iron-rich, humus-poor, reddish material that is firm or very firm when moist and that hardens irreversibly when exposed to the atmosphere). Not used in A horizons.	
r	C	Weathered or soft bedrock, including saprolite or dense till, that roots can easily penetrate along joint planes. Sufficiently incoherent to permit hand-digging with a spade.	
d	A, B, C	"Dense" horizon with high bulk density and physical root restriction. Low numbers of connected pores. Examples: dense basal till (Cd) or plow pans (Ad).	
m	B	Continuous or nearly continuous cementation or induration of the soil matrix by, for example, carbonates (km), silica (qm), sesquioxides (sm), or carbonates and silica (kqm).	c
b	A, E, B	Buried horizon of a mineral soil. Implies that the horizon is part of a paleosol.	b

Sources: Guthrie and Witty (1982); Canadian nomenclature after the Expert Committee on Soil Science (1987).

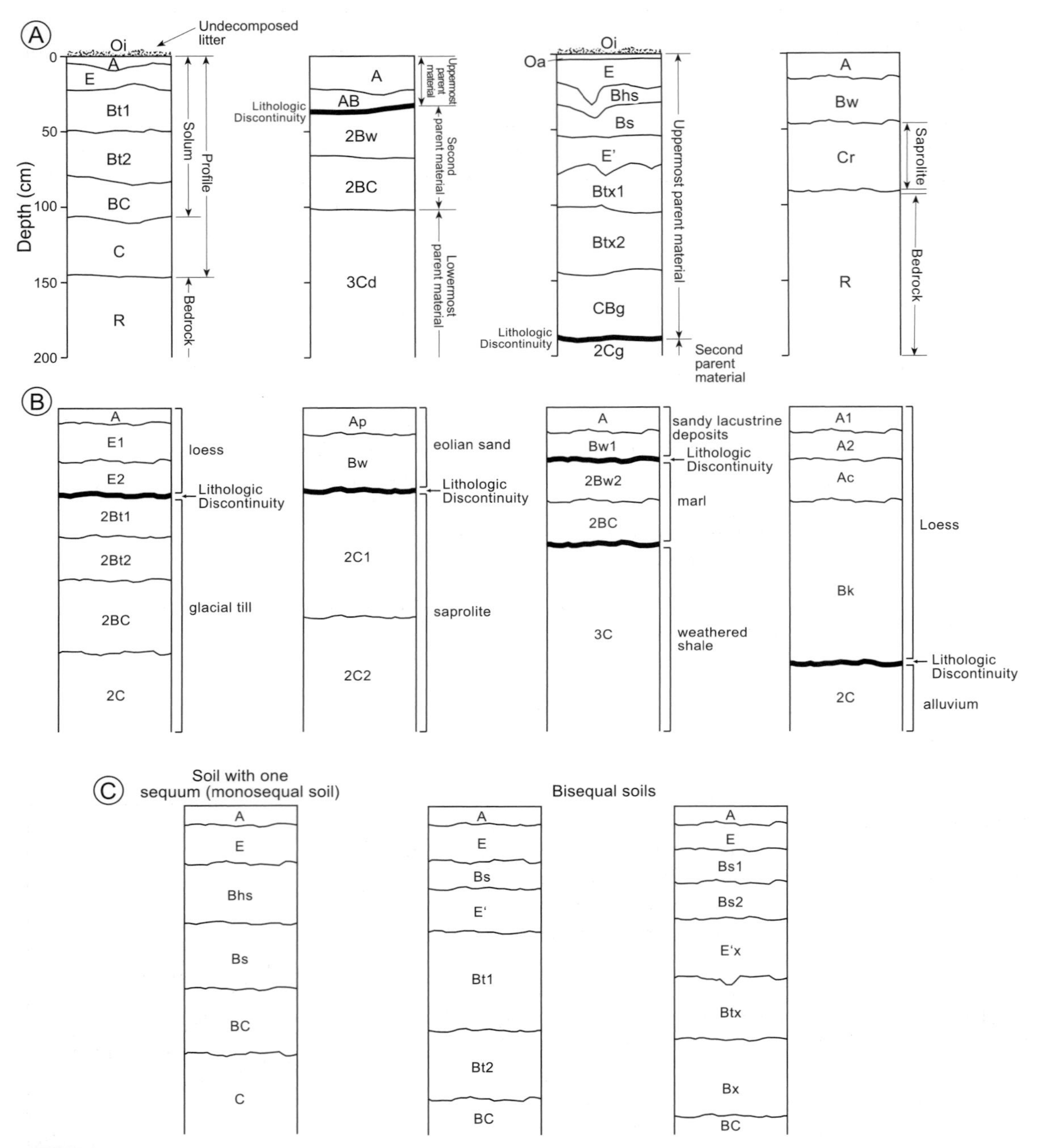

Fig. 3.5 Various combinations of soil profile horizonation, illustrating how soil horizon terminology functions in different types of soils.

O horizons are usually missing because they have been incorporated into the plow (Ap) horizon. In areas where organic production is low or decomposition rates are high, O horizons may not be present on the soil surface at all times of the year because they have decomposed. O horizons provide a buffer between the mineral soil and the atmosphere. They insulate the soil below from extremes of temperature and moisture, and provide mechanical protection from raindrop impact, runoff and other erosional forces. When present, they facilitate infiltration.

Table 3.3 Transitional soil horizons

Type of transitional horizon	Characteristics	Examples	Possible types
Gradational	Used where one horizon grades or transitions into another. The transitional horizon is characteristically more like the horizon designated first.	AE: Transitional between A and E, but more like the A horizon. BC: Transitional between the B and C, but more like the B horizon.	AB, AE, AC EA, EB BA, BE, BC CA, CB
Mixed/interrupted	Used where discrete, intermingled bodies (individual parts) of two horizons exist within one layer. Intermingling is so "congested" that separation of the bodies into individual horizons is not justified.	E/B: composed of individual parts of E and B horizon components in which the E component is dominant and surrounds the B materials.	A/E, A/C, A/B E/A, E/B B/A, B/E, B/C C/B

Source: Guthrie and Witty (1982).

O horizons form as organic debris falls to the surface and accumulates. Although this debris or litter is produced most rapidly in warm and wet climates (Fig. 3.6), the amount in the O horizon at any one time is actually in a state of dynamic equilibrium (Sharpe *et al.* 1980). Inputs of litter may be continuous throughout the year, although in most ecosystems the main influx of dead organic debris to the soil surface precedes a dry or cold season (Fig. 3.7). For this reason, O horizons usually attain their greatest thickness immediately after leaf fall. Although leaf fall is temporally predictable, additions of twigs and other woody debris are not, and occur randomly throughout the year. Wind and ice storms are responsible for major additions of non-leafy litter; these inputs vary greatly in time and space. The "inputs side" of the equation describing the amount of organic matter in O horizons is balanced by decomposition, or turnover, which also has seasonal biorhythms. Because decomposition is also high in warm, wet climates, O horizons tend to be thin there. They are thickest in cool, dry climates, or locally where

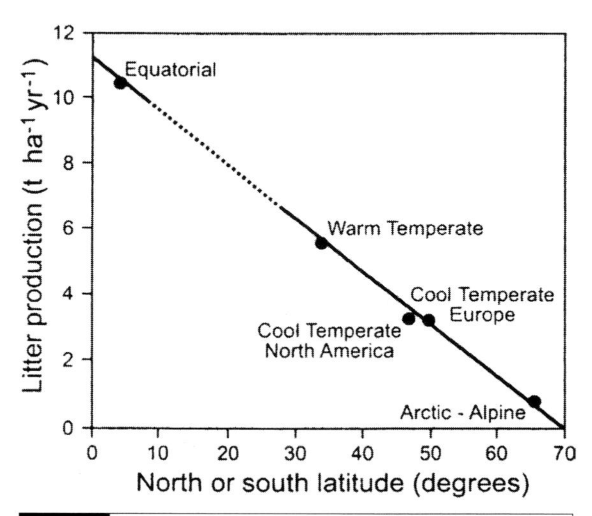

Fig. 3.6 Annual production of litter in relation to latitude (humid climate areas only). Area where the line is dotted reflects the latitude where many deserts exist, and litter production is not directly reflected by the line. After Bray and Gorham (1964).

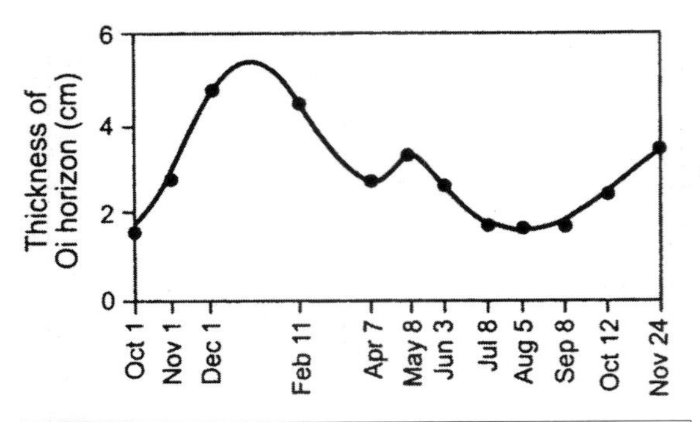

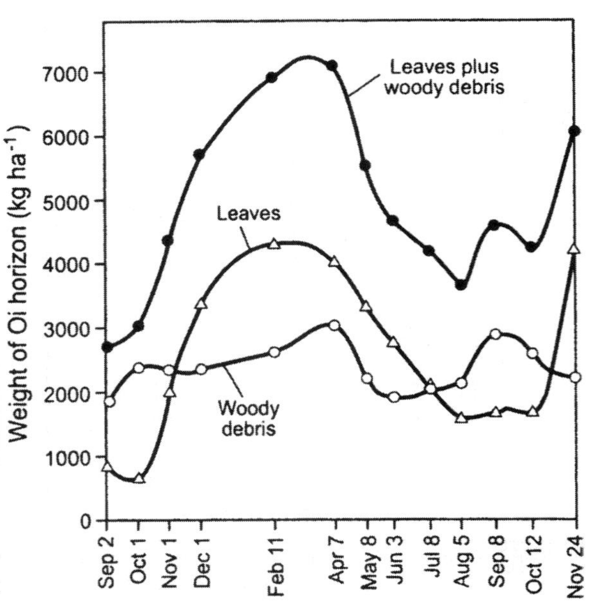

Fig. 3.7 Thickness and mass data for Oi horizons in a well-drained, silt loam soil (fine-silty, mixed, mesic Typic Hapludalf) in Wisconsin, throughout the year. After Nielsen and Hole (1964).

standing water and acidic litter conditions inhibit decomposition. This latter situation is best epitomized in bogs of mid and high latitudes, where O horizons are meters thick, and the soils are mapped as deep Histosols (see Chapters 8 and 9).

O horizons form quickly and can be lost just as quickly. It is not unusual for O horizons on fertile forested sites with abundant moisture and microorganisms (especially worms; see Kevan 1968) to be completely decomposed and incorporated into the mineral soil below by midsummer of the following year. On a longer timescale, Albert and Barnes (1987) found that the O horizon of a Michigan forest had reformed 50 years after the forest had been clear-cut. The O horizon in the virgin stand was thinner and weighed less per unit area than did the O horizon in the stand that had been cut 50 years previous. Schaetzl (1994) studied a chronosequence of sites in the Great Lakes region, where thick O horizons develop on sandy Spodosols. His study design centered around a series of plots that had been burned; after the burn the forest and the O horizon regenerated. His data indicated that O horizons reform very quickly after fire; some sort of steady state is achieved in little over a century (Fig. 3.8). Thus, at least for midlatitude sites, O horizons appear to be able to recover from disturbance relatively quickly. Similar data are available for O horizons in different ecological settings (Peet 1971, Fox *et al.* 1979, Jacobson and Birks 1980, O'Connell 1987). Nonetheless, few data exist on endpoints of O horizon development: (1) the potential thickness that an O horizon will attain, given freedom from disturbance and (2) the thickness that the O horizon would be at that steady-state endpoint. O horizons in cooler climates probably take longer to reach a steady state and attain greater thicknesses than in tropical or subtropical areas (cf. Raison *et al.* 1986, O'Connell 1987, Schaetzl 1994).

Types and subdivisions of O horizons

O horizon suffixes refer to the degree of decomposition of the organic material within (Table 3.2). Raw or nearly raw organic material is called *fibric*, for the dominance of plant fibers. Oi horizons are composed of predominantly fibric material (*i* is the second letter in fibric). Highly decomposed O horizon material is given the adjective sapric and horizons dominated by sapric materials are designated Oa. Note again that the second letter in sapric is *a*. Those with intermediate decomposition levels (hemic) are Oe horizons. In old literature, Oi, Oe and Oa horizons were referred to as L, F and H layers, respectively, referring to Litter, Fermentation (or *Formultningsskiktet*) and Humus (or *Humusamneskiktet*), or A_O and A_{OO} horizons. Raw organic material on the soil

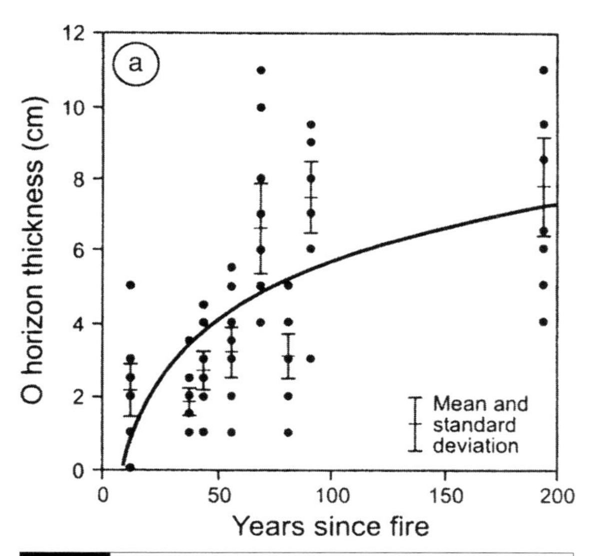

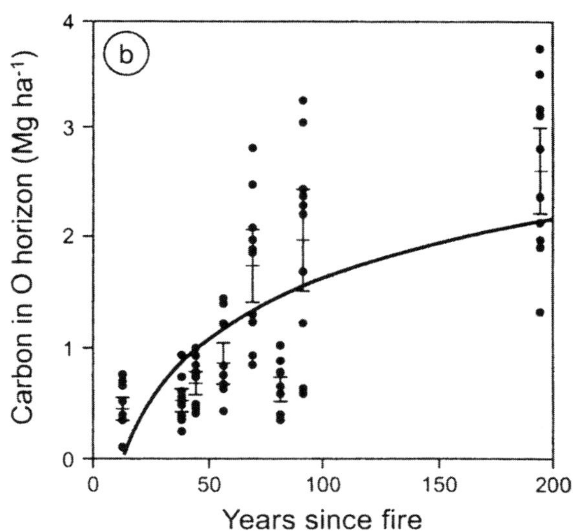

Fig. 3.8 Changes in O horizon character on some burned but previously forested plots in northern Michigan. Ten samples, represented as dots, were taken for each plot; mean and (standard deviation/2) values are also shown as error bars. After Schaetzl (1994). (a) Thickness of the Oi plus Oe horizons for plots of different ages. (b) Carbon stored in the O horizon for plots of different ages.

surface is still referred to as *litter*. And the O horizon is still, collectively, referred to as the *forest floor* by foresters and biologists (Pritchett 1979, Fox *et al.* 1987, Melillo *et al.* 1989). This classification has its problems, however, since some foresters include the A horizon in with the forest floor and the humus layer.

The terminology used to describe O horizons has a long and storied history (Chertov 1966). Terms such as raw humus, peat, acid humus, leaf mold, duff, mor, moder and mull are but a few of the terms that have been used to describe organic material on the surface of mineral soils (Romell and Heiberg 1931, Heiberg and Chandler 1941, Wilde 1950, 1966), beginning as far back as the late nineteenth century (Muller 1879). The earliest classification, still in use today in some parts of the world, divides litter layers into mull (Danish *muld*) and mor (Danish *mor* or *maar*) types. *Mull* humus is dominated by bacterial decomposition at the microscale and intense worm activity at macroscales. It has a high (>5.0) pH, with a crumb-like structure and diffuse lower boundary (Koshel'kov 1961). Worms help to form the character of the mull O horizon by mixing raw organic materials with mineral material from the A horizon below, and as this mixed material is excreted it helps form the crumb-like structure typical of mull (Bernier 1998). The more acidic, slower to decompose, *mor* type of humus is associated with fungi-dominated decomposition and a general lack of macrofauna. Mor usually forms in coniferous litter, with its matted roots, needles and litter fragments (Heiberg and Chandler 1941). Some mors have pH values below 4.0. Phenolic substances, common in vegetation of low base saturation, may form protective coatings on plant cellulose materials, further retarding decomposition (Kevan 1968). The large amounts of phenolic substances also render the litter unpalatable to earthworms, limiting their role in the decomposition process. Rather, fungi, protozoa, collembola and mites are responsible for much of the (slower) breakdown of acid, needleleaf litter into mor humus.

Earthworms and arthropods are seen as vital to the development of mull humus from calcium-rich, broadleaf litter. Thus, mull humus has significant incorporations of mineral matter, while mor humus is essentially all organic matter, with much less mixing between the O horizon and the mineral soil. In fact, thick mats of mor humus are often stratified into layers of varying degrees of decomposition (Pritchett 1979). In sum, mull humus is typical of higher pH, base-rich soils

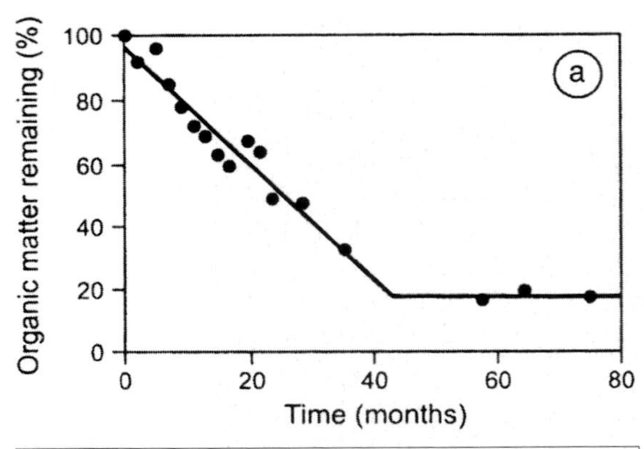

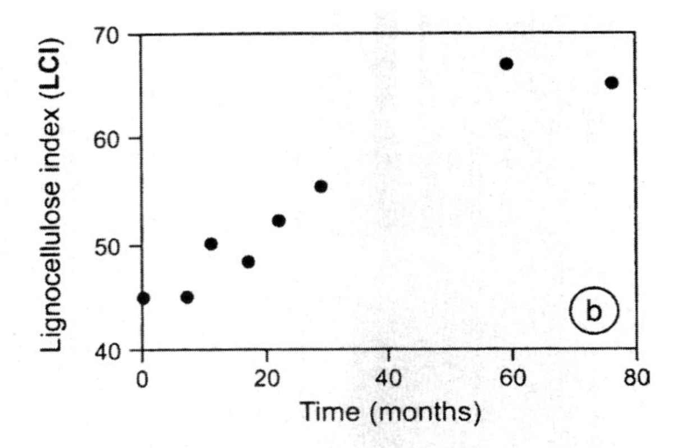

Fig. 3.9 Changes in an O horizon dominated by red pine (*Pinus resinosa*) litter over an 80-month timespan. After Melillo *et al.* (1989). (a) Percentage of the original mass of litter vs. time. (b) Lignocellulose index (LCI) vs. time. The LCI is the percentage fraction of lignin in the lignocellulose component of the decaying litter.

under broadleaf forests while mor humus characterizes the more acidic environments of coniferous stands. Litter of intermediate character is termed *moder* or duff mull (Hoover and Lunt 1952).

Biological degradation of raw litter in O horizons

The process whereby fresh litter is changed, fragmented and decomposed to humus, i.e., *humification*, is continuously ongoing in O horizons, unless they are extremely dry or frozen (see Chapter 12). Breakdown and decomposition are perfomed by soil macro and microorganisms, whose activity is in turn a function of temperature and moisture. Warm, moist environments favor litter decomposition and humus formation, and also facilitate high levels of productivity of the terrestrial plants that produce the litter. A useful way to envision this breakdown/decomposition process is to realize that both macro- and microorganisms are feeding on it simultaneously, with the most palatable and easily decomposed materials going first, while some of the larger and more resistant materials must first be fragmented, perhaps several times, by macrofauna such as millipedes. With time, the mass of organic material becomes more and more decomposed, until it is completely humified.

After the raw organic debris (fibric material) falls to the surface, it immediately begins to be consumed by organisms of all sorts and sizes as a food (energy) source (Burges 1968) (Fig. 3.9); fungi are the first microscopic invaders. Soil macrofauna also ingest the litter and in so doing increasingly fragment it (Schulmann and Tiunov 1999). The fragmentation process conditions the litter for further decomposition by microbial populations in that it increases its surface area and usually mixes it with soil materials and other microorganisms (Gunnarsson *et al.* 1988). Earthworms are particularly good at fragmentation; litter decomposes much faster in the presence of earthworms, particularly those in the family Lumbricidae. The worms comminute the organic matter they ingest and mix it with mineral soil; their casts are fertile ground for microorganisms and are well known as sites of enhanced microbial activity within the upper profile (Kevan 1968). Actinomycetes and bacteria multiply within earthworms, and this growth continues in the feces (Kevan 1968). Earthworm casts are also important because, if excreted at the surface or within the O horizons, they bury other litter and create a more favorable microenvironment for its further decomposition.

Equally important in fragmentation and mixing processes are millipedes, collembola, mites and isopods (Cárcamo *et al.* 2000) (see Chapter 6). Fragmented and partially digested litter that has passed through the guts of the larger soil fauna is a favored substrate for soil bacteria. Collembola are also important in this regard,

Fig. 3.10 Overthickened O horizons, composed of leaves and moss, on top of limestone boulders in the upper peninsula of Michigan. Isolation of this organic matter from soil macrofauna and the high pH of the rock have impeded its decomposition. Photo by RJS.

as they readily consume the fecal pellets of larger soil arthropods and break this material down further; *coprogenic* materials make up a large part of Oa horizons. In tropical areas where termites are abundant (see Chapter 10), soils have low organic matter content. One reason for this is that termites are so efficient at digesting cellulose that little is left for the soil. This situation contrasts with midlatitude environments where macrofauna like millipedes, mites and earthworms pass most of the organic materials through their guts with most of the material remaining in organic form; O and A horizons are well stocked with humus.

Eventually, deposits of excrement from soil meso- and macrofauna accumulate. According to Kevan (1968), the fecal matter is, at first, chiefly pellets of collembola and mites. The fecal pellets are again fed upon by thecamoebae and enchytraeid worms, as well as collembola (again), which reduce it to an amorphous, black material called *humus*. Thus, multiple cycles of ingestion, fragmentation, mixing and excretion must occur for all the litter to be in a finely divided and well-decomposed state.

Regardless of how much physical fragmentation occurs, however, eventually litter must be decomposed down biochemically, so that the C—H bonds within get broken. In this regard, hemicelluloses and celluloses get broken down by microbes, while substances such as lignin, plant cuticle, animal chitin and related substances persist (Kevan 1968). Lignin and chitin are eventually broken down by fungi, but some soil fauna can assist through the action of intestinal symbionts.

In situations where the litter is isolated from most soil macrofauna, as happens where the soil is extremely shallow to bedrock, the decomposition process is much slower and is dominated by microfauna. Litter will simply accumulate and thicken to the point that Folists (organic soils over bedrock or gravel) will develop (Fig. 3.10).

Like any organic residue, under favorable conditions soil organisms will utilize it as a source of carbon and energy (Kevan 1968, Martin and

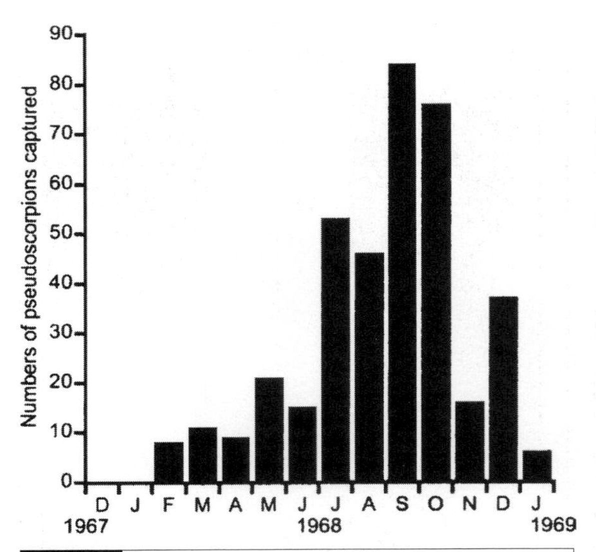

Fig. 3.11 Numbers of *Chelonethi* pseudoscorpions captured monthly, from an O horizon in an ash–oak forest in Great Britain. Pseudoscorpions are carnivorous soil fauna, smaller than 4 mm in length, which feed primarily on insects, spiders, springtails and mites. After Jones (1970).

Haider 1971). In a Michigan soil, for every gram of live weight of earthworm, 8–27 mg of leaves (dry weight) were consumed per day (Knollenberg *et al.* 1985). At this rate, Knollenberg *et al.* (1985) hypothesized that the worm population at their research site could consume almost 95% of the leaves that fell within 4 weeks. Decomposition is hastened by periodic additions of moisture, by active soil fauna that ingest the litter and excrete it in a partially or wholly digested form, and by warmer temperatures. Likewise, ingestion of litter by soil fauna increases when the soil is warmer and wetter. Raw (1962) calculated that leaf consumption by worms increased as a function of temperature:

$$\mathrm{LC} = 2.61\,(T - 2) \qquad r^2 = 0.98$$

where LC is the rate of leaf consumption in mg dry weight of leaves per gram of live earthworm weight per day, and T is the soil temperature (in °C) at 30.5 cm depth.

Since litter is a primary source of food for many soil organisms, which are in turn a food source for higher organisms, the numbers and mass of litter-dwelling organisms varies systematically throughout the year (Fig. 3.11). Often, litter-dwellers are in greatest numbers in the warm season, such that by autumn (in the mildly acidic and well-drained soils of midlatitudes) the previous year's litter has been mostly consumed. Most litter-feeders will therefore migrate into the mineral soil to estivate, or lay eggs and die, before the onset of a dry or cold season.

A horizons

The A horizon, colloquially referred to as topsoil, is the dark, uppermost, *mineral* soil horizon (Table 3.1). Mineral particles in the A horizon have dark coatings of humus and organic matter in advanced stages of decay. Isolated fragments of organic matter, dead plant and animal fragments, fecal pellets, seeds and pollen grains, as well as opal phytoliths are also commonly found within the A horizon. Most A horizons form beneath an O horizon; the latter contributes partially decomposed organic matter to the A horizon (Fig. 3.12). Organic materials mixed into the A horizon then further decompose into humus, which is a brown to black material that coats the soil particles. Mixing of humus and organic materials into the mineral soil can occur as soil animals drag plant

Fig. 3.12 Soil from northern Michigan, showing a thick O horizon overlying a thin and somewhat wavy A horizon. The light-colored horizon at the base of the photo is the E horizon. Many small sugar maple seedlings have sprouted in the O horizon; most will soon die because of the dense shade of the forest. Photo by RJS.

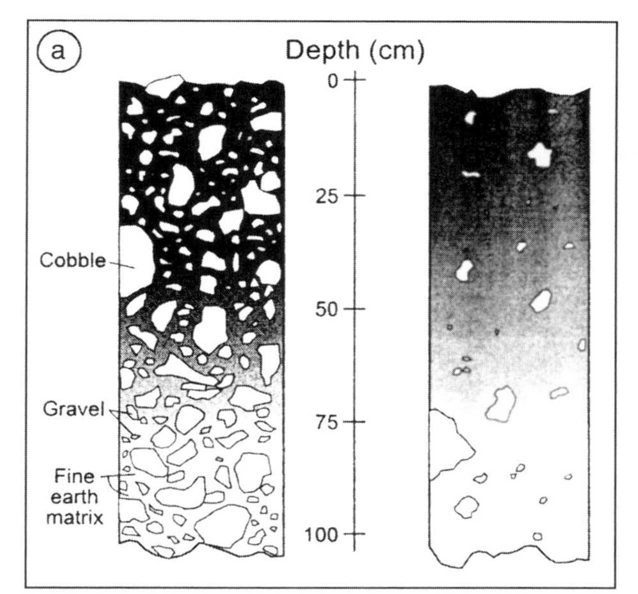

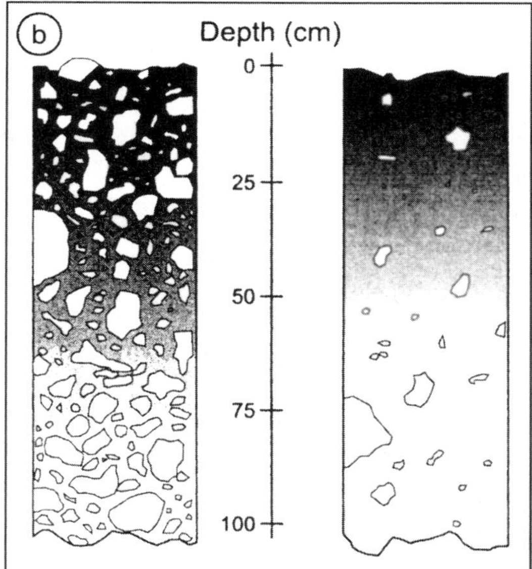

Fig. 3.13 Influence of large amounts of coarse clasts on the formation of A horizons in soils. In gravelly and rocky soils, the A horizon has less "space" in which to form, causing it to (a) accumulate more organic materials in the limited volume than it normally would, and hence be darker, or (b), or develop deeper than it normally would. After Schaetzl (1991b).

matter into their burrows (Nielsen and Hole 1964) or as small plant fragments wash down into pores and cracks and along root channels (Graham and Southard 1983). By depositing casts of mineral soil on and in the O horizon, burrowing animals such as ants and worms incorporate parts of the O horizon into the A, and mix the two; ultimately, this mixed zone will revert to A horizon as the organic materials decompose further. Darkening of the mineral material by additions of organic matter and humus is called *melanization* (Schaetzl 1991b) (see Chapter 12).

In general, the amount of organic matter in the A horizon is dependent on the balance between additions and losses. Additions occur from both above ground as leaf fall, and below ground as roots and soil fauna die and decay in place. Losses include translocation of organic matter to deeper horizons, erosion of organic-rich materials and oxidation/mineralization of organic matter to CO_2. Because the processes of additions and losses vary considerably in space and time, A horizons also vary in characteristics both spatially and temporally. They tend to be darkest where organic matter additions far exceed losses, such as cold or wet sites where oxidation of organics is slow. Grassland soils also have thick dark A horizons, because of the large, ongoing additions of organic matter by roots (Ponomareva 1974). Some forest soils, however, can also exhibit thick, dark A horizons, as for example where the forest is thin and grasses occur in the understory (Afanas'yeva 1966, Nimlos and Tomer 1982), where forest has recently replaced grassland vegetation (Geis *et al.* 1970) or where the forest soil is very rocky (Gaikawad and Hole 1965, Small *et al.* 1990). High amounts of coarse fragments cause soils to have thicker and/or darker A horizons, as the large volume of coarse clasts in the soil limits the volume of soil in which organic matter can be distributed (Schaetzl 1991b) (Fig. 3.13).

In a forested soil, Nielsen and Hole (1964) removed litter from plots for 5 years. Organic matter in the upper part of the soil on this plot decreased, and the number of earthworm middens decreased from 64 000 to 22 000 ha^{-1}. On another site, leaf litter was imported, resulting in a doubling of O horizon thickness. On this plot, organic matter increased, as did midden densities (from 64 000 to 125 000 ha^{-1}). This experiment illustrates the comparatively ephemeral nature of the organic coatings in A horizons, and hence the horizon itself.

Plowed A horizons are designated Ap (Table 3.2). Ap horizons usually have a sharp lower boundary, occurring at the base of the plow zone. In areas where the A horizon is thin, the Ap horizon has incorporated some parts of the lower horizons within it.

E horizons

E horizons are light-colored mineral horizons that show evidence of loss of organic matter, clay, oxides, iron and/or aluminum, usually due to downward translocation of these substances by infiltrating water. The loss of the above constituents, *eluviation*, usually implies that grains coated with these substances have been stripped clean and that many of the dark, weatherable minerals, e.g. biotite, pyroxene, amphibole, have decomposed (see Chapter 12). Thus, E horizons are usually light-colored because they are dominated by uncoated, sand- and silt-sized quartz and feldspar grains. In older texts, E horizons are referred to as A2 horizons. Because the E horizon exhibits more total eluviation than perhaps any other horizon, its symbol was changed from A2 to E in 1982 (Guthrie and Witty 1982). The E designation, proposed by Whiteside as early as 1959, makes it the only major horizon with a definite and specific, genetic connotation.

E horizons form as vertically or laterally moving water strips sand and silt grains of their coatings and weathers many of the dark minerals. Thus, E horizons form primarily in humid climates, on freely draining (no high water tables, no water-impeding layers) sites or where water perches and moves laterally above an aquitard or aquiclude. In soils where water perches on a slowly permeable layer, chemical reduction of Fe and Mn occurs (Kemp *et al.* 1998). The reduced forms of these two elements are easily translocated in laterally flowing soil water (see Chapter 13), leading to an Fe- and Mn-depleted horizon with many of the morphological characteristics of a typical clay-impoverished E horizon.

E horizons could be thought of as A horizons that have been stripped of all coatings except humus. Thus E horizons can be conceptually considered the downward, or lower, extension of A horizons. Where A horizons are thick, as in grassland soils, E horizons are often absent. Conversely, E horizons are commonplace in forested soils in humid climates, where translocation is frequent and intense. In some soils with minimal pedoturbation, such as acidic sandy Spodosols, the E immediately underlies the O horizon. Here, decaying organic matter is only very slowly (if at all) mixed into the mineral soil. As the litter decomposes within the O horizon, the small organic molecules wash into the mineral soil and completely bypass the E; deposition occurs in the B horizon below.

With time, E horizons can grow downward at the expense of the B horizon below. This form of E horizon is often considered a degraded B and can be taken as clear evidence of continued profile deepening and development (Bullock *et al.* 1974) (see Chapter 12). Degraded B horizons often have the outward appearance of an E but can contain fragments of B material within (Payton 1993b).

B horizons

Just as the A and E horizons are zones within the profile where eluviation (losses) are dominant, B horizons below often show evidence of *illuviation* (Latin *il*, in, and *luv*, washed), or translocation of materials *in*. Often, illuvial substances or weathering by-products are carried downward by percolating water, and become deposited within the B as the water soaks into the peds. In so doing, the illuvial substances get plastered onto the ped face as coatings, or cutans (Table 3.4). Illuvial materials are also deposited in B horizons due to other processes, such as desiccation, adsorption, filtering and precipitation, although not all of these will result in obvious cutans. In illuvial B horizons, suffixes are used to convey information about the nature of the illuvial materials, which might include clay, iron, aluminum, carbonates, sodium, humus, gypsum, sulfur and silica, alone or in combination (Table 3.2). Placing a suffix on a B horizon, or indeed on any horizon, as an indicator of illuvial materials is a judgement

Table 3.4 Types of illuvial cutans (ped coatings) and cemented zones

Component	Cutan name[a]	Cemented name
Clay	Argillan, clay skin, clay film, Tonhautchen	
Silt	Silan, siltan	
Humus	Organ, organan	
Manganese	Mangan	
Crystalline salts, e.g., carbonates, chlorides and sulfates	Soluan	
Skeleton grains such as silt and very fine sand	Skeletan, neoskeletan	
Sesquioxides of iron and aluminum	Sesquan	Ortstein, placic
Calcium and magnesium carbonates	Calcan, calcitan	Caliche, calcrete, nari, kankar, croute calcaire
Iron	Ferran	Iron pan, laterite, plinthite, ferricrete, ironstone
Goethite	Goethan	
Allophane	Allan	
Sodium salts, esp. halite	Halan	Salcrete
Gypsum	Gypsan	Gypcrete
Silica		Silcrete, duripan
Various combinations of materials	E.g., ferri-argillan, organo-argillan	

[a] Cutan names are also occasionally used when referring to pore linings. Prefix modifiers such as "neo-" and "quasi-" are, respectively, used when the cutans are minimally developed or "almost like" the root cutan name. For example, neo-argillan or quasi-sesquan.

call. If the observer believes they see evidence for illuvial coatings and feel that such a notation is significant and important to the soil description, they are justified in adding the suffix as a B horizon descriptor. There is no quantitative rule that must be met or some minimum that must be exceeded. Thus, as are many aspects of soil description, choosing whether to add a descriptive suffix or not is based on experience, and is sometimes more art than science.

With time, illuvial coatings can thicken. Thicker cutans are often cited as evidence of increased soil development. Conversely, processes are always operative that work to destroy soil order and break up cutans.

Weak B horizons (termed Bw) are primarily expressed not by their illuvial products but by slight reddening associated with weathering and the accumulation of residual weathering products such as oxides, or by the development of soil structure and the concomitant loss of rock structure. In soils that have formed in carbonate-rich parent materials, like some glacial sediments, the lower limit of the B horizon is taken as the depth of carbonate leaching. The many types of B horizons, and how they form, will be explained in later chapters.

B horizons are sometimes referred to as subsoil materials. Because B horizons take longer to form and to be destroyed, and by virtue of their location in the subsoil are less likely to be eroded than are A and E horizons above, many soil classification schemes place great emphasis on B horizon characteristics (Expert Committee on Soil

Science 1987, Soil Survey Staff 1999). Similarly, paleopedologists often realize that the burial process itself can concomitantly erode the uppermost horizons, leaving only the B horizon as evidence of the soil that existed prior to burial (Sorenson 1977, Olson 1997, Olson and Nettleton 1998).

The sequum concept

A sequum is a couplet of an eluvial horizon above an illuvial horizon, usually an E and an underlying B horizon. Many soil profiles in humid regions have an E–B sequum. Those that have two sequa are termed *bisequal soils* (Schaetzl 1996) (Fig. 3.5). Often, bisequal soils form where two different sets of eluvial–illuvial processes are ongoing in the same pedon, e.g., clay translocation at depth, forming a E–Bt sequum, and iron/aluminum translocation above, forming a E–Bs sequum. The second, or lower, E horizon in a bisequal soil is usually denoted with the prime (E′). Although it is the second E horizon in the profile, an intervening horizon occurs between it and the E of the upper sequum, necessitating the use of the prime (Fig. 3.5).

The solum concept

The solum is an old concept that still is in use today. It refers to the soil horizons above the C horizon, which means it usually includes the O, A, E and B horizons, if they are present. The Soil Survey Division Staff (1993) defined it as a set of horizons that are related through the same cycle of pedogenic processes; it includes all horizons now forming, so as to exclude buried sola. Thus, the solum can be thought of as the entire, genetically formed soil profile. If the C horizon is taken to represent unaltered parent material (see below), then the solum would by default consist of all horizons that have been pedogenically altered.

Determining the lower limit of the solum is often difficult, and in many instances may be meaningless. In soils developed on calcareous parent materials, the solum is often taken to be the leached zone, while the C horizon, below the solum, remains calcareous. In soils developed on weathered rock (saprolite), however, the lower boundary of the solum is very diffuse and defining it may be very difficult. Clay-rich Bt horizons may gradually yield to unaltered saprolite, although the boundary is nearly impossible to define since illuvial clay commonly extends well below the solum, along vertical fissures in the saprolite. Douglas *et al.* (1967) found that the lower boundary of the solum was often placed too shallowly. They noted that plant roots and illuvial clay extend well into the C horizon in Illinois soils. Buol *et al.* (1997) suggested that the maximum lower limit of the solum must be the maximum depth of perennial roots, implying that if the solum is soil, it must have been influenced by biotic activity. Perhaps the best way to delimit the lower boundary of the solum is to examine the layer just below; if it shows little or no sign of pedogenic alteration, it is not within the solum proper. In sum, the solum is a theoretical construct and, while useful in some instances, is often a source of confusion more than enlightenment (Figs. 3.2, 3.4).

C horizons

Traditionally, the material between the solum and bedrock has been designated as the C horizon. Defined early as "parent material that has been unaltered by soil-building forces" (Kellogg 1930), C horizons are, today, viewed as the parent material from which the mineral soil horizons above have formed. Shallow profiles that rest directly on bedrock often do not have a C horizon (Willimott and Shirlaw 1960). When the parent material is calcareous, most soil scientists assume that the lower limit of leaching corresponds to the upper limit of unaltered parent material, or the C horizon. This is not always true. Many, if not most, C horizons have been altered somewhat by surficial (pedogenic and geogenic) processes (Godfrey and Riecken 1954, Douglas *et al.* 1967). If the original parent material was calcareous, many C horizons are slightly leached of carbonates. Still others that are calcareous show evidence of illuviation of carbonates from above (Schaetzl *et al.* 1996). Indeed, Douglas *et al.* (1967) stressed that the depth to detectable carbonates is not suitable as a criterion for determining the C horizon. Yet, this remains the main criterion used to identify the C horizon (solum) boundary in many soils.

Although defined as "unaltered" material by Kellogg in 1930, the weathered nature of the C horizon was soon recognized (Kellogg 1936). Indeed, many C horizons are slightly weathered or altered by pedogenic processes. Many have thin, patchy illuvial deposits, such as argillans or other secondary minerals, along joints and major ped faces. Others often show evidence of oxidation, gleying and jointing, and can be well within the depth of rooting. Such minor alteration and illuviation is difficult to exclude from the C horizon concept, since water that enters that part of the soil is not without solids and solutes, regardless of where it originated within the profile. Depth sequential sampling can be used to illustrate these subtle but real depth trends. For these reasons, the B–C (and hence, lower solum) boundary is often diffuse and somewhat arbitrary (Simonson and Gardner 1960), and criteria used to determine it are often contradictory (Fig. 3.14). Most studies that have examined the depth of the solum concluded that the upper boundary of the C horizon is, in actuality, deeper than might be observed based on field criteria (e.g., Douglas *et al.* 1967). Tandarich *et al.* (2002) review the various concepts used to define the C horizon, showing that several different types of C horizons have been proposed to describe the altered parts of the C, such as zones that are weathered, oxidized, leached, illuvial and/or gleyed.

Many soil profiles have formed in more than one parent material (Schaetzl 1998). The contact between two unlike parent materials is called a *lithologic discontinuity* (Fig. 3.5) (see Chapter 8). Some profiles could have more than one lithologic discontinuity, where more than two distinct parent materials are present. In soils with a lithologic discontinuity, the C horizon could be in the upper material and/or in the second (next lower) parent material. In the former case, the C horizon would represent parent material for the horizons above. That is, the C could be viewed as material from which the horizons above have formed. In the latter instance, the 2C horizon material could be quite different than was the material above, in its original state. Thus, one could not examine 2C material for clues as to the original characteristics of the upper, overlying sediment.

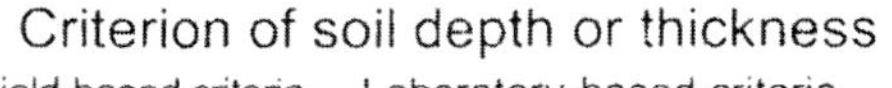

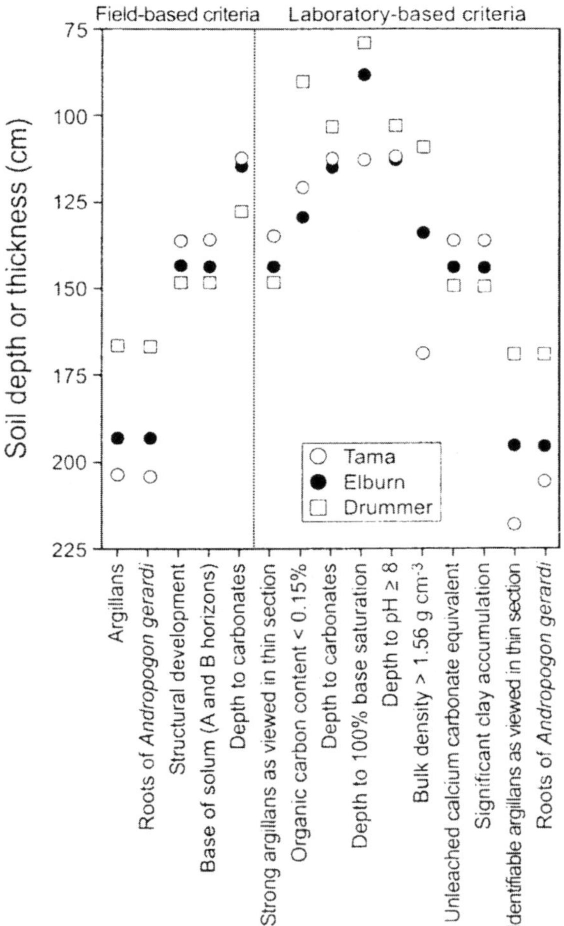

Fig. 3.14 Thickness of the solum of three different soils, based on several criteria. Note that criteria used to determine the depth to the bottom of the solum (i.e., the B–C boundary) are seldom in agreement. After Douglas *et al.* (1967).

D horizons

The C horizon concept, as used today, was originally devised by soil scientists who often did not have information on deep subsurface stratigraphy. Because the C horizon often contains significant amounts of chemically and physically altered material (i.e., leached of carbonates, jointed or oxidized), it is often viewed as a flawed concept.

To clear up this confusion, Tandarich *et al.* (1994) reintroduced the *D horizon* concept, not so much in the context of the *soil* profile, but the *pedo-weathering* profile. We use the term "reintroduce" because the D horizon concept was actually *introduced* by Dokuchaev in 1900 for a horizon below the C (Tandarich *et al.* 2002). (Earlier use of the "D" had applied it to bedrock.) Ninety-four years later, Tandarich *et al.* chose to limit the C horizon to the modified part of the traditional C. The part of the traditional C horizon that is *unaltered* by pedogenic processes, but is not hard bedrock, is therefore recognized as the D horizon (Tandarich *et al.* 1994). Others, too, have noted that the C horizon alone was insufficient to connote its modified character and have variously referred to this material as Cu or C4 (Follmer 1979b). Tandarich *et al.* (1994) defined the D horizon as geogenic or non-pedogenic horizons (unaltered zones) of fresh sediment, excluding consolidated bedrock, characterized by original rock or sedimentary fabric, lack of tension joints, and lack of alteration features of bio-oxidation origin. In areas that are comparatively shallow to bedrock, the D horizon may not have ever existed, or may quickly have been converted to C or solum material. On old landscapes that have been deeply weathered, the D horizon will be difficult, if not impossible, to identify. In short, D horizon material is unaltered whereas C horizon material *is* slightly chemically and/or physically altered. A, E and B horizons are all biologically, chemically *and* physically altered.

Tandarich *et al.* (1994) embraced the concept of the pedo-weathering profile as one that merges traditional soil profile concepts (above) with those of the geologic weathering profile. The latter have traditionally been used by Quaternary geologists (e.g., Kay and Pierce 1920, Frye *et al.* 1960b, Willman *et al.* 1966). According to Tandarich *et al.* (1994), an unwritten agreement between Curtis Marbut, then head of the soil survey program in the US Bureau of Soils, and Morris Leighton, a Quaternary geologist in Illinois, created the schism between the soil profile, which was to remain in the realm of soil scientists, and the weathering profile, which was given over to those more geologically oriented. This made sense at the time (1950s), since Quaternary geologists, while interested in soils, were more inclined to focus on sediments below the profile. Similarly, pedologists at the time were trying to diminish their dependence on geological theories. To this end, the pedo-weathering profile concept of Tandarich *et al.* (1994) re-merges the geologic and pedologic views of the weathered portion of the Earth's crust into a coherent theme. The pedo-weathering profile retains the concept of unaltered parent material, which exists in parts of the midwestern USA (Follmer 1984), but labels it D horizon.

While the D horizon concept has not yet been fully accepted, its usage seems to be a logical extension of horizon nomenclature into the deep subsurface, reflective of a greater interest in processes and materials that exist below the solum. As of this writing, no national-level soils agency, such as the Natural Resources Conservation Service, has adopted the D horizon concept. Nonetheless, we support the concept, and will use it accordingly.

Buried soils

Soils that are or have been buried are referred to as paleosols, geosols or fossil soils (see Chapter 15). They are introduced here because, once recognized as such, they may also be described using traditional soil horizon nomenclature. Horizons in buried soils are described just like horizons in surface soils, except that b follows the horizon designation. For example, a buried soil might have Ab, Bab, Btb and C horizons. (Normally, the paleosol C horizon does not receive the b suffix.) If a buried paleosol later becomes exhumed, the b suffix is no longer applied.

Whether the b suffix is used or not depends on whether the soil is actually, by definition, buried. The Soil Survey Staff (1999) defines a buried soil as one that is covered with a surface mantle of new soil material that either is $\geq$50 cm thick or is 30–50 cm thick and is at least half as thick as the buried soil. Thus, the 50 cm minimum limit has precedence. It is assumed that the surface mantle is minimally altered by pedogenesis. However, Schaetzl and Sorenson (1987) argued that any covering of sediment, if it is clearly identifiable as such, regardless of thickness, is enough to bury the soil below. Obviously, in situations where the covering is thin, pedogenic processes will quickly act to incorporate that new mantle into the buried soil, and in effect "unbury" it. This process is called soil welding (Ruhe and Olson 1980) (see Chapter 15). Our position is as follows: if it can be demonstrated that sediment exists on top of the profile and is not genetically a part of that profile, the soil below should be considered buried and the b used on its horizons.

Chapter 4

Basic concepts: soil mineralogy

Minerals are naturally occurring inorganic compounds that have a characteristic chemical composition and a regular, repeating three-dimensional array of atoms in a crystal structure. Minerals can be classified according to their chemical composition and crystal structure, or according to whether they are primary (inherited from the parent material without chemical alteration) or secondary (formed by chemical weathering of other, pre-existing minerals). Primary minerals tend to dominate in coarser size fractions, whereas secondary minerals are most abundant in the clay and fine silt fractions (Fig. 2.3). Knowing the structure and properties of soil minerals is essential for understanding the mineral transformation (including weathering) and transport processes that are important in soil genesis. This chapter emphasizes mineral structure, classification, properties, occurrence and identification, whereas later chapters (especially Chapters 13 and 15) focus on the role of minerals in soil genesis and geomorphology, e.g., as indicators of soil age or paleoenvironments.

Bonding and crystal structures

Understanding the chemical and structural classification of minerals requires knowledge of bonding and crystal structure in inorganic solids, particularly solids in which O^{2-} is the anion. Oxygen (in the form of the oxyanion O^{2-}) makes up about 47% (by mass) of the Earth's crust. It is by far the most abundant element in the crust (Fig. 4.1). Consequently, nearly all major groups of soil minerals, including silicates, oxides, phosphates, carbonates and sulfates, are ionic solids in which O^{2-} is the primary anion. The most common cations in soils are Si^{4+}, Al^{3+} and Fe^{3+}, which reflects their crustal abundance as well (Fig. 4.1). The bonds between these cations and O^{2-} are predominantly ionic, and mineral structures consist of large O^{2-} anions in close packing around a small polyvalent cation such as Si^{4+} or Al^{3+} (Fig. 4.2). The number of O^{2-} anions surrounding each cation is determined by ratio of the cation

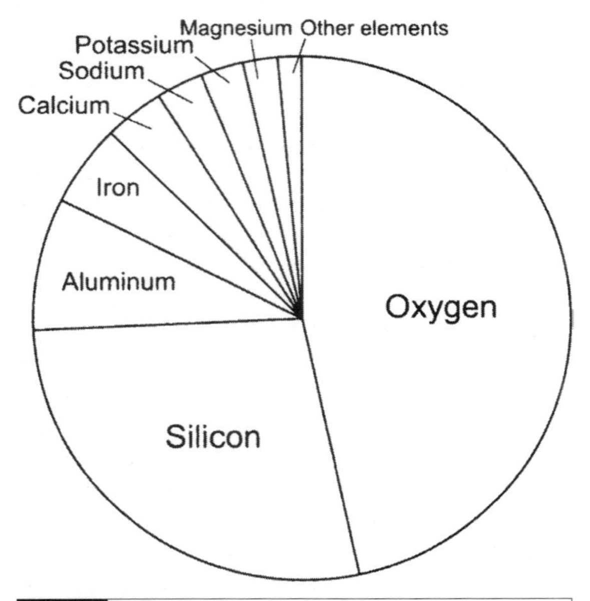

Fig. 4.1 Relative abundance (*mass* %) of the most common elements in the Earth's crust. When examined on a *volumetric* basis, oxygen comprises about 94% of the crust, because of its comparatively large size.

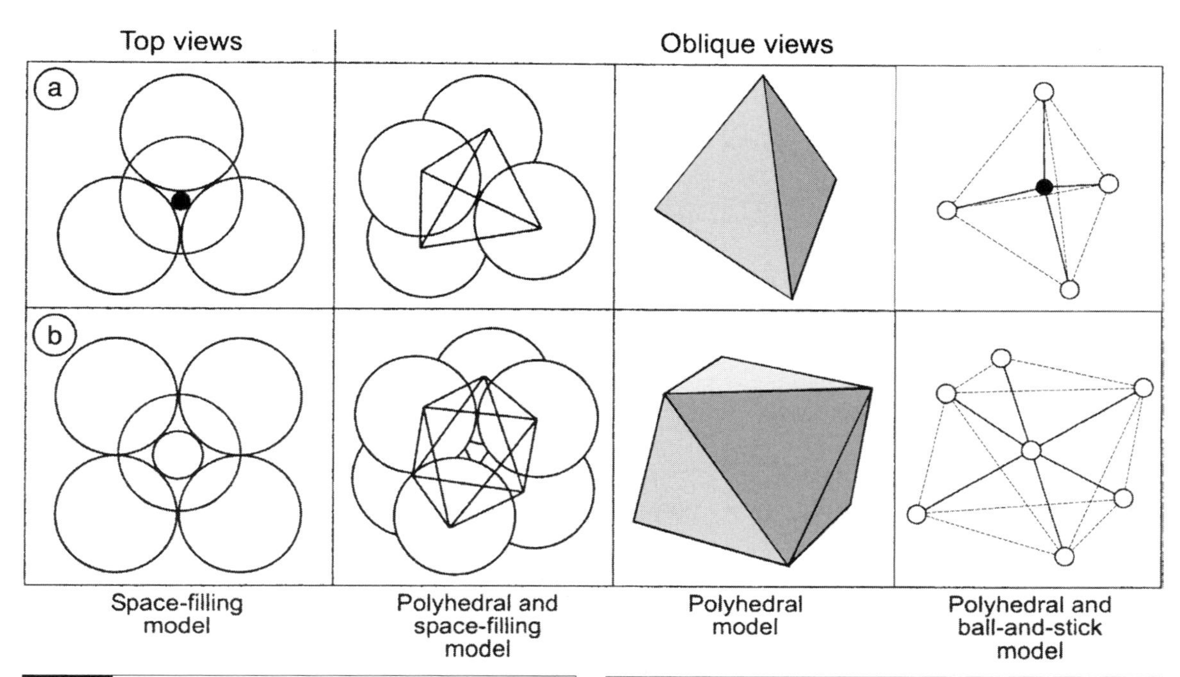

Fig. 4.2 Different views and models of silica tetrahedra and octahedra. (a) Si–O tetrathedra. The small silicon cation is surrounded by four 0^{2-} anions. The corners of the tetrahedra represent the centers of 0^{2-} anions surrounding the tetrahedral cation. (b) Octahedra. The central cation, e.g., Fe^{3+}, Al^{3+} or Mg^{2+}, is surrounded by six 0^{2-} anions.

radius to the O^{2-} radius, as well as the number of O^{2-} anions that can fit around the cation in a close-packed arrangement (Fig. 4.2). Table 4.1 lists the ionic radii, the radius ratio ($r_{cation}/r_{O^{2-}}$) and the resulting coordination number of O^{2-} anions that surround the cation in soil minerals. Because of its small size, Si^{4+} is coordinated to four O^{2-} anions to form silica tetrahedra (SiO_4^{4-}), whereas the larger Al^{3+}, Fe^{3+}, Fe^{2+} and Mg^{2+} cations are coordinated to six O^{2-} anions to make an octahedral structure (Fig. 4.2). The sphere-packing or space-filling models of mineral structures (left half of Fig. 4.2) are the most accurate representations of mineral structures, but other schematic representations provide more information about the three-dimensional arrangement of ions in a mineral. Ball-and-stick diagrams (far right of Fig. 4.2) use small balls to represent ions and lines to depict the bonds between atoms. Polyhedral models (middle of Fig. 4.2) show lines that connect O^- ions at the corners of tetrahedra and octahedra but do not show ions or bonds. Polyhedral models are most useful for showing the sharing of corners, edges or faces (i.e., one, two or three O^- ions) between adjacent polyhedra.

Table 4.1 Oxygen anion, common cations, their radii, and coordination numbers with O^{2-} in common minerals

Ion	Ionic radius (nm)	Radius ratio with oxygen[a]	Coordination number in minerals[b]
O^{2-}	0.140	–	
Si^{4+}	0.042	0.34	4
Al^{3+}	0.051	0.41	4,6
Fe^{3+}	0.064	0.51	6
Fe^{2+}	0.074	0.63	6
Mg^{2+}	0.066	0.47	6

[a] Ratio of cation radius to O^{2-} radius.
[b] Refers to the number of cations that surround a single metal atom in the closest-packed arrangement.

Oxides

Metal oxides, which are ubiquitous in soils, have relatively simple chemical formulas and crystal

Table 4.2 Characteristics of aluminum oxides in soils

Mineral	Frequency of Coccurrence in soils	Usual crystal shape	Intense X-ray diffraction peaks (nm)	Thermal analysis peaks (°C)
Gibbsite γ-$Al(OH)_3$	Most common Al oxide in soils and bauxite deposits	Pseudo-hexagonal plates (or rods)	0.485 0.437 0.432 0.245 0.239	Endotherm 300–330
$Al(OH)_3$(am)	Common in soils; intermediate in the formation of gibbsite	Amorphous	None	Endotherm 150–200
Bayerite α-$Al(OH)_3$	Uncommon in soils; common in bauxite deposits	Triangular pyramids	0.222 0.471 0.435 0.173 0.320	Endotherm 300–330
Boehmite γ-AlOOH	Not common in soils; occurs in bauxite deposits	Tiny particles	0.611 0.316 0.235 0.186 0.185	Endotherm 450–580
Diaspore α-AlOOH	Not common in soils; common in bauxite deposits	Tiny particles	0.399 0.232 0.208 0.246	Endotherm 540

Source: Hsu (1989).

structures (Tables 4.2, 4.3). Most metal oxides in soils are secondary minerals, formed during the weathering of primary minerals that contain iron or aluminum; only a few oxides are primary minerals inherited from the parent material. Secondary iron and aluminum oxides are major components of the clay fraction of highly weathered soils such as Oxisols and Ultisols (Fig. 4.3) (see Chapter 13). In younger soils, these same oxides do not make up a significant fraction of the soil mass but are typically present as coatings on mineral grains, where they influence aggregation and retention of certain ions (Hsu 1989).

Iron and aluminum oxides are composed of Al^{3+}, Fe^{3+} or other cations that are octahedrally coordinated to oxygen (Fig. 4.4). They are classified according to the type of metal cation and the manner in which adjacent octahedra bond to one another and are arranged in space. Octahedra can share corners, edges or faces, with edge-sharing being most common. In some cases, two or more minerals have identical crystal structures but differ in the arrangement of atoms in space, i.e., the manner in which adjacent octahedra share corners or edges.

Aluminum oxides

The most abundant aluminum oxide in soils is *gibbsite*, $Al(OH)_3$, sometimes referred to as *hydroargillite* in European literature. When gibbsite

Table 4.3 Characteristics of iron oxides in soils

Mineral	Soil environment	Crystal shape	Density (g cm^{-3})	Maximum Al substitution	Intense X-ray diffraction peaks (nm)	Thermal analysis peaks (°C)
Goethite α-FeOOH	Most common Fe oxide in soils	Needles, laths	4.26	1/3 of Fe^{3+} may be substituted	0.418[a] 0.245 0.269	Endotherm 280–400
Lepidocrocite γ-FeOOH	Poorly drained, non-calcareous soils	Laths	4.09	Little	0.626 0.329 0.247 0.194	Endotherm 300–350 Exotherm 370–500
Ferrihydrite (Fe_2O_3) $5.9H_2O$	Precursor to hematite; forms when Fe^{2+} is rapidly oxidized in presence of organic matter or silicates	Spherical or aggregated	3.96	Not known	0.254 0.224 0.197 0.173 0.147	Endotherm 150
Hematite α-Fe_2O_3	Highly weathered soils	Hexagonal plates	5.26	Up to 1/6 of Fe^{3+} may be substituted	0.270 0.368 0.252	None
Magnetite Fe_3O_4	Primary mineral persists in coarse grains; contains Fe^{2+} and Fe^{3+}	Cubes	5.18	Possible	0.253 0.297	None
Maghemite γ-Fe_2O_3	Formed from magnetite by oxidation of Fe^{2+} to Fe^{3+}; may form under burning vegetation	Cubes	4.87	Possible	0.253 0.297	None

[a] Goethite's 0.418 nm peak may be obscured by the 0.426 nm peak of quartz; size fractionation to remove quartz is essential.

Source: Schwertmann and Taylor (1989). See also Table 2.3.

forms discrete particles rather than coatings on mineral grains, it typically occurs as hexagonal plates (Hsu 1989). Gibbsite is composed of sheets of $Al(OH)_6$ octahedra that share edges (Fig. 4.4). In the octahedral sheets, 2/3 of the cation sites in the center of the octahedra are vacant; 2/3 are occupied by Al^{3+} (Fig. 4.4). Gibbsite formation is promoted by high rainfall in freely draining settings that allow for silica to be leached, and by warm temperatures that promote weathering of primary minerals (Furian *et al.* 2002). Thus, gibbsite is most common in soils

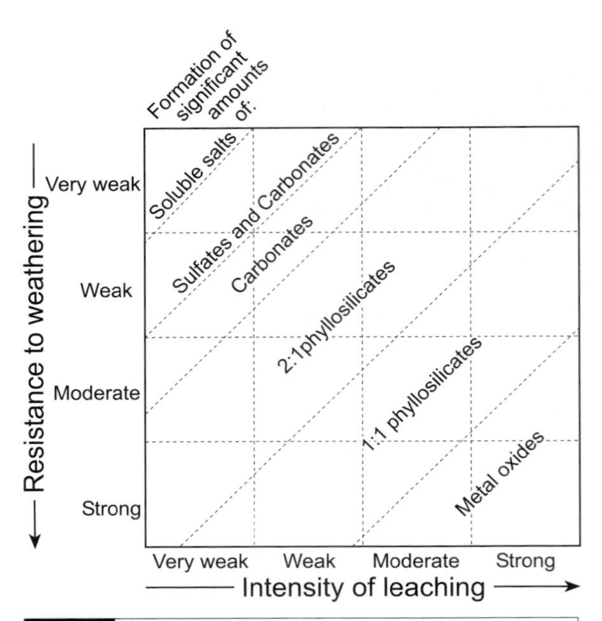

Fig. 4.3 Occurrence of various types (groups) of secondary minerals in soils, as a function of resistance to weathering and leaching intensity (see also Chapters 10 and 13). After Crompton (1962).

of warm, tropical climates; amorphous $Al(OH)_3$ is more common in less weathered soils. An abundance of gibbsite in midlatitude settings may suggest that the soil is quite old (Wang *et al.* 1981, Ogg and Baker 1999).

Gibbsite is believed to be the dominant crystalline form of $Al(OH)_3$ in soils because its formation is favored by the presence of organic matter and anions that inhibit crystallization (Hsu 1989). Other forms of $Al(OH)_3$ differ in their stacking of adjacent Al octahedral sheets. Bayerite is found in bauxite deposits but is rare in soils; nordstrandite rarely occurs in soils (Table 4.2). In addition to $Al(OH)_3$, two forms of AlOOH occur in highly weathered soils: diaspore (α-AlOOH) and boehmite (γ-AlOOH) (Gilkes *et al.* 1973, Sadleir and Gilkes 1976). These two polymorphs of AlOOH differ in the arrangement of double chains of octahedra (Fig. 4.4).

Iron oxides

Iron oxides are more abundant in soils than are aluminum oxides. Several different Fe oxide minerals can be found in soils, depending on the weathering environment (Table 4.3). Iron oxides form when Fe^{2+} in primary minerals is oxidized, released from the mineral structure and hydrolyzed to form hydroxy iron polymers that crystallize into iron oxides. Iron oxides can occur both as coatings on mineral grains in mild and moderate weathering environments, and as discrete particles in stronger weathering environments (Schwertmann and Taylor 1989). Because iron oxides have a high surface area, even a small amount of Fe oxides can greatly enhance aggregation and affect soil color (Table 2.3). Iron oxides also promote aggregation; Fe oxides have positively charged surface groups that can bind to negatively charged sites on clays and organic matter, and also possess negatively charged sites that adsorb cations (Borgaard 1983).

Most iron oxides contain Fe^{3+} that is octahedrally coordinated to six O^{2-}, OH^- or H_2O groups. Structurally, the various forms of iron oxide differ in the number of O atoms shared between octahedra and the distribution of "vacant" octahedra with no central Fe^{3+} cation. In most Fe oxides, 2/3 of the octahedral sites are occupied and 1/3 are vacant. In goethite and hematite, Al^{3+} can substitute for Fe^{3+} in the crystal structure (Schwertmann and Taylor 1989) (Table 4.3, Fig. 4.4). Aluminum substitution is more important in soils derived from aluminosilicate-rich parent material (Allen and Hajek 1989).

Goethite (α-FeOOH), the most widely distributed iron oxide in soils, can be found in temperate to tropical, and semi-arid to humid climates. Goethite imparts a brown to yellowish brown color to soils, although when hematite is present the goethite hue may be masked (Table 2.3). Goethite particles are small ($<0.1\ \mu m$). When goethite forms discrete particles, it occurs as needles, although in highly weathered soils it may assume a more rounded appearance (Schwertmann 1984). Goethite consists of double chains of Fe–O octahedra that run parallel to the c-dimension. The octahedra form double chains (Fig. 4.4) that share corners or edges; the double chains are staggered, with H bonds between O and OH of neighboring octahedra that do not share O atoms. Goethite can have up to 1/3 of the Fe^{3+} ions substituted by Al^{3+} (Schulze 1984).

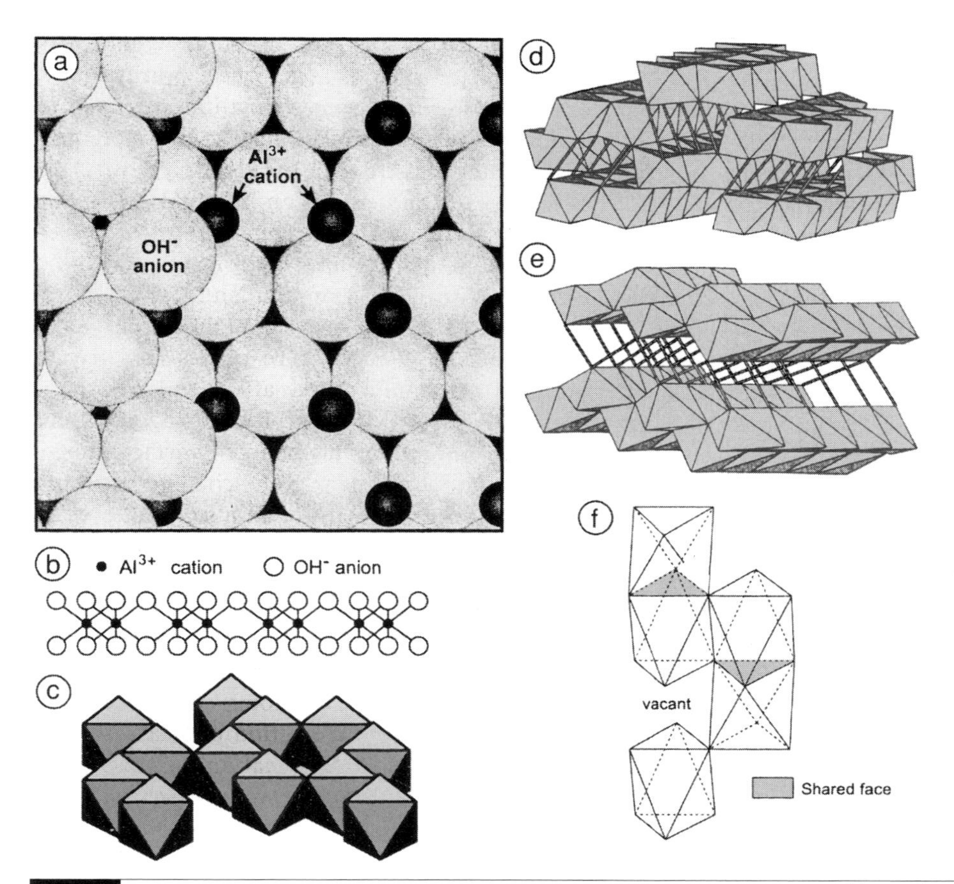

Fig. 4.4 Schematic diagrams of various oxide clay minerals. (a) Schematic diagram of the gibbsite ($Al(OH)_3$) structure showing small Al^{3+} cations surrounded by six larger OH^- ions —three below the cation and three above. The Al^{3+} ions occupy 2/3 of the octahedral positions. (b) Ball-and-stick model of side view of a gibbsite octahedral sheet consisting of lower OH plane, Al^{3+} plane and upper OH plane. (c) Schematic, polyhedral representation of the gibbsite structure showing octahedra sharing edges. (d) Schematic diagram of either diaspore (α-AlOOH) or goethite (α-FeOOH), depending on whether Al^{3+} or Fe^{3+} is the ocathedral cation. The figure shows staggered double chains of octahedra, with corners of octahedra shared between adjacent double chains. (e) Schematic diagram of either boehmite (γ-AlOOH) or lepidocrocite (γ-FeOOH), depending on whether Al^{3+} or Fe^{3+} is the ocathedral cation. The figure shows zigzagging double chains of octahedra, with edges shared between adjacent double chains. (f) Schematic diagram of hematite (α-Fe_2O_3), showing shared crystal faces and vacant parts of the crystal structure.

Lepidocrocite (γ-FeOOH) is a polymorph of goethite in which double chains of Fe octahedra form zigzag layers, with hydrogen bonding between the layers (Fig. 4.4). Lepidocrocite is orange and is typically found in concretions or mottles, i.e., as redoxymorphic features in soils that are subject to alternating oxidizing and reducing conditions (Schwertmann and Taylor 1989) (see Chapter 13).

Hematite (α-Fe_2O_3) forms in strongly weathered soils but it can also be inherited from the parent material. The name "hematite" is loosely translated as "blood rock" – because even small amounts of hematite will give soils a strong, almost blood red coloration (Table 2.3). Pedogenic hematite forms hexagonal plates up to 50 nm in diameter (Schwertmann and Taylor 1989). In the hematite structure, 2/3 of the octahedral sites are occupied by Fe^{3+}, and the Fe–O octahedra are arranged such that each Fe-containing octahedron shares one face with another. Hematite can have up to 1/6 of the Fe^{3+} ions substituted by Al^{3+} (Bigham *et al.* 1978).

Ferrihydrite is a poorly crystalline, fine-grained Fe oxide with the formula $5Fe_2O_3 \cdot 9H_2O$. Its reddish brown color is intermediate between the red of hematite and the yellow-brown of goethite (Table 2.3). Ferrihydrite grains or crystals can form aggregates, with particle sizes of 2 to 5 nm. The small size, high surface area, and poor crystallinity (with only short-range ordering that produces very broad X-ray peaks) cause ferrihydrite to be very reactive and susceptible to dissolution. Thus, it dissolves in selective dissolution treatments that do not affect goethite and other crystalline Fe oxides. Consequently, many of the "amorphous Fe oxides" described in older literature are now presumed to be ferrihydrite. Ferrihydrite is believed to be structurally similar to hematite, with more vacant Fe sites (fewer Fe^{3+} ions) and some of the O^{2-} anions replaced by OH^- and H_2O. It is an intermediate in the formation of hematite (and also goethite) and forms in environments where Fe^{2+} becomes oxidized in the presence of organic matter or silicate. Ferrihydrite has been reported in subtropical to cold climates, and is also found in bogs and spodic horizons (Schwertmann and Taylor 1989).

Magnetite (γ-Fe_3O_4) is a primary mineral inherited from parent material; it can be found as black, magnetic particles in the coarse fraction of many soils. Magnetite contains 1/3 Fe^{2+} and 2/3 Fe^{3+}. It may form when Fe^{2+} is oxidized to Fe^{3+}. Magnetite and its weathering product maghemite are found mainly in soils derived from basalt or other mafic igneous rocks (Allen and Hajek 1989). Because magnetite in soils can become aligned with the Earth's magnetic field, its presence in soils, especially in buried soils, is valuable as a dating tool (see Chapter 14).

Manganese oxides

Manganese oxides are more complex and less understood than are Al or Fe oxides. One of the difficulties in studying pedogenic Mn oxides is that their poor crystallinity makes it difficult to distinguish one Mn oxide from another, especially in mixtures.

Manganese oxides can contain Mn^{2+}, Mn^{3+} and Mn^{4+} octahedrally coordinated to O^{2-}. They can form either "tunnel" or "layer" structures of Mn–O octahedra, with varying degrees of hydration. The Mn octahedra share edges to form single, double or triple chains, or flat layers. Tunnel structures form when the chains share corners to form the edges of a tunnel (similar to the tunnels of goethite) (Fig. 4.4). The tunnels differ in height and width depending on whether a single, double or triple chain forms the tunnel boundary. Both the tunnels and the interlayer spaces can hold water or adsorbed cations. Of the Mn oxides, birnessite and vernadite (both layer structures) seem to be the most common, though todorokite (tunnel) and lithiophorite (layer) also have been reported in soils (McKenzie 1989). Manganese oxides are commonly found as discrete nodules, resembling fragments of black pepper, in wet soils that are undergoing frequent and alternating periods of oxidation and reduction (see Chapter 13).

Chlorides, carbonates, sulfates, sulfides and phosphates

Chlorides (mainly halite, NaCl) have a simple structure, with each Na^+ ion octahedrally coordinated to six Cl^- anions, and each Cl^- anion surrounded by six Na^+ anions. Chlorides are extrememly soluble and occur mainly as salt crusts on the surface of arid soils, particularly soils derived from saline parent material or influenced by saline waters or aerosols (Doner and Lynn 1989) (see Chapter 12).

Carbonates such as calcite ($CaCO_3$) or dolomite ($CaMg(CO_3)_2$ are less soluble than chlorides, but are nonetheless still typically found only in dry soils or in young soils or soils with calcareous parent material (see Chapter 12). Calcite is the most common soil carbonate and can be either pedogenic (forming in root zones where CO_2 concentrations are high) or inherited from calcareous parent materials. Soil dolomite is believed to be inherited from calcareous sediments or eolian dust. Structurally, each Ca^{2+} or Mg^{2+} is octahedrally coordinated to six O^{2-}, one from each of six CO_3^{2-} anions. The octahedra are distorted because of ion sizes, so calcite and dolomite have rhombohedral structures.

Sulfates that contain Ca^{2+}, Na^+ or Mg^{2+} are relatively soluble and occur predominantly in dry regions. Gypsum ($CaSO_4 \cdot 2H_2O$) is the most common sulfate mineral in dry soils and can be either inherited or pedogenic (see Chapter 12). In contrast, jarosite ($KFe_3(SO_4)_2(OH)_6$) forms in acid environments where FeS_2 from pyrite or mine spoils is exposed to air and oxidized. FeS_2 is the most common sulfide mineral in soils and is rapidly oxidized when exposed to air or oxygen-rich water (Doner and Lynn 1989).

Phosphate minerals are not abundant in soils, though apatite ($Ca_5(F,Cl,OH)(PO_4)_3$) has been identified in young soils. In general, phosphate minerals are less soluble than either carbonates or sulfates (Lindsay *et al.* 1989).

Silicates

Silicates are minerals composed of silica tetrahedra (Fig. 4.2), with Si^{4+} cations surrounded by four O^- anions, with different silicates having different crystal structures, chemical formulas, (Table 4.4, Fig. 4.5) and temperature of formation, all of which affect their stability in soils (Figs. 4.3, 9.1). Most of the important primary soil minerals are silicates, including quartz, feldspars, micas, pyroxenes and amphiboles. Primary silicate minerals are most common in the sand fraction (Fig. 2.3), with the relative abundance of each depending upon the composition of the parent material and the extent of weathering. Secondary silicates, such as kaolinite and smectite, form by the weathering of primary silicates and are most abundant in the clay fraction.

Silicate minerals are classified according to how many corner O^{2-} anions each silica tetrahedron shares with other tetrahedra, as well as on the geometric arrangement of neighboring tetrahedra (Fig. 4.5). Additional classification criteria will be addressed for the two groups of silicates that are most common in soils: tectosilicates and phyllosilicates.

Nesosilicates, sorosilicates and cyclosilicates

Silicates that consist of independent SiO_4^{4-} tetrahedra that are not bonded to other SiO_4^{4-} tetrahedra are known as nesosilicates or orthosilicates (Fig. 4.5). Olivine $(Mg,Fe)SiO_4$ is a common nesosilicate in which neighboring silica tetrahedra are held together by electrostatic attraction between the SiO_4^{4-} tetrahedra and interstitial Mg^{2+} or Fe^{2+} cations. Olivine forms at high temperatures and weathers readily in soils (Allen and Hajek 1989).

Sorosilicates such as epidote contain pairs of tetrahedra that share one corner O atom; they have the formula $Si_2O_7^{6-}$ (Fig. 4.5). Cyclosilicates (Fig. 4.5) typically have six tetrahedra arranged in a ring, each sharing two corner O atoms, to give the general formula $Si_6O_{18}^{12-}$. Sorosilicates and cyclosilicates such as beryl and tourmaline are not common in soils (Allen and Hajek 1989).

Inosilicates (chain silicates)

The inosilicate or chain silicate group (Fig. 4.5) includes pyroxenes and amphiboles, which generally originate in mafic and intermediate rocks. Single-chain silicates such as pyroxenes have long chains of silica tetrahedra that share two corner O^{2-} ions and have the general formula SiO_3^{2-}. Double-chain silicates such as amphiboles are composed of parallel chains of silica tetrahedra in which half the tetrahedra are cross-linked to an adjacent chain. This gives the general double-chain formula of $Si_4O_{11}^{6-}$, and provides for greater resistance to weathering than with single chains. Amphiboles such as hornblende are more stable than pyroxenes, e.g., augite or diopside, but less stable than tectosilicates such as feldspars or quartz (Table 9.2). The most common cations that balance the negative charge of the silicate chains in amphiboles and pyroxenes are Ca^{2+}, Mg^{2+} and Fe^{2+} (Allen and Hajek 1989).

Tectosilicates (framework silicates)

Tectosilicates (framework silicates), of which quartz is the best known and most common, have the general formula SiO_2. The silica tetrahedra in quartz and other framework silicates are each bonded to four other tetrahedra; all O^{2-} anions in each tetrahedron are shared between two tetrahedra to give a cross-linked three-dimensional network of silica tetrahedra (Fig. 4.5). Because quartz has a three-dimensional network of bonded tetrahedra rather than a linear or planar structure like other silicates, it has no planes of weakness and is quite resistant to both physical and chemical weathering (Fig. 9.1, Table 9.2). There are other

Table 4.4 Structural classification of primary silicates (other than phyllosilicates) in soils

Silicate class and Si : O ratio	General formula	Number of shared oxygens in each tetrahedra	Examples of minerals
Nesosilicates (orthosilicates), isolated tetrahedra	SiO_4^-	None	Olivine Garnet Zircon
Sorosilicates, double tetrahedra	$Si_2O_7^-$	1	Epidote
Cyclosilicates	$Si_6O_{18}^{12-}$	2	Beryl, tourmaline
Hexagonal ring formed by six tetrahedra, each sharing two corners			
Inosilicates			
single chain	SiO_3^{2-}	2	Pyroxenes: augite, diopside, enstatite, hypersthene
double chain	$Si_4O_{11}^{6-}$	3	Amphiboles: hornblende, actinolite, tremolite
Phyllosilicates, layer sheet silicates	$Si_2O_5^{2-}$	3	Kaolin group: serpentines, talc, micas, vermiculite, smectites, chlorite
Tectosilicates, structural silicates; share all four apical oxygens of each tetrahedron	SiO_2	4	Feldspars: orthoclase, albite, and anorthite Quartz Zeolites

polymorphs of SiO_2, including biogenic opals or phytoliths, and polymorphs that form at high temperature, but quartz is the most stable at the Earth's surface and the most common mineral in soils. It is also the second most abundant mineral in the Earth's crust, second only to feldspars (Allen and Hajek 1989).

Feldspars are tectosilicates in which one to two out of every four Si^{4+} ions are substituted by Al^{3+}, which gives the framework structure a net negative charge and a general formula of $MAlSi_3O_8$, where M represents a metal cation (e.g., Ca^{2+}, Na^+ or K^+) that resides between silica tetrahedra and balances the negative charge of the silicate structure. Feldspars in which Ca^{2+} or Na^+ balance the negative structural charge are known as *plagioclases*. The plagioclase series or family includes two end members – one in which 100% of the charge is satisfied by Ca^{2+} cations (anorthite) and one in which in which 100% of the charge is satisfied by Na^+ (albite) (Fig. 4.5). Plagioclase feldspars can have any composition,

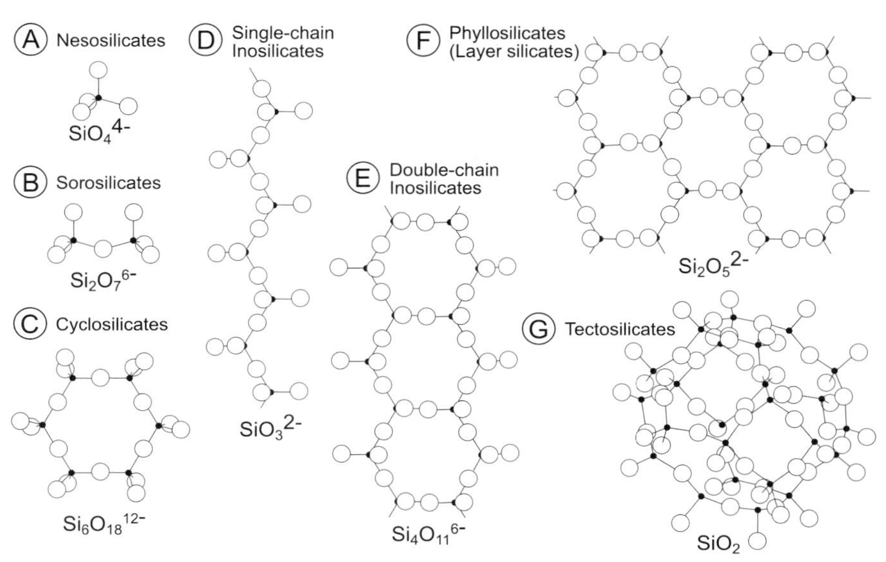

Fig. 4.5 Schematic (ball-and-stick) representations of silicates, arranged according to increasing sharing of O^{2-} at the corners of SiO_4^{4-} tetrahedron. (a) Nesosilicates or island silicates. (b) Sorosilicates, which share one corner of each tetrahedron. (c) Cyclosilicates, which share two corners of each tetrahedron. (d) Single-chain inosilicates, which share two corners of each tetrahedron. (e) Double-chain inosilicates, which share two or three corners of each tetrahedron. (f) Phyllosilicates or layer silicates, which share three corners of each tetrahedron, all in same plane. (g) Tectosilicates, such as quartz, which share all four corners of each tetrahedron in a three-dimensional framework.

ranging from 100% Ca^{2+} to 100% Na^+ cations, but those that are nearer the albite end of the continuum are more weathering-resistant (Fig. 9.1). Because of aluminum substitution and interstitial cations, plagioclase feldspars are considerably more susceptible to weathering than is quartz; this information can be used to assess the degree of weathering in soils by comparing the loss of feldspar to that of quartz (Table 14.5). Feldspars weather when their interstitial cations are removed and leached, which promotes subsequent dissolution of the silicate structure. Potassium feldspars ($KAlSi_3O_8$) cannot form a solid solution with plagioclases because K^+ is a much larger cation. Potassium feldspars (K-spars) are much more resistant to weathering than are plagioclases (Fig. 9.1). Microcline, orthoclase and sanidine are the most common forms of K-spar.

Zeolites are framework silicates in which the silica tetrahedra form a more open and less uniform structure than is observed in quartz or feldspars. In zeolites, tetrahedra are linked to form open, cage-like structures with more interstitial space than in quartz or feldspars. Zeolites differ in the number of linked tetrahedra and the size of the interstitial cavities. Like feldspars, zeolites have Al^{3+} substitution for some of the Si^{4+}, which produces a negative structural charge that is balanced by Na^+, K^+, Ca^{2+} and Mg^{2+}, and sometimes Ba^{2+} or Sr^{2+}. Zeolites can form in basalts and other mafic igneous rocks; they also form authigenically in sedimentary rocks, particularly those derived from volcanic rocks, and can be found in a variety of soil environments (Ming and Mumpton 1989).

Phyllosilicates (layer silicates)

Structure and classification

Phyllosilicates or layer silicates contain silica tetrahedra in which all three O^{2-} ions at the base of each silica tetrahedron are shared between two tetrahedra, and the linked tetrahedra are arranged to form a "sheet" of pseudo-hexagonal rings with the general formula $Si_2O_5^{2-}$ (Fig. 4.5). The O^- ions at the base of the tetrahedra are called the "basal plane" or "siloxane surface." The apical O^{2-} ions of each tetrahedron are not shared with other silica tetrahedra, but rather are shared or bonded to a metal hydroxide octahedral

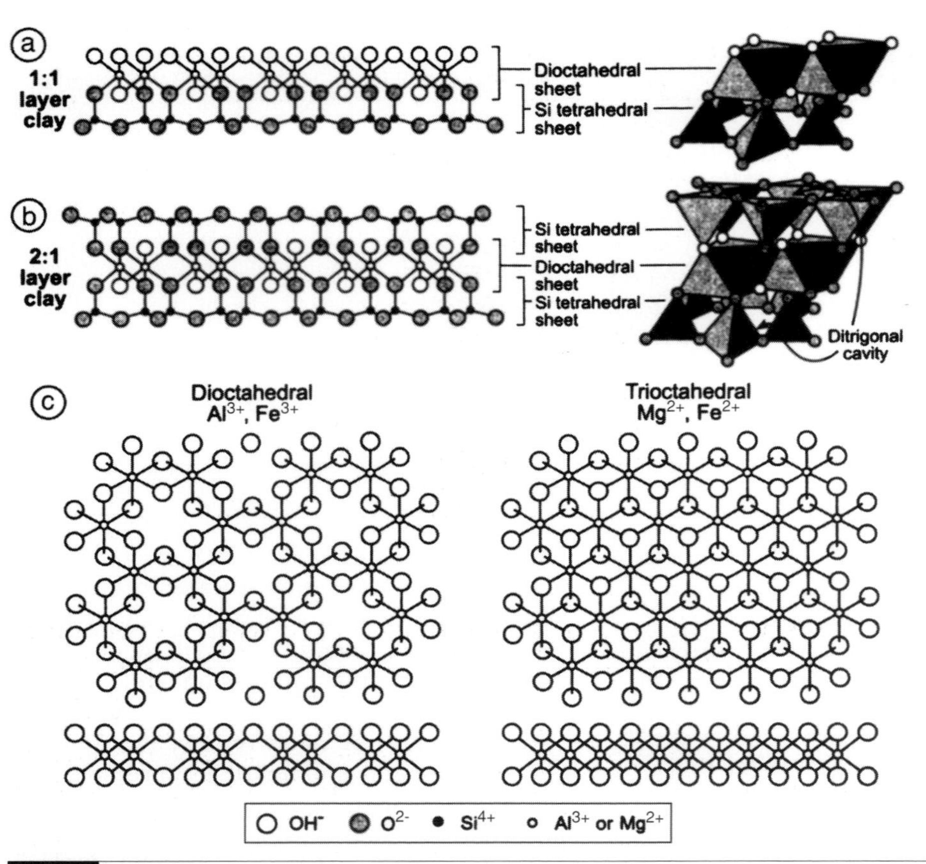

Fig. 4.6 Schematic diagrams of layer silicates, or phyllosilicates. (a) Ball-and-stick diagram of a 1 : 1 layer phyllosilicate, composed of one silica tetrahedral sheet composed of basal plane of O atoms, Si^{4+} cations, apical oxygens shared with an octahedral sheet, OH groups in the octahedral sheet (not shared); a trivalent octahedral cation occupying 2/3 octahedral sites, and an octahedral, outer OH plane. A polyhedral model of the same phyllosilicate structure is located to the right. In this truncated polyhedral structure, not all the connections between tetrahedra and octathedra can be shown. (b) Ball-and-stick diagram of a 2 : 1 layer phyllosilicate, showing upper and lower silica tetrahedral sheets sharing apical oxygens with the dioctahedral sheet within the 2 : 1 layer. The octahedral OH groups in a 2 : 1 layer silicate are inside the crystal structure. A polyhedral model of the same phyllosilicate structure is located to the right. In this truncated polyhedral structure, not all the connections between tetrahedra and octathedra can be shown. (c) Comparison of a dioctahedral sheet (2/3 cation sites occupied by 3+ cations) and a trioctahedral sheet (all cation sites occupied by 2+ cations), as seen from above.

sheet, e.g., $Al(OH)_3$ or $Mg(OH)_2$ (Fig. 4.6). The combination of an $Si_2O_5^{2-}$ silica tetrahedral sheet with an $Al(OH)_3$ octahedral sheet creates an mineral with the overall "unit cell" formula[1] of $Al_2Si_2O_5(OH)_4$. Phyllosilicates where one silica tetrahedral sheet is bonded to one octahedral sheet are called 1 : 1 phyllosilicates, signifying that the unit cell consists of one tetrahedral sheet and one octahedral sheet. Kaolinite and serpentine are examples of 1 : 1 layer silicates (Table 4.5). The term 'layer' refers to the combined tetrahedral and octahedral sheets, with most phyllosilicates classified as being 1 : 1 or 2 : 1 layer minerals.

[1] The "unit cell" is the smallest set of atoms in the crystal structure that contains a complete sample of the crystal pattern that is repeated in space, to form the mineral.

Table 4.5 Classification of 1:1 phyllosilicates according to the type of octahedral sheet (dioctahedral or trioctahedral) and octahedral cation, layer charge, formula and interlayer forces

Octahedral sheet and cation	Layer charge	Mineral or group name	Ideal formula	Interlayer bonding
Kaolin group				H-bonding
Dioctahedral	0	Kaolinite	$Al_2Si_2O_5(OH)_4$	
Al^{3+}	0	Halloysite	$Al_2Si_2O_5(OH)_4 \cdot 2H_2O$	
	0	Dickite	$Al_2Si_2O_5(OH)_4$	
	0	Nacrite	$Al_2Si_2O_5(OH)_4$	
Serpentinite group				
Trioctahedral	0	Chrysotile	$Mg_3Si_2O_5(OH)_4$	H-bonding
Mg^{2+} and some Fe^{2+}	0	Antigorite	$Mg_{3-3/m}Si_2O_5(OH)_{4-6/m}$	
	0	Lizardite	$[(Mg,Fe)_{3-x}Al_x][Si_{2-x}Al_x]O_5(OH)_4$	

In 2:1 layer silicates, two silica tetrahedral sheets surround or "sandwich" one octahedral sheet (Fig. 4.6). In 2:1 layers, the basal plane of O atoms, i.e., the siloxane surface, of each tetrahedral sheet is on the outside of each 2:1 layer, creating an upper and lower siloxane surface. The apical oxygens are bonded to the octahedral sheet on the inside of the layer (Fig. 4.6). A 2:1 layer that contains two $Si_2O_5^{2-}$ tetrahedral sheets sandwiching an $Al(OH)_3$ octahedral sheet has a unit cell formula of $Al_2Si_4O_{10}(OH)_2$. The formulas of 2:1 phyllosilicates have twice as many Si tetrahedra and only half as many OH groups per unit cell compared with the formula of 1:1 layer silicates because the octahedral sheet is shared between two different tetrahedral sheets. Note that mineral formulas are written with the octahedral cations (Al^{3+} in these examples) listed before the tetrahedral cations (Si^{4+}); the O^{2-} and OH^- groups that balance the cation charge are listed last. Examples of phyllosilicates with 2:1 layers include micas, smectites and vermiculite (Table 4.6).

In addition to being classified as 1:1 or 2:1 layers, phyllosilicates are also classified according to the *type* of octahedral sheet they contain. As shown previously for gibbsite $Al(OH)_3$ (Fig. 4.4), trivalent octahedral cations only occupy 2/3 of the cation sites in the octahedral sheet; 1/3 of the sites are vacant (Fig. 4.6). Octahedral sheets that contain predominantly trivalent cations such as Al^{3+} or Fe^{3+} are called "*di*octahedral sheets" because only *two* out of every three of the octahedral sites are filled (Fig. 4.6), giving a 1:1 layer formula of $Al_2Si_2O_5(OH)_4$. In contrast, divalent cations like Mg^{2+} or Fe^{2+} will occupy all three octahedral sites (Fig. 4.6); a 1:1 layer composed of a $Mg(OH)_2$ has the formula $Mg_3Si_2O_5(OH)_4$. Thus, the term "trioctahedral" can be remembered as signifying that there are three divalent octahedral cations in the unit cell formula; dioctahedral sheets have two trivalent octahedral cations per unit cell (Tables 4.5, 4.6).

Additional criteria for phyllosilicate classification include the amount of isomorphous substitution in the mineral structure and whether the substitution occurs in the tetrahedral or octahedral sheet. Isomorphous substitution is the replacement of one ion in the crystal structure by another ion of similar charge and radius (Table 4.1) without altering the crystal form. In clays, isomorphous substitution typically involves substitution of lower-valence cations for higher valence cations, i.e., Al^{3+} for Si^{4+}, in the crystal structure. Each substitution by a lower-charge cation creates a net *negative charge* in the layer structure. The amount of charge per unit cell or formula unit is called the "layer charge." Isomorphous substitution can occur in either the

Table 4.6 Classification of 2:1 phyllosilicates according to octahedral cation and sheet type (dioctahedral or trioctahedral), layer charge, charge location and formula

Octahedral sheet and cation	Layer charge	Mineral or group name	Ideal formula	Location of isomorphous substitution
		Pyrophyllite–talc group		
Dioctahedral				
Al^{3+}	0.01	Pyrophyllite	$Al_2Si_4O_{10}(OH)_2$	
Fe^{3+}		Ferripyrophyllite	$Fe_2Si_4O_{10}(OH)_2$	
Trioctahedral				
Mg^{3+}	0.02	Talc	$Mg_3Si_4O_{10}(OH)_2$	
Fe^{2+}	0.01	Minnesotaite	$Fe^{II}{}_3Si_4O_{10}(OH)_2$	
Ni^{3+}	0.04	Willemseite	$Ni_3Si_4O_{10}(OH)_2$	
		True micas		
Dioctahedral				
Al^{3+}	1.0	Muscovite	$KAl_2(Si_3Al)O_{10}(OH)_2$	Tet sheet
Al^{3+} and Fe^{3+}	1.0	Celadonite	$K_{0.99}(Al_{0.2}Fe^{3+}_{0.9}M^{3+}_{0.9})(Si_{3.96}Al_{0.04})O_{10}(OH)_2$	Oct sheet
Al^{3+} and Fe^{3+}	0.6–0.85	Illite (hydrous mica)	$K_{0.75}(Al_{1.3}Fe^{3+}_{0.4}M^{2+}{}_{0.2})(Si_{3.4}Al_{0.6})O_{10}(OH)_2$	Tet (oct) sheet
Al^{3+} and Fe^{3+}	0.6– 0.8	Glauconite	$K_{0.75}(Al_{0.5}Fe^{3+}0.9M^{2+}{}_{0.6})(Si_{3.8}Al_{0.2})O_{10}(OH)_2$	Oct (tet) sheet
Trioctahedral				
Fe^{2+} and Mg^{2+}	1.0	Biotite	$K(Mg,Fe)_3(Si_3Al)O_{10}(OH)_2$	Tet sheet
Fe^{2+} and Mg^{2+}		Phlogopite	$K(Mg,Fe)_3(Si_3Al)O_{10}(OH)_2$	Tet sheet
Al^{3+} and Li^{+}		Lepidolite	$K(Li,Al)_{2.5-3}(Si_3Al)O_{10}(OH)_2$	Tet sheet
Mainly trioctahedral; some dioctahedral				
	0.6–0.9	Vermiculite	$Mg_x/2(Mg,Fe^{3+})(Si_4\text{-}xAl_x)O_{10}(OH)_2$	Either
		Smectites		
Dioctahedral				
	0.2–0.6	Beidellite	$Na_{0.33}Al_2(Si_{3.67}Al_{0.33})O_{10}(OH)_2$	Tet sheet
	0.2–0.6	Nontronite	$Na_{0.33}Al_2(Si_{3.67}Al_{0.33})O_{10}(OH)_2$	Tet sheet
	0.2–0.6	Montmorillonite	$Na_{0.33}(Al_{1.67}Mg_{0.33})Si_4O_{10}(OH)_2$	Oct sheet
Trioctahedral				
	0.2–0.6	Saponite	$Na_{0.33}Mg_3(Si_{3.67}Al_{0.33})O_{10}(OH)_2$	Tet sheet
	0.2–0.6	Hectorite	$Na_{0.33}(Mg_{2.67}Li_{0.33})Si_4O_{10}(OH)_2$	Oct sheet
		Chlorite		
Trioctahedral/ trioctahedral	1.0			Mainly tet sheet

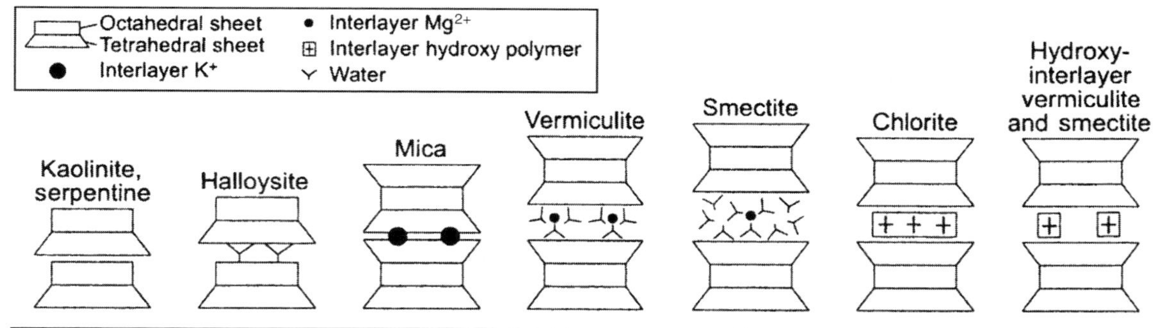

Fig. 4.7 Schematic diagram of common soil phyllosilicates showing octahedral and tetrahedral sheets and type of interlayer material. From left: 1 : 1 clays (kaolinite and serpentine); halloysite with interlayer water; mica group with interlayer potassium residing within the ditrigonal cavity in its basal plane of O atoms, yielding collapsed 2 : 1 layers; vermiculite (2 : 1) with hydrated interlayer cations such as Mg^{2+}; smectite group (2 : 1) with expanded interlayers and layers of water around each interlayer cation; chlorite (2 : 1 : 1) with a continuous interlayer octahedral metal hydroxide sheet; hydroxy-interlayered vermiculite or smectite with clusters of octahedral metal hydroxides.

tetrahedral sheet (Al^{3+} for Si^{4+}) or in the octahedral sheet, e.g., Mg^{2+} for Al^{3+} or Fe^{3+}. Clays can be categorized based upon the amount of tetrahedral charge or octahedral charge, i.e., isomorphous substitution in the tetrahedral or octahedral sheet, as well as on the basis of the total layer charge. Isomorphous substitution is only important in 2:1 clays; 1:1 layers have negligible isomorphous substitution. The layer charge created by isomorphous substitution is neutralized by cations (or cationic polymers) that are adsorbed between two adjacent 2:1 phyllosilicate layers (Fig. 4.7). The type of interlayer cation or material is also used as a basis of phyllosilicate classification.

Each of these classification criteria will be described more fully below, using specific examples. As shown in Tables 4.5 and 4.6, the type of layer (1:1 or 2:1) is the primary criterion for classifying phyllosilicates. The next most important criteria are total layer charge, whether the layer charge is in the tetrahedral or octahedral sheet, and whether the octahedral sheet is dioctahedral or trioctahedral. The type of interlayer cation or material is also important in some cases.

1:1 Layer silicates: kaolin group

The kaolin group of minerals are 1:1 dioctahedral minerals in which Al^{3+} is the octahedral cation (Table 4.5). Kaolinite, dickite and nacrite all have the formula $Al_2Si_2O_5(OH)_4$. The kaolin group of minerals have little or no isomorphous substitution, and hence a negligible layer charge. Instead, adjacent 1:1 layers are held together by hydrogen bonds between the OH of the octahedral sheet and basal O of the tetrahedral sheet (Fig. 4.7). The three kaolin minerals differ in the stacking arrangement of the neighboring 1:1 layers (Bailey 1980). Dickite and nacrite form in hydrothermal environments and are not usually found in soils. Kaolinite, which is very common in soils (Fig. 15.12), can either have a hydrothermal origin or can be pedogenic. Pedogenic kaolinite particles are small (<1 μm) and typically are pseudo-hexagonal. Kaolinite can form by weathering of other aluminosilicates in slightly acidic, well-drained and highly weathered soils in which silica and relatively mobile base cations such as Ca^{2+} and Mg^{2+} have been removed by leaching (Dixon 1989).

Halloysite ($Al_2Si_2O_5(OH)_4{\cdot}2H_2O$) has the same general formula as kaolinite except that there are two interlayer water molecules per unit cell. Halloysite minerals can be tubular, lath-shaped, or even spheroidal. Halloysite that has been completely dehydrated is platy. It is poorly ordered, probably because of the interlayer water and effect of drying on the interlayer bonding. Halloysite forms by weathering in acidic, volcanic sediment and soils (Dixon 1989).

1:1 Layer silicates: serpentine group

The serpentine minerals are 1:1 trioctahedral minerals that usually have Mg^{2+} in the octahedral sheet, although Co, Cr, Ni and Al also can substitute (Table 4.5). The serpentine minerals are found in soils derived from ultramafic rocks (serpentinites) that originate in the ocean crust (see Chapter 8). Chrysotile ($Mg_3Si_2O_5(OH)_4$), which produces asbestos fibers, is composed of 1 : 1 layers that curl into cylindrical or spiral fibers; the curling occurs to maximize hydrogen bonding between layers. Antigorite and lizardite (Table 4.5) both have overall platy morphology (Bailey 1980).

Serpentines form by hydrothermal alteration of olivine, pyroxene and peridotite, and are also found in ultramafic igneous and metamorphic rocks. Serpentines weather easily, rendering them uncommon in the clay fraction, except in very young, unweathered soils. Serpentine-derived soils tend to be high in pedogenic chlorite, which is normally an unstable clay mineral, and smectite, especially iron-rich smectite, as well as more uncommon minerals such as talc and lizardite (Wildman *et al.* 1968, Parisio 1981). High Fe contents lead to the formation of iron oxides, especially goethite, and the development of reddish-brown soil colors. Impeded drainage may allow Fe mottles and concretions to form. Some authors report that initial clay mineral suites are preferentially altered to vermiculite and smectite clays (Wildman *et al.* 1968, Ducloux *et al.* 1976).

Overview of 2 : 1 layer silicates

All 2 : 1 phyllosilicates contain two $Si_2O_5^-$ tetrahedral sheets that share apical O atoms with a either a dioctahedral sheet (containing Al^{3+} or Fe^{3+}) or a trioctahedral sheet (containing Mg^{2+} or Fe^{2+}) (Fig. 4.6). There are many more types of 2 : 1 phyllosilicates than 1 : 1 phyllosilicates because 2 : 1 layer silicates have differing amounts of isomorphous substitution in either the tetrahedral or octahedral sheet. Thus, 2 : 1 clays are classified according to their layer charge (the net charge on the 2 : 1 layer caused by isomorphous substitution) *and* on whether the isomorphous substitution occurs in the tetrahedral sheet (Al^{3+} substituting for Si^{4+}) or the octahedral sheet. The groupings shown in Table 4.6 reflect both the overall layer charge as well as its location. The classification of some common 2 : 1 phyllosilicates, based on total layer charge and the source of the layer charge, is also shown schematically in Fig. 4.8. Another criterion used to classify 2 : 1 phyllosilicates is the type of interlayer cation or material. Interlayer cations (or cationic hydroxide sheets or polymers) balance net negative charge in the 2 : 1 layer; the interlayer material is part of the mineral's overall chemical formula.

2 : 1 Layer silicates: talc–pyrophyllite group

Talc and pyrophyllite are the simplest 2 : 1 clay minerals in terms of chemical composition, consisting of an octahedral sheet that is sandwiched between tetrahedral sheets. Talc and pyrophyllite are the prototype 2 : 1 minerals because they have no isomorphous substitution in either the tetrahedral sheet or the octahedral sheet and therefore have relatively simple formulas. Pyrophyllite ($Al_2Si_4O_{10}(OH)_2$) has an Al dioctahedral sheet; talc is trioctahedral and has an Mg octahedral sheet ($Mg_3Si_4O_{10}(OH)_2$). Because there is no isomorphous substitution and the 2 : 1 layers are uncharged, adjacent 2 : 1 layers are held together only by weak dispersion forces. The weak interlayer forces are responsible for the slippery or greasy feel of talc and for its lubricating properties; the layers readily slide past each other. Talc forms from ultramafic parent material in metamorphic and hydrothermal environments; it also forms in soils from weathering of pyroxenes and amphiboles. Pyrophyllite is rare in soils (Zelazny and White 1989).

2 : 1 Layer silicates: mica group

Micas are primary 2 : 1 layer silicates found in many soil environments. Although there are many types of micas (Table 4.6), only muscovite and biotite are common in soils. In micas, the 2 : 1 layers have isomorphous substitution of lower-valence cations for higher-valence cations in the tetrahedral and/or octahedral sheet to yield a layer charge of about 1.0 mol per formula unit. The high negative charge on the 2 : 1 layers is balanced by K^+ held tightly between adjacent layers by strong electrostatic forces (Fig. 4.7).

Muscovite, which is the most common soil mica, has an $Al(OH)_3$ dioctahedral sheet. In addition to the Al^{3+} in the octahedral sheet, Al^{3+} also

Landscapes: serpentine barrens

Soils derived from materials weathered from serpentine-rich rocks tend to be highly infertile. Many such landscapes are covered with sparse and distinctive plant communities. These serpentine "barrens" or savannas stand out against the usually more dense vegetation on nearby soils formed from other parent materials (Marye 1955, Proctor 1971, Proctor and Woodell 1975, Rabenhorst *et al.* 1982, Graham *et al.* 1990b). To live on such soils, plants must have special adaptations, such as the ability to absorb calcium, which offsets the negative effects of magnesium, or the ability to withstand dry conditions and periodic soil erosion. The paucity of vegetation cover sets up a feedback; it allows erosion to occur, which makes the soil more droughty, making it still harder for vegetation, especially woody vegetation, to survive.

The grass cover on serpentine areas also leads to different profile morphologies, as compared to surrounding soils. For example, Mollisols may form on serpentine areas because only grasses and small shrubs can tolerate the extreme edaphic environment (Graham *et al.* 1990b), and because the neutral soil reaction favors high biological activity (Parisio 1981). The high pH values also inhibit lessivage and the high Fe contents promote brunification (see Chapter 12).

Soil profile formed in serpentine residuum, in the Soldier's Delight Natural Area, near Baltimore, Maryland. The knife is resting on bedrock. Photo by B. N. Weisenborn.

substitutes for 1/4 of the Si^{4+} ions in the tetrahedral sheets. Because Al^{3+} has a smaller positive charge than does Si^{4+}, each Al^{3+} that substitutes for Si^{4+} creates a net negative charge in the 2 : 1 layer. Comparison of the formula of muscovite, $KAl_2(Si_3Al)O_{10}(OH)_2$, with that of pyrophyllite, $Al_2Si_4O_{10}(OH)_2$, reveals that Al^{3+} substitutes for one out of every four Si^{4+} cations in the tetrahedral sheets, and that the negative layer charge is balanced by interlayer K^+.

Biotite is the second most common soil mica, although it can be locally important in young soils derived from mafic parent material. Like muscovite, biotite has a layer charge of 1.0 caused by Al^{3+} substituting for Si^{4+} in the tetrahedral sheet. Biotite differs from muscovite, however, by

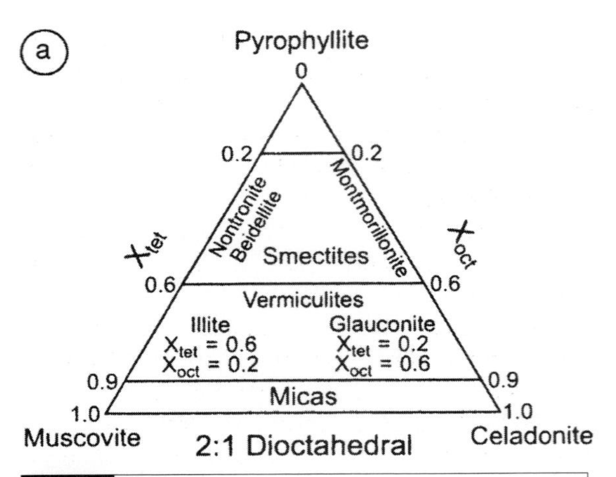

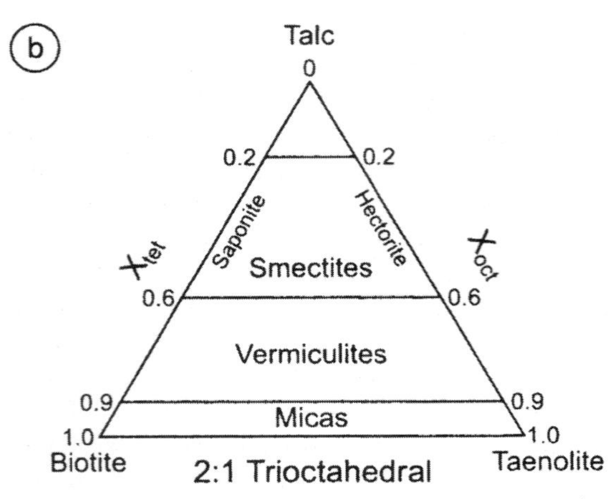

Fig. 4.8 Triangular diagrams of 2 : 1 clays showing their total layer charge on the vertical axis and location of layer charge on the horizontal axis. After Zelazny and White (1989). (a) Dioctahedral 2 : 1 clays. (b) Trioctahedral 2 : 1 clays.

having a trioctahedral sheet in which Fe^{2+} is the dominant octahedral cation. The Fe^{2+} gives biotite a darker color than muscovite, and also allows biotite to weather more rapidly.

Micas are platy minerals with pronounced cleavage; soil micas are primary minerals that are inherited via physical weathering of the parent material; they do not form *in situ*. Muscovite is found in granitic igneous rocks, as well as in metamorphic rocks from medium and high-grade metamorphism. It is also found in sericite, as a retrograde metamorphic alteration product of feldspar and other aluminosilicates. Biotite has a similar origin, but is also found in rocks that are more mafic, consistent with biotite's high Fe and Mg content. Biotite tends to weather up to 100 times more rapidly than muscovite (Fig. 9.1) and is less common in soils, especially weathered soils. The difference in weathering rates can be attributed to different weathering mechanisms. In muscovite, the primary weathering mechanism is loss of interlayer potassium, which begins at particle edges. Thus, small muscovite particles weather faster than larger particles. In biotite, electrostatic repulsion between interlayer K^+ and the OH^- groups of the trioctahedral sheet enhances the weathering rate, as does oxidation of Fe^{2+} to Fe^{3+} (Harris *et al.* 1985, Fanning *et al.* 1989).

Other micas (Table 4.6) are uncommon in soils, except locally where they are abundant in the parent material. Celadonite has isomorphous substitution mainly in its dioctahedral sheet; taenolite has isomorphous substitution in its trioctahedral sheet (Table 4.6, Fig. 4.8). Illite and glauconite, both of which form in marine environments, are dioctahedral micas that have isomorphous substitution in both the octahedral and dioctahedral sheets (Table 4.6, Fig. 4.8). In illite there is more substitution in the tetrahedral than octahedral sheet, whereas glauconite has more octahedral than tetrahedral substitution. In older soil literature, the term illite is sometimes used to refer to fine-grained, partially weathered soil muscovite (also called "hydrous mica"), but there have been efforts by mineralogists to reserve the term illite for micas formed via diagenesis. Although most micas have a layer charge close to 1.0, illite and glauconite, which form in marine environments, have a somewhat lower layer charge (0.6 to 0.8), which overlaps somewhat with vermiculite, described below.

Most micas have K^+ as the interlayer cation. Interlayer K^+ is held tightly or "fixed" between adjacent mica layers because the properties of micas and of the K^+ ion provide ideal conditions for K^+ "fixation." Because muscovite and biotite have high layer charges in the tetrahedral sheet, the electrostatic forces between the 2 : 1 layers and interlayer K^+ are very strong. Because K^+ is a large, low-charge cation that is only weakly hydrated by water, it is attracted so strongly to the highly charged mica surface that it releases its "hydration shell" of water molecules. The size of the K^+ ion provides a nearly perfect fit with the pseudo-hexagonal or "ditrigonal" ring formed by the basal-plane O atoms (Figs. 4.5, 4.6b), so interlayer K^+ in micas sits partially inside the ditrigonal "cavities," allowing the mica layers to collapse together around the interlayer K^+ ions (Fig. 4.7). Thus, the separation between adjacent 2:1 layers of micas is too small to allow entry of water molecules or other cations into the interlayer region, and interlayer K^+ is "fixed" or "nonexchangeable." The fixation of interlayer K^+ in micas causes micas to have low cation exchange capacity, because cation exchange is restricted to external surfaces of "packets" of mica layers. This characteristic also plays an important role in identification of micas by X-ray diffraction, as described later.

2:1 Layer silicates: vermiculite

Vermiculites have a layer charge of 0.6 to 0.9 moles per unit cell, and can be trioctahedral or dioctahedral (Table 4.6). The layer charge of vermiculite is similar to that of the low-charge micas illite and glauconite; vermiculite is distinguished from those micas by having Mg^{2+} or other readily exchangeable cations in the interlayer region to balance the layer charge, whereas illite and glauconite have interlayer K^+. The Mg^{2+} interlayer cation in vermiculite is strongly hydrated, meaning that water molecules are strongly attracted to and retained by Mg^{2+} ions. The attraction between Mg^{2+} and H_2O is greater than the attraction between Mg^{2+} and the surface of vermiculite, so the adsorbed Mg^{2+} cations are surrounded by one or two shells of water molecules when in the interlayer region of vermiculite. Because the separation between adjacent vermiculite layers is large enough for Mg^{2+} ions and their hydration shells, other ions can enter the interlayer region and replace the Mg^{2+}. As long as the replacing cations are small and strongly hydrated, the interlayer region of vermiculite will remain open and accessible (Douglas 1989). Because vermiculite has a moderately high layer charge and has Mg^{2+} as a readily exchangeable interlayer cation, vermiculite has a high cation exchange capacity (CEC) (1600–2000 $mmol_c\ kg^{-1}$) (Barshad and Kishk 1969). However, if cations such as K^+ or NH_4^+ enter the interlayer region, these large, low-charge cations cause vermiculite to collapse like mica, and the K^+ and NH_4^+ are fixed in the vermiculite interlayer. This behavior is used to help identify vermiculite by X-ray diffraction.

Vermiculites typically obtain their negative charge from isomorphous substitution in the tetrahedral sheet; in some instances, isomorphous substitution in the octahedral sheet can actually produce a net positive charge, particularly with trioctahedral vermiculites. Vermiculites are reported to be unstable intermediates in the mica weathering process (Kittrick 1973). Trioctahedral vermiculite can form from biotite if the rate of interlayer K^+ loss exceeds the rate of iron release from the octahedral sheet. This transformation is favored at pH values greater than about 7.5, because lower pH promotes dissolution of the 2:1 layers rather than simple transformation via loss of interlayer K^+ (Fanning *et al.* 1989).

2:1 Layer silicates: smectite group

Smectites are a group of 2:1 phyllosilicates characterized by a low layer charge of 0.2 to 0.6 moles per unit cell. They can be either trioctahedral or dioctahedral, and can have either tetrahedral or octahedral charge (Table 4.6, Fig. 4.8). The type of octahedral sheet and the source of layer charge is related to their mode of formation.

Soil smectites tend to form (and persist) in soils rich in silica, Mg^{2+} and Ca^{2+}. These mobile ions are found in poorly drained environments and in low-leaching environments with intense, long dry seasons (Folkoff and Meentemeyer 1985). Most soil smectite is pedogenic, and it is most abundant in the very fine clay fraction. Smectites can form either by partial weathering of other 2:1 clays such as mica, vermiculite and chlorite, or be "neoformed' by dissolution of other minerals with subsequent precipitation/recrystallization of dissolved ions. The later process is referred to as "dissolution and synthesis." Dioctahedral smectites can form by either mechanism, whereas trioctahedral smectites are either inherited from the sedimentary material or form by "simple transformation" (partial weathering without dissolution of 2:1 layers) of other trioctahedral 2:1 clays. Dioctahedral smectites are more stable and more abundant in soils than are trioctahedral smectites (Borchardt 1989).

Montmorillonite, with its $Al(OH)_3$ dioctahedral sheet with isomorphous substitution of Mg^{2+} and other divalent cations for octahedral Al^{3+}, is common in poorly drained or dry soils derived from Mg-rich intermediate and mafic rocks. Precipitation of montmorillonite from solution is favored by slightly acidic pH (<6.7) and low concentrations of organic ligands. Beidellite, which has also has an $Al(OH)_3$ dioctahedral sheet but has isomorphous substitution of Al^{3+} for Si^{4+} in the tetrahedral sheet, can precipitate from solution when the organic matter content is higher, since organic ligands complex Al^{3+} and promote Al^{3+} incorporation into the tetrahedral sheet (Kittrick 1971, Weaver *et al.* 1971). Beidellite can also form by direct transformation of other 2:1 clays, i.e., the 2:1 layers remain intact during weathering. Nontronite is an Fe^{3+}-rich smectite with tetrahedral charge that forms by simple transformation of iron-rich chlorite and vermiculite or glauconite (Robert 1973). Although the ideal formulas for smectites (Table 4.6) are relatively simple, smectites that form by precipitation in poorly drained environments contain a much wider range of octahedral cations and exchangeable cations, incorporating whatever ions are most abundant in the soil solution at the time of precipitation.

The low layer charge of smectite leads to relatively weak electrostatic forces between the 2:1 layers and the interlayer cations; smectite layers do not collapse completely even when K^+ is the interlayer cation. When Na^+ is the interlayer cation, the low layer charge of smectite and the small size and low charge of Na^+ produce only weak attraction between Na^+ and the smectite surface. The weak forces permit water to enter the interlayer region between the 2:1 layers and force the layers apart (Fig. 4.7); interlayer separations in Na^+ montmorillonite can be hundreds of times greater than when K^+ or Mg^{2+} is the interlayer cation. The extent of smectite swelling also depends on the total layer charge, and on whether the charge is in the octahedral or tetrahedral sheet. Montmorillonite has octahedral layer charge and swells more in water and other polar solvents than does beidellite, which has tetrahedral charge. When isomorphous substitution occurs in the octahedral sheet, the charge is farther away from the mineral surface than when the charge is in the tetrahedral sheet. The greater separation between the cations and the negative layer charge produces weaker electrostatic attraction between the clay and the interlayer cations, allowing more water molecules to enter the interlayer space, which in turn enables greater swelling between clay layers.

Natural shrink–swell processes in montmorillonitic soils during seasonal wet–dry cycles play an important role in soil genesis. Shrink–swell cycles cause a form of soil mixing called argilliturbation (Dudal 1965, Jayawardane and Greacen 1987) (Table 10.1). In many smectite-rich soils of wet-and-dry climates, frequent and strong argilliturbation processes produce the defining morphological characteristics of the soil order Vertisols (Wilding and Tessier 1988, Coulombe *et al.* 1996) (see Chapter 10). Swelling in smectite-rich soils weakens cohesive forces in the soil and contributes to soil creep and landslides (Borchardt 1989). Soils with small amounts of smectite do not swell extensively, yet their physical properties can be strongly influenced by smectite, particularly if soil pore waters are rich in Na^+.

Sodium weakens the interparticle forces between montmorillonite particles and deflocculates (disperses) fine-grained smectite particles. The dispersed montmorillonite particles, many of which are <0.2 μm in diameter, can be translocated through the soil profile but eventually clog soil pores and may inhibit drainage. Dispersion of montmorillonite at the soil surface can produce surface crusts that dramatically decrease infiltration rates. Thus, the small size of montmorillonite particles, coupled with the presence of Na^+ in shallow water tables or irrigation water, often create conditions of poor tilth in dry climates where this setting is most likely to occur.

2:1 Layer silicates: chlorite and hydroxy-interlayered clays

True chlorites are primary trioctahedral 2:1 clay minerals with a layer charge similar to that of mica, but with a positively charged octahedral sheet of metal hydroxide between adjacent 2:1 layers to balance the layer charge (Table 4.6, Fig. 4.7). The octahedral sheet within each 2:1 layer of chlorite is usually trioctahedral, dominated by either Mg^{2+} or Fe^{2+} (Barnhisel and Bertsch 1989) (Table 4.6). Chlorite is sometimes called a 2:1:1 phyllosilicate because of the interlayer octahedral sheet. In chlorite, the interlayer hydroxide sheet is continuous and blocks the entire interlayer region; other ions cannot enter the interlayer region unless the interlayer hydroxide sheet dissolves. True chlorites form in mafic igneous and metamorphic environments; when present in soils, they are inherited from the parent material. Chlorite is generally an unstable clay mineral, usually limited to areas with weakly developed soils derived from chlorite-bearing parent materials, or in cold climates where weathering is inhibited (Gao and Chen 1983, Yemane *et al.* 1996).

The term "pedogenic chlorite" is sometimes used to refer to 2:1 layer silicates (vermiculite or smectite) with extensive metal hydroxide polymers in the interlayer region. The terms hydroxy-interlayered vermiculite (HIV) and hydroxy-interlayered smectite (HIS) encompass vermiculite and smectites with positively charged Al^{3+} or Fe^{3+} hydroxide polymers in the interlayers. The most common interlayer material is aluminum hydroxide, though ferruginous parent materials (particularly chlorite) can produce iron hydroxide polymers in the interlayers (Barnhisel and Bertsch 1989). In some literature, the terms chlorite–vermiculite intergrade and chlorite–smectite intergrade are used to refer to HIV and HIS. The interlayer hydroxide polymers in HIV and HIS can range from nearly complete interlayers (as in "pedogenic chlorite") that completely block ion adsorption sites and prevent smectite swelling, to discrete islands that only slightly modify the ion adsorption and swelling behavior of vermiculite and smectite. Although other ions can enter the interlayer region of HIV and HIS when the hydroxy interlayers are not extensive, the metal hydroxide polymers are strongly adsorbed and not readily replaced. Thus, HIV and HIS have lower cation exchange capacities than do vermiculite or smectite. Because hydroxide interlayers tend to minimize swelling of smectite, such interlayers can have a favorable effect on soil physical properties in smectite-rich soils.

HIV and HIS can form either by weathering of chlorite or by precipitation of hydroxide interlayers during weathering of mica and vermiculite; hydroxide interlayers form as silica is leached. The presence of hydroxide interlayers can stabilize vermiculite and smectite and slow the rate of silica loss. However, when the pH is above about 7.2, the interlayers tend to dissolve (Barnhisel and Bertsch 1989).

Identification of phyllosilicates by X-ray diffraction

Soil clay fractions contain mixtures of phyllosilicates, typically aggregated by oxides, organic matter and occasionally carbonates. Phyllosilicates in these mixtures can be identified using X-ray diffraction (XRD) techniques after a series of pretreatment steps (described below) that remove aggregating agents and isolate the clay fraction (Kunze and Dixon 1986, Moore and Reynolds 1989).

Theory of XRD

X-ray diffraction methods generally allow for the distinction among different clays based upon the expansion or contraction of the interlayer space – the region between adjacent 2:1 or 1:1 layers – in the presence of different cations and solvents (Brown and Brindley 1980, Whittig and Allardice 1986). X-ray diffraction not only allows specific clay minerals to be identified from complex mixtures, but also can provide semi-quantitative estimates of their abundance (Brindley 1980, Brown and Brindley 1980, Whittig and Allardice 1986, Wilson 1987, McManus 1991, Courchesne and Gobran 1997). Most researchers focus is on the *qualitative* use of XRD for mineral identification, because the poor crystallinity, small particle size and diverse chemical composition of pedogenic clays greatly complicate quantitative XRD techniques (Whittig and Allardice 1986).

The basic principle underlying mineral identification by XRD is that minerals are crystalline materials with perioidic, repeating planes of atoms with uniform distances between these planes. The distance between a plane of atoms in a 2:1 or 1:1 layer (for example the plane of

Landscapes: geophagy in the tropics

Geophagy, the purposeful consumption of soil, is a nearly universal, transcultural phenomenon, which is particularly widespread in tropical nations, especially in Africa (Abrahams and Parsons 1996, 1997). In general, most geophagy involves eating clay, whether dried to a powder or shaped into disks, lozenges or other forms. The source of the clay may be B or C horizon material weathered in place, although some peoples mine clay from insect mounds, such as those of termites or wasps (Hunter 1984a, 1993) (see figure). Once a good source of geophagic materials is located, most cultures go back to the same soil pits again and again; it is not chosen indiscriminantly. Termite mounds are often preferred sites for geophagical materials.

(a)

(b)

Two areas where soil is being mined for geophagical clays. Photos by John Hunter. (a) Termite (macroterme) mound in Zambia. (b) Pit in clay-rich soils in Guatemala.

Why do peoples purposefully "eat dirt"? Physiologically, geophagy is most common among pregnant and lactating women, giving rise to the notion that the clay is ingested as a supplemental source of nutrients, minerals in particular. Persons with intestinal parasites such as hookworm continually suffer gastric irritation, and often resort to clay-eating to assuage this condition (Hunter 1973). Ingesting clay can sooth heartburn and diarrhea, alleviate hunger pangs, or simply satisfy a craving. This is the primary reason that children are given clays to eat (Vermeer 1966). When adults, these children are likely also to give clays to their children, perpetuating the culture.

By far the most common physiological association of geophagy is with pregnancy, but it also is common during menstration and lactation (Hunter 1984a, b, 1993, Abrahams and Parsons 1997). Because pregnancy exterts a physiological cost in terms of mineral nutrition, the resultant increase in clay-eating may be a behavioral response to a physiological need. The cravings of pregnant women are often associated with a nutritional need, and in many Third World cultures this craving is satiated by eating clay. Hunter (1984a) analyzed clay tablets eaten by the people of Central America, and found that they are rich in iron, calcium and manganese, and contain not insignificant amounts of copper, potassium and zinc (see also Hunter 1984b, 1993). Pregnant women in parts of Nigeria do not drink milk and their food is grown on calcium-deficient Oxisols; hence, the calcium they obtain via geophagy is vital to the health of their fetus (Vermeer 1966). Women in Guatemala overwhelmingly stated that they stop eating clay after giving birth. Thus, geophagy can be seen in the context of a nutritional supplement during a period of increased need, and if so, it fulfills a physiological need (Abrahams and Parsons 1997). Hunter (1984b) described a 21-year-old woman in Sierra Leone who was raising four children in a mud hut. She was pregnant again, and eating an exceptionally large amount of mud (about 290 g daily) from a wasp nest. But in so doing she was providing herself with 160% of the recommended pregnancy supplementation of zinc, 62% of calcium, 56% of manganese, 47% of iron and 30% of phosphorous. This was truly remarkable, and might help explain how she had given birth to 4 healthy children.

The roots of geophagy, however, lie in spiritual and religious beliefs (Hunter 1984a). Religious geophagy almost always involves the ingestion of clay tablets, formed by craftspersons and later blessed by monks, priests or holy persons. Clay tablets come in many forms, shapes and sizes; they are eaten in many parts of the world, fulfilling spiritual as well as physiological needs (Hunter 1984a, Hunter *et al.* 1989). The are sold and traded as commodities (see figure).

With these two main motivations in mind, Hunter (1973) developed a culture–nutrition hypothesis to explain geophagy in humans. He suggested that, over long periods of time, the cultural practice of geophagy subconsciously responds to the physiological needs of the body under stress. Pregnancy, for example, puts the woman in physiological stress, demanding a behavioral response. Long ago, the responses varied from person to person and place to place. Eventually, trial-and-error led to a subconscious knowledge that eating clay is beneficial in these situations. This knowledge is then passed on as a cultural behavior, which may become institutionalized and eventually develop into a cottage industry in which clay tablets are manufactured and sold (Hunter *et al.* 1989).

Geophagy is not limited to humans (French 1945, Setz *et al.* 1999, Krishnamani and Mahaney 2000). Many animals eat soil, especially as a source of salt,– at "salt licks." In the Tambopata–Candamo Reserve Zone of southeast Peru, macaws eat and lick exposed clays in cliffs, almost daily. It is thought that the clay acts as a type of "jungle antacid." Macaws eat a variety of seeds and fruit depending on the season. Many of the items they eat are high in various types of toxins. The minerals in the clay may help detoxify the seeds and fruit.

Tablets made from geophagal clays. Photos by John Hunter. (a) Tablets drying in the sun, Guatemala. (b) Marketplace display of geophagal clays in various shapes.

O^{2-} ions that form the siloxane surface) and that same plane of atoms in an adjacent layer is called the *d*-spacing and is characteristic of the layer type and type of interlayer material (Fig. 4.9). Phyllosilicate clay samples can be prepared so that clay particles are oriented with their 1 : 1 and 2 : 1 layers parallel to one another. Then, after being treated with different cations, solvents, and heat treatments, the *d*-spacings of the various clays in the samples can be determined and used to facilitate their identification.

X-ray diffraction requires an X-ray *diffractometer*, which consists of an X-ray generator, a filter to remove all but one wavelength of X-rays, slits to collimate the X-ray beam, a sample chamber (which holds a glass slide or porcelain plate that contains the oriented clay sample), a goniometer that rotates the sample (and/or the detector) and measures the angle between the two and a detector to measure the intensity of diffracted X-rays at different angles (Fig. 4.10a). An X-ray diffractogram (Fig. 4.11) is a recording of the intensity of diffracted X-rays as a function of diffraction angle Θ. Additional details about choosing the correct components of an X-ray diffractometer system, e.g., voltage, target metal, slits and filters, are reported elsewhere (Cullity 1978, Whittig and Allardice 1986, Wilson 1987).

When monochromatic X-rays of wavelength *l* reach a sample of oriented clay particles at an angle of incidence Θ (Fig. 4.10b), the X-rays pass through the solid until they strike atoms in the crystal structure and are reflected or diffracted. Because X-rays are diffracted from many atoms at different locations in the crystal, they are scattered in all directions, and many of the scattered X-ray waves are out of phase with one another and cancel each other out. However, scattered X-rays that are *in phase* reinforce one another at angles that depend upon the separation between parallel plans of atoms. The condition for constructive interference (X-ray waves in phase when they reach the detector) and the production of an intense X-ray signal at the detector is quantitatively described by *Bragg's law*, which relates the wavelength λ of the X-rays to the distance *d* between parallel planes of atoms in different unit cells (the *d*-spacing of adjacent clay layers), and the sine of the angle of incidence Θ between the X-ray beam and the plane of atoms (Fig. 4.10b):

$$\text{Bragg's Law: } n\lambda = 2d\sin(\Theta).$$

When the difference in path length between X-ray 1 and X-ray 2, i.e., $2r$ in Fig. 4.10b, is equal to an integral number n of wavelengths, i.e., $2r = n\lambda$,

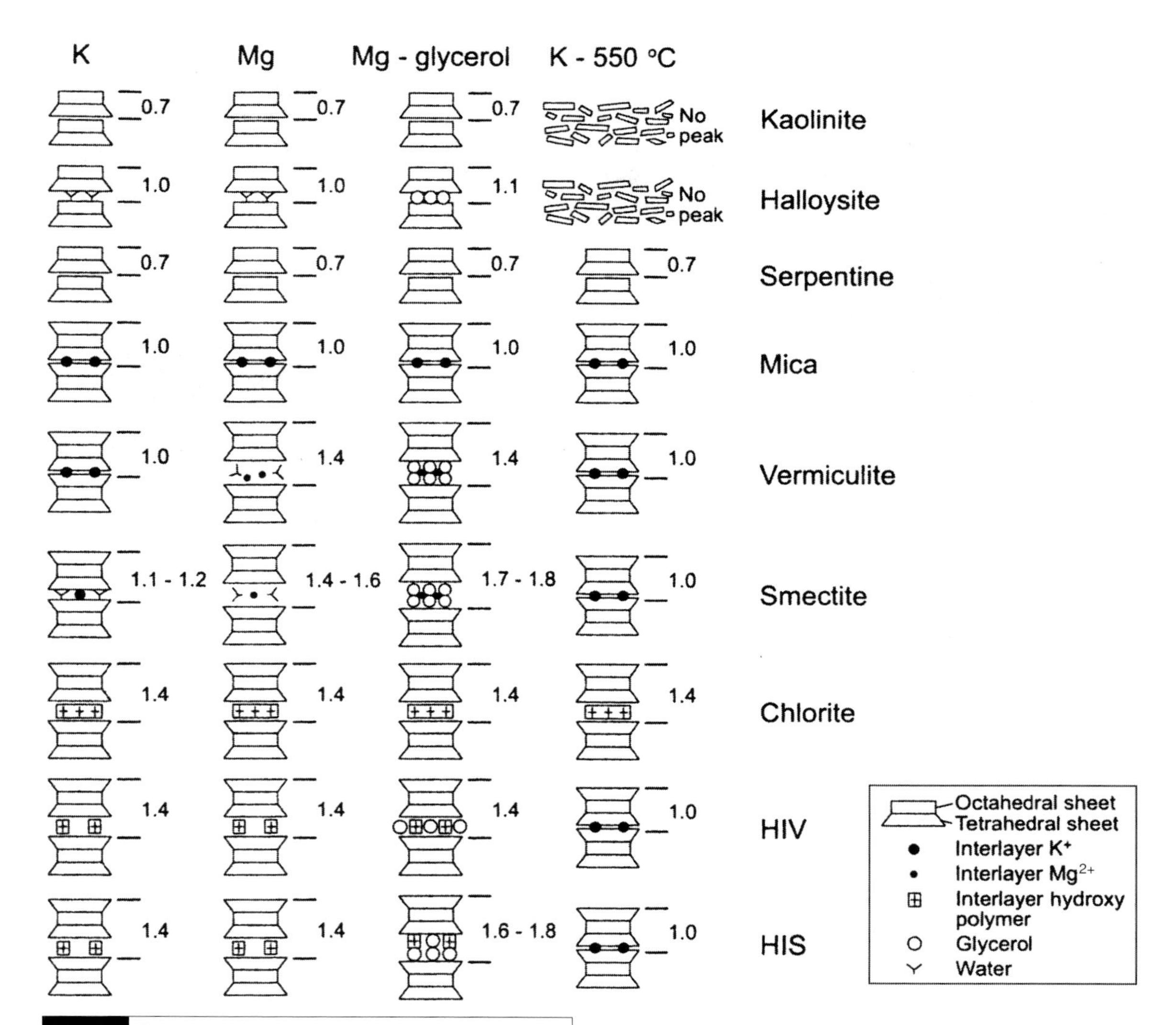

Fig. 4.9 Schematic diagram showing how clay mineral 001 *d*-spacings (in angstroms) change in response to various X-ray diffraction treatments: K^+, Mg^{2+}, Mg^{2+} and glycerol, and heat treatment to 550 °C. Collapse or expansion depends on layer charge and on the type of interlayer material and interlayer forces in the clay.

the X-rays interfere constructively. The geometric relationships shown in Fig. 4.10b show that $r = d \sin(\Theta)$. Thus, constructive interference (diffraction) occurs when $n\lambda = 2r = 2d \sin(\Theta)$, or when $d = n\lambda/2\sin(\Theta)$. Identification of clays involves measuring the angle Θ at which scattered X-rays interfere constructively and produce a peak in the intensity of X-rays reaching the detector, using Bragg's law to calculate the *d*-spacing corresponding to each peak, and then comparing the measured *d*-spacings with the known *d*-spacings of various clay minerals (Fig. 4.9). First-order diffraction peaks correspond to $n = 1$ in Bragg's law, and give the true *d*-spacing of the mineral. Second-order diffraction peaks occur at angles where $n = 2$ in Bragg's law, and give an apparent *d*-spacing equal to 1/2 the true *d*-spacing; third-order peaks give apparent *d*-spacings of 1/3 the true *d*-spacing, and so on. Thus, a mineral with a first-order or true *d*-spacing of 1.42 nm will give a second-order peak at 0.71 nm, a third-order peak at 0.474 nm and a fourth-order peak at 0.356 nm. First-order peaks are normally used to identify clay minerals, but occasionally the second-order peak from one mineral can overlap with the

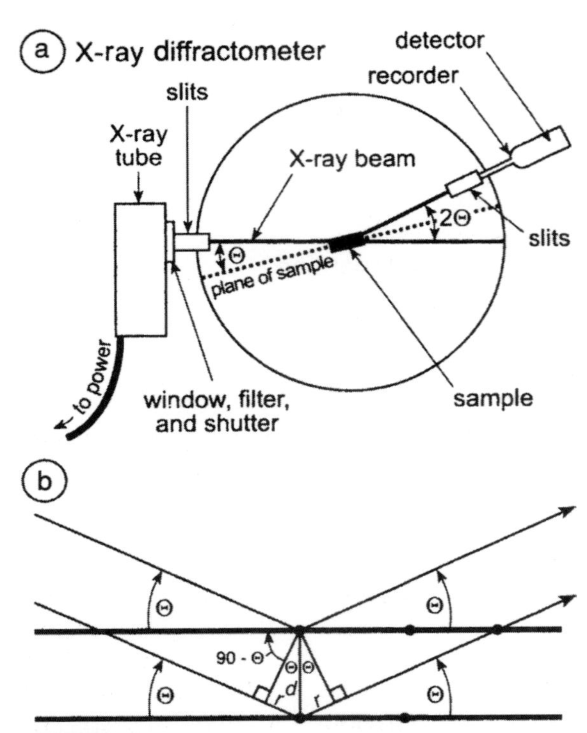

Fig. 4.10 (a) Schematic diagram of X-ray diffractometer showing the X-ray source that generates X-rays from a metal target, the shutter that opens and closes for analysis and sample insertion, the metal filter to filter all but desired wavelengths of X-rays, slits to collimate the radiation, sample holder, recorder and detector. The angle Θ is the angle between the incident radiation and the plane of the sample. 2Θ is the angle between the scattered radiation reaching the detector and the incident radiation. (b) Schematic diagram showing how incident X-ray radiation is scattered by atoms in the crystal structure of a mineral. Scattered X-rays interfere constructively at the detector when the path difference ($2r$) between X-rays hitting one plane of atoms and X-rays hitting a parallel plane of atoms is given by $2r = n\lambda$ where λ is the wavelength of the radiation.

first-order peak of another mineral and complicate mineral identification. This situation will discussed later in the context of identifying 1 : 1 clays in the presence of chlorite.

Application of XRD

Before clays can be identified by XRD, the soil must be pretreated to remove aggregating agents so that the clay size fraction can be separated from coarser particles by sedimentation (Brown and Brindley 1980, Kunze and Dixon 1986, Whittig and Allardice 1986). The first pretreatment step is typically to remove carbonates and soluble salts. Organic matter is then removed by treatment with H_2O_2. If necessary, iron oxides are also removed by using various chemical reducing agents. The resulting soil sample is Na-saturated and essentially free of aggregating agents. It is then washed with distilled water to remove excess salt and to promote dispersion. The sand and silt fractions are separated by wet-sieving and sedimentation, and the resulting clay suspension is concentrated by centrifugation.

The clay suspension is next subjected to a series of cation, solvent and heat treatments, each designed to *change the* d*-spacing* of certain layer silicates and facilitate their identification within the sample (Fig. 4.9). A portion of the clay suspension is reacted with a solution containing small, strongly hydrated cations such as Na^+ (Brindley and Brown 1980) or Mg^{2+} (Whittig and Allardice 1986) to saturate interlayer adsorption sites with Mg^{2+} or Na^+. Another part of the clay suspension is reacted with a KCl solution to saturate the interlayer cation adsorption sites with K^+. The cation-treated clay slurries are then either pipetted onto a glass slide or filtered through a porcelain plate to orient the 2 : 1 and 1 : 1 clay layers and make them lie flat. The samples are then X-rayed to determine the various *d*-spacings in each sample. The Mg-saturated sample is subsequently equilibrated with either glycerol or ethylene glycol, whereas the K-saturated sample is heated to 550 °C. Both the heat-treated and Mg–glycerol-treated samples are then X-rayed again. The treatments with K^+, Mg^{2+} (or Na^+), Mg–glycerol and heat cause predictable changes in the *d*-spacings of different phyllosilicates depending each mineral's layer type, layer charge, type of interlayer material and swelling characteristics (Fig. 4.9).

As shown in Fig. 4.9, the 1 : 1 kaolin and serpentine minerals have *d*-spacings of about 0.7 nm in all cation and solvent treatments; the *d*-spacing is unaffected by cation and solvent treatment because 1 : 1 clays have no isomorphous substitution and no interlayer cations to affect the *d*-spacing. However, heating to 550 °C causes kaolinite to decompose and its 0.7 nm peak to disappear. Serpentines do not decompose when

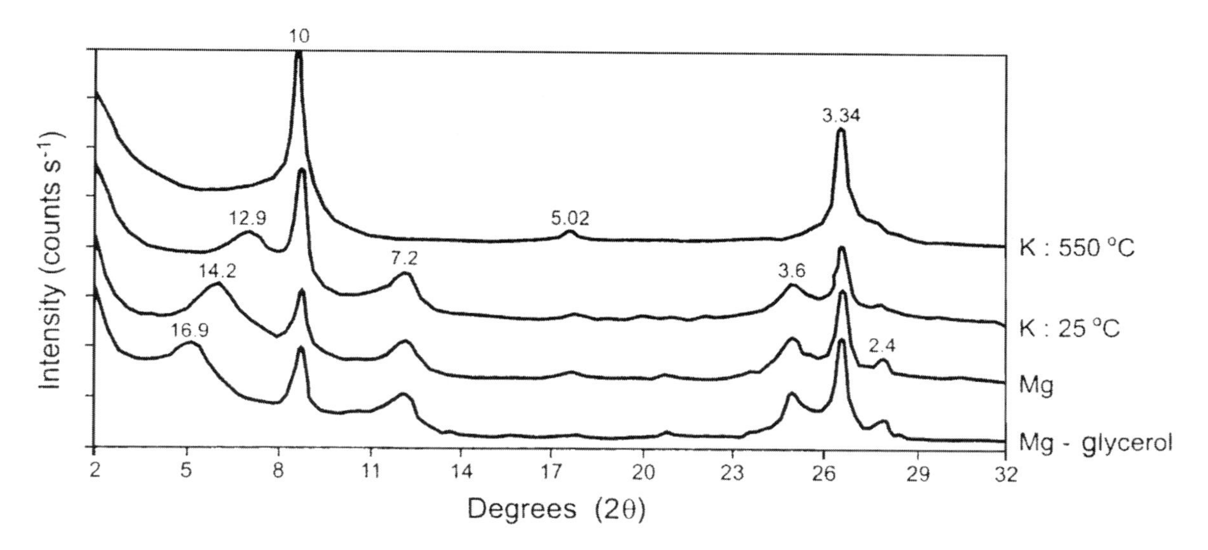

Fig. 4.11 Typical X -ray diffractograms of a soil sample that contains smectite (as shown by expansion to 1.67 nm when treated with ethylene glycol and collapse to 1.2 nm when treated with K^+), mica (as shown by 1.0 nm and 0.5 nm peaks in Mg^{2+} and Mg-ethylene glycol treatments), kaolinite (as shown by 0.72 nm and 0.357 nm peaks that disappear upon heating to 550 °C). The intense peak at 0.334 nm is clay-size quartz.

heated and can be more difficult to identify, but their sharper X-ray peaks (due to geologic, not pedogenic origin) facilitate their identification. Halloysite is a 1 : 1 clay with interlayer water. When treated with glycerol, it replaces that interlayer water, allowing halloysite to expand from 1.0 to 1.1 nm.

In Mg^{2+}-treated clays, all 2 : 1 layer silicates except mica, i.e., chlorite, vermiculite, smectite, HIV and HIS, have *d*-spacings of 1.4 nm. Micas have a *d*-spacing of 1.0 nm because they hold K^+ so tightly that they remain K-saturated even when reacted with NaCl or $MgCl_2$. The smaller *d*-spacing of micas reflects the collapse of mica layers around interlayer K^+. When Mg-treated clays are treated with glycerol or ethylene glycol, low-charge 2 : 1 clays (smectite and weakly interlayered HIS) expand to 1.6–1.8 nm (Fig. 4.9). Higher-charge clays do not expand and have the same *d*-spacings before and after glycerol treatment (Fig. 4.9). K^+ treatment causes vermiculite to collapse to 1.0 nm because of its higher layer charge, while smectites collapse only to about 1.2 nm when treated with K^+. Chlorite, HIV and HIS maintain a 1.4 nm *d*-spacing when K-treated because their interlayer hydroxide polymers prevent collapse of the layers. Heating the sample to 550 °C not only causes kaolinite to decompose but also provides a means of distinguishing chlorite from HIV and HIS. When the sample is heated to 550 °C, chlorite retains its interlayer metal hydroxide sheet and keeps its 1.4 nm *d*-spacing, whereas the discontinuous hydroxide interlayers in HIV and HIS decompose, causing at least partial collapse of those layers to lesser *d*-spacings (Fig. 4.9).

Clay mineral identification strategies

Specific strategies used to identify important soil phyllosilicates are described below and can be inferred from Fig. 4.9. Mica is identified by the presence of a 1.0 nm peak in Mg–glycerol-treated clays. Micas give a 1.0 nm peak in every treatment because the layers are collapsed around interlayer K^+; only micas give a 1.0 nm peak in Mg–glycerol samples. In some cases, biotite can be distinguished from muscovite by comparing the ratios of the first-order peak at 1.0 nm with the intensity of the second-order peak at 0.5 nm. In muscovite, the 0.5 nm peak is about 50% as intense as the 1.0 nm peak; biotite's high iron content causes the second-order peak to be only 20% as intense as the 1.0 nm first-order peak (Brindley 1980, Fanning *et al.* 1989). Another means of distinguishing muscovite from biotite is based on the difference in unit-cell dimensions between dioctahedral and trioctahedral micas, requiring either a special sample chamber or a random powder sample (Brindley and Brown 1980).

Smectite can be distinguished from other 2 : 1 phyllosilicates by the presence of a 1.2 nm peak in K-treated samples and a peak at 1.6 to 1.8 nm when Mg-saturated samples are treated with glycerol. Although HIS may also expand in the presence of glycerol, HIS does not collapse with K-treatment.

Vermiculite is characterized by 1.4 nm peaks with Mg and Mg–glycerol treatments, and a 1.0 nm peak after K treatment. Vermiculite can be difficult to identify in samples that also contain mica, smectite and chlorite; the best approach is to determine the peak/area ratios for the 1.4 nm peak and the 1.0 nm peak. Collapse of vermiculite in K treatment would cause the (1.4 nm/1.0 nm) peak/area ratio to be greater in Mg-treated than K-treated samples.

Chlorite has a 1.4 nm *d*-spacing that persists in all treatments. It is the only mineral that retains a 1.4 nm peak when heated to 550 °C.

HIV and HIS exhibit 1.4 nm *d*-spacings with K and Mg treatments. When treated with glycerol, HIS may expand slightly to 1.6 nm; HIV does not expand. Both HIV and HIS collapse from 1.4 nm to smaller *d*-spacings when heated to 550 °C (Fig. 4.9). The less extensive the hydroxy interlayers, the greater the collapse. In samples that also contain chlorite and vermiculite, HIS can be identified by expansion with glycerol; HIV can be inferred from the ratios of the 1.4 nm and 1.0 nm peaks in different treatments: If HIV is present, the 1.4 nm/1.0 nm peak ratio will be greater in the room-temperature K-saturated sample than in the sample that was heated to 550 °C.

Kaolinite or serpentine can be identified easily in the absence of chlorite and HIV by the presence of a relatively broad 0.7 nm peak in K-saturated samples. Although no other minerals have interlayer spacings of 0.7 nm, kaolinite's first-order peak at 0.7 nm coincides with the second-order peak for chlorite, making it difficult to identify kaolinite when chlorite is present. Because kaolinite decomposes at 550 °C but chlorite retains its 1.4 nm *d*-spacing (and 0.7 nm second-order peak) at 550 °C, kaolinite can be positively identified if the 0.7 nm/1.4 nm ratio decreases when the K-treated sample is heated to 550 °C. In some cases, kaolinite can be identified in the presence of chlorite on the basis of peak width. True chlorites have very sharp X-ray diffraction peaks, whereas pedogenic kaolinite gives a broad peak at 0.7 nm. A sharp 1.4 nm peak and a broad 0.7 nm peak may provide evidence that both chlorite and kaolinite are present.

Identification of iron and aluminum oxides

Iron and aluminum oxides in soils can be identified by XRD only in highly weathered soils, where they are present in sufficiently high concentrations. Even in highly weathered soils such as Oxisols or Ultisols, XRD identification of Fe and Al oxides usually requires particle-size separation to isolate the clay fraction, where concentrations of pedogenic Al and Fe oxides are greatest. Iron oxides can also be separated magnetically. Another approach is differential XRD, in which the diffractograms are obtained of a sample that contains oxides and a sample that has been treated to remove oxides; the difference between the treated and untreated diffractograms yields the diffraction pattern attributable to the oxide minerals. Yet another approach involves treating the sample with concentrated NaOH to dissolve silicates, and then X-raying the remaining solid phase (Schwertmann and Taylor 1989). The most intense X-ray peaks of common aluminum and iron oxides are reported in Tables 4.2 and 4.3, respectively.

Mössbauer spectroscopy, which gives information about the chemical environment of Fe^{2+} and Fe^{3+} in iron oxides, can also be used to identify iron oxides (Bowen and Weed 1981, Parfitt and Childs 1988). The method is based on measurement of magnetic fields at different temperatures; the field of Fe in different oxides responds differently to changing temperature, depending on the specific chemical environment of Fe in the mineral.

Iron and aluminum oxides can also be identified by thermal analysis (Buhmann and Grubb 1991). Differential thermal analysis involves measuring the difference in temperature between the sample and a reference material as the sample is heated. When a mineral undergoes an exothermic reaction (energy given off), the temperature of the sample is greater than the temperature of the reference material. An endothermic transition causes the sample to have lower temperature than the reference material. Dehydroxylation (loss of H_2O when an OH^- and another H^+ are lost from the mineral structure during heating) gives endotherms; recrystallization can produce exotherms. Thermal analysis is more often used for quantitative analysis when oxide minerals have already been identified by other methods, because many different Fe and Al oxides give overlapping endotherms in the 300–400 °C range.

Selective chemical dissolution can also be used to quantify different groups of Fe and Al oxides based on their dissolution by different reagents. Because dissolution rates depend on factors such as crystallinity and surface area, selective dissolution cannot be used to identify specific oxide minerals, which influence their dissolution rates.

Chapter 5

Basic concepts: soil physics

Soil physics is the branch of soil science that deals with the physical properties of soils. Included in this arena are the measurement, prediction and control of these physical properties, as well as the ways in which such knowledge facilitates various applications, e.g., irrigation scheduling. Just as physics deals with matter and energy, soil physics is generally concerned with the state and movement of matter and energy in soils.

Soil physical properties such as water content, texture and structure, as well as soil physical processes such as water retention and transport, soil temperature and heat flow, and the composition of the soil atmosphere (mainly O_2 and CO_2), all affect the rates of weathering and soil genesis. In this chapter we discuss the basic concepts of soil physics that are necessary as background to the more in-depth discussions of soil genesis and geomorphology that follow.

Soil water retention and energy

Many pedogenic process begin and end with the flow of water in soils (see Chapters 12 and 13). Water is the main agent by which solids and ions are transported within soils; knowledge of the forces acting upon water flow is, therefore, important to an understanding of soil genesis, not to mention soil use and management.

Water is retained in soils in two ways – in adsorbed and absorbed forms. *Adsorbed* water is retained on the *surfaces* of soil particles by chemical or physical binding, due to the natural adhesive attraction between water and solids. Water that is *absorbed*, however, is taken *into* the solid (soil mineral, rock particle or organic substance). Water retention in soils, in both adsorbed and absorbed forms, increases with increasing contents of clay and organic matter because of the affinity of water for those solids.

Liquid water has extremely high surface tension and viscosity because the H atoms in each H_2O molecule form strong hydrogen bonds to the O atoms of neighboring water molecules. The polar, hydrogen-bonding nature of water also creates strong adhesive attraction between water molecules and soil particles. In particular, water molecules can form H-bonds with OH groups on oxide minerals and edges of clay minerals, and with NH and OH groups on soil organic matter. In addition, water molecules also strongly solvate cations that are adsorbed by soil minerals. Because the H-bonds and ion-dipole bonds between water and soil particles are relatively strong, the first few layers of soil water get tightly adsorbed by soil particles (Fig. 5.1). Because of the strong attraction between water and charged and polar sites in clays and organic matter, water retention increases as the surface area of a soil increases and as the amount of clay and organic matter increase. Adsorbed water can range from one or two layers of water molecules (a few Å thick) to tens of water molecules thick (about 80 Å) in smectite-rich soils; swelling soils that are rich in smectite retain more water than other soils (see Chapter 4). Soil water is retained not only by adhesive forces between water and soil solids, but also by H-bonds among water molecules, i.e., cohesive forces. Water

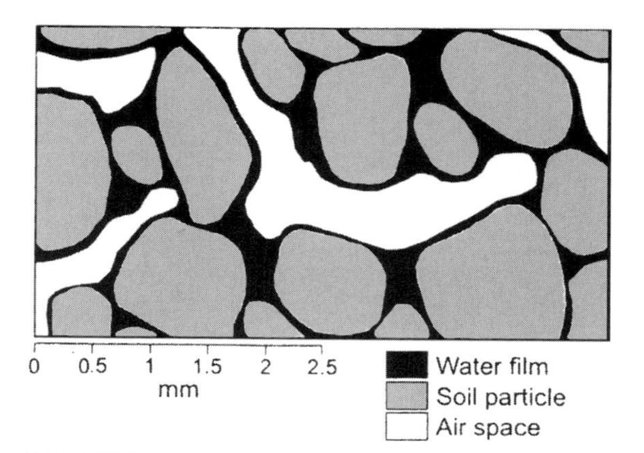

Fig. 5.1 Schematic diagram of soil pores showing water adsorbed in small pores and surface films; large pores remain air-filled.

held in small pores above the saturated zone, sometimes called "capillary water," is retained there by a combination of adhesive and cohesive forces. More water is held in fine-grained soils, therefore, because of their greater surface area and the greater number of smaller pores.

Soil water content can be expressed as either gravimetric water content q_g (mass of water per mass of dry soil), or as volumetric water content q_v (volume of water per unit volume of soil). Gravimetric water content is readily measured by weighing a sample of moist soil, oven-drying it at 105 °C for 24 h (or until the mass stops decreasing), and then reweighing the oven-dry soil (Gardner 1986). Gravimetric water content is then calculated as:

$$q_g = (\text{mass moist soil} - \text{mass oven dry soil})/\text{mass oven dry soil.}$$

Oven-drying at 105 °C removes most, but not all water; some is retained as tightly adsorbed molecules. Nevertheless, oven-drying is a widely used (but destructive) method to determine q_g that requires little in the way of equipment other than an oven and a balance.

Volumetric water content (q_V) is difficult to measure directly, and is usually calculated from measurements of q_g and the soil's bulk density (r_b), with the following equation:

$$q_v = r_b q_g / r_w$$

where r_w, the density of water, is equal to 1.0 g cm^{-3}.

Volumetric water content can be determined non-destructively with specialized equipment that measures moisture-dependent properties of soils (Topp and Davis 1985, Gardner 1986). One of the most widely used non-destructive methods is time domain reflectometry (TDR), in which soil water content is estimated by measuring the dielectric number (e) of the soil based on the travel time of a magnetic pulse between two probes. In TDR, two probes are inserted into the soil, typically vertically from the soil surface, although horizontal probes can be inserted into the wall of a soil pit. An electromagnetic pulse is then emitted from one probe and reflected back from the second probe to the source, at which the travel time and velocity are determined. The velocity of the electromagnetic pulse in the soil depends on soil permissivity, which depends in turn on soil type, solute concentration, and water content. Calibration equations appropriate for the soil type are then used to calculated q_v from e (Topp *et al.* 1980, Roth *et al.* 1990). TDR is best suited for long-term monitoring of soil moisture at one site, because calibration curves must be developed for each different soil. Other non-destructive methods of measuring water content are ground penetrating radar, neutron attenuation and gamma-ray attenuation (Gardner 1986).

Although water content is an important soil property, the potential energy of soil water governs its retention and movement. Like other fluids, water flows from regions where it has greater potential energy to regions where it has less. The potential energy of soil water is expressed relative to the potential energy of pure liquid water in a reservoir where interactions of water with the edges are negligible. The total potential energy of soil water (y_{tw}) is the sum of all of the forces acting on it, including gravity (y_g, gravitational potential), the attraction between the soil particles and water (y_m, matric potential) and the attraction between dissolved ions and water (y_o, osmotic potential). Soil water potential is typically reported in units of pressure (bars, atmospheres or pascals), in large part because water potential typically is measured by applying pressure (or suction) to the soil, and measuring the

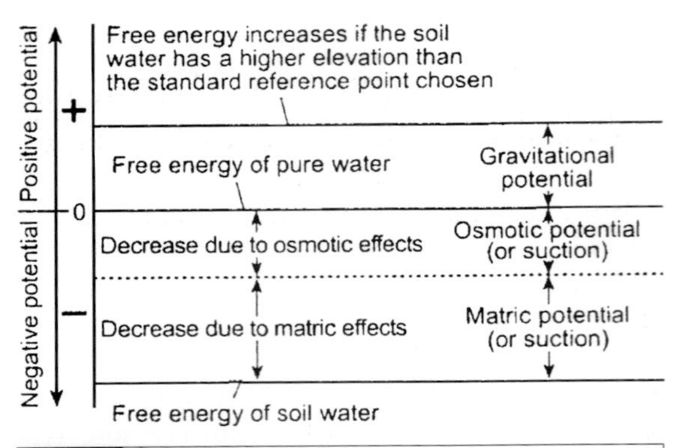

Fig. 5.2 Representation of water potential, with free, pure water having a potential of zero. Water ponded above the soil surface yields a positive potential, while the attraction of water for dissolved ions (osmotic forces) and for particle surfaces (matric forces) decrease the potential energy of soil water. After Brady (1974).

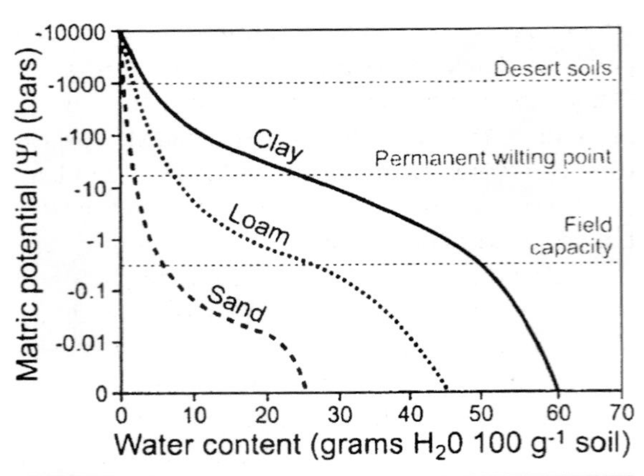

Fig. 5.3 Water characteristic curve (water potential vs. water content) for sand, loam and clay soils. Field capacity is shown at −0.3 bar, and permanent wilting point is shown at −15 bar. After Brady and Weil (2001).

pressure (or tension) necessary to extract it from the soil. Thus, the total soil water potential is given by

$$y_{tw} = y_g + y_m + y_o.$$

Although the osmotic potential in saline soils can affect the availability of water to plants, it is a much less important component of y_{tw} in unsaturated soils than is matric potential, or suction (y_m). In fact, in unsaturated, non-saline soils, y_m is typically about 95% of y_{tw} (Donahue *et al.* 1983). In situations where the soil air pressure is higher than the atmospheric pressure, or where the overlying soil exerts pressure on the soil water, additional water potential terms would be needed to reflect soil air pressure and overburden pressure.

At points below the water table, y_g is zero or positive and the total potential y_{tw} is also positive (Fig. 5.2), because the overlying water exerts pressure on the soil water at the depth where y_{tw} is measured. In a saturated soil (all pore spaces are filled with water), the matric potential y_m is equal to zero, because the water in the middle of large pores is not attracted to the soil matrix.

Water that drains freely from a soil under the influence of gravity is often referred to as *gravitational water*, and is the water for which cohesive forces between neighboring water molecules exceed the matric forces (attraction to the soil matrix). As water drains out of a soil, matric potential y_m and hence the total water potential y_{tw} decrease and become negative. The negative values of y reflect the fact that water is held in the soil by matric forces that exceed the force of gravity; removing this water from the soil requires energy inputs, i.e., gravity, oven-drying or suction. As the soil dries and drains, weakly held water in large pores (greater y_m) is removed first, and water adjacent to particle surfaces held by stronger matricforces (more negative y_m) remains. The first layer of adsorbed water is estimated to have a water potential of −8000 bars, whereas the outer layers of adsorbed water films have y_m ranging from −0.3 to −0.1, depending on soil texture. Drainage of a saturated soil under the influence of gravity removes water from large pores (Fig. 5.1), in which the downward pull of gravity is stronger than the adhesive forces between the soil particles and neighboring water molecules. The water potential measured after gravitational drainage stops is typically around −0.3 bars of suction. The water content of a soil after gravitational drainage ceases, called *field capacity*, is often equated with the water content when $y_w = -0.3$ bar. As can be seen in Fig. 5.3, the water content at field capacity is greatest in clayey soils, because of their greater surface area and stronger matric forces, and least in sands. Another important water potential benchmark is the *permanent wilting point*, which is the water potential below which plants are unable to extract soil water. The permanent wilting point varies with plant type. Many crop plants have a permanent wilting point of about −15 bar, whereas desert plants can extract soil water at potentials as low as −100 bar.

The relationship between soil water potential and water content, often called the moisture characteristic function, depends largely on soil texture. Moisture characteristic curves, which are plots of y_w or y_m vs. q, differ greatly between sandy soils and clay-rich soils (Fig. 5.3). Laboratory methods for determining the relationship between water content and matric potential are described in Klute (1986); field methods are detailed in Bruce and Luxmoore (1986). At any given matric potential, clays have higher water contents than do coarser soils because of the greater surface area and porosity of clays. Conversely, at a particular water content, clayey soils have a more negative y_m than do coarser-textured soils, because of the greater surface area of the clays and the smaller pore sizes.

The moisture characteristic curve (plot of y_m vs. q) for a given soil is not a unique function; it depends on whether the soil is being wetted or dried. Figure 5.4 illustrates this *hysteresis effect* in soils. Note that the moisture content in a drying soil is different, at the same matric potential, than it is in a wetting soil. Hysteresis in the soil moisture characteristic curve is caused by pores emptying in a different order than they fill, and by entrapment of air during wetting. During wetting, large pores may fill first, and trap air in small pores or in small cavities between pores. If wetting/drying were to occur slowly enough for equilibrium to occur, then the smallest pores would pull the water into them as air escaped from them, and the wetting and drying curves would be identical.

Different components of soil water potential can be measured with different methods. Tensiometers can be used to measure soil water potentials of 0 to −1 bar by allowing an airtight, porous ceramic tube (connected to a monometer) to equilibrate with soil water (Cassell and Klute 1986). Pressure plates can be used to measure water potentials down to almost −15 bar. Piezometer tubes can be used to measure hydrostatic and overburden pressure in saturated soils

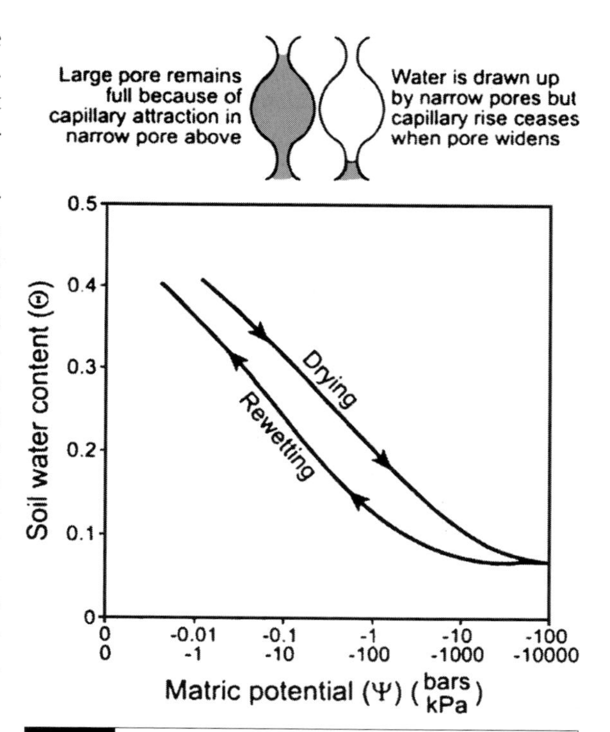

Fig. 5.4 Example of the hysteresis effect in soils. The water content of the soil, at the same matric potential, varies depending on whether the soil is being dried or wetted. After Brady and Weil (2001).

(Reeve 1986; see also Campbell and Gee 1986, Rawlins and Campbell 1986).

Soil water movement

Water flow in soils is vitally important in soil genesis, because percolating water dissolves minerals, and transports ions, colloids and metal–organic complexes downward in the soil profile. In addition, upward movement of water in response to evapotranspiration is vital to plant growth, and in some areas can lead to salt accumulations at the surface (see Chapter 12).

The first step in soil water transport is *infiltration*, the process by which water moves into the soil from above the surface. *Percolation* is the process by which water moves through the soil, vertically and laterally. The infiltration rate of a soil, measured in length units per unit of time, e.g., 4.2 cm h^{-1}, depends upon factors that affect the permeability of the soil, such as soil texture and pore size, the presence of swelling clays (which can plug soil pores as they expand), as well as on the organic matter content at the surface. Hydrophobic organic matter repels water, and limits infiltration until water is ponded on the surface. Infiltration rate is also affected by the initial soil water content in the soil below the leading edge of the percolation (wetting) front, as well as by the depth to the water table and to any impermeable subsurface layers. In addition, soil temperature affects infiltration; decreasing temperatures cause a decrease in infiltration due to the greater viscosity of water at low temperatures. Of course, infiltration is negligible in most frozen soils.

Infiltration and percolation rates diminish with time after initial wetting. The reasons for this include: (1) clay minerals and other colloids go into suspension, plugging pores as they infiltrate, (2) surface crusts, of low permeability, can form, especially in bare soils and (3) clay minerals swell as they wet, further decreasing pore space, size and connectivity. Entrapment of air in soil pores during infiltration can also inhibit percolation. Mathematical descriptions of infiltration are a focus of soil physics, but are not described in detail here (see Jury *et al.* 1991).

Once water has entered the profile, percolation or redistribution of water in the soil occurs under *saturated* and/or *unsaturated* conditions. Under saturated conditions, i.e., saturated flow, the steady-state vertical flux of water, J_{sat}, past depth z in the soil profile can be described by *Darcy's law*:

$$J_{sat} = K_{sat} H/z$$

where K_{sat} is the saturated hydraulic conductivity of the soil, and the H is the hydraulic head difference. Hydraulic head is the distance between the upper surface of the water table and depth z in the soil (Amoozegar and Warrick 1986). Water flows from areas of greater total water potential (y_{tw}) to areas of lesser y_w. In saturated soils, the hydraulic head or gravitational potential is the driving force for downward flow. Hydraulic conductivity of the soil, under saturated flow conditions, depends on porosity and pore size distribution; it is greater (faster) in sandy soils than in fine-textured soils. Although field conditions rarely if ever conform to Darcy's law because of heterogeneity and the fact that steady-state conditions seldom occur in the field where percolation and evaporation are occurring, this simple equation for saturated flow can provide insight into the variables that control water flux through soils.

Water flow in unsaturated conditions, i.e., unsaturated flow, still depends on the hydraulic conductivity and the driving force for water flow, but differences in water content or matric potential provide the primary driving forces for unsaturated water flow. When the soil is not saturated, only the water-filled pores are accessible for water flow, and hydraulic conductivity decreases as the water content and matric potential decrease. In addition, because the larger pores drain and fill with air first, the hydraulic conductivity initially declines very rapidly as water content decreases. Also, the air-filled pore spaces act as barriers to water flow; causing water flow to follow more tortuous paths along thin water films on soil particles and in small pores. Thus, water flux declines rapidly with decreasing water content. Methods for measuring unsaturated flow properties are described in Green *et al.* (1986).

In unsaturated conditions, where y_m dominates the total water potential, water flows from regions of higher water content (less negative y_m) to regions of lower (more negative) matric potential, i.e., from wet to dry areas in the soil. Gravitational drainage can occur under unsaturated conditions only when the matric potential is greater than about -0.3 bar (field capacity). However, as noted above, at matric potentials below -0.3 bar, the attraction of water for particle surfaces exceeds the pull of gravity, so flow is governed by differences in matric potential, not by gravitational drainage. Consequently, unsaturated flow can be upward in response to evaporation at the surface, downward as water is redistributed after a wetting event, or laterally.

In soils in which there are abrupt changes in soil texture at horizon boundaries, such as at lithologic discontinuities (see Chapters 3 and 9), rates of percolation can change markedly. When coarse-textured soil horizons overlie fine-grained layers, the lower hydraulic conductivity of the smaller pores in the fine-grained soil can cause water moving as saturated flow to stop (perch) at the boundary. However, the larger matric suction that is typical of the finer-textured soil below the discontinuity may eventually facilitate water flow into it, much like a sponge. In contrast, when highly porous, fine-grained horizons overlie coarse-textured layers, the flux of water through the fine-grained layer to the interface is insufficient to displace air from the larger pores in the coarse material until the hydraulic head over the layer is sufficient to displace air from the large pores. In short, water often will not flow into the coarser material below until the finer soil above is saturated, or nearly so. Once flow channels are established in the coarse layer below, fingered flow occurs through the preferential flow paths that are already wetted, as water from the fine-grained overlying layer flows to the regions where larger pores in the sandy underlying horizon have been filled. The importance of water flow to soil genesis is illustrated with examples in later chapters, especially Chapter 12.

Soil temperature

Soil temperature affects many aspects of soils, especially plant growth on and biological activity and water movement within them (Post and Dreibelbis 1942, Baker 1971, Berry and Radke 1995, Sharratt *et al.* 1995). Temperature data are important for estimating evaporation rates, mineral weathering rates, freeze–thaw processes, and frost development within soils. Because they reflect longer-term trends than do air temperatures, soil temperatures might be better used to gauge trends in global climate (Gilichinsky *et al.* 1998). Temperatures also influence the movement of moisture through the soil matrix, e.g., in cold regions soil frost can lower permeabilities and lead to runoff and erosion (Pierce *et al.* 1958, Zuzel and Pikul 1987, Todhunter 2001). In general, a 10 °C increase in temperature causes a doubling in the rates of many biological and biochemical processes.

Soil temperatures are largely determined by the rate at which heat is exchanged between the soil and the soil surface, i.e., soil heat flux (Oliver *et al.* 1987). Soil heat flux is influenced by many different internal factors, e.g., heat capacity, soil moisture content and its effect on latent heat capacity and thermal conductivity. External factors that affect heat flux include air temperature and the presence of insulating materials at the air–soil boundary, e.g., crop and vegetative cover, snowpack and forest leaf litter (Pierce *et al.* 1958, Johnsson and Lundin 1991, Schaetzl and Isard 1996).

An important component of the soil temperature equation is the amount of solar and long-wave radiation that reaches the soil surface, and those properties that govern its absorption and reflection (Fig. 5.5). Clouds can reflect up to 70% of the incident radiation. The amount of solar energy absorbed by atmospheric gases and particulate matter depends on altitude, latitude and time of year, as well as the time of day. At higher altitudes, the atmosphere is less dense, so incoming radiation interacts with fewer molecules before reaching the Earth's surface, and a higher proportion of the energy reaches the Earth's surface. The effects of latitude and season of the year are related to the angle of incidence of sunlight, which controls the distance the energy travels through the atmosphere before reaching the surface, as well as the amount of surface reflection vs. absorption. At high latitudes and in winter months when the sun is low in the sky, the sun's energy hits the Earth at low angles of incidence, and as much as half of the UV energy can be reflected.

Once UV radiation passes through the atmosphere and reaches the Earth, it is either reflected or absorbed. The fraction of incident energy reflected by the surface is known as *albedo*. Albedo depends not only on the angle at which the sun's energy strikes the Earth, but also on the angle and direction of any ground slope, and on ground cover, soil color and soil moisture. At the poles, much of the radiation that reaches the Earth is diffuse radiation that has been scattered, so slope direction and angle have less effect on albedo. Snow both insulates the soil surface and reflects over 75% of the energy back into the atmosphere. Light-colored sand also has a very high albedo. The albedo of vegetation ranges from 5% to 30% because of variations in leaf color and ground

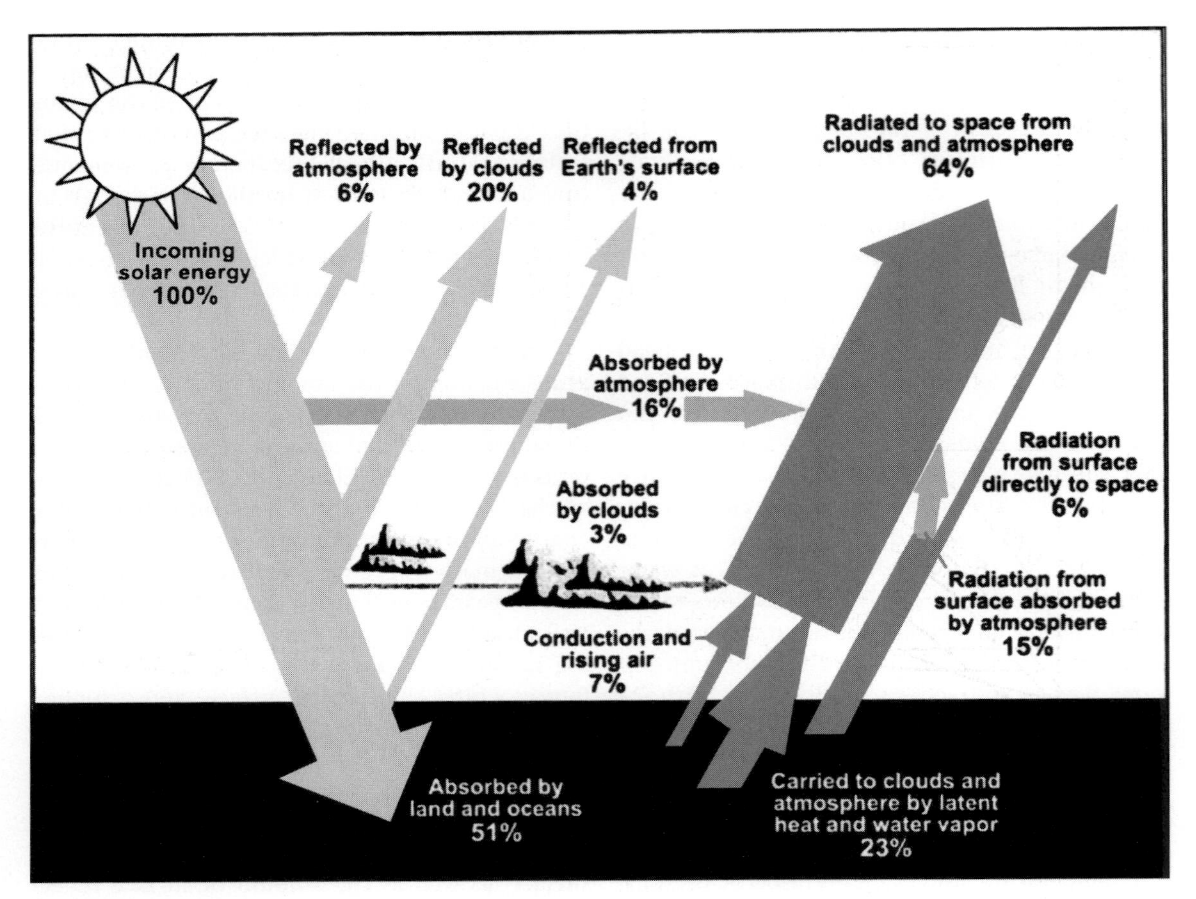

Fig. 5.5 Schematic diagram of the Earth's energy budget.

cover. In contrast with the high albedo of snow, water reflects less than 10% of the incident radiation, and is an excellent absorber of the sun's energy, except at very low solar incidence angles. Increasing soil moisture increases energy absorption and decreases reflection because wet soils are darker than dry soils.

Heat flow in soil

Soil temperature is controlled by energy inputs to the soil (described above), as well as by the thermal properties of the soil (described in this section). Energy absorbed by the soil can either be (1) used to warm it, (2) radiated back to the atmosphere, (3) conducted to some depth below the soil surface or (4) used to evaporate water. Because soil is a heterogeneous, porous medium, heat flow through it depends on the relative proportions of solid, liquid and gas in the soil, on particle size, bulk density, and on the physical arrangement of the soil solids, and on the specific heat capacity and thermal conductivity of each soil constituent. The *specific heat capacity* of a substance is the heat energy required to heat 1 g of that substance 1 °C. Water has a high heat capacity (1.0), which gives it the ability to absorb or lose a large amount of energy with relatively little change in temperature. In contrast, air has an extremely low heat capacity and warms up rapidly in the day and cools down rapidly at night. The specific heat capacity of most soil minerals is about 0.2, or about 1/5 that of water, whereas organic matter has a heat capacity intermediate between that of minerals and water. Because the heat capacity of water is high and because higher water contents promote absorption rather than reflection of the sun's energy, variation in soil water content has a much greater effect on soil temperature than does variation in

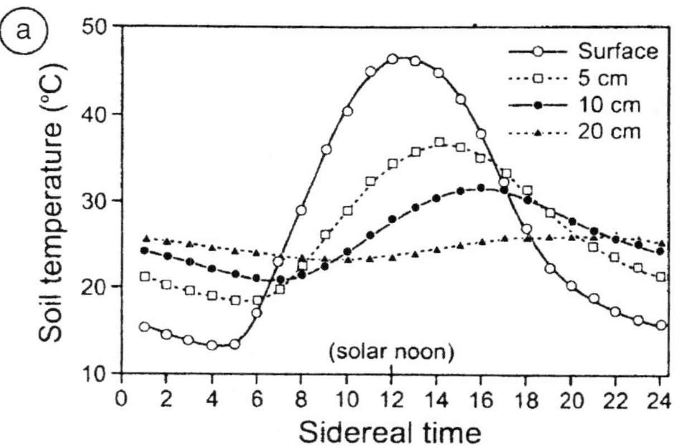

Fig. 5.6 Soil temperature patterns, through time and with depth, for well-drained soils. (a) Changes in daily soil temperature, theoretical and generalized, as a function of depth and time of day. (b) Annual temperature curves, theoretical and generalized, for various depths. (c) Actual, once-daily soil temperatures at an upland, forested site in northwest lower Michigan, taken at 50 cm depth. Data from RJS.

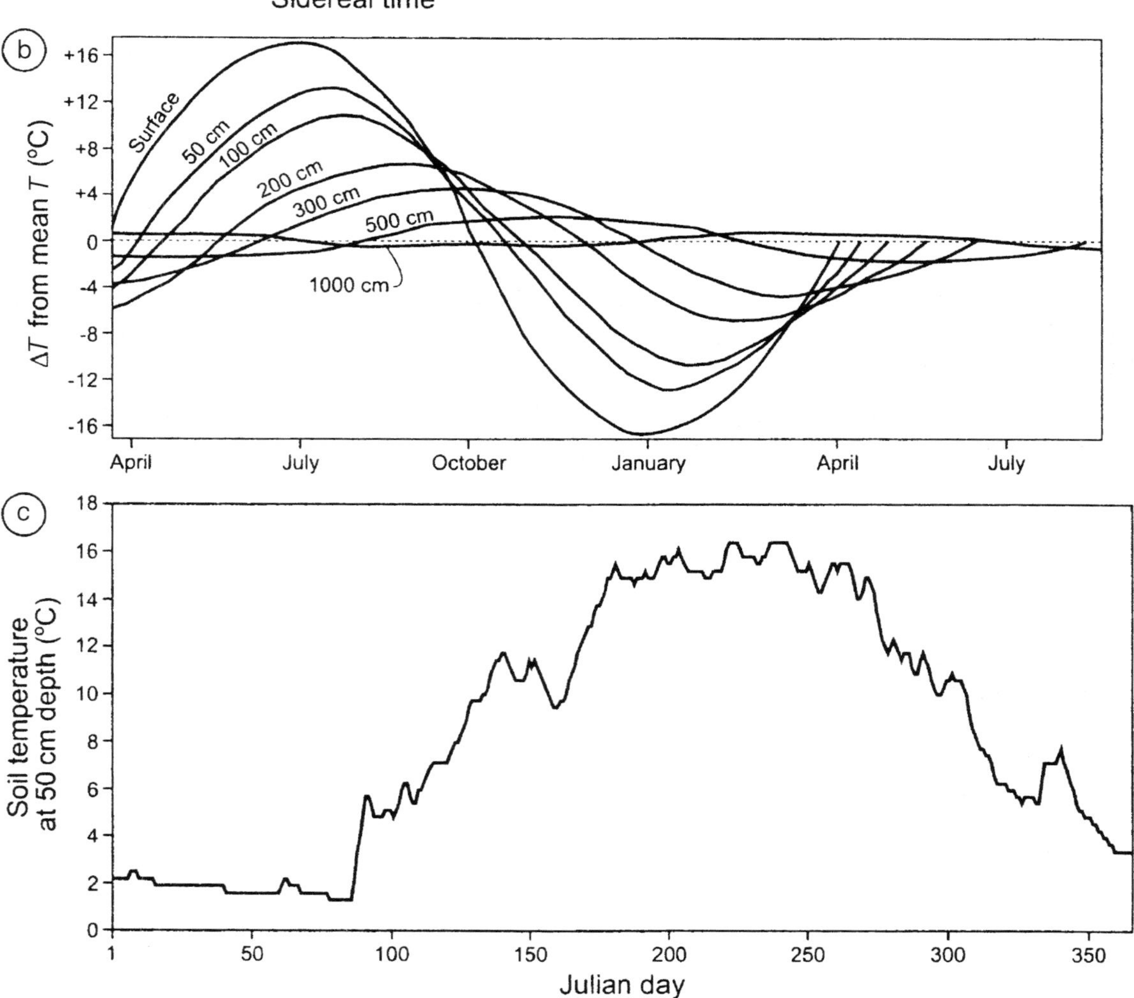

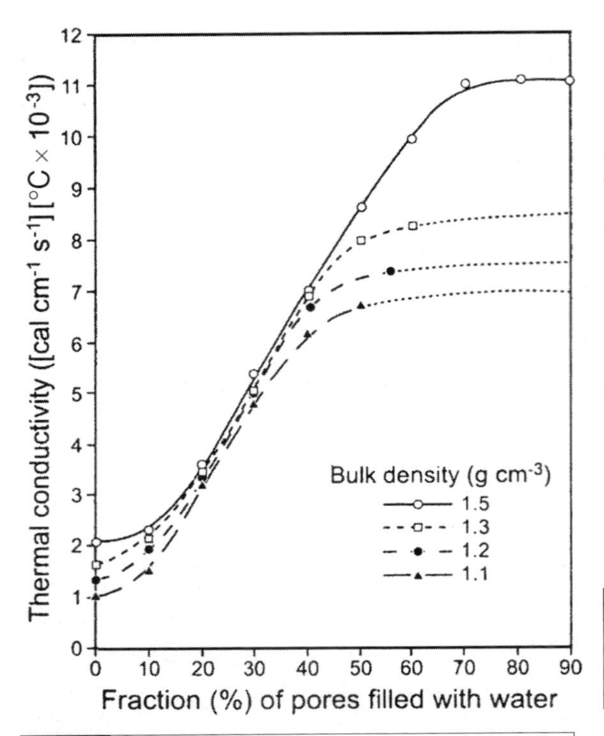

Fig. 5.7 Relationship between thermal conductivity and water content for soils with different bulk densities.

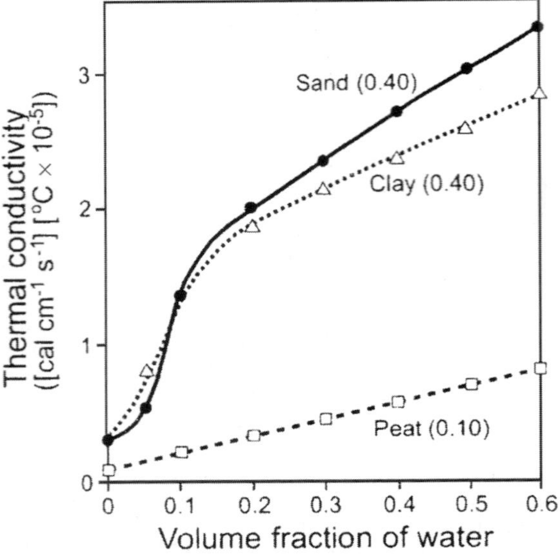

Fig. 5.8 Relationship between thermal conductivity and water content for soils with different textures, showing that sandy soils are better thermal conductors than are finer-textured soils.

either mineral content, bulk density or organic matter content. Clearly, a soil's heat capacity and hence its temperature depend largely on the relative proportions of water, air and solids within it at any one time.

Heat transport also plays a key role in determining soil temperature, since the soil profile is constantly sandwiched between two "boundary layers" that are of different temperature – the atmosphere and the deep subsurface. Heat flow in soils can occur via *conduction* through solids and water as well as by *convection* and *diffusion* in air-filled pores. Soils may warm considerably when a warm rain falls onto an otherwise cool soil, or vice versa (Fig. 5.6c). In addition, water and water vapor transport is a form of *latent heat transport*. Water must absorb energy (540 cal g^{-1}) to evaporate; when water vapor moves from one region of the soil to the other, or leaves the soil to return to the atmosphere, the vapor is transporting energy equal to the heat of vaporization. When the water condenses at another location, the heat is released. Thus, evapotranspiration by plants and water vapor transport are also types of energy transport mechanisms.

Conduction is an important heat-flow process, and is governed by the thermal conductivity (in units of cal cm^{-1} s^{-1} °C) of the soil. The thermal conductivity of quartz is about 15 times greater than that of water, and about 300 times greater than that of air. Consequently, heat transfer in soils occurs primarily via conduction through solids and water, and more slowly via diffusion, convection and water vapor transport in the air-filled pores. Because minerals are the best heat conductors in soil and air is a poor heat conductor, the thermal conductivity of soil increases as its bulk density increases (Fig. 5.7), due to increased particle–particle contact. Similarly, thermal conductivity is greater in coarse-textured soils than in fine-textured soils (Fig. 5.8) because of better solid-phase continuity. Because heat conduction is much greater in water films than in air-filled pores, increasing the water content of a dry soil can cause a tremendous increase in heat conduction – water fills small pores first and creates inter-grain pathways for conduction.

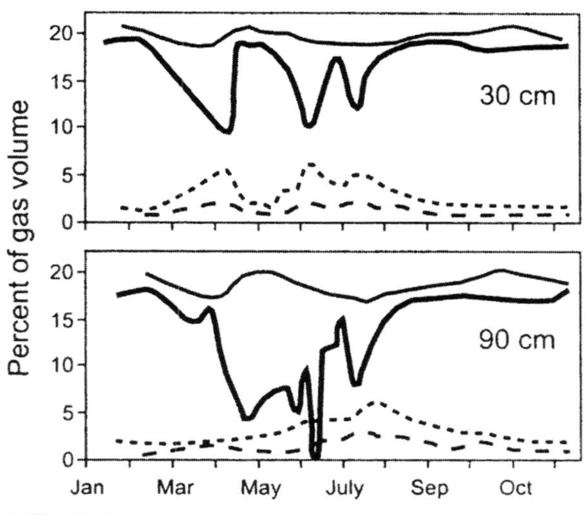

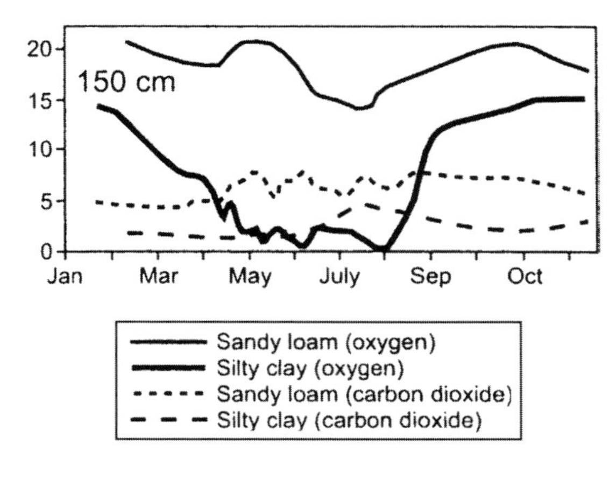

Fig. 5.9 Typical composition of soil gases, with respect to O_2 and CO_2, at various periods of the year, by texture and at various depths.

Models and mathematical equations to describe heat flow near the surface are complex because conduction, convection, diffusion, water vapor transport and evapotranspiration, as well as incident radiation, must all be taken into consideration, and these are rarely at steady state. However, soil temperature measurements at different depths throughout a year provide a great deal of information about heat flow in soils. Figure 5.6a shows soil temperature variations in a sandy soil at depths down to 1 m. The graph is based on average daily soil temperature measurements over many months, so only long-term trends are shown. Note that the mean soil temperature is the same at all depths, and that annual temperature changes can be fit with a sine function that has a period of 1 year. The amplitude of the sine wave, i.e., the difference between the maximum or minimum temperature and the average temperature, increases nearer the surface. At depth, soil temperature varies less and less. Annual soil temperature variations are observable to 1000 cm depth, although the temperature only varies about the average by 0.2 °C at this depth.

Another important feature, shown in Fig. 5.6b, is that there is a lag between heating or cooling at the surface and heat transfer to or from the deeper soil. The maximum soil temperature was recorded in July at the soil surface, in September at 200 cm, and in December at 500 cm. The increasing lag time with increasing depth reflects the time required for energy to be transfered to depth. Because of this lag time for heat transfer, heat flow is downward in the summer, when the soil surface is warmer than at depth, whereas in winter, heat flow is upward because the surface soil is colder than the deep soil, which experiences its maximum temperature in the winter (Fig. 5.6b).

In contrast with the annual temperature cycle, diurnal variations in soil temperature typically penetrate only the upper few centimeters of soil (Fig. 5.6c). At a depth of 20 cm, the amplitude of temperature varation is only 2 °C, with a lag of 8 hs. Daily oscillations in temperature are completely "damped" at 50 cm, which is one reason why mean annual soil temperatures, used to determine soil temperature regime (see Chapter 8), are based on data taken at 50 cm depth (Carter and Ciolkosz 1980, Schmidlin *et al.* 1983, Isard and Schaetzl 1995).

Soil gas composition and transport

One must always remember that the main source of gases in soil air is the atmosphere above the soil. Gases diffuse from the atmosphere into the soil, and vice versa; few gases are actually formed within the soil itself. The composition of gases in the soil atmosphere is, therefore, generally similar to the composition of air above the soil surface. Differences occur because root and microbial respiration cause soil CO_2 concentrations to be much higher and O_2 concentrations to be typically much lower in soils than in the atmosphere (Fig. 5.9).

Gases move through soils mainly via diffusion; convection is negligible except in the uppermost centimeter. Oxygen and CO_2 can dissolve in water and diffuse through water films and water-filled pores, yet diffusion in water is much slower than is gas diffusion, so the primary mode of transport is gas-phase diffusion. Diffusion is governed by Fick's law of diffusion, which specifies that diffusion depends on the concentration gradient between two points and on the diffusion coefficient of the gas. The latter can be quite low, given that soil is a medium in which gas molecules have to travel through narrow pores and along tortuous pathways. Thus, the effective diffusion coefficient of a gas in soil is retarded by a factor known as the gas tortuosity factor, which is determined experimentally but depends on soil texture and pore size distribution in the gas-filled pores.

Fine-textured soils with small pores have the most limited gas exchange with the air, especially when moist. Thus, in warm, clayey, moist soils, CO_2 concentrations may exceed 5% by volume, in contrast to about 0.035% CO_2 in the atmosphere (Fig. 5.9). Conversely, CO_2 is produced as oxygen is consumed, so O_2 decreases under conditions where CO_2 increases (Fig. 5.9). In clayey soils, under anaerobic conditions, it is possible for O_2 concentrations to fall to near zero as CO_2 concentrations climb to over 5%. Conversely, in dry sandy soils, the soil atmosphere is closer to equilibrium with the atmosphere, so oxygen in soil pores is rapidly replaced via diffusion from the atmosphere as it is consumed by roots and microbes.

The high CO_2 concentrations in soils create alkalinity in the soil solution, precipitation of carbonate minerals, and assist in the the development of Bk horizons (see Chapter 12). Oxygen depletion below the soil surface can also lead to chemically reduced zones in the soil, whereas oxidizing conditions promote dissolution of Fe^{2+} and Mn^{2+} minerals and the formation of iron (and manganese) oxides (see Chapter 12).

Although water vapor is a gas, water vapor movement in soils is different than flow of CO_2 and O_2. Water vapor transport can occur through air-filled pores, as with transport of other gases, although the driving force for water vapor transport is differences in relative humidity, which is largely a function of temperature of the soil air, as long as soil particles are covered by a water film. Thus, water vapor transport depends on the gradients in soil air temperature (or on dr_v/dT) as well as the temperature gradient of the solid phase. In addition, water-filled pores are accessible to the flow of water vapor, because water vapor can condense on one side of a water "plug" and other water molecules can evaporate on the other side, with the net effect being that water vapor appears to flow through water-filled pores.

In general, water vapor flows from areas of lower temperature to areas of higher temperature, where the relative humidity is less. Thus, in the day, when the soil surface is hot and water evaporates from the surface water vapor flows upward in the soil toward the drier surface. At night, vapor flow can either be downwards (if the soil surface cools sufficiently), or it may continue to flow upward, if the relative humidity is still lower at the surface.

Chapter 6

Basic concepts: soil organisms

Organisms can be classified according to a variety of different criteria. Functional classification schemes group organisms according to their role in the food web (Table 6.1; Fig. 6.1), whereas phylogenetic classification (the basis for the binomial system of scientific names) is based upon organisms' morphology, physiology, habitat and genetic relationships. Living things were originally divided between the plant and animal kingdoms, but organisms are now classified into three domains: Eucarya, Bacteria and Archaea, and seven kingdoms (Woese *et al.* 1990). Eukaryotes have their genetic material organized inside a nuclear membrane. They can be single-celled (algae, yeasts, most protozoa) or multicellular (most fungi, and all the various plants and animals), and range from primary producers at the bottom of the food web to predatory animals at the top. In contrast, the cells of prokaryotes (bacteria and archaea) lack a nuclear membrane and an organized nucleus. Although important in swamps and the initial colonization of extreme environments, archaea do not play a major role in soil genesis. In this chapter, each of the major groups of soil biota will be introduced, with emphasis on their role in the soil ecosystem. The importance of organisms in soil genesis is discussed more fully elsewhere, especially Chapters 10 and 12.

Primary producers

Primary producers are photosynthetic organisms at the bottom of the food web. Through the process of photosynthesis, they use the sun's energy to convert CO_2 into organic compounds and molecular oxygen, both of which are essential for other organisms' survival. Algae, cyanobacteria and plants are all primary producers (Bold and Wynne 1979). Cyanobacteria, which formerly were called "blue green algae," are single-celled photsynthetic bacteria that are also capable of N-fixation. Algae are single-celled eukaryotes. Both cyanobacteria and algae can grow on the surfaces of rocks and soils and can contribute considerable organic matter to moist, fertile soils (Donahue *et al.* 1983). Cyanobacteria and algae can also form mutualistic or symbiotic associations with fungi to make lichen (see Chapter 14). In lichen, the algae or cyanobacteria phtosynthesize to provide organic C for the fungus; the fungus provides nutrients to the photosynthetic partner, and can also regulate light (Ahmadjian 1993). Cyanobacteria can also fix N for the fungus. Lichens play a role in colonizing rocks and harsh environments by attaching to rocks and facilitating the early stages of weathering (Fig. 6.2) (see Chapters 9 and 14). In coastal environments with frequent fog, lichens that hang from trees, e.g., "Spanish moss," cause the fog to condense and drip onto the soil. This condensed moisture can greatly increase soil moisture and affect the distribution of plants and the rate of soil weathering.

Most primary producers are within the plant kingdom. Here we emphasize the relationship between plants and other soil organisms, in particular the effects of plant roots and litter. Most biological activity in soils is concentrated at the soil surface just below the litter layer – in the root

Table 6.1 Classification of organisms based on functionality in the soil system

Types of organisms	Main functional role	Trophic level	Details
Bacteria, fungi	Decomposers	Second	
	Mutualists and symbionts	Second	Rhizobia form symbiotic relationship with legume roots and fix N_2 for plants; fungi form mycorrhizal associations with roots and provide nutrients and water to plants; can dissolve P-containing minerals
Bacteria, fungi, nematodes, small arthropods	Pathogens	Second and third	Pathogens cause disease
	Parasites	Second and third	Parasites consume host organism
Nematodes, microarthropods, herbivorous burrowing mammals	Root-feeders	Second	Eat roots, sometimes killing the plants
Protozoa (eat bacteria), nematodes (eat fungi and bacteria), microarthropods (eat fungi)	Bacterial feeders and fungal feeders (grazers)	Third	Consume bacteria, typically releasing excess NH^{4+} and other nutrients into soil; eat beneficial and harmful bacteria
Earthworms, macroarthropods	Shredders	Third	Shred plant litter and ingest bacteria and fungi, which live in digestive tract; improve soil aggregation
Large nematodes (eat small nematodes), large arthropods, mice, shrews, birds, voles, etc.	Higher-level predators		Control populations of lower-level predators; burrowing animals transport soil and organic matter in burrows; transport small organisms

zone and A horizon. The litter layer, or O horizon, is an important source of organic carbon for soil organisms and for soil development (see Chapter 3). Dead root cells add organic matter to the soil and provide channels that influence aeration and infiltration.

Roots can comprise from 30% to 50% of plant biomass (Donahue *et al.* 1983). Some plants have shallow, fibrous root systems, whereas others have deep taproots. Root cell walls are composed of cellulose and lignin and are often surrounded by a protective layer of mucilage

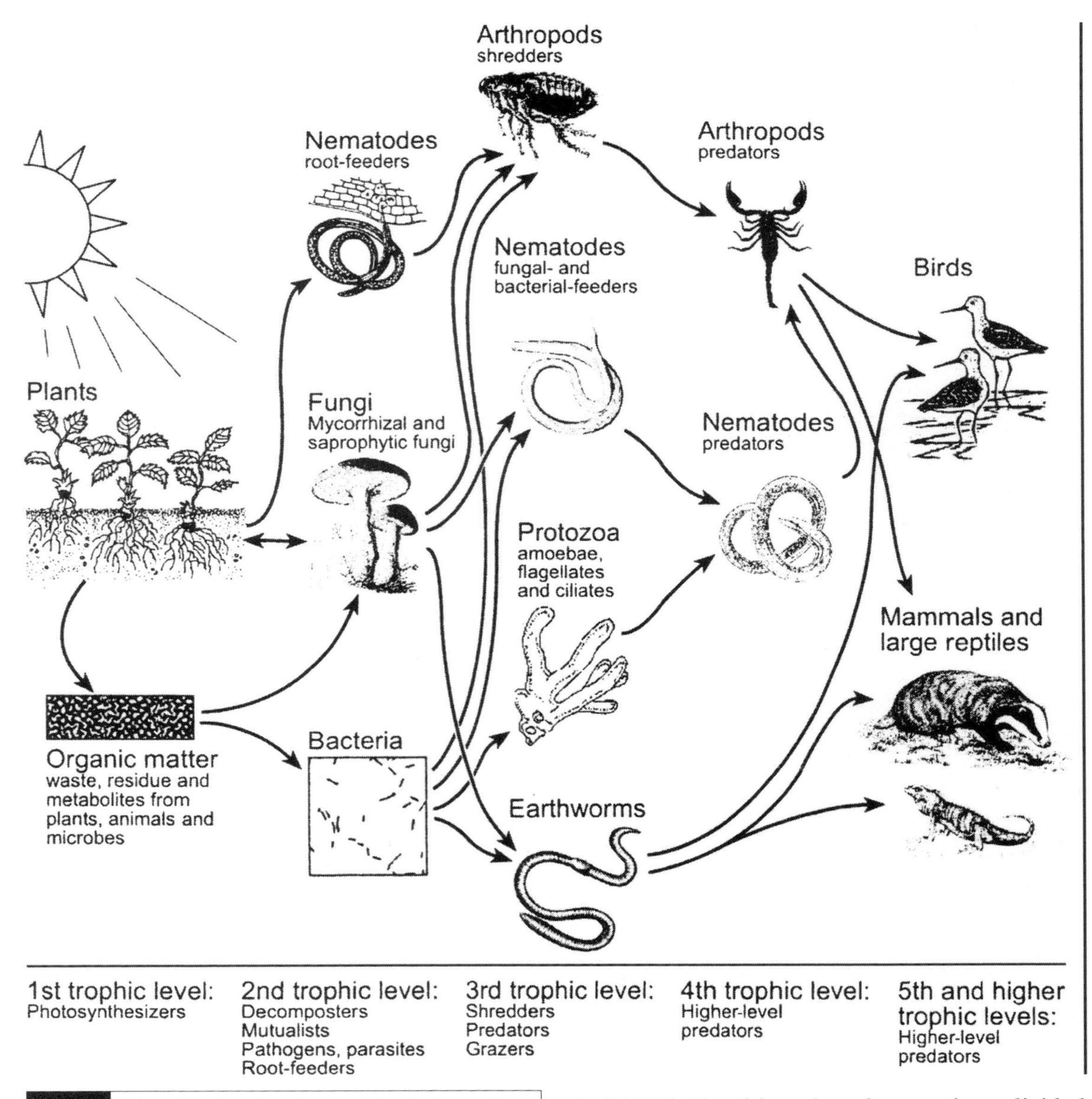

Fig. 6.1 The soil food web and the functional relationships among soil organisms. After Ingham (1999d).

(polysaccharides) (Fig. 6.3). The thin zone of soil immediately around and influenced by living roots is known as the *rhizosphere*. It is rich with biological activity. The boundary of the rhizosphere cannot be defined precisely because its extent depends on the relative rates of root exudation of soluble compounds and microbial utilization of those compounds. However, roots typically influence about 1 to 2 mm of surrounding soil (Pinton *et al.* 2001). The rhizosphere is sometimes divided between the ectorhizosphere (around the outside of the root), the rhizoplane (the root surface) and the endorhizosphere (the root cells that can be colonized by bacteria or fungi).

Living roots release compounds known as exudates, which can dramatically stimulate microbial activity. Root exudates can be divided into two categories: low-molecular-weight water-soluble compounds that are easily metabolized by bacteria (sugars, amino acid, organic acids, hormones and vitamins) and complex compounds

(a)

(b)

Fig. 6.2 Lichen (*Xanthoparmelia lineola*) thalli on rocks on a basaltic ridge at 2200 m elevation, near Ranchos De Taos, New Mexico. *Xanthoparmelia lineola* grows slowly here because of the dry climate. This lichen species has been studied by Benedict (1967) in the Colorado Front Range. The knife is 8.5 cm in length. Photos by P. J. Webber. (a) *Xanthoparmelia lineola* thalli on a basalt boulder. (b) The same lichen on a rock that was carved, presumably in the late 1940s, by Henry Trujillo, a local shepherd. The thallus that covers the carved letters could provide an estimate of growth rates for this species within its great period (see Chapter 14).

that are insoluble in water and need to be hydrolyzed by extracellular enzymes before they can be assimilated by bacteria (cell walls, mucilage) (Cheng *et al.* 1993, Brimecombe *et al.* 2001). About 80% of the C lost from living roots is in the form of polysaccharides known as mucigel (Hale *et al.* 1978). Soluble root exudates stimulate microbial growth; the numbers of bacteria in rhizosphere soil are up to 1000 times greater than in bulk soil, ranging from 10^9 to 10^{12} per gram of soil (Foster 1988). The numbers of soil invertebrates in the rhizosphere can be up to 100 times greater than in bulk soil.

Soil fauna

Soil fauna includes all the living organisms that are not primary producers. We will refer to this large body of organisms as "soil animals" as our

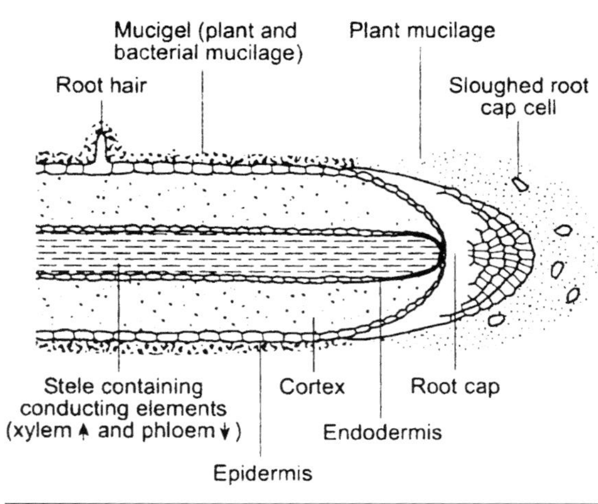

Fig. 6.3 Schematic diagram of the structure of a typical root, and its surroundings within the soil.

focus is on the soil. Soil animals that live mainly in the soil are termed *soil infauna*. Ranging in size from microscopic to many centimeters in length (Fig. 6.4), they play vital roles in the soil system. Our focus here is on the smaller forms of these organisms; the roles of the meso- and mega- soil fauna are covered in Chapter 10.

Bacteria

Although the rhizosphere can contain up to 10^{12} bacteria per gram, bacteria usually occupy only a tiny fraction of the soil volume. Bulk soil, by way of comparison, contains 10^8 to 10^9 bacterial cells per gram (Ingham 1999d). There are more than 10^4 bacterial species, and it is estimated that about half of the bacteria in soils have not been identified (Paul and Clark 1996). Bacteria can be categorized according to phylogenetic relationships, according to their energy and nutritional requirements, or according to their functional role in soils. In the context of soils it is most relevant to focus on the carbon and energy requirements of bacteria and their functional roles in the soil ecosystem.

Bacteria that require *organic* carbon compounds for cell growth are known as *heterotrophs*, whereas *autotrophs* utilize CO_2 as their soil C source. *Photoautotrophs* are primary producers such as cyanobacteria that use the sun's energy to convert CO_2 into cellular organic C. The majority of soil bacteria are *heterotrophs* and require a source of organic C; heterotrophs can obtain energy by using either organic compouds (*organotrophs*) or inorganic compounds (*lithotrophs*) as electron donors. Lithotrophs get their energy by reducing inorganic compounds other than oxygen, e.g., nitrate, sulfate, ferric iron and manganese, and are active in waterlogged soils where oxygen is unavailable. Essentially all oxidation and reduction (redox) reactions in soil are biochemical reactions mediated by bacteria (see Chapter 13).

Heterotrophic bacteria can be divided into three functional groups based on the *source* of organic C they use for cell growth. *Pathogens* derive their C from a host plant or animal, and cause disease. *Mutualists* form a mutualistic or symbiotic relationship with another organism, such as the association between cyanobacteria and fungi to create lichen. Another example is the symbiotic assocation of N-fixing bacteria (Rhizobia) with legume roots, in which the bacteria get organic C from the plant and the bacteria provide fixed N to plants. *Decomposers* are the most abundant, and arguably most important, functional group of soil bacteria. They live primarily on plant litter, dead roots and root exudates. Some decomposers utilize only simple, water-soluble organic compounds, whereas other decomposer bacteria can either metabolize large, complex compounds or release extracellular enzymes that hydrolyze complex molecules into simple compounds that can then be more readily assimilated. Decomposers depend upon plant-derived organic C in the litter layer, root zone and rhizosphere. In turn, they can promote plant growth by the release of plant hormones. They also release polysaccharides that promote aggregation and aeration, and assimilate and cycle nutrients in plant litter.

Actinomycetes, which are multicelled, filamentous bacteria, are also a type of decomposer. They produce the characteristic musty odor of wet soil. About 90% of the actinomycetes in soil are streptomycetes. They live in soil and plant litter, and can degrade complex substances such as lignin, chitin, humin, aromatic compounds, keratin and cellulose (Paul and Clark 1996). Actinomycetes also produce a range of antibiotics, including erythromycin, neomycin and tetracycline (Paul and Clark 1996).

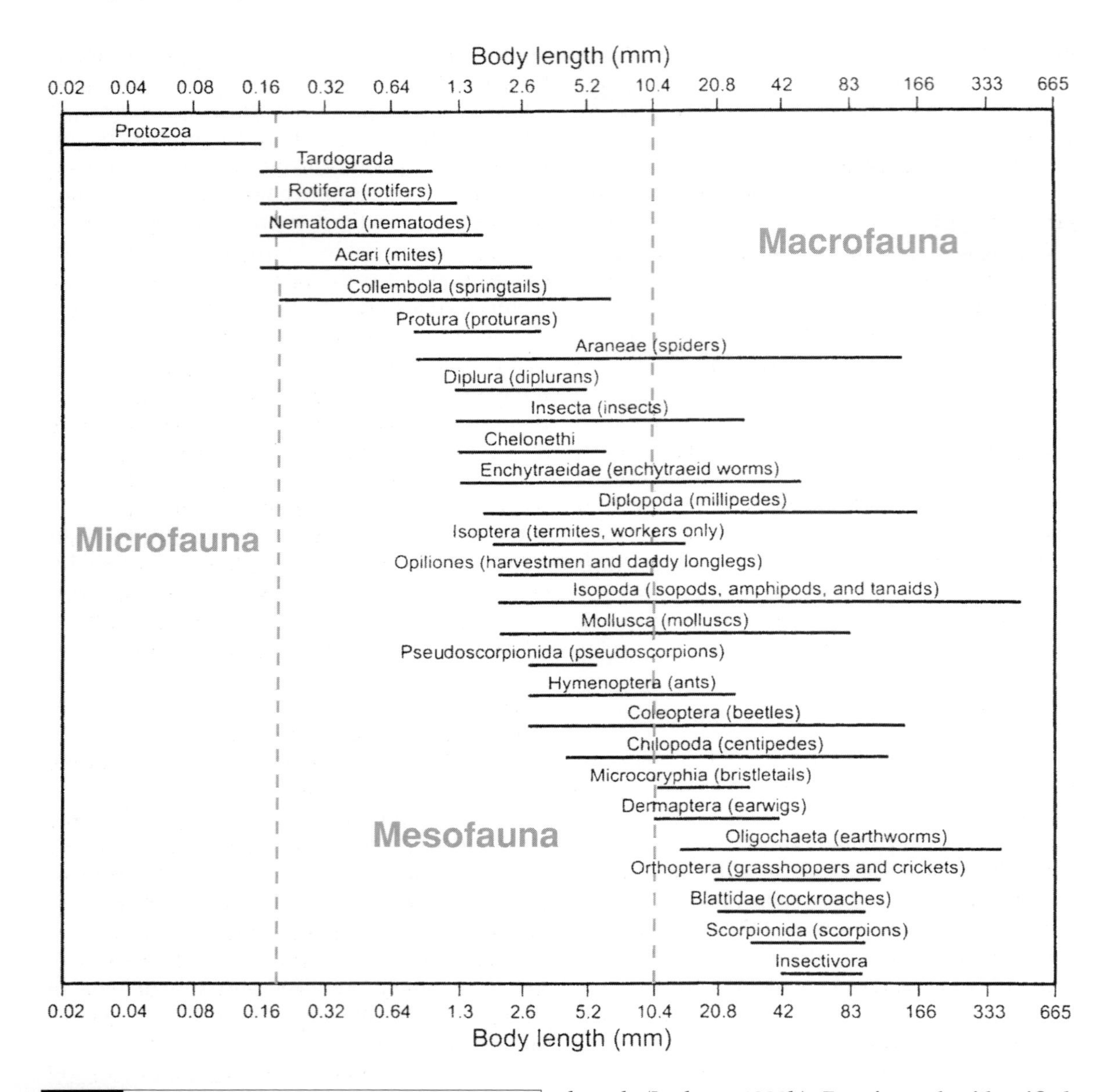

Fig. 6.4 Dimensions of soil biota in relation to other soil materials. Compiled from various sources.

Fungi

Fungi are aerobic, heterotrophic organisms that obtain their C and energy from organic compounds in either organic litter or living plant and animal tissue (Fig 6.1). Fungi include one-celled yeasts as well as a range of multicellular filamentous organisms such as molds, mildews, smuts, rusts and mushrooms that range in size from a few micrometers to several meters in length (Ingham 1999b). Fungi can be identified by the growth of *hyphae*, which are filaments or threads that sometimes form mats called *mycelia*. Fungi can also be divided into the same three functional groups described above for bacteria. Within each functional group, however, fungi perform slightly different roles than do bacteria.

Fungal decomposers, also known as saprophytic fungi, are the most numerous and visible fungi in soils. Decomposer fungi convert complex organic substances into simpler molecules like organic acids that bacteria can use. Fungi are similar to actinomycetes in their ability to degrade compounds like cellulose, chitin and aromatic rings, but they are more active at lower pH values than are actinomycetes, and are more prevalent in litter derived from wood. Fungi contain much less N than do bacteria and can thus grow on organic compounds with much wider C:N ratios than those that are favored by bacteria (Ingham 1999a). Fungi also play key roles in the formation of soil humic substances.

Fungi that colonize plant roots in a mutualistic or symbiotic relationship are known as *mycorrhizae* (literally, "fungus root"). Nearly all plants have mycorrhizal fungi associated with their roots, although plants with fibrous roots rely less on mycorrhizae for successful growth than do plants with taproots. Many noxious weeds found in nutrient-rich environments are non-mycorrhizal. Most mycorrhizae obtain their C from organic compounds produced by the host plant. In exchange, the fungi provide the plant with nutrients by accessing parts of the soil solution that are inaccessible to the roots and by releasing extracellular enzymes that solubilize nutrients such as P.

There are seven different types of mycorrhizal associations, but about 90% of mycorrhizae fall into three of the categories: ectomycorrhizae, arbuscular mycorhizae and ericoid mycorrhizae. Ectomycorrhizae, which include edible wild mushrooms such as truffles, infect trees and shrubs, mainly in temperate regions. They form a mantle or sheath around the root surface, and penetrate between, rather than through, root cortex cells (Fig. 6.5). Many different fungal species can form ectomycorrhizae, and sometimes a single tree can have several different species of mycorrhizae simultaneously. Fungal mycelia can produce enzymes that promote the release of phosphate and inorganic N for the plant. A second type of mycorrhizal association is arbuscular mycorrhizae (previously known as vesicular arbuscular mycorrhizae). They can infect nearly all agricultural crops, many grasses and numerous trees and shrubs, including legumes. They occur in soils that have near-neutral pH and typically are P-limited but not limited by available N (Paul and Clark 1996). Arbuscular mycorrhizaae penetrate the root cortex, where they form arbuscules composed of finely branched hyphae that fill the cytoplasmic space of root cortical cells (Fig. 6.5) and transfer nutrients to the roots (Paul and Clark 1996). Some, though not all, also have hyphal extensions that form internal storage vesicles between cortical cells. In addition, some hyphae extend out from the root into the soil, where they can take up nutrients that would not be accessible to an uninfected root. Some of the external hyphae have asexual spores that were once thought to be storage vesicles. Ericoid mycorrhizae form coils inside the root cells, with fungal hyphae extending a few millimeters beyond the roots. They are found mainly found on acidic, nutrient-poor soils.

Fungal hyphae are important in soil genesis because they can penetrate cracks in rocks and even mineral grains, promoting physical and chemical weathering. In addition to their role in weathering, hyphae also play an important role in soil aggregation, and can increase the infiltration and water-holding capacities of soils.

Protozoa

Protozoa are single-celled organisms in the Protista kingdom. They are the largest soil organisms not in the animal kingdom.

Most are "grazers" that feed or graze mainly on bacteria, though some can also eat fungi, and large predatory protozoa can eat small protozoa. There are three groups of protozoa that differ in their size, morphology and method of locomotion. The relationships among different protozoa, their prey and their predators are shown schematically in Fig. 6.1. Ciliates are the largest protozoa (up to 500 μm) and can be either grazers or predators. Juvenile ciliates move by coordinated motions of tiny hair-like cilia, whereas adults lose their cilia and use tentacles to reach their food (Paul and Clark 1996). Because of their relatively large size, ciliates are most abundant in coarse-grained soils.

Amoebae are intermediate in size and move about by the use of a pseudopod – a temporary foot that is also used to trap bacteria and fungi. Some amoebae have a protective covering (testate amoebae); naked amoebae do not. Most ameoebae are grazers, though some are disease-causing parasites (Ingham 1999c).

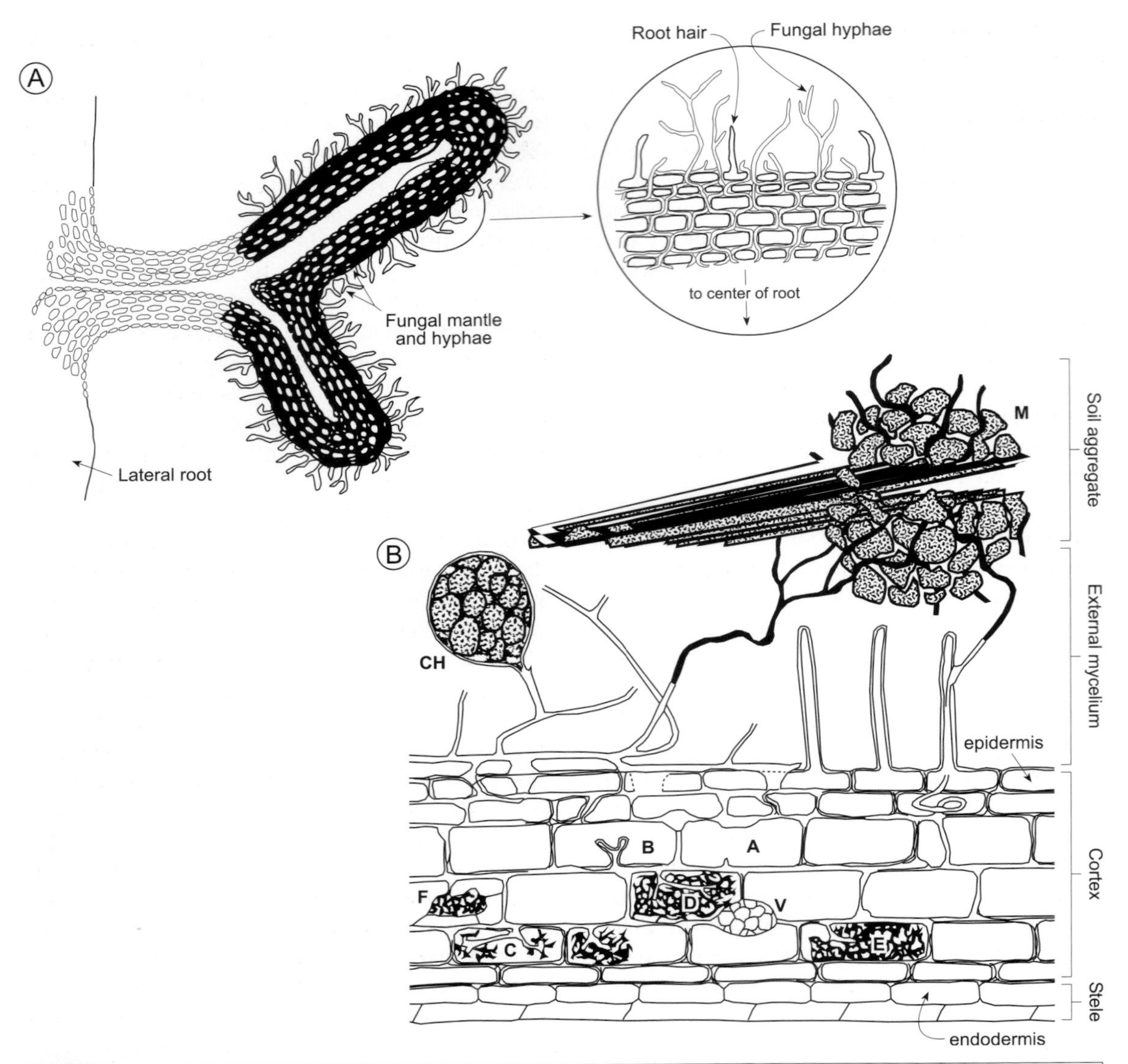

Fig. 6.5 Diagrams of two different types of mycorrhizae. (a) A mantle of ectomycorrhizae around a root, with hyphae penetrating between epidermal cells. After Donahue *et al.* (1983). (b) Arbuscular mycorrhizae, showing progressive stages in the development and digestion of fungal arbuscules (A–F) inside cortical cells. The arbuscules survive for 4 to 10 days, and provide nutrients to roots. A lipid storage vesicle (V) and chlamydospores (M and CH) are also shown. After Paul and Clark (1996).

Zooflagellates are the smallest protozoa and move using a whip-like flagellum. Although some zooflagellates are free-living in the soil, most form either commensal, symbiotic or parasitic associations with insects and vertebrate animals. Termites contain symbiotic zooflagellates in their guts and rely on them to digest wood (Paul and Clark 1996) (see Chapter 10).

Protozoa are most abundant in the rhizosphere because most protozoa feed on bacteria. They are much more numerous in rhizosphere than in the bulk soil. Protozoa play an important role in the soil N cycle by digesting bacteria, which have a higher N content than other soil biota. When protozoa digest bacteria, excess N is released as NH_4^+. Protozoa also play a key role in the soil ecosystem by regulating bacterial populations (Ingham 1999c).

Larger soil animals

Soil animals, of the kingdom Animalia, range in size from tiny nematodes to earthworms to large mammals and reptiles that burrow in the soil and tread upon it (Fig. 6.4). Though individual animals and their burrows are large, their numbers and total biomass per hectare are small compared to microbes and small invertebrates (Ingham 1999a).

Nematodes

Nematodes are non-segmented worms that range in size from about 0.1 to over 1 mm (Fig. 6.4). Because of their large size, they are more common/abundant in coarse-grained, porous soils, moving through the soil in water films on particle surfaces. Different nematodes operate at different trophic levels (Fig. 6.1). Some are plant parasites that feed on roots, whereas others are grazers that feed on bacteria or fungi (Paul and Clark 1996). Some larger nematodes are predators that eat other nematodes as well as other small organisms, and some nematodes are omnivores that eat a range of organisms or eat different organisms at different stages of life. Predatory nematodes attach to their prey and suck out the contents of their prey, leaving only the exoskeleton (Paul and Clark 1996).

Nematodes that graze on bacteria and fungi fulfill the same ecological role as do protozoa, cycling nutrients and regulating microbial populations. They also disperse bacteria and fungi to different parts of the soil, transporting them either in their digestive system or attached to their exterior (Ingham 1999d).

Arthropods

The word "arthropod" means "jointed foot," and arthropods include insects (springtails, beetles, ants and termites), crustaceans (sow bugs), arachnids (mites and spiders) and myriapods (centipedes, millipedes and scorpions). Arthropods are both diverse and numerous in soils; there can be several thousand different species of arthropods per hectare. The range of sizes for many soil arthropods are shown is shown in Fig. 6.4. Springtails and mites are dominant in many soils, though ants and termites can prevail in certain sites (Fig. 6.6). Arthropods occupy a range of niches in the soil food web, including root feeders (herbivores), grazers, shredders and predators. Herbivores such as cicadas, root maggots and root worms eat roots and can damage plants. Springtails, mites and silverfish are grazers that eat bacteria and fungi that are on or near roots. A related and very important functional group of soil fauna is the shredders, which live near the soil surface and shred plant litter while eating bacteria and fungi that adhere to the litter (see Chapter 3). Shredders will also eat roots and the microbes associated with roots if no litter is available. Shredders include millipedes, sow bugs, termites, some types of mites and roaches (Fig. 6.7).

Larger arthropods such as spiders, ground beetles, scorpions, centipedes, pseudoscorpions, ants and some mites are best viewed as predators. They are beneficial when they eat pests such as some nematodes, but certain predators also eat beneficial organisms. For example, centipedes will prey on earthworms. Most predators and shredders live near the soil surface; those that live deeper than about 5 cm are usually less then 2 mm long and blind (Moldenke 1999).

All arthropods provide the beneficial service of mixing and aerating the soil as they feed (Lyford 1963). Soil particles get cycled repeatedly through the digestive tract of many types of soil organisms, where the soil is mixed with mucus and organic matter to make fecal pellets (see Chapter 3). Fecal pellets are an important type of aggregate in many soils. Arthropods that feed on bacteria and fungi cycle and release nutrients, mixing them throughout the soil. Burrowing arthropods also dramatically affect porosity and infiltration (Moldenke 1999).

Arthropod fauna such as ants (Hymenoptera) and termites (Isoptera) are found worldwide and move vast amounts of soil material (Mandel and Sorenson 1982, Levan and Stone 1983). Termites

Fig. 6.6 Termite mounds in the Jalapao region of east–central Brazil. Photo by A. F. Arbogast.

Fig. 6.7 A common millipede of midlatitude forests (*Narceus americanus*, in the Class Diplopoda and Order Spirobolida), with leaves for scale. The majority of the 1400 known species of North American millipedes eat decaying vegetable matter and fungi and play a crucial role in decomposition of leaf litter and nutrient cycling. Photo by RJS.

are particularly abundant in Australia, Africa and throughout the tropics, whereas earthworms are the dominant burrower in more humid locations like Europe and parts of North America (Lee and Wood 1971a, Lobry de Bruyn and Conacher 1990).

Termites are, arguably, the most widespread and important of the soil arthropods. The major differences between termites and many other soil fauna are: (1) they can digest cellulose, which most other infauna cannot, (2) they use their excreta, which is highly resistant to further decomposition, to line the mound rather than allowing it to immediately biocycle, leading to (3) a long period of organic matter immobilization within the mounds, which themselves can take decades to fully deteriorate (Lee and Wood 1968, Grube 2001). In short, they extract large quantities of organic matter and store it in intractable form in their mounds, dramatically affecting the carbon cycle (Lee and Wood 1968). Other soil animals, especially worms and millipedes, are most effective at comminuting soil organic matter, passing it through their guts and mixing it with soil, *facilitating* the rapid breakdown and biocycling of litter. Termites, therefore, should *not* be considered, as some have said, the "tropical analogs of earthworms."

Earthworms

Earthworms (annelid worms) are among the largest soil invertebrates and typically contribute more to soil biomass than do other soil invertebrates (Lee 1985); they are most abundant in medium-textured soils that are moist and not too hot. Earthworms tend to avoid sandy soils. There are over 700 genera and 7000 species of earthworms. They burrow through soil or litter and ingest soil, organic matter and attached microorganisms. As this material moves through their guts, the ingested organic matter is broken into smaller pieces, inoculated with microorganisms, and mixed with mineral particles. For this reason, they are considered shredders in the soil ecosystem. Earthworms excrete small aggregates (casts) that are enriched in microorganisms, organic matter and nutrients. Earthworm casts improve soil structure and enhance aeration, infiltration and water-holding capacity (Donahue *et al.* 1983). Litter-feeding earthworms and beetles burrow within and mix surface litter into the upper mineral soil, thereby assisting in the formation of A horizons. They also assist in the decomposition and compaction of Oa horizon material from Oi material. Slugs, woodlice, snails and millipedes are also important in the breakdown of raw organic matter into humified substances (see Chapter 3).

By ingesting raw or slightly decomposed organic matter from the O horizon, earthworms facilitate its incorporation into the mineral soil in two ways: (1) directly, by depositing humus-rich casts in the subsurface and (2) by increasing the amount of macropores, especially vertically oriented macropores, thereby facilitating translocation of small fractions of organic matter into the mineral soil (Stout and Goh 1980). For this reason, they are extremely important to *melanization* (Table 13.1), and are a major reason why the A horizons in many soils are thicker than expected; soils with abnormally "thin" A horizons are often worm-free.

Earthworms are categorized according to where they live in the soil and the types of burrows they form. Epigeic species, called "red wigglers" by fishermen, are small earthworms that live at the soil surface or in the litter layer. They require high organic matter concentrations but are adapted to a broad range of moisture conditions. Endogeic earthworms, in contrast, ingest organic matter in the upper layers of the mineral soil. As they burrow, they fill in their burrows with casts. Anecic species primarily burrow vertically, deep in the soil, and move organic matter downward and move soil from lower to upper horizons. "Nightcrawlers" are the best-known anecic earthworms. They create permanent burrows that can be several meters deep. This type of earthworm feeds on organic matter found at the soil surface, bringing it deep into their burrows and enhancing its decomposition (Edwards 1999) (Fig. 6.8). If the openings of their burrows are not blocked with casts or plugs of organic matter, they can facilitate the downward migration of water, pesticides and nutrients, and the high organic-matter content of burrow walls may

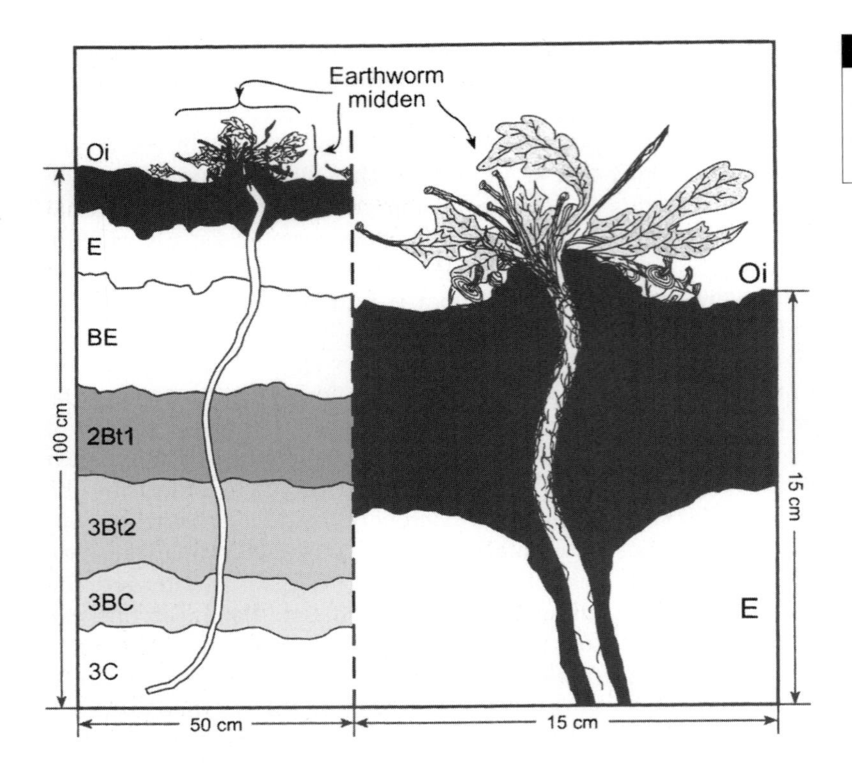

Fig. 6.8 Drawing of an earthworm (*Lumbricus terrestris* – a "nightcrawler") midden and associated soil profile. After Nielsen and Hole (1964).

adsorb dissolved ions and pesticides (Edwards 1999).

Mammals, birds, reptiles and amphibians

Most soil biology books or chapters do not devote a great deal of space to the roles of the largest fauna – the vertebrates – to soils. Perhaps this is because because many, e.g., birds, reptiles and larger mammals, primarily live outside of the soil and their influence is seen as "minimal." Nonetheless, these fauna can compact or loosen the soil, and all contribute feces and thus function within various nutrient cycles. Indirectly, they can denude the vegetation and facilitate soil erosion. Soil compaction and denudation are particularly noteworthy around waterholes in grasslands and deserts where, for example, ungulates congregate in the dry season.

Mammals primarily influence the soil by burrowing and living within it. Many rodent and mole species are particularly adept at burrowing; their influence is primarily seen as associated with soil mixing (see Chapter 10). Künelt (1961) noted that, in California, gophers may turn over as much soil in 5 months as earthworms do in 5 years. Many amphibians also perform similar functions, although they are more geographically restricted to wet soils. Many snakes and skinks (*Scincus* spp.) are also proficient burrowers.

Although no bird species is considered a true soil animal, many build their nests on or within the soil and feed there. Many feed upon soil animals, such as worms and beetles, and in certain areas they can decimate the ground cover. In Antarctica, penguins and sea birds have profound influence on soils, primarily through their excrement. Recent work has focussed on the effects of penguin guano as an important source of organic acids, driving podzolization in soils of penguin rookeries that otherwise would have little soil development (Beyer *et al.* 1997, 2000). Bird guano-influenced soils are referred to as *ornithogenic soils* (Tatur 1989, Myrcha and Tatur 1991, Beyer *et al.* 1997).

Landscapes: invading vs. migrating worms

Anglers, take note: that nightcrawler on your hook is an invader. According to recent research by Sam James, an earthworm ecologist, most bait worms in North America are European. Canada, New England, some mid-Atlantic states and much of the upper Midwest are entirely populated by non-native earthworms (see Samuels 2000). The rest of the continent contains a mixture of non-natives and native earthworms. While the ancestry of one's local earthworm might not seem like a big deal, scientists say otherwise; the encroachment of the non-native species has a profound effect on soil structure and genesis.

James has concluded that the dividing line between native and non-native species roughly coincides with the furthest advance of the Wisconsin ice sheet, about 22 000 years ago. North of that limit, there are few native earthworm species. Why? During glaciation, earthworms were obliterated from the landscapes that were covered with ice, and permafrost pushed many others even farther south. In the 180–220 centuries that followed, they simply have not retaken their original territory; migration of earthworms is a very slow process! Support for Dr James' hypothesis comes from an unpublished Dutch study that measured the rate at which earthworm populations expand their range. Dutch researchers placed earthworm colonies in a field free of worms. After 1 year, the colonies of European worms, a species that is an especially good colonizer, had expanded by ≈10 m. If you take 10 m per year as the maximum rate of range expansion and multiply that by 22 000 years, that yields only about 220 km (137 miles). In other words, native North American earthworms have expanded their territory to the north of the glacial border by the length of a small state since the ice sheets left. Many earthworms to the north are, therefore, not native and have migrated there. Instead they are European and Asian immigrants, brought there inadvertently by humans in the horticultural trade. As people moved plants, probably in the colonial period, worms were inadvertent hitchhikers.

Why worry about whether alien worms are on American soil? In a wooded area with worms, soils have thinner O horizons and thicker A horizons, while in worm-free zones, like those in New England, northern Michigan or Minnesota, leaf litter is much thicker and lower horizon boundaries are more diffuse (Stout and Goh 1980). Nutrient cycling is also more dramatic when worms are present. And the invasion is still in progress, with worms moving into areas that have long been worm-free.

Hawaii provides another example. It was historically earthworm-free, until it started drawing settlers from all parts of the globe. It now has a vast diversity of introduced tropical and European earthworms. Humans also introduced pigs to Hawaii; these pigs now run wild. The pigs preceded the worms and made an easy living in the Hawaiian woods, finding enough food on the ground that they did little rooting below the surface. Once the worms arrived, though, the pigs developed a taste for them and are now tearing up the remaining natural forest in search of the non-native worms.

Chapter 7

Soil classification, mapping and maps

The essence of any classification is to place a name on an definable entity, in this case a basic soil unit. Once this is done, it is possible to arrange them in an orderly system and establish interrelationships among them (Beckmann 1984). Most importantly, though, it allows for communication about the entities being classified.

Users of classification schemes agree that a name should carry with it a unique and defined range of characteristics, whether it be a plant (e.g., *Pinus sylvestris*), rock (e.g., arkose sandstone), or soil (e.g., a Lithic Fragiaquept). For soils, that range of characteristics may involve physical properties such as clay content of the B horizon, thickness of the A horizon, or color, among others. Over time, these ranges are changed as new and more complete information about the universe of soils (or plants or rocks) accrues. Thus, most classification systems are open-ended, and easily amended.

Before we try to classify a soil, we must first define it. We agree with Johnson (1998a), who defined *soil* as organic or lithic material at the surface of planets and similar bodies that has been altered by biological, chemical and/or physical agents. But to classify soils, it is essential that we also agree on some sort of singular or *basic unit* (Johnson 1963). In many disciplines, it is relatively easy to determine the basic unit that is being classified: a plant, a rock, a virus. In soils, however, the elementary or basic unit of classification is not always clearly differentiable. How deep should it go? How areally extensive should it be? How much internal variability should be allowed? For example, Ollier *et al.* (1971) found that people of the Baruya tribe in New Guinea had many names for *soil material*, but not for *profiles* per se. Based on their use of soil, the classification scheme of the Baruya people worked well, which makes the point that all classification schemes must fill a specified need.

The basic soil unit, like the basic units of all classification schemes, must be a small, arbitrary volume which must not be able to be further subdivided, and which is of a convenient size for study and measurement, like the hand specimen in geology (Van Wambeke 1966, Arnold 1983, Beckmann 1984). Ideally, it should also be able to be sampled, i.e., it should be three-dimensional and have volume, with clear boundaries. It should include the entire vertical thickness of the soil. Some early classification systems used as the basic soil unit the soil profile, which is a two-dimensional entity and in theory cannot be sampled. Later, the *pedon* was chosen as the basic soil unit (see Chapter 3), largely because it *could* be sampled; in Soil Taxonomy it continues to be used as the basic unit of soil.

Soil geography, mapping and classification

As Arnold (1983) noted, there is an apparent dichotomy of thought when considering soils as a part of a continuous landscape, and the concept of the basic soil unit (see also King *et al.* 1983). Soils vary both gradually and abruptly across the landscape, since the five soil-forming factors (climate, organisms, relief, parent material and

Fig. 7.1 An abrupt soil boundary. The pedon at the left has formed in sandy glaciofluvial sediment, while the one at the right has formed in clayey glaciolacustrine sediment. This abrupt contact formed as meltwater traversing the lake plain ripped gullies into it, and later filled them with sand. Presque Isle County, Michigan. Photo by RJS.

time), and therefore the pedogenic processes, also vary across that same landscape (Johnson 1963) (see Chapter 12). At places where one of the factors changes abruptly, such as at a bedrock or parent material interface, the soil may change character abruptly as well (Fig. 7.1). More often, pedologic gradations are gradual.

A taxonomic class is a segment of the soil continuum which has predetermined limits. The challenge of soil classification is to set these taxonomic limits, lines and boundaries in the most appropriate way, within the constraints of this soil continuum, so that they can actually be mapped (Cline 1963). Regardless of the class limits chosen, they will not fit all landscapes equally well. A class limit might work well on one landscape, where it nicely delimits soils formed from Pleistocene deposits from those formed in Holocene alluvium, but on other landscapes these limits may not work, because perhaps all the soils exist so close to the class limit that they cannot be easily differentiated. The challenge of the mapper or soil geographer, hopefully with help from the soil classifier, is to define class limits which make taxonomic sense *and* are mappable. In theory, the soil mapper tries to delineate pure, contiguous units of pedons, called polypedons. A *polypedon* is a soil volume, spread across the landscape, that consists of several like and contiguous pedons; it forms the real-world link between the pedon (i.e., the basic soil unit) and the taxonomic class. Just as a human being cannot possibly exhibit all the possible range of characteristics of the race, neither can a pedon exhibit all the possible characteristics of a taxonomic class, e.g., one pedon cannot exhibit all possible ranges of slope. Thus, the polypedon was developed as the link between reality (i.e., the landscape, with all its complexities) and taxonomy (in its purest sense). Johnson (1963) defined polypedons as contiguous pedons that fall within a defined (taxonomic) range. They are bounded by areas of not-soil, or by pedons of unlike character, i.e., those that are outside of the taxonomic range of the polypedon. Polypedons, not pedons, are the stuff of Soil Taxonomy (see Chapter 3). Thus, the soil mapper, in attempting to map taxonomic units on the landscape, tries to separate polypedons from each other.

The system of Soil Taxonomy

In this book, we focus on one of many systems used worldwide to classify soils. Buol *et al.* (1997) provided a splendid review of the many other systems, such as those of New Zealand, France, Russia and Canada. Soil Taxonomy (Soil Survey Staff 1999) grew out of the work of Guy D. Smith, a pioneer in the early soil survey of the United States (Smith 1979, 1983, 1986).

An historical overview

In order to best understand the rationale for Soil Taxonomy, it is best to examine it in light of the historical events that led to its development. In the United States, the earliest mapping was done under the jurisdiction of Milton Whitney, head of the US Soil Survey (Simonson 1986). From its inception in 1899, the Soil Survey program has existed to map the soil resources of the nation and provide land-use interpretations, especially as related to crop yield.

Early concepts of pedogenesis in the United States were based on soils as products of rock weathering (Simonson 1986). It made sense, therefore, to key the soil classification system to geology. The basic cartographic unit was the *soil type*, which was the precursor to the current concept of the *soil series* (Simonson 1952, Smith 1983). The series concept in the early 1900s was very different than today's. The series was viewed as spanning vast physiographic regions, such as the Miami series which was mapped on glacial deposits of all kinds, or Cecil which was widespread on the old, weathered landscapes of the Piedmont (Gibbs and Perkins 1966). Each series could span a variety of texture classes; Miami silt loam, Miami loam, Miami clay loam, etc. were all *phases* of the Miami series. Because soils were regarded at this time as essentially weathered rock, parent material considerations far outweighed all others.

Earlier, in Russia, V. V. Dokuchaev was espousing a different approach (see Chapter 1). His work, done largely between 1870 and 1900, viewed soil as a natural body with properties that resulted from the influence of climate and living organisms acting on parent material over time, as conditioned by relief (Smith 1983). These concepts were translated from Russian to German by Glinka (1914). The then-head of the US Soil Survey, Curtis Marbut, was so influenced by these ideas that he changed the way soils were classified in the United States, emphasizing that they should be classified based on characteristics of their *horizons*. Marbut considered the following horizon characteristics to be important for distinguishing one series from another: number, arrangement, thickness, texture, color, chemical composition and structure, as well as the geology of the parent material. He eventually developed a multicategoric classification system (Marbut 1928, 1935), one of the more important advances of which was the concept of the *normal soil*: a freely draining, mature soil having clearly expressed horizons (see Chapter 12). The idea of a mature soil, probably based on the ideas of the geographer William Morris Davis and his geographical cycle in which landscapes are seen to move from youth through maturity and old age, was a major part of the system. It continues to be used today by people who see soils as evolving toward some profile endpoint, rather than continuously evolving both progressively and regressively (Johnson and Watson-Stegner 1987). The concept also remains popular in Russia. Soils that did not fit the normal soil concept, e.g., organic soils, alkali soils and wet soils, were not included in the higher categories of the system, again emphasizing the importance of the upland, well-developed soil; all others were an aberration of sorts. Normal soils were placed within two categories at the highest level: *Pedocals* and *Pedalfers* (Fig. 7.2). Pedo*cal*s are soils of dry regions that accumulate *calcium* in the profile. Ped*al*f*e*rs are soils of humid regions, enriched in *Al* and *Fe* in their subsurface horizons. These names continue to be used loosely and do retain some descriptive value (Table 7.1).

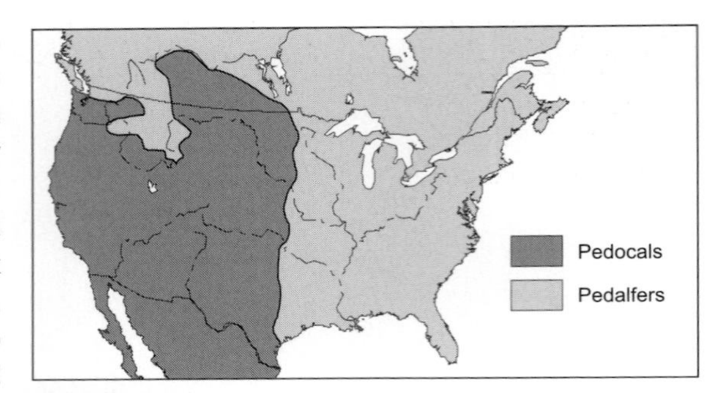

Fig. 7.2 Distribution of Pedocals and Pedalfers (old terms) in the USA and adjacent regions.

Shortly on the heels of the 1935 Marbut classification came the long-standing classification of 1938 (Baldwin *et al.* 1938). This system de-emphasized the Pedocal/Pedalfer scheme, and in its place classified all soils into one of three major categories, or *orders*: Zonal (previously normal soils), Azonal (soils with weakly expressed or incomplete profiles, i.e., young soils) and Intrazonal (soils whose profiles are dominated by local or particular factors of soil formation, such as ash, salts or high water tables). The three categories

Table 7.1 Outline of the 1938 soil classification system

Category VI (Order)	Category V (Suborder)	Category IV (Great soil groups)
Zonal Soils: Pedocals	Soils of the cold zone	Tundra soils
	Light-colored soils of arid regions	Desert soils
		Red Desert soils
		Sierozem
		Brown soils
		Reddish Brown soils
	Dark-colored soils of the semi-arid, subhumid and humid grasslands	Chestnut soils
		Reddish Chestnut soils
		Chernozem soils
		Prairie soils
		Reddish Prairie soils
Zonal soils: Pedalfers	Soils of the forest–grassland transition	Degraded Chernozem soils
		Noncalcic Brown or Shantung Brown soils
	Light-colored podzolized soils of the timbered regions	Podzol soils
		Brown Podzolic soils
		Gray-Brown Podzolic soils
	Lateritic soils of forested warm-temperate and tropical regions	Yellow Podzolic soils
		Red Podzolic soils (and Terra Rossa)
		Yellowish Brown Lateritic soils
		Reddish Brown Lateritic soils
		Laterite soils
Intrazonal soils	Halomorphic (saline and alkali soils)	Solonchak or saline soils
		Solonetz soils
		Soloth soils
	Hydromorphic soils of marshes, swamps and seep areas	Wiesenböden (meadow soils)
		Alpine Meadow soils
		Bog soils
		Half Bog soils
		Planosols
		Groundwater Podzol soils
		Groundwater Laterite soils
	Calcomorphic	Brown Forest soils (Braunerde)
		Rendzina soils
Azonal soils		Lithosols
		Alluvial soils
		Dry sands

Source: Modified from Baldwin *et al.* (1938).

were an outgrowth of Dokuchaev's normal, transitional and abnormal soil orders. Again, zonal soils were considered typical of broad geographic regions, or belts. Soil orders were broken into suborders, and then into great soil groups (Tables 7.1, 7.2).

This first serious attempt at a US soil classification scheme was not without problems. Many

Table 7.2 General characteristics of zonal soils of the 1938 system of classification

Soil type	Profile characteristics	Native vegetation	Soil-development processes
Zonal soils			
Tundra	Dark brown peaty layers over grayish, mottled profile; with permafrost	Lichens, moss, flowering plants and shrubs	Gleization and mechanical mixing
Desert	Light gray or light brownish gray solum, low in organic matter and shallow to calcareous material	Scattered shrubby desert plants	Calcification
Red Desert	Light reddish brown A horizon, brownish red or red B horizon with some illuvial clay; shallow to secondary carbonates	Desert plants	Calcification
Sierozem	Pale grayish soil grading into secondary carbonates at $\leq$0.3 m	Desert plants	Calcification
Brown	Brown soil grading into a whitish Bk horizon at 0.3–1 m	Shortgrass and bunchgrass prairie	Calcification
Reddish Brown	Reddish brown A horizon, grading into a red or dull red, clay-rich B horizon, and then into a zone of secondary carbonates	Tall bunchgrass and shrubs	Calcification
Chestnut	Dark brown, friable A horizon over a brown prismatic B horizon with secondary carbonates at 0.5–1 m	Mixed tall and shortgrass prairie	Calcification
Reddish Chestnut	Dark reddish brown A horizon; reddish brown or red B; secondary carbonates at $\geq$0.7 m	Mixed grasses and shrubs	Calcification
Chernozem	Black or very dark grayish brown A horizon, 1–1.5 m thick, grading through a lighter B, to secondary carbonates	Tallgrass and mixed grass prairie	Calcification

(cont.)

Table 7.2 *(cont.)*

Soil type	Profile characteristics	Native vegetation	Soil-development processes
Prairie	Very dark brown or grayish brown A horizon, grading through brown to lighter-colored parent material	Tallgrass prairie	Calcification with weak podzolization
Reddish Prairie	Dark brown or reddish brown A horizon, grading through reddish brown, clay-rich Bt; moderately acid	Tallgrass and mixed grass prairie	Calcification with weak podzolization
Degraded Chernozem	Nearly black A horizon, somewhat bleached, grayish E, incipient Bt horizon; vestiges of secondary carbonates	Forest encroaching on tallgrass prairie	Calcification followed by podzolization
Non-calcic Brown (Shantung Brown)	Brown or light brown, friable A horizon over a pale reddish brown or dull red B	Open deciduous forest with brush and grasses	Weak podzolization and some calcification
Podzol	Thin O horizon over a very thin, dark gray A horizon, a thin whitish gray E, a dark or coffee brown Bhs horizon and a yellowish brown BC; strongly acid	Coniferous, or mixed coniferous–deciduous forest	Podzolization
Brown Podzolic	Acidic O horizon over a thin, gray-brown or yellowish brown E, over a brown Bw horizon; solum is seldom more than 60 cm thick	Deciduous or mixed coniferous–deciduous forest	Podzolization
Gray-Brown Podzolic	Thin O horizon over dark A horizon 5–10 cm thick, over a grayish brown E, over a brown, clay-rich Bt horizon; less acid than Podzols	Mostly deciduous forest with mixture of conifers	Podzolization

(cont.)

Table 7.2 (*cont.*)

Soil type	Profile characteristics	Native vegetation	Soil-development processes
Yellow Podzolic	Thin, dark-colored O horizon over a pale yellowish gray E, 15–100 cm thick, over a clay-rich, yellow Bt horizon, over yellow, red and gray mottled parent material; acid	Coniferous or mixed coniferous and deciduous forest	Podzolization with some laterization
Red Podzolic	Thin O horizon over a yellowish brown or grayish brown A horizon, over a deep red B; acid	Deciduous forest with some conifers	Podzolization and laterization
Yellowish Brown Lateritic	Brown friable clay and clay loam horizons over yellowish brown, heavy but friable clays; acid to neutral	Evergreen and deciduous broadleaved trees	Laterization and some podzolization
Reddish Brown Lateritic	Reddish brown or dark reddish brown, friable, clayey soil over deep red, granular clay; C horizon reticulately mottled	Tropical rain forest to edge of savanna	Laterization with little or no podzolization
Laterite	Red-brown surface soil with a thick, red B horizon and a red or mottled C; very deeply weathered	Tropical selva and savanna	Laterization and a little podzolization
Intrazonal soils			
Solonchak	Thin, gray salty surface crust with a fine granular mulch just below, more salts in lower solum	Sparse growth of halophytic grasses and shrubs	Salinization
Solonetz	Very thin, friable A horizon underlain by dark, hard columns, usually highly alkaline	Halophytic plants; much bare surface	Solonization (desalinization and alkalization)
Soloth	Thin grayish brown, friable A horizon over whitish, leached E horizon, underlain by dark brown, clayey Btn or Btz horizon	Mixed prairie or shrub	Solodization (dealkalization)

(*cont.*)

Table 7.2 (*cont.*)

Soil type	Profile characteristics	Native vegetation	Soil-development processes
Wiesenböden and Alpine Meadow	Dark brown or black A horizon grading to a gray, mottled subsoil	Grasses and sedges	Gleization and some calcification
Bog	Brown, dark brown or black organic materials over brown peat	Swamp forest or sedges and grasses	Gleization
Half Bog	Dark brown or black peat over gray or mottled mineral materials	Swamp forest or sedges and grasses	Gleization
Planosols	Strongly leached, acid A and E horizons over a compact or cemented Bt "claypan"	Grass or forest	Podzolization and gleization; laterization in the tropics
Groundwater Podzols	O horizon over a very thin, acid A horizon, over a whitish gray E horizon up to 100 cm thick; Bsm horizon is brown or very dark brown; gleyed, gray C horizon	Forest	Podzolization and gleization
Groundwater Laterites	Gray or grayish brown A horizon over a leached, yellowish gray E, over a thick, reticulately mottled and cemented "hardpan" at ≥30 cm; pan can be over 1 m thick	Tropical forest	Podzolization and laterization
Brown Forest (Braunerde)	Very dark brown, friable A horizon, grading through a lighter-colored solum; high amounts of calcium	Forest, usually broadleaf	Calcification with very little podzolization
Rendzina	Dark, grayish brown to black, granular A horizon underlain by gray or yellowish, usually soft, calcareous material	Usually grasses; some broadleaf forest	Calcification

(*cont.*)

Table 7.2 (cont.)

Soil type	Profile characteristics	Native vegetation	Soil-development processes
Azonal soils			
Lithosols	Thin, stony A horizon; little or no illuviation; stony parent materials	Depends on climate	Varies
Alluvial soils	Little profile development; some organic matter accumulation; stratified parent materials	Depends on climate	Varies
Sands, dry	Essentially no profile; loose sands	Scanty grass, scrubby forest or bare sand	Varies

Source: Modified from Baldwin *et al.* (1938).

soils did not fit into a category; some fit into several. Most great soil groups were weakly defined, and most definitions of class limits were qualitative (Riecken 1945). The system was based on the virgin soil profile, and broke down by forcing the classifier to pigeon-hole cultivated or eroded soils into categories based on what they would have been like prior to cultivation. In short, the 1938 soil classification system left much to subjective interpretation. A more quantitative system was needed (Smith 1983).

Major advances in soil chemistry and soil physics, better understanding of soil landscapes after years of mapping, the introduction of the catena concept (Bushnell 1942) and with it a better understanding of local-scale (within-series) soil variability, and the need for increasingly quantitative soil interpretations for agricultural purposes all led to the call for a new, objective system of soil classification. James Thorp and Guy Smith, in 1949, declared their intent to depart from the zonal soil concepts that were the backbone of the 1938 system, "in favor of terms based on soil characteristics" (Thorp and Smith 1949: 125). This was to be a major turning point in soil classification.

This new classification, eventually called *Soil Taxonomy*, had to have certain characteristics (Cline 1949a). First, it could not abandon already entrenched series concepts or names. Second, the primary objective of the new classification was to facilitate soil survey/mapping operations; Soil Taxonomy was designed to improve use and management of the soil. Let no one be confused; the new system was not put in place to better categorize soils based on *genesis* alone, although the categories needed to make genetic sense (Cline 1963, Bockheim and Gennadiyev 2000). The categories of the new system must have pedogenic ties, or they could not be mapped, but the criteria that governed the system was based on quantifiable soil *properties*. The move away from an emphasis on soil processes was based on the assumption that soil properties were the result of processes, and properties were more quantifiable (Bockheim and Gennadiyev 2000). Thus, soil genesis would not overtly appear in the new system, but it lay behind it (Cline and Johnson 1963, Smith 1983).

The new system, spearheaded by Guy Smith, was published as a series of working drafts or approximations (Thorp and Smith 1949). Each approximation was circulated to a number of practicing soil scientists and scholars and amended as per their comments. Advice was sought at specially organized meetings, and by circulating the approximations at national and international soils conferences. Hundreds of people were involved in the process. Finally, in 1960, the new 7th Approximation was deemed sound enough to be published widely (Soil Survey Staff 1960). Fifteen years later, Soil Taxonomy was again published in a thick, green, hardcover book; it was essentially the 8th Approximation, but now it would become known simply as Soil Taxonomy or, to many, "the green book" (Soil Survey Staff 1975). The National Soil Survey Center staff continues to update the system, as new knowledge comes forth, especially with respect to soils outside the United States (e.g., Soil Survey Staff 1999).

How the system works

There are basically two philosophies in classifying soils: genetic and morphologic (Beckmann 1984). The genetic approach is based on knowledge of pedogenic processes or on correlations of inferred pedogenic factors (see Chapter 12). The purely morphologic approach excludes some of the genetic interpretation in favor of an emphasis on properties of individual soils, i.e., pedons. Soil Taxonomy (Soil Survey Staff 1999) is based on definable, measurable and sampleable soil characteristics; it is thus primarily a morphologic taxonomic system but with strong genetic underpinnings. The pedon is the basic soil unit; it is what is sampled. The system draws heavily upon experience of soil survey teams in the United States, although it can, theoretically, be applied globally.

Soil Taxonomy is hierarchical, with the six major categories (from largest to smallest) being: order, suborder, great group, subgroup, family and series. With each smaller category, more information is provided about the soil. Soil orders are broadly defined and general, but small (12) in terms of members. Soil series, at the other end of the spectrum, are smallest in terms of specificity, but large in terms of members (>15 000 and growing!). Thus, the system is arranged from the general to the specific, just like the Linnaean system in biology.

Taxa within the system, e.g., orders, suborders, etc., are often defined based on the presence or absence of quantitatively defined *diagnostic horizons*, which can be at the surface or in the subsurface (Table 7.3). Surface (typically A) horizons are called *epipedons*. Not all horizons are diagnostic. For example, a B horizon enriched in illuvial clay would be designated Bt. If it was strongly developed and also met certain defined criteria, it would be a diagnostic *argillic* horizon. But if it lacked a minimum amount of illuvial clay it might be a *cambic* horizon (Table 7.3). All argillic horizons have a t suffix; *not* all Bt horizons, however, qualify as argillic.

Soil names, at the highest four levels of the system, are constructed by assembling formative elements or syllables. Many formative elements come from Greek and Latin, while others are fragments of words commonly used today (Heller 1963). For the 12 soil orders, the formative elements are listed in Table 7.4. Each order name contains a formative element and a connecting vowel (*o* or *i*), ending with *sol* (Latin *solum*, soil). In the Gelisol order the formative element is *el*; in Aridisols it is *id*. These formative elements continue to be used as endings for the suborders, great groups and subgroups. Thus, all the names of all taxa (higher than series) that are in the Entisol order end in *ent*. Each order is strictly defined; many require the presence of at least one diagnostic horizon (Tables 7.4, 7.5). Many of the major order concepts were taken directly or in large part from the 1938 system of classification (Baldwin *et al.* 1938) (Table 7.6). Many of the orders also correspond, roughly, to a sequence of development: Entisol – Inceptisol – Spodosol/Alfisol – Ultisol – Oxisol (Fig. 7.3), although soil development can take many detours (see Chapter 12). For some orders, the influence of lithology, climate or local site characteristics are dominant (Figs. 7.3, 7.4).

Soils are further categorized into suborders. Suborder names are two syllables long; the first syllable connotes something about an important property or properties of the soil, whereas the second is the formative element of the order (Table 7.7). There are currently 64 suborders. The Fibrist suborder is constructed by adding the formative element *Fibr* (meaning fibrous) (Table 7.7) to the *ist* for the Histosol order. Fibrists are Histosols (organic soils) in which the organic matter is minimally decomposed, or fibrous. Likewise, a Psamment is a sandy Entisol (*Psamm* means sandy and *ent* is for Entisol). Psamments are weakly developed soils (i.e., Entisols) dominated by sandy particle sizes. In this way, the formative elements are strung together to make words that have

Table 7.3 Common diagnostic horizons used in Soil Taxonomy

Diagnostic horizon[a]	Defining criteria[b]	Minimum thickness (cm)	Common field equivalence	Required for which orders?	Commonly found in which orders?
Histic epipedon	Organic soil material (peat or muck), characterized by saturation and reduction	20	Oi, Oe, Oa	Histosols, if thickness criteria are met	Histosols
Folistic epipedon	Same as histic, but not wet; can exist in a non-Histosol (i.e., mineral soil); commonly overlies rock	15	Oi, Oe, Oa		Histosols
Mollic epipedon	Dark (Munsell color value ≤3 when moist and ≤5 when dry; chroma ≤3 when moist), humus-rich (≥0.6% organic carbon) horizon in which bivalent cations, especially Ca, are dominant on the exchange complex; base saturation >50%; granular structure is common; usually forms under grasses	25	A, Ap	Mollisols, except some Albolls	Vertisols
Umbric epipedon	Similar to mollic but base saturation is <50%; usually formed in cool, humid climates, under forest		A, Ap		Ultisols
Anthropic epipedon	Generally similar to mollic but has high P content due to long-term human occupation		Ap		

Plaggen epipedon	Dark surface horizon created by long-term manuring	50	Ap	
Melanic epipedon	Thick, dark horizon, rich in organic carbon, usually formed by continual, thin additions of volcanic ash		A, Ap	Some Andisols
Ochric epipedon	All other epipedons; typically thin, formed under forest	None	A, Ap	Entisols, Alfisols, Ultisols, Oxisols, Andisols, Inceptisols, Aridisols, Gelisols, Spodosols, some Mollisols (Albolls)
Albic horizon	Bright (Munsell color chroma ≤ 3 when moist and ≥ 5 when dry), leached E horizon; most grains have few or no coatings of clay or Fe oxides; often coarse-textured		E	Spodosols, Alfisols, Andisols, Ultisols, some Mollisols (Albolls)
Argillic horizon	Horizon of illuvial clay; contains ≥ 1.2 times as much clay as overlying horizons; has coatings of oriented clay (argillans) on ped faces; fine : coarse clay ratio is greater than in overlying horizons; can exist as clay bands (lamellae)	7.5	Bt	Alfisols, Ultisols, Mollisols, Aridisols, Spodosols
Kandic horizon	Weathered horizon with more clay than an overlying horizon; most of the (not necessarily illuvial) clay has low CEC and is either 1 : 1 phyllosilicates or oxide clays; may not exist as lamellae	30	Bt	Ultisols

(cont.)

Table 7.3 (cont.)

Diagnostic horizon[a]	Defining criteria[b]	Minimum thickness (cm)	Common field equivalence	Required for which orders?	Commonly found in which orders?
Natric horizon	Similar to argillic, but has more sodium (which accelerates and facilitates clay translocation); dense and slowly permeable, in parts; columnar structure is common		Btn		Mollisols, Aridisols, Alfisols, Vertisols
Glossic horizon	Degraded argillic, kandic or natric horizon; Albic (E) material interfingers into Bt or Bn horizon		E/Bt, E/Btn, E/Btx		Alfisols
Agric horizon	Illuvial horizon, below the Ap, in which long-term cultivation has led to translocation of silt, humus and clay		Varies		
Calcic horizon	Horizon of secondary carbonate accumulation, commonly in dry climates but can occur as precipitation from shallow groundwater	15	Bk		Aridisols, Vertisols, Mollisols, Inceptisols
Petrocalcic horizon	Horizon with cemented, secondary carbonates	10	Bkm		Aridisols
Gypsic horizon	Horizon of secondary gypsum accumulation, in arid climates on gypsum-rich parent materials	15	By		Aridisols, Vertisols, Mollisols

Petrogypsic horizon	Horizon with cemented, secondary gypsum; dry fragments do not slake in water	10	Bym		Aridisols
Duripan	Subsoil pan cemented by illuvial silica; most fragments will not slake in weak acid; common to dry regions, where they often form in association with Bkm horizons; parent materials commonly have volcanic origins		Bqm		Mollisols, Inceptisols, Alfisols, Andisols
Fragipan	Horizon that restricts entry of water and roots; air-dry fragments slake in water; not geogenic; shows pedogenic features; commonly has coarse prismatic structure with E material between prisms; brittle and dense	15	Bx, Ex, Btx		Alfisols, Inceptisols, Ultisols, Spodosols
Salic horizon	Horizon in which salts more soluble than gypsum, such as halite, have accumulated; high electrical conductivity	15	Bz		Aridisols, Vertisols, Mollisols
Spodic horizon	Horizon dominated by illuvial, amorphous compounds of organic matter and Al, with or without Fe; typically underlies albic horizons; dark and/or reddish colors; low ($\leq$5.9) pH	2.5	Bs, Bh, Bhs, Bsm, Bhsm	Spodosols	Spodosols

(cont.)

Table 7.3 (cont.)

Diagnostic horizon[a]	Defining criteria[b]	Minimum thickness (cm)	Common field equivalence	Required for which orders?	Commonly found in which orders?
Placic horizon	Thin (<10 mm thick), black to dark reddish pan, cemented by Fe and organic matter; commonly wavy or convoluted	None	Bsm		Spodosols
Sombric horizon	Acidic horizons dominated by illuvial humus, not due to cultivation; humus is not associated with aluminum		Bh		
Oxic horizon	Sandy B horizon dominated by kaolinite, oxide clays and other resistant minerals; highly weathered; low CEC; clay content increases with depth; on old, stable, weathered surfaces; porous	30	Bo	Oxisols	Oxisols
Cambic horizon	Minimally developed B horizon; cannot be sandy; often called "color B"		Bw, Bg		Inceptisols, Gelisols, Vertisols, Aridisols

[a] The diagnostic horizons listed are all subsurface horizons unless the word "epipedon" follows the name.
[b] Greatly simplified from Soil Survey Staff (1999). The reader is referred to that source for precise definitions.

Table 7.4 Names, formative elements and generalized descriptions of soils in the 12 taxonomic orders

Order	Formative element	Derivation[a] of formative element	Pronunciation	General characteristics and descriptors[b]
Gelisol	-el	L. *gelare*, to freeze	Gel	Soils with permafrost within 100 cm of the surface
Histosol	-ist	Gr. *histos*, tissue	Histology	Organic soils without shallow permafrost, dominated by decomposing organic matter; most are saturated at times
Spodosol	-od	Gr. *spodos*, wood ash	Odd	Soils in which translocation of compounds of Fe, humus and Al is dominant
Andisol	-and	Modified from *ando*	And	Soils that have often formed in parent materials with a large component of volcanic ash
Oxisol	-ox	F. *oxide*, oxide	Oxides	Highly weathered, relatively infertile soils dominated by oxide, low-activity clays
Vertisol	-ert	L. *verto*, turn, mix	Invert	Dark soils of semi-arid grasslands and savannas which develop deep cracks in the dry season; cracks swell shut in wet season as the shrink–swell clays rehydrate and expand
Aridisol	-id	L. *aridus*, dry	Arid	Soils of dry climates with some development in the B horizon, often as precipitated compounds of calcium or other salts
Ultisol	-ult	L. *ultimus*, last	Ultimate	Acid, leached soils of warm, humid climates that have a B horizon enriched in clay, usually 1:1 and oxide clays
Mollisol	-oll	L. *mollis*, soft	Mollify	Base-rich soils that have a thick, dark A horizon, often formed under grasslands or savanna
Alfisol	-alf	Aluminum (Al) and iron (Fe)	Ralph	Soils that are less acidic than Ultisols, and which have a B horizon enriched in silicate clays
Inceptisol	-ept	L. *inceptum*, beginning	Inception	Soils with weak B horizon development
Entisol	-ent	Meaningless syllable from "recent"	Recent	All other soils, usually very weakly developed, on young surfaces or eroded/disturbed sites or forming on difficultly weatherable materials

[a] F., French; Gr., Greek; L., Latin.

[b] Greatly simplified from Soil Survey Staff (1999). The reader is referred to that source for precise definitions.

Source: After Soil Survey Staff (1975).

Table 7.5 Diagnostic horizons and properties, arranged by taxonomic order

Soil order	Commonly observed diagnostic horizons and properties[a]	Representative horizon sequence(s)
Alfisol	Argillic horizon (high base status)	A/E/Bt/C
	Fragipan	A/E/Btx/C
Andisol	Melanic epipedon	A/Bw/C or A/Bs/C
	Andic properties	
Aridisol	Natric horizon	A/Eg/Btn/Bk/By/C
	Calcic, petrocalcic horizon	A/Bkm/Ck
	Gypsic, petrogypsic horizon	A/Cym/Cy
	Argillic horizon	A/E/Bt/Ck
	Duripan	A/B/Cqm
	Salic horizon (aridic soil moisture regime)	Az/Cz
Histosol	Histic materials	Oi/Oa/Oe
Mollisol	Mollic epipedon	A/Bt/C or A/Bk/C
Oxisol	Oxic horizon	A/Bo/Cr
Spodosol	Spodic materials, albic horizon	Oa/E/Bhs/Bs/C or Oe/A/E/Bs/C
	Fragipan	A/E/Bs/Ex/Bx/C
Ultisol	Argillic horizon (low base status)	A/E/Bt/C or E/Bt/C or A/E/Bt/Bv/C
Vertisol	(Slickensides, cracks)	A/Css or A/Bss/Ck
Gelisol	Gelic materials	O/Bgjj/Cf
Inceptisol	Cambic horizon (and others)	A/Bw/C
Entisol	(None)	A/C or A/Bs/C or A/Bw/C (if sandy)
All orders except Aridisols	Redoximorphic features (aquic soil moisture regime)	A/Bg/Cg

[a] These are the diaganostic horizons that the soils of each order usually do/can have; not all are required for a soil to classify within that order.

Source: Bockheim and Gennadiyev (2000).

meaning and connote an image. Because each of the taxa formed by stringing together formative elements is quantitatively defined, the words have a strict meaning that is measurable. The little room for subjective interpretation is usually a good thing, but can sometimes be restrictive.

Soil moisture regimes

At the suborder level, one of the most common descriptors refers to the soil moisture or wetness status, which is primarily a function of climate (water supply) and topography/relief as it affects the local water table. Because of the importance of water in soil genesis, knowing the *soil moisture regime* of a soil is vital to understanding its genesis and, more importantly, use and management. Soils dry out and wet up on various timescales (daily, seasonally and over long timescales of climate change) and spatial scales (upper horizons may be much drier than lower ones, a perched water table may exist), complicating the assignment of a soil moisture regime. Similarly, across landscapes soil moisture can vary tremendously, even among pedons only a few meters apart. In theory, there are three basic types of soil moisture regimes: saturated, leaching (soil water is adequate to move materials downward and completely through in the profile) and dry (most water moving in the soil eventually evaporates, leaving behind various precipitates). Many soils span two of these regimes, depending on the climate of any particular year.

Table 7.6 Taxonomic equivalencies between the 1938 soil classification system[a] and Soil Taxonomy[b]

1938 classification	Soil Taxonomy (1999)	
Great soil groups	Great groups or other taxa mostly included	Great groups or other taxa partly included
Alluvial soils	Fluvaquentic and fluventic Inceptisols and Mollisols; Fluvents	Cumulic Mollisols; Cryorthents, Psammaquents, Endoaquents and Endoaquepts
Alpine Meadow soils	Cryaquods and Cryands	Cryaquolls and Cryaquepts
Ando soils	Andisols	
Bog soils	Histosols	
Brown soils	Aridic Argiustolls, Argixerolls, Haploxerolls and Haplustolls; Ustollic and Xerollic Argids and Cambids	Frigid families of Xerollic Argids, Cambids and Xerolls; frigid families of Aridic Haplustolls and Hapludolls
Brown Forest soils	Eutrudepts and Haplustepts	Haploxerolls and Hapludolls
Brown Podzolic soils	Cryudands; mesic families of Entic Orthods	Dystrudrepts, Eutrudepts and Fragiudepts
Calcisols	Calcids, Calciargids, Petrocalcids, Calcigypsids and Calcicryids	Haplocambids, Haplodurids, Haplustepts and Haploxerepts
Calcium Carbonate Solonchaks	Calciaquolls	Aquic Calciustolls
Chernozems	Cryolls; mesic families Argiustolls and Haplustolls	Mesic and frigid families of Haploxerolls
Chestnut soils	Frigid families of Argixerolls, Durixerolls, Haploxerolls and Palexerolls; mesic families of Aridic Argiustolls, Argixerolls, Haploxerolls and Haplustolls	Mesic and frigid families of Calcic and Calcidic Argixerolls and Haploxerolls
Degraded Chernozems	Mollic Hapludalfs and Endoaqualfs	Alfic Argiudolls and Argiustolls
Desert soils	Mesic families of Argidurids, Haplargids and Paleargids	Mesic families of Haplocambids and Haplodurids
Gray-Brown Podzolic soils	Hapludalfs and Glossudalfs	Aeric Endoaqualfs; Fragiudalfs and mesic families of Hapludults
Gray Wooded soils	Eutrudepts and Dystrudepts	Mesic and frigid families of Udalfs and Ustalfs; Glossudalfs
Groundwater Laterite soils	Plinthaquults, Plinthudults, Plinthustalfs and Plinthustults	
Groundwater Podzols	Aquods	Haplohumods; Aquands; Spodic Psammaquents
Grumusols	Vertisols	Vertic Endoaquepts, Epiaquepts and Endoaquolls; Ustertic, Vertic and Xerertic Haplargids and Haplocambids

(cont.)

Table 7.6 *(cont.)*

1938 classification	Soil Taxonomy (1999)	
Great soil groups	Great groups or other taxa mostly included	Great groups or other taxa partly included
Half-Bog soils	Histic Cryaquepts and Humaquepts	Histic Cryaquolls, Endoaquolls and Epiaquolls; Histic Glossaqualfs
Humic Ferruginous Latosols	Humults	Dystrudands; Oxisols
Humic Gley soils	Argiaquolls, Cryaquolls, Haplaquolls and Humaquepts	Aquults and Aquands; Calciaquolls, Fluvaquents; Mollic Endoaquepts and Endoaqualfs
Humic Latosols	Humults	Dystrudands and Hydrudands; Humic Ustox, Perox and Udox
Hydrol Humic Latosols	Hydrudands	
Laterite soils	Acrotorrox, Acroperox and Acrudox	
Latosolic Brown Forest soils	Dystrudands	Udivitrands
Latosols	Humults	Oxisols; Hydrudands, Rhodustults, Kandiustults and Dystrudepts
Lithosols	Lithic extragrades	"Shallow" familes of any soil series
Low Humic Gley soils	Aquults, Endoaquents, Endoaquepts and Endoaqualfs	Aquepts, Aqualfs and Aquods; Cryaquents, Fluvaquents and Psammaquents
Low Humic Latosols	Oxic Dystrustepts and Haplustepts	Haplustox and Kandiustox
Non-calcic Brown soils	Durixeralfs, Haploxeralfs and Palexeralfs	Haploxerepts
Planosols	Albaqualfs, Argialbolls and Fragiaqualfs	Glossaqualfs, Fragiaquults and Albaquults
Podzols	Spodosols	Cryands and Aquands; Psammoturbels and Psammorthels
Prairie soils and Brunizems	Argiudolls, Haplaquolls, Argiaquolls, Hapludolls, Argixerolls and Haploxerolls	Haplocryolls, Durixerolls, Palexerolls and Udic Argiustolls
Red Desert soils	Haplocambids, Argidurids, Haplargids and Paleargids	Haplocalcids

(cont.)

Table 7.6 *(cont.)*

1938 classification	Soil Taxonomy (1999)	
Great soil groups	Great groups or other taxa mostly included	Great groups or other taxa partly included
Reddish Brown Lateritic soils	Humults and Rhodudults; Rhodic Paleudalfs and Paleudults	
Reddish Brown soils	Haplustalfs and Paleustalfs; Typic and Ustollic Haplocalcids, Haplocambids and Haplargids	Haplotorrands; Aridic Argiustolls, Calciustolls and Haplustolls
Reddish Chestnut soils	Haplustalfs and Paleustalfs; Aridic and Typic Argiustolls and Paleustolls	Haplustolls
Reddish Prairie soils	Thermic families of udic subgroups of Argiustolls and Paleustolls	Torrands; thermic families of Paleudolls
Red-Yellow Podzolic soils	Thermic families of Fragiudults, Hapludults, Kandiudults, Kanhapludults and Paleudults	Mesic families of Fragiudults, Hapludults, Kandiudults, Kanhapludults and Paleudults; thermic families of Paleudalfs; some Haplustalfs and Paleustalfs
Regosols	Psamments, Orthents	Entic subgroups of Haploxerolls, Hapludolls and Haplustolls; Psammaquents, Xerochrepts and Cryands
Rendzina soils	Rendolls	Xerolls and Ustolls
Sierozems	Haplocambids, Argidurids, Haplargids and Paleargids	
Solonchak soils	Salorthids	Halaquepts
Solonetz soils	Natric great groups of Alfisols, Aridisols and Mollisols	
Soloth soils		Natraqualfs; Natric subgroups of Cryolls; Duraquolls, Durixeralfs and Haploxeralfs
Sols Bruns Acides	Dystrochrepts, Fragiochrepts and Haplumbrepts	Udorthents and Xerumbrepts
Subarctic Brown Forest soils	Cryepts	Cryolls
Tundra soils	Aquiturbels	Vitricryands, Cryochrepts and Cryumbrepts

[a] After a variety of sources, including Baldwin *et al.* (1938), Cline (1955), Harper (1957), Kellogg (1941, 1950), Kellogg and Nygard (1951), Marbut (1935), McClelland *et al.* (1959), Oakes and Thorp (1950), Simonson *et al.* (1952), Tavernier and Smith (1957) and Thorp and Smith (1949).
[b] Modified from Soil Survey Staff (1975).

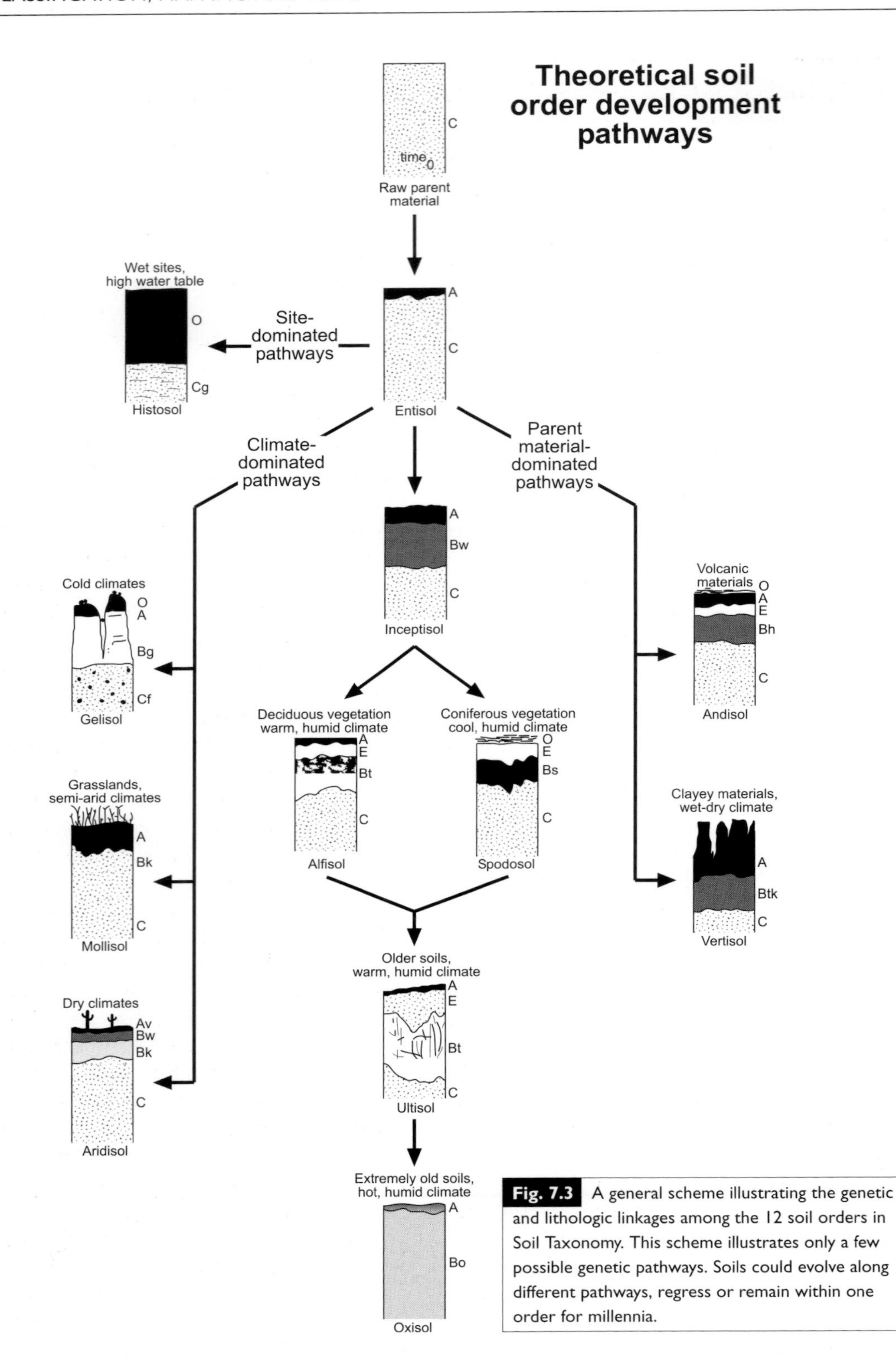

Fig. 7.3 A general scheme illustrating the genetic and lithologic linkages among the 12 soil orders in Soil Taxonomy. This scheme illustrates only a few possible genetic pathways. Soils could evolve along different pathways, regress or remain within one order for millennia.

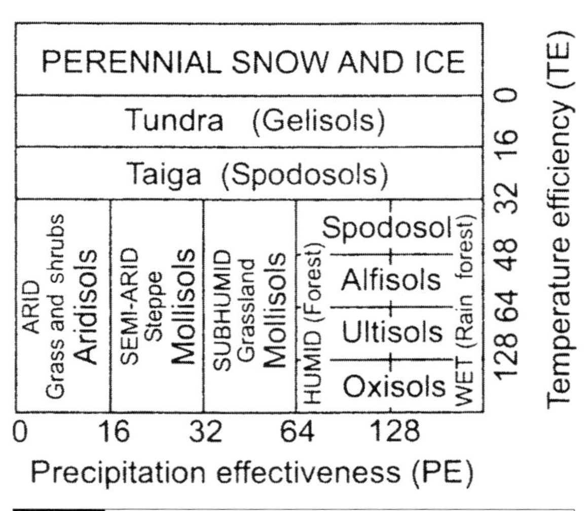

Fig. 7.4 Matrix showing how several major soil orders relate with each other, and how they vary along two climatic axes. PE values correlate to moisture availability and are: >128 = superhumid, 64–128 = humid, 32–64 = subhumid, 16–32 = semi-arid, and <16 = arid. TE values correlate to temperature and are: >128 = very hot, 64–128 = hot, 32–64 = warm, 16–32 = cool and <16 = cold. After Thornthwaite (1931) and Ruhe (1975a).

Soil moisture regime refers specifically to either groundwater or water held at a tension of less than 1500 kPa (15 bar) in the soil. Water held at tensions >1500 kPa is retained so tightly that most plants cannot remove it. When a soil horizon is saturated, the water within is retained at tensions near zero, or even at negative pressures. Consequently, a horizon is considered dry when the water within is held at tensions exceeding 1500 kPa, and is considered moist if the water within is held at tensions <1500 kPa but more than zero. Soils saturated with salty or saline water are considered salty rather than dry.

Soil moisture regimes are loosely related to macroclimate (Paetzold 1990). For example, permeable soils in humid climates have water that is available to plants most of the time. Conversely, one might think that soils in deserts are necessarily dry. They may be dry, moist *or* saturated, however, depending on their position on the landscape, because they may receive water from sources other than precipitation, such as groundwater. Extra water such as runoff is available to soils in any climate, provided they occupy certain landscape positions. In the northern hemisphere, precipitation is more effective in soils on north aspects than in soils on south aspects (Franzmeier *et al.* 1969, Hunckler and Schaetzl 1997, King *et al.* 1999), possibly leading to different soil moisture regimes on opposing slopes. Soils on steep slopes or slowly permeable materials may lose part or most of the precipitation that impinges upon them as runoff, making them drier than they would otherwise be. Thus, across vast areas soils may have similar soil moisture regimes, due to climate, but on certain landscape positions soils may exist that occupy the drier or wetter regimes, based on local influences.

Determining the soil moisture regime of a soil requires knowledge of inputs of water to it, throughputs of water vapor and liquid within it and losses of water from it (Fig. 7.5). We usually only infer soil-moisture status over longer timescales, based on climatic data for which we have good areal coverage and long temporal runs. Data from drought and abnormally wet years must be avoided. Soil Taxonomy uses probabilities to determine the climate of the normal year (Soil Survey Staff 1999) and bases its assessment of soil moisture regimes on that.

The moisture characteristics of a soil vary with depth and throughout the year. A soil may be continuously moist in some or all horizons either throughout the year, or for some part of the year. It may be either moist in winter and dry in summer, or the reverse (Soil Survey Staff 1999). Soil moisture regimes are defined such that significant variations in soil moisture with depth and over time are factored in. But what *part* of the soil profile must be considered? Soil Taxonomy uses its best estimate of the water in the *soil-moisture control section*. The upper boundary of this section is the depth to which a dry soil will be wetted by 2.5 cm of water within 24 h. Its lower boundary is the depth to which the same soil will be moistened by 7.5 cm of water within 48 h. The intent of these rather arbitrary limits is that the soil-moisture control section roughly corresponds to the rooting depths for many crops. To put some numbers on these concepts, one can assume that the soil-moisture control section of a finer-textured soil is between 10 and 30 cm depth. In sandy soils, the section might be 30 to 90 cm deep (Soil Survey Staff 1999).

Table 7.7 Derivations, mnemonics and generalized meanings of the formative elements for suborders

Formative element	Derivation[a]	Mnemonic	Connotation/meaning[b]
Alb	L. *albus*, white	Albino	Has an albic (bright E) horizon
Anthr	Gr. *anthropos*, human	Anthropology	Has an anthropic epipedon, modified by humans, usually implying long-term human occupancy
Aqu	L. *aqua*, water	Aquarium	Aquic or peraquic soil moisture regime; wet, saturated at parts of the year as indicated by redoximorphic features
Ar	L. *arare*, to plow	Arable	Mixed horizons, by agricultural plowing
Arg	L. *argilla*, white clay	Argillite	Has an argillic (Bt) horizon, enriched in illuvial clay
Calc	L. *calcis*, lime	Calcite	Has a calcic (Bk) horizon, enriched in secondary $CaCO_3$
Camb	L. *cambiare*, to exchange		Has a cambic (Bw) horizon, slightly weathered
Cry	Gr. *kryos*, icy cold	Crystal	Exists in a cold, cryic soil temperature regime (<8 °C)
Dur	L. *durus*, hard	Durable	Has a duripan (Bqm), dense and cemented with silica
Fibr	L. *fibra*, fiber	Fibrous	Organic soil dominated by raw or virtually raw (fibric) materials
Fluv	L. *fluvius*, river	Fluvial	Soils on floodplains or developed in alluvium
Fol	L. *folia*, leaf	Foliage	Organic horizon, not necessarily wet, with abundant leaves; often lying on bedrock
Gyps	L. *gypsum*, gypsum	Gypsum	Has a gypsic (By) horizon, enriched in secondary gypsum
Hem	Gr. *hemi*, half	Hemisphere	Organic soil dominated by by partially decomposed (hemic) materials
Hist	Gr. *histos*, tissue	Histology	Presence of organic soil materials
Hum	L. *humus*, earth	Humus	Abundant organic matter, usually in the B horizon (Bh)
Orth	Gr. *orthos*, true	Orthographic	Common type of soil for this order
Per	L. *per*, throughout time		Little seasonality with respect to temperature or moisture; often implies tropical climate
Psamm	Gr. *psammos*, sand	Psammite	Sandy textures dominate
Rend	Modified from Rendzina		Abundant carbonates in soil and parent material
Sal	L. base of *sal*, salt	Saline	Has salic (salty) horizons (Bn, Bz), with excess soluble salts
Sapr	Gr. *saprose*, rotten	Saprophyte	Organic soil dominated by highly decomposed (sapric) materials
Torr	L. *torridus*, hot and dry	Torrid	Torric (aridic) soil moisture regime; dry

(*cont.*)

Table 7.7 (*cont.*)

Formative element	Derivation[a]	Mnemonic	Connotation/meaning[b]
Turb	L. *turbidis*, disturbed	Turbulent	Mixed by cryoturbation (soil ice)
Ud	L. *udus*, humid	Udometer	Udic soil moisture regime (humid, moist)
Ust	L. *ustus*, burnt	Combustion	Ustic soil moisture regime (semi-arid to subhumid)
Vitr	L. *vitrum*, glass	Vitreous	Presence of glass or volcanic ash
Xer	Gr. *xeros*, dry	Xerophyte	Xeric (Mediterranean) soil moisture regime; dry summers and moist winters

[a] Gr., Greek; L., Latin.
[b] Simplified from Soil Survey Staff (1999). The reader is referred to that source for precise definitons.
Source: Soil Survey Staff (1975).

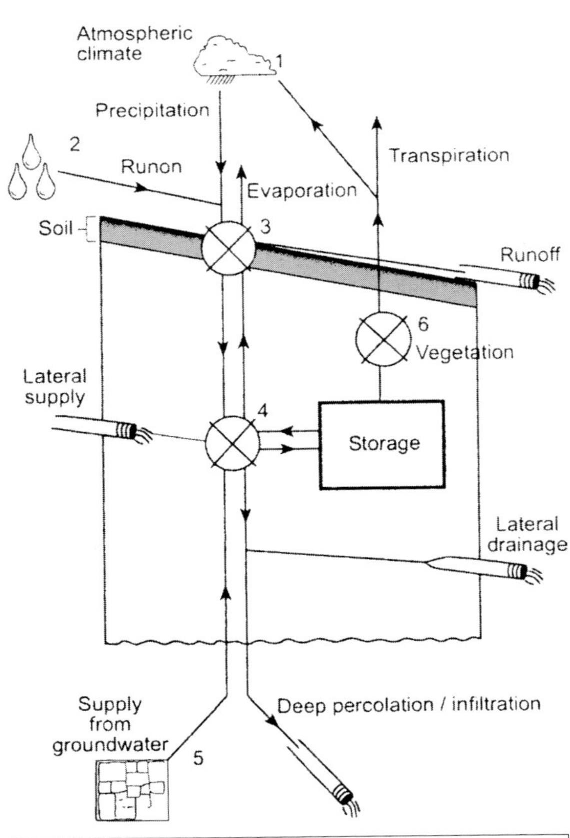

Fig. 7.5 Factors influencing the soil moisture regime of a soil. After Souirji (1991).

Most soil moisture regimes correspond approximately to major climate regions, such as arid (aridic), Mediterranean (xeric), semi-arid (ustic) and humid (udic) climates (Table 7.8). The wettest, *aquic*, is better envisioned as occurring locally, in low, wet spots where the water table is high. The concept of aquic soils involves a periodically reducing (anoxic) regime due to periodic saturation (Soil Survey Staff 1999). Thus, many landscapes have two types of soil moisture regimes: upland soils are in the regime that corresponds to the macroclimate (e.g., ustic, udic, xeric), while aquic soils occur at various locations where high (or perched) water tables impinge upon the profile. If groundwater is always close to the surface, as in salt marshes or deltas, the term *peraquic* is used (Table 7.8).

Soils that have an *aridic (torric)* moisture regime normally occur in dry climates, although a few occur in semi-arid climates where soil properties inhibit infiltration. Leaching occurs after storms, but much of this water later moves upward along matric tension gradients, as unsaturated flow. The result is a soil with a B horizon that has precipitates, such as $CaCO_3$ or soluble salts (see Chapter 12). Soil moisture regimes of semi-arid climates, in which the concept is one of soils that are usually dry but at some times are moistened for days or weeks at a time, are split into two groups. Those with a wet season that

Table 7.8 Classes and definitions of soil moisture regimes

Soil moisture regime	Simplified definition and limits
Aquic	Reducing regime in periodically saturated soils, due to either a perched water table (episaturation) or a regional groundwater table (endosaturation). During dry seasons the soil is not always saturated.
Peraquic	Reducing regime in continuously saturated soils, due to a high regional groundwater table.
Aridic and torric	Typical of upland soils in dry climates. In the warm season of normal years the soil-moisture control section is dry more than half of the time, or is moist for <90 consecutive days.
Ustic	Typical of semi-arid grasslands and savannas. In normal years, the soil-moisture control section is dry for ≥90 cumulative days, but moist for >180 cumulative days or >90 consecutive days.
Xeric	Typical of Mediterranean climates. In normal years, the soil-moisture control section is dry for ≥45 consecutive days in the summer and fall, and moist for ≥45 consecutive days in winter.
Udic	Typical of humid climates where precipitation is greater than evapotranspiration. In normal years, the soil-moisture control section is not dry for 90 cumulative days. Extended dryness (>45 days) in summer is not allowed.
Perudic	Typical of extremely wet climates where leaching occurs in all months; the soil water tension rarely reaches 100 kPa.

Source: Simplified from Soil Survey Staff (1999). The reader is referred to that source for precise definitons.

corresponds to the warm season are considered *ustic*; those where wintertime rainfall is the rule are *xeric* (Engel *et al.* 1997). In ustic soils, moisture is usually limited but is present when conditions are suitable for plant growth, as in grasslands and savannas (Soil Survey Staff 1999). Some leaching occurs here – more than in aridic regions – although subsoil accumulations of secondary carbonates are commonplace (see Chapter 12). In xeric soil moisture regimes, precipitation is concentrated in the cool season when plants are dormant, and thus it is more effective at leaching. To qualify as xeric, a soil must be wet for a period of time in winter, and dry in summer (Table 7.8). Most humid climates have adequate precipitation to wet and thoroughly leach soils annually. This concept is embodied within the *udic* soil moisture regime, in which water moves entirely through the profile in most years. Thus, accumulations of secondary carbonates and salts are generally, though not always, absent (e.g., Schaetzl *et al.* 1996). In climates where every month is wet and deep leaching is more or less continuous, the term *perudic* is used (Table 7.8).

The name of a *great group* consists of its suborder name and an additional prefix descriptor that consists of one or two formative elements that provide more information about soil properties (Table 7.9). Great group names, therefore, have three or four syllables and end with the name of a suborder. As an example, a *Fragiaquod* is a soil that is in the Spodosol order (-od) and has an aquic soil moisture regime (-aqu-) and a fragipan (Frag-). A fragipan is a diagnostic subsurface horizon that is dense and often brittle (see Chapter 12). Likewise, a *Natrargid* is an Aridisol (-id) that contains argillic (-arg-) and natric (Natr-) diagnostic horizons (Table 7.3).

The Soil Survey Staff stopped the process of adding onto soil taxonomic names with great groups (Table 7.9). *Subgroup* names, therefore, consist of the name of a great group modified by one or more adjectives that precede the great group name as a separate word or words (Table 7.10).

Table 7.9 Derivations, mnemonics and generalized meanings of the formative elements for great groups

Formative element	Derivation[a]	Mnemonic	Connotation[b]
Acr	Gr. *akros*, at the end	Acrolith	Extremely weathered, low CEC values
Al	Modified from aluminum	Aluminum	Illuvial horizons dominated by aluminum
Alb	L. *albus*, white	Albino	Has an albic (bright E) horizon
Anhy	Gr. *anydros*, waterless	Anhydrous	Very dry
Anthr	Gr. *anthropos*, human	Anthropology	Has an anthropic epipedon, usually implying long-term human occupancy
Aqu	L. *aqua*, water	Aquarium	Aquic or peraquic soil moisture regime; wet, saturated at parts of the year as evidenced by redoximorphic features
Argi	L. *argilla*, white clay	Argillite	Has an argillic (Bt) horizon, enriched in illuvial clay
Calci, calc	L. *calcis*, lime	Calcium	Has a calcic (Bk) horizon, enriched in secondary $CaCO_3$
Cry	Gr. *kryos*, icy cold	Crystal	Exists in a cold, cryic soil temperature regime (<8 °C)
Dur	L. *durus*, hard	Durable	Has a duripan (Bqm), dense and cemented with silica
Dystr, dys	Gr. *dys*, ill; dystrophic, infertile	Dystrophic	Low base saturation, acid
Endo	Gr. *endon*, *endo*, within	Endoderm	Regional groundwater table present, "endosaturation," as in "Endoaquods"
Epi	Gr. *epi*, on, above	Epiderm	Perched water table present, "episaturation," as in "Epiaquods"
Eutr	Gr. *eu*, good; eutrophic, fertile	Eutrophic	High base saturation, high (non-acid) pH, alkaline, fertile
Ferr	L. *ferrum*, iron	Ferruginous	Abundant iron, usually implies "within B horizon"
Fibr	L. *fibra*, fiber	Fibrous	Organic soil dominated by by raw or virtually raw (fibric) materials
Fluv	L. *fluvius*, river	Fluvial	Soils on floodplains or developed in alluvium
Fol	L. *folia*, leaf	Foliage	Organic soil, not necessarily wet, with abundant leaves; often lying on bedrock
Fragi	L. *fragilis*, brittle	Fragile	Has a fragipan (Bx or Btx horizon), brittle and dense
Fragloss	Compound of frag- and gloss		Has fragipan (Bx or Btx horizon) and glossic (tonguing E or Ex) horizon
Fulv	L. *fulvus*, dull brownish yellow	Fulvous	Dark brown color, presence of dissolved organic carbon; acid
Glac	L. *glacialis*, icy	Glacial	Presence of ice lenses or wedges
Gyps	L. *gypsum*, gypsum	Gypsum	Has a gypsic (By) horizon, enriched in secondary gypsum
Gloss	Gr. *glossa*, tongue	Glossary	Has a glossic horizon (tonguing of E into Bt horizon)
Hal	Gr. *hals*, salt	Halite	Salty, saline; salts present in the soil profile
Hapl	Gr. *haplous*, simple	Haploid	Minimum horizonation, not a "complicated" profile

(cont.)

Table 7.9 (*cont.*)

Formative element	Derivation[a]	Mnemonic	Connotation[b]
Hem	Gr. *hemi*, half	Hemisphere	Organic soil dominated by by partially decomposed (hemic) materials
Hist	Gr. *histos*, tissue	Histology	Presence of organic soil materials
Hum	L. *humus*, earth	Humus	Abundant organic matter, usually in the B horizon (Bh)
Hydr	Gr. *hydor*, water	Hydrolysis	Presence of water, wet
Kand, kan	Modified from kandite		Dominated by 1 : 1 phyllosilicate clays
Luv	Gr. *louo*, to wash	Alluvium	Illuvial materials in B horizon
Melan	Gr. *melas*, black	Melanic	Black or dark in color; abundant organic carbon
Moll	L. *mollis*, soft	Mollify	Has mollic (dark, thick, base-rich A horizon) epipedon
Natr	from *natrium*, sodium	Natron	Has natric (Btn) horizon, rich in sodium
Pale	Gr. *palaeos*, old	Paleosol	Extreme profile development, usually of B horizon; old soils or paleosols
Petr	Gr. *petra*, rock	Petrology	Has a rock-like or a cemented horizon near the surface
Plac	Gr. base of *plax*, flat stone		Has a thin ironpan (placic horizon)
Plagg	Ger. *plaggen*, sod		Has a plaggen epipedon, produced by long-time manuring
Plinth	Gr. *plinthos*, brick	Plinth	Presence of plinthite
Psamm	Gr. *psammos*, sand	Psammite	Sandy textures dominate
Quartz	Ger. *quarz*, quartz	Quartz	Extremely high quartz content
Rhod	Gr. base of *rhodon*, rose	Rhododendron	Dark red color
Sal	L. base of *sal*, salt	Saline	Has salic (Bz) horizon, dominated by salts more soluble than gypsum
Sapr	Gr. *saprose*, rotten	Saprolite	Extremely weathered or decomposed
Somb	F. *sombre*, dark	Somber	Has a sombric (acid, Bh) horizon
Sphagn	Gr. *sphagnos*, bog	Sphagnum	Dominated by fibers of *Sphagnum* moss
Sulf	L. *sulfur*, sulfur	Sulfur	Presence of sulfides or their oxidation products; may have a sulfuric horizon
Torr	L. *torridus*, hot and dry	Torrid	Torric (aridic) soil moisture regime; dry
Ud	L. *udus*, humid	Udometer	Udic soil moisture regime (humid, moist)
Umbr	L. *umbra*, shade	Umbrella	Has umbric (dark, thick, low base saturation) epipedon
Ust	L. *ustus*, burnt	Combustion	Ustic soil moisture regime (semi-arid to subhumid)
Verm	L. *vermes*, worm	Vermiform	Abundant worm activity, or mixed by animals
Vitr	L. *vitrum*, glass	Vitreous	Presence of glass or volcanic ash
Xer	Gr. *xeros*, dry	Xerophyte	Xeric (Mediterranean) soil moisture regime; dry summers and moist winters

[a] F., French; Ger., German; Gr., Greek; L., Latin.
[b] Simplified from Soil Survey Staff (1999). The reader is referred to that source for precise definitons.
Source: After Soil Survey Staff (1975).

Subgroup adjectives normally take on one of three functions: (1) denoting the soils as the *central concept* for that great group, (2) indicating that the soil is an *intergrade* between its great group and another or (3) noting that the soil has properties that are not representative of the great group, but neither are they transitional to another kind of soil (*extragrade* soils). The adjective *Typic* is used when a soil represents the central concept of the great group, e.g., Typic Haplocalcid. In other cases, the Typic subgroup is simply used for all the soils of a great group that don't fit into the possible intergrade or extragrade subgroups; the Typics are the leftovers. Typic subgroups have all the diagnostic properties of the order, suborder and great group to which they belong (Soil Survey Staff 1999). *Intergrade* subgroups belong to one great group but have some properties of (i.e., are transitional to) another order, suborder or great group. For example, *Vertic Torrifluvents* have some properties of Vertisols, but display the complete set of diagnostic properties of Torrifluvents – Entisols on floodplains (-fluv-) in a hot (Torric) soil-temperature regime. *Extragrade* subgroups do not necessarily show properties of another order, but rather, they display important properties that may affect their use and management. They are not transitional to another order or suborder. They are named by modifying the great group name with an adjective that connotes something about the nature of the aberrant properties (Soil Survey Staff 1999). For example, a *Typic Haplorthod* is the central concept of Haplorthods, an *Andic Haplorthod* is an intergrade Haplorthod to an Andisol, and a *Lithic Haplorthod* is an extragrade whose noteworthy characteristic is the presence of shallow bedrock. In certain circumstances, multiple adjectives may be used in a subgroup name, e.g., Abruptic Haplic, or Cumulic Vertic (Table 7.11).

Preceding the subgroup name is that of the family, within which additional detail is provided. Family differentia include terms related to particle-size class or classes, mineralogy, cation-exchange activity, calcareous and reaction class, soil temperature, and, in a few families, thickness, rupture resistance and classes of coatings and cracks. The names of most families have three to five of these descriptive terms that modify the subgroup name (Soil Survey Staff 1999). Table 7.12 provides a listing of a variety of family names, as examples of the diversity of soils; it also shows how the system works.

Soil-temperature regimes

At the family (and sometimes higher) level, soils are split out according to their long-term temperature (Paetzold 1990). Soil temperature is an important component of a soil, for it regulates biological activity in and on the soil, and geologic and pedologic processes within and below the soil, and can dramatically affect the state of soil water (Schmidlin *et al.* 1983, Jensen 1984, Alexander 1991, Isard and Schaetzl 1995, Mount 1998). Frozen soils behave quite differently than do soils where the water is in liquid form (Cary *et al.* 1978). Soil temperatures are variable between horizons, and among various parts of the landscape (Carter and Ciolkosz 1980, Nullet *et al.* 1990). Near-surface temperatures fluctuate daily and seasonally (Beckel 1957). Daily soil-temperature fluctuations may be very large, especially in soils of dry climates; at the other extreme, under melting snow, the temperature at the soil surface may be isothermal with depth. Wet soils fluctuate less in temperature than do dry soils, since the specific heat of water is nearly four times that of dry soil (see Chapter 5).

Soil Taxonomy recognizes (essentially) five *soil-temperature regimes* (Table 7.13). The temperature regime of a soil is often only estimated from long-term atmospheric data from nearby sites (Bocock *et al.* 1977, Reimer and Shaykewich 1980). This approach is taken because real soil-temperature data that can be gathered today are reflective only of contemporary conditions. Hardly ever do long-term temperature data exist from soils, requiring that we determine long-term soil temperatures based on statistical relationships between the soil and the atmosphere (Isard and Schaetzl 1995). Soil-temperature regimes are difficult to map because, like atmospheric climate, soil climate changes on short and long timescales (Isard and Schaetzl 1995, Mokma and Sprecher 1995). What was mesic this year may have been frigid in five of the last 10 years.

Table 7.10 Derivations, mnemonics and generalized meanings of the formative elements for subgroups

Adjective	Derivation[a]	Connotation
Albaquic		Combination of *Albic* and *Aquic*
Abruptic	L. *abruptum*, torn off	Abrupt textural change, usually from A (or E) to B horizon
Acraquoxic		Combination of *Acr-*, *Aquic* and *Oxic*; intergrade to an Acraquox
Acrudoxic		Combination of *Acr-*, *Udic* and *Oxic*; intergrade to an Acruduox
Acrustoxic		Combination of *Acr-*, *Ustic* and *Oxic*; intergrade to an Acrustox
Aeric	Gr. *aerios*, air	Aeration; slightly drier, better drained, than is normal for the great group
Albaquultic		Combination of *Albic*, *Aquic* and *Ultic*; intergrade to an Albaquult
Albic		Has an albic (bright E) horizon
Alfic	*Al* for aluminum, and *Fe* for iron	Intergrade to an Alfisol, often implying that the soil has a Bt horizon
Alic	Modified from aluminum	High amounts of Al^{3+}
Andic		Intergrade to an Andisol, often implying that the soil has had additions of volcanic glass, ash or pyroclastic materials
Anionic	Gr. *anion*, a thing going up	Negatively charged colloids
Anthraquic		Has an anthropic epipedon
Anthropic		Has an anthropic epipedon, modified by humans, usually implying long-term human occupancy
Aqualfic		Combination of *Aquic* and *Alfic*; intergrade to an Aqualf
Aquandic		Combination of *Aquic* and *Andic*; intergrade to an Aquand
Aquentic		Combination of *Aquic* and *Entic*; intergrade to an Aquent
Aquertic		Combination of *Aquic* and *Vertic*; intergrade to an Aquert
Aquic		*Trending toward* an aquic soil moisture regime; often implies that the soil is moderately well drained (if no -aqu- in suborder position)
Aquicambidic		Combination of *Aquic*, *Cambic* (implying a cambic (Bw) horizon, slightly weathered from the parent material) and *Aridic*; intergrade to an Aquicambid
Aquodic		Combination of *Aquic* and *Spodic*; intergrade to an Aquod
Aquollic		Combination of *Aquic* and *Mollic*; intergrade to an Aquoll

(*cont.*)

Table 7.10 (*cont.*)

Adjective	Derivation[a]	Connotation
Aquultic		Combination of *Aquic* and *Ultic*; intergrade to an Aquult
Arenic	L. *arena,* sand	Has thick, sandy epipedon
Argiaquic		Combination of *Argic* and *Aquic*
Argic		Has an argillic (Bt) horizon, enriched in illuvial clay
Argidic		Combination of *Argic* and *Aridic*; intergrade to an Argid
Argiduridic		Combination of *Argic*, *Duric* and *Aridic*; intergrade to an Argidurid
Aridic		Intergrade to an Aridisol, often implying that the soil moisture regime is borderline Aridic or Torric
Calciargidic		Combination of *Calcic*, *Argic* and *Aridic*; intergrade to a Calciargid
Calcic		Has a calcic horizon
Calcidic		Combination of *Calcic* and *Aridic*; intergrade to a Calcid
Cambidic		Combination of *Cambic* and *Aridic*; intergrade to a Cambid
Chromic	Gr. *chroma*, color	High chroma (bright-colored, not gray); usually imples "low in organic matter"
Cumulic	L. *cumulus,* heap	Overthickened, dark epipedon
Duric		Has a duripan or a mostly- cemented, duripan-like layer
Duridic		Combination of *Duric* and *Aridic*; intergrade to a Durid
Durinodic	L. *durabilis*, lasting or enduring, and L. *nodus,* knot	Has significant amounts of durinodes or is brittle
Dystric		Low base saturation, acid
Entic		Intergrade to an Entisol, often implying that some part of the soil is more weakly developed than is typical for the great group
Epiaquic		Combination of *epi-* and *aquic*; surface wetness, usually implying a perched water table within the profile
Eutric		High base status or pH
Fibric		Contains some organic materials with a low degree (fibric) of decomposition
Fluvaquentic		Combination of *Fluv-*, *Aquic* and *Entic*; intergrade to a Fluvaquent
Fluventic		Combination of *Fluv-* and *Entic*; intergrade to a Fluvent
Fragiaquic		Combination of *Fragic* and *Aquic*

(*cont.*)

Table 7.10 (*cont.*)

Adjective	Derivation[a]	Connotation
Fragic		Has a fragipan (Bx or Btx horizon) or a fragic-like horizon, brittle and dense
Glacic		Presence of ice lenses or wedges
Glossaquic		Combination of *Glossic* and *Aquic*
Glossic		Has a glossic horizon, tonguing horizon boundaries
Grossarenic	L. *grossus*, thick, and L. *arena*, sand	Has thick, sandy epipedon
Gypsic		Has a gypsic horizon
Halic		Has high amounts of soluble salts
Haplargdic		Combination of *Haplic*, *Argic* and *Aridic*; intergrade to a Haplargid
Haplic		Minimum horizonation, not a "complicated" profile
Haplocalcidic		Combination of *Haplic*, *Calcidic* and *Aridic*; intergrade to a Haplocalcid
Haploduridic		Combination of *Haplic*, *Duridic* and *Aridic*; intergrade to a Haplodurid
Haploxeralfic		Combination of *Haplic*, *Xeric* and *Alfic*; intergrade to a Haploxeralf
Haploxerollic		Combination of *Haplic*, *Xeric* and *Mollic*; intergrade to a Haploxeroll
Hemic		Contains some organic materials of intermediate degree (hemic) of decomposition
Histic		Intergrade to a Histosol; can imply that the soil has an O horizon thinner than is required for a Histic epipedon
Humaqueptic		Combination of *Humic*, *Aquic* and *Inceptic*; intergrade to a Humaquept
Humic		Has abundant organic matter
Hydraquentic		Combination of *Hydric*, *Aquic* and *Entic*; intergrade to a Hydraquent
Hydric		Presence of water, often implying a layer of water in inorganic clays and humus, or simply a layer of water within the profile
Inceptic		Not strongly developed
Kandic		Dominated by 1:1 phyllosilicate clays; probably has a Bt horizon that nearly qualifies as a kandic horizon
Kandiudalfic		Combination of *Kandic*, *Udic* and *Alfic*; intergrade to a Kandiudalf
Kandiustalfic		Combination of *Kandic*, *Ustic* and *Alfic*; intergrade to a Kandiustalf
Kanhaplic		Combination of *Kandic* and *Haplic*
Lamellic	L. *lamina*, thin plate	Argillic horizon that consists entirely of lamellae (clay bands)

(*cont.*)

Table 7.10 (*cont.*)

Adjective	Derivation[a]	Connotation
Leptic	Gr. *leptos*, thin	Has thin solum
Limnic	Modified from Gr. *limne*, lake	Presence of limnic materials (lake sediments) at depth
Lithic	Gr. *lithos*, stone	Shallow to hard bedrock (lithic contract)
Mollic		Intergrade to a Mollisol, often implying that A horizon is not quite dark and/or thick enough to be a mollic epipedon
Natrargidic		Combination of *Natric*, *Argic* and *Aridic*; intergrade to a Natrargid
Natric		Has a natric (Btn) horizon, rich in sodium
Natrixeralfic		Combination of *Natric*; *Xeric* and *Alfic*; intergrade to a Natrixeralf
Ochreptic		Combination of *Ochr-* and *Inceptic*; intergrade to an Ochrept
Ombroaquic		Water on the soil surface; saturated in the upper solum at times of the year
Oxic		Intergrade to an Oxisol, often implying a low CEC or a dominantly oxide clay mineralogy
Oxyaquic		Oxygenated water occupies saturated soil horizons for at least some part of the year
Pachic	Gr. *pachys*, thick	Has a thick epipedon
Paleargidic		Combination of *Pale-*, *Argic* and *Aridic*; intergrade to a Paleargid
Palexerollic		Combination of *Pale-*, *Xeric* and *Mollic*; intergrade to a Palexeroll
Paralithic	Gr. *para*, beside and *lithic* (stone)	Shallow to soft and/or weathered bedrock (paralithic contact)
Petrocalcic		Has a petrocalcic horizon
Petrocalcidic		Combination of *Petro-*, *Calcic* and *Aridic*; intergrade to a Petrocalcid
Petroferric	Gr. *petra*, rock, and L. *ferrum*, iron	Shallow to ironstone (petroferric contact)
Petrogypsic		Has a petrogypsic horizon
Petronodic		Has significant amounts of nodules or concretions
Placic		Has a thin ironpan (placic horizon)
Plagganthreptic		Combination of *Plaggen*, for plaggen epipedon, produced by long-time manuring, *Anthropic* and *Inceptic*; intergrade to a Plagganthrept
Plinthaquic		Combination of *Plinthic* and *Aquic*
Plinthic		Presence of plinthite
Psammentic		Combination of *Psamm-* and *Entic*; intergrade to a Psamment
Rhodic		Dark red color
Ruptic	L. *ruptum*, broken	Intermittent or broken horizons, often over bedrock

(*cont.*)

Table 7.10 (*cont.*)

Adjective	Derivation[a]	Connotation
Salic		Has a salic (Bz) horizon, dominated by salts more soluble than gypsum
Salidic		Combination of *Salic* and *Aridic*; intergrade to a Salid
Sapric		Contains some organic materials of advanced degree (sapric) of decomposition
Sodic	Modified from sodium	Abundant sodium in the profile
Sombric		Has a sombric (acid, Bh) horizon
Sphagnic		Dominated by fibers of *Sphagnum* moss
Spodic		Intergrade to a Spodosol, often implying that the soil has a Bs or Bh horizon that is spodic-like but fails to meet one or more criteria for a spodic horizon
Sulfaqueptic		Combination of *Sulfic*, *Aquic* and *Inceptic*; Intergrade to a Sulfaquept
Sulfic or Sulfuric		Presence of sulfides or their oxidation products; may have a sulfuric horizon
Terric	L. *terra*, earth	Organic soils with a mineral layer within 1.5 m of the surface
Thaptic	Gr. *thapto*, buried	Soils profile contains a buried soil or horizon
Thapto-Histic	Gr. *thapto*, buried, and *histos*, tissue	Combination of *Thapto* and *Histic*; soil profile contains a buried organic soil or soil horizon
Torrertic		Combination of *Torric* and *Vertic*; intergrade to a Torrert
Torrifluventic		Combination of *Torric*, *Fluv-* and *Entic*; intergrade to a Torrifluvent
Torriorthentic		Combination of *Torric*, *Orthic* and *Entic*; intergrade to a Torriorthent
Torripsammentic		Combination of *Torric*, *Psamm-* and *Entic*; intergrade to a Torripsamment
Torroxic		Combination of *Torric* and *Oxic*; intergrade to a Torrox
Typic	Typical, normal	Central concept; not normally grading toward another type of profile
Udandic		Combination of *Udic* and *Andic*; intergrade to a Udand
Udertic		Combination of *Udic* and *Vertic*; intergrade to a Udert
Udic		Grading toward a udic soil moisture regime
Udifluventic		Combination of *Udic*, *Fluv-*, and *Entic*; intergrade to a Udifluvent
Udollic		Combination of *Udic* and *Mollic*; intergrade to a Udoll
Udorthentic		Combination of *Udic*, *Orth-* and *Entic*; intergrade to a Udorthent
Udoxic		Combination of *Udic* and *Oxic*; intergrade to a Udox

(*cont.*)

Table 7.10 *(cont.)*

Adjective	Derivation[a]	Connotation
Ultic		Intergrade to an Ultisol, often implying that the soil is neither weathered enough nor acidiic enough to be an Ultisol
Umbric		Has an umbric epipedon
Ustalfic		Combination of *Ustic* and *Alfic*; intergrade to a Ustalf
Ustandic		Combination of *Ustic* and *Andic*; intergrade to an Ustand
Ustertic		Combination of *Ustic* and *Vertic*; intergrade to an Ustert
Ustic		Has a soil moisture regime that borders on ustic
Ustifluventic		Combination of *Ustic*, *Fluv-* and *Entic*; intergrade to an Ustifluvent
Ustollic		Combination of *Ustic* and *Mollic*; intergrade to an Ustoll
Ustoxic		Combination of *Ustic* and *Oxic*; intergrade to an Ustox
Vermic		Worms very abundant, or mixed by animals
Vertic		Intergrade to a Vertisol, often implying that the soil cracks in a dry season, but not deep enough or for long enough periods of time to classify as a Vertisol
Vitrandic		Has volcanic ash, glass, cinders, pumice and pumice-like fragments
Vitric		Has some volcanic ash and/or glass influence
Vitritorrandic		Combination of *Vitric*, *Torric* and *Andic*; intergrade to a Vitritorrand
Vitrixerandic		Combination of *Vitric*, *Xerric* and *Andic*; intergrade to a Vitrixerand
Xanthic	Gr. *xanthos*, yellow	Yellow colors
Xeralfic		Combination of *Xeric* and *Alfic*; intergrade to a Xeralf
Xereptic		Combination of *Xeric* and *Inceptic*; intergrade to a Xerept
Xerertic		Combination of *Xeric* and *Vertic*; intergrade to a Xerert
Xeric		Has a xeric (Mediterranean) soil moisture regime; dry summers and moist winters
Xerofluventic		Combination of *Xeric*, *Fluv-* and *Entic*; intergrade to a Xerofluvent
Xerollic		Combination of *Xeric* and *Mollic*; intergrade to a Xeroll

[a] Derivation is included here only if not previously provided in an earlier Table.

Source: Soil Survey Staff (1975, 1999). The reader is referred to those sources for precise definitions.

Table 7.11 Compound formative element names for subgroups

Abruptic Xeric, Abruptic Haplic, Abruptic Argiduridic
Acrudoxic Vitric, Acrudoxic Ultic, Acrudoxic Hydric, Acrudoxic Thaptic
Aeric Fragic, Aeric Humic, Aeric Umbric, Aeric Vertic, Aeric Chromic
Albic Glossic
Alfic Vertic, Alfic Lithic, Alfic Humic, Alfic Oxyaquic, Alfic Arenic
Anionic Aquic
Aquertic Chromic
Aquic Durinodic, Aquic Arenic, Aquic Dystric, Aquic Cumulic, Aquic Natrargidic, Aquic Petroferric, Aquic Duric, Aquic Lithic
Arenic Plinthaquic, Arenic Umbric, Arenic Rhodic, Arenic Aridic, Arenic Plinthic, Arenic Ustic, Arenic Ultic
Argiaquic Xeric
Aridic Leptic, Aridic Lithic
Calcic Lithic, Calcic Pachic, Calcic Udic
Chromic Udic, Chromic Vertic
Cumulic Vertic, Cumulic Ultic
Duric Xeric, Duric Histic
Durinodic Xeric
Dystric Vitric, Dystric Fluventic
Entic Grossarenic, Entic Udic
Eutric Thaptic, Eutric Pachic, Eutric Lithic
Fluvaquentic Vertic
Fluventic Humic
Fragic Oxyaquic
Glossic Ustic, Glossic Vertic
Grossarenic Plinthic
Halic Terric
Haplic Ustic, Haplic Haploxerollic, Haplic Palexerollic
Histic Lithic, Histic Placic, Histic Plinthic
Humic Lithic, Humic Inceptic, Humic Psammentic, Humic Rhodic, Humic Xanthic, Humic Xeric, Humic Pachic
Hydric Thaptic, Hydric Pachic
Lamellic Ustic
Leptic Vertic, Leptic Udic, Leptic Torrertic
Lithic Xeric, Lithic Mollic, Lithic Ustic, Lithic Petrocalcic, Lithic Ruptic, Lithic Ultic, Lithic Ruptic-Inceptic, Lithic Ruptic-Entic
Oxyaquic Vertic, Oxyaquic Ultic
Pachic Ultic, Pachic Vertic, Pachic Vitric, Pachic Udertic
Petrogypsic Ustic
Petronodic Xeric, Petronodic Ustic
Ruptic-Histic, Ruptic-Lithic, Ruptic-Ultic
Sodic Xeric, Sodic Ustic
Thapto-Histic
Umbric Xeric

Source: Soil Survey Staff (1975, 1999). The reader is referred to those sources for precise definitions.

Table 7.12 | Examples of soil families within the system of Soil Taxonomy, with generalized characteristics discernible from the name

Family names	Generalized description of the main characteristics of that family
Central concept suborders	
Fine-loamy, mixed, non-acid, mesic Typic Udifluvent	Central concept for a weakly developed soil (Entisol), probably on a floodplain (fluv-) in a udic (humid) soil moisture regime. Ochric A horizon unless materials are stratified. No diagnostic B horizon. Mixed mineralogy, non-acid soil pH, and fine-loamy textures.
Fine, mixed, acid, isomesic Typic Placaquept	Central concept for an Inceptisol with placic horizon and an ochric epipedon (A horizon). Mesic soil temperature regime with little seasonality (iso-). High water table results in aquic (aqu-) soil moisture regime. Acid soil pH, mixed mineralogy and fine textures (silts and clays).
Clayey, oxidic, isohyperthermic Typic Rhodudult	Central concept for a very red (Rhod-), clay-rich Ultisol in a udic soil moisture regime. Dominated by oxide clays such as goethite and hematite. Mean annual soil temperature is >22 °C, with little seasonality.
Fine-silty, mixed, mesic Typic Fragiaqualf	Central concept for an Alfisol with an argillic Bt horizon and a fragipan. Probably has a Btx horizon. High water table gives the soil an aquic soil moisture regime. The soil temperature regime is mesic, mineralogy is mixed, and the textures are fine-silty, the latter characteristic indicating that the likely parent material is loess.
Coarse-loamy, mixed, superactive, subgelic Typic Aquiturbel	Central concept for a Gelisol with cryoturbated horizons. Water perches on the permafrost below, leading to aquic conditions. Mixed mineralogy and coarse-loamy textures. High organic matter content gives it a high (superactive) CEC/clay ratio. This is a typical Gelisol on permafrost.
Intergrade suborders	
Loamy-skeletal, oxidic, mesic Ultic Palexeralf	Thin (ochric) A horizon in an Alfisol. Has well-developed (Pale-) argillic Bt horizon (all Alfisols have an argillic horizon). In xeric (Mediterranean) soil moisture regime. Intergrading to an Ultisol. Mesic soil temperature regime, mineralogy dominated by oxide clays (hence the intergrade status), loamy textural family but with many coarse clasts (skeletal).
Fine, montmorillonitic, frigid Argiaquic Argialboll	Dark, thick A horizon (Mollisol) with noteworthy E (albic) and argillic (Bt) diagnostic horizons. Intergrade to an Argiaquoll, suggesting wetness within the profile (grading into an aquic soil moisture regime, aqu-). Frigid soil temperature regime, montmorillonite clay mineralogy dominates, fine textures (clays and silts).
Coarse-loamy, mixed, thermic Ustorthentic Haplocambid	Aridisol (in aridic soil moisture regime) with a cambic Bw horizon and an ochric A horizon. Intergrade to an Ustorthent, which is a weakly developed (Entisol) soil in an Ustic soil moisture regime. Thus, the soil moisture regime of this soil is aridic, but close to ustic. Thermic soil-temperature regime, mixed mineralogy, coarse-loamy textures.

(cont.)

Table 7.12 (cont.)

Family names	Generalized description of the main characteristics of that family
Loamy-skeletal, mixed, mesic Durixerollic Lithic Haplocambid	This Aridisol is both an intergrade (to a Durixeroll) and an extragrade (lithic). It has a weakly-developed Bt, Bk or Bw (cambic) horizon and is shallow to bedrock. It has a thicker and darker A horizon than is typical of Cambids, as it intergrades to a Mollisol. The aridic soil moisture regime is grading to xeric, so summers must be drier than winters. It has a duripan, a mesic soil-temperature regime, mixed mineralogy, and loamy textures amid many rock fragments (skeletal).
Very fine, montmorillonitic (calcareous), mesic Vertic Ustifluvent	A weakly developed soil that has a tendency to develop cracks in the dry season (Vertic intergrade), but the cracks do not stay open long or often enough to qualify as a Vertisol. Formed in very fine, clayey alluvium, rich in montmorillonite clays. Ustic soil moisture regime tends to promote cracking allowing the soil to get very dry at times of the year. Mesic soil temperature regime. Minimal leaching leads to carbonates within the profile.
Euic Fluvaquentic Cryohemist	Organic soil (Histosol) in which most of the organic material is partially (Hemic) decomposed. In a cryic soil temperature regime. This soil intergrades with Fluvaquents, which are wet mineral soils that form on floodplains. Thus, it probably has strata of mineral (i.e., not organic) alluvium among the organic materials. Organics are not highly acid (Euic).
Extragrade suborders	
Fine-loamy, mixed, mesic Cumulic Hapludoll	Soil with a dark, thick A horizon (Mollisol) in udic (humid) soil moisture regime with simple horizonation (Hapl-). Overthickened (cumulic) A horizon. Mesic soil temperature regime, mixed mineralogy, fine-loamy textural family.
Loamy-skeletal, serpentinitic, mesic Lithic Haploxeroll	Soil with a dark, thick A horizon (Mollisol) in xeric (Mediterranean) soil moisture regime with simple horizonation (Hapl-). Shallow to serpentine bedrock. Mesic soil temperature regime, serpentinitic mineralogy, loamy textural family but with many clasts (skeletal).
Medial-skeletal over fragmental or cindery, amorphic over mixed, isomesic Humic Ustivitrand	Andisol dominated by relatively unweathered volcanic materials (cinders and ash), which have a high content of glass-like materials (vitr-). Ustic soil moisture regime with a dark, thick A horizon. Probably formed in two distinct parent materials: the upper one with ash amidst many clasts (cinders, skeletal), the lower one being mostly cinders (fragmental) of mixed mineralogy.
Sandy, siliceous, thermic Grossarenic Haplohumod	Spodosol with Bh horizon (hum-). Ochric A horizon, and probably has an albic E horizon. Overthickened (Gross-) sandy (aren-) surface, or upper profile, dominated by quartz sand (siliceous). Thermic (subtropical) soil-temperature regime.
Fine-loamy, siliceous, semiactive, thermic Plinthaquic Paleudult	Weathered, upland Ultisol in a humid (udic) climate. Old, strongly developed soil (Pale-) with a thick profile. Has plinthite (secondary iron compounds) and is slightly wetter than typical Paleudults; gradational to an aquic soil moisture regime but still is within udic. Dominated by silica minerals (sands).

(cont.)

Table 7.12 (*cont.*)

Family names	Generalized description of the main characteristics of that family
Fine, mixed, superactive, mesic Chromic Gypsitorrert	Vertisol that develops wide, deep cracks during the dry season. In a torric (dry) soil moisture regime, implying a short wet season; the cracks are usually open. Cracks may actually open not so much due to rain but due to run-on from precipitation in nearby uplands. Has a subsurface horizon of accumulated gypsum. Light-colored (Chromic) epipedon. Mesic soil-temperature regime. Fine textures (high clay content), but mixed mineralogy. High organic matter content gives it a high (superactive) CEC/clay ratio.
Mixed, mesic Aquic Xeropsamment	Weakly developed sandy (Psamm-) Entisol with minimal B horizon development. In a Xeric soil moisture regime, but with indications of wetness in the profile, from a water table, hence the extragrade to an aquic soil moisture regime. Mesic soil temperature regime, and the sands are not strongly dominated by quartz.
Fine, ferruginous, isothermic Xanthic Acroperox	Yellow-colored (Xanth-) Oxisol (highly weathered soil dominated by oxide and 1:1 clays) in a continuously wet climate, leading to a perudic soil moisture regime. Very leached, even for Oxisols, with a low CEC value (Acr-). Isothermic soil temperature regime implies a continuously warm, tropical climate. Fine (clayey and silty) soil textures. High iron content (ferruginous).

For classification purposes, Soil Taxonomy attempts to determine the mean, long-term temperature of soils at 50 cm, chosen because daily temperature fluctuations are dampened out below that depth. At 50 cm or deeper, soil temperatures change more in response to short-term weather patterns and seasonal trends than they do to diurnal patterns (Soil Survey Staff 1999).

What, exactly, are the relationships between air and soil temperatures? In theory, the long-term mean soil temperature should change little below a certain depth, perhaps about 9 m (Soil Survey Staff 1999). According to the Soil Survey Staff (1999), a single temperature reading at 10 m is usually within 0.1 °C of the mean annual air temperature (MAAT), while a reading at 6 m is assumedly within 1 °C of the MAAT, regardless of when it is taken. With increasing depth, temperatures will gradually increase as geothermal sources of energy are approached. Mean annual soil temperature (MAST) is often simply estimated to be MAAT plus 1° or 2 °C. This relationship breaks down in areas that have thick snowpacks, as snow is an effective insulator from cold air temperatures (Hart and Lull 1963, Ping 1987, Sharratt *et al.* 1992, Isard and Schaetzl 1995). MAST is also affected by the amount and distribution of precipitation, the protection provided by shade and by O horizons in forests, slope aspect and gradient, and irrigation (MacKinney 1929, Smith *et al.* 1964, Soil Survey Staff 1999). Factors such as soil color, texture and content of organic matter have negligible effects (Smith *et al.* 1964). In the tropics, where seasonal differences in air temperature are negligible, variation in soil temperature responds most to elevation and to wet vs. dry seasons.

An overview and assessment of the system

Soil Taxonomy, like most other soil classification systems worldwide, is a highly logical and useful system. In short, it works. One shortcoming, in our opinion, of the otherwise logical system of Soil Taxonomy is the limited number of descriptors that can be used for extragrades. Generally, only one or two extragrade descriptors are allowed, where many may be applicable.

Table 7.13 Classes and definitions of soil-temperature regimes

Soil-temperature regime	Simplified definition[a] and limits
Hypergelic	MAST is ≤ −10 °C. Gelisols only
Pergelic	MAST is between −4 °C and −10 °C. Gelisols only
Subgelic	MAST is between 1 °C and −4 °C. Gelisols only
Cryic	MAST is <8 °C, but the soil does not have permafrost[b]
Frigid	MAST is <8 °C, but warmer in summer than Cryic soils; the summer minus winter soil temperature is >6 °C (i.e., there is seasonality)
Mesic	MAST is between 8 °C and 15 °C and the summer minus winter soil temperature is >6 °C (i.e., there is seasonality)
Thermic	MAST is between 15 °C and 22 °C and the summer minus winter soil temperature is >6 °C (i.e., there is seasonality)
Hyperthermic	MAST is >22 °C and the summer minus winter soil temperature is >6 °C (i.e., there is seasonality)
Isofrigid	MAST is <8 °C and the summer minus winter soil temperature is <6 °C (i.e., there is little seasonality)
Isomesic	MAST is between 8 °C and 15 °C and the summer minus winter soil temperature is <6 °C (i.e., there is little seasonality)
Isothermic	MAST is between 15 °C and 22 °C and the summer minus winter soil temperature is <6 °C (i.e., there is little seasonality)
Isohyperthermic	MAST is >22 °C and the summer minus winter soil temperature is <6 °C (i.e., there is little seasonality)

[a] MAST, mean annual soil temperature.
[b] By definition, soils with permafrost within 100 cm are within the Gelisol order (Table 7.3). The soil temperature regime for Gelisols is, therefore, not calculated.

Source: Simplified from Soil Survey Staff (1999). The reader is referred to that source for precise definitons.

Those that have problems applying Soil Taxonomy are often too close to it and get bogged down in the details. To be sure, classifying a soil, i.e., going from a pedon to a name, when done strictly by the book, is often cumbersome and difficult. However, using the names, i.e., going from the name to a pedon concept in one's mind, given just a little practice, can be simple and straightforward.

Critics argue that Soil Taxonomy is too quantitative – that it relies too much on laboratory data which can take months to acquire. In the process, the classification of a pedon is in limbo until those data are available. However, this problem is inherent in any highly quantitative system. Others (e.g., Birkeland 1999) noted that it may focus too much on surface horizons and not enough on subsoil development. The latter often provides more information about the age of the surface the soil may have formed on, thereby rendering it more useful from a geomorphic standpoint. Soil quality researchers, on the other hand, feel that not enough emphasis is placed on surface horizons, which are of most direct utility to human use and are the first to be eroded, polluted, etc. Campbell and Edmonds (1984) felt that Soil Taxonomy lacks a strong geographic focus. That is, many soil taxa, as they exist on the Earth's surface, have boundaries that do not coincide with natural features, making them difficult to map.

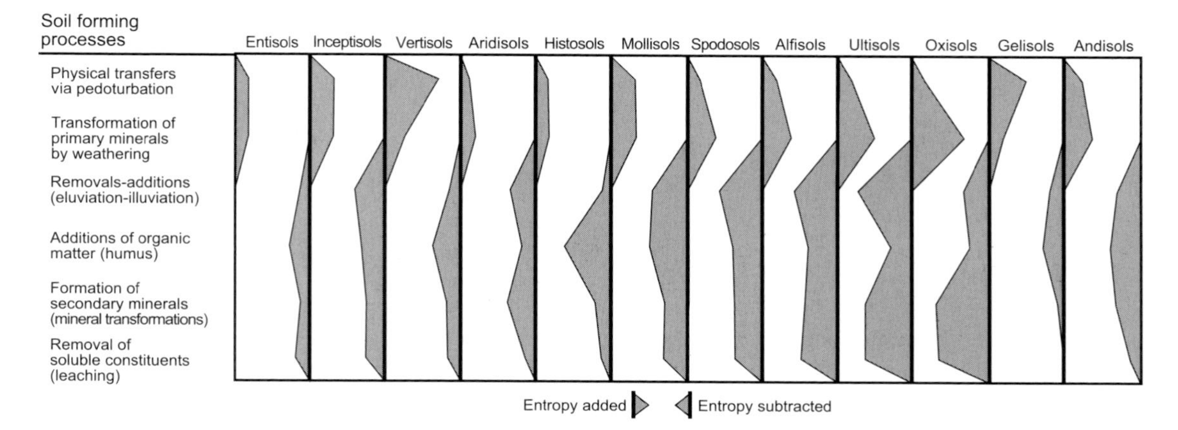

Fig. 7.6 Degree of expression of some of the major process bundles, for the 12 soil orders, as expressed in the degree of entropy added or subtracted from the soil. Significantly changed from Smeck *et al.* (1983); we apologize if the changes distort the original intent.

They argue that, because Soil Taxonomy is designed to assist soil surveys, it should, by its very nature, consider geographic features of soil distributions. Also, the abstract, taxonomic entities defined by Soil Taxonomy do not always coincide with genetic soil units. Along these same lines, many have stated that Soil Taxonomy is not genetic enough – that it relies on observable characteristics more than interpretation of processes. Processes that have left little measurable imprint on the soil are not often considered in the classification. This is true. But as Guy Smith (1983: 43) noted,

> Processes that go on can rarely be observed or measured. They vary with the season and with the year. They change if climate changes or as the result of new cycles of landscape dissection. They leave marks in the soil, but the marks may persist long after the processes that produced them have ceased to act.

Thus, Soil Taxonomy assumes that if a given set of processes has been dominant for a long period of time, they will have left their mark morphologically or chemically in such a way as to be measurable; their genesis can often only be inferred. The problem with placing too much emphasis on genetic interpretations of soils lies in the fact that genetic interpretations of soils differ among researchers, and change as knowledge advances. Recall also that Soil Taxonomy was designed to further our *use and management* of soil, knowledge of which generally outpaces knowledge of genesis. Soil Taxonomy tries to insure that different people will arrive at the same classification of a soil, even if they disagree on its genesis (Smith 1983). This trait insures that the system will be somewhat independent of the day-to-day vagaries of pedogenic theory, or the views of any particular soil scientist. (Note that the system *is* amended frequently, in light of advances in pedogenic understanding and theory.) Genesis does not appear so much in the *definitions* of the taxa in Soil Taxonomy, but it lies everywhere *behind* them (Smith 1983, Ciolkosz *et al.* 1989) (Fig. 7.6). The founders of Soil Taxonomy intended the system to keep soils of similar genesis in the same taxa, although this is not always possible because many properties can be developed by different processes, i.e., they exhibit equifinality. The further one proceeds through the system, the more similarity will the soils have with regard to genesis. Soils of a given suborder are somewhat alike genetically, but at the series level the similarities are as close as knowledge permits.

Other criticisms center on class limits and the inherent inflexibility of the system (Hallberg 1984). Beckmann (1984) felt the class limits of Soil Taxonomy were often arbitrary (see also Webster

1968). To that end, most diagnostic horizons are a quantitative reflection of our understanding of pedogenic processes, taken beyond a certain minimum point. For example, an argillic horizon is designed to reflect the process of clay illuviation, that has gone on for a certain period of time. Although most, if not all, soils have technically experienced some clay illuviation, making note of it within the taxa of a soil is simply not worthwhile unless the process has proceeded to the point where it has dramatically impacted the profile. How dramatically is enough? That answer is found in the quantitative limits outlined by the Soil Survey Staff (1999). For a more thorough discussion of how genetic thought factored into the making of Soil Taxonomy, consult the Guy Smith interviews (Smith 1986).

The Canadian system of soil classification

Unlike Soil Taxonomy, which is a global system, the Canadian system of soil classification is intended primarily for use in Canada (Table 7.14). Provision is not made for soils in thermic and warmer soil-temperature regimes, and for soils in torric and xeric soil moisture regimes (Expert Committee on Soil Science 1987, Soil Classification Working Group 1998). However, a strength of the system lies in its classification of cold soils and salt-affected soils, of which Canada has many millions of hectares.

The two classification systems have many similarities. The structure and logic are essentially the same. Some of the soil orders are essentially equivalent in the two systems: Spodosols and Podzols, Alfisols and Luvisols, Entisols and Regosols, Inceptisols and Brunisols. The main difference between the two systems lies in how wet soils and salt-affected soils are classified. The Canadian system breaks these soils out at the highest level of classification: Gleysols and Solonetzic soils, respectively. In Soil Taxonomy, on the other hand, drainage class is not a defining criterion until the suborder level, at which various -aqu- suborders appear, e.g., Aqualfs, Aquels. Likewise, the presence of salts in a soil horizon does not enter into Soil Taxonomy until the great group level, e.g., Natrudalfs, Natrustolls.

Soil mapping and soil maps

The most direct application of soil classification is in soil mapping and soil survey. The fundamental unit of a soil map is the *map unit* (Soil Survey Division Staff 1993). Each map unit is called a *delineation*, which generally contains the dominant components in the map unit name, e.g., Drummer silty clay loam. A Drummer map unit would be dominated by soils that fall within the limits of the Drummer series, along with some inclusions of other soils. Almost all soil map units have inclusions of similar and dissimilar soils.

What can be done to minimize the area or number of inclusions on smaller-scale maps? One solution is to define the map units differently. On a large-scale map of a pasture, for example, units of Drummer silty clay loam are appropriate and readily mappable with minimum inclusions. On a smaller-scale county soil map, the soil scientist might choose to delineate units of Aquolls, gently sloping or Flanagan–Drummer soils. That is, broadening the definition of the map unit enables the mapper to delineate areas that are still reasonably pure, based on the map legend.

The amount of map unit inclusions varies from place to place; in some complex landscapes inclusions may comprise half or more of the map unit (Fig. 7.7). A map unit with no inclusions would be equivalent to a polypedon – taxonomically pure. Such map units are rare. A polypedon may be split into two or more map units, each with a different phase, such as slope. Thus, one Drummer polypedon may be split into adjoining two Drummer map units, one with a 0–2% slope and one with 2–4% slope.

The amount of detail that can be shown on the soil map is a function of scale; large-scale maps, like county soil maps, can show a great deal of detail, while small-scale maps of states or regions must generalize (Lyford 1974). Even on the largest-scale map, some degree of generalization

Table 7.14 Approximate equivalents between the Canadian system of classification and Soil Taxonomy

Canadian system	Soil Taxonomy
Order and great group	Great groups or other taxa partly or mostly included
1 Chernozemic soils	Ustolls and Cryolls
1.1 Brown (Light Chestnut)	Aquic and Vertic Ustolls; Calciustolls; Cryolls other than Natric
1.2 Dark Brown (Dark Chestnut)	Ustolls other than Aridic and Udic; Cryolls other than Natric
1.3 Black	Frigid and Cryic Argiudolls and Hapludolls; Aquic and Vertic Argiustolls and Haplustolls; Calciustolls other than Typic and Aridic; Cryolls other than Natric
1.4 Dark Gray	Cryic and Frigid Ustolls and Udolls
2 Solonetz soils	Alfisols and Mollisols with a natric horizon
2.1 Solonetz	Natrustalfs, Natrustolls, Natraquolls, Natricryolls and Natralbolls; Frigid Natrargids
2.2 Solod	Glossic Natraqualfs, Natrudalfs, Natrustalfs, Natrudolls, Natrustolls and Natralbolls; Frigid Natraquolls
3 Luvisolic soils	Udalfs
3.1 Gray Brown Luvisol	Hapludalfs and Glossudalfs
3.2 Gray Luvisol	Cryalfs and Udalfs that do not have a kandic or natric horizon
4 Podzolic soils	Spodosols not within an aquic soil moisture regime
4.1 Humic Podzol	Humods and Humicryods
4.2 Ferro-Humic Podzol	Cryods and Placorthods
4.3 Humo-Ferric Podzol	Cryods Placorthods and Frigid Orthods
5 Brunisolic soils	Inceptisols not within an aquic soil moisture regime
5.1 Melanic Brunisol	Eutrocryepts, Eutrudepts and Hapludolls
5.2 Eutric Brunisol	Cryepts and Eutrudepts
5.3 Sombric Brunisol	Dystrudepts and Eutrudepts
5.4 Dystric Brunisol	Dystrudepts and Dystrocryepts
6 Regosolic soils	Entisols
6.1 Regosol	Entisols not within an aquic soil moisture regime
7 Gleysolic soils	Soils of aquic suborders that are Frigid or Cryic
7.1 Humic Gleysol	Cryaquolls, Calciaquolls and Humaquepts; Frigid Endoaquolls
7.2 Gleysol	Aquents, Aquepts and Fluvents
7.3 Eluviated Gleysol	Albolls, Aquolls and Aqualfs
8 Organic soils	Histosols except Folists
8.1 Fibrisol	Frigid and Cryic Fibrists
8.2 Mesisol	Frigid and Cryic Hemists
8.3 Humisol	Frigid and Cryic Saprists
8.4 Folisol	Frigid and Cryic Folists

Source: Simplified and combined from Leahy (1963), Soil Survey Staff (1975, 1999), the Expert Committee on Soil Science (1987) and the Soil Classification Working Group (1998).

Table 7.15 The five orders (scales) of soil survey maps

Order of the map	Minimum map unit size (ha)	General uses
First order	≤1	Intensive uses, building lots, experimental farm plots
Second order	0.6–4	Intensive, general agriculture and zoning, urban planning; typical of most county-level soil surveys
Third order	1.6–16	Extensive, range land determinations, broad-scale zoning
Fourth order	16–252	Extensive, for broad land-use potential
Fifth order	252–4000	Very extensive, regional soil maps for general inventory

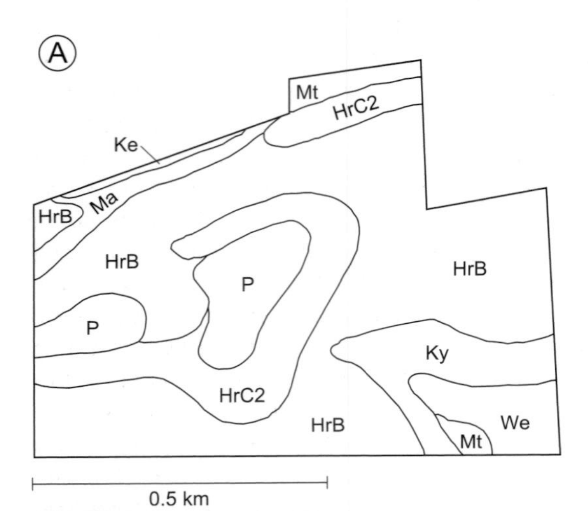

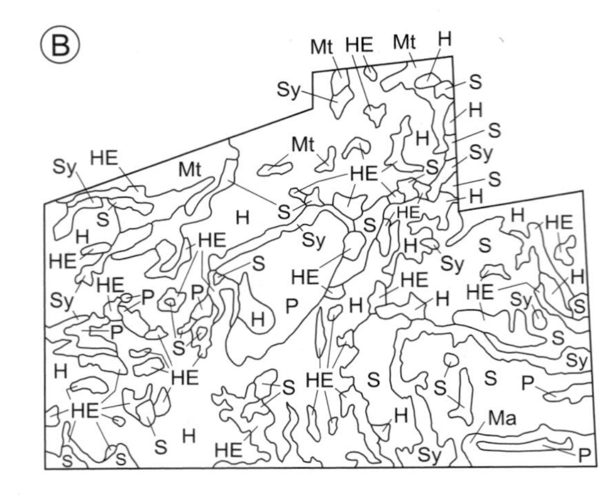

Fig. 7.7 Two soil maps of the same 50-ha area in Winnebago County, Wisconsin. After Hole and Campbell (1985). (a) Published county soil map (Mitchell 1980). (b) Soil map used for research, based upon intensive field mapping and air photo interpretation. H, Hortonville silt loam, eroded; HrB, Hortonville silt loam, 2–6% slopes; HrC2, Hortonville silt loam, 6–12% slopes; HRE, Hortonville silt loam, severely eroded; Ke, Keowns silt loam; Ky, Korobago silt loam; Ma, Manawa silty clay loam, 0–3% slopes; Mt, Mosel silt loam, 0–3% slopes; P, Poygan silty clay loam; Sy, Symco silt loam; S, Symco silt loam, eroded; S+, Symco silt loam with overwash deposit; We, Wauseon silt loam.

is required; it is not possible to map every pedon. Generally, five types or orders of soil maps, based on scale and detail, are recognized (Table 7.15). Second-order maps, like those of county soil surveys, keep errors and generalization to a minimum. As the area mapped gets larger, the amount of detail that can be shown decreases, such that similar (but not identical) map units must be grouped together.

Purpose and utility of soil maps

Soil maps were originally created as a mechanism with which to assess land value, for taxation purposes. Today, soil maps are still used for that purpose, but they have a multitude of other functions, such as land-use zoning and planning, determining septic tank filter field suitability, environmental quality protection and management, siting highway routes, in agriculture and horticulture, etc. (Lammers and Johnson 1991). An accurate soil map is an excellent surficial geology map as well, for quite frequently the initial parent material of the soil can be determined for a soil map unit (Schaetzl *et al.* 2000) (see Chapter 14). Soil maps can be also used as a proxy for natural landscape wetness. Large-scale maps are often used to determine the location and extent of wetlands and floodplains. Schaetzl (1986a) developed a natural soil drainage index, which ranges from 0 to 99, with wetter soils having larger numbers. Once converted to this index, soil map units can then be examined and remapped; the resultant maps are a spatial representation of landscape wetness. Groffman and Tiedje (1989) applied such data to identify areas likely to undergo denitrification. In a similar way, wetness-mediated processes like weathering and leaching, not to mention biomass accumulation, could be mapped using spatial data on soils and soil wetness. One could also produce maps of other soil attributes, such a N and organic matter content, erodability and runoff potential, soil attenuation potential, etc. (Cates and Madison 1993).

In the past, soil maps were produced as paper products only. Today, most are available as digital files, which allow for a multitude of additional uses in, for example, a geographic information system (GIS) (Schaetzl 2002, Schaetzl *et al.* 2002).

Types of map units

There are three main types of map units: *consociations*, *complexes* and *associations*. Consociations are map units delineated for, or dominated by, one taxon or series. At least half of the pedons in that map unit should belong to the soil for which the map unit is named, e.g., Milaca loam (Campbell and Edmonds 1984). Dissimilar soil inclusions are permitted, but only within predetermined limits (Soil Survey Division Staff 1993). Map unit *complexes* and *associations* consist of two or more dissimilar soil taxa occurring in a pattern that is so complex that it cannot be resolved at the map scale. In a soil association, the pattern is coarse enough that it can be resolved (and hence, portrayed) on a 1 : 24 000 soil map. Thus, a map unit *complex* might be named Milaca–Mora complex or the Grayling–Graycalm association. Usually, the estimated percentages of each series within the complex are provided in the map legend, and sometimes the degree of complexity is explained. For example, Mora soils may occur as isolated patches (low spots) within the Milaca background, they may occur randomly, or in a linear pattern. Fridland (1965) noted that regularly occurring soil complexes (or as he called them, soil combinations) can be divided into three types: (1) microcombinations, often due to microrelief such as treethrow pits or mounds, (2) mesocombinations, linked to patterns of mesorelief like swell-and-swale topography, and (3) macrocombinations, which relate to macrorelief, and thus are similar to catenas. Soil combinations of type 1 and 2 are usually so complex that they would be mapped as map unit complexes. Type 2 and 3 combinations are sufficiently large that they may be delineated as consociations.

When attempting to map a consociation, the mapper is, in theory, trying to delineate polypedons. This is a Herculean task, given the great amount of spatial heterogeneity that exists on the landscape. Thus, the mapper tries to delineate landscape segments that are as taxonomically pure as possible, within the constraints of the mapping exercise they are undertaking (Amos and Whiteside 1975). For example, for a soil map of an agronomy research plot, where high amounts of accuracy are required, most map units would need to be perhaps as much as 95% pure. That is, only 5% of the pedons in the map units would be allowed to be inclusions. Small-scale maps, like those used for general planning purposes, allow higher percentages of soils that are not of the same series (Fig. 7.7). In such a map, the legend should mention not only what percentage of the map unit is a different series, but it should note how much of that map unit comprises pedons of similar vs. dissimilar soil series. *Similar* series may differ only in subtle ways, such as slight changes in texture or depth to carbonates, while *dissimilar* soil series may be in another order, as for example when pockets of Histosols occur on a till plain in Minnesota, or when Natraquolls dot the shortgrass landscape of Canada. Amos and Whiteside (1975) took this exercise even further, determining not only the number of map unit inclusions in soils of a Michigan landscape, but also the degree of contrast those inclusions had with the soil the map unit was named

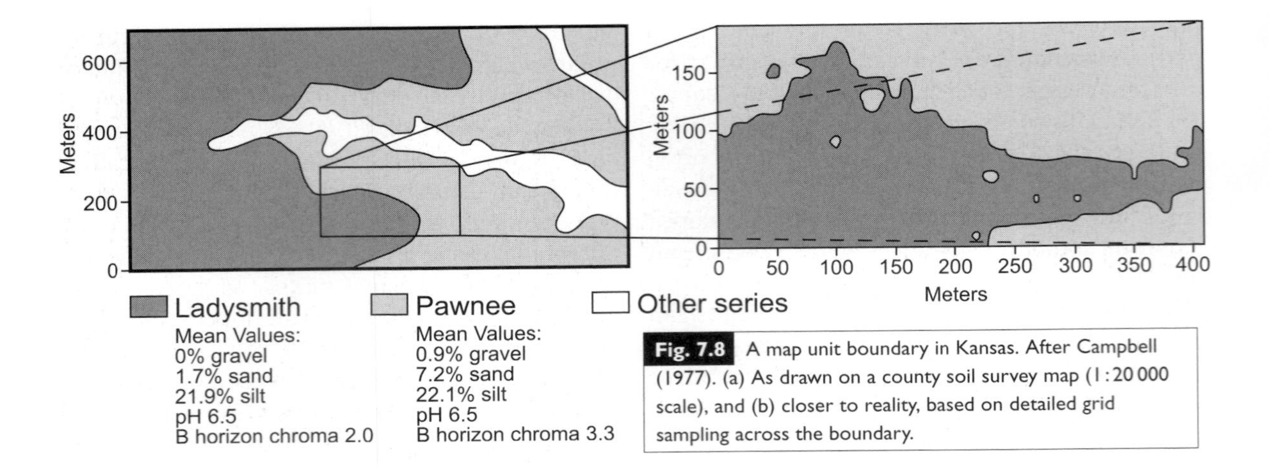

Fig. 7.8 A map unit boundary in Kansas. After Campbell (1977). (a) As drawn on a county soil survey map (1 : 20 000 scale), and (b) closer to reality, based on detailed grid sampling across the boundary.

for. Their work pointed out that it is not just the amount of inclusions that is important in assessing soil map quality, but how *different* these are from the soil that the map shows on the landscape.

Map units are only subdivisions defined for the purpose of depicting, on maps, landscape units observed in the field (Campbell and Edmonds 1984). The mapper must locate the place on the landscape where the *rate of change* from one polypedon to another is most rapid and draw a boundary there. A map unit boundary does not provide an exact location where soil series A yields to soil series B (Campbell 1977). Rather, the boundary is an estimate of the place where, on the landscape, the change from polypedon A to polypedon B is best shown or most likely. Pedons of soil A will occur as impurities in map unit B, and vice versa (Fig. 7.8).

Making a soil map: fieldwork

Most mapping today, unless done by contract for a specific use such as a for a research farm or housing subdivision, is based on aerial photography. The mapper uses certain preconceived notions of what types of soils to expect, what parent materials might exist in the landscape and the degree of variability it may contain. This knowledge comes from office-work done prior to entering the field. The mapper then examines numerous aerial photographs of the area, looks for places where vegetation, slope or some other landscape phenomenon changes abruptly, and draws a line there (Fig. 7.1). Because these lines *could* correspond to soil boundaries, they still need to be field-checked. Thus, the mapper rarely goes into the field "blind." To do so would be a learning experience, but little more.

The best place to start mapping in the field, of course, is at the beginning. The first pedon encountered, like so many others that will follow, must be exposed and classified, usually to series. Exposing the soil/pedon is often done with a hand auger, shovel or a push probe (Fig. 7.9). If more time or more detail is needed, a pit may be opened by hand or with a backhoe. In rocky areas where bucket augers are not optimal or where soils are shallow to bedrock, the mapper may simply use a shovel or spade to open a small, shallow pit and determine the soil series based on the upper few decimeters of the profile. In any event, a decision must eventually be made on the pedon – usually to series.

The mapper uses morphologic clues to determine the soil series present at the site. All soil series are strictly defined, and ranges have been determined for each major property, such as types and arrangement of horizons, thickness, texture and color or horizons, and depth to redoximorphic features (mottles), among others (Lyford 1938) (Fig. 7.10). Thus, the mapper examines these and other major clues, e.g., probable parent material and evidence for lithologic discontinuities, to determine the series at the site. Usually, the mapper knows the major series of their

(a)

(b)

Fig. 7.9 Bruce Knapp, a soil mapper of the Natural Resources Conservation Service using a bucket auger to determine the soil series at that site. He augers (a) and then examines the auger shavings (b) for color, texture, consistence, etc. Photos by RJS.

region so well that only one to three clues are enough to ascertain what soil series exists at that site. For example, on upland landscapes the mapper may know that most of the time only one of three series occurs there. One may typically be found in association with a very gravelly and rocky surface, while the others lack coarse fragments. Each may be associated with a particular vegetation type. Land use can also help. One series may be more erodible than the others; the list of possible clues is long but, after many hours of mapping, reliable.

Of the possible mapping clues, the presence or absence of major horizon types is certainly the most important. The mapper must determine, for example, the type of epipedon (mollic, ochric, etc.), and the type of subsurface horizons (e.g., Bt, Bh, Bg) and if any of these meet the criteria for a particular type of diagnostic horizon. For example, Milaca loam, a coarse-loamy, mixed, superactive, frigid Typic Hapludalf, must have an ochric epipedon, an argillic horizon, and no other unique or extragrade features such as shallow bedrock or a fragipan (although it is allowed to have an albic horizon). It must also have coarse-loamy textures. If the pedon examined is similar but has, for example, a fragipan, the pedon would classify as a Fragiudalf. Alternatively, if the pedon were similar but had carbonates at a shallow depth, it would still classify as a Typic Hapludalf, but within a series called Emmet. Each series has a suite of similar series, all of which differ by one or more aspects – texture, thickness, presence or absence of a diagnostic horizon, drainage class, etc. The mapper must be aware of these similar soils, and be able to eliminate all of them so as to settle in on the correct series for that pit or auger site.

The natural soil drainage class is determined by first locating oneself on the landscape; poorly drained soils are not likely on steep slopes or on the tops of sandy hills. This knowledge gives the mapper clues about the soil's natural drainage class well before it is even examined. The information that the soil provides about

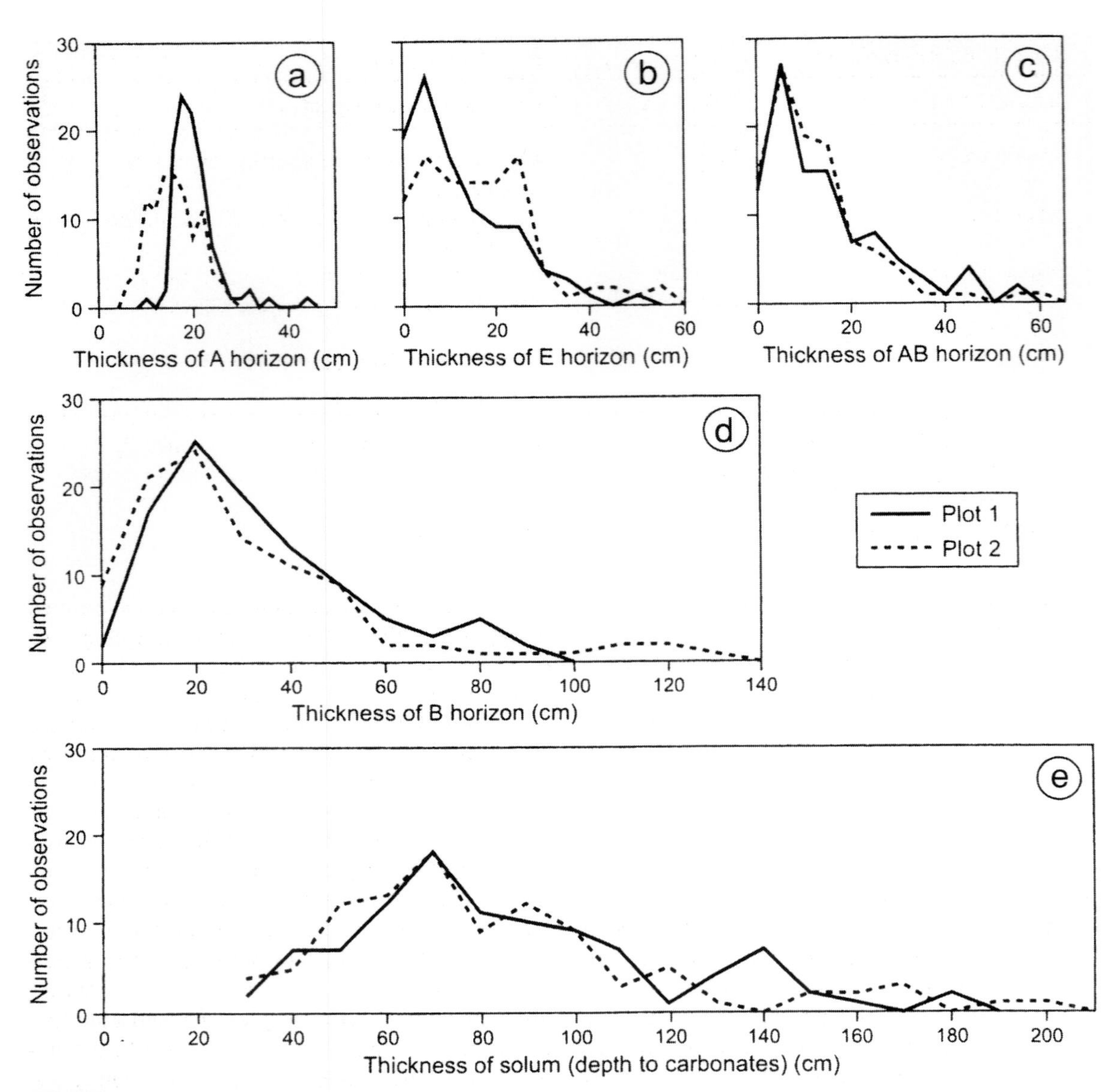

Fig. 7.10 Variability in horizon thickness and depth to carbonates for a Typic Hapludalf in Ontario, Canada. After Protz *et al.* (1968).

its natural drainage class includes presence or absence of a water table, and most importantly, presence, character and depth to mottles, or redoximorphic features, or gleying. Although we discuss the major types of natural soil drainage classes in Chapter 14, we note here that the wetter the soil, with respect to a water table, the more likely it is to undergo oxidation–reduction processes, which leave behind a tell-tale morphology of gray and reddish markings. Generally, the wetter the soil the higher in the profile these features will occur. For most soil series, a complete suite of series counterparts exist, all of which have formed in similar parent materials and exist as a drainage sequence (or catena) (Table 7.16). An experienced mapper can interpret the morphology of a soil and determine, rather quickly, the natural drainage class, thereby eliminating many soils from the list of possibilities.

Once the soil has been identified to series, the mapper examines the landscape and attempts to visualize the extent of that polypedon, i.e., locate

Table 7.16 Examples of soils series from two soil catenas or drainage sequences

Natural drainage class	Soil series[a]	Family taxonomic designation
Well drained[b]	Milaca	Coarse-loamy, mixed, superactive, frigid Typic Hapludalfs
Moderately well drained	Mora	Coarse-loamy, mixed, superactive, frigid Aquic Hapludalfs
Somewhat poorly drained	Ronneby	Coarse-loamy, mixed, frigid Udollic Epiaqualfs
Poorly drained and very poorly drained	Parent	Coarse-loamy, mixed, frigid Typic Epiaquolls
Very poorly drained	Twig	Coarse-loamy, mixed, acid, frigid Histic Humaquepts

[a] All the soils (1) have developed in materials exposed by deglaciation in the Late Pleistocene, (2) have the same parent material (dense loamy glacial till on drumlins or moraines) and (3) have formed under forest vegetation.
[b] "Drier" drainage classes are defined, but only in sandy parent materials.

the map unit boundary. The boundary should be drawn where the rate of change of soil properties is the greatest (King *et al.* 1983). Determining the position of the map unit boundary and the correct taxonomic identification of the unit are the most important parts of soil mapping. Mapping really boils down to how well (and rapidly) one can identify map units and determine their extent.

There are many tricks to discerning possible soil boundaries in the field (Scull *et al.* 2003). Recall that the mapper has a sense of where possible soil boundaries are, based on interpretation of aerial photographs of vegetation and terrain, prior to entering the field. But these possible boundary locations must be field-checked. Usually, the mapper is armed with a knowledge of the five soil-forming factors (Jenny 1941b), and knows that as one factor changes on the landscape, the soil is also likely to change (Muckenhirn *et al.* 1949, Gile 1975a, b). Thus, the mapper looks for changes in vegetation, topography or parent material as clues to a possible change in soil, and thus a soil map unit boundary. Usually, one of the most reliable of these genetic factor clues is topography. Subtle differences in slope curvature (concave vs. convex) or steepness can affect which soil exists at a site (King *et al.* 1983) (Fig. 7.11). Downslope, the soil mapper may be able to find where the soil changes to a series with a shallower water table. This type of pattern is so predictable that one could almost map soils or at least some attributes of them, based on terrain alone (Klingbiel *et al.* 1987, Moore *et al.* 1988, 1993, Bell *et al.* 1992). In the office, the mapper will judge where polypedon edges are and draw lines on the map (air photo) in the appropriate locations. Across small parts of the landscape, at this scale, subtleties in topography, as reflected in the natural drainage class of a soil, are often the major factor that affects map unit boundaries. In Fig. 7.12a, where most of the map unit boundaries are caused by changes in drainage class, spodic horizons are shown as having formed in soils that are well drained; drier and wetter soils lack this diagnostic horizon. In very poorly drained landscape positions, muck has accumulated and a Histosol is mapped.

In dry climates, most of the landscape is well drained. Soil map unit boundaries may then be associated with changes in slope gradient or aspect, or surface age, since the different slopes will reflect younger surfaces or will have affected the genesis of the soils (Gile 1975a, b). For example, on steeper slopes on a North Dakota prairie, more water runs off and argillic horizons do not form, while argillic horizons *have* formed on more gently sloping parts of the landscape (Fig. 7.12b). Likewise, at the base of slopes, A horizons may be overthickened, as shown in the Bowbells series.

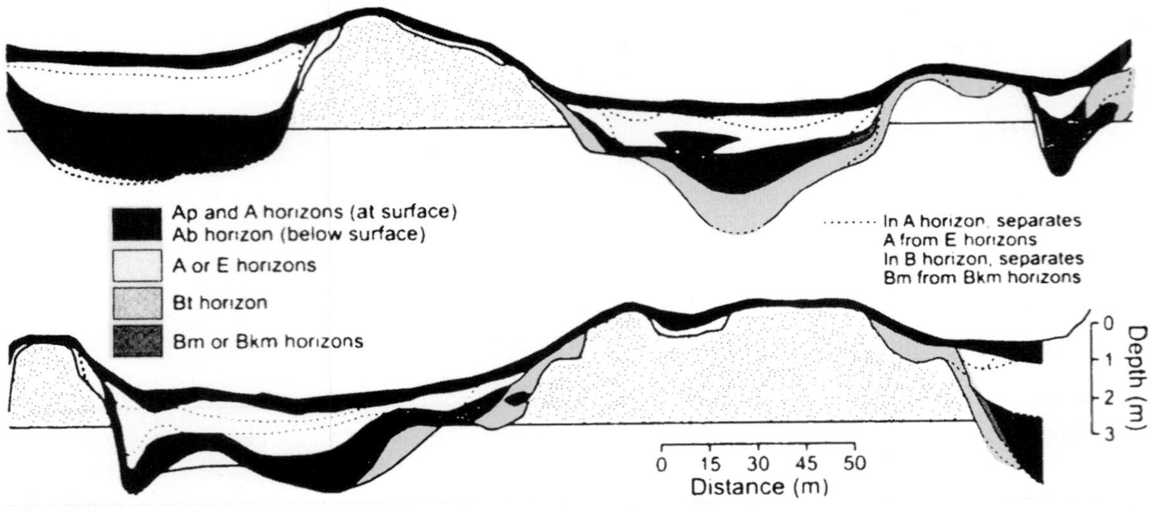

Fig. 7.11 Transects showing soil profile variability across a gently rolling, Saskatchewan landscape. After King *et al.* (1983).

On many landscapes, changes in topography also accompany changes in parent material, or perhaps signal a change to a surface of different age, e.g., a lower, eroded surface or a higher, constructional surface like a dune (Gile 1975a, Nettleton and Chadwick 1991). This is particularly true in "rock country," where certain types of rock may stand up as ridges or hold steeper slopes than other kinds of rock (Fig. 7.12c). But this system also works in glaciated terrain; outwash plains tend to be flat while moraines and dunes are rolling. Floodplains tend to be flat while uplands tend to be undulating, and the contact between the two (i.e., the valley wall) is often sharp. Moving from a plain onto a sloping surface, therefore, often indicates a change in parent material and drainage class, and hence, soil series. On older landscapes, deposition of a new parent material may lead to a soil boundary at the contact of the old and new sediment (Gile 1975a). Soils vary as a function of slope concavity and convexity, at sediment-accumulating sites vs. sites where sediment is lost (Walker *et al.* 1968) (see Chapter 14). Similarly, and often very importantly, the mapper looks for changes in vegetation, which may signal a different parent material or drainage class (Host and Pregitzer 1992). Landscapes of low relief and uniform parent materials have very subtle soil boundaries (Gile 1975b). On such landscapes, map unit boundaries are nonetheless often associated with subtle changes in relief, manifested as a change in drainage class. On some flat lacustrine plains all soils are in the same drainage class but parent materials may change laterally.

Once the boundary between two soil map units has been established, the mapper continues looking for places where the soil's natural drainage class changes so a new series can be delineated. Other types of map unit boundaries can occur as slopes change from steep to gentle, or as aspect changes from cool, northeast-facing slopes to warm, dry slopes on the southwest. In these ways, soil mappers use information about the factors of soil formation (climate, organisms, relief, parent material and time) (see Chapter 12) to estimate how soils may change across the landscape, and soil samples provided via their shovels or augers support or reject their hypotheses. In the end, the degree to which reality can be represented (how much error there is) on the map depends on the time the mapper has to do the job, the scale at which the map is being drawn, the complexity of the landscape and the skill of the mapper.

Making a soil map: after the field

Making a soil map obviously does not stop in the field. In the office, the mapper will review

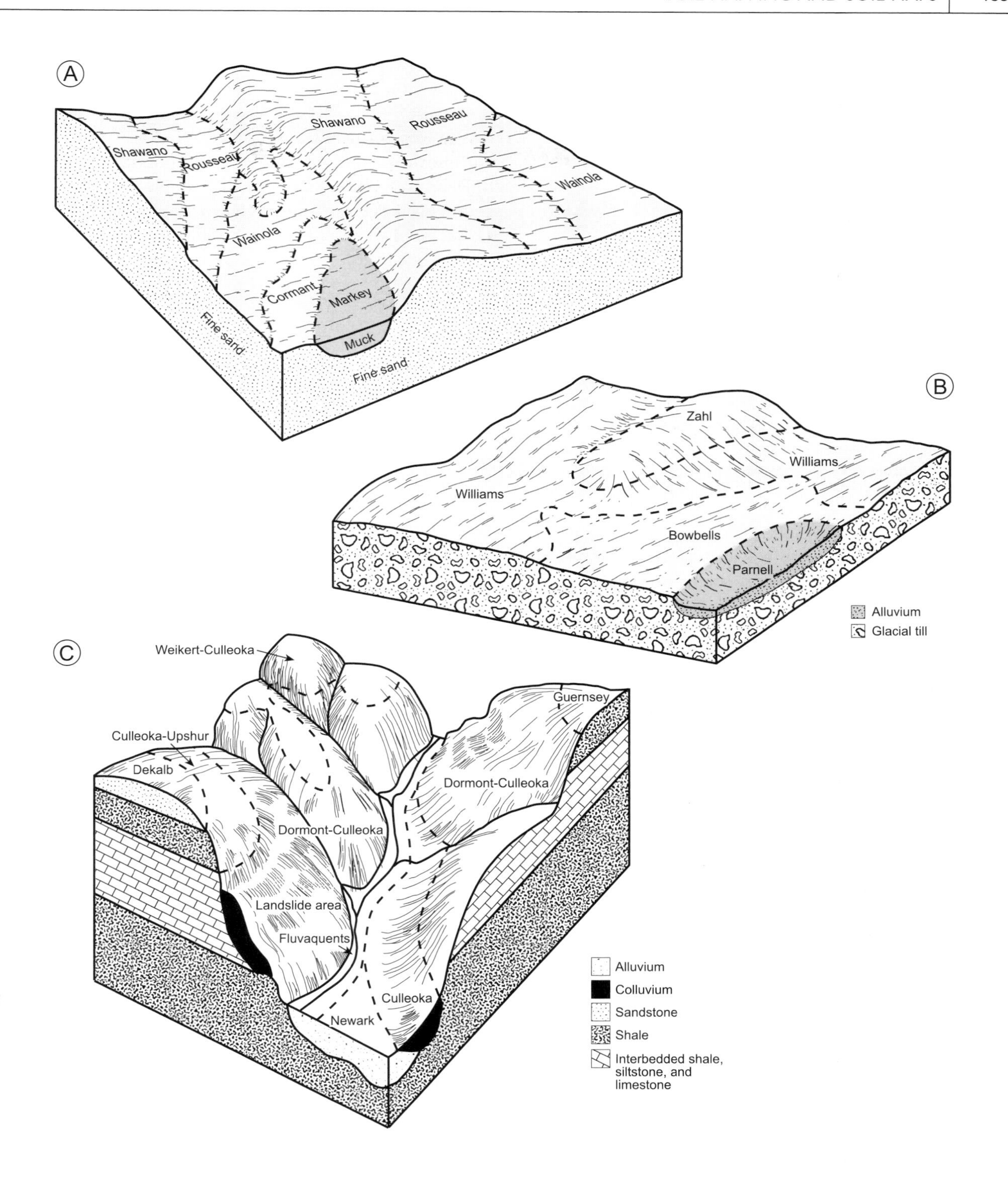
A
Shawano
Rousseau
Shawano
Rousseau
Wainola
Wainola
Cormant
Markey
Muck
Fine sand
Fine sand
B
Zahl
Williams
Williams
Bowbells
Parnell
Alluvium
Glacial till
C
Weikert-Culleoka
Guernsey
Culleoka-Upshur
Dekalb
Dormont-Culleoka
Dormont-Culleoka
Landslide area
Fluvaquents
Culleoka
Newark
Alluvium
Colluvium
Sandstone
Shale
Interbedded shale, siltstone, and limestone

Fig. 7.12 Typical patterns of soils on various landscapes, as shown on block diagrams. (a) A glaciated landscape (udic, frigid) in Wisconsin. After Gundlach *et al.* (1982). Most of the soil map unit boundaries area associated with changes in natural soil drainage class, in this landscape dominated by a sandy parent material. Shawano, mixed, frigid Typic Udipsamments (excessively well drained); Rousseau, Sandy, mixed, frigid Entic Haplorthods (well drained); Wainola, Sandy, mixed, frigid Typic Endoaquods (somewhat poorly drained); Cormant, Mixed, frigid Mollic Psammaquents (poorly drained); Markey, Sandy or sandy-skeletal, mixed, euic, frigid Terric Haplosaprists (very poorly drained). (b) A glaciated landscape (ustic, frigid) in North Dakota. After Seelig and Gulsvig (1988). Most of the soil map unit boundaries are associated with changes in slope and topography, although the Parnell series also includes a change in parent material. Zahl, Fine-loamy, superactive, frigid Typic Calciustolls (well drained); Williams, Fine-loamy, mixed, superactive, frigid Typic Argiustolls (well drained); Bowbells, Fine-loamy, mixed, superactive, frigid Pachic Argiustolls (well drained) Parnell, Fine, smectitic, frigid Vertic Argiaquolls (poorly drained). (c) A mountainous, unglaciated landscape (udic, mesic) in southwestern Pennsylvania. After Seibert *et al.* (1983). Most of the soil map unit boundaries are associated with changes in parent material or, less commonly, natural soil drainage class.

Residuum: Dekalb, Loamy-skeletal, siliceous, active, mesic Typic Dystrudepts (acid sandstone); Guernsey, Fine, mixed, superactive, mesic Aquic Hapludalfs (shale); Weikert, Loamy-skeletal, mixed, active, mesic Lithic Dystrudepts (interbedded shale, siltstone, and sandstone); Upshur, Fine, mixed, superactive, mesic Typic Hapludalfs (shale); Dormont, Fine-loamy, mixed, superactive, mesic Oxyaquic Hapludalfs (shale and siltstone).

Colluvium: Culleoka, Fine-loamy, mixed, active, mesic Ultic Hapludalfs.

Alluvium: Newark, Fine-silty, active, nonacid, mesic Aeric Fluvaquents.

the map unit boundaries that were drawn in the field and transpose them onto an orthophotoquad. Each map unit will be checked for accuracy and proper labeling. After that, the lines will be digitized and sent to a cartographer as a digital line file with attributes for each map unit (series, slope, etc.).

Most county-scale soil maps have map sheets that span 6–8 square miles (15.5–20.7 km^2). In general, a *team* of soil mappers works on a county soil map. As a means of quality control, the soils on a given map sheet are not mapped solely by one person. Rather, the sheet is divided into sections and different people map different sections. Then, the edges of the sections (or sheets) are matched or joined. Soil map units should edge-match perfectly, but rarely do. The mappers are left to work out their joins and in so doing, an error check is made of each sheet. Sometimes two *different* soil series will meet at a map sheet edge, requiring a reconciliation of sorts. Often, additional fieldwork will be required to rectify such a situation, which in the end leads to a better map.

Error and variability in soils and on soil maps

Landscapes with high amounts of soil variability provide a significant challenge to the mapper. To produce an effective soil map, the surveyor must delineate uniform mapping unit consociations, or pedologically complex units that are consistently and explicity described (Edmonds *et al.* 1985). This is seldom possible, leading to error. Soil map error occurs due to poor mapping and/or highly complex landscapes, or because the scale of the map does not allow for all the known complexities to be mapped (Ball and Williams 1968, Walker *et al.* 1968, Campbell 1979). Error varies as a function of many sedimentologic and geomorphic properties. For example, variability might be much greater on some landscape positions than on others (Lepsch *et al.* 1977a, Ovalles and Collins 1986), or in some types of parent materials (Drees and Wilding 1973). Variability can occur *between* map units, as on complex landscapes of undulating relief where parent materials change frequently over short distances, and it may increase or decrease with the age of the soil landscape (Barrett and Schaetzl 1993, Phillips 2001a). In theory, the complexity of such a landscape could still be mapped without error. *Within*-map unit variability is usually so great, however, that the map cannot account for all of it, leading to inclusions (Lark and Beckett 1995).

Soil spatial variability takes two forms: (1) continuous, natural variations due to spatial variability in the soil-forming factors, and (2) discontinuous variation, often manifested as inclusions. The latter type of variation is stochastic, in that measured values of soil properties cannot be precisely predicted from values at neighboring locations without large numbers of samples (Crosson and Protz 1974, Campbell 1979). Many of

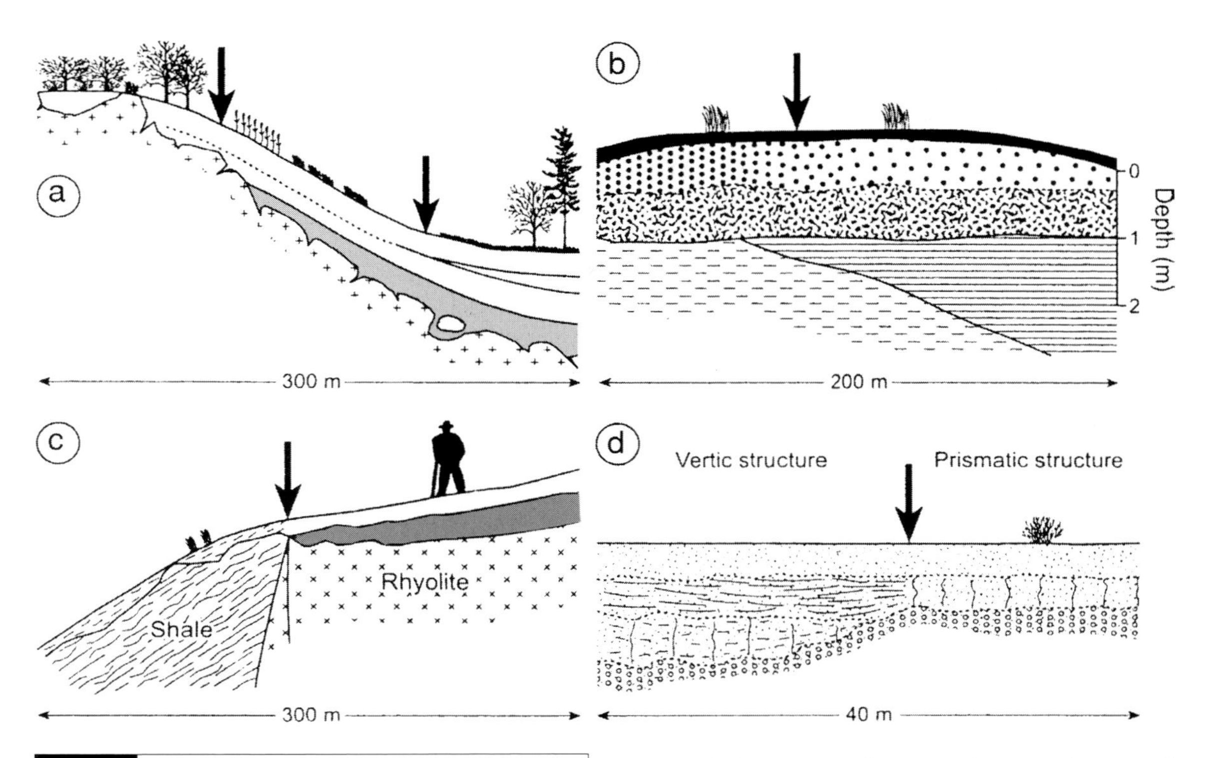

Fig. 7.13 Fuzziness and uncertainty in soil boundaries. After Lagacherie *et al.* (1996). (a) High fuzziness, low uncertainty. Gradual variation in soils along a catena, but this variation is expressed nicely as changes in vegetation. (b) High fuzziness, high uncertainty. Gradual variation in stoniness of a horizon in the lower profile, which is not expressed on the surface. (c) Low fuzziness, low uncertainty. Abrupt variation of soils due to a subsurface geological boundary is manifested at the surface as a slope break and a change in surface stoniness. (d) Low fuzziness, high uncertainty. Undetectable but abrupt variation in subsoil structure which is not manifested on the surface.

these inclusions would be extragrades, e.g., Lithic Hapludult inclusions within a Typic Hapludult map unit (similar soils are not considered inclusions; only dissimilar soils are listed as inclusions in mapping unit descriptions). Obviously, the mapper tries to define map units that will contain as few inclusions as possible, with a commonly set goal of 85% purity for a consociation. In glaciated areas, the 85% goal is extremely difficult to meet (Wilding *et al.* 1965, Beckett and Webster 1971).

Another type of error on soil maps involves the location of map unit boundaries (Lagacherie *et al.* 1996). Map unit boundaries misrepresented spatially are an example of *uncertainty*. That is, the map unit location only approximates the actual location of the boundary between the two unlike polypedons. Someone standing on the landscape near a map unit boundary would express the uncertainty as, "I am *probably* in map unit A." The spot was mapped as soil A but there exists some degree of uncertainty in the map. Uncertainty may or may not be manifested on the surface (Fig. 7.13). Soil boundaries also have a degree of *fuzziness*, i.e., they are not abrupt, but rather one mapping unit grades into the other. Fuzziness indicates ill-defined soil attributes close to the boundary. Soil boundaries along a catena, in which all the soils have formed in uniform parent materials, are usually fuzzy (Fig. 7.13a). Explaining your location vis-à-vis a fuzzy soil boundary could go like this: "The soil here, at the edge of mapping unit A, is more like series A than series B." Interpreting soil maps, especially digital maps in a GIS, is done more accurately when the degrees of uncertainty and fuzziness are taken into account. Lastly, knowing the *nature*, gradual or sharp, of the map unit boundary is also critical to correct interpretation of the soil landscape.

Because map units are not taxonomically pure, and because important land-use decisions are constantly being made, based on more or less

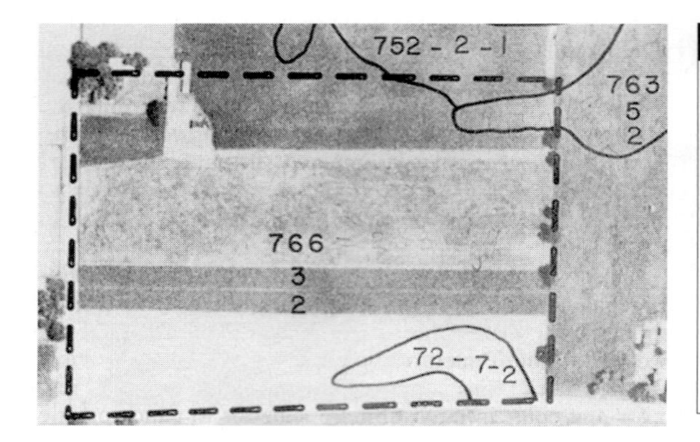

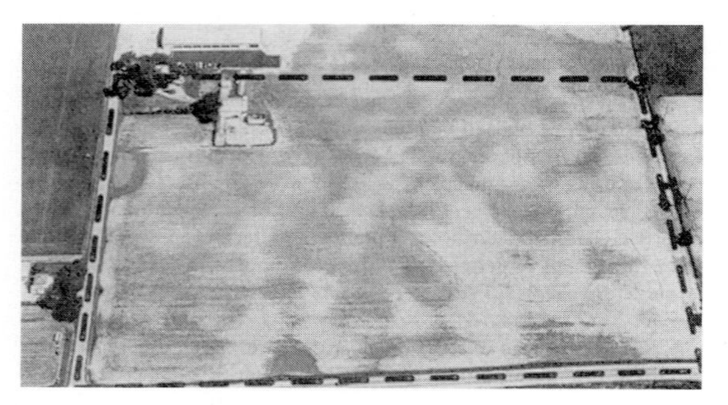

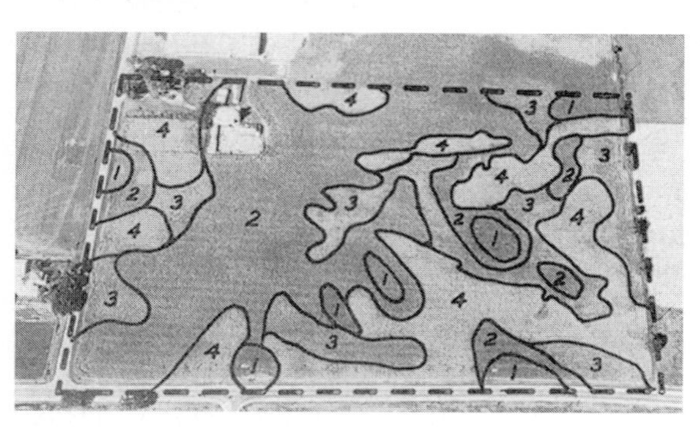

Fig. 7.14 Four oblique aerial photographs of a field in Dane County, Wisconsin. After Milfred and Kiefer (1976). (a) The field with soil boundaries overlain on it, as depicted in the Dane County soil survey. (b) The same field in July, 3 days after receiving 2.5 cm of rain. Soil patterns primarily reflect soil moisture. (c) The same field in September, showing patterns in the corn crop, which are indicative of soil series and soil moisture differences within the field. (d) The same field, mapped in more detail based on data obtained from repetitive aerial photographs. This level of detail is not possible with most county soil surveys, but does illustrate the complexity of soils that is not always captured on maps.

pure units, knowing the amount of error or variability *within* a map unit is very important (Mader 1963, Cipra *et al.* 1972, Campbell 1978, Campbell and Edmonds 1984) (Fig. 7.14). Parent materials like loess or eolian sand have low within-unit variability, while glacial deposits and some types of lacustrine sediments are highly variable from place to place (Campbell 1979).

Soil landscape analysis

Geographers and soil scientists are interested in understanding the pattern of soil map units, or the distribution of soils on landscapes (Fridland 1974). Many decades ago, Dokuchaev noted that soils exist on the landscape in distinct geographic combinations, later called soil combinations or areals (Neustruyev 1915, Prasolov 1965). Closely related are the concepts of soil catenas and soil associations, all very geographic in nature, dealing with the spatial relationships of soils to each other (Fridland 1965). The science of *soil landscape analysis* explores such issues (Hole 1978, Hole and Campbell 1985). Francis Hole (1953) was perhaps the first to draw attention to the fact that soils can and should be studied as three-dimensional bodies, at a time when the focus was primarily on the two-dimensional profile.

Soil landscape analysis places soils within their landscape context, and in so doing provides a basis for characterizing and explaining soil patterns. Soil landscape analysis uses as its basic unit the soil body – an abstract concept that is more analogous to a polypedon than a map unit. Fridland (1965) refers to a similar concept

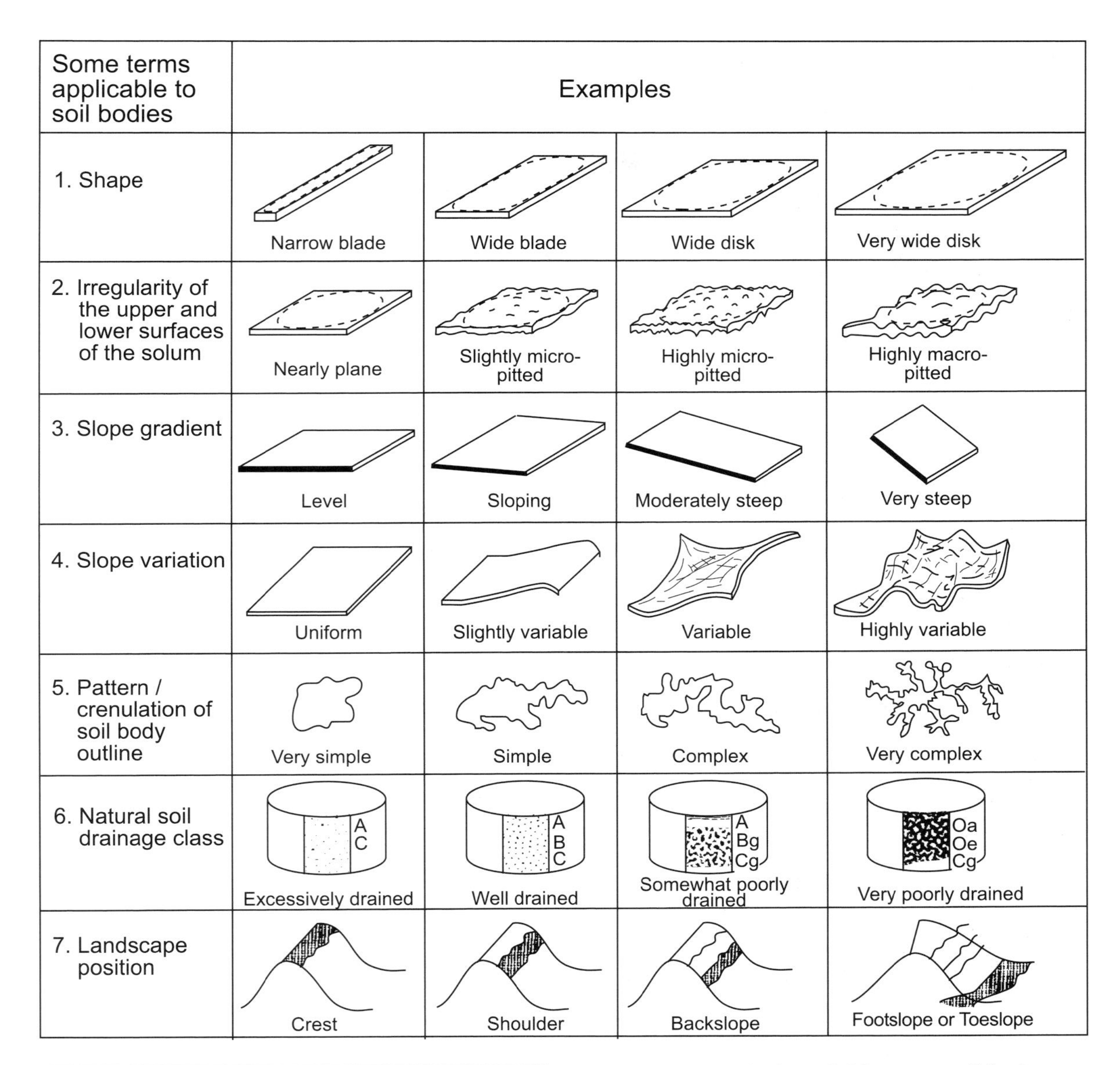

Some terms applicable to soil bodies
Examples
1. Shape
Narrow blade
Wide blade
Wide disk
Very wide disk
2. Irregularity of the upper and lower surfaces of the solum
Nearly plane
Slightly micro-pitted
Highly micro-pitted
Highly macro-pitted
3. Slope gradient
Level
Sloping
Moderately steep
Very steep
4. Slope variation
Uniform
Slightly variable
Variable
Highly variable
5. Pattern / crenulation of soil body outline
Very simple
Simple
Complex
Very complex
6. Natural soil drainage class
A
C
Excessively drained
A
B
C
Well drained
A
Bg
Cg
Somewhat poorly drained
Oa
Oe
Cg
Very poorly drained
7. Landscape position
Crest
Shoulder
Backslope
Footslope or Toeslope

Fig. 7.15 Examples of terms used to describe the various aspects of soil map units, when examined from a soil landscape analysis perspective. After Hole (1953).

as that of the *elementary soil areal*. Think about soil bodies, the basic units of study in soilscape analysis, as having distinct boundaries. The upper boundary is discrete – where the soil surface meets the atmosphere. The lower limit of the soil is sometimes obvious but sometimes diffuse and unclear; nonetheless we are forced to define the lower limit of a soil, or a pedon (see Chapter 3). The lateral boundaries of this same soil body are determined by its taxonomic description, i.e., by humans, except where the soil body abuts non-soil material like a lake or bare bedrock (Hole 1953). Thus, the extent of a soil body is definable. Others abut it, fitting together like a jigsaw puzzle. Explaining the nature of this soil puzzle is the realm of soil landscape analysis. How each soil varies vertically falls in the realm of pedology.

Soil landscape analysis takes the emphasis away from describing the profile, and places more weight on describing and explaining the *patterns*

Table 7.17 A selection of methods than can be used to characterize the soil landscape

Type of soilscape measure[a]	Description of method and what it portrays
Soil body shape	Shapes (random, elongate, circular, anastomosing, complex, etc.) of soil bodies on the landscape (see Fig. 7.17)
Alignment of soil boundaries	Compass orientation of the lines that separate soil map units (see Fig. 7.17)
Number and arragement of nodes	Distribution of the points on the landscape where map unit boundaries join another boundary; also of interest is the sheer numbers of nodes, as a measure of soilscape complexity
Density of soil bodies	An indirect measure of the size of soil map units (mean, median, range, standard deviation, etc.) (see Fig. 7.19)
Composition	Proportionate amounts of the various types of soil taxa (series, subgroups, great groups, etc.) on a landscape
Taxonomic contrast	Relative numbers of taxa on a landscape
Size of soil bodies	Quantitative assessment of the varying sizes of soil map units
Index of soil heterogeneity	Number of soil bodies per unit area (i.e., the density) multiplied by the taxonomic contrast; provides an indication of the taxonomic complexity of a given area
Number of landscape positions	Assessment of which of the five major landscape positions are represented on the soil map under question
Soil moisture regime	Assignment of a value to each mapping unit, based on its natural soil drainage class, and performing various mathematical manipulations with these data (see Fig. 7.18)
Distinctness of soil boundaries	Estimation of the pedogenic sharpness of boundaries between map units
Taxonomic rank of soil boundaries	Providing a taxonomic rank to each soil boundary, as a measure of landscape complexity
Degree of landscape openness	Topographic, hydrologic and pedogeomorphic information used to determine the amount of integrated vs. closed (internal), surface drainage; corollary to this index is a measure of the length of drainageway per unit area (i.e., drainage density)

[a] Most measures presented are readily quantifiable.

Source: Hole and Campbell (1985).

that soil bodies make on the landscape, so as to better see how the soil fits into that landscape. In this way, soil landscape analysis is seen as being a core part of soil geography. The basic premise of soilscape analysis is that the *geographic distributions of soil characteristics* are the criteria that can be used to distinguish one *landscape* from another, as a profile description can be used to distinguish one *pedon* from another (Pavlik and Hole 1977). One tries to determine the nature and patterns governing the distribution of soils as reflected in the make-up of the soil cover (Fridland 1965).

Describing soil bodies and soil landscapes

In soil landscape analysis, each soil body is described by a number of semi-quantitative and quantitative terms – shape, surface and subsurface irregularity, slope gradient and variation, outline crenulation/pattern, wetness and position on the landscape (Pavlik and Hole 1977) (Fig. 7.15). For example, one could quantify the nature of the soil boundary (sharp, gradual, etc.) or the soil body itself in several ways (Table 7.17, Fig. 7.15). The shape, slope, pattern, surface irregularity and natural drainage class of a soil body can all be

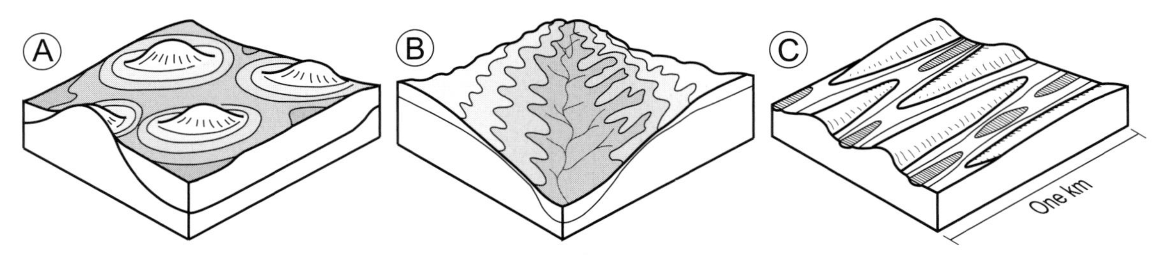

Fig. 7.16 Examples of three different types of landscapes and the shapes of the soil bodies (or map units) that might be observed on them. After Hole (1978). (a) Isolated, coppice dunes on a plain. (b) Fluvially dissected plateau. (c) Drumlin field.

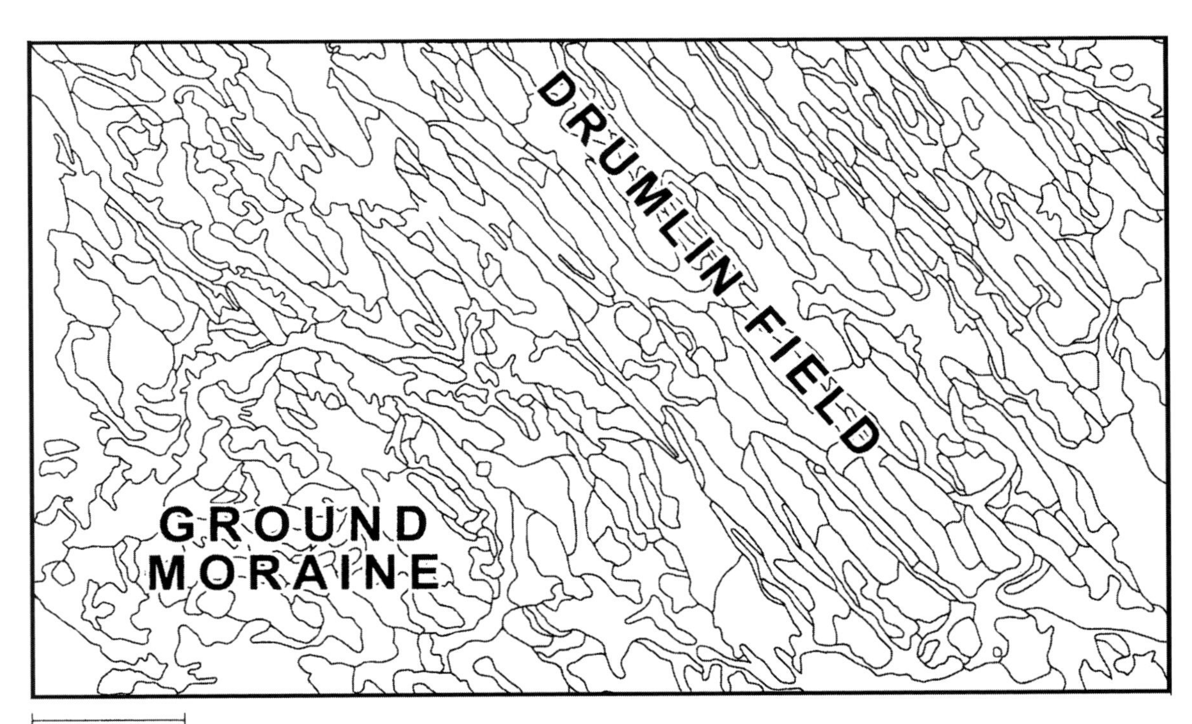

Fig. 7.17 Soil map unit boundaries in a part of Presque Isle County, Michigan (Knapp 1993). The area with a high degree of linearity is a drumlin field, while the more random pattern in the lower left is a ground moraine.

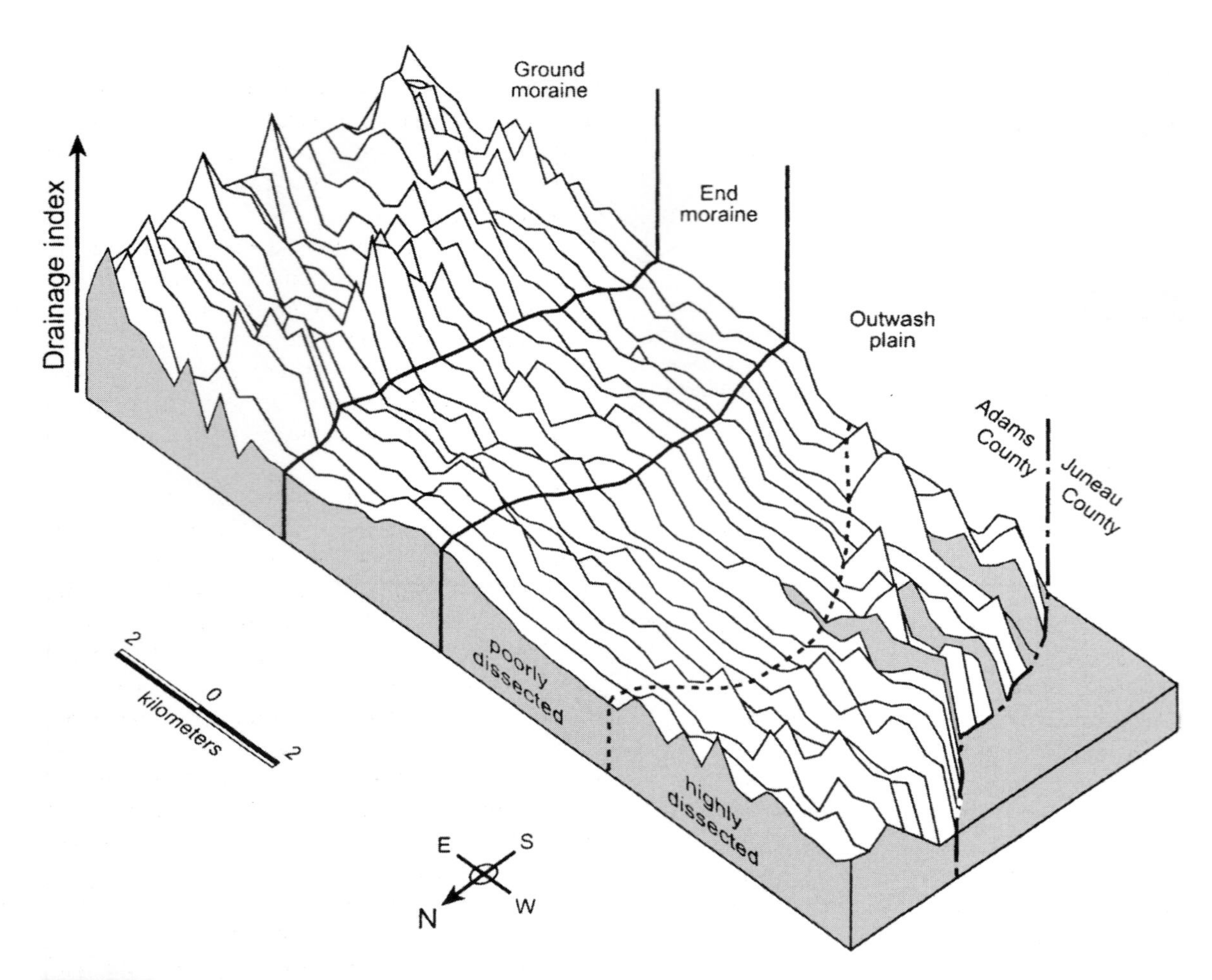

Fig. 7.18 Mesh diagram of natural soil drainage index values for part of Adams County in central Wisconsin. Higher peaks indicate wetter soils. Note how the landscape is much more variable, with respect to wetness, in the part of the outwash plain that is highly dissected. The high, dry end moraine also contrasts well with the lower ground moraine, which has many wet soils in depressions. After Schaetzl (1986a).

determined, and after a number of soil bodies on a landscape have been studied, generalizations made about the spatial character of the soilscape (Hole 1953, Schaetzl 1986a). In sum, these properties are a landscape's *pedogeomorphic fabric* (Habermann and Hole 1980).

Shapes of soil bodies on a landscape can be a very useful indicator of the physiography of that landscape and how it may have evolved (Table 7.17, Fig. 7.16). One can describe the general shapes of soil map units in qualitative terms such as disks, spots, stripes and dendritic patterns, or they can be quantified by using terms such as length : width ratios, area : perimeter ratios, degree of boundary crenulation, etc. (Fridland 1965, Hole 1978). The *orientation* of *soil boundaries*, equivalent to the arrangment of the soil bodies on the landscape, can be examined, as a way of determining the degree of linearity that exists on the landscape. Fluted landscapes with drumlins would exhibit high linearity, while an end or ground moraine might not (Fig. 7.17). Density of soil bodies per unit area can also be examined, although this will vary according to map scale and the mappers themselves. Some mappers are "splitters" while others are "lumpers." Likewise, the size of map units, which determines map unit density, is affected by access (forested areas are difficult to access and hence may contain

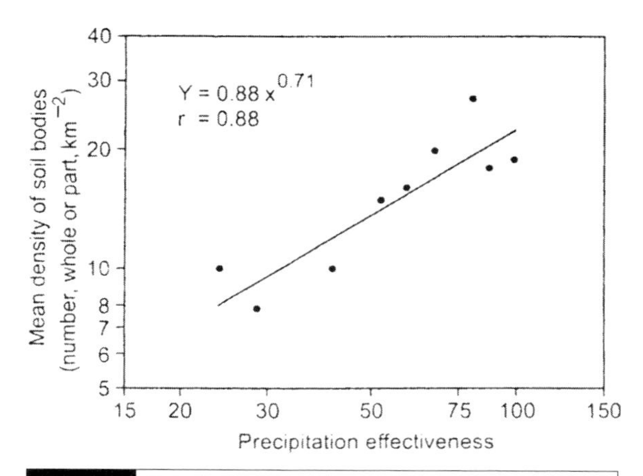

Fig. 7.19 Relation between the density of soil bodies and precipitation effectiveness, a measure of leaching, from the dry climate of Colorado to the humid climate of Indiana. After Hole (1978).

larger map units) and whether the mappers chose to combine some consociations into map unit complexes. Thus, map unit size is a difficult and potentially misleading soilscape parameter, and must be used with caution. One possible way to avoid this problem is to examine patterns that are more geomorphic than pedologic in origin, like slope or drainage class. In this way, the patterns will emerge regardless of the soil classification scheme used, accessibility of the landscape, or the biases of the mapper (Habermann and Hole 1980).

Information on *pedologic diversity*, a measure of the heterogeneity of a soilscape, may assist biologists interested in *biodiversity* and provide important geological/geomorphological information as well (Hole 1978). It may be quantified by examining the suite of different soil taxa on a landscape, or by examining the taxonomic contrast across a number of soil boundaries (Schaetzl 1986a). Contrast may be defined based on natural soil drainage class, slope and productivity, or taxonomically. Taxonomic contrast can take all forms, from high-level contrast (a Mollisol abutting an Entisol or an Andisol) to contrast at the series level (two different soil series, both coarse-loamy, mixed, mesic Typic Hydrudands).

Although difficult to generalize, many soilscapes, especially "young" ones and those with high local relief, have great pedologic diversity, while old, flat landscapes may be dominated by one or two taxa. Schaetzl (1986a) has shown that pedologic diversity is not necessarily related to readily observable landscape qualities such as local relief. Landscapes of high relief can be more pedologically diverse and internally contrasting than more subdued ones. At the very lowest level of taxonomic contrast are soil boundaries where the same series is mapped on each side of the boundary, but the slope of each map unit is different. This map unit boundary exhibits no taxonomic contrast whatsoever, only slope contrast.

Some soilscapes may be primarily composed of well and somewhat poorly drained soils, while others are mainly wet, and still others have nearly equal proportions of soils from four or more drainage classes – this is landscape contrast according to soil wetness. With respect to natural soil drainage class and soilscape diversity, Hole (1953) developed, and Schaetzl (1986a) expanded upon, a type of soil drainage index designed to capture the essence of the soilscape's diversity with respect to wetness. Variability in drainage index (0–99) across various types of landscapes can then plotted (Fig. 7.18).

As a logical next step, data on soil bodies from entire landscapes can be combined and interregional comparisons made (Habermann and Hole 1980). For example, the soil landscape of the plains of Kazakhstan is one of large soil bodies, while areas of similar Mollisols in North Dakota may be more intricate, with patches of small soil bodies interspersed in glacial kettles or along drainageways. Hole (1978) reported on one such comparison, in which the density of soil bodies increased in direct relation to precipitation effectiveness, i.e., leaching potential (Fig. 7.19). Pavlik and Hole (1977) illustrated how soil information can be used to discriminate one type of landscape from another, and to determine the degree of difference between adjoining soilscapes. In sum, soilscape analysis provides a refreshing way to examine landscapes and in so doing, to glean landscape-based information that might otherwise have remained hidden.

Part II

Soil genesis: from parent material to soil

Chapter 8

Soil parent materials

The influence of parent materials on soil properties has long been recognized. Early pedologists and soil geographers based their concepts of soils largely on its presumed parent material. Later, parent material was viewed simply as a factor that influences soil development – an influence that diminishes in importance with time.

No one disputes the fact that geologic controls and factors are important to a complete understanding of soil development and soil patterns (Fig. 8.1). Parent material is the framework for the developing soil profile. In fact, the more holistically one examines a soil, including not just the upper 2 m or simply the solum, but also the many deep horizons, the more important parent material seems.

One goal of soil geomorphology is to identify the types and origins of a soil's parent material; any study of soil genesis must ascertain what the parent material of the soil is. Information on parent material is paramount to the study of soils, especially soils on Pleistocene or younger surfaces (Roy *et al.* 1967, Schaetzl *et al.* 2000). Many times, the role of the soil geomorphologist is to produce a surficial geology map from a soils map. From this surficial sediment map comes a better understanding of the sedimentological and surficial processes that were operative in the past. This exercise may be relatively easy for young soils on young landscapes, but it is an extreme challenge on older landscapes. Conversely, one almost cannot make a soil survey without detailed information about surficial geology, illustrating the interrelatedness of soils and parent materials in almost all landscapes.

Several problems may arise when trying to positively identify the parent material of a soil. First, it may have been highly weathered and altered prior to the current period of pedogenesis. In many such cases, the parent material is actually a previous soil and the soil is, by definition, polygenetic (e.g., Waltman *et al.* 1990) (see Chapter 11). Second, there may be more than one parent material, either as discrete or intimately mixed layers (e.g., Frolking *et al.* 1983, Rabenhorst and Wilding 1986a). And some of these layers may be so thin that pedogenesis has blurred them.

This chapter focusses on the discernment and characteristics of the major types of soil parent materials, and what they might be able to tell us about the geomorphic and sedimentologic history of a site.

Effects of parent material on soils

Parent material is one of the five state factors (Jenny 1941b) (see Chapter 11). Soils, per se, develop as various pedogenic, geomorphic and biologic processes act on initial material(s) changing it into a soil that is unique to that location in time. Completely unaltered parent material, which occasionally occurs below the C horizon, is infomally called the D horizon (see Chapter 3). Not always obvious and easily recognized, D horizon material may be buried deeply beneath water, sediment or rock, frozen, or it may have just formed, such as volcanic ash or fresh beach sand. Buol *et al.* (1997: 136) considered parent material

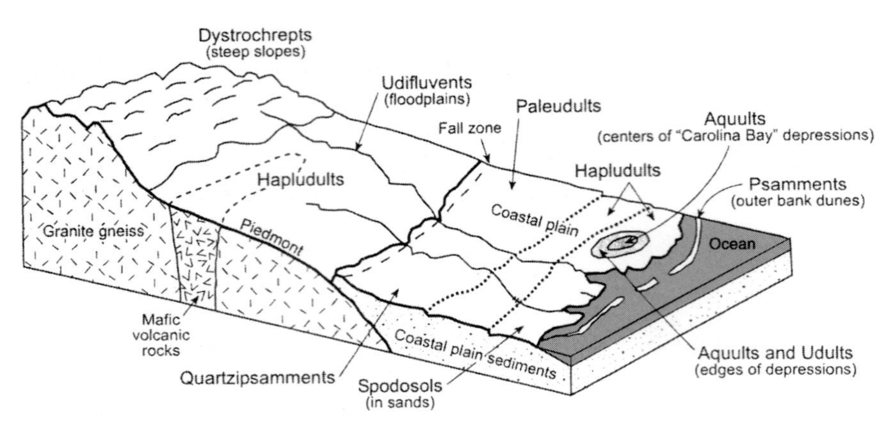

Fig. 8.1 Block diagram of the major soils and parent materials in the Piedmont and Coastal Plain regions of the southeastern United States, illustrating the strong association between parent material and soil development at the highest levels of classification. After Markewich and Pavich (1991).

to be "that portion of a C or R horizon which is easily obtainable but reasonable distant" below the solum. This approach will work for soils developed in deep, uniform parent materials, but will not suffice for soils that have formed in multiple layers of sediment; the upper materials may be so altered by pedogenesis that it is impossible to characterize the original parent material. For this reason and many more, ambiguities as to what exactly constitutes a given soil's parent material are commonplace.

For many soils, especially weakly developed Entisols, Gelisols and Inceptisols, discernment of their parent material may be relatively easy (e.g., Ehrlich *et al.* 1955, Joseph 1968, Schaetzl 1991a, Mason and Nater 1994). The solum may be thin enough and the soil young enough that a pit a few meters deep exposes the C horizon. In such soils, one can surmise some characteristics of the parent material, e.g., color and texture, by examining the profile only (Oganesyan and Susekova 1995). Mineralogical characteristics also may not have changed markedly. For thicker profiles and older soils, the solum may be so different from the original parent material that only generalizations can be made about the original C horizon. With time, soils increasingly change with respect to their parent materials (Muckenhirn *et al.* 1949, Yaalon 1971, Chesworth 1973). The impacts of climate and relief often become dominant over that of parent material (Short 1961), such that adjacent or neighboring soils formed from different materials will seem to converge, morphologically. Color is quickly changed under different oxidizing or reducing conditions, by translocation of clay–humus complexes, or salts, and by melanization and rubefaction (see Chapter 12). Pedoturbation can quickly destroy stratification, or create layering where none had existed before (Johnson 1989, 1990) (see Chapter 10). Leaching, weathering and/or additions of material on the top of the soil surface may change soil pH, such that the original pH may not be discernible.

Conversely, many properties of parent material are persistent and continue to exert an impact on the soil for long periods of time (Ehrlich *et al.* 1955). Clayey parent materials will develop into clayey soils and sands generally produce sandy soils; texture has a way of persisting (Yaalon 1971). Some sandy parent materials may weather into a loamy soil, however, if many of the sand grains are weatherable minerals, rather than quartz. A soil horizon may have more clay than it inherited from the parent material either because its larger particles have weathered to clay, or because of translocation of clay into that horizon. Conversely, clay can be lost from not only the horizon but also the profile, leaving a sandier profile behind. Similar statements could be made for a suite of soluble substances common to soils.

Parent material, through its impacts on texture and surface area, also affects rates of pedogenesis. In general, pedogenesis proceeds faster in sandier materials, because of its lower surface area. Color can change more rapidly in sandy materials (see Chapter 12). Similarly, sola may be thicker on coarser-textured parent materials.

The mutability of time$_{zero}$

In the simplest case, a soil parent material is deposited, or exposed by catastrophic erosion, and pedogenic processes immediately begin to "work" on the raw parent material: time$_{zero}$. Rarely, however, is pedogenesis that simple (Johnson 1985). Two complicating scenarios can occur.

First, erosion may slowly or rapidly remove material as soil forms in it, lowering the surface through time. In this situation, the lower boundary of the soil/solum continually encounters fresh parent material (Pavich 1989). If the soil contained coarse fragments, the soil surface may become enriched in them as a *lag concentrate* or *carpetolith* (stone line), perhaps including some *pedisediment* (see Chapter 13). Horizon boundaries migrate downward into fresh sediment as the soil surface erodes. In this scenario, the eroding (unstable) surface lessens the utility of the time$_{zero}$ concept.

Second, soils can receive slow, steady or intermittent, surficial additions of fresh mineral material, slowly aggrading the soil surface (Johnson 1985, McDonald and Busacca 1990). (Catastrophic additions of material, as with floods or mass movements, essentially "reset" the clock and do not apply here; see Chapter 13.) Slow additions of sediment allow soil processes to adjust; eventually the soil profile may thicken (Riecken and Poetsch 1960, Wang and Follmer 1998). Johnson (1985) called this process "developmental upbuilding"; it is also known as cumulization (see Chapters 13 and 14). Cumulization can be defined as eolian, hydrologic or human-induced additions of mineral particles to the surface of a soil. It often occurs on alluvial surfaces, e.g., floodplains, although it can also occur when loess or eolian sand is slowly added to upland soils. Small, slow inputs of sediment such as loess epitomize this process, and because loess is deposited across the entire landscape, cumulization would then not be spatially restricted to low-lying, sediment-receiving areas. In cumulization, therefore, the surface of the profile gradually "grows upward" as sediment is slowly added, and the new sediment plus the existing profile become the parent material for the "new" soil.

All soils must have at least one time$_{zero}$, or a starting point. At time$_{zero}$, pedogenic processes begin their work on some sort of loose surficial material or weathered rock. For many soils, that starting point represents only one of many such time$_{zeros}$, as the soil may form, get eroded and begin forming again in the remaining parent material. Rapid, deep burial may essentially isolate the soil, allowing pedogenic processes to start anew in the overlying sediment. For other soils, such as those in mountainous settings, there may be little loose parent material initially, and the most important processes may be the alteration of bedrock into saprolite and then soil.

Polygenetic soils are those that have evolved along more than one pedogenic pathway; thus, many have more than one time$_{zero}$ (see Chapter 11). When an Aridisol encounters a suddenly wetter climate, the new, humid-climate pedogenic pathway uses the old soil as parent material (Bryan and Albritton 1943, Chadwick *et al.* 1995). Similarly, a soil may encounter a change in vegetation, forcing modifications to its pedogenic processes. In both cases, the pre-existing soil is viewed as the parent material for the one that will henceforth develop, i.e., parent material may not necessarily be pedogenically *unaltered* sediment.

In sum, soil parent material should be evaluated in light of surface stability and pedogenic pathways. Stable geomorphic surfaces may be attained in the short run but "stability" is never retained over geologic timescales (see Chapter 13). Thus, the concept of parent material is one that also has short-term meaning. What is "parent material" to a soil today will be part of the solum, eroded or buried at some time in the geologic future. This line of reasoning also reinforces the importance of being able to determine what the parent material of a soil is/was, for in so doing we

Fig. 8.2 Mining bauxite-rich limestone "residuum" in Jamaica. The ore here is, essentially, the lower profile of an Oxisol. Photo by RJS.

can learn a lot about surface stability, soil development and landscape evolution. For this reason, obtaining accurate information about soil parent material is often one of the most important and challenging tasks of the soil geomorphologist (e.g., Chadwick and Davis 1990, Dahms 1993, Schaetzl *et al.* 2000).

A classification of parent materials

Early approaches to soil survey and classification relied heavily on interpretations of a soil's parent material, largely because soils were thought of as disintegrated rock mixed with decaying organic matter (Simonson 1952, 1959). Terms such as "granite soils" and "loessial soils" were common. Emphasis was on parent material, in part by necessity since pedogenesis was still poorly understood. The more we learned about soils, however, the more we realized that soils formed in a similar parent material may have that one thing in common, but still can vary considerably spatially and temporally (Phillips 2001a). It was no longer acceptable to simply refer to a soil as a product of its parent material. Today, we acknowledge that understanding the parent material of a soil goes a long way toward knowing its initial state, or $time_{zero}$ condition, and this information provides certain limits to what the soil can eventually develop into. Thus, although we seldom use a parent material type as a soil name modifier, knowledge of soil parent materials continues to be important. Like any broad topic, understanding all of its complexities may be best done by categorization and classification, which is what we attempt below.

Residual parent materials

All loose, unconsolidated materials that overlie bedrock are referred to as *regolith* (see Chapter 3). Regolith may be either *residual* (formed in place) or it may have been *transported* there by a surficial vector, like water, gravity or wind. Residual regolith is called *residuum* (Short 1961). Often, it is impossible to know with certainty that the residuum has not been transported at some point in the past. For example, on limestone terrain in Mediterranean regions, a red sediment often interpreted to be residuum overlies carbonaceous bedrock. However, this sediment is often overthickened in depressional areas and on footslopes (Wieder and Yaalon 1972, Yassoglou *et al.* 1997), and thinner on steeper slopes (Benac and Durn 1997). Thus, what may initially have been residuum has undoubtedly been transported to some extent. Similar relationships exist in the residual bauxite areas of Arkansas. Saprolite rich in aluminum but relatively poor in iron is termed *bauxite* (Sherman *et al.* 1967) (Fig. 8.2). Here, long-term weathering of nepheline syenite bedrock

Fig. 8.3 Residual soil over sandstone bedrock in the Appalachian Plateau of West Virginia. Photo by D. L. Cremeens.

on uplands has led to thick, residual deposits of bauxite ore, rich in gibbsite and kaolinite (Gordon *et al.* 1958). Thick bauxite deposits on lower slopes exist because of transportational (colluvial and alluvial) processes.

The thickness of the residual cover on a slope is often a function of the interplay between the weathering processes that produce it and the transport processes that erode and carry it away. On slopes with thick residual accumulations, weathering rates exceed transport rates and the slope is said to be *transport-limited*. On other slopes, residuum is thin; these may be *weathering-limited* slopes (Fig. 8.3). The terms do not imply that either process is, in and of itself, proceeding slolwy or rapidly. They simply refer to the balance between the two opposing processes.

Saprolite

Regolith with a sharp bedrock contact often suggests that it (the regolith) has an erosional/transportational origin, by which the contact was eroded and sharpened. Conversely, where regolith overlies a thick weathered zone of soft rock that retains the geologic structure (i.e., *saprolite*), the possibility of a purely *residual origin* is strengthened. Saprolite is *in situ*, isovolumetrically and geochemically weathered rock that still retains some of the original rock structure, such as strata, veins or dikes (Aleva 1983, Whittecar 1985, Stolt and Baker 1994). It contains both primary minerals and their weathering products

Fig. 8.4 Gneiss saprolite exposed in a roadcut in western North Carolina. Photo by RJS.

(Hurst 1977) and is usually soft enough that it can be penetrated with a sharp shovel blade or knife (Fig. 8.4). More often than not, thick saprolite implies fairly deep weathering dominated by chemical, rather than physical, processes. Often, the boundary between saprolite and the soil profile is gradual (Stolt *et al.* 1991). Weathering processes dominate in saprolite, whereas soil processes dominate above, in the pedological profile. Weathering of rock into saprolite has been called *saprolitization* or *arenization*: the physical disintegration of the host rock, induced by the chemical weathering of some weatherable minerals (Eswaran and Bin 1978b). Saprolitization increases porosity, promoting processes that act to further break up the rock, such as shrink–swell (Frazier and Graham 2000).

Saprolite is best expressed and thickest on old, gently sloping, transport-limited surfaces. Therefore, saprolite is usually thickest on stable bedrock uplands, especially those that have gentle slopes (Aleva 1983, Whittecar 1985, Pavich 1989, Stolt *et al.* 1992, Ohnuki *et al.* 1997). Here, weathering by-products cannot be flushed from the system except in fractures, of which there may only be a few and these may be filled with illuvial clay. On steep and incised landscapes, weathering by-products are removed so rapidly (in solution and by mechanical erosion) that the landscape tends to become denuded and the bedrock–saprolite weathering front continues to deepen. Colluvium is often thickest at the bases of slopes and in small valleys or microtopographic lows (Cremeens and Lothrop 2001). On older surfaces, such as in parts of the Appalachian Mountains, there has been ample time to produce a thick layer of saprolite, sometimes more than 30 m in thickness (Bouchard and Pavich 1989, Cremeens 2000). Saprolite can form quickly and can persist for several thousands of years, with some examples dating to the Tertiary.

Depending on their degree of weathering and composition of the parent rock, saprolites can either have weathered isovolumetrically or they can have "collapsed" from the original rock. Collapse occurs as minerals are weathered and their weathering products are removed in solution. Therefore, it is common in rocks with high amounts of soluble minerals, such as limestone. Pavich (1989) reported a mass loss of about 1/3, and still more mass is lost when saprolite is converted to soil. The total mass lost in the saprolite-to-soil changeover can be another 75%, leaving behind mostly clay-sized material (Pavich 1989). Alternatively, the formation of saprolite from crystalline rocks is commonly isovolumetric, i.e., there is a loss of mass but little or no volume change (Stolt and Baker 2000). Velbel (1990) listed the three mechanisms by which isovolumetric weathering can occur:

(1) Minerals weather in different sequences, and "take turns" maintaining the rock structure. For example, hornblende weathers while plagioclase feldspar alters to clay, while later the feldspar alteration products (clay) maintain the structure while the hornblende weathers.

(2) A rigid framework that resists collapse is produced during weathering, maintained largely by resistant minerals such as quartz and muscovite.

(3) Etch pits form as portions of minerals weather at different rates. Material is removed in solution from etch pits, but because the general outline and bulk volume of the mineral remains unaltered, the rock does not collapse.

Another process at work to inhibit or counteract collapse involves soil biota. They, along with roots, facilitate the "opening up" of the saprolite, forming the pores that lead to the decreased bulk density of soil vis–vis saprolite.

In order for saprolite to continue weathering, it must be porous enough to allow water to penetrate and remove weathering byproducts, e.g., CaO, Na_2O and SiO_2 (Pavich 1989). Permeability also facilitates the translocation of clay and dissolved substances; hence, saprolites may have argillans, even at depth. Many saprolites are more clay-rich near the surface than at depth, however, because clay particles are arrested before they can move to great depths. A follow-up to this involves the examination of stream chemistry in landscapes where bedrock weathering is the main producer of ions, for such knowledge allows for calculation of rates of saprolite production (Velbel 1990). Pavich (1989) and Cleaves *et al.* (1970) both reported that it can take as long as 1 million years to form 4 m of saprolite in the Appalachians.

Many studies of saprolite approach the problem because of a geomorphic interest. By knowing the rates of saprolite formation and its thickness on the landscape, one can achieve some degree of understanding of landscape age and evolution.

Hunt (1972) described the typical saprolite profile in the southern Appalachians, on crystalline rocks such as schist, gneiss and granite (see also Calvert *et al.* 1980a, b and Pavich 1989). From the hard bedrock upward, one first encounters discolored, partially weathered bedrock with a bulk density 10–20% lower than the bedrock. The weathered rock does not ring when hit with a rock hammer, but instead produces a dull thud. Cracks are coated with yellow and brown hydrated iron oxides. *Structured saprolite*, which makes up most of the regolith volume, overlies the weathered bedrock (Pavich *et al.* 1989). Argillans and iron stains are common, and its density is only half of the bedrock; the original structure and fabric are preserved. Weathering and translocation has resulted in losses of many basic cations and any soluble salts, especially Na_2O and CaO (Pavich 1989). In places where erosion has been slow and weathering rapid, structured saprolite is overlain by *massive saprolite*, implying that rock structure has been lost, mostly due to bioturbation, wetting and drying, creep and other near-surface processes. Thus, quartz veins found below, in the structured saprolite, would end abruptly at the base of the massive saprolite, and in its place one might find only scattered quartz fragments. To many pedologists, massive saprolite is simply another term for the lower solum, in which pedogenic and pedoturbative processes have destroyed the original saprolite structure. That is, massive saprolite is often better described as soil material. The transition from saprolite to the soil profile, which can be abrupt or gradual, is largely controlled by the following processes: (1) mixing and formation of structure by fauna and plant roots, (2) accumulation of organic matter, (3) additions of eolian materials and (4) translocation of clay (De Villiers 1965, Brimhall *et al.* 1991b). The change from rock structure with very small pores, to the blocky and granular pedogenic structure with much larger pores, is a fundamental part of the saprolite–soil transition (Fig. 8.5). This usually occurs at the maximum depth of rooting, about 1 m according to Fölster *et al.* (1971), but often deeper. Formation of soil from saprolite is not isovolumetric. The volume *increase* (dilation) that occurs in the bedrock-to-soil morphogenesis can be 200–300% (Brimhall *et al.* 1991b).

Flach *et al.* (1968) coined the term *pedoplasmation* for the collection of physical and chemical changes that alter saprolite to soil. It is particularly important in many Oxisols and Ultisols. Literally, this term means the formation of soil plasma or what Fölster *et al.* (1971) referred to as a *homogenization horizon*. The processes responsible for pedoplasmation include bioturbation and other mechanical disturbances such as shrink–swell and root action. Brimhall *et al.* (1988) were

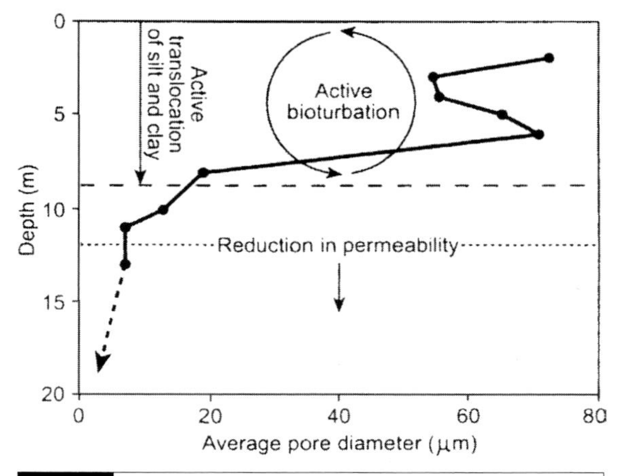

Fig. 8.5 Average pore diameter in a core from an Oxisol and its subjacent saprolite in Mali. After Brimhall *et al.* (1991b).

able to distinguish this homogenized zone from the saprolite below and noted that zircon minerals in the saprolite resembled those in the bedrock, while zircons in the soil above were smaller and more rounded. Because the upper zircons had presumably been deposited by wind, they must have been mixed to depth by homogenization processes such as pedoturbation and root growth. Zircons in the saprolite, below the homogenization horizon, had not yet been exposed to these processes.

In sum, perhaps the best way to summarize bedrock-to-soil morphogenesis is to quote Brimhall *et al.* (1991b: 54–55):

> Lateritic weathering is simultaneously induced both from below and above by advance of two very different processes; one at depth affecting incipient alteration of bedrock apparently under water-saturated reducing conditions, and the other completing a progressive evolution in chemistry, structure, and physical properties, which has proceeded downward by replacement of each zone by its superadjacent and pedogenically more mature counterpart leading towards complete oxidative equilibrium with the atmosphere. Thus, the zonal sequence is a series of dynamic partial equilibrium states between the solid earth and the atmosphere, mitigated by the migration and reactivity of percolating groundwater and ground gasses.

Saprolite formation in the humid tropics

The main differences between saprolitization in the humid tropics and elsewhere revolves around the much greater rates at which it forms, the types of minerals affected by weathering and the different mobilities of the ions that result from weathering. Given the intense heat, the availability of water (at least during parts of the year) and the great age of many tropical landscapes, residual parent materials there are highly weathered and frequently underlain by thick saprolite. During saprolitization in this environment, chemical weathering is especially intense, leaving many soils with a high proportion of resistant materials and ions, such as Ti, Zr and various sesquioxides and metal oxyhdroxides, mostly as fine clay (Soileau and McCracken 1967, Bigham *et al.* 1978). In tropical soils, bases and silica are translocated from the profile (see Chapter 12), leaving behind large amounts of free Fe and Al; what happens to these elements markedly impacts the evolution of the soil. Some may stay in the profile to form nodules, concretions and crusts, while others may be transported laterally into lower, wet areas at the bases of slopes where they can form massive ferruginous crusts (Fölster *et al.* 1971) (see Chapter 12).

Kaolinite, gibbsite and other sesquioxides are generally regarded as end products of weathering in the humid tropics, with gibbsite often seen as the ulitimate end product (Aleva 1983). Gibbsite is most common in dry, hot soils and may be almost absent in soils with high water tables because it cannot crystallize there (Cady and Daniels 1968). Where the rock is aluminum-rich, boehmite and gibbsite are dominant minerals. Bauxite (Al-rich saprolite) tends to form in more humid climates than do many Fe-rich saprolites. In order to concentrate Al at the expense of Fe, two factors must also occur: (1) the parent rock must be iron-poor, such as in basic volcanics and phyllites, and (2) the site must have high biotic productivity so that the Fe can be chelated by organic acids and removed from the system (Thomas 1974). Desilication must also be a strong process in order to concentrate aluminum to the extent that it is considered bauxite.

In the humid tropics, saprolites may exceed 100 m in thickness (Aleva 1983), with the thickness of the saprolite dependent upon many factors, primarily age and stability of the surface, type of rock, location on the landscape and climate. On many basic and ultrabasic rocks, there may be no real saprolitic zone, as the base of the soil rests directly on rock or on a thin weathering crust (Soil Survey Staff 1999) (Fig. 8.6).

Using a mass balance approach to soil formation and saprolitization on an Oxisol in Mali, Brimhall *et al.* (1991b) were able to show that much of the Fe and Al in these soils was actually not residual, but was brought into the soil system from upslope or in eolian dust. The eolian dust can penetrate only to a certain depth, depending on pore size and depth of pedoturbation. Thus, it appears that purely residual accumulations of sesquioxides may be restricted to the basal parts of the saprolite. Many of the "chemically mature" sediments in the tropical soils above may be brought in as dust from nearby weathered landscapes. Much of that dust ends up in oceanic areas immediately downwind (Fig. 8.7).

Saprolite often displays dicrete zones of varying weatheredness and pedogenic imprinting. Acworth (1987) described four main saprolite zones below the solum; Jones (1985) described five (Fig. 8.6). The bedrock–saprolite interface, often tens of meters below the surface, is usually diffuse and shaped mostly by fracture patterns in the bedrock. This is the region where soluble constituents and weathering by-products are removed in groundwater, but rock structure is preserved. Corestones of bedrock are common well up into the saprolite (Berry and Ruxton 1959). The permeability of this zone is high; solutes are able to move between corestones and within fractures. In a description of the typical rock–saprolite–soil sequence in the Gold Coast (Africa), Brückner (1955) noted that in many sections it is possible to distinguish a lower layer of saprolite from an upper layer (Fig. 8.8). Roots proliferate only in the upper, redder layer. Eswaran and Bin (1978a) described six soil-saprolite zones in humid Malaysia – the uppermost of which was considered soil. Below this zone is a stone line or stone zone; the stones may be either resistant quartz/chert fragments or sequioxidic nodules (see below). Further below are three zones of

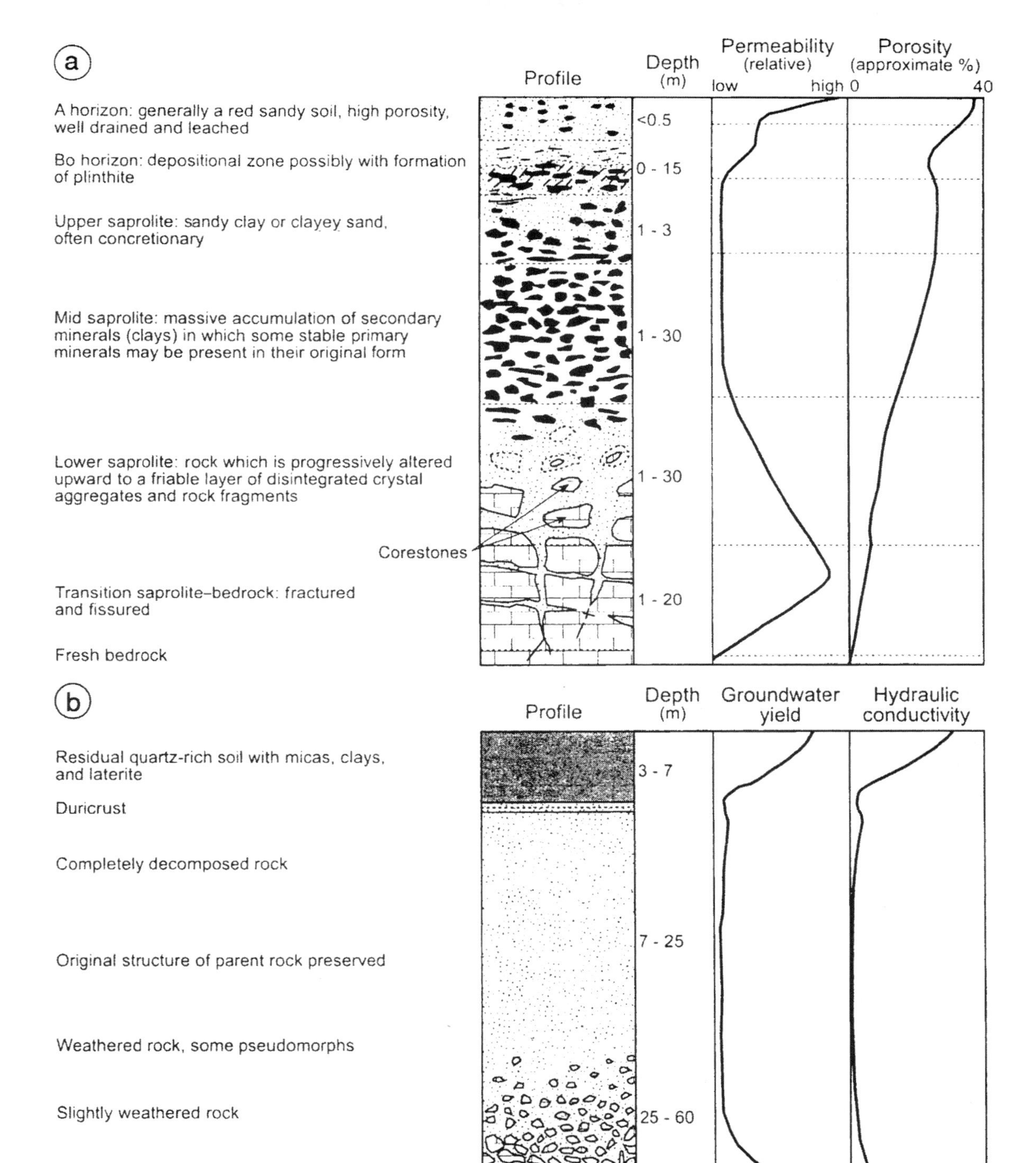

Fig. 8.6 Typical weathering profiles developed on crystalline bedrock in the humid tropics. (a) After Acworth (1987). (b) After Jones (1985).

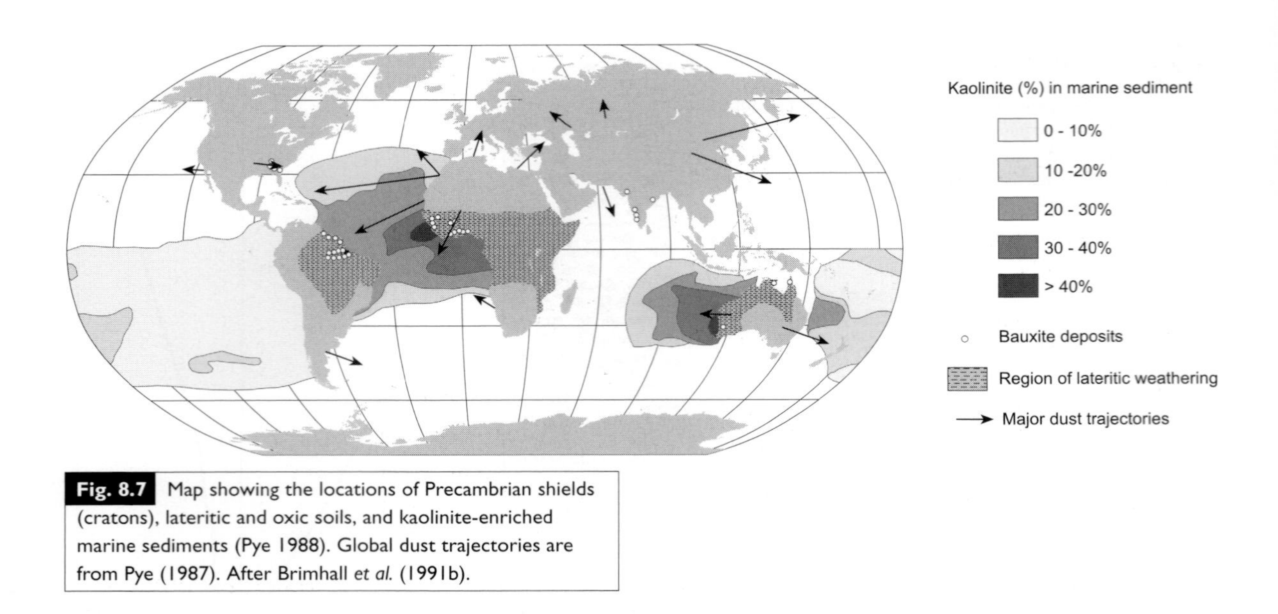

Fig. 8.7 Map showing the locations of Precambrian shields (cratons), lateritic and oxic soils, and kaolinite-enriched marine sediments (Pye 1988). Global dust trajectories are from Pye (1987). After Brimhall *et al.* (1991b).

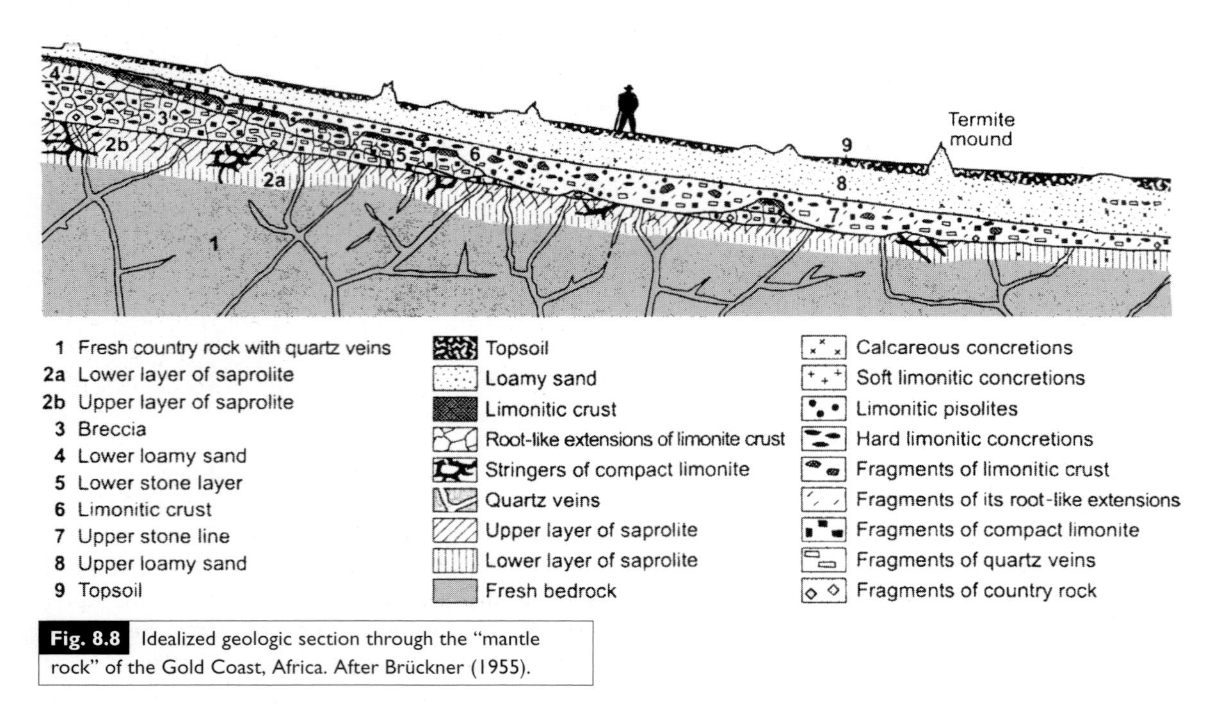

Fig. 8.8 Idealized geologic section through the "mantle rock" of the Gold Coast, Africa. After Brückner (1955).

decreasing amounts of weathering and increasing rock structure.

Weathering intensity, which increases toward the surface, follows the pH gradient; soils are often most acidic at the surface and less so at depth, due to accumulations of weathering by-products (van Wambeke 1962) (Fig. 8.9). The increase in weathering may be either gradational or abrupt, largely dependent on water flow pathways and parent material mineralogy. Fingers of

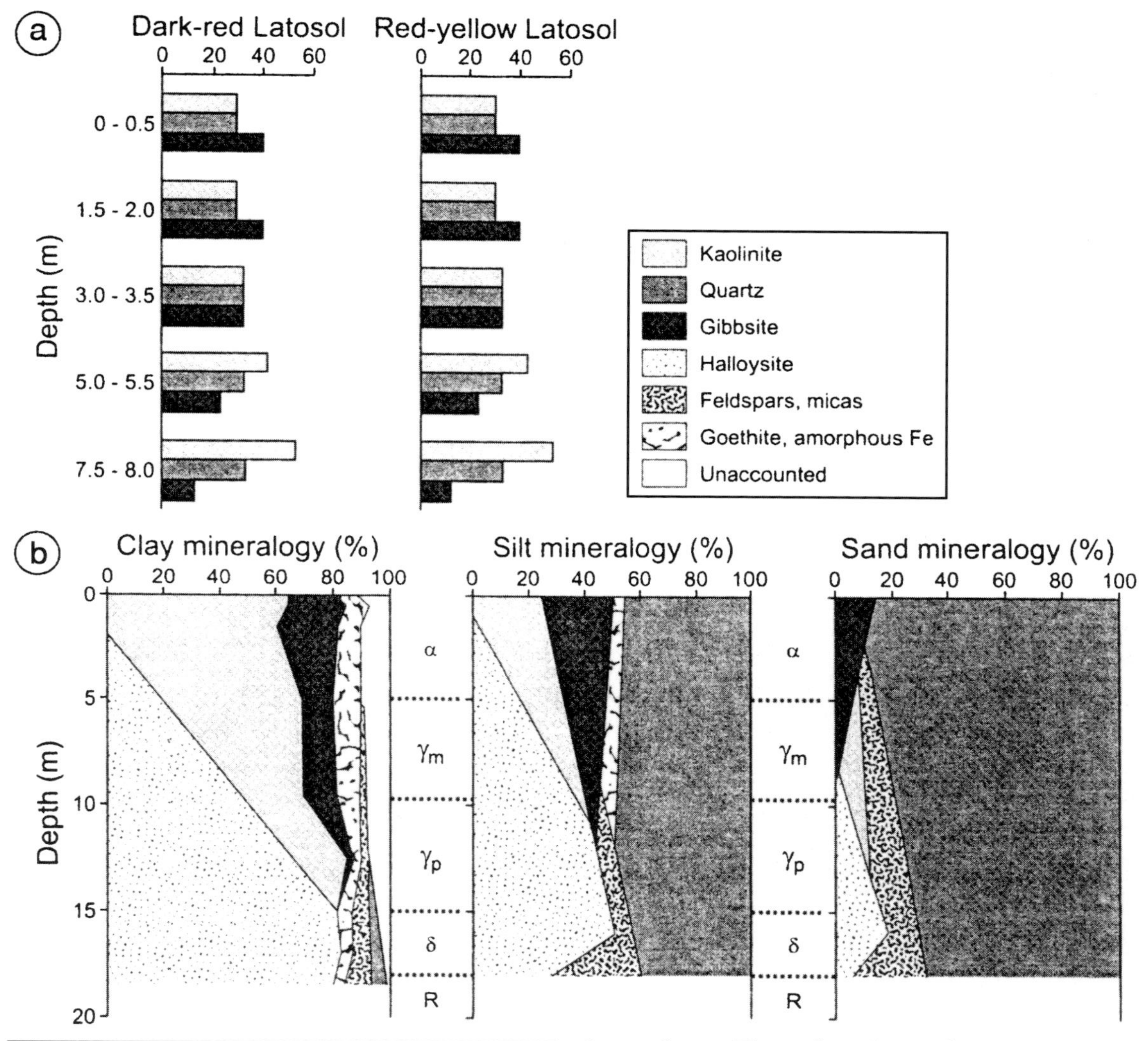

Fig. 8.9 Variation in mineralogy, with depth, for some weathering profiles in the humid tropics. (a) Mineralogy of the fine earth fraction of two Acrustox in Brazil. After Macedo and Bryant (1987). (b) Mineralogy of the sand, silt and clay fractions of a deep weathering profile in Malaysia. After Eswaran and Bin (1978b).

more highly weathered materials protrude down into less weathered rock, reflective of infiltration conduits. Likewise, protuberances and diapir-like structures of less weathered rock may commonly be found higher in the weathering profile than one would expect; they are preserved because infiltrating water is steered around, rather than into, them. The mineralogy changes that are associated with increased near-surface chemical weathering can be observed in the rock-to-deep-saprolite-to-soil transition (Fig. 8.9). Conditions also become less and less oxidixing with depth, leading to mottled or reduced zones and some degree of iron segregation into mottles. If large amounts of free iron are present, ironstone fragments or plinthite will form (Fölster *et al.* 1971, Eswaran and Bin 1978a).

Although it is difficult to make general statements about the changes in mineralogy from the base of tropical saprolites to the soil profile, it is clear that weatherable minerals decrease in near-surface layers. One of the first secondary minerals

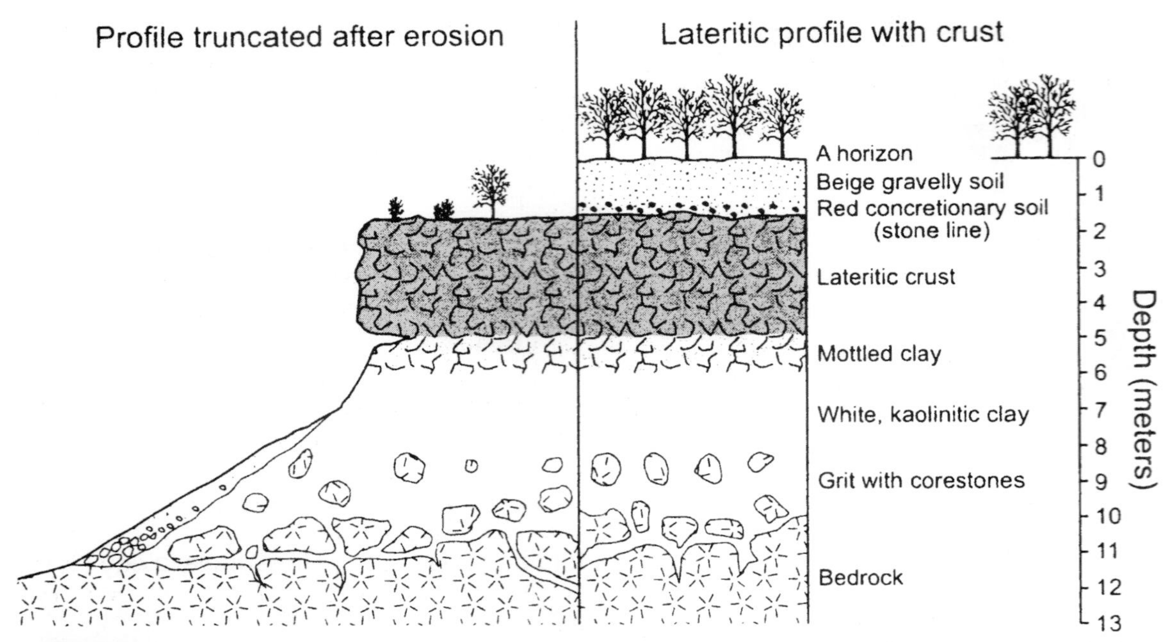

Fig. 8.10 Generalized bedrock–saprolite–soil sequence under forest and savanna, Sudan. After Thomas (1974).

observed in deep saprolite in the humid tropics is kaolinite, which weathers from feldspars. The abundance of kaolinite is often (though not always) expressed as whitish layers in the upper saprolite, below the soil per se (Fig. 8.10). Moss (1965) and Thomas (1974) referred to this zone as the "pallid zone." Kaolinite is then replaced by (weathered to) gibbsite in nearer surface zones, as the soil color changes to reds and browns (Aleva 1983, Macedo and Bryant 1987) (Fig. 8.9). Gibbsite can also alter to kaolinite by desilication (Sivarajasingham *et al.* 1962). The decrease in kaolinite in near-surface layers may therefore be due to weathering, desilication or eluviation in particulate form, to greater depth (Aleva 1987). In highly weathered soils, however, kaolinite is probably weathering out, or has mostly weathered out, of near-surface layers. In some saprolites, pseudomorphic alteration of feldspar phenocrysts occurs directly to halloysite and/or gibbsite, the aggregates of which retain the original fabric. In this environment, Al is essentially immobile while Fe can be reduced and move out of, or into, preferred areas in solution. Thus, in some highly leaching environments, the amount of aluminum-rich weathering products, i.e., boehmite, is greater near the surface because more Fe has been leached (Aleva 1965).

Another feature typical of humid tropical weathering profiles is the presence of secondary minerals in the sand and silt fractions – as small ironstone fragments or parts of partially altered primary minerals surrounded by an alteration product. According to Curi and Franzmeier (1984), high hematite and gibbsite contents reflect a leached, dry soil environment with high Fe^{3+}concentrations in solution, while moist sites typically have more goethite and kaolinite, reflective of lower levels of desilication. The predisposition for goethite to form in wetter horizons is supported by the work of Fölster *et al.* (1971), who noted that while goethite and hemitite can form in the same horizon, in wetter horizons goethite is dominant.

The red, brown and white masses of kaolinite and gibbsite, so common in saprolite, become mixed and merged into a more uniform, almost isotropic, reddish brown material in the solum (Flach *et al.* 1968). In the upper B horizon, this uniform brown material will have developed strong blocky or granular structure,

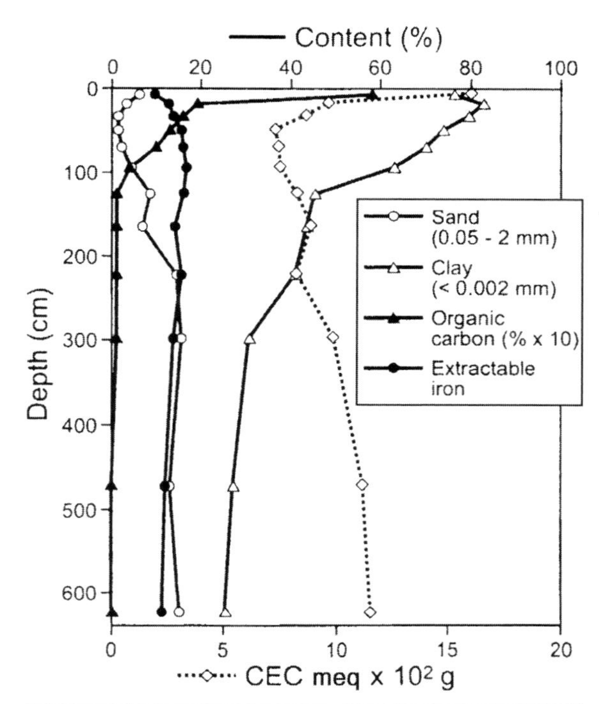

Fig. 8.11 Depth functions for a Tropeptic Hapludox from Puerto Rico, formed in andesite breccia. The mean annual rainfall exceeds 200 cm, and there is no dry season. Data from Flach *et al.* (1968).

clay contents will have increased almost threefold and bulk densities will have dropped significantly (Fig. 8.11). Silt and even sand-sized masses of kaolinite, common in the saprolite, are weathered in near-surface horizons such that in the upper solum all the kaolinite is in the clay fraction. Thus, clay in the saprolite becomes increasingly disaggregated and "finer" as weathering proceeds in near-surface layers.

Effects of rock type on residuum

The character (mineralogy, texture, porosity, base status, etc.) of the rock largely determines the nature of its residuum (Stephen 1952, Plaster and Sherwood 1971). In many instances, this relationship is intuitive and obvious. For example, sandstone will produce sandy, porous residuum; shale leads to clayey residuum. Quartzite is so difficult to weather that it produces little residuum and soils on quartzite tend to be shallow. Limestone weathers so completely that its residuum mainly forms as insoluble components that had been within the parent rock, such as clay, are released by weathering (Moresi and Mongelli 1988). Basic rocks, like basalt, diorite and gabbro, weather to a dark, clay-rich residuum that is high (initially) in pH. Soils formed in this type of residuum tend to be more fertile and "balanced" with respect to cations than are the soils formed from acidic rocks. Smectite may form from magnesium and sodium-rich, acidic rocks, if the climate is not so humid that these ions are leached from the profile. Ultramafic rocks like peridotite, which are very high in magnesium and iron, and low in aluminum, potassium, calcium and sodium, produce silty and clayey residuum (Rabenhorst *et al.* 1982). Table 8.1 and Figs. 8.12 and 8.13 provide baseline information on some common rocks and minerals.

The discussion below focuses on residuum from different rock types, but the emphasis is on those that do not develop thick sequences of saprolite, e.g., many sedimentary rocks, gabbros and basalts. Rocks that do form saprolitic residuum, such as granite, gneiss and schist, are discussed only briefly below, since saprolitization has already been covered.

Residuum from siliceous crystalline rocks

Granite and granitic gneiss are common coarse-grained, acid crystalline rocks; other examples include tonalite, quartz monzanite and granodiorite. These rocks form deep under the Earth's crust, where slow cooling allows for the formation/development of large mineral crystals. Quartz is a common mineral in these rocks, as are various types of feldspars (Fig. 8.13). Biotite, distinctive because of its shiny appearance and plate-like cleavage, may be a "weak link" in these rocks, from a weathering standpoint (Isherwood and Street 1976, Dixon and Young 1981). It tends to be the least resistant to chemical attack, weathering to vermiculite in humid, leaching environments (Borchardt *et al.* 1968).

Because of their high quartz content, acid igneous rocks tend to weather, primarily by

Table 8.1 Classification and composition of sedimentary rocks

	Texture	Composition	Rock name
Clastic	Coarse-grained	Rounded fragments of any rock type; quartz, quartzite and chert dominant	Conglomerate
		Angular fragments of any rock type; quartz, quartzite and chert dominant	Breccia
	Medium-grained	Quartz and rock fragments	Sandstone
		Quartz and feldspar fragments	Arkose sandstone
	Fine-grained	Quartz and clay minerals	Siltstone
	Very fine-grained	Quartz and clay minerals	Shale, siltstone or mudstone

(*cont.*)

physical means, to sandy and gravelly residuum, often with low base status and poor nutrient reserves (Muckenhirn *et al.* 1949, Stephen 1952, Plaster and Sherwood 1971, Eswaran and Bin 1978a, Wang *et al.* 1981, Evans and Bothner 1993, Oganesyan and Susekova 1995). Dixon and Young (1981) referred to the weathering of granite to coarse-textured soils as "arenization," or the formation of sandy mantles. Weathered granite saprolite has been called *grus*, and its formation from bedrock, *grusification* (Isherwood and Street 1976, Dixon and Young 1981). Sand grains in grus might contain a significant amount of weatherable minerals, such as mica, chlorite and feldspar, while in the resulting soil profile these minerals have often been weathered away. Clays will be primarily derived from weathering of micas and feldspars (Rebertus and Buol 1985). Si and Fe are common exchangeable cations (Short 1961). Acid, light-colored or reddish soils which have lost silica often form in this type of residuum (Ehrlich *et al.* 1955, Eswaran and Bin 1978a).

Gneiss and schist, metamorphic equivalents of granite, often have high amounts of mica (Fig. 8.12). Their saprolite tends to be silty, but coarser-textured and more permeable than is residuum from other metamorphic rocks. Illite and vermiculite may be common clay minerals in the saprolite. Residuum from chlorite schist may be more clayey, with high amounts of magnesium and perhaps smectite.

Residuum from base-rich crystalline rocks

Subsilicic rocks such as basalt, diabase, dolerite, gabbro and metagabbro are dark-colored and rich in base cations, especially Fe and Mg. Other rocks included in this category are andesites, diorites and hornblende gneisses (Fig. 8.12). These rocks are dominated by "ferromagnesian" minerals

Table 8.1 *(cont.)*

	Texture	Composition	Rock name
Chemical or Organic	Medium- to coarse-grained	Calcite ($CaCO_3$)	Limestone
	Microcrystalline, conchoidal fracture		Micrite
	Aggregates of oolites		Oolitic limestone
	Fossils and fossil fragments, loosely cemented		Coquina
	Abundant fossils in calcareous matrix		Fossiliferous limestone
	Shells of microscopic organisms, clay-soft		Chalk
	Banded calcite (cave or "hot spring" deposits)		Travertine
	Textural varieties similar to limestone	Dolomite ($CaMg(CO_3)_2$)	Dolomite
	Cryptocrystalline, dense, conchoidal fracture	Chalcedony (SiO_2)	Chert
	Fine to coarse crystalline	Gypsum ($CaSO_4 \cdot 2H_2O$)	Gypsum
	Fine to coarse crystalline	Halite (NaCl)	Rock salt

such as amphiboles, pyroxenes, biotite, olivine and chlorite (Rice *et al.* 1985). Hornblende may comprise a significant component of the sand fraction of the C horizon; in the solum it may have been weathered away. Plagioclase feldspars in these rocks are rich in calcium, and like many of the other minerals that comprise these rocks, weather rapidly. Generally speaking, minerals that crystallize early from molten materials under high temperatures and pressures tend to

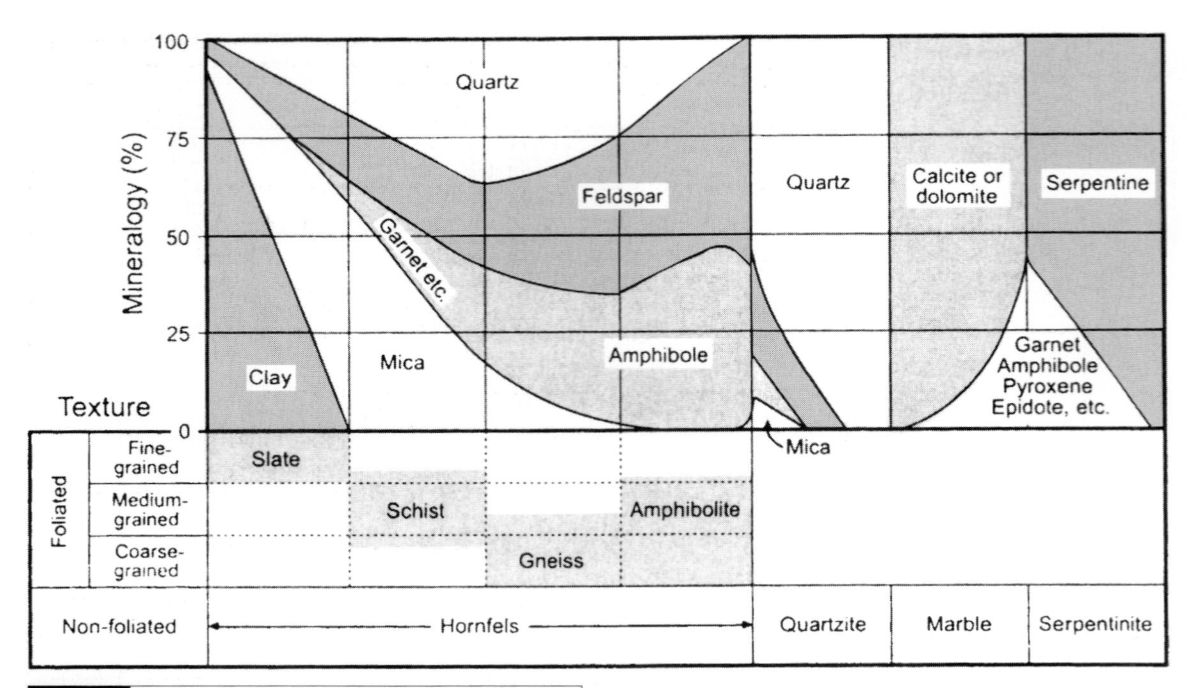

Fig. 8.12 A textural/mineralogical classification of the common metamorphic rocks. After Foster (1971).

weather more rapidly than many others, as they are well out of equilibrium with the cool, often wet, surficial environment.

Soils formed from these types of rocks tend to be dark, clayey and base-rich (Fig. 8.14). The low quartz content in these rocks is the reason for the general lack of sand in the saprolite and the soil profile, although exceptions occur (Jackson *et al.* 1971, Graham and Franco-Vizcaino 1992). As long as calcium, magnesium and other basic cations persist in the residuum, due either to low leaching regimes or eolian influxes of calcium-rich dust, a high base status can be maintained (Graham and Franco-Vizcaino 1992). In humid climates where base cycling is not strong, leaching may quickly remove these bases from the soil, leading to more acidic soil conditions. High amounts of bases and organic matter can, if combined with a wet–dry climate or an intermittently perched water table, lead to the formation of smectites and Fe–Mn concretions (Plaster and Sherwood 1971, Rice *et al.* 1985). Otherwise, kaolinite, vermiculite and halloysite are common secondary clay minerals; smectites are most common when the soils are weathering in dry climates. Rice *et al.* (1985) provided a weathering sequence for a soil developed on metagabbro (Fig. 8.15). Red and dark-colored soils develop on base-rich rocks, with the red colors coming from oxidized iron minerals and the darker colors from organic matter which strongly binds to the Ca ions released from the rock. Dark primary minerals such as hornblende and olivine may also exist in the residuum and the soil.

Ultramafic rocks are characterized by low (<40%) silica contents and relatively high amounts of magnesium, iron, chromium and nickel (Mansberg and Wentworth 1984, Lee *et al.* 2001). The most common ultramafic rocks are peridotite and serpentinite, both of which are rich in the 1:1 clay mineral serpentine; others are dunite and soapstone. According to folklore, serpentine gets its name from the resemblance of the soils formed on it to a mottled, greenish-brown snake.

Soils formed in ultramafic residuum are extremely rich in magnesium, iron, nickel and chromium, and poor in calcium, aluminum,

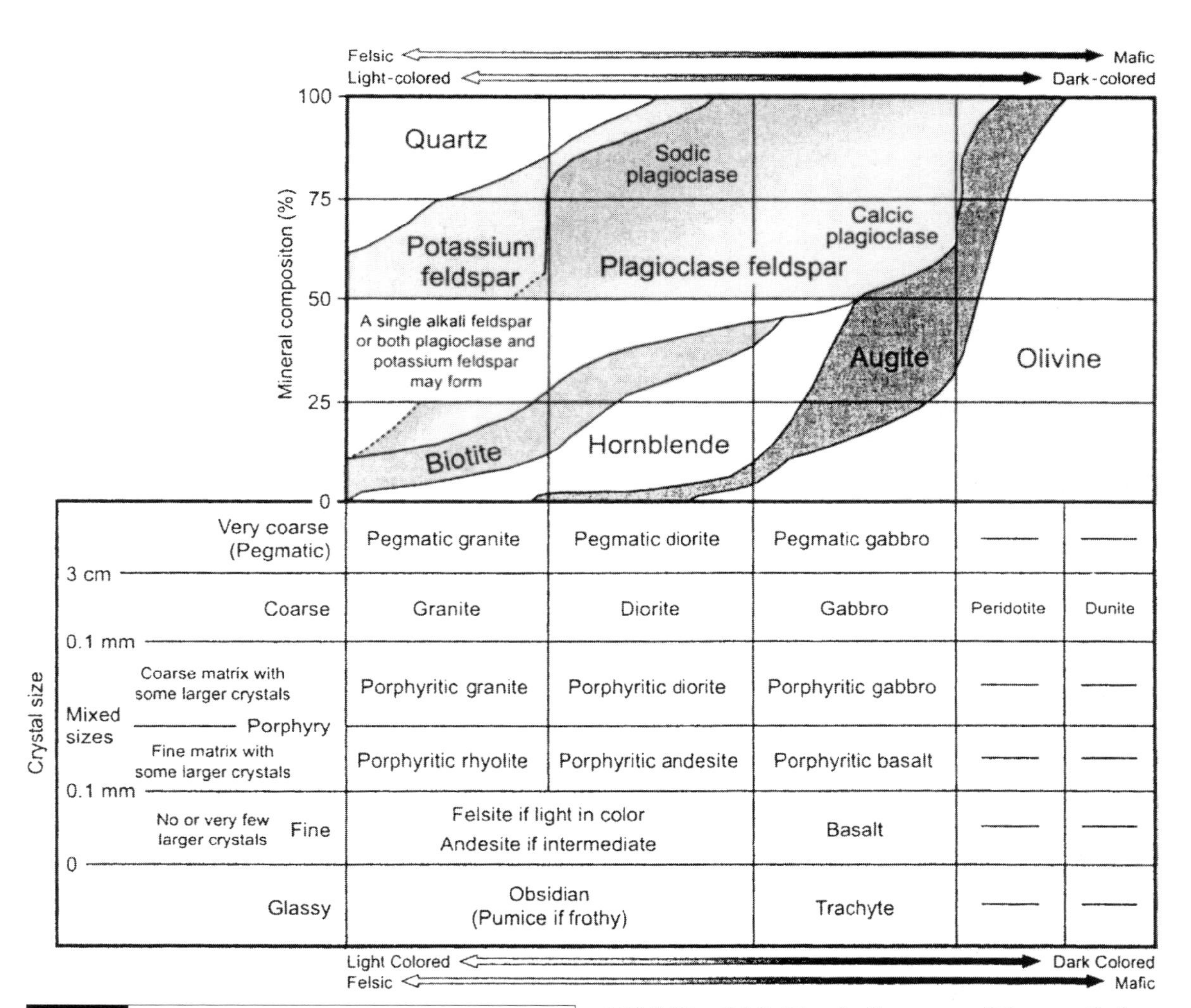

Fig. 8.13 Classification of common igneous rocks, based on mineral composition and crystal size. Minor accessory minerals are omitted for brevity. The name "granite" is used in the widest sense; several names are, however, in common usage for this type of rock. The modifier "porphyritic" is used only when crystals are very prominent in the rock. After Foster (1971).

potassium and sodium. Quartz, feldspar and mica minerals are rare, while normally uncommon minerals such as talc, magnetite and olivine are found in abundance. Residual soils and their ultramafic residuum tend to be shallow, droughty, nutrient-poor, erodible, stony and fine-textured (Parisio 1981, Bulmer and Lavkulich 1994) (Fig. 8.16). The shallowness of these soils has been attributed to (1) the lack of resistant minerals such as quartz and feldspar, (2) limited clay mineral formation due to a lack of aluminum, (3) accelerated erosion due to sparse vegetative cover and (4) removal of mass by magnesium leaching (Parisio 1981).

Because serpentine weathers to release high amounts of Mg and metal ions, its residuum and soils have very low Ca : Mg ratios (Walker *et al.* 1955, Rabenhorst and Foss 1981). Roberts (1980) reported Ca : Mg ratios (based on total elemental content) for some serpentine soils in Newfoundland that typically were <0.012. Nickel, cobalt and chromium levels are commonly high as well (Rabenhorst *et al.* 1982), and can even reach toxic levels. Soils formed on serpentine are therefore

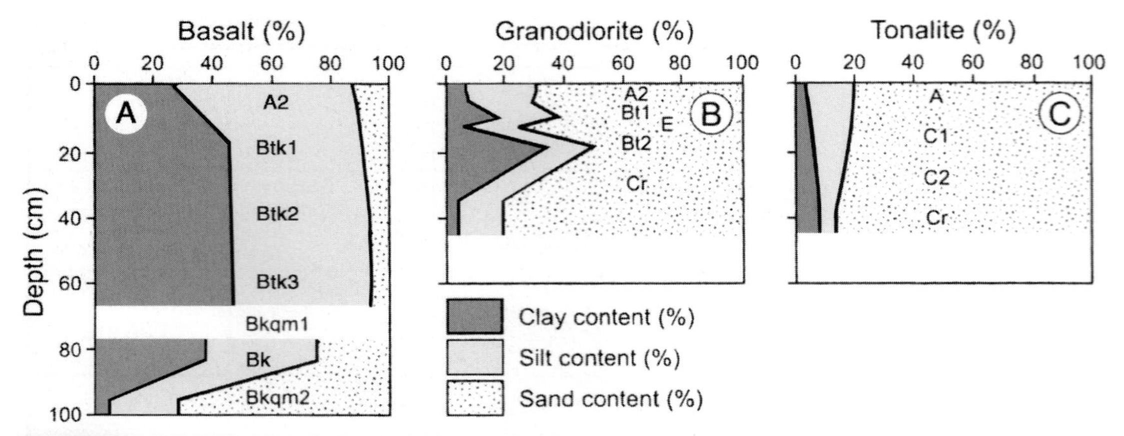

Fig. 8.14 Particle size distribution, with depth, in three residual soils developed in Baja California. Basalt is a fine-grained, base-rich rock. Tonalite is a coarse-grained, acid igneous rock, while granodiorite is intermediate between the two in many respects. After Graham and Franco-Vizcaino (1992).

notoriously infertile because of the very low Ca : Mg ratios, metal toxicities or macronutrient deficiencies (Gordon and Lipman 1926, Spence 1957). In most horizons, ≥80% of the exchange sites are occupied by Mg^{2+}. The high magnesium contents block the plant's ability to take in nutrients, especially calcium. With time and continued leaching, Mg and Ni are depleted from serpentine soils (Fig. 8.16). Fe, Si and Al, which were present in very low amounts in the parent material, then become relatively enriched in secondary Fe oxides (Wildman *et al.* 1968, Rabenhorst *et al.* 1982, Bulmer and Lavkulich 1994). Despite the leaching, the soils retain high amounts of exchangeable Mg, especially in the lower profile, keeping pH values high.

Residuum from sedimentary rocks

For the most part, residuum formed from clastic sedimentary rocks is texturally much like the clastic particles within the rock, e.g., sandstones weather to sandy sediment. Clay mineral assemblages, however, can vary from one rock unit to another.

Shale is perhaps the most easily weathered of the common sedimentary rocks. For the most part, weathering of shales (and sandstones) tends to be dominated by physical (disintegration) processes, rather than chemical weathering. Shales tend to be high in layer silicates and feldspars and the soils on them tend to be clayey, thin, dark and slowly permeable (Oganesyan and Susekova 1995). Some shales contain pyrite and other sulfurous minerals, leading to very acidic soils that are dominated by kaolinite. Other shales are high in carbonates and bases, with illite and smectite clays. Hence, a soil that develops on acid shale may be quite different from one on calcareous shale (Table 8.1). Weathering, lessivage and other pedogenic processes may initiate earlier in acid residuum, prompting thicker sola and more clay formation and translocation.

Sandstones, commonly >60% sand and usually dominated by quartz, weather to sandy residuum. The rate of weathering largely depends on the nature of the cementing agent. Most commonly, sandstone residuum is acidic and deep, although erosion may, in places, prevent thick residuum from accumulating. Arkose has a high feldspar content, and weathers to a more clayey residuum. Greywacke is a sedimentary rock consisting of angular fragments of quartz, feldspar and other minerals set in a fine-textured base. Soils formed in greywacke residuum are often loamy. Quartz and feldspars are common minerals in residual soils on "pure" sandstone.

Landscapes: the mesa-and-butte desert landscape of the Colorado Plateau

In deeply dissected, arid landscapes, bedrock outcrops are often a pronounced part of the landscape. The Grand Canyon area of the Colorado plateau is a classic example of this type of landscape. In humid regions, weathered rock and soils may occur on all but the steepest parts of such a landscape. In dry climates, however, there is little vegetation to protect slopes from erosion by wind and water and bare bedrock is common.

Sandstone and limestone tend to be resistant rock units in dry climates, and thus they tend to stand up as the more nearly vertical walls of the "canyons". Shale is a weaker rock that retains more moisture after precipitation events and weathers quickly by wetting and drying processes. Thus, it crops out as gently sloping surfaces with shallow Entisols. For example, the solum of the Rizno soil in the figure below is typically about 20 cm thick, above an R horizon. Farther from the main river valleys, in "mesa-and-butte country" the landscape commonly has lower relief, and while it still has rock outcrops, many surfaces have a thin covering of residuum. Much of it is washed down from higher elevations during rainstorms. Any of a number of different profiles can form in this setting. If moist enough for grasses, Mollisols can develop. On older surfaces argillic horizons are common. Clayey materials may have developed into Vertisols.

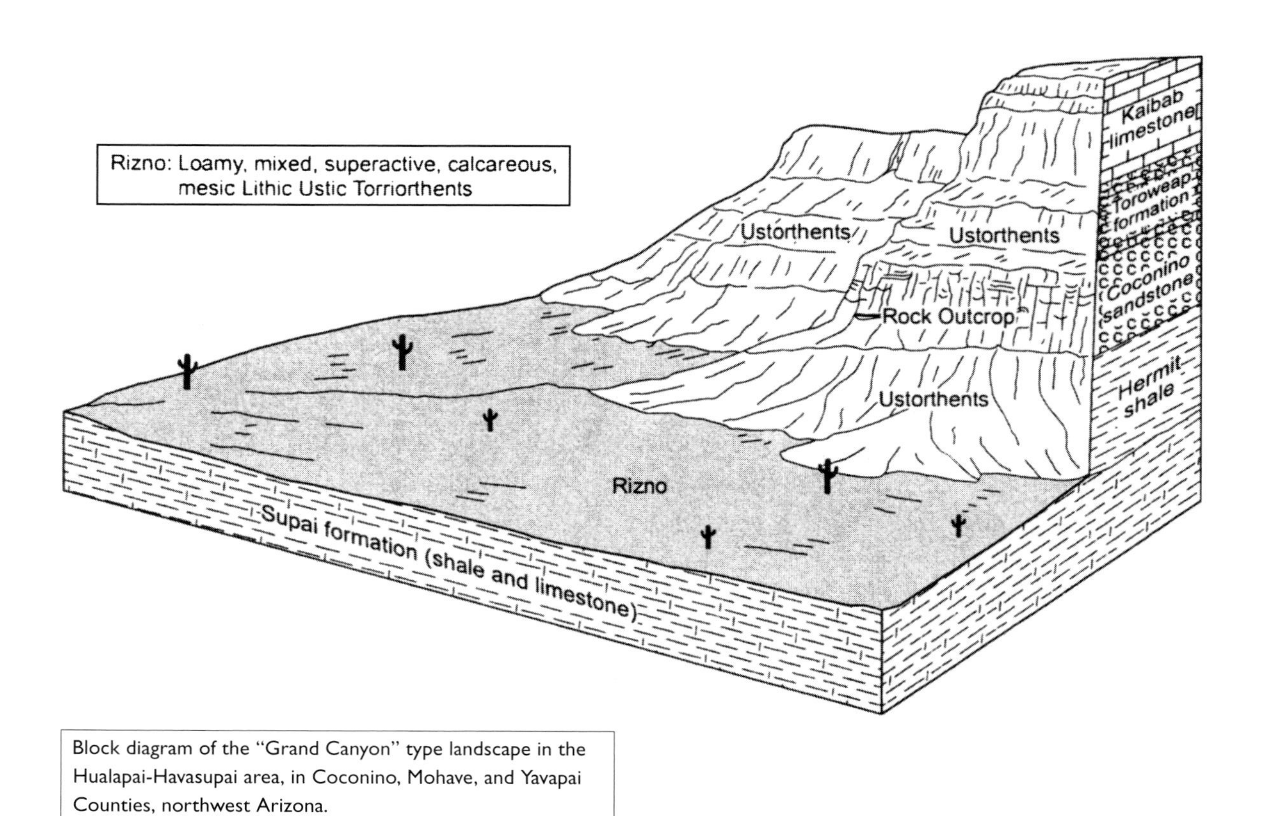

Block diagram of the "Grand Canyon" type landscape in the Hualapai-Havasupai area, in Coconino, Mohave, and Yavapai Counties, northwest Arizona.

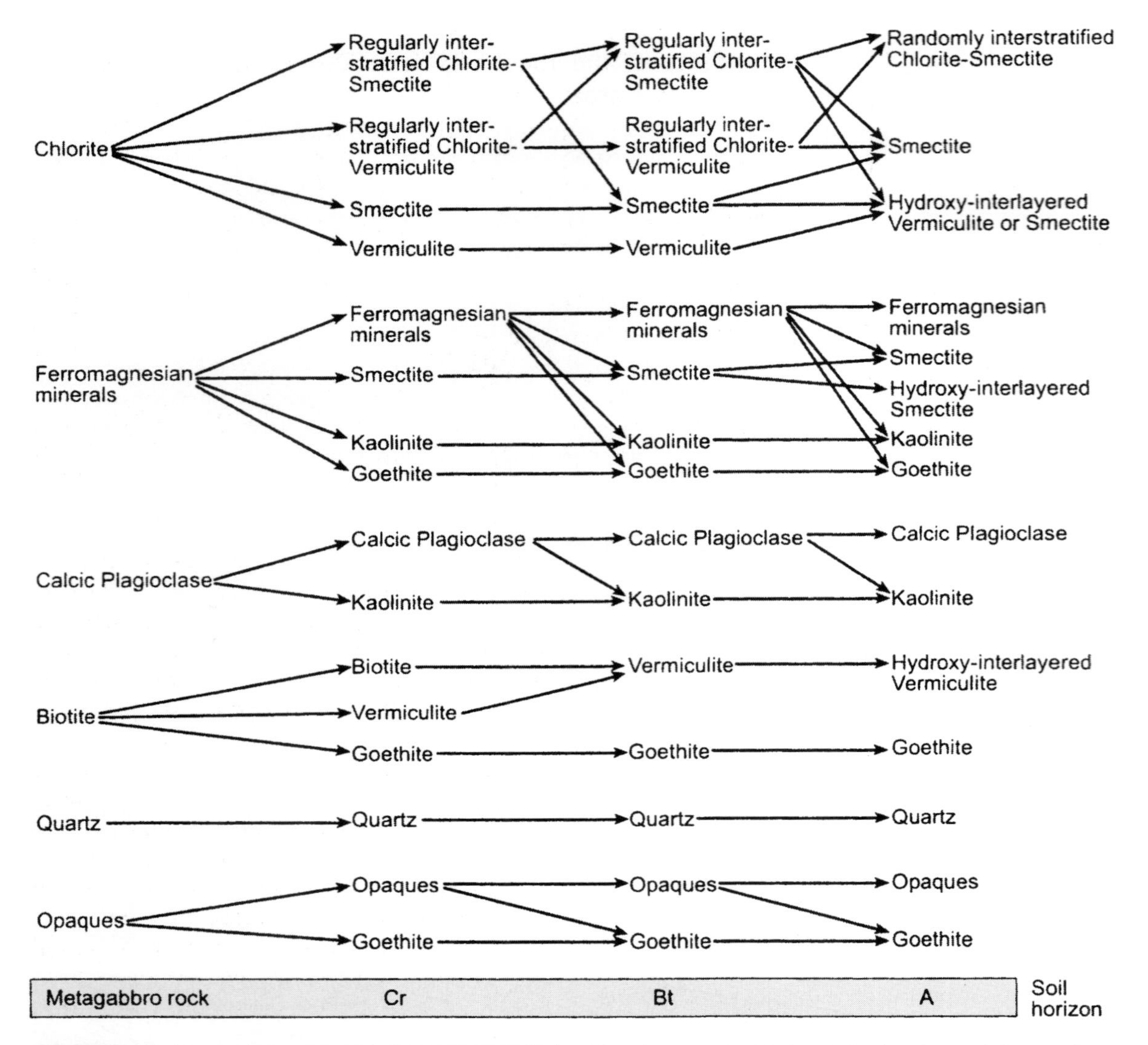

Fig. 8.15 Idealized mineral transformations in a soil profile formed on metagabbro saprolite, in North Carolina. After Rice *et al.* (1985).

Within the carbonaceous family of sedimentary rocks (Table 8.1), chemical weathering (carbonation and hydrolysis) is dominant over physical weathering. Oxidation of Fe, especially sulfides like pyrite, if present in the parent rock, can enhance carbonate dissolution (Atalay 1997). Limestone is more weatherable and soluble than dolomite, but both produce little residuum. Insoluble residue contents of 1% or less are common (MacLeod 1980, Danin *et al.* 1983). Most, if not all, of the $CaCO_3$ in the limestone leaves as soluble bicarbonate (HCO_3^-), with the Ca and Mg cations being biocycled into the soil, or leached. Usually, these rocks contain some chert fragments and smaller amounts of insoluble residue, which end up being a component of the residuum (Barshad *et al.* 1956, Rabenhorst and Wilding 1984, Moresi and Mongelli 1988). Residue from limestone and dolomite is often clay- and silt-rich, having been a component of the sea floor when the carbonates were precipitated or deposited (Plaster and Sherwood 1971, Rapp 1984, Crownover *et al.* 1994). Limestone residuum is typically dominated by illite clays, although aragonite, siderite and quartz clays are also common (Plaster and Sherwood 1971, Oganesyan and Susekova 1995). Some of the

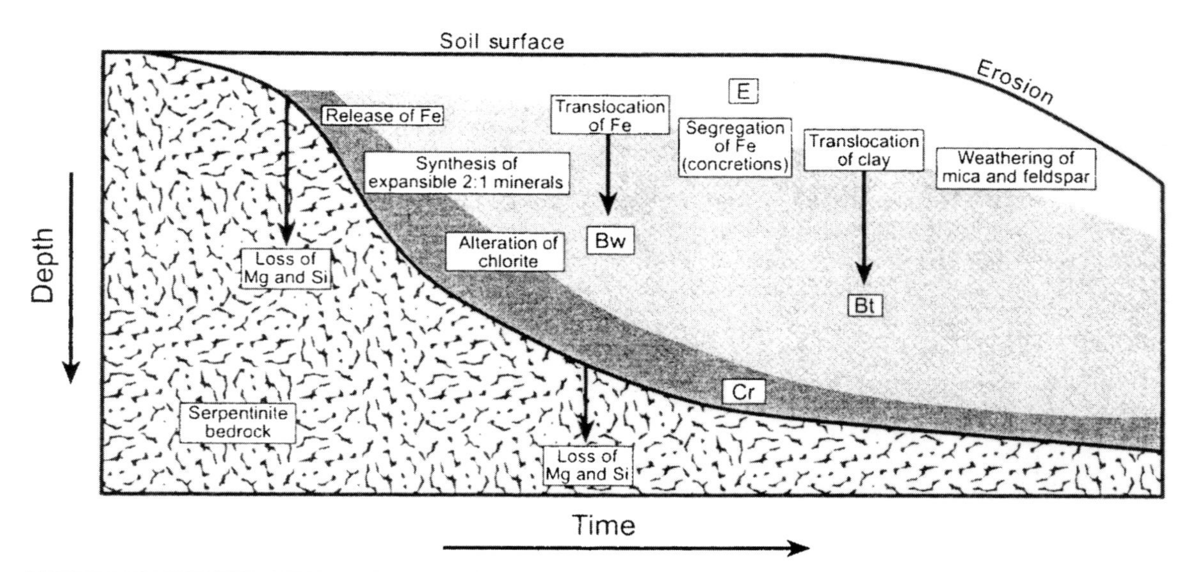

Fig. 8.16 Generalized model of soil genesis in serpentinite residuum. After Rabenhorst *et al.* (1982), and modified with data from Bulmer and Lavkulich (1994).

quartz may have been deposited as a silica gel, to later dehydrate to chert, jasper and flint (Buol *et al.* 1997). Goethite also has been reported in the sand fraction of limestone residuum (Plaster and Sherwood 1971).

Limestone weathers along preferential seams or pockets, where water and various acids can easily penetrate. Once developed, these pathways become preferred sites for additional weathering. Thus, the classic red residuum of limestone typically occurs in pockets, deep cracks and fissures, often with bare rock between (Atalay 1997). In Europe, the rocks that stand up, between wide, deep cracks in limestone, are called lapis (Danin *et al.* 1983). Frolking *et al.* (1983) described a brown clay zone, several millimeters thick, that lies between the red clay and the fresh bedrock. In other cases, the contact is clean and abrupt (Atalay 1997). It is often assumed that clayey, red, weathered material washes off the rock into the fissures, keeping the rock relatively fresh and free of residue, while at the same time facilitating the accumulation of residuum in pockets (Fig. 8.17). On marls and other carbonaceous rocks that have few fissures and cracks, more water runs off, leading to erosion and thin residual soil profiles (Verheye 1974).

The red color of limestone residuum is attributed to *rubefaction* (*rubification*), or reddening, of iron-bearing minerals within (Benac and Durn 1997). Rubefaction is enhanced by aeration, warm temperatures, a wet season sufficient to permit growth of microorganisms (specifically iron bacteria) and rapid turnover of organic matter (Glazovskaya and Parfenova 1974, Boero and Schwertmann 1989). Fractured limestone uplands in Mediterranean climates are well suited to provide these factors. Iron compounds (e.g., oxyhydroxides) released by weathering are precipitated as iron hydroxides, ferrihydrite and hematite, coating clays and imparting red color to the soil and residuum (Glazovskaya and Parfenova 1974, Bech *et al.* 1997). Once formed, these compounds tend to persist, even in the solum.

Many studies of the red, clay-rich residuum that overlies limestone and dolomite focus on the extent to which it is "pure" residuum vs. outside "contaminants," most notably from eolian influx. To resolve this question, researchers have examined the silt and clay mineralogy, and texture, of the insoluble fraction in the parent rock and compared it to that of the residuum (Glazovskaya and Parfenova 1974, MacLeod 1980, Frolking *et al.* 1983, Muhs *et al.* 1987) (Table 8.2). Questions asked include:

- Do certain minerals increase or decrease in the residuum vis-á-vis the rock?

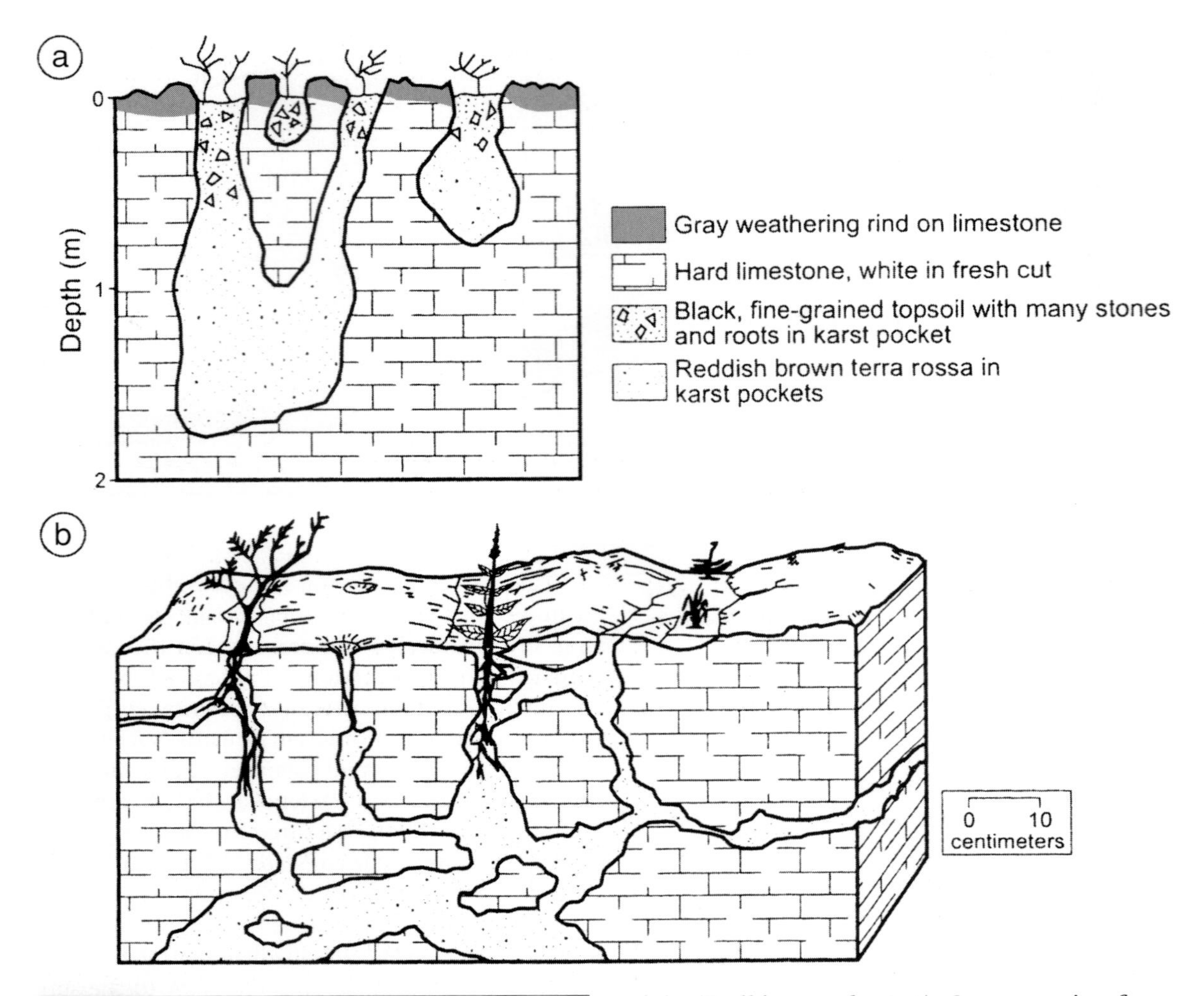

Fig. 8.17 The pattern of soil–residuum–rock in fissured outcrops of limestone. (a) After Rapp (1987). (b) After Atalay (1997).

- Are some minerals in the regolith not present in the rock?
- If not, could they be weathering products of minerals in the rock or must they have come from outside the system?

For example, Moresi and Mongelli (1988) noted that the main effect of chemical weathering on limestone in Italy was the transformation of illite to kaolinite in the red residue. They also reported on SiO_2 contamination of the residuum from biologic (plant opal) sources. Chert-rich stone lines in the rocks that continue into the regolith are also strong evidence for a purely residual origin (Frolking *et al.* 1983). Quartz grains from eolian sources may have different shapes and surface textures than those taken directly from the bedrock; these data can also assist in identifying the source of the sediment (Frolking *et al.* 1983, Rabenhorst and Wilding 1986a, Levine *et al.* 1989). Figure 8.18 illustrates this point with SEM photomicrographs of silt grains removed directly from limestone and from the A horizon of the same profile which has formed in limestone residuum *and* loess. The former grains have an obvious euhedral shape, typical of authigenic quartz crystals formed in voids in limestone (Rabenhorst and Wilding 1986a). The latter, however, are pitted and rounded, presumably due to eolian transport (see also Frolking *et al.* 1983).

Limestone residuum, like that of calcareous shales and marls, is saturated with bases when it

Table 8.2 Characteristics of dolomite residuum, overlying red clay and loess in Wisconsin uplands

Characteristic (units)	Dolomite residuum[a]	Red clay	Loess
Sand content (mean %)	10.1	5.2	1.4
Silt content (mean %)	43.0	14.7	70.0
Clay content (mean %)	47.0	82.0	28.0
Fine clay content (mean %)	34.0	70.0[b]	16.0
Smectite content[c]	1.1	4.5	4.7
Vermiculite content[c]	1.1	1.9	2.7
Mica content[c]	3.6	1.5	1.5
Kaolinite content[c]	1.0	2.4	1.7
Quartz content[c]	4.2	2.0	1.7
$I_{100}:I_{101}$ ratio[d]	0.70	1.02	0.23

[a]Dolomite residuum represents material taken directly from the rock.
[b]Much of the fine clay could have been illuvial.
[c]Mineralogy is an average of the fine and coarse clay fractions, based on XRD peak areas. 5 = >50%, 4 = 31–50%, 3 = 16–30%, 2 = 5–15 and 1 = <5%.
[d]Ratio of the 100 and 101 XRD peaks for the 1–10 μm quartz fraction. This ratio measures the relative concentration of euhedral quartz grains in a sample. Assumedly, chert weathers to euhedral grains, whereas eolian loess has primarily angular quartz grains with pitted surfaces. Higher $I_{100}:I_{101}$ ratios indicate more chert weathered from the dolomite, and less loessial influence.

Source: Frolking *et al.* (1983).

weathers out of the parent rock. This residuum may eventually become leached, depending on climatic and biotic conditions at the site. Certain pedogenic processes, e.g., lessivage, will be inhibited until bases are removed and the pH drops below a certain point (see Chapter 12). Melanization will be more restricted to the upper profile, due to the strong Ca–humus bonds that develop, rendering organics relatively immobile. For these reasons, soils formed from limestone residuum often are less developed and have thinner sola than comparable soils developed on other parent materials (Ehrlich *et al.* 1955). Although limestone is a highly weatherable and unstable rock in humid climates, fragments may persist for long periods of time in the residuum as infiltrating water takes preferred pathways around these relicts. The decarbonation effect created by infiltrating, acidic water is counteracted by the continual release of bases at the limestone–residuum contact and from limestone fragments within the soil, as the rock weathers (Ehrlich *et al.* 1955). If clay translocation does occur, much of the illuviated clay may be located in a thin layer near to, or immediately on top of, the limestone.

Organic parent materials

Organic materials released from vegetation accumulate faster than they can decompose where decomposition rates are slow and where they cannot be mixed with mineral matter. When organic materials accumulate to great thicknesses the soil classifies as a Histosol (Soil Survey Staff 1999). Organic soil materials occur in various stages of decomposition, from raw vegetable matter, through peat, mucky peat, peaty muck and muck (Malterer *et al.* 1992) (Table 3.2). The degree of decomposition is, ultimately, dependent upon a number of factors, including (1) supply of oxygen to the organic materials, (2) water chemistry, especially pH, (3) temperature, (4) biodegradability of the plant matter, (5) amount of faunal activity, such as worms and beetles and (6) the time available for decomposition.

Histosols can form on upland or lowlands (Ciolkosz 1965, Daniels *et al.* 1977). They are most

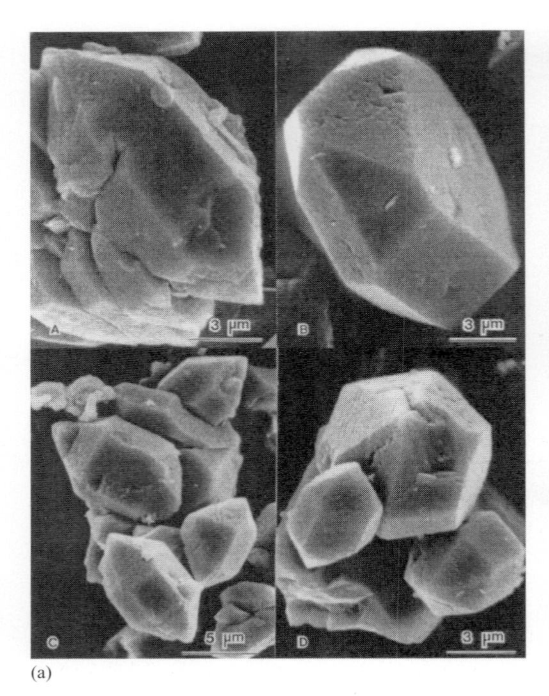

(a)

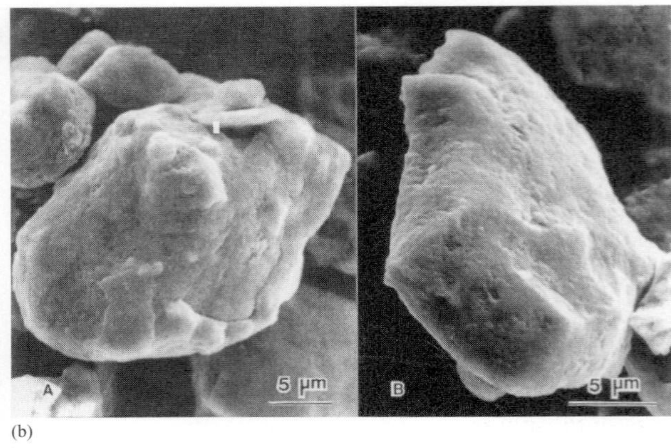

(b)

Fig. 8.18 Photomicrographs of quartz grains from a Texas soil. After Rabenhorst and Wilding (1986a). Photomicrographs by M. C. Rabenhorst. (a) Euhedral prismatic quartz grains from the medium silt (5–20 µm) fraction of the non-carbonate residue of a Texas limestone. (b) Silt-sized quartz grains from the upper solum, showing rounded edges and pitted surfaces, possibly indicating a loessial origin.

common in wet, cool sites such as bogs, fens, swamps, marshes and tidal backwaters (Gates 1942, Heinselman 1970, Bockheim *et al.* 1999) because the rate of decomposition is slowed by wet, waterlogged, and sometimes anoxic (anaerobic) conditions (Rabenhorst and Swanson 2000). Although many wet sites are highly productive from a biological perspective they also tend to have low rates of decomposition, ultimately spelling out the conditions for organic matter accumulation. The organic materials in Histosols may be a mix of many different kinds of plant material, or they may be dominated by woody debris, *Sphagnum* moss, heather, grasses and sedges, depending on the site. Over time, organic materials may continue to accumulate and thicken or they may slowly decompose as site conditions change. The latter situation is likely to occur as the climate warms or dries, and water tables fall. Many Histosols burn when they are dry, retaining layers of charcoal indicative of dry paleoclimatic periods. Most organic soil materials preserve within them even more detailed records of past plant occupation, fire history and environment, which impacts the soil forming in them for (potentially) millennia, in the form of plant macrofossils and pollen (see Chapter 15).

On bare bedrock surfaces, even on uplands, organic materials can accumulate and thicken, providing that the climate is cool (Fig. 3.10). Folists (organic soils that are rarely saturated with water under natural conditions) can occur in cool, high mountains, where they form directly on bedrock or fragments of bedrock (Witty and Arnold 1970). They are essentially an overthickened O horizon composed of leaves, twigs, root residues and other organic materials (Bochter and Zech 1985).

An historian once described the efforts of men fighting fires on Isle Royale in Lake Superior, in 1935, where Folists are common on basalt bedrock, "The soil is leaf mold and humus, lying in a shallow layer over clay and rock. The soil itself burns." The lack of mineral soil material below the O horizon precludes mixing with sands, silts and clays and in places keeps the litter mat drier than it would otherwise be, thereby slowing decomposition. On surfaces that are not bare rock, Folists can occur as organic drapes on large boulders that stand above the mineral soil (Fig. 3.10). Schaetzl (1991a) reported on Folists on Bois Blanc Island, a cool site in Lake Huron. Here, acid organic materials have accumulated in and on top of coarse beach gravel dominated by limestone. The high pH assists in the formation of Ca–humus precipitates (specifically, Ca precipitates the polyphenolic precursors of humic and fulvic substances), slowing decomposition of the organic materials (Bochter and Zech 1985). Although organic materials are, technically, the parent material of Folists, most plants are affected by and are able to utilize the mineral material below, especially if it is fractured bedrock or skeletal materials (Alexander 1990).

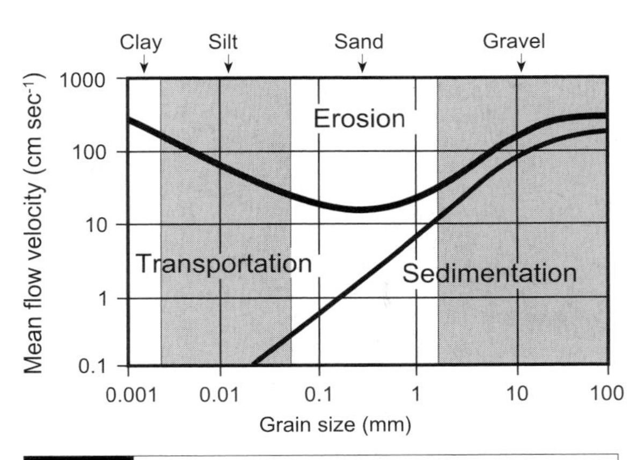

Fig. 8.19 The Hjulstrom diagram. This diagram depicts water/current speeds necessary for transportation or deposition of various grain sizes. The upper portion depicts velocities needed to erode stationary deposited particles.

Transported parent materials

Worldwide, more parent materials are transported than are residual (Hunt 1972). For transported parent materials, their origin and mode of transport have altered their character from where they originally were. Many become intimately mixed as a result, while others are nicely sorted by particle size and/or mineralogy. The origins of transported materials are often revealed through an understanding of their sedimentology, mineralogy and geomorphology. For example, eolian sand has a distinctive texture and sorting, and is typically found in dunes. Alluvium tends to be stratified. The categorization of transported parent materials below illustrates one possible grouping.

Transported by water

Materials transported and deposited by running water in streams are called *alluvium*. All alluvium is first entrained (picked up) by the stream, transported and then deposited. Understanding these processes can be of benefit in the interpretation of alluvial parent materials (Friedman 1961). Other sediment that has some connection to water includes marine sediment (oceans) and lacustrine deposits (lakes). Most alluvium is deposited within stream channels, on floodplains and in deltas and alluvial fans. Thin, scattered alluvial deposits may also occur at the base of slopes, the result of sheetflow (non-channelized water).

Transport of sediment by water is primarily a function of the sediment size and the energy (velocity) of the stream, although other factors such as the water's viscosity and turbulence are also important. Coarse sand, gravel and larger stones are difficult to pick up (entrain), as are most clays, because their plate-like structure creates strong cohesive forces (Fig. 8.19). Thus, it is easier for a stream to entrain silt than either sand or clay. Transportation, on the other hand, is more directly proportional to the size of the sediment than is entrainment. Most clays will settle out of suspension only in calm water, whereas even the finest sand requires velocities of 1 cm s^{-1} to avoid deposition.

Most alluvial deposits are stratified (layered) to some extent. Lack of stratification implies either deposition in turbid, swirling waters, and subsequent re-entrainment, or commonly, some form of post-depositional mixing. Coarse-textured sandy deposits, such as mid-channel bars, are often stratified but cross-bedded (Fig. 8.20). Finer, overbank deposits are silty to clayey, and finely stratified in more horizontally lying beds. Levee sediments are also stratified, but somewhat coarser than overbank sediments on the floodplain proper. In oxbow lakes and backswamp areas on floodplains, clays and silts can settle out of calm water. Point bar deposits are formed as the stream migrates across the floodplain and, like so many fluvial materials, often possess a fining upward sequence. This occurs because larger particles are deposited first, and as the energy of the river decreases during

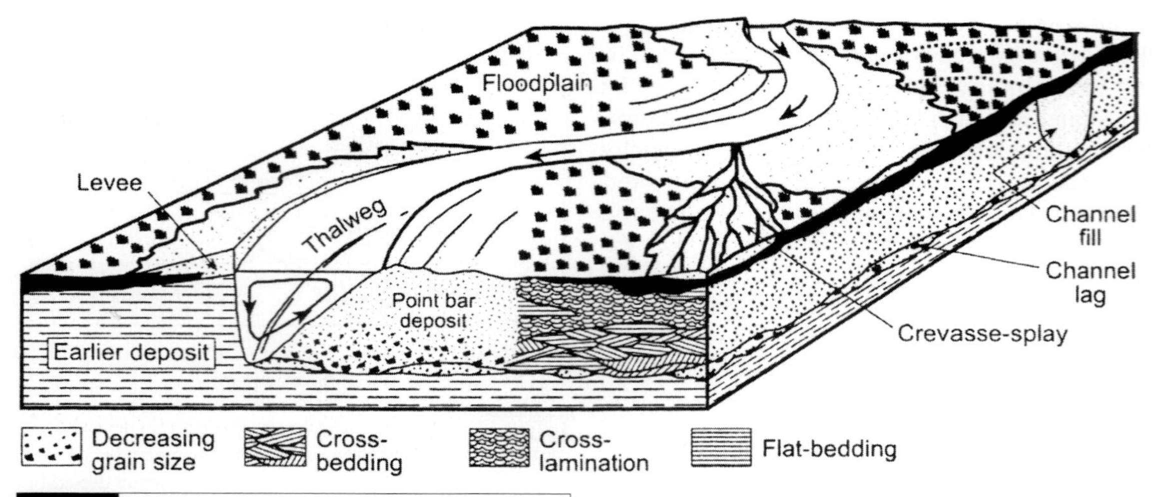

Fig. 8.20 Types of depositional systems and deposits in a meandering river. After Allen (1970).

a depositional episode, continually smaller and smaller particles are deposited until at the top of the sedimentary sequence, only the finest materials remain (Baker *et al.* 1991). Coarsening upward sequences are rare, for they would imply that the river was gaining in energy and velocity while building its channel. The young age of most upwardly aggrading alluvial landforms such as floodplains points to their classification as Entisols (Fig. 8.21).

Because of their stratification, alluvial deposits get progressively younger towards the top of the deposit. They also can exhibit many sudden changes or breaks in sediment texture, packing or bedding, along strata. These sedimentologic "breaks" can dramatically affect pedogenesis. Water (wetting fronts) carrying dissolved and suspended materials tends to stop or "hang" at such discontinuities, especially where coarse material overlies finer sediment (see Chapter 12). In such a situation, the infiltrating water may not enter the coarser material below until the fine material above is saturated, or nearly so. Thus, the contact (discontinuity) point and the sediment above are preferred sites for deposition of materials translocated in infiltrating water.

Alluvial facies can be continuous for many kilometers, or they may be highly discontinuous.

Fig. 8.21 An alluvial valley in Monroe County, West Virginia, in the Ridge and Valley Province. Fluvents occupy the floodplain. Photo by D. L. Cremeens.

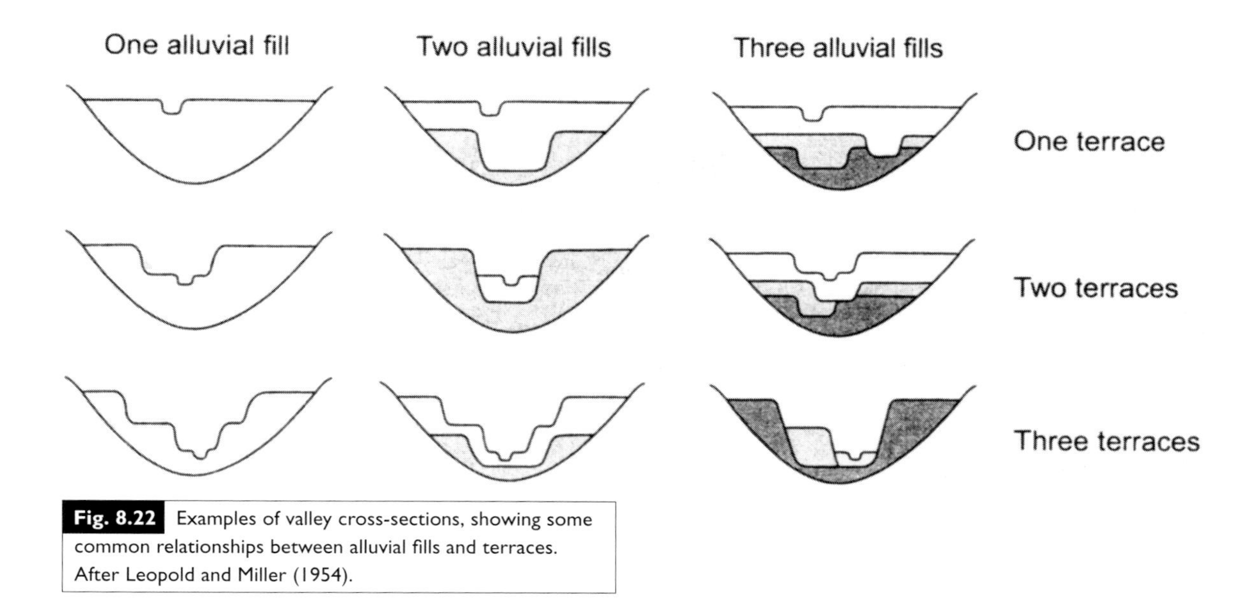

Fig. 8.22 Examples of valley cross-sections, showing some common relationships between alluvial fills and terraces. After Leopold and Miller (1954).

Overbank flood deposits could extend for many kilometers, while a mid-channel bar sand deposit might only be a few meters wide. Alluvial fan deposits are also notorious for being highly variable across space, as the river channel migrates laterally with every depositional event.

Many streams in North America and Europe carried glacial meltwater during the waning phase of the last glaciation. Once large valleys choked with coarse sediment, these streams today are *underfit*, carrying much less water than their large valleys would suggest. Many of these streams filled their valleys with sediment, only to later deepen and incise them when the large influx of glacial debris no longer was a factor. Post-glacial incision of some of these meltwater (and many other) valleys left high terraces behind (Fig. 8.22). Soil development on these terraces can provide clues to the timing of terrace abandonment (Howard *et al.* 1993).

Lacustrine (lake) deposits can be highly variable in texture but often show excellent stratification (Fig. 8.23). In deep lakes, clayey sediments can be deposited. This was and is especially common in proglacial lakes, which had a constant supply of silt and clay-sized material from meltwater. In glacial lakes and even some non-glacial lakes, freezing in winter coupled with a lack of sediment influx allows even the finest clays to settle out of suspension. In summer, the deeper waters of the lakes are more turbid and have slow currents, allowing only silts and coarser materials to be deposited. Thus, an annual "couplet" of fine, dark clay (the dark colors originate from finely comminuted organic matter) and lighter silts (with some quartz) get deposited annually. These annual couplets are called *varves* (Bradbury *et al.* 1993, Schaetzl *et al.* 2000).

In the shallower parts of lakes, wave action can be a major erosive force, beveling off any high spots on the bottom of the lake. Eventually, the bottoms of many lakes can become quite flat, due to wave beveling in shallow parts of the lake, coupled with infilling of the deeper sections by lacustrine sediments (Clayton and Attig 1989).

Offshore sand bars are an important component of many lacustrine environments. In glacial Lake Saginaw in southern Michigan, for example, these sandy deposits represent some of the only coarse-textured sediment on the lake plain. Many of these sandy deposits have since been deflated and later deposited as dunes (Arbogast *et al.* 1997). These sandy soils are often the primary sites of well-drained, upland soils on lake plains.

Lakes and certain parts of floodplains rich in carbonate rocks may develop thick deposits of marl, a light-colored, carbonate-rich form

Fig. 8.23 Clayey lacustrine sediments in the C horizon of a soil in northern Michigan. These sediments were deposited in glacial Lake Algonquin. The light areas are carbonates. Knife for scale. Photo by RJS.

of organic–limnic sediment (Haile-Mariam and Mokma 1990). Shells and shell fragments are common indicators of marl, which is finely divided calcite (Johnston *et al.* 1984).

Desert lakes (playas) typically are also very flat, and usually contain salt deposits and saline soils in their centers (Peterson 1980). Eolian dust that blows out of these playas during dry periods is important to soil development on the nearby uplands (Chadwick and Davis 1990).

One group of widespread parent materials associated with coasts is coastal plain deposits. These sediments form in association with current and former coastlines, thereby including former barrier islands, deltas, beaches and bars. Most coastal plain sediments are sandy, and the landscapes that result after the sea level falls and exposes the sediments subaerially are of low relief.

In the coastal system, many additional sedimentary environments can be isolated, which have saline and brackish conditions; tidal activity adds additional complications. Estuaries are drowned river valleys usually caused by Holocene rises in sea level. In order to keep estuaries from filling with alluvium, rivers must deposit sediment at rates that do not exceed the transportational abilities of waves and currents. Otherwise, the estuary will fill and a delta will slowly form. Estuarine facies include muds, silts and sands; farther seaward, tidal deposits are common. At the heads of estuaries, fluvial sands can be found, which grade into finer-textured sediments in the estuary proper.

In certain areas along the sea coast, sediments can become rich in sulfur and the sulfur-bearing mineral pyrite. Through a process called *sulfidization*, these sediments develop into acid sulfate soils, with pH values below 3.0 being common (Fanning and Fanning 1989) (Fig. 12.76). The process is initiated as sulfur-bearing sea water comes into contact with tidal marsh sediments, usually in "backwater" areas and not on the "open" coast proper (Darmody and Foss 1979) (see Chapter 12).

Deltas are perhaps some of the more complicated fluvial depositional systems. Their character can vary from sandy through clayey, not only across the delta but vertically within it. Ages of the sediment also vary greatly, as the major distributary channels shift from one side of the delta to another over time. In general, delta sediments are deposited in wedges that get thinner and finer-textured farther from the delta head (or river mouth), to the prodelta. Nonetheless, splays of coarser material can occur at any place, leading to sandy deposits overlying silts, or any imaginable combination.

Oceanic coasts are complex systems of deposition and erosion. On actively eroding coasts, sea cliffs and stacks may provide unmistakable evidence of the power of the sea. Other coasts show evidence of ongoing deposition and reworking

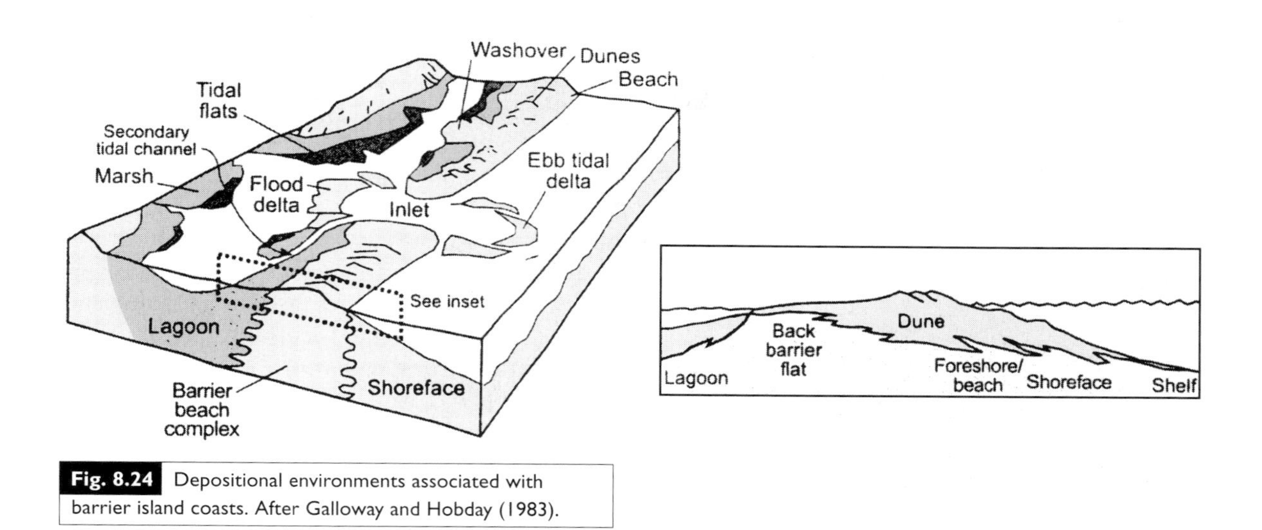

Fig. 8.24 Depositional environments associated with barrier island coasts. After Galloway and Hobday (1983).

of sands by waves and currents. The one variable common to all sea coasts and nearby environments is the recent geologic changes in sea level. Eustatic (global) sea level rise following the melting of the last ice sheets was on the order of 150 m. These changes forced coasts to retreat, and with them offshore barrier islands composed of clean sand, often with dunes at the landward side (Fig. 8.24). Barrier islands separate the ocean from the brackish lagoon. Lagoons are important sedimentation systems for not only fine sediment but also for organics. They often contain organic-rich muds over (deep) sands, which may be interbedded with sands washed over the barrier during storms. On the landward side of the lagoon, brackish marshes and tidal flats accumulate various types of sediments from streams that feed into the coastal system, and acid sulfate soils may form (Fig. 12.76). Position in the marsh generally controls grain size. On tidal flats, sediments tend to be coarser near low-tide lines, and fine near high-tide lines (Daniels and Hammer 1992).

Transported by wind

A variety of soil parent materials are transported through the atmosphere. Some, like eolian sand and silt, are picked up by the wind and moved a certain distance, depending on wind velocity and grain size. Others, like pyroclastic materials (fragments ejected during a volcanic explosion), either settle out of the air or are moved still farther by wind (Kieffer 1981).

Wind is best capable of entraining and moving small particles, typically sand size or less. Strong winds, coupled with dry sediment of the appropriate size which is not stabilized by vegetation, are optimal for eolian entrainment and transport (Fig. 8.25). Once entrained, larger particles will bounce along (saltate on) the surface, while smaller particles may be lifted up and move in suspension. As saltating grains impact the

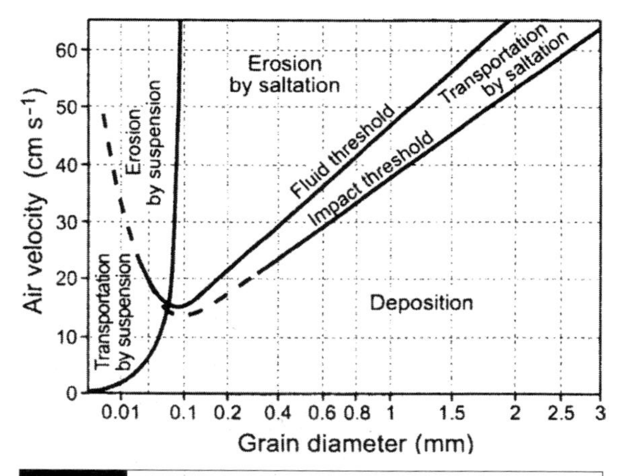

Fig. 8.25 Various threshold velocities needed for wind to erode/entrain, transport and deposit dry particles. After Pye and Tsoar (1990).

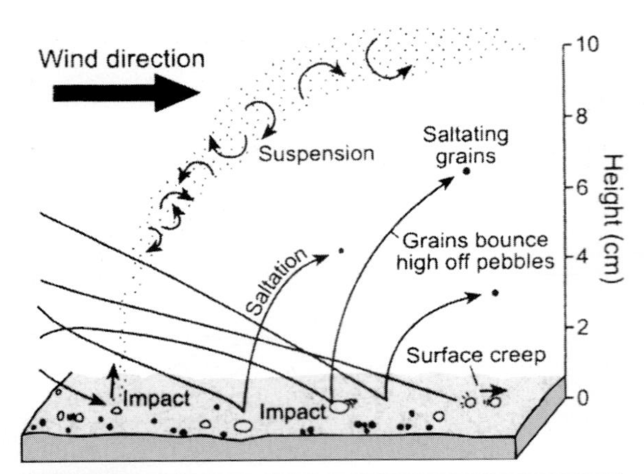

Fig. 8.26 Idealized example of particle movement by the wind: creep, saltation and suspension. After Christopherson (2000).

surface, they dislodge others, which may then begin to move by creep, saltation or in suspension (Fig. 8.26). Erosion by wind (i.e., *deflation*) may remove most of the fines, sometimes resulting in a lag concentrate, or pavement, at the surface (Parsons *et al.* 1992). Stones eroded, grooved and shaped by this type of sand blasting are called *ventifacts* (Sharp 1949, Whitney and Splettstoesser 1982) (Fig. 8.27). When wind velocities drop below a critical value, deposition occurs. The coarsest particles will settle out of suspension (or saltation) first, and as wind speeds drop still more the smaller particles will also be deposited. For this reason, eolian (wind) deposits are usually fairly well sorted.

Fig. 8.27 Ventifacts on an alluvial fan, Death Valley National Park, California. Photo by RJS.

Isolated landforms made of eolian sand are called dunes, or sand sheets or coversand if the sand resembles a "blanket" on the landscape (Cailleux 1942, Muhs and Wolfe 1999, Holliday 2001). Dune fields form when winds above a threshold velocity impact sandy surfaces that lack adequate stabilizing vegetation. The instability is usually caused either by fresh deposition of sand (and hence no vegetative cover) or by fire, drought or other factors destroying stabilizing vegetation. Many coastal dunes are associated with the first scenario.

Like slopes, entire landscapes can either be considered sediment-limited or transport-limited (Kocurek and Lancaster 1999). Sediment-limited situations occur where either (1) the available sediment is of the wrong size for eolian transport, or (2) the grains are not susceptible to entrainment and transport, as may happen when the sand is wet, cemented, covered with lag gravel or vegetated. Transport limitations primarily involve wind speed, which may also be affected by a cover of vegetation. Many of the world's dune fields are not active today because of limitations with respect to sediment supply or, more commonly, transport vectors. Thus, they represent a time in the past when conditions were different, and sand was mobile (Muhs and Zárate 2001). Small shifts in climate (moisture balance, wind strength or land use) might be enough to initiate or reinitiate sand transport (Keen and Shane 1990, Muhs and Holliday 2001). Soils that are forming on stabilized dunes can provide information about when the dunes stabilized and pedogenesis began (e.g., Arbogast *et al.* 1997, Arbogast 2000).

Medium and fine, quartz-rich sands dominate many dunes (Ahlbrandt and Fryberger 1980, Dutta *et al.* 1993, Arbogast 1996, Muhs *et al.* 1999) (Fig. 8.28). Often there is no measurable clay fraction (Arbogast *et al.* 1997). Eolian deposits are often yellow, buff or brown in color, due to thin coatings of oxidized iron and clay on the quartz grains. Sand grains may appear "frosted" and chipped due to impact in transit (Fig.8.18b). Many are well rounded. Experimental evidence suggests that sand grains are continually comminuted as they saltate, becoming more "mineralogically mature" over time (Dutta *et al.* 1993).

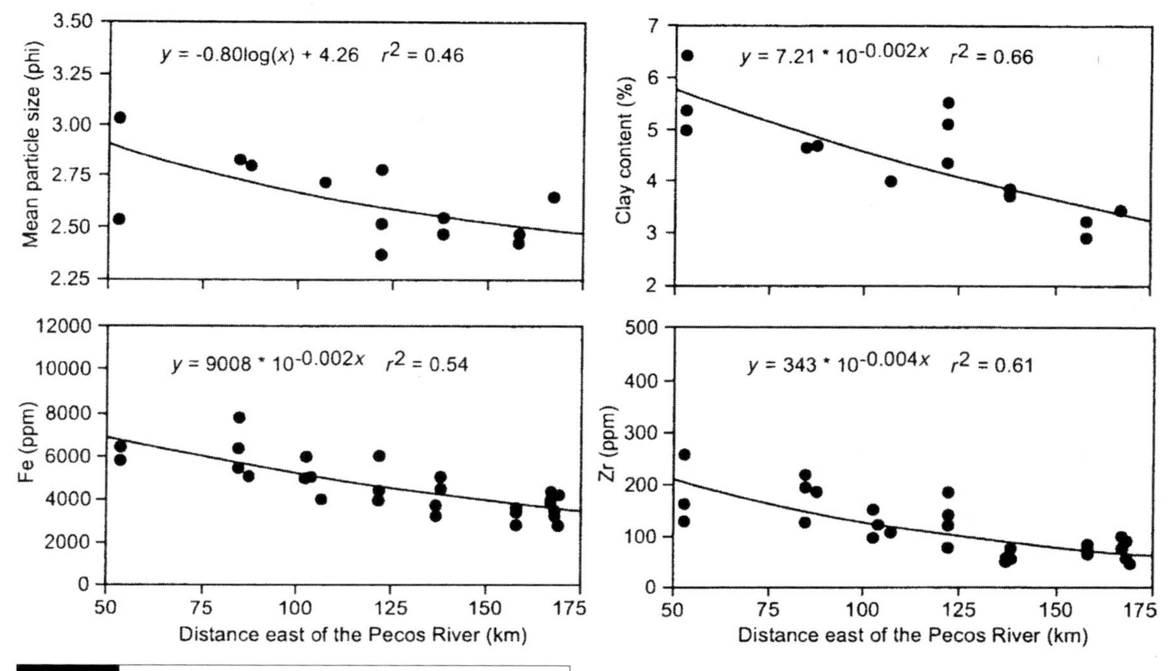

Fig. 8.28 Variation in dune sand characteristics away from the Pecos River, Texas, the sand source. After Muhs and Holliday (2001).

The type of soils that form in dune sand, and the rate at which various ions and minerals are released to the soil by weathering, are largely a function of climate and initial dune sand lithology, age of the deposits notwithstanding (Fig. 8.29). Most soils formed in dune sand are quartz-rich. Determining whether they are low in weatherable minerals because of long-term weathering or because the parent material is quartz-rich is often an important question for dune soils (Dutta *et al.* 1993, Muhs *et al.* 1997a) (Fig. 8.30). Such information is also useful in identifying the ultimate source material and age of the dunes (Muhs *et al.* 1997a, b, Arbogast and Muhs 2000).

Many soils have small amounts of eolian sand in their upper horizons. The evidence for such additions is, however, often elusive. Crawford *et al.* (1983) were able to discriminate eolian sand grains in soils from those that had weathered out of bedrock *in situ*; eolian sand grains were more rounded (Fig. 8.31). Another indicator of eolian materials (sands) in soils is the particle size distribution within the sand fraction; eolian sand is often quite well sorted while sand from other sources may not be as sorted. Similar statements could be made for loessial influence in soils.

Sandy soils tend to form rapidly and their pH can adjust quickly to various state factor inputs, largely because of sand's low surface area (see Chapter 12). Podzolization is a common process on dunes in cool, humid climates; in grasslands melanization is dominant in dune soils. Because dune sand is so quartz-rich and clay-poor, lessivage is seldom a dominant process, although if the dune has some small amounts of clay, textural clay bands known as *lamellae* can form (Oliver 1978, Gile 1979, Berg 1984, Schaetzl 1992b, 2001, Rawling 2000).

Clay-sized particles, if flocculated, can move as sand and form clay dunes (Butler 1956, Bowler 1973, Munday *et al.* 2000). *Parna*, an eolian sediment of dominantly clay size, is common in parts of Australia, where it can attain thicknesses in excess of 8 m (Dare-Edwards 1984).

One of the most widespread types of eolian sediment is wind-blown silt, or loess. Loess is unconsolidated and dominated by silt-size, usually

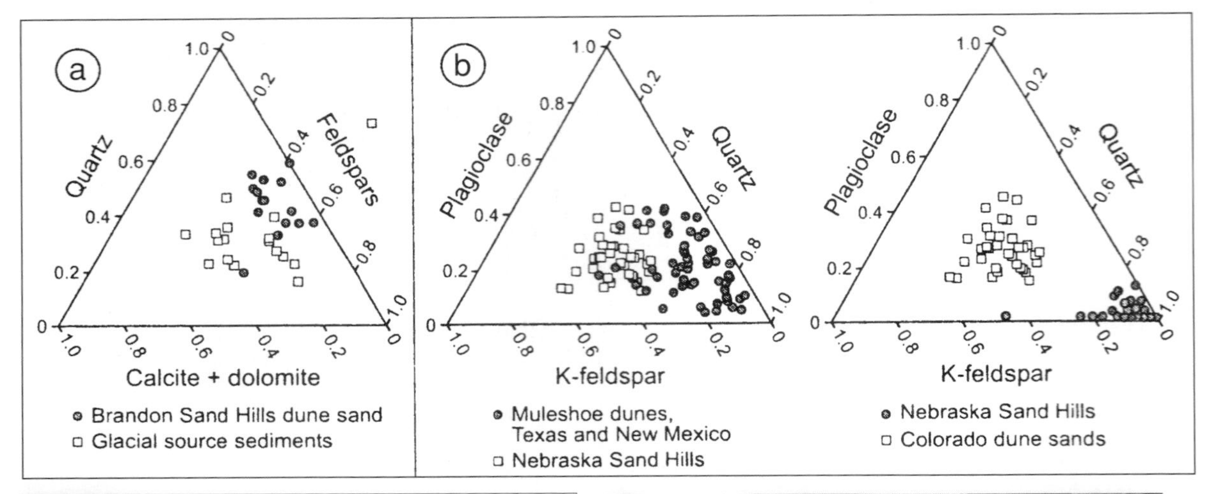

Fig. 8.29 Characteristics of dune sand. (a) Ternary plot of the mineral composition of dune sand vs. the presumed glacial source material. The dune data are from the Brandon Sand Hills, Manitoba. After Muhs *et al.* (1997a). (b) Ternary plots of the mineral composition of dune sand from different dune fields. After Muhs *et al.* (1997b) and Muhs and Holliday (2001).

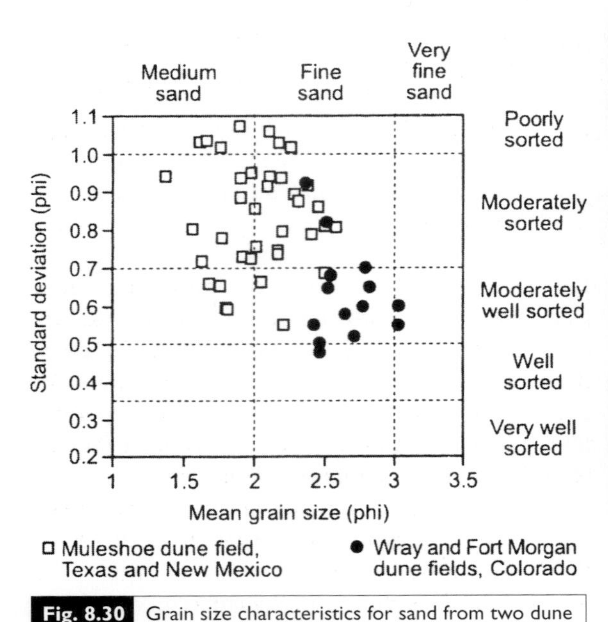

Fig. 8.30 Grain size characteristics for sand from two dune fields in the Great Plains, USA. After Muhs *et al.* (1999).

Fig. 8.31 Photomicrographs of coarse sand from a soil near Point Reyes, California. After Crawford *et al.* (1983). (a) Rounded grains brought to the soil by eolian transport. (b) Angular grains weathered from bedrock, present in the residuum.

quartz, particles. In "raw" form it is usually tan or light brown in color, with 10YR hues dominating. Thick loess sheets are widespread in the mid-continent USA, China, Argentina and eastern Europe (Eden *et al.* 1994) (Fig. 8.32). In the central USA, loess sheets can exceed 50 m in thickness. In exposed faces of thick deposits, loess stands vertically, with pronounced vertical cleavage

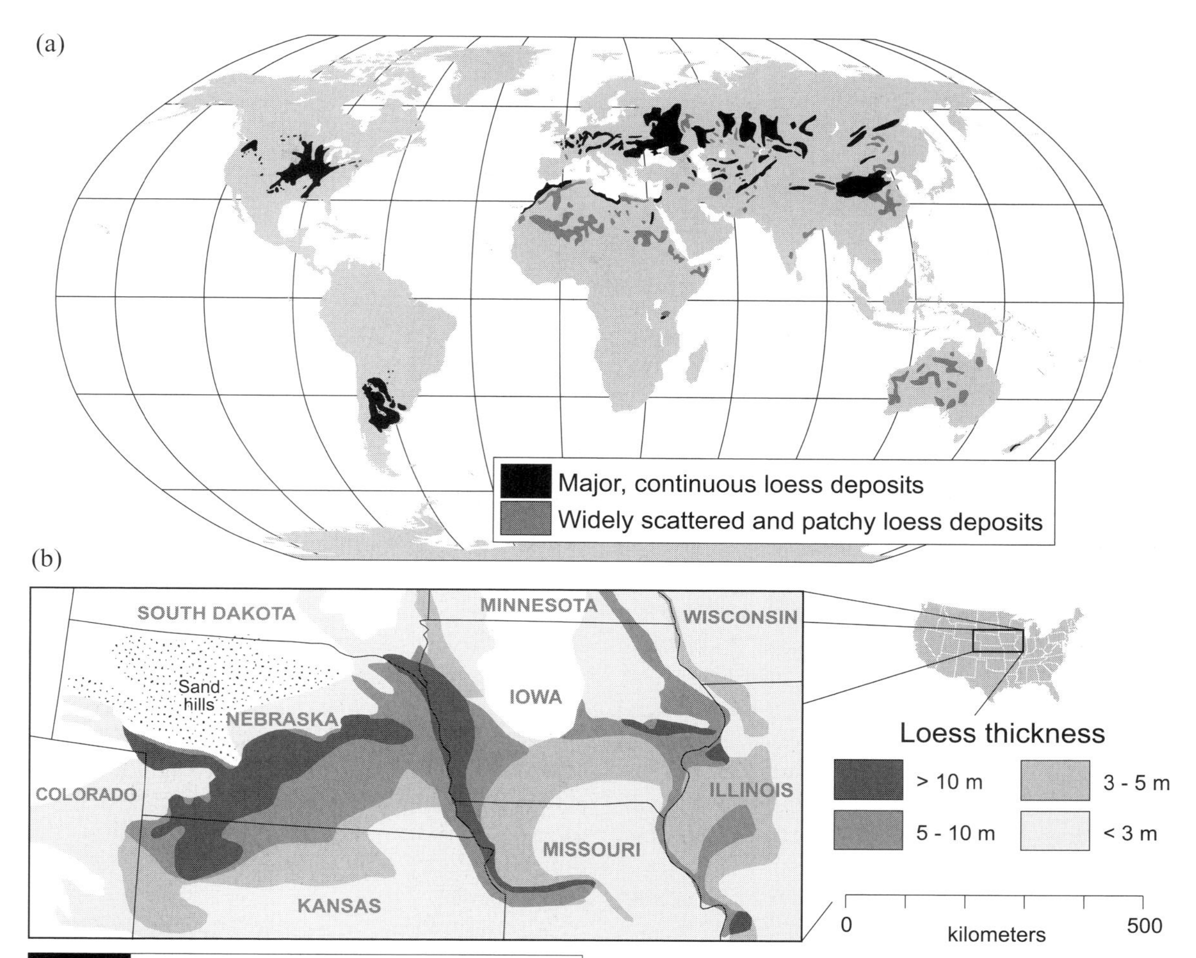

Fig. 8.32 (a) World distribution of loess deposits. After Snead (1972) and Muhs and Zárate (2001). (b) Loess distribution and thickness in the central United States. After Muhs and Bettis (2000).

(Ruhe 1975b) (Fig. 8.33). Coarse loess, deposited just slightly downwind from the source, may be slightly stratified, but most loess is essentially massive and initially calcareous. Loess deposits are easily erodible by water and wind; gullying is not uncommon. Of course, all manner of gradational loess–sand deposits exist where loess sheets interfinger with sand, and vice versa (Mason 2001, Muhs and Zárate 2001).

Loess has two main origins: non-glacigenic (desert) loess and glacial loess. Non-glacigenic loess forms downwind of dry, somewhat devegetated areas, where sand may be blowing into dunes, such as the Nebraska Sand Hills (Fig. 8.32). The silt grains are winnowed and deposited farther downwind, as loess (Yaalon 1969, 1987, 1997b). Non-glacigenic loess in parts of Colorado and Nebraska is presumed to have been eroded from shales and siltstones during cold, dry periods in the Holocene (Aleinikoff *et al.* 1999, Muhs *et al.* 1999). Even deposition rates and transport directions of non-glacigenic loess, however, are impacted by glacial–interglacial climate cycles (Mason 2001). The largest accumulation of loess in the world is non-glacigenic. Loess Plateau sections commonly attain thicknesses of 150 m or more (Huayu and Zhisheng 1998), with over 300 m of loess being reported. The China loess exists downwind from the deserts of Mongolia and China, in the Chinese Loess Plateau (Kukla and An 1989, Ding *et al.* 1992, 1999). Here loess has been brought in by the dry, west winds of the winter Asian monsoon (Lu and

Fig. 8.33 An exposure of loess near Natchez, Mississippi. As shown here, loess typically will stand in vertical faces. Photo by D. L. Cremeens.

Sun 2000) from the dry deserts to the northwest of the Loess Plateau, over the past 2.5 million years (Ding *et al.* 1993, Eden *et al.* 1994) (Fig. 8.34). The non-glacigenic China loess also has a glacial origin, of sorts. During glacial periods increased aridity and expansion of the deserts led to dust influxes onto the Plateau; loess input was much less during cool, moist glacial periods (Ding *et al.* 1995).

In the midlatitudes, equatorward of the margins of the various Quaternary ice sheets, loess is commonplace (Markewich *et al.* 1998). This loess originated with meltwater deposits in river valleys. Silty glacial debris was carried out from the ice sheets, whose grinding was very effective at comminuting larger rocks and sand into silt-sized particles (Assallay *et al.* 1998). Loess is often called glacial flour, because, like wheat, it is physically ground (by ice) into smaller sizes, and because it has the same consistency when dry. The meltwater valleys were filled with milk-colored, silt-rich meltwater in summer. The floodplains of meltwater rivers such as the Mississippi, Missouri, Wabash and Illinois would have been dry and unvegetated in winter. Strong winds associated with the ice sheet blew the silt out of the dry valleys and onto the adjoining uplands. Areas not covered by loess were either under the ice sheet or are distant from source areas.

Five major factors affected the glacial loess sedimentary system: (1) initial coarseness and abundance of the source sediments, (2) width or size of the valley, (3) distance between source area and accumulation site, (4) strength and direction of winds and (5) bluff height and angle (Putman *et al.* 1989). Sand, too large to be moved from the valleys, occasionally accumulated as dunes on the valley floors, only to be destroyed by the next year's floods.

Although loess can be preferentially deposited in some areas to form hills (loess dunes), usually loess simply covers the landscape evenly, like a blanket. These loess sheets are thickest near the source, whether it be a desert margin or a meltwater valley (Smith 1942, Frazee *et al.* 1970, Kleiss 1973, Fehrenbacher *et al.* 1986), and on interfluves (Ruhe 1954). The thinning of loess from the source areas is so predictable that it has often been described mathematically (Simonson and Hutton 1942) (Fig. 8.35). In the northern midlatitudes, loess typically thins to the east and southeast of a source, although in the case of a meltwater valley, there is a secondary but more subtle thinning trend to the west and northwest (Frazee *et al.* 1970, 1986, Ruhe *et al.* 1971, Ruhe 1973, Muhs and Bettis 2000) (Fig. 8.35). The secondary thinning pattern to the west and northwest is usually ascribed to infrequent easterly winds, while

Landscapes: terra rossa soils of the Mediterranean

Terra rossa is an Italian term for soils that are red, shallow, clayey and undifferentiated, usually above calcareous bedrock (Stace 1956). Terra rossa soils are often called "red Mediterranean soils" because they dominate that area, although most "limestone soils" in this region are, in fact, not red (Verheye 1974, Benac and Durn 1997). Terra rossa has also been used to describe the red, residual material that overlies limestone, chalk, marble or dolomite (Darwish and Zurayk 1997). Most terra rossa soils classify as Xeralfs, Ochrepts or Xerolls; many are Rhodoxeralfs (Soil Survey Staff 1999). The red or brown colors are due to iron oxides and hematite (Jackson and Frolking 1982). Many have a significant clay increase at the bedrock interface, where bases released from the weathering rock cause clay to flocculate (Durn *et al.* 1999).

It has long been presupposed that these soils formed in residuum from carbonaceous bedrock, i.e., the clay and other silicate minerals are acid-insoluble residue (Barshad *et al.* 1956, Moresi and Mongelli 1988). This theory would imply that the soils are very old and on exceedingly stable surfaces. The presence of many soils in limestone that are less developed and closely match the mineralogy of the residuum tends to support this model (Glazovskaya and Parfenova 1974). Verheye (1974) noted that terra rossa soils are primarily found on stable uplands, where erosion is minimal; on other landscape positions less-developed, brown and yellow soils occur. Some, however, observed that the red residual sediments are also accumulating in karst depressions as thick deposits and on accretionary slope positions (Benac and Durn 1997).

The residuum theory of terra rossa development has, however, been recently rejected. The arguments against it are:

(1) The sharp contact between terra rossa and bedrock, usually but not always observed (Glazovskaya and Parfenova 1974), is inconsistent with a residual origin (Olson *et al.* 1980).

(2) There is typically more "residuum" in terra rossa soils than can be accounted for by weathering alone (Olson *et al.* 1980, Muhs *et al.* 1987, Durn *et al.* 1999). (For example, at a site in Greece about 130 m of limestone would have to have weathered (and all the residuum stay *in situ*) to produce only 40 cm of red clay (Macleod 1980).).

(3) Cyanobacterial weathering patterns on the limestone indicate that weathering essentially ceases after residuum gets a few decimeters thick (Danin *et al.* 1983).

(4) The clay content and silt : clay ratio in the soils are often different than those of the acid-insoluble limestone residue (Macleod 1980, Durn *et al.* 1999).

(5) The amount of iron in the limestone residuum is often much lower than would be expected, given the red color and high iron contents of the terra rossa soils.

(6) Clay minerals and other elemental oxides in the rock and residue do not match those in the red sediment (Olson *et al.* 1980, Moresi and Mongelli 1988).

(7) The angularity of the silt and sand grains in the residuum match classical eolian particles more closely than those encased in the limestone (Rapp 1984).

An additional line of evidence comes from geomorphology. Overthickened silty sediments in depressional areas and on footslopes attest to the instability of terra rossa landscapes (Wieder and Yaalon 1972, Yassoglou *et al.* 1997); residuum does not accumulate to great thicknesses on such weathering-limited slopes. Yassoglou *et al.* (1997) noted that most terra rossa soils in the Mediterranean region have undergone such severe erosion that the complete profile is rarely found (see also Boero and Schwertmann 1989). Even in the lower-relief landscapes of Midwestern US, terra rossa is primarily debris that has been transported by slope processes (Olson *et al.* 1980).

It has now been determined that much of the clay and silt in terra rossa soils has eolian origins (Rapp 1984). In the Mediterranean region, eolian clay and silt from the Sahara and Ukraine fall out as "mud" or "blood" rain and snow (Lundqvist and Bengtsson 1970, Macleod 1980, Yaalon 1987, 1997a). Dust from the Sahara impacts locations as far north as Great Britain (Pitty 1968, Prodi and Fea 1979). Limestone weathering contributes *some* residuum to the terra rossa soils, but the majority of the rob parent material is eolian.

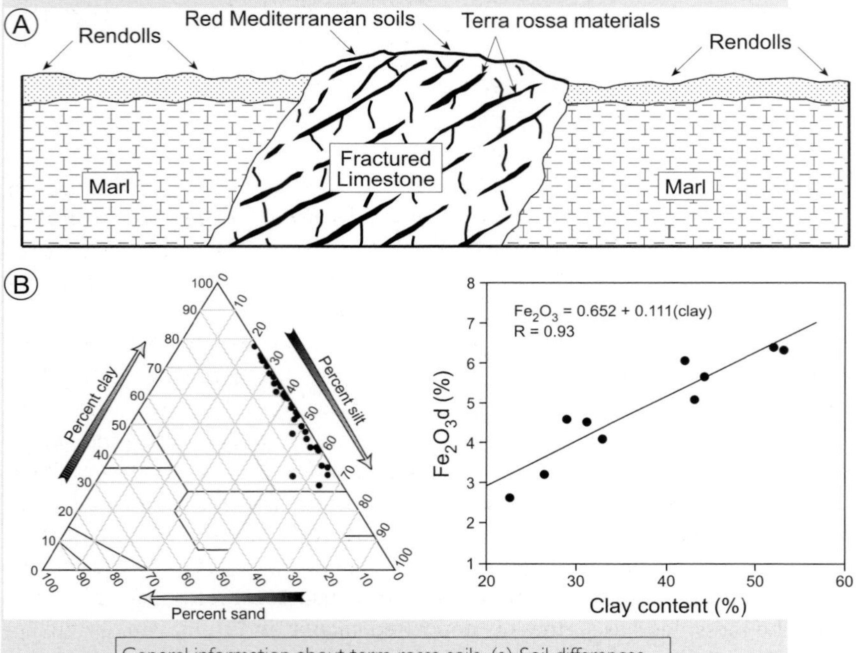

General information about terra rossa soils. (a) Soil differences on marl vs. fractured, hard limestone, in the Mediterranean. After Atalay (1997). (b) Particle size data, plotted on a standard textural triangle, for some terra rossa soils of Croatia and Indiana, and the relationship between clay content and dithionite-extractable Fe content of some terra rossa soils in Greece. After Durn *et al.* (1999), Olson *et al.* (1980) and Yassoglou *et al.* (1997).

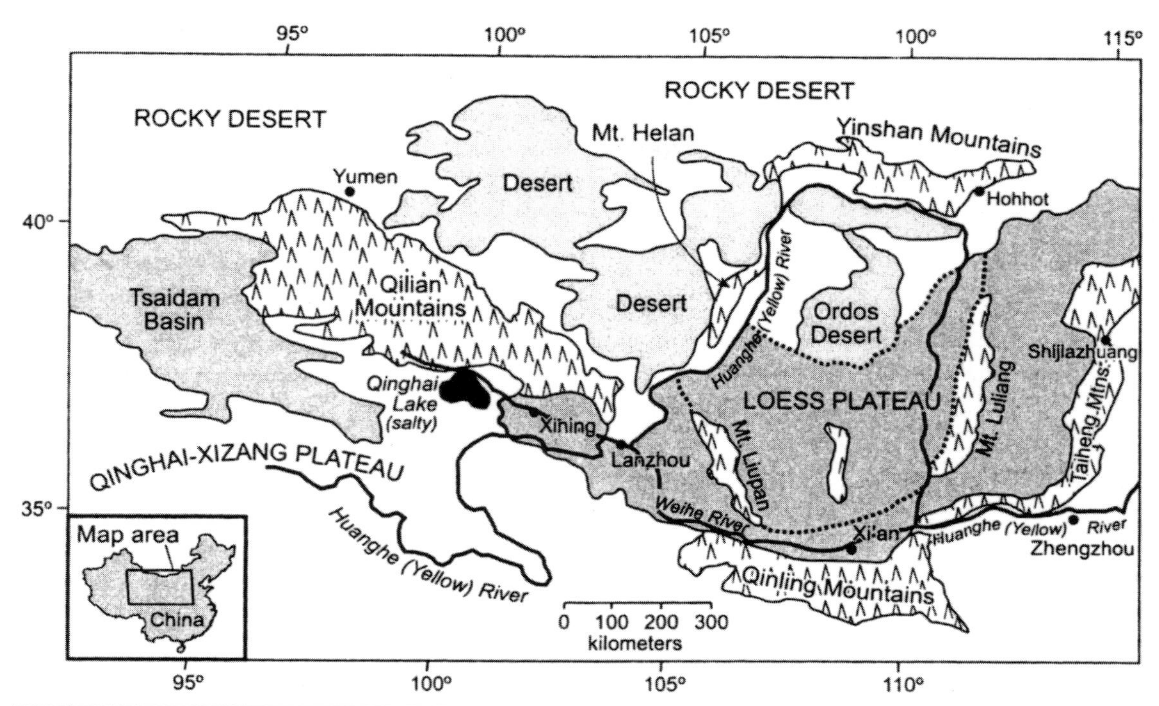

Fig. 8.34 The Chinese Loess Plateau and nearby areas. After Zhongli *et al.* (1993).

the dominant west-to-east thinning pattern of the loess clearly indicates that the dominant winds were from the west (Fehrenbacher *et al.* 1965). Not only does the thickness of loess decrease farther from its source, but it also gets finer and better sorted; the contents of clay and fine silt increase while coarse silt and fine sand contents decrease farther from the source area (Frazee *et al.* 1970) (Fig. 8.36). There is also some evidence that carbonates and other elements in the loess decrease as a function of distance from the source, all of which point to the exceptional sorting abilities of wind (Fig. 8.36).

Complexities associated with the shape/topography of the underlying surface and with source area geography may be reflected in the distribution patterns of loess. For example, Ruhe (1954) showed that loess is least sorted when it overlies summit landscape positions, and is better sorted on slopes. Some evidence has also shown that loess is slightly thicker on lee sides of hills, rather than being a uniform blanket (Simonson and Hutton 1942). Lastly, data from Fehrenbacher *et al.* (1965) and Putman *et al.* (1989) suggest that loess is both thicker and coarser near valley sections that are wider.

Loess deposition rates vary greatly in space and time (McDonald and Busacca 1990). Presumably, loess accumulation rates were maximal when the ice sheet was rapidly melting and when winds were strong (Muhs and Bettis 2000). Thus, loess sheets in many areas are associated with the retreat of glaciers from the headwaters of major drainage basins (McKay 1979, Johnson and Follmer 1989). It is also commonly assumed that loess is deposited earlier and more rapidly on uplands nearer to the source area than on those farther away. For example, in Illinois and Iowa, it is postulated that the earliest additions of loess occurred near the rivers (Ruhe *et al.* 1971). Later, loess deposition enveloped more and more of the landscape, so that at the time of maximal deposition, sites far away were also receiving silt inputs (Ruhe 1969). As loess deposition slowed near the end of the glacial period, sites farthest from the source ceased receiving loessial additions and soils began to form. The last sites to receive loess

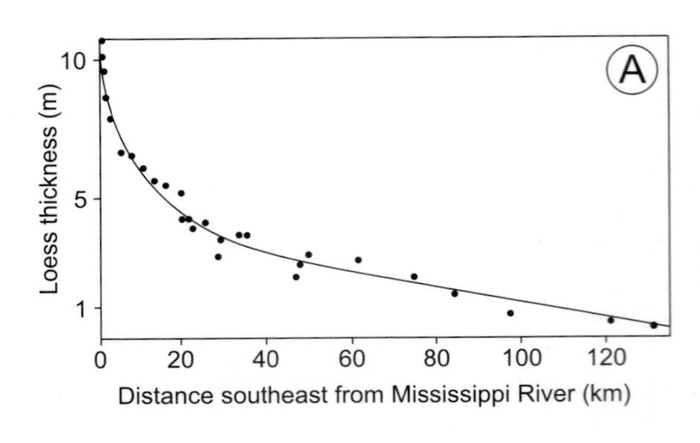

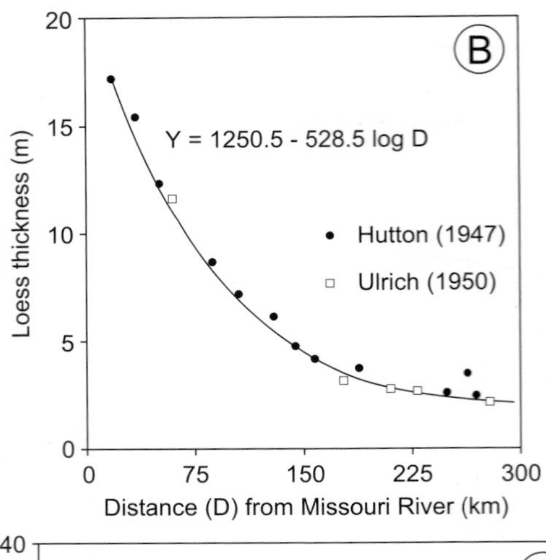

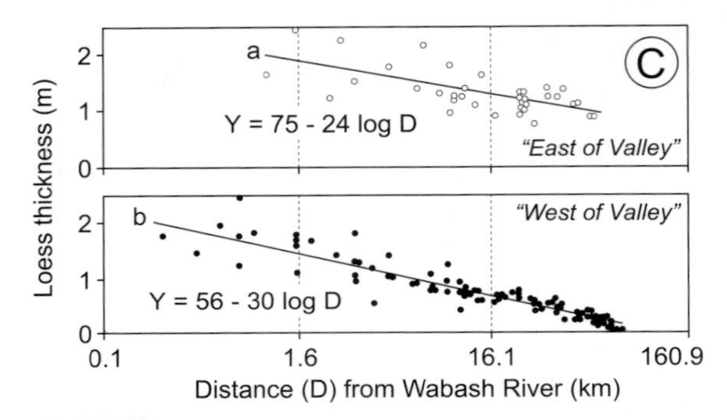

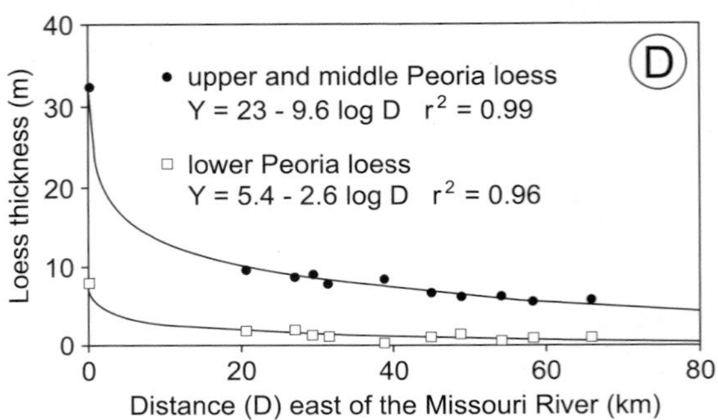

Fig. 8.35 Various mathematical relationships between certain loess properties and distance from the loess source. (a) Loess thickness from the Mississippi River in southern Illinois, east toward southern Indiana. After Frazee *et al.* (1970). (b) Loess thickness from the Missouri River in western Iowa, east toward south–central Iowa. After Hutton (1947), Ulrich (1950) and Ruhe (1969). (c) Peoria loess thickness from the Wabash River in southern Illinois and Indiana, to the east and the west. After Fehrenbacher *et al.* (1965). (d) Peoria loess thickness from the Missouri River, east into western Iowa. After Muhs and Bettis (2000).

were those nearest the river. Thus, the age of the *base* of the loess gets progressively younger as distance from the river increases, and the window within which the loess was deposited gets progressively shorter farther from the river (Ruhe 1969). Slower loess deposition rates farther from the source areas presumably allowed for syndepositional weathering of the loess, facilitating increased leaching and pedogenesis (Ruhe 1973, Muhs and Bettis 2000).

Loess-derived landscapes are time transgressive, with those surfaces farthest from the source rivers being the oldest. Near the rivers, soils develop on younger surfaces and in coarser loess, creating a complex space–time set-up for soil formation (Ruhe 1969). Post-depositional processes on the steep slopes near the major river valleys were particularly effective at eroding the coarser loess, leaving behind gullied and eroded landscapes, with steep slopes and minimally developed soils. Farther from the source areas, where erosion is slower and landscapes are flatter and more stable, the loess is thinner, finer-textured and more weathered. Also, if the loess covered a slowly permeable surface containing a paleosol, this surface would get buried progressively shallower in thin loess areas, facilitating weathering and soil development there, for two main reasons: (1) a "wetness effect" in which the buried soil perches water and facilitates weathering in the loess via ferrolysis, and (2) an acidity effect, because the buried soils are commonly leached

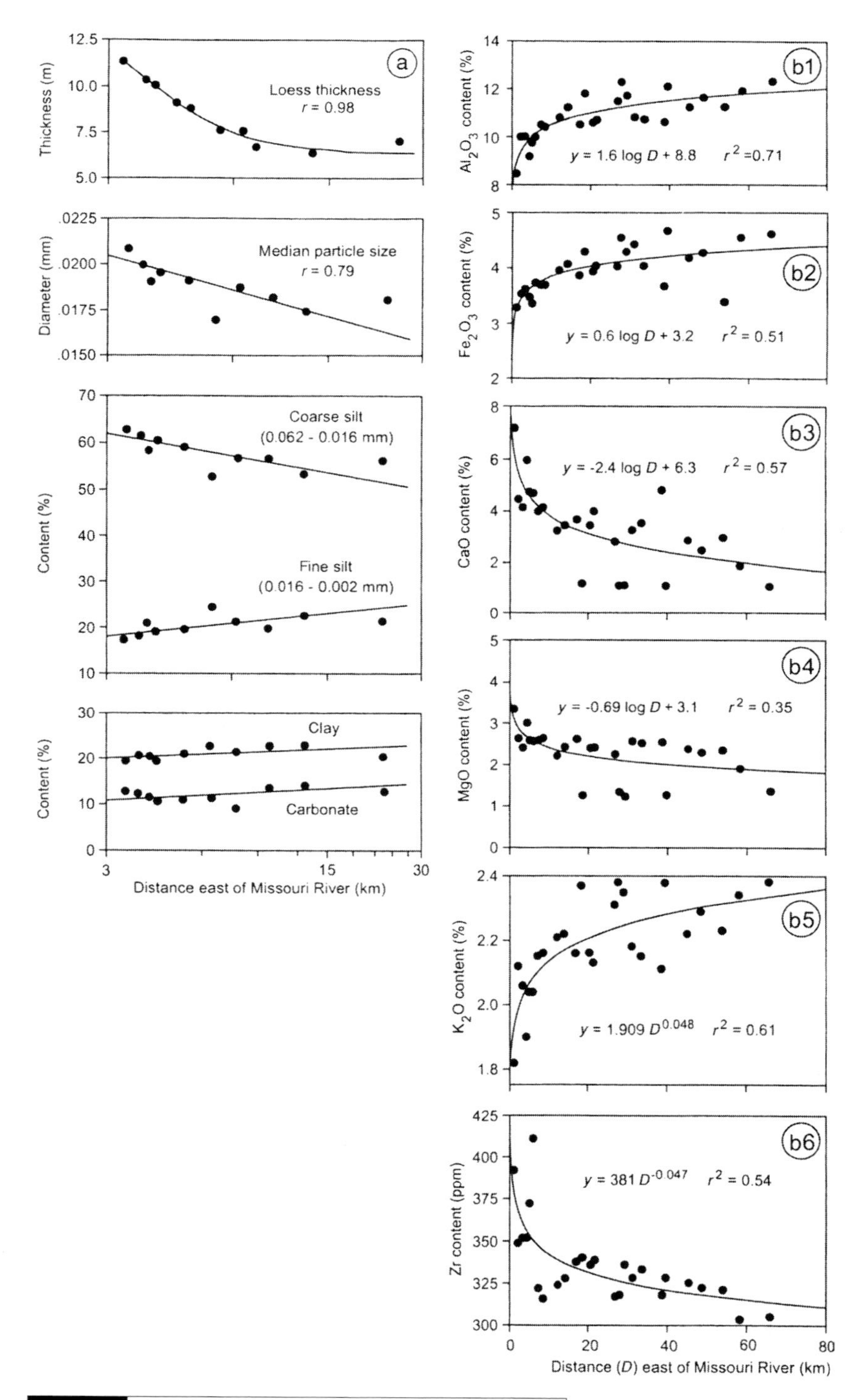

Fig. 8.36 Variation in loess characteristics, eastward from the Missouri River in southern Iowa. (a) After Ruhe (1954). (b) After Muhs and Bettis (2000).

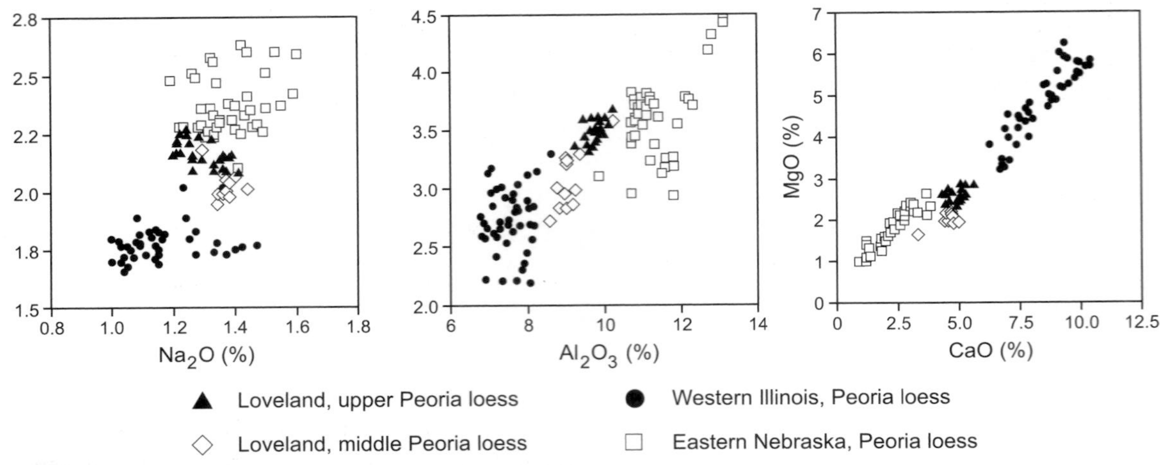

Fig. 8.37 Concentrations of various elemental oxides in different loess deposits from eastern Nebraska and western Illinois. The clustering of the loesses indicates different source regions and environments of deposition. After Muhs and Bettis (2000).

and acidic, facilitating loss of bases from the (often) calcareous loess above. In sum, soils tend to be better developed in thin loess areas than they are nearer the loess source (Ruhe 1973, Harlan and Franzmeier 1977).

Trends in loess thickness have great impact on soil development (Ruhe 1954). If loess deposition is slow, the new sediment is readily incorporated into the soil profile. Under this scenario, the soil profile, especially the A horizon, slowly thickens as the loess is mixed into it (McDonald and Busacca 1990). This process is called cumulization (see Chapter 13). In cases where loess deposition is rapid, sedimentation outstrips the soil's ability to incorporate the loess into the profile, and the soil becomes effectively buried (Fig. 12.78). Such buried soils, or paleosols, are quite common in thick loess columns (Ruhe *et al.* 1971, Gerasimov 1973, Valentine and Dalrymple 1976, Wintle *et al.* 1984, Olson 1989, McDonald and Busacca 1992, Feng *et al.* 1994a, Maher and Thompson 1995). The length of the pause in sedimentation and the rate of soil formation during that time interval will determine whether the buried soil is a strongly developed one or just a dark band enriched in organic matter (Daniels *et al.* 1960).

Silt loam and silty clay loam textures dominate loess soils. These soils are also quite porous, generally fertile and high in pH, unless underlain by a shallow, acid paleosol. Most loess-derived soils are relatively young, having formed in the Late Pleistocene or early Holocene. Thus, weathering of primary minerals is typically not advanced, and their mineralogy is related closely to the mineral suite inherited from the parent material (Jones *et al.* 1967). The clay mineralogy of loessial soils varies, depending on the clay minerals in the parent material, which again is primarily a function of source area.

Grain size, mineralogy, carbonate content and other characteristics are often used to differentiate loess deposits from each other, or to determine whether a thin cap of loess exists on another parent material (Dixon 1991, Dahms 1993, Mason and Jacobs 1998, Muhs *et al.* 1999) (Fig. 8.37). Sponge spicules, shells of terrestrial snails (mollusks), opal phytoliths and various other microfossils are common in loess (Jones and Beavers 1963, Jones *et al.* 1963, Fehrenbacher *et al.* 1965, Rousseau and Kukla 1994, Rousseau *et al.* 2000). Sponge spicules are blown out of the wet environments of the meltwater valleys. Snails live within the upland loess soils; their shells are well preserved in calcareous loess (Conkin and Conkin 1961). Opal phytoliths are biogenic silica fragments produced by plant roots (see Chapter 15). Magnetic susceptibility and paleomagnetism of the loess are only two of several types of paleoenvironmental proxy data that can be gleaned from loess (Hayward and Lowell 1993, Shaw 1994, Feng and Johnson 1995, Liu *et al.* 1995, Maher and Thompson 1995, Meng *et al.* 1997,

Table 8.3 Grain-size based nomenclature for some common pyroclastic deposits

Grain size	Unconsolidated tephra	Consolidated pyroclastic rock
<1/16 mm	Fine ash	Fine tuff
1/16–2 mm	Coarse ash	Coarse tuff
2–64 mm	Lapilli tephra	Lapillistone
>64 mm	Bomb (fluidal shape) and block (angular) tephra	Agglomerate or pyroclastic tephra

Source: Orton (1996).

Grimley 2000). Also included in this proxy data list are various geochemical, elemental and mineral data and their ratios (Muhs and Bettis 2000) and contents of various isotopes (Beer *et al.* 1993, Frakes and Jianzhong 1994, Thompson and Maher 1995) (see Chapter 15).

"Dust" is a term that refers to silt-size or smaller sediment that has an eolian origin (Simonson 1995). The importance of eolian dust to soil genesis in arid landscapes is now unquestioned (Jenkins and Bower 1974, Chester *et al.* 1977, Rabenhorst *et al.* 1984, Amit and Gerson 1986, Yaalon 1987, Levine *et al.* 1989, Reheis *et al.* 1995, Blank *et al.* 1996, Naiman *et al.* 2000) (see Chapter 12). Muhs *et al.* (1990), for example, provided evidence for dust transport from the Sahara Desert to islands in the Caribbean Sea (Fig. 8.38). Cores from the ocean basins and ice sheets, direct observations of dust storms and their trajectories, as well as soils data confirm that dust transport is global in scale and that it has been ongoing for millenia (Pitty 1968, Biscaye *et al.* 1974, Prodi and Fea 1979, Glaccum and Prospero 1980, Pokras and Mix 1985, Dansgard *et al.* 1989, Rea 1990).

Volcanic parent materials: Andisols

Volcanic materials are typically classified by their size and elemental content (Table 8.3). Low-viscosity magmatic lavas tend to effuse slowly out of volcanic vents and harden to basalt.

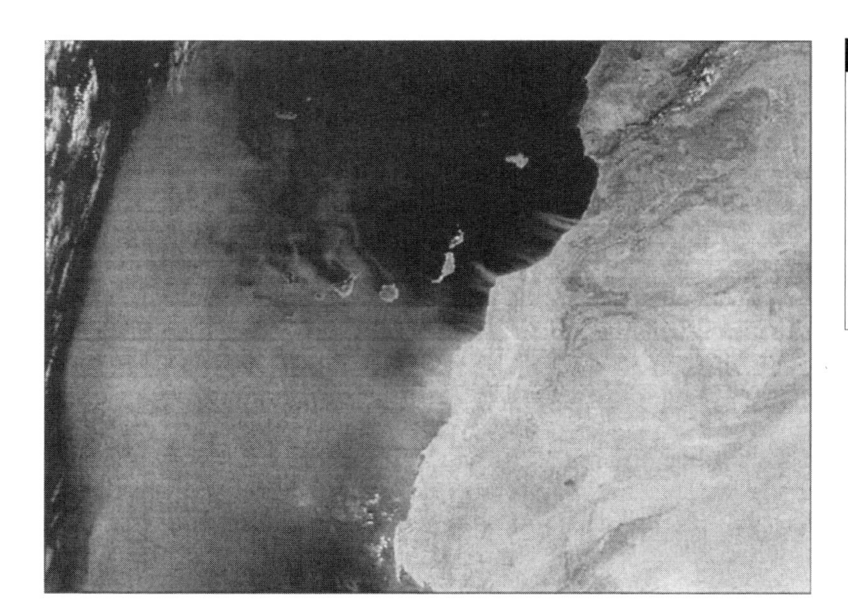

Fig. 8.38 A major dust storm over northwest Africa and the Sahara Desert on February 11, 2001. Of particular interest in this satellite image are the eddies and waves on the lee side of the mountainous Canary Islands (center). Image provided by the SeaWiFS Project, NASA/Goddard Space Flight Center, and ORBIMAGE.

Explosive eruptions tend to be associated with silicic lavas that are more viscous and slightly cooler. On the far end of the silicic spectrum are various airborne, ejected, pyroclastic materials, classified primarily on size: dust (<63 μm), ash (63 μm–2 mm), lapilli (2–64 mm), and blocks and bombs (>64 mm). Scoria and pumice are ill-defined terms for glassy, vesicular fragments in the lapilli and bomb size range (Orton 1996). Ashfall deposits tend to cover the entire landscape, with the exception of perhaps the steepest slopes.

The distribution and spread of pyroclastic materials is largely determined by their size and the direction of the prevailing winds. Large particles (e.g., lapilli and bombs) follow trajectories that are more or less ballistic, little affected by wind. Few travel farther than a few kilometers from the vent. Pyroclastic materials, aka *tephra*, are significantly impacted by local and regional winds. Particles smaller than 250 μm that also penetrate the tropopause may circle the earth many times before falling (Orton 1996). Because winds vary in speed and direction at different levels of the atmosphere, dispersal of the ash cloud can be complex. Nonetheless, ash deposits usually thin with distance, and at any one site ash deposits exhibit a fining-upwards sedimentology. Obviously, many sites that have soils developed in ash also record a history of episodic additions of ash, such that layers of "fresh" ash often overlie paleosols developed in an older ash, and so on (Hay 1960). Great thicknesses of ash deposits may exist at a site, each one comprising a distinct layer. Many of the volcanic materials associated with Andisols have undergone subsequent transport as mudflows, filling canyons and low-lying areas, or as eolian sediments (Arnalds *et al.* 1995). The dating of ash deposits and placing them within their proper chronostratigraphic context is called *tephrochronology* (see Chapter 14).

Glass and amorphous, glass-like materials dominate tephra. Their lack of crystallinity is caused by the rapid cooling of the molten ejecta; there is not enough time for crystals to form before cooling hardens the molten material. Light-colored ash is acidic, with quartz, hornblende and rhyolite being the main crystalline minerals. Darker ashes tend to be basic (mafic) and enriched in olivine and other ferromagnesian minerals. Amorphous materials dominate the mineralogy of younger soils developed in ash deposits. With time, however, the amorphous glassy minerals weather and neoform into other, crystalline minerals, such that the impact of the ash as the soil's parent material is slowly diminished.

Because volcanic parent materials and the soils that develop from them are so unique, the Soil Survey Staff (1999) established the Andisol order for them (Kimble *et al.* 2000). The central concept of Andisols is that of soils developing in volcanic ejecta (such as volcanic ash, pumice, cinders and lava) and/or in volcaniclastic materials. After only a minimal amount of weathering of these types of parent materials, the colloidal fraction tends to become dominated by amorphous materials and short-range-order minerals dominated by Al and Si, such as allophane, imogolite and ferrihydrite (Yoshinaga and Aomine 1962a, b). Secondary Al-, Fe- and Si-humus complexes and chloritized 2:1 clay minerals are also common, depending on the specific nature of the parent material (Shoji and Ono 1978, Parfitt and Saigusa 1985). Weathering of parent minerals that are *not* volcanic can also lead to the formation of short-range-order minerals; some of these soils also are included in the Andisol order (Soil Survey Staff 1999). However, for Andisols, weathering processes in tephra tend to dominate. The Andisol order is defined based on characteritsics inherited from weathering products of volcanic materials, not on the assumed presence of volcanic ejecta – a critical distinction.

Volcanic parent materials and their weathering products are somewhat unique. For this reason, acid soils formed in volcanic materials, with their thick, dark A horizons and abundant organic acids, have been distinguished with names such as "Ando soils" (Japanese *an*, dark, and *do*, soil), Humic allophane soils and "Andosols" (Thorp and Smith 1949, Simonson and Rieger 1967, Wada and Aomine 1973, Simonson 1979, Tan 1984). The dark A horizon that is characteristic of these soils was also deemed unique enough that it was eventually separated out as a melanic epipedon (Table 8.3).

Andisols are usually found in humid, volcanically active areas – typically in or near volcanic mountain ranges. Areas with abundant Andisols include Iceland, Alaska, Japan, Indonesia, the Andes Mountains of South America and Russia's Kamchatka peninsula (Seki 1932, Ulrich 1947, Tan 1965, Simonson and Rieger 1967, Arnalds *et al.* 1995, Arnalds and Kimble 2001). In these regions, the likelihood of encountering Andisols tends to increase with altitude.

In Andisols, weathering and mineral transformations provide a prominent pedogenic signature. Assisted by the non-crystalline nature of the primary minerals and the high porosities of the soils, weathering is often exceedingly rapid (Hay 1960). Many Andisols are pedogenically young, because they have formed either in recently deposited ash or in parent materials with thin increments of geologically young ash embedded within (Tan and Van Schuylenborgh 1961). They are frequently pedogenically "rejuvenated" and highly polygenetic (Bakker *et al.* 1996). Therefore, weathering of primary aluminosilicates in most Andisols has proceeded only to the point of formation of short-range-order minerals (Dahlgren and Ugolini 1991).

Most Andisol-like profiles have formed in *acidic* volcanic deposits, and because these deposits contain Al- and Si-rich materials they form allophane and other short-range-order minerals when they weather. In humid climates, silicic tephra weathers to Andisols rich in allophane, ferrihydrite, imogolite, halloysite and hydroxy-interlayered vermiculite (Vacca *et al.* 2003). Basic materials tend to weather to soils richer in halloysite. Volcanic materials in dry climates weather more slowly and soils may take much longer to develop into Andisols (Dubroeucq *et al.* 1998). Non-volcanic Andisols can form where parent material rich in acidic silicate minerals and a warm, more or less continuously moist climate combine to promote rapid weathering. Hydrous oxides of Fe, Al and Si, along with short-range Fe- and Al-organic matter complexes, can then form (in either environment) as these cations are released from primary minerals (Shoji *et al.* 1985, Hunter *et al.* 1987). Once weathered, the cations (primarily Si, Al and to a lesser extent Fe) released are usually quickly complexed by organic acids, producing stable organo-metallic complexes when the soil pH is low (Parfitt and Saigusa 1985, Shoji *et al.* 1993). In the end, the formation of immobile Al–organic matter complexes leads to high organic matter contents in near-surface horizons, dark soil colors and even potential aluminum toxicity (Wada and Aomine 1973).

Andisols have many unique characteristics that merits their distinction as a separate order: thick, dark A horizons, high porosity, low bulk density, low permanent charge, an exchange complex dominated by variable charge surfaces and high anion adsorption and moisture retention capacities (Shoji and Ono 1978, Wada 1985, Ping *et al.* 2000). Other common but not requisite characteristics include low pH and base saturation values, and high amounts of exhangeable aluminum, much of which is complexed with organic matter (Aran *et al.* 2001). Many Andisols fix high amounts of phosphorous. When wet they are greasy and smeary. One of the outstanding features of Andisols is their generally high natural productivity, despite the fact that many are on steep slopes that preclude intensive agriculture.

Andisol profiles often display geological and pedogenic layering (Fig. 8.39). They are unlike almost all other soils, in that their surface horizons are often less weathered than the B horizons below because new parent materials are frequently added to the surface (McDaniel *et al.* 1993, Bakker *et al.* 1996). Obviously, lithologic discontinuities are common, at breaks between ash layers. They are so common, in fact, that Arnalds *et al.* (1995) did not recommend distinguishing individual ashfall layers as separate parent materials. Thin ashfalls may lead to cumulization of the A horizon of the profiles, while thick ashfalls bury the existing surface soils (Shoji and Ono 1978, Smith *et al.* 1999).

Weathering of primary minerals in Andisols can provide ample aluminum to complex large amounts of organic matter. Unlike podzolization (see Chapter 12), where metal–cation complexes are mobile and become translocated to form E–Bhs sequa, little such translocation occurs in

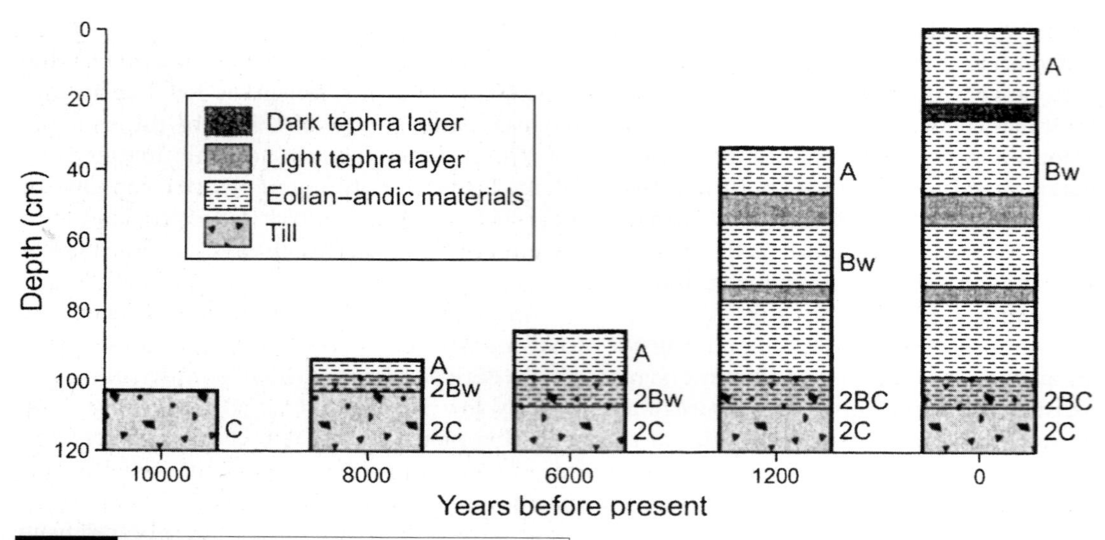

Fig. 8.39 Chronostratigraphic development of a typical Andisol in Iceland that has formed in tephra and volcanic materials reworked by wind. After Arnalds *et al.* (1995).

Andisols. Why? One hypothesis presupposes that the organic matter immobilization is caused by such high aluminum contents in the soil that the humic compounds become saturated in the A horizon and cannot be readily translocated (Aran *et al.* 2001). Weathering of volcanic materials in Andisols can indeed provide large amounts of Al cations to the soil solution. Another likely cause has to do with the types of organic acids involved: larger, humic-type organic acids are not as likely to become mobile when chelated with Al cations.

Andisols often abut Spodosols on the landscape; they genetically have much in common. Both orders are dominated by soil horizons whose active components consist mainly of amorphous materials (Flach *et al.* 1980). Whereas Andisols are associated with the *immobilization* of Al and Si by organic matter complexation, in podzolization the organic acids assist in the *mobilization* of metal cations. Large humic acid molecules tend to dominate allophane-rich Andisols (Wada and Aomine 1973, Flach *et al.* 1980). Podzolization is triggered by organic acids of a different type: fulvic acids (Aran *et al.* 2001). Unlike humic acids, fulvic acids render Al and Fe cations mobile by forming chelate complexes. Thus, as one progresses from wamer, drier environments to cooler, humid, forested conditions, Andisols may yield to soils with spodic-like profiles (Simonson and Rieger 1967, Shoji *et al.* 1988). Along such an Andisol–Spodosol transect the amounts of organic materials, Fe, Al and Si that have been translocated to the B horizon increases, E horizons get more acidic and smectite, rather than allophane, comes to dominate the E horizon mineralogy (McDaniel *et al.* 1993). In drier pedogenic environments, organic matter production is less. Al is more likely to be polymerized, forming short-range-order aluminosilicates, allpohane and imogolite in near-surface horizons. With time, additions of more free silica, dehydration and structural rearrangement of the allophane and imogolite can lead to the formation of halloysite (Dubroeucq *et al.* 1998). Under strong leaching conditions, desilication will facilitate the formation of gibbsite.

Transported by ice

Geologists use the term *drift* to refer to material carried by glacial ice. The glacial sedimentary environment is unique and complex because sediments are often transported both by solid ice and by, later, meltwater (Ashley *et al.* 1985). Indeed, according to Ashley *et al.* (1985: 2), "If any generalization can be made concerning the distribution of glacial sedimentary facies, it is that few generalizations can be made." Glacial sediments vary from unsorted, boulder-rich till to stratified lacustrine silts and clays. Transitions from one deposit to another may occur abruptly, or they may imperceptibly grade together. Multiple deposits, stacked one upon another and all dating to the same ice advance, are common. It is little wonder that perhaps the most complex soil landscapes are those that have been recently deglaciated.

Material deposited directly by the ice, glacial *till*, is often unstratified and unsorted (Dreimanis 1989). Till tends to run the gamut from very clay-rich material to coarse, sandy sediment, depending upon the rocks and landscapes the ice most recently overrode. Large clasts may be concentrated in some parts of a till deposit while others may be nearly gravel and stone-free. Perhaps the most distinguishing characteristic is its unpredictability, from a sedimentological perspective (Attig and Clayton 1993). Parts may be crudely sorted and stratified, interfingered with glaciofluvial sediments, while other tills may be totally unstratified (Hartshorn 1958, Khakural *et al.* 1993) (Fig. 8.40). Till can be sandy, clayey, silty, cobbly, dense or porous, acidic or calcareous (Willman *et al.* 1966, Levine and Ciolkosz 1983, Mohanty *et al.* 1994, McBurnett and Franzmeier 1997). In many areas, however, there is a critical distinction made between basal till (deposited under the ice sheet) and ablation or superglacial till. The former is often dense and compacted, while the latter is more porous and often shows evidence of the influence of running water, as stringers of sorted sand lenses. In some soils with dense fragipans in the subsurface, the close-packed nature of the fragipan has been ascribed to inheritance from compact till (Lyford and Troedsson 1973).

Most till soils are fairly young, having formed in till that was deposited by the last (Wisconsin, Pinedale, Würm) or next to last (Illinoian, Bull

Fig. 8.40 Glacial till overlying glaciofluvial sediment (outwash), southern Wisconsin. Photo by RJS.

Lake) major ice advance in North America (Allan and Hole 1968, Dalsgaard *et al.* 1981, Franzmeier *et al.* 1985, Berry 1987, McCahon and Munn 1991, Applegarth and Dahms 2001). Another common aspect of till soils and landscapes is their "layer-cake" stratigraphy, whereby one till sheet is deposited upon another. Buried soils (paleosols) are, therefore, routinely found in the older till deposits (Bushue *et al.* 1974, Griffey and Ellis 1979). In many landscapes, the till itself does not contain a soil, but is buried by loess that contains the modern soil at its surface.

Glacial meltwater is capable of entraining and sorting parts of the drift, carrying away the silt and clay (some of which will eventually become loess) and leaving behind the larger sand and gravel particles as glaciofluvial sediment, or outwash. *Outwash* is stratified, sandy and gravelly sediment that is impacted by pedogenesis in many of the same ways as sandy alluvium or dune sand (Fig. 8.40). It is unlike dune sand in that outwash (1) contains coarse fragments and (2) is usually sorted *and* stratified; the strata affect water flow in the soils that are to follow. Contacts between strata are preferred sites for illuviation and weathering. This has been demonstrated numerous times, e.g., illuvial clay tends to accumulate at the contact between a sandy and an underlying gravelly zone (Bartelli and Odell 1960a).

Transported by gravity

Slumps, slides, flows and soil creep are all the result of the influence of gravity on regolith (Moser and Hohensinn 1983, Vaughn 1997). Material moved downslope, primarily under the influence of gravity, is called *colluvium*. If this movement is restricted to channelized flow, *alluvium* would be better. Some refer to the *process* as *colluviation* (Goswami *et al.* 1996). Colluvium that has been reworked by running water or slopewash is called *co-alluvium* (Fig. 8.41). Colluvium can range from fine-textured to extremely cobbly and bouldery material, depending on the source material composition, and this variability can occur in both the lateral and vertical dimensions (Fig. 8.42). It is usually quite heterogeneous but often does have crude textural zones or strata within, which can impact pedogenesis at a later point (Ciolkosz *et al.* 1979, Kleber 1997). The contacts between such zones can be quite abrupt.

Slope failure occurs because the shear strength of the material is compromised, or because shear stresses become too great. Increased shear stress can occur as the slope is overloaded with heavy, wet snow or by tectonic stresses and faulting (Forman *et al.* 1991, Amit *et al.* 1995). Undercutting the slope by rivers, jointing in the parent rocks, devegetation and saturated soil conditions after long rain events all act to weaken the cohesiveness (shear strength) of the sediment/soil and initiate failure (Eschner and Patric 1982, Riestenberg and Sovonick-Dunford 1983, Graham *et al.* 1990a). Other situations where slope failure is common include thin soils over bedrock, areas of convergent subsurface flow, and the presence of shallow-rooted vegetation or loss/lack of vegetation and impermeable layers within 1–3 m of the surface (Neary *et al.* 1986) (Fig. 8.43). In the Appalachian Mountains, many debris flow deposits date back to hurricanes in the geologic past (Liebens and Schaetzl 1997) or to infrequent, intense storms in recent history (Williams and Guy 1973, Neary *et al.* 1986). Some colluvial materials, in climates that today are warm, date to times of periglacial activity when surficial materials often were saturated because of the slow permeability of the underlying permafrost (Ciolkosz *et al.* 1979, Mills 1981).

Perhaps the two factors that contribute most to slope instability and are most related to colluvial thickness are slope gradient and vegetative cover (Salleh 1994, Goswami *et al.* 1996). Steep slopes with minimal vegetation are the most prone to failure, other things being equal. The effect of slope gradient is obvious, but vegetation cover may affect colluvial thickness in two ways: (1) less vegetation leads to more slope failures and (2) vegetation helps to trap colluvium or hold slope mateirals in place, so that they do not end up as colluvium. Most areas of intense and rapid slope failure have *initiation zones*, which are often steep, shallow-to-bedrock areas, *runout zones*, valleys where colluvium moves rapidly downslope, and the fan-shaped *deposition zones*, where it comes to rest.

Colluvial soils, usually most common on the lower parts of slopes (Fig. 8.41), have primarily been studied in mountainous areas (Hoover and Ciolkosz 1988, Liebens 1999, Ogg and Baker 1999). Colluvium and colluvial soils tend to be thickest in the low-gradient, deposition zones at the bases of slopes, in depressions and at the ends of small valleys, in steeply sloping terrain, and thinnest on steeply inclined upslope areas (Goswami *et al.* 1996, Ohnuki *et al.* 1997) (Figs. 8.43, 8.44). Thus, landscapes with frequent slope failure often are a complex of deep, colluvial soils in coves and valleys, while upland areas contain residual soils, saprolite and many "scars" where bedrock is exposed or soil is thin at best (Whittecar 1985, Mew and Ross 1994). Scar areas (initiation zones) are often prone to further slope failure, because the thin soils over hard bedrock are only weakly retained on the slope (Fig. 8.45). Thus, a cycle often

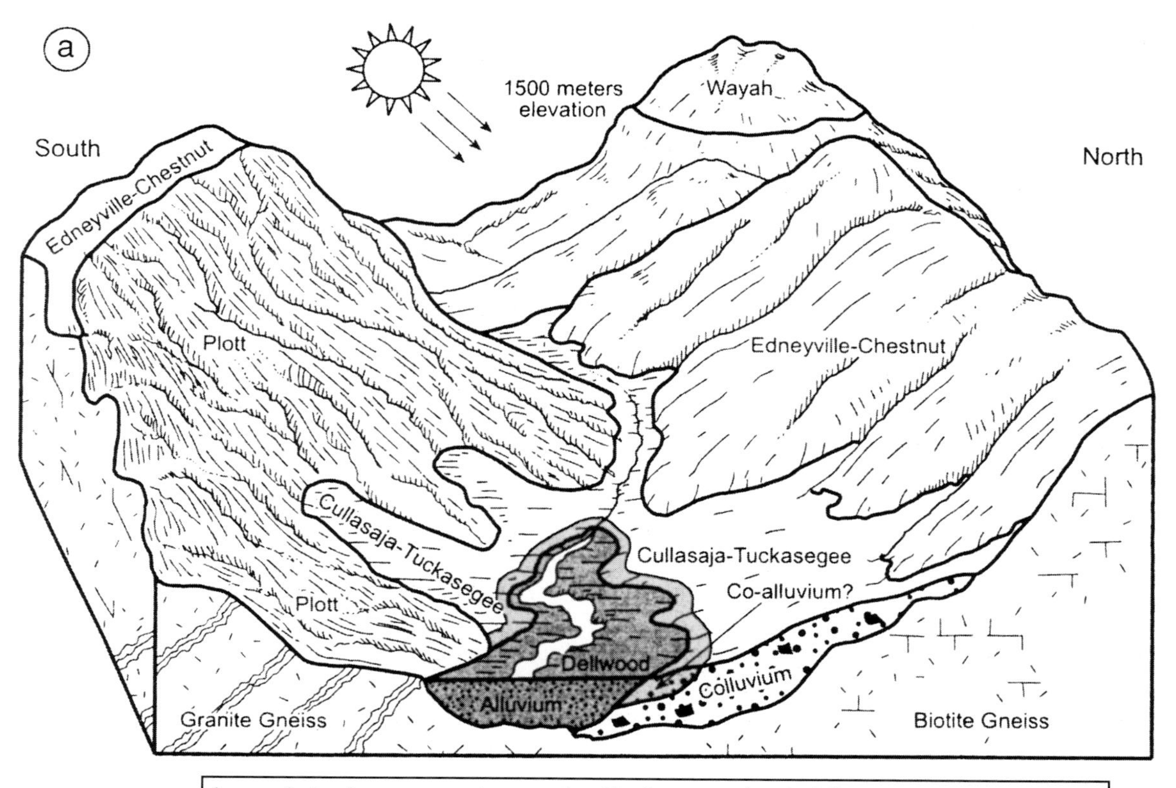

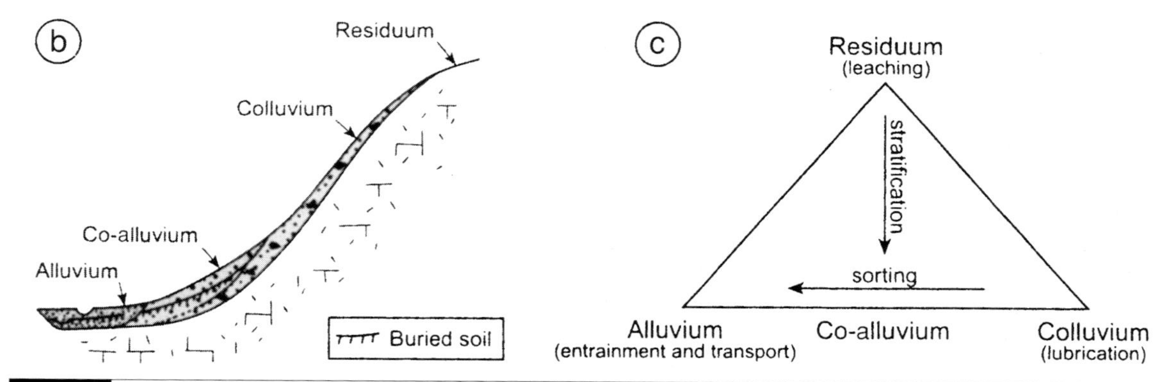

Fig. 8.41 Types of parent materials and landscapes in the Appalachian region, USA. (a) Block diagram of the soil–landscape–parent material relationships in Macon County, North Carolina. Residuum occurs on upper slopes while colluvium and alluvium are at the bases of the slopes. After Thomas (1996). (b) Schematic valley cross-section in a first-order drainage basin, illustrating the typical distribution of regolith types along the slope. After Cremeens and Lothrop (2001). (c) Ternary chart showing the varying degrees of sorting, and interactions of water with regolith, colluvium and alluvium. After Cremeens and Lothrop (2001).

a

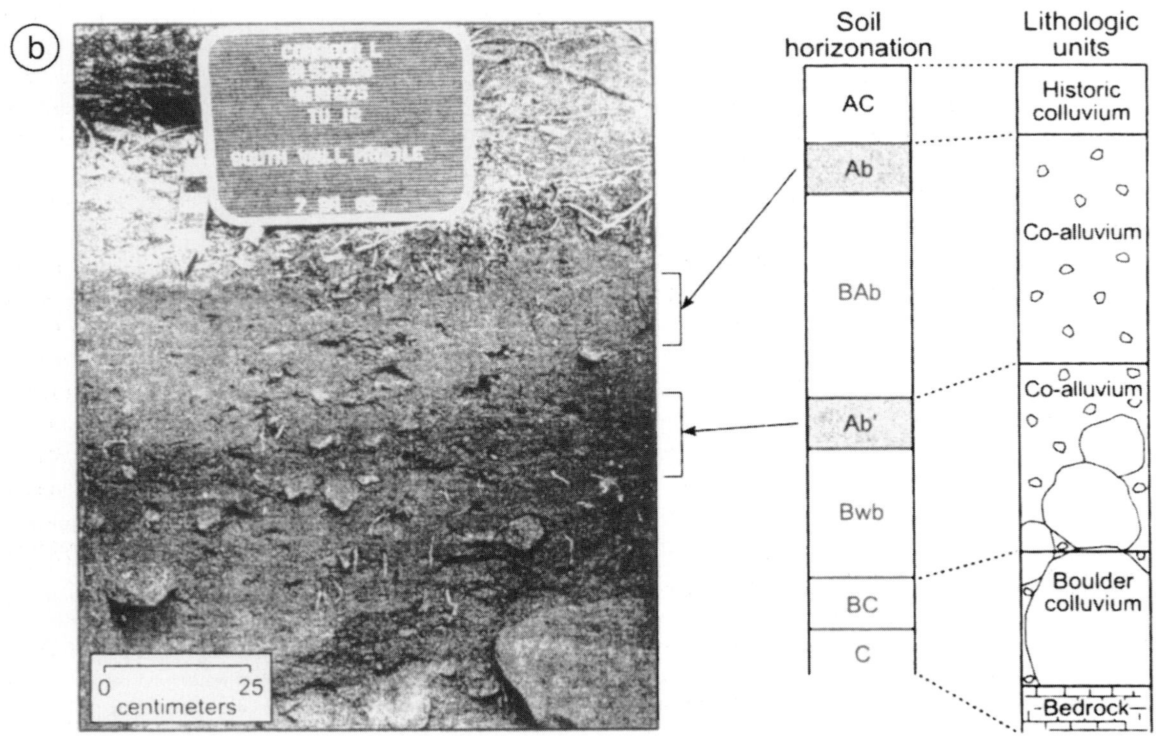
b
0
25
centimeters
Soil horizonation
Lithologic units
AC
Ab
BAb
Ab'
Bwb
BC
C
Historic colluvium
Co-alluvium
Co-alluvium
Boulder colluvium
Bedrock

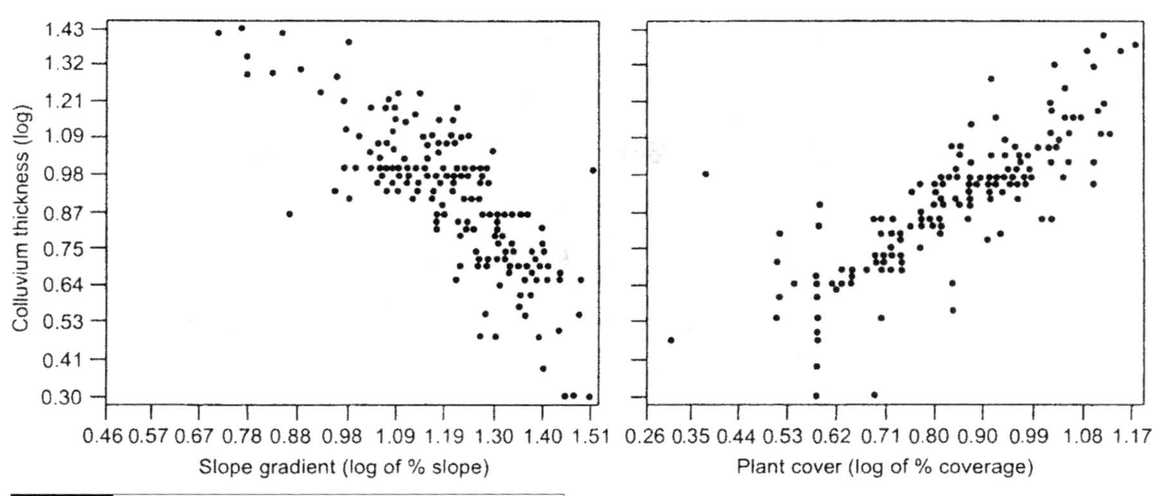

Fig. 8.43 Statistical relationships between thickness of colluvium and two factors that greatly affect slope stability: slope gradient and plant cover. Both slope gradient and plant cover increase to the right, along the *X*-axis. After Salleh (1994).

ensues, where areas of slope failure tend to repeat this process (Liebens and Schaetzl 1997). Unlike many other soil "parent materials," colluvium contains pre-existing weathered and pedogeneticized material (buried soils), reflecting the episodic nature of its deposition.

Also of concern in settings where colluvial soils are common is the issue of slope stability. Colluvium is created as a result of unstable slopes, and the soils that form on the ensuing colluvium also may be developing on geomorphic surfaces that are periodically unstable and prone to erosion. Many colluvial deposits bury or truncate pre-existing soils (buried paleosols), or bury still older colluvium (Waltman *et al.* 1990, Crownover *et al.* 1994, Kleber 1997, Yassoglou *et al.* 1997). These contacts represent lithologic breaks, which also can occur within the same colluvial mass (Whittecar 1985).

Lithologic discontinuities in soil parent materials

Lithologic discontinuities represent zones of change in the lithology of soil parent material that are interpreted to be primarily lithologic, rather than pedogenic. Lithologic discontinuities provide important information on parent material origins and subsequent pathways of soil development (e.g., Kuzila 1995). Knowing whether a soil has a lithologic discontinuity, and its nature (abruptness, topography, etc.) is an essential starting point for any pedogenic study (Parsons and Balster 1966, Chapman and Horn 1968, Raad and Protz 1971, Evans 1978, Meixner and Singer 1981, Norton and Hall 1985) and required for any quantification of soil development (Haseman and Marshall 1945, Evans and Adams 1975a, Santos *et al.* 1986, Chadwick *et al.* 1990). One must also be able to discriminate between "true" lithologic discontinuities, which are inherited and which have formed geologically, from similar "breaks" within soil profiles that are strictly pedogenic (Paton *et al.* 1995). Given enough time, soils can develop

Fig. 8.42 Examples of colluvial soils and paleosols in the unglaciated Allegheny Plateau of West Virginia. (a) Colluvium and a colluvial soil (Typic Hapludult) above sandstone bedrock. Photo by D. L. Cremeens. (b) Exacvation into co-alluvium (colluvium with an alluvial component) on a toeslope, showing two buried paleosols on older, previously stabilized surfaces. The lower paleosol contains charcoal that dates between 2470 and 3450 BP, and is buried by co-alluvium. The upper paleosol is buried by more recent alluvium and slopewash deposits, and dates to about 600 BP. After Cremeens and Lothrop (2001).

(a)

(b)

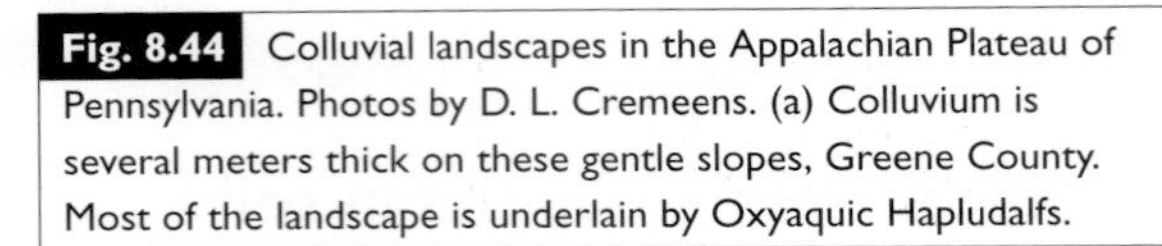

Fig. 8.44 Colluvial landscapes in the Appalachian Plateau of Pennsylvania. Photos by D. L. Cremeens. (a) Colluvium is several meters thick on these gentle slopes, Greene County. Most of the landscape is underlain by Oxyaquic Hapludalfs. (b) Colluvial material creeping into the base of this small stream valley, while the stream attempts to remove it, Nicholas County. The soils on the slopes are Typic Haplumbrepts.

abrupt textural breaks at horizon boundaries; not being inherited from the parent material they would not be considered lithologic discontinuities. Many of these are referred to as "abrupt textural changes" by the Soil Survey Staff (1999). Although the focus in this section is on *inherited geologic discontinuities*, the principles put forth are applicable in either case. Schaetzl (1998) reviewed the methods by which lithologic discontinuities can be detected, and their importance to pedogenesis; the discussion that follows is based largely on his work.

Lithologic discontinuities are common; Schaetzl (1998) statistically sampled a listing of US soil series and determined that about a third (33.5%) of them have lithologic discontinuities, and nearly 7% have two or more. Thus, although soils with discontinuities are common, detection of subtle discontinuities is often difficult because their pedogenic impacts are difficult to ascertain and pedogenesis often blurs them. Additionally, many sedimentary deposits assumed to be homogenous throughout often are crudely stratified (e.g., till), forcing the investigator to decide to what extent contacts between individual strata should be considered discontinuities (Asamoa and Protz 1972, Cremeens and Mokma 1986, Schaetzl *et al.* 1996, Liebens 1999). In the end, determining whether or not a sediment or soil has a lithologic discontinuity often has a bit of subjectivity to it.

Types of discontinuities

Distinct breaks in the lithology of soils can occur in two ways: geologically–sedimentologically or pedologically (Raad and Protz 1971, Follmer 1982, Paton *et al.* 1995). In theory, a geologic lithologic discontinuity (LD) is the physical manifestation of either (1) a break/change in sedimentation or (2) an erosion surface. The interruption in sedimentation may involve a period of non-deposition, with perhaps with a paleosol developed at the break point, or period of erosion. Similarly, it could involve a change in sedimentation system, such as from deposition of sands by running water to deposition of silts by wind (Kuzila 1995, Soil Science Society of America 1997). Follmer (1982) felt that LDs are typically expressed by a departure in depth trends between zones of otherwise relative uniformity. Arnold (1968) used the term "abrupt changes" to describe loci within depth functions where LDs might be indicated (see also Raad and Protz 1971).

Pedologically formed discontinuities, such as biomantles, usually involve processes that, in essence, sort near-surface sediments. They are common but often conceptually ignored (D. L.

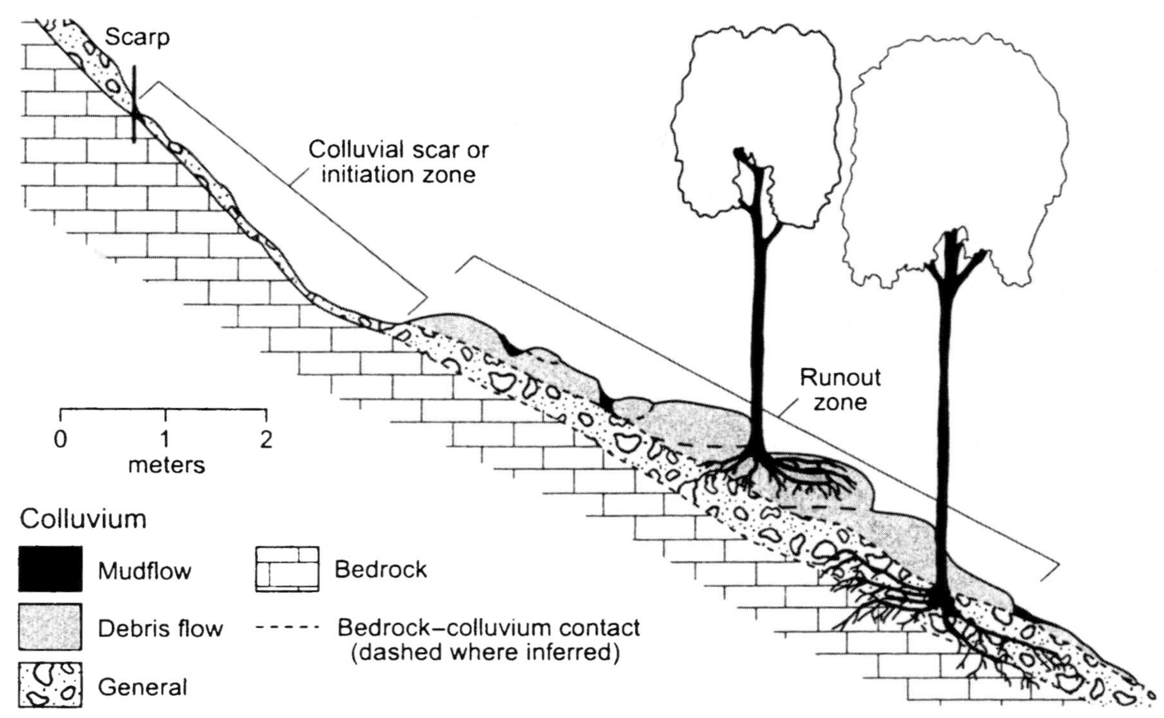

Fig. 8.45 Cross-section through a wooded slope near Cincinnati, Ohio, showing various colluvial deposits and their relationship to standing trees. After Riestenberg and Sovonick-Dunford (1983).

Johnson 1990, Humphreys 1994, Nooren *et al.* 1995). Two or more layers are then formed within the solum, perhaps with the discontinuity near the depth at which the influence of the surficially driven sorting process, e.g., splash erosion, bioturbation or tree uprooting, diminishes (D. L. Johnson 1990). Pedogenically formed depth functions usually exhibit more gradual changes than do geologically formed discontinuities (Smeck *et al.* 1968). In tropical regions, where discontinuities and stone lines exist in soils on old landscapes, their detection and accurate interpretation has enhanced our understanding of soil geomorphic systems and their evolution (Bishop *et al.* 1980, Johnson 1993b, Paton *et al.* 1995) (see Chapter 13). In fact, it is likely that pedologically formed discontinuities are more common on older landscapes, where the processes of surface wash, creep, eolian transport and bioturbation have acted in concert for many years to form relatively stone-free biomantles and texture-contrast soils, many of which exhibit stone lines at depth (Johnson 1993b, 1994, Schwartz 1996).

Because lithologic discontinuities reflect changes in past sedimentation/erosion systems, their existence is highly applicable to studies of near-surface sedimentary history (Fig. 8.40). Applications include detecting and distinguishing among various deposits of loess, colluvium, alluvium, till and lacustrine sediments (Raukas *et al.* 1978, Bigham *et al.* 1991, Karathanasis and Golrick 1991, Schaetzl 1998). Identification of a paleosol in the upper part of a sedimentary layer implies not only that a discontinuity exists, but also that a soil forming interval has occurred between the depositional events (Ruhe 1956, 1974a, Foss and Rust 1968, Schaetzl 1986c, Ransom *et al.* 1987, Olson 1989, Tremocoldi *et al.* 1994, Anderton and Loope 1995). Identification of a stable geomorphic surface buried within the near-surface stratigraphic column also has implications for climatic change, landscape stability and archeology, among many others (Holliday 1988, Cremeens and Hart 1995, Curry and Pavich 1996).

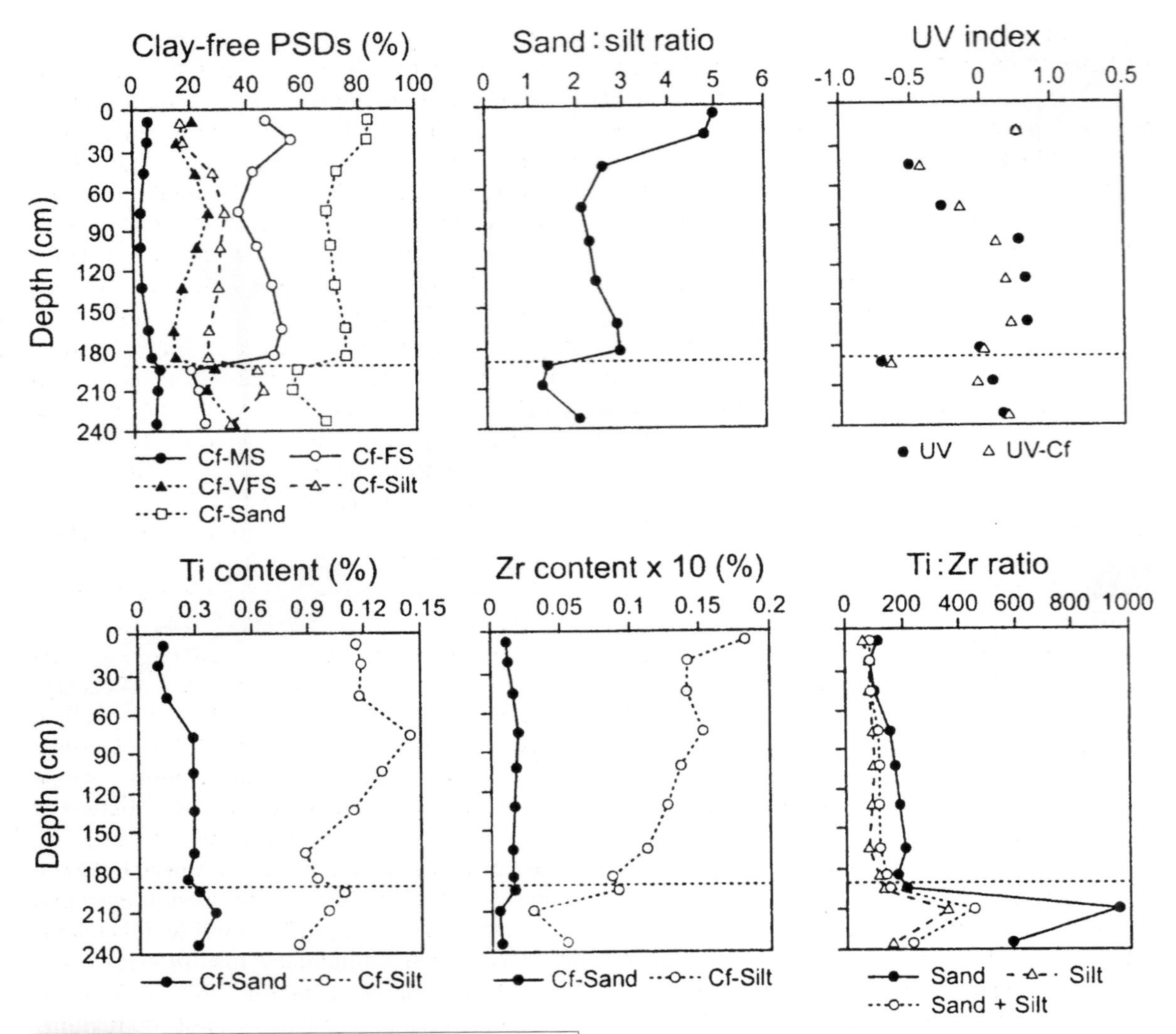

Fig. 8.46 Various depth functions for a upland Ultisol in Taiwan that has a lithologic discontinuity at 190 cm. After Tsai and Chen (2000).

Detection of discontinuities

Examination and sampling of the soil from the surface downward, at numerous, closely spaced intervals, is the first step in detecting lithologic breaks. If the data are then plotted as depth functions, "breaks" in certain, but not all, depth trends often suggest the presence of lithologic discontinuities (Fig. 8.46). This method is perhaps the most commonly used and probably the best to identify discontinuities.

Depth functions for the various pedogenic parameters may not always be in agreement or not coincide with the "real" discontinuity, because of (1) the subtlety of the discontinuity or (2) large amounts of variability within an apparently homogeneous, lithologic unit, such as till (Oertel and Giles 1966, Busacca and Singer 1989). Alternatively, the depth function being examined may not be appropriate for the detection of LDs, such as the depth distribution of iron in a Spodosol. In this case, an LD may be indicated where, in reality, there is none, because of the choice of poor evaluative tools. Thus, it is clear that knowing which parameter to use for the identification of an LD is important. Oertel and Giles (1966) observed that the disagreement among three studies on the locations and/or existence of LDs within the same soil was due to the use of different criteria for their detection, some of which were probably inappropriate. Beshay and Sallam (1995) could not reconcile depth function data designed to detect LDs because some of their parameters were developmental (i.e., acquired via pedogenesis) and some were sedimentological.

Which parameters, then, should be used to detect lithologic discontinuities, from depth function data? Table 8.4 lists some of the many parameters that have been successfully used to detect lithologic discontinuities in soils (Schaetzl 1998). The most important point to make about proper vs. improper parameters for detecting LDs is: use of any soil constituent that is pedologically mobile, i.e., the soil plasma, which is ions and particles <2 μm or <4 μm (cf. Meixner and Singer 1981, Asadu and Akamigbo 1987), can produce erroneous results. Therefore, avoid the use of pH, clay content or mineralogy, organic matter, CEC, base saturation, various iron minerals, certain forms of magnetic susceptibility, and total or partial elemental data, as well as morphological properties dependent upon the above, such as color, structure, consistence, bulk density, electrical conductivity or horizon boundary characteristics (Table 8.5). Although depth functions for one or more of these parameters can and often do change abruptly at an LD (e.g., Karathanasis and Macneal 1994, Kuzila 1995), they should not be used to detect one. Soil development indices, such as Bilzi and Ciolkosz's (1977) relative horizon distinctness (RHD) index, also should not be used to detect LDs. At best, pedogenic data or indexed compilations of acquired pedogenic characteristics should be used only to *infer* the presence of LDs, not to definitively *identify* them. In most environments that lack permafrost, particles coarser than about 30 μm in diameter can be considered immobile (Karathanasis and Macneal 1994). Caution must be exercised, however, in situations where large "particles," e.g., concretions, could have formed pedogenically, or more commonly, where they could have been moved by pedoturbation (Wood and Johnson 1978, Johnson *et al.* 1987) (see Chapter 10). Biomantles (Johnson 1990, Humphreys 1994) could be mistaken for a second parent material, since they commonly overlie a stone line that itself may be mistaken for an LD (Johnson 1989). Because the shape and sphericity of sand grains can be diagnostic of their depositional environments (Patro and Sahu 1974), this sort of data, again on an immobile fraction, could be useful for detecting LDs (Schaetzl 1998). Of the sand grain shape parameters examined by Schaetzl (1998), however, only mean feret diameter proved to be useful. Subtle differences in feret diameters occurred near discontinuities, and in one pedon they increased below the discontinuity. The *standard deviation* of sand grain roundness and sphericity data also may be more efficacious in differentiating parent materials than mean values alone (Patro and Sahu 1974) (Fig. 8.47).

To be even more discriminatory, identification of a geologic LD should be based on data from an immobile and (preferably) slowly weatherable soil fraction (Langohr *et al.* 1976, Schaetzl 1998). Immobile elements, as mentioned above, are excellent indicators of LDs because they reflect sedimentology better than do mobile, or plasma, elements (Wascher and Collins 1988). However, some of these "immobile" constituents may be weatherable, and thus may have have been lost from parts of the solum, when in fact the entire solum has formed in one parent material. Weathering processes can potentially alter these data, especially in old soils (Nikiforoff and Drosdoff 1943). In a profile formed from uniform materials, depth functions of immobile *and* slowly weatherable parameters should plot along a line that has no major "breaks" with depth (Santos *et al.* 1986). Therefore, because surficial weathering can reduce the amount of larger particles, many researchers have used both immobile *and* inert mineral grains, such as beryl or zircon, to assess LDs (Table 8.4). Although this is a time-tested method for ascertaining the location of LDs, there is one drawback: many soils, especially highly weathered ones, contain very low amounts of such minerals (Tsai and Chen 2000).

Most importantly and finally, calculation of depth functions for immobile and/or inert components is best done on a *clay-free basis*, as this removes the effects of the mobile element, clay (Rutledge *et al.* 1975a, Asady and Whiteside 1982, Bigham *et al.* 1991). Clay-free silt and clay-free

Table 8.4 Parameters that have been successfully used to detect (or refute) the presence of lithologic discontinuities in soils[a]

Indicator	References
Presence, absence or change in the content of a mineral	Barnhisel *et al.* 1971, Raad and Protz 1971, Karathanasis and Macneal 1994, Kuzila 1995
Presence or absence of a detrital fossil	Karathanasis and Macneal 1994
Content of one or more resistant minerals in a silt fraction	Rutledge *et al.* 1975b, Chapman and Horn 1968, Tsai and Chen 2000
Content of one or more resistant minerals in a sand fraction	Chapman and Horn 1968, Wascher and Collins 1988, Tsai and Chen 2000
Elemental composition or abundance in a sand fraction	Arnold 1968
Elemental composition or abundance in a silt fraction	Alexander *et al.* 1962, Barnhisel *et al.* 1971, Foss *et al.* 1978, Norton and Hall 1985, Karathanasis and Macneal 1994
Elemental composition of the non-clay fraction	Oertel and Giles 1966
Heavy mineral content	Chapman and Horn 1968, Khangarot *et al.* 1971, Cabrera-Martinez *et al.* 1989
Clay mineralogy	Follmer 1982, Dixon 1991, Kuzila 1995
Oxygen isotope composition ($\delta^{18}O$) of quartz	Mizota *et al.* 1992
Magnetic susceptibility	Singer and Fine 1989, Fine *et al.* 1992
Content of coarse fragments	Follmer 1982, Arnold 1968, Raad and Protz 1971, Asamoa and Protz 1972, Meixner and Singer 1981, Schaetzl 1996, 1998, Liebens 1999
Shape of sand particles	Schaetzl 1998
Total sand content, content of a sand fraction or mean sand size	Oertel and Giles 1966, Arnold 1968, Borchardt *et al.* 1968, Gamble *et al.* 1969, Caldwell and Pourzad 1974, Meixner and Singer 1981, Follmer 1982, Schaetzl 1992b
Total silt content or content of a silt fraction	Oertel and Giles 1966, Caldwell and Pourzad 1974, Price *et al.* 1975, Meixner and Singer 1981, Follmer 1982
Clay-free total sand or a clay-free sand fraction	Wascher and Collins 1988, Busacca 1987, Busacca and Singer 1989, Karathanasis and Macneal 1994, Schaetzl 1996, 1998, Tsai and Chen 2000
A clay-free and carbonate-free sand fraction	Raad and Protz 1971, Schaetzl 1998
Clayfree silt or a clay-free silt fraction	Chapman and Horn 1968, Barnhisel *et al.* 1971, Asamoa and Protz 1972, Price *et al.* 1975, Rutledge *et al.* 1975a, Wascher and Collins 1988, Busacca 1987, Busacca and Singer 1989, Karathanasis and Macneal 1994, Schaetzl 1996
A clay-free and carbonate-free silt fraction	Raad and Protz 1971
Any one of a number of fine earth fractions between 20 and 500 μm	Langohr *et al.* 1976, Santos *et al.* 1986

(*cont.*)

Table 8.4 *(cont.)*

Indicator	References
Ratio of one sand fraction to another	Fiskell and Carlisle 1963, Oertel and Giles 1966, Wascher and Collins 1988, Cabrera-Martinez *et al.* 1989, Hartgrove *et al.* 1993, Beshay and Sallam 1995, Liebens 1999
Ratio of one silt fraction to another (in some cases, clay-free)	Follmer 1982, Kuzila 1995
Ratio of sand : silt or silt : sand (in some cases, clay-free)	Chapman and Horn 1968, Raad and Protz 1971, Smith and Wilding 1972, Asady and Whiteside 1982, Busacca 1987, Busacca and Singer 1989, Tsai and Chen 2000
Ratio of two minerals in a sand fraction	Chapman and Horn 1968, Beshay and Sallam 1995
Ratio of two minerals in a silt fraction	Price *et al.* 1975, Follmer 1982, Busacca and Singer 1989, Bigham *et al.* 1991
Ratio of an element to a resistant mineral in the silt + sand fraction	Santos *et al.* 1986
Ratio of two or more elements in a sand fraction	Smith and Wilding 1972, Tsai and Chen 2000
Ratio of two or more elements in a silt fraction	Smith and Wilding 1972, Foss *et al.* 1978, Amba *et al.* 1990, Bigham *et al.* 1991, Karathanasis and Macneal 1994, Tsai and Chen 2000
Formulas involving particle size fractions:	
Uniformity value	Cremeens and Mokma 1986, Schaetzl 1998, Tsai and Chen 2000
Comparative particle size distribution index	Langohr *et al.* 1976

[a]Parameters listed are of the kind that can be displayed as depth functions; qualitative and/or morphological indicators of discontinuities are not addressed in this table. See Raukas *et al.* (1978) for a list of parameters used to study tills, some of which could have application within pedology.

Source: Modified from Schaetzl (1998).

sand contents are routinely used to detect LDs (Table 8.4, Fig. 8.47). Their use is highly recommended. Using skeletal (sand and silt) data from Table 8.6, one might mistakenly assume that an LD exists between the E and Bt1 horizons; each decreases >10% in the Bt1 horizon. However, clay-free silt data show that no LD exists, and that the large decrease in sand and silt between the E and Bt1 horizons is an artifact of the large *increase* in clay content. If possible, one might take this precaution one step further and use clay-free *and* carbonate-free data (Raad and Protz 1971).

Ratios of some of the above parameters are also widely used in the detection of LDs (Santos *et al.* 1986) (Table 8.4). The ratio of two sedimentologic parameters has an advantage over data of just one because more data are incorporated, implying that the magnitude of the between-horizon differences might be increased (Foss *et al.* 1978). Ratios also have, however, a distinct mathematical

Table 8.5 General utility of various parameters used to detect lithologic discontinuities in soils

Very useful in most instances[a]	Useful in some instances	Rarely useful; should be used only in conjunction with parameters	Not generally useful
Amount of a clay-free, immobile particle size fraction	Amount of a particle size fraction larger than clay	Clay mineralogy	pH and electrical conductivity
Elemental composition of an immobile and difficultly weatherable particle size fraction larger than clay	Mineral presence or absence	Magnetic susceptibility	Organic matter content
Content of a resistant mineral in a particle size fraction larger than clay	Heavy or light mineral content	Paleobotanical or paleontological evidence	Clay content and mineralogy
	Shape (roundness, sphericity, etc.) of a particle size fraction larger than clay		Morphological indicators such as structure, consistence, color, bulk density and horizon boundary characteristics
	Presence of a stone line or zone		CEC and base saturation
			Certain kinds of magnetic susceptibility
			Contents of most elements, except those associated with difficultly weatherable minerals (Zr and Ti)
			Soil "development indices"

[a] Application of ratios of and between useful parameters is generally encouraged.

Source: Schaetzl (1998).

disadvantage over other parameters. Extremely small values, when positioned as the denominator, can inflate the ratio and make interpretation difficult. Many ratios of resistant mineral contents encounter this problem, since the values are small to begin with and can even be zero (Beshay and Sallam 1995). Molar ratios of elements, and resistant/weatherable mineral ratios, have been used to determine the age or weathering status of soils (e.g., Beavers *et al.* 1963, Busacca and Singer 1989, Bockheim *et al.* 1996) and often do change abruptly at discontinuities (Foss *et al.* 1978). But

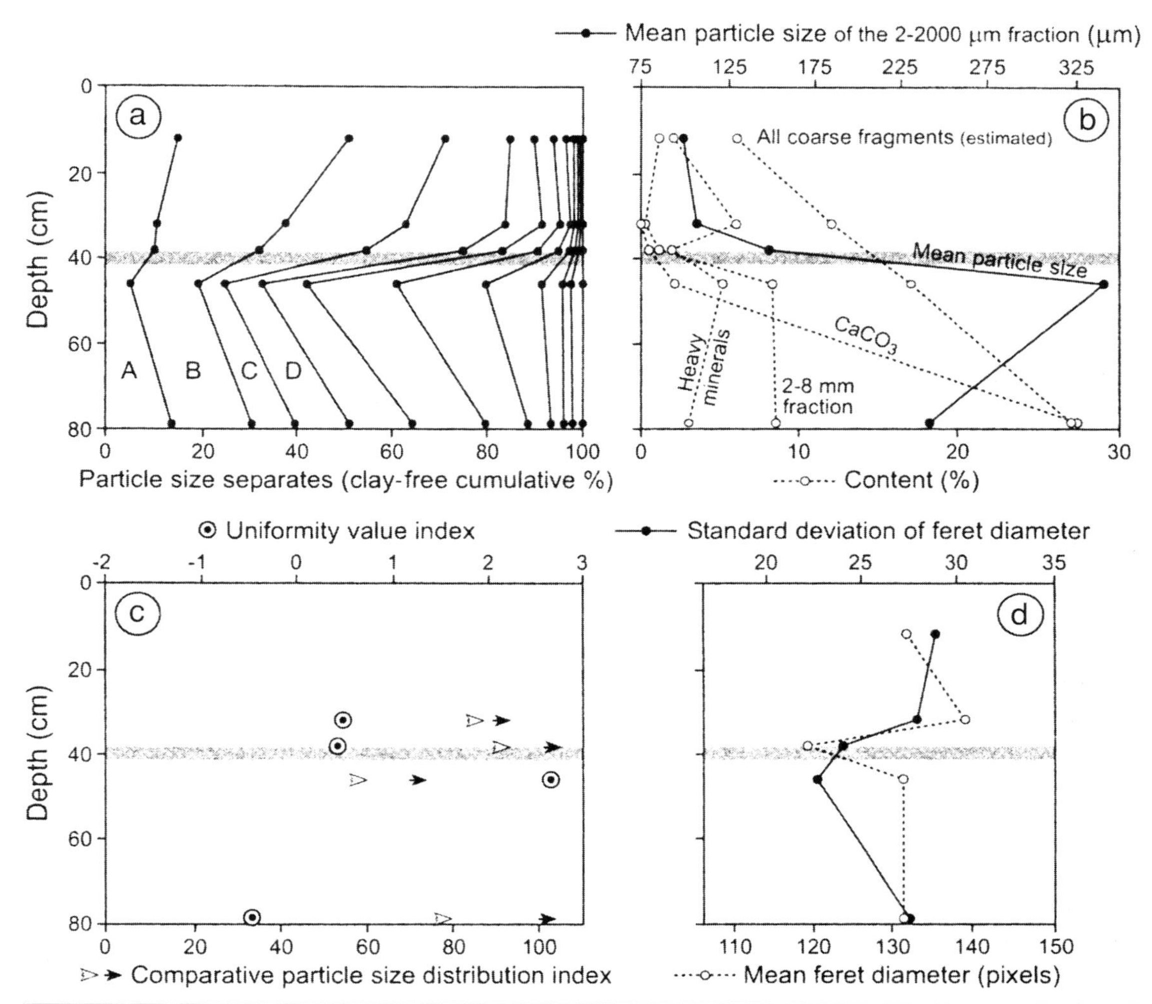

Fig. 8.47 Depth functions for an Alfisol in northern Michigan (Omena-2 site of Schaetzl (1998)). (a) Clay-free, cumulative percentages of particle size separates of the fine earth (<2 mm dia.) fraction. The particle size breaks begin with fine silt (0.002 to 0.016 mm, labeled "A"), and get coarser to the right (breaks at 0.053, 0.106, 0.18, 0.25, 0.355, 0.5, 0.71, 1.0, 1.4 and 2.0 mm dia). The lithologic discontinuity, as judged in the field, is shown by the broad, stippled band. (b) Depth functions of various pedologic and sedimentologic data. All are in units of percentages, except for mean particle size, which is in micrometers. Mean particle size was calculated for the clay-free size fraction only. (c) Data from three indices designed to show lithologic discontinuities, based on particle size data (comparative particle size distribution (CPSD) Index: light arrows, CPSD including coarse fragment data: dark arrows, uniformity value (UV) index: circles). In each case, the value plotted shows the index value when compared to the horizon *immediately above*. For the UV index, greater deviations, either way from zero, indicate a stronger likelihood that the two horizons are separated by a discontinuity. Higher values for the CPSD index indicate a lower likelihood of a discontinuity between the two horizons, hence the use of arrows pointing to the right – the farther to the right the value plots, the lower the likelihood of a lithologic discontinuity between that horizon and the horizon above. (d) Mean and standard deviation data for feret diameter of coarse sand grains.

Table 8.6 Sand, silt and clay data for a soil that illustrate the utility of calculating clay-free data for identifying lithologic discontinuities

Horizon	Sand	Percentage of fine earth fraction: Silt	Clay	Totals	Clay-free silt (%)
A	45	45	10	100	50
E	50	46	4	100	48
Bt1	38	36	26	100	49
Bt2	42	35	23	100	45
Bt3	42	35	23	100	45
BC	45	35	20	100	44
C	43	35	22	100	45

because they also reflect pedologic development, they should not be relied upon solely for detecting LDs.

In sum, when *many lines of evidence* point to a discontinuity at a given depth, one can have confidence in concluding that a discontinuity exists (Evans 1978, Raukas *et al.* 1978, Beshay and Sallam 1995). Taking off on this generalization, several researchers have developed mathematical indices that incorporate several parameters, so as to discern the presence or absence of a lithologic discontinuity. These indices usually compare data between the horizon in question and the one immediately above. The Comparative Particle Size Distribution (CPSD) Index of Langohr *et al.* (1976) uses data from the particle size fractions between 2 μm and 2 mm, i.e., it is "clay-free." The CPSD determines the similarity of particle size fractions of two samples (in this case, two horizons), such that the higher the CPSD the more similar are the samples. Perfect similarity will result in a CPSD of 100. Rindfleisch and Schaetzl (2001) developed a modified version of the CPSD, including data on the 2–4 mm, 4–8 mm and total coarse fragment contents, again all expressed in percentages of the entire sample (not just the fine earth fraction). In its original form, perfect similarity will result in CPSD values of 100; in this "modified Index" values can exceed 100.

Another index that has been used successfully to detect lithologic discontinuities is the uniformity value (UV) of Cremeens and Mokma (1986) (Tsai and Chen 2000). As with other indices, the UV compares particle size data with data from the horizon above. It is calculated as:

$$\mathrm{UV} = \frac{[(\%\text{silt} + \%\text{very fine sand})/(\%\text{sand} - \%\text{very fine sand})] \text{ in upper horizon}}{[(\%\text{silt} + \%\text{very fine sand})/(\%\text{sand} - \%\text{very fine sand})] \text{ in lower horizon}} \text{ minus } 1.0.$$

The closer the UV is to zero, the more likely that the two horizons formed from similar parent materials; Cremeens and Mokma (1986) assumed that UV values >0.60 indicated nonuniformity.

Morphological indicators, if chosen carefully, might also help locate LDs, or they may substantiate LDs that were suggested by other methods. For example, horizon boundaries often coincide with LDs (Bartelli and Odell 1960a, b). Schaetzl (1996, 1998) noted that the depth of leaching in some Udalfs in Michigan is usually coincident with an LD (Fig. 8.47). Unusually thick or thin horizons may be indicative of a discontinuity. Depth functions of physical parameters such as bulk density or porosity have also been used to detect LDs (e.g., Arnold 1968), although their use should be done with caution. Stone lines within a profile may indicate the presence of an erosional episode and thus can be at or near a discontinuity (Ruhe 1958, Parsons and Balster 1966). Care must be taken, however, since many stone lines are biogenic (Johnson 1989, 1990, Johnson and Balek 1991).

Fig. 8.48 An Inceptic Hapludalf from northern Michigan where the 2Bt horizon occurs at a lithologic discontinuity between calcareous till (below) and a sandy loam surface of unknown origin. The Bt horizon is evident as a dark, wavy layer 30–60 cm below the surface. Note the large number of coarse clasts at or near the discontinuity. The Bt horizon exists as a thin illuvial layer immediately above the calcareous C horizon. As carbonates are leached from the calcareous part of the soil, clay is presumably remobilized and translocated to the contact again. Tape increments are 10 cm. Photo by RJS.

Detection of LDs becomes increasingly more difficult as soils develop (Fine *et al.* 1992). Translocation of mobile soil components, mixing processes and transformations within the profile may blur and confound the sedimentologic evidence of the discontinuity. Thus, in situations where an LD is suspected but the soil is old, it becomes increasingly imperative that any data on mobile elements, such as pedogenic iron (Singer and Fine 1989) or translocated clay, be removed from the data set.

Although most LDs can be detected by the examination of depth functions of carefully selected parameters, this method has the disadvantage of being somewhat subjective. More quantitative methods, involving discriminant analysis, have been reported (Norton and Hall 1985, Litaor 1987). However, these often require some sort of a priori reasoning regarding the possible location and/or existence of discontinuities. Like depth functions, discriminant functions are best calculated using data on an immobile and slowly weatherable soil fraction.

Pedogenic importance of discontinuities

Pedogenic processes that involve translocation are variously impacted and altered at and by the discontinuity (see Chapter 12). Discontinuities where fine material overlies coarse layers dramatically affect eluviation–illuviation processes because of increased hydraulic tensions in the upper, fine-textured material. Unsaturated flow cannot enter the lower, coarser-textured sand until it is nearly saturated (Bartelli and Odell 1960a, b, Clothier *et al.* 1978). Infiltrating water may "hang" at the fine-over-coarse discontinuity (Asamoa and Protz 1972, Busacca and Singer 1989, Khakural *et al.* 1993), leading to less frequent wetting of the materials below the discontinuity and deposition of illuvial materials immediately above it and at the contact (Fig. 8.48). When water does "break through" the contact, it often moves through the lower material rapidly and as finger flow (Liu *et al.* 1991, Boll *et al.* 1996, Dekker and Ritsema 1996). In these situations, LDs essentially cause the zone of illuviation to be vertically compressed. Soil morphology and horizonation are dramatically affected.

Where coarser materials overlie fine-textured sediment, infiltrating water is "pulled" into the lower material, and again, deposition is affected. In situations where the soil is saturated, water flow is again impacted by finer materials beneath the coarser sediment, as water may perch at the discontinuity.

Chapter 9

Weathering

An important step in the formation of *soil* from *rock* involves *weathering* of the rock into smaller and/or chemically altered parts (Yatsu 1988). Weathering is the physical and chemical alteration of rocks and minerals at or near the Earth's surface, produced by biological, chemical and physical agents or their combination, as they adjust toward an equilibrium state in the surface environment (Pope *et al.* 2002). Few soils form directly from bedrock. Usually there is an intermediate step in which weathering processes form residuum from the rock, or geomorphic processes comminute and erode it, forming various types of regolith. Both are then acted upon by pedogenic processes to form soil (see Chapter 12). In the end, rocks become discolored, structurally altered, acquire precipitates of weathering by-products and experience surface recession as a result of weathering. In this section we briefly examine the main components and processes of weathering; it logically follows on our treatise of soil parent materials in Chapter 8.

Rocks are variously physically and chemically unstable at the Earth's surface, and hence they *weather*, because this environment is far different from the one in which they formed. For most rocks, the surface (soil) environment is colder, with less pressure and increased amounts of oxygen, water and biota, than their formative environment, be it below the sea floor or deep within the crust. As noted in Chapter 4, minerals formed in such environments are often primary minerals. Many of these are unstable within soils and weather to secondary minerals, commonly clay minerals (see Chapter 4).

Geomorphically and pedogenically, several processes are associated with weathering, such as erosion (the wearing away of rocks or sediments/soils) and transport (the movement of those same materials), which collectively are termed *denudation*. Weathering, erosion, mass wasting and transportation are all included within denudational processes. During transport, sediment may also be *corraded* (eroded by friction) from the underlying rock or sediment.

Rocks (Table 9.1) and minerals (Fig. 9.1) vary in their resistance to weathering. Minerals that weather slowly or difficultly are termed "resistant" minerals, whereas those that weather more quickly are variously referred to as "weatherable" or "weak" minerals. The susceptibility of various minerals to weathering is largely captured by Bowen's series, which illustrates that minerals that crystallize first from a molten state are most susceptible to weathering (White and Brantley 1995) (Chapter 4). The structure of minerals also dramatically affects not only the susceptbility of a mineral to weathering, but the manner in which it weathers (Fig. 9.2).

Rocks with many weathering-susceptible minerals are vulnerable to weathering. However, rocks with many individual minerals that are only loosely held together will also be highly weatherable, regardless of their mineralogy. Weathering susceptibility of rocks is also a function of the size of the individual minerals as well as the degree to which they are cemented together. In general, coarse-textured rocks, with larger mineral crystals, such as granite, weather more rapidly than do finer-textured intrusive

Table 9.1 A classification of the physical strength of some rock types

Description	Examples
Very weak rocks	Chalk, rock salt, lignite
Weak rocks	Coal, siltstone, schist
Moderately strong rocks	Slate, shale, sandstone, ignimbrite
Strong rocks	Marble, limestone, dolomite, andesite, granite, gneiss
Very strong rocks	Quartzite, dolerite, gabbro

Source: Selby (1980).

rocks. It is easier for physical weathering processes to separate the mineral particles in the intrusive igneous and acidic metamorphic rocks (e.g., gneiss) than in a finer-textured rock like basalt. For many clastic sedimentary rocks, e.g., sandstone, the degree of cementation of the grains is a primary factor determining their resistance to weathering.

Physical weathering

In physical weathering, also known as *mechanical weathering* or *disintegration*, physical stresses combine to break rock into smaller pieces. Commonly, these breaks or fractures occur along joints or bedding planes, or between mineral grains as in a sandstone or granite. Little or no chemical alteration is implied. The sizes of fragments created by physical weathering depends on the degree of fracturing and the spacing of fractures and grain-to-grain contacts within the rock. A granite with large phenocrysts will necessarily weather into larger mineral fragments than will a schist composed of small mineral grains.

The primary importance of physical weathering centers on the formation of additional surface area and pore space. As rocks get fractured, their surface area increases geometrically (Fig. 9.3). Chemical weathering (see below) reactions operate on *surfaces*, and thus as a rock gets increasingly fractured, the surface area upon which chemical reactions can occur increases. Likewise, fractured rocks with more pores allow more oxygen and water to come into direct contact, again facilitating chemical weathering.

Water and oxygen are important to many chemical weathering reactions. Highly fractured rock tends to have increased amounts of porosity and permeability, which when combined with abundant water facilitate the transport of byproducts of chemical weathering reactions. If those by-products are not flushed from the system, the weathering reactions slow and eventually stop. Pore spaces also facilitate the growth of roots, which in turn can cause further physical break-up. Roots also facilitate chemical break-up by their exudates and various fungal hyphae associated with them. They also release CO_2, which combines with water to form carbonic acid, further enhancing weathering.

Forms of physical weathering

Physical weathering occurs as stress is *added to* a rock (usually from within an existing plane or point of weakness) or as pressure is *released from* the rock, allowing it to expand and crack. The latter process is called *unloading* or *dilatation*. Unloading occurs when weight is gradually or suddenly removed from dense rock masses, causing the rock to expand and sometimes resulting in sheets (sheeting) or flakes (flaking) that appear to peel from the rock. Dilatation can be up to 1% in some granites. Rock corners and edges are the most susceptible to sheeting and flaking, since a three-face corner has even more surface area to attack, and therefore potential for rounding. Thus, one major impact of these processes is to round the edges of rocks.

Rocks undergoing unloading typically expand along planes that are at right angles to the direction of pressure release. When this occurs on a large scale and the sheets are thick and/or

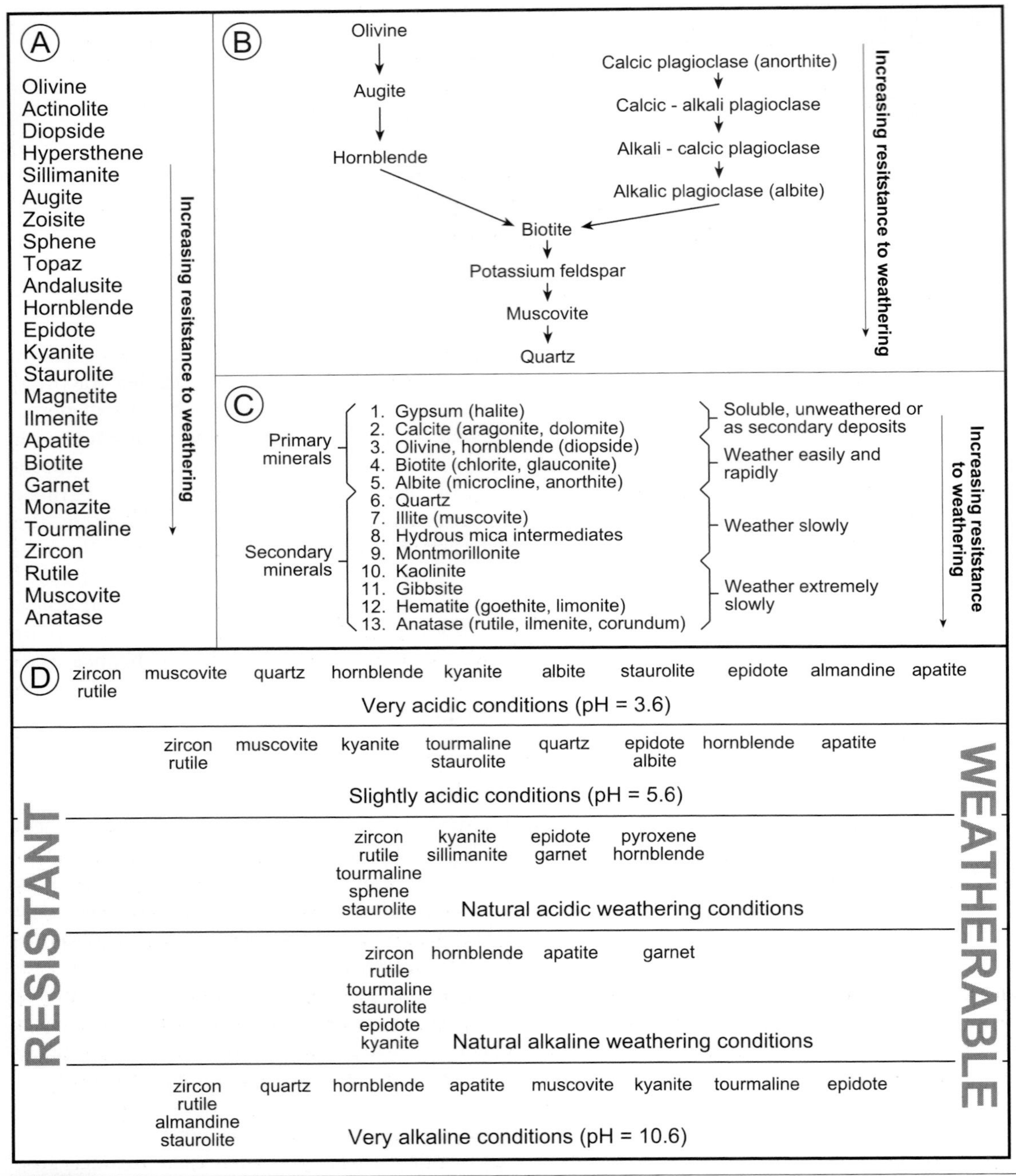

Fig. 9.1 Various rankings of common rock-forming minerals with respect to their resistance to weathering in general. Exact correspondence among the sequences is not implied. (a) After Pettijohn (1941), based on their frequency in sedimentary rocks of increasing age, which is interpreted as degree of stability or resistance to weathering. (b) After Goldich (1938), which is roughly the inverse sequence described for minerals as they crystallize from a melt. This sequence is often referred to as Bowen's series. (c) After Jackson *et al.* (1948). (d) Resistance and weatherability at various pH levels. After Weyl (1950) and Nickel (1973).

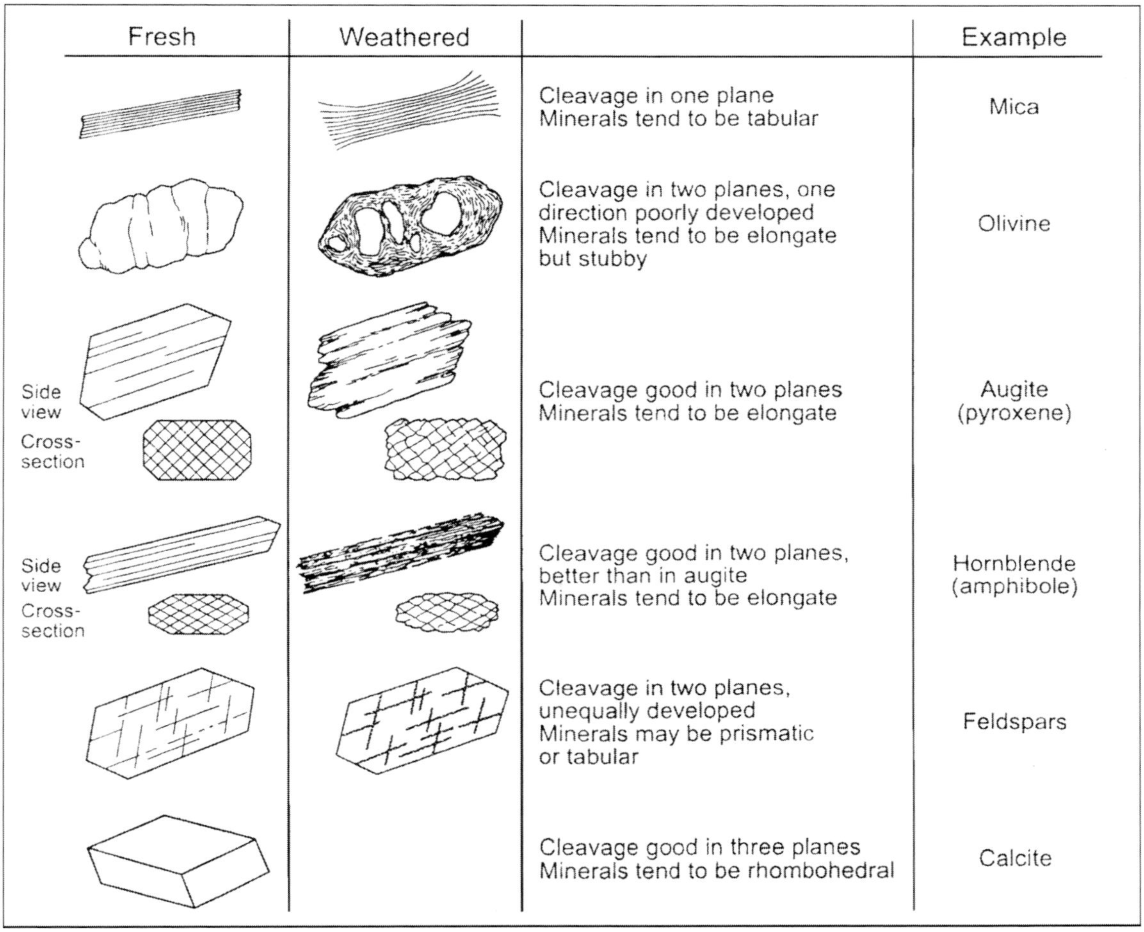

Fresh	Weathered		Example
		Cleavage in one plane Minerals tend to be tabular	Mica
		Cleavage in two planes, one direction poorly developed Minerals tend to be elongate but stubby	Olivine
Side view Cross-section		Cleavage good in two planes Minerals tend to be elongate	Augite (pyroxene)
Side view Cross-section		Cleavage good in two planes, better than in augite Minerals tend to be elongate	Hornblende (amphibole)
		Cleavage in two planes, unequally developed Minerals may be prismatic or tabular	Feldspars
		Cleavage good in three planes Minerals tend to be rhombohedral	Calcite

Fig. 9.2 Examples of how mineral structure and cleavage affect the way minerals weather. After Hunt (1972).

curved, the term *exfoliation* is used (Bradley 1963). Exfoliation is often restricted to massive, otherwise unjointed rocks; slaking or sheeting is used for smaller scale versions of the same. Continued sheeting and exfoliation wear away the sharp edges of the rock, rendering it round or spheroidal in shape, hence the name spheroidal weathering (Fig. 9.4).

Most forms of physical weathering occur when internal stresses overcome the strength of the rock, causing it to disintegrate. These stresses can be caused by a number of processes: (1) salt crystal growth as salty water within cracks in the rock evaporates, (2) ice crystal growth as water within cracks freezes, (3) root growth within cracks and (4) expansion of the rock as it is wetted or heated. All of these stresses are more effective if they occur in numerous cycles. Salt crystals, ice and roots are all agents that are added from *outside* the weathering "boundary layer," expanding into the volume.

Salt and gypsum weathering is a common form of rock disintegration in arid regions and near sea coasts where salty water is abundant (Reheis 1987b, Doornkamp and Ibrahim 1990). In urban areas there is also sometimes an overabundance of salts due to ice removal, fly ash dust, general traffic dust, etc. Water with dissolved substances penetrates crevices in rocks and upon drying, salt crystals grow and force apart the rock (Holmer 1998). Water is later drawn to the dry salt, continuing to enhance salt weathering (Amit *et al.* 1993). In some salty soils, the lack of rocks in the upper solum has been attributed in part to physical shattering of clasts by salts and other forms of mechanical weathering (Dan

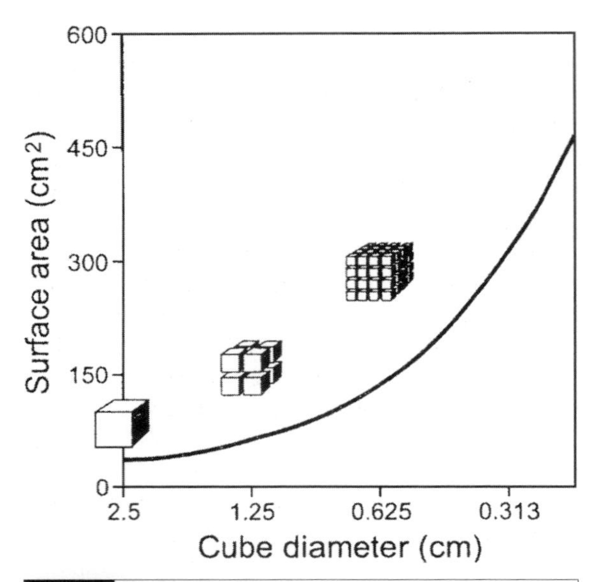

Fig. 9.3 Change in surface area as a cube-shaped rock is repeatedly fractured.

et al. 1982, Gerson and Amit 1987) (Fig. 12.62). Amit *et al.* (1993) stressed, however, that salt concentration alone does not cause shattering of clasts, for within petrosalic (Bzm) horizons in the lower solum, shattering does not occur. Rather, changes in moisture content and temperature in the upper solum, coupled with the presence of salts and rocks with microcrystalline fractures, lead to the rapid and complete rock shattering. The process is as follows: salt crystal shattering relies upon the formation of a salt crust on the rocks. These crusts plug the openings of the cracks, confining saline solutions to the interiors. Upon further drying and/or heating, the pressures inside the rocks can result in shattering.

Salt weathering is not restricted to deserts (Goudie and Viles 1997). Physical weathering, initiated by additions of rain rich in sulfuric acid (H_2SO_4) attributed to the burning of sulfur-rich coal, can weather rocks, especially carbonate rocks, as follows:

$$H_2SO_4 + CaCO_3 + H_2O \longrightarrow CaSO_4 \cdot 2H_2O + CO_2$$

Gypsum ($CaSO_4 \cdot 2H_2O$) crystals form just below the rock surface, occupying 38% more volume than the original calcite crystals (Feddema and Meierding 1987). The newly formed gypsum crystals can flake the limestone away at rates as high as 0.08 mm yr^{-1}.

The role of freeze–thaw activity and ice crystals in weathering is undeniable and yet, exactly how it functions is debatable. In theory, as water in cracks in rocks freezes, it expands by 9%, potentially forcing the rock apart and

Fig. 9.4 Spheroidal weathering of granite caused by sheeting and flaking, Joshua Tree National Monument, California. Photo by RJS.

shattering it. This process, also called *frost shattering* or *gelifraction*, works best in rocks with deep, narrow cracks where the pressure is contained wholly within the rock and not released as the developing ice mass pushes outside it. Obviously, this process can only operate in climates that at least occasionally have freezing temperatures, and the more cyclicity there is, e.g., in the high alpine or high midlatitudes, the better (Thorn 1979, Dredge 1992). Optimal conditions for frost shattering involve rapid freezing of water-filled cracks from the top downward, allowing the ice to act as a seal. Recent work on this process has, however, suggested that the main agent of disintegration may not be expansion by ice but the migration of thin films of water along microfractures. These films may be adsorbed so tightly that they cannot freeze, but they can exert pressure on the rock and cause it to fracture (White 1976, Walder and Hallet 1985, Bland and Rolls 1998).

Thermal expansion of rocks is driven by insolation (solar heating) and by fire. In theory, expansion occurs as the outer layers of a rock are heated by the sun or a fire, only to contract as they cool (Peel 1974). The expansion and contraction lead to compressive and tensile stresses, i.e., thermal fatigue (Halsey *et al.* 1998). Fatigue effects reduce the cohesive strength of intergranular bonds in the rock and initiate microfracture development (Warke *et al.* 1996, Hall and Andre 2001). The degree to which minerals in the rock expand with temperature is a function of their coefficient of thermal expansion, and in sunlight, their albedo. Minerals vary in these properties, setting up a condition in which the parts of the rock expand and contract at different rates, contributing to stresses and breakdown. For breakage to finally happen, however, the elastic limit of the rock must be exceeded, and this also varies among rock types (Bland and Rolls 1998).

Fire has long been known to be an agent of rock weathering (Blackwelder 1926, Ollier and Ash 1983, Allison and Bristow 1999). During fire, rock temperatures can approach 500 °C. "Fire weathering" is particularly effective if the fire is immediately followed by a cooling rain. Shattered rocks on the surface of hot deserts and in recently burned areas attest to the efficacy of thermal expansion as a weathering process, in addition to shattering by salt crystals.

Biotic forms of physical weathering have long centered on the role of plants. Roots penetrate fissures in rock and can exert enough pressure to break them apart; many of us have observed boulders that have been moved or split by tree roots. However, fauna also play a large role in disintegration processes, but usually function at the smaller end of the scale where their impacts are less noticed (Avakyan *et al.* 1981). Many burrowing animals ingest sand and gravel and as this material passes through their gut it gets physically comminuted (Suzuki *et al.* 2003). Large mammals like badgers and wombats are more than able to break apart rock or cemented soil horizons as they tunnel.

Lastly, wetting and drying cycles are also an effective agent of physical weathering, especially in rocks that are weak initially or are able to absorb large amounts of water (Hall and Hall 1996, Gokceoglu *et al.* 2000, Paradise 2002).

Chemical and biotic weathering

Chemical weathering, also known as *decomposition*, involves the biochemical breakdown of rocks and minerals into different products and compounds. Common end products of chemical weathering are clay minerals, soluble acidic compounds and various ionic species. The intensity of chemical weathering is largely governed by climatic conditions which govern temperature and the movement (or lack thereof) of water in the soil and rock, water quality and the mineral and chemical composition of the rocks themselves. As noted above, rock permeability and porosity, and surface area are also important, as they facilitate or retard flushing of reaction by-products. Indeed, in areas where chemical weathering is abnormally active or inactive, based on what would be climatically "expected," one can usually look to parts of the chemical weathering environment not governed by climate, e.g., permeability, rock chemistry, strength of weathering agents, etc. for an explanation (Pope *et al.* 1995).

Chemical weathering can be subdivided into congruent and incongruent dissolution. Congruent dissolution occurs when a mineral goes completely into solution and nothing is immediately precipitated from it. Many highly soluble

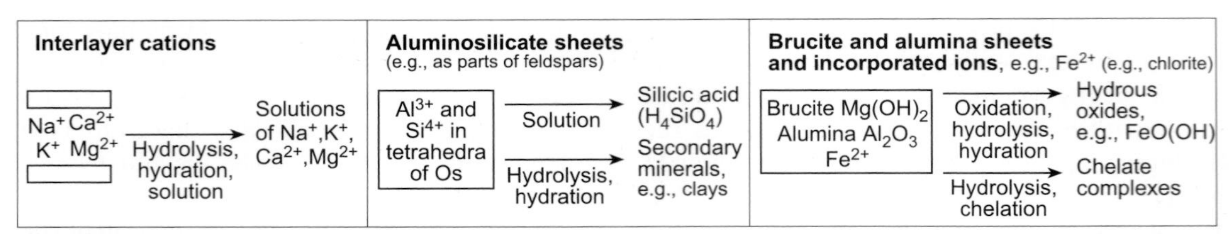

Fig. 9.5 Diagrammatic representation of the three main categories of chemical weathering, using a silicate mineral as an example. After Bland and Rolls (1998).

minerals, e.g., halite or marble, weather congruently, i.e., dissolve, leaving virtually nothing behind and forming a type of "collapsed" landscape (*karst* is the name given to this landscape type when the rock is limestone). In incongruent dissolution, all or some of the ions released by weathering precipitate at or near the site of weathering, to form new minerals such as phyllosilicates or oxyhydroxides.

Following on this two-fold generalization, Bland and Rolls (1998) divided chemical weathering into three categories (Fig. 9.5):

(1) Solution of ions and molecules.
(2) Production of new minerals, e.g., clay minerals, oxides and hydroxides.
(3) Residual accumulation of insoluble or otherwise unweatherable materials.

Any or all of these categories results in a chemical change in a preexisting mineral and the formation of new minerals or parts (ions) that could be used to "construct" them. Hence, chemical weathering involves decomposition, but an equally important part of the process is the synthesis of new compounds that occurs after the main reaction is done.

Some of the more important and commonly mentioned subsets of chemical weathering are discussed below.

Hydration

Many minerals hydrate, or take water *into* their chemical structure, readily. In hydration, water is added to the crystal lattice of a mineral but the original material does not change chemical composition. One example is:

$$\underset{\text{annhydrite}}{CaSO_4} + 2H_2O \longrightarrow \underset{\text{gypsum}}{CaSO_4 \cdot 2H_2O}$$

The hydration of annhydrite to gypsum results in expansion and mechanical deformation. Likewise, hematite can be hydrated to goethite:

$$Fe_2O_3 + H_2O \longrightarrow Fe_2O_3 \cdot H_2O$$

Secondary clay minerals, e.g., smectites, also hydrate readily (see Chapter 4).

Related to hydration are the processes of *oxidation–reduction*. These processes primarily affect minerals rich in Fe, Mn or S (see Chapters 13 and 14).

Solution

All minerals are soluble; some are much more soluble than others. Some of the more soluble of the common minerals are halite, potash (KCl), gypsum and, to a lesser extent, $CaCO_3$. The factors which affect solubility are myriad and beyond the scope of this discussion. However, the concept of ionic potential is central. Defined as ionic charge divided by ionic radius, ionic potential refers to the ease with which an ion can be removed, in solution, from a compound such as a mineral. Ions with low ionic potential typically are the main components of weathering solutions, implying that they are more easily leached, and hence, more soluble (Bland and Rolls 1998) (Fig. 9.6). Temperature affects solubility; most minerals are more soluble at higher temperatures. Most solubility reactions are also highly pH dependent. Many minerals, such as silica and aluminum oxides, are most soluble at high *and* low pH values (Fig. 9.7).

Unlike many other chemical weathering reactions, solution processes are usually reversible. When the saturation point of a solution is exceeded, it becomes supersaturated and some of the dissolved ions precipitate. These precipitates may, however, be different from those that were involved in the initial dissolution reaction.

Hydrolysis

Hydrolysis, (Greek, *hydro,* water, *lysis,* split) is the most important chemical weathering process

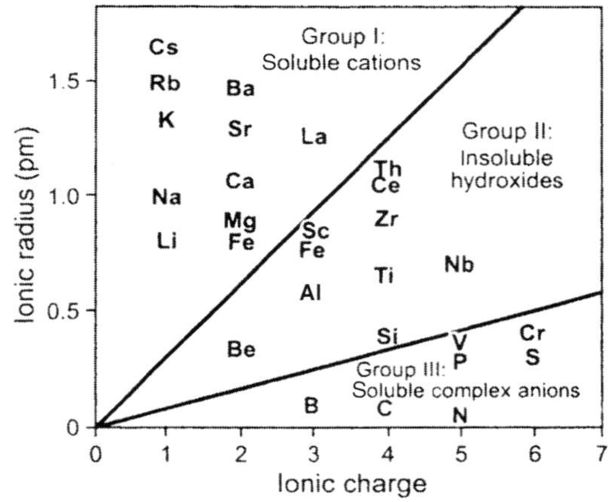

Cation	Ionic potential (charge / ionic radius)
Na^+	1.0
K^+	0.8
Ca^{2+}	2.1
Mn^{2+}	1.6
Fe^{2+}	2.6
Mg^{2+}	3.1
Fe^{3+}	5.6
Al^{3+}	6.6
Ti^{4+}	5.9
Mn^{4+}	7.4
Si^{4+}	9.8

Fig. 9.6 The ionic potentials of various elements common to soils and regolith. The ion of each element can be placed into one of three categories that roughly coincide with their relative potential to be leached from soils. After Paton (1978).

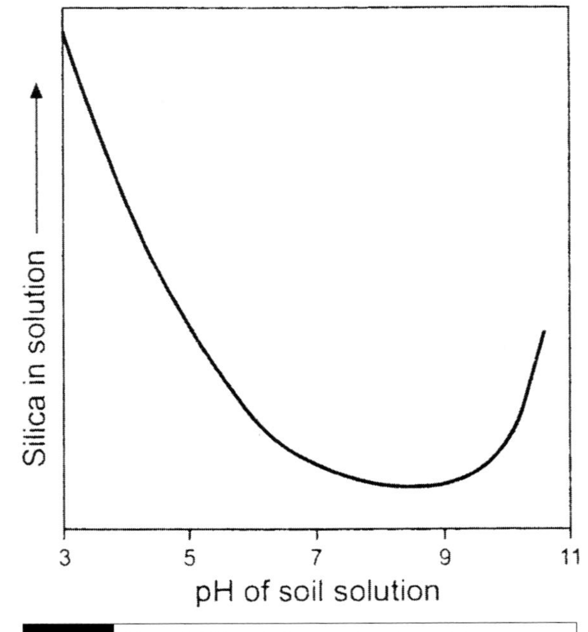

Fig. 9.7 The solubility of silica as a function of pH.

for silicate minerals. In hydrolysis, the water molecule is split into H^+ (a proton) and OH^- (a hydroxyl) ions, and is the main agent that drives mineral dissolution. The H^+ proton reacts with silicate minerals in a form of cation exchange. The cation that is released from the mineral then reacts with the OH^- anion to form a new substance. The H^+ ions also derive from acid dissociation, which is why hydrolysis increases at low pHs. Examples of this process abound, such as the one reported by Ahnert (1996):

$$2KAlSi_3O_8 + 2(H + OH) \longleftrightarrow 2HAlSi_3O_8 + 2KOH$$

(potassium feldspar + water ⟷ aluminosilicic acid + potassium hydroxide)

and then a series of reactions ensue, involving the byproducts of the hydrolysis, both of which are unstable:

$$2KOH + H_2CO_3 \longrightarrow K_2CO_3 + 2H_2O$$

(potassium hydroxide + carbonic acid ⟶ potassium carbonate)

The KOH is soluble and is carried away in solution, whereas the aluminosilicic acid reacts further:

$$2HAlSi_3O_8 + 9H_2O \longrightarrow Al_2Si_2O_5(OH)_4 + 4H_4SiO_4$$

(aluminosilicic acid ⟶ kaolinite + silicic acid)

Because of the importance of hydrolysis in silicate weathering, we provide another example reaction:

$$8NaAlSi_3O_8 + 6H^+ + 28H_2O \longrightarrow 3Na_{0.66}Al_{2.66}Si_{3.33}O10(OH)_2 + 14H_4SiO_4 + 6Na^+$$

(albite ⟶ smectite + silicic acid)

Both these reactions are typical of hydrolysis in that: (1) a primary silicate mineral is weathered using only water and commonly occurring acids, (2) a clay mineral forms as a result, (3) most of the reaction by-products are soluble and (4) in the end, the pH of the soil solution is lowered (an acid is produced). Because many of the by-products are soluble, much of the original mass is lost. In some environments, e.g., the humid tropics, the silica remains soluble and leaves the system via desilication (see Chapter 12), while in others it may precipitate as quartz or some other microcrystalline silicate mineral.

Carbonation

Carbonation is a general term for weathering processes associated with CO_2 in aqueous solution. In soils and regolith, CO_2 quickly combines with water to form a weak acid – *carbonic acid*:

$$CO_2 + H_2O \longleftrightarrow H_2CO_3.$$

Water that is in equilibrium with the atmosphere contains about 0.03% CO_2 and has a pH of 5.6 because of dissolved carbonic acid. However, soil air has much higher concentrations of CO_2 due to respiration by soil fauna and plant roots. Thus, carbonic acid concentrations in soil water can, naturally, be quite high. Carbonation impacts rocks at the surface, but because soil water is more aggressive than rainwater, it is particularly effective within the profile.

Carbonic acid, through its bicarbonate anion, is particularly effective at weathering calcareous rocks like limestone and marble:

$$\underset{}{H_2CO_3} \longrightarrow H^+ + \underset{\text{bicarbonate}}{HCO_3^-}$$

$$H^+ + HCO_3^- + \underset{\text{limestone.}}{CaCO_3} \longrightarrow Ca^{2+} + 2HCO_3^-$$

Both the Ca^{2+} and the bicarbonate are soluble and able to be leached, which means that, in theory, carbonation of calcareous rock is congruent. Generally, when limestone or marble weather via carbonation, all that remains are the insoluble components of the rock that were trapped in it as clastic impurities.

It should also be noted that carbonation is also effective on other types of minerals, as illustrated below:

$$\underset{\text{K-feldspar}}{2KAlSi_3O_8} + \underset{\text{carbonic acid}}{2H_2CO_3} + 9H_2O \updownarrow \underset{\text{kaolinite}}{Al_2Si_2O_5(OH)_4} + 2K^+ + \underset{\text{bicarbonate}}{2HCO_3^-} + \underset{\text{silicic acid.}}{4H_4SiO_4}$$

Weathering by carbonation creates a plethora of weathered forms (Fig. 9.8). Landscapes dominated by dissolution of carbonate rocks are termed *karst*. In such landscapes, carbonation has enlarged the natural fracture patterns in the rocks, leading to high levels of porosity and permeability. Thus, surface waters infiltrate rapidly, and there exists few surface streams. Underground rivers, collapsed caves, swallowholes and sinkholes are common features of karst landscapes (Fig. 9.8b).

Biological agents

Plants and their associated microbiota directly impact weathering by generating chelating compounds (see Chapter 12), by modifying pH through production of CO_2 and organic acids, by altering the exposed surface areas of minerals, via nitrification and by affecting the residence time of water (Drever 1994). Of these, the production of organic acids may be the most important contribution to biochemical weathering in soils. These acids can either be released directly from the organism or derive as by-products of organic decomposition. Bland and Rolls (1998) described two ways in which organic acids contribute to chemical weathering:

(1) Monovalent cations in the mineral are replaced by protons (H^+ cations) and then removed in solution.
(2) Di- and trivalent cations are made soluble either after redox reactions have occurred or by chelation by organic acids (Schatz 1963).

In either event, they are made readily soluble (Fig. 9.9).

Obviously, plants also play a critical role in the fate of weathering by-products, for if they are not biocycled, leaching or loss from the system in other ways becomes more likely.

(a)

(b)

Fig. 9.8 Some features associated with dissolution of limestone rocks and karst. Photos by RJS. (a) Small-scale dissolution features, sometimes called *karren*, on a surface limestone boulder, Indiana. (b) A swallowhole in a karst landscape – a location where surface water enters the subsurface hydrologic system.

(a) HOOC–(benzene ring)–OH, OH + M^{3+} → HOOC–(benzene ring)–O, O–M^{+} + $2H^{+}$

(b) $2H_2O + M^{2+}$ + oxalic acid → chelate complex + $2H^{+}$

metal ion + oxalic acid → chelate complex

Fig. 9.9 Examples of chelation of metal cations by organic acids. (a) Chemical structure of a typical phenolic compound (in this case, protocatechuic acid) before and after it has formed a chelate complex with a metal cation. After Vance *et al.* (1986). (b) A generic example of chelation. After Bland and Rolls (1998).

Recently, research on rock weathering has increasingly focussed on the role of fungi, which often form symbiotic relationships with plant roots. These *ectomycorrhizal fungi* mobilize essential plant nutrients directly from minerals through excretion of organic acids (Landeweert *et al.* 2001). Direct weathering by fungi can, however, occur (Etienne and DuPont 2002, Hoffland *et al.* 2002). Through their use of fungi in their mounds, termites have even been shown to chemically weather clay minerals (Jouquet *et al.* 2002).

Lichens, which are composite organisms whose body (thallus) consists of a fungus and an alga growing symbiotically, are early colonizers on bare rock surfaces and are therefore instrumental in weathering. They weather the rock both physically and chemically (Williams and Rudolph 1974, Chen *et al.* 2000, Aghamiri and Schwartzman 2002). Their nutrient-absorbing and transmitting bodies, *mycelia*, are composed of a complex web of fine thread-like *hyphae*. The hyphae can penetrate the rock surface along microcracks and generate enough stress to fracture parts of it. Likewise, as water is absorbed into the hyphae they expand and create additional pressure. The swelling action of the organic and inorganic salts originating from lichens also adds to physical break-up of rocks. On bare rock surfaces, lichens enhance the water-holding capacity locally, which may increase chemical dissolution and, climate permitting, frost wedging (Bjelland and Thorseth 2002). Lichens also emit organic acids, especially oxalic acid, which are very effective chemical weathering agents (Ascaso and Galvan 1976). The exact nature of the chemical weathering process associated with lichens remains unclear, but it is suspected that the acids from lichens (1) donate protons (H^+) that cause hydrolysis of the minerals and (2) act as agents of chelation. Lastly, the CO_2 emitted by the lichen plays a role in weathering via its generation of carbonic acid.

Products of weathering

The fate of the ions released by chemical weathering is highly variable. Some of these cations and anions are biocycled, while others, especially those that are not plant nutrients and those that are highly soluble, are leached and may end up in the groundwater. Some cations may be temporarily trapped within mineral and grain voids as secondary crystalline or amorphous deposits. In this state, they are at least temporarily unavailable to plants and not flushed into the groundwater.

Using this knowledge, a large literature has emerged that is concerned with measuring the rate and type of weathering, usually on a watershed scale. The technique applies data on the concentrations and types of cations in ground and surface waters exiting the basin to determine the types and rates of weathering occurring within (Velbel 1985, Pecher 1994, Richards and Kump 1997, Furman *et al.* 1998). The assumption in these studies is that the soluble cations released by weathering reactions eventually leave the system via flux from springs, streams and groundwater, and if the geology is known, a mass balance approach can provide information of which minerals are weathering and at what rates (Duan *et al.* 2002, West *et al.* 2002). Rates of landscape denudation due to chemical weathering can even be calculated using such data (Pope *et al.* 1995, Yuretich *et al.* 1996).

Controls on physical and chemical weathering

Many factors control and impact weathering rates and govern the reactions, not the least of which is macroclimate, because of its impact on moisture availability and temperature (Büdel 1982). Temperature impacts physical weathering in that it can affect freeze–thaw processes and, in hot deserts, salt weathering. It may be equally or more important for chemical weathering because temperature controls the rate of the reaction(s); reaction rates increase at higher temperatures (White *et al.* 1999). This concept is captured in the general rule of thumb that, for every 10 °C increase in temperature, the reaction rate is doubled. Obviously, this "rule" only holds within certain temperature limits, usually taken to be 5–35 °C. Nonetheless, the implications of this rule are huge; in the tropics many chemical reactions operate more rapidly than in midlatitudes, and if water is adequate, deep weathering profiles of highly weathered regolith can result (Fig. 9.10).

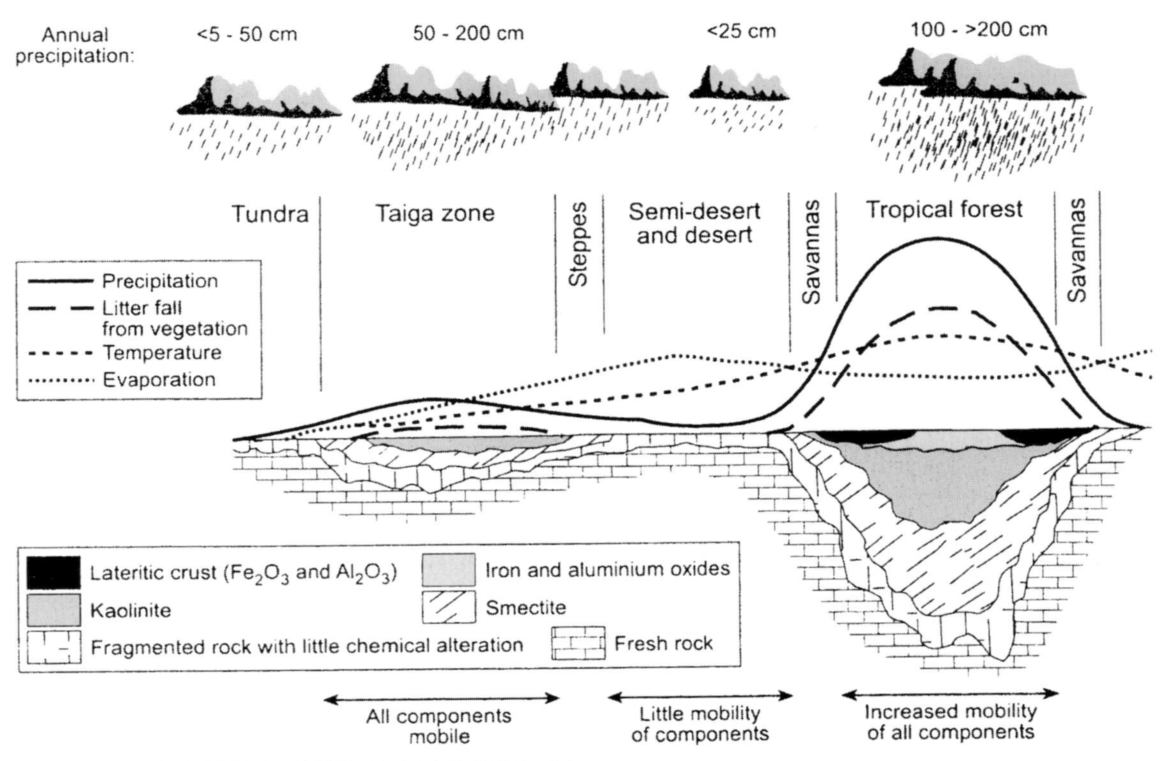

Fig. 9.10 Depth of weathering (= regolith thickness) and soils along a latitudinal transect from the equator to the pole, showing the major climatic belts. Although generally correct, this simplistic view of weathering across the globe creates misconceptions that ignore denudation rate adjustment and climate change. After Strakhov (1967).

We stress that temperature affects *rates*, not the *type* of reactions or processes; there is no reason to postulate a special type of "tropical weathering" as has been done in the past.

Physical weathering and its products have traditionally been viewed as being dominant in cold and/or dry climates. Peltier's (1950) classic work, though now dated, retains its value in pointing out that physical weathering is best expressed in cold, dry climates, especially those like alpine areas with their frequent freeze–thaw cycles (Fig. 9.11). Although not shown in Fig. 9.11, physical weathering is also common in deserts and along sea coasts, where salts and water commingle.

A major control on chemical weathering involves the degree to which water can participate in the reaction, including the ability to flush the system of weathering by-products. In stagnant but wet settings, chemical weathering quickly grinds to a halt, as by-products accumulate. Since water availability is a function of not only local factors but also macroclimate, it has long been assumed that chemical weathering is dominant in hot, wet (humid) climates where the reactions involved are assisted by warm temperatures. Conversely, physical weathering is assumed to be dominant over chemical weathering in cold, dry environments, where freeze–thaw processes are active and where cold temperatures inhibit chemical reactions (Fig. 9.11). Several versions of the diagram shown in Fig. 9.10 have been published for decades, illustrating and (to some) confirming this point. This notion, although largely correct, has recently been challenged (Pope *et al.* 1995). Hall *et al.* (2002) pointed out that, contrary to long-standing belief, weathering, including chemical weathering, is not strictly temperature-limited but is, instead, limited more by moisture availability. Even in polar regions, rock temperatures are more than adequate to support physical *and* chemical weathering, providing that water is present. Wherever water is available, chemical weathering can be a major component of the weathering regime (Thorn *et al.* 2001). Pope *et al.* (1995) argued that the macroclimatically

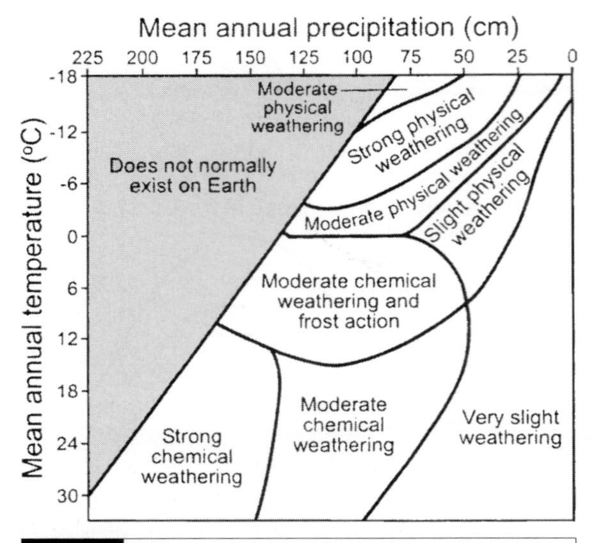

Fig. 9.11 Diagram suggesting the relative importance of the various types of physical and chemical weathering in different environments on Earth. After Peltier (1950).

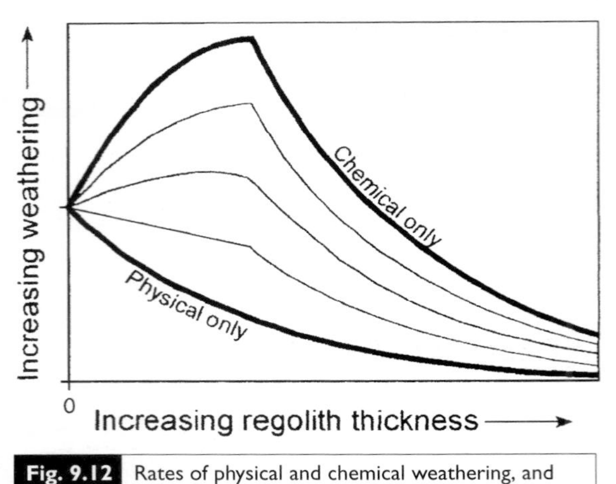

Fig. 9.12 Rates of physical and chemical weathering, and combinations thereof, as a function of regolith thickness. After Ahnert (1987).

driven approach to weathering is flawed because it is based on visual observations, and that the real mechanisms of weathering, observable only under a microscope, do not support the broad-brush, macroclimate–zonal categorization of Peltier (1950). The climatic models of weathering fall apart because the process link between tropospheric climate and boundary layer weathering processes is often disjunct.

As regolith thickness increases, the rocks below get exposed to fewer and fewer wet–dry and freeze–thaw cycles, and the influence of biota and macroclimate gets diminished. Thus, regolith thickness tends to be inversely proportional to the intensity of physical weathering (at the top of the bedrock) (Fig. 9.12). This is an example of a negative feedback mechanism that operates between weathering and denudation (Ahnert 1987). When regolith is thin enough, weathering can operate, but as it thickens, the impact of regolith is to force weathering reactions to slow and possibly cease. Because of this interaction, weathering and denudation rates tend to adjust mutually in a type of dynamic equilibrium, resulting in a more constant thickness of regolith than would otherwise occur. This line of reasoning can help explain the deep regoliths that lie outside of climatic "expectations," such as in Australia, the Rocky Mountains, Spain, or the Appalachian Mountains.

Assessing weathering intensity

For a number of reasons – applicable in soil geomorphology and genesis – it is often useful to know the degree to which a soil or regolith is weathered. Often, degree of weathering parallels degree of soil development. In theory, the older the sediment or landform (surface) is, the more weathered it should be. Climate, however, has a strong impact as well. Thus, weathering data tend to focus (minimally) on *relative* degrees of weathering at different sites, but if possible, more rigorous numerical treatments of weathering intensity can be achieved. Actual *rates* of weathering are normally left to geochemists, who commonly derive such data based on mass balance equations of fluxes of weathering by-products exiting a section of landscape in surface or groundwater (see above).

The methods used to determine degree of weathering of rocks range from qualitative ones involving color or hardness, to highly quantitative measures involving mineralogy or chemistry. Many of these rock weathering methods are discussed in Chapter 14, where their application is fit into the context of using weathering to assess the age of the sediment or a landform they reside within. Examples, particularly those associated with Robert Ruhe's work in Iowa, are provided in Chapter 13.

Chapter 10

Pedoturbation

Pedoturbation, popularized by Francis D. Hole (1961), is synonymous with soil mixing. The mechanisms and vectors by which this physical mixing is accomplished are many, and function from microscopic scales at which crystals grow and deteriorate, to larger mixing associated with uprooted trees, massive termite mounds and debris flows. The importance of pedoturbation has traditionally not been emphasized by most soil and earth scientists. Nonetheless, it is ubiquitous. Not only is it a regressive (mixing) process but it can also enhance soil genesis, horizonation and order. In short, although it is a form of *mixing*, pedoturbation is not, as we shall see, always synonymous with *homogenization*.

Pedoturbation affects soil genesis and its developmental pathways almost continually, but it often goes little noticed. Knowledge of pedoturbation is vital for the study of pre-existing stratification, such as in archeology (Wood and Johnson 1978, Rolfsen 1980, Stein 1983, Bocek 1986, McBrearty 1990, Balek 2002) and sedimentology, as well as for those who study pedogenic layering, e.g., soil scientists or geomorphologists (Johnson *et al.* 1987). For example, geological lithologic discontinuities can be blurred or completely masked by pedoturbation processes. Pedoturbation is in large part responsible for maintaining macroporosity in soils, which in turn aids in infiltration and retards runoff and erosion. Physical mixing of organic matter into soils is largely accomplished via pedoturbation (Fig. 12.2). In short, there is hardly a single pedogenic pathway that is not affected or altered by pedoturbation.

In order to understand pedoturbation, as with many other pedogenic processes, a starting point must be determined to analyze the effects of that process on the soil. This raises potential problems of circularity (Van Nest, pers. comm. 2002), i.e., the researcher must ascertain what the pre-existing state of the soil was, in order to fully document/quantify/understand pedoturbation at that site or in that soil. This is tough work, largely because pedoturbation varies spatially and temporally, besides being highly scale-dependent.

Many traditional soil genesis studies have proceeded by ascertaining how the soil profile (an ordered state) has developed. Comparatively few studies, however, have stressed or examined the converse: the formation of disorder (haploidization) from an otherwise pedologically ordered (horizonated) soil. Still fewer have examined the preservation of disorder, by pedoturbative processes, or the formation of soil order by pedoturbation. This oversight, perceived or real, is the impetus for this chapter.

Classifying pedoturbation: proisotropic vs. proanisotropic

Pedology is replete with studies that examine how *pedologic order* (anisotropy, soil horizonation, layering) evolves from sediments with *geologic disorder* (isotropic parent materials such as loess or dune sand) or *geologic order*, such as stratified alluvium or saprolite. This mindset has proliferated, deliberately or inadvertently, the notion that pedoturbation is a regressive soil process – one that blurs soil horizons or prevents them from forming (see Chapter 11). Fewer studies observe that pedoturbation can actually create, or preserve at best, anisotropy that results from pedogenesis.

With this in mind, we follow the classification of others (Hole 1961, Johnson *et al.* 1987), who placed pedoturbations into one of two categories: *proisotropic* or *proanisotropic*. The former term implies a condition tending toward ("pro") isotropy (disorder, randomness), while proanisotropic pedoturbation means a tendency toward order, nonrandomness and sorting (Fig. 10.1). More rigorously defined (Johnson *et al.* 1987; see also Hole 1961: 375), *proisotropic pedoturbations* are processes that *disrupt*, *blend* or *destroy* horizons, subhorizons, or genetic layers and/or *impede their formation*, and cause morphologically simplified profiles to evolve from more ordered ones. Proanisotropic pedotrubations are processes that *form* or *aid in forming/maintaining* horizons, subhorizons, or genetic layers and/or cause an overall *increase in profile order*. Soil mixing should be envisioned as having components of each of these two sets of interacting processes and factors.

Classifying a pedoturbative process as proanisotropic or proisotropic depends on whether it promotes a more simple or more complex soil profile. Proisotropic pedoturbation is akin to homogenization, although the homogenization may only occur in part of the profile and only for a brief period of time. The literature contains many examples of proisotropic pedoturbation, such as the destruction of soil horizons by burrowing animals, or tree uprooting (Schaetzl *et al.* 1990). Borst (1968) described how ground squirrel burrowing activities had destroyed incipient horizons and prevented the development of argillic horizons. By continually mixing soils in the same area, these squirrels are capable of impacting the soils to a depth of 75 cm every 360 years. Proanisotropic processes act in the opposite fashion; they create order (horizonation) in soils and/or strengthen existing order.

Seldom do pedotubation processes act entirely in either of these two ways; rather, they usually have elements of both, with one form of pedoturbation more strongly expressed than the other (Fig. 10.1). For example, earthworms may mix O horizon material into the A horizon, and in so doing isotropically blur the two horizons (Nielsen and Hole 1964). But in so doing the worms thicken the A horizon at the expense of the O, thereby promoting horizonation. Alternatively (and probably concurrently) they may mix small amounts of mineral matter from the A horizon with raw litter and deposit organic-rich casts on top of the O horizon; the O horizon thus thickens – a form of anisotropic mixing (Bernier 1998).

Archeology provides examples of how both forms of pedoturbation can operate synchronously. Human artifacts dropped on the soil surface are lowered into the soil as worms bring casts from the A horizon and deposit them above the artifacts (Cremeens 2003). After a few thousand years, the artifacts are lowered into the soil, but their layering and even their spatial organization, can be preserved (Darwin 1881, Atkinson 1957, Van Nest 2002). To most observers, a "before" and "after" snapshot of the upper horizons in such a soil would be indistinguishable. The main difference lies not in the fine earth fraction, but in the coarse materials – the artifacts. Likewise, the burrowing of cicada nymphs in grassland soils can be viewed as almost entirely proanisotropic. Hugie and Passey (1963: 79) observed mixing of soil horizons by cicadas only at abrupt and clear soil boundaries, suggesting that the cicada nymphs continually backfill their burrows with local soil materials, keeping open only a short lead tunnel.

These examples illustrate the point that pedoturbation is rarely wholly isotropic. It may be mostly anisotropic, or it could be partly anisotropic and partly isotropic, depending on such factors as time of year, which part of the profile the activity occurs in, which size fraction is being examined, etc. Think of it another way: isn't it true that for many, if not most soils, the processes subsumed under the various types of pedoturbation are ongoing from the start? Are they not integral to the formation of the soil, rather than upsetting some other idealized state? For example, one definition of a B horizon involves the obliteration of original rock structure (including primary sedimentary

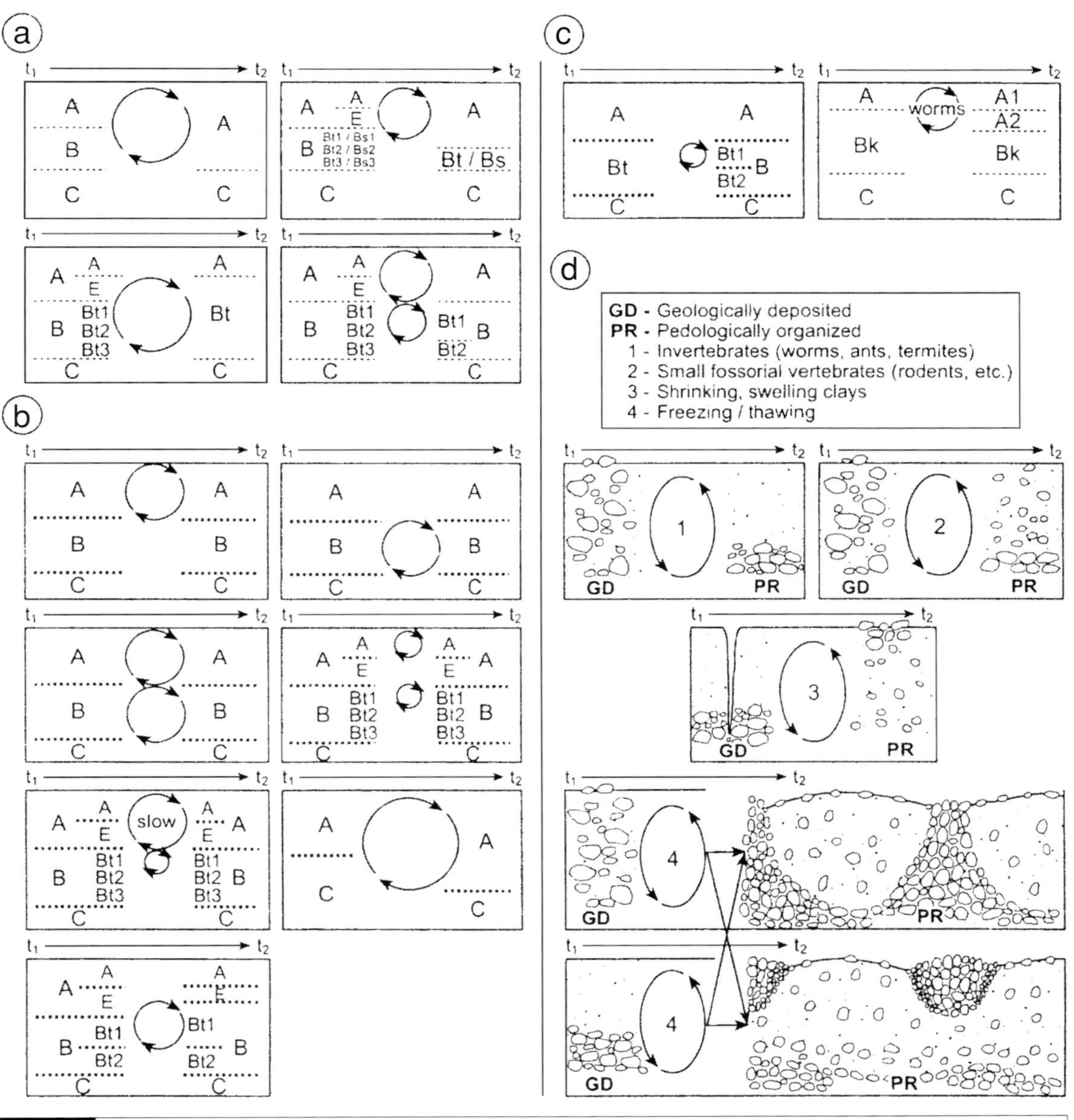

Fig. 10.1 Examples of proisotropic and proanisotropic pedoturbation. After Johnson *et al.* (1987). (a) Hypothetical examples of proisotropic pedoturbation, in which a horizon or horizons are simplified or destroyed by pedoturbation, during the time period t_1 to t_2. (b) Hypothetical examples of proanisotropic pedoturbation, in which a horizon or horizons are maintained despite the mixing, during the time period t_1 to t_2.
(c) Hypothetical examples of proanisotropic pedoturbation, in which a new horizon is produced by pedoturbation, during the time period t_1 to t_2. In the right diagram a Vermdudoll with two distinct A horizons is formed by intense faunalturbation. (d) Hypothetical examples of proanisotropic pedoturbation, in which geologically organized and/or unorganized sediment is reorganized by various types of pedoturbators.

structures). At least some of the major "players" in this process are roots and soil fauna. Thus, few would argue that biota are helping to form pedogenic structure in this instance, but how much of this work would also fall under the rubric of bioturbation? And if the processes subsumed under pedoturbation may also lead to increased order and anisotropy, should not these processes be accorded a primary (formative) rather than secondary (de-formative) role?

Pedoturbation, particle size and biomantles

Whether a pedoturbative process is proisotropic or proanisotropic, as well as the spatial expression of that process, often depends on not only the form of the process (mixing by tree uprooting vs. ant mounding) and its intensity and duration, but also on the *particle size fraction* that is being examined. That is, pedoturbation can be regressive or progressive, isotropic or anisotropic, depending on one's point of view – is the focus the fine earth fraction, or gravel? Many forms of pedoturbation mix the fine earth fraction while concomitantly sorting and creating order in the larger size fractions, like gravel or stones. For example, many insects, arthropods and small mammals burrow in and through the soil profile, bringing soil material to the surface and/or reorganizing it within the soil profile proper (Wiken *et al.* 1976, Munn 1993). For these soil *infauna*, some fragments are too large to move. These fragments cannot be carried to the surface or even moved upward in the profile by infauna. Rather, in the process of burrowing around such fragments, they eventually settle downward, as tunnels collapse and smaller materials nearby are moved to the surface. In time, these fragments become concentrated at the approximate maximum depth of burrowing. The coarse fragment zone thus formed is often called a *stone line*, stone zone or stone layer (Williams 1968, Johnson 1989, 2002). Above the stone line is a covering of relatively stone-free material which, because it was formed by bioturbation, is called a *biomantle*. A biomantle is a layer of material sorted and brought to the surface by animals (Soil Survey Staff 1975: 21, Johnson 1990, Humphreys 1994, Balek 2002).

Because they have a distinctive mode of formation, many biomantles also have a characteristic *biofabric*. For example, a worm-worked biomantle will have a strong crumb or granular fabric associated with worm casts and voids (Peacock and Fant 2002). Horizons within the biomantle can be quite thoroughly mixed or they can retain some original stratification. Van Nest (2002) discussed how artifact patterns, rock foundations and brick patio arrangements can be preserved beneath a biomantle, as long as the burial processes are slow and spatially uniform. The large clasts are lowered so slowly and evenly that their original spatial arrangement is preserved. In such a case, the principle of superposition (see Chapter 14) obviously does not apply, as the age of the stone line and biomantle are synchronous. This fact *must* be taken into consideration whenever interpretations are being made in soils, based on stratigraphy and superposition (Piperno and Becker 1996, Cremeens 2003).

Charles Darwin, in his classic (1881) book *On the Formation of Vegetable Mould through the Action of Worms*, showed how bioturbation texturally differentiates soils. In particular, Darwin described how a loamy, worm-worked topsoil is formed above a subsurface layer of flints and cinders that had been spread on the soil surface years earlier. These stones had migrated downward due to worm bioturbation to form a subsurface stone line. Such stone lines are omnipresent in areas where the soil has coarse and fine fractions, and where burrowing infauna have occupied the soil for a long time. If the stone line contains human artifacts, a false impression of a paleo land surface can be created by these artifacts at a consistent depth. Knowing of these processes and their importance, Atkinson (1957: 222) stated that many archeological materials have been displaced downwards, and that in some cases the amount of displacement may have been sufficient to alter their apparent stratigraphic relationships (Erlandson 1984). Van Nest (2002) used this knowledge to explain why artifacts >3500 years old are often found several centimeters beneath the soil surface, and thus often overlooked by routine archeological surveys. The upper limit on the size of materials or structures that can

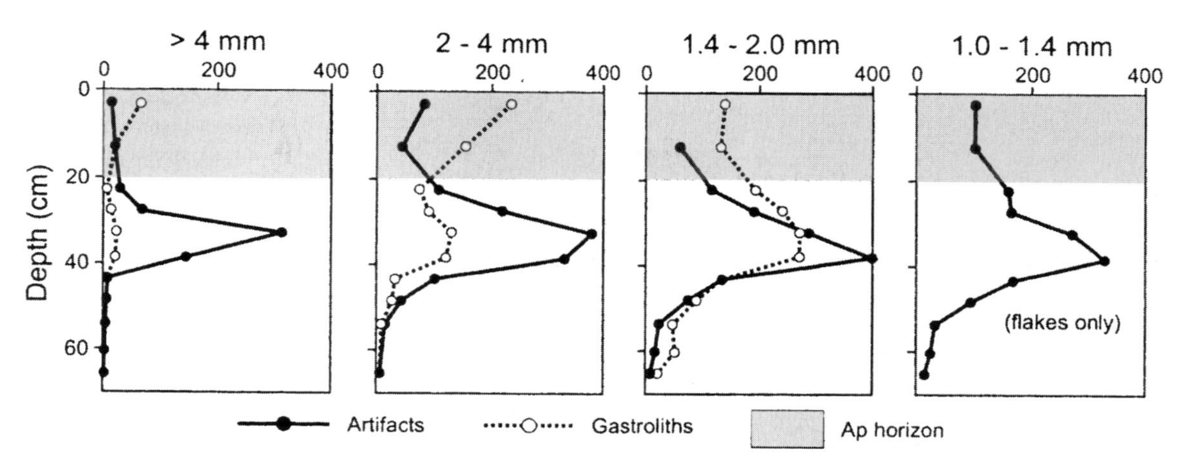

Fig. 10.2 Distributions of artifacts and gastroliths with depth in an upland soil in Warren County, western Illinois. Note the presence of a developing stone zone at depth. After Van Nest (2002).

be lowered by bioturbation is quite large. Darwin showed, for example, that some of the fallen stones of Stonehenge have come to be buried to depths of several centimeters since they originally toppled.

The depth of the stone line and the size of fragments within it depends on the burrower (how deep it burrows), the composition of the soil and the amount of time that the process has been ongoing. For example, ants may form a stone zone at 2 m, composed primarily of particle larger than 3 mm in diameter. For gophers, the stone zone may be deeper and limited to stones longer than 15 cm. Johnson (1989) noted that a certain species of pocket gopher in California does not transport stones that are larger than ≈6–7 cm in long diameter. It follows that if the largest clasts in the soil are smaller than the maximum size clasts the infauna are capable of moving, no stone line or stone zone will be formed at depth.

Often, contained within the stone line or zone are *gastroliths* – gravel-sized fragments that are ingested by birds to facilitate digestion. After they pass through the bird's gut or are coughed out, their fate is similar to that of any other stone, human artifact or coarse clast (Fig. 10.2).

In cases where bioturbation is intense and widespread, most of the original soil horizonation is destroyed (but the coarse fragment part of the soil either retains its stratification or develops stratification from a less-ordered state). Where stone-free sediments, like loess, overlie sandy and gravelly sediment and the gravels are all capable of being moved by infauna, mixing may be mainly anisotropic and no stone line may form. For example, Munn (1993) discussed how prairie dog burrowing in soils developed in alluvium over gravel thoroughly mixes the two zones. In other instances, where the burrowing activities are scattered in time and space, soils take on a form of patchiness and overall landscape complexity increases; vestiges of previously formed soil horizons can therefore be maintained. Indeed, the patchiness or pattern of entire landscapes is often due simply to long-term pedoturbation, usually by mounders such as prairie dogs and the South American vizcacha (Löffler and Margules 1980, Branch 1993).

While many forms of pedoturbation create a stone-free biomantle, other forms of pedoturbation can bring large rocks, including archeological artifacts, to the surface (Holmes 1893). Tree uprooting is a major cause of large rocks and stones being thrust upward (Lutz 1960, Daniels *et al.* 1987, Schaetzl *et al.* 1990, Small *et al.* 1990). Then, when fine materials are washed off the root + soil mass, coarse fragments can become concentrated as a surface lag (Denny and Goodlett 1956, 1968, Beatty and Stone 1986) (Fig. 10.3). Freeze–thaw processes, i.e., cryoturbation,

Fig. 10.3 An eroding root mass from a tree that was uprooted about 2–3 years previously. Erosion by water and wind has led to the formation of rock pedestals on the surface, a form of lag concentrate. Thus, a form of pedoturbation has also led to sedimentologic sorting. Photo by RJS.

bring large clasts to the surface and sort them (see below). Argilliturbation can function similarly (Johnson and Hester 1972, Muhs 1982), as can some larger mammals and reptiles. Thus, pedoturbation can be viewed as both proisotropic (destroying surface horizons and mixing the fine earth and small gravel fraction) *and* proanisotropic (forming a gravel and coarse-fragment-enriched stone zone at depth and a stone-free biomantle). It simply depends on the particle size fraction that is being considered (Johnson *et al.* 1987).

Baxter and Hole (1967) studied ant mounds on a prairie remnant in Wisconsin, where the mounds occupy 1.7% of the surface. They assumed a 12-year occupancy for each mound, implying that each point on the landscape might be occupied by a mound every 600 years. At this intensity of faunalturbation, Baxter and Hole (1967) argued that strongly horizonated (anisotropic) Alfisols had been changed into a more isotropic Mollisol in the 3500 years since prairie replaced forest. One of the more interesting ways in which the ants had accomplished this was their selective "mining" of B horizon material, leading to the formation of a humus- *and* clay-rich A horizon in the resulting Mollisols. Thus, as with worms, many soils with diffuse horizon boundaries can ascribe this morphology to long-term faunal activity.

Expressions of pedoturbation

Pedoturbation, in all its various forms (Table 10.1) is studied either by observing the process, such as ants moving soil particles (Perez 1987b), or by examining and interpreting the end products of such processes (Baxter and Hole 1967, Schaetzl 1986b, Cox *et al.* 1987a, Johnson 2002). In the former case, morphological or chemical signatures of the pedoturbation are observed and measured. And in order to identify past pedoturbation evidence, similar soils on a younger surface might need to be examined. Nearby soils may provide clues about pedoturbative processes that are spatially isolated, such as tree uprooting. Uprooting may affect a landscape continually for millennia, but may affect a given pedon only once every 5000 years. Thus, evidence of active pedoturbation in nearby pedons can be used to infer that the process is operative but temporally and spatially discontinuous.

Signatures of previous pedoturbation are expressed as within-profile, morphological imprints and as surface topographic features. Within-profile imprints include slickensides, faunal fecal pellets, reorganization and reorientation of the soil microfabric, subsurface or surficial stone lines, broken and disrupted horizons and mixing in general (Wang *et al.* 1995). Surface expressions of pedoturbation occur as microrelief;

Table 10.1 Major types of pedoturbation vectors

Form of pedoturbation	Soil mixing vectors
Aeroturbation	Gas, air, wind
Anthroturbation	Humans
Aquaturbation	Water
Argilliturbation	Shrinking and swelling of clays, such as smectite
Cryoturbation	Freeze–thaw activity, ice crystals
Crystalturbation	Crystals, such as ice and various salts
Faunalturbation[a]	Animals, including insects
Floralturbation[a]	Plants
Graviturbation	Mass movements, such as creep
Impacturbation	Extraterrestrial impacts such as comets and meterorites, and human-generated impacts, e.g., artillery shells and bombs
Seismiturbation	Earthquakes

[a]Collectively, floralturbation and faunalturbation are referred to as *bioturbation*.

Source: Hole (1961) and Johnson *et al.* (1987).

examples include gilgai, treethrow mounds and pits, anthills, termite mounds, patterned ground, and in-slumped animal burrows of all kinds (Haantjens 1965). The scale at which these features are manifested ranges from microscopic to those observable with the unaided eye.

Forms of pedoturbation

Hole (1961), Wood and Johnson (1978) and Johnson *et al.* (1987) variously classified the vectors or forms of pedoturbation (Table 10.1). These are discussed below, in no particular order.

Aeroturbation

Little work has been done on pedoturbation by gases. Mixing of soils by gases occurs on a macro scale as soil particles are transported by wind, and on a micro scale as gases, emitted as by-products of organic or inorganic reactions, move soil particles micromillimeters at a time. Erosion and transportation of sediments by *wind* is not, strictly speaking, pedoturbation, although it clearly can and will affect pedogenesis (Johnson 1985) and soil productivity. Only when the particles are translocated within (more likely, from place to place on top of) the same pedon can wind transport be considered aeroturbation *sensu stricto*. Aeroturbation includes minuscule forms of mixing as gases are given off from various microscopic life forms, roots and small faunal inhabitants. In general, only a few soil particles are moved, giving this process more in common with Brownian motion than with the macroscopic forms of pedoturbation.

Aquaturbation

In most instances, water flow in soils is viewed as an organizing, rather than an mixing vector (see Chapter 12). Examples include the translocation of colloids and ions to form illuvial B horizons. Aquaturbation, mixing by water, is nonetheless ubiquitous in all soils. All water that moves into a soil is capable of dissolving soluble substances and entraining clastic particles. Water usually moves clastic materials downward in the profile; if infiltrating water moves clastic particles to consistent depths, such as clay translocation from the near-surface zone to depths of 60–100 cm, aquaturbation could be viewed as proanisotropic (clays are being sorted and delivered to a similar, clay-enriched zone, while at the same time

Table 10.2 Pedoturbation processes that can move coarse fragments

Process	Representative sources
Faunalturbation: burrowing, tunneling or mounding	
Earthworms	Darwin 1881, Webster 1965b, Johnson *et al.* 1987, Ponomarenko 1988
Pocket gophers	Murray 1967, Cox and Allen 1987a, Johnson *et al.* 1987, Johnson 1989, Cox and Scheffer 1991
Ground squirrels	Borst 1968, Johnson *et al.* 1987
Mole rats	Cox *et al.* 1987b, 1989, Cox and Gakahu 1987
Prairie dogs	Carlson and White 1987, 1988
Termites	Lee and Wood 1971a, b, Gillman *et al.* 1972, Holt *et al.* 1980, Bagine 1984, Nutting *et al.* 1987, Lobry de Bruyn and Conacher 1990, Grube 2001
Ants	Baxter and Hole 1967, Salem and Hole 1968, Humphreys 1981, Mandel and Sorenson 1982, Levan and Stone 1983
Floralturbation: tree uprooting	Johnson 1990, Schaetzl 1990, Schaetzl *et al.* 1990, Small 1997, Small *et al.* 1990
Cryoturbation (also congelliturbation): freezing and thawing	Bryan 1946, Inglis 1965, Pissart 1969, Manikowska 1982, Mackay 1984, Van Vliet-Lanöe 1985, Pérez 1987a
Argilliturbation: swelling and shrinking of clays	Springer 1958, Johnson and Hester 1972, Cooke *et al.* 1993
Graviturbation: mass wasting	
Soil settling	Moeyersons 1978
Creep	Schumm 1967, Williams 1974, Moeyersons 1978, 1989
Debris flow	Van Steijn *et al.* 1988
Aeroturbation	
Wind erosion	Wilshire *et al.* 1981
Eolian accumulation	McFadden *et al.* 1987
Aquaturbation	
Splash creep	Moeyersons 1975
Runoff creep	De Ploey *et al.* 1976
Surface wash	Kirkby and Kirkby 1974
Hydraulic erosion	Abrahams *et al.* 1984, Poesen 1987
Crystalturbation: growth and wasting of crystals	Cooke *et al.* 1993
Seismiturbation: earthquakes and vibrations	Clark 1972, Hole 1988
Anthroturbation	
Trampling	Barton 1987
Tillage	Kouwenhoven and Terpstra 1979
Off-road vehicle traffic	Elvidge and Iverson 1983, Pérez 1991

Source: Modified from Poesen and Lavee (1994).

forming an eluvial horizon above). However, the movement of each individual particle, if viewed separately, could be construed to be proisotropic mixing. It is the concert of particles, all moved to a similar position within the profile, that acts to make this form of pedoturbation generally proanisotropic.

Aquaturbation usually occurs on microscopic or otherwise very small scales. As water tables rise and fall, and as wetting fronts penetrate the soil, some soil particles are indisputably moved, if only a few micrometers. Water can also inhibit organization. A spring seep, for example, where water is slowly flowing upward, can inhibit the formation of some types of subsurface horizons.

Faunalturbation

Soil fauna, termed *infauna*, are probably the most commonly observed, frequently mentioned and intensively studied agents of pedoturbation (Thorp 1949, Heath 1965, Hole 1981, Stein 1983, Lobry de Bruyn and Conacher 1990). Fauna such as earthworms, mammals such as wombats, badgers, gophers and moles, and many species of insects such as ants and termites burrow in the soil as a means of finding food and providing for shelter, or for hibernation, estivation or reproduction (Thorp 1949, Hole 1981, Carpenter 1953, Van Nest 2002). Many pass soil particles through their guts as they burrow, while others simply push soil aside or move it to a preferred location.

The main pedogenic activities of soil fauna include mounding, mixing, forming voids, forming and destroying peds, regulating soil erosion, impacting plant and animal litter, facilitating the movement of water and air within the soil and regulating nutrient cycles (Hole 1981, Lobry de Bruyn and Conacher 1990). When fauna burrow through the soil, their actions result in both proanisotropic and proisotropic mixing (Wiken *et al.* 1976). The degree of selectivity with which the soil animals move particles, and the locations of their burrows, often determine whether the faunalturbation is predominantly proanisotropic or proisotropic (Humphreys and Mitchell 1983, Johnson 1989, 1990).

Sediment moved by soil fauna often has distinct textural characteristics which are often different from the native soil (Table 10.3). Mound material may become slightly sorted as the mound collapses and water washes over it (Wiken *et al.* 1976). Burrowing, whether it be by ants (Lokaby and Adams 1985), other insects or mammals, creates large voids in the soil (Fig. 10.4). These voids provide conduits for rapid passage of gases and fluids, and at a later time, the soils may have lower bulk densities (Salem and Hole 1968, Lobry de Bruyn and Conacher 1990). Many plants rely on recently pedoturbated soils as seedbeds; without this type of disturbance their germination successes would be greatly diminished (Dean and Milton 1991, Schiffman 1994).

Generally, one envisions soil fauna as mixing and homogenizing the soil to some depth, usually to the mean or maximum depth of burrowing. However, soil conditions may also play a role, which may force a feedback situation. For example, Hole and Nielsen (1970) observed that nests of the ant *Formica cinerea* are shallow and mounds as much as 3 m wide on poorly drained soils, while the same ant burrows almost 2 m deep and builds narrow, tall mounds on well-drained soils. They suggested that water table limitations on the wetter soils impact burrowing activities, and thereby limit whole-profile mixing there. The contrast in clay contents between the upper and lower solum, which is typically higher in wetter soils (Shrader 1950), could be attributed to the mixing effects of ants on the drier soils, which counteracts lessivage there (Hole and Nielsen 1970).

Infilled animal burrows, termed *krotovinas* or crotovinas (Borst 1968), often exhibit contrasting color to their surrounding matrix, making for easy recognition. They provide unmistakeable evidence of bioturbation. When they become filled with A horizon material, as is often the case, krotovinas are darker than the surrounding matrix. In some Aridisols, krotovinas are easily recognized, since they fill in with darker material from horizons above, while the matrix is rich in light-colored, secondary carbonate (Fig. 10.5).

Table 10.3 Characteristics of material transported by ants and termites vs. the native soil

Location	Rate of accumulation (mm yr^{-1})	Soil mass moved to surface	Soil turnover rate (t ha^{-1} yr^{-1})	Mound longevity (yr)	References
Termites					
Northern Australia	0.0125	–	–	10	Williams 1968
Northern Australia	0.02–0.10	65	470 g m^{-2} yr^{-1}	–	Lee and Wood 1971a
Northern Australia	0.05–0.025	20 in 25–50 years	–	–	Holt *et al.* 1980
Northern Australia	–	11–26	–	–	Spain and McIvor 1988
West Africa	0.02	–	1.25	80	Nye 1955c
Africa	–	17.6	–	–	Maldague 1964
Africa	–	20–25 m^3 ha^{-1}	–	–	Lepage 1972, 1973
Africa	–	–	0.35	–	Nel and Malan 1974
Africa	0.04–0.115	–	–	–	Pomeroy 1976
Africa	–	–	0.3	–	Wood and Sands 1978
Africa	–	–	4.0	–	Aloni *et al.* 1983
Africa	–	3.7	–	–	Akamigbo 1984
Africa	0.06	–	1.06	–	Bagine 1984
Africa	1.5–2[A]	–	–	20–25	Lepage 1984
India	–	–	15.9 g m^{-2} day^{-1}	–	Gupta *et al.* 1981
South America	–	–	0.78	–	Salick *et al.* 1983
Africa	0.19	3.0 t ha^{-1} yr^{-1}	3	8	Aloni and Soyer 1987
Africa		–	–	15–20	Janeau and Valentin 1987
North America	–	–	0.07	–	Nutting *et al.* 1987

(cont.)

Table 10.3 (*cont.*)

Location	Rate of accumulation (mm yr^{-1})	Soil mass moved to surface	Soil turnover rate (t ha^{-1} yr^{-1})	Mound longevity (yr)	References
Ants					
South Australia	–	–	400 cm^3 ha^{-1} yr^{-1}	100	Greenslade 1974
Eastern Australia	–	–	8.41	months	Humphreys 1981
Southeast Australia	0.03	–	0.35–0.42	–	Briese 1982
Eastern Australia	–	–	<0.05	–	Cowan *et al.* 1985
Michigan	–	85.5 g m^{-2}	–	–	Talbot 1953
Wisconsin	–	2500 kg in 1040 m^2	11.36	–	Salem and Hole 1968
USA	–	0.28–0.8 g m^{-2}	–	–	Rogers 1972
USA	–		11.36	–	Wiken *et al.* 1976
Wisconsin	–	948 kg ha^{-1}	–	–	Levan and Stone 1983
England	–	–	8.24	–	Waloff and Blackith 1962
England	–	40–400 mg $nest^{-1}$ day^{-1}	–	–	Sudd 1969
Argentina	0.085	300	11	–	Bucher and Zuccardi 1967

Source: Lobry de Bruyn and Conacher (1990).

Nielsen and Hole (1964) studied the middens formed as earthworms drag raw litter into their vertically oriented burrows. Worm burrows were filled with litter to an average depth of 11 cm, which was also the average A horizon thickness. Annually, 46 leaves are gathered into each midden, meaning that 2.8 million leaves per hectare are dragged into the subsurface by earthworms in this Wisconsin forest. Likewise, deep burrowers may bring materials from below the solum into the near-surface environment (Thorp 1949). Burrowing activities of pocket gophers in the Siskiyou Mountains of California have led to thicker A horizons than in adjacent forested areas where the gophers are absent (Laurent *et al.* 1994). At the same time, forested soils have O horizons, while the areas populated by gophers generally do not.

Mounders and non-mounders

When infauna are discussed in light of their effects on soil, two groups are immediately recognized: mounders and non-mounders. Ants, termites, certain large earthworms and insects, rodents, gophers and foxes are usually mounders (Wood and Sands 1978, Reichman and Smith

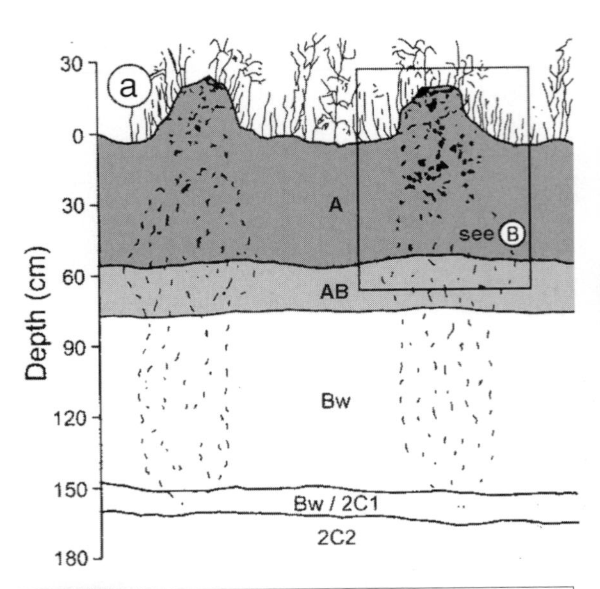

Fig. 10.4 Ant mounds formed by a species of the genus *Formica*. (a) Diagram of mounds formed in a Mollisol in Wisconsin. After Baxter and Hole (1967). (b) Cut-away view of a mound, probably from the same ant, but in a sandy Entisol in Michigan. Note the large krotovinas (infilled animal turnous) and the overthickened A horizon within the mound. Photo by RJS.

Fig. 10.5 Cicada krotovinas (dark-colored infillings, from clay-rich Bt horizon above) in an otherwise light-colored, carbonate-impregnated Bkm horizon of an Aridisol, New Mexico. Photo by RJS.

1991, Whitford and Kay 1999) (Table 10.4, Fig. 10.6). Their pedogenic role has a proisotropic component; they bring soil materials to the surface and deposit them in mixed piles. Nonmounders, such as the *Citellus* ground squirrel of California (Borst 1968) tend to simply redistribute material in the subsoil. Both types of burrowing activity are capable of lowering objects within the soil. Abandoned mounds provide homes for other fauna, which invade and establish residence there (Anduaga and Halffter 1991), while others become permeable depressions which act as loci for infiltration. Thus, the increased amounts of water that may move through those isolated sites may, in time, counteract the proisotropic tendencies of the pre-existing pedoturbation. Mounds, sometimes built as an adaptation to high water tables or shallow-to-bedrock conditions, are sometimes occupied for only a short period of time, up to a few years, and then abandoned (Watson 1962). Other mound-builders never really occupy their mounds; they are simply waste piles of soil formed during burrowing. Eventually the mounds, which are usually richer in smaller-size particles and certain ions than is the soil below and next to them, collapse and form a biomantle.

Mounds of many soil fauna represent deliberate or inadvertent concentrations of soil constituents, whether they be a certain clastic size fraction, organic matter or a chemical fraction (Lee and Butler 1977, Levan and Stone 1983, Coventry *et al.* 1988, Moorhead *et al.* 1988). For example, many ant and termite mounds have higher pH values and base concentrations than do the surrounding soil (Wiken *et al.* 1976,

Table 10.4 Characteristics of termite mounds built by the three Australian termite species at "Redlands," North Queensland, Australia

	Density (mounds ha^{-1})	Bulk density (Mg m^{-3})	Maximum (and mean) height (m)	Mean basal diameter (m)	Mass of soil in mounds (Mg ha^{-1})	Total basal area (m^2 ha^{-1})	Mean depth of termite activity (m)
Amitermes vitiosus	173.2	1.45	1.35 (0.54)	0.50	15.3	59	0.4
Drepanotermes perniger	25.6	1.36	0.90 (0.53)	0.73	3.0	15	0.4
Tumulitermes pastinator	28.1	1.02	0.93 (0.57)	0.81	2.4	16	0.2

Source: Holt *et al.* (1980).

Fig. 10.6 Several large ant (*Formica* spp.) mounds, near Grayling, Michigan. Photo by RJS.

Salik *et al.* 1983, Nutting *et al.* 1987). Most have higher N and organic matter contents (Lobry de Bruyn and Conacher 1990). Many mounds are loci of organic matter enrichment, the sources being feces, sailvary secretions, corpses and dead predators, as well as organic material deliberately transported into the mound (Lee and Wood 1971a, Lee and Butler 1977). In any event, the nutrients in most mounds are relatively inaccessible to plants while the mound is active, as most insects keep the mounds plant-free. However, these nutrients are released to the soil after the mounds are abandoned, enriching the soil and making mound occupancy times important from a nutrient-cycling perspective. Indeed, in many ecosystems, former mounds represent nutrient pools and loci of more porous soil that can be exploited by plants as preferred germination sites, leading to a patchiness that is a direct result of bioturbation. Termite mounds are occasionally exploited by indigenous peoples as a food supplement; this practice is called geophagy (see Chapter 4).

Earthworms

Earthworms are a very important group of infauna (Darwin 1881, Atkinson 1957) (Fig. 10.7). Most are non-mounders. They are very important in the production of soil structure (by casts) and macroporosity (by biopores), and in the consumption and mixing of organic materials into the mineral soil (Kladivko and Timmenga 1990, Zhang and Schrader 1993). Large tropical earthworms can even create microrelief with their casts on the order of decimeters (Haantjens 1965). The mixing activities of earthworms have been shown to be beneficial to crops (Mackay *et al.* 1982, Stockdill 1982) and very efficient at sorting finer from larger, i.e., too large to pass through their gut, materials. Vermudolls and Vermustolls have an A horizon dominated by "wormholes, wormcasts, or filled animal burrows" (Soil Survey Staff 1992). Many other types of Mollisols that lack discrete horizon boundaries often owe such morphologies to long-term mixing of their upper

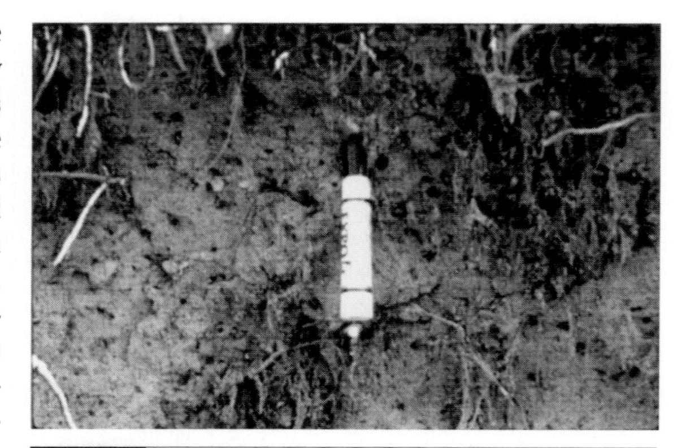

Fig. 10.7 Earthworm krotovinas in a Michigan Alfisol. Photo by RJS.

layers by infauna, especially worms and ants (see also Langmaid 1964). Worms are generally rare or absent from the drier and colder climates of the world, and from sandy soils.

Worm casting on the surface and at depth dilutes any illuvial clay concentration that may have existed, resulting in more isotropic textures with depth. One of the explanations given by Buol *et al.* (1997) for the occurrence of nearly equal amounts of clay in A and B horizons is that B horizon clay is continually being translocated upward by ants, worms and other biota. This hypothesis is substantiated by Baxter and Hole (1967), who found that clay contents of ant mounds in a prairie soil were similar to that of the Bt horizon below.

Arthropods

Arthropod fauna such as ants and termites occur worldwide. They mix and impact soil differently, however, based on their feeding and burrowing habits. Termites feed on wood and plant material containing cellulose, such as seeds and grasses, whereas ants ingest seeds and fluids from other insects. Most ants excavate subterranean nests and build mounds, but do less to alter the soil than do termites. Most ant activity involves moving soil "out of the way" as they dig new tunnels; many ant mounds are small and some ants do not build mounds at all. On the other hand, many species of termites deliberately move soil *into* large mounds, or termitaria. These structures contain runways, shelters and sheeting, and are cemented together with feces, saliva and undigested wood fibers (Gillman *et al.* 1972, Nutting *et al.* 1987). Conversely, ants are more likely to enrich their mounds in *raw* organic matter, like twigs, seeds and grass shards.

The amount of soil mixed and otherwise influenced by mound-building arthropods is a function of the size of the mounds, as well as their density and longevity (Table 10.5). Densities upwards of 500 termite mounds per hectare have been reported in Australia (Spain *et al.* 1983); Wiken *et al.* (1976) reported ant mound densities exceeding 1100 ha^{-1}. In time, as burrows, mounds and hibernacula are created and destroyed, a larger and larger proportion of the surface and subsurface comes to exhibit evidence of bioturbation. On landscapes where the mounds are inhabited and maintained for long periods of time, as with some species of termites, the impact of pedoturbation is very intense in these localized spots (Fig. 6.6). The impact on the soils at the site of the mound may be so large that it may take many centuries for it to be obliterated. Likewise, some ants move to a new mound annually, allowing them to impact more of the landscape, but the net impact of each mound is less (Greenslade 1974).

The effect of arthropod burrowing on soils could fill volumes; only a few examples are provided here. Changes in soils brought about by animals include nutrient content, pH, bulk density and organic matter content (Watson 1962, Mandel and Sorenson 1982, Levan and Stone 1983, Litaor *et al.* 1996). Baxter and Hole (1967) found that ant mounds had much higher amounts of available P and K than did the soil below, or next to, the mounds. Termite mounds, and the soils immediately below, are commonly enriched in many nutrients, even precious metals (Fig. 10.8). Termite mounds studied by West (1970) had anomalously high zinc, gold and silver contents. They were underlain at 10–25 m by rocks rich in those metals, pointing not only to the enriching tendencies of termites but also to their extreme depth of burrowing. Some mounds are so nutrient-rich, and the nearby soils so infertile, that some have even suggested they could be crushed and used as fertilizer (Watson 1977). This type of surface enrichment is vital to the long-term health of ecosystems that would otherwise have low productivity, e.g., the dry tropics (Coventry *et al.* 1988). Additionally, faunalturbation may indirectly result in changes in flora (King 1977, Mandel and Sorenson 1982).

The effects of faunalturbation on the formation and maintenance of soil porosity cannot be overestimated, as the burrows (aka galleries) formed provide rapid conduits for infiltrating water (Green and Askew 1965, Stone 1993), can "wick up" water from deep below the soil (Watson 1962) and can provide entryways for gases. Soil infauna have also been shown to dramatically

Landscapes: worm faunalturbation of artifacts on the hillslopes of western Illinois

For most of his life Charles Darwin was fascinated by earthworms, and his last book (1881) was devoted to synthesizing his life-long study of them. Darwin had become well acquainted with the fact that continued bioturbation by earthworms resulted in the burial of objects larger than the worms were capable of moving (2.1 mm as measured by Darwin). Included among these objects were ancient artifacts and architectural features, such as coins and paving stones, and the remains of stone foundations. The process Darwin described involved earthworms bringing fine sediment to the surface, facilitating the slow, downward settling of objects that were originally deposited there. Given sufficient time, a fine-grained "vegetable mould" (A horizon) would form, over a subsurface layer of stones a few decimeters in thickness. For Darwin, the known age of the archeological materials provided important insights into the rates at which this process might operate over geological timescales – at first relatively rapid in shallow portions of the soil where the worms were most active, but much slower for objects being buried to greater depths, eventually reaching a maximal depth to which the worms would burrow.

Many potential archeological sites buried in this way have been overlooked or undervalued. One archeological landscape where this process has been important is the loess-mantled uplands of the Midwestern United States, where for generations archeologists have relied almost exclusively on surface reconnaissance surveys to locate sites. Recent research shows that burial of artifacts to sub-plow zone depths in biomantles has occurred even on these relatively young (Late Pleistocene) surfaces. Without a full appreciation of bioturbation processes and a theoretical framework within which to study them (Johnson 1993a, 2002), archeologists struggled to explain why so many Holocene artifacts were buried on these upland surfaces. In a case study from western Illinois, Van Nest (2002) showed how Archaic Period (>2500 years), but not younger, artifacts had become buried in stone zones 30–40 cm beneath the surface. In these prairie-influenced soils, burial was accomplished mainly by earthworm faunalturbation.

Van Nest (2002) observed that archeological sites with artifacts at the surface often occur in ribbon-like patterns on shoulder and upper backslopes along headwater valleys. If present, artifacts found upslope or downslope from these sites are usually buried. Using biomantle–stone-zone theory, Van Nest explained this type of distribution. On hilltops, soil loss by creep was effectively nil, allowing artifacts to be buried beneath a worm-worked biomantle, ultimately settling to the depth of bioturbation. On steep backslopes, artifacts were commonly found near the surface and within the plow zone. Here, the rate of soil creep exceeds the rate of burial by soil fauna; as the artifacts are buried, erosion exposes them and they remain at the surface. These sites are readily found by survey archeology teams. At the bases of the slopes, artifacts may be buried by sediment washed from upslope, or by alluvium. These findings counter the once-prevalent assumption that upper slopes, where cultural remains occur at the surface, are eroded and degraded. Upland archeology can begin anew, with its new "conceptual bag of tricks."

Table 10.5 The effects of termites on physical soil properties

Termite species	Sample location	Coarse sand content (%)	Fine sand content (%)	Silt content (%)	Clay content (%)	References
Macrotermes bellicosus	Mound	30.0	38.0	15.0	17.0	Watson 1969
	Nearby soil	67.0	23.0	5.0	5.0	
Amitermes laurensis	Mound	19.0	49.0	5.0	24.0	Lee and Wood 1971a
	Nearby soil	26.0	65.0	4.0	4.0	
Nasutitermes exitiosus	Mound	23.0	37.0	5.0	33.0	Lee and Wood 1971a
	Nearby soil	36.0	50.0	6.0	7.0	
Macrotermes bellicosus	Mound	8.5	26.4	9.7	54.4	Boyer 1973
	Nearby A horizon	26.1	40.2	12.6	18.1	
	Nearby deep subsoil	21.1	23.3	6.4	47.1	
Macrotermes bellicosus	Mound	5.3	23.3	14.3	55.3	Boyer 1975
	Nearby soil	30.5	25.8	5.9	35.5	
Macrotermes falciger	Mound	22.0	37.0	12.0	29.0	Watson 1977
	Nearby A horizon	52.0	38.0	5.0	5.0	
Amitermes vitiosus	Mound	30.6	34.1	7.8	27.5	Holt *et al.* 1980
	Nearby soil	39.3	35.4	7.7	17.6	
Trinervitermes trinervoides	Mound	2.4	46.2	14.3	37.1	Laker *et al.* 1982
	Nearby soil	6.2	67.5	5.9	20.0	
Cubitermes oculatus	Mound	61.0		19.6	19.8	Wood *et al.* 1983
	Nearby A horizon	77.0		12.0	10.9	
Nasutitermes sp.	Mound	24.7	25.6	12.6	37.0	Akamigbo 1984
	Nearby soil	31.0	37.4	7.4	24.2	
Macrotermes michaelseni	Mound	48.0		14.0	38.0	Arshad *et al.* 1988
	Crust Nearby A horizon	67.0		15.0	18.0	
Iridomyrmex purpureus	Mound	29.0	56.0	9.0	5.0	Greenslade 1974
	Nearby soil	30.0	52.0	10.0	6.0	
Paltothyreus tarsatus	Ant spoil	55.0	14.5	7.6	20.1	Lévieux 1976
	Nearby A horizon	44.0	33.0	12.6	7.6	
	Nearby soil at 25 cm	42.0	34.0	14.5	8.0	
	Nearby soil at 50 cm	49.0	30.0	11.6	8.9	

Source: Lobry de Bruyn and Conacher (1990).

influence the development of soil structure, from the granular structure formed by worm casts to the subangular blocky structure of ant hills to the cylindrical peds formed as cicada krotovinas are infilled (Buntley and Papendick 1960, Hugie and Passey 1963, Wiken *et al.* 1976).

Termite activity in the tropics

Because of their abundance and near ubiquity, ants, earthworms and termites are dominant global faunalturbators (Lee 1967, Humphreys 1981, Martius 1994). In this section we focus on the role of *termites* in the soil systems of the

Landscapes: the Mima mound mystery . . . solved

Mima mounds, named for the Mima Prairie in Washington State, are dome-shaped earth mounds, often >2 m high and 10–50 m in diameter, that are widespread on many grassland landscapes (Cox 1984, Cox and Roig 1986, Lovegrove and Siegfried 1986). Also called pimple mounds or prairie mounds, they sometimes cover vast acreages at densities exceeding 100 ha^{-1}. The mounds occur only on soils shallow to bedrock or a subsurface pan, such as a duripan, or in soils that are periodically waterlogged (Cox and Scheffer 1991). The origins of these features has long been controversial (Scheffer 1947). One hypothesis had them as erosional features – relict landforms left behind after wind and water eroded areas between (Waters and Flagler 1929, Melton 1935, Ritchie 1953). Others provided evidence that the mounds form as vegetation (shrubs or clumps of grass) trap eolian or fluvial sediment, making them purely constructional features that date back to a past time when wind erosion and sediment transport were more active (Barnes 1879). A third hypothesis gave them a paleo-periglacial origin, like Arctic stone circles, in large part because areas between the mounds were often bare of vegetation and strewn with large rocks (Masson 1949, Newcomb 1952, Malde 1964). Berg (1990) proposed a seismic hypothesis.

The fossorial rodent hypothesis of Mima mound origin, accepted today as fact, was once ridiculed. To quote Ritchie (1953: 41), "This novel idea left the geologist in a position of arguing on what a gopher might or might not do (to form such mounds)." Its main proponent is George Cox, a biologist at San Diego State University (Cox and Gakahu 1986, Cox and Allen 1987a, Cox 1990, Cox and Hunt 1990, Cox and Scheffer 1991). This hypothesis postulates that Mima mounds are formed as pocket gophers, moles or tuco-tucos translocate soil material centripetally from their subsurface homes (Price 1949). As the gophers tunnel outward they push soil material behind them, building up a mound. The mounds serve as nest chambers in the thin or poorly drained soils, providing the increased soil thickness necessary to protect them from predation, winter cold or high water tables (Cox and Scheffer 1991). Where soils are thick, gophers are not restricted in siting their nests and thus move from place to place, and mounds are not formed. The fossorial rodent hypothesis is strongly supported by association (wherever mounds are found, rodents are always present in or near the mound fields) and by various types of soils data from mound soils. Between the mounds is commonly a zone where soil is thin and rocks are numerous, largely because the soil has been removed from these areas by the rodents (Cox and Allen 1987b). The mounds contain small stones only (Cox and Gakahu 1986); stone lines containing stones too large for gophers to move (> ≈6 cm dia.) commonly underlie and ring the mounds (Ritchie 1953). Small stones are not concentrated on mound tops, as would occur if the mounds were erosional. Gophers are territorial animals, explaining the regular pattern of mounds on the landscape. The limited amount of soil available to form mounds explains why larger mounds are generally located farther from their nearest neighbors than are smaller mounds (Cox and Gakahu 1986). Finally, many mounds continue to be occupied by gophers (Cox and Gakahu 1986) and those that lack gophers contain krotovina, and collapsed nest chambers. Mima mounds provide a unique look into the world of biogeomorphology – the study of landscapes that result from the interaction of abiotic and biotic processes.

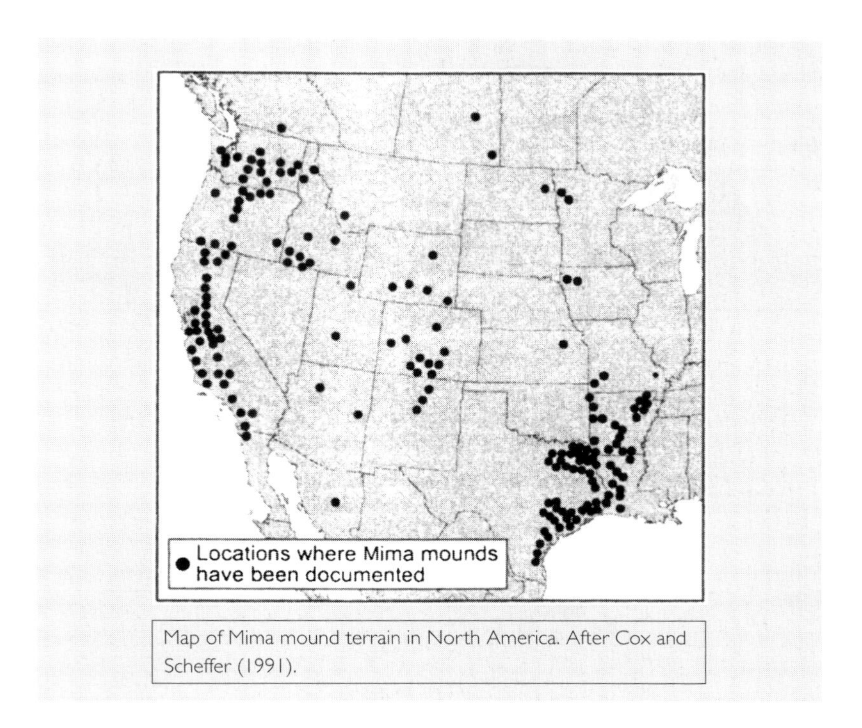

Map of Mima mound terrain in North America. After Cox and Scheffer (1991).

humid tropics, recognizing that they are also important in other warm climates.

Most termites are social insects that occur in discrete mounds or nests, and access remote food resources via covered runways or galleries (Martius 1994). Some species of termites build tall mounds while others create numerous covered runways (sheetings) on the soil surface that can extend over 50 m beyond the central nest (Bagine 1984). These "forage soil sheetings" are a very effective way of bringing stone-poor, fine-textured soil materials to the surface, and unlike mounds, are widely dispersed and fairly ephemeral. Termite mounds can be extremely numerous and over 10 m in height (Pullan 1979) (Fig. 10.9). By bringing up only sand-size and smaller-size fractions, termites create a stone-free mound which will eventually become a texturally distinct biomantle of less than 1 m to more than 10 m in thickness (Pomeroy 1976, Holt *et al.* 1980). Most stone-free biomantles in the tropics are generally ascribed to the actions of termites, ants, worms and similar infauna (Nye 1955c, D. L. Johnson 1990, Paton *et al.* 1995) (Figs. 10.10, 10.11).

Termite nests consist of soil and wood particles cemented together with feces and saliva (Wood *et al.* 1983, Martius 1994). A typical mound is assembled from sand-sized clay pellets and has a hard, compact, clay-rich casing that is resistant to erosion (Nye 1955c) (Fig. 10.10a). Evaporation of Ca-rich groundwater can lead to carbonate precipitation in the casing, further hardening it (Fig. 10.12). Some species cover the casing with a thin layer of loose soil (Lee and Wood 1968). Inside the casing is the nest framework, which is essentially a mass of tunnels, nurseries and the royal chamber – all composed of fine materials (Matsumoto 1976). Nye (1955c) suggested that the nest is composed mainly of material that has passed through the bodies of the termites, necessitating that there be no particles in it larger than fine sand size.

Termites consume large amounts of plant material (Pomeroy 1976, Wood and Sands 1978, Brener and Silva 1995). However, even when termite populations are high, they still consume less than 3% of the dead biomass (Matsumoto 1976). Much of the biomass they ingest and redeposit

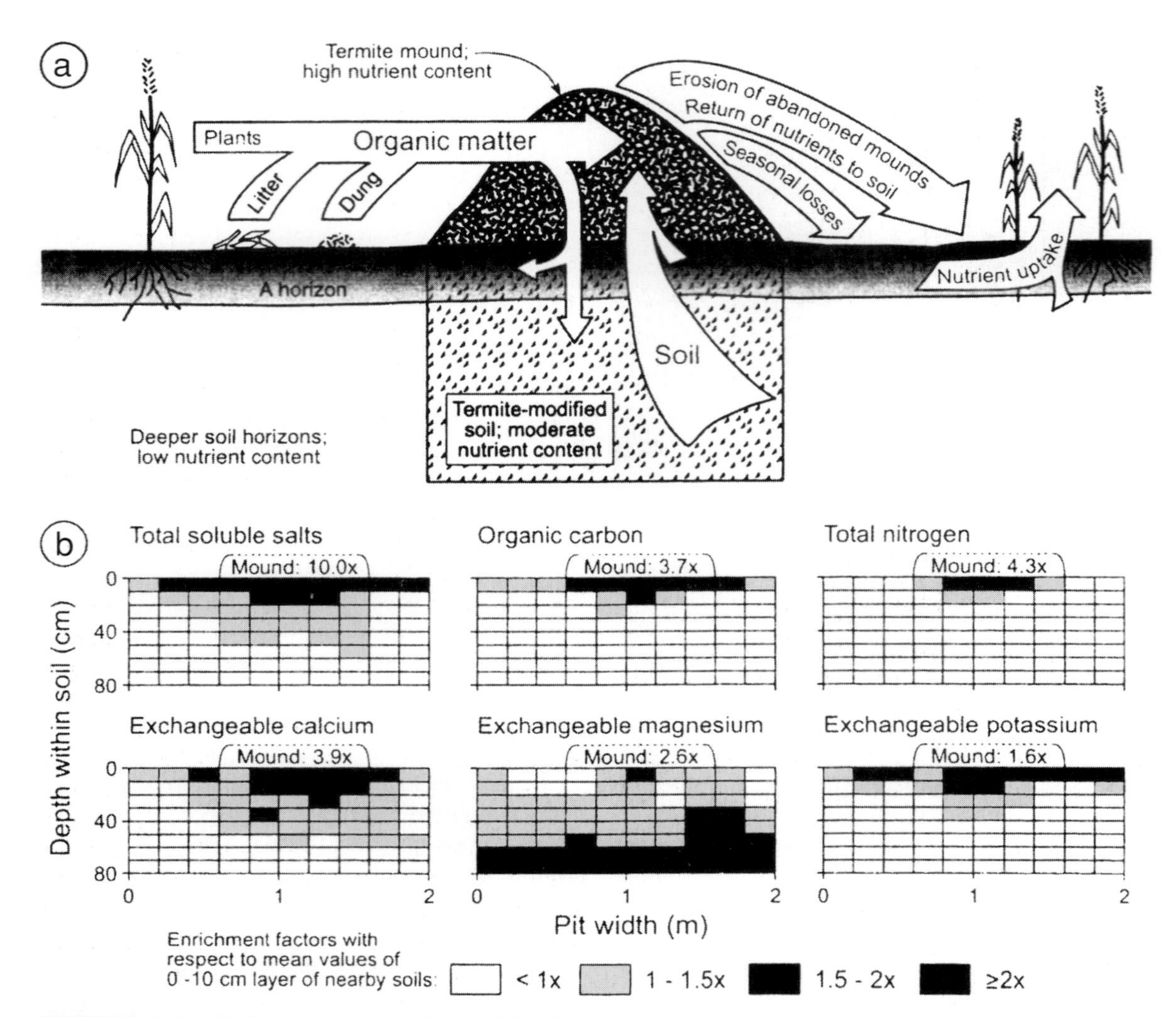

Fig. 10.8 Efects of termite mounds on soil fertility below and near the mound. After Coventry *et al.* (1988).
(a) Generalized flow patterns of nutrients through termite mounds. Similar flow paths could be fashioned for ant mounds. (b) Distribution patterns of various soil chemical components within and beneath termite (*Amitermes vitiosus*) mounds.

in the mound is so throroughly consumed and reconsumed by cannibalism and consumption of feces, however, that little is often left behind to enrich the soil. In fact, one of the reasons that many tropical soils have so little organic matter centers on the fact that much of it is so efficiently consumed by termites (Fig. 10.13). In the nest many species of termite cultivate fungus gardens or "combs." Biomass is fed to the fungi by the termites; this is another way that organic materials associated with termite mounds are completely digested. Lastly, many termites eat their own excreta and dead, keeping the nutrient cycle tight and "within-mound" (Hesse 1955). In sum, the soils in and near termite mounds may be nutrient-rich but are notoriously humus-poor and on a slow nutrient cycle (Lee and Wood 1971b, Pomeroy 1976) (Table 10.6, Figs. 10.13, 10.14). When one considers that mounds can remain viable and occupied for over a century (Watson 1972), and most are occupied for 10–20 years (Holt *et al.* 1980), the potential for nutrients to be tied up in mounds is considerable (Pomeroy 1976). Even when abandoned, it may take from

(a)

(b)

(c)

Fig. 10.9 Examples of termite mounds. (a) Mound near Niamey, Niger. Photo by Jennifer Olson. (b) Mound in Ghana that has been mined for geophagical clays. Photo by J. Olson. (c) Another mound in Ghana that also has been mined for geophagal clays. Photo by John Hunter.

a few years to a few decades before the mound disintegrates and the nutrients become plant accessible (Watson and Gay 1970). In Brazil, Salik *et al.* (1983) reported that an average of 165 nests per hectare are abandoned annually in a tropical rain forest where over 1500 nests occur per hectare, pointing to a perhaps a more rapid turnover of soil and nutrients than is typical.

With a few exceptions (Holt *et al.* 1998), termitaria are finer-textured, higher pH, nutrient-rich patches in the otherwise nutrient-poor soils of the tropics (Salik *et al.* 1983). In a part of Queensland, Australia, termite mounds contained only about 1% of the A horizon soil mass but 5–7% of the nutrients (Coventry *et al.* 1988). Coventry *et al.* (1988) calculated "enrichment factors" for various nutrients, where the factors are the relative enrichment over and above that of the upper 10 cm of surface soil. Mounds sometimes showed enrichment factors exceeding two times (Fig. 10.8).

Archeologists are concerned about the effects of termites on soils not only because they can lower artifacts to the depth of burrowing but because they change the pH and alter the aeration of the soil, which affects bone preservation (Watson 1967, McBrearty 1990). In general, the effects of increased aeration more than offset the higher pH values created within mounds, and thus termites do more to destroy bones than preserve them (Lee and Wood 1971b).

Floralturbation

Floralturbation or *phytoturbation* (Vasenev and Targul'yan 1995) is the mixing of soil by plants, usually via (1) root expansion, (2) decay and infilling of root channels, (3) root movement caused by agitation by wind (Hintikka 1972, Schaetzl *et al.* 1990) and (4) uprooting, especially of larger plants such as trees.

Root growth may cause upward expansion of the soil surface. Later, as the roots die and decay, the infilling of the root channels may allow materials from upper horizons to penetrate deeply into the subsoil (Roberts 1961). Agitation of roots has been studied theoretically (Hole 1988) but

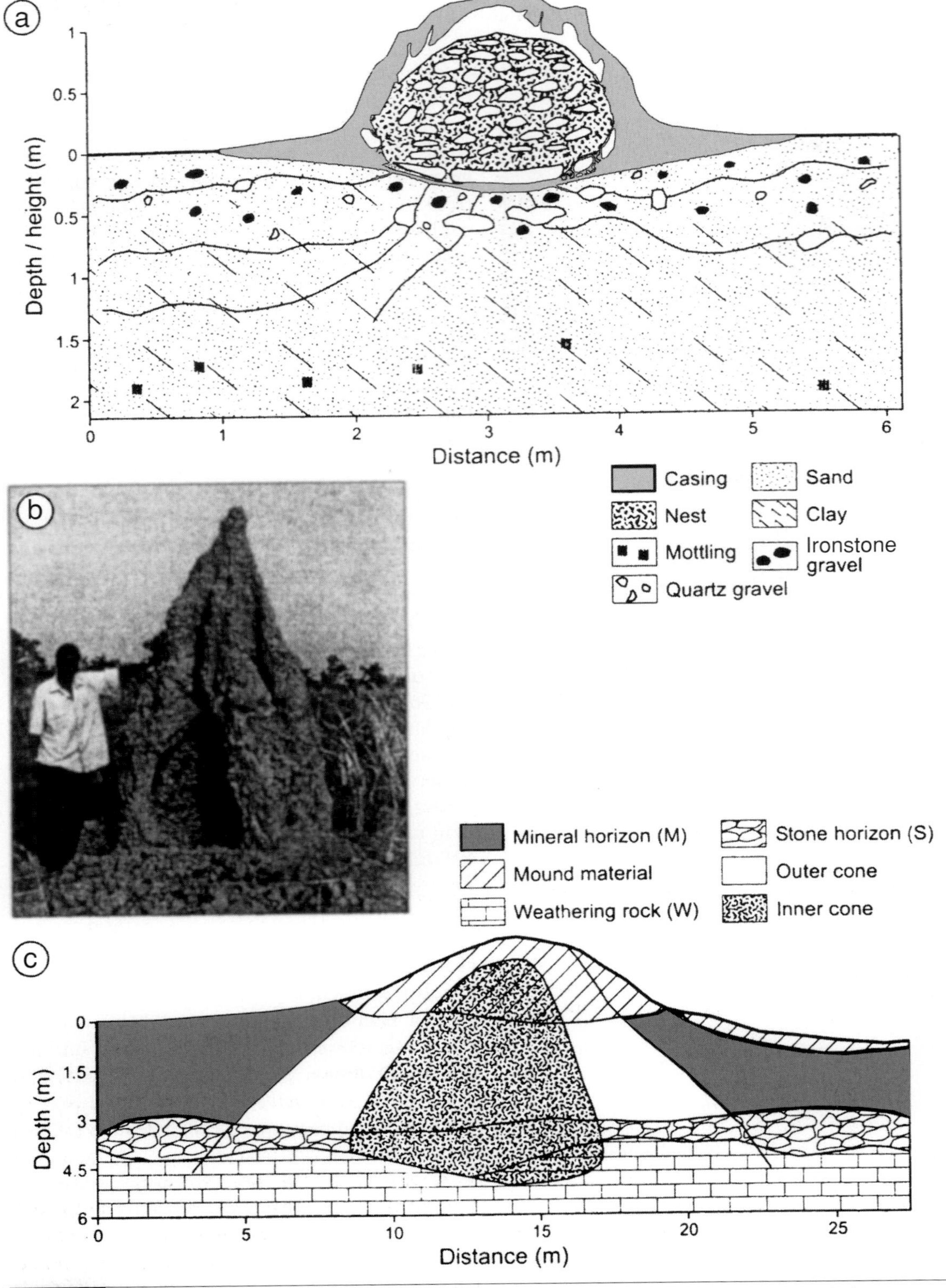

Fig. 10.10 Termite mound structures. (a) Diagram of a typical mound. After Nye (1955b). (b) Termite mound in Ghana, showing the casing and interior galleries. Photo by J. M. Hunter. (c) Diagram of a termite mound, showing its components and relationship to M, S and W horizons. After Watson (1962).

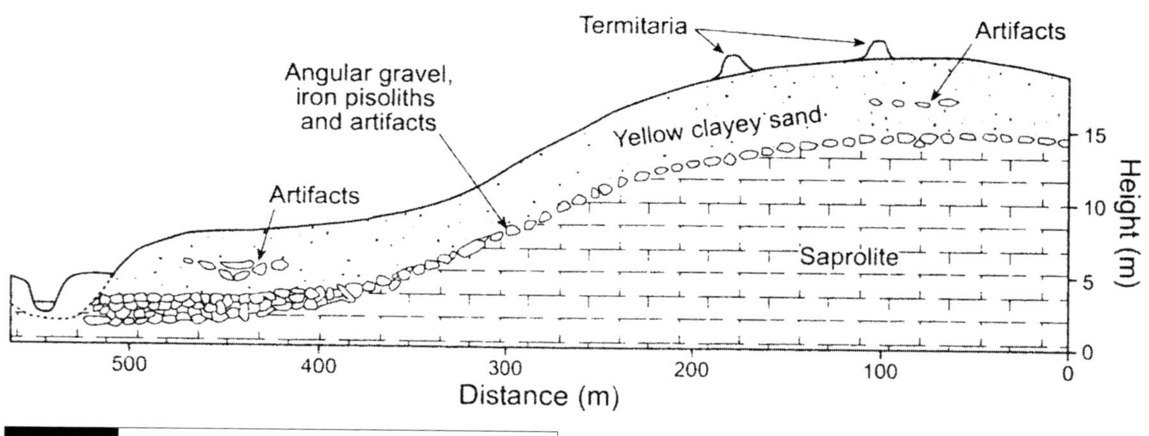

Fig. 10.11 Cross-section through a fluvial terrace in Zaire, showing a stone line at the base of a biomantle created by termites. After De Ploey (1964).

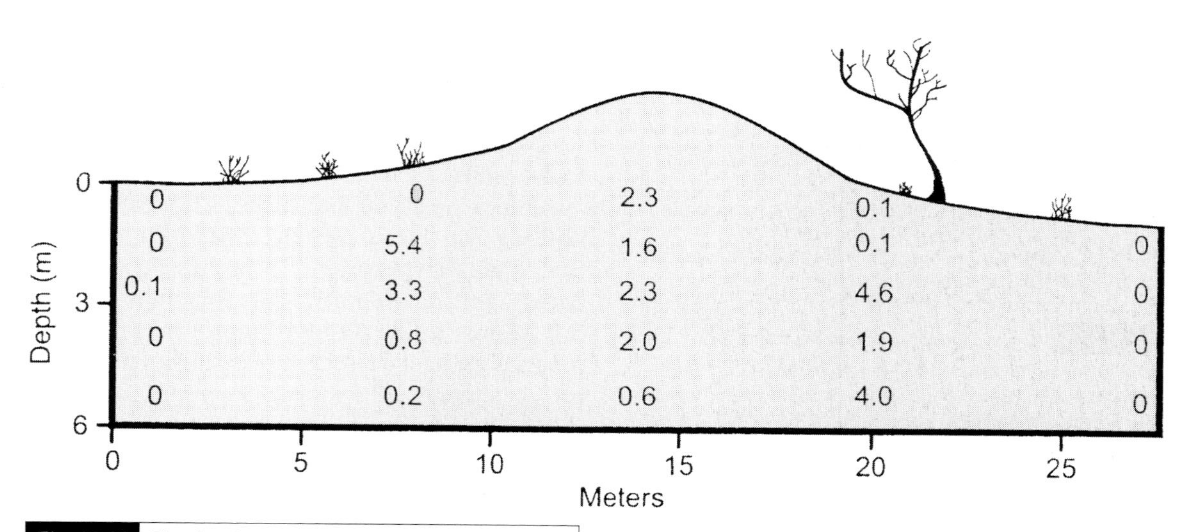

Fig. 10.12 The distribution of carbonates below a termite mound. Units shown are percentages of $CaCO_3$ in the fine earth fraction. After Watson (1962).

little work has been done on the field measurement of this process, or its implications for soil mixing (however, see Stone 1977). Thus, almost by elimination, uprooting has been the most studied form of floralturbation. It is most common in forested areas (Schaetzl *et al.* 1990). Because it is best manifested as a form of microtopography, the effects of tree uprooting are less noticeable after cultivation.

The term uprooting implies that a tree has fallen with most of its larger roots intact (Fig. 10.15). Brown (1979) used the term *arboturbation* to describe this process. Treefall or tree-tip are also synonymous terms, but they may also refer to trees whose trunks are broken near the base, with little or no soil disruption. Uprooting may disrupt considerable amounts of soil and often results in the formation of a pit (where the roots once were) and an adjacent mound, where soil slumps off the roots (Fig. 10.16). It contorts, mixes and may even overturn soil horizons. Floralturbated soil slumps off deteriorating

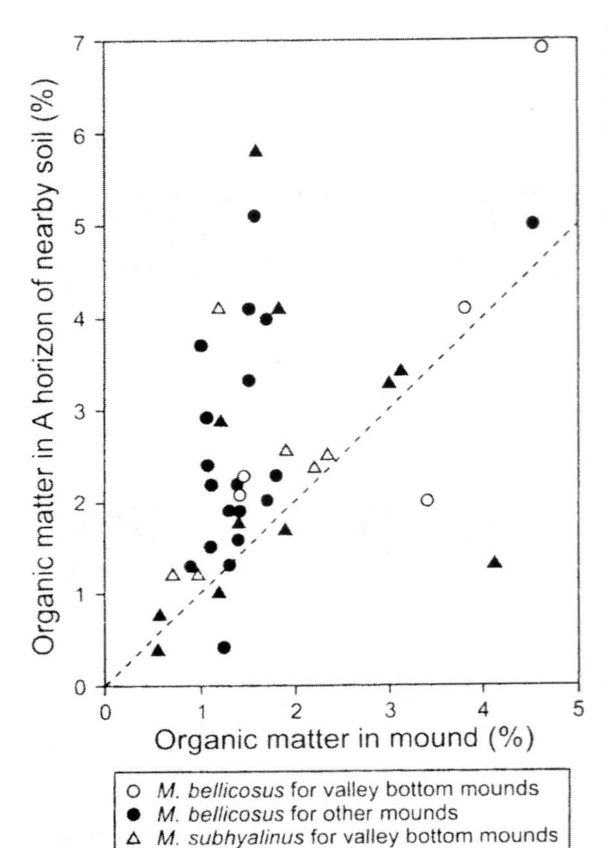

Fig. 10.13 Organic matter contents of termite (*Macrotermes*) mounds in Uganda, compared to the A horizon of nearby soils. Points above the line indicate sites in which the nearby soil contained more organic matter than did the mound. After Pomeroy (1976).

Table 10.6 Nutrient contents and other parameters associated with termitaria and nearby lateritic soils in Brazil

Parameter	Termitaria	"Laterite" soil
pH	3.9	3.9
Carbon (%)	22	2.9
Total N (%)	1.5	0.3
Total P (%)	0.08	0.02
Exchangeable K (%)	0.03	0.003
Exchangeable Ca (%)	0.02	0.002
Exchangeable Mg (%)	0.1	0.001
Exchangeable Na (%)	0.004	0.002

Source: Salik *et al.* (1983).

root plates and tends to create irregular patches of mixed and discontinuous horizons in the mound (Pawluk and Dudas 1982, Vasenev and Targul'yan 1995). Intact masses of soil may slump off the root plate and horizons may fold over each other, doubling their thickness (Veneman *et al.* 1984, Schaetzl 1986b). Schenck (1924) likened this process to natural plowing of the soil; it may be important because base-rich and nutrient-rich subsoil material is brought to the surface. Certain pedogenic characteristics, such as the interrupted and cyclic horizon character of many Spodosols and Alfisols, imply past floralturbation. E/B and B/E horizons, common to many forest soils, may be a direct result of floralturbation.

Treethrow initially regresses well-horizonated forest soils such as Spodosols or Alfisols to lesser-developed orders such as Inceptisols or Entisols, or keeps weakly developed soils in that state by persistent, proisotropic pedoturbation. Soil development subsequent to uprooting seems to be favored more in pits than mounds (Schaetzl 1990), because the treethrow pit is a zone of stronger leaching (Veneman *et al.* 1984, Price and Bauer 1984), presumably enhancing pedogenesis if infiltration is not inhibited (see Chapter 13). Coarse clasts are readily brought to the surface by uprooting, and thus gravelly biomantles or lag concentrates may be formed on uprooting mounds (Small *et al.* 1990) (Fig. 10.17).

Many midlatitude forests are replete with thousands of mounds and pits per hectare. In these landscapes, this type of pedoturbation literally is responsible for structuring the surface and the resultant vegetation patterns. In these areas, uprooting has led to so much spatial distortion of soil horizonation that the soils are very difficult to map at small scales (Alban 1974, Meyers and McSweeney 1995, Kabrick *et al.* 1997).

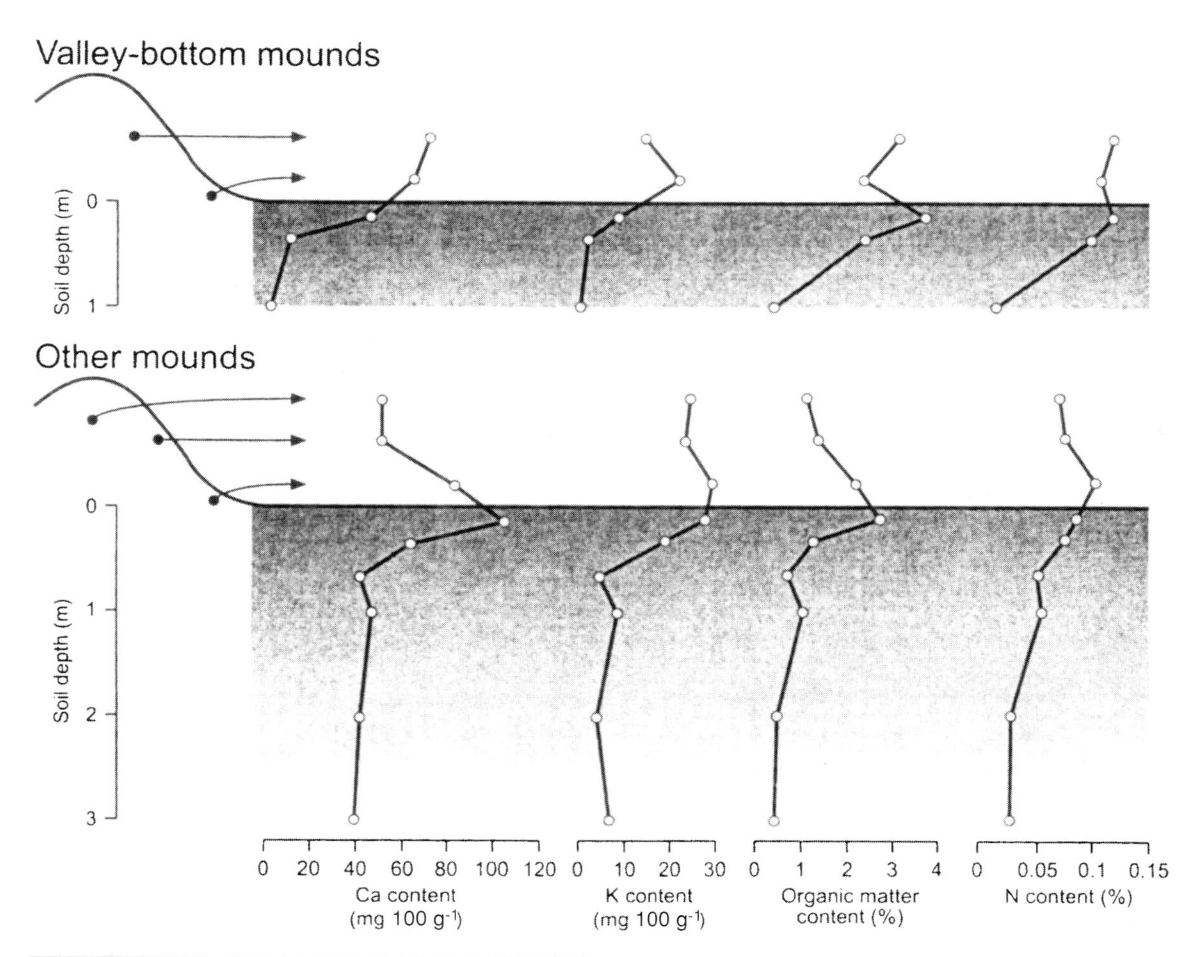

Fig. 10.14 Depth functions of soil constituents within and below termite mounds in Uganda. Data for soils "below" the mound are actually from soils adjacent to the mound, so as to reflect the native soil prior to mound building. After Pomeroy (1976).

Cryoturbation and Gelisols

Approximately 8×10^6 km^2 of soils in the Arctic, Subarctic and Boreal ecozones and in high mountain areas of the globe are associated with permafrost and undergo annual or more frequent freeze–thaw cycles (Tedrow 1962, 1977). Cryoturbation, a subset of the larger term for frost action in soils (*geliturbation* or *congelliturbation*), is an important pedogenic and geomorphic process in such landscapes (Bryan 1946). Bryan (1946) suggested the term *cryopedology* for science that deals with cryogenic processes in soils and near-surface sediments. Eluvial–illuvial soil processes, so typical of mid- and low-latitude soils, are slowed and inhibited by long cold seasons and frozen substrates, as the surface and the soil below heaves, mixes, flows and slumps under the influence of cryoturbation and graviturbation.

Soil freezing and cryoturbation

Water expands about 9% in volume upon freezing; this expansion can create significant amounts of heaving, downslope transport and mixing in soils if repeated frequently (Konrad 1989). In some soils, the expansion caused by freezing is vital – it creates voids that enhance infiltration and hydraulic conductivity (Chamberlain and Gow 1979, Schaetzl and Tomczak 2002). The permeability of frozen soils depends

Fig. 10.15 A stand of red pine (*Pinus resinosa*) trees that have been recently uprooted, Potter County, Pennsylvania. Photo by RJS.

on the type of frost present. Four main types of soil frost are generally recognized: (1) concrete, (2) honeycomb, (3) stalactite and (4) granular (Post and Dreibelbis 1942, Pierce *et al.* 1958). Of these, concrete frost, which forms in wet sands and soils with weak structure grade, is the least permeable.

Freezing fronts develop in soils just as wetting fronts do – from the top down, as the soil loses heat to the cold atmosphere. If permafrost (permanently frozen ground) exists, ice and subfreezing conditions may grow upward from it, while the freezing front from the surface grows downward as well (Fig. 10.18).

Freezing in moist soils is accentuated as water in vapor and liquid forms moves upward or laterally from unfrozen soil nearby, towards the freezing front. Here it can form ice lenses (Radke and Berry 1998). This process is called *ice segregation*, and is manifested as ice crystals and ice lenses, vein ice and ice wedges. Ice segregation features are almost pure ice; soil and other debris are pushed aside as they grow, heaving the soil surface (Van Vliet-Lanoë 1988, Konrad 1989, Chamberlain *et al.* 1990). Over time, these structures may grow into areally extensive ice lenses. They are more prevalent in wet soils than in dry soils because of the greater water supply.

A second type of freezing-related displacement occurs in clay-rich soils, below the freezing front. Here, large negative pore water pressures can form vertical *shrinkage cracks* (Chamberlain *et al.* 1990) that create additional pore space within the soil. Soils may even collapse in on themselves due to this shrinkage.

The primary effect of periglacial and/or intense frost action in soils is one of sorting: coarse fragments are often pushed or pulled to the surface and concentrated in various patterns (Corte 1963, Inglis 1965). This process is termed *upfreezing* (Vilborg 1955, Anderson 1988). Perhaps the core component of cryoturbation, upfreezing is of special importance in engineering and archeology. Engineers are concerned with foundation instability created by cryoturbation processes, and archeologists are aware that upfreezing can displace artifacts and destroy stratigraphy (Wood and Johnson 1978). Upfreezing begins as the soil develops a freezing front that advances from the surface downward (Fig. 10.19). Because the thermal conductivity of rocks in the soil is greater than that of soil matrix, the freezing front advances more rapidly through the rocks. The soil expands (heaves) as it freezes and develops ice lenses. At some point, each rock is frozen to the frozen soil above it, and as that soil heaves, the rock is *pulled* upward, since there is nothing

(a)

(b)

(c)

Fig. 10.16 A sequence of uprooted trees whose root plates are in various stages of disintegration. Photos by RJS. (a) An uprooted red pine (*Pinus resinosa*) tree. The event is less than 4 weeks old. Potter County, Pennsylvania. (b) An uprooted tree in Brownfield Woods, near Champaign, Illinois. The event probably happened about 1–3 years previously. (c) A root plate in an advanced state of disintegration. Near L'Anse, Michigan.

that holds it to the unfrozen soil below (Inglis 1965). (Competing hypotheses that involve ice lenses forming below rocks and pushing them upward, have since been rejected (Taber 1943).) As one might envision, upfreezing is greatest when the rate of frost penetration is slow. During thaw, the rock remains anchored in the frozen soil above, while thawed soil around it settles and slumps into the cavities below. The partially infilled cavities prevent the rock from returning to its original location, resulting in a small increment of upward movement. The cavity may also narrow and infill slightly by lateral frost heave (Anderson 1988), causing fine particles such as silt to be translocated downward in the soil. For this reason, silt flows and concentrations are common in areas of frequent frost. Factors that maximize ice lens growth, such as high moisture contents or slow advance of the freezing front, favor high rates of upfreezing. Therefore, upfreezing is potentially greatest in fine-textured soils because they hold more water than do coarse-textured soils, and because water has greater mobility in fine pores, when it is near freezing (Anderson 1988).

When the ground surface is distorted and mixed, freezing fronts penetrate the soil at angles and therefore pull clasts at an acute angle to the surface (Vilborg 1955). Movement of clasts in a non-vertical dimension by upfreezing can actually yield larger net displacements, because the clast cannot return as easily to its original position. On smaller scales, needle-ice formation may cause significant heaving and mixing in the uppermost O and A horizons, and can move large fragments within and on top of the surface (Pérez 1986, 1987a, b).

Upfreezing should not be confused with frost heave, which is simply the expansion of wet soil during freezing. Heave is facilitated by ice lenses, which grow and accumulate in the lower parts of the active layer, providing that temperature and moisture conditions allow. When these lenses melt, the large pores/cavities that remain allow for the translocation, in suspension, of silt and clay. Silt is not normally translocable in soils because pore sizes are not large enough. *Pervection* is the name given to the translocation of

Fig. 10.17 Large rocks caught in the root plate of an uprooted tree in northern Michigan. Uprooting is one the the few forms of pedoturbation that can effectively move large clasts upward in the soil profile. Photo by RJS.

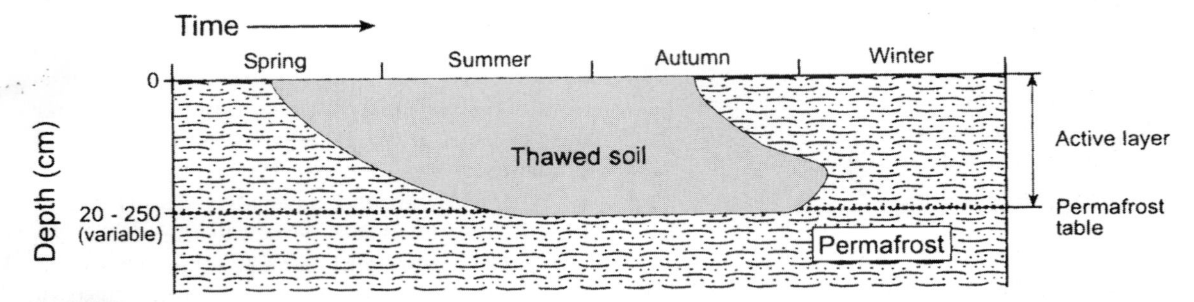

Fig. 10.18 Schematic representation of the annual, spatio-temporal cycle of freezing and thawing in soils with permafrost at depth. After Van Vliet and Langhor (1981).

silt grains in this manner (Table 12.1). It is only common in soils that undergo frequent freezing and thawing, and is typically manifested as silt caps on stones and as silt-enriched horizons (Fedorova and Yarilova 1972, Frenot *et al.* 1995). Silt caps form as silt-rich, percolating meltwater replaces ice sheaths around stones by infilling the voids which appear above them as they settle into cavities left after melting (Payton 1993a).

Permafrost

Permafrost is rock or soil that remains at or below freezing for at least two consecutive years; it is defined on temperature, not composition. The top of the permafrost zone is the permafrost table (Fig. 10.18). Permafrost can extend to great depths and areas with it usually have a surface zone, i.e., an *active layer*, that thaws seasonally (Miller *et al.* 1998). Some permafrost is pure ice; other types may include layered ice and clastic or humic materials. Dry permafrost contains insufficient moisture to be cemented, and is common to the very driest of sites, especially Antarctica (Ferrians *et al.* 1969, Bockheim 1997). Cryoturbation processes are minimal in soils with dry permafrost (Bockheim 1980a). According to Bockheim and Tarnocai (2000) permafrost may underlie over one-fourth of the Earth's land surface. Near the margins of continuous permafrost its distribution becomes discontinuous or sporadic, usually being found only within organic soils with high insulating abilities (Zoltai and Tarnocai 1975).

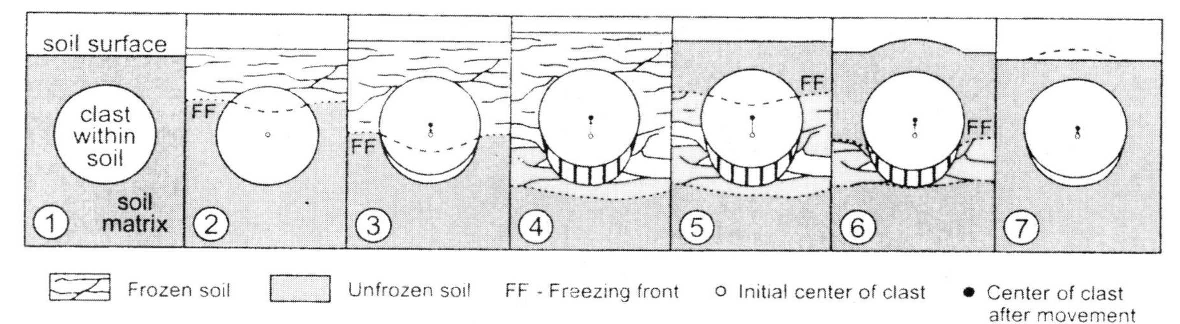

Fig. 10.19 The frost-pull process, which results in upfreezing of a clast within the soil. Ice lenses in the soil cause frost heave (expansion of the soil) in panels 1–4. By panel 3, the clast is frozen into the overlying soil and pulled upward as the soil is heaved, forming a cavity below the clast. Settling in the soil around the clast (panels 5–7) produces a bulge at the surface and partial infilling of the cavity. The clast cannot return to its original position because of the infilled material below it, resulting in a net upward displacement. After Anderson (1988).

The active layer is the compartment within which pedogenesis occurs. It is strongly influenced by the permafrost below (Miller *et al.* 1998). If the permafrost is dry, water can infiltrate it. If it consists of solid layers, infiltration is restricted, placing the soil on a much different pedogenic pathway. Because most permafrost is impermeable, soils within the active zone tend to have perched water tables. Water cannot infiltrate below the permafrost, and little evaporates. Only on sloping surfaces and in the driest polar deserts is the active zone not saturated for much of the warm season.

In polar regions, the thickness of the active layer generally increases as latitude decreases and on preferred aspects. On a more local scale, the thickness of the active layer is a function of the nature of the substrate, insulating cover, e.g., a histic epipedon, vegetative cover, relief, aspect, etc. Well-drained sites tend to thaw to greater depths than do wetter sites. Sites with thick histic epipedons and those that have late-lying snow patches may thaw only a few centimeters into the actual mineral soil, because of the insulating effects of the peat.

On sloping surfaces, the active zone will commonly "flow" slowly downslope (Benedict 1976). This process, due to both cryoturbation and graviturbation, is called *gelifluction* or *solifluction*. Usually involving saturated masses (lobes) of soil whose cohesive forces have been weakened by repeated freeze–thaw activity, solifluction can occur on slopes of even very gentle inclination, at rates up to 1 m yr^{-1}. It is also an important mechanism by which organic matter gets incorporated into soils, literally as it is buried by gelifluctinig materials.

Patterned ground

A very common morphological outcome of long-term cryoturbation is *patterned ground* – sorted and unsorted circles, nets, steps, stone stripes and polygons (Washburn 1956, 1980, Black 1969, Fitzpatrick 1975, Tedrow 1977, Thorn 1976) (Fig. 10.20). Although the genesis of patterned ground falls within the realm of geomorphology, the forms are largely due to a *pedo*turbative process. In fact, the reason that patterned ground is so common and persistent on the polar landscapes is because cryoturbation is omnipresent while other forms of pedoturbation *are not*. If faunalturbators were common in this landscape, patterned ground would be less dramatic and in some areas, absent (Smith 1956). Patterned ground has long been called "structure soils," just as one might use the term Mollisol. This term was eventually abandoned, but its use highlights the genetic link between cryoturbation and pedogenesis in these environments.

Patterned ground results from repeated freezing and thawing in the active layer. The details of the type and formation of patterned ground are varied and, suffice it say, more than one mechanism can lead to its formation (Washburn 1956,

Landscapes: the boreal forest of interior Alaska

Near the boreal forest–tundra ecotone in central Alaska, permafrost is an important component of the soil landscape (Swanson 1996). Here, like many other sites where the landscape is not a uniform plain, meso- and microtopography have a highly important influence on pedogenesis, largely because they impact snow patch thickness and soil wetness, which in turn impacts soil temperature (Mueller *et al.* 1999).

On a typical slope, gravel highs on bedrock knobs yield the driest soils, usually Haplorthels, with Oe-Bw1-Bw2-2Bc-2Cr horizonation beneath spruce–birch forest. Permafrost is thick. The wettest, coldest sites are within microtopographic lows; Hemistels (Oi–Oe–Oe/Ag–Oe/Agf) here typically have permafrost in their lower sola. Intermediate sites have Oi–Oe–Bw/A–Bg/A–2BC–2Cf horizonation and some permafrost.

Feedback mechanisms fuel the genetic linkages between cold, wet sites with thick O horizons and the shallow permafrost that underlies them. Summer thaws penetrate less deeply in such soils due to the high heat capacity of water. Once frozen in winter, these soils give off more heat than do dry soils, since ice has a high thermal conductivity. This leads to colder soil temperatures and shallower permafrost, which enhances organic matter accumulation, setting up a positive feedback loop. The shallower permafrost causes water to perch and perpetuates the soil wetness. Dry sites on bedrock knobs warm quickly in summer, facilitating the mineralization of any organic matter that accumulates there. Dry conditions inhibit peat formation and warm temperatures drive the permafrost table deeper. These soils have oxidized Bw horizons and less evidence of cryoturbation. Although many site factors influence depth to permafrost, at many locations subtle microtopographic variation is enough to cause major differences in soils across very short distances.

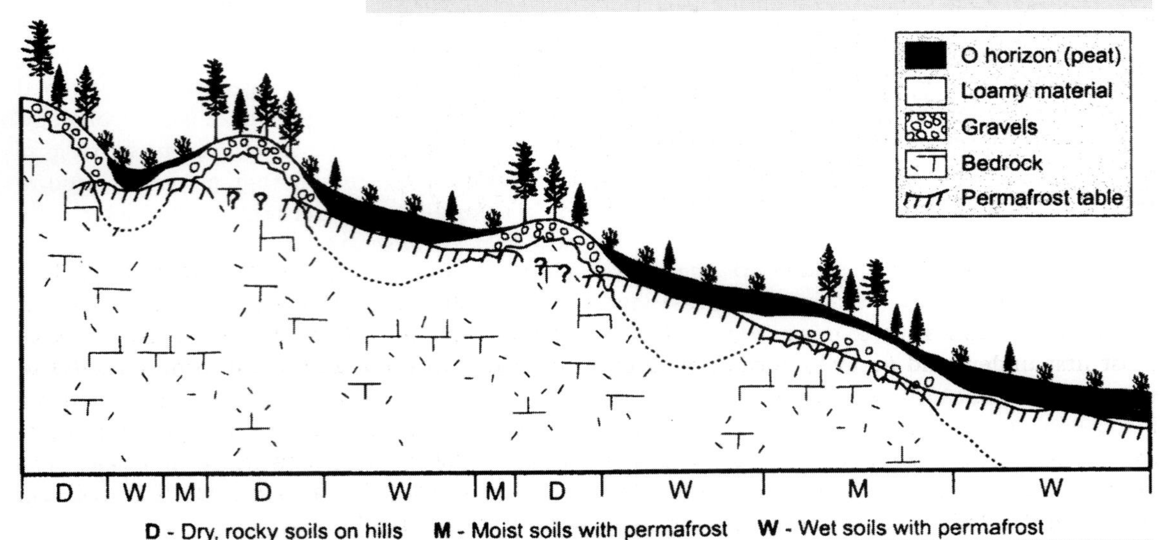

Soil-landscape diagram for the area near Hughes, central Alaska. Soil horizonation is vertically exaggerrated, relative to topography. After Swanson (1996).

Fig. 10.20 Examples of patterned ground. Photos by F. E. Nelson. (a) Sorted patterned ground, Cathedral Massif, northwestern British Columbia. Atlin Lake is in background. Gelifluction has been documented in the areas of fine material between the coarse stripes. (b) Low-centered ice wedge polygons, North Slope of Alaska. Active ice wedges underlie the troughs (edges) of the polygons.

Fig. 10.21 An exceptionally large, exposed ice wedge along the eroding Beaufort Sea coast. The ice wedge is associated with high centered lowland peat polygons. Note the cryoturbated mineral soil material on both sides of the ice wedge resulting from the expanding ice wedge. Mackenzie Delta area, North West Territories, Canada. Photo by C. Tarnocai.

1980, Goldthwait 1976, Hallet and Prestrud 1986, Van Vliet-Lanoë 1988). But, in order to understand the formation of patterned ground, one must first understand *ice wedge* formation. Frost or ice wedges are V-shaped bodies of ground ice, usually less than 4 m in depth and 2 m wide, that typically form in areas of continuous permafrost, where mean annual air temperatures are below about −6 °C (Péwé 1966, Romanovskij 1973) (Fig. 10.21). Ice wedges begin forming in winter when the permafrost extends to the surface, fractures and cracks, usually by thermal contraction. The cracks form in a somewhat random, polygonal pattern (Fig. 10.22). Indeed, most patterned ground is formed by intersecting frost wedges (Fig. 10.22b). Repeated cracking in the same location may cause a frost wedge to grow in size as soil water, in vapor and liquid form, migrates to the site. Growth can also occur as meltwater seeps into the open crack. Later, as this water freezes, the wedge expands and grows, and nearby soil material is upturned, displaced and distorted (Fig. 10.22).

Some wedges, called sand wedges or ground wedges, do not fill with ice; the open cracks are filled with blowing sand and snow. Sand wedges presumably form in cold, dry landscapes with strong winds, minimal snow cover and an

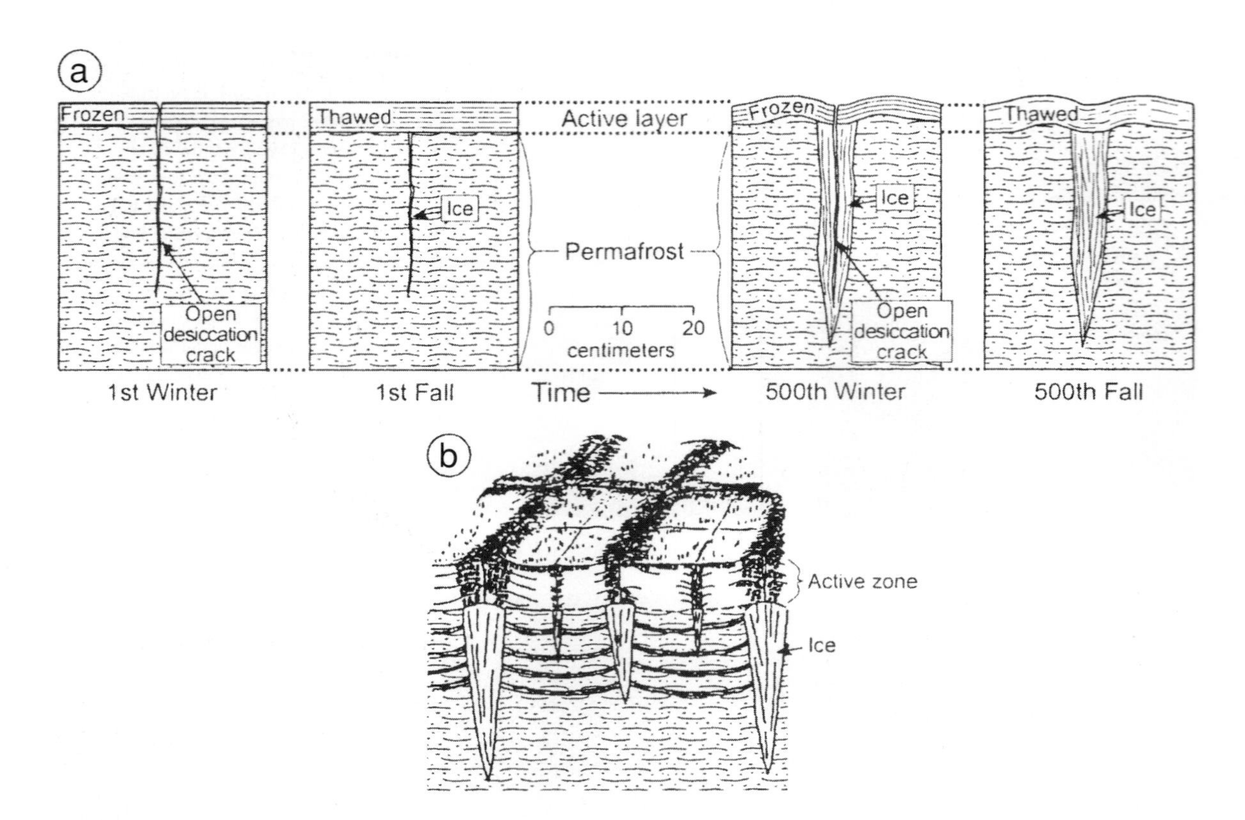

Fig. 10.22 Frost wedge morphology and genesis. (a) Idealized diagram showing the formation of a frost wedge. After Tedrow (1977). (b) Frost wedges and patterned ground, in three β dimensions. After Romanovskij (1973).

abundant supply of sand. They eventually take on the same form as ice wedges (Fig. 10.23), but because they are infilled with sandy material, exhibit minimal slumping when the wedge becomes relic (Mackay and Matthews 1983, Wayne 1991).

When temperatures are no longer cold enough, ice wedges become fossil and inactive. They are then variously called frost wedges, fossil ice wedges, ice wedge casts or ice wedge pseudomorphs (Billings and Peterson 1980) (Fig. 10.24). Frost wedges undergo morphological change when they thaw, making their unequivocal identification somewhat subjective (Harry and Gozdzik 1988). Evidence from frost wedges is, nonetheless, retained after the cold conditions have gone and can be used to infer a great deal about paleoenvironments (Sharp 1942, Smith 1942, Smith 1962, Fosberg 1965, Péwé 1966, Kozarski 1974, Walters 1994, Mader and Ciolkosz 1997). Ice wedge casts fill with sediment from nearby and overlying materials, and usually retain an irregular wedge shape (Péwé 1966, W. H. Johnson 1990). The sediment in the cast typically shows synclinally curved stratification and normal faulting, which occurs as the sediment is let down when the ice in the wedge melts. (Thaw is usually from the top down.) Sediment near the cast then gets even more distorted. The end structure is usually wider than the original ice wedge (Black 1976). Sediment in ice wedge casts may not be distorted, however, since there would have been little ice to melt and create additional volume.

Both upfreezing and outfreezing from polygon centers are thought to be the main drivers behind the formation of patterned ground, manifested as a network of *linear or curvilinear depressions* or *shallow ridges* separated by troughs (Goldthwait 1976, Harry and Gozdzik 1988). Driven by frost heave within wedges, the polygons

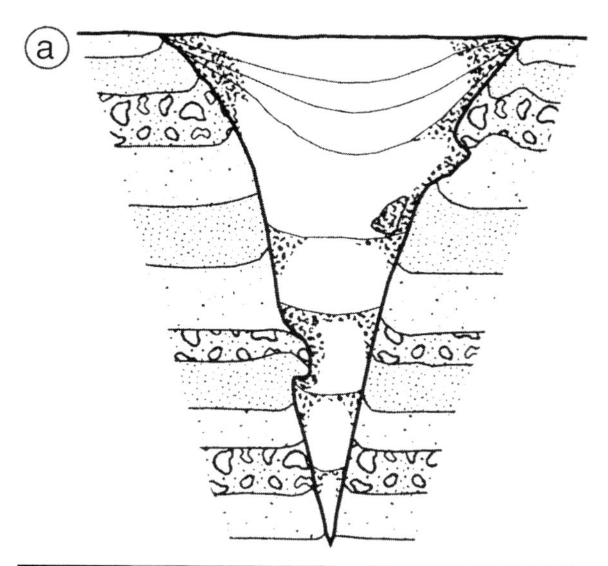

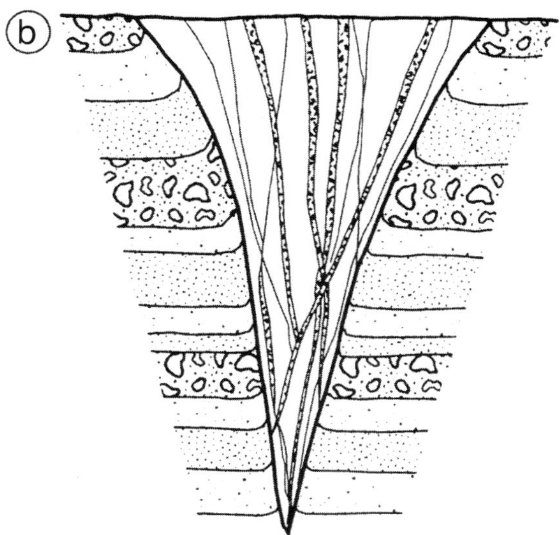

Fig. 10.23 Diagrammatic sketches of ice wedge casts. After Black (1976). (a) Ice wedge cast originally formed of ice, showing upturned strata and slumpage within the fossil cast. (b) Ground (or sand) cast, showing upturned strata, but vertical fabric within the cast itself.

associated with the frost wedges develop a type of crude circulation that brings coarse material to the surface along the cracks much like a diapir (Fig. 10.25). At the same time, organic matter is subducted at the edges of the polygons, producing organic matter maxima at depth (Hallet and Prestrud 1986). If the upthrust material is fine-textured, it may later subside to form a network of troughs, as may the fine material within the polygon center. Troughs/cracks can accumulate organic materials, leading to a polygonal pattern of soils with overthickened O and A horizons (Ugolini 1966, Lev and King 1999). Often, the channel between the polygons contains a mat of organics over waterlogged gravel (Tedrow 1977), while the intrapolygon soils are better drained (Fig. 10.26). In some low-centered ice wedge polygons, however, the center may be the wettest part of the landscape and thick deposits of organic materials may accumulate on the soil surface (Ping 1997) (Fig. 10.27). In coarser-textured materials, polygonal ground may be manifested as ridges of rock and gravel (Zoltai and Tarnocai 1975).

Gelisols and pedogenesis in the presence of cryoturbation

Cryoturbation and the pedogenic impact of ice lenses and frozen sediment are central to Gelisols. Pedogenic processes in Gelisols are driven primarily by the volume changes associated with ice formation and thermal contraction (Bockheim *et al.* 1997). Ping (1997) and Höfle *et al.* (1998) listed three major process suites associated with Gelisol genesis: cryoturbation, oxidation–reduction and organic matter accumulation. *Cryopedogenic processes*, defined by Britton (1957: 30) include

> all mechanisms operative in the substrate that relate to the formation and degradation of permafrost, as well as frost-cracking, -splitting, -churning, -stirring, -thrusting, and -heaving actions, induced either by expansion or contraction of frozen materials exposed to large temperature stresses, or to a repeated alternate freezing and thawing of surface materials.

Bockheim and Tarnocai (2000) included freezing and thawing, frost heaving, cryogenic sorting, thermal cracking and ice segregation as subsets of cryopedogenic processes. Canadian soil scientists have long recognized the importance of cryoturbation as a pedogenic process (Soil Classification Working Group 1998).

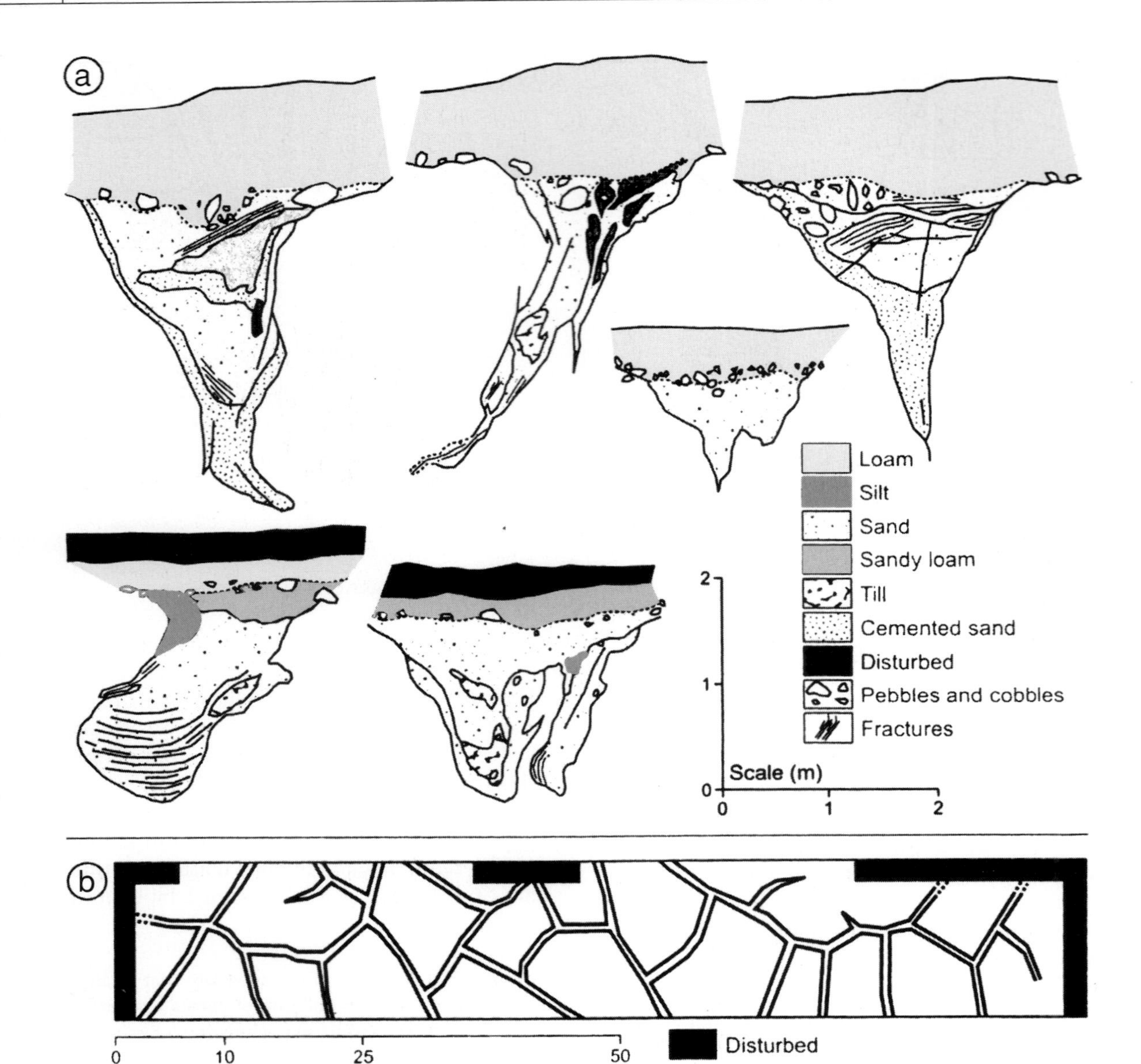

Fig. 10.24 Patterns of ice wedges and ice wedge casts. After Walters (1994). (a) Cross-sections of ice wedge casts from the Iowan erosion surface, northeast Iowa. (b) Plan view of the wedges as they appear, interconnectedly, on the landscape.

Long-term cryoturbation in Gelisols is manifested as irregular and broken horizons, involutions, oriented coarse fragements, silt-enriched layers and caps on stones, and (sometimes large) organic matter accumulations on top of and within the permafrost layer (Frenot *et al.* 1995, Michaelson *et al.* 1996, Bockheim *et al.* 1997) (Fig. 10.28).

Most Gelisols contain permafrost within 2 m of the surface. They are roughly equivalent to Cryosols (Soil Classification Working Group 1998). Pedogenesis, in the traditional, midlatitude mindset, is restricted to the active layer, and is severely slowed by the cold soil temperatures (Fig. 10.29). Thus, one should simply think of Gelisol landscapes as just another pedogenic province or regime, but in this case one that is

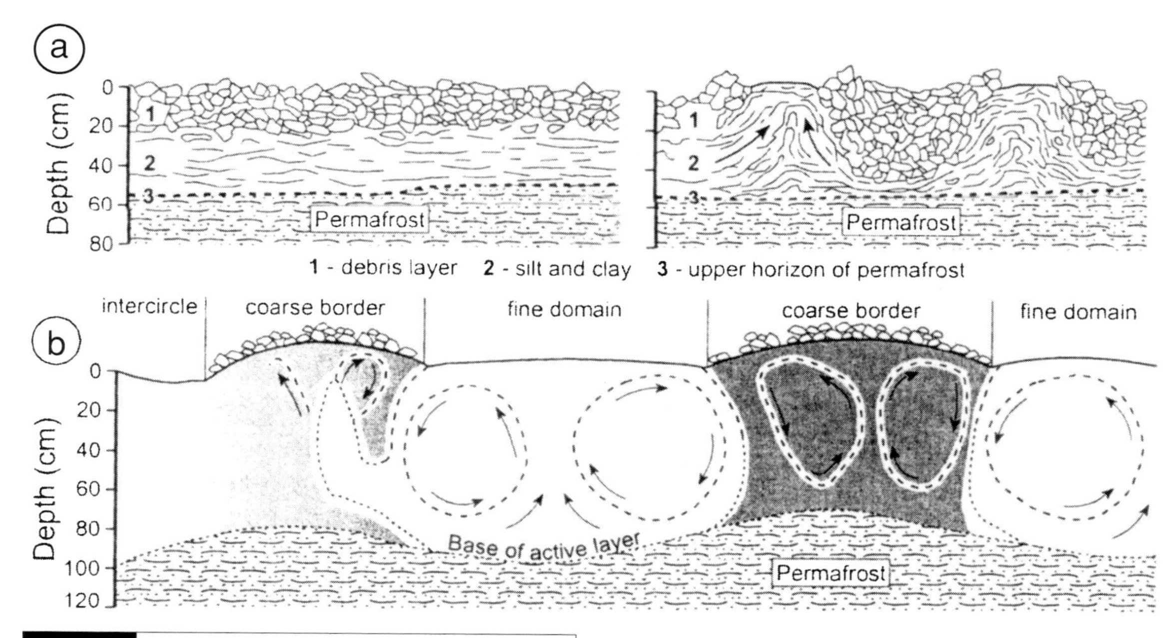

Fig. 10.25 Models of soil movement by cryoturbation, and the formation of patterned ground. (a) After Tedrow (1977). (b) After Hallet *et al.* (1988).

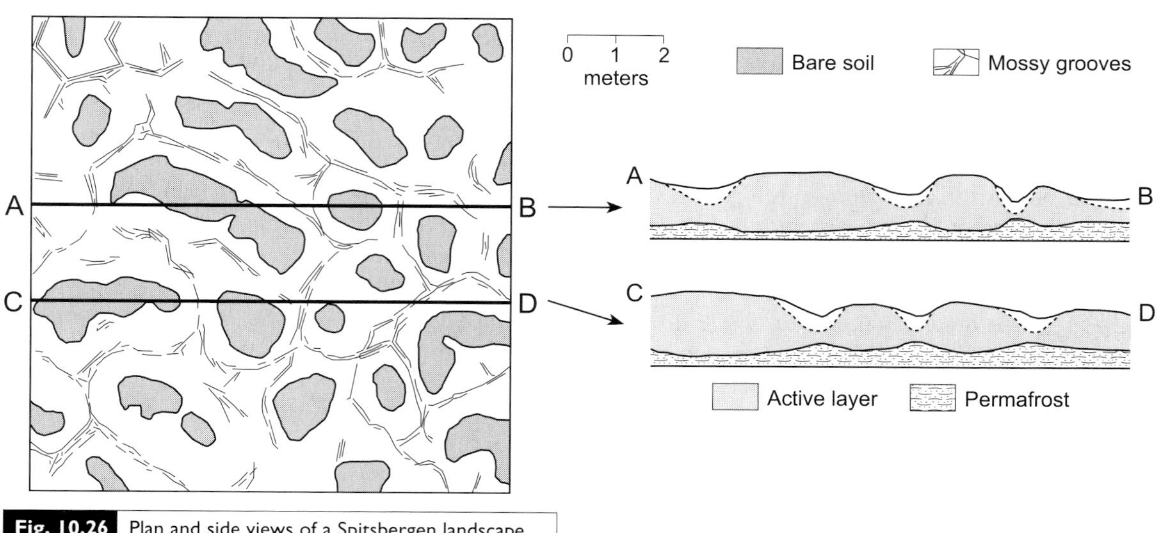

Fig. 10.26 Plan and side views of a Spitsbergen landscape with patterned ground. After Smith (1956).

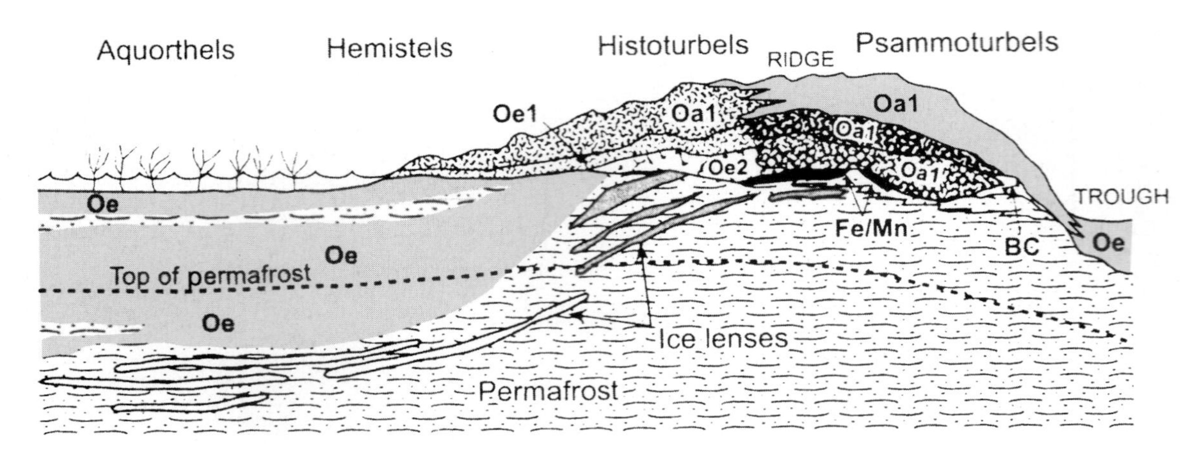

Fig. 10.27 Schematic cross-section of the soils in a low-centered ice wedge polygon near Barrow, Alaska. After Everett (1979).

dominated by cryogenic processes and one that is particularly sensitive to human activities.

Gelisols have three suborders: Histels (dominated by organic materials), Turbels (with strong evidence of cryoturbation, e.g., irregular, broken, or distorted horizon boundaries, involutions, the accumulation of organic matter on top of the permafrost, ice or sand wedges, and oriented rock fragments) and Orthels (with less evidence of cryoturbation) (Soil Survey Staff 1999). The effects of cryoturbation are most evident in mineral, as opposed to organic, soils. Gelisols are found mainly in polar regions, the subarctic and boreal regions, as well as localized high mountain areas. Low soil temperatures cause pedogenic processes, especially weathering, humification and mineralization, to proceed very slowly, making the impact of parent material and natural soil drainage stronger than in most landscapes (Oganesyan and Susekova 1995). Because Gelisols are frozen for much of the year they have the potential to preserve pedogenic features that developed at some time in the past (Gubin 1994). Wet conditions prevail in many Gelisols because of perched water above the permafrost (Tarnocai 1994). Organic matter production is higher in polar landscapes than one might assume, because the soils get warmer in summer than does the atmosphere (Blume *et al.* 1997). Stagnant, perched water and shallow permafrost, coupled with abnormally high organic matter production allows for the accumulation of thick peat layers at the surface, which are subsequently cryoturbated (Zoltai and Tarnocai 1975, Swanson 1996). Decomposition of organic matter is impeded not only by standing water but also by cold temperatures and decomposition-resistant plant material (Höfle *et al.* 1998). The high organic matter contents of most Gelisols have potentially important long-term, global climate implications. Being a storehouse for large amounts of carbon, they may release large amounts of CO_2 to the atmosphere under a climate warmed by greenhouse gases, possibly setting off a complex positive feedback loop that further accentuates global warming.

Gelisols occur on geomorphic surfaces that are, despite their low slope, inherently unstable. Lobes, stripes, terraces and steps are common, as the sediment in the active zone continues, albeit episodically, to move downslope (Benedict 1976, French 1993). As this material is slowly transported, mixing at many different levels and scales occurs. Horizon boundaries are, therefore, highly contorted, broken and often vertically aligned. O horizon material is frequently incorporated into the subsurface, where it can remain for many years (Ugolini 1966, Brown 1969). Much of the incorporated histic material eventually comes to reside at the top of the permafrost or in its upper few decimeters (Michaelson *et al.* 1996). In short, mixing and burial are slow but

(a)

(b)

Fig. 10.28 Gelisols with involuted and contorted horizonation. Small knife for scale. Photos by C. Tarnocai. (a) Strongly cryoturbated Gelisol associated with a non-sorted circle. The lighter material in the soil profile is parent material pushed up by cryoturbation. The dark colored B horizons are strongly contorted by cryoturbation. Bathurst Island, Nunavut, Canada. (b) Cryoturbated Gelisol associated with earth hummocks. The soil has a thin organic layer at the top of the hummock and thicker organic accumulation in the inter-hummock depression. The upper part of the soil profile has a B horizon with granular structure. The lower part of the profile is perennially frozen. Yukon coast, Canada.

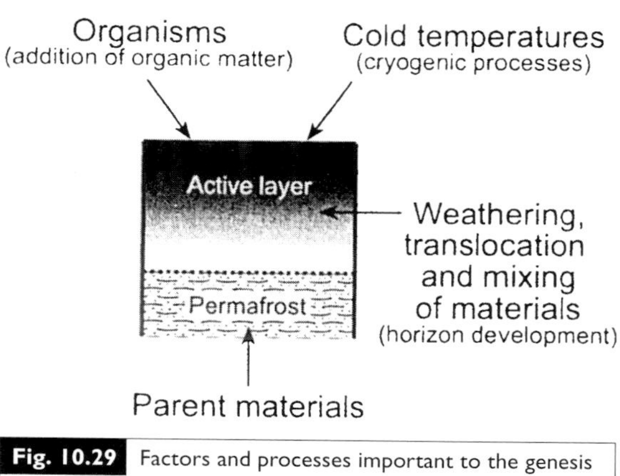

Fig. 10.29 Factors and processes important to the genesis of Gelisols. After Tarnocai (1994).

omnipresent in soil landscapes undergoing cryoturbation.

Cryoturbation leads to repeating, complex and seemingly random patterns of microtopography, which when combined with variable patterns of parent materials and vegetation produce highly variable soils at small and medium scales (Everett 1980, Lev and King 1999, Mueller *et al.* 1999). Earth hummocks and frost boils create specialized niches and microsites for plant growth, peat formation and oxidation–reduction (Ping 1997). Soil horizons, already mixed and contorted by cryoturbation, are further pushed toward non-horizontality by microtopography, which creates areas of sedimentation, saturated areas where peat can form, and highs that are prone to wind erosion. Thus, Gelisols can be highly spatially complex on scales of a few meters, but across large tracts of landscape the overall soil pattern may actually be monotonous.

Freeze–thaw activity, typical of spring and fall, and vein ice formation create granular and platy structure in surface horizons. The growth of the freezing front, downward, into the soil each autumn leads to extreme desiccation and high cryostatic pressures in the soil between it and the permafrost (Bockheim and Tarnocai 2000). The cryostatic pressure placed on the soil that is trapped between the two freezing fronts causes it to be squeezed, creating lower horizons that are often dense and massive.

Gelisol pedogenesis is perhaps best described by Tarnocai (1994) (Fig. 10.30). Permafrost develops quickly after $\text{time}_{\text{zero}}$. Structure forms quickly, aided by heave and shrinkage processes promoted by ice lens formation. Establishment of vegetation leads, eventually, to the formation of O and A horizons, but this material may be cryoturbated at any time. Soil horizon boundaries are contorted. Eventually, a dense subsurface

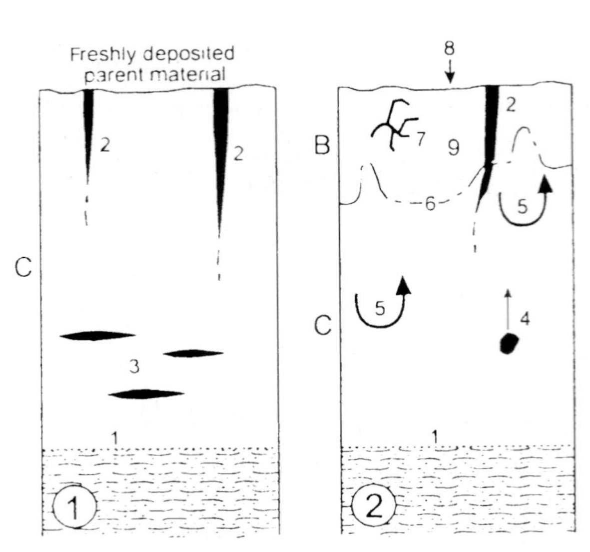

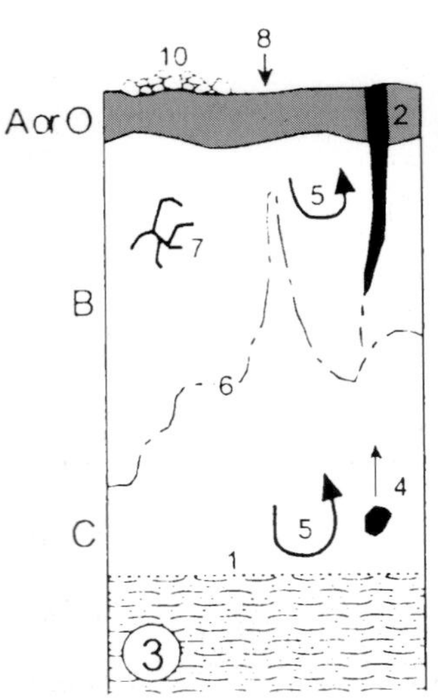

Fig. 10.30 Stages in the development of a Gelisol (Turbel). After Tarnocai (1994).

forms due to freezing front pressures and desiccation. Patterned ground develops and the B horizon, although distorted, shows the effects of podzolization, gleization or brunification.

Cryopedogenic processes are not restricted to Gelisols; they occur in all soils that freeze. Likewise, processes associated with warmer soils, such as podzolization, calcification, salinization, weathering and humification (see Chapter 12) also occur in Gelisols, although usually at reduced rates and intensities (Bockheim 1982, Claridge and Campbell 1982, Forman and Miller 1984, Ugolini 1986a, Ugolini *et al.* 1987, Sokolov and Konyushkov 1998). Lessivage is rare in polar landscapes (Ugolini 1986a, Bockheim and Tarnocai 2000), although pervection is a signature process in many Turbels.

Bear in mind that the polar landscape is highly diverse, especially hydrologically, containing some of the driest and wettest soils on the planet. Parts of the flat, permafrost-laden landscapes are wet and waterlogged for much of the short warm season. Conversely, large expanses of Anarctica are extremely dry; soils accumulate salts; oxidation and salt weathering are dominant processes (MacNamara 1969, Bockheim 1982, Beyer *et al.* 2000). Despite this diversity, several studies have tried to classify, map and otherwise subdivide the polar landscapes into pedogenically homogeneous subregions. There does exist a certain degree of climatically driven pedogenic zonation in the polar zone (Tedrow 1968, Tuhkanen 1986, Ugolini 1986a, Bockheim and Ugolini 1990, Campbell and Claridge 1990, Blume *et al.* 1997, Sokolov and Konyushkov 1998). Soil distribution in the highest latitudes is primarily dependent upon terrain and parent material (Fig. 10.31). Goryachkin *et al.* (1999) subdivided the polar pedogenic landscape into three soil zones: High Arctic barren, Mid-Arctic tundra and Low Arctic and Subarctic tundra (Fig. 10.32). Pedogenic processes change markedly across these regions (Fig. 10.33). Little information exists on the importance or geographic variation in the strength or intensity of cryoturbation across this region of Gelisols. In his review of the major pedogenic processes in polar regions, Ugolini (1986a) scarcely mentions cryoturbation.

We suspect that cryoturbation is at a maximum in the mid-Arctic. Equatorward of the Low Arctic and Subarctic tundra region one gradually encounters less permafrost and more trees, and podzolization becomes increasingly dominant. Wet, gleyed soils, many with histic (O) epipedons, are common. Near the pole, increasingly larger parts of the landscape are dry, windswept islands where inputs of salt aerosols are high and where rocky parent materials have been little pedogeneticized and leached. Here, high

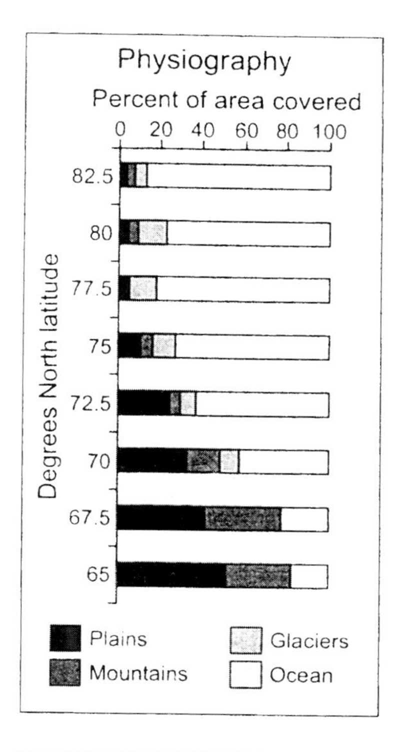

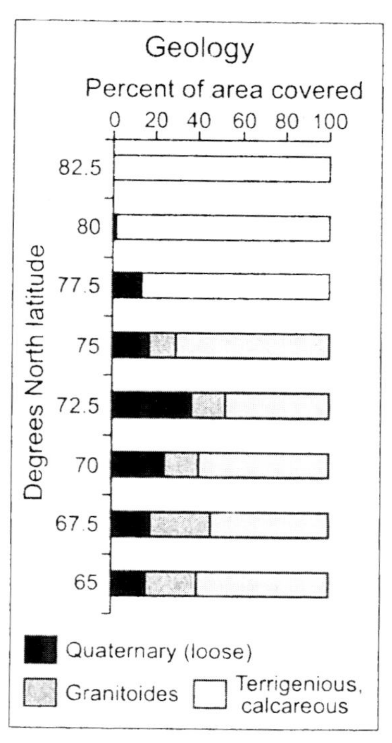

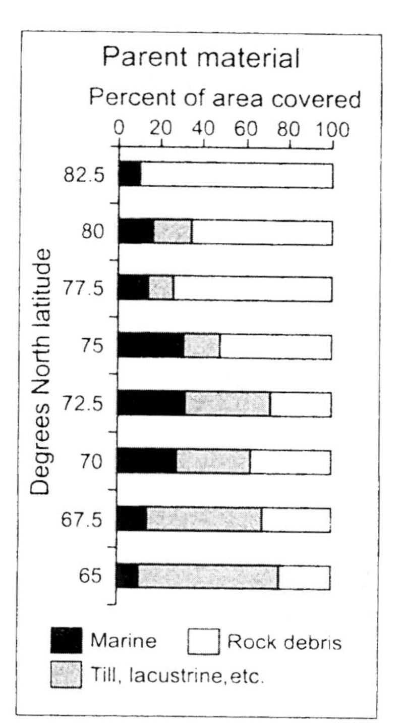

Fig. 10.31 The distribution of terrain and parent material features in the Arctic, as a function of latitude. After Goryachkin *et al.* (1999).

winds and scarce precipitation create conditions where cryoturbation is subsidiary to salinization and calcification. In short, soil development, but not necessarily cryoturbation, decreases poleward (Goryachkin *et al.* 1999). Indeed, many of the very coldest, driest (ultraxerous) soils in Antarctica and in the polar deserts of the northern hemisphere have virtually no pedogenic signature, except for salt concentrations at the surface or shallow depths, and a cover of desert pavement (Claridge and Campbell 1968, Blume *et al.* 1997). Their lack of even an appreciable amount of organic matter has earned them the name *ahumic soils* (Ugolini and Bull 1965, Claridge and Campbell 1982, Bockheim and Ugolini 1990).

Argilliturbation and Vertisols

Some 2:1 silicate clays, e.g., smectite, have high coefficient of linear extensibility (COLE) values; they expand upon wetting and shrink when dry. Soils dominated by these types of clays undergo significant volume changes upon wetting (Jayawardane and Greacen 1987). Because the formation and preservation of smectite clay is exacerbated in wet–dry climates (Folkoff and Meentemeyer 1987), soils rich in smectite commonly occur in wet–dry climates. In the dry season, these soils shrink and develop deep cracks, while in the wet season the clays swell and the cracks close.

Vertisols are defined based on the dominance of argilliturbation. They are high in smectite clay content and are often dominated by fine clay (<0.2 μm dia.) (Nelson *et al.* 1960). Plastic and sticky when wet, Vertisols become very hard when dry (Jayawardane and Greacen 1987). Clay contents can approach 90% or more (Hallsworth and Beckmann 1969). The high COLE values of Vertisols typically increase with increasing amounts of clay, particularly montmorillonite and beidellite (Nelson *et al.* 1960, Roy and Barde 1962, Ahmad and Jones 1969, Fadl 1971, Graham and Southard 1983, Newman 1983, Hussain *et al.* 1984, Yerima *et al.* 1985) (Fig. 10.34), although exceptions do occur, e.g., Isbell (1991).

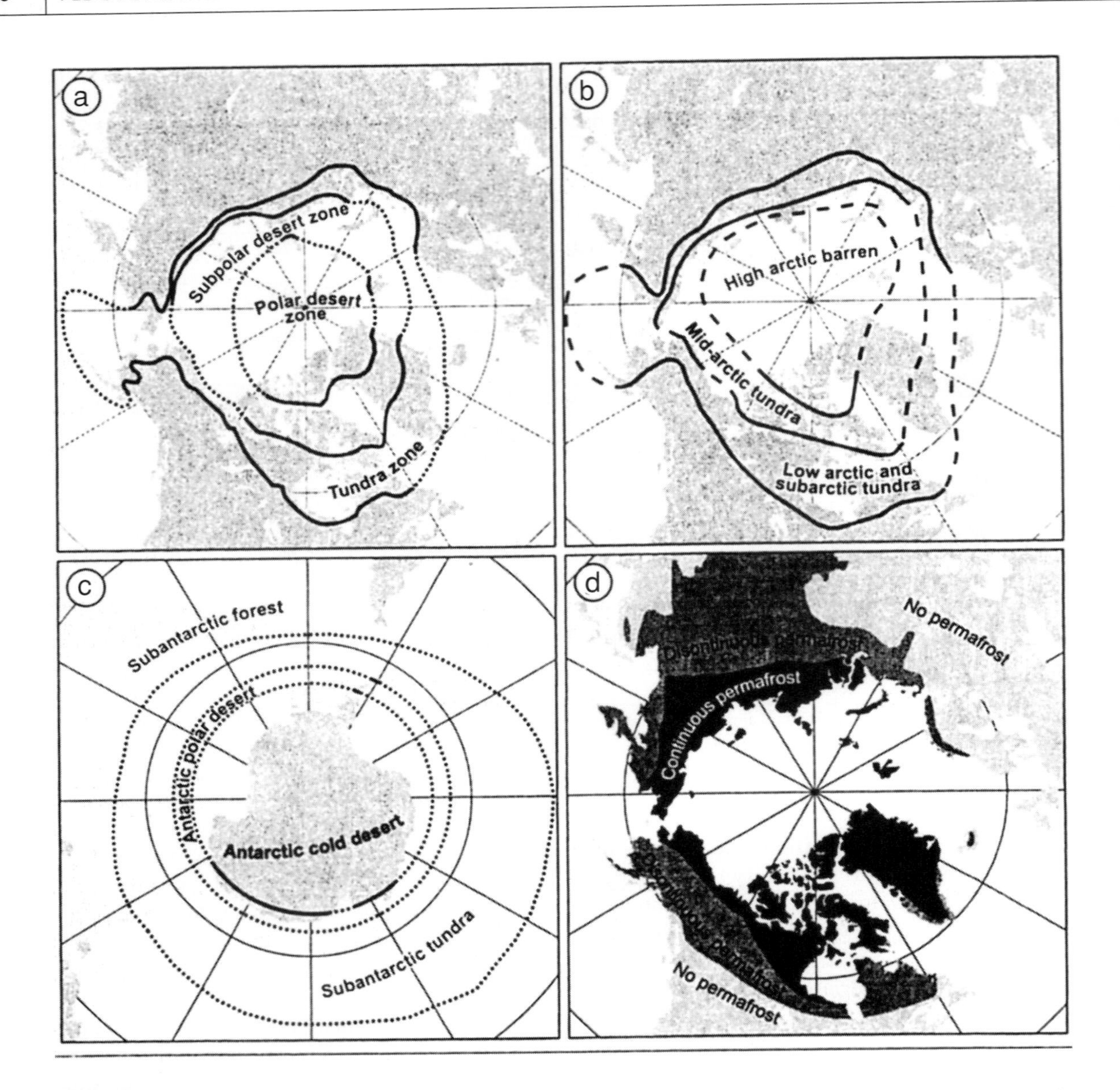

Fig. 10.32 Zonation in the Arctic and Antarctic, as it pertains to soils and permafrost. See also Tedrow (1968). (a) Soil zonation in the Arctic. After Tedrow (1977). (b) Soil zonation in the Arctic. After Goryachkin *et al.* (1999). (c) Soil zonation in the Antarctic. After Bockheim and Ugolini (1990). (d) Permafrost zone in the Arctic. After Péwé (1969).

Dudal (1965) estimated that there are about 257 million ha of dark clay soils in the world – a minimum estimate of the area of Vertisols. Vertisols dominate parts of the Deccan Plateau of India, where they have developed on gently sloping plains underlain by weathered basalt (Simonson 1954b, Roy and Barde 1962, Sehgal and Bhattacharjee 1988). In Morocco and Algeria they are known as Tirs (Del Villar 1944). In Sudan and other parts of semi-arid Africa, Vertisols and similar soils are widespread (Stephen *et al.* 1956, El Abedine *et al.* 1971, Acquaye *et al.* 1992, Favre *et al.* 1997). They are also dominant on the prairies in eastern Texas (Nelson *et al.* 1960, Kunze *et al.* 1963, Newman 1983; see below). Australia is home to >80 million ha

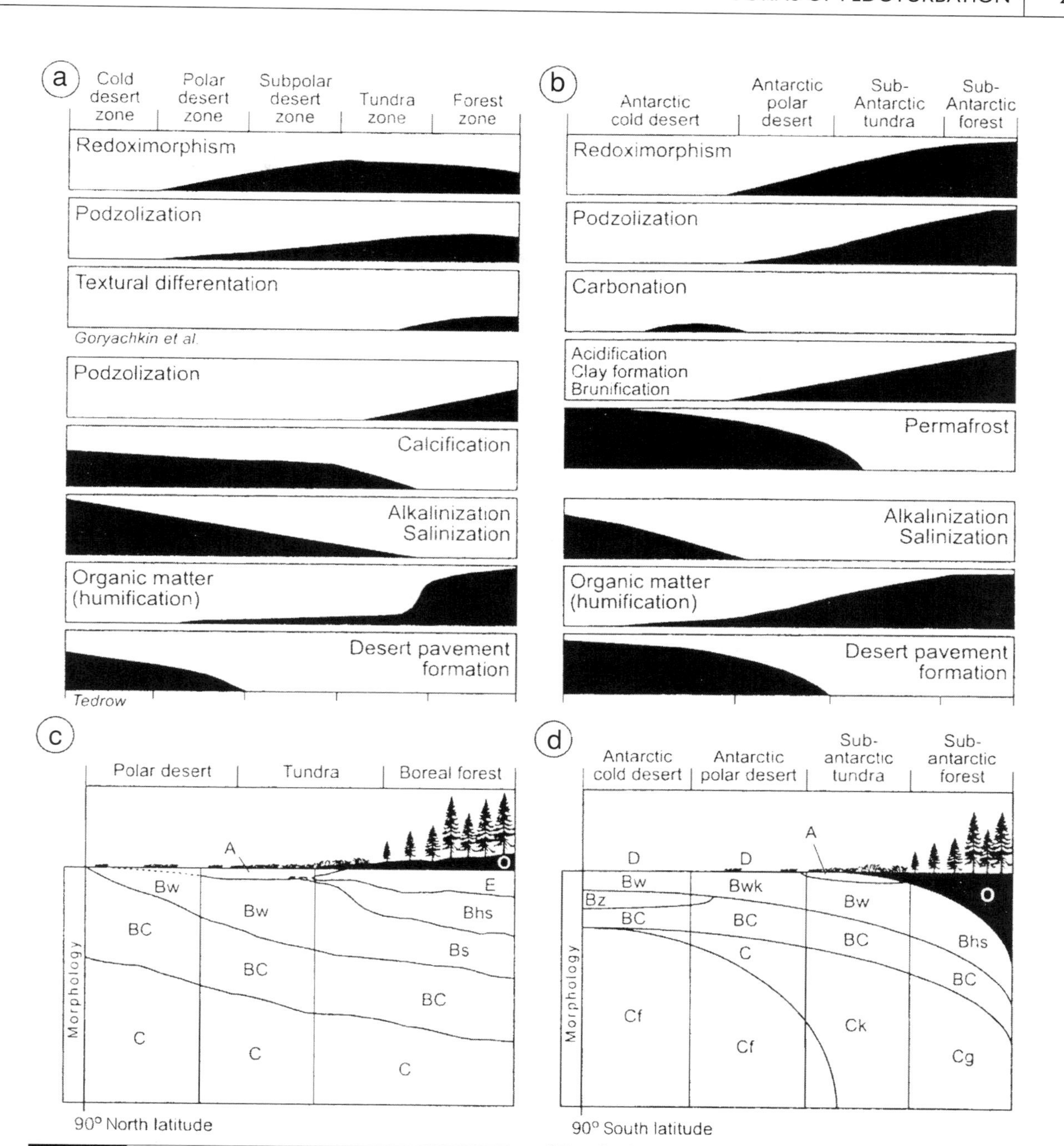

Fig. 10.33 Change in strength/intensity of pedogenic processes and soil morphology in the various polar ecozones. (a) Pedogenic processes in the Arctic. After Goryachkin *et al.* (1999) and Tedrow (1977). (b) Pedogenic processes in the Antarctic. After Blume *et al.* (1997), who modified the original pedogenic zonation figure from Bockheim and Ugolini (1990). (c) Soil morphologies in the Arctic. After Ugolini (1986a). (d) Soil morphologies in the Antarctic. After Bockheim and Ugolini (1990).

of Vertisols (Costin 1955a, Hallsworth *et al.* 1955), giving it the largest acreage of Vertisols of any nation (Isbell 1991).

Vertisols are usually found on flat to gently rolling surfaces (Roy and Barde 1962, Ahmad 1983, Isbell 1991). On steep slopes erosion may outpace profile formation, leading to shallow sola and weakly developed Entisols, while Vertisols occupy adjoining lowlands (Ahmad 1983). Because of the inherent slow permeabilities of Vertisols, low slopes tend to assist them in wetting up (De Vos and Virgo 1969). Many of the best and deepest Vertisols are in shallow depressions, which act as settling basins for clays eroded nearby, and where water can pond. Extended periods of ponding further assists in smectite neoformation, and virtually assures complete wetting of the profile, accentuating the wet–dry seasonality of the site.

Smectites contribute to shrink–swell processes in three ways. First, they can absorb large amounts of water in their interlayer spaces, causing expansion (Anderson *et al.* 1973). When they lose that water upon drying, they shrink. Second, smectites acquire a different structure when wet. Edge-to-edge patterns are common in wet, swollen smectites, while when dry, they often acquire a plate-like arrangement. Third, smectites are usually some of the smaller clay minerals in soils, often being dominant in the fine clay fraction ($<0.2\ \mu m$). COLE is more strongly correlated with fine clay than with coarse clay content (Yerima *et al.* 1989). Soils with abundant fine clay also have much higher *specific surface areas* and greater numbers of small, interparticle pores than do soils with more coarse clay and silt (Coulombe *et al.* 1996). Some have also suggested that the attraction of water for the *surfaces* of clays, rather than their interlayer spaces,

Landscapes: the Blackland and Grand Prairies of east Texas

The Blackland and Grand Prairies of east Texas occupy a nearly level to gently rolling, dissected plain. They are dominated by Vertisols that have developed on calcareous marine shales, marls and clays (Diamond and Smeins 1985). The Grand Prairie has shallower soils and is climatically drier, hence grasses are shorter than on the Blackland Prairie. Flat and gently sloping uplands merge into small narrow valleys on the Grand Prairie. Large rivers crossing the area have broad but shallow valleys. The Blackland Prairie once supported a tallgrass prairie dominated by bluestem, sideoats and switchgrass. Deciduous bottomland woodland and forest were common along rivers and creeks. The high soil fertility is ideal for row crop agriculture, although some hay meadows and a few ranches remain. This area receives 700 to 1200 mm of precipitation during the year, but is known for extreme heat and dryness during midsummer, when wide, deep cracks open in the smectite-rich Vertisols. Precipitation is most common in spring and in fall.

The overwhelming importance of clay-rich, smectitic parent material in a semi-arid climate cannot be missed when examining this landscape. In McLennan County, most of the soils are classified within the ustic soil moisture regime. Only the Tinn series, in stream valleys, does not develop wide enough cracks to classify as an Ustert. The other soils are Udic intergrades to Usterts. The chromic modifer for the Ferris series implies a lower organic matter content, probably due to its position on steeper slopes, where erosion exposes the lighter colors of the B horizon. The Houston Black soil series (Udic Haplustert) is one of the first Vertisols to be studied anywhere (Templin *et al.* 1956) and remains today a classic Vertisol. It dominates the lowlands in Williamson County. Uplands here are shallow to soft chalk bedrock, and thus underlie shallow Inceptisols and Entic Mollisols. Weathering of the chalk coupled with the semi-arid climate keep soil pHs high. Weathered materials from the chalk bedrock are rich in expansible clays and provide the basic cations necessary for their continued neoformation in the chalk saprolite. In short, this landscape is not likely to lack for smectite clays in the near future!

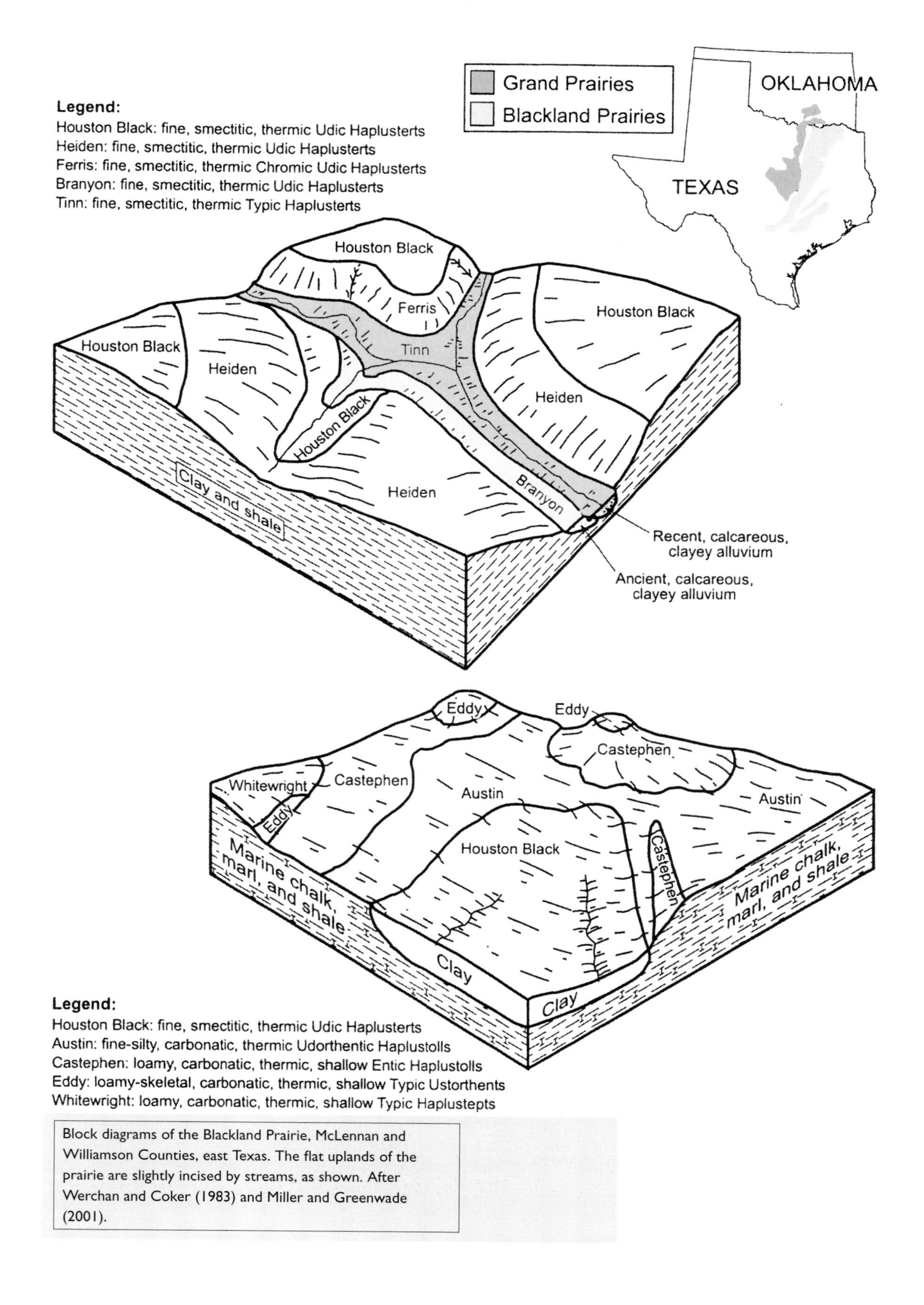

Block diagrams of the Blackland Prairie, McLennan and Williamson Counties, east Texas. The flat uplands of the prairie are slightly incised by streams, as shown. After Werchan and Coker (1983) and Miller and Greenwade (2001).

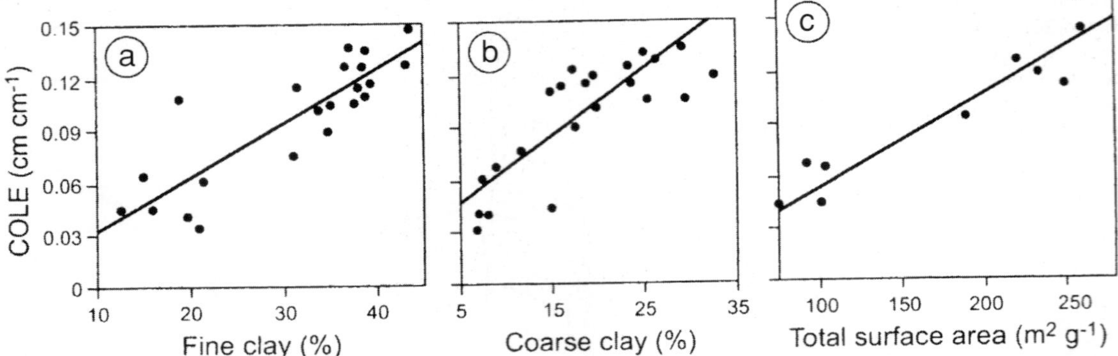

Fig. 10.34 Relationships between coefficient of linear extensibility (COLE), clay content and surface area, for some Vertisols and Alfisols in Cameroon. After Yerima *et al.* (1989).

causes expansion upon wetting. The major proportion of volume changes in soils, as they wet and dry, has also been attributed to changes in microstructure and porosity at submicroscopic levels, rather than interlayer expansion of smectites alone (Wilding and Tessier 1988, Coulombe *et al.* 1996). This view is quite different from the view that most shrinking and swelling is due to interlayer expansion and contraction of smectite clays as they wet and dry. The fact that smectites that have higher specific surface area appear also to have higher shrink–swell potential supports this hypothesis (Coulombe *et al.* 1996). Williams *et al.* (1996) sought out a middle ground and suggested that it is the combination of smectite mineralogy *and* high clay content that leads to high COLE values. In sum, many now believe that the shrink–swell processes in Vertisols are more coincidental than requisite, since soils with mixed mineralogy can also have high COLE values. The real causation may be the *size* of the clay in Vertisols more than the clay *mineralogy*; the coincidence lies in the fact that smectites tend to occur mostly as very fine clay *and* have high expansibility.

Early researchers described Vertisols as self-plowing, self-swallowing or self-churning soils; the name comes from the Latin *vertere*, to turn over or invert. The self-plowing concept has traditionally involved: (1) the shrinking of the soil matrix and the formation of desiccation cracks, (2) the partial infilling of the cracks with material from outside the profile, or from higher up within the same profile, (3) the wetting and expansion of the soil matrix, leading to a space problem, resulting in (4) soil material moving (churning) horizontally and vertically to accommodate the volume of the infilled material (Simonson 1954b, Bronswijk 1991).

In Vertisols, cracks form at the surface and expand downward. Many cracks reopen in the same place year after year (El Abedine *et al.* 1971). Deepening cracks facilitate the drying of the lower solum. The largest cracks tend to be in surface microbasins, sometimes with one primary crack in the center (Newman 1983). Soil material sloughs off the sides of open cracks and fills them; this is expected because the subsoil often shrinks more than the surface soil, resulting in cracks that widen with depth and overhang near the surface (Hallsworth *et al.* 1955). We cannot overemphasize this point, since cracks that are widest at the top will have less soil material fall into them than those that widen with depth. Soil material can also be blown in or knocked in by biota, or washed in by rain. Any cracks that remain open after swelling are soon filled as surface soil washes in (Hallsworth *et al.* 1955). Direct evidence of this process is present in most Vertisols as darker, humus-rich masses in lower or mid-profile positions (El Abedine *et al.* 1971, Hussain *et al.* 1984). Rewetting causes the cracks to close; it can occur from the bottom up, from the top down, or from side to side (Graham and Southard 1983, Coulombe *et al.* 1996). Heavy rains or flooding can cause the soil to wet from the bottom up, while moderate and gentle rains tend to cause upper horizons to wet first (Dasog *et al.* 1987).

Fig. 10.35 Gilgai microtopography on a Texas prairie. Photo by W. Lynn.

Soil structure within Vertisol A horizons, often described as a "surface mulch," is typically strongly granular (Dasog *et al.* 1987). These clay- and humus-rich horizons undergo even more frequent cracking than the Bss horizons below, fracturing peds into small granules. The hard, nut-like structure resembles loose gravel on the surface (Dudal and Eswaran 1988). Additionally, grass roots contribute to the integrity of granular peds in the A horizons, and worm casts are also abundant.

Gilgai

Argilliturbation is manifested within Vertisol profiles and on the surface as gilgai microtopography, with relief exceeding commonly 15 cm (Fig. 10.35). *Gilgai* is an Australian aboriginal term meaning small water hole, referring to microtopographic depressions (Paton 1974). Australia has some of the most extensive and varied gilgai forms anywhere (Hallsworth *et al.* 1955, Isbell 1991). The term gilgai has since come to mean an assemblage of small mounds *and* depressions (microknolls and microbasins) produced by argilliturbation (Knight 1980, Soil Survey Staff 1999). Related terms include melonhole, crabhole, puff and shelf (Hallsworth *et al.* 1955).

The variability of gilgai is somewhat predictable, and quite considerable; gilgai are not just random mounds and depressions, and they are rarely the same from one landscape to the next. Newman (1983) felt that each Vertisol series has its own signature gilgai, much like a fingerprint. Hallsworth and Beckmann (1969) described six distinct kinds of gilgai (Paton 1974, Bhattacharyya *et al.* 1999). Regularity of mounds and depressions can be such that amplitudes can be readily determined (Jensen 1911, Hallsworth *et al.* 1955, Stephen *et al.* 1956). White and Bonestell (1960) reported that the respective surface area occupied by gilgai in South Dakota varies between 1/8 and 1/1 (microridge/microvalley). On steep slopes gilgai may align down the slope (White and Bonestell 1960).

Gilgai form by any of a number of stress/heave configurations (Knight 1980). Most common is the assumption that they form in the areas between cracks, i.e., the stresses induced by additional soil material within cracks is transferred to ridges between. Depressions are most likely associated with cracks or former crack locations (Graham and Southard 1983). Establishment of a gilgai pattern, especially depressions which occasionally pond water, can lead to additional pedoturbation when crayfish invade these wet sites and begin burrowing; ant mounds tend to be concentrated on knolls (Newman 1983).

Slickensides

Expansion stresses within Vertisols force masses of soil to slide past each other, forming a subsoil structure composed of wedge-shaped aggregates, also called *parallelepipeds* or *lentils* (Blokhuis 1982). These structural units, found only in Vertisols, are most common in the B (Bss) horizon at depths of 25–125 cm. De Vos and Virgo (1969: 199) described these features: "Individual peds have triangular or trapezoidal faces tapering to points at each end in the form of a 'double wedge,' but forming an obtuse-angled dome at the upper and lower sides." They argued that the term bicuneate should be used to describe these double

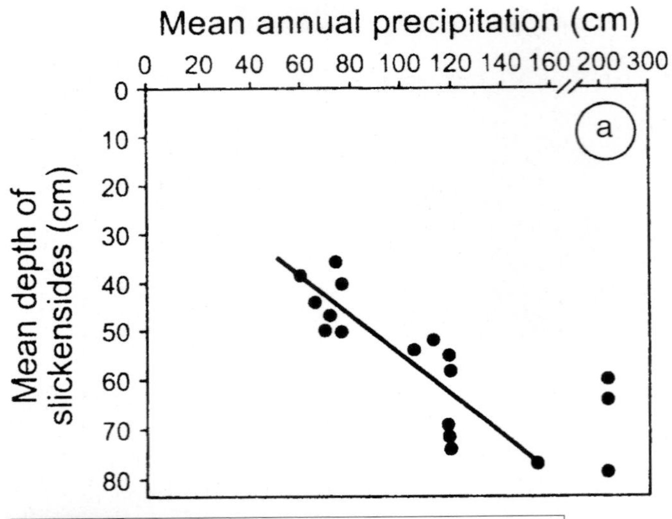

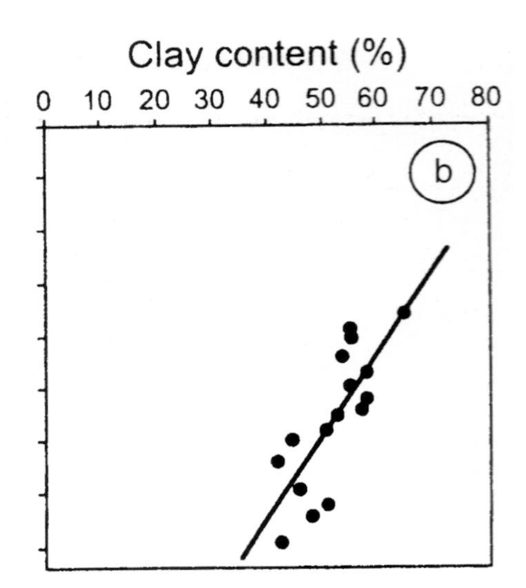

Fig. 10.36 Mean depth of slickensides in some Vertisols of India, as a function of (a) mean annual precipitation, and (b) clay content. After Vadivelu and Challa (1985).

(a)

(b)

Fig. 10.37 Slickensides in Vertisols. (a) Large slickenside in a Texas Vertisol. Photo by W. Lynn. (b) Close-up of a slickenside in a New Mexico Vertisol. Photo by RJS.

wedge-shaped aggregates in Vertisol B horizons (Latin *bi*, two, *cuneus*, wedge). Found closer to the surface in drier climates, they are better expressed where the wet–dry seasonality is extreme (Ahmad 1983, Vadivelu and Challa 1985) (Fig. 10.36). As the soil expands, the structural units are slightly recast and remolded. In time, they acquire blocky or wedge shapes, with smooth edges and facets. Most are wider than they are tall (Ahmad 1983).

Slickensides, which are highly characteristic of Vertisols, are the shear planes formed along wedge-shaped aggregate edges, in the subsurface. They represent the plane along which the expanding soil matrix has moved, much like a thrust fault. Often shiny when moist, slickensides appear polished or grooved when freshly exposed, or striated if the aggregate slid past another which had gravel or sand grains protruding from it (Lynn and Williams 1992) (Fig. 10.37).

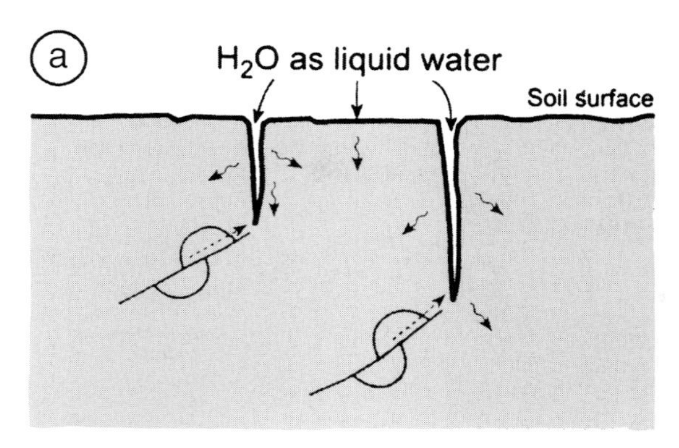

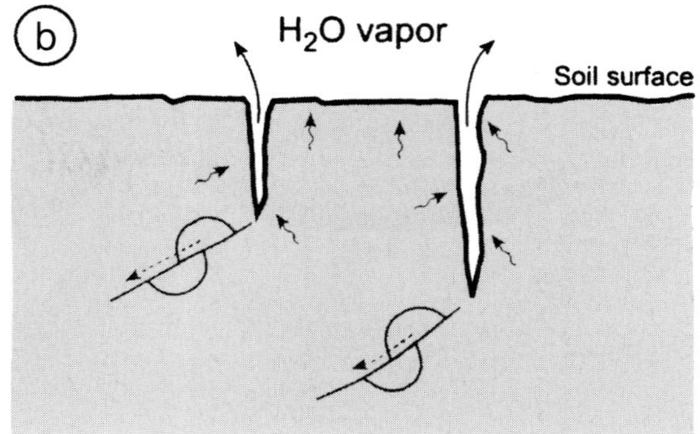

Fig. 10.38 Movement within a Vertisol upon wetting (a) and drying (b). After Lynn and Williams (1992).

Larger slickensides are often fluted (Fig. 10.37a). The shear surface is seldom planar, but rather is slightly curved like a flattened spoon (White and Bonestell 1960, Lynn and Williams 1992). Once formed, a slickenside is the surface of least resistance to shear for the next wetting/expansion event; thus they persist. *In situ* slickensides often align along a plane that is from 20° to 60° from the horizontal, indicating that most of the shear is occurring out and up, rather than vertically or entirely horizontally (Knight 1980). Lynn and Willams (1992) noted that many slickensides represent planes of movement in two directions, upward upon wetting, and downward upon drying, and that many vertical, surface cracks lead down to a slickenside in the subsurface (Fig. 10.38).

Models of argilliturbation

Argilliturbation in Vertisols occurs in two ways: (1) by swelling pressures and mixing within the solum, and (2) by self-swallowing processes that lead to volume disruption because surface materials have accumulated in the lower profile. Although self-swallowing is ongoing in Vertisols (El Abedine *et al.* 1971, Graham and Southard 1983), recent research has de-emphasized its role, instead emphasizing simple expansion–shrinkage processes (Beckmann *et al.* 1984). Evidence for this comes from Vertisols with pronounced gilgai that have protrusions of subsoil and subjacent material into near-surface locations (Paton 1974) (Figs. 10.39, 10.40). Newman (1983) called these structures "chimneys," while other have used the terms *diapir* and *mukkara*. These features may extend completely to the surface as gilgai mounds or tuffs, although most do not (Figs. 10.39). Vertisol sola are much shallower above chimneys (Kunze *et al.* 1963). Shallow depressions between chimneys are often referred to as *bowls*; they range from 2–5 m across and 1–2 m deep (Williams *et al.* 1996).

The presence of chimneys and bowls has forced a rethinking of the traditional self-swallowing model of Vertisol genesis, for if cracks formed in *random* locations and material was deposited in them evenly each year, chimneys would not exist. The self-swallowing model also does not explain certain features of Vertisols, such as systematic depth functions of organic carbon, salts and carbonates, and the presence of E and Bt horizons, which should be destroyed by repetitive cracking and argilliturbation (White and Bonestell 1960). Material in chimneys often has different textures *and* colors than the surrounding material; the areas between chimneys are traditional black clay while diapirs/chimneys consist of lighter-colored, brown clay, like subsoil material (Hallsworth *et al.* 1955, Hallsworth and Beckmann 1969, Paton 1974, Coulombe *et al.* 1996). Paton (1974) hypothesized that, in chimneys, subsoil material has been thrust upward due to differential loading pressures on wet clay, which can flow upward through overlying soft sediments, as in salt domes. Hirschfield and Hirschfield (1937) showed that the subsoil under gilgai chimneys expanded considerably more, upon wetting, than did the surface layer. These

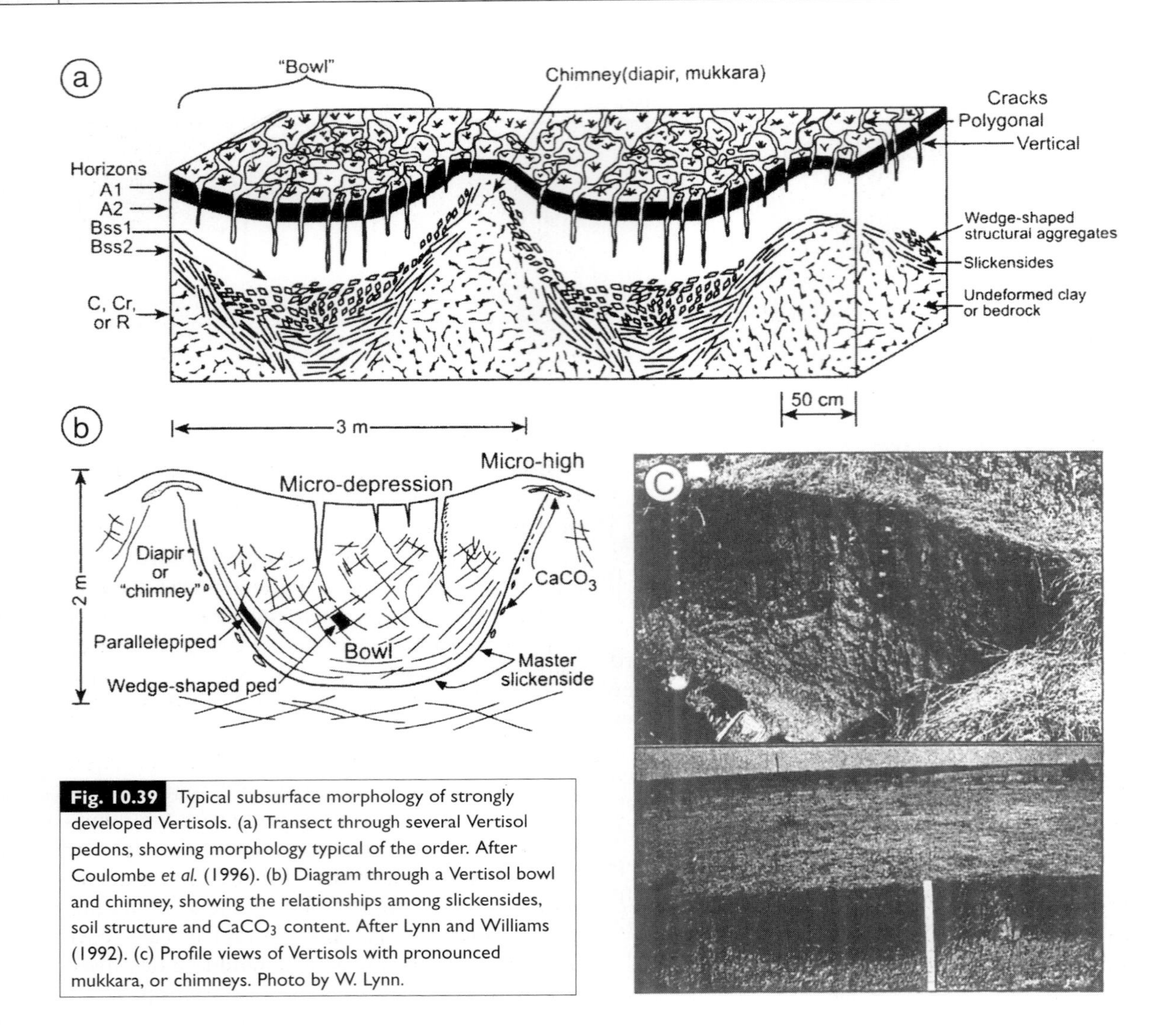

Fig. 10.39 Typical subsurface morphology of strongly developed Vertisols. (a) Transect through several Vertisol pedons, showing morphology typical of the order. After Coulombe *et al.* (1996). (b) Diagram through a Vertisol bowl and chimney, showing the relationships among slickensides, soil structure and $CaCO_3$ content. After Lynn and Williams (1992). (c) Profile views of Vertisols with pronounced mukkara, or chimneys. Photo by W. Lynn.

data support Paton's (1974) model, which involves chimney materials pushing upward through a darker, thin surface mantle that periodically cracks and swells. This hypothesis goes a long way toward explaining why material in the chimneys is lighter in color. Annual cracking, which efficiently incorporates organic matter into the soil to the depth of cracking (Knight 1980), would not theoretically affect chimney pedons. Cracking would be mostly restricted to the bowls between chimneys, which obtain much drier states and crack to greater depths (Fig. 10.39). Also in support of this model is the observation by Lynn and Williams (1992) that the outer surfaces of bowls often have large, master slickensides that essentially form the boundary between chimneys and bowls (Knight 1980, Wilding *et al.* 1991) (Fig. 10.37). These can be traced from nearly vertical orientations near the surface (next to the chimney) to nearly horizontal orientations at the base of the bowl, about 1–2 m below the surface.

Wilding and Tessier (1988) expanded upon Paton's model. They noted that the area between the chimneys forms a type of syncline, or synclinorum, with the chimneys occupying the area between (Beckmann *et al.* 1984, Williams *et al.* 1996) (Fig. 10.39). In the Wilding and Tessier (1988)

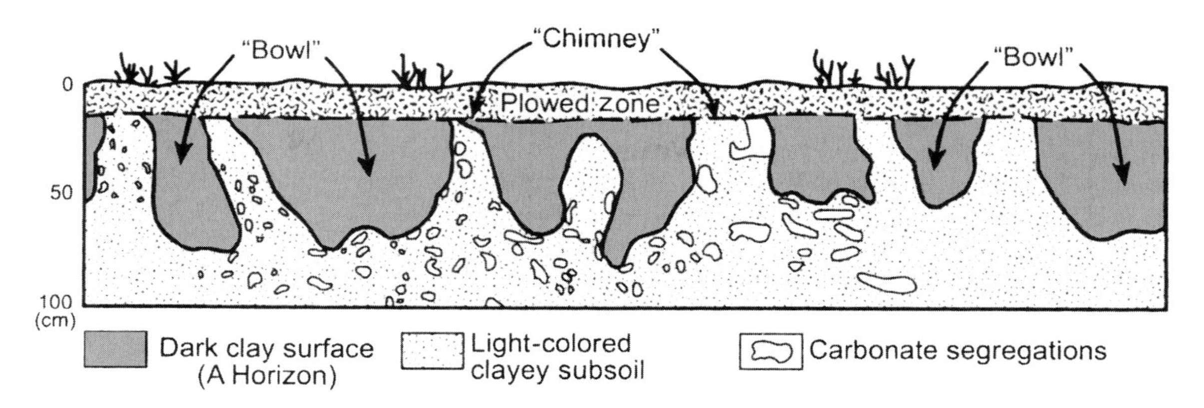

Fig. 10.40 Cross-section (true to scale) along an exposure of Vertisols in Queensland, Australia. Chimneys and bowl structures are evident. After Beckmann *et al.* (1984).

model, dry, cracked soil rewets from the base of the cracks upward. The lighter-colored subsoil, which may have higher COLE values, expands in all directions (Beckmann *et al.* 1984). The only way for the increased volume to be accommodated is by buckling of the surface, similar to the way folds develop in a carpet (Fig. 10.40). Once the folds are initiated, further expansion can be accommodated by additional vertical motion of the subsoil. Slickensides form at this point and induce the formation of thrust cones or bowls, between which are chimneys. Within the thrust cones, shrink–swell processes, slickensides, self-churning and argilliturbation are maximal, while between the bowls (in the chimneys) only slow upward movement of subsoil material is occurring. This set-up is enhanced when COLE values in the subsoil exceed those of the upper sola, a situation common to landscapes with gilgai (Hallsworth and Beckmann 1969, Beckmann *et al.* 1984). The plastic limits of the lighter subsoil material vs. the darker bowl material are such that, at certain moisture contents, the subsoil might still be deformable while the dark soil above would be dry and rigid (Beckmann *et al.* 1984). This juxtaposition leads to chimney formation as the subsoil buckles upward. At the base of long slopes, downslope creep pressures force the entire sequence to be compressed, forcing chimneys to fold over each other in the downslope direction, like waves breaking on a beach (Fig. 10.41). Some of the subsoil light brown clay of the chimney gets washed into the bowls, essentially completing the circuit. Leaching is maximal within bowls, which are microdepressions and crack the most, and minimal on chimneys (Table 10.7). Therefore, carbonates tend to be much more deeply translocated within bowls, and often within chimneys secondary carbonates and gypsum occur very near the surface (Hallsworth *et al.* 1955, Lynn and Williams 1992) (Figs. 10.39, 10.40, 10.42).

All in all, argilliturbation and shrink–swell processes in Vertisols lead to intimately mixed horizons and short-range spatial variability. Horizonation is faint and boundaries get blurred, especially in bowls. And yet, because these processes lead to pedons (bowls) with overthickened A horizon material between chimneys that are very different and much shallower, they should probably be considered proisotropic *and* proanisotropic (Johnson *et al.* 1987).

The development of bowls/thrust cones potentially leads to a series of interesting feedback mechanisms. First, we must assume that the finest clays have the most shrink–swell potential, regardless of their mineralogy. Agents that flocculate those clays into larger masses with lower shrink–swell capacity include organic matter, gypsum and carbonates. In the lower, wetter bowls, organic matter tends to accumulate, leading to flocculation of clays and potentially less shrink–swell activity. Working against this trend, however, is the tendency for bowls and microtopographic lows to be more intensively leached, allowing clays near the surface to be less flocculated and thereby enhancing shrink–swell

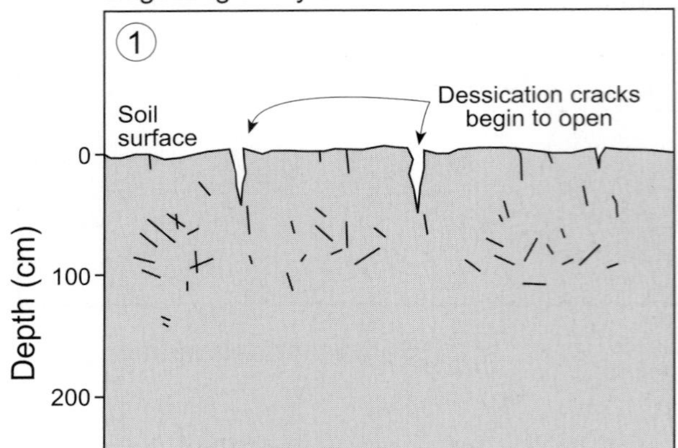

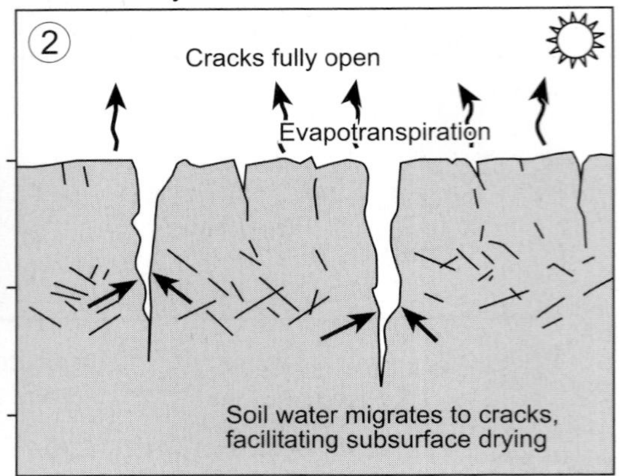

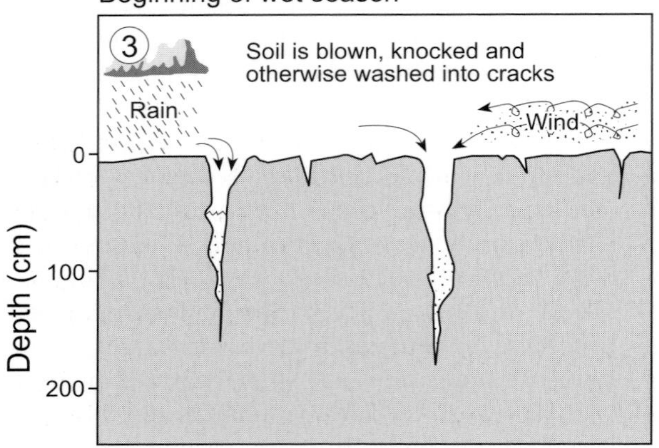

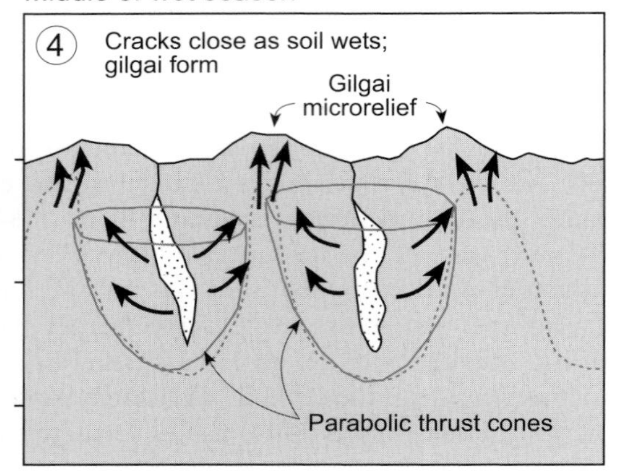

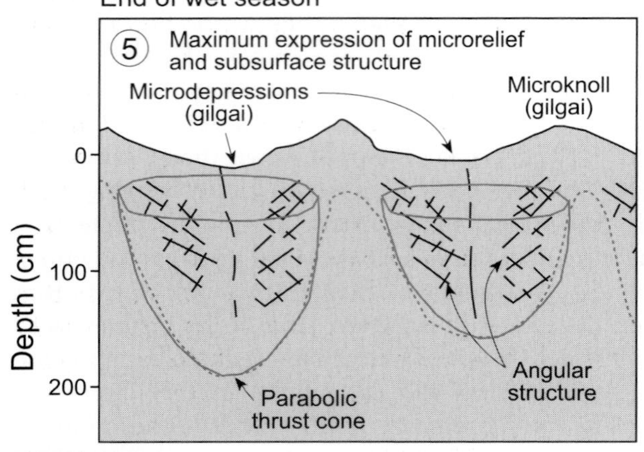

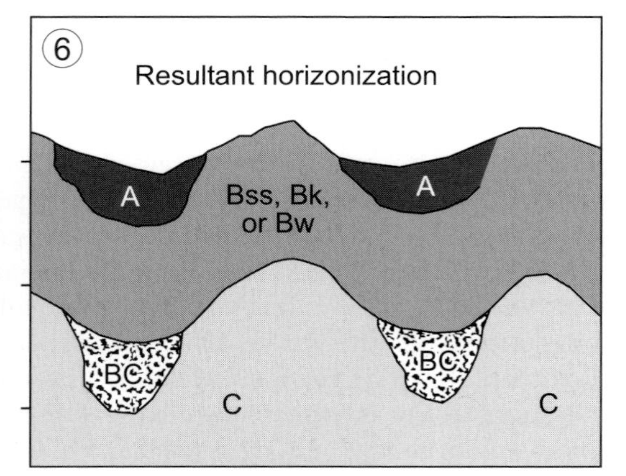

Fig. 10.41 Current thinking, compiled from several sources, of the formation of Vertisols by argilliturbation and shrink–swell processes, into the traditional bowl-and-chimney subsurface morphology.

Table 10.7 General morphological trends between microbasin (bowl) and microhigh (chimney) landscape positions in Vertisols

Property	Microbasin (bowl)	Microhigh (chimney, knoll)
A horizon thickness	Thicker	Thinner
Solum thickness	Thicker	Thinner
Color	Darker	Lighter
Organic carbon content	Greater	Lower
pH	Lower	Higher
Structure grade	Finer	Coarser
Consistence	Very firm	Extremely firm
Cracking characteristics	Many, wide, deep	Fewer, narrower, shallower
Shrinkage	Higher	Lower
Slickenside expression	Anticlinal	Synclinal
Moisture content (mean)	Higher	Lower
Depth of leaching	Greater	Lesser
Carbonates	Less, deeper	More, higher in profile

Source: Wilding *et al.* (1991).

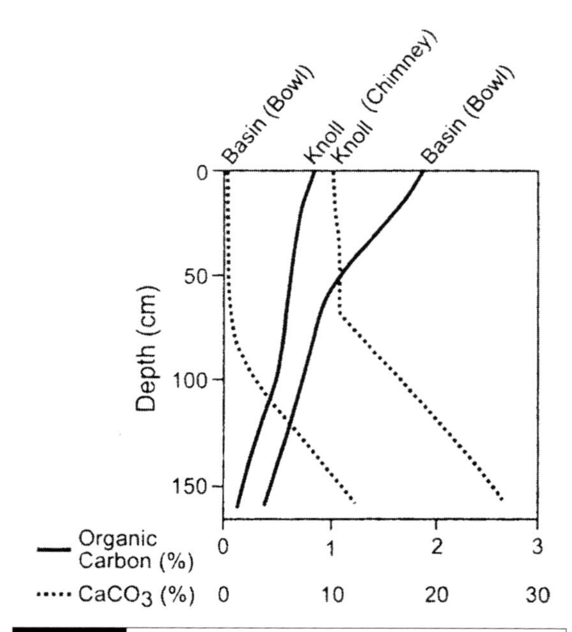

Fig. 10.42 Depth distributions of organic carbon and $CaCO_3$ for a microbasin and microknoll in a Texas Vertisol. After Newman (1983).

processes. Although these two process bundles might offset each other, Williams *et al.* (1996) suggested that the leaching of carbonates and gypsum has more influence on the system than does the accumulation of organic matter. In carbonate-rich chimneys, argilliturbation processes will be discouraged by the tendency for smectites there to be flocculated. The end result is a system that tends to perpetuate argilliturbation in the bowls, opening up the soil with cracks and facilitating leaching. At the same time, carbonates and gypsum (which are not translocated from the chimneys) work to chemically de-emphasize cracking there, which in turn minimizes wetting and leaching processes.

Despite the two viable theories of Vertisol genesis outlined above, the genesis of diapirs, chimneys and mukkara, as well as thrust cones, is still somewhat unclear. They can occur on some landscapes but be inexplicably absent on adjoining ones (Fig. 10.43). They can occur without gilgai.

Although the general processes within Vertisols have been known for decades, details of the process are still working themselves out. However, current research appears to be indicating that Vertisols are not as isotropic as once thought and that *thrusting* due to swelling may be more important than *mixing*.

Rates of argilliturbation

The rates of argilliturbation and soil formation in Vertisols largely depends on the dynamics of surficial materials being incorporated into cracks, and exactly *where* one examines it, i.e., bowl or chimney.

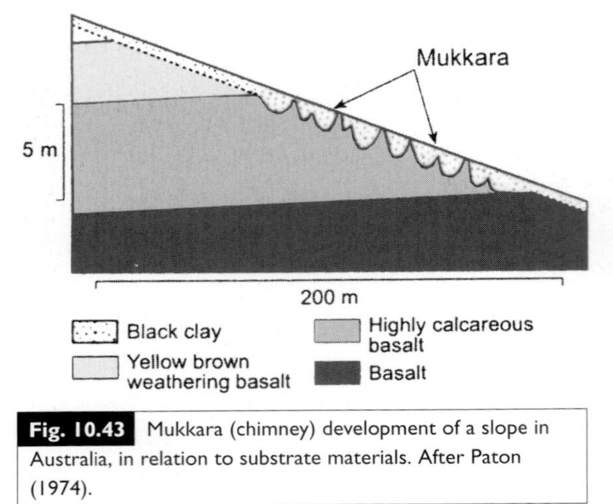

Fig. 10.43 Mukkara (chimney) development of a slope in Australia, in relation to substrate materials. After Paton (1974).

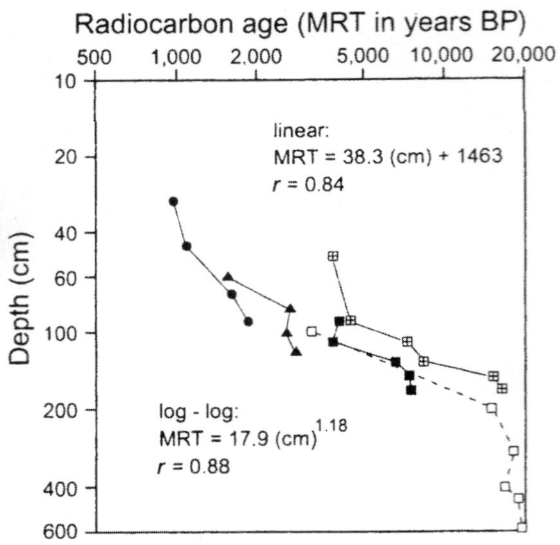

Fig. 10.44 Mean residence time (MRT) (see Chapter 14) of organic matter in five Vertisols and Vertisol-like profiles from Israel. After Yaalon and Kalmar (1978).

Simonson (1954b: 201) postulated that mixing of a Vertisol could occur "in a matter of centuries." El Abedine *et al.* (1971) measured the volume of infilled soil material in closed cracks in a Vertisol in Sudan and determined that the entire profile could be overturned in 8700 years. Yaalon and Kalmar (1978) recalculated mixing rates, using the data of El Abedine *et al.*, and determined turnover times of 700–1250 years for the *upper* solum. Radiocarbon ages of soil organic matter tend to confirm that rates of mixing are less than the rates of organic matter additions. In Vertisols with continuous, deep mixing, ^{14}C ages should be similar with depth, but they seldom are (Scharpenseel 1972c). Rather, ^{14}C ages increase with depth, meaning that organic matter additions procede at a more rapid rate than does mixing. Southard and Graham (1992) used an isotopic tracer, ^{137}Cs, produced from open-air nuclear testing and which is strongly adsorbed onto clays, to estimate the rates at which surficial materials had become incorporated into the subsoil of a Vertisol. Within 25 years since the peak ^{137}Cs fallout, some mixing had occurred to the maximum depth of cracking, but most mixing was limited to the upper solum. Their work indicated that sloughing of soil into the bottoms of *deep* cracks occurs only in the very driest years, and that argilliturbation decreases markedly with depth. This work may help explain why the Bss horizon and its slickensides are maximally developed at the *mean*, rather than the *maximal*, depth of cracking. Over time, more soil material is likely to fall to the mean depth of cracking than to any other depth, and therefore expansion pressures will, on average, be greatest at that depth.

Yaalon and Kalmar (1978: 36) noted that "the process of turbation is apparently much less significant than commonly believed, and is essentially a moderate rate process." They concluded (1978: 41) that intensive pedoturbation is not required to produce the typical characteristics of Vertisols, citing mean residence time (MRT) ^{14}C dates on soil organic matter, which should be uniform with depth if mixing was complete and thorough (Fig. 10.44). Their data show that MRT dates increase significantly with depth, as they do in soils that lack large amounts of pedoturbation, suggesting that most mixing in Vertisols is confined to the upper solum.

Argilliturbation and coarse fragments

Argilliturbation is one of the few forms of pedoturbation that can transport gravel, gastroliths, carbonate nodules and large rocks to the surface, making it proanisotropic (Blokhuis 1982). Some Vertisols even have sand maxima at the surface (Yaalon and Kalmar 1978). For fine particles like

(a) Advancing wetting front

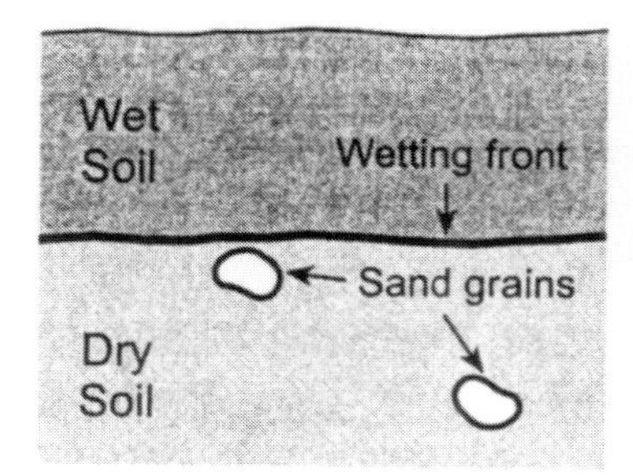

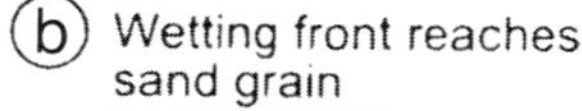

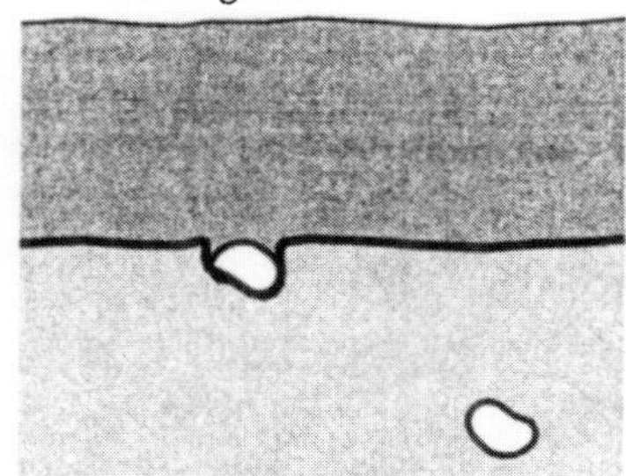

(c) Sand grain uplifted by swelling clay

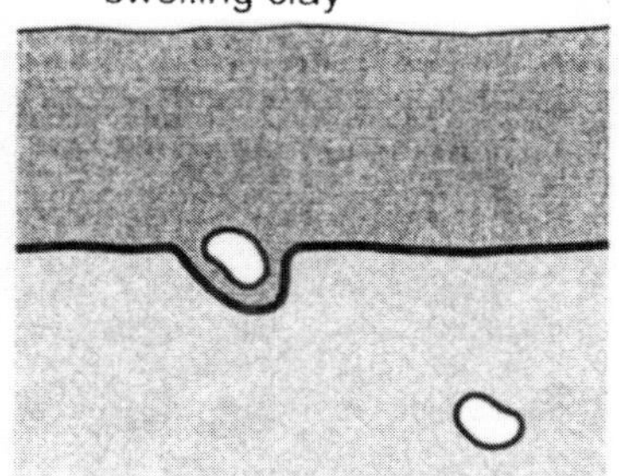

Fig. 10.45 Diagrammatic representation of how argilliturbation and expanding clays can move small objects upward in a soil profile. After Yaalon and Kalmar (1978).

sand the process functions as shown in Fig. 10.45. As a wetting front advances into a soil with abundant smectites, water rapidly flows around mineral/sand grains, ahead of the front. Clays below the grain expand, pushing it upward. Movement of the grain down, into dry, hard clay, is simply not as likely as is upward thrusting. Drying of the wet soil occurs rather uniformly, causing the grain to remain at its new position, or to fall back only a fraction of the distance it was lifted.

For larger clasts (larger than many cracks), another factor is at work (White and Bonestell 1960). As larger clasts are moved upward by thrusting, they are unable to fall back into cracks because the shrinkage cracks are too small, and thus become concentrated at the surface (Springer 1958, Mabbutt 1965). In the end, coarser material comes to overlie finer material (see also Jessup 1960). Because of this process, the bedrock–soil contact in Vertisols can be abrupt; as soon as a rock is dislodged from the bedrock it is swept up and into the solum (Johnson *et al.* 1962: 393).

In extreme situations, argilliturbation can result in a surface stone pavement (Johnson and Hester 1972, Muhs 1982). Johnson and Hester (1972) pointed out how all clasts can be brought to the soil surface by upward heave, but only the largest clasts remain there, as others fall into cracks and are cycled through the profile (Fig. 10.46). Vertisols and vertic intergrade Mollisols in Iraq have limestone fragments in their upper sola, but are stone-free below, suggesting that argilliturbation has moved them upward (Hussain *et al.* 1984). Similarly, in Australia, Mabbutt (1965) described Aridisols with pronounced stone pavements, stone-free B horizons and silcrete duricrusts at depth (Fig. 10.47). He hypothesized that these stones had been moved upward from the duricrust below by argilliturbation. Since that time, the climate has dried, argilliturbation has ceased, and an ordered A–E–B–C profile has developed, while retaining the inherited rock depth trend. A second possibility that Mabbutt (1965) considered is that, after a certain number of rocks are moved to the surface, an intrinsic threshold is crossed. The rocks act as a mulch, effectively minimizing surface cracking by limiting evaporation.

Graviturbation

Downslope transport of soil and sediments under the influence of gravity is termed *mass wasting* or *mass movement*. Usually, interstitial water aids in this movement by reducing the shear strength of the materials. Turbation of regolith while in transport is typically a function of the category of mass movement, of which there are four: slides, flows, falls and heaves. In slides, cohesive blocks of material move downslope along a distinct plane of failure. Mixing within the sliding material may be minimal. Potential for mixing within falls and flows, however, is considerable, either while the soil material is in transit or due to settling and compaction/rearrangement immediately upon cessation of movement. Nonetheless, because most of the aforementioned mass movements are generally catastrophic or of moderate size, they are usually not considered within the

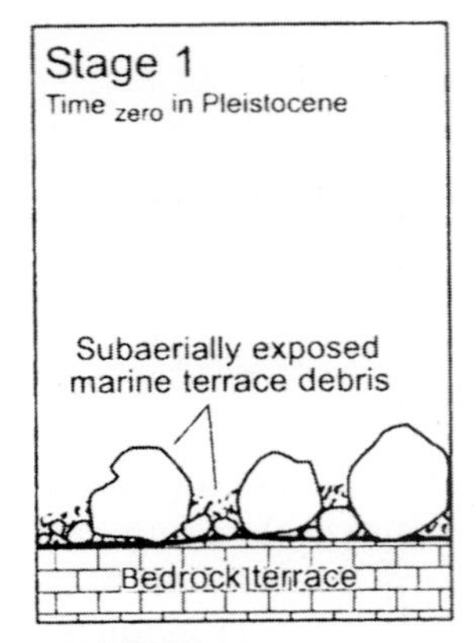

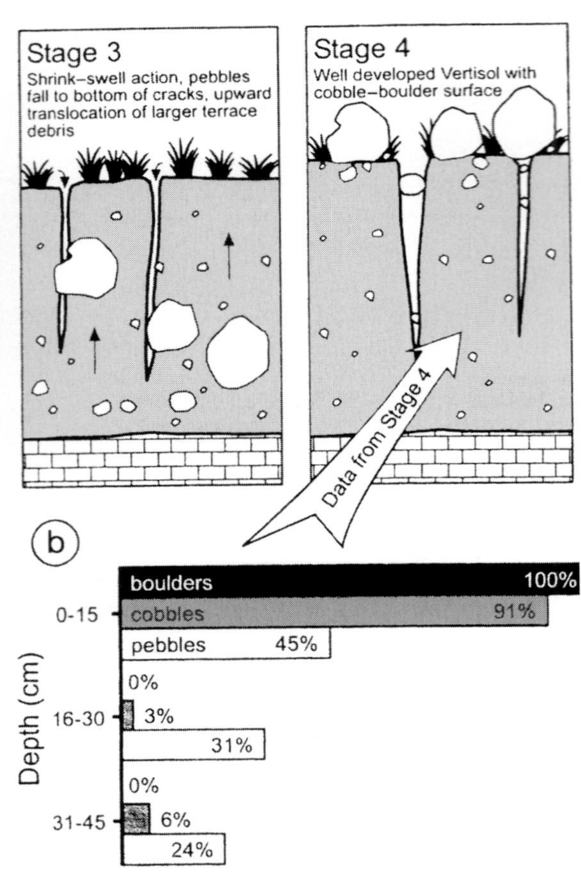

Fig. 10.46 Illustrations of the mixing effects of argilliturbation on coarse fragments. After Johnson and Hester (1972). (a) Hypothetical model showing how clasts on the surface of a marine terrace on San Clemente Island can be argilliturbated to the surface. (b) Content of coarse fragments of varying sizes in a Vertisol on San Clemente Island, California.

rubric of pedoturbation. Indeed, most of the soil may be lost and destroyed during falls, slides or flows, rather than mixed in place.

Turbation via heave is due primarily to ice crystal growth (cryoturbation), although slight expansion does occur due to wetting and drying. Heave contributes in large part to the slow but persistent process of seasonal creep or soil creep, which is generally confined to the upper 20–40 cm of soil and diminishes rapidly in intensity with depth (Young 1960, Finlayson 1981, Heimsath *et al.* 2002). Frost heaving is most efficient in silty materials.

Graviturbation processes are more active as slope angle increases and as the effects of stabilizing vegetation decrease (Schumm 1967). Slope aspect, in that it affects microclimate, may also affect slope movement processes. Since graviturbation has been little studied from a pedological perspective (but see Pérez 1984, 1987a, c), it is difficult to be specific about the type and amount of mixing caused by this vector of pedoturbation.

Anthroturbation

Humans are perhaps the most obvious contemporary soil mixers, and yet our activities are often ignored. Anthroturbation includes all forms of soil mixing by humans, including agricultural activities. Humans plow the soil and in so doing quickly destroy all vestiges of previous horizonation in the plow layer, or Ap horizon. The mixing induced by plowing and tillage can lead to the formation of an *agric horizon* immediately below the mixed zone, which is enriched in illuvial silt, clay and humus (Soil Survey Staff 1999). Plowing changes the character of the plowed horizon and impacts pedogenic processes deeper in the profile. For example, large pores are formed in the plow layer, which when combined with the absence of vegetation immediately after plowing, permit turbulent flow of muddy water to the base of the plow layer (Soil Survey Staff 1999). The water can enter wormholes or fine cracks below the Ap horizon, depositing suspended materials there. Eventually, characteristic features of the agric horizon form: worm channels, root channels and surfaces of peds become coated with a dark-colored mixture of organic matter, silt and

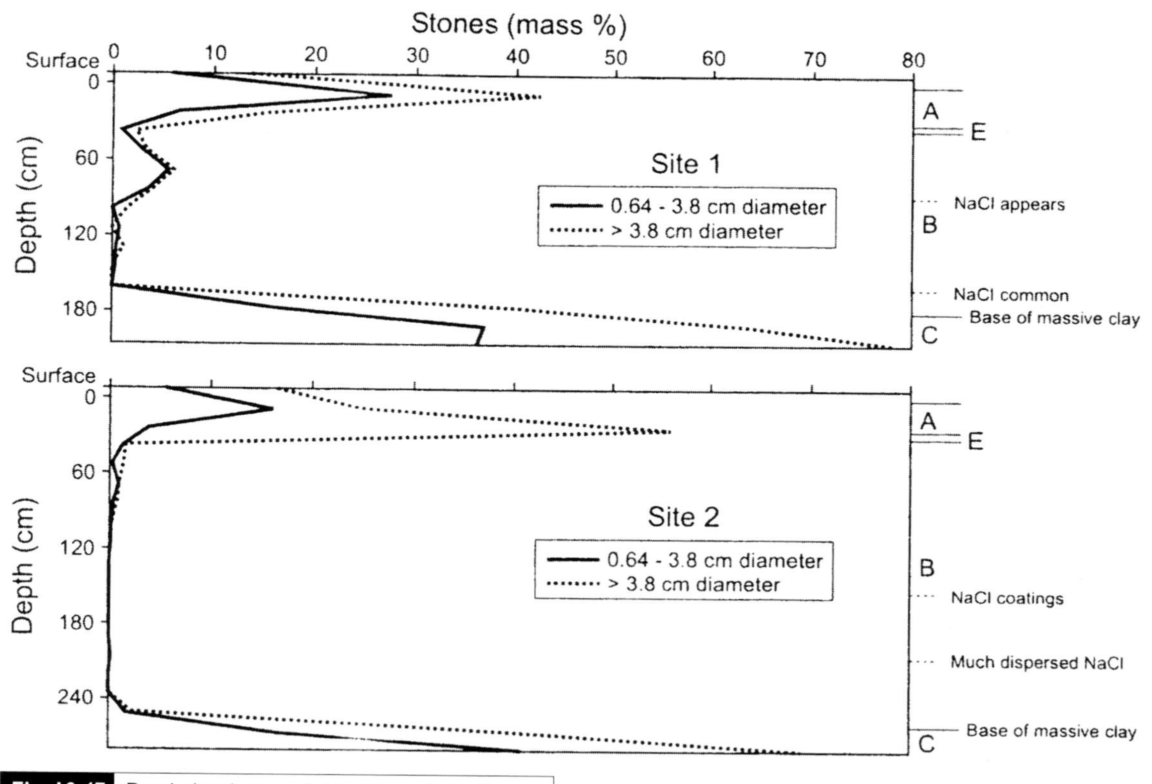

Fig. 10.47 Depth distribution of stones in an Australian Aridisol. After Mabbutt (1965).

clay which can become so thick that it fills the holes.

Effland and Pouyat (1997) defined a subset of anthroturbation, *urbanthroturbation*, which *excludes* agriculture-related mixing. It includes activities one might find in cities, at mines and wherever soils are scalped, filled or otherwise modified for human use (Fanning and Fanning 1989). Names given to soils that have been markedly impacted by urbanthroturbation include anthrosols, urban earths and urban soils (Sjöberg 1976, Woods 1977, Stroganova and Agarkova 1993) (see Chapter 11).

Lesser-studied forms of pedoturbation

Crystal growth (other than ice) has been shown to be an effective agent of weathering (Winkler and Wilhelm 1970, Ugolini 1986b), but has been less studied as an agent of pedoturbation, wherein it is referred to as crystalturbation. It commonly occurs as the water in an ion-rich soil solution is lost to evaporation or root uptake. The resultant evaporite crystals grow and expand into the surrounding matrix, disupting some soil material. Some of the more common crystals that form authigenically in soils include sodium chloride (NaCl), gypsum ($CaSO_4 \cdot H_2O$), calcite ($CaCO_3$), dolomite $CaMg(CO_3)_2$, silica (SiO_2) and jarosite ($KFe_3(OH)_6(SO_4)_2$) (Kittrick *et al.* 1982, Dixon and Weed 1989, Nettleton 1991). Most of these minerals are so small as to be insignificant as pedoturbation vectors. However, when the volume of minerals is great relative to the soil matrix, or when their formation/dissolution frequencies are high relative to horizonation processes, they can be disruptive. Salt playas in desert regions seldom have any appreciable horizon development, due primarily to their unique hydrology but also due to salt crystal growth during repeated periods of hydration and drying.

Seismic activity is capable of moving and crushing solid rock; the soil that exists above that rock also is exposed to mixing processes as the seismic waves traverse through the crust. Seismiturbation is usually associated with small-scale forms of graviturbation and mass wasting, as the disrupted soil moves, slumps, slides and flows into more stable positions.

The term impacturbation was only recently coined by Johnson *et al.* (1987), and has been little studied. At least 140 terrestrial astroblemes are now known, and several new ones are discovered annually. Impacts of various sorts continually affect the Earth, and many occur on land which retains a cover of soil. Their effect on soils is spatially minuscule in comparison to other forms of pedoturbation, but nonetheless is Herculean in magnitude at the impact site. Finally, we point out a new form of human-induced impacturbation that has yet to be studied in detail: bombturbation (explosive munitions).

Chapter 11

Models and concepts of soil formation

Soils are very complex natural phenomena. They exist at the interface of the lithosphere, atmosphere and biosphere, and perform vital functions within the hydrologic cycle. They inherit, react to and impact (seemingly simultaneously) spatial and temporal variability from all three of these realms. Erosion, burial, climate change, biomechanical movement and mixing processes, water table effects, inputs of eolian dust, microclimatic effects of aspect and topography, and innumerable other nuances of the soil-forming environment all interact to form the most complex of natural systems. To top it off, there is no agreed-on endpoint toward which the soils are developing; there is no equifinality. Earth's soils are each on their own individual journey, to an end that may be impossible to envision or obtain.

How can one possibly make sense of this noise and sensory overload? By using conceptual models which help us understand the soil system and soil formation and distinguish a signal from all the noise. Conceptual models are essential tools of science; they are simplified descriptions of natural systems. Rather than being precise mathematical descriptions that can be solved, given ample data, they are used to help put soils information into perspective and provide insight into the system interrelationships, process linkages and the nuances of pedogenesis and soil geomorphology (Dijkerman 1974). They are a way to organize, simplify and enumerate the factors that affect soil systems or the processes occurring in those systems (Smeck *et al.* 1983). They help to view things in unique and ordered ways, organize our thoughts and provide a conceptual framework within which to view facts (Johnson and Watson-Stegner 1987). Some models are complex but most try to take the complex and make it simple. By their very nature they are simplifications of reality. In fact, perhaps the simpler the model, the better.

Models have allowed soil science and soil geomorphology to progress. History is full of examples by which the application of certain pedogenic models influenced perceptions of soil genesis. It is equally full of examples in which certain models were ignored, despite being published, or not known. In these situations, had these models been used, the way in which the soils were perceived might have been very different. Models can also serve as mental blinders, keeping us from seeing aspects of reality that don't necesarily fit with the model. They can constrain our viewpoint on the world, and force us to see it with and through a certain type of conceptual lens.

In this chapter we present, apply and evaluate several of the major, conceptual pedogenic models and explain how each provides a unique perspective on soil genesis and geomorphology. By devoting a chapter to this topic, we emphasize that it forms the basis and foundation for soils research and study. It is our view that, as Eldredge and Gould (1972) stated, science progresses more by the introduction of new world-views than by the steady accumulation of information. For this reason, it is appropriate to present the various major pedogenic models and to devote an entire chapter to that task. The models are presented in their historical sequence, so as to observe how one model often was inspired by another or developed due to a shortcoming of a predecessor, although we emphasize that many models did more than just build on a predecessor. Many were almost reactionary, taking soil genesis theory in a vastly different direction. Thus, just as soil development pathways are not always progressive, the models used to describe them, if examined temporally, also ebb and flow in their approach toward pedogenesis.

It is our hope that the useful models discussed below will, as suggested by Smeck *et al.* (1983), lead to the formulation and development of even more advanced and testable models. Because no model is perfect, we must always strive to develop better models – ones that help us to better organize and understand our observations about the complex reality that is the soil landscape.

Dokuchaev and Jenny: functional–factorial models

When it comes to pedogenic models, the "granddaddy of them all" is the functional–factorial or state factor model, often ascribed to Hans Jenny but actually first developed by the Russians. The functional–factorial categorization is used because the model (or models) envision soils as forming due to the interplay of several state factors and because, generally, the model is written much like an algebraic function,

$$S = f(\text{factor}_1, \text{factor}_2, \text{factor}_3, \ldots).$$

Hence the name.

Each factor is assumed to impact the soil through a variety of pedogenic processes. A functional–factorial model does not, however, include processes directly within its framework. It leaves this link to the person applying the model.

During the 1870–80s, Russian geologist/agricultural chemist V. V. Dokuchaev, largely under an agronomic and tax mandate, attempted to unravel and make sense of the genetic pathways of soils on the loess-mantled Russian plains. Dokuchaev viewed soils not just as an amalgamation of weathered rock debris, as did others of his time, but as natural bodies worthy of study in their own right. His effort culminated in his treatise on Russian Chernozem soils (Dokuchaev 1883, Johnson 2000). To do this, Dokuchaev, along with his colleagues N. M. Sibirtsev and K. D. Glinka, formulated a factorial model of soil formation (Afanasiev 1927). The people who influenced him have been meticulously reviewed by Tandarich and Sprecher (1994) and are not discussed here. His model is formulated thus,

$$P = f(k, \Phi, g, v)$$

where P is soil (*pochva*) or soil properties, k is climate (*klimat*), Φ is organisms (*organism*), g is subsoil (*gornaya poroda*) and v is age (*vosrast*). Later, he added relief as a fifth factor (Nikiforoff 1949). Soil is the dependent variable both in this and most models that we will discuss. Thus, soil properties and the processes that interact to form them are viewed as being a function of four interacting factors. These concepts were included in a German soils text by Glinka (1914), and were then brought to the Western world in Marbut's (1927) translation.

Wilde (1946) expanded on Dokuchaev's model and equated the soil (S) as an integral of three, somewhat reorganized soil forming factors,

$$S = \int (g, e, b)\, dt$$

where g is geologic parent material, e is environmental influences, b is biological activity and t is time.

Factors of soil formation

Dokuchaev's model set the stage for what Hans Jenny would develop: the most influential of all soil genesis models. A Swiss-born soil chemist, Jenny spent much of his career first at the University of Missouri and then at the University of California. He was influenced, though not directly, by the Dokuchaev functional–factorial school of thought, developing his state factor model in the 1930s and 1940s (Tandarich *et al.* 1988). Jenny's ideas about soils as functions of various factors had been championed in the United States for decades by E. W. Hilgard, G. N. Coffey, C. E. Kellogg and others. It was Jenny, however, who adopted it to *study* how soils varied as a function of various state factors

(Wilding 1994). Jenny's publications spoke to his early interest in factorial analysis of soil development (Jenny and Leonard 1934, Jenny 1935). For historical background on Jenny and the development of this model, see Tandarich *et al.* (1988), Tandarich and Sprecher (1994) and Arnold (1994).

Jenny attempted to further quantify Dokuchaev's "five factors model" (Johnson 2000). Jenny's model, eloquently expressed in his 1941 book, *Factors of Soil Formation*, had a huge theoretical impact on the pedologic community. But, as Johnson and Hole (1994) observed, the intellectual ambiance that prevailed in 1941 was probably optimal for the "formalization" of the model that many had been using informally for decades. By 1941, the five factors approach had been installed as *the* principle paradigm of soil genesis. Its rise to prominence was, in essence, based on its utility in soil mapping, classification and human use (Johnson *et al.* 1990). It had appeared in a number of major American and international publications (Dokuchaev 1893, 1899, Glinka 1914, Neustruev 1927, Marbut 1928, 1935, Joffe 1936, Byers *et al.* 1938, Thorp 1948). Thus, the largely Russian pedological approach had, by then, been accepted, endorsed and widely promulgated in the Western world. To top it off, no competing theory existed (Johnson and Hole 1994). Thus, the stage was set for Jenny's 1941 book on the state factor model to be a success, and the clear and simple language and elegant illustrations within did not disappoint.

Jenny's model represented a final "settling in" on *five* factors out of the great complexity that was the soil environment; others before (and after, see Stephens (1947)) had tried various numbers and definitions of factors. The five state factors seemed to resonate, and like many "successful" models, it was "sold" well in Jenny's 1941 treatise. Each factor is meant to define the state and history of the soil system (Wilding 1994), and are thus referred to as state factors. They are not *forces* or *causes* but rather *factors* – independent variables that define the soil system. The now-famous model, often referred to as the *clorpt* model is,

$$S = f(cl, o, r, p, t, \ldots)$$

where S is the soil or a soil property, cl is the climate factor, o is the organisms factor, r is the relief factor, p is the parent material factor, t is the time factor and the string of dots represents other, unspecified factors that may be important locally but not universally, such as inputs of eolian dust, sulfate deposition in acid rain (Phillips 1999) or fire (Ulery and Graham 1993). The factors define the soil and the environmental system in terms of the *controls* on pedogenesis, pedogenic processes and soil distribution (Wilding 1994). Jenny's *clorpt* model was his way of conveying the idea that if we could specify the state of the entire soil system precisely, we would be able to predict its exact soil properties. In order to precisely define the state of the system, however, we would need to address and define at least five aspects of it. These five factors define the state of the system, hence they are referred to as *state factors*. The factors were not meant to explain *how* these particular conditions influenced soil properties, i.e., this was not a process model, only that a given set of environmental conditions would result in a particular soil property.

Climate and organisms are considered the more "active" factors while relief, parent material and time are "passive," i.e., they are being "acted upon" by active factors and pedogenic processes.[1] In this light, the parent material factor should be viewed as the initial state of the system, including its physical, chemical and mineralogical characteristics as well as all other inorganic and organic components. The model defines the soil in terms of the *controls* on pedogenesis and soil distribution factors, which Jenny called an "environmental formula." It demonstrated that the soil system could be quantitatively investigated, at a time when the scientific community was hungry for any quantification it could get.

[1] Johnson *et al.* (1990) discussed in some detail a very complete listing of passive and active "vectors" in pedogenesis. Vectors are different than factors, but the distinction between active and passive is nowhere drawn more clearly than in the aforementioned paper.

The "five factors model," as it has come to be known, remains today the primary pedogenic model throughout the world. It is the standard against which all other pedogenic models are still judged – the main model used to explain soil distributions at most scales, especially on small-scale maps (Ciolkosz *et al.* 1989). It has allowed us to view the soil as part of a larger environmental system. This viewpoint facilitates predictions and connections of soils with the rest of the physical world, providing a conceptual framework on which to hang all this information, that might otherwise seem impossibly complex and unrelated.

The model has perhaps its most utility at intermediate scales, e.g., in soil mapping. Field mappers will often be able to explain and predict soil variation as a function of the five factors (Jenny 1946, Johnson and Hole 1994). Indeed, soil mapping may be viewed as a field solution to the state-factor equation. As Birkeland (1999: 142) put it,

> In the field, coring and mapping the soils, one has to wonder why the soils differ. The differences in soil may be due to differences in parent material, topographic position, slope steepness, redistribution of moisture, vegetation, age of the associated landscape, etc. Because all of these can be seen or visualized in the field, in time a correlation of factors with mapping boundaries develops. Thereafter, the mapping proceeds at a more rapid pace and one can predict the location of contacts better.

The state-factor model is also appropriately used as a teaching tool, still used in most introductory courses, since it is elegant and easy to understand. The value, utility and comprehensibility of the factorial approach to the understanding of natural systems is underscored by the fact that it is used in related disciplines such as geoarcheology (Holliday 1994), Quaternary geology (Birkeland 1974) and Earth surface systems (Huggett 1991).

An advantage of the state factor approach is that the factors/controls are generally observable and measureable and vary spatially, making this model a favorite of soil mappers and soil geographers. The major contributions of the model center around its ease of comprehension and the fact that it, being a factorial model, attempts to explain the soil system characteristics in terms of external variables. In so doing it reveals little about soil system dynamics and pedogenic processes, although that was never Jenny's aim (Yaalon 1975, Huggett 1976a, Wilding 1994). The conceptual link between the factors and pedogenic processes, however, must be made by the person using the model; this is a shortcoming of Jenny's and every other factorial-type model. For example, one might use the functional–factorial model to surmise correctly that soils at the base of a hillslope will be different than those at the summit because the state factor of relief (at least) is different. However, only through a *knowledge* of oxidation–reduction *processes* will one be able to correctly hypothesize what those differences should be, how extensive they might be and why they occur.

Many have attempted to "solve" the state-factor equation, and have made great pedogenic advances in the process. A "solution" would imply the ability to predict or describe the state of the soil system or a property thereof in terms of the state factors (Phillips 1989). Mathematically *solving* the equation involves determining what the numerical coefficients of each factor in the equation are and using them to more quantitatively define the nature of the environmental system and its interrelationships. The equation cannot be solved, however, partly because the soil system is so complex that we cannot ever "define" it well enough in terms of state factors, but primarily because the factors are *not* independent, e.g., organisms and soil covary (Stephens 1947, Yaalon 1975, Phillips 1993c). Time may be the only truly independent state factor (Chesworth 1973). Jenny (1941b: 16) even acknowledged their lack of independence in his initial model formulation. Many view this as a conceptual shortcoming of *the model*, when it may, in fact, be a shortcoming of *our* abilities to gather enough of the right kinds of data! Since one might never be able to collect all the data necessary to actually solve the equation, what practical value would solving it have? As Phillips (1993c) put it, the inability to solve the state-factor model does not diminish its utility as a conceptual framework.

Johnson and Hole (1994) point out that the state-factor model falls short on two additional, theoretical elements, both of which had been mentioned in pre-1941 publications. One element is the notion that soil morphological properties and conditions can evolve such that they, by their very presence, can cause profiles to change more or less independently of any external environmental factors. These *pedogenic accessions*, important aspects of soil development, are not captured in the model (see below). An example might be a soil horizon that continues to accumulate illuvial clay and develops to the point where it becomes an aquitard. The pedogenic accession, the slowly permeable Bt horizon, dramatically changes soil development by perching water and facilitating oxidation–reduction cycles in the overlying horizon. Ferrolysis acts on the clays and weathers them, eventually causing the soil to become clay-poor, acidic and sandy (see Chapter 12). All these changes occur because of intrinsic factors within the soil, not the extrinsic environmental factors captured in the state factor model. Soils can and do change radically without any changes in the soil-forming factors.

The second missing element is the absence of biomechanical processes as originally defined in the *o* factor. Jenny's *o* factor concept was primarily centered on plants and their *biochemical* impacts on soil; this is understandable as Jenny was a soil chemist. To sum up Johnson's (2000) concerns,

> The five factors model is fine for students, and is useful as an explanatory-mapping model in soil survey work. Certainly scholars should be aware of the 'whys' and 'wherefores' of its impacts on the field. But as a major theoretical tool in research and graduate training, it is visionally restrictive.

We do not imply that the state-factor model *cannot* incorporate biomechanical processes, but rather that it *did not*, and its practitioners generally also *chose not* to include them in their application of the model. Thus, the lack of biomechanical processes in the myriad of studies that have used it is not due to a major conceptual flaw, but rather one of omission on the part of the scientists applying it.

It is important to note that Jenny spent some effort trying to solve the state-factor equation, even going so far as to observe that "the fundamental equation of soil formation is of little value unless it is solved" (Jenny 1941b: 17). Although it cannot be "solved" in a strict mathematical manner, and this is viewed by some as a serious shortcoming (Stephens 1947), each factor or variable can nonetheless be isolated and its effect on the soil system evaluated in a semi-quantitative manner (Richardson and Edmonds 1987). These types of equations (there are five) are called pedogenic functions (Stephens 1947) and the functional–factorial model has been applied mainly through the use of pedogenic functions. Jenny (1941b: 17) noted that, to ascertain the role or impact of a soil-forming factor X on pedogenesis, several soils must be examined in which factor X is allowed to vary, while all the remaining factors must be held constant. Thus, he obtained the following set of equations to describe the five possible scenarios and called them "functions":

$S = f_{cl}(\text{climate})_{o,r,p,t}\ldots$ — climofunction

$S = f_{o}(\text{organisms})_{cl,r,p,t}\ldots$ — biofunction or floralfunction (plants only)

$S = f_{r}(\text{relief})_{cl,o,p,t}\ldots$ — topofunction

$S = f_{p}(\text{parent material})_{cl,o,r,t}\ldots$ — lithofunction

$S = f_{t}(\text{time})_{cl,o,r,p}\ldots$ — chronofunction

Combinational functions are also possible, though uncommon, e.g., a topolithofunction: $S = f_{r,p}(\text{relief, parent material})_{cl,o,t}\ldots$. Each state factor "function" thus provides a mechanism or formulation within which to evaluate and determine the effects of that factor on soil development, or theoretically, to predict S on the basis of one or more of the independent variables (Muckenhirn *et al.* 1949, Phillips 1993c). To do so, the investigator routinely examines a group of soils in which four (or less commonly, three) of the state factors are held constant, and allows one (or two) to vary. The series of soils so examined is called a "sequence," while the equation(s) derived from the soils is the "function." For example, a series of soils that have all formed in the same parent material, are of the same age and formed (presumably) under the same climate and vegetation would comprise a *toposequence*. Only the factor of relief has been allowed to vary within this group or set of soils. The *equation* that describes clay content in the B horizon as a function of, for example, feldspar content of the parent material would be the *lithofunction*. The functional–factorial model of Jenny (1941b) has been used many times to examine climosequences, biosequences, toposequences, lithosequences and chronosequences. In fact, that is the primary research utility of the model. Examples of each are briefly presented below.[2] Obviously, holding the other four factors constant, as much as is possible, is paramount to the formulation of any and all "functions."

The time factor

Perhaps the most common of the sequences derived from the state-factor equation is the chronosequence. In a chronosequence, a series of soils of varying age are examined, providing valuable information about soil development, particularly the developmental stages soils may go through. Although chronosequences are discussed in more detail in Chapter 14, some important points will be made here.

Critical to the development of a chronosequence is the notion of $\text{time}_{\text{zero}}$, or the time when soil formation first started. For many soils, this can be ascertained with reasonable certainty, e.g., those developing on a mid-Holocene ashfall deposit. For others, such as an old plateau in the humid tropics, it might be acceptable to simply get within an order of magnitude of the age (in years). Confounding all this is the notion that many soils are polygenetic (see later sections of this chapter for definitions) (Bryan and Albritton 1943, Johnson *et al.* 1990, Fuller and Anderson 1993, Mossin *et al.* 2001). In others, development is arrested at a certain moment in time by an erosion or burial event, resetting the clock and forcing a new $\text{time}_{\text{zero}}$ to begin (Mausbach *et al.* 1982). Finally, the accuracy of the chronosequence is always at the mercy of the accuracy and availability of information about the ages of the soils or surfaces they have formed on (Yaalon 1975). In many instances, relative ages of soils are known, but without accurate absolute dating a chronofunction cannot be constructed and the rates of soil development cannot be accurately ascertained. Phillips (1993b, 2001a) brought up another problem: soil development can be so sensitive to initial conditions and can be so affected by small perturbations in the system that variability in development increases dramatically through time. Thus, soils of similar age can be radically different, even on the same site, further reducing the predictability of the *t* factor (see below). Another problem with chronosequences is that a certain amount of constancy of the other four factors must be assumed, and while this may be acceptable for the *p* and *r* factors, it is highly unlikely that *cl* and *o* have been immutable over the length of even the shortest chronosequences (Stevens and Walker 1970).

Despite all these potential complications, chronofunctions have been highly valuable in determining rates of soil development and how these rates may vary through time, as well as the absolute lengths of time necessary for certain pedologic features or characteristics to develop (Machette 1985, Mellor 1987). In a chronofunction, time is the independent variable and some soil property or index is taken as the dependent variable (Fig. 11.1). The chronofunction is therefore usually presented as a best-fit regression line, fit to a scatter of points where each point represents a soil or soil property at a particular time. This tendency has led to a high degree of quantification of pedogenic properties as a function of time; chronofunctions tend to be presented statistically. Depending on the soil property, the data may be best fit to linear, logarithmic, exponential or other types of functions (Schaetzl *et al.* 1994).

Eventually, after enough study and compilation of chronofunctions, synthesizing data can be presented on the general *amount of time necessary* for certain pedogenic features or horizons to form (Fig. 11.2). This is a goal (#1) of much of chronofunction research. Another goal or group of goals (#2) centers on determining how soil

---

[2] The discussion of the various state factors, below, may seem uneven because others are discussed in more detail in other chapters. For example, relief and water table effects are discussed in Chapter 13 and paleoclimate is a focus in Chapter 15.

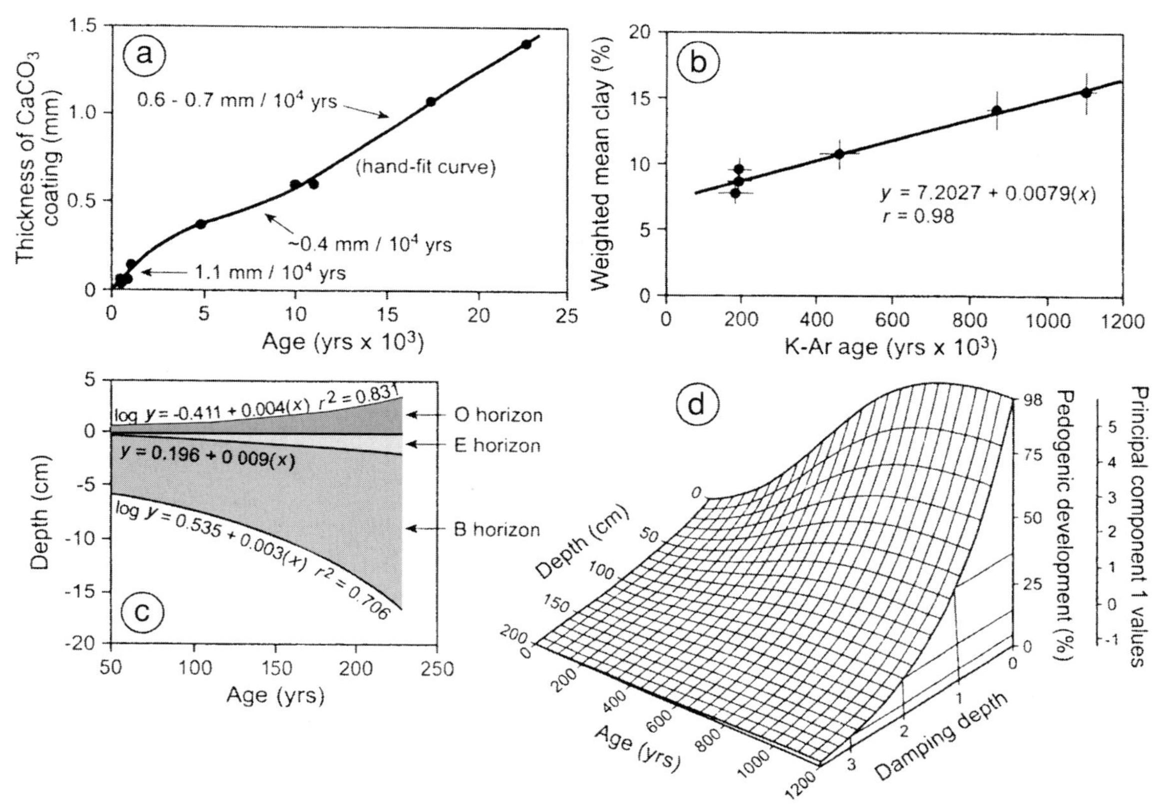

Fig. 11.1 Examples of soil chronofunctions, in graphic and equation form. (a) Mean thickness of pedogenic carbonate coatings on the bottoms of pebbles in some soils in Idaho. The chronosequence was developed for soils on a series of fluvial terraces. Note that the chronofunction indicates that the rate of accumulation changes through time. After Vincent *et al.* (1994). (b) Profile-weighted mean clay contents in some soils on a volcanic field in Sonora, Mexico. The chronosequence was developed for soils on a series of lava flows. After Slate *et al.* (1991). (c) Excellent graphic display of chronofunction data on horizon thicknesses for some soils in Norway. The chronosequence was developed for soils on a series of glacial moraines. After Mellor (1985). (d) Three-dimensional plot of percent pedogenic development, as indicated by principal component scores, vs. age and depth. After Sondheim *et al.* (1981).

development rates *change through time* rather than just the amount of change from a certain time in the past, as in goal #1. For example, one might be able to determine if soils reach some steady state after a period of time, or if the processes reach a threshold beyond which the rate of change increases or decreases markedly (Fig. 11.1).

As will be noted later in this chapter, chronofunctions more or less assume smooth pedogenic change through time. They often do not or cannot take into account perturbations in development that may have occurred in the past. For example, soils can regress entirely or briefly, and morphologic evidence for this may be lacking.

The parent material factor

Lithosequences represent a series of soils that have developed in parent materials that sequentially change along some gradient. Parent material differences could be related to a number of factors, such as texture, mineralogy (of clay or coarser materials), coarse fragment content, and even parent material layering/stratification (Schaetzl 1991a). A traditional type of lithosequence occurs in residual soils, for they form a natural laboratory of soil variation along a sequence of rocks (parent materials) that vary in mineralogy or texture (Cady 1950, Barnhisel and Rich 1967, Parsons and Herriman 1975) (Fig. 15.12). Parent material is the focus of Chapter 8.

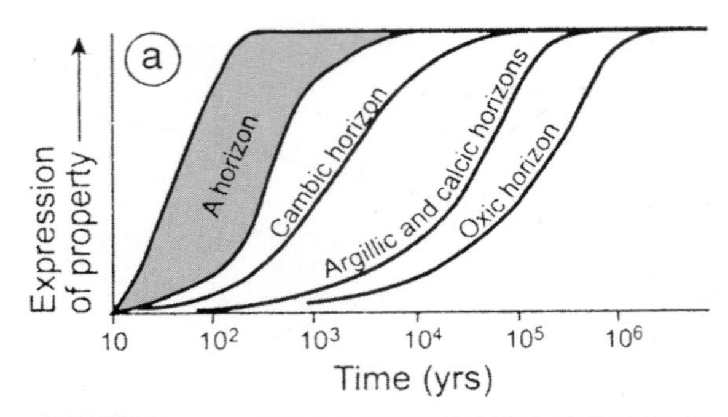

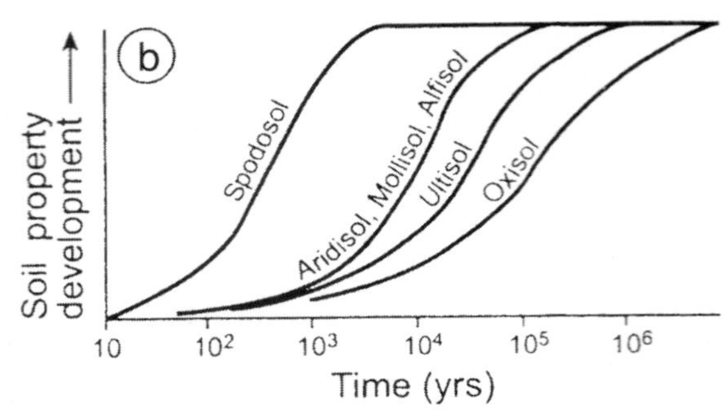

Fig. 11.2 General curves, compiled from many chronosequence studies, illustrating the time necessary for certain horizon types or soil orders to form, assuming that the pedologic "clock" is not reset. In part, after Birkeland (1999).

Parent material changes can affect many soil properties. For example, coarser-textured soils generally develop faster and to greater depths because they have less "soil volume" that needs to be pedogeneticized and less surface area that needs to be "painted" with coloring agents such as humus or iron (Schaetzl 1991b) (Figs. 3.13, 11.3). In order for lithofunctions to be pedologically meaningful, however, we must be able to assign numerical values to different parent materials, e.g., percent sand, percent gravel, $CaCO_3$ content, weathering potential or porosity (Jenny 1941b, Pedro 1966). Nonetheless, many lithofunctions are non-parametric, with data sets that are minimal and not continuous (Yaalon 1975) (Fig. 11.4). Therefore, unlike chronofunctions, lithosequences are commonly portrayed graphically, rather than as a scatter of points fit to a regression line (Fig. 11.5).

The relief factor

Soil functions examining the effects of relief are termed toposequences or catenas, emphasizing the impact of topography (see Chapter 13). The term hydrosequence is also used for this type of soil function, especially when the effect of depth to the water table is of concern.

The factor "relief" carries with it a number of subfactors, among which Jenny singled out slope and water table effects. Corresponding to water tables are such subfactors as soil moisture, degree of oxidation within the groundwater, and vegetation differences (Boersma *et al.* 1972, Knuteson *et al.* 1989, Daniels and Buol 1992).

ASPECT AND SOIL DEVELOPMENT

Aspect – the compass direction that a slope faces – is a relief subfactor which primarily impacts soils indirectly, through its effect on microclimate (Lee and Baumgartner 1966, Franzmeier *et al.* 1969, Small 1973, Nullet *et al.* 1990, Kutiel 1992, Hunckler and Schaetzl 1997). However, exactly how this microclimate effect is manifested varies greatly from landscape to landscape.

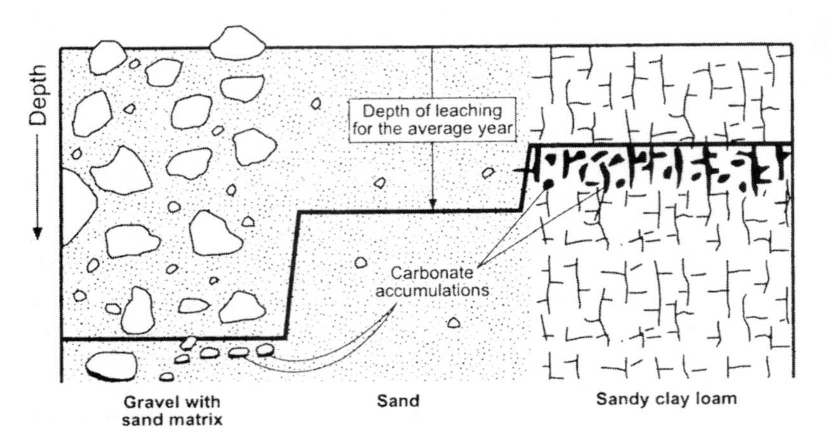

Fig. 11.3 Hypothetical depth of leaching and soil development as a function of parent material porosity and surface area, which is related to texture and coarse fragment content. After Birkeland (1999).

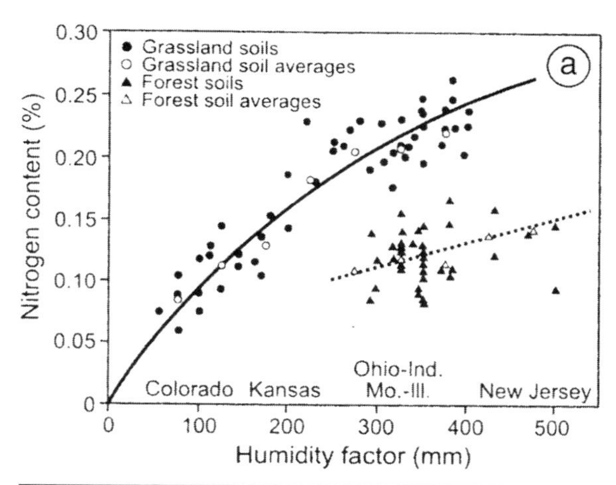

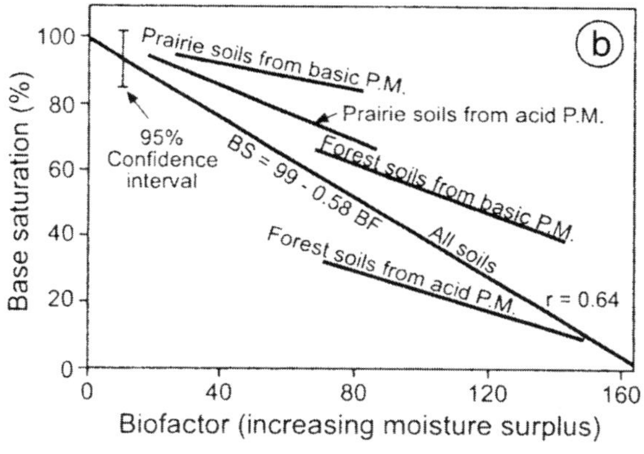

Fig. 11.4 Quantitative relationships between an integrative climatic parameter and a soil property. (a) Humidity factor (mean annual precipitation (mm) / absolute saturation deficit (mm)) vs. nitrogen content of the A horizon. After Jenny (1941b). (b) Biofactor (a measure of precipitation excess, over and above environmental water demand) vs. base saturation. P.M., parent material. After Kohnke *et al.* (1968).

The incidence angle of the sun varies from one slope aspect to another, due to differences in local topography and the subsequent partial shading of the landscape (Lee and Baumgartner 1966). Equatorial latitudes are virtually unaffected by this shading phenomenon because solar radiation is received from angles both north and south of (and always very close to) celestial zenith during the entire year. At other latitudes aspect is a highly important subfactor to soil development because solar angle is lower and the sun is always in one hemisphere (southern or northern). In high latitudes, the summer sun circumscribes a near circle in the sky, potentially negating the effect of aspect, but because in the northern hemisphere[3] many north-facing slopes retain snowpacks for much longer into the spring, the north–south slope differences in soil temperature and growing season length are significant, even at these latitudes.

North of the Tropic of Cancer (23.5° N latitude), the sun is always in the southern part of the sky. That is, south-facing slopes get more direct sunlight than do north-facing slopes. This is called insolation receipt (Lee and Baumgartner 1966). For this reason, atmospheric and soil temperatures tend to be higher on south-facing slopes (Cantlon 1953, Whittaker *et al.* 1968, Franzmeier *et al.* 1969, Hutchins *et al.* 1976, Macyk *et al.* 1978, Hairston and Grigal 1994) and soils on south-facing slopes tend to be drier (Finney *et al.* 1962, Carter and Ciolkosz 1991).

In theory, slopes that face due south should be the warmest and driest, although this is rarely the case. The coolest and most mesic slopes face north or northeast while the warmer, drier slopes face south and southwest. A major reason for this "offset" centers on early morning sunlight, much of which is used to evaporate water (dew) and thus has less influence as a source of heat. Later afternoon sun is used more for direct heating, making southwest-facing slopes the warmest and driest. North- and northeast-facing slopes are also kept cool because they remain in shadow at the warmest time of day (late afternoon). Thus, soils on southwest-facing slopes should be the least leached, and pedogenic and weathering processes that are energy-limited should be maximal there. Likewise, soils on northeast-facing slopes should be more leached and processes that are less energy-dependent (or which are stronger in cooler climates) should be dominant.

Another aspect-related component centers on the effect of *shadowing*. Shadowing, like aspect, is not really an issue in equatorial latitudes where the sun is nearly always directly overhead. But in

[3] Our discussion will be confined to northern hemisphere examples; the opposite situation holds in the southern hemisphere.

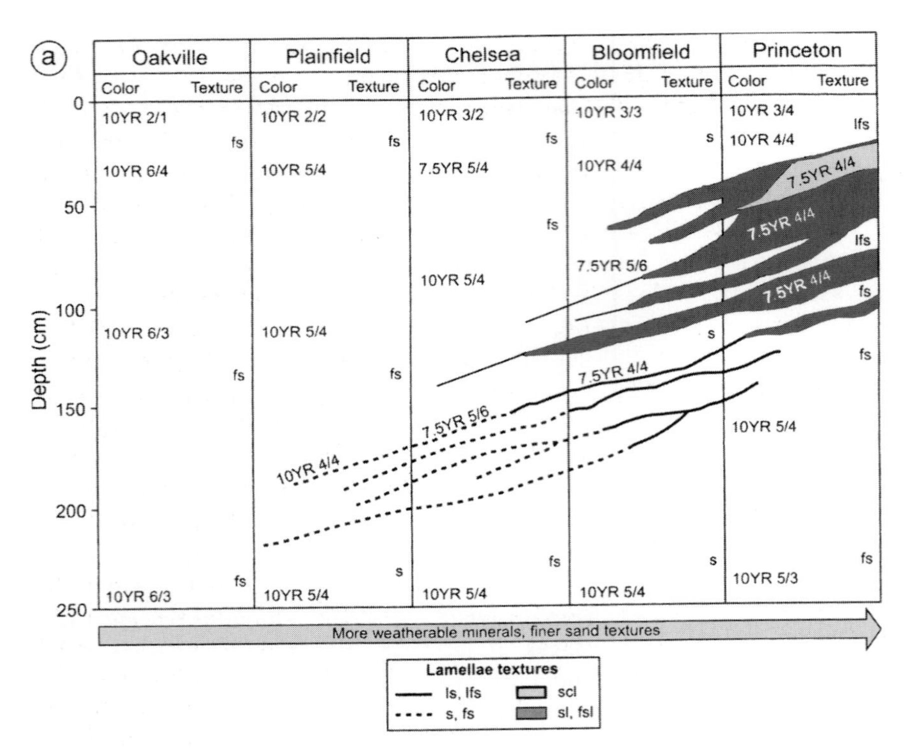

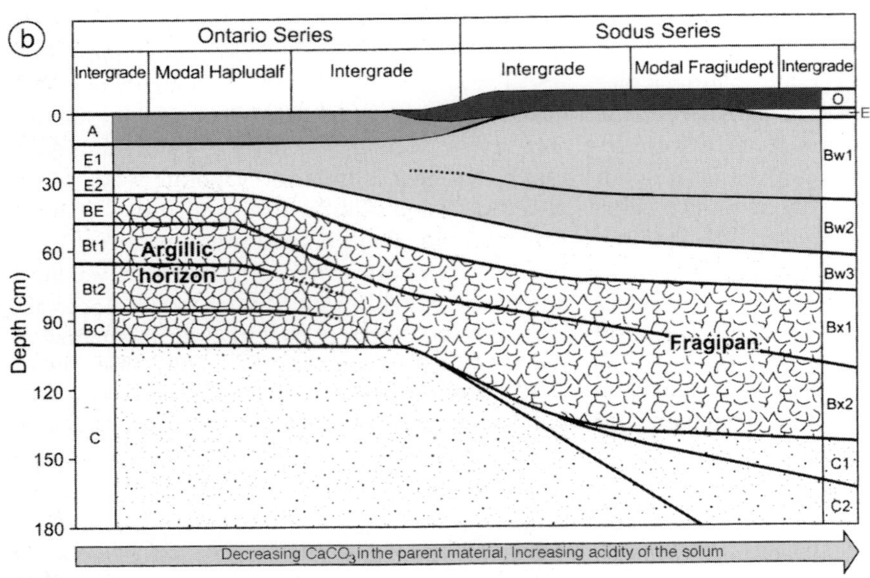

Fig. 11.5 Diagrams illustrating the changes in soil profiles that occur as a function of lithology. (a) Soil morphology along a lithochronosequence in Indiana. The soils vary in age but also in sand texture and mineralogy. After Miles and Franzmeier (1981). (b) Soil morphology along a lithosequence in New York state. The parent materials of the soils vary in $CaCO_3$ content. After Cline (1949b).

high latitudes, where the sun stays low in the sky, and in areas of deep, narrow valleys, much of the area of north-facing slopes stays in near-perpetual shadow and may retain snowpacks long into the spring. However, the real importance of shadowing for such locations is that the lower parts of south-facing slopes may also be in perpetual shadow, i.e., they may never receive direct solar radiation. Soils on lower slope positions may also be cooler because of night-time cold-air drainage (Franzmeier *et al.* 1969). In such cases, only the soils on the upper parts of the slopes will exhibit the traditional, aspect-related differences.

The pedogenic impact of aspect is not always clear. Some studies have indicated that soil development is greater on north- and northeast-facing slopes (Hunckler and Schaetzl 1997) (Table 11.1); others have found the opposite (Finney *et al.* 1962, Franzmeier *et al.* 1969, Losche *et al.* 1970, Macyk *et al.* 1978) (Fig. 11.6). In many instances, it is not even correct to state that soil development is "better" on one slope or the other. Rather, certain energy-related aspects of soil development, such as weathering, clay formation, acidification, illuviation and solum thickness may be better expressed on south-facing slopes (Finney *et al.* 1962, Small 1973), while other, moisture-related, aspects that are often better expressed under cooler conditions, such as organic matter accumulation, melanization, eluviation and leaching, may be stronger on north-facing slopes (Krause *et al.* 1959, Franzmeier *et al.* 1969, Zech *et al.* 1990, Carter and Ciolkosz 1991). Leaching-related measures, such as solum and E horizon thickness, should be better expressed on the cooler, wetter northeastern facing slopes (Finney *et al.* 1962, Hunckler and Schaetzl 1997). Weathering-related soil properties such as clay content, clay mineralogy and rubification tend to be better expressed on south-facing slopes (Losche *et al.* 1970).

Still, the effects of aspect are not straightforward. In southeast Michigan, for example, Cooper (1960) found shallow, intensely developed sola on south-facing slopes, and deeper, less intensely developed sola on north-facing slopes. Marron and Popenoe (1986) found that a greater degree of soil development existed on north-facing slopes in California, as indicated by redder and more clay-rich B horizons. To resolve this apparent confusion, Hunckler and Schaetzl (1997) proposed that aspect must be viewed in the context of macroclimate and the dominant pedogenic processes in the area. If one assumes that aspect affects primarily the energy and moisture status of soils, then the overall effect of aspect will be dependent on the energy and moisture needs and limitations of the pedogenic processes on those slopes. For mesic sites with an abundance of moisture, energy may be limiting for certain processes like weathering and humification; soils on such sites may exhibit increased development on the warmer, south-facing slopes. In these same mesic areas, pedogenic processes that are not as dependent upon energy but upon moisture, such as eluviation, podzolization and melanization, may be better expressed on northern aspects (Table 11.1). Conversely, in "pedogenically dry" areas (dry with respect to the amount of water that could be utilized by the dominant pedogenic process), moisture may be limiting. Here, the moister (poleward-facing) slope may contain better developed soils. But again, in cold, dry regions, energy may be more pedogenically limiting than moisture; in these areas southern aspects may have the best-developed soils.

Another complication with regard to aspect centers on paleoclimate. Many soils have undergone several climatic cycles during the Pleistocene, with the current one being a warm, dry cycle. Soils on these sites may reflect the interaction of paleoclimate and aspect as much or more than they do modern climate aspect (Carter and Ciolkosz 1991). For example, soils on cinder cones in arid and semi-arid eastern Arizona should be best developed on north- and northeast-facing slopes, since this is clearly a moisture-limiting environment. However, Rech *et al.* (2001) found a mix of aspect-related soil development. Weathering and solum thicknesses were greater on south-facing slopes, presumably because they continued to develop during cold glacial climates, while north-facing slopes were frozen for long periods of time. Moisture-related, pedogenic parameters that may have been able to operate even during glacial climates, such as leaching and acidification, were greater on north-facing slopes.

Landscapes: the Palouse Hills, Washington state

Parts of eastern Washington state are covered with thick deposits of Palouse loess. The silty soils developed in this loess are Xerolls, well known for their productivity. Wheat is the main crop. This region, in the rainshadow of the Cascade Mountains, receives only about 50 cm of annual precipitation; most falls during the cool winters. The silty Palouse soils can retain a great deal of water, but little of it remains beyond one growing season because much of the wintertime precipitation runs off the frozen ground.

The loess hills of the Palouse resemble ocean waves because almost all of the landscape is in slope. The rolling topography of the region, with relief upwards of 60 m, has many steep slopes. The loess hills are distinctly asymmetrical, almost crescentic in shape. Most ridges are oriented in a southeast–northwest direction; the cooler northeast-facing slopes are the steepest (Lotspeich and Smith 1953). Because the dominant wind direction is from the southwest, the northeast slopes accumulate thick snow drifts, while moderate to little snow is retained on the gentle southeast-facing slopes (Rockie 1934). Snow is the primary source of soil moisture during spring.

The deep Palouse–Thatuna soils are dominant in the loess hills of this region. Athena soils are found on hill crests and summits. According to Lotspeich and Smith (1953), soil development in this landscape is a function of the amount of infiltration. The main source of water is snow, which drifts onto preferred leeward sites. Thatuna soils are on north slopes and other slopes that receive runoff from higher positions, or additional moisture from snowdrifts that accumulate on the north slopes. They are the recipients of snowmelt, and have argillic horizons. The additional snow that the northeast slopes receive leads to headward erosion and continual steepening of the slopes (Kirkham *et al.* 1931). In effect, this is a positive-feedback mechanism that keeps these slopes "sheltered" enough to continue to be snow-receiving sites. Athena soils are the driest soils in this "catena" because they receive very litte water from snowmelt – the snow blows off the ridge crests. They have Bk horizons. Soils within the Palouse series, on gentle windward slopes, are intermediate in morphology and development.

The Palouse catena, not traditional in that it contains only soils that are well-drained, illustrates the importance of topography in soil development. In this case, the role of topography is to redistribute energy provided by infiltrating snowmelt water, and in so doing it affects the degree of soil development. The driest sites are on the ridges, not so much because of runoff but because snow blows off these sites in winter. Lee sites, where large snowdrifts develop, are the most thoroughly leached and soils here are the best developed. Thus, within a few meters, one can go from soils with Bk horizons to leached soils with Bt horizons.

Aspect also affects soil development through its impact on slope and surficial processes. For example, in high latitudes the incidence of freeze–thaw activity may be greater on north-facing slopes, while the opposite situation may set up in the midlatitudes. Greater freezing and thawing may lead to more mass movement and gentler slopes, and create distinct slope asymmetry (Fig. 11.7). Additional snow accumulations, usually on lee slopes, also can greatly impact slope processes (French 1971) and soil development. In some regions, the increased snow accumulations on east and northeast slopes causes great slope instability but this effect may be offset by increased

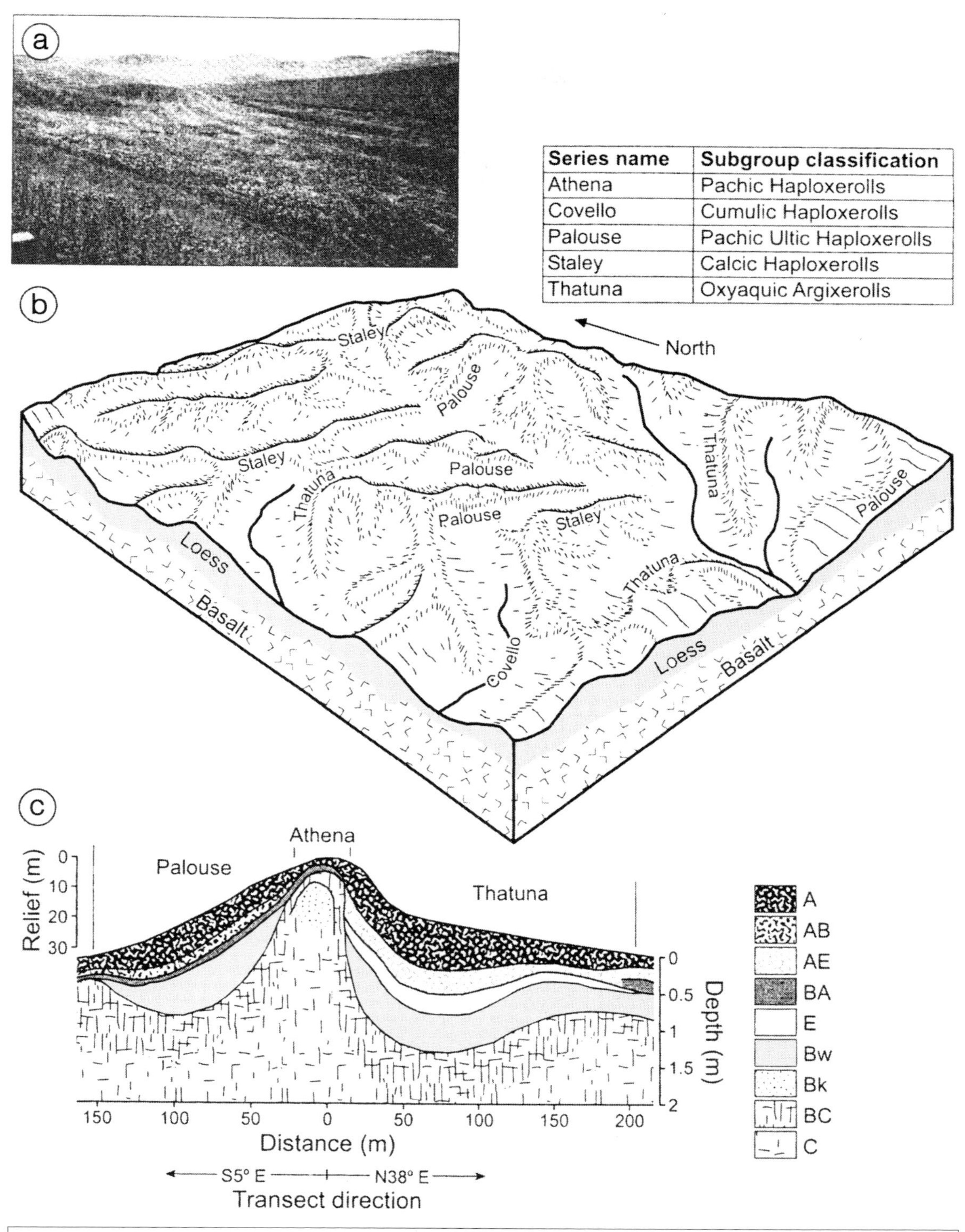

Series name	Subgroup classification
Athena	Pachic Haploxerolls
Covello	Cumulic Haploxerolls
Palouse	Pachic Ultic Haploxerolls
Staley	Calcic Haploxerolls
Thatuna	Oxyaquic Argixerolls

Soils and landscapes of the Palouse Hills of Washington state. (a) Oblique aerial image of the landscape. Photo by Robert Ashworth. (b) Block diagram of the soils and landforms in Whitman County, Washington. After Donaldson (1980). (c) Typical soil pattern across one of the loess ridges. After Lotspeich and Smith (1953).

Table 11.1 Data from soils on north–northeast and south–southeast aspects in northern Michigan

Variable (units)	Value for soils on N–NE slopes (Mean (SD))	Value for soils on S–SW slopes (Mean (SD))	Indicates stronger podzolization on which slope?	Level of significance[a,b]
Data indicative of eluviation				
Depth to top of E (cm)	8.0 (3.1)	7.4 (3.5)	Neither	ns[3]
Depth to top of B (cm)	24.6 (7.2)	17.1 (5.2)	N–NE	0.02
E thickness (cm)	16.6 (6.1)	9.7 (4.0)	N–NE	0.02
E pH (H_2O)	4.7 (0.4)	4.5 (0.3)	Neither	ns
E hue (Munsell YR)	5.8 (1.2)	6.0 (1.3)	Neither	ns
E value (Munsell units)	4.6 (0.5)	4.8 (0.4)	Neither	ns
Fe_o in E (g kg^{-1})	0.2 (0.2)	0.4 (0.2)	N–NE	0.05
Al_o in E (g kg^{-1})	0.2 (0.1)	0.3 (0.1)	N–NE	0.01
Fe_p in E (g kg^{-1})	0.1 (0.0)	0.2 (0.1)	N–NE	0.05
Al_p in E (g kg^{-1})	0.1 (0.0)	0.2 (0.1)	N–NE	0.01
$Fe_o - Fe_p$ in E (g kg^{-1})	0.1 (0.1)	0.3 (0.2)	Neither	ns
$Al_o - Al_p$ in E (g kg^{-1})	0.1 (0.1)	0.2 (0.1)	N–NE	0.01
ODOE[c] of E	0.0 (0.0)	0.1 (0.1)	N–NE	0.02
Data indicative of illuviation				
B thickness (cm)	34.4 (10.0)	30.6 (4.1)	Neither	ns
Uppermost B pH (H_2O)	5.4 (0.2)	5.5 (0.2)	Neither	ns
Second B pH (H_2O)	5.6 (0.3)	5.7 (0.2)	Neither	ns
Uppermost B hue (Munsell YR)	4.5 (1.1)	6.0 (1.3)	N–NE	0.02
Uppermost B value (Munsell units)	3.1 (0.4)	3.9 (0.3)	N–NE	0.01
Uppermost B chroma (Munsell units)	4.3 (1.0)	5.6 (0.8)	N–NE	0.02
Fe_o in uppermost B (g kg^{-1})	5.1 (2.0)	3.5 (1.1)	N–NE	0.04
Al_o in uppermost B (g kg^{-1})	4.7 (2.2)	3.5 (1.1)	Neither	ns
Al_o in second B (g kg^{-1})	3.5 (1.7)	2.1 (0.7)	N–NE	0.01
Fe_p in uppermost B (g kg^{-1})	2.5 (1.5)	1.1 (0.5)	N–NE	0.01
Al_p in uppermost B (g kg^{-1})	2.5 (1.0)	1.4 (0.4)	N–NE	0.01
Fe_p in second B (g kg^{-1})	1.1 (0.7)	0.6 (0.5)	N–NE	0.03
$Fe_o - Fe_p$ in uppermost B (g kg^{-1})	2.4 (1.5)	2.4 (1.2)	Neither	ns
$Fe_o - Fe_p$ in second B (g kg^{-1})	1.3 (0.7)	1.3 (0.6)	Neither	ns
$Al_o - Al_p$ in second B (g kg^{-1})	2.0 (1.3)	1.3 (0.7)	N–NE	0.02
ODOE[c] of uppermost B	0.3 (0.2)	0.1 (0.1)	N–NE	0.01

[a] One-tailed test using Wilcoxon test.
[b] ns: not significant at $\alpha = 0.05$.
[c] Optical density of the oxalate extract, an indicator of organic acids.

Source: Hunckler and Schaetzl (1997).

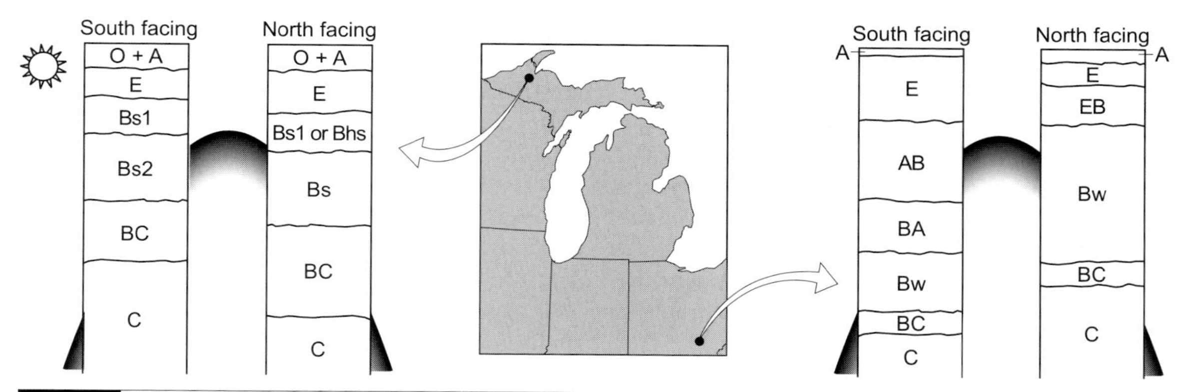

Fig. 11.6 Effects of slope aspect on soil morphology at two midlatitude sites in the United States. After Finney *et al.* (1962) and Hunckler and Schaetzl (1997).

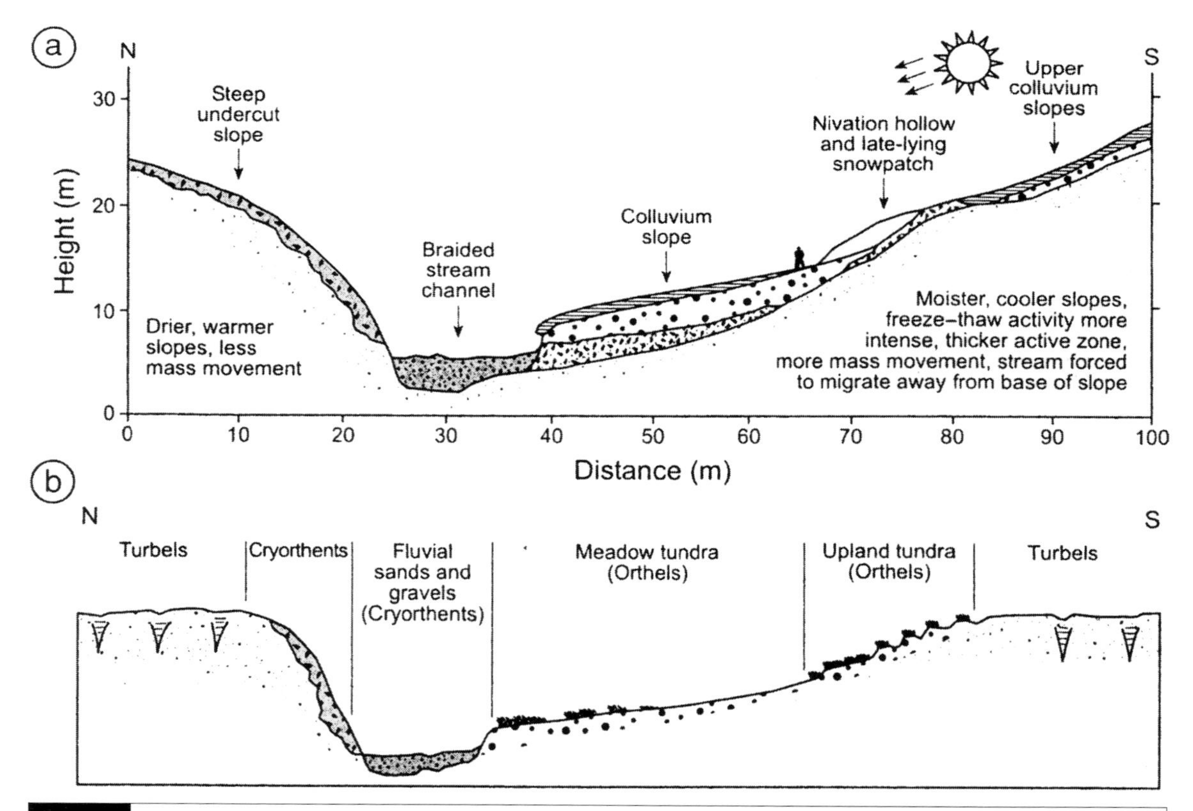

Fig. 11.7 Slope asymmetry on the Beaufort Plain, northern Canada. The moister northeast-facing slopes accumulate more snow in winter. Solifluction and nivation processes are stronger on these slopes, transporting more sediment to the base of the slope. This forces the river to shift position and undercut and oversteepen the southwest-facing slope, the soils of which, via feedback processes, get still drier because of its steepness. After French (1971). (a) Geomorphology and sediments. (b) Soils and vegetation.

infiltration and biomass production. Slope instability also impacts soils, which are less developed on unstable (and younger) slopes.

In sum, slope aspect is a dynamic factor that impacts soils on a number of levels, all of which revolve around microclimate. Insolation receipt (which includes shadowing) and winds impact other processes and characteristics such as snow accumulation, soil moisture and temperature. These, in turn, impact slope processes and soil development.

Also included within the relief factor are the concepts of elevation, important in mountainous terrain, and slope gradient or steepness (Acton 1965). Finally, the three-dimensional slope curvature of the surface must also be considered, for certain slopes are water-concentrating slopes while others are divergent, e.g., nose slopes (see Chapter 13).

Variability in soils across short distances is usually best explained by the relief factor. Within a few meters horizontally soils can be radically different due to higher or lower water tables, being located on steeply sloping erodible sites or on cumulic sites at the base of the slope where sediment accumulates from upslope. The portrayal of toposequences can take many forms, although the most useful is perhaps the kind where variations in soil morphology or chemistry are portrayed along them (Fig. 11.8).

The biotic factor

Both Jenny (1958) and Crocker (1952) defined the biotic factor as the potential floristic list, or the potential natural vegetation of the site. What *actually grows there*, according to Jenny, is dependent on the other state factors. As mentioned above, this factor had, in Jenny's (1941b) original interpretation, primarily a biochemical meaning and was more or less restricted to plants. Jenny's research emphasis with regard to the *o* factor was on the nitrogen and organic matter contents of soils as a function of plant type and amount, and on the parallelism between plant succession and soil development. His research on biochemical functions, as well as the paucity of information on the "faunal" factor as of 1941, may have led to this interpretation. The biomechanical component, which can occur as animals and plants move and rearrange soil, was largely ignored by Jenny and is seldom studied within the context of the model (Johnson and Hole 1994). Jenny even referred to the *o* factor as the "plant factor" (Jenny 1958). It is not that Jenny did not consider animals as being part of the *o* factor, but, as he stated, "because of lack of sufficient observational data . . . , the discussion of animal life is omitted" from the book (Jenny 1958: 203). Although many studies on the effects of animals on soil development had been published in the early years of pedology and soil science, they were by and large ignored by Dokuchaev and Jenny. Only later would the literature "catch up." Eventually, scientific papers on the effects of animals would become increasingly prominent, e.g., Thorp (1949), Hole (1981), Johnson (1989, 1990, 1993a) and Cox and Scheffer (1991).

The biochemical/biotic factor is one of the more difficult to isolate and tease out within the functional–factorial framework, across small or medium-sized areas, because plants (and animals) rarely function independently of the other factors. In fact, it is more likely that vegetation is a function of soil, rather than vegetation being an independent variable as stated in Jenny's equation (Noy-Meir 1974). As a result of these interdependencies, Yaalon (1975) felt that true biotic functions were rare and suggested that biotic attributes should be treated more like a dependent, rather than an independent, variable.

Plant communities change through time, naturally by succession and through disturbance, adding another level of complexity to the biofunction. These interdependencies again make it difficult to generate a true biosequence (Crocker 1952). Even when climate, relief and time can be controlled (usually within a small area), it is difficult to find vastly different plant communities while holding parent material constant (Curtis 1959). Thus, most biosequences cover large expanses, often from forest into grassland, and are rarely "pure" biosequences because climate covaries with biota (Shrader 1950, White 1955, White and Riecken 1955, Bailey *et al.* 1964, St. Arnaud and Whiteside 1964, Severson and Arneman 1973).

Another way to "skin this cat" commonly involves an analysis of the impacts of vegetation

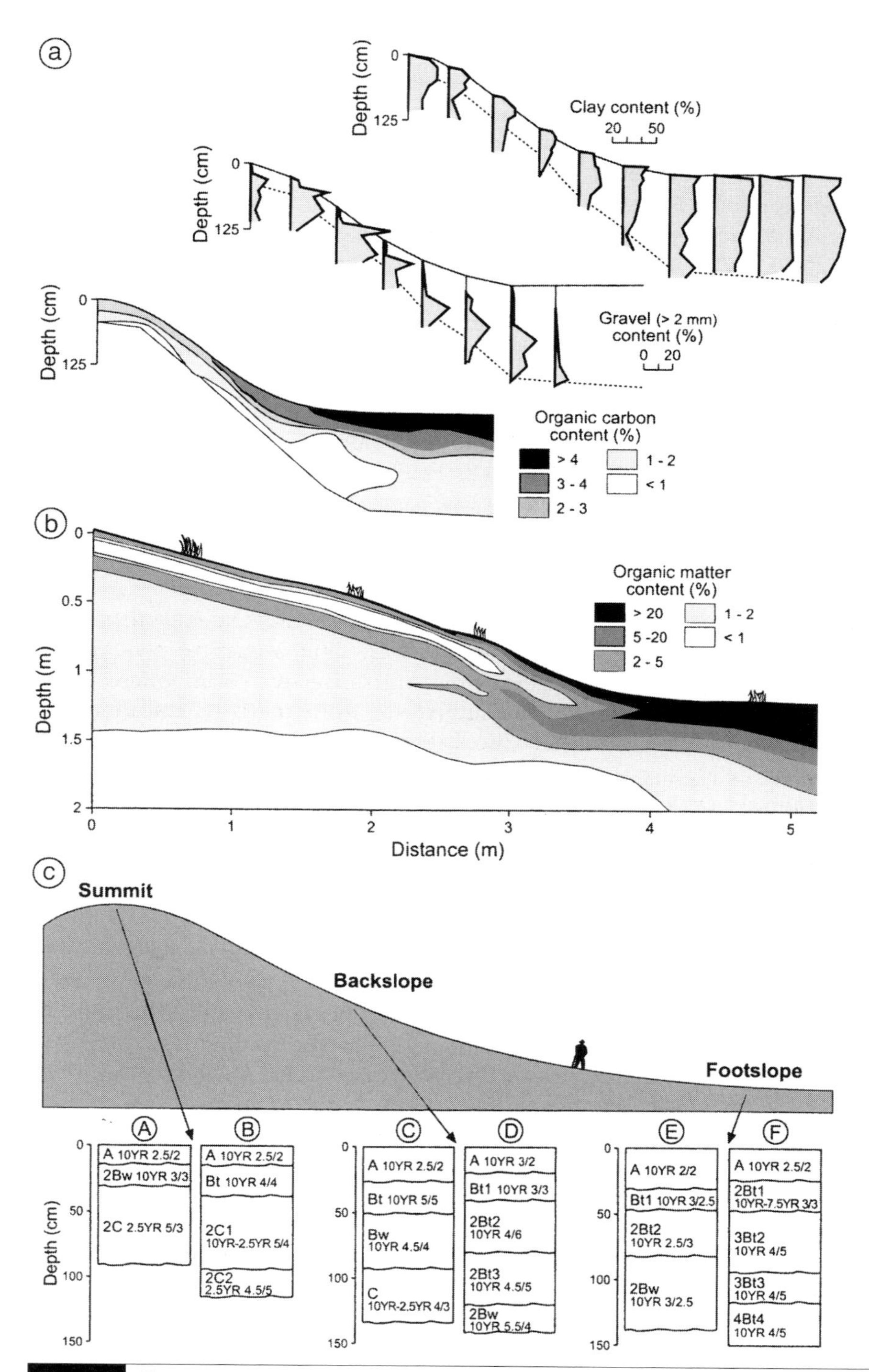

Fig. 11.8 Changes in soil properties along hillslopes. (a) Depth functions of clay content, gravel content and organic carbon content for a series of soils on a slope in Iowa. After Walker (1966). (b) Organic matter in soils along a hillslope in Sweden. Large data sets provide the ability to draw isolines and therefore present a somewhat continuous surface. After Mattson and Lönnemark (1939). (c) Horizonation and soil color of representative profiles along two catenas on glacial moraines in the Ruby Mountains, Nevada. After Birkeland *et al.* (1991).

change on existing soils (Fuller and Anderson 1993). That is, it is assumed that the biosequence did not begin at $time_{zero}$, but that at some $time_{zero+1}$ there was a vegetation change in some areas but not in others. Analysis of the effects of these varying biotic changes presents a useful type of biosequence (Geis *et al.* 1970, Barrett and Schaetzl 1998, Mossin *et al.* 2001).

The climate factor

Once early soil scientists began to realize that soil was more than just weathered rock, climate replaced parent material as the most important soil-forming factor. Russia has a suite of east–west trending climate zones that are uninterrupted by mountains or oceans, such that any early pedologist traveling on a north–south transect could not help being swayed by the impact of climate on soils (Kohnke *et al.* 1968). Hilgard's (1906) early classification of US soils used as a first "cut" the division between soils of arid and humid climates.

The *cl* factor was originally defined as the regional climate, although clearly the climate that the soil reacts to is influenced by the biotic cover that lies between the soil and the atmosphere, by slope aspect, snow cover, and many other factors. Thus, *soil climate* is often quite a different thing than is regional atmospheric climate (Schaetzl and Isard 1991).

Although climate is clearly one of the most important state factors (Yaalon 1983), climosequences generally, by their very nature, are difficult to isolate. First of all, climate is seldom independent of biota, and thus a climo-biosequence is much more common and easier to find than is a pure climosequence. Second, by its very nature climate is always changing. Thus it is difficult to really know how the aspects of climate variation along the "modern climosequence" existed in space and magnitude in the geological past. What is a dry tropical to humid tropical sequence today may have been a very different one in the past. Deserts may have been wetter, the seasonal timing of precipitation, so important to pedogenesis, may have changed even if the annual precipitation totals did not. Thus, although climosequence data must be viewed cautiously, where a good climosequence has been teased out or modeled, the results can be very useful in the interpretation of pedogenic properties (Arkley 1963).

In most climosequences, the primary climatic variables (independent variables in a regression equation) are mean annual precipitation, mean annual temperature or some index that measures water need or evapotranspirative demand, of which there are many (Lang 1915, Thornthwaite 1931, Yaalon 1975, 1983). Only one of a myriad of possible climatic inputs is regressed against soil properties such as nitrogen content of the A horizon, depth of leaching, clay content and mineralogy, and translocation depths of any of a number of soil components (Kohnke *et al.* 1968) (Figs. 11.4, 11.9). One can immediately see the difficulty of capturing the essence of "climate" in one or even a few variables, which is why well-crafted climofunctions are a challenge to set up.

ZONATION OF SOILS IN ALPINE REGIONS

Like biosequences, climosequences tend to be studied over large areas, e.g., from a desert to a humid climate or from the core of a desert to its fringes (Buntley and Westin 1965, Yair and Berkowitz 1989). However, a common form of climosequence is one that examines soils (recognizing, as always, that vegetation often parallels climate) on a mountainside or within a mountain range (Retzer 1956, Whittaker *et al.* 1968, Hanawalt and Whittaker 1976, Nettleton *et al.* 1986, Dahlgren *et al.* 1997, Darwish and Zurayk 1997, Trifonova 1999). Generally, air temperatures get predictably cooler and evapotranspirative demand diminishes as elevation increases, and moisture/precipitation changes as well, providing an areally "tight" climosequence that often has good control on lithology and relief (well-drained soils can be chosen at all sites) (Cortes and Franzmeier 1972). The environmental lapse rate – the change in air temperature with elevation – varies as a function of water vapor content, but is generally about 6.4 °C per 1000 m. Thus, one can assume that a location at 3000 m will be, on average, about 19 °C cooler than one at the base of the mountain at sea level. As air temperatures get cooler with height, precipitation usually increases as well, although in very high mountains there is a drop-off in precipitation at the highest elevations, because the cold

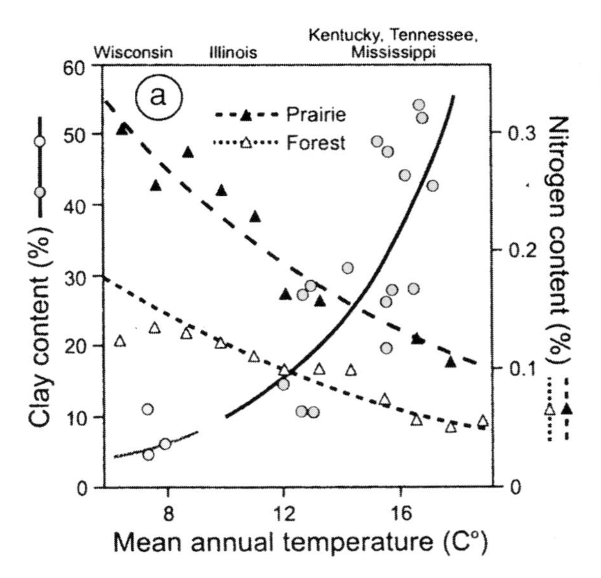

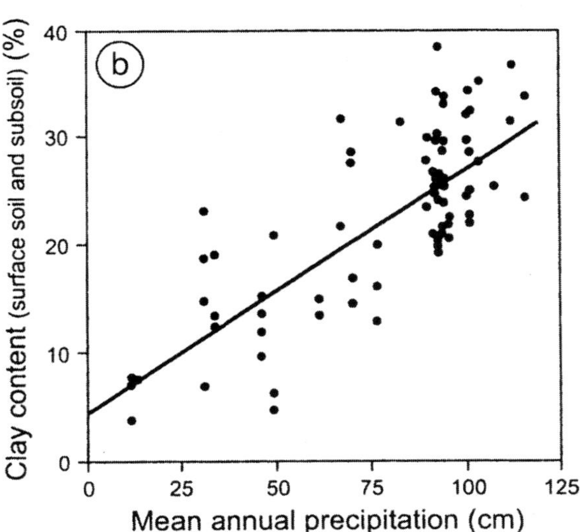

Fig. 11.9 Quantitative relationships between a climatic variable and a soil property. After Jenny (1941b). (a) Mean annual temperature vs. clay content (probably of the most clay-rich horizon), and vs. nitrogen content of the A horizon. (b) Mean annual precipitation vs. mean clay content to a depth of 100 cm.

air at this height cannot hold as much moisture. However, the continued lowering of evapotranspirative demand with height may act to offset the drier conditions in the high alpine areas, rendering even them climatically humid.

The sequence of soils that occurs along an alpine climosequence (or climo-biosequence) vary greatly from region to region. It is difficult to generalize. The soils and vegetation at the base of the sequence are dependent upon the general macroclimate of the region; the soils could be Aridisols, Mollisols, etc. With height, climate and vegetation change, and are usually expressed as discrete vegetation *zones* that wrap around the mountain but are lowest on the warmer, drier southwest slopes (northern hemisphere) (Fig. 11.10). In all but the lowest and highest mountain ranges there is a forested zone somewhere along the sequence (Amundson *et al.* 1989a). In mountain ranges with multiple forested zones, the uppermost one is usually a coniferous zone, which gives way to the treeless *alpine zone* at height (Fig. 11.10). The ecotone between the uppermost forest and the alpine zone is called the alpine treeline; it is usually not a distinct line but a series of disjunct patches and outliers/inliers of stunted trees. As the climate gets more and more severe with increasing height, trees become limited to sites that are sheltered, appearing only in patches or groups.

Soils change along with vegetation and climate in alpine climosequences. Generally, "effective moisture" and organic matter content increase with elevation (Chadwick *et al.* 1995) (Fig. 11.11). This trend occurs because biomass production increases (to a point) and then decreases along the elevation transect, but concomitant with that, litter mineralization decreases steadily as temperatures decrease with height. The ebb and flow of the intensity of the various pedogenic processes along alpine transects generally corresponds to climate. For example, calcification in soils at the base of mountains (in desert or dry grassland areas) diminishes with height as leaching increases. Leaching, however, then becomes less pronounced in the high alpine zones.

For the remainder of this section, our focus will be on the soils of the high alpine, i.e., the areas above treeline. Soil patterns here are easier to generalize than they are along the entire mountain slope.

The soils of the high alpine zone are generally thin, well drained and coarse-textured (Grieve 2000). Physical weathering processes dominate over chemical weathering (Gao and Chen 1983).

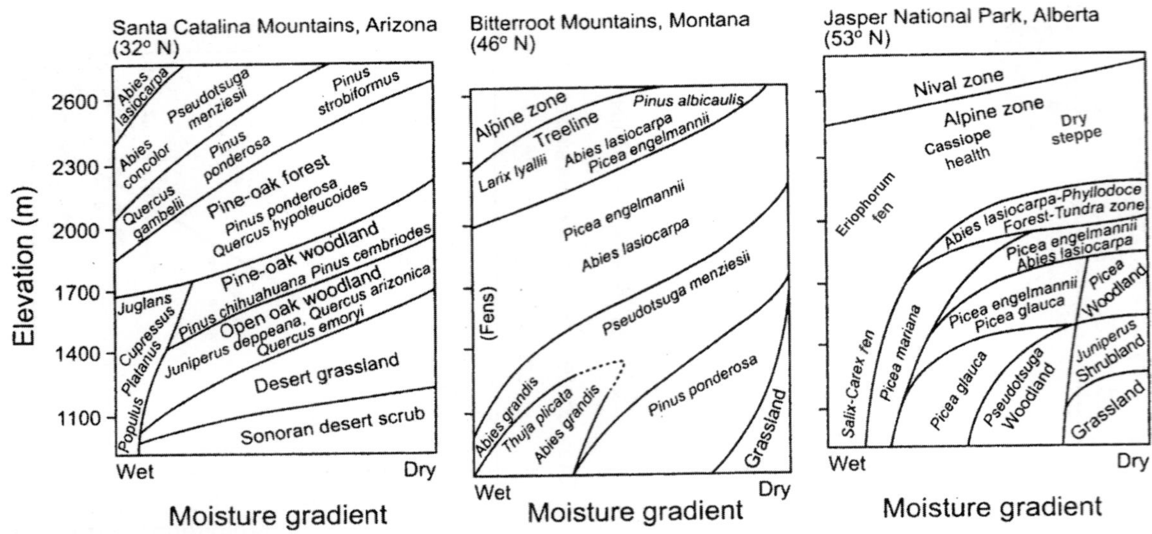

Fig. 11.10 Vegetation–elevation mosaics for three mountain ranges in western North America. *Wet* and *Dry* sides imply the northeast and southwest slopes of the ranges, respectively. (a) Whittaker and Niering (1965). (b) Arno (1979). (c) La Roi and Hnatiuk (1980).

The little water, necessary for chemical weathering, that does exist here is frozen for much of the year. In summer, however, minimal amounts of hydration and chelation are operative (Ellis 1980a). Sola get increasingly thinner and soils get less developed with height within the alpine zone, until they are little more than raw parent material (with interspersed patches of bedrock) covered for much of the year with snow and ice (Fig. 11.12). This highest area of the alpine (except for the areas of perennial snow and ice) is often called the shatter zone or blockfield zone (Rudberg 1972). Cryoturbation is dominant in the soils of the high alpine, not just because this is a cold place but also because the diurnal temperature cyclicity is greater in the alpine than anywhere else on Earth (Harris 1982). The thin air and high elevations promote rapid nocturnal cooling, and at the same time the intense insolation at high altitudes allows for rapid heating during the day.

The soils of the high alpine are different enough that they are often studied and discussed as a unique group: alpine soils (Retzer 1956, 1965). Costin (1955b) defined alpine soils as occurring above treeline where the ground is snow-covered for more than half of the year. Soils of the alpine zone are highly varied in form and type, from Histosols to Gelisols to Entisols and Inceptisols. Topography is the most important soil-forming factor here. Slopes are generally unstable; solifluction and processes associated with patterned ground are common (Benedict 1970, Grieve 2000). The alpine zone is a very windy place; many fine materials, like silt and clay, are brought in and removed by wind, which is of course, impacted by topography (Thorn and Darmody 1980, Dahms 1993). Thus, thin eolian caps are common in alpine soils (Litaor 1987, Dixon 1991). Many alpine soils are shallow to bedrock, and most are coarse-textured and rocky/gravelly, reflecting the dominance of physical weathering and freeze–thaw activity. Although pedogenesis is typically slower here than in warmer climates, many of the same processes are operative, providing that a similar suite of soil-forming factors sets up (Stützer 1999).

One way to make "geographic sense" of the alpine is to use geomorphic models to subdivide the area into areas of relative unformity. Burns and Tonkin (1982) developed a two-level classification of the alpine, based on work done in the southern Rockies but which has found widespread use among those who study alpine soils and landforms around the globe (Fig. 11.13). On a gross scale, the alpine can be divided into ridge tops, valley sides and valley

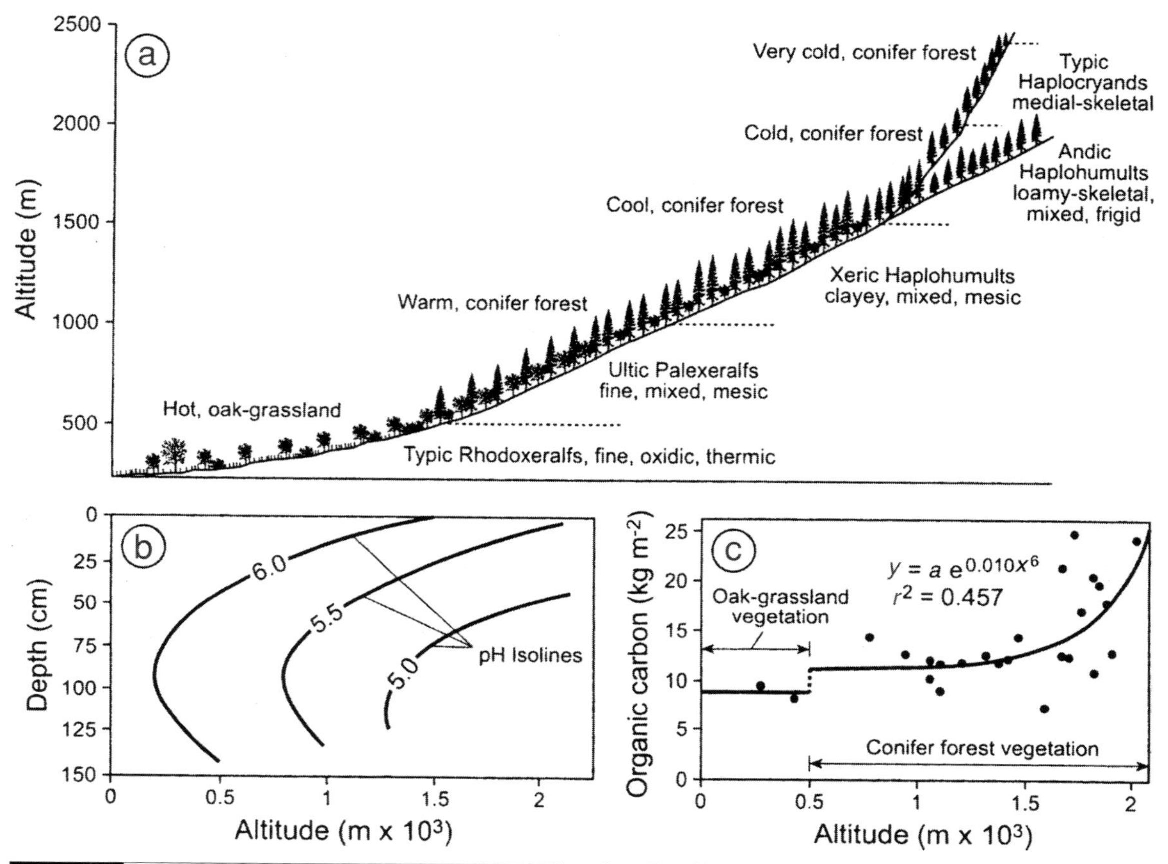

Fig. 11.11 Trends in vegetation and soil properties with elevation, in the Cascade Mountains, California. After Alexander *et al.* (1993). (a) Dominant soil types and plant communities, idealized. (b) pH trends with depth and altitude. (c) Organic carbon contents in the upper meter of soil. Note the change due to vegetation.

bottoms. Periglacial activity tends to dominate on the ridge tops, which generally escaped Late Pleistocene glaciation. The slopes here are fairly stable and, while heavily cryoturbated, have thicker profiles. The steep slopes of the valley side tundra areas, coupled with recent glacial activity, interact to create slope instability and thin, rocky soils interspersed with rock outcrops. Glacial and fluvial processes dominate the valley bottoms.

Burns and Tonkin (1982) developed an even more detailed assessment the windswept ridge top areas: the synthetic alpine slope (SAS) model. It incorporates the major factors that control the distribution of alpine soils: aspect, topography, seasonal snow accumulation, distribution of plants, and alpine eolian sediments (Fig. 11.13b). All of the components are rarely seen on one slope, but form a mosaic within the alpine zone. The SAS model is based on the assumption that spatial variations in soil characteristics follow topographically controlled variations in snow cover, the distribution of which also affects soil development. Because wind and topography impact snow cover and melting more than other local-scale factors, the SAS is largely based on how topography impacts wind and snow drifts. The SAS illustrates that the highest alpine sites are often the most windblown and snow-free, and hence the driest and most cryoturbated. Areas with long-lasting snowpacks tend to have minimal vegetation and rocky soils. Armed with this soil-geomorphic model, alpine soil patterns and characteristics, especially A horizon thickness and degree of eolian contribution, become more predictable and understandable.

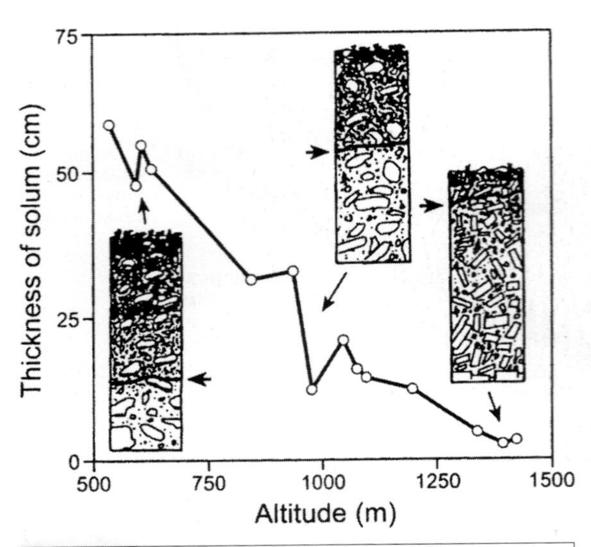

Fig. 11.12 Changes in soil thickness and morphology with elevation in the Brooks Range of Alaska. Most of the soils are Cryepts. The line represents maximum solum thickness observed at that altitude, while the arrows on the soil profile drawings show the bottom of the solum. After Tedrow and Brown (1962).

Largely because of the persistent winds, soil patterns also get extremely "patchy" within the alpine zone and within each of its subzones. Long-lasting snow patches form in the lee of wind obstructions, e.g., tree islands or large rocks. These features, both the snow patches and the tree islands, have a dramatic effect on soil development (Stepanov 1962, Ellis 1979, Grieve 2000) (Fig. 11.14). In protected areas, snow and litter can accumulate; soils there are some of the best developed and leached anywhere in the alpine zone (Stepanov 1962). Other soils, in exposed locations, are desiccated and eroded by the strong winds. Snowmelt water facilitates weathering, and any eolian material within the snow is added to the soil (Thorn and Darmody 1980), eventually building up a fine-textured mantle. Then, after the fine mantle has been established, the long-lasting snow patches help to reduce its subsequent erosion by wind and water.

In conclusion, the primary research effort with regard to climate–soil relationships has not

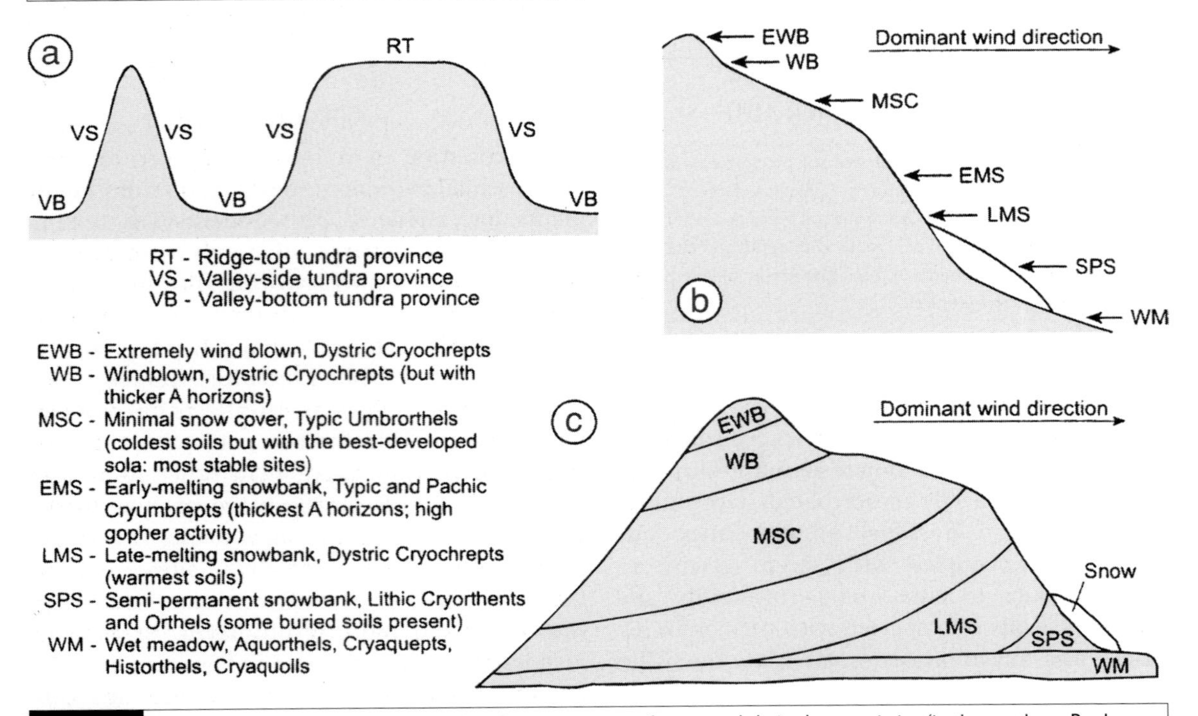

Fig. 11.13 Alpine geomorphic/soil provinces, after Burns and Tonkin (1982). (a) The major alpine geomorphic provinces. (b) Microenvironmental sites on the lee slopes, based on the synthetic alpine slope (SAS) model. Dominant soil types and their characteristics (in the southern Rocky Mountains) are listed. (c) The SAS model for both windward and lee slopes in alpine environments.

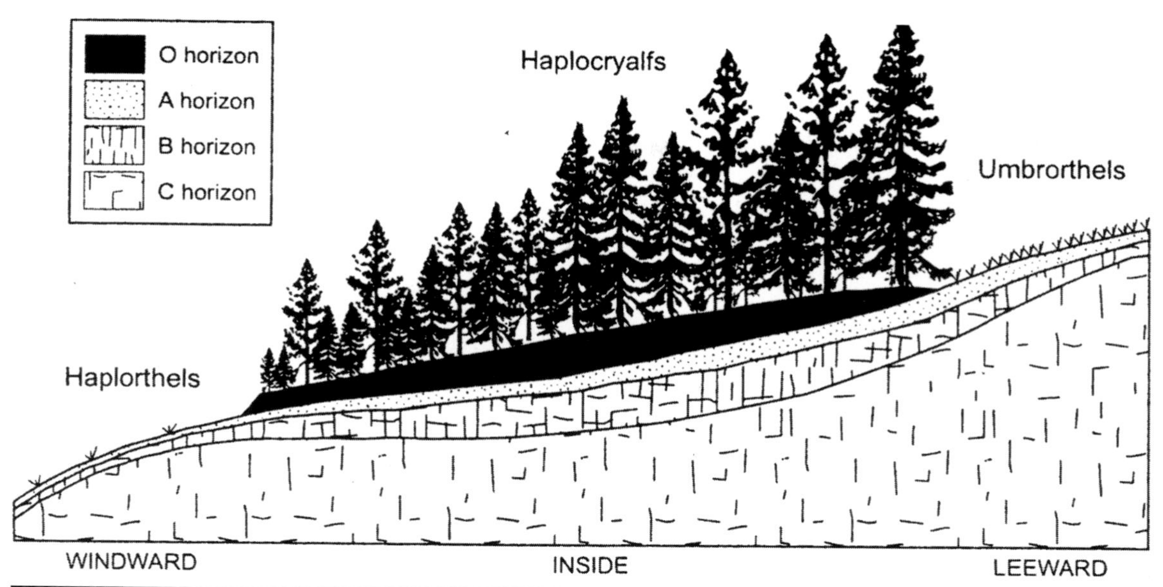

Fig. 11.14 Soil horizonation and classification as affected by a tree island. Snow accumulates on the leeward side of the island, while windward soils get eroded and have thin sola. Soil development is strongest within the tree islands, where the thick O horizons can form, as these sites are protected from the wind. After Holtmeier and Broll (1992).

so much been in the arena of climosequences, but rather it has been centered on (1) examining the effects of paleoclimate on modern soil properties and (2) using soils to provide information on paleoclimates (see Chapter 15). Challenges left to the researcher include unraveling the effects of (1) seasonal and temporal variability of precipitation and (2) local spatial variations due to topography, slope and aspect (Yaalon 1975).

Humans as a soil-forming factor: anthrosols

Human agency affects soils through any and all of the state factors; it is not just another *o*. Humans already influence the *t* factor by "resetting the clock" in the same way that geologic events might do so, and by forming new materials and surfaces for pedogenesis. Jenny (1941b), however, recognized that humans have an impact on soils, as part of the *o* factor (Bidwell and Hole 1965). However, he also noted that "a considerable number of human influences on soil appear to stand in no direct relationship to soil-forming factors" (Jenny 1941: 203). Effland and Pouyat (1997) noted that human impacts can be incorporated into the state factor model in either of two ways: (1) they can be considered on an equal basis with other organisms and hence included within the *o* factor, or (2) they can be given "factor status" as:

$$S = f(a, cl, o, r, p, t, \ldots)$$

where *a* is the anthropogenic factor, or as Yaalon and Yaron (1966) called it, the *metapedogenic factor*. Certainly, adding the sixth "*a* factor" increases the interdependence of the factors, since humans can and do alter four of the remaining five factors (and maybe we will even alter time someday!). Another problem associated with the *a* factor has to do with the length of time it has been operative. Because of the relatively brief time-span over which humans have altered soils, the *a* factor can rarely be given equal status with the other five factors. Instead, humans usually act as soil *modifiers* rather than soil *formers* (Table 11.2). Lastly, when the human influence is strong, the other state factors are usually forced to change, e.g., to cultivate a soil one must remove the natural vegetation. Thus, no matter how hard one may try, it is almost impossible (though neither may it be necessary) to treat the human soil-forming factor on equal ground with the five given originally by Jenny.

Because the effects of humans are somewhat atypical and not "natural" by some pedological

Table 11.2 Examples of metapedogenetic (human-caused) soil modifications

Manipulation	Principal processes observed in the soil
Topographical features	
Terracing or land leveling	Reduction of erosion; humus content increase; rejuvenation of pedogenic processes; catenary slope differentiation altered
Dam construction	Stopping of sedimentation and leaching; water table rise; salt accumulation
Draining of wetlands and mining	Surface subsidence
Hydrological factors	
Drainage, lowering of water table	Improved oxidation; structure formation; permeability change
Planting of windbreaks	Change of moisture regime; base saturation altered; carbonate leaching
Dredging of bays, harbors and wetlands	Additions of dredged materials onto otherwise undisturbed soils
Flooding of paddy fields	Hydromorphic water regime; reduced oxidation; gleying
Changing the hydrologic cycle by cloud seeding and irrigation	Potentially increased leaching or translocation; additional soluble salt accumulations
Chemical factors	
Irrigation with sodic water	Adsorption of sodium; structural degradation; decrease in permeability
Clay marling or warping	Textural change in upper horizons; moisture regime change; base saturation altered
Cultivation and cropping factors	
Deforestation and plowing in temperature areas	Mixing of upper horizons; change in pH; retardation of podzolization or depodzolization
Deforestation and shifting cultivation in tropical areas	Erosion; dehydration of iron oxides
Overgrazing	Erosion of surface horizons; reduction in infiltration

Source: Yaalon and Yaron (1966) and Bidwell and Hole (1965).

standards, they are treated only minimally in many pedology books. Although we examine some aspects of the *a* factor, we discuss little, for example, on the pedogenic effects of agricultural practices, even though long-standing agriculture dramatically affects soils and can even lead to the formation of plaggen and mollic epipedons.

Records of human impact on soils, dating back over 7000 years (Zi-Tong 1983), could fill volumes, but it can be summarized according to the main direct and indirect impacts that humans have (Bidwell and Hole 1965). Directly, humans till and terrace the soil, facilitating erosion and/or leading to overthickened profiles (Chang 1950, Sandor *et al.* 1986a, Smith and Price 1994). Burial and ceremonial mounds are constructed (Bettis 1988, Cremeens 1995) (Fig. 11.15). Amendments, such as manure and fertilizers, are added (Gaikawad and Hole 1965). Long-term manuring can lead to an overthickened, organic-rich A horizon that may qualify as a plaggen epipedon (Table 7.3) while long-term cultivation without large additions of organic materials can lead to decreased organic matter contents (Fig. 11.16).

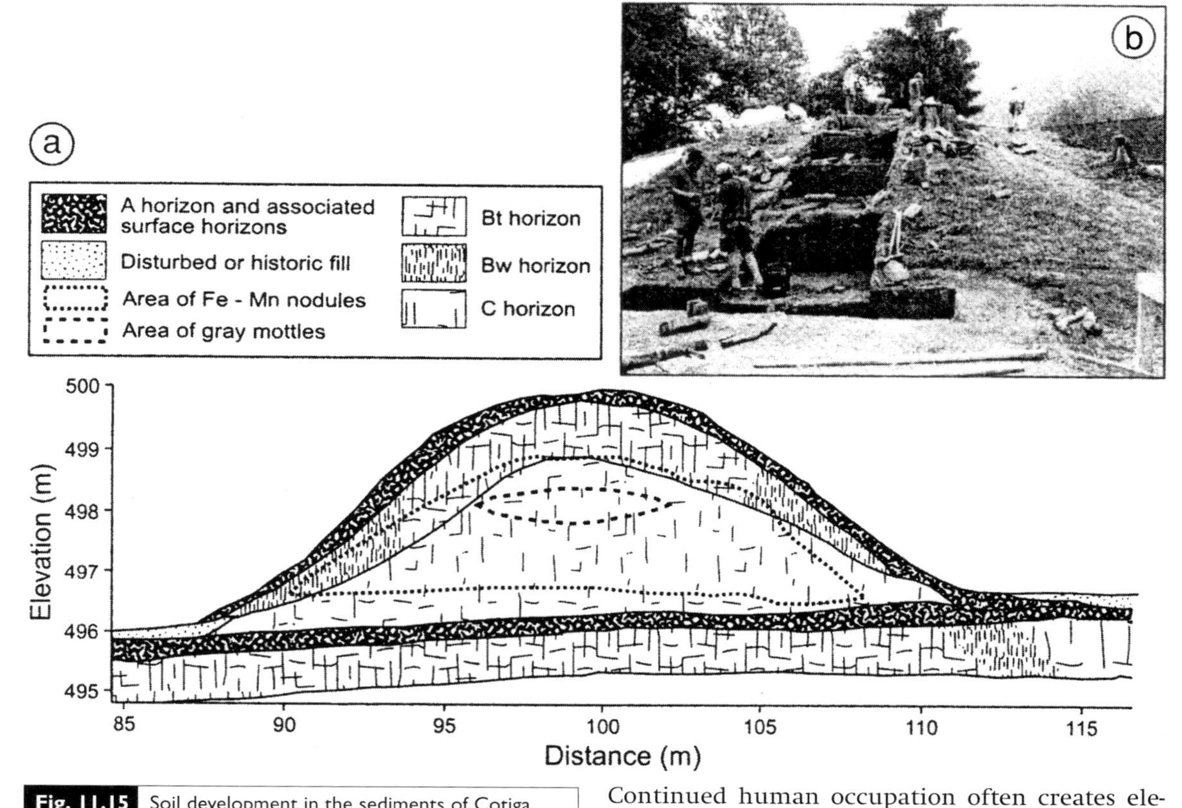

Fig. 11.15 Soil development in the sediments of Cotiga Mound, a mortuary mound from the Early Woodland period, West Virginia. (a) Soil development in Cotiga mound, as seen in cross-section. After Cremeens (1995). (b) Excavation of the mound. Photo by D. L. Cremeens.

Fig. 11.16 Schematic diagram of organic matter changes in soils, under natural and cultivated conditions, for a site in New Mexico. After Sandor *et al.* (1986b).

Continued human occupation often creates elevated P levels in the soil, which if high enough qualify the epipedon as anthropic (Kaufman and James 1991, Leonardi and Miglavacca 1999, Soil Survey Staff 1999) (Table 7.3). The sources of the P are commonly associated with bone and kitchen middens – residue related to mealtime activities. For this reason, elevated P levels in soils are commonly used by archeologists as being diagnostic for (paleo)human habitation (Eidt 1977, Sandor *et al.* 1986c, Schlezinger and Howes 2000).

Anthropologists and archeologists have long referred to human-impacted soils as anthropogenic soils or Anthrosols (Sjöberg 1976, Woods 1977). The term has caught on with soil scientists as well, although it has not yet received formal recognition as a taxonomic class. Another, emerging application of Anthrosols centers on urban soils, urban earths and other forms of disturbed land (Indorante and Jansen 1984, Stroganova and Agarkova 1993). In cities, the soilscape is often a mix of "natural" soils and those affected in some way by human actions (Alexandrovskaya and Alexandrovskaya 2000). Effland and Pouyat

(1997) described an "anthroposequence" – a series of soils, often from the city center to the countryside, in which the effects of humans change along a predictable gradient (McDonnell and Pickett 1990). They went on to coin the term "anthropedogenesis" for the role of humans in changing the pathway of soil development.

Urban Anthrosols are directly impacted by human activity in many ways. They are modified by compaction and additions and removals of chemicals, and they become overthickened where spoil or "fill dirt" is brought onto a site to raise its elevation or to provide foundation materials (Parker *et al.* 1978, Kahle 2000). They intercept pesticides, contaminants and other toxic substances and often contain fragments of human activity such as glass, bricks, wood, nails and ash (Effland and Pouyat 1997). Large parts of many cities actually rest and are built on spoil, rather than natural soils and sediments (Short *et al.* 1986a,b, Buondonno *et al.* 1998). Humans also mine, excavate and backfill areas outside of cities, creating a form of disturbed land, and in so doing, reset the pedogenic clock (Shafer 1979, Ciolkosz *et al.* 1985, Strain and Evans 1994).

Indirectly, human alteration of soils is ongoing and can reach well into the hinterland, beyond direct human habitation (Stottlemyer 1987, Burghardt 1994). Exhaust from motor vehicles and emissions from factories add metals to soil (Parker *et al.* 1978, Coester *et al.* 1997, Hiller 2000, Perkins *et al.* 2000). Acids from coal-burning factories bring inordinate amounts of sulfur into the soil system, far from the cities that are the sources. The list could go on, but suffice it to say that disturbed, urban and anthropogenic soils are increasing in area and are rapidly becoming an important field within pedology (DeKimpe and Morel 2000).

Simonson's process-systems model

Roy Simonson developed and presented a process-systems model of soil formation in a classic 1959 paper, and refined it 19 years later (Simonson 1978). Unlike the state-factor model of Jenny (1941b) that was long on factors that affected soil development but said nothing directly about pro-

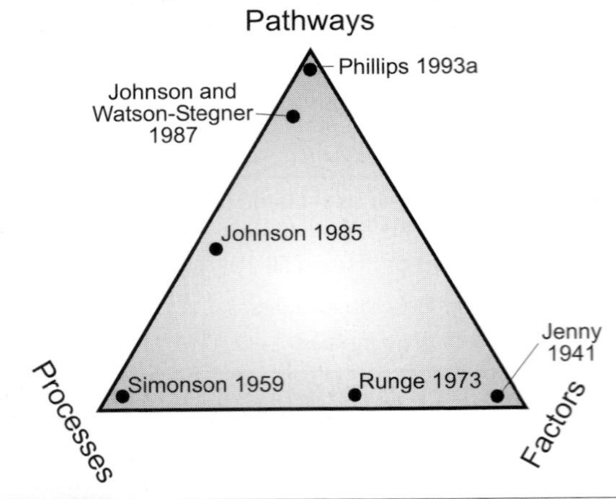

Fig. 11.17 Schematic representation of the emphases of a few pedogenic models. Interpretation is that of the authors and does not imply any type of judgement or value.

cesses that actually formed the soil, Simonson's model was entirely process-based (Fig. 11.17).

The background for Simonson's model is worth exploring. In his 1959 paper, Simonson drew attention to the paradigm shifts that had been occurring in soil science. He noted that the functional–factorial model, born in Russia 75 years earlier, was yielding to a point of view that held that soils evolve continuously. Each soil "stage" may appear, disappear, recur and fade away, leading to changing soil types and patterns. This ebb-and-flow viewpoint placed an emphasis on process and therefore set the stage for the process-systems model. Simonson noted that soils all have similarities and differences to each other. This observation was important for it reinforced the commonality that all soils had; the differences they exhibit, however, are due to different strengths of the *same types of processes* operating on similar materials.

In the process-systems model, soil genesis is viewed as consisting of two steps: (1) the accumulation of parent material and (2) the differentiation of that parent material into horizons. Yaalon (1975) noted that the general operation of any process-response model, like Simonson's or Runge's (below), can be characterized in terms of three determinants: (1) the initial state of the system (#1 above), (2) the processes it is subjected

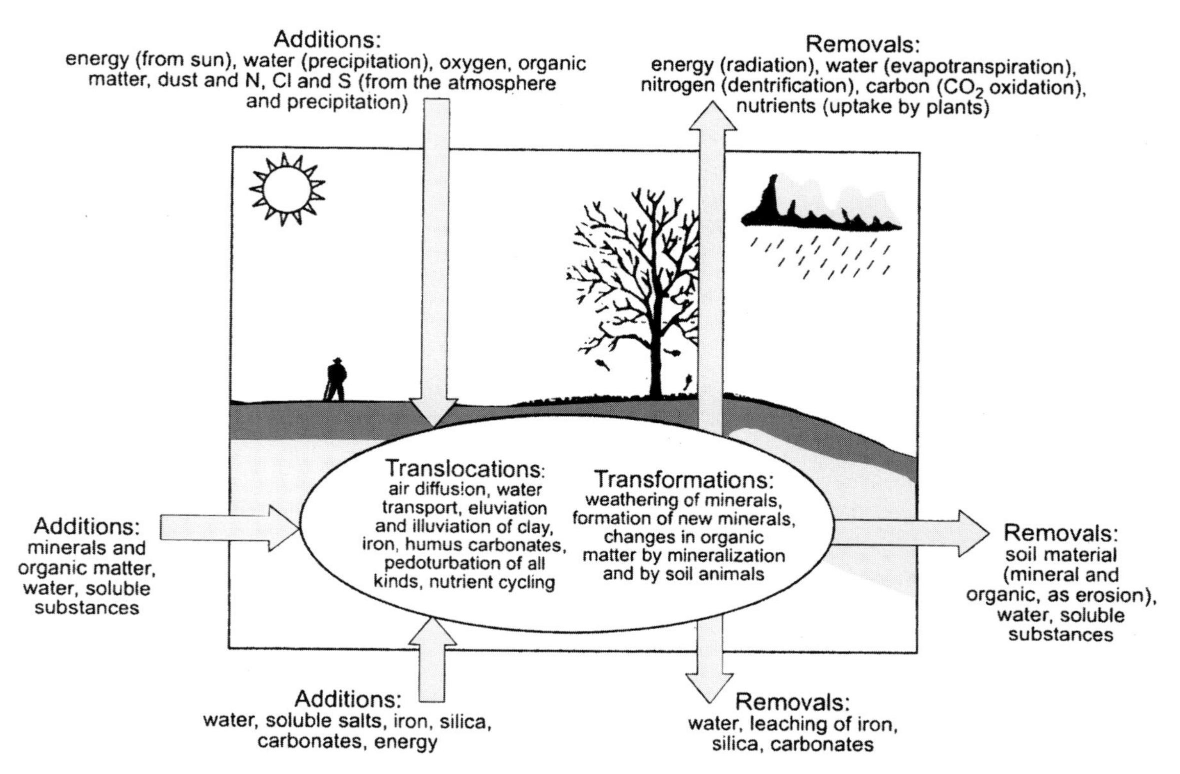

Fig. 11.18 Diagrams illustrating some common additions, removals, translocations and transformations in soils, and how these are useful in interpreting soil development. After Dijkerman (1974) and Simonson (1978).

to (#2 above) and (3) stage or duration of the processes (time). The focus of Simonson's model is on the second step. Simonson felt that soils differed because the processes they shared varied in *degree*, not *kind* (Fig. 11.18). The four major kinds, or bundles, of processes were designed, by necessity, to be very general, in order to cover the full range of pedogenic processes (Smeck *et al.* 1983). Although not originally conceived as an equation, the model can be written as

$$S = f(a, r, t_1, t_2)$$

where S is the soil, a is additions, r is removals or losses, t_1 is transfers/translocations and t_2 is transformations. Simonson (1978) envisioned that losses and additions are to the soil as a whole, while translocations are losses and additions, i.e., movements, between horizons within a single pedon. These four sets of processes occur simultaneously in all soils. Their *balance* and *character* governs the actual ultimate nature of the soil (Simonson 1978). Simonson's model might be considered part of a broader class of mass balance/budget models, to which sediment, water, nutrient and geochemical balance models also belong.

Horizons and the profile were viewed as developing their distinctive character because of a unique set of additions, removals, etc. The development of each and every horizon involves processes that fit within these process categories (Table 11.3). Many A horizons, for example, are dominated by additions of organic material, whereas salts, clay and other materials are removed. Weathering of primary minerals and alteration of raw organic matter into humus are also dominant processes.

The process-systems model of Simonson was not developed for, and has not found strong usage in, the interpretation of soil variability; for this purpose the functional–factorial model of Jenny (1941b) is more appropriate. Users of Simonson's model have to have some knowledge of processes in order to apply it effectively to explain soil

Table 11.3 Major soil horizons as they pertain to components of Simonson's (1959) process-systems model[a]

Horizon	Additions	Removals	Transformations
O	1. Organic matter (OM) 2. Atmospheric fallout (ions, dust) 3. Mineral matter from below (by burrowing organisms)	1. Decomposed OM and humus	1. Raw OM to humus (humification)
A	1. OM and humus	1. Bases (Ca, Mg, K, Na) (unless biocyled) 2. Clay 3. Fe and Al	1. Primary mineral weathering 2. Humus to a more decomposed state
E	None permanently	1. Bases 2. Clay 3. Fe and Al	1. Primary mineral weathering 2. Clays degraded
B	1. t = clay 2. s = Fe and Al 3. h = humus 4. y = gypsum 5. z = soluble salts 6. n = sodium (Na) 7. k = $CaCO_3$ 8. w = few or none 9. g = None	1. g = ferrous iron	1. w = primary mineral weathering 2. g = Fe^{3+} to Fe^{2+}
C	1. None, technically, though clay and bases are common additions, in small amounts	1. None	1. None, technically, although some weathering is expected

[a]Compiled as representative examples. Transfers are omitted because they are captured by the essence of "removals + additions" when the soil is evaluated on a horizon-by-horizon basis. If examined on a whole-soil basis, transfers become horizon-to-horizon movements of materials and energy, while additions and removals are considered to be from (or to) the soil as a whole.

spatial variability. Phillips (1989: 167) summed it up best when he said:

> The process and state-factor approaches are not exclusive. The process approach is more useful and appropriate for analyzing and describing the properties and development of a particular soil profile. However, the state-factor approach is often more useful for understanding the geographical variation of soils and ecosystem-level relationships with other environmental components.

The model continues to be used (successfully) as a framework for many studies that center on pedogenic processes (e.g., Cutler 1981, Vreeken 1984b, Hoosbeek and Bryant 1992, Gessler *et al.* 2000) and theory (e.g., Conacher and Dalrymple 1977, Johnson and Watson-Stegner 1987). Because of the way it compartmentalizes each of the four sets of processes, the model is also adaptable to computer simulations of soil genesis (Kline 1973, Levine and Ciolkosz 1986, Gaston *et al.* 1992, Hoosbeek and Bryant 1992, 1994). However, the model has proven most useful when evaluating the gains and losses of materials from profiles and horizons, as a means of piecing together their genesis. In this regard, the process-systems model has formed the conceptual backbone of an emerging branch of pedology that is concerned with mass balance calculations (see Chapter 14). In this field, gains and losses of various constituents from soil horizons (and the profile) are determined by comparing their amounts with those of an immobile, slowly weatherable mineral, usually zircon, tourmaline or quartz (Brimhall *et al.* 1988, 1991b, Sohet *et al.* 1988, Olsson and Melkerud 1989, Chadwick *et al.* 1990, Merritts *et al.* 1992, Jersak *et al.* 1995).

Runge's energy model

Ed Runge, a soil scientist then at the University of Illinois, developed a factorial model of soil development that is somewhat of a hybrid between the state factor model of Jenny and Simonson's process-systems model (Runge 1973) (Fig. 11.17). In formulating the model, Runge found a way to merge a considerable amount of process into Jenny's factorial framework, although Yaalon (1975: 199) called it nothing more than a "new verbal dress for the same model." Runge emphasized two *priority factors* from Jenny's model, climate and relief, which he felt were of most importance. He combined them into a single *intensity factor* that he defined as the amount of water available for leaching (w), which was governed by climate and topography, because certain sites are run-on sites and others are runoff sites. Thus, when examined together, the two factors produce a process vector that is roughly comparable to the potential for water to enter and percolate through the profile. Runge saw the w factor as an *organizing vector* that utilized gravitational energy to organize the profile, decrease profile entropy and create horizons, i.e., make it more anisotropic. Water that ran off the surface, i.e., not available for leaching, represented gravitational energy that was forever lost to the system. He then combined parent material and organisms (again assuming that the organisms factor was primarily concerned with plants and hence biochemical in nature) into a single intensity factor called organic matter production (o), or lack of mineralization. The rationale for the o factor is more difficult to tease out. Basically, Runge knew that plants were the source of organic matter (humus) in the soil and assumed that their ability to grow was governed largely by parent material. For example, if the parent material is infertile, little organic matter can be produced. Thus, while a number of environmental factors govern the amount of organic matter production, or primary biomass production, of plants, Runge felt that many of them were captured by the parent material factor. Unlike water available for leaching (w), the o factor was seen as a renewing or rejuvenating vector. How? Humus coats mineral soil particles (melanization) and thus inhibits weathering. The prairie grasses near Runge's home (Illinois) are also excellent base cyclers, and thus the better they grow the higher the pH of the soil remains, further inhibiting weathering and lessivage (clay translocation). As a result, the more humus-rich the profile, the less weathered and the more isotropic it often is. Runge, consequently, viewed the o factor as offsetting the w factor. He also included a time factor (t).

The model, which relies heavily on gravitational energy that drives infiltrating water and in turn causes horizonation, and relies (indirectly) on radiant solar energy for organic matter production, has come to be known as the energy model (Smeck *et al.* 1983):

$$S = f(o, w, t)$$

where S is the soil, o is organic matter production, w is water available for leaching and t is time. Each of the two intensity factors is conditioned by a number of capacity factors. W is conditioned by such factors as duration and intensity of rainfall, runoff vs. run-on, soil permeability, evapotranspirative demand, etc. Factor o is conditioned by nutrient (especially P), air and water availability, soil fertility, available seed sources, fire, etc.

The energy model is a hybrid between the factorial model of Jenny (1941b) and the pure process-based model of Simonson (1959) (Fig. 11.17), combining many of the positive attributes of factor-based models and process-based models. Nonetheless, it is just as difficult to quantify (Huggett 1975). It also does not quantify energy, though it is an "energy model." The model is simple, easily comprehended, and process-oriented.

The energy model is applicable in a number of settings where sites with excess water due to runon tend to have more strongly developed and horizonated soils, i.e., it works best in strongly leaching environments. Runge noted that it is probably limited in application to soils on unconsolidated surficial deposits like loess or till, in which leaching can operate freely. The model may not work well on residual soils. On the prairies of Saskatchewan, Miller *et al.* (1985) reported on "depression focused" recharge by snowmelt that lead to greater leaching in those areas. Sola were thicker in these sites, where water available for leaching was greatest (see also Pennock *et al.* 1987) (Fig. 13.21). In areas where a high water table inhibits the water available for leaching, soil development in both prairie landscapes is also inhibited, illustrating the efficacy of the energy model. These and similar studies, e.g., Anderson and Burt (1978), Sinai *et al.* (1981), Donald *et al.* (1993), Fuller and Anderson (1993) and Manning *et al.* (2001), continue to show that infiltrating water is a source of organizational pedogenic energy, and in so doing validate Runge's energy model. Applications of this general body of knowledge to landscape-scale patterns of denitrification are also myriad, e.g., Elliott and DeJong (1992).

Perhaps the strongest criticism of the model centers on the *o* factor. In Runge's surroundings at Illinois, soils are mostly Mollisols in which humic-acid rich, high pH organic matter is clearly a retarding vector. Large amounts of humus retard weathering and inhibit horizon differentiation. However, in forested regions, especially forests with a coniferous component, organic matter is acidic and via the chelating action of the fulvic acids formed by their decomposition, horizonation is promoted. Even in such landscapes, however, the model has been successfully applied. Schaetzl (1990) reported accelerated soil development in pits formed by tree uprooting. Treethrow pits are loci of greatly increased infiltration. They also have higher organic matter accumulations, which in the forests of Michigan also act as an organizing vector. Thus, it cannot be completely ascertained whether the increased soil development in pits is due more to increased infiltration or higher contents of organic acids. In conclusion, whereas the model is most applicable to grassland environments developed on unconsolidated parent materials, as originally stated by Runge (1973), it is still useful for forested sites.

Johnson's soil thickness model

Pedogenic models generally focus on the formation of the profile, the development of horizons, the loss/degradation of those horizons and similar topics. Most assume, for ease of comprehension, that the soil surface is static and that parent material has already been in place. Almond and Tonkin (1999) called this concept "top down pedogenesis" in which the depth of alteration increases with time, but one where the main impacts come from the external, subaerial environment "on top of" the soil.

This situation, however, does not hold for aggrading surfaces where soils get buried (rapid aggradation) or upbuild slowly (slow aggradation) by additions of loess, alluvium, tephra, etc. (Nikiforoff 1949, Phillips *et al.* 1999). Almond and Tonkin's study highlighted upbuilding and soil thickness concepts that were addressed by Donald Johnson more than a decade earlier: soil thickness is an important soil geomorphic component, and additions to and removals from the soil surface are an integral part of pedogenesis, not to mention classification. Johnson (1985) isolated soil thickness and set about examining the processes that affect it. It is an interesting twist on previous pedogenic models which focused on the totality of soil development as the dependent variable. In effect, one could argue that the soil thickness model is an outgrowth of Simonson's process-systems model, but focuses on additions and removals from the soil surface. Although such a focus may sound trivial, it is not.

In the model, the thickness (T) of a mineral soil is viewed as a dynamic interplay involving processes of profile deepening (D), upbuilding (U) and removals (R):

$$T = D + U + R.$$

Soils get thinner when $D + U < R$, and they get thicker when $D + U > R$, $D > U - R$ or $U > D - R$. Deepening refers to the downward migration of the lower soil boundary, generally accomplished via leaching and weathering. Upbuilding refers to allochthonous surficial additions of mineral and organic materials. This is generally accomplished by eolian additions, additions of organic material from plants, slopewash, etc. It also includes slight subsurface additions due to root growth or in-migration of fauna that die *in situ*. Upbuilding can take two forms: developmental and retardant. *Developmental upbuilding* implies that the surface additions are slow enough that pedogenesis can keep pace (McDonald and Busacca 1990, Almond and Tonkin 1999). The new materials are pedogeneticized as fast as they are added. This concept is captured in the term "cumulization" as well. Examples of developmental upbuilding include slow additions of loess or overbank alluvium. In *retardant upbuilding* materials are added to the surface faster than pedogenic processes can effectively incorporate them, and the soil, momentarily at least, gets buried (Schaetzl and Sorenson 1987). In this case, the soil per se gets no thicker, only buried by sediment that is considered not a part of the soil profile.

Removals refer mainly to losses of material from the surface through erosion and mass wasting, although subsurface removals by throughflow, leaching and biomechanical processes are also included. Oxidation and mineralization of organic matter are also considered removals. Phillips *et al.* (1999) defined truncation as soil thinning and surface lowering that is not of a magnitude to completely remove horizons.

The conceptual value of this model is threefold. It (1) draws attention to a concept (soil thickness) that had heretofore not been examined per se, (2) emphasizes the dynamic nature of the soil surface as one that is constantly experiencing erosion and/or deposition (Haynes 1982, Sunico *et al.* 1996) and (3) it introduces concepts that will eventually lead to Johnson and Watson-Stegner's soil evolution model two years later. Just as soil thickness, as a soil property, is shown to be either static, regressive (getting thinner) or progressive (getting thicker), so can soil development be viewed in the same manner. For most soils, thickness is a dynamic condition that ebbs and flows through time (Fig. 11.19); so it is with soil development.

Johnson and Watson-Stegner's soil evolution model

The thrust of Johnson and Watson-Stegner's (1987) soil evolution model is that soils *evolve*, ebb and flow, rather than unidirectionally develop and progress from "not soil" to some theoretical, steady state endpoint. The background for their model is complicated, but it was certainly influenced by the Russians, especially Nikiforoff (1949), whose lesser-known model or concept of soil evolution was based on two assumptions: (1) that soil development is continuously affected by certain progressive processes and (2) that these processes do not operate steadily. Each process begins, peaks and then fades over time. Each soil experiences successive waves of these beginnings, peaks and endings, each for a different type of process. For example, decalcification may be followed by lessivage, which may then be followed by podzolization and fragipan formation. Although the Russians, Nikiforoff in particular, conceived of some overlap among these "waves" the process waves nonetheless maintained their identity throughout the soil development timescale. Thus, the Russian concept of soil evolution was like the human evolution – generally progressive, occurring in waves of distinct species (soils). Although it uses the same name (evolution), the Nikiforoff model is not quite the same as the concept embodied in the soil evolution model of Johnson and Watson-Stegner (1987).

Johnson and Watson-Stegner, two geographers then at the University of Illinois, argued that the term "soil genesis" had, by the 1980s, come to be synonymous with soil formation and that soil formation implies better soil organization, or a loss of entropy and isotropism. Thus, many came

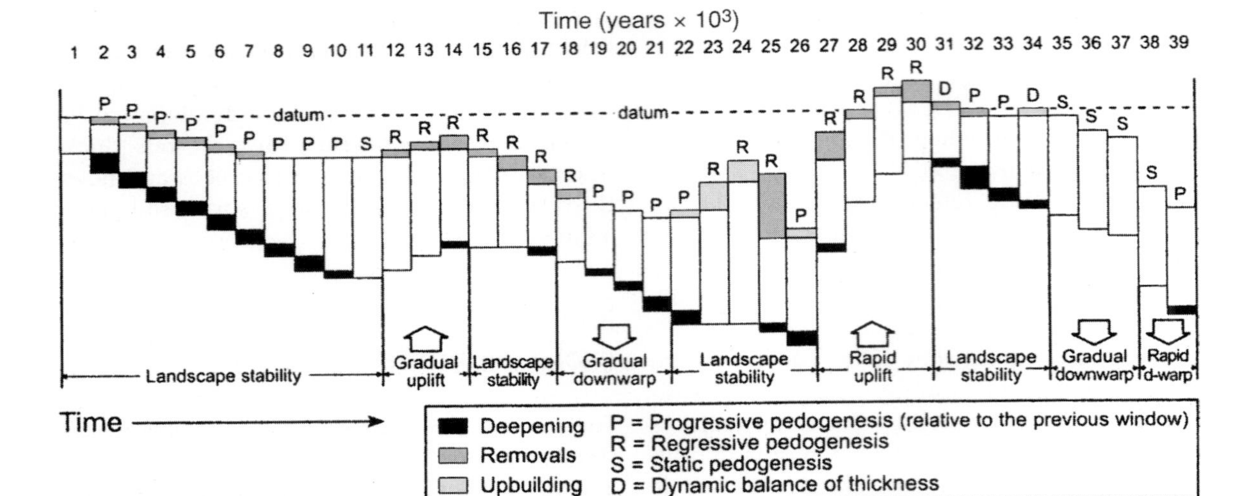

Fig. 11.19 Conceptual diagram showing the relationships of various soil thickness processes in a hypothetical soil in a dynamic landscape, illustrating progressive, regressive and static pedogenesis. The diagram shows various permutations of soil thickness and elevation changes relative to a constant datum, by means of 39 pedon time windows, each arbitrarily specified at 1000 years. $Time_{zero}$ was 39 000 years ago, implying that pedon 39 is the same soil after 39 000 years of pedogenesis. Pedons 1 to 7 show the changing soil thickness relationships from $time_{zero}$ to 32 000 years. Although surface removals occur, they lag behind the pace of soil deepening; surface additions are negligible. The soil becomes thicker via deepening and, in the absence of appreciable haploidization, experiences progressive pedogenesis. During the period 8000 to 10 000 years (pedons 8 to 10), removals and additions are negligible but deepening continues, as the soil continues to reflect progressive pedogenesis. Pedon 11 shows no change in thickness from pedon 10, as no appreciable additions or removals occur, and deepening processes are negligible. Static pedogenesis characterizes this stage of soil evolution. Pedons 12 to18 represent a 7000-year period of regressive pedogenesis in which removals are greater than deepening due to uplift and increased erosion. Uplift ceases at the end of pedon 14, and subsequent landscape stability is reflected in pedons 15 to 17; however, pedons 15 to 17 continue regressing because surface removals are greater than subsurface deepening. Landscape downwarping commences with pedon 18, and the soil continues regressing due to removals and negligible deepening. Pedons 19 to 30 are self-explanatory as the thickness of the soil continues to change. Pedon 31 is characterized by structural stability of the landscape. Removals are balanced by profile deepening processes, so that while the soil actively lowers relative to the datum, it neither thins nor thickens. This thickness condition is a case of dynamic balance. Pedons 32 and 33 reflect a return to progressive pedogenesis with modest or negligible removals and active deepening. In Pedon 34 rates of upbuilding match those of deepening, which thickens the soil and causes it to rise relative to the datum: another example of dynamic balance of thickness. Static pedogenesis typifies pedons 35 to 38. Negligible removals, additions or deepening occur during this 4000-year period of slow downwarping that culminates with rapid downfaulting (pedon 38). The thickness of the soil remains constant, and by the end of pedon 38 time the soil surface is again below datum. Pedon 39 marks a return to progressive pedogenesis with some deepening in the absence of appreciable removals or additions. Rapid episodic downfaulting of the landscape continues during this last (39th) millennium of pedogenesis. After Johnson (1985).

to perceive soils, by virtue of the language that was/is being used, to be a system that progressed along a pathway of continued development, despite many position pieces and empirical research products that pointed to the importance of soil regression, mixing and even erosion and burial as "normal" pedogenic processes (Hole 1961, Runge 1973, Vreeken 1975, Wood and Johnson 1978, Humphreys and Mitchell 1983, Johnson *et al.* 1987).

Why had so many scholars come to view soil genesis as strictly the "organizing" half of the equation, despite a significant body of evidence for contrary processes (Nikiforoff 1949)? Much of the answer has to do with the application of the state-factor model to better understand soil *development*. Jenny's state-factor model (paraphrased and quoted from Johnson and Watson-Stegner 1987: 349),

> encouraged viewing soil genesis through a formational/developmental filter. Its main message is that soils form and progressively develop under the influence of environmental factors. The time factor became implicitly linked to the formational/developmental model, in that soils were seen as progressing from simple to complex and ultimately toward a stable "mature" state.

Regressive processes that simplified or regressed soils were known, but essentially ignored. Studies had pointed out that the morphology of many soil profiles reflected not just organization and progressive formation, but disorder, mixing, regression and haploidization (simplification). Indeed, an entire soil order (Vertisols) was devoted to the concept. Processes that might lead to profile simplification included not just pedoturbation but also melanization, biocycling of nutrients, erosion and high water tables, to mention a few. The pedogenic implications of all these sorts of processes had been more or less dismissed. Because the state-factor model held sway for decades (and still does), the pedogenic community came to focus on soil development as a progressive process. Additionally, existing models of soil development did not focus on soil thickness. Should a "developing" soil get thicker with time, and if so, should it thicken indefinitely?

A model that could see both sides was clearly needed. Such a model needed to address progressive and regressive soil development, as well as thickness concepts. In this light, Johnson and Watson-Stegner (1987) presented their soil evolution model, as

$$S = f(P, R)$$

where S is the soil or a soil property, P is progressive pedogenesis and R is regressive pedogenesis. The soil evolution model stresses that soils proceed along two *interacting genetic pathways* that reflect variable exogenic/endogenic processes, factors and conditions (Fig. 11.17). Every soil has a progressive pathway along which the soil "moves forward or develops" and a regressive pathway that typifies a reversion to an earlier or simpler form. Whether a soil develops or regresses depends on which pathway is stronger at the moment or in the recent past. Even though a soil may display morphologic and/or physicochemical order and stability, and thereby reflect the predominating strength of the horizonation (progressive) vectors, the subsidiary haploidization vectors continue to operate. Static pedogenesis, a minor pathway less often realized, is possible when the two major pathways are essentially equal in strength.

Each pathway has three components, each of which consists of two opposing vectors or sets of processes: (1) horizonation/haploidization vectors or processes, (2) retardant or developmental upbuilding vectors or processes and (3) profile deepening/thinning vectors or processes (Table 11.4). The *progressive pathway* is composed of horizonation processes, developmental upbuilding and soil deepening/thickening (Tables 11.4, 11.5, Fig. 11.20). Progressive pedogenesis, or soil progression, is synonymous with soil formation, development and organization. It includes those processes and factors that singularly or collectively lead to organized and differentiated (more anisotropic) profiles. When progressive pedogenesis dominates, a soil develops more, thicker and better-expressed genetic horizons. The *regressive pathway* includes haploidization (simplification) processes, retardant upbuilding and soil thinning (Tables 11.4, 11.5). Regressive pedogenesis, or soil regression, reverses, stops or slows soil progression and development. It includes those processes and factors which singularly or collectively lead to simpler and less differentiated (more isotropic) profiles. When regressive pedogenesis dominates, soil horizons become thinner, blurred and/or mixed, and even eroded. Figure 11.20 provides examples of how the two soil evolution pathways interact, including an breakdown of the three components included in each pathway.

Table 11.4 Vector/process components of soil genesis and its two major soil evolution pathways

Progressive pedogenesis	Regressive pedogenesis
Horizonation Proanisotropic conditions and processes that promote organized profiles; the profile differentiating aspects of additions, removals, transfers, transformations, intrinsic feedbacks and proanisotropic pedoturbations	**Haploidization** Proisotropic conditions and processes that promote simplified profiles; the profile rejuvenating aspects of additions, removals, transfers, transformations, melanization, nutrient biocycling, enrichment, intrinsic feedbacks and proisotropic pedoturbations
Developmental upbuilding Pedogenic assimilation of surface-accreted materials	**Retardant upbuilding** Pedogenic impedance produced by surface-accreted materials
Soil deepening Downward migration of the lower soil boundary into fresh, relatively unweathered material	**Soil thinning** Surface erosion and mass wasting

Source: After Johnson and Watson-Stegner (1987).

An important aspect of the model is that mixing (pedoturbation) processes are not exclusively haploidizing; some may promote solum order (proanisotropic pedoturbations) (Hole 1961, Johnson *et al.* 1987). Much older literature focussed only on the "blurring" or proisotropic aspects of pedoturbation (Blum and Ganssen 1972). As discussed in Chapter 10, pedoturbation processes are typically thought of, even today, as being proisotropic or tending to promote simplification. This is certainly the case when the mixing is inter-horizon, such as mixing by ants that burrow through and move soil within the entire profile. But intra-horizon mixing need not destroy horizons, and it could even promote their identity. Johnson *et al.* (1987) showed how, in theory, pedoturbation can be anisotropic and help to create or maintain horizons.

The soil evolution model places an emphasis on *vectors* or *pathways* (plural!) of soil development: regressive and progressive. The continual "give and take" between the two pedogenic pathways is the essence of soil evolution (Fig. 11.21). As soils evolve, their morphologic and physicochemical profiles progress and regress through time; they do not simply advance or develop unidimensionally.

Changes in pedogenic pathways: how and why?

Knowing that pedogenic pathways are variably progressive or regressive, we next explore how and why these pathways ebb and flow. Pedogenic pathways are affected by many intrinsic and extrinsic factors. Extrinsically, the universe of possible pathways a soil can follow are "set" initially by parent material, topography and climate, e.g., Nettleton *et al.* (1975). For example, in the humid southeastern United States, on sandy parent materials the pedogenic pathway usually involves podzolization, leading to Spodosols or similar profiles (Markewich and Pavich 1991). Ultisols are unlikely to develop within this parent material "constraint." Parent material and topography, therefore, *precondition* and *constrain* the pedogenic system to only a certain number of initial pedogenic pathways. As topography changes and as parent material is eroded or added to, however, other options may open up.

Pedogenic pathways can be altered and changed due to externally or internally driven, active or developmental *forces*. These changes vary along a continuum from extremely subtle to major, e.g., as shown in Fig. 11.22. Subtle *external* change such as a slightly drier climate for a

Table 11.5 Components of the progressive and regressive pathways of pedogenesis

Major components	Examples
Horizonation vectors (progressive pathway)	
Additions (to one or more horizons)	Energy fluxes Insolation Heat transfers (black body; adiabatic and advection processes; endothermic–exothermic reactions) Oxidation (slow–biochemical; fast–fires) Gravity Mass fluxes Water (on-site rain and snow or ice melt; fog condensation and leaf drip; surface water run-on; subsurface water runthrough; water table) Gases (diffusion of H_2O, CO_2, O_2, N_2, etc; mass flow via wind, pressure gradients) Solids (eolian dust–ash, loess; ions in rain and snow, released by weathering, fluvial and mass wasting additions; organic matter production in place; additions of feces, urine, pollen, seeds, microbes, decaying plant and animal bodies and parts)
Removals (from one or more horizons)	Leaching of soluble constituents Eluviation and lessivage of solids in suspension
Translocations (from one horizon or portion of the profile to another)	Humus (A to B and C) Clay (A to B and C) Silt (A to B and C) Al and Fe (A to B and C) Soluble salts, silica and gypsum (A to B and C) Carbonates (A to B and C)
Transformations (of materials in place and of the soil chemical environment)	Weathering of primary and secondary minerals Organic matter (synthesis and resynthesis of humus, other compounds) Oxidation–reduction reactions Alkaline to acid pH changes with time
Developmental feedback processes induced by intrinsic thresholds and pedogenic accessions	Clay illuviation following leaching of carbonates and exchangeable Ca^{2+} Dispersion and translocation of clay in the presence of Na^+ Time-delayed formation of Bhs horizons consequent upon some amount of accumulation of Fe^{3+} and Al^{3+} in Bs horizon to effect immobilization of humus (−)
Proanisotropic pedoturbations (mixing that acts to form and/or maintain horizons)	Faunalturbation (animals) – a factor in formation of some A horizons, stone lines, stone zones, and biomantles Argilliturbation – a factor in blending some Bt horizons and in promoting surface stone pavements in Vertisols Crystalturbation – a factor in formation of calcic, petrocalcic, salic and gypsic horizons

(cont.)

Table 11.5 (*cont.*)

	Aeroturbation – a factor in promoting vesicular horizons, desert pavements
	Cryoturbation – a factor in promoting surface stone pavements, polygonal ground, stone circles, garlands, etc., in cold lands
Deepening	Downward migration of lower soil boundary
Developmental upbuilding	Pedogenic assimilation of relatively small amounts of eolian and slope additions; occurs on upland soils or high-energy, well-drained lowland sites
Haploidization vectors (regressive pathway)	
Additions	Same as above
Removals	Same as above
Translocations	Same as above
Transformations	Same as above
Nutrient cycling (soil enrichment and rejuvenation)	Biocycling, eolian enrichment, clay flocculation, melanization, and other simplifying processes (owing to renewing effects of nutrient inputs, such as Ca^{2+}, Mg^{2+}, humus, N, P, etc.)
Simplifying feedback processes induced by intrinsic thresholds and pedogenic accessions	Retardation or cessation of clay illuviation and vertical water percolation in B horizon due to pore reduction and clogging, which promotes increased runoff and erosion A smectite-rich Bt horizon becomes vertic, engulfs its A horizon, and evolves to a Vertisol with concomitant profile simplification
Proisotropic pedoturbations (horizon disruption and/or destruction processes)	Aeroturbation (soil gases) Aquaturbation (water) Argilliturbation (shrinking and swelling clays) Cryoturbation (freezing and thawing) Crystalturbation (salt crystal growth and wastage) Faunalturbation (animals) Floralturbation (plants) Graviturbation (mass wasting) Impacturbation (comets, meteoroids, munitions) Seismiturbation (earthquake disturbance) Anthroturbation (humans)
Removals	Episodes of erosion via deflation, sheetwash, stripping, and piping (owing to destructional effects of winds, thunderstorms, torrential rains, fire, removal of vegetation, animal population pressures, slope steepening, etc.), or normal mass movements
Retardant upbuilding	Impeding or retarding effects of relatively large amounts eolian- and slope-derived materials added to the soil surface. Occurs on some uplands (e.g., bluff sites of sand and loess accretions) and many floodplains, catenary footslopes and toeslopes, depressions and poorly drained sites.

Source: Johnson and Watson-Stegner (1987).

few decades might slow clay translocation, but not change the pedogenic pathway; it still is ongoing but it has simply changed its rate, and probably decreased in strength. Plant succession from aspen forest to hemlock–pine may accelerate podzolization in a soil that was already on that pathway. Similarly, subtle *internal* changes in pedogenic pathway or vector may occur as a soil exhausts its primary minerals, slowing the production of clay. Pedogenic pathway changes can occur when a Bs horizon of a Spodosol is continually enriched in iron and humus, root proliferation within it is promoted, and thus via enhanced root decay the rate at which that horizon gains humus is ever increasing. Before, roots had sought out the illuvial B horizon as a source of nutrients and soil water and now this horizon is more "lucrative" in that regard. The change was a matter of degree, not kind.

In order to examine fully what factors, internal and external, affect pedogenic pathways to a greater degree and how this fits into the soil evolution model, we must briefly digress into geomorphology where the concept of *thresholds* is entrenched and has been increasingly utilized, e.g., Langbein and Schumm (1958), Coates and Vitek (1980), Chappel (1983), Carson (1984) and Wescott (1993). Schumm (1979) noted that perhaps "threshold" is not the best word for this concept, suggesting the term "critical zone." As an example of a threshold, sediment will remain stationary in a stream bed until a critical (threshold) flow velocity is reached, after which it will freely move. The threshold is a value, unique to the system, beyond which the system adjusts or changes, not just in rate but in kind. Similarly, sediment yield per unit area in dry climates was found to be directly proportional to precipitation and runoff up to a point, a threshold point, beyond which sediment yield drops significantly, even with increased precipitation (Langbein and Schumm 1958). Why? Because vegetation reaches a threshold coverage level at some value of precipitation, holding the soil in place and reducing surface erosion and sediment yield.

Schumm's classic paper (1979) called attention to the concept of thresholds in geomorphology and defined two main types of thresholds: *extrinsic* and *intrinsic*. In an extrinsic threshold, an external variable or factor changes progressively, triggering abrupt changes or failure within the affected system (in this case, either geomorphic or pedologic). The system responds to the external change at what is called the extrinsic threshold level; below the threshold level no (or minimal) change occurs (Schumm 1979). Many extrinsic thresholds are caused by climatic or geomorphic change. Intrinsic thresholds, on the other hand, occur when a system changes without a change in an external variable. Schumm (1979) provided an example from geomorphology: long-term weathering of rock that reduces the strength of slope materials until eventually the slope fails. The critical shear strength value of the rock, in this case, sets the threshold level, below which the system is forced to readjust or change its "pathway." Another commonly referenced example of an intrinsic threshold comes from fluvial systems (Schumm and Parker 1973). Sediment storage in small stream catchments, brought in by tributaries with steeper gradients, accumulates because the trunk stream is unable to remove it all. This steepens the valley floor but there is no real system change until a threshold steepness (gradient) is achieved (Fig. 11.23). After this point, the trunk stream, realizing the increase in stream power when its gradient exceeded a threshold value, begins incising its valley and removing the accumulated sediment. The system has gone from an aggrading one to one dominated by downcutting (Patton and Schumm 1975). That is the "pathway change" initiated because of a threshold crossing.

Soil systems, like all others, have intrinsic and extrinsic thresholds. Muhs (1984: 100) defined a pedologic threshold as:

> the limit of soil morphologic stability that is exceeded either by *intrinsic change* of the soil morphology, chemistry, or mineralogy, or by a subtle but progressive change in one of the *external* soil-forming factors parent material, climate, relief and organisms. [Emphasis added.]

An extrinsic threshold may involve soil texture, almost like a lithosequence; as parent material

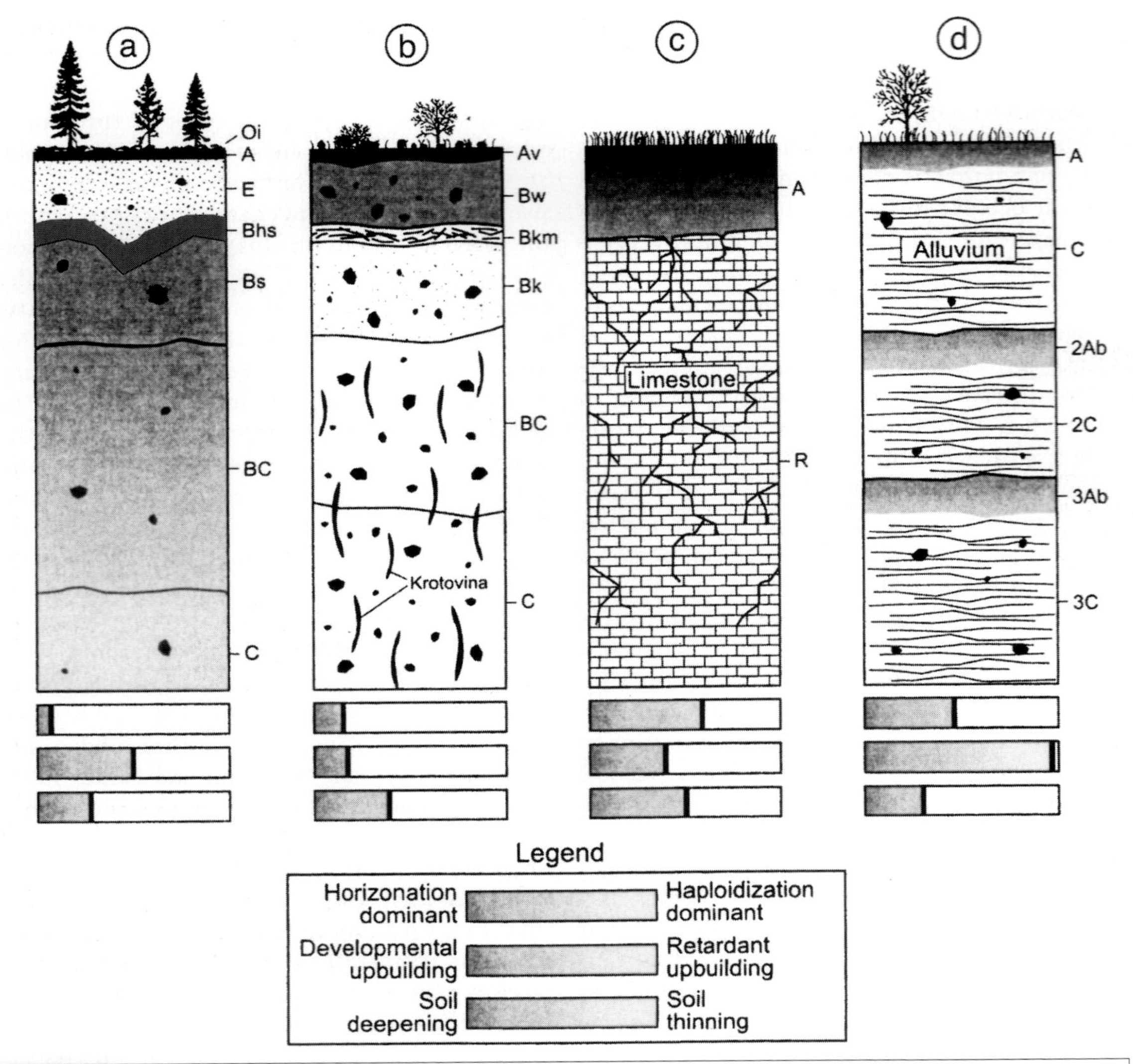

Fig. 11.20 Examples of hypothetical soil profiles operating under various permutations of horizonation–haploidization vectors. (a) A Typic Haplorthod developed in sandy glacial outwash in a coniferous forest in Manitoba. Horizonation vectors dominate, as almost no soil fauna are present to mix the soil. Runoff is minimal and thus soil thinning processes are weak while concomitantly the B–C boundary continues to deepen. The only major regressive pathway is provided by infrequent tree uprooting (Schaetzl 1986b, Schaetzl and Follmer 1990). (b) A Typic Petrocalcic developed on an alluvial fan in a desert in Nevada. Horizonation vectors dominate as translocation of $CaCO_3$ brought in by eolian dustfall has led to a thick petrocalcic horizon. After the Bkm horizon developed a pedogenic threshold occurred. Less and less water reaches the lower solum because of the almost impermeable nature of the Bkm horizon, and hence profile deepening is slowed. Additions of eolian dust has led to some profile thickening, as the dust is quickly incorporated into the Av horizon (McFadden *et al.* 1987), although slight deflation losses are also occurring. Generally vertically oriented cicada burrows in the subsoils attest to some mixing at depth (Hugie and Passey 1963), although this process subsided after the laminar cap on the petrocalcic horizon formed. (c) A Typic Rendoll developed on limestone bedrock, on a steep slope in Greece. Dissolution of the pure limestone bedrock is slow in this climate, and even when it does weather it provides little mineral material to the soil; the main process affecting soil thickness, therefore, centers around episodic inputs of dust from the Sahara (Prodi and Fea 1979). However, the steep slope leads to runoff that offsets the eolian additions. Melanization is dominant; given the high pH values of the soil, clay translocation is minimal. The limited profile "space" keeps worm and small insect bioturbation concentrated and active, leading to a homogenized but thin profile. The soil approaches static pedogenesis. (d) A Typic Ustifluvent on the floodplain of a medium-sized river in Kazakhstan, which drains a rugged, wetter upland area. Thus, the stream frequently floods and during these floods small increments of alluvium are added to the surface. In the intervals between floods horizonation processes operate and form a thin A horizon which, however, gets buried by the next flood event, hence the buried paleosols at depth. The alluvium is base-rich, which prevents clay translocation, and sandy, which limits worm faunalturbation.

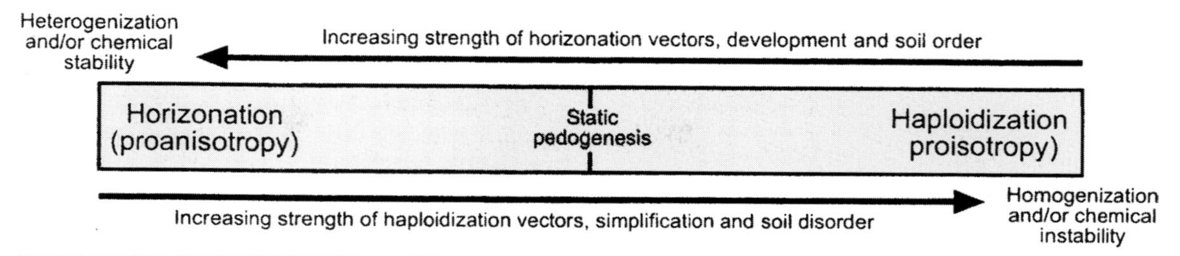

Fig. 11.21 Theoretical relationships between the horizonation and haploidization vectors, part of the progressive and regressive soil evolution pathways (see Table 11.4). The diagram illustrates how each vector is opposing the other and yet is usually operative in a soil at any given time, although their relative strength determines whether the soil progresses or regresses. After Johnson and Watson-Stegner (1987).

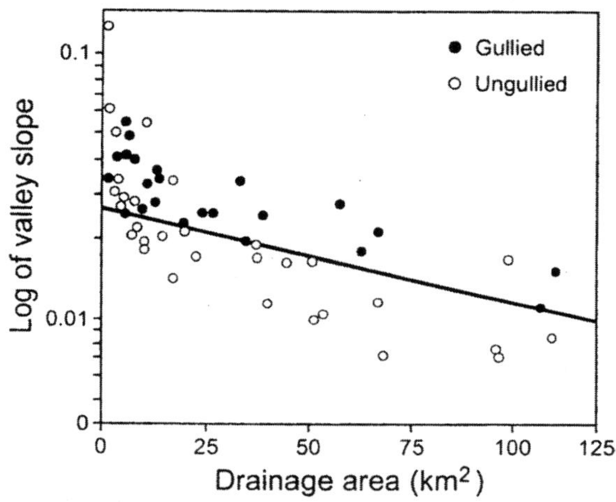

Fig. 11.23 Relation between valley slope and drainage area for the Piceance Creek Basin, Colorado. The line defines the intrinsic threshold gradient, beyond which the fluvial system begins to incise and becomes gullied. After Patton and Schumm (1975).

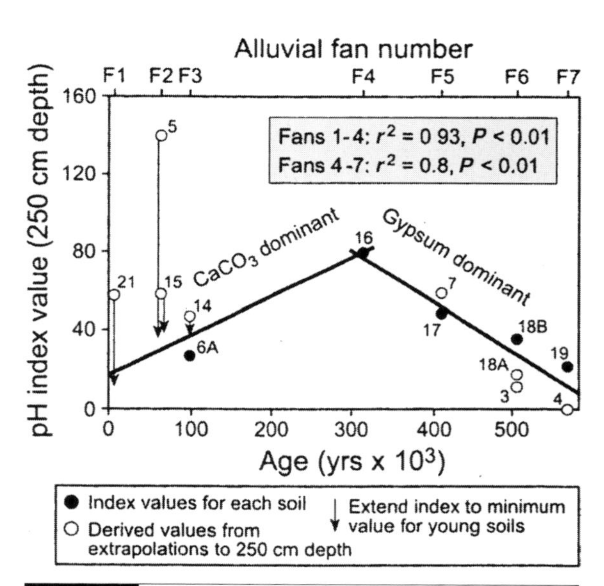

Fig. 11.22 A chronosequence of soils that on alluvial fans in Wyoming that show evidence for a change in pedogenic pathway at about 300 000 years ago. Early in these soils' development, pedogenic carbonate accumulated from eolian sources. Later, despite continued additions of carbonate, accumulation of secondary gypsum has been the main pedogenic process. This change in pathway is illustrated by a pH index (y-axis). After Reheis (1987b).

texture changes the pedogenic processes may also abruptly change at a certain point (Pedro *et al.* 1978, Duchaufour 1982). Likewise, climate or vegetation may shift, which they continually do, forcing a pathway change. Or a group of burrowing animals moves onto a site, promoting mixing and regressive pedogenesis. These are all external drivers that can cause a change in the pedogenic pathway (see also Graham and Southard 1983, Reheis 1987a).

Internal drivers that can cause pedogenic pathways to change can also occur; these changes can be (but often are not) manifested completely independently of the state factors. For example, clay translocation (lessivage) cannot begin until a threshold amount of base cations has been removed from exchange sites (Muhs 1984). As soils in wet–dry climates weather and attain a threshold level of smectite clay, argilliturbation begins and offsets lessivage that may have been operative (Muhs 1982).

Landscapes: pedogenic thresholds in the Buckskin Range, Nevada

One might expect thin soils on the arid, talus slopes of the Buckskin Range in Nevada, but many soils here are rocky only in near-surface horizons (Blank *et al.* 1996). Typical talus pedons (Xeralfic Paleargids) consist of large rocks with the interstices filled with vesicular sandy loam or loam material. Below this material lies clayey, prismatic- and blocky-structured horizons that are nearly rock-free. As one might expect in a desert, the development of these arid soils begins as raw talus acts as a trap for eolian dust. (Sites on these mountains that lack talus do not trap dust and have very thin-to-bedrock soils.) Rainfall washes the dust deep into the rock crevices. After a minimum amount of dust has accumulated, an intrinsic pedogenic threshold is passed and primary minerals in the dust begin to weather to a smectitic clay. This does not occur in near-surface horizons, which are too dry. Continued additions of dust coupled with its transformation to smectite pushes the soil past yet another threshold, and argilliturbation begins. The talus gets rafted upward in the incipient Vertisols, producing a soil that has a boulder zone near the surface over a clay-rich, relatively rock-free subsoil. Muhs (1982) and Johnson and Hester (1972) described a similar sequence of development on the Channel Islands of California. The talus protects the underlying soil from erosion.

Landscape images of the Buckskin Range, Douglas County, Nevada. Widespread occurrence of rocky talus is obvious. Photos by R. Blank.

Understanding intrinsic thresholds is important so that we do not always ascribe changes in soil morphology and pedogenic pathway to external forcing (McFadden and Weldon 1987). Intrinsic changes within the soil system can also result in pedogenic shifts (Muhs 1984). In a similar vein, Torrent and Nettleton (1978) introduced to the soil community the importance of pedogenic feedback mechanisms. They defend feedback as the returning of a part of the effects of a given process to its beginning or to a preceding stage, so as to reinforce that process. Thus, many feedback mechanisms or processes are self-accelerating to a point beyond which they decelerate and possibly terminate as some aspect of the process becomes limiting. Examples include the accumulation of organic matter in A horizons, clay accumulation in Bt horizons and the development of windows in Bkm horizons. Each of these processes tends to accelerate to a point, due to positive feedbacks set up within the soil, and then decelerate and stop changing after a time.

Vertisols provide excellent examples of pedogenic thresholds. They can develop directly from pre-existing soils, and similarly other soils can develop from Vertisols. Graham and Southard (1983) described a situation where eolian materials overlie the smectite-rich argillic horizon (2Bt?) of a Mollisol. The Mollisol was in a dry enough climate to develop subsoil cracks. Erosion of the upper solum of the Mollisol exhumed the clayey subsoil and allowed cracks to extend to the surface; eventually the profile developed into a Vertisol. Nettleton *et al.* (1969) had previously showed how argillic horizons could become so clay-rich and develop sufficient shrink–swell tendencies (and cracks) that they could engulf overlying materials. Although Graham and Southard's 2Bt horizon did not entirely "swallow" the overlying horizons into its subsoil cracks, this example nonetheless provided an example of one possible Vertisol development pathway. Graham and Southard's (1983) study provided an excellent example of polygenesis, and illustrates how erosion (or any form of geomorphic instability), as an external forcing vector, can lead to a shift in pedogenic pathway. In this case, some threshold of erosion had to have been crossed before the pedogenic pathway could shift to one where argilliturbation overwhelmed the other pedogenic organizing processes and vectors. Additionally, it should be noted that this "pathway shift" may be difficult to reverse, since it is unlikely that the Vertisol could develop back into a Mollisol without some sort of external forcing, such as deposition of smectite-poor sediment on top of the profile, or perhaps a change in climate.

Another Vertisol example, on San Clemente Island off the California coast, illustrates how Vertisols can develop from pre-existing soils without an external forcing mechanism. The Mediterranean climate here has dry summers and wet winters, and the parent materials are coarse marine sands and gravels on uplifted marine terraces. Muhs (1982) described a chronosequence of soils that develop from Mollisols to Alfisols to Vertisols. Older soils contained more and more smectite, presumably added by eolian influx, such that eventually, when older than $\approx$200 000 years, cracking and vertic properties become evident. Eventually, on San Clemente, Vertisol morphology is attained.

One of the many factors that causes intrinsic thresholds to be crossed centers on *pedogenic accessions* – pedogenically acquired/evolved features – that alter the genesis of the soil in some way and can force feedback processes. Accessions include pedogenically acquired pans, fabric, concretions, pH and various plasma features (see Johnson *et al.* (1990) for more examples). The feedback processes regulated by these acquired accessions may be more or less independent of any external conditions or vectors. For example, many desert soils accumulate secondary $CaCO_3$ in their B horizons. Eventually, the carbonates plug the soil pores, creating an aquiclude. This pedogenic accession – the Bkm horizon – forces the soil to cross an intrinsic threshold and the pedogenic pathway drastically changes from one in which water flows primarily downward and thus vertically oriented processes prevail, to one where water infiltrates and the runs laterally. Pedogenic accessions and the feedback mechanisms they promote (Yaalon 1971, Torrent and Nettleton 1978) are captured within the soil evolution model of Johnson and Watson-Stegner (1987) but are not included, or at least directly mentioned, in any of the three major pedogenic models previously discussed, i.e., Jenny (1941b), Simonson (1959) and Runge (1973).

The soil evolution concept casts some doubt on the validity of long-term, strictly linear chronofunctions. Linearity can occur in some chronofunctions which capture only a fragment of the soil's development. At the very least, the slope of the line, i.e., rate of development, can be assumed to change at times that lie outside of the limits of the chronofunction, in the future and the past. These changes simply cannot be discerned from the evidence presented in the soil's

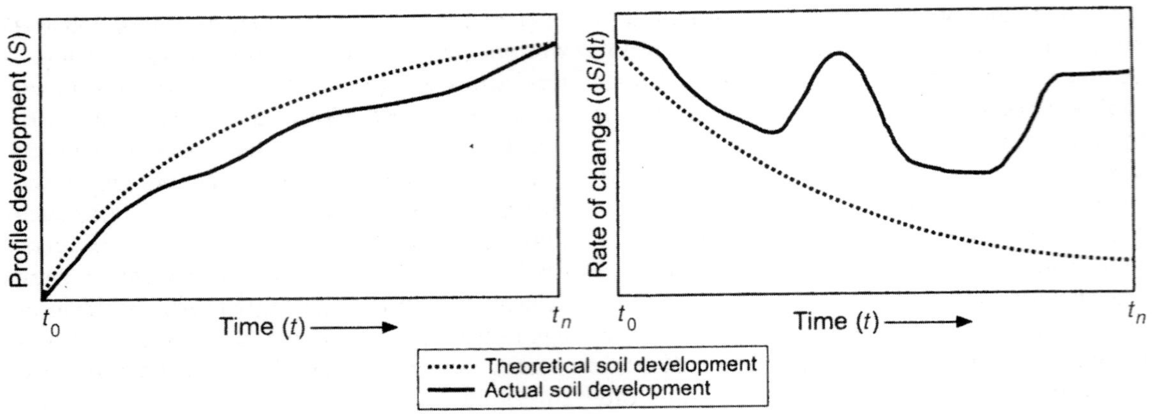

Fig. 11.24 Examples illustrating the difference between soil development vs. rates of development. After Johnson *et al.* (1990).

morphology and chemistry (Fig. 11.24). Also, we might not be able to discern the non-linearities and changes in development that do exist because we normally are operating on incomplete data; any real-world data set is of necessity incomplete. In short, the soil evolution model forces us to look at soil chronofunctions in a richer, different light.

Chief among the shortcomings of the soil evolution model is the ambiguity of the terms, especially "pathway." Intuitively easy to understand for some, it is nonetheless, a difficult concept to define rigorously. The model, while theoretically correct, is also difficult to actually apply, making it less useful to the field soil mapper than does the state-factor model. Most would agree that the state-factor model is better for soil science (mapping, classification, etc.), while the soil evolution model is perhaps equally or more useful for pedogenic and soil geomorphic research and higher-level theoretical training. Thus, while it is conceptually rigorous, use of the soil evolution model is generally restricted to scholars and those with a thorough understanding of the pedogenic system, for it requires that the user be able to elucidate possible scenarios to explain soil evolution on site. Without a solid pedogenic background, it is difficult to do that. The soil evolution model also focusses on soil as a whole, rather than its properties. This is probably on purpose, since it is difficult to imagine that soil properties regress or progress; the soil as whole regresses or progresses. Thus, the model takes a larger view of soil genesis, perhaps at the expense of the "pieces." Finally, because the soil evolution model is phrased in generalities, it is much more difficult to test than are some others. Designing a study to test the pathway shifts that a soil may or may not have taken, its degree of regression and/or progression, is a Herculean task. Perhaps it is one model we are destined to speak of in generalities only.

In sum, the soil evolution model redefines the terms pedogenesis and soil genesis, giving equal definitional rank to regressive or simplifying soil processes (Table 11.4). It recognizes and emphasizes that soils are complex, open, process and response systems that are continually adjusting, by various degrees, scales and rates to changing energy and mass fluxes, thermodynamic gradients and environmental conditions. Disturbance and change within the soil system are, at last, viewed as natural and predictable.

The normal soil concept

The soil evolution model, the 1990 winner of the Association of American Geographers' Geomorphology Specialty Group G. K. Gilbert Award for excellence in geomorphic research, was a reaction to the flaws in the normal soil concept of unidirectional soil development. Normal soils, synonymous with zonal or mature soils, were thought to be the endpoint of soil development with each kind of normal or zonal soil to its own particular climate (Marbut 1923, Baldwin *et al.* 1938, Cline 1949b, Frei and Cline 1949, McCaleb 1954, Johnson *et al.* 1990). Because both

Dokuchaev and Jenny emphasized climate as a dominant, active soil-forming factor, it came to be assumed that in each climate there existed a profile morphology that all soils would develop toward, unless thwarted by excess wetness, rockiness, sandy or clayey textures, etc. The "climax" soil morphology for a climate was captured in the "normal soil" concept, not unlike the climax forest concept that dominated ecological theory at the time (Clements 1936). Indeed, the first major treatise on soil development, Dokuchaev's (1883) *Russkil Chernozem*, was essentially a mandate for the zonal soil concept and the dominant influence of climate on soil formation. Thus, from its very beginnings, the science of pedology has been influenced by the notion that each climate or bioclimate has a dominant, almost climax-like, soil type. The "one climate – one soil" idea was reinforced by Bryan and Albritton's (1943) paper, influential though conceptually flawed, on "monogenetic" soils. In a way, our book follows some of those same lines of reasoning in Part III, although we approach it with a wider conceptual and theoretical "bag of tricks."

Belief in the normal soil concept during the early and mid twentieth century was almost a doctrine (Johnson and Watson-Stegner 1987), espoused and supported by all the major soil scientists of the day: Marbut, Jenny, Thorp, Baldwin and Kellogg. The science of geomorphology was, at that time, conceptually no different; the zonal implications of climatic geomorphology (1950–90) initiated by Peltier (1950) were showcased by Büdel in his 1982 book *Climatic Geomorphology* (Johnson 2000). Ecologists were also focussing on the progression of vegetation communities through a series of stages, to a stable and unchanging climax community (Cooper 1913, Clements 1936, Graham 1941, McComb and Lomis 1944, Drury and Nisbet 1971). Thus, all the major soil, landscape and vegetation successional-climax models and theories came to emphasize "normal", "linear" and "progressive" development (Johnson 2000).

Even in Russia, where the concept of "soil evolution" was being expressed decades earlier (Nikiforoff 1949), it was essentially developmental. The Russians viewed soil evolution "as a succession of continuous changes of the matter which make impossible stabilization of any particular stage of the process... (They abandoned) the concept of stable soil types in a fixed equilibrium with the environment" (Nikiforoff 1949: 220). Thus, their concept of soil development was one that was progressive; there was little opportunity for "going back" although neither did they entirely buy into the normal soil concept. The reason they stuck to the idea of a series of process waves or a progression of soil development centered on the importance they ascribed to primary mineral weathering in pedogenesis (Nikiforoff 1949: 221):

> The concept of the inevitability of continuous or progressive changes of the soil, whether along an evolutional spiral or otherwise, is based largely on the assumption that certain irreversible processes continuously affect soil formation. . . These processes consist especially, if not entirely, of the decomposition of primary minerals which are unstable in the thermodynamic field of the pedosphere. . . Their irreversibility refers to the fact that resynthesis of the original materials from the products of their decomposition in the soil . . . is impossible.

Thus, although the Russians had not held fast to the idea of a steady-state end member of soil formation, "in a fixed equilibrium with the environment," they had nonetheless insisted (and correctly so) that there are some aspects of soil formation that you simply cannot "go back" on. Weathering is one of them. Rode, a prominent Russian pedologist, supported the idea that soils can evolve – some processes are cyclic – but some, like weathering, are not.

Emphasis within pedology was, therefore, placed on determining the processes (developmental of course) that led to the formation of the normal soil for its given climate. The literature was full of terms such as "mature soil," "normal soil" and "zonal soil," all of which revolve around the concept that soils in a particular climate developed toward some end member, beyond which continued development was minimal. Once formed, such soils were thought to reflect an equilibrium with their climate and were inherently buffered against small or even modest environmental changes (Johnson *et al.* 1990). This mentality is understandable, given the widespread adoption of similar concepts in

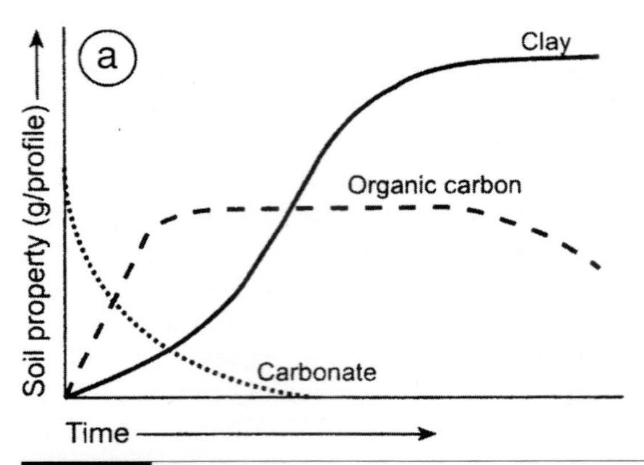

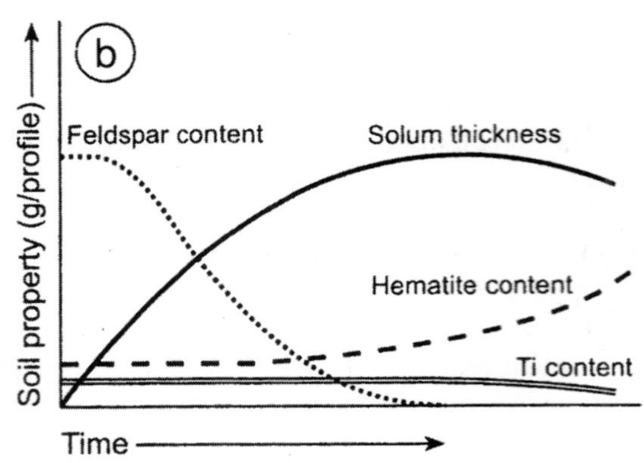

Fig. 11.25 Hypothetical variation in soil properties with time. A humid climate is assumed. Eventually, carbonates are leached from the soil. Organic carbon contents increase to a steady-state value, as does clay. The curves shown fit many, but not all, soils. After Birkeland (1999).

related fields. Although this pervasive view may have been imbalanced with respect to soil genesis, it was probably necessary at the time, in that pedologists needed to know how horizons became differentiated and how soils "formed" before they could focus on how they regressed or evolved.

Steady-state conditions

Although soils do evolve and progress/regress through time, individual soil properties can eventually achieve steady-state conditions, at least for a given length of time (Fig. 11.25). In a steady-state condition, the soil or certain of its properties are assumed to be in a sort of quasi-equilibrium at a given moment in time. Vidic and Lobnik (1997) suggested that steady state occurs when the rate of progressive soil development equals that of regressive soil development. Any soil can be in a steady state but they all eventually change over time. The "determination" of steady-state or not steady-state conditions within soils is dependent to a large degree on the timescale over which the change is being evaluated (Howard *et al.* 1993).

When a soil is in a steady-state condition, *inputs* of matter and energy to the system are ongoing but the *reactions*, as best as can be ascertained by measurement and observation of the resultant *properties*, are minimal. If the soil or its properties *are* reacting and changing over time, their rate of change is immeasurably slow. That is, the value (however expressed or determined) of that soil or property changes little over time, and the small changes that occur do not reflect a shift in pedogenic pathway. An example of a steady-state condition in soils is the organic matter content of most A horizons. Even though the horizon continually has varying inputs of raw litter and losses via mineralization, the overall amount of organic matter in the horizon remains roughly the same (over medium to long timescales).

As should be obvious, the judicious use of chronofunctions is an appropriate way to isolate and identify potential steady-state conditions and assess how long they may take to form (Fig. 11.25). The soil evolution model, as now understood, does not preclude the attainment of steady-state conditions. However, it may dictate that the condition will either be ephemeral or that all aspects of the soil cannot be in a steady-state condition at the same time. That is, some aspects of the soil will always be changing even if some are deemed to be in a quasi-equilibrium (Howard *et al.* 1993).

Polygenesis and polygenetic soils

The original concept of a monogenetic soil coincided well with that of the climatically equilibrated, normal soil. The monogenetic soil was one in which the soil continued developing toward some endpoint that was in equilibrium with the present climate (Bryan and Albritton 1943). Thus, climatic stability was the basis and underlying tenet of monogenesis (Johnson *et al.*

1990). Monogenetic profiles were viewed as being the result of pedogenic processes that can be described by a set of variables that had constant relationships with each other for an interval of time sufficient to cause the soil attributes to form.

By definition, then, a polygenetic soil must have experienced at least two such episodes of monogenesis and exhibit a complex assemblage of attributes that developed in response to changes in pedogenic variables. Thus, if it could be demonstrated that the climate or some other external factor/variable had changed, the soil could be considered polygenetic (Reheis 1987a). Beckmann (1984) said it best when he opined that a soil has a heritage, rather than an origin. Bryan and Albritton's polygenetic concept was, however, quite restrictive in that *major* change, i.e., glacial–interglacial magnitude, was envisioned (Watson 1968). How should minor climatic oscillations be handled? Johnson and Watson-Stegner (1987) (and Johnson *et al.* 1990) had an easy answer. Because pedogenic pathways are always being impacted to varying degrees by changes in external and internal forcings, all soils should be considered polygenetic. Johnson and Watson-Stegner's (1987) definition of the concept lies at the other end of the continuum from that of Bryan and Albritton (1943), suggesting that very subtle and minor changes force enough of a change to render a soil polygenetic. Thus, because all soils have certainly undergone subtle process changes, they must all be polygenetic. In this vein, Johnson *et al.* (1990: 309) offered this definition of polygenesis:

> Polygenetic soil (and polygenesis) connotes multiple genetic linkages of exogenous and endogenous processes, factors and conditions, including evolved accessions, thresholds, and feedbacks, that vary with time. Therefore, all soils are polygenetic, and the older the soil, other things being equal, the more polygenetic it is.

Using this mindset, because pedogenic pathways have always been subtly shifting and evolving, the notion of the monogenetic soil is moot.

We take the middle ground by affirming that polygenetic soils have demonstrably undergone some sort of major environmental change (that could be intrinsically or extrinsically driven). Many soils, especially latest Holocene soils, might still be best considered monogenetic. For if we throw out monogenesis, then we are back to where we were previously and will simply have to adopt another term for soils that have undergone major vs. subtle pathways change (Follmer 1998b). There is good reason for being able to discriminate between the two, and by regarding *all* soils as polygenetic there is no way to communicate the distinction. Furthermore, under the Johnson and Watson-Stegner (1987) definition soils that are polygenetic would not need to have any measurable expression of that polygenesis. Birkeland (1999) asserted, and we agree, that the pedogenic impacts that confirm polygenesis, i.e., the chemical and morphologic signatures within the profile, must be *detectable*; if they are not then the polygenetic impact may be so small that it is not worth noting. And if polygenesis need not be physically or chemically expressed, the meaning of the term is devalued because it cannot be scientifically verified.

Phillips' deterministic chaos and uncertainty concepts

Soil evolution taken one step further epitomizes the deterministic chaos concepts espoused by Jonathan Phillips, a soil geographer and geomorphologist at the University of Kentucky. In several papers Phillips (1993a, b, c) supports and expands upon the idea of soil evolution *sensu* Johnson and Watson-Stegner (1987). Phillips' ideas focus on a mathematical and theoretical explanation of the extreme soilscape complexity that is "out there," especially with respect to soil pattern, utilizing the soil evolution model and chaos and non-linear dynamical system theory for support.

The soil evolution model, with its pedogenic ebbs and flows, suggests that soil development is inherently non-linear (Phillips 1993c) (Fig. 11.24). Pachepsky (1998) described non-linear dynamic soil systems as those in which "almost the same" environmental conditions can render different developmental pathways. Deterministic chaos theory adds that patterns, in this case soil patterns, arise from the complex interactions of many elements behaving in an apparently random yet deterministic manner, via iterative non-linear systems. Deterministic uncertainty is a perspective on soil spatial variability that uses two fundamental axioms: (1) the reductionist view that variability can be explained with more and better measurements of the soil system and (2) the non-linear dynamics view that variability may be an irresolvable outcome of complex system dynamics (Phillips *et al.* 1996). It includes concepts such as dynamical instability, chaos and divergent self-organization (Phillips 2001c). Deterministic uncertainty takes these two fundamental axioms, that many view as being irreconcilable, and attempts to reconcile them into a unifying theory.

Chaos is characterized by sensitive dependence of a system on initial conditions and small perturbations, such that even minute differences at some point along a soil's evolution can lead to large and increasing differences as the soil system evolves (Huggett 1998a). Chaotic behavior is more likely when soil regression is strong, although it can occur in any scenario. The theory suggests that increasingly divergent soil development can occur over time, i.e., soils on a surface can become increasingly variable with time, even if they have all been subject to the same suite of pedogenic factors and exogenous inputs (Ibanez *et al.* 1990, Barrett and Schaetzl 1993, Huggett 1998a). Why? Minor perturbations emerge into complex patterns that can be used to explain the soil landscape even when these processes cannot be explicitly modeled (Phillips *et al.* 1996). In short, the effects of minor perturbations are large and long-lived (even if the perturbations themselves are not) because the soil system is presumably *unstable*. At least some of this "persistence" is due to feedbacks that are created when they first occur (Phillips 2001c). If these perturbations did not persist and grow, Phillips (1993a) argued that soil variability probably could not increase over time. Thus, the complexity of the soil fabric does not arise so much due to forcings from complex external forcing vectors, such as vegetation or microtopography; this type of complexity is called stochastic complexity (Phillips 1993b, 2001c). Rather, it assumes that underlying constraints or influences tend to persist and grow with time until they are disproportionately large compared to the magnitude of the original perturbation. External forcings are *significant* to soil evolution, but Phillips' work asserts that they are not always *necessary* and that in situations where divergence occurs in the absence of variations in environmental controls and external forcings, then nothing but dynamical instability, i.e., chaos, can explain it.

Minuscule local perturbations or soil variations at an indeterminate earlier time, such as any hypothesized $\text{time}_{\text{zero}}$, become magnified and lead to significant spatial variability. While these patterns can appear to be chaotic and random under deterministic chaos, Phillips argues for a modeling and theoretical paradigm that can address these inherently non-linear processes. Much of Phillips' work centers on unraveling and disentangling which parts of soil patterns have been caused by deterministic uncertainty vs. the natural heterogeneity of the external pedogenic factors.

To explain and model this concept, Phillips (1993b) restated the soil evolution model

$$S = f(P, R)$$

as

$$dS/dt = dP/dt - dR/dt$$

expressing all the variables as a function of time, or as a rate. By solving for this equation, Phillips (1993a) gets to

$$St = St - 1 + \Delta P - \Delta R$$

suggesting that the state of the soil at any given time_t is dependent on its condition or state at time_{t-1} and on any changes in the progressive or regressive pathway. By running this model a number of times in various numerical simulations, an unpredictable, though stochastic, complex pattern of development is produced. Think of the simulation as running the model once for each pedon on a landscape and then observing the degree of development of each pedon (Fig. 11.26). The resulting pattern is highly variable, i.e., not all pedons will have attained similar degrees of development.

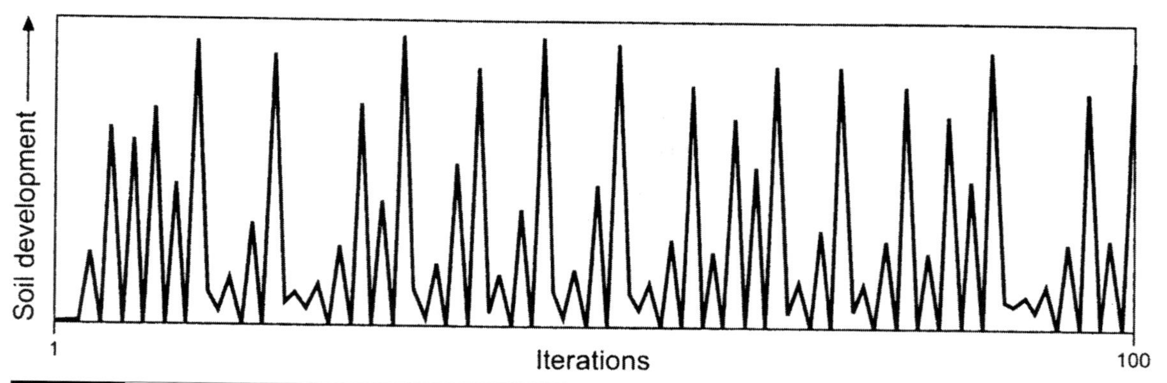

Fig. 11.26 Degrees of soil development produced by running a number of numerical simulations of a simple deterministic (mathematical) model of soil development. One could think of the simulation as running the model once for each pedon on a landscape and then seeing what the development of each pedon is. The resulting pattern is obviously highly random and variable. After Phillips (1993b).

In sum, Phillips' work shows that rich and complex patterns of soil development can ensue even if environmental inputs are reasonably uniform, due to *deterministic uncertainty*. Thus, the apparent chaos that one often "sees" on the soil landscape may be unrelated to the age of the surface, as even subtle variations in initial conditions can lead to vast differences in soils on the same surface. To quote Phillips (1993b: 175):

> In the case of dramatic local spatial variability where no variation in soil-forming factors can be discerned, the chaos model offers a plausible explanation where no other explanation is available.

One way to understand the implications of Phillips' model is to take one of three perspectives on the variable soils of a place. The chaos theorist might throw his hands in the air and attribute the pattern, at least in part, to inherent, irresolvable complex dynamics within the soil system. The traditional pedologist might argue that the pattern is understandable and decipherable if the pedologic processes were better understood, and to do that we need more data. The deterministic uncertainty theorist would take the middle ground (Phillips *et al.* 1996) and say that known and measurable causes account for the variability, but unstable, chaotic dynamics must be invoked to explain the magnitude and persistence of the variability.

Obviously, the implications of this model cast a shadow on the use of soil chronofunctions as a surface age dating tool, especially when the number of observations is low. Chronofunctions tend to be based on surfaces of known age with the data based on the larger-scale soil "order" while "anomalous" pedons are ignored. Chaos theory assists in interpreting these spatially complex soil landscapes and in so doing, provides a powerful theoretical framework for the understanding of order and disorder in soils.

Drawbacks of the model, however, lie in the fact that it cannot discern which part of the variation in development is due to initial conditions vs. random noise, or indeed, if any initial perturbations even existed (Amundson 1998). Pachepsky (1998) also noted that if the soil system is truly chaos-like, then our ability to predict soil development is limited. Thus, it has yet to be seen whether deterministic uncertainty theory will, eventually, allow for any kind of predictive ability, as the state-factor model did right from the start. As Phillips (1993b) notes, deterministic uncertainty raises more questions than it can, as yet, answer.

Non-linear dynamical systems theory, rather than diminishing the possibility of predictability, provides a "new context for predictability." A chaotic system is unpredictable in the traditional sense of a deterministic model that can predict exact outcomes forever. However, in some cases predictability can actually be improved, e.g., chaos-based models of sediment transport, river flows, etc., by accounting for the chaotic "memory" in apparently random time series. In

Table 11.6 | Major geochronologic units of the Cenozoic era

Era	Period	Epoch	Duration of unit (Ma[a])	Unit began (Ma ago)
Cenozoic	Quaternary[b]	Holocene	0.01 (10 000 years)	0.01 (10 000 years ago)
		Pleistocene	≈1.65	≈1.65
	Tertiary	Pliocene	5	7
		Miocene	19	26
		Oligocene	12	38
		Eocene	16	54
		Paleocene	11	65

[a] Ma: millions of years
[b] The status of the Quaternary Period is, at the time of this writing, being re-evaluated.

pedology it has been demonstrated that these ideas have explanatory power and help generate field-testable hypotheses that would not otherwise arise. Nonetheless, the approach may not ever have the same predictive power of the state-factor model. Rather, non-linear systems theory is probably going to be most helpful in addressing/explaining the genesis of those soils that the state factor model does not fully explain or predict.

Other models

It is not our intention to suggest that all of pedogenic theory is captured in the six models outlined above. It most certainly is not. Because of space limitations we have had to restrict our discussion to these six. Others are certainly worthy of note, including Wilde (1946), Stephens (1947), Hole (1961), Chesworth (1973), Huggett (1975), Johnson *et al.* (1987, 1990) and Sommer and Schlichting (1997). Models by Butler (1959) and Paton *et al.* (1999) are discussed in later chapters.

The geologic timescale and paleoclimates as applied to soils

In this chapter devoted to pedogenesis it might seem out of place to discuss the details of geologic timescales. However, consider that soils are always forming within the constraints of time. Time is the one factor that affects all soils equally. It plays no favorites.

The past 65 million years lie within the Cenozoic era (Table 11.6). Almost no soils currently at the Earth's surface began forming before the Cenozoic because the surfaces that soils form on are rarely stable for more than a few million years. In all likelihood, the oldest surface soils probably date back to only the Pliocene, and most soils have their $time_{zero}$ somewhere in the Quaternary period – the last 1.65 million years of the Cenozoic. The Quaternary has been a time of repeated climatic shifts and glacial advances and retreats. Continental ice sheets invaded the upper midlatitudes of the northern hemisphere several times, moving out from accumulation zones in central Canada and northern Eurasia (Bowen 1978, Andrews 1987). These advances and retreats were brought on by climatic forcings that were probably associated with cyclicity in insolation (solar radiation) amounts that the Earth was receiving, due to orbital geometry oscillations (Martinson *et al.* 1987, Ruddiman and Wright 1987) (Fig. 11.27). This cyclicity is referred to as Milankovitch cycles, after the Serbian mathemetician Milutin Milankovitch who first described them in the 1930s.

Milankovitch cycles have three components (Berger 1988). The *eccentricity of the orbit* occurs because the Earth's orbit is elliptical and the elongation of the ellipse, i.e., the eccentricity, changes on an approximate 100 000-year cycle.

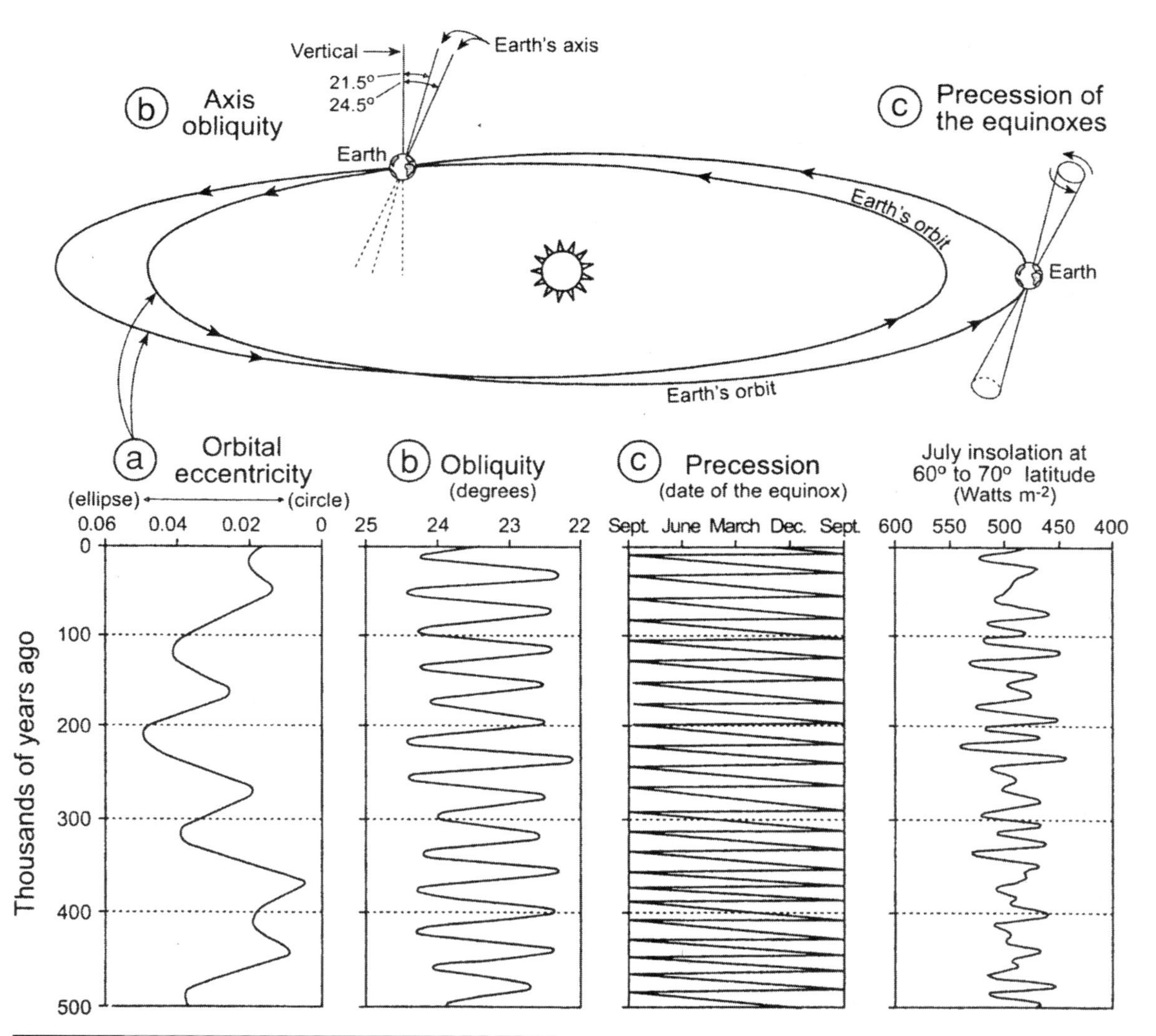

Fig. 11.27 The cyclicity of the three astronomic factors (eccentricty, obliquity and precession of the equinoxes) that have created cyclicity in the amount of insolation received on Earth. The cycles are called Milankovitch cycles, after Milutin Milankovitch, the mathematician who discovered them. After Covey (1984).

Orbital obliquity refers to the tilt of the Earth's axis with respect to its orbital plane. Today the tilt, or obliquity, is 23.5°, but it varies between about 22.5 and 24.5° on a 41 000-year cycle. Greater tilt generally leads to greater seasonality, especially in high latitudes. *Precession of the equinoxes* is in reference to the time of year when the seasons occur in relation to the Earth's position in its orbit. Currently, the Earth is farthest from the sun in the northern hemisphere winter, but this date gradually changes because the axis of the Earth wobbles like a slowly rotating top. The precession cycle varies between 19 000 and 23 000 years. In the end, the intricate combinations of the three cycles produce great cyclical variations in the amount of insolation received at the Earth's surface, which presumably drives the glacial cycles (Fig. 11.27). Obviously, other inputs and factors such as volcanic activity, continental alignments, changes in sea ice coverage and albedo of the land surface also affect glacial advances and retreats.

Major glacial advances are called glaciations or glacials, while the period of warmer climate between, when the ice sheets retreated (melted), are called interglaciations or interglacials.

Table 11.7 Correlations between oxygen isotope stages and various glacial advances or retreats

Oxygen isotope stage (from core V28–238)	Glacier characteristics	Midwestern US	Rocky Mountain	Sierra Nevada	Alps	Northern Europe
1	Interglacial	Holocene	Holocene	Holocene	Holocene	Holocene
2–4	Glacial	Wisconsin	Pinedale (varies)	Tioga, Tenaya and Tahoe	Würm	Weischel
5	Interglacial	Sangamon			Riss–Würm	Eem
6	Glacial	Illinoian	Bull Lake	Mono Basin	Riss	Saale
7–10[a]	Interglacial, but deteriorating climate	Post-Yarmouth	?	?	?	?
11[a]	Interglacial	Main Yarmouth	?	?	Mindel–Riss?	Holstein?
12+	Glacials and Interglacials	Pre-Illinoian (Kansan, Nebraskan, etc.)	Sacagawea Ridge	Sherwin	Mindel, Gunz, etc.	Elster, Menap, etc.

[a] Based on personal correspondence from Leon Follmer (Illinois State Geological Survey), as well as data from Sharp and Birman (1963), Richmond (1986) and Chadwick *et al.* (1997).

Smaller retreats within a larger glacial advance are called interstadials, while small readvances are called stadials (Table 11.7). Each glacial advance brought with it new parent materials while it eroded or buried others, the glacial sediments coming largely as the result of erosion of soils and rocks farther up ice. Sea levels fell and rose in response to the accumulation of glacier ice, which had as its ultimate source evaporation off the oceans. In mountainous areas, alpine glaciers advanced onto the piedmont lowlands and retreated back, in step with the cycles of the larger continental ice sheets. In the tropics and subtropics, the climatic cycles were often manifested as wet or dry periods, with the wet periods coinciding with the cold, glacial periods in the high and midlatitudes. In many deserts, a change to a cool, wet climate, called a *pluvial* period, accompanied the glacial advances (Benson *et al.* 1990).

Until about the 1950s there were thought to have been only four major glacial advances based on a suite of four fluvial terraces in river valleys that had carried glacial outwash (Cooke 1973). The assumption was that each glacial cycle filled the valleys with outwash, forming a high terrace, while during interglacial periods the terraces were incised, leaving a suite of four terraces from, hence, four major glaciations. However, by the mid twentieth century, data from the sea floors had begun to emerge, revealing a pattern of glacial cycles that was much more complex (Emiliani 1955, 1966, Shackleton 1968, 1977).

Data on past climates and ice volumes were obtained from the shells of one-celled plankton (mostly foraminifera (forams) but also coccoliths and diatoms) that had accumulated in layers on the ocean floors. These plankton secrete silica- and $CaCO_3$-rich shells, using sea water as the oxygen source. Thus, the isotopic composition of the shells varies in direct proportion to that of sea water (Shackleton and Opdyke 1976). Most oxygen is ^{16}O, but about 0.2% is a heavier isotope, ^{18}O. During glacial periods, the oceans are depleted in the lighter ^{16}O, because it preferentially evaporates and gets tied up in ice sheets. Thus, the oceans get enriched in ^{18}O during glacials, and the isotopic composition of the foram shells reflects this change in the seawater. During interglacials, foram shells become proportionately re-enriched in ^{16}O. Armed with this knowledge, paleoclimatologists pulled cores from of deep-sea sediments and extracted forams from them. The mean isotopic composition of the shells for any particular layer was taken as an indicator of

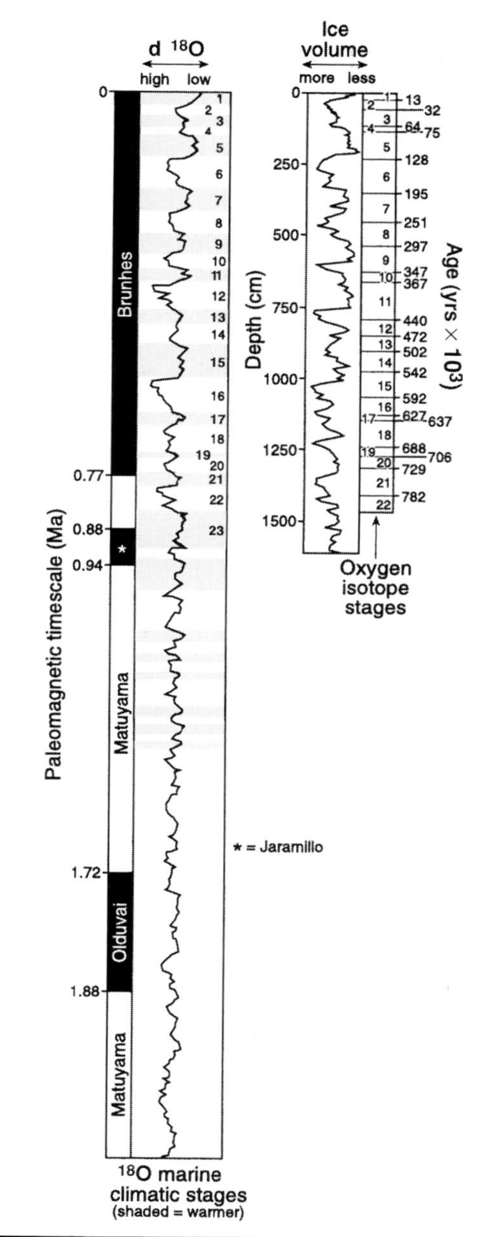

Fig. 11.28 Oxygen isotope composition of deep-sea core V28–238, Pacific Ocean, and approximate ages of each numbered oxygen isotope stage. For more detail on the upper portion of this sequence, see Fig. 11.29. Paleomagnetism timescale is also included. After Shackleton and Opdyke (1973) and Harland *et al.* (1982).

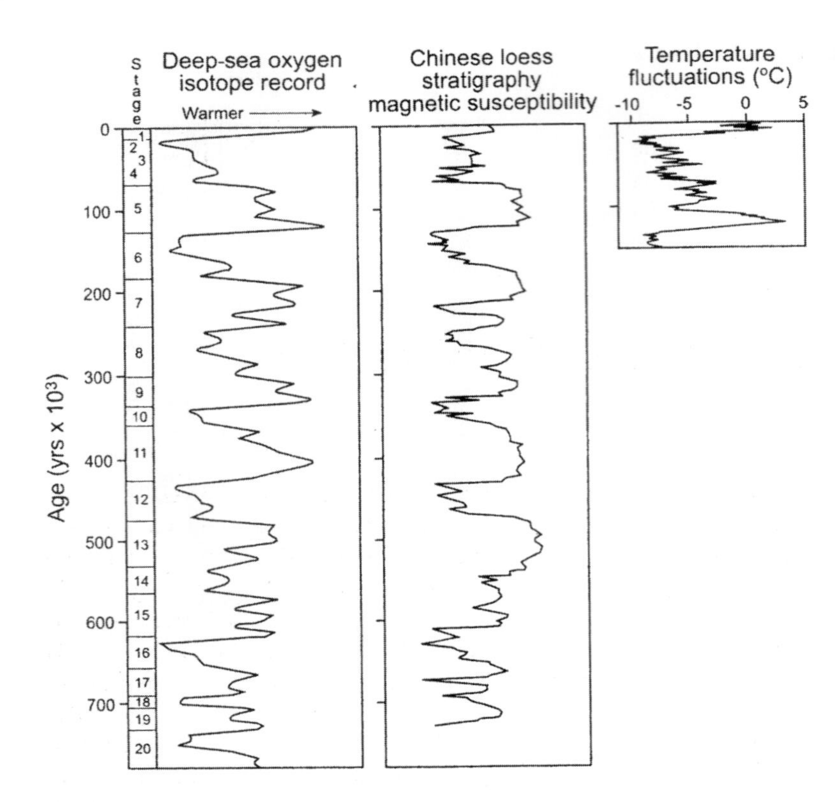

Fig. 11.29 Correlations among various paleoclimate indicators and the marine isotope record (see Chapter 15). After Jouzel *et al.* (1987) and Kukla (1987).

planetary ice volume, although the exact location of the ice sheets could not be determined from these data (Shackleton and Opdyke 1973). Analysis of the "type" deep-sea core, V28–238 (Shackleton and Opdyke 1973), which extended back to almost 800 000 years, revealed that many more than four cycles of ice build-up and retreat had occurred (Fig. 11.28). The cycles were numbered from the top of the cores to the bottom. Thus, oxygen isotope stage 1 was considered the current interglacial (Holocene). Stage 2 was the last major glaciation, etc. Odd-numbered stages were assigned to interglacials, while even-numbered stages correspond to glacials. Stages were identified as far back as number 23 (Shackleton and Opdyke 1973, 1976). The age of these sediments is primarily based on correlations of their paleomagnetism to the paleomagnetic timescale (Fig. 14.49). Most scholars today use the term marine isotope stage (MIS) to refer to these types of subdivisions of geologic time.

The use of oxygen isotope stages has become the conventional way of referring to climatic intervals during the Quaternary, while in different parts of the world the advance or retreat associated with a particular oxygen isotope stage is often given a name unique to that region (Table 11.7). Therefore, the convention of using oxygen isotope stages is one that most can recognize and adapt to. Data from other paleoclimatic indicators continue to be highly supportive of the oxygen isotope data as the standard for climate and ice volume fluctuations during the Quaternary (Fig. 11.29).

Little is known about the earliest glacial advances; there may have been a dozen or more. We really only know a significant amount about the last two or three major ice advances, and the minor readvances that are a part of each. The information about the last glacial advances comes primarily from the stratigraphic record contained within the glacial sediments and the loess deposits that indirectly resulted from them. Thus, it is a land-based record (Table 11.7). Our understanding of soil formation, burial and erosion is, necessarily, placed within this climatic record.

Chapter 12

Soil genesis and profile differentiation

The formation of a soil from raw parent material or from a pre-existing soil encompasses the concept of soil genesis, or *pedogenesis*. Soil genesis involves both progressive and regressive processes (Johnson and Watson-Stegner 1987). It includes all processes operative within the soil, whether they act to promote horizonation, preserve it, or even destroy it. And it is the underlying principle behind many soil classification systems which endeavor to group soils of similar genesis, based on observable morphology or chemical properties (Cline and Johnson 1963, Smith 1986). The link between soil genesis and classification, shown in one respect in Fig. 7.6, can also be envisioned as follows: soil-forming factors → soil-forming processes → diagnostic horizons and materials → soil taxonomic system (Bockheim and Gennadiyev 2000).

We emphasize pedogenic processes in this book for several reasons First, we must understand at least some things about soil genesis if we are to classify, and indeed manage, soils in a logical manner. Second, only by knowing the processes that formed soils can we ever hope to predict how they may change, as inputs (e.g., precipitation, heat, ions) and human uses change over time (Bockheim and Gennadiyev 2000). Lastly, only by knowing *how* soils formed can we map them or explain their past, present or future distributions. Thus, *soil genesis* is an integral part of *soil geography*.

Forming maintaining or destroying a soil involves an extremely complex set of processes, much akin to building, maintaining or destroying a house. Both can be compartmentalized into discrete, subcontracted blocks or bundles like (in the house example) (1) excavating the basement, (2) pouring the cement walls, (3) framing in the above-ground walls and roof, (4) installing plumbing and electrical components, (5) painting and finish work, etc. When examined in a soil genesis context, the project might be viewed as a series of process bundles or macroprocesses, e.g., weathering, dealkalinization, lessivage, podzolization and ferrolysis (Arnold 1983, Bockheim and Gennadiyev 2000) (Table 12.1). Each of these bundles or macroprocesses, in turn, has a certain degree of internal cohesiveness, but is composed of more precise and detailed parts (microprocesses). For the home project, the names given to the more detailed jobs might include (1) hanging dry (gypsum) wallboard, (2) installing windows, (3) adding trim to window frames and (4) painting. For the soil, the discrete process names might include (1) feldspar weathering, (2) chelation and (3) translocation of amorphous silica. Our point is that soil genesis, like building a house, can be viewed at a number of nested scales and processes (Zonn 1995).

Pedro (1983), like Simonson (1959) before him, viewed pedological evolution as consisting of two types: (1) processes associated with *weathering* of primary minerals, the *release* of various elements and their possible *recombinations* to produce new, more stable compounds and structures and (2) processes relating to the *arrangement* or *redistribution* of these soil constituents into horizons. We recognize that weathering is an important precursor to many pedogenic processes, and that it is ongoing in all soils (see Chapter 9).

Table 12.1 Processes of soil formation: a synopsis

Term/Process	Definition, and some references to the process
Soil genesis; pedogenesis	Surficial processes that aid in the formation and/or destruction of a soil body. Processes developing during and after weathering, related to the arrangement and reorganization of soil plasma and other constituents (Pedro 1983). Progressive and regressive processes are included.
Subsets thereof (Pedro 1983): Associative pedogenesis	*In situ* weathering and pedogenesis. Little disruption of plasma or skeleton fabric. Results in the formation of Bw horizons.
Dissociative pedogenesis	Dissociation between skeleton and plasma. Translocation.
Soil development; soil formation	Surficial processes that aid in the formation of, or otherwise do not destroy or weaken, horizons. Only progressive processes are included.
Soil evolution	The concept that soils change, both progressively and regressively, and all combinations thereof, through time (Johnson and Watson-Stegner 1987).
Horizonation	Processes involved in the formation and/or differentiation of parent material into discrete soil horizons. Formation of pedogenic anisotropy (Johnson and Watson-Stegner 1987).
Haploidization	Processes involved in the destruction or blending of existing soil horizons. Profile simplification (Johnson and Watson-Stegner 1987).
Eluviation	Downward or lateral movement of material out of a portion of a soil profile.
Illuviation	Movement of material into a portion of a soil profile, usually from overlying horizons.
Biocycling; photocycling; phytocycling; biogeochemical cycling	The movement of ions from the soil to the biosphere, via root uptake, and back to the soil, via littering, humification and mineralization. Biogeochemical cycling involves broader scales of cycling, including the hydrosphere and atmosphere.
Leaching; depletion; base cation leaching	Washing out (eluviating) of soluble materials completely out of the solum.
Decarbonation	Loss of carbonates from a soil, usually by leaching.
Recarbonation	Adding carbonates to a soil that had previously been leached, usually by increased Ca and Mg cycling or as additions of calcareous dust (Fuller *et al.* 1999).
Enrichment	General term for additions of material to a soil.
Surficial erosion; soil removals; superficial impoverishment	Removal of material from the surface layer(s) of a soil (Roose 1980, Johnson 1985).
Alluviation	Loss of clay from A and E horizons by lateral transport by water (Jackson 1965).
Upbuilding	Allochthonous surficial additions of mineral and organic materials to a soil (Johnson 1985).
Developmental upbuilding	Additions to a soil that are slow enough that pedogenesis can keep pace and incorporate them into the profile (Johnson 1985, McDonald and Busacca 1990, Almond and Tonkin 1999).

(continued)

Table 12.1 *(continued)*

Term/Process	Definition, and some references to the process
Retardant upbuilding	Additions to a soil that are faster than pedogenic processes can assimilate them into the profile, so that the soil, momentarily at least, gets buried (Johnson 1985, Schaetzl and Sorenson 1987).
Cumulization	Overthickening of the A horizon, caused when the rate of surface additions exceeds that of pedogenic assimilation. The surface of the soil to grows upwards (Riecken and Poetsch 1960, Hole and Nielsen 1970, McFadden *et al.* 1987).
Non-cumulative genesis; strain	Soil collapse as soluble materials, e.g., carbonates, are removed. Includes subsurface lateral and vertical removals through pervection, lessivage and leaching (Hole and Nielsen 1970, Johnson 1985, Chadwick *et al.* 1990).
Loosening;dilation	Increase in volume of voids by the activity of plants, animals and humans, by freeze–thaw or other physical processes and by removal of material by leaching.
Hardening	Decrease in volume of voids (porosity) by collapse and compaction, and by in-filling of voids with fine earth, carbonates, silica and other materials.
Soil deepening; profile deepening	The slow, downward migration of the lower soil boundary into fresh, relatively unweathered material below (Johnson 1985).
Decalcification; acidification	Removal of calcium ions or $CaCO_3$ from one or more soil horizons (van Breeman and Protz 1988).
Calcification; calcosiallitization	Processes leading to the accumulation of secondary calcium carbonate in soils (Gile *et al.* 1966, Sobecki and Wilding 1983, Machette 1985, Schaetzl *et al.* 1996).
Sparmicritization	Dissolution of sparry calcite by carbonation and its recrystallization into the mineral micrite, usually facilitated by soil organisms (Kahle 1977).
Lixivation	General term for the movement of soluble salts within soils (Duchaufour 1982).
Salinization	The accumulation of soluble salts such as sulfates and chlorides of calcium, magnesium, sodium and potassium in soils.
Desalinization; solonization; alkalization	The leaching of salts from the upper solum to the lower solum, resulting in a leached upper profile above a Btn horizon with columnar structure (Kellogg 1934, Munn and Boehm 1983).
Dealkalization; solodization	The leaching of sodium ions and salts from (out of) natric horizons (Pedro 1983).
Podzolization; spodsolization; spodosolization	The migration of Al and organic matter, with or without Fe, to the B horizon, resulting in the relative concentration of silica (i.e., silication) in an eluviated layer (DeConinck 1980, Ugolini and Dahlgren 1987). See also acidocomplexolysis.
Depodzolization	The gradual erasure of podzolic morphology from a soil, usually because of a change in climate or vegetation that weakens or stops the podzolization process (Hole 1975, Barrett and Schaetzl 1998).

(continued)

Table 12.1 *(continued)*

Term/Process	Definition, and some references to the process
Andisolization	Processes operative in soils that contain a large proportion of volcanic materials such as ash. Similar to podzolization, in which the fine earth fraction of the soils comes to be be dominated by amorphous compounds.
Latosolization	Processes that lead to the residual accumulation of sesquioxides in soils, as bases (especially Ca, Mg, Na and possibly K) and silica are leached, under long-term weathering in a hot, humid climate. There is little in-migration or translocation of Fe, as occurs in laterization.
Desilication; ferrallitization; allitization; siallitization	The chemical migration of silica out of the solum, leading to the concentration of sesquioxides, with or without formation of ironstone (laterite; hardened plinthite) and concretions (Eswaran and Bin 1978a, Latham 1980, Pedro 1983).
Degrees of desilication	(Pedro 1983)
Allitization	Total desilication and dealkalinization, forming gibbsite and ferric hydrates.
Monosiallitization	Partial desilication. Total dealkalinization, forming 1 : 1 phyllosilicates and ferric hydrates.
Bisiallitization	Partial desilication. Partial dealkalinization, forming 1 : 1 and 2 : 1 phyllosilicates.
Laterization	Chemical migration of Fe compounds into and out of the soil, leading to the formation of Fe concentrations at preferred sites. Much of the Fe probably comes from outside of the profile.
Silicification	Accumulation of secondary silica, either neoformed or translocated, into a soil or soil horizon (Milnes and Twidale 1983, Chadwick *et al.* 1987, 1989, Evans and Bothner 1993).
Lessivage; illimerization; argilluviation	The mechanical migration of clay particles from the A and E horizons to the B horizon, producing Bt horizons enriched in clay (Fridland 1958, Smith and Wilding 1972).
Argillation; neoformation	The formation of clay *in situ* (Fridland 1958, Alekseyev 1983).
Pervection	The mechanical migration of silt (but also including some clay) particles from the A and E horizons to the B horizon (or lower) of a soil (Paton 1978, Bockheim and Ugolini 1990).
Self-weight collapse	Process whereby a wet soil or sediment collapses, increasing its overall density and reducing pore space (Bryant 1989).
Pedoturbation	Biological and physical mixing, churning and cycling of soil materials (Hole 1961, Lee and Wood 1971b, Wood and Johnson 1978, Johnson *et al.* 1987, Poesen and Lavee 1994).
Proisotropic pedoturbation	Regressive pedoturbative processes that disrupt, blend, destroy or prevent the formation of horizons or subhorizons. Results in a simplification of the profile or the inhibition of further horizon development (Johnson *et al.* 1987).
Proanisotropic pedoturbation	Progressive pedoturbative processes that form or aid in the formation and maintenance of horizons or subhorizons. Results in, or promotes, increased profile order (Johnson *et al.* 1987).

(continued)

Table 12.1 *(continued)*

Term/Process	Definition, and some references to the process
Pedoturbation processes: mixing and cycling by:	
Aerotubation	Gas, air, wind (McFadden *et al.* 1987).
Anthroturbation; anthrosolization	Human activity (Eidt 1977, Short *et al.* 1986).
Aquaturbation	Water (Abrahams *et al.* 1984).
Argilliturbation; vertization	Swelling and shrinking of clays (Graham and Southard 1983).
Bombturbation	Exploding ordnance, such as bombs or missiles.
Cryoturbation	Freezing and thawing; ice crystal growth (Benedict 1976, Manikowska 1982, Van Vliet-Lanoë 1985, Tarnocai 1994).
Crystalturbation	Growth and wasting of salts.
Faunalturbation	Animals (burrowers especially) (Webster 1965a, Langmaid 1964, Cox *et al.* 1987a, Johnson 1989).
Floralturbation	Plants (treefall, root growth) (Schaetzl 1986b, Small *et al.* 1990).
Graviturbation	Mass wasting (creep, solifluction, etc.) (Moeyersons 1978).
Impacturbation	Comets, meteoroids.
Seismiturbation	Earthquakes.
Disintegration; physical weathering	The physical breakdown (weathering) of mineral and organic compounds into smaller pieces, with little or no change in composition (Jackson 1965).
Arenization	Physical disintegration of a rock induced by the chemical weathering of some of its weatherable minerals (Eswaran and Bin 1978b).
Decomposition; chemical weathering	The chemical breakdown of mineral and organic materials, with a concomitant change in chemical composition and structure.
Synthesis	The formation of new particles of mineral and organic species from ionic components (Jackson 1965).
Main pedoweathering pathways:	(Pedro 1983)
Acidocomplexolysis	Strongly acid and complexing attack on minerals by organic acids, affecting the extraction and elimination of Al, Fe and basic cations from the soil (see podzolization).
Alkalinolysis	Strongly alkaline attack due to the presence of Na carbonates in solution, affecting alumina and silica (see solodization).
Xerolysis	Attack on clays due to alternating periods of wetness and dryness (see ferrolysis).
Littering	The accumulation of organic litter on the mineral soil surface (Hart *et al.* 1962).
Humification; maturation	The slow transformation of raw organic material into highly decomposed humus (Anderson 1979, Zech *et al.* 1990, Malterer *et al.* 1992).
Paludification; paludization	Processes, dominant in wet soils, regarded by some as geogenic rather than pedogenic, including the accumulation of deep (>30 cm) deposits of organic matter, as in Histosols.
Ripening	Chemical, biological and physical changes in organic soil material after air (oxygen) penetrates previously waterlogged material. Usually accompanied by humification.

(continued)

Table 12.1 *(continued)*

Term/Process	Definition, and some references to the process
Mineralization	The release of oxide solids due to the decomposition of organic matter.
Melanization	The darkening of light-colored, unconsolidated mineral materials by the admixture of organic matter and humus (Fenton 1983, Schaetzl 1991b).
Leucinization	The paling or lightening (in Munsell color value) of soil horizons by the loss of humus either through transformation to light-colored ones or through removal (eluviation).
Braunification; brunification; rubification; ferrugination; humosiallitization	Release of iron from primary minerals, followed by the dispersion of particles of iron oxide in increasing amounts. Their progressive oxidation or hydration gives the soil mass brownish, reddish-brown and red colors (Glazovskaya and Parfenova 1974, McFadden and Hendricks 1985, Boero and Schwertmann 1989, Schwertmann and Taylor 1989, Blume *et al.* 1997).
Gleization; gleyzation; gleyification; hydromorphism	The chemical reduction of iron under anaerobic (waterlogged) conditions, with the production of bluish, greenish-gray or whitish-gray matrix colors (Schlichting and Schwertmann 1973, Vepraskas 1999).
Sulfidization; sulfidation; pyritization	The process of accumulation of sulfides in soils and sediments, as sulfur-reducing bacteria change sulfate SO_4^{2-} in water and sediment to secondary ferrous sulfide (FeS) or pyrite (FeS_2). Occurs primarily in wet, anaerobic soils along sea coasts, where a source of sulfur exists (Fanning and Fanning 1989, van Breeman 1982).
Sulfuricization	The oxidation of sulfide-bearing minerals, producing sulfuric acid, which lowers soil pH and promotes additional weathering (Fanning and Fanning 1989).
Ferritization	Process whereby dissolved ferrous sulfates produced by sulfuricization are oxidized and recrystallize to form ferrihydrite and ferric sulfates.
Chelation	Chemical process whereby the ligand of an organic acid reacts (combines) with a metal cation. The ion is removed from the mineral and rendered chemically mobile in the soil system (Atkinson and Wright 1957, Buurman and van Reeuwijk 1984).
Cheluviation	Combination of chelation and eluviation. Dominant subprocess of podzolization.
Ferrolysis; xerolysis	Process whereby reduced Fe (in a wet season) replaces base cations on clay mineral exchange sites. Bases may then be leached or moved out laterally in groundwater, by combining with HCO_3^- anions. During the dry season, Fe is oxidized, producing red mottles and exchangeable H^+. The H^+ attacks clay mineral structures and liberates silica and Al (Brinkman 1970, 1977b, Ransom *et al.* 1987a, McDaniel *et al.* 2001).
Chloritization	Subset of ferrolysis in which the Al^{3+} cations that have been released from clay minerals remain, to form chlorite and Al-interlayers in existing clays that have very low CECs and high amounts of exchangeable Al (Dijkerman and Miedema 1988).
Gypsification	Accumulation of secondary gypsum, either neoformed or translocated, in a soil or soil horizon (Carter and Inskeep 1988).

Source: Compiled by Schaetzl and Anderson, but patterned after Buol *et al.* (1997).

Weathering alone can form recognizable soil horizons. For example, Bw horizons (Soil Survey Staff 1999) are slightly weathered or reddened horizons that have formed primarily due to weathering, with little gain or loss of constituents. Pedro (1983) called this type of pedogenesis "associative pedogenesis without plasm transfer into the solid phase" (see brunification, below). In this chapter we also include those pedogenic processes that are primarily redistributive in nature. Pedro (1983) referred to these as dissociative pedogenic processes: those *with* plasma transfer into the solid phase. Within the context of soil genesis, we also give equal weight to pedoturbation processes, emphasizing that they are not necessarily regressive in nature (see Chapter 10).

Many pedogenic processes involve the mobilization and transport (and usually the eventual deposition within the profile) of soil components that can be referred to as *plasma*. Plasmic components such as ionic species, organic compounds, etc. are readily mobile within soils, while the parts that remain behind are referred to as *skeleton* (Pedro 1983). Over time, even the skeleton can be weathered and its ionic components translocated.

Eluviation–illuviation

An important first step in the discussion of pedogenic processes is their organization into a logical scheme. In this regard, we adopt the process-systems approach of Simonson (1959) (see Chapter 11): the formation of soil horizons can be ascribed to additions to and removals from the soil system, as well as transfers and/or transformations within it. Most transformations that affect soils involve primary and secondary mineral weathering (see Chapters 3 and 10). This chapter focusses on such transformations from a pedogenic perspective, and on inter-horizon transfers. In certain instances, additions to the soil surface, e.g., dust, organic matter, are also discussed where they are viewed as being pedogenically important.

The loss of material from a horizon, whether it be in solution or suspension, is called *eluviation*. It is unclear whether the term could also apply to the mass transfer of solid material, e.g., peds of A horizon material falling into the open cracks of a Vertisol, out of a horizon. *Illuviation* refers to the gain of material by a horizon, from an overlying horizon (vertical illuviation) or from a horizon upslope (lateral illuviation). Thus, eluviation and illuviation often occur as couplets or bundles. What is lost from one horizon is usually gained by another, unless it exits the profile entirely. Commonly observed eluvial–illuvial couplets include those associated with clay, Fe, Al, humus, carbonates and silica. In order for eluviation and illuviation to occur, there must also be a transfer. An eluvial–illuvial couplet, if considered within the Simonson (1959) framework, involves vertical or lateral transfers.

Impact of surface area

The strength or intensity of many eluvial–illuvial processes is assessed by determining the actual amount of material translocated, e.g., clay or carbonate, or, if that cannot be done, by using a semi-qualitative assessment of illuviation based on coatings, color or some other easily measured variable (see Chapter 14). An example of the latter might be iron (hematite) coatings that make a soil red, organic coatings that color it black, or a determination of the volume of argillans based on a number of thin-section point counts. Determination of coatings, their relative abundance and/or thickness, can be done in a variety of ways.

The impact of eluvial or illuvial processes is often dependent not only on their strength and duration, but on the amount of surface area the soil presents. It is often only the *surfaces* that get coated, whether these surfaces are mineral grains, coarse fragments or the more ephemeral ped surfaces. Coarse fragments have much less surface area than does the fine earth fraction. Soils with abundant clay and silt have more surface area to become coated (often an indication of illuviation) or stripped of existing coatings (an indication of eluviation) than do sands and gravels (Fig. 2.6). Thus, eluvial–illuvial processes take more time to become morphologically evident in finer-textured soils. The effect of surface area is most obvious in cobbly and rocky soils; rocks and other coarse fragments occupy considerable volume, forcing the remaining processes to be compressed into less space than if the same soil had no coarse fragments (Schaetzl 1991b). Surface area is also important to soils because particle surfaces retain water, cations, anions and nutrients.

Examples of the impact of surface area abound (Broersma and Lavkulich 1980). Secondary carbonates accumulate more rapidly in gravelly parent materials, as compared to gravel-poor sediment (Gile *et al.* 1966). E horizons form more quickly in coarse-textured soils than in finer-textured soils. For this reason, the Soil Survey Staff (1999) does not allow diagnostic cambic horizons to be sandy, as they can form too quickly in such materials. The intention of the cambic horizon concept is one of slight to very strong degree of alteration of primary minerals but some weatherable minerals are present. It is not intended to be used for quickly formed horizons in sandy materials. Schaetzl (1991b) observed that melanization and the accumulation of organic matter with depth in some gravelly soils is promoted by high contents of coarse fragments, as indicated by the strong correlation between organic carbon content and coarse fragment content ($r = 0.93$; $p < 0.001$). Holding other factors constant, as the amount of coarse fragments increases, the soil *volume* in which melanization can operate, and the surface area on which a limited number of organs can form, decreases. Thus if similar amounts of organic matter are incorporated into two soils, one with abundant coarse fragments and gravel, and one with few gravels or cobbles, two hypothetical scenarios are envisoned (Fig. 3.13):

(1) Organic matter will accumulate to approximately similar depths in both soils, but in the gravelly soil (with less surface area) the amount of organic matter in inter-gravel voids and the thickness of organs on skeletal grains will be greater than in the low gravel soil, resulting in *higher concentrations* of organic matter in the gravelly soil.

(2) The amount of organic matter in inter-gravel voids of the upper horizons of the gravelly soil will not differ markedly from that of the low gravel soil, but the organic matter will be *incorporated more deeply* into the gravelly soil.

Process bundles

In this section, we examine the major pedogenic process suites or bundles.

Processes associated with organic matter

O horizons are formed as organic materials such as leaves, needles and wood accumulate on the soil surface (Bray and Gorham 1964). *Littering* is the term applied to this process (Table 12.1). Litter accumulates as it falls from living and dead vegetation, is blown onto the site and/or washes in from upslope (Mitchell and Humphreys 1987).

As litter decomposes it forms a substance called *humus*. The transformation of raw organic material into humus is *humification* (Ganzhara 1974, Zech *et al.* 1990, Malterer *et al.* 1992) or *maturation* (Duchaufour 1982). Litter decomposition begins as litter is fragmented and broken up by larger soil organisms such as millipedes, earthworms, collembola and isopods. As the litter becomes more and more fragmented, the surface area made available to soil microorganisms for humification is greatly increased. It may be useful to envision this breakdown/decomposition process as one wherein both macro- and microorganisms are feeding on the litter simultaneously, with the most palatable and easily decomposed materials going first, while some of the larger and more resistant materials need to first be variously fragmented by macrofauna. For example, hemicelluloses and celluloses in the litter are broken down rapidly by microbes, while substances such as lignin, plant cuticle and animal chitin persist (Kevan 1968). With time, the mass of organic material becomes increasingly decomposed, with larger amounts of fecal material with its fragmented and humified organic materials encased within. Eventually, the litter is completely humified, and its C : N ratio is concomitantly lowered. At pH > 5.0, bacteria and actinomycetes are the main decomposers, producing a type of mull humus, whereas at pH < 5.0, fungi are the main decomposers, leading to a "greasy" type of mor humus (see Chapter 3).

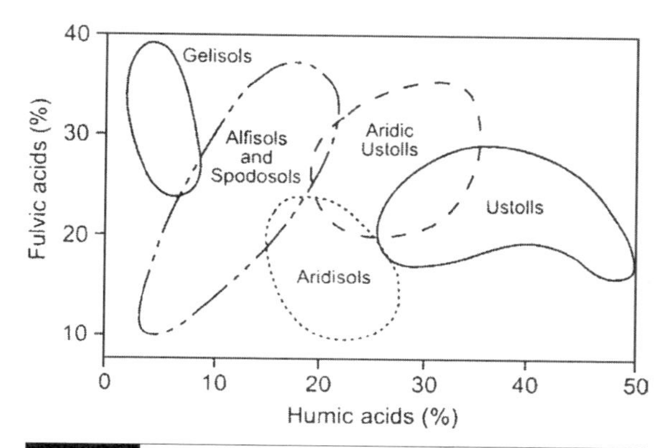

Fig. 12.1 Variation in humic and fulvic acid contents in soils of various types. After Volobuyev (1962).

A variety of factors affect the rate of litter decomposition. It is fastest in warm, moist, nutrient-rich environments. Humification is accentuated in rocky materials, under grass vegetation, in wet soils, and in parent materials with high carbonate contents (Smith *et al.* 1950, Gaikawad and Hole 1965, Anderson *et al.* 1975, Schaetzl 1991b). Adsorption of organic substances and humus by clays generally slows decomposition and humification. Clay–humic acid complexes are especially common in grassland soils (Fig. 12.1), in part explaining why decomposition of humus in these soils, beyond a certain point, is slow. For example, after cultivation, grassland soils quickly lose organic matter as the less decomposed humic fraction is fractionated and otherwise destroyed (Martel and Paul 1974, Tiessen and Stewart 1983, Gregorich and Anderson 1985, Zhang *et al.* 1988). However, after long periods of cultivation, the organic matter content of these soils eventually stabilizes, as the inputs of organic matter (small though they may be) balance out the losses via decomposition, and because the difficult-to-decompose fraction continues to increase in proportion to the more raw humus. In the end, soil particles become coated with only the most resistant ligno-protein residues, which impart dark colors even in small quantities. Particularly important to the humification/maturation process is the C:N ratio of the raw litter; litter with higher C:N ratios decomposes more slowly. In the end, highly stable humus compounds are composed of a combination of the less palatable parts of the original organic matter, and compounds synthesized by soil biota (Buol *et al.* 1997).

In soils with high pH values and abundant exchangeable calcium, many of the organic molecules form Ca-humates, which renders them resistant to further decomposition. Calcium is a highly efficacious stabilizing agent for organic matter, regardless of its stage of decay, sequestering organic compounds in a carbonate film (Duchaufour 1976, Zech *et al.* 1990). Additionally, many of the humic fraction molecules and compounds bind onto the surfaces of fine clay, which serves the same purpose – the resulting clay–humus complexes are highly stable. Clay–humus complexes of grassland soils reside mostly in the fine-clay fraction (Dudas and Pawluk 1969, Oades 1989). Once formed, these complexes can greatly lengthen the turnover time of soil organic matter. The calcium and fine clays bind with aromatic humic substances; for this reason the term humic acids is generally given to the suite of organic matter in organic matter- and base-rich soils.

Humus eventually gets incorporated into the mineral soil, as organs (Table 3.4) or humus-rich fecal pellets. There are at least three pathways from the O horizon, where organic matter exists as litter in varying states of decomposition, to the mineral soil (Fig. 12.2):

(1) The litter may decompose (humify) within the O horizon and, because of its colloidal size, be translocated into the A horizon by percolating water.
(2) The litter (in a more raw state, although it need not be entirely raw) may be translocated into the soil and decompose there. The translocation may occur as soil fauna drag the litter into the soil (Fig. 6.8) or as it washes or falls into open krotovinas or cracks.
(3) Organic matter can be added to the mineral soil by decomposition of roots and dead soil organisms *in situ*; this pathway is most important in soils where root density and turnover are high, such as grasslands. Ponomareva (1974) provided evidence that humus accumulation in Chernozems (Ustolls) can also be contributed as water-soluble root excretions, directly from plants.

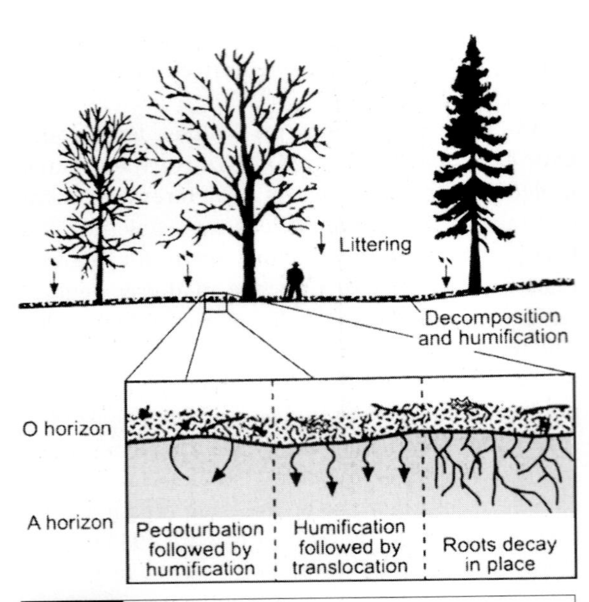

Fig. 12.2 The three pathways whereby litter in the O horizon eventually becomes incorporated into the A horizon as humus.

In most soils all three routes are operational, although in acid soils forming under coniferous litter, the second pathway is less effective because of the relative paucity of infauna. Where the humification–translocation pathway is dominant, e.g., sandy Spodosols, humus is incorporated into the soil more shallowly than in soils where pedoturbation is active, e.g., in silty Udolls. Thus, in many acid Spodosols the A horizon is thin or almost non-existent, and the lower O horizon boundary is sharp. Where bioturbation–humification is the dominant process bundle, the A horizon tends to be thicker, the rate of humification is increased and the O–A boundary is blurred.

In most soils, the rates of littering and humification eventually achieve a steady state. Warm, moist climates tend to have thinner O horizons because humification is rapid, despite the fact that litter production is generally high there. Disturbances such as fire will occasionally impact almost all ecosystems, temporarily upsetting this equilibrium but in the case of O horizons, it can be re-established in a few decades (Fox *et al.* 1979, Jacobson and Birks 1980, O'Connell 1987, Schaetzl 1994). On wet and cold sites, litter can accumulate to great thicknesses; this process is called *paludification* if the build-up is due to a high water table or cold conditions that inhibit decomposition (Gates 1942, Krause *et al.* 1959, Frazier and Lee 1971, Miller and Futyma 1987, Rabenhorst and Swanson 2000). In Histosols associated with a high water table, the degree of decomposition varies as a function of temperature and oxygen supply. Generally, the more oxygen that comes into contact with litter the faster it will decay. Because the main source of oxygen to soils is the atmosphere, decomposition will often proceed more rapidly in near-surface, though saturated, organic layers. Thus, one can often find sapric muck near the surface, while at depth the organic materials are in hemic or fibric states of decomposition (Table 3.2).

The endpoint of humification occurs when humus gets increasingly decomposed, eventually becoming fully mineralized. *Mineralization* (Table 12.1) results in the release of C, H, N and other ions that were contained in the humus, to the soil solution, making them available for leaching, translocation, neoformation and biocycling.

The darkening of a soil horizon or soil material, usually due to the addition of humus, is called *melanization* (Table 12.1). This bundle of processes involves the development of dark, humus-rich coatings on ped faces and mineral grains, rendering the horizon a dark brown or black color. Successive coats leave a more lasting and deeper color. Obviously, humification must precede melanization. The degree to which a soil becomes melanized is a function of the rate and duration of humus production, the types of humus produced, as well as its surface area. Minimal weathering of primary minerals in some soils can be explained by humus coatings on the mineral particles, as the Ca-humus acts like a protective coating on the soil minerals, especially clay minerals. Think of the humus coating as a type of protective paint on mineral grains (Anderson *et al.* 1974).

Melanization can occur in any horizon where organic matter is added and retained, e.g., Bh horizons, although it usually dominates A horizons. It is, essentially, the hallmark and distinguishing process of A horizons, especially those that have formed under grassland vegetation. Documented cases of mollic epipedons beneath forest vegetation, where a period of grassland dominance is not inferred, generally occur in frigid or colder soil temperature regimes (Nimlos and Tomer 1982; Anderson *et al.* 1975), or on the cool side of mesic temperature regimes (Gaikawad and Hole 1965) where mineralization rates are slow. Dark, thick epipedons also occur beneath forest vegetation where the sites are wet; the wetness inhibits decomposition.

Formation of granular structure

A horizons of all but the most sandy soils typically have granular structure (Fig. 2.11a). Granular peds are approximately spherical or polyhedral and are bounded by curved or very irregular faces that do not exactly match those of adjoining peds (see Chapter 3). The peds have strong internal cohesiveness, yet are fairly porous. They are formed, in large part, as soil passes through earthworms and exit their bodies as casts (Buntley and Papendick 1960). The mixing action that occurs in the guts of the worms multiplies the chemical bonds between organic and inorganic materials, and mixes the entire mass with microbial gums, resulting in crumb-shaped fecal pellets (Duchaufour 1976). Vermudolls and Vermustolls are dominated by earthworm activity and have very strong granular structure. Formation of this type of structure is also aided by wet–dry and freeze–thaw activity. Once formed, fine plant roots tend to seek out inter-ped voids, enhancing the integrity of the peds as they grow and expand. Root growth on the outsides of the peds slightly constricts the soil material within the ped, making it even stronger relative to forces that might destroy it. Hole and Nielsen (1970) refer to this phenomenon as interlacing roots. Roots that pierce granular peds add to their porosity.

Once formed, humus and humic substances bind together as well as coat these peds. Much of this humus "glue" is simply water-soluble, root exudates (Ponomareva 1974). Also contributing to cohesiveness of the peds are microbial gums and other decomposition products of plant and animal biomass (Hole and Nielsen 1970).

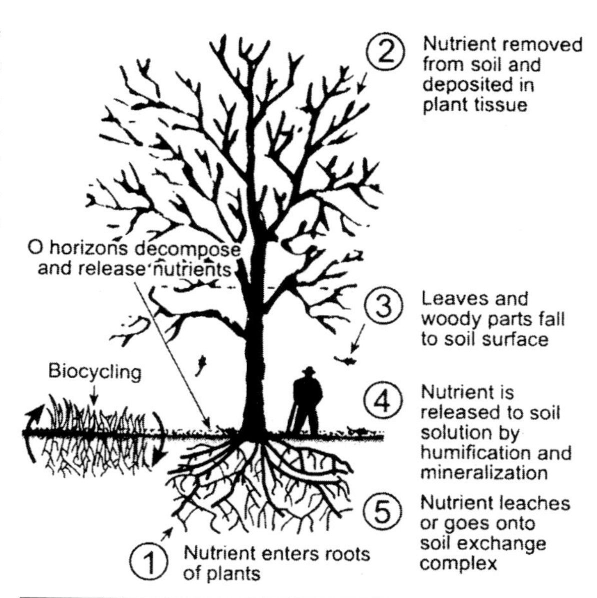

2 Nutrient removed from soil and deposited in plant tissue
O horizons decompose and release nutrients
3 Leaves and woody parts fall to soil surface
Biocycling
4 Nutrient is released to soil solution by humification and mineralization
5 Nutrient leaches or goes onto soil exchange complex
1 Nutrient enters roots of plants

Fig. 12.3 Generalized diagram of some of the processes involved in nutrient biocycling. After Fanning and Fanning (1989).

Acidification and base cycling

Many soil processes involve downward translocation of pedogenic plasma materials. Upward translocation is generally only possible via biocycling, pedoturbation and capillary movement of soluble materials in the soil water. In biocycling, nutrients (although theoretically any ion) are taken up by plants, used for growth and returned to the soil in litter or as roots decay (Bormann *et al.* 1970, Gosz *et al.* 1976, Helmisaari 1995). It is one of the main ways by which downward transfers of soil materials are compensated (Duchafour 1982). A discussion of biocycling is appropriate here, as almost all organic matter in the soil system is cycled; it forms within plants, decomposes and becomes mineralized, and along the way can contribute to melanization (Fig. 12.3).

Nutrients can be returned to the soil in various ways, as throughfall, stemflow, root exudates, litter fall and directly as the plant decomposes, either above ground or below (Eaton *et al.* 1973, Sanborn and Pawluk 1983). Animals biocycle materials as well, passing them back to the soil as feces and when their bodies decompose. In this regard, Francis Hole often referred to things in nature, whether they were trees, people or houses

Table 12.2 | Measured amounts of biocycled Ca, P, Mg and K in some forests in the southern Appalachian Mountains

Nutrient	Amount contributed by litter fall (kg ha^{-1}yr^{-1})	Amount in the Oi horizon (kg ha^{-1}yr^{-1})	Relative turnover rate
Ca	34	46	Low
P	4	5	Medium
Mg	6	3	High
K	13	4	Highest

Source: Sharpe *et al.* (1980).

as either "soil" or "not yet soil." His point was that, in the end, parts of almost everything have come from the soil and will end up as soil. The text below, displayed in front of a display case at the Soils Building on the University of Wisconsin campus, where Hole taught, was probably written by him:

> Structure built by an avian engineer using solid waste (plastic, paper, tin foil); organic debris from vegetation; and mineral soil. The soil was compacted into a platy, stratified deposit which is essentially a series of crusts of reduced hydraulic conductivity. The structure, a segment of a concretion, is formed on a tree branch close to the canopy. It is an epiphytic pedological feature, whose fate is to be translocated by free fall to the soil surface, where it will eventually be incorporated into the soil, except that part which decomposes first.

Displayed in the glass case was a robin's nest.

Biogeochemical cycling includes elemental transfers among not only the soil and biota, but also the atmosphere, hydrosphere and rock below (Duchaufour 1982, Likens *et al.* 1998, Likens 2001). Some of the more common ions that are biocyled include the macronutrients (N, P, K, Mg, S, Ca, H, O, C, Fe) and the micronutrients (Cl, Mn, Zn, B, Cu, Si, Mo, V, Co, Na), as well as others that are more plant-specific (Al, Pb, Ni). Of these, O, C, N and S also occur in the gaseous state and are therefore readily exchanged between soil, plant and air. The most commonly biocycled ions are probably (in order) Ca, K, Mg and P; additionally, N and Mn are strongly biocycled. Grasses and some species of trees (e.g., sugar maple, yellow birch, aspen) are strong base-cyclers while many of the oaks and coniferous tree species are not.

The *rate* of biocycling of bases is also a critical part of the pedogenesis equation. Rates of biocycling are dependent upon several factors: (1) the rate at which bases are taken up by the plants (this is species-dependent but also is a function of availability), (2) the rate at which bases are returned to the soil (deciduous species return more foliage annually than do coniferous trees; grass litter and roots are almost completely turned over annually) and (3) the decomposition or mineralization rates of the litter (low in cold climates, rapid in warm, wet climates, and variable as a function of litter type and C:N ratio). Fire can rapidly release bases otherwise trapped in O horizons, although the bases (in the ashes) are then prone to removal by wind and water. Across a variety of forest types in the southeastern United States, Sharpe *et al.* (1980) determined the amounts of four elements that are biocycled and returned to the soil as litter. Their data indicate that P and K are more rapidly biocycled (higher turnover rate) and that the litter is a large storehouse of bases that biocycle more slowly (Table 12.2). The lower turnover rate for Ca may be due to its immobilization by fungi in the litter (Lawrey 1977, Lousier and Parkinson 1978).

In the low leaching regimes of many grasslands, base cycling of Ca facilitates the formation of stable Ca-humates (Rubilin 1962). These humates bind to mineral grains, forming clay–humus complexes which not only protect the mineral particle from many forms of weathering, but also render the humus fraction resistant to further decomposition by microbes. Clays coated with humus are easily aggregated and become essentially immobilized, inhibiting lessivage (Sanborn and Pawluk 1983).

Most grasses are particularly adept at biocycling silicon, much of which gets deposited as opal within their leafy tissues, giving the leaves a certain amount of rigidity and forming knife-like edges. Certain species of sedges and horsetails have so much silica that Native Americans used them to scrub dishes. The amorphous, opal secretions in plant roots and leaves, usually silt-sized, are called *phytoliths*. They get incorporated into the soil as leaves and stems fall to the surface. Some soils can accumulate thousands of kilograms of phytoliths per hectare (Fanning and Fanning 1989). The shape, size and surface textures of opal phytoliths are reasonably unique to each type of plant and can reside in the soil for long periods of time, especially in acid soils, making them a paleoecological indicator (see Chapter 15).

Plants, especially grasses and certain tree species, are well-known base cyclers. The biocycling of calcium is often cited as a mechanism whereby the pH of udic grassland soils is maintained at high levels, despite continued profile leaching. Because many trees do not readily biocycle bases, and indeed some like hemlock, hickory and beech actually cycle Al (Chenery 1951), bases in many forest soils are more freely removed by percolating water. The loss of bases is one mechanism whereby soils get acidifed. In this instance, *acidification*, an important precursor to many other pedogenic processes, is a direct result of *decalcification* and low amounts of base cycling (Table 12.1, Fig. 12.4). In acidification, the exchange complex of the soil comes to be increasingly occupied by H^+ and Al^{3+} cations, while bases (Ca^{2+}, Mg^{2+}, K^+ and Na^+) get removed. Bases are removed from the profile in two ways: either they are leached from the soil (in percolating water, either vertically or laterally) or plants take them up and the biomass is later removed from the system (by animals that eat the plant matter and defecate elsewhere, by fires that burn the plant matter and then the ash is blown or washed away, etc.). Thus, the underlying cause of acidification is the loss of base cations from the soil; they are usually replaced by hydrogen protons (H^+) from H_2O. *Recarbonation* (Fuller *et al.* 1999) (Table 12.1) occurs when carbonates are added to a leached soil, sometimes as a deep-rooted, base-cycling plant species invades a site that had previously been covered with a non-base-cycling plant.

Organic acids also play a key role in acidification (de Vries and Breeuwsma 1984). Organic acids and H_2CO_3 are weak acids that are able to dissociate protons as a function of pH. Organic acids are most important in acidic soils, while carbonic acid is relatively more important in calcareous soils. Both are capable of donating protons (H^+) to the soil solution.

Leaching and leucinization

Leaching is the primary way that bases and other soluble compounds are removed from the profile. The term leaching is often misapplied; it should be used only for the *complete removal of soluble constituents* from the profile. Colloids are not leached, they are translocated. Ions in solution are not, technically, leached if they are redeposited in the lower solum.

Humid climate soils, where water periodically wets the entire profile, are variously leached. As soil water percolates, soluble constituents are removed and translocated to subprofile locations; many end up in the groundwater. Those that are highly mobile are sometimes quickly removed from the profile. Eventually, a leached zone develops, from which the most soluble ions have been removed. In parent materials rich in carbonates, the leached zone is easily determined; unleached materials below effervesce upon exposure to a weak acid, usually 10% HCl. In most soils on calcareous, Pleistocene-aged deposits, the leached zone is equivalent to the thickness of the solum.

The thickness of the leached zone is a function of several factors. Sites that have more infiltration, perhaps due to a wetter climate or a wetter (run-on) site, are typically more leached than are drier sites. Coarser-textured parent materials are more quickly and deeply leached because they have less surface area to strip of soluble ions and because they usually have more quartz, which is relatively inert (Fig. 12.5). Soils

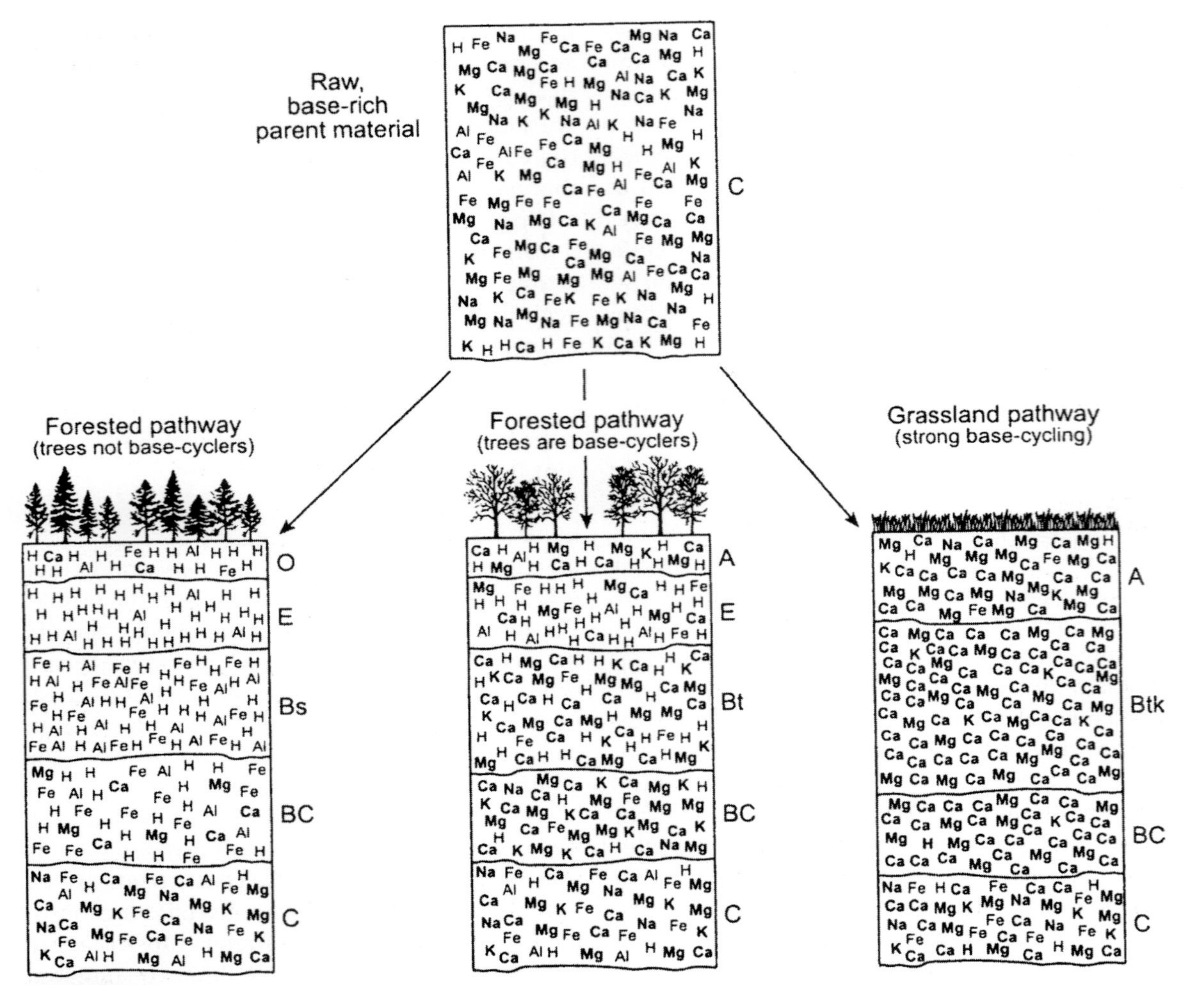

Fig. 12.4 Theoretical pedogenic pathways associated with a base-rich (but probably not calcareous) parent material. Base cations are bolded. The first soil develops under a non-base-cycling, leaching pathway associated with coniferous trees. The second soil develops under base-cycling maple trees, while the third forms in association with strongly base-cycling tallgrass prairie. Note the distributions of cations within and between the soils, as an indication of the degree of acidification and base-cycling.

with less base-cycling are also likely to be more intensively leached. Depth of leaching commonly increases with time, making it a potential relative dating tool (see Chapter 14).

All ions are mobile in soils; the ones that are very slowly mobile, such as Zr and Ti, are considered, for the sake of communication, immobile. Through the use of mass balance pedogenesis studies (see Chapter 14), the relative mobilities of elements within soils have been established. While there are site-to-site and climate-to-climate variations, the *general* mobility sequence for semi-arid climates, based on Busacca and Singer (1989) and Harden (1987), is:

$$Mg \gg Na \geq Ca \geq Fe > Al > Ti > Si > K > Zr.$$

In a humid climate, Bain *et al.* (1993) reported:

$$Mg > Na > Ca > K$$

for bases only.

The degree to which a cation or anion is mobile in soils is largely dependent upon their ability to form stable ions in aqueous solutions (Paton 1978). Whether or not these ions will persist in the soil solution is in turn dependent on their

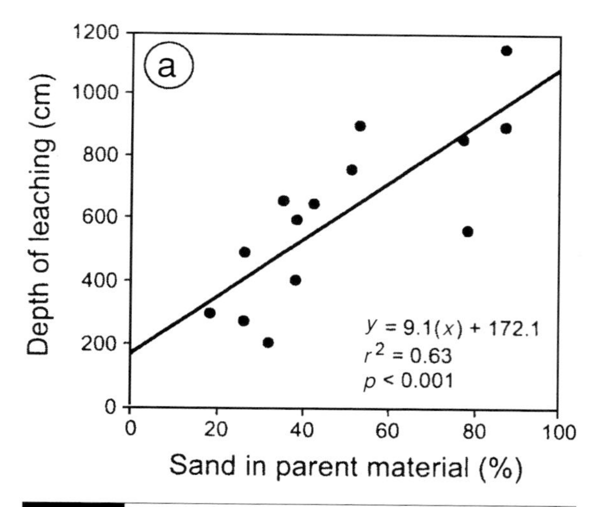

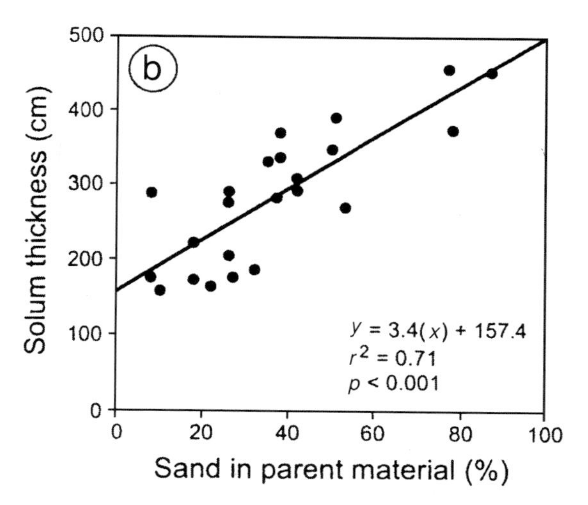

Fig. 12.5 Scatterplots of (a) depth of leaching and (b) solum thickness for the Sangamon Geosol in Indiana. The Sangamon Geosol is strongly developed in this area, having formed over a period in excess of 100 000 years. After Jacobs (1998).

reaction and relation to the hydroxyl anion (OH^-), which involves a complex interplay between ionic size and valency (Fig. 9.6). In the end, it comes down to a balance between geometry and charge, since ionic size determines how many hydroxyl anions can be accommodated around it, while its valency determines the number of positive charges that must be neutralized.

The primary morphologic manifestation of leaching is the formation of light-colored E horizons, a process called *leucinization* (Table 12.1). The light colors are due to the predominance of clean quartz grains and the preferential weathering of dark-colored primary minerals. Some of the most easily weathered primary minerals are dark in color, while quartz and potassium feldspar, both resistant minerals, are light (Fig. 8.13). E horizons tend to be more acidic than lower horizons, often because of loss of bases. Where eluviation is strong, E horizons can form rather quickly, perhaps in as little as a few decades.

Leucinization operates in opposition to melanization. Only when the eluvial zone gets thick enough to outpace melanization, or outdistance it with depth, can a distinct E horizon develop below the A. This process interplay (Fig. 12.6) shows that eluviation alone cannot form E horizons, for in the uppermost profile it is often effectively counteracted by melanization. Therefore, another way that E horizon formation can be enhanced is to minimize the depth of melanization, e.g., in dry, sandy soils beneath acid forest (mor) litter. The soil organism population is low in such soils, and the litter decays so slowly that sometimes O horizons rest directly on E horizons; there is virtually no melanization.

Lessivage

One of the most common processes associated with leucinization is the translocation of clay-sized particles from E to Bt horizons (McKeague and St. Arnaud 1969, Dixit 1978). Translocation of clay-sized particles in suspension is called lessivage (French *lessive*, washing) or argilluviation (L. *argilla*, white clay, *luv*, washed) (Table 12.1). Formation of clay *in situ* is called *argillation*. Although most lessivage occurs from the upper profile to the lower (from A and E horizons to B horizons), the process does occur laterally as well. Most translocated clays are silicate clays; oxide clays tend to occur in environments that are less conducive to lessivage.

Soil components that are able to be translocated, e.g., clay, are collectively called plasma (see Chapter 2). Gravel and sand grains are typically considered skeletal, while silt can be plasmic or skeletal, depending on the circumstance. Translocation of silt grains in suspension is called *pervection*, and is typically a significant process only in cold climates (Table 12.1; see Chapter 10).

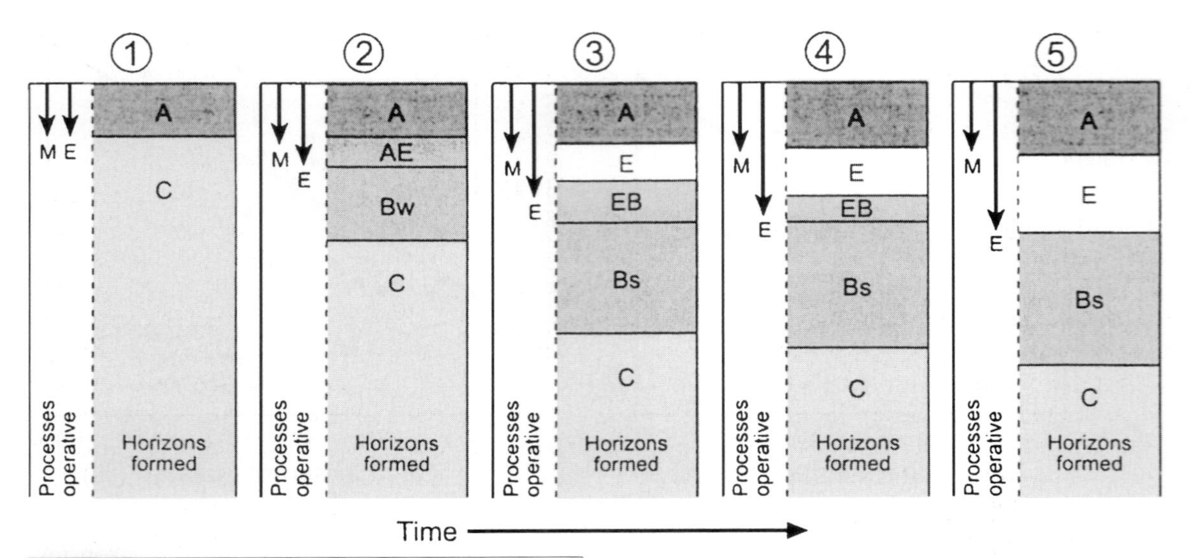

Fig. 12.6 Interplay between melanization (M), a process bundle that darkens soils by the addition of humus, and eluviation/leaching (E), a process bundle that (in this case) strips soil particles of their coatings and makes them lighter. In the example shown, melanization forms dark A horizons while leaching/eluviation promotes leucinization, or the lightening of horizons, thereby forming E horizons.

Associated with, and sometimes mistaken for, lessivage are two companion processes: *decomposition* and *synthesis* of clays (Table 12.1). Clays can be chemically weathered (decomposition) in the upper profile of acidic soils (see Chapter 9). As clays weather, their soluble by-products are translocated, in solution, to the B horizon, where they can reprecipitate to form new minerals (synthesis). Clay neoformation in the lower solum is common for several reasons: (1) pH increases with depth, (2) wetting fronts strip weathering by-products from the upper profile and as they accumulate in the lower profile, supersaturation drives neoformation and (3) loss of water at the wetting front favors precipitation and synthesis, also because of supersaturation. When lessivage is dominant, the clay mineralogy in the E and Bt horizons is usually quite similar, whereas under a decomposition and synthesis regime the clay mineralogy of the E and Bt horizons can be quite different.

Because lessivage is so widespread, terms that refer to illuvial clay and horizons rich in illuvial clay permeate various soil taxonomies. Soil Taxonomy (Soil Survey Staff 1999) defines an *argillic horizon* as a Bt horizon that contains a defined quantity of illuvial clay. B horizons are given a *t* suffix to indicate illuvial clay, for the German word for clay, *der Ton*. Alfisols and Ultisols must, by definition, have an argillic horizon; many Inceptisols and Entisols are developing one. A minority of Aridisols, Spodosols and Andisols have Bt horizons, while many Oxisols and Vertisols probably have since lost their Bt horizons because of weathering and argilliturbation, respectively. In Aridisols, the Bt horizon usually dates back to when the climate was more humid. In Canada, Luvisolic soils have illuvial clay-rich B horizons. The French use the term *Sols Lessives* to refer to similar soils. In Australia, soils with a clay-impoverished horizon above a clay-rich horizon are called *duplex* and *texture-contrast soils* (Gunn 1967, Koppi and Williams 1980, Chittleborough 1992).

Not all profiles with clay-enriched B horizons are due to lessivage or decomposition/synthesis (Chittleborough 1992). Relative clay enrichment in the B horizon can be caused by (1) sand and silt destruction (weathering) in the horizons above, (2) preferential erosion of finer materials from, or additions of coarser materials to, the upper profile and/or (3) comminution of silt and coarse clay to fine clay in the B horizon (Oertel 1968, Smeck *et al.* 1981). Bishop *et al.* (1980) concluded that texture-contrast soils in parts of Australia are due to the slow downslope movement of a clay-impoverished layer above a more sedentary, clay-rich layer, rather than to lessivage. The upper layer is assumed to be a biomantle that has had most of its fines removed by water and wind; its mobility can be confirmed by the presence of a stone line at the contact with the B or Cr horizon (see Chapter 13). This type of genetic interpretation is not without merit on sloping landscapes that have been stable for long periods of time (Paton *et al.* 1995). On younger landscapes, such as those that date to Quaternary glaciations, the more traditional interpretation of vertical clay translocation via lessivage is more plausible (Johnson 2000), and certainly is supported by the development of argillans, glossic tongues and lamellae (see below).

Because Bt horizons can form in more than one way, criteria must be in place to verify that lessivage has been active. The primary criterion used to infer lessivage, either in the past or ongoing, is cutans of illuvial clay (argillans) on ped faces and grain surfaces (Soil Survey Staff 1999), bearing in mind that pedoturbation can destroy them (Nettleton *et al.* 1969). Argillans with multiple layers of oriented clay form on ped faces in loamy or finer-textured soils that maintain relatively stable aggregates. In sandy soils illuvial clay coats individual sand grains and is also manifested as bridges between them (Buol and Hole 1961) (Fig. 2.17e). In thin section, the laminated coatings of argillans on ped faces are readily identifiable, often within pores and at pore–grain contacts, as laminated, birefringent layers. Eventually, argillans can completely fill the inter-ped pores (Fig. 2.21) while most, less well-developed argillans occupy only the edges of pores (Fig. 12.7). Gradual filling of the pores in a Bt horizon by illuvial clay causes it to become an aquitard or even an aquiclude. When viewed macroscopically, argillans usually appear as smooth surfaces on ped faces, and if wet they can have an almost glassy sheen.

Argillan surfaces develop as layer silicates, moving as suspended particles in the soil solution, are deposited in thin layers on the surface of the ped or skeletal grain (Fig. 12.8). The depositional process occurs as the clay–soil water suspension is absorbed into peds when the wetting front slows (Buol and Hole 1961). Silicate clay platelets then get "plastered" onto ped surfaces in ever-thickening layers, forming a smooth, shiny argillan.

As would be expected, the ease with which clay can be translocated is a function of mineralogy and size. Fine clay (<0.2 μm) is much more mobile than is coarse clay (0.2–2.0 μm), regardless of mineralogy (Dixit 1978). Thus, the fine/coarse clay ratio is commonly used to identify whether a horizon has been undergoing lessivage; fine/coarse clay ratios must be higher in argillic than in overlying eluvial horizons (Oertel 1968, Smith and Wilding 1972, Soil Survey Staff 1999). This ratio is typically lowest in E horizons. Smectite tends to be the most readily translocated clay mineral, usually because it is one of the smallest clay particles, while the relatively larger kaolinite and oxide clays are more difficult to translocate (Grossman *et al.* 1964).

Another criterion used to classify a Bt horizon as argillic is an overall increase in clay, relative to the eluvial horizons above. Cremeens and Mokma (1986) developed the I/E index, which reflects the degree of development of Bt horizons:

$$\text{I/E index} = (\text{fine clay}_{\text{Bt}}/\text{total clay}_{\text{Bt}})/(\text{fine clay}_{\text{E}}/\text{total clay}_{\text{E}}).$$

Most Bt horizons form in freely draining, upland soils in humid (but not perhumid) climates in which acidification is ongoing (Cremeens and Mokma 1986) and which undergo frequent wet–dry cycles. Although lessivage is not an important process in arid climates, in every climate there is some rain and with each infiltration event the possibility exists for clay translocation.

The process

Like all eluvial–illuvial couplets, lessivage has three components: mobilization, transport and deposition. First, clay must be dispersed and made to go into suspension. Then, there must be percolating water to transport it, and finally there must exist a means to deposit it. This section is devoted to this three-fold suite of processes.

Clay platelets exist in two states in soil. When acting independently, they are said to be *dispersed*. When stuck together electrochemically

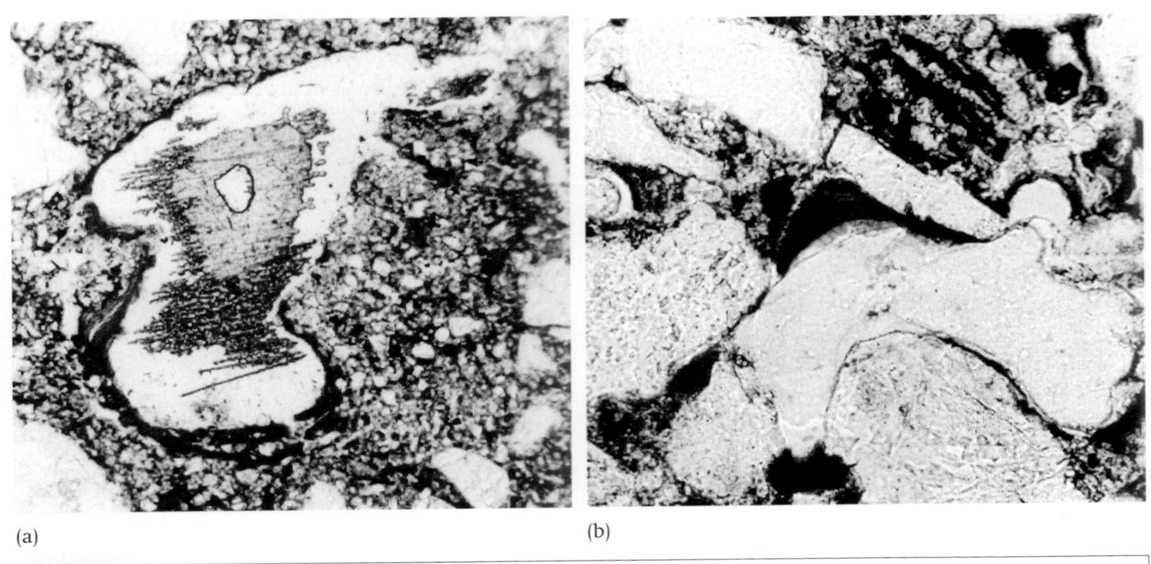

(a) (b)

Fig. 12.7 Thin-section photomicrographs from Alfisols that show argillans. Images by B. N. Weisenborn. (a) Small argillan in a pore. The large, light-colored grains are the quartz sand skeleton. (b) The void formed by a weathered mineral, being filled with illuvial clay.

or physically, they are *flocculated*. Clays that are flocculated behave as larger masses and are much less easily translocated. Groups of flocculated clay minerals are called floccules or microaggregates.

Clay can be dispersed both physically and chemically. If a cementing agent such as sesquioxides or silica is causing the flocculation, that agent must be removed chemically before the clay can be dispersed. Physically, wetting of a dry soil can sometimes cause dispersion and disruption of fabric (McKeague and St. Arnaud 1969, Soil Survey Staff 1999). When heavy rain impacts a dry, clay-rich soil, some platelets are physically extracted from floccules and translocated. Clay that is being translocated rapidly or through large pores can move significant distances as long as

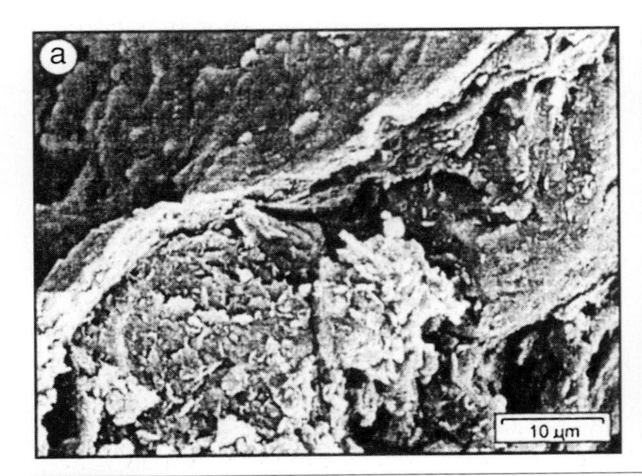

Fig. 12.8 Scanning electron photomicrographs of argillans in Bt and Btx horizons of Alfisols from northern Michigan. Images by B. N. Weisenborn. (a) Argillan from the Bt horizon of a Glossudalf in central Michigan. (b) Argillan on a pore wall in a Btx horizon of a Haplorthod from northern Michigan.

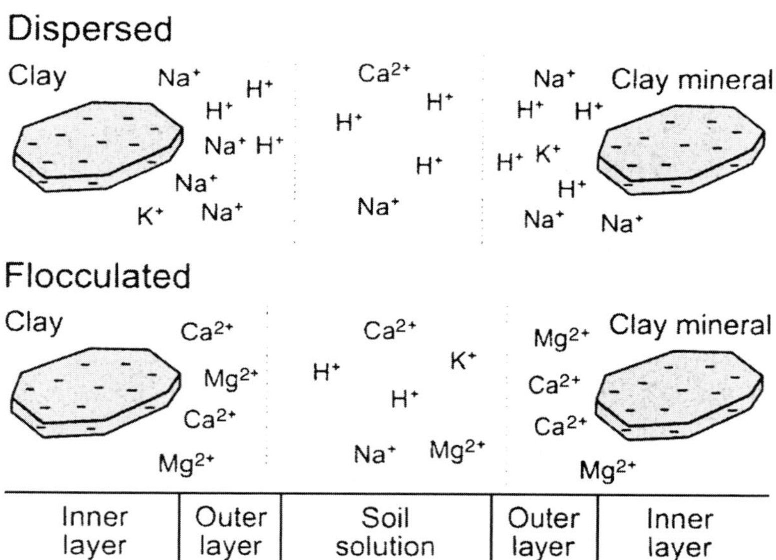

Fig. 12.9 Diagrammatic representation of clay minerals in flocculated and dispersed states. The thickness of the outer and inner layers, and the cations in those layers and in the soil solution, are also shown.

the translocating agent, i.e., percolating water, remains active (Johnson *et al.* 2001). Examples include clay films deep within limestone fissures and in calcareous soils, which are not conducive, chemically, to dispersion (Gile 1970, Goss *et al.* 1973). However, if conditions do not chemically favor dispersion, the clay will soon flocculate and cease moving. Thus, dispersion is most effective chemically, occurring as base cations, primarily, K, Ca and Mg, in the soil solution and on exchange sites are replaced by H (or Na). Removal of any free iron oxides also assists in dispersion and translocation (Lepsch *et al.* 1977b). Lessivage is a weak or almost non-existent process in calcareous materials, because the clay exchange sites in such material remain occupied by Ca and Mg cations. According to Ciolkosz *et al.* (1989), dispersion is favored at base saturation values $<40\%$. Thus, a period of weathering, leaching and acidification, i.e., *decarbonation*, must usually precede lessivage (Allen and Hole 1968, Gile and Hawley 1972, Vidic and Lobnik 1997). In leaching regimes, the upper solum eventually becomes acidic enough and low enough in electrolytes (salts), that clay platelets can be dispersed. Decarbonation is, therefore, an important prerequisite for lessivage.

Dispersion vs. flocculation is driven by the thickness of the outer layer, i.e., the cation swarm, near the negatively charged clay mineral edges. The clay mineral itself is considered the inner layer (Fig. 12.9). In the acidic upper profile, most of the outer layer is occupied by small H^+ cations. Large numbers of these +1 valence cations are required to satisfy the negative charges of the inner layer. Although the entire inner–outer layer system is always electrostatically neutral, the large assemblage of positive charges in the outer layers repel each other, causing it to thicken. In this situation, the clay platelets are unable to get close enough to allow the natural attraction between two neutral objects (van der Waals forces) to take over and flocculate them. Thus, the clay platelets remain separate, or dispersed. Acidic, base-poor conditions favor dispersal because many of the cations in the outer layer are H^+. The upper (eluvial) parts of soil profiles are also favored sites for dispersal because any potentially flocculating cations are often chelated by organic acids (Dixit 1978). In most situations, Na^+ cations also favor dispersal.

Dispersion is also favored in leached zones because of low electrolyte concentrations (Dixit 1978). Electrolytes are salts within the soil solution; their concentration is usually lower in leached horizons. Where electrolytes are concentrated in the soil solution (in base-rich and calcareous horizons), they repel the cations in the outer layer, making it thinner and favoring flocculation. Thus, the more water that can be flushed through a soil, the more likely clays within are to disperse, and consequently, to be translocated.

In base-rich or unleached materials, flocculation is favored. In such materials the outer layer, while still electrostatically neutral, is thinner because it is populated with fewer cations that have +2 and +3 valency, like Ca and Mg (Fig. 12.9). The outer layers do not repel each other as strongly, allowing van der Waals forces to operate and cause flocculation. In low concentrations, organic compounds are effective agents of dispersion, although at high concentrations they can promote flocculation (Visser and Caillier 1988). Flocculation of this sort can repeat itself until a small stack of clay platelets, called a *clay domain*, forms.

Lessivage requires percolating water as a vector of transport, and generally, the more rapid the percolating water the more capable it is of translocating clay. Rapidly percolating water is facilitated in two ways: (1) large, highly interconnected and vertically aligned pores, and (2) intense rainfall or rapid snowmelt. If these two mechanisms operate simultaneously, clay can even be translocated through calcareous material (Goss *et al.* 1973).

There are three general mechanisms that can lead to the deposition of clay in Bt horizons: (1) cessation of the wetting front due to (a) desiccation, (b) loss of energy (as at a water table), (c) the presence of a lithologic discontinuity or an aquitard/aquiclude, (2) flocculation within a base-rich material and/or (3) deposition by filtering (pores get too small). Clay in suspension must stop when and where the wetting front stops and water is pulled into peds by matric forces. Many Bt horizons are best developed, therefore, at the modal depth of wetting front penetration. In perhumid climates where soils are constantly wet, such as in parts of southeastern Alaska, the Olympic Peninsula of Washington, and the British highlands, where soils are continually moist, Bt horizons are absent (Soil Survey Staff 1999); there is no mechanism to get migrating clay to stop in a preferred zone. On the other hand, soils with a seasonal moisture deficit are dry at depth but wet up after rainfall events. There are three reasons why soils that occasionally dry out at depth often develop Bt horizons: (1) wetting a dry soil favors dispersion, (2) dry soils have larger pores that facilitate translocation of the clay–soil water suspension and/or (3) the cessation of the wetting front, and with it the capillary withdrawal of water into peds and clay *onto* them, is favored when the soil dries. As water penetrates these soils, clay (most of which is carried near the wetting front) gets deposited at the depth of wetting front penetration. Clay is also deposited onto peds at the top of a water table, because the percolating water loses its energy there and stops. However, Bt horizons are usually better expressed in well-drained soils than in soils with high water tables (Cremeens and Mokma 1986), perhaps because the wetness inhibits weathering or because water table fluctuations cause illuviation to occur at varying depths, diluting the illuvial materials. Also, the likelihood of the dry soil getting wetted by an infiltration event, a process which favors lessivage, is more common in upland soils and less common in soils with high water tables.

The most common reason why clay is deposited in a preferred zone is because of an increase in base cations; often this coincides with the top of a calcareous layer (Fig. 8.48). Most Bt horizons on young landscapes where the parent materials are calcareous are associated with the leached–calcareous boundary. Increased electrolyte concentrations within the calcareous materials also assist in flocculation. When that depth is also the modal depth of wetting front penetration, the situation exists for depositing a large amount of clay within a thin depth increment, and forming a Bt horizon very quickly. For soils on calcareous parent materials, however, formation of the Bt horizon is aided not only because the carbonates help flocculate illuvial clay in the lower solum, but also because carbonate dissolution in the upper solum is a *source* of clay. Much soil carbonate is found in association with clay minerals, and thus as the carbonates are leached, clay is free to be translocated, and typically will move to the top of a calcareous zone, or deeper if the percolating water is moving rapidly. It should now be obvious why Bt horizons tend to form at the leached–unleached (acidic–base-rich)

Fig. 12.10 A soil from Michigan with a pronounced concentration of illuvial clay and sesquioxides at a sand-over-gravel lithologic discontinuity. Units on tape are 10 cm. Photo by RJS.

contact in soils forming on calcareous parent materials (Fig. 8.48).

Wetting fronts also stop because matric forces cause water to hang at a fine-over-coarse contact, such as a loess over gravel lithologic discontinuity (Fig. 12.10; see Chapter 8). If this discontinuity occurs below the solum, any illuvial clay there may develop into a second clay maximum for the profile. Such horizons are called *beta horizons* because they sometimes occur below a similar (alpha) illuvial horizon. Beta Bt horizons tend to be composed of finer clay than their alpha equivalents. They are extremely common in soils that have lithologic discontinuities (Bartelli and Odell 1960a, b, Ranney and Beatty 1969), shallow bedrock or aquicludes (Schaetzl 1992a). The deposition of the clay within the upper (finer) material enhances the textural contrast across the discontinuity, initiating a positive feedback that will facilitate the development of the Bt horizon. In some soils, the beta horizon corresponds to the modal depth of wetting fronts associated with the wet season ("winter rainfall"), while the upper (alpha) Bt horizon corresponds to dry-season wetting fronts.

Clay-rich and/or thick Bt horizons can become aquitards. This thickening and fining of the Bt horizon is facilitated as pores get plugged with illuvial clay, creating a filter effect which might explain why so many Bt horizons have their maximum clay contents near the top of the horizon, with gradually diminishing contents below. More and more clay is captured by the Bt horizon, and the positive-feedback loop quickly leads to the formation of a nearly impermeable Bt horizon. Water may begin to perch on the Bt, leading to ferrolysis in the horizons above (Table 12.1). The main reasons that some Bt horizons do not become aquicludes center on pedoturbation and the formation of strong structure; water can usually percolate between peds.

Argillans are "incontrovertible evidence" of clay migration (Brewer 1976, Bronger 1991, Chittleborough 1992). The lack of argillans in soils with Bt horizons, however, should not be taken as evidence of lack of lessivage (Oertel 1961, Nettleton *et al.* 1969, Chittleborough *et al.* 1984). Many argillans are destroyed by pedoturbation, especially argilliturbation. Because most argillans are enriched in fine clay, and fine clay has more smectite than other clay fractions, one would expect shrink–swell processes to be operative in horizons with thick argillans. For this reason, some soils retain the strongest evidence of illuviation, as argillans, in the lower, not upper, Bt horizon (Buol and Hole 1961).

Argillic horizons can be overprinted with other illuvial materials as pedogenic pathways change. This is particularly common in soils on the humid–arid climate boundary. Here, small changes in climate can cause a shift from lessivage to calcification. Cutans can then provide

Fig. 12.11 Lamellae in a sandy soil in Michigan. The clay in the lamellae provides additional cohesion, allowing them to stand out as positive features. Dry sand from above has fallen onto each protruding lamella. Shovel for scale. Photo by RJS.

a record of such climatic changes (see Chapter 15).

LAMELLAE

In sandy soils, Bt horizons are often in the form of thin (<5 mm thick) layers of clay, associated with Fe oxides, called *lamellae* (singular: lamella). Rather than being deposited throughout the horizon, as occurs in finer-textured soils, lamellae are thin, sheet-like zones of preferred illuviation in sand (Thorp *et al.* 1959). They are often contorted, wavy and discontinuous (Fig. 12.11). Most lamellae are more clay- and Fe-rich than the surrounding matrix, and are therefore redder in color (Schaetzl 1992b, Rawling 2000). When viewed in thin section, the clay within lamellae usually forms bridges between sand grains. Lamellae are often located in the lower solum, usually overlain by an A–E–Bs, A–E–Bw, A–Bw or comparable sequum. In sandy landscapes the presence of lamellae at depth is ecologically important, as these thin bands dramatically increase the water-holding capacity of the soil and, hence, site productivity (Host *et al.* 1988, McFadden *et al.* 1994). Although the clay in lamellae has, presumably, come from the entire profile above it, for the sake of communication lamellae are called Bt horizons and the areas between are referred to as E horizons. Since it would be impractical to describe each lamella individually, zones of lamellae are grouped and named E&Bt or Bt&E horizons, depending on whether the clean sand (inter-lamellae areas) or the bands are dominant, respectively (Soil Survey Division Staff 1993). If sufficiently thick, the lamellae-rich horizon may qualify as argillic (Soil Survey Staff 1999).

The formation of lamellae is understood at one level, but because they can apparently form in different ways, detailed knowledge of the pedogenic processes involved in their formation is lacking (Rawling 2000). Pedogenic lamellae must first be differentiated from geologic lamellae. Many clay bands that look like lamellae, especially the thicker ones, are probably geogenic in origin (Hannah and Zahner 1970), conforming to geologic bedding planes (Robinson and Rich 1960, Liegh 1998, Schaetzl 2001). Many lamellae of this ilk occur so deep that they could not possibly be pedogenic. Pedogenic lamellae are usually thinner and cross geologic bedding planes (Schaetzl 2001). They also are found only within the lower solum, and usually have clear, sharp upper boundaries and ragged, diffuse lower boundaries.

Pedogenic lamellae form as percolating water, carrying very small amounts of clay in suspension, deposits clay in thin bands (Gile 1979, Bond 1986, Kemp and McIntosh 1989). The edges of wetting fronts in sandy soils are often contorted, and if the percolating water was carrying most of the clay near the edge of the wetting front, cessation of the wetting front will lead to a contorted zone of deposition that resembles an incipient lamella. Once deposited, this zone may act as a filter for future wetting fronts, stripping them of some clay and thickening as it does so. A related question centers on why the wetting fronts stop where they do. Since most sand has varying degrees of bedding and stratification, it has been assumed that lamellae begin forming at these subtle pore size discontinuities (Van Reeuwijk and de Villiers 1985, Bouabid *et al.* 1992, Soil Survey Staff 1999). Nonetheless, many lamellae are much too contorted to be aligned entirely with pore size discontinuities, and may simply reflect random wetting front locations.

It is also possible that clay illuviation occurs while the wetting front is moving, and not at its static margin. The explanation for this is as follows. As the wetting front moves through the soil,

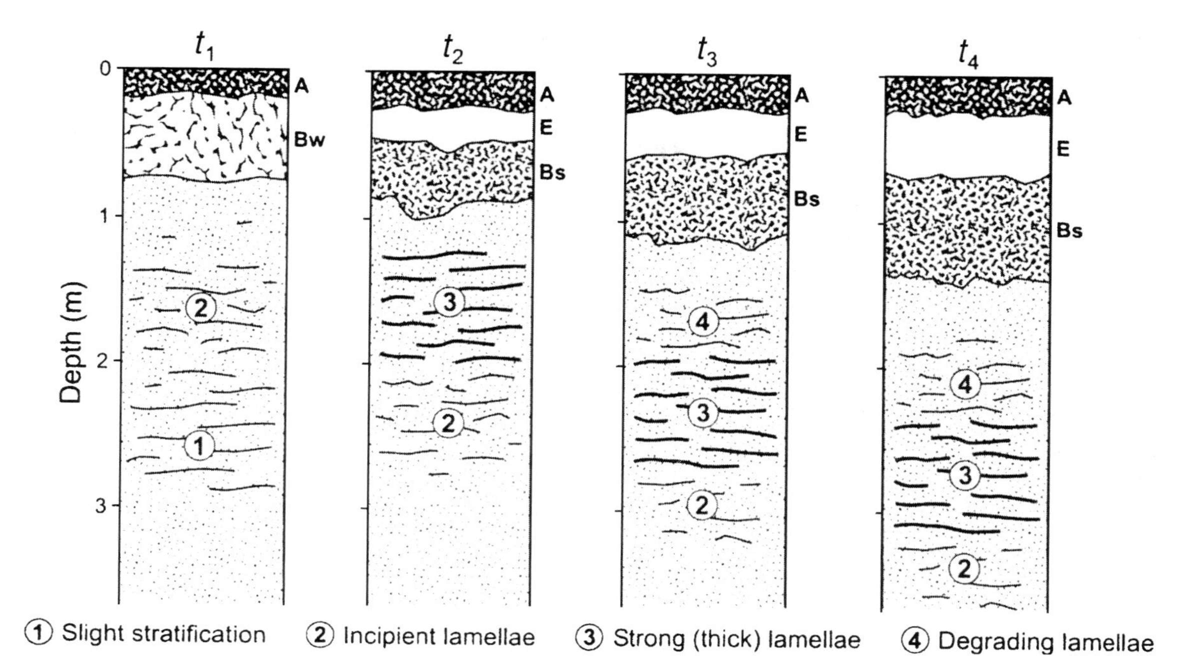

Fig. 12.12 Schematic diagram showing how a sandy soil in a udic soil moisture regime and frigid soil temperature regime might develop over time. Emphasis is on how clay lamellae develop and "migrate" through the profile.

it continually picks up colloids. At some point, a maximum colloidal concentration threshold is crossed, forcing clay to be deposited even as the wetting front continues to move (Bond 1986). Flocculation, leading to the formation of a clay band, can also occur at a zone where pH or free iron oxide content increases (Miles and Franzmeier 1981, Berg 1984).

Small amounts of clay are easily and quickly translocated in sands, enabling lamellae to form quickly (Berg 1984). Most researchers assume that lamellae increase in thickness and clay content with time, although they can degrade as well (Fig. 12.12). Lamellae are assumed to form initially in the upper solum but as the soil gets increasingly acidic, they degrade and the clay within them is translocated deeper (Berg 1984). Along these lines, we paraphrase the Soil Survey Staff's view of lamellae genesis (1999: 82–83):

> Lamellae nearer the soil surface have the least concentration of clay and the faintest color contrast from the overlying E horizon. These may be degrading. They are generally more wavy and discontinuous than those below. Lamellae in the mid-solum appear to have the highest concentration of clay near the top of each lamella. In the lower part of these lamellae, some sand grains are devoid of clay and some have only thin clay coatings, similar to lamellae above. The deepest lamellae are commonly very thin and not very wavy. Parallel to each other, they may be, in fact, sedimentary strata or bedding planes.
>
> Clay is episodically moved from upper to lower lamellae. Clay stripped from the lower part of one lamella is redeposited in the top of the next lower lamella; lamellae thicken as clay is added to their tops. Loss of clay from its base, coupled with additions of clay to the top of a lamella, cause it to move away from any bedding plane from which it may have originated. The stripping of clay from the lower part of a lamella and redeposition in the top of the next lower lamella continues this upward movement. Because the movement of each lamella upward is not uniform throughout its extent, lamellae are wavy rather than smooth. Branching in lamellae occurs where a part of a lamella has moved up more rapidly than the overlying part of the next higher lamella and they become joined in this part.

The above view of lamellae genesis is only one possible scenario. Schaetzl (2001) has shown that lamellae can form when thin, geogenic clay bands at bedding planes are moved *deeper* by

(a) (b)

Fig. 12.13 Glossic horizons in a Michigan Glossudalf. Photos by RJS. The scales of the photos are not exactly the same.

percolating water. Until we know more about the genesis of lamellae, it is best to assume that they represent a good example of pedogenic equifinality (Rawling 2000) – the same *form* may have developed in different *ways*.

DEGRADATION OF BT, BX AND BTN HORIZONS
In the early stages of soil development, argillic horizons thicken, redden and get more clay-rich with time, i.e., they continue to develop along one pedogenic pathway. In warm (thermic), humid environments, the Bt horizon clay mineralogy can gradually become dominated by low-activity clays, even as the clay content of the Bt horizon increases. When that happens, the horizon is called a kandic horizon (Soil Survey Staff 1999). Across the midlatitudes, most Bt horizons develop to a point and then begin to degrade, as the upper solum gets increasingly acidic and leached (Cady and Daniels 1968). The increased acidity, often in conjunction with ferrolysis (see below) causes the clay minerals, especially 2:1 clays, in the upper Bt horizon to become unstable. The unstable clays and their weathering byproducts get translocated to the lower profile (DeConinck *et al.* 1968). Here, they may be more stable or they may neoform into other clay minerals. This type of Bt horizon degradation can be viewed as a localized change in pedogenic pathway from lessivage to clay destruction, since some of the clay that is lost from the upper Bt is destroyed by chemical weathering. Alternatively, clay from the degrading Bt horizon may simply be remobilizing and moving in suspension to the lower solum, where pH and base saturation conditions are more favorable for flocculation and deposition (Ranney and Beatty 1969, Bullock *et al.* 1974). Under this scenario, the soil might be seen as continuing its pre-existing pedogenic pathway, through which E horizons thicken and Bt horizons grow downward in clay-rich sola via lessivage.

Degradation of Bt horizons is similar to the degradation of fragipans and natric horizons (see below). All of these horizons occasionally function as aquitards, forcing perched water to flow across the top of the horizon and percolate along preferred pathways (Langohr and Vermeire 1982, Ransom *et al.* 1987b). Chartres (1987) noted that degradation features are most common in soils that have a marked textural contrast between the E and B horizons, and which have slowly permeable B horizons. Lindbo *et al.* (2000) observed that Fe–Mn nodules, indicative of perched water, are common in the E horizons of soils with degraded Bt horizons. Because perched water is probably involved in their degradation, processes related to

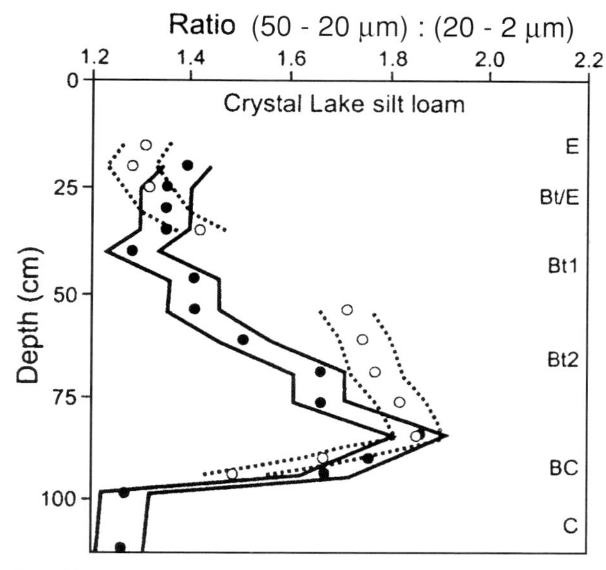

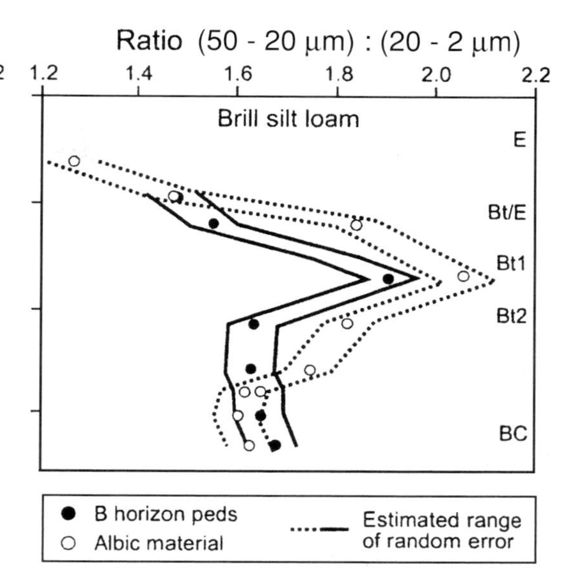

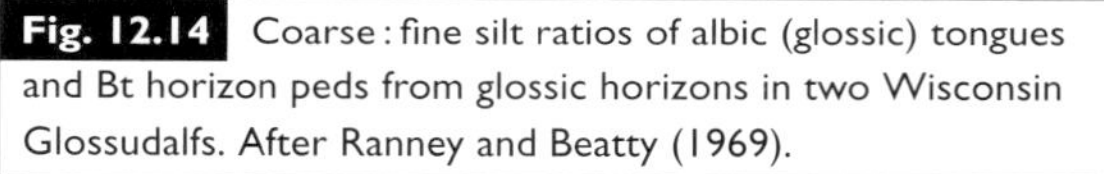
Fig. 12.14 Coarse : fine silt ratios of albic (glossic) tongues and Bt horizon peds from glossic horizons in two Wisconsin Glossudalfs. After Ranney and Beatty (1969).

ferrolysis (Brinkman 1970, 1977b) and oxidation–reduction (Lindbo *et al.* 2000) must be considered.

Bt, Btn and Bx horizons degrade from the top down, as percolating water strips away bases and clay, leaving behind light-colored, acid, siliceous, skeletal material composed mainly of quartz, muscovite and free silica on ped faces (Bullock *et al.* 1974, Chartres 1987, Payton 1993b). Daniels *et al.* (1968) referred to these areas as eluvial bodies. These albic bodies, which are almost pure sand and silt near the top of the B horizon but become more clay-rich with depth, have also been referred to as grainy gray ped coatings, silans, skeletans, albans and silica powder (Figs. 12.13, 12.14). Even if clay minerals are present in these eluvial bodies, they are usually coarse, low-CEC clays (kaolinite and Al-chlorite). With time, the thickness of the albic zone increases, as the degradational processes continue farther into the ped centers.

The Bt horizon degrades initially at the top and along permeable conduits, i.e., ped faces and pores (Bouma and Schuylenborgh 1969) (Fig. 12.15). Roots also preferentially locate at the sites; their secretions help to degrade the clays by acidifying these areas while their uptake of oxygen creates reducing conditions. Percolating, acidic water moves preferentially along these ped faces (prism faces in fragipans), making chemical weathering most intense there, and increasing the likelihood of clay translocation. Perched water on top of the B horizon helps to force translocation and degradation (Bullock *et al.* 1974, Payton 1993b), since reducing conditions can quickly be established and any iron-rich materials can be reduced and translocated as Fe^{2+} (see below). Indeed, many of the whitish areas on ped faces are simply redox depletions, from which Fe and Mn has been reduced and/or removed (Bouma and Schuylenborgh 1969). Thus, the degrading conduit areas lose Fe and gain Si (Payton 1993b). Reducing conditions allow Fe and Mn to migrate from the conduit (tongues) to ped interiors, where the oxidizing conditions facilitate precipitation as Fe–Mn nodules or neocutans (Bouma and Schuylenborgh 1969), Mn near the edge and Fe farther inside. With time, even these nodules and cutans are "not safe," as degradation continues.

Normally, a degrading Bt horizon has an irregular upper boundary that is marked by narrow to broad penetrations (tongues, fingers) of E (albic) material into the Bt horizon (Jha and Cline 1963, Langohr and Vermeire 1982) (Figs. 12.13, 12.15). They get finer and occur at wider spacings with depth. This tonguing, degrading horizon, designated E/Bt or Bt/E, is called a *glossic horizon*

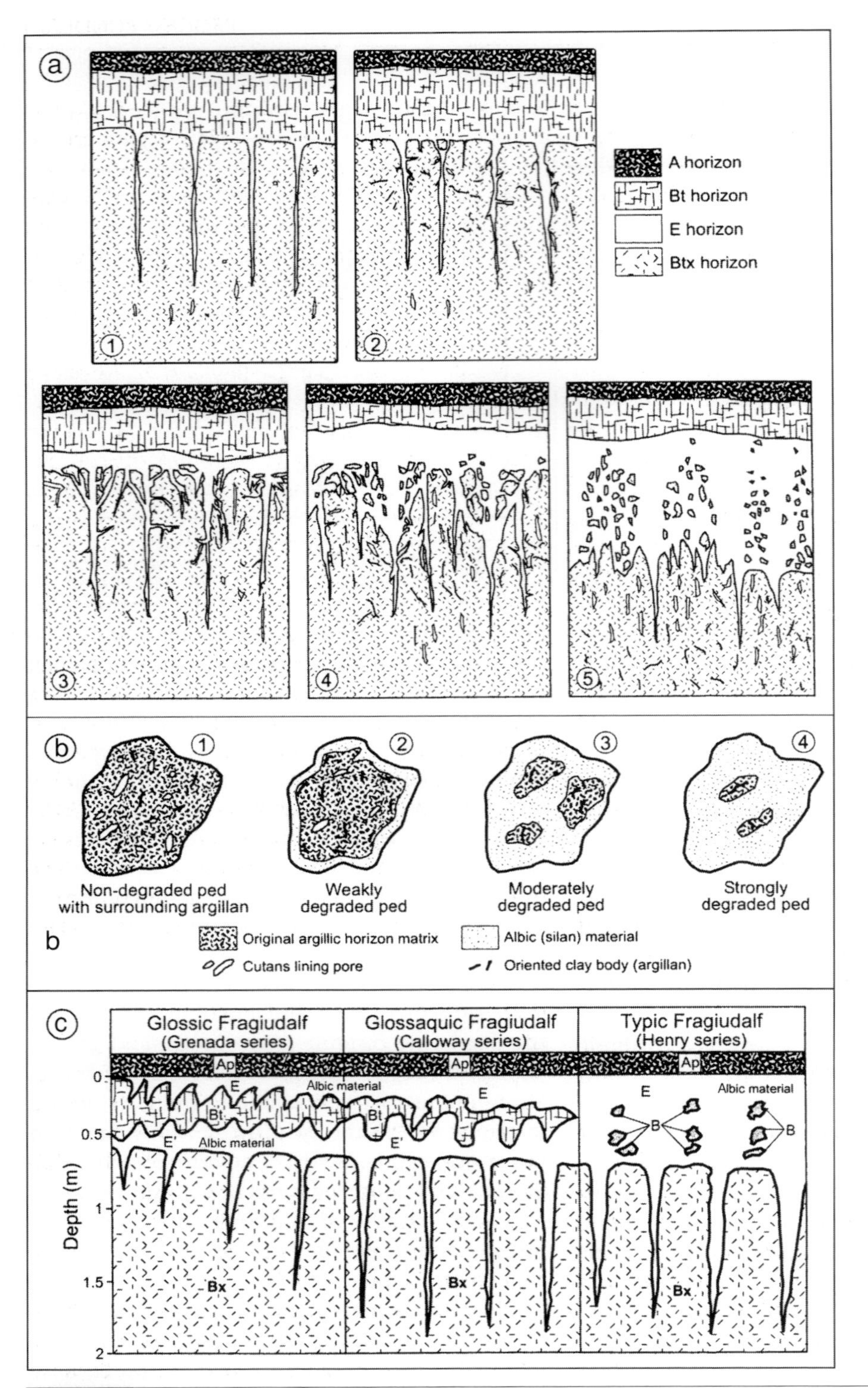

Fig. 12.15 Schematic representations of the degradation Bt and Btx horizons. Although each study examined different types of degradation (Bt vs. Btx), many of the same principles apply. (a) Profile-scale degradation of a fragipan (Btx horizon). After Lindbo *et al.* (2000). (b) Degradation of a ped from a Bt horizon. After Bullock *et al.* (1974). (c) A development/degradation sequence of fragipan soils developed in loess in the lower Mississippi Valley. After Bartelli (1973).

in Soil Taxonomy (Soil Survey Staff 1999). Glossic tongues form first as thin, whitish coats composed of quartz-rich, silt and fine sand grains that are too large to translocate but which have been stripped of Fe and clay by the percolating, often anoxic and acidic, soil water. Organic acids and aluminum cations may assist in the weathering and translocation of these materials, as the tongues appear to be in the most acidic parts of the horizon (McKeague *et al.* 1967, Duchaufour 1982). As the peds degrade, they become like a "house of cards," retaining their overall structure because their skeletal grains remain in place, but the glue that holds them together, i.e., the clay and Fe oxides, is slowly removed. As the clays and cementing agents degrade, silica and other cations are released; these may precipitate to form allophane or other amorphous materials, which can act as a binding agent for the albic materials and partially offset the loss of Fe and clay. Peds so cemented tend to retain this original fabric but, when dry, will shatter easily under only pressure. Near the end of this process, small, Fe- and Mn-rich nodules and remnants of the B horizon exist amidst a sea of albic material. The argillans and small, nodular remnants of the argillic horizon remain as isolated peds in the lower part of the E horizon, at sites that were farthest from the original albic tongues. In advanced degradational stages, all argillans and Bt horizon remnants will have been destroyed.

While the mechanisms of B horizon degradation are somewhat understood, the causative agents for glossic horizon formation have intrigued scientists for years. What are the external drivers, if any, of this process? Cline (1949b) assumed that this process was part of a natural pedogenic progression associated with long-term base depletion and acidification. Whether the acidification was reinforced by a vegetation change from base-cycling deciduous forest vegetation to mixed forest or conifers, with their acidic litter, or not, is debatable. Bartelli (1973) essentially stated that increased water flow through soils, as occurs in wetter climates and at lower slope positions, leads to increased ferrolysis – the driving force behind Bt horizon degradation. In Europe, a theory emerged that the degradation was human-induced, by land clearing, deforestation and the loss of bases from the system that accompanied it (Tavernier and Louis 1971). Langohr and Vermiere (1982) debunked this theory, however. They suggested that Bt horizons degrade when a permafrost-derived, dense layer exists at depth; soils without such a layer tend to not develop glossic features.

As Bt horizons degrade and E horizons thicken, the acid, sandy eluvial zone formed can become parent material to another pedogenic process (podzolization) and therefore, the soil can develop a second sequum. Bisequal soils (Gardner and Whiteside 1952, Allen and Whiteside 1954, DeConinck *et al.* 1968, Schaetzl 1996) typically have a E–Bs sequum that is developing in the E part of the thicker, older E–Bt sequum. The Bs horizon that forms is reflective of podzolization, which is best expressed in coarser-textured, acidic materials, essentially making lessivage a preconditioning agent for podzolization in fine-textured soils. In coarse-textured soils in certain settings, however, lessivage can operate simultaneously with podzolization (Guillet *et al.* 1975). Nonetheless, in finer-textured soils there usually exists a period of lessivage followed by podzolization after the eluvial part of the profile has gotten sufficiently acidic and coarse-textured due to loss of clay.

Fragipans

Characteristics

A fragipan (Latin *fragilis*, brittle) is a genetic, dense and usually brittle subsurface horizon (Soil Survey Staff 1999). It is a diagnostic horizon in Soil Taxonomy (Soil Survey Staff 1999). The high bulk density and non-connective porosity make fragipans water- and root-restricting; most are aquitards. Most fragipans are in the lower solum, sometimes below a spodic upper sequum (Fig. 3.5); they are commonly associated with Bt horizons. Fragipans have low organic matter contents, relative to the horizons above them. Because other types of horizons can also be dense, root-restricting and low in organic matter, the tendency for moist fragipan peds to be *brittle* is commonly used as a distinguishing characteristic.

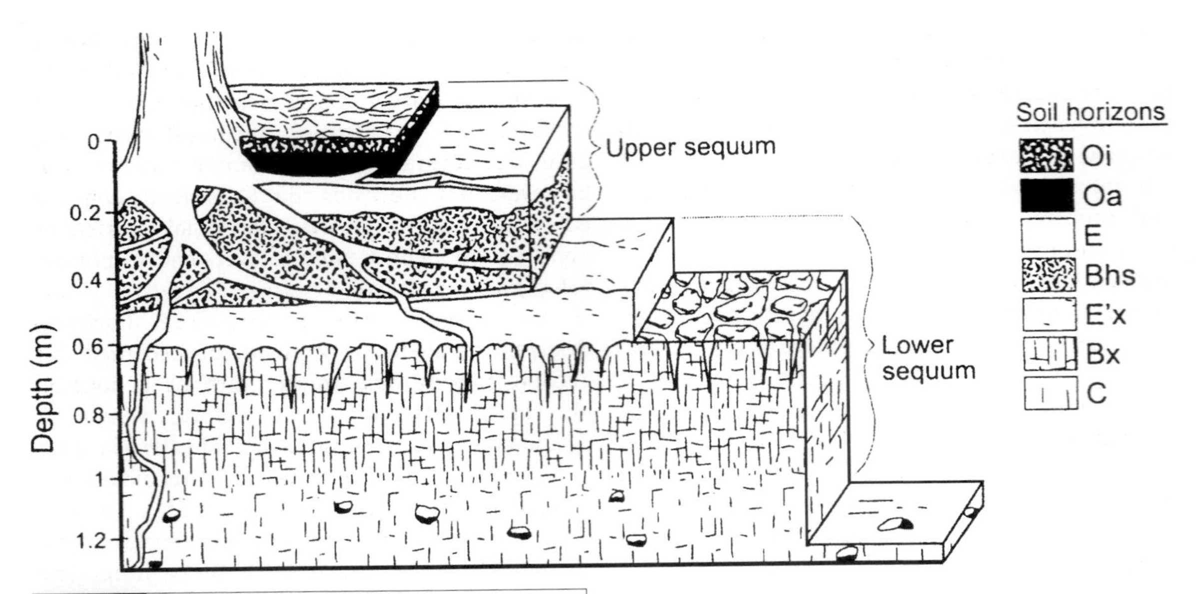

Fig. 12.16 Typical expression of a bisequal soil, probably an Oxyaquic Fragiorthod, with a fragipan in the lower sequum. After Hole (1976).

Brittleness is defined as the tendency for a ped to rupture suddenly when pressure is applied, as opposed to the slow, plastic deformation that many other peds undergo. Similarly, air-dry fragments of fragipans tend to slake when placed in water; slaking can best be described as a slow "crumbling like a pile of table sugar." Fragipans are typically loamy or silty in texture, and usually form in udic soil moisture regimes, beneath forest vegetation. Fragipan horizons are given an x suffix: Bx, Btx, Btgx, Ex and Egx. Many fragipans perch water at some time of the year and exhibit redoximorphic features. Many also show evidence of translocated clay. The top of the fragipan is usually quite abrupt, whereas the lower boundary is gradual or diffuse. Well-developed fragipans have coarse prismatic structure. The exteriors of these large (5–50 cm dia.) prisms typically are light-colored and sandier than are the interiors, and form a polygonal pattern in plan view (Fig. 12.16). Because they are the most permeable parts of the pan, they have been stripped of clay and iron oxides by percolating water. These Ex (albic) seams are also the primary locations where roots are found in the pan proper.

Genesis and evolution

There have been several major reviews of fragipans and their genesis (Grossman and Carlisle 1969, Smalley and Davin 1982, Smeck and Ciolkosz 1989, Ciolkosz *et al.* 1995) and dozens of empirical studies of fragipans. The result of all this work can be summed as follows: (1) fragipan genesis is still poorly understood, and (2) Bx horizons of similar morphology may form in different ways, i.e., *equifinality* may be involved (Buntley *et al.* 1977). We examine fragipan genesis by highlighting the main points of concern and processes that may be involved, and end with a model that incorporates much of the current genetic thinking about fragipans.

Two main issues must be addressed to understand fragipan and fragic property genesis:

(1) How does the pan become so dense? Most studies attribute the high bulk density of fragipans to dense packing of individual grains and inter-grain bridging by various compounds. Many fragipans have high concentrations of silt and very fine sand – particle sizes that are inherently prone to close packing. This characteristic has taken on a "brick and mortar" analogy, where sand and silt particles are the brick, and clay or some other cementing agent is the mortar (Knox 1957, Hutcheson and Bailey 1964).

(2) What is the cause of brittleness and cementation in the pan? Most studies attribute the brittleness to pedogenesis, particularly to illuviated cementation agents. Any of a number may be invoked: aluminosilicates, amorphous compounds, hydrous oxides and gels of Si and Al, clay minerals and sesquioxides. These materials may occur in the horizon as coatings/cutans, void infillings or grain-to-grain bridges.

FORMATION

Because they parallel the soil surface and display varying degrees of expression as a function of topography and time, it is assumed that fragipans are pedogenic features. One school of thought regarding their genesis ascribes their dense nature to *relic processes*, either geogenic or pedogenic. That is, if the parent material was dense, all that is needed to form the pan is a cementing agent, and maybe not even that if the original cementation was strong enough. Antoine (1970) was perhaps the first to suggest that the dense nature of fragipans may be inherited from dense basal till. His finding was supported by subsequent work (Lyford and Troedsson 1973) and by observations of fragipans above Cd horizons. Obviously, this hypothesis does not apply to fragipans formed in loess. In other studies, the coincidence between a buried paleosol or weathering zone and the fragipan horizon was strong enough to suggest that the pan is actually a polygenetic feature, but with many of its properties inherited from a pre-existing period of pedogenesis (Buntley *et al.* 1977, Bruce 1996).

Other scholars have suggested that the fragipan (1) is relict from a period when the soil had permafrost, (2) developed many of its characteristics while under the influence of permafrost or (3) has formed due to contemporary freeze–thaw processes (FitzPatrick 1956, Collins and O'Dubhain 1980, Van Vliet and Langhor 1981, Bridges and Bull 1983, Habecker *et al.* 1990, Payton 1992, 1993a). This family of hypotheses focusses on several key morphologic properties: (1) the depth and abruptness of the upper boundary of the fragipan and its correlation to the paleo-permafrost table, (2) its polygonal/prismatic structure, which is remarkably similar, though on a smaller scale, to patterns developed by ice segregation features, (3) the subhorizontal platy structures of fragipans which resemble the platy structure formed when ice lenses form in soils (Van Vliet and Langohr 1981). Other morphological features of fragipans, such as silt concentrations or caps on peds and stones, clay flows, platy structure and vesicular porosity, are also explainable by invoking frost activity. Freezing conditions *can* create dense fabric as the freezing front advances downward toward a permafrost layer below (see Chapter 10).

Instead of the weight of a glacier or the pressures exerted by freezing water, others suggested that dense fabric can result simply from the self-weight collapse of a sediment that has been appropriately preconditioned. The main proponent of this theory is Ray Bryant (1989). He suggested that certain types of sediment can undergo physical ripening and/or self-weight collapse, creating a dense condition that may persist. Others had suggested that collapse of this sort may occur after a period of weathering (Yassoglou and Whiteside 1960, Nettleton *et al.* 1968), but Bryant was the first to propose that this process could occur in the early stages of soil development. Desiccation, alone, may be enough to initiate a type of collapse or densification (Parfitt *et al.* 1984).

Many fragipans occur at a lithologic discontinuity, prompting study of how the interruption of percolating water that occurs there might be related to their genesis (Harlan and Franzmeier 1977, Habecker *et al.* 1990). Of particular importance is the possibility that percolating water might deposit dissolved or suspended substances at the discontinuity, and that these might act as cementing agents (Karathanasis 1987b, Smeck *et al.* 1989). Roots that proliferate at the discontinuity might cause that area to experience more wet–dry cycles, and may enhance precipitation of solutes by absorbing soil water. Some of the discontinuities also coincide with the top of a buried paleosol.

A primary genetic focus, one that is necessary to resolve before fragipan genesis can be fully understood, centers on the nature of the brittleness and any cementing agents that may

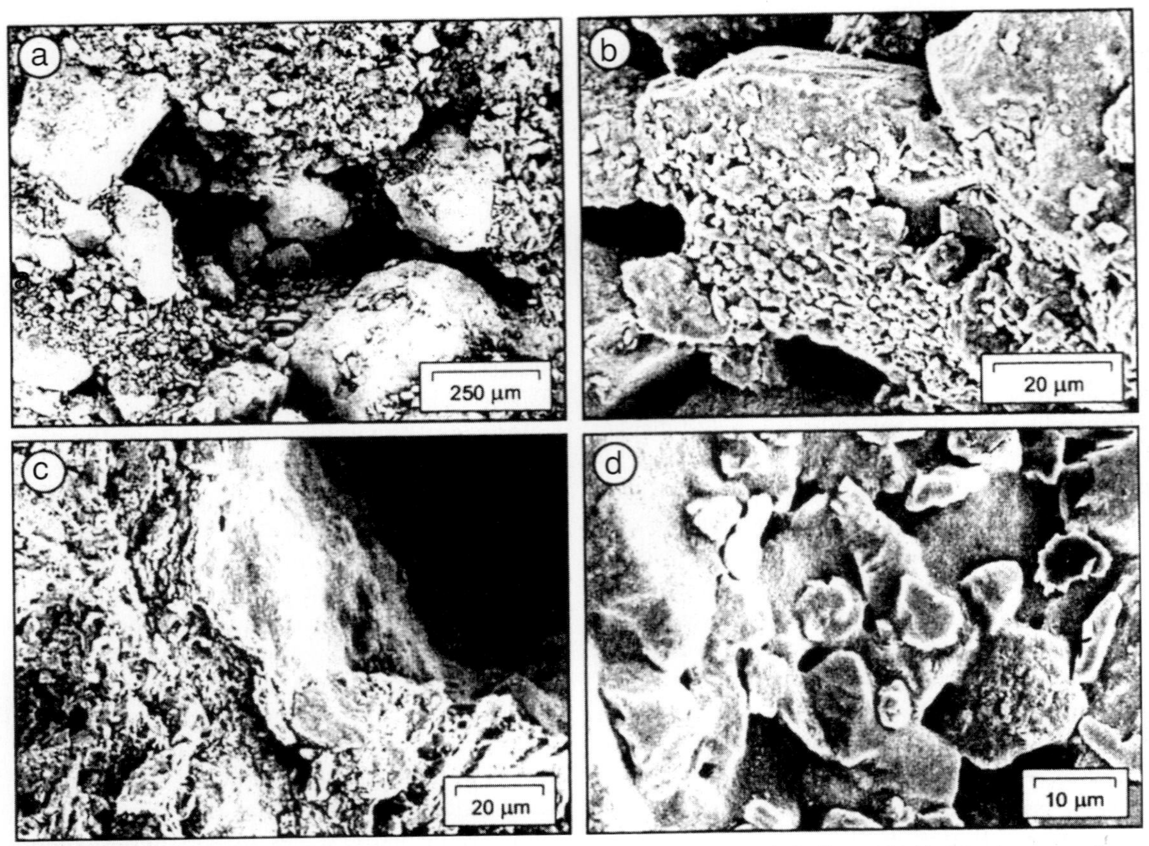

Fig. 12.17 Scanning electron photomicrographs images of properties characteristic of fragipans. Images by B. N. Weisenborn. (a) Close packing of clay and silt grains around an architecture of sand particles. The image was deliberately selected to show the large void; the majority of this Btx horizon is densely packed as shown around the periphery. (b) Clay bridging between skeletal sand grains. (c) A thick, laminated ferriargillan (and sesquan?), including some amorphous material, on a ped face in a Bt horizon of a Glossudalf from northern Michigan. (d) Close-up of an amorphous coating draping (and cementing) some silt particles.

be responsible for it. This question has been difficult to resolve, as the cementing agent in fragipans is not overtly visible, e.g., $CaCO_3$ or organic matter. Some have suggested that there really is no cementing agent, instead noting that the structure, packing or internal arrangement of the clay and silt particles within a skeleton of sand grains is adequate to explain the brittleness (Hutcheson and Bailey 1964). Most researchers, however, feel that a cementing agent *is* a part of fragipan genesis. They fall into one of two camps (or into both camps simultaneously):

(1) Cementation occurs due to amorphous materials, such as silica or iron/aluminum compounds, which have either been illuviated or released by *in situ* weathering (Bridges and Bull 1983). Deposition occurs when the soil solution becomes supersaturated via desiccation (De Kimpe 1970, Harlan *et al.* 1977, Steinhardt *et al.* 1982, Ajmone Marsan and Torrent 1989, Karathanasis 1987a, b, Norfleet and Karathanasis 1996) (Fig. 12.17). Cementation occurs after a physical mechanism creates *close packing*.
(2) Silicate clay minerals form bridges that hold the brittle matrix together (Grossman and Cline 1957, Knox 1957, Wang *et al.* 1974, Lindbo and Veneman 1993). Or perhaps close packing *and* the correct architecture are

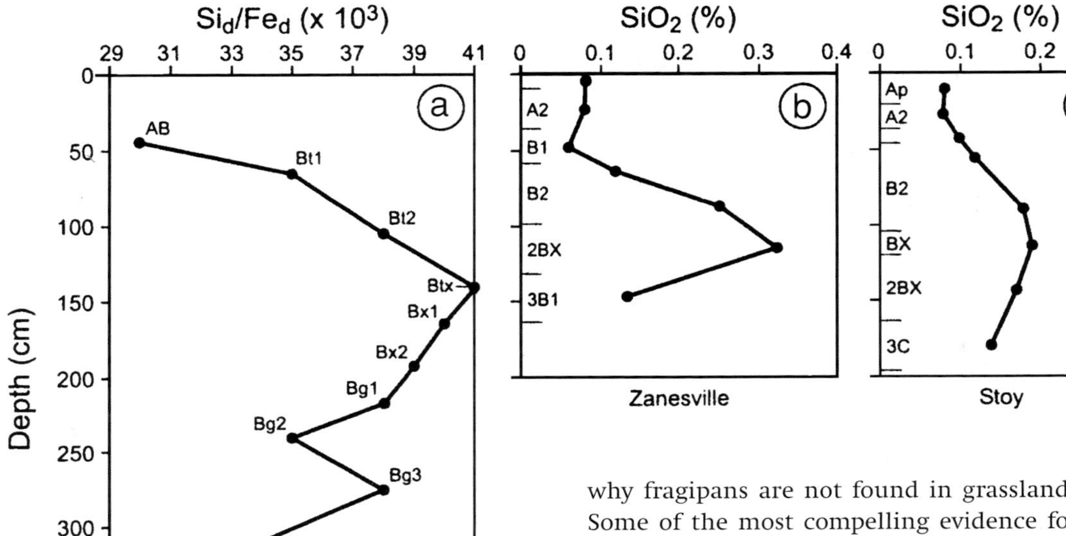

Fig. 12.18 Depth functions illustrating the coincidence of high silica content and fragipan expression. (a) Free silica/free iron ratio in Fragiudalf from Italy. After Ajmone Marsan and Torrent (1989). (b) and (c) Total SiO_2 in Fragiudalfs from Indiana. After Harlan *et al.* (1977).

enough to create brittleness and high bulk density, especially in Ex horizons that may have been stripped of amorphous materials (Miller *et al.* 1993).

Cementation by an amorphous silica compound or a Si-rich aluminosilicate seems to be favored by many researchers and confirmed by depth plots (Bridges and Bull 1983, Tremocoldi *et al.* 1994, Norfleet and Karathanasis 1996) (Fig. 12.18). Steinhardt and Franzmeier (1979) found that natural fragipan clods broke down after silica was extracted from them. Extraction of Fe and Al in the same way, however, had little effect. SEM and EDXRA analyses have identified Si-rich bridges between silt grains in fragipans (Norton *et al.* 1984). The silica is thought to coat and form bridges between grains, plug pores and cause rigidity and brittleness. Because grasses biocycle silica, less is available as a bonding agent, perhaps explaining why fragipans are not found in grassland areas. Some of the most compelling evidence for a silica or aluminosilicate cementing agent in fragipans comes from southern Indiana, where Harlan *et al.* (Harlan and Franzmeier 1977, Harlan *et al.* 1977) observed that free SiO_2 normally increases in fragipan horizons. The soils here are formed in silty loess over sandy loess, above a buried paleosol. The fragipans tend to form in the lower part of the silty loess. Harlan *et al.* (1977) proposed that silica-enriched soil water, generated in the weathered, upper solum, percolates and tends to hang in the silty loess because of its greater matric tension, and not move into the sandy loess below. When the base of the silty loess is within the rooting zone, it desiccates during the summer, precipitating the silica and creating conditions that can lead to a fragipan. In thicker loess, the base of the silty loess is below the rooting zone. It does not dry out as often, and less silica is precipitated; in these settings the soils lack a fragipan. Because some fragipans do not exhibit high amounts of extractable silica (Yassoglou and Whiteside 1960, De Kimpe *et al.* 1972), it is likely that it (silica) is an important, but not the only, player in the formation of fragipan brittleness.

With this background in mind, we turn to the conditions that must exist for a soil to develop a fragipan, the processes that must occur to precondition the soil for its formation and the processes that interact to form the Bx or Btx horizon. Most fragipans form in acidic parent materials, or at least in parent materials that are readily leached of carbonates (Yassoglou and Whiteside 1960, Miller *et al.* 1993). Percolating water is necessary to translocate the weathering by-products to the lower solum, where they can precipitate as cementation agents. Illuviation of silicate clay to this depth is also a plus. Illuviation of soluble materials and clay can be enhanced at a particular depth if a lithologic or weathering discontinuity occurs there. Smeck *et al.* (1989) defined a *weathering discontinuity* as the lower limit of the weathering front (more of a "zone" than a "front"). In calcareous soils it may coincide with the lower limit of leaching. It also may coincide with the top of a buried soil. Weathering by-products, particularly hydrous oxides of Si and Al, are likely to accumulate at weathering discontinuities – they are generated *above* it, translocated *to* it and preferentially deposited *at* it because of changes in soil chemistry. A dense, closely packed parent material, rich in very fine sand and silt, also enhances the likelihood of fragipan formation, since the effectiveness of a bonding agent is partially dependent upon a conducive architecture in which there are many grain-to-grain contacts. Few, if any, fragipans are formed in sands or clays. Clays have so much surface area that any bonding agent is spread too thinly, and the shrink–swell propensities of clayey soils tend to be stronger than the bonding agents (Ciolkosz *et al.* 1979, Smeck *et al.* 1989).

Bryant (1989) illustrated how otherwise porous parent materials can become dense by processes he called physical ripening and self-weight collapse, individually or in coordination. In parent materials of the proper texture, Bryant suggested that, once saturated, they will physically ripen and develop close packing, prismatic structure and high bulk densities after an initial desiccation event (Jha and Cline 1963). Subsequent desiccation events may further ripen/densify the material only if they are more intense, i.e., the soil desiccates further. Additional wet–dry cycles will not disrupt dense, ripened material unless it has a high COLE value. In short, a slurried sediment will densify (ripen) upon desiccation. These desiccation events may normally occur at the very earliest stages of pedogenesis, as a wet sediment dewaters (dries out). Physical ripening is most efficient in subsoil horizons that do not undergo subsequent pedoturbation and wet–dry cycles. Thus, it works well as a prerequisite for the development of many fragipan soils. In addition to physical ripening is the concept of self-weight collapse, wherein fabric rearrangement and densification occur as a sediment is wetted to saturation (Bryant 1989). Eolian sediments are especially prone to self-weight collapse; not coincidentally many fragipans occur in loess. Thus, Bryant has set up two ways in which open, porous sediment can become denser and more closely packed, and both involve saturation. In physical ripening, a saturation event is followed by desiccation, but the concept of self-weight collapse illustrates that desiccation may not even be necessary in certain types of sediments. The two processes can act independently or together.

Using the above theoretical guidance and some data from some fragipan soils in Michigan, Weisenborn (2001) developed a comprehensive model of fragipan evolution (Fig. 12.19). Although no one model may ever be able to explain fragipan formation for all soils, her model works for Fragiorthods and Fragiudalfs in the Great Lakes region, as it addresses the three main components of fragipan genesis. It *assumes* that the parent material is *collapsible* or of the correct texture, open structure, COLE, thickness and wetness. It *accommodates* and explains how the requisite fragipan physical properties can be attained, by invoking *physical ripening* and/or *self-weight collapse*. And it *accounts for* fragipan binding and brittleness by invoking the *precipitation of amorphous materials* (bonding agents) at a pedogenic *weathering front* or *lithologic discontinuity*. The model does not exclude other processes from having an important influence; we note particularly the potential for frost action in fragipan genesis (Van Vliet and Langhor 1981).

DEGRADATION

Fragipans eventually degrade, as illustrated in the northeastern United States where soils on young glacial tills have fragipans while those on older tills do not. Like argillic horizons, their degradation stems mainly from the fact that they are aquitards and occasionally form perched zones of saturation (Payton 1993b). Some of the acidic,

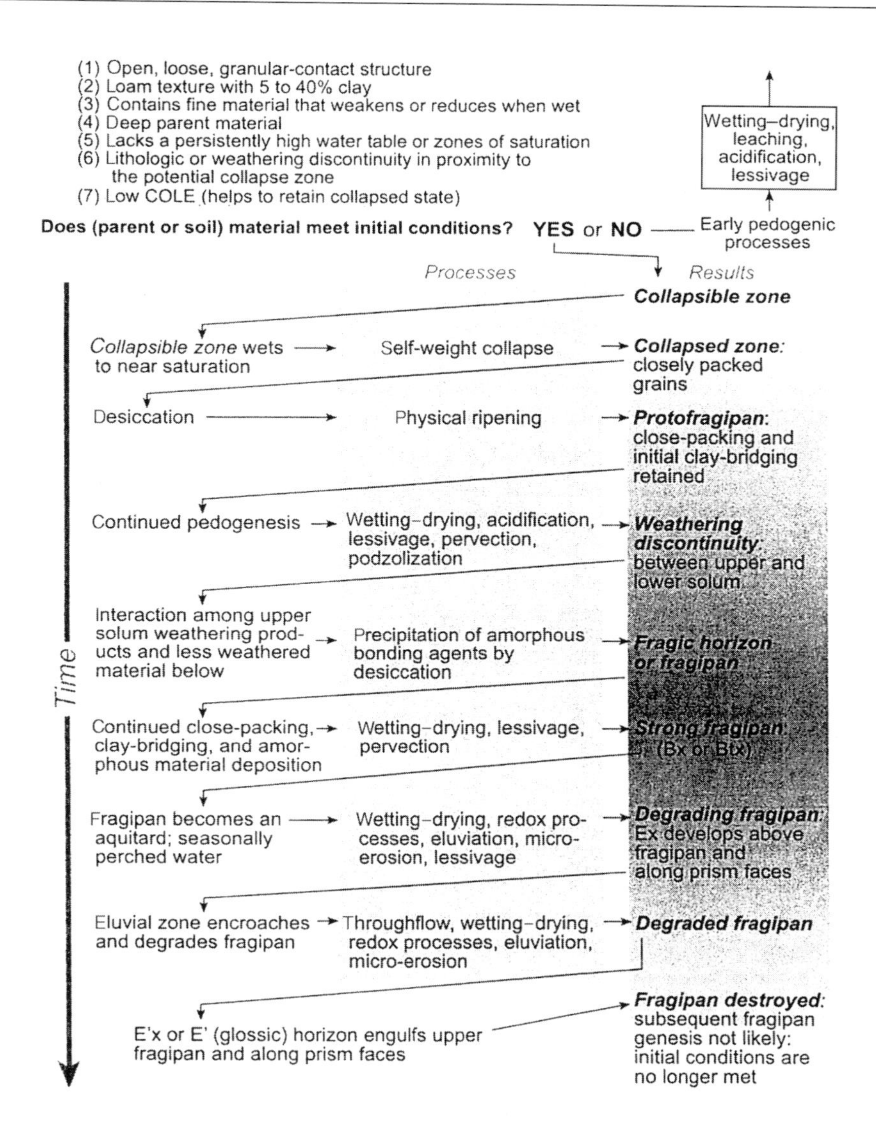

Fig. 12.19 The Michigan model of fragipan evolution. After Weisenborn (2001).

base-poor, perched water percolates across the top of the pan or vertically through it. Vertically percolating water tends to move between the large Bx prisms, forming light-colored, leached zones between. Simultaneously, light-colored zones develop in the uppermost pan; these are labeled as Ex horizons as long as they retain their brittleness. With continued leaching, they may become stripped of their bonding agents, including clay and Fe, and develop into non-brittle and continuous E horizons (Steele *et al.* 1969).

The processes that form the albic streaks within the pan and the Ex horizon itself are similar to those that form glossic horizons (see above). Acidic, chemically aggressive water is forced to pass through the few pores in the pan, removing clay particles in suspension (Miller *et al.* 1993) and chemically weathering others. Payton (1993b) linked this form of "micro-erosion" of clay coatings and void walls to destabilization of the clay–Fe system after the Fe is reduced and eluviated. Weathering by-products, many of which are potential amorphous bonding materials, are remobilized and moved downward, leaving the fragic architecture but removing the "glue." For this reason, many Ex horizons are more porous,

Table 12.3 Factors involved in the supply vs. demand for oxygen in soils

"Supply side" factors	"Demand side" factors
Water table depth	Roots
Oxygen content of soil water	Microbial activity
Soil texture	Soil temperature
Soil porosity	Soil texture
Soil permeability	
Soil profile thickness	

Table 12.4 Order of utilization of principle electron acceptors in soils, and associated reactions

Process	Reaction
O_2 disappearance	$1/2\ O_2 + 2e^- + 2H^+ = H_2O$
NO_3^- disappearance	$NO_3^- + 2e^- + 2H^+ = NO_2^- + H_2O$
Mn^{2+} formation (Mn reduction)	$MnO_2 + 2e^- + 4H^+ = Mn^{2+} + 2H_2O$
Fe^{2+} formation (Fe reduction)	$FeOOH + e^- + 3H^+ = Fe^{2+} + 2H_2O$
HS^- formation	$SO_4^{2-} + 6e^- + 9H^+ = HS^- + 4H_2O$
H_2 formation	$H^+ + e^- = ½\ H_2$
CH_4 formation (fermentation)	$(CH_2O)_n = n/2\ CH_4$

Source: Mausbach and Richardson (1994).

of lower density and more fragile than the Bx horizons below.

Fragipan degradation is slightly different, and perhaps more intense, than Bt horizon degradation because it almost always involves oxidation–reduction processes. Reduced conditions tend to be more common in fragipan soils, leading to increased iron and manganese mobility. Thus, many Ex horizons are Egx horizons, and have Fe–Mn nodules and mottles within (Lindbo *et al.* 2000). Eventually, the fragipan may be converted to a thick eluvial horizon or, if its illuvial clay can be retained and some additional clay brought in from the upper solum, it could develop into a Bt horizon.

Oxidation–reduction and gleization

Soils are generally either oxidized or reduced – terms used to describe whether the soil system has oxygen available to microbes (oxidizing) or whether it does not (reducing). When the demand for oxygen exceeds the ability of the atmosphere to supply it, available O_2 is used up and the soil becomes anaerobic or anoxic (Ponnamperuma *et al.* 1967, Patrick and Mahapatra 1968, Patrick and Henderson 1981, van Breeman 1987) (Table 12.3). The atmosphere is the main *source* of oxygen in soils, into which it moves by diffusion. Microbes, along with roots, are the main oxygen *sinks*, utilizing the most easily accessible sources of oxygen first, such as entrapped soil air (Table 12.4). When that source is no longer available, as in a saturated soil, they utilize the dissolved oxygen in soil water. When *that* O_2 supply is gone, anaerobic conditions develop and a series of biochemical reactions occur, the endpoint of which is the gain of an electron, i.e., reduction, for certain vulnerable cations. Nitrate is the first of a series of soil components to be reduced in the anaerobic soil system. After the nitrate is used up, the sequence continues as manganese is reduced from Mn^{3+} (or Mn^{4+}) to Mn^{2+}. Next, ferric iron compounds (Fe^{3+}) are reduced to ferrous iron (Fe^{2+}), and finally sulfates are reduced

(Vepraskas 1999). Most soils have so many iron compounds and so little sulfate that reduction of sulfates is rare.

As soil conditions change from oxidizing to reducing and back again, cations change states and the suite of secondary minerals associated with them also changes. Under oxidizing conditions, Fe and Mn cations (and their associated minerals) tend to be immobile and lend a brown or red color to the soil. Thus, dry upland soils that allow free entry of oxygen into the soil tend to be brown or red (Table 2.2). Reducing conditions force a new suite of soil minerals to form, thereby changing the soil matrix to shades of gray, green or other pale hues. The condition represented by long-term reduction is referred to as *gleying*, and the processes associated with long-term gleying are called *gleization* (Table 12.1). In older literature reduction associated with groundwater is called *groundwater gley*. An alternate model of gleying is the *pseudogley model*, in which the brief periods of reduction occur below a perched water table. Gleization occurs in permanently or near-permanently saturated soils that have relatively high organic matter contents – a microbial food source, fueling their demand for oxygen. Gleyed soils are wet soils, as the water is the primary mechanism by which atmospheric oxygen is excluded from the soil. If the soil is fine-textured, it is even more likely to be gleyed, since the diffusion rate of oxygen is lower in clays than in sands, and because fine-textured soils often have higher microbial populations. Nonetheless, a sandy soil can *become* gleyed, i.e., develop gray colors, more rapidly than can a fine-textured soil, because the sand has much less surface area to "color." Gleyed horizons are typically given the suffix g, as in Bg or Cg horizons (Table 3.2).

Gleying is a pedogenic process that can also be considered a geochemical process. The reactions involved are primarily geochemical and any translocation often occurs along flowlines of equal hydrological potential. The results of gleying can occur within the profile or well below it, although deeper horizons are less likely to be gleyed, even if they are saturated, because the demand for oxygen decreases with depth.

Reduced iron is mobile and is capable of diffusing through and within the soil solution; oxidized iron is not. Fe and Mn compounds can eventually migrate out of gleyed horizons, rendering many of them iron-poor. Regardless, the lack of Fe and Mn, or the presence of reduced Fe and Mn, both result in the same morphologic expression: gray, muted colors. Obviously, the loss of reduced cations implies that the net flow of the water in the saturated soil is out of the soil. In a recharge wetland where water is entering a wet soil from outside, materials are being added to the soil and gleyed colors may be more difficult to achieve (Mausbach and Richardson 1994).

In many soils, the soil atmosphere undergoes alternating periods of oxidation and reduction, often referred to as *redox conditions* (Vepraskas 1999). Morphologic features associated with redox conditions take many forms within soils (nodules, concentrations, masses and pore/ped linings) and are generally referred to as *redoximorphic features* (Schwertmann and Fanning 1976, Vepraskas 1999) (see Chapter 13). While saturated, Fe and Mn may move along concentration gradients or with currents in the soil solution, i.e., they are mobile and free to move within the soil. Later, under oxidizing conditions, the reduced Fe can oxidize to insoluble ferrihydrite or ferric hydroxide ($Fe(OH)_3$), forming red or orange Fe concentrations, or blackish Fe–Mn concentrations (Table 2.3). Over time, continued oxidation–reduction cycles cause Fe and Mn to become concentrated in these red mottles; surrounding areas have lower chromas and grayer hues (Richardson and Hole 1979). Brown or red redox concentrations were called red mottles or high chroma mottles in older literature.

Equally likely in such a soil is the formation of areas of Fe depletions, or gray mottles. Gray areas generally have low iron and manganese contents; the gray colors reflect uncoated sand and silt grains.

Many soil horizons are mottled, i.e., they have variegated patterns of soil colors due to redox depletions and concentrations (see Chapter 13). Like gleying, the redox processes that cause soil horizons to be mottled are primarily geochemical. However, mottling is perhaps more important pedologically than is gleying, for not only is mottling more common but it is a useful indicator of water table fluctuations. Thus, mottling can provide information on the duration of saturation in a soil, what time of year it may occur, and the dissolved oxygen status of the water. Mottle patterns can even be used to identify ephemeral features, such as a perched water table.

Ferrolysis

Brinkman (1970) coined the term ferrolysis to refer to the process whereby the CEC of a soil is lowered and many of its clay minerals are destroyed due to exchange reactions involving iron (Table 12.1). Ferrolysis occurs in soil horizons that are alternatively saturated and dry, usually due to a perched water table on a slowly permeable B horizon, during a wet season (Bartelli 1973). The B horizon, is an aquiclude, usually a strong Bt, Btn or Btx horizon (Fig. 12.20). In the rice paddies of southeast Asia that have been under cultivation for centuries, the aquiclude can also be a plowpan (Brinkman 1970, 1977a). Many paddy soils have a base-poor, bleached upper profile from which clay has been destroyed by ferrolysis (Zi-Tong 1983). Long-term ferrolysis will cause the soil horizons in the upper solum to be sandier and more acidic because clay decomposition by ferrolysis is, according to Brinkman (1970), 10 times more efficient than weathering by organic acids. Soils undergoing ferrolysis have upper sola with gray matrix colors, indicating generally reducing conditions, with red or brown goethite or ferric hydroxide mottles within the matrix but mostly along cracks and ped faces. The mottles form during the dry season when conditions become oxidizing. The soil is not saturated long enough, however, for true gleying to occur.

Ferrolysis requires cycles of oxidation and reduction in a wet–dry climate. During the *anaerobic stage*, soils are saturated and Fe^{3+} is reduced to Fe^{2+} by soil organisms using organic matter as an energy source. The organic matter is concurrently oxidized, forming OH^- anions (Brinkman 1977b). Because it is the *upper* solum that is saturated, which is near to a large source of organic matter (the A horizon), microbial activity is high in the saturated horizon and reduction can occur quickly. The resulting Fe^{2+} ions displace +2 and +1 base cations and Al^{3+} from clay mineral exchange sites. This exchange reaction increases soil pH (Ransom and Smeck 1986). Bases and aluminum are then free to be removed from the soil system with soil water; many precipitate. Removal is often accomplished as lateral flow, with soil water slowly flowing along the top of the aquiclude (Fig. 12.20).

During the dry season *aerobic stage*, the Fe^{2+} cations adsorbed to the clay surfaces are oxidized to Fe^{3+}; some precipitate as minerals containing Fe^{3+}. Some of the adsorbed Fe oxidizes and combines with oxygen and water to form reddish mottles of goethite (FeOOH) and/or ferric hydroxide ($Fe(OH)_3$) (Bartelli 1973, Eaqub and Blume 1982):

$$4Fe^{2+} + O_2 + 10H_2O \rightarrow 4Fe(OH)_3 + 8H^+$$

Thus, overall, the soil develops $Fe(OH)_3$ mottles (redox concentrations) in the aerobic phase and loses some bases (due to leaching) in the anaerobic phase. In return, the soil solution gains H^+ cations, since the formation of the ferric hydroxide during the dry season releases H^+ cations. Via hydrolysis, any adsorbed H^+ cations attack the octahedral sheets of the clay from their edges inward, liberating silicic acid, Al^{3+} and some Mg^{2+}. Clays with higher CEC values are attacked most efficiently because they can adsorb more Fe cations; 1 : 1 clays are the least susceptible. Some of the silica that is released from the clays as they weather gets reprecipitated as amorphous material in gray silans or skeletans (Brinkman 1977b). The Mg^{2+} cations will eventually be leached, while the Al^{3+} cations stay in the soil to further acidify it (Fig. 12.20). Thus, in every wet–dry cycle base cations are leached (or potentially leached) and a part of the clay lattice is destroyed. Weathering by-products are removed laterally in soil water, as it flows on top of the B horizon.

Some of the free Al may form Al interlayers in existing clays and a low CEC version of chlorite (with high amounts of exchangeable aluminum), a process Brinkman called *chloritization* (Dijkerman and Miedema 1988). Other Al cations may combine with silica to form amorphous compounds. Release of Al^{3+} cations is, therefore, another means by which the soil's CEC is lowered (Reuter 1965).

The outcome of every wet–dry cycle is predictable – base cations are leached, H^+ cations become more prevalent, clay minerals are

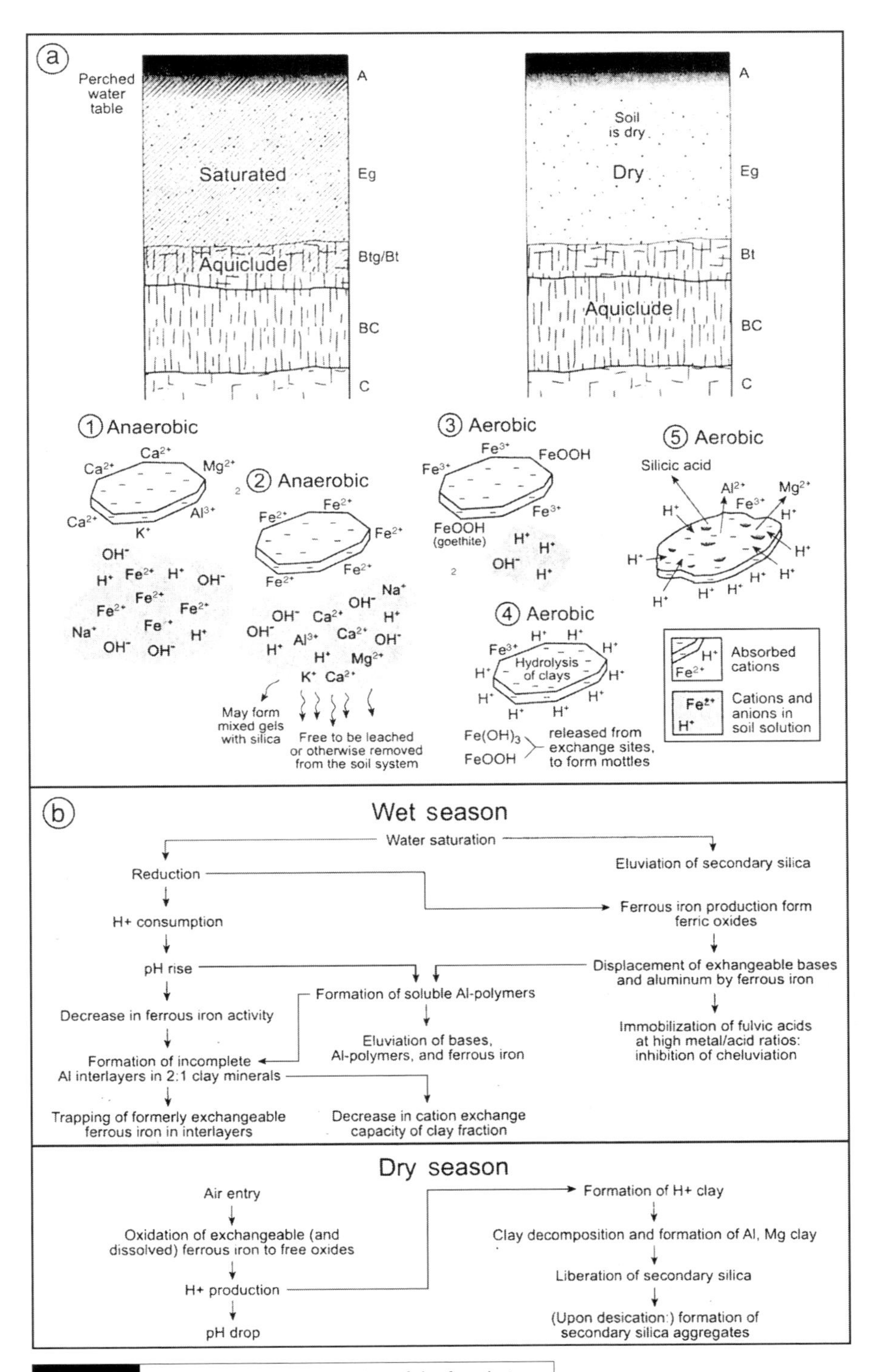

Fig. 12.20 Diagrammatic representation of the ferrolysis process, and the stages in its annual cycle. Part (b) after Mausbach and Richardson (1994).

Landscapes: Planosols. . .ferrolysis gone amuck

Planosols (usually Mollic or Vertic Albaqualfs) are considered by many the classic example of ferrolyzed soils. These acidic soils have a dense, clay-rich Bt horizon ("claypan") which perches water (Jarvis *et al.* 1959). Above the Bt is a leached, chemically weathered, silty/sandy E or Eg horizon, which commonly has abundant Fe–Mn accumulations, indicative of occasional saturated, reducing conditions. Planosols tend to be found on older, flat landscapes and in slight depressions (Culver and Gray 1968a, b). It is precisely these soils that led Brinkman (1970) to develop his theory of ferrolysis.

Planosols can be viewed as ferrolysis taken to the *n*th degree. Normally productive Mollisols evolve to Albaqualfs (Planosols) through a number of processes that take considerable time to gestate fully. Initially, melanization and base cycling dominate, as in most Mollisols. Because of climate or vegetation changes, or simply due to the slight depressional site that the soil is located on, the soil gets more intensively leached/acidified, and develops a contrasting E–Bt sequum. On flat landscapes, water moves primarily downward, rather than laterally, facilitating translocation of clay and silt to the B horizon. Eventually, the Bt horizon gets so clay-rich that it passes an intrinsic threshold and begins to accumulate clay by sieving, as attested by clay contents that are highest at the top. The A and E horizons become leached of most of their bases, while the lower solum stays base-rich. Organic matter also is translocated, leading to a profile with a secondary organic carbon maximum in the B horizon (Culver and Gray 1968a). The Bt horizon becomes less and less permeable, perching water in spring and fall and promoting ferrolysis. Acidification and destruction of the clays in the upper solum results, forming sandy, acid A and Eg horizons over a clay-rich Btg horizon. Some of the silt in the upper solum is translocated to the Bt, especially during dry periods when deep cracks open, forming silans on ped faces (Culver and Gray 1968a). Vegetation gets more and more impoverished as ferrolysis continues, and the overall productivity of the landscape declines. The only hope for these soils comes as landscape dissection leads to steeper slopes, deeper water tables and better internal drainage. Drier conditions may also facilitate enough lateral flow to break through the Bt horizon, allowing base-cycling and pedoturbation to introduce base-rich materials from below. It should also be noted that some Planosols develop as salt-affected soils are leached of sodium in solodization (White 1961, 1964).

The development sequence discussed above and shown in the figure illustrates several key points about grassland soils. First, their organic matter contents are not stable through time, and may decline in the oldest soils. Second, as with all soils, they pass through thresholds and exhibit changes in their pedogenic pathways.

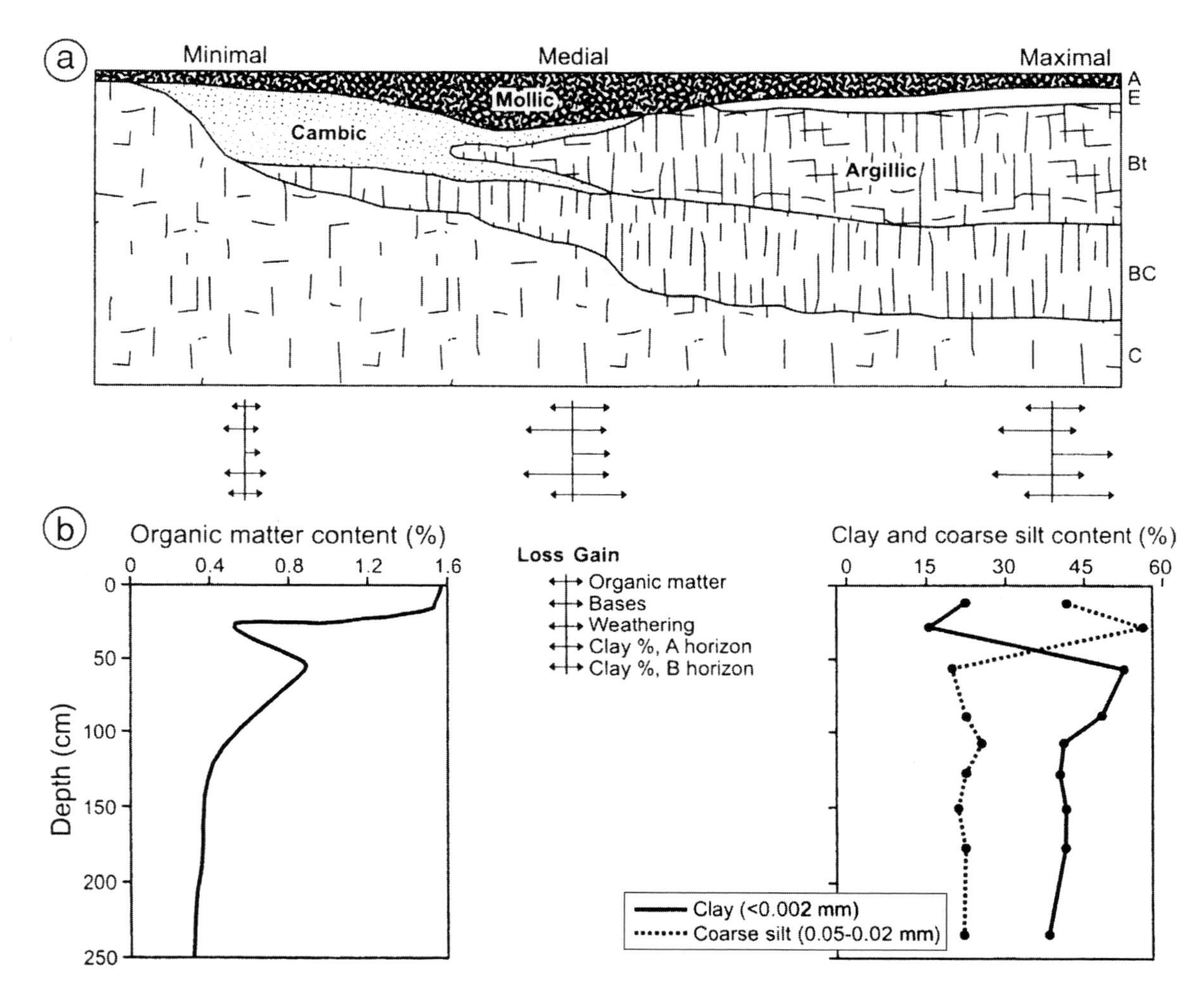

Planosol genesis and data. (a) Model of Planosol development in Oklahoma. (b) Some depth functions for the Nardin (end-member) soil (Mollic Albaqualf). After Culver and Gray (1968).

destroyed and some Al is added to the soil solution. The release of silicic acid, coupled with the destruction of phyllosilicate clays, can cause soils to have an abundance of quartz (probably neoformed from the silicic acid) in the clay fraction of wet–dry zone (Hardy 1993). Some Al-rich clays may form due to chloritization. Hydrolysis of easily weatherable minerals such as biotite or augite in the sand and silt fractions will replenish the supply of bases and slow, but not stop, the process. The pH continues to drop and clay minerals continue to weather, resulting in an acid, sandy horizon, usually an Eg or E horizon, above a mottled aquiclude (B horizon). Eventually, ferrolysis will work its way down into the aquiclude and degrade it, creating windows that will lessen the duration of saturation and cause the process to slow down, and perhaps even stop.

It should be noted that soils with similar morphologies can form without invoking ferrolysis. Eaqub and Blume (1982) described two soils in Bangladesh that had morphologies typical of soils undergoing ferrolysis. They argued that weathering and lessivage alone can produce the clay-poor epipedons; ferrolysis is not required.

Processes associated with the humid tropics

Most humid tropical soils are highly weathered. The process bundles associated with the formation of these soils, many of which are Oxisols,

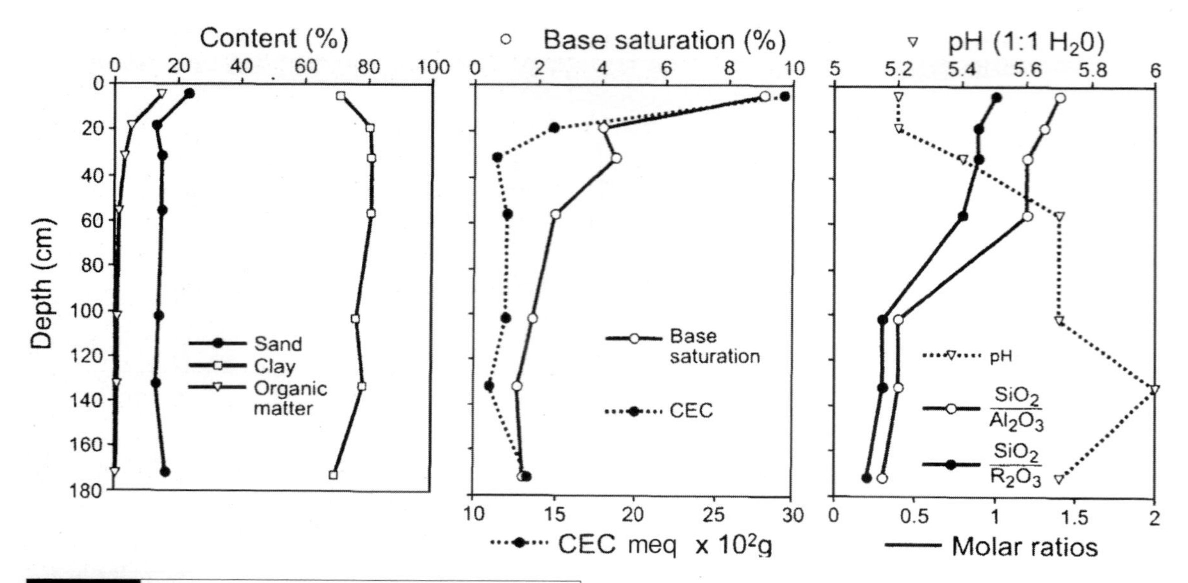

Fig. 12.21 Depth functions for a Typic Hapludox from Costa Rica, formed in volcanic materials on an old alluvial terrace. The current vegetation is savanna and the mean annual rainfall is about 260 cm, with a 2–3-month dry season. Data from Martini and Macias (1974).

have traditionally been termed *laterization* or *ferrallitization* (Table 12.1). Fanning and Fanning (1989), however, divided this process bundle into two subtly different processes: *laterization* and *latosolization*. In this section we point out the similarities and differences between these process suites, and discuss a companion process: *desilication*.

Oxisols have an *oxic horizon* – a mineral subsurface horizon of sandy loam or finer texture, with low CEC and few weatherable minerals (Soil Survey Staff 1999). The genetic concept of the oxic horizon involves extreme weathering and long-term soil formation under a hot, humid climate. Intense weathering and leaching are evident on many fronts. Values of CEC and effective CEC, amounts of exchangeable bases, divalent/monovalent cation ratios and extractable P values are all very low (Tanaka *et al.* 1989), while the amounts of sesquioxide clays are high (Fig. 12.21). In easily weathered parent materials and when climatic conditions are favorable, oxic horizons can form in soils on young surfaces and over a relatively short period (Soil Survey Staff 1999), but generally they require great lengths of time to develop. Thus, most are highly polygenetic (Muggler and Buurman 2000).

The clay mineralogy of oxic horizons is dominated by 1 : 1 clays such as kaolinite, and oxide clays such as goethite, hematite and gibbsite. Kaolinite is usually the only phyllosilicate mineral of any abundance. Smectites are only observed in waterlogged sites where the potential to remove weathering by-products is low. Maghemite is present in soils derived from basic or ultrabasic rocks (Buol and Eswaran 2000). The silt and sand fractions are generally dominated by quartz along with small amounts of other resistant minerals such as anatase, zircon and rutile, depending on the parent material. Most coarse fragments are coated with sesquioxides.

Oxisols are reddish and clay rich. Horizon boundaries are gradual, especially in the subsurface where the only differentiating criteria may be a change in structure or consistence. A horizons are the most acid and tend to be low in organic matter. Bennema (1974) found that organic carbon contents decrease exponentially with depth in many Oxisols. Thus, deep horizons below have much less organic matter than even the A horizons, making it difficult to differentiate one from another (Martini and Macias 1974). Another important attribute is that, in many soils with oxic horizons, the clay content is relatively

Landscapes: Red, upland soils of the humid tropics

We know comparatively little about tropical soils. Although it is commonly assumed that tropical landscapes are monotonously red, weathered and underlain by laterite (see below), these landscapes are actually spatially diverse and variable (Sanchez 1976, Richter and Babbar 1991). Indeed, perhaps their main commonality is a lack of seasonal temperature cycles (Buol 1973). Other commonalities worthy of note include the highly weathered nature of the minerals in the red and yellow tropical soils, and their low ionic activity, base saturation status and CEC. Many tropical soils are very clay-rich and acidic (Martini and Macias 1974). That these clay-rich soils also have such low CEC values is a testament to the low exchange capabilities of the clay minerals within. Another unifying attribute of tropical soil landscapes, at least upland landscapes, is their red color (Buol and Cook 1998). The red colors (Ardiuno *et al.* 1984, Rebertus and Buol 1985, Bech *et al.* 1997) are due to an abundance of Fe_2O_3 compounds and minerals (Al_2O_3 is colorless) (Table 12.5). In wet, low-lying areas, the iron mineralogy changes and the soils take on yellow-orange colors.

Many humid tropical soils are infertile (Buol 1973). The nutritive status of many tropical soils can be characterized as one where relatively large concentrations of organically bound nutrients exist in the A horizon, while very low concentrations occur in the subsoil (Stark and Jordan 1978, Buol and Cook 1998). Especially problematic for these soils are the low levels of phosphorous, not only because they have low P amounts naturally but also because the sesquioxides are capable of strongly fixing PO_4^- (Kamprath 1973, Bigham *et al.* 1978). Additions of fertilizer or organic matter is only a short-term solution, for soon they are again infertile. Fertility for most cropping applications is, therefore, dependent more on nutrient contents than on other factors such as rainfall, past soil erosion, compaction, etc. However, where additions of fresh parent material, such as alluvium, colluvium or volcanic ash, are present, fertility can be high.

Oxisol profiles are usually very thick. A horizon organic matter contents are low, probably due to the rapid decomposition and consumption (by fauna) of organic materials in the hot environments. Thus, A horizons are thin and light-colored. Only in tropical grasslands and savannas are high amounts of soil organic matter common. Often, very thick sequences of saprolite underlie the profiles; thicknesses over 200 m have been recorded (Kroonenberg and Melitz 1983). Many tropical soils on old, stable landscapes also have one or more stone lines at depth (Dijkerman and Miedema 1988).

The clay mineralogy often gives rise to a uniquely tropical type of friable and porous soil structure, which Buol (1973) called fine granular oxic structure (Soil Survey Staff 1999). The fine crumbs, less than 2 mm in size, are composed of sand grains held together by clay and sesquioxides (Escolar and López 1968). These crumbs can adhere to each other and form stable conglomerate-like masses that approach 5 mm in diameter and facilitate high infiltration capacities and minimal runoff (Van Wambeke 1962) (Table 12.6). Bulk densities are generally low, commonly near 1 g cm^{-3}.

Table 12.5 Common iron and aluminum compounds and minerals in soils

Mineral or compound	Chemical formula
Goethite	α-FeO(OH)
Hematite	α-Fe_2O_3
Ferrihydrite	$Fe^{3+}{}_2O_3 \cdot 0.5(H_2O)$
Maghemite	γ-$Fe^{3+}{}_2O_3$
Lepidocrocite	γ-FeO(OH)
Boehmite	AlO(OH)
Amorphous ferric iron	$Fe(OH)_3$

Table 12.6 Ranges of infiltration rates in some soils of Puerto Rico

Soil order	Minimum infiltration rate (cm h^{-1})	Maximum infiltration rate (cm h^{-1})
Oxisols	8.4	15.4
Ultisols	7.4	23.6
Mollisols	8.2	19.5
Alfisols	2.7	11.5
Inceptisols	2.7	13.2
Entisols	2.3	27.5
Vertisols	0.1	9.5

Source: Sanchez (1976).

constant with depth, indicating little or no clay mobility. Lessivage is not a dominant process in many humid tropical soils, and where it does occur the translocated clay does not illuviate into any one preferential depth zone, as it does in other soils (Eswaran and Bin 1978a, Soil Survey Staff 1999). Argillans are almost impossible to identify in the field in soils with kaolinitic or oxic mineralogy (Beinroth 1982). If observed in thin section, they tend to be thin and degrading. Oxide clays are not readily dispersed, probably due to cementation by sesquioxides, and thus tend to form flocculated aggregates that are almost immobile within the profile. This tendency makes the soils feel and appear to be much less clayey than they really are. Some actually feel sandy. Another artifact of this unique clay mineralogy and soil structure centers on plant-available water capacity. Oxisols often have low water-holding capacities because most of their pores are either very large *between* the granules (and thus do not retain much water against the forces of gravity) and are very small *within* the granules (and thus retain water at too great a tension to be plant-available). However, the stability of the clay granules with respect to water provides the soils with the ability to resist erosion better than soils with similar amounts of clay, but different mineralogies.

Pedogenic processes in the humid and subhumid tropics are influenced by: (1) abundant moisture (at least at some time in the year), enhancing *chemical weathering* and vigorous plant growth, (2) high temperatures, providing *energy* for inorganic chemical reactions and biological activity, (3) old landscapes, providing ample *time* for many persistent but slow processes to run to completion and commonly (4) active *biomechanical agents* such as termites and ants (see Chapter 10). These factors combine to create a situation involving rapid mineralization of organic matter and intense and nearly complete weathering of primary minerals (De Villiers 1965). Aluminum and ferric iron cations are relatively insoluble and tend to remain in soils, while soluble weathering products can be translocated out of the profile. Due to the high rainfall in the humid tropics, there is ample potential to translocate all but the most insoluble cations from the profile. Even silica is leached from the profile: a process called *desilication* (Scholten and Andriesse 1986, Buol and Eswaran 2000). Residual Fe and Al tend to combine with oxygen to form oxide clays in such high amounts that parts become cementated into a brick-like substance of various forms and names such as laterite, ferruginous nodules and concretions (Livens 1949).

Laterite

Tropical soils are still poorly understood. Confounding this problem is a plethora of confusing and inaccurate terminology that has been perpetuated for years, much based on inference rather than science. Many older textbooks and publications on tropical soils simply refer to them as red soils, red loams or red earths (Greene 1945). Later, terms such as laterite and latosols came

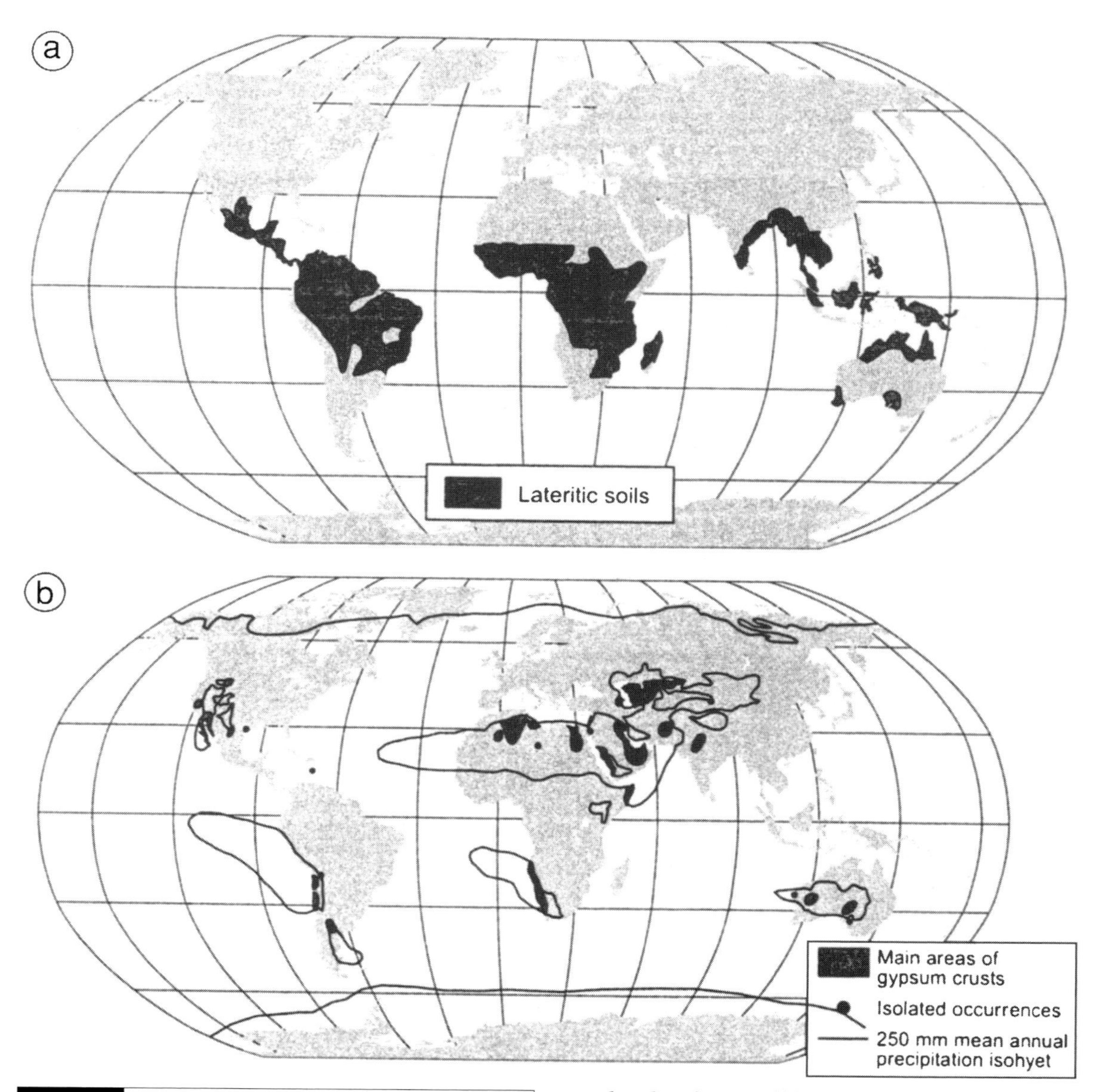

Fig. 12.22 Global distributions of some soil types. (a) An old map (source unknown) showing the supposed global extent of laterite and lateritic soils. (b) Generalized distribution of soils with gypsum accumulations, including some playas and saline sinks. After Watson (1985).

into vogue, based on the travels of a few scientists to tropical regions (Sherman and Alexander 1959, Martini and Macias 1974, Cline 1975) (Fig. 12.22). These terms, however, meant different things to different audiences. Quickly, the term *tropical soils* was taken to mean red soils high in iron that can harden irreversibly, which is essentially the definition of laterite. For a long time (e.g., McNeil 1964), it was thought that tropical lateritic soils would, if cleared of vegetation, harden irreversibly and become worthless brick pavement in a few years. This notion was later shown to be inaccurate.

Part of this confusion stemmed from the fact that soil science began in the *midlatitudes*; information about *tropical soils* was minimal. Upon visiting the tropics, midlatitude soil scientists were shown extreme and unique examples of soils, e.g., laterites, and carried these ideas back to

Fig. 12.23 Brick quarry in a clay-rich Oxisol, near Kano, Nigeria. Photo by J. Olson.

their countries (Sanchez 1976). Misguided ideas that laterites and red, brick-like soils dominated the tropics were promulgated, while vast areas of the tropics that were covered with soils not unlike those found in midlatitudes were largely ignored. For example, in the 1938 system of soil classification (Baldwin *et al.* 1938), only one zonal soil order was defined for the forested tropical regions: lateritic soils of forested warm-temperate and tropical regions. By 1955, this grouping had changed name (but not number) to Latosols, which meant soils that are developing or can develop lateritic characteristics (Kellogg 1949, Cline 1955, Sherman and Alexander 1959). Laterization became known as a suite of processes whereby soils developed into latosols. By the early twentieth century, more soil scientists were conducting research in the humid tropics and reporting that laterite was relatively uncommon (Hardy 1933). In 1949, Charles Kellogg advocated abandoning the term laterite within soil classification.

By now it should be obvious that perhaps no other group of soils has such an ill-conceived, persistent terminology as have laterites. The origin of the term laterite can be traced to an 1807 report by Francis Buchanan, a Scottish scholar (Beinroth *et al.* 2000). In his travels to India, Buchanan described soft, red, Fe-rich soil material which hardened upon drying, rendering it useful as a construction material (Fig. 12.23). He called this material brickstone or laterite, from the Latin *later*, brick (Sivarajasingham *et al.* 1962). Buchanan's original definition referred to "an iron-oxide-rich, indurated, quarryable slag-like or pisolitic illuvial horizon, developed in the soil profile" (van der Merwe 1949). The laterite observed by Buchanan was soft enough to be cut into blocks with a knife, but upon exposure to air it quickly became hard. This laterite stone was found mainly as a cap on the summit of basaltic hills and plateaus in the Kerala state of India, where even today it is exploited for construction materials. Not all tropical soils had laterites of widespread, sheet-like morphology. Often, lateritic materials were present as hard gravel pisoliths, blocks or ferralitic concretions which stood out as low humps in the terrain. They consisted of gravel-sized concretionary nodules set within a matrix of silt and clay and obtained thicknesses between 1 and 5 m (Alexander *et al.* 1956). Thus, the more study that occurred, the greater the variety of soils and laterites that was discovered.

Confusion evolved as to whether laterite was a type of *soil* or a type of *material* (Kellogg 1950). In the original meaning of the term, Laterite soils contained a laterite horizon as a *part of the soil profile* (van der Merwe 1949). Thus, Laterite soils, or at least parts thereof, came to be characterized by forming hard, impenetrable and often irreversible pans when dried. Confusion arose

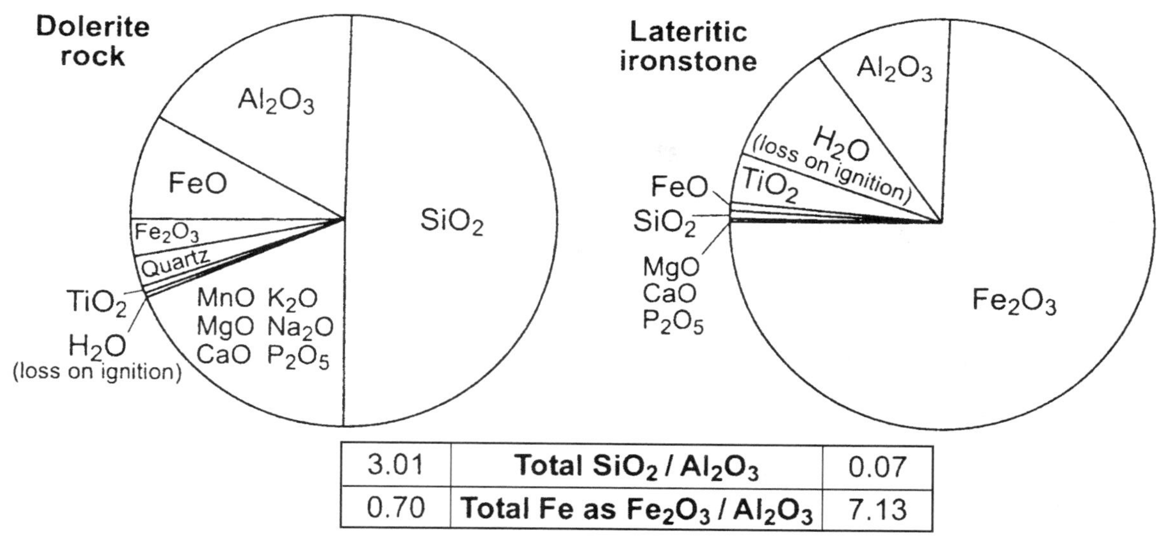

3.01	**Total SiO_2 / Al_2O_3**	0.07
0.70	**Total Fe as Fe_2O_3 / Al_2O_3**	7.13

Fig. 12.24 Charts showing the chemical composition of a dolerite rock parent material and the lateritic ironstone that has formed from it, via laterization, Eagle Mountain, Guyana. All values are mass percent. The increase in water lost on ignition, in the ironstone, occurs because of the hydroxyls in the mineral structures of gibbsite, goethite and kaolinite. These are converted to water during heating, which is necessary for the chemical analysis: two structural OH groups → one structural O + H_2O. After Alexander and Cady (1962) and Fanning and Fanning (1989).

because a variety of *materials* with many types of compositions and various origins *also* had this lateritic tendency, and had been called laterites (Kellogg 1949, Carter and Pendleton 1956). These materials range from the pedogenic iron cappings on the plateaus of southern India to whole soils in the humid tropics, to iron-rich breccias (rocks) and slopewash accumulations (colluvium).

Another source of confusion was the abundance of names for the same, cemented, iron-rich material; whereas most called it laterite, others have referred to this material as ironstone, duricrust or ferricrete. Laterite is highly variable in hardness; where should the line be drawn? Using a hardness criterion, unfortunately, made it nearly impossible for soils to be classified as lateritic or as having a laterite horizon, at the time of exposure, i.e., in a freshly dug soil pit. Field soil scientists and engineers were required to make a field call about whether or not the soil contained laterite based on their experience, not on the morphology exposed in the pit. To avoid this problem, some relied on definitions based on chemical criteria. Laterites contain mixtures of hydrated oxides of aluminum and iron, often with a small percentage of other oxides, chief among them being manganese and titanium oxides (Fig. 12.24). Van der Merwe (1949) suggested that the term laterite should be further restricted to materials with free alumina, while most definitions stressed the ratio (SiO_2 / (Fe_2O_3 + Al_2O_3)), also written as SiO_2/R_2O_3, as a key chemical criterion. Soils with ratios between 1.33 and 2.0 were taken as indicative of lateritic soils, while those with ratios >2.0 were indicative of non-lateritic soils. (Care must be taken to exclude data on quartz silica from the SiO_2 part of the ratio.)

The term laterite has now fallen into disfavor. Most now prefer to use other definitions for hard, cemented soil materials, such as ferricrete or ironstone for iron-rich cemented crusts, alcrete or bauxite for aluminum-rich cemented crusts, calcrete for calcium carbonate-rich crusts, and silcrete for silica-rich cemented crusts (Fookes, 1997). In Soil Taxonomy, the term *plinthite* (Greek *plinthos*, brick) is now used in place of laterite (Sivarajasingham *et al.* 1962, Daniels *et al.* 1978b, Soil Survey Staff 1999). Plinthite is an iron-rich, humus-poor mixture of clay with quartz and other minerals that commonly occurs as dark red redox concentrations in platy, polygonal or reticulate patterns. Plinthite changes irreversibly to an ironstone hardpan or to irregular aggregates when exposed to wet–dry cycles and heat from the sun. Like laterite, moist plinthite is soft enough to be cut with a spade; after irreversible hardening it is no longer considered plinthite but is called ironstone. Horizons with plinthite, usually B horizons, are given the suffix v (Table 3.2). In this book, we use the terms laterite and plinthite almost interchangeably.

Plinthite is common in many Oxisols and even some Paleudults on old (Miocene to Pliocene) surfaces (Cady and Daniels 1968) and in Ultisols of the humid subtropics. It forms by segregation of iron species, much of which has probably been translocated into the site/horizon from overlying horizons or laterally. Generally, plinthite is thought to form in horizons that are saturated for some time during the year. Initially, iron is segregated in the form of soft, more or less clayey, red or dark red redox concentrations (see Chapter 13), but these concentrations are not considered plinthite unless there has been enough segregation of iron to permit their irreversible hardening.

Alexander and Cady (1962) hypothesized that laterite (plinthite) develops in one of three ways: (1) condensation from an iron-rich gel, (2) formation from rock pseudomorphs or (3) infilling of voids. It is usually dominated by goethite and hematite, with some kaolinite, although bauxitic laterites have abundant gibbsite and boehmite. Many laterites contain virtually no free aluminum minerals. Geomorphically, laterite is found in two settings: (1) as continuous crusts on flat uplands, where they act as hard caps on the table-like landforms (Alexander and Cady 1962) and (2) at footslopes, in seepage areas where reduced Fe in soil solutions encounters oxidizing conditions and precipitates.

An interesting and useful corollary to laterization is a discussion of the mechanisms that interact to harden plinthite into ironstone. That lateritic materials do harden irreversibly is important to millions of people in the humid tropics, for they depend on the laterite bricks as roadbed material, for home construction, and for many other uses. It is also vital that the soil's capacity to harden be known before it is cultivated. Thus, people in the tropics will frequently remove soil material and force it to undergo several wet–dry cycles, to see if it might be a useful roadbed material (Buol 1973).

At a minimum, two mechanisms occur when soft laterite (plinthite) hardens: crystallization of the Fe minerals, and dehydration of the soil mass. Drying of the iron-rich soil matrix causes crystallization of much of the previously amorphous iron compounds. Sesquioxides alone, if present in large quantities, can cause cementation. Alexander and Cady (1962) noted that residual iron accumulations are usually not adequate to develop hard laterite; additions of Fe in solution are usually necessary to provide enough Fe to form hard laterites. That is, Fe enrichment beyond just residual appears to be prerequisite to the formation of hard laterite; this process is now known as *laterization*.

Many tropical soils, especially those that either had or were considered as possibly developing laterite, were once classified as Latosols. Kellogg (1949, 1950) defined them as zonal soils whose dominant characteristics are (1) low SiO_2 : R_2O_3 ratios in the clay fraction, (2) low base exchange capacities, (3) an abundance of low-activity clays, (4) low contents of weatherable minerals, (5) low amounts of soluble constituents, (6) a high degree of aggregate stability, (7) no essential horizons of accumulation or illuviation, (8) thin A horizons, (9) low silt contents and usually (10) red colors. Latosols could contain laterite. The Latosol concept was later largely subsumed into the Oxisol order and the oxic horizon. Latosol is, however, still a popular and useful taxonomic term in parts of the world, especially Brazil (Tanaka *et al.* 1989).

Laterization

Laterization and *latosolization* are older terms used to describe processes in soils of the humid tropics, usually beneath broadleaf evergreen forest (Kellogg 1936). Both concepts revolve around long-term weathering of primary minerals, including phyllosilicate clays. Silica is leached (e.g., Fig. 12.24), leaving mainly iron and aluminum cations which combine with oxygen to form iron- and aluminum-rich clay minerals such as hematite, boehmite and goethite (Fig. 8.9). Formation of thick residual accumulations of oxides and metals is facilitated by the fact that these landscapes are old and flat and therefore clastic weathering products are not as rapidly carried away by streams as they are in the midlatitudes. Additionally, these landscapes have not been rejuvenated by glaciation.

Accumulations of Fe in soils are due both to residual accumulation (latosolization) and to processes that involve transfers of Fe (laterization). In *laterization*, Fe compounds *migrate* into and out of the soil, leading to concentrations at preferred sites; it places a premium on Fe mobility (Fanning and Fanning 1989). Much of the Fe probably came from sites outside of the profile, such as the saprolite below or from upslope. Therefore, processes that lead to Fe mobility are paramount, primarily movement in wet, waterlogged soils as ferrous Fe (Fe^{2+}) and in all soils, due to complexation (chelation) with organic matter. Aluminum is relatively immobile, which is why some tropical soils have minimal amounts of aluminum and why tropical soil classification systems stress Fe. Evidence for translocation of Fe comes from many sources. Alexander and Cady (1962) point to iron-cemented quartz, weathered out of a sandstone, as evidence that the Fe was brought to the site in solution, as Fe^{2+}. Accumulated Fe may eventually develop into plinthite.

The exact nature of the processes by which Fe accumulates in the soil profile remain poorly understood. Preferred sites of Fe accumulation may occur in certain parts of the profile where, for instance, oxidizing conditions exist. Many lateritic concretions are found in association with quartz gravels, implying that the Fe in the groundwater is oxidized and precipitated when it encounters these gravel layers, which are more oxygenated (Livens 1949). This form of laterite supports the redox mode of formation. Loss of the ability to translocate chelated Fe also occurs at water tables and lithologic discontinuities. Some Fe may move by capillarity, as water and dissolved ions in wet soil below is wicked into dry soil above. The type of laterite formed in this manner is called groundwater laterite. It may *form in lowlands* such as alluvial plains and swamps, but later, as the landscape is dissected and the lowlands become plateaus, it may be widespread across *upland surfaces*, forming a resistant caprock (Fig. 8.10).

The Fe in laterite does not accumulate evenly throughout the soil matrix, leading to the reticulate patterns so typical of plinthite. When overlying horizons are eroded, the plinthite may take the form of nodules or pisoliths, as described long ago by Buchanan (Beinroth *et al.* 2000).

Latosolization

Latosolization includes processes that lead to the formation of oxic horizons, or soils formerly known as Latosols (Fanning and Fanning 1989). In latosolization, *residual* sesquioxides accrue, as bases and silica are leached from the profile, under long-term weathering in a hot, humid climate; there is little assumed in-migration or translocation of Fe, as occurs in laterization. Some of the bases that are released by weathering are biocycled, leading to slightly higher concentration in the A horizon than at depth. The residual material that remains behind is rich in sesquioxides of Fe and Al, Ti and Mn oxides, oxyhydroxides and hydroxide minerals, as well as heavy elements like Ni, Zr and Cr (Beinroth 1982). Clay minerals such as kaolinite neoform from these residual ions. Perhaps because of the large amounts of residual ions present or because of the long time periods involved, the Fe and Al compounds in these soils are typically highly crystalline. That is, they have neoformed into minerals with discrete ordered structures that are difficultly weatherable. This is, in essence, the concept of the oxic (Bo) horizon genesis. An important side effect of latosolization, with its loss of soluble materials, is collapse; "the land shrinks like a rotten apple" (McFarlane and Bowden 1992).

Evidence for purely residual Fe accumulation in soils is given by Alexander and Cady (1962). They reported that, in West Africa, upland soils developed on residuum from basalt had much thicker laterite accumulations than soils developed on granite, because granite has fewer Fe-bearing minerals. A significant modification to this genetic model was provided by Brimhall *et al.* (1988, 1991b). Their data support the hypothesis that many upland tropical soils are enriched in mineralogically mature sediment from *eolian sources*, suggesting that only in the lowermost part of many Oxisol profiles is most of the Fe truly residual. Faure and Volkoff (1998) reported similar findings and suggested that the neutral pH values of some upland tropical soils may be due to inputs of Ca-rich dust from nearby deserts. Thus, it appears likely that not all of the Fe and Al in many Oxisols is truly residual.

Associated with laterization and latosolization is the almost complete destruction of organic matter. It has long been known that Oxisols contain low amounts of organic matter, although many contain more than their colors suggest (Gobin *et al.* 2000). The red colors, brought about by high amounts of Fe_2O_3 compounds, mask organic matter. Buol (1973) even suggested that the organic matter of some Oxisols may be almost colorless. Additions of litter from vegetation does little to add organic matter to the soil, because most of what falls is quickly decomposed and the constituent components are either taken up by flora, washed from the soil or leached. Few are retained in the soil proper because of the low CEC values of the oxide clays.

Latosolization often leads to the neoformation of oxide clays. These clay minerals are the dominant pigmenting agents in saprolites, producing the browns and reds so typical of tropical soils and saprolites. They form in soils where there is abundant free Fe^{3+}. Iron compounds are released by weathering as amorphous $Fe(OH)_3$ (in oxidizing situations) or as dissolved $Fe(OH)_2$ (rarely, in reducing conditions) (Hurst 1977). Eventually in moist soils, and immediately in the dry ones, the two compounds listed above dehydrate and crystallize as goethite, lepidocrocite, maghemite or hematite (Fig. 12.25). Hematite formation tends to be favored in environments that are warmer and drier, and in soils with lower organic matter contents because organic acids chemically complex with Fe, discouraging inorganic crystallization to hematite (Kämpf and Schwertmann 1982, Schwertmann *et al.* 1982) (Fig. 12.26). Hematite is more likely to form in warm, dry conditions because dehydration is a necessary process in its formation, and because at warmer sites soil organic matter contents are more likely to be low. Goethite is more stable in humid, cooler environments, in organic-rich soils (Fig. 12.26), and in more acidic soils where organic acids are prevalent (Kämpf and Schwertmann 1982).

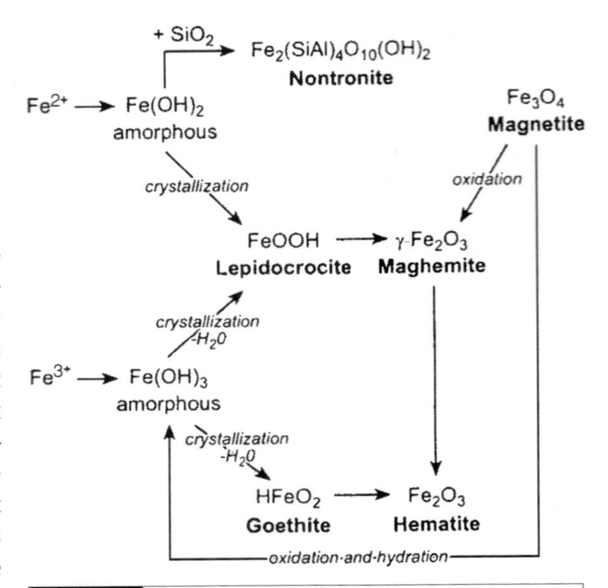

Fig. 12.25 Schematic diagram of the mineral synthesis pathways dominated by iron oxides. After Sherman *et al.* (1967).

Desilication

Central to the concept of latosolization, and necessary to the process of laterization, is the loss of silica from the profile. *Desilication* occurs when water that is undersaturated with silica percolates through a soil at the correct pH; it is facilitated by freely draining conditions in a hot, humid environment. Because the solubility of silica increases in warmer temperatures, on the order of about 1 ppm solubility per °C, soil solutions in the tropics are more likely to be undersaturated than are those elsewhere. As silica is lost from the profile, residual aluminum and iron remain. Desilication decreases in intensity with depth, as percolating water picks up silica and becomes less aggressive. Often desilication occurs concomitantly with the loss of bases (Curi and Franzmeier 1984) and some monomeric silicic acid: $Si(OH)_4$. The amount of silica removed depends, first, on residence time of the water around the silicate mineral, and the type of mineral (Buol and Eswaran 2000). Even in wet soils the groundwater may be so low in bases and

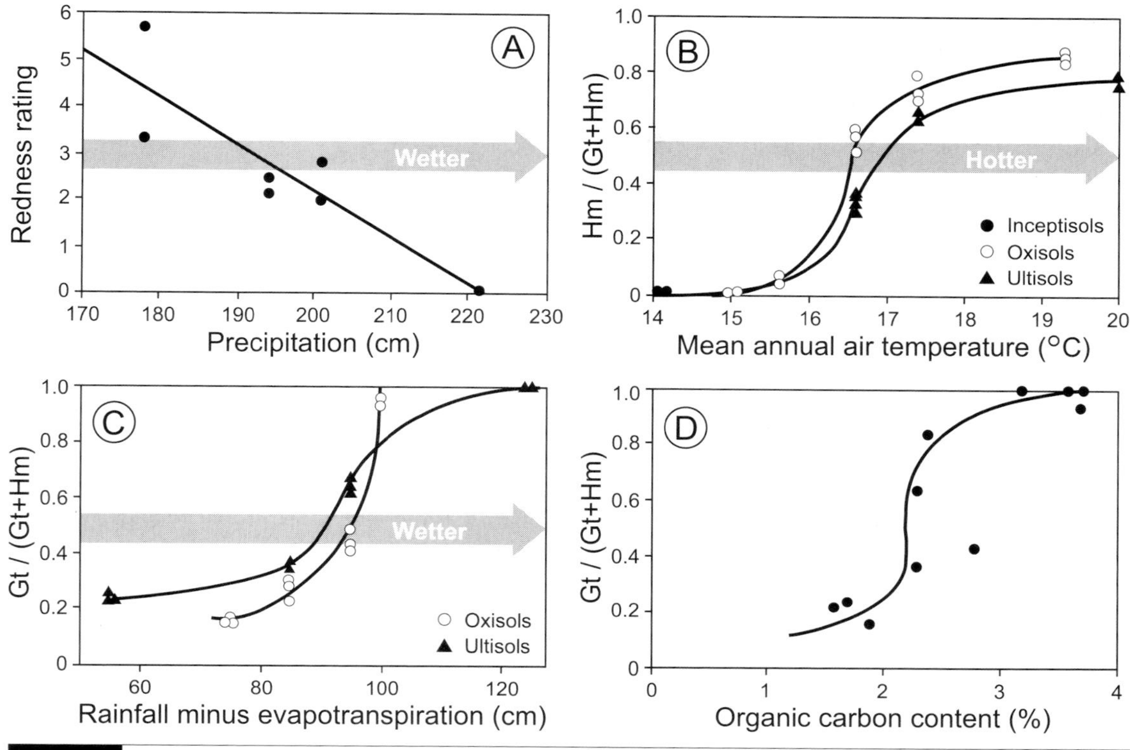

Fig. 12.26 Relationships between iron oxide mineralogy and various climate/soil factors. (a) Redness rating (Torrent *et al.* 1983) of soils vs. precipitation, Blue Ridge Mountains of western North Carolina. After Bryan (1994). (b) Hematite content (as a percentage of hematite + goethite) vs. air temperature, Brazil. After Kämpf and Schwertmann (1982). (c) Goethite content (as a percentage of hematite + goethite) vs. an index of excess moisture availability, Brazil. After Kämpf and Schwertmann (1982). (d) Goethite content (as a percentage of hematite + goethite) vs. soil organic carbon content, Brazil. After Kämpf and Schwertmann (1982).

silica that, despite the high water table, the soils are continuously flushed and leached, and 2 : 1 clay minerals cannot form.

The mobility/loss of silica from upland soil profiles is manifested in two ways: (1) collapse of saprolite and soils, and (2) precipitation as evaporites in seasonally waterlogged bottomlands where silica-rich vadose zone waters accumulate (McFarlane and Bowden 1992). Another important effect of desilication is a lowering of the Si : Al ratio and a concomitant enrichment of 1 : 1 and oxide clays (Buol and Eswaran 2000). When the silica potential is as low as it is in these soils, there is virtually no possibility for synthesis of 2 : 1 clay minerals (Soil Survey Staff 1999). Given the strong desilication that occurs in well-drained soils, however, even kaolinite is unstable and will weather. Evidence that desilication is strong and ongoing in many humid tropical soils is provided by $SiO_2 : R_2O_3$ ratios (Martini and Marcias 1974). In Fig. 12.21, it is evident that the $SiO_2 : R_2O_3$ ratio of the Udox is extremely low. Related soils all had $SiO_2 : R_2O_3$ ratios less than 3.0, indicating that they contain high amounts of kaolinite and gibbsite.

Desilication and long-term weathering can lead to high amounts of aluminum-rich minerals, although in some tropical soils even aluminum is soluble (McFarlane and Bowden 1992). Aluminum can lead to low pH values and if taken to extremes, biotoxicity. Liming is not always a solution to the low pH values, because the Al is such a strong buffer (Martini and Marcias 1974). Soil and subsoil residue that are especially high in Al are often mined as bauxite (Ahmad *et al.* 1966, Scholten and Andriesse 1986) (Fig. 8.2).

Landscapes: Soil geomorphology in old, tropical landscapes

Buol and Eswaran (2000) described three main types of Oxisol landscapes:

(1) On stable slopes, acid igneous rock is slowly weathered into soils rich in low-activity clays, over thick saprolite. Little lateral transport of material occurs and thus stone lines are uncommon. If Bt horizons do occur, the soils are Kandiudults or Kandiudalfs. The mineralogy of the soils largely reflects the bedrock parent materials (Scholten and Andriesse 1986).

(2) On younger landscapes formed from easily weatherable basic rocks, Oxisols are on the most stable surfaces, or sometimes in weathered alluvium. Andisols and Inceptisols are associated with volcanic materials.

(3) The most common landscape has a late-Tertiary, high plateau with deep weathering profiles (Du Preez 1949, Wright 1963). Oxisols on the plateau typically have a resistant plinthite cap 2–12 m in thickness (Thomas 1974) (Table 12.7). Nearer the dry edges of the plateaus, the cap is thicker and harder despite the increased erosive pressures, because the edge experiences drying most often and ferrous iron may migrate laterally to the edge, oxidize and precipitate (Hunter 1961). The laterite cap is like bedrock; soils at the edge are shallow and rubbly. Thick saprolite underlies the plateau center. On sideslopes, slow and continuous translocation of sediments is a central soil geomorphic process; stone lines are ubiquitous there. In many tropical landscapes, the laterite upland is all that remains of an old plain which has since been dissected (Kroonenberg and Melitz 1983, Dijkerman and Miedema 1988).

In parts of Australia and Nigeria, uplands are underlain by remnants of old weathering surfaces (plinthite or silcrete) (Muckenhirn *et al.* 1949, Gobin *et al.* 2000). Much of the geologic literature simply refers to these caps as duricrusts – a more generic name for a hardened, near-surface crust, whatever the cementing agent may be. In Australia's Northern Territory, Wright (1963) described a similar situation. Beneath continuous laterite are great thicknesses of sediment that contains fragments and nodules of ironstone; these grade to saprolite at depth. Occasional higher knolls on the uplands have Alfisols and Ultisols. Beneath the hard plinthite of the plateau surface one encounters, in order, kaolinite-rich oxic horizons, saprolite and finally hard bedrock. The soils on dissected edges of this landscape have formed in the kaolinite-rich material or the saprolite. These parent materials occur in broad bands along the dissected edges of the uplands. The soils formed in the kaolinitic materials also contain colluvium from the plateau surface. The residual soils located on the broad band associated with the bedrock are thin. Finally, at the base of the slope, alluvium and deep colluvium are the parent materials.

In the Ivory Coast, a hard laterite cap on the hilltops is also physically broken up at the margins, as softer material beneath is undercut. Fragments of broken laterite litter the side slopes. As forests get established on the side slopes, their organic acids help to chelate the ferric iron, allowing much of it to be translocated downslope to footslope positions, where it may recement laterite rubble (Alexander and Cady 1962). Likewise, iron-rich groundwater may penetrate the laterite rubble on footslopes and assist in recementation (Sivarajasingham *et al.* 1962).

In Hawaii, Oxisols dominate the flat uplands and Ultisols with argillic horizons (but otherwise similar mineralogical and chemically) are on side slopes (Beinroth *et al.* 1974) (see Figure). Clay translocation occurs on the side slopes because the geomorphic instability that occurs there. Soil creep leads to shear planes within the soils, causing the soil structure to break down and reactivate clay for illuviation. Argillans attest to this rejuvenation. In the bottoms of the valleys, the surfaces are young and Inceptisols are mapped.

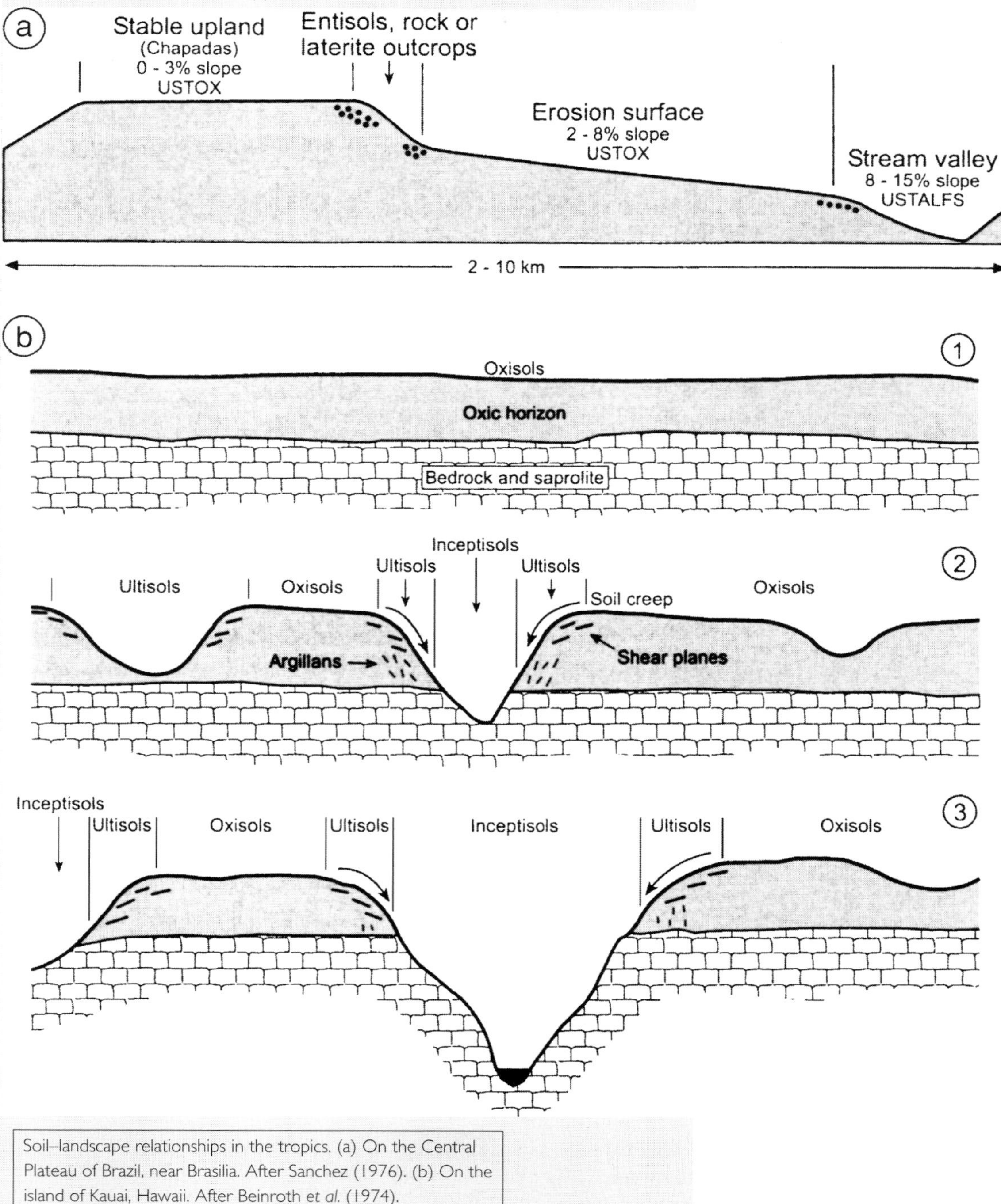

Soil–landscape relationships in the tropics. (a) On the Central Plateau of Brazil, near Brasilia. After Sanchez (1976). (b) On the island of Kauai, Hawaii. After Beinroth *et al.* (1974).

Table 12.7 Geomorphic position and soil types on a landscape in Nigeria with a wet–dry climate

Landform (landscape position)	Dominant soils (subgroup classification)
Plateau top	Typic and Rhodic Kandiudox
Sandstone escarpment, plateau edge	Ustoxic Quartzipsamments
Lower, eroded hills and uplands (off the plateau)	Ustic Kandihumults, Typic Kanhaplustults, Typic Plinthaquults and Plinthustults, Typic Plinthohumults
Floodplains	Aeric Endoaquents
River's edge	Oxyaquic Quartzipsamments

Source: Gobin *et al.* (2000).

Latosolization involves more than just weathering and desilication, however, since these processes alone cannot convert rock to soil. Rather, they convert it to *saprolite*. Saprolite is unlike soil in that it retains *rock structure* (see Chapter 8). The soil profile has *soil structure* and *soil horizons*, which in tropical regions are formed not just through desilication and accumulation of residual oxides, but also through the chemical and biomechanical actions of plants and fauna. Soil fauna disrupt and mix the weathered rock, moving some of it to the surface. Termites, in particular, mine parts of the subsurface and move those components to the surface, creating a biomantle (see Chapter 10). Additionally, lessivage and melanization are at work to form clay-rich kandic horizons and organic matter-enriched A horizons (Table 7.3). Where bioturbation is strong, any existing kandic horizons rich in illuvial clay are destroyed and converted to red oxic horizons.

Three-phase tropical pedogenesis

A widely accepted paradigm for tropical soil genesis was proposed by Philippe Duchaufour (1982). His three phases of weathering for tropical climates are assumed to occur in the general absence of organic matter, and although not technically a development series, they can be viewed as such (Table 12.1, Fig. 12.27). From phase 1 to phase 3, soils show increased (1) weathering of primary minerals, (2) loss of silica and (3) neoformation of clay minerals, formed from weathering by-products (Singh *et al.* 1998). The three-phase pedo-weathering sequence ties together many of the concepts that have been discussed earlier in this section.

Phase 1 is *fersiallitization*. These soils, typical of Mediterranean regions and hot semi-arid climates, are not strongly weathered. Although this process suite can occur on any parent material, it is best expressed on porous, calcareous substrates, e.g., terra rossa soils (see Chapter 8). The red colors indicate that these soils contain significant amounts of iron oxides, but little silica has been lost. Base saturation remains high, and Bt horizons are enriched in 2 : 1 clays. Many of these soils would classify as Ultisols with argillic horizons. As the name implies, these soils retain high amounts of Fe, Si and Al.

Phase 2 is *ferrugination*. In the transitional phase 2, weathering increases in intensity, but some primary, weatherable minerals can persist. Desilication increases in intensity and some 1 : 1

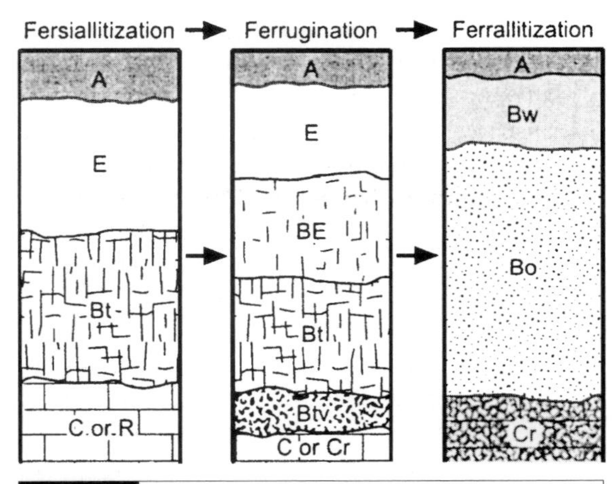

Fig. 12.27 Duchaufour's (1982) three-phase weathering/pedogenesis sequence for warm and hot climates. Typical profiles are shown (not to scale), but many other morphologies are possible.

clays, such as kaolinite, neoform. Conditions are not right, however, for the formation of gibbsite. Base saturation is lower, but lessivage is still active. Plinthite and kandic horizons can form. These soils are typical of subtropical savanna landscapes.

Phase 3 is *ferrallitization*. These acid soils are strongly weathered; kaolinite and oxide clays dominate (Latham 1980). Of the primary minerals from the parent material, only quartz remains. The soil is enriched in Fe and Al oxides and desilication has resulted in a loss of silica. Gibbsite and oxide clays neoform. Lessivage is not strong, as the clays have become increasingly resistant to dispersion. Many of these soils are Oxisols on thick sequences of saprolite.

Rubification

Rubification is the reddening of soils or regolith, usually by various forms of oxidized iron minerals such as hematite (Schwertmann *et al.* 1982) (Table 12.1). Red soils are almost always on uplands. From a global perspective, they are most common in xeric soil moisture regimes and the humid tropics.

Factors important to rubification include: good aeration, warm temperatures, parent material rich in Fe-bearing minerals, a wet season sufficient to permit growth of microorganisms, specifically iron bacteria, and rapid turnover of organic matter (Glazovskaya and Parfenova 1974, Boero and Schwertmann 1989). Fractured limestone uplands in Mediterranean climates are well suited to provide many of the above factors. Yaalon (1997b) stressed that rubification is a corollary feature resulting from the long summer dry period. This fact seems to be substantiated by Yassouglou *et al.* (1997), who noted that soils in the Mediterranean region tend to be redder on south-facing slopes, which would experience a hotter, drier summer than soils on north-facing slopes. Fractured limestone, which underlies terra rossa soils (see Chapter 8), also facilitates drainage and drying. On soft, less permeable marls, dark-brown Rendolls typically form, while adjoining, fractured limestones usually are host to redder soils with more hematite, probably because they have a drier soil microclimate (Boero and Schwertmann 1989, Atalay 1997).

The red and yellow colors in tropical soils come from an abundance of Fe_2O_3 and Al_2O_3 minerals (Eswaran and Sys 1970). Even though Al_2O_3 is colorless, it is also correlated with red colors, since Fe_2O_3 and Al_2O_3 contents are often positively correlated. Red tropical soils contain mostly ferric iron in the form of fine-grained hydroxides and oxides, as most of the Fe-rich silicate minerals have been weathered away. The hue of these soils is more a function of which ferric pigment (mineral) predominates (Hurst 1977) than it is of the *amount* of iron (Plice 1942) (Table 2.3). For example, Soileau and McCracken (1967) found that the redness of Ultisols was not as much due to the *thickness of iron coatings* on soil particles as it was due to the *form* of their iron oxides. Bigham *et al.* (1978) pointed out, however, that the issue of color and Fe oxide content is complicated; soil color is also affected by the concentration, particle size and distribution of the iron oxides.

The actual degree of redness or yellowness is best related to the amount of each particular iron mineral, for each has a unique hue and intensity of coloration (Waegemans and Henry 1954, Buol and Cook 1998). The yellowish-brown soils of the midlatitudes tend to have high amounts of lepidocrocite and goethite, although Davey *et al.* (1975) suggested that high amounts of aluminum oxides can impart a yellowish hue to soils that otherwise would have been red due to high Fe_2O_3 contents. In tropical soils, goethite is more common in brownish horizons, kaolinite is most common in the white and yellow horizons; deep red colors are associated with hematite (Table 2.3). Soils with 7.5–10YR hues have at least some goethite and little or no hematite (Fig. 12.28). Hues between 2.5YR and 10YR generally indicate that hematite content exceeds goethite content (Eswaran and Sys 1970, Bigham *et al.* 1978). Deep, beet-red colors are usually associated with submicron-size hematite (Hurst 1977). Because color is a good indicator of iron oxide mineralogy, therefore, it is used at high categorical levels in several tropical soil classification systems (Soil Survey Staff 1999).

At different landscape positions, water table relations and wetness affect the types of minerals that are likely to form, and hence soil colors.

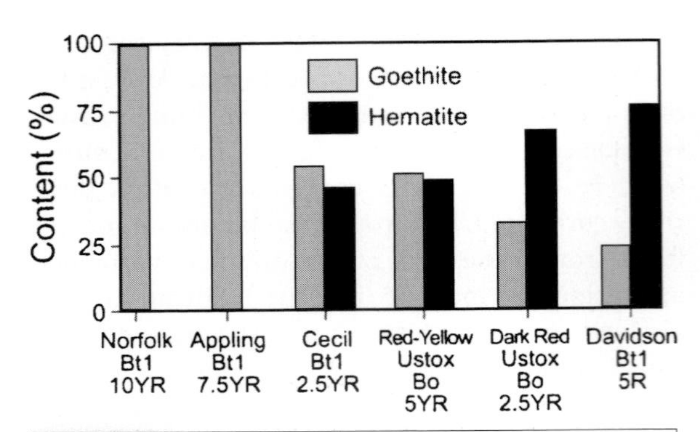

Fig. 12.28 Mineralogical composition of the iron oxide fractions of selected soil horizons from some Ultisols from North Carolina and some Oxisols from Brazil. After Bigham *et al.* (1978).

Redder colors, associated with hematite, are more common on dry uplands (Curi and Franzmeier 1984) and in dry climates (Fig. 12.29). Wetter soils with higher water tables will generally have yellower hues and more goethite (Macedo and Bryant 1987). Even short periods of reduction, for example around decaying roots, will lead to the removal of hematite, leaving yellow-brown goethite as the dominant mineral (Buol and Eswaran 2000). Under prolonged wet conditions, gleying can and will occur in Oxisols as in other soils, leading to gray colors and a lack of oxidized iron (see Chapter 13).

Because determining Fe_2O_3 content in the laboratory is a fairly time-consuming process, several studies have attempted to develop quantitative relationships between it and color (see Chapter 14). Early attempts (Soileau and McCracken 1967) found no consistent relationship between hue and free iron oxide content, probably because hue, alone, does not capture the essence of coloration provided by iron oxides. However, Hurst's (1977) redness rating index, although derived for use in saprolite, has been successfully applied to soils (Liebens and Schaetzl 1997). In a soils application, Torrent *et al.* (1983) determined that there is a strong relationship between the redness rating (modified from Hurst's, see Torrent *et al.* (1980b)) and hematite content, both for tropical Oxisols as well as for a variety of red soils from Europe (Fig. 12.29).

Processes associated with moisture deficits

In this section, we focus on soil genesis in landscapes characterized by a marked deficiency of soil water in some or most of the year, e.g., from deserts to mid-grass grasslands (Fig. 12.30). The suite of dryland pedogenic processes primarily involves varying degrees of melanization and weathering, coupled with translocation of soluble materials into and within, but not necessarily *out of*, the profile. After each rainfall event, some of the more readily soluble compounds are brought into solution and translocated downward. At some depth the wetting front stops and these compounds are deposited. Because the soils seldom get completely flushed by deep wetting events, the B and C horizons continue to accumulate soluble compounds. As with clay, continued illuviation can eventually result in the plugging of pores and the formation of an aquiclude. Because the pedogenic imprint in dryland soils is in the lower profile, soils in aridic soil moisture regimes typically have their diagnostic horizons in the subsurface; many are Aridisols (Table 7.4).

The type of soluble materials that are translocated in these soils is a function of availability (Table 12.8). How deep they get translocated is a function of solubility, precipitation and permeability, i.e., how deep the wetting fronts typically move. Salts and soluble compounds have different mobilities and solubilities in soils:

$$Cl^- > SO_4^{2-} > HCO_3^- > CaSO_4\ 2H_2O > CaCO_3$$

(chlorides > sulfates > bicarbonate > gypsum > carbonates).

In upland soils, the most soluble materials are translocated the deepest. In lowland soils that are shallow to a water table, soluble materials get wicked upward by capillarity and deposited in the soil profile, meaning that the most soluble materials are found in the *highest* parts of the profile. Therefore, many possibilities exist for varying salt contents and distributions within desert soils and landscapes. In upland soils on the more humid end of dry climates, only $CaCO_3$ may remain in the profile, as other materials will have been leached. In soils that are a bit drier, carbonates and gypsum may still remain within the profile,

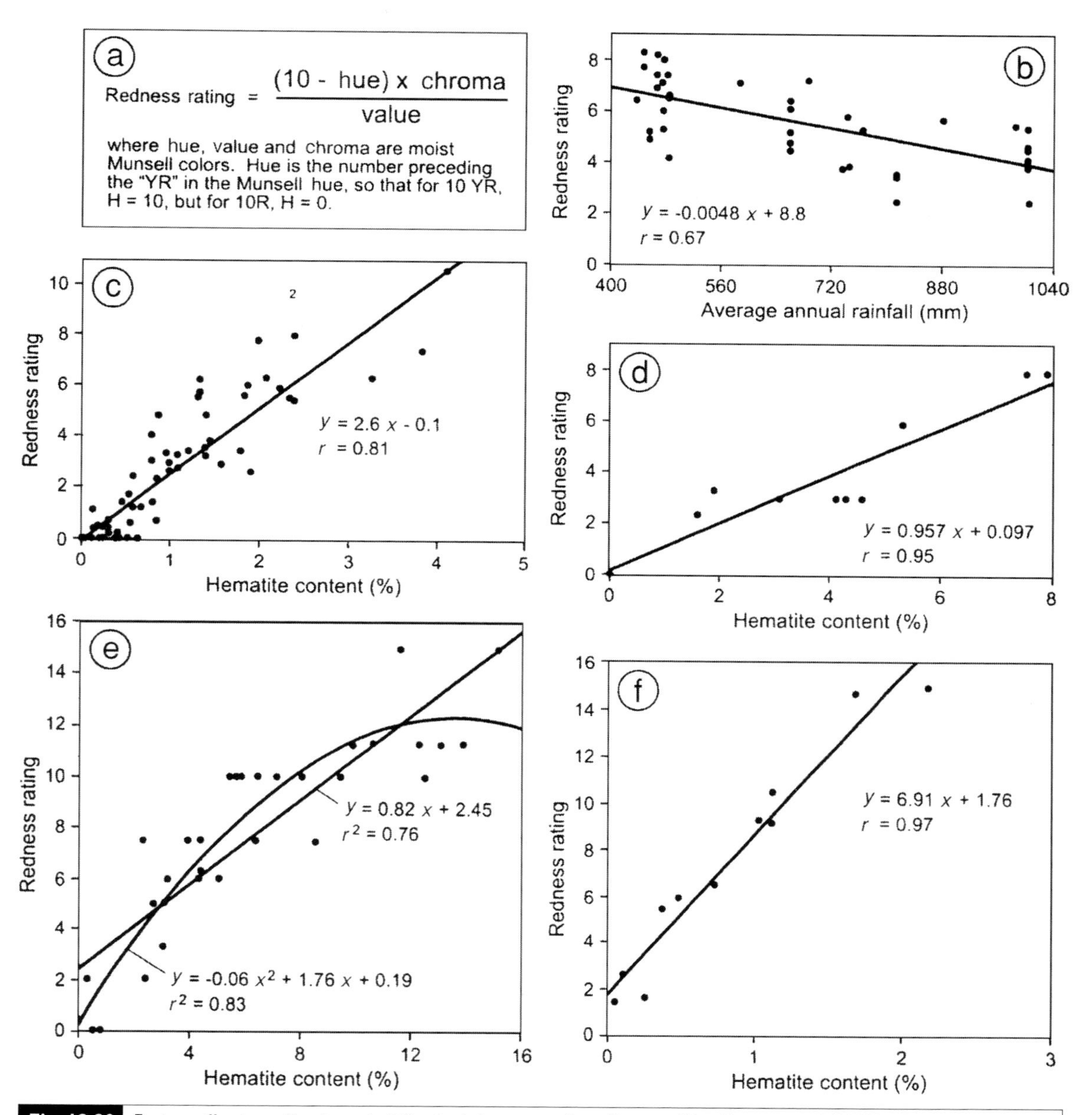

Fig. 12.29 Factors affecting soil redness (rubification). In most studies, the B horizon color is examined because the organic matter contents of the A horizon mask colors. (a) The Redness rating formula of Torrent *et al.* (1983). (b) Relationship between mean annual precipitation and the redness rating index of Torrent *et al.* (1983), for some soils in Greece. After Yassoglou *et al.* (1997). (c) Relationship between hematite content and the redness rating index of Torrent *et al.* (1983), for some Alfisols, Inceptisols and Ultisols from Europe. After Torrent *et al.* (1983). (d) Relationship between hematite content and the redness rating index of Torrent *et al.* (1983), for some Ultisols and Inceptisols from North Carolina. After Graham *et al.* (1989). (e) Relationship between hematite content and the redness rating index of Torrent *et al.* (1983), for some Oxisols and Ultisols from Brazil. After Torrent *et al.* (1983). (f) Relationship between hematite content and the redness index of Hurst (1977), for some Xeralfs in Spain. After Torrent *et al.* (1980).

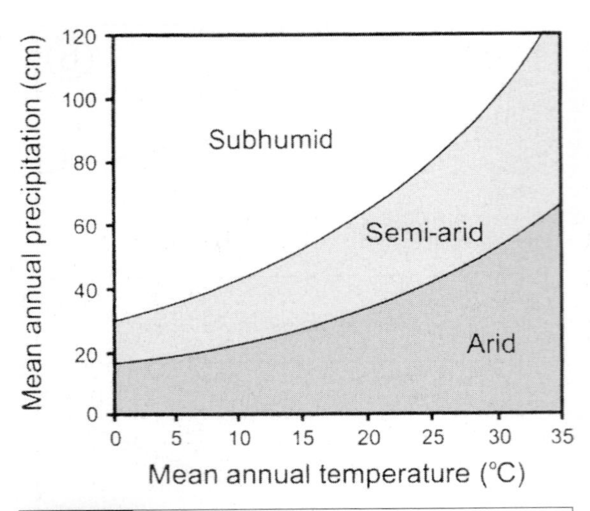

Fig. 12.30 Generalized limits of arid, semi-arid and subhumid climates, based solely on mean annual temperature and precipitation. Hyperarid climates, not shown, would lie in the extreme lower right corner of the graph. After Bailey (1979).

but the gypsum max will typically be deeper, as gypsum is more soluble. Chlorides are found in upland soils only in the driest deserts, and then only if there is a source of chlorides, such as sea water or saline groundwater, nearby. The rule of thumb is: as the climate gets drier, salts in upland desert soils become more common and closer to the surface.

Calcification

Calcification refers to the accumulation of secondary calcium and magnesium carbonates in soils. It is a dominant process in many dry soils, because aridity coupled with the low solubility of $CaCO_3$ make them difficult to leach (Southard 2000). Calcification is common in subhumid grassland soils (Sobecki and Wilding 1983) (Figs. 12.30, 12.31) and in humid climates where soil horizons are clay-rich and slowly permeable (Wenner *et al.* 1961, Schaetzl *et al.* 1996). In short, wherever a source of calcium exists and there is inadequate water (energy) to translocate it from the profile, secondary carbonates can accumulate (Stuart and Dixon 1973).

Because the dominant form of secondary carbonates is $CaCO_3$, soils with abundant subsurface

Table 12.8 Common chemical species brought to desert soils via precipitation, and their origins

Constituents	Major origins
Chlorides (Cl^-)	Sea water; dissolution of particulate chlorides
Sulfates (SO_4^{2-}, HSO_4^-)	Sea water; lake waters; dissolution of sulfate particulates; condensation of gaseous H_2SO_4; sulfite oxidation by O_2, O_3, H_2O_2 and metal catalysts
Sulfites (H_2SO_3, HSO_3^-, SO_3^{2-})	SO_2 dissolution (particularly important near sites of industrial emissions)
Carbonates (CO_3^{2-}, HCO_3^-, H_2CO_3)	CO_2 dissolution; dissolution of particulate carbonates; sea water; lake water
Nitrites (HNO_2, NO_2^-)	Dissolution of NO, NO_2 and HNO_2
Nitrates (NO_3^-)	Nitrate oxidation by O_3; dissolution of gaseous NO_2^- and HNO_3; dissolution of particulate $NaNO_3$ from soils
Ammonium (NH_4OH, NH_4^+)	Dissolution of gaseous NH_3; dissolution of particulate $(NH_4)_2SO_4$ and NH_4NO_3
Hydroxyls (OH^-)	Water dissociation
Active hydrogen (H^+)	Water dissociation
Metal cations (Na^+, K^+, Ca^{2+}, Mg^{2+}, Fe^{3+})	Sea water; lake water; dissolution of particulates

Source: Pye and Tsoar (1987).

Fig. 12.31 A Typic Argiustoll from Texas, with accumulations of secondary carbonates in the subsoil.

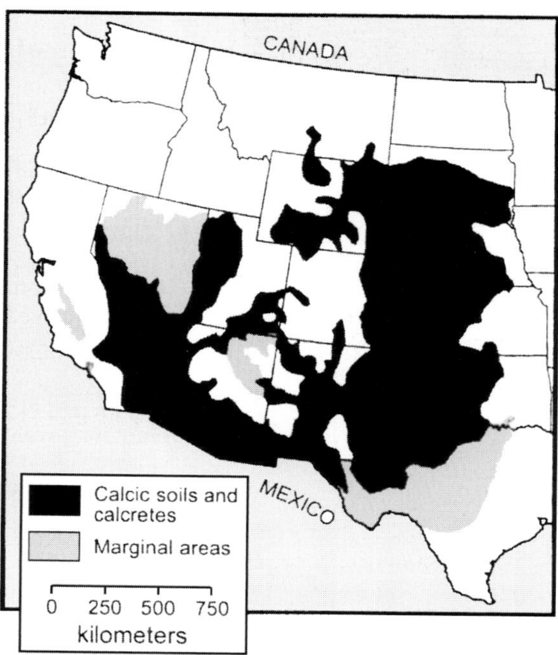

Fig. 12.32 Distribution of calcic soils in the western United States. Marginal areas have discontinuous or poorly preserved calcic soils. After Machette (1985).

carbonates are referred to as *calcic soils* – roughly equivalent to Pedocals (Marbut 1935, Jenny 1941a) (Figs. 7.2, 12.32). If the $CaCO_3$-enriched B horizon meets certain criteria, it is a diagnostic *calcic* (Bk) *horizon* (Soil Survey Staff 1999). Calcic horizons are characterized by layers of carbonate-coated pebbles or thin filaments of carbonate, but can also develop thick, massive, indurated horizons that resemble limestone (Machette 1985). The latter horizons are called *petrocalcic* (Bkm) (Soil Survey Staff 1999). Other, older terms for petrocalcic horizons include *caliche* and *calcrete* (abbreviated from "caliche concrete") (Aristarain 1970, Blümel 1982, Dixon 1994). Birkeland (1999) and other pedologists working in the western United States prefer the term *K horizon* (German *Kalk*), first proposed by Gile *et al.* (1965) because there was no horizon designation in 1965 for horizons dominated by cemented $CaCO_3$. We acknowledge the use of the K horizon term in the literature but use the more accepted Bk and Bkm horizon nomenclature because (1) there currently *exists* a term for such horizons (petrocalcic), (2) the K horizon is not officially recognized by the Soil Survey Staff (1999), and (3) it can be confused with Butler's (1959) K-cycle terminology (Chapter 13).

The process

Salomons *et al.* (1978) and Monger (2002) reviewed the main models by which the carbonates get deposited in the vadose zone of soils: (1) *in situ* dissolution and reprecipitation (Blank and Tynes 1965, Treadwell-Steitz and McFadden 2000), (2) upward capillary flow from shallow groundwater, aka the *per ascensum* model (Nikiforoff 1937), (3) various biogenic models, and (4) the *per descensum* model which involves carbonate-rich solutions descending in percolating water, to be deposited as the wetting front stops (Gile *et al.* 1966). The *per descensum* model has garnered the most attention and is our focus (Durand 1963, Reeves 1970, Blümel 1982) (Fig. 12.33). The last three models involve three stages: (1) provision of a carbonate-rich solution, (2) movement of that solution within the soil or near-surface environment and (3) precipitation of carbonates from solution. Carbonates are rendered soluble through a process called *carbonation* (see Chapter 9). Carbonic acid (H_2CO_3), formed as CO_2 and water

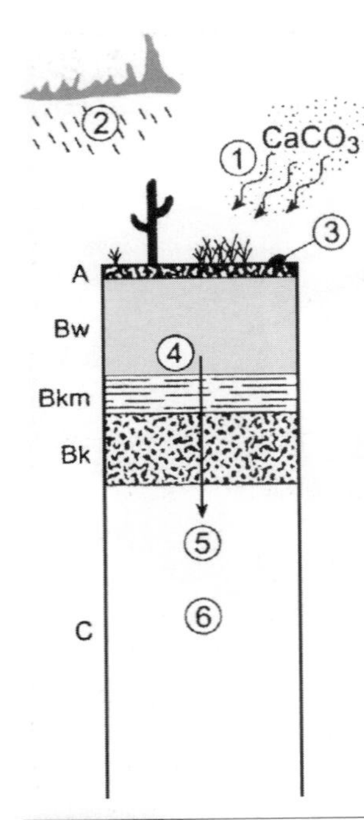

2
1 CaCO3
3
A
Bw
4
Bkm
Bk
5
6
C
Formation of calcic horizons
1 Carbonate (CaCO3)-rich dust blows onto soil surface
2 Water in precipitation combines with atmospheric and soil carbon dioxide (CO2) to make weak carbonic acid (H2CO3)
$H_2O(l) + CO_2(g) \rightleftharpoons H_2CO_3(aq)$
3 Calcium carbonate in (and on top of) the soil reacts with carbonic acid, rendering it mobile
$H_2CO_3 + CaCO_3 \rightleftharpoons Ca^{2+}(aq) + 2HCO_3^-(aq)$
4 Calcium and bicarbonate ions are translocated in the soil with percolating water
5 Dry conditions at depth lead to precipitation of secondary carbonates
$Ca^{2+}(aq) + 2HCO_3^-(aq) - H_2O(l) \rightleftharpoons CaCO_3 + CO_2(g)$
6 Deep wetting events may cause resolution of secondary carbonates and translocation / precipitation at still greater depths

Fig. 12.33 Chemical reactions involved in the dissolution and reprecipitation of soil carbonate, and a general model of the formation of calcic horizons.

combine, reacts with $CaCO_3$ and renders it mobile, as Ca^{2+} and HCO_3^- ions:

$$CaCO_3 + H_2CO_3 \leftrightarrow Ca^{2+} + 2HCO_3^-$$

The Ca^{2+} ions move in soil water. When at some point in the soil the reaction is driven to the left, $CaCO_3$ is precipitated as secondary carbonate. Conditions that drive carbonate dissolution include a moister soil environment (as long as the water is not saturated with $CaCO_3$ and its by-products), lower pH values, higher CO_2 contents in the soil air and cooler temperatures (Brook *et al.* 1983). Although temperature affects carbonate chemistry to a lesser extent than do the other factors and conditions, cold water *is* able to dissolve more carbonate than warm water (Fig. 12.34). The temperature effect is only important when comparing calcification between regions, not locally. This discussion assumes that the permeability of the soil exceeds the rate at which wetting fronts move through it. If, for example, a soil has a clayey, slowly permeable layer, water with dissolved carbonates may tend to perch there, and as roots take up the water the carbonates will be precipitated (Schaetzl *et al.* 1996).

Carbonate precipitation is driven by decreases in soil moisture and carbon dioxide partial pressure (pCO_2) and by increases in pH (Salomons *et al.* 1978). Desiccation occurs as the wetting front simply stops due to lack of energy (i.e., there is no more water to push it downward, and neither are there enough matric forces to pull it downward). Uptake by plant roots also may force the wetting front to stop. Wetting fronts may also stop if they hit an aquiclude or aquitard. For many desert soils, the aquitard is an existing petrocalcic horizon, forcing additional precipitation of carbonates on top, further thickening it and setting up a positive feedback mechanism in which the petrocalcic horizon grows upward through time. The importance of percolating water to the process of carbonate deposition is illustrated by studies which document soils with carbonate deposits immediately above a lithologic discontinuity (e.g., Buol and Yesilsoy 1964, Stuart and Dixon 1973). Gypsum also depresses the solubility of $CaCO_3$ (Reheis 1987b). In theory, rainfall will dissolve both gypsum and carbonate present at the soil surface (from dustfall) and translocate them into the soil. However, as soon as the soil solution accumulates a significant amount of gypsum, the

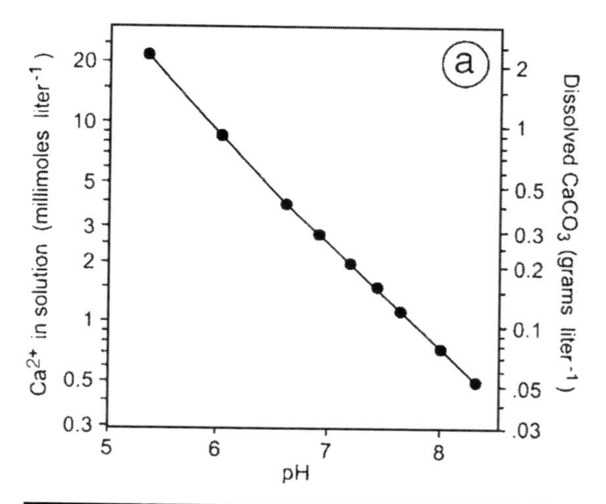

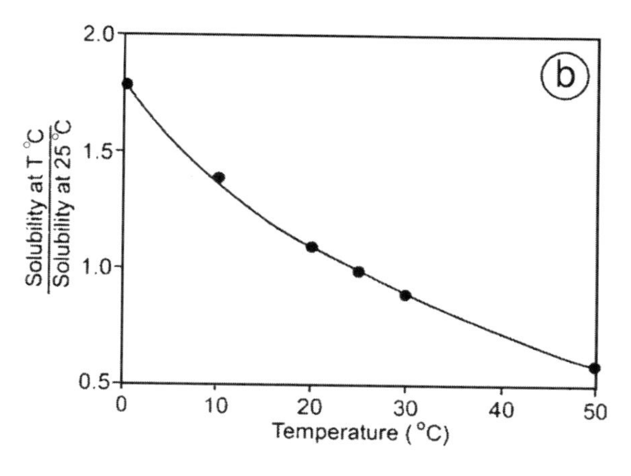

Fig. 12.34 Relationships pertaining to the solubility of $CaCO_3$ in soils. (a) Solubility of $CaCO_3$ in relation to the pH of the equilibrium solution. (b) Solubility of $CaCO_3$ at different temperatures. After Arkley (1963).

addition of Ca^{2+} and SO_4^{2-} ions will depress the solubility of $CaCO_3$ and it will precipitate (Reheis 1987b). Thus, the more gypsum in a soil, the less soluble $CaCO_3$ will become, providing a feedback mechanism to keep the gypsum max lower than the carbonate max.

The main sources of CO_2 in soil air are respiration from plant roots and decaying organic material, in addition to that which diffuses in from the atmosphere. As a result of respiration, the pCO_2 of soil air is much greater than atmospheric levels, driving the carbonate reaction toward dissolution (Rabenhorst *et al.* 1984). Because of the diffusion of atmospheric air (with its low CO_2 contents), soils tend to exhibit an increase in CO_2 concentration with depth (Boynton and Reuther 1938). One might then postulate that the increased pCO_2 with depth facilitates increased amounts of carbonate dissolution in the lower profile, leading to a lack of carbonate deposition. However, most researchers now agree that the main reason that carbonates are deposited at depth centers on desiccation (which can include desiccation by freezing) (Cox and Lawrence 1983), rather than changes in pCO_2 content.

Soil CO_2 contents also respond to climate and vegetation forcings. When conditions are moist, e.g., during a wet climatic interval, soil organic matter contents rise as vegetation responds to the increased precipitation. Increased CO_2 emissions from higher numbers of roots, coupled with larger populations of CO_2-emitting microbes, will in turn increase CO_2 partial pressures, pushing the reaction to the right and dissolving more carbonates. Higher CO_2 partial pressures also will lower soil pH. Likewise, enhanced precipitation will force wetting fronts and carbonates to deeper into the profile.

Because carbonates are translocated more deeply in wet than in dry years, the depth to secondary carbonates in many soils is thought to reflect, regionally, mean annual precipitation (Arkley 1963, Sehgal and Stoops 1972) (Fig. 12.35). Jenny (1941b) and Jenny and Leonard (1934) found that the depth to the top of the Bk horizon (D), in cm, is directly proportional to mean annual precipitation (P), in cm:

$$D = 6.35(P - 30.5).$$

Extrapolating this relationship to $D = 0$ would, however, imply that the Bk horizon should be at the surface (i.e., no leaching) in climates where $P \leq 30$ cm (Fig. 12.35a). This is not the case because the actual depth to the calcic horizon also depends on local factors such as slope, position on the landscape, permeability, presence or absence of a surface crust, vegetation cover, and the temporal distribution and intensity of precipitation. For example, runoff is common in deserts because of the intensity with which precipitation often falls and because the soil surface often has

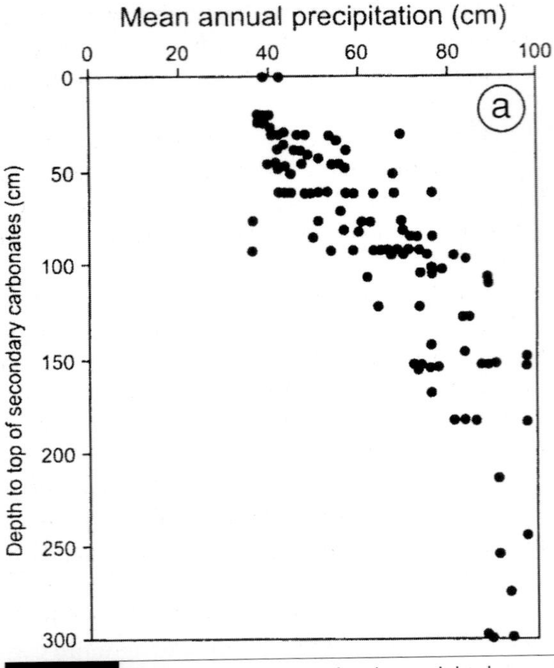

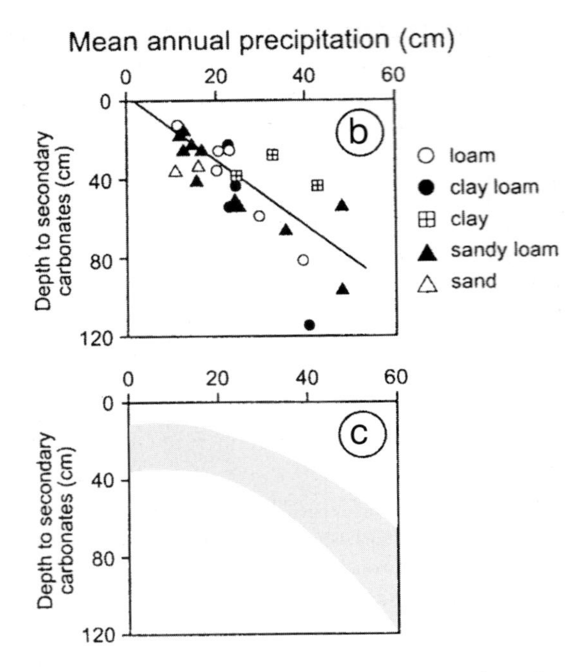

Fig. 12.35 Relationships between the observed depth to the top of the zone of secondary carbonates and mean annual precipitation. (a) After Jenny (1941b). (b) After Arkley (1963). (c) After Yaalon (1983).

a slowly permeable crust or is dotted with gravel and rocks. Runoff from steeper slopes leads to pedogenically drier conditions there, while sites at the base of slopes are pedogenically wetter due to run-on (Yair *et al.* 1978). Thus, thickness of, and depth to, B horizons may be greater at the bases of slopes that on level or steep surfaces. Additionally, in arid and hyperarid climates, rainfall is often intense and concentrated in only a few, large storms. Thus, there is almost always *some* leaching; Bk horizons do not form at the surface. Also, the depth to carbonates may be more closely related to the few, large precipitation events that such soils receive, than to some sort of annual mean (Amundson *et al.* 1997). Yaalon (1983) described a curvilinear relationship between rainfall and depth to carbonates which may be closer to reality (Fig. 12.35C). Thus, it is clear that *soil climate* is not the same as *atmospheric climate*.

Because calcic horizons form slowly and can persist in landscapes for long time intervals, the depth to the calcic horizon is also a function of paleoclimate, making the relationships in Fig. 12.35 even more problematic (Netterberg 1969a, McDonald *et al.* 1996, Khokhlova *et al.* 2001). Carbonates that had been precipitated at depth in a previous, dry climatic interval may be redissolved and driven deeper in the profile. During wet climatic intervals, e.g., the latest Pleistocene in the deserts of the southwestern United States (Phillips 1994), deep Bk horizons may form. If followed by a drier period, e.g., the Holocene, soils may develop a Bk horizon higher in the profile, and thereby exhibit two Bk horizons (McDonald *et al.* 1996). A modeling study by McDonald *et al.* (1996) pointed out that processes operative in "wet years" in dry soils today are good analogs to "normal years" during Pleistocene pluvial climatic intervals.

Computer models point to the complexity of calcification. They show that the depth of Bk horizons in dryland soils is primarily a function of climate, but texture, coarse clast content, dust influx, soil CO_2 contents, soil age, presence/absence of gypsum, and type and density of vegetation cover are also important (Marion *et al.* 1985, Mayer *et al.* 1988, McFadden *et al.* 1991). Most importantly, the depth of carbonates is dramatically influenced by eolian inputs of carbonate-rich dust (see below). Dust not only provides a source of carbonate but dramatically affects the way in which water moves through the profile (Treadwell-Steitz and McFadden 2000).

SOURCES OF CARBONATE

There are three possible sources of pedogenic carbonate: groundwater, soil parent material or external sources which deliver carbonate directly to the soil surface. Plants are capable of biocycling Ca^{2+} (Goudie 1996) but only in rare situations do they actually *add* Ca to the profile. Machette (1985) pointed out that, in most dry environments, groundwater is too deep and the overlying sediment too coarse for capillary action to bring carbonate-rich water into the profile. Where this process does occur, e.g., in lowlands and on floodplains, the carbonate zones can be indurated to depths of 10 m or more (Machette 1985).

Although chemical weathering is slow in dry environments, carbonaceous rocks like limestone and other rocks rich in Ca^{2+}, e.g., basalt and granites rich in Ca-feldspars, can release significant amounts of Ca via weathering (Blank and Tynes 1965, Kahle 1977, Rabenhorst and Wilding 1986b, Boettinger and Southard 1991). However, the release of cations by weathering is exceedingly slow in dry environments, and thus is not considered a *major* source of carbonate (Lattman 1973). If rock weathering *were* a major carbonate source, trace elements such as Al and Ti that are found in the rock would also be present in the secondary carbonate; they usually are not (Aristarain 1970).

Many Aridisols, especially those on surfaces of Pleistocene age or older, have accumulated pedogenic carbonate within parent materials that are essentially carbonate-free. For example, soils at the top of the Mormon Mesa, Nevada, have Bkm horizons over 1 m in thickness, but they have almost no carbonate in their parent material (Gardner 1972). If parent material alone had been the source of the pedogenic carbonate in this, and similar, desert soils with Bk and Bkm horizons, the insoluble residue left behind would be much thicker than what is presently observed. In the case of the Mormon Mesa soil, Gardner (1972) estimated that over 36 m of parent material would have to have weathered *in situ* to produce the soil carbonate present today. In cases like this, an external carbonate source must be invoked.

Landscapes: Southern Arizona Basin and Range

Centered on the state of Nevada and extending from southern Oregon to western Texas, the Basin and Range is an immense physiographic region of north–south-trending, faulted mountains separated by wide, dry bolsons (basins of interior drainage). In Southern Arizona this landscape is part of the Sonoran Desert with its trademark cactus, the stately saguaro, with its branched, tree-like form (see Figure). What little rainfall that does fall occurs mostly in late summer. The basin and range was formed about 20 million years ago as the Earth's crust stretched, thinned, and then broke into some 400 mountain blocks that partly rotated from their originally horizontal positions. To add to the complexity, Miocene volcanoes in what is now Arizona and Mexico emitted silicic lava and ash. Since the Pliocene, mass wasting and erosion of adjacent mountain ranges have gradually filled the basins with thousands of meters of sediment. Massive alluvial fans and bajadas have formed at the contact between basins and ranges.

Lithic Entisols and Aridisols dominate the weathered bedrock uplands. Soil development in the basins depends on the age of the geomorphic surface and sediment type. Many of the older surfaces consist of alluvial fans and pediment surfaces covered with alluvium. Here, soil development ranges from very strong, e.g., the Hickiwan series with its thick petrocalcic horizon at 35 cm, to Calcids and Argids of moderate development (see Figure). Downslope, these landforms may be incised, and younger surfaces exposed. Arroyo valleys incised into the gravelly and sandy alluvium have Torriorthents. Some soils, like the Casa Grande series with its Btknz horizons, have accumulated an abundance of soluble materials. In short, this is a landscape with a wide variety of soils typical of warm deserts, and one in which the degree of soil development is roughly coincident with surface age.

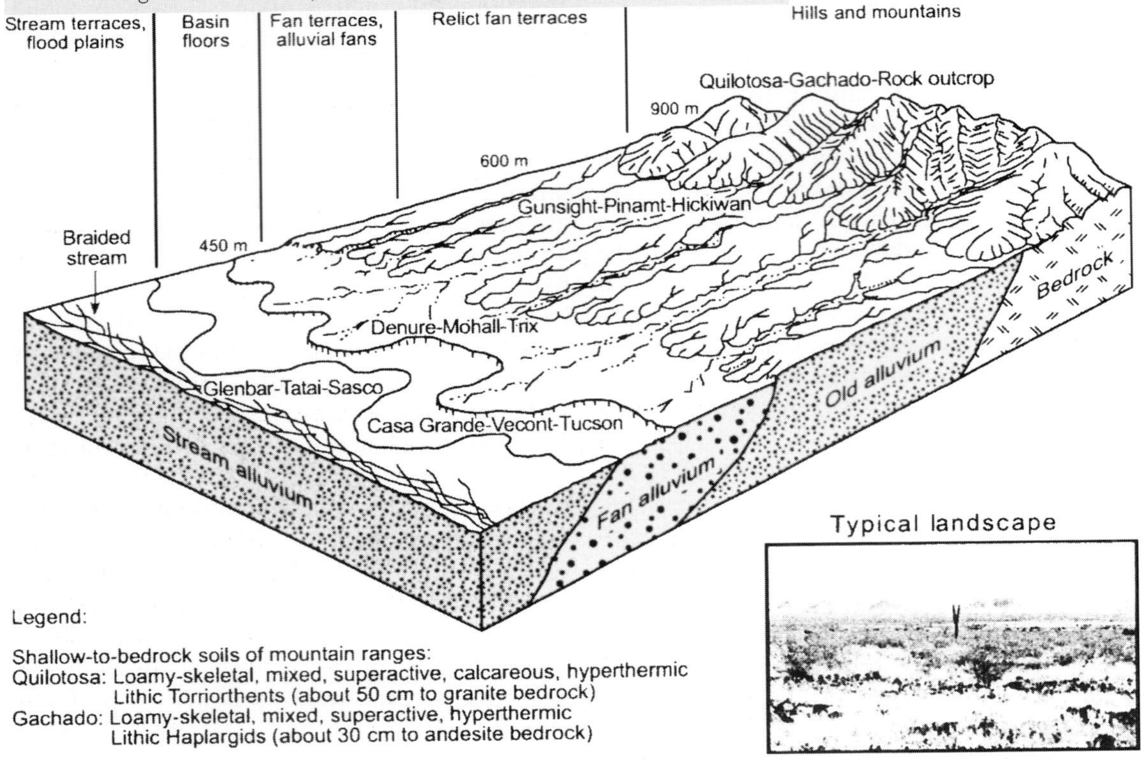

Legend:

Shallow-to-bedrock soils of mountain ranges:
Quilotosa: Loamy-skeletal, mixed, superactive, calcareous, hyperthermic Lithic Torriorthents (about 50 cm to granite bedrock)
Gachado: Loamy-skeletal, mixed, superactive, hyperthermic Lithic Haplargids (about 30 cm to andesite bedrock)

Well-developed soils of older alluvium:
Gunsight: Loamy-skeletal, mixed, superactive, hyperthermic Typic Haplocalcids
Pinamt: Loamy-skeletal, mixed, superactive, hyperthermic Typic Calciargids
Hickiwan: Loamy-skeletal, mixed, superactive, hyperthermic, shallow Calcic Petrocalcids

Soils of varying development on incised alluvium:
Denure: Coarse-loamy, mixed, superactive, hyperthermic Typic Haplocambids
Mohall: Fine-loamy, mixed, superactive, hyperthermic Typic Calciargids
Trix: Fine-loamy, mixed, superactive, calcareous, hyperthermic Typic Torrifluvents

Soils of basin floors and modern stream valleys:
Casa Grande: Fine-loamy, mixed, superactive, hyperthermic Typic Natrargids
Vecont: Fine, mixed, superactive, hyperthermic Typic Haplargids
Tucson: Fine-loamy, mixed, superactive, hyperthermic Typic Calciargids
Glenbar: Fine-silty, mixed, superactive, calcareous, hyperthermic Typic Torrifluvents
Tatai: Fine-loamy, mixed, superactive, hyperthermic Sodic Haplocambids
Sasco: Coarse-silty, mixed, superactive, hyperthermic Typic Haplocambids

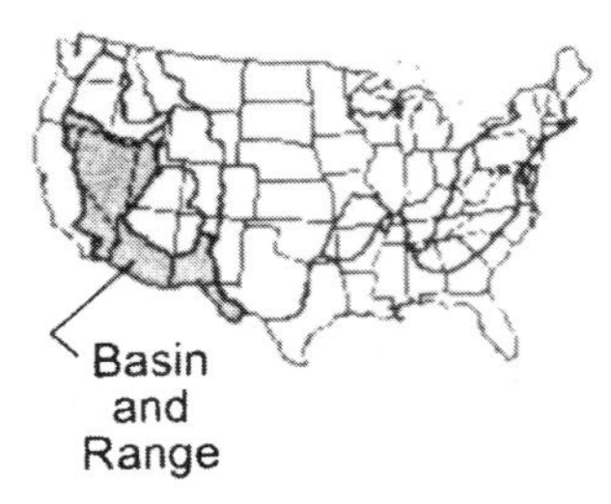

Block diagram of the soils and landscapes of a part of the Basin and Range province of the Tohono O'odham nation, in the Sonoran Desert of south–central Arizona. After Breckenfield (1999).

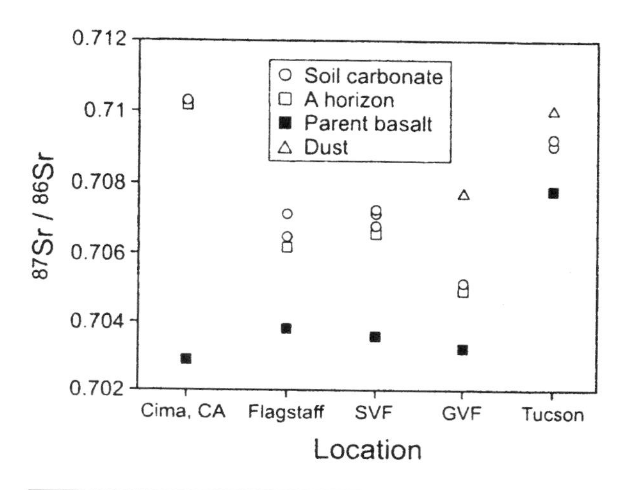

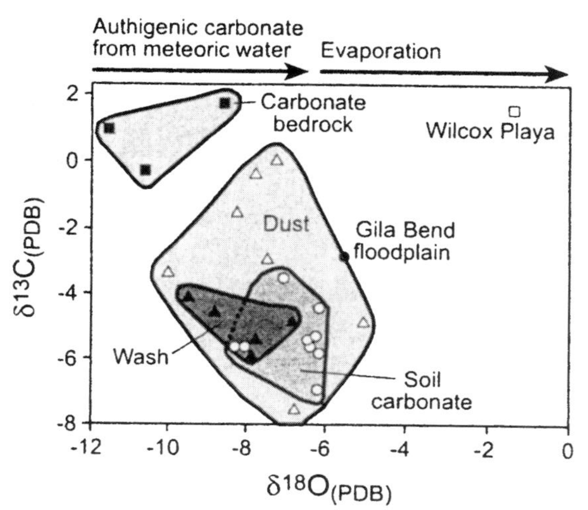

Fig. 12.36 Isotopic composition of soil carbonate, carbonate bedrock, dust, wash (fluvial material), floodplain and playa materials, illustrating that the main source material for dust is eroded soil carbonate itself. The data are from southern Arizona landscapes. After Naiman *et al.* (2000).

Most researchers today agree with the early conclusions of Brown (1956) and Ruhe (1967) that the main source of carbonate in many Aridisols is external. The primary external source is carbonate-rich eolian dust, blown onto the site and later dissolved and washed in by infiltrating water, although in coastal areas sea spray can contribute significant amounts of calcium (Quade *et al.* 1995, Offer and Goossens 2001). Some carbonate is brought in, dissolved, with rain or snow while still other carbonates enter the soil as coatings on silt- and clay-sized dust particles. Whether the coating is dissolved by water and translocated into the soil or not, the clastic particle could be removed by wind and transported yet again, suggesting that soils do not necessarily need to get markedly thicker even if they show signs of long-term inputs of dust (Gile *et al.* 1966). Inputs of gypsum ($CaSO_4 \cdot H_2O$) dust can also be an important Ca^{2+} source. Machette (1985) outlined four lines of evidence to support the hypothesis that airborne materials are the primary source of pedogenic carbonate: (1) most calcic soils are well above any groundwater influence, (2) the relationship between soil age and development would not be present if the carbonate came from groundwater, (3) abundant sources of eolian carbonate exist in (and upwind of) many dry areas and (4) rainfall can contain high amounts of calcium (Junge and Werby 1958). Copious data have been generated on the rates, timing and mineralogy of dust additions, what proportion is associated with silt- and clay-sized clastic minerals, and how this process has impacted pedogenesis. For soils in the desert southwest of the United States, dust was confirmed as the source of carbonate, using C and Sr isotopes (Naiman *et al.* 2000). In this case, the main source of dust was previously existing soil carbonate that had been eroded, not weathered bedrock (Fig. 12.36).

Dust commonly blows out of dry lake beds such as playas, alluvial fans and other alluvial flats (Young and Evans 1986, Gunatilaka and Mwango 1987, Chadwick and Davis 1990). Dust downwind from carbonate-rich areas is much higher in carbonates and gypsum than dust from other sources, including playas (Reheis and Kihl 1995), and soils immediately downwind from dust source areas tend to be better developed than soils farther away (Lattman 1973). Another significant source of dust is weathering bedrock escarpments (Reheis and Kihl 1995). Influx rates vary as climate varies; moist conditions during Pleistocene glacial periods (pluvial conditions in the western United States) greatly reduced influx rates of dust influx, and hence, soil development (Chadwick and Davis 1990).

Research is continuing on the rates at which dust is assimilated into desert soils, and the factors that influence these rates. For example, dust falling onto a hyperarid site (no matter what the rate of influx) is likely to blow away without ever impacting the soil; this situation is described as *moisture-limited* (Machette 1985). Moisture-limited sites have a greater Ca^{2+} influx than can be accommodated by local rainfall. Conversely, *influx-limited* sites are, theoretically, wet enough to accumulate more carbonate if more dust fell onto the site. Some influx-limited sites lie in desert areas that have high amounts of snowmelt infiltration (Machette 1985). Thus, the amount of carbonate that a soil accumulates is a function of not only climate (primarily precipitation amount and character) but also dust influx rate and carbonate content, as well as surface age (Machette 1985, Harden *et al.* 1991a). Slate *et al.* (1991) observed that soils near major dust sources develop slowly and have carbonates evenly distributed through the profile, presumably because they accumulate eolian sediment faster than translocation can occur. Sites farther from the dust source, with slower influx rates, have carbonates concentrated within a few horizons. They interpreted this relationship to mean that soils with slower rates of dust input have time to allow for carbonate translocation into the subsoil. Determining long-term rates of carbonate accumulation is complicated by the fact that dust influx rates and climate have varied over geologic time. Topography also enters into the equation: dust that falls onto a knob or steep slope could be blown or washed away.

Dust influx over geologic time is often manifested as secondary accumulations of carbonate, gypsum, salts and other soluble material, over and above that of the parent material (Fig. 12.37). For example, Machette (1985) reported $CaCO_3$ accumulation rates of 2–10 g m^{-2} yr^{-1} for latest Pleistocene soils in New Mexico. Accumulation rates were less for older soils, possibly because of erosion episodes, or because during wet climatic intervals some of the carbonates were leached from the profile, or due to varying rates of influx. Scott *et al.* (1983) reported carbonate accumulation rates of 5 g m^{-2} yr^{-1} in Utah. In the Mojave Desert, Schlesinger (1985) reported 1–3 g m^{-2} yr^{-1}.

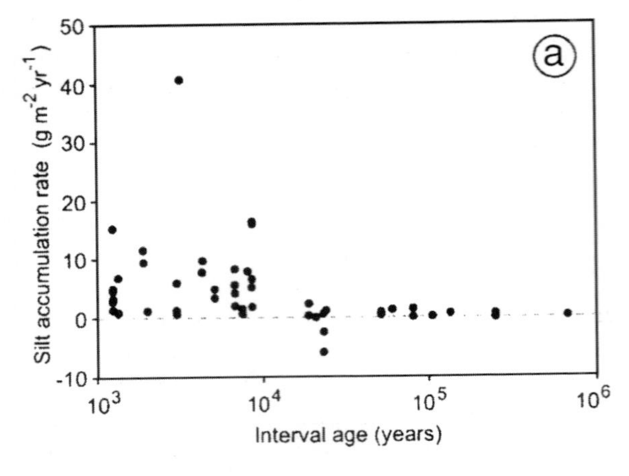

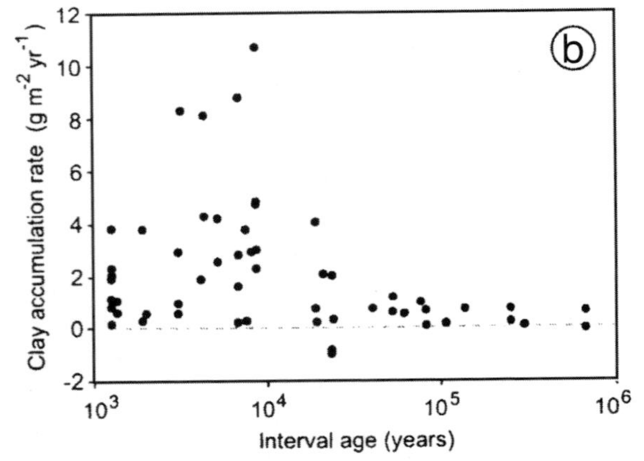

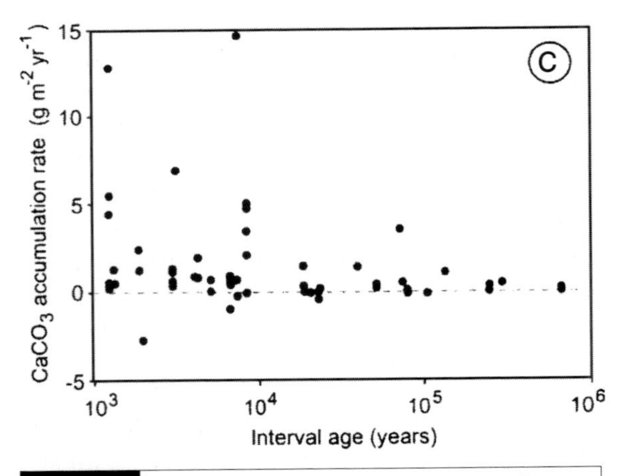

Fig. 12.37 Accumulation rates of (a) silt, (b) clay and (c) $CaCO_3$, in soils from southern Nevada and California. Negative numbers imply losses. After Reheis *et al.* (1995).

Finally, we note that gypsum can have an influence on carbonate accumulation. Lattman and Lauffenburger (1974) proposed that bacteria may reduce the sulfate in gypsum, creating a significant source of Ca^{2+}:

$$2CaSO_4 \cdot H_2O + \text{energy} \rightarrow 2Ca^{2+} + 2H_2S + 5O_2.$$

Presumably the energy for the organic reaction comes from soil organic matter. Evidence for this model comes from areas in Nevada where eolian dust contains significant amounts of gypsum. Carbonate horizons here are exceptionally thick (Gardner 1972); these areas are often near to, or overlie, gypsum sources, regardless of the nature of the rock detritus from which the soil formed. Many Bkm and Ckm horizons smell of H_2S gas when crushed, suggesting that sulfur is involved in their formation.

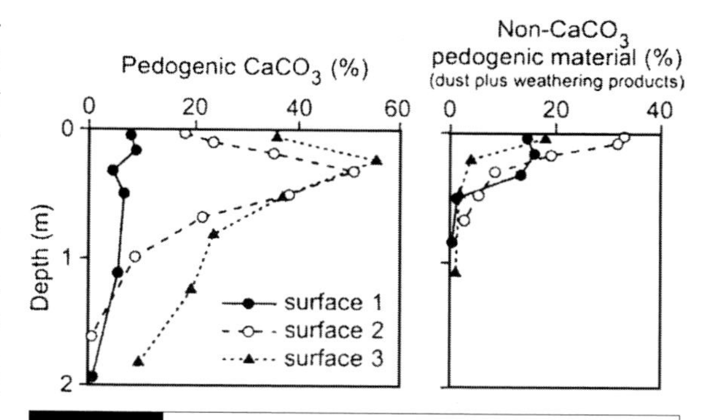

Fig. 12.38 Variations in pedogenic carbonate vs. other pedogenic materials, with depth, in three soils on the Kyle Canyon fan, Nevada. For details on the methods by which these data were calculated, see Reheis *et al.* (1992).

Physicochemical models of $CaCO_3$ accumulation

It was recognized decades ago that soils do not accumulate $CaCO_3$ in a clear developmental sequence (Hawker 1927). Soon after a semi-quantitative model of carbonate accumulation was established, it began to be used to interpret pedogenesis and soil/surface age in drylands (Gile and Grossman 1979). Most of the traditional, time-tested models for carbonate accumulation depend heavily on physicochemical processes (Abtahi 1980). These will be discussed first, followed by models which also include biogenic processes.

All the various models describe accumulation of *secondary* carbonate. Differentiating the amount of secondary carbonate from primary (inherited) carbonate is a difficult but necessary first step (Fig. 12.38). Rates of carbonate accumulation and dust influx are best calculated on sites where the slopes are stable and where the bedrock does not contribute significant amounts of calcium to the soil. Slate *et al.* (1991) described such a setting in New Mexico. Their data illustrate the slow, steady increase in carbonates and clay that is thought to occur in soils with a reliable, upwind source of carbonate-rich dust (Fig. 12.39).

The most widely cited of the quantitative, physicochemical models for describing secondary carbonate accumulation is that of Gile *et al.* (1966), although others exist (Alonso-Zarza *et al.* 1998). The model relies on physicochemical processes for the accumulation of carbonates, and describes a sequence of carbonate accumulation leading ultimately to carbonate-plugged Bkm horizons (Reeves 1970). It is called the *per descensum* model because it relies heavily on inputs of carbonates "descending" into the soil via percolating water. This model, developed for the southwestern United States, has generally (though not entirely; see Lattman (1973)) withstood the test of time. For this reason it is a powerful tool in the correlation of various deposits and geomorphic surfaces (Wells *et al.* 1985, Vincent *et al.* 1994). The four stages of the model depict the various widely observed morphologies of $CaCO_3$ accumulation; many intermediate stages also occur (Table 12.9, Fig. 12.40).Monger *et al.* (1991a) provided micromorphological data for soils in the various stages.

The carbonate accumulation patterns and morphology of the stages are affected by age and parent material. Two distinct sequences of accumulation were modeled: for gravelly (>50% gravel) and for less gravelly (<20%) parent materials. Materials with intermediate amounts of gravel have intermediate morphologies. In Stage I, if the soil is not gravelly, carbonates occur as thin, white filaments. Filaments may represent old root traces where carbonate-rich water once accumulated; the roots took up only pure H_2O,

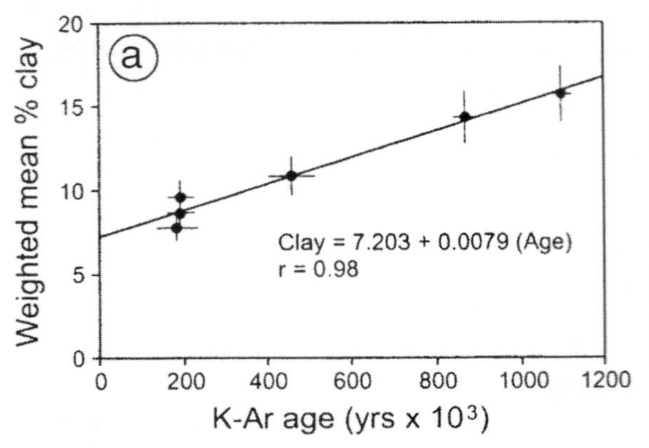

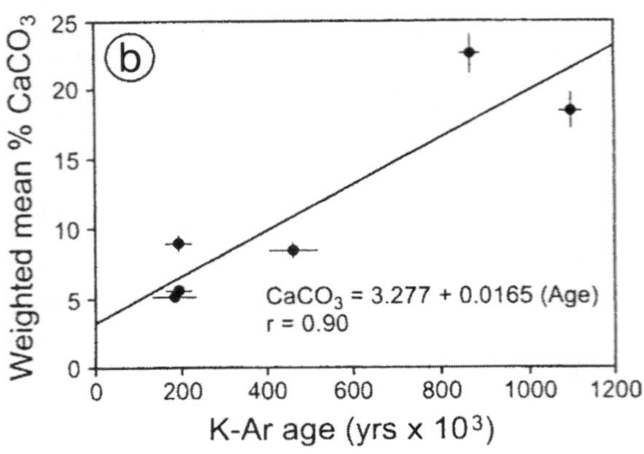

Fig. 12.39 Relationships between surface age and soil properties (weighted mean clay (a) and $CaCO_3$ (b) contents) for the Pinacate volcanic field, New Mexico. K–Ar ages on the underlying lava flows should be interpreted as maximum ages. The rate of carbonate accumulation at this site is slower than at many other sites in the western United States. After Slate *et al.* (1991).

leaving $CaCO_3$ behind. In gravelly Stage I soils, secondary carbonates accumulate as thin, discontinuous carbonate coatings, primarily on the undersides of pebbles and gravel. They are generally not visible unless the soil is being examined closely, e.g., with a hand lens (Fig. 12.41). Carbonates tend to accumulate on the bottoms of clasts since percolating, carbonate-rich water will tend to persist there, and when it later evaporates or is taken up by roots, carbonates are deposited (Treadwell-Steitz and McFadden 2000). The larger the rock, the more carbonate-rich water it can intercept, and thus the faster its coating will grow (Treadwell-Steitz and McFadden 2000). A unique situation was reported by Amundson *et al.* (1997), however, for soils in southern Baja California. Here, carbonate coatings are found on the *tops* of large clasts. The explanation for this geographically isolated phenomenon may be connected with soil thermal and hydraulic gradients. In Stage I, carbonates are most common at about 15–60 cm depth.

By Stage II, pebbles show continuous and thicker coatings of carbonate, and some voids between pebbles become filled with white, non-indurated carbonate. In Stage II there is continued infilling of voids and as Gile *et al.* (1966) noted, closing of gaps between flakes and filaments. If the soil is gravel-poor, the carbonates accumulate as small nodules. Some nodules take on cylindrical form or may resemble baby rattles, perhaps because they cement infilled cicada burrows, or krotovinas (Hugie and Passey 1963, Suprychev 1963) (Fig. 10.5).

Stage III carbonate morphology has many inter-pebble fillings. All or nearly all of the skeletal grains in the upper Bk horizon are coated with carbonate (Fig. 12.42). Carbonates coat the tops and sides of clasts in gravelly Stage III soils because eolian additions of clay and silt will, by now, have caused the soil to be much finer-textured. These fine materials, most common in the upper profile, fill macropores and allow percolating water to be retained in close contact with all sides of large clasts; when this water dries, carbonates are deposited around the clasts (Treadwell-Steitz and McFadden 2000). Soils in dust-poor areas may take considerably longer to reach Stage III. Late in Stage III, whether in gravelly or non-gravelly materials, carbonates fully impregnate the entire Bk horizon and voids get plugged, causing the horizon to become an aquitard. Part or all of it is indurated. If bioturbation is prominent during late Stage II and early Stage III, plugging may not occur (Sehgal and Stoops 1972). Plugging of pores, like most of the other carbonate accumulation processes, is strongest at the top of the Bkm horizon and decreases in intensity with depth. Stage III and greater B horizons have traditionally been referred to as K, or petrocalcic, horizons.

Table 12.9 Morphologic stages[a] of $CaCO_3$ accumulation for soils developed in non-calcareous parent materials

Stage	Gravelly parent material	Non-gravelly parent material
I	Thin discontinuous coatings on pebbles, usually on undersides; some filaments; matrix can be calcareous next to stones; about 4% $CaCO_3$	Few filaments or coatings on sand grains; <10% $CaCO_3$
I+	Many or all coatings on pebbles are thin and continuous	Filaments are common
II	Continuous coatings on all sides of pebbles; local cementation of few to several clasts; matrix is loose and calcareous enough to give whitened appearance	Few to common carbonate nodules, 0.5–4 cm diameter; matrix between nodules is slightly whitened, carbonates occur in veinlets and as filaments; parts of the matrix can be non-calcareous; about 10–15% $CaCO_3$ in whole sample, 15–75% in nodules
II+	Same as Stage II, except carbonate in matrix is more pervasive	Carbonate nodules are common; 50–90% of matrix is whitened; about 15% $CaCO_3$ in whole sample
III	Horizon is 50–90% continuously filled with carbonate (K fabric); color mostly white; carbonate-rich layers more common in upper part; about 20–25% $CaCO_3$	Nodules coalesce; there are so many nodules, internodule fillings and carbonate coats on grains that over 90% of horizon is white; matrix is cemented; carbonate-rich layers more common in upper part; about 20% $CaCO_3$
III+	Most pebbles have thick carbonate coats; matrix particles continuously coated with carbonate or pores plugged by carbonate; nearly continuous cementation; >40% $CaCO_3$	Most grains coated with carbonate; most pores plugged; >40% $CaCO_3$
IV	Upper part of is nearly pure cemented carbonate (75–90% $CaCO_3$) consisting of laminar carbonate layers that are <1 cm thick in total; the rest (lower part) of the horizon is plugged with carbonate (50–75% $CaCO_3$)	
V	Platy, laminar layer(>1 cm thick) is strongly expressed, with signs of incipient brecciation and pisolith (thin, multiple layers of carbonate surrounding particles) formation; vertical faces and fractures are coated with laminated carbonate	
VI	Brecciation and recementation, as well as pisoliths, are common; multiple generations of laminae, brecciation and pisoliths are evident	

[a]Developed for soils in the southwestern United States.

Sources: Gile *et al.* (1966), Bachmann and Machette (1977), Machette (1985) and Birkeland (1999). Recognition should be made, however, of the early contribution of Hawker (1927).

The Bkm horizon of Stage IV perches water, allowing for carbonate deposition directly on top (Reeves 1970). Eventually, a laminar, indurated horizon forms and grows upward, engulfing the overlying horizons while at the same time filling the voids below (Alonso-Zarza *et al.* 1998). This laminar layer exhibits a succession of thin (≈1 mm), low porosity, $CaCO_3$-enriched layers; it consists of almost pure $CaCO_3$, with little other allogenic clastic materials. Upper

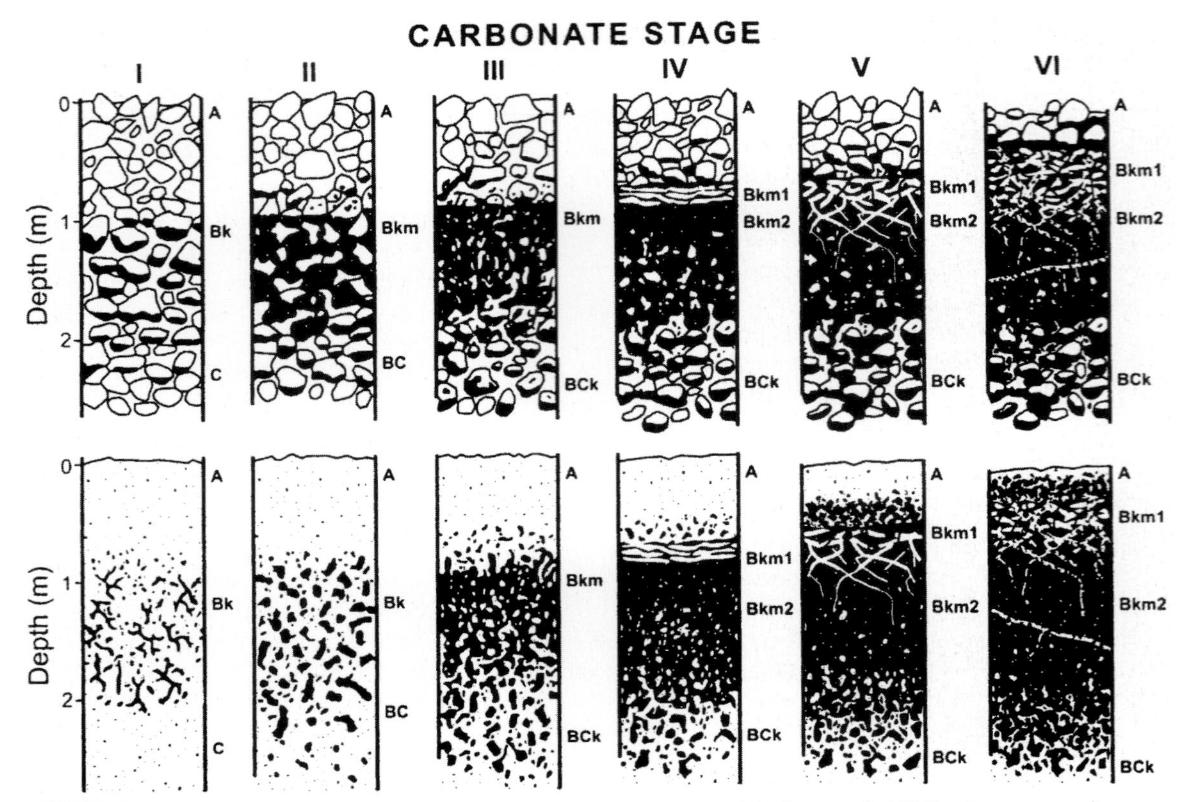

Fig. 12.40 Model of carbonate accumulation in soils. Developed primarily based on observations of soils in the southwestern United States. After Gile *et al.* (1966) and Machette (1985).

laminae are younger than lower ones. As the laminar Bkm horizon thickens, the soil surface will, presumably, grow upward. At this point there is convergence between gravelly and non-gravelly stages (Gile *et al.* 1966). Irregularities in the top of the Bkm horizon are infilled more rapidly, as percolating water preferentially ponds there. In time, the laminar horizon attains smoothness and horizontality (Aristarain 1970). Nonetheless, the petrocalcic horizon is not totally impermeable; it has windows – sites of preferential flow. These windows are kept free of carbonate accumulation as percolating water is funneled rapidly through them after heavy rainstorms (Fig. 12.43). Many are formed by mammal bioturbation. Because they have less void space and surface area, gravelly soils reach Stage IV more rapidly than do non-gravelly soils (Gile *et al.* 1966, 1981, Marion *et al.* 1985); the process is even more rapid in soils with larger gravels (Treadwell-Steitz and McFadden 2000). Thus, petrocalcic horizons may be able to form in gravelly parent materials while in clast-poor materials of the same age, cementation may not yet have taken place.

Machette (1985), working with data from an earlier paper (Bachman and Machette 1965), proposed two additional stages of carbonate accumulation (Table 12.9, Fig. 12.40). He suggested that Stage IV be restricted to soils with laminar horizons thinner than 1 cm. Stage V was then defined to include soils with laminar horizons >1 cm in thickness. The laminar Stage V and VI horizons are like limestone (Fig. 12.44). Some Stage V soils have concentrically banded pisolitic structures, suggestive of brecciated Stage II and IV materials. Pisoliths are subangular to spherical features, 0.5 to 10 cm in diameter, surrounded by thin layers of secondary carbonate (Birkeland 1999). Presumably, they grow by carbonate accretion around a core, which could be, e.g., a pebble or a fragment of Bkm material.

Fig. 12.41 A Typic Haplocambid near Las Cruces, New Mexico, with Stage I carbonate accumulation. Units on the stick are decimeters. Photo by D. L. Cremeens.

Stage VI soils show evidence of multiple episodes of brecciation and pisolith formation, suggesting extreme age and polygenesis (Bryan and Albritton 1943); some may date to the Pliocene (Fig. 12.45). Disoriented, brecciated fragments (also found in Stage V) are evident in and above the laminar horizon, but many may be recemented. The reason for this is clear: as carbonate infills the lower solum, it acts *displacively*, literally moving skeletal grains out of the way due to crystallization pressures, filling fractures and voids with secondary carbonate (Reheis *et al.* 1992). The carbonate content exceeds the original pore volume, forcing expansion and causing clastic grains seemingly to float within a carbonate matrix. Machette (1985) estimated that expansion could reach 400–700%. Secondary carbonate can also act *replacively*, wherein $CaCO_3$ replaces primary silicate grains (Reheis *et al.* 1992). Replacive carbonate is a relatively new concept, although it has been documented in Europe (Millot *et al.* 1977) and Africa (Watts 1980). The plugged Bkm horizons of Stages III through VI can lead to erosion of upper horizons, as water from intense rainfall or snowmelt events perches on top and runs laterally, facilitating the brecciation process that begins in Stage V (Aristarain 1970).

Although a particular form or type of secondary carbonate can be ascribed to a certain climate, the correspondence is not always assured (Netterberg 1969b). Many calcretes have been formed, reworked, eroded and reformed many times over, and thus are very difficult to interpret from a paleoenvironmental point of view (Alonso-Zarza *et al.* 1998). Stage V and VI morphologies confirm that many calcretes, especially the thick ones that have many different forms and morphologies, are a product of multiple climatic cycles (Sanz and Wright 1994).

Using radiocarbon and other dating methods (see Chapter 14), it is possible to establish the age of secondary carbonates and put some age limits on the carbonate stages shown in Table 12.9. Because some of the CO_2 involved in the formation of the carbonic acid is derived from plant (root) and microbial respiration, a portion of the carbon in the reprecipitated $CaCO_3$ is biologic and can be dated by the radiocarbon method (Amundson *et al.* 1994). Likewise, the stable isotope contents of soil carbonate can provide information about the paleoenvironment during which they were formed (Quade *et al.* 1989, Quade and Cerling 1990) (see Chapter 15). Dating by ^{14}C on soil carbonate have established that the youngest carbonate occurs nearest the top of the Bkm horizon, and ^{14}C ages increase with depth (Buol and Yesilsoy 1964, Reeves 1970).

Applying this dating information, rates of carbonate accumulation have been established, allowing soil development to be correlated to surface age (Fig. 12.46). Rates vary from region to region (Machette 1985, Sehgal and Stoops 1972). Figure 12.45 shows how the attainment of carbonate Stages I through VI varies across the southwestern United States. In favored locations, Stage VI can be attained on surfaces dating back to the Early Pleistocene, while others of presumably similar age remain in Stage III. Holocene

(a)

(b)

Fig. 12.42 Aridisols near Las Cruces, New Mexico, with late Stage III carbonate accumulation. Both soils have an indurated Bkm horizon, but lack the laminar cap that defines Stage IV development. (a) A Petrocalcic Ustollic Paleargid. Photo by RJS. (b) An Ustollic Haplargid. Units on the stick are in decimeters. Photo by D. L. Cremeens.

Fig. 12.43 A long trench through a Petrocalcid in southern New Mexico. Note the windows (lacunae) in the Bkm horizon, and the possible krotovinas below and within, probably attributable to badgers. Photo by D. L. Cremeens.

soils have formed in a dry, interglacial period and thus exhibit rapid rates of carbonate accumulation. Older soils would have been exposed to at least one wetter, cooler, glacial climatic interval, in which carbonates would not have accumulated rapidly, or perhaps even been partially leached. Their long-term carbonate accumulation rates would not be as rapid as the Holocene soils. Sehgal and Stoops (1972) examined a sequence of soils from dry to subhumid climates, and found

(a)

(b)

Fig. 12.44 A Stage VI Aridisol near Las Cruces, New Mexico. The soil is estimated to be 1.6 million years old. Photo by RJS. (a) Leland Gile discusses the evolution of the soil. (b) Close-up of the upper profile. Pisoliths in and above the Bkm horizon are evident. Knife for scale.

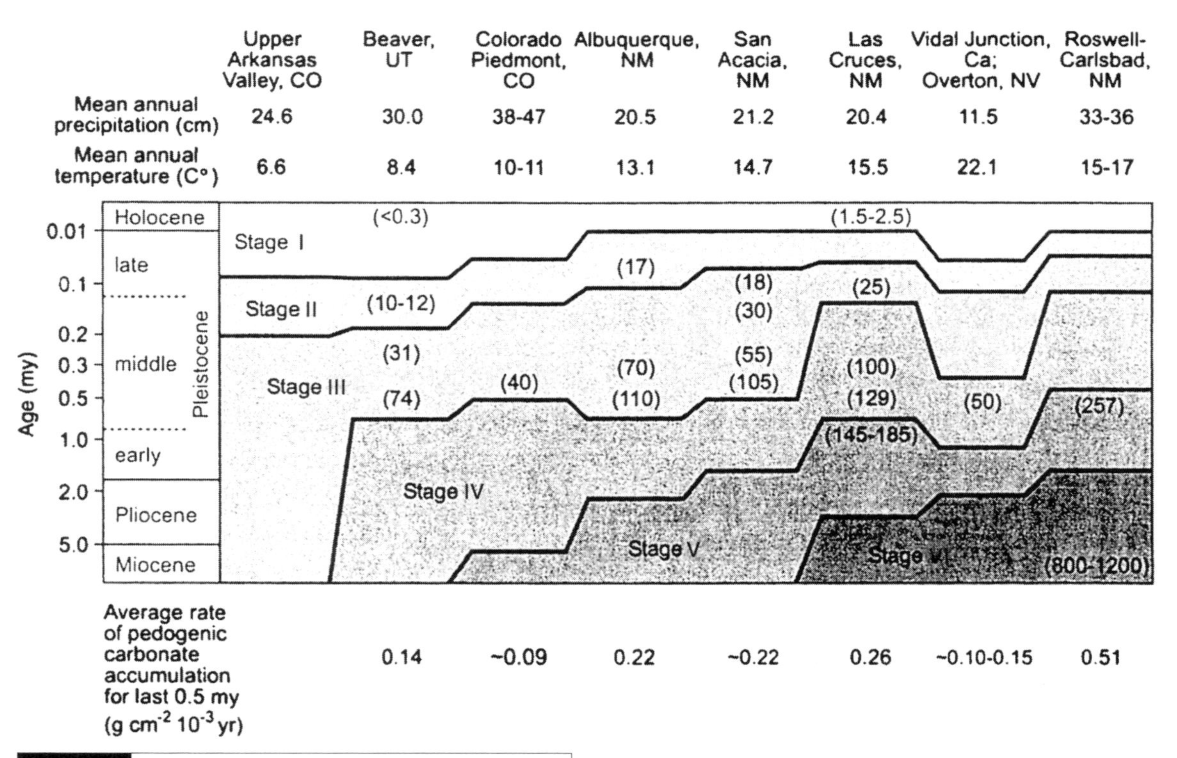

Fig. 12.45 Maximum stages of carbonate morphology in gravelly alluvium in the southwestern United States. Numbers in parentheses are the mean g of $CaCO_3$ cm^2 column. After Machette (1985) and Birkeland (1999).

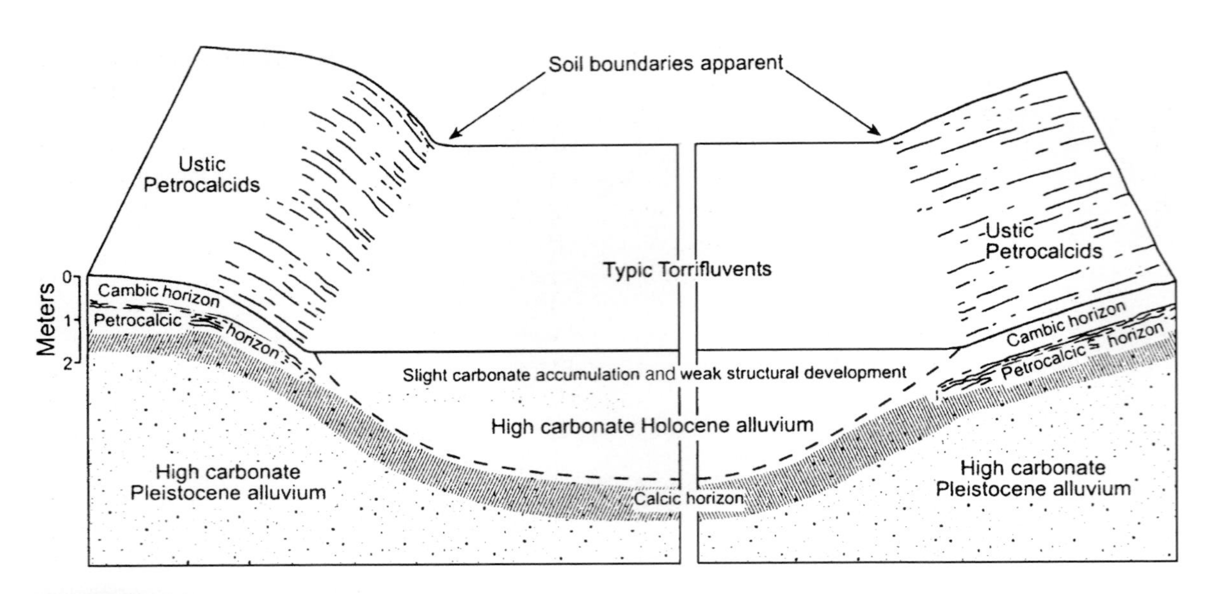

Fig. 12.46 Block diagram illustrating the difference in calcic horizon development in high carbonate parent materials of different ages. After Gile (1975a).

that the forms of secondary carbonates changed along this transect. As would be expected, the amounts of secondary carbonates in the soils also decreased from dry to subhumid climates. Rates of carbonate accumulation also appear to vary as a function of the types of rock that dominate the gravel fraction (Lattman 1973). The extent and development of secondary carbonates is greatest where carbonates and basic igneous rocks dominate, intermediate in soils with large amounts of siliceous sedimentary detritus and least where acid igneous gravels are common.

Rabenhorst *et al.* (1991) and Rabenhorst and Wilding (1986b) presented an alternative physicochemical model of petrocalcic horizon formation, for soils underlain shallowly by limestone, in dry, western Texas (see also Blank and Tynes 1965). They envisioned that the carbonate horizons in these soils developed from *in situ* dissolution of limestone, followed by reprecipitation of soil carbonate, often at a lithologic discontinuity within the limestone (West *et al.* 1988b). In the Rabenhorst–Wilding model, percolating water containing carbonic acid and organic compounds dissolves some of the porous limestone, enlarging some of the pores (Stage 2). It is known that limestone dissolution is enhanced by the acid secretions of algae and fungi, eventually leading to precipitation of micrite crystals in pores, a process called *sparmicritization* (Kahle 1977). (*Micrite* is a term used for calcite crystals less than 4 μm in size.) Pores continue to enlarge and micrite linings coalesce. Ca^{2+} ions are added in precipitation and from continued dissolution of bedrock, leading to a plugged condition, similar to Stage III of Gile *et al.* (1966). A laminar cap of secondary carbonates forms above the plugged horizon (Stage 5), similar to Stage IV of Gile *et al.* (1966). Evidence in support of the Rabenhorst–Wilding model includes (1) the low non-carbonate residue in the Bkm horizon, i.e., the horizons were nearly pure carbonate and (2) the presence of limestone fragments within the Bkm horizon (West *et al.* 1988a, b). Grain displacement by growing calcite crystals could not explain the low (≈2%) non-carbonate residue levels. In contrast, soils in New Mexico, where Gile and his colleagues developed their model, often contain 25–54% non-carbonate residue in the Bkm horizon. The Rabenhorst–Wilding model also explains the somewhat anomalous presence of fluorite (CaF_2) in Bkm horizons in west Texas Calcids, since it is found in the bedrock as well and would not have been leached from the soil system.

A related model by West *et al.* (1988b) incorporates many of the same components, but relies more heavily on the formation of the Bkm horizon in conjunction with a dense, weathering-resistant layer in the limestone. This more resistant limestone bed remains after softer layers above and below have weathered, and forms the locus of secondary carbonate deposition. Bkm horizons in Texas often embed fragments of limestone. Eventually, geologic erosion lowers the soil surface, placing the Bkm horizon within the leached zone, causing it to break down.

BIOGENIC MODELS OF CARBONATE ACCUMULATION

Recent research has repeatedly pointed to the influence, be it subtle or dominant, of biological factors in calcification (Verrecchia 1994, Monger 2002). Most of these *biogenic models* also include a physicochemical component. Organisms may act as *catalysts* for carbonate precipitation, either passively or actively, and may function as *sources* of carbonate materials (Goudie 1996). Obviously, by removing CO_2 from soil air they also mediate the carbonate precipitation process (Krumbein and Giele 1979). Goudie (1996) argued that soil biota can also have a much more active role in carbonate precipitation, e.g., fungi may trigger carbonate precipitation by dumping their excess Ca^{2+} (Verrecchia 1990).

Secondary carbonates have long been known to form as Ca^{2+} ions precipitate on, or thoroughly permeate, former biological substrates such as root hairs, fungi and actinomycetes (Calvet *et al.* 1975, Klappa 1979, Phillips and Self 1987, Vaniman *et al.* 1994). Like any substance, the ions often precipitate passively onto an existing surface – in this case a biological one. If that substance/substrate is a root hair, the carbonate feature produced can be variously referred to as a rhizolith, root tubule or root cast (Goudie 1996, Wright *et al.* 1995). Monger *et al.* (1991b), however, documented a more *active* role of soil microorganisms, suggesting that they can *directly precipitate* calcite. Cited as evidence are fossilized remains of calcified fungal hyphae, among other biogenic features, in Bk horizons (Kahle 1977, Phillips and Self 1987). These hyphae resemble thin filaments (Phillips *et al.* 1987) (Fig. 12.47). In an interesting experiment in which soil columns were irrigated with calcium-rich water, calcite formed only in the soils that contained microorganisms; none formed in sterile soils (Phillips *et al.* 1987).

Amit and Harrison (1995) proposed a carbonate accumulation model that incorporates physicochemical and biogenic processes, and thus may be the most holistic and widely applicable model to date. It was developed in a sand dune landscape in a hyperarid region of Israel – but one with surprisingly high biological activity. Here, carbonate accumulations show distinct biogenic origins; many are calcified fungal hyphae. All of the secondary, micritic carbonates in these young (1425 ^{14}C years old) soils occurred in conjunction with roots, bacteria or fungi. Even the nodules of calcite contain high amounts of biogenic material. Over time, the accumulation of secondary carbonates, along with dust additions, decreases the permeability of the soil. At some point, the soil crosses an intrinsic threshold and physicochemical processes of carbonate accumulation begin. Thus, biogenic processes initiate calcification, and at some later point are joined by physicochemical processes. Additions of dust are an important component of this model because it reduces moisture loss between precipitation events, thereby enhancing carbonate precipitation. This model works best on highly permeable parent materials with a high degree of biogenic activity (Amit and Harrison 1995).

Another calcification model (Alonso-Zarza *et al.* 1998) involves biogenic carbonate accumulation, along with surface evolution and stability. This model highlights the complex interaction between pedogenesis and surface stability in arid regions, integrates biophysical processes and points out that disruptions to the classic Gile *et al.* (1966) model are not only possible, but likely (Verrecchia 1987, Sancho and Meléndez 1992).

To summarize, secondary carbonates in soils have multiple origins. Certainly, as carbonate-rich water in a soil is lost to root uptake or evaporation, the dissolved carbonate must precipitate.

This is a physicochemical reaction. However, the reaction could be facilitated or even initiated biochemically. Fungal hyphae and fruiting bodies, algae, bacteria, plant roots and even pupal cases might provide optimal sites for initial precipitation and continued growth of calcite and micrite crystals (Phillips *et al.* 1987). There remains much to learn of the intricacies of the calcification process, even though the general sequence and process have been known for decades.

Gypsification

In addition to carbonates, desert soils accumulate other soluble compounds, for similar genetic reasons – there is not enough water movement through the soil to translocate them out of the profile (Veenenbos and Ghaith 1964, Dan and Yaalon 1982). The process whereby soils accumulate secondary *gypsum* (or calcium sulfate: $CaSO_4 \cdot 2H_2O$) is *gypsification*. *Gypsic* (By and Cy) horizons contain significant amounts of secondary gypsum. If cemented, they are referred to as *petrogypsic* (Bym) horizons, gypsum crusts or gypcrete (Watson 1985, Dixon 1994).

Whether soils accumulate gypsum depends as much on a *source* as on the leaching regime. Many dry areas of the world lack sources of gypsum and are devoid of gypsic soils (Fig. 12.22b). The primary source of gypsum for most gypsic soils is eolian dust derived from gypsum-bearing rocks (Reheis 1987b, Dixon 1994). Near oceans, gypsum impacts soils as dry deposition of evaporated sea spray; for this reason many gypsic soils are near ocean coasts (Fig. 12.48). Hydrogen sulfide (H_2S), the ultimate source of oceanic gypsum, is derived during oceanic upwelling of gases produced by

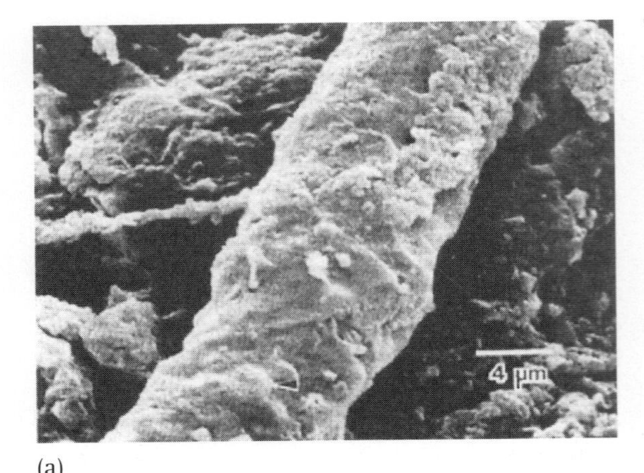

(a)

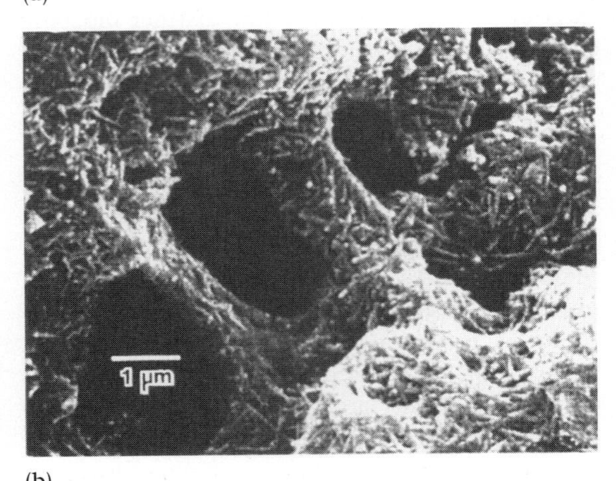

(b)

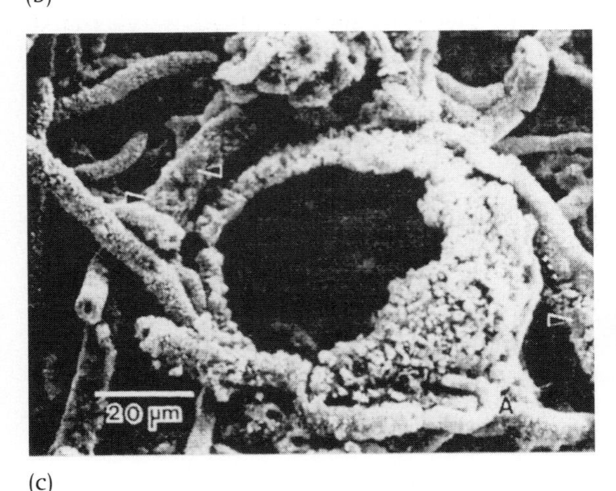

(c)

Fig. 12.47 Scanning electron photomicrographs of calcified filaments from calcareous soils in South Australia. After Phillips *et al.* (1987). (a) Calcified filament with a relatively smooth surface, with calcite rhombohedra beneath a possible organic sheath. (b) Submicron-size, calcified rods of possible bacterial origin. (c) Images of fruiting bodies associated with filaments. The large hollow sphere may be an oogonium. The small curved filament in the lower right (at A) may be an antheridium attached to the oogonium. Pustular structures at the arrows may be bacteria.

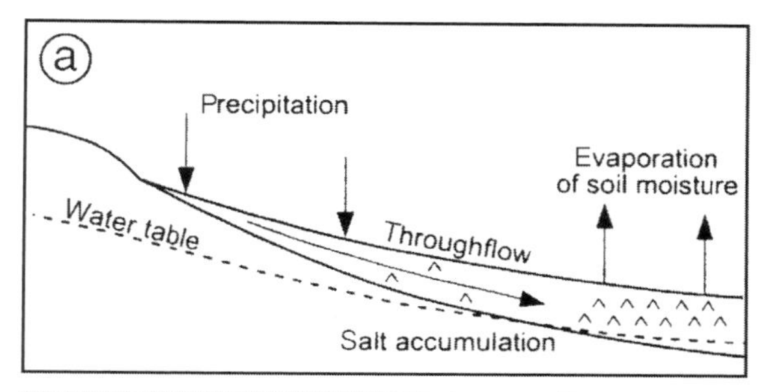

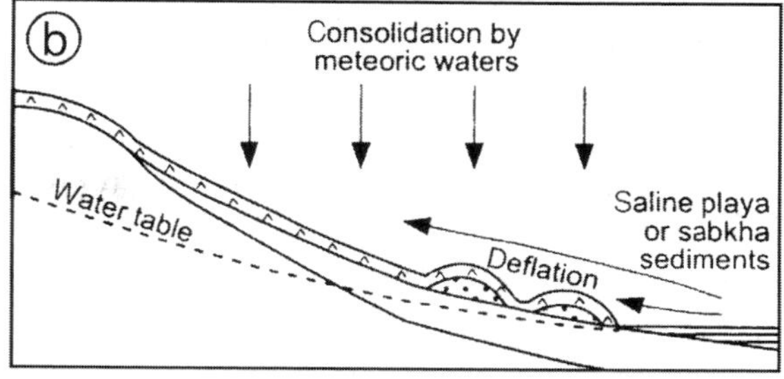

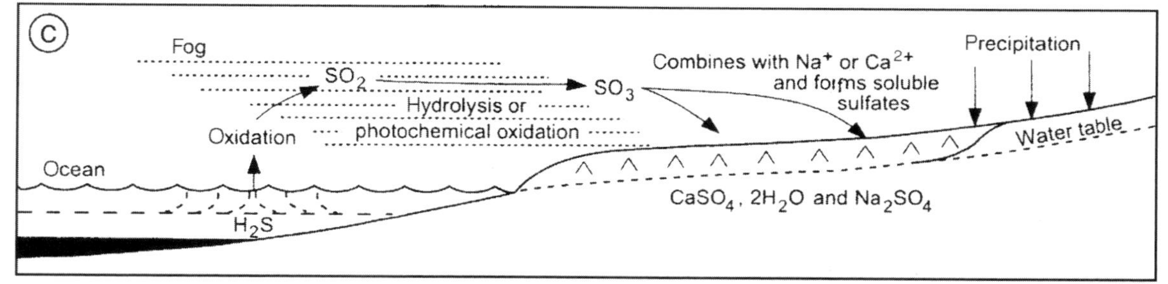

Fig. 12.48 Three ways in which gypsum sources can be delivered to soils. After Watson (1985). (a) Downslope accumulation as a result of leaching of hillslope sediments rich in gypsum, coupled with throughflow. Gypsum accumulates in lowlands, often as a surface crust. (b) Deposition of gypsiferous dust onto nearby uplands. Dust is later translocated into the soil by precipitation events. (c) Gypsum brought to coastal areas by fog or as dry deposition.

sea floor bacteria. When dissolved in fog, it is oxidized to SO_2.

Inland, gypsic horizons occur where parent materials are rich in gypsum, such as gypsiferous shale, either at the site or upwind (Eswaran and Zi-Tong 1991). One of the most common sources of gypsum in dry climates are playas underlain by gypsum-rich groundwater (Nash *et al.* 1994). After the groundwater rises into the soil by capillarity and gypsum precipitates, that gypsum can later be blown onto nearby soils (Watson 1985, 1988). If gypsiferous soils are eroded and the By or Bym horizon exposed, they can deflate and become a gypsum source for soils downwind. In soils with large amounts of salts and gypsum, these substances will come to resemble secondary carbonate – white-colored deposits in the lower profile, or sometimes occurring as a surface crust (Watson 1979, 1985).

Of all the common soluble materials that accumulate in soils, only $CaCO_3$ is less soluble than gypsum, meaning that the bulk of accumulated carbonate is usually located shallower in the profile than the gypsum max in freely draining desert soils; By horizons should be below Bk horizons. If gypsum-rich horizons *overlie* carbonate-rich horizons, it is a tell-tale sign that the gypsum has accumulated via capillary rise. Where capillary flow leads to surface gypsum deposits, the crust formed is often referred to as *croûte de nappe*.

Because carbonates and gypsum often co-occur in soils, differentiating them from each other is critical to the accurate identification and naming of the horizon and to the interpretation of the genesis of the soil. Usually, secondary gypsum will not effervesce when exposed to weak HCl, while carbonates will, although the definitive test for pedogenic gypsum in soils is the identification of euhedral gypsum crystals with a hand lens (Carter and Inskeep 1988). Indeed, close inspection of pedogenic gypsum crystals within the soil fabric may be the only way to differentiate it from gypsum originally in the parent material (Amit and Yaalon 1996).

In most other respects, gypsification is similar to calcification. Indeed, in many desert and semi-arid soils, gypsum and carbonates are intermingled within one or more horizons. Gypsum may accumulate uniformly throughout sandy soils, while in finer-textured or gravelly material it may be more concentrated in masses or clusters, sometimes called snowballs. In gravelly or stony material, it also may accumulate in pendants below the rock fragments, as do carbonates. The differences lie in gypsum's increased solubility (Carter and Inskeep 1988) and in the fact that it depresses the solubility of $CaCO_3$ (Reheis 1987b). Gypsum also commonly accumulates as surface crusts when groundwater evaporates in shallow lakes, playas, chotts or sabkhas (Busson and Perthuisot 1977, Schwenk 1977, Watson 1979, 1988). Carbonates, however, seldom occur as surface evaporite deposits because of their lower solubility (Watson 1979). Like calcic horizons, gypsic horizons have discrete developmental stages.

Silicification

Accumulation of secondary silica in soils is called *silicification*. Silica is abundant in all soils – in silicate minerals and/or as easily weathered tephra and volcanic ash/glass. Soils that accumulate sufficient amounts of soluble silica may eventually develop a Bqm horizon – a silica hardpan or *duripan* (Southard *et al.* 1990, Soil Survey Staff 1999). Duripans, also called *silcretes* or *siliceous duricrusts*, are often firm and brittle, even when wet (Stephens 1964, Summerfield 1982, Twidale and Milnes 1983, Dixon 1994). Many duripans contain >90% silica, along with other illuvial materials, especially $CaCO_3$ (Watts 1977, Dixon 1994). Cementation of soil fabric by silica is verified if fragments do not slake in water or after prolonged soaking in acid (HCl).

Silicification of soils occurs in much the same way as the *per descensum* model of calcification. Silica-enriched horizons can also form *per ascensum*, as silica-rich groundwater moves upward via capillarity (Summerfield 1982). Thus, duripans (and gypcretes) have many morphologic features in common with calcic and petrocalcic horizons (Callen 1983, Harden *et al.* 1991b, Reheis *et al.* 1992, Vaniman *et al.* 1994). In many parts of the world duripans or silcretes stand up as resistant layers, even when the overlying soil horizons have been eroded, reflecting the great age of the geomorphic surface (Milnes and Twidale 1983)

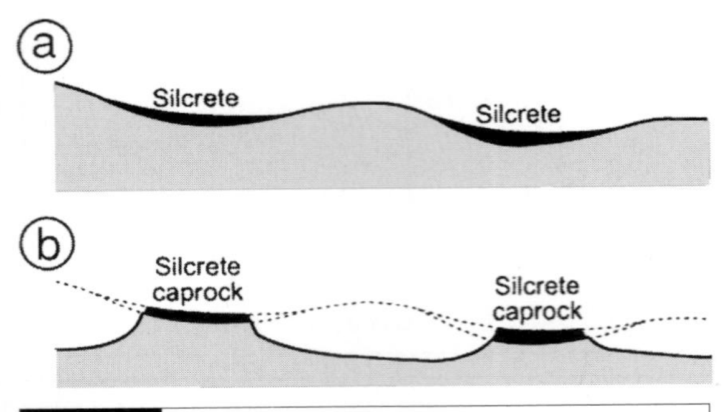

Fig. 12.49 Inversion of topography in arid parts of Australia, as induced by hard, siliceous duricrusts. Window (a) is $time_1$, window (b) is $time_2$. After Twidale and Milnes (1983).

(Fig. 12.49). For example, most Australian silcretes appear to be of Tertiary age (Callen 1983, Milnes and Twidale 1983). Duripans are best developed in Mediterranean climates, under xeric moisture regimes with a winter rainfall peak.

Duripans commonly are broken into very coarse prisms with coatings of opal lining the prism faces and large pores. The prisms presumably form by slight volume changes that result from wetting and drying, since they are absent in duripans of arid regions (Soil Survey Staff 1999). Weak duripans can occur in humid climates, where the soils have (often) formed in volcanic materials.

Like petrocalcic horizons, indurated duripans have an abrupt upper boundary, often with a laminar top consisting of a nearly continuous layer of secondary silica (Boettinger and Southard 1991). Water often perches on top of the pan during the rainy winter season. Strong duripans are dominated by spherical and ellipsoidal nodules of microcrystalline and opaline silica, with primary silicate minerals observable in partially weathered states (Boettinger and Southard 1991). These nodules can agglomerate into microscopic glaebules or microagglomerates (Chadwick *et al.* 1989a). Composed of silt and clay held together by poorly crystalline silica (or carbonate) cement, microagglomerates can eventually grow to become durinodes (Latin *durus*, hard, *nodus*, knot) – weakly cemented to indurated silica nodules ≥1 cm in diameter.

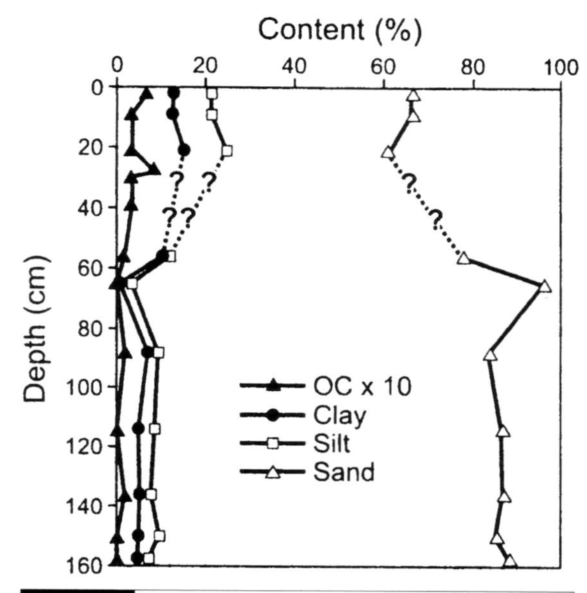

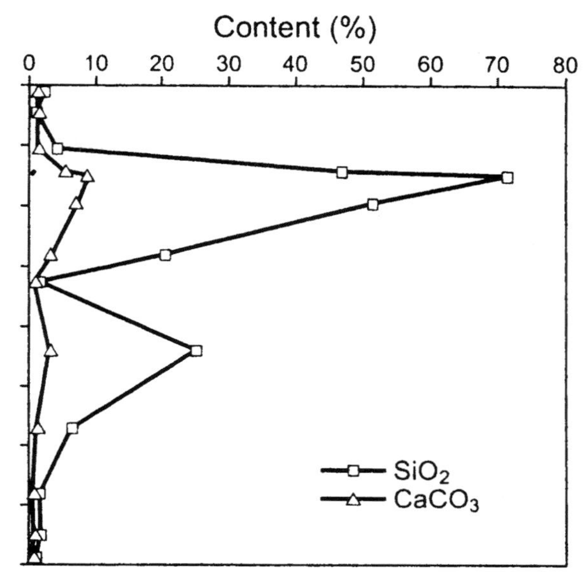

Fig. 12.50 Depth plots for a Xeric Haplodurid in Southern California's Mojave Desert. Data from a thin saprolite (Crk horizon) seam at 47–75 cm have been omitted for clarity of presentation. Data from Boettinger and Southard (1991).

The sources of silica are varied. Many soils have at least some silicate minerals like feldspar and quartz, which can serve as a silica source (Boettinger and Southard 1991). Quartz is difficult to weather and minimally soluble, although Summerfield (1982, 1983) has argued that quartz *dust* is characterized by disordered surfaces and smaller sizes, both of which render it more soluble (Siever 1962). Other silica sources include volcanic ash and pyroclastic materials, as well as amorphous forms: opal phytoliths and diatoms (Jones and Beavers 1963, Jones and Handreck 1967, Scurfield *et al.* 1974, Summerfield 1982) (see Chapter 15). The geographic association of duripans with areas affected by volcanism lends support to the notion that volcanic materials such as tuffs, ignimbrites and volcanic glass are excellent silica sources (Soil Survey Staff 1999). Glass tends to weather readily and liberate a great deal of silica (Chadwick *et al.* 1989a). Indeed, many soils with duripans may show evidence of recent ash deposition.

A primary issue about *silicification* centers on silica mobility. Silica is not generally thought of as a soluble substance in soils in the same sense as are salts and carbonates. But it *can* be soluble in the wetter, higher pH environment of the upper solum, only to be precipitated in deeper, drier horizons with near-neutral pH values. The solubility of silica is highest at pH values >9 and <4 (Fig. 9.7). Thus, the presence of Na^+ ions in soils with duripans is, therefore, important because it can raise the pH. The chemical behavior of silica in soils is also influenced by mineralogy, surface chemistry, moisture regime, leaching potential and biochemical activity (Milnes and Twidale 1983). Silica is translocated either as monosilicic acid (H_4SiO_4) or as an aqueous sol (Beckwith and Reeve 1964).

One important process relating to sources of silica centers around the differences between silica and carbonate solubilities. Calcite and quartz have inverse solubility relationships in alkaline environments (Summerfield 1982). The presence of soil carbonate may itself be responsible for raising the pH to a point where quartz is more soluble and silica more mobile. Like soils with illuvial carbonates, salts and gypsum, soils with duripans occur where the climate is just wet enough to weather minerals and translocate dissolved silica, but not out of the profile (Callen 1983, Southard *et al.* 1990, Thiry and Milnes 1991). Thus, Bq and Bqm horizons are often overlain by horizons that have been depleted in silica (Torrent *et al.* 1980a) (Fig. 12.50). If the parent material contains abundant calcium, a calcic or petrocalcic horizon tends to co-occur with the duripan, or it may occur immediately below the duripan, attesting to the slightly greater solubility of $CaCO_3$. Because of its greater solubility and abundance within dry lands, horizons of secondary carbonate often form more rapidly than do horizons of secondary silica (Reheis *et al.* 1992). Also, secondary calcite is more readily *redissolved* than silica, facilitating the translocation of carbonates to lower positions within the profile (Chadwick *et al.* 1987).

In soils that are acidic in the upper solum, silica is released into solution by weathering processes and is precipitated in deeper, higher pH horizons. Silica precipitation in a pH-friendly soil environment is primarily due to supersaturation of the soil solution. Thus, if present in the soil solution, silica tends to be deposited in horizons that have near-neutral pH values, either as amorphous opal or microcrystalline forms such as quartz and chalcedony; opal is most common (Summerfield 1983). The amorphous forms can later be transformed to quartz (Siever 1962). Repeated dissolution and reprecipitation is commonplace, as evidenced by etched and embayed borders of quartz crystals and opal fragments in soils. In the initial stages of duripan formation, much of the silica released by weathering and redeposited as a cementing agent is translocated only short distances. Monosilicic acid, which forms as silica is released into the soil solution, combines with water, is adsorbed on soil grains, polymerizes and precipitates primarily onto clay, sesquioxide and primary mineral surfaces as the soil dries. Secondary carbonates, however, preferentially precipitate in large voids on existing calcite crystals, more quickly plugging the horizon. The adsorbed silica can provide a template for further precipitation. The silica deposition process, therefore, tends to predominate in small voids and at grain–grain contacts, cementing the grains without completely plugging the voids between them (Chadwick *et al.* 1987). Although active *biogenic processes* involved in duripan formation have not yet been identified (Callen 1983), passive silicification of plants (replacement of plant parts with secondary silica) has been reported (Ambrose and Flint 1981).

Blank and Fosberg (1991) proposed a two-step model for the formation of agglomerates, involving *encapsulation* of quartz grains by silica, followed by its *alteration* to opaline silica. They suggested that agglomerates can form when loess, rich in quartz (and volcanic glass), is deposited on the surface of a soil with a duripan. Loess particles are translocated, intact, to the top of the pan, where they become coated (encapsulated) with carbonates and silica. Encapsulation may occur when perched water is lost to evaporation and deposits carbonates and silica. Then, in response to high pH values, the encapsulated loess agglomerates are altered to opal and other forms of amorphous silica. This model suggests that silica in duripans need not all have been translocated in solution.

Soluble salts

Salts continually enter soil profiles from dust, dissolved in rainwater and as by-products of chemical weathering (Table 12.8). Eolian processes may contribute salts as dried sea spray, as salt coatings on clastic particles and in pyroclastic materials (Yaalon and Lomas 1970, Dan and Yaalon 1982, Berger and Cooke 1997). Indeed, sea spray and fog are major sources of airborne salts (Watson 1985, Pye and Tsoar 1987, Schemenauer and Cereceda 1992). For upland soils, the sources of salts are atmospheric, especially dry deposition from oceans and salty lakes (Gerson and Amit 1987), and from weathering bedrock.

Under hyperarid conditions and at low temperatures, evaporites of calcium and magnesium chlorides, nitrates and other soluble salts can accumulate in soils (Dan *et al.* 1982, Soil Survey Staff 1999). Pariente (2001) found that an abiotic threshold exists at about 200 mm of annual precipitation; soils receiving less than that amount of rainfall have high soluble salt contents, while soils receiving more than 200 mm have very low contents. Our focus here is on the most soluble salts, e.g., sodium chloride (halite), soluble sulfates (thernadite, hexahydrite, epsomite and mirabilite) and bicarbonates (trona and natron). Because these salts are so readily dissolved in soil water, their location within the profile is an indicator of atmospheric

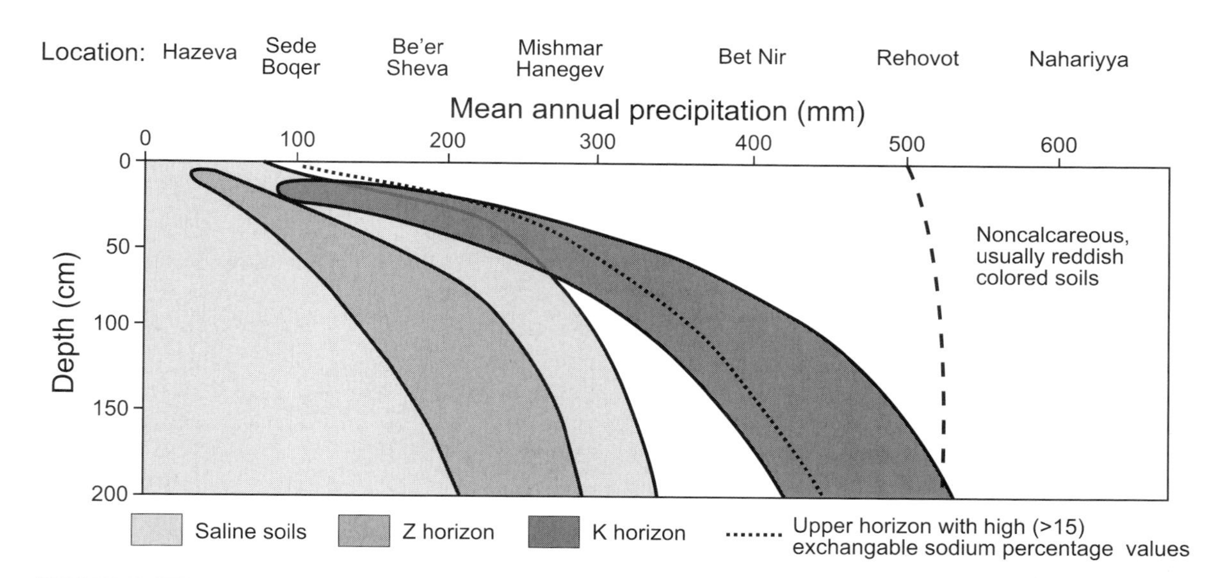

Fig. 12.51 Schematic diagram illustrating the depth of translocation of soluble salts, gypsum and carbonates along a climate sequence in Israel. After Dan and Yaalon (1982).

macroclimate or nearness to salty groundwater (Fig. 12.51).

Salts tend to accumulate in slowly permeable soils, from which they cannot be leached, or in dry climates where leaching is minimal and where a salt source is available (Fehrenbacher *et al.* 1963, Duchaufour 1982, Munn and Boehm 1983, Reid *et al.* 1993, Bockheim 1997). Among Earth's dry climates, those that have distinct wet and dry seasons tend to favor salt accumulation (Lewis and Drew 1973). During wet seasons, salts weather out of minerals, are washed into the soil, and are translocated, so as to become concentrated in certain horizons or landscape positions.

Providing that sources of the salts exist, they tend to accumulate according to their mobilities (Berger and Cooke 1997). In an upland, leaching regime, the salts with the highest mobilities will be lowest in the profile. In some arid climates, however, salts may not accumulate in upland soils due to extremely deep water tables, or insufficient weathering to release them from rocks. In a lowland setting where soils are shallow to groundwater, as in shallow depressions, on valley floors and in playas, the most mobile salts will move to the highest part of the profile by capillarity (Abtahi 1980, Young and Evans 1986) (Fig. 12.52a,b). Likewise, soils at the bases of slopes may have salts translocated *deeper* in the profile, because of additional run-on water, or they may have *more* salts because the run-on water carries additional salts with it. In sum, to understand how salts get concentrated in soils we must know the forces acting on soil and groundwater, and how these vary spatially and temporally (Arshad and Pawluk 1966, Seelig *et al.* 1990).

In upland soils, the depth to which salts are translocated will depend on precipitation regime, the pore space and permeability of the soil and the local topography. Wetting fronts usually do not penetrate as deeply in older soils that have accumulated a silty mantle of eolian sediment for two reasons: (1) the soil has less pore space, especially large pores, that facilitate deep wetting and (2) the surface may be plugged or have a crust, making it less permeable and leading to runoff (Amit and Yaalon 1996).

Salt content is typified/measured in the laboratory by the electrical conductivity of a saturated soil paste (Pariente 2001). Perhaps the only *field test* for gypsum and salts in soils is taste, because they impart no color to the soil at low concentrations. Horizons enriched in salts more soluble than gypsum are given the z suffix. Accumulation of soluble salts in soils is called *salinization* (Table 12.1).

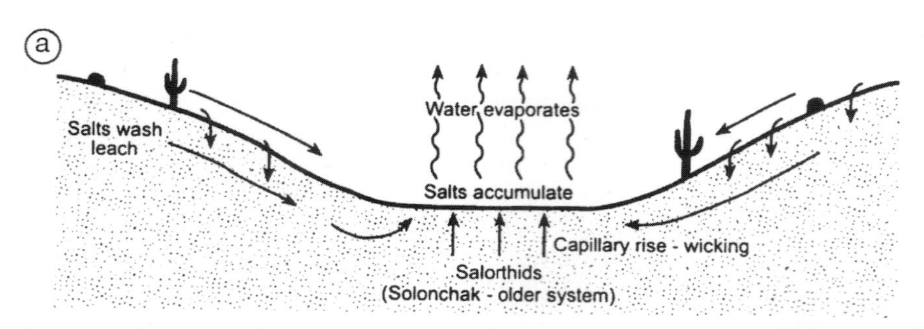

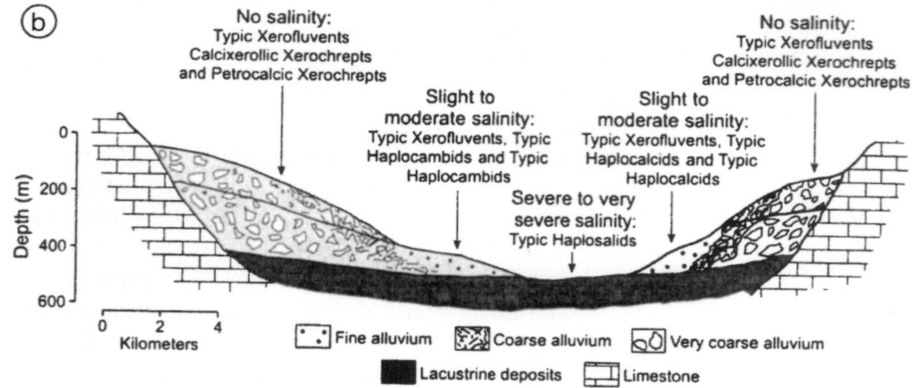

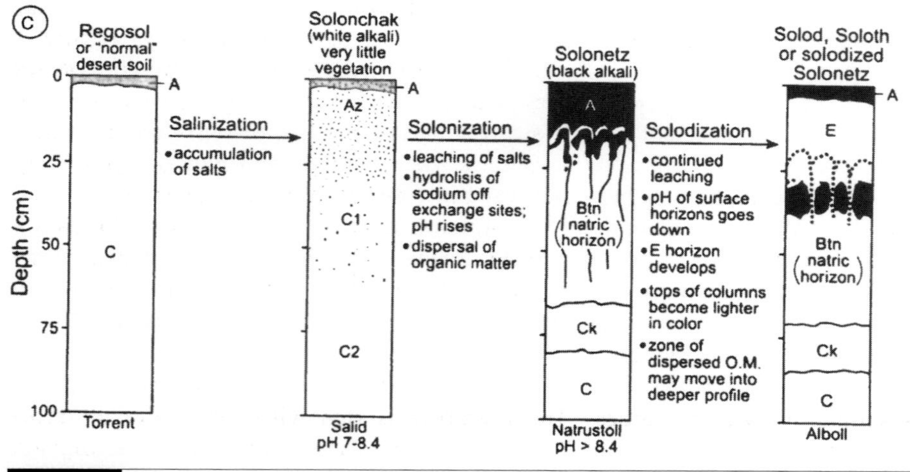

Fig. 12.52 Processes of accumulation of salts in dryland soils. (a) Idealized diagram showing how salts can accumulate in lowlands in dry climates. After Fanning and Fanning (1989). (b) Cross-section of a valley in Iran, showing the relationships among parent materials, topography and soil development. Note that the soils in the valley bottom are influenced by saline groundwater, while upland soils have accumulated secondary carbonate. After Abtahi (1980). (c) Diagrammatic illustration of the classical salination–solonization–solodization sequence for salt- and sodium-affected soils. After Kellogg (1936).

Sodium salts

Some of the most common salts in dryland soils are compounds of sodium, and to a lesser extent, magnesium. Ultimately, sodium salts originate from chemical weathering of rocks. Soils get exposed to sodium salts from initially Na-rich parent materials, from slightly saline groundwater, or from salty water running onto the soil surface

(Henry *et al.* 1985, Miller *et al.* 1985). Secondary sources may also be atmospheric (as rainout of aerosols derived from oceanic sources and playas) (Downes 1954, Wilding *et al.* 1963, Gunn 1967, Dimmock *et al.* 1974, Ballantyne 1978, Gunn and Richardson 1979, Peterson 1980, Bockheim 1997). B horizons enriched in sodium salts are partly due to (1) lower permeabilities brought on by increased amounts of clay, and/or (2) leaching that is only sufficient to translocate them to that part of the profile (Parsons *et al.* 1968).

Both Na and Mg cations are normally quite mobile in soils, but the depth at which they accumulate in soils varies greatly. Of the two cations, most emphasis is placed on Na^+ and its salts, because even a small amount of sodium can cause dramatic morphological changes in soils. Sodium, like hydrogen, another small monovalent cation, deflocculates (disperses) clay colloids, allowing them to be easily translocated. When enough illuvial clay (with adsorbed sodium) accumulates in a Bt or Btn horizon, it forms a sticky, jelly-like mass that, upon drying, is nearly impervious, structureless and clay-rich, possibly classifying as a *natric* horizon (Soil Survey Staff 1999). Thus, sodium salts can be considered a type of backdoor cementing agent, for they create a situation whereby soils become swollen, hard and impermeable. Salts, like carbonates, can also form impermeable pans (Bzm horizons) if all the soil pores are plugged with secondary precipitates.

The family of so-called neutral salts include not only sodium chloride (NaCl) but also gypsum, sodium sulfate (Na_2SO_4) and various other sulfates, bicarbonates and chlorides. If present in high enough amounts, saline soils with pH values <8.5 will result, with their white, salty surface crusts. Soils in which so-called alkali salts, such as sodium and potassium carbonates, are dominant have high levels of exchangeable Na^+ and pH values that exceed 8.5 (Duchaufour 1982). In these alkali or black alkali soils organic matter is dispersed through the soil matrix, rendering the soil black (Byers *et al.* 1938).

Sodium ions are highly soluble and readily leached from permeable, humid climate soils. Accumulation of sodium in the soil must, therefore, be due to some factor that either impedes the loss of sodium from the profile, or adds it to the profile faster than it can be removed. Leaching could be restricted by a slowly permeable layer at depth, such as bedrock or a clay-rich paleosol (Arshad and Pawluk 1966). Alternatively, in swales or at the bases of slopes, sodium-rich groundwater can be wicked upward (Sandoval and Shoesmith 1961, Miller *et al.* 1985) while upslope catena members are leached of sodium (Westin 1953, Lewis and Drew 1973) (Fig. 12.52b). Salts of Ca and Mg are less soluble and, in this situation, would get precipitated below sodium salts (Munn and Boehm 1983, Sobecki and Wilding 1983, Reid *et al.* 1993).

Understanding the processes that rid/flush sodium from the soil vs. those that replenish it, and seeing how they vary from place to place, is tantamount to explaining the genesis and distribution of sodium-enriched soils. This balance changes with time, season and landscape position; sodium-affected soils seldom cover the entire landscape (Arshad and Pawluk 1966). Rather, they exist at preferred spots where the balance is shifted toward sodium accumulation, rather than leaching (MacLean and Pawluk 1975) (Fig. 12.53). Typically, sodium accumulations occur in slight depressions on the landscape or in areas of low slope where sodium-enriched groundwater is close to the surface. Pedoturbation, such as by burrowing mammals, can counteract sodium and salt accumulations in soils, by mixing in C horizon materials and by increasing porosity, which facilitates leaching and flushing of salts (Munn 1993).

TYPES OF SODIUM-AFFECTED SOILS

The literature is rich with terms that describe sodium-affected soils (Byers *et al.* 1938, Johnson *et al.* 1985). Many agree at least some salt-affected soils go through three evolutionary phases: (1) salinization, (2) solonization, and (3) solodization (Fig. 12.52c). These old Russian terms refer to pedogenic morphologies associated with the accumulation of sodium (salinization), followed by its gradual loss (solonization and solodization). This "salt cycle" concept has long been accepted in Russia and the United States (Kellogg 1934, Byers *et al.* 1938). Hence the modifier that many still use to refer to salt-affected soils is *Solonetzic* (Russian, little salt), of which three kinds are

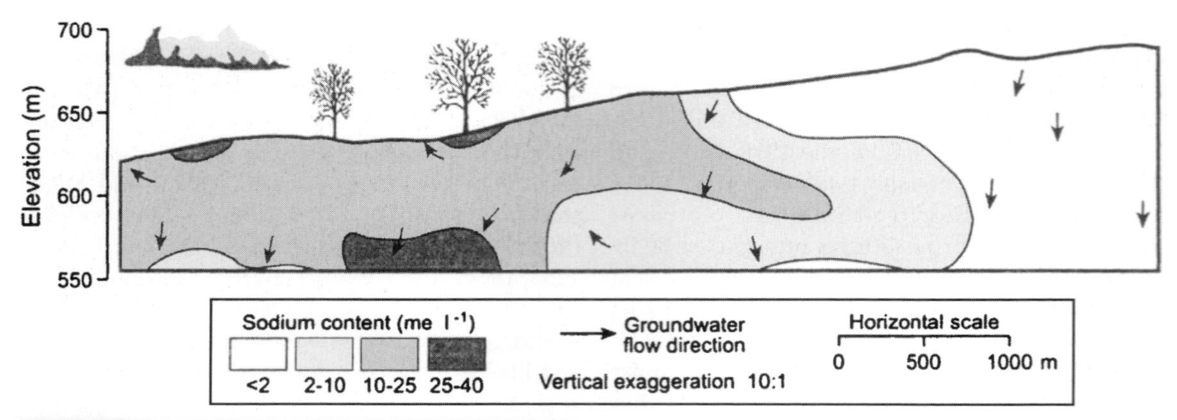

Fig. 12.53 Groundwater flow paths and sodium content of groundwater at a rolling, semi-arid steppe in Alberta, Canada. After Maclean and Pawluk (1975).

commonly recognized: the *Solonetz*, *Solodized Solonetz* and the *Solod*.

Salinization, the first stage, is the accumulation of neutral Na and Mg salts (McClelland *et al.* 1959, Parakshin 1974, Peterson 1980). It is often manifested as whitish salt accumulations on the surface. Soils undergoing salinization may develop a *salic* horizon (Soil Survey Staff 1999). Soils that result from this process have been called *Solonchaks* (Russian, much salt) (Joffe 1949); in Soil Taxonomy they usually classify as Salids. The terms *Calcium Carbonate Solonchak* or *Calcic Solonchak* are used if calcium and magnesium salts, rather than sodium salts, are dominant (Redmond and McClelland 1959). Gypsum is common in these soils. Solonchaks tend to be well flocculated, due to the presence of some divalent cations (Ca, Mg) (Fig. 12.52c). In extremely sodic soils, however, soil structure is absent, as the sodium salts disrupt the structure and the dispersed clays make the soil very slowly permeable (Reid *et al.* 1993). The dispersed horizon acts as a barrier to sodium accumulation (from below) in horizons above it and perches water. Thus, over time, a leached, slightly acidic eluvial zone can develop over the Btn horizon, much like a glossic horizon.

Solonization (or *desalinization*) is the removal/leaching of salts from the upper solum, possibly due to a change in climate, additions of irrigation water, or a change in the surface or subsurface hydrology. Solonization results in the formation of *Solonetz* soils. In this second stage of saline soil genesis, *salts* are leached but *sodium ions* still remain on the exchange complex (White 1978). Thus, accompanying solonization is the *alkalization* of the soil, in which more and more of the exchange complex comes to be dominated by sodium, with pH values >8.5 in part of the profile. With even as little as 15% sodium saturation, clays will disperse and be free to migrate downward in the profile, eventually forming a Bt or Btn horizon with a coarser-textured, poorly structured eluvial horizon above (White 1978) (Fig. 12.54). The Btn horizon can become

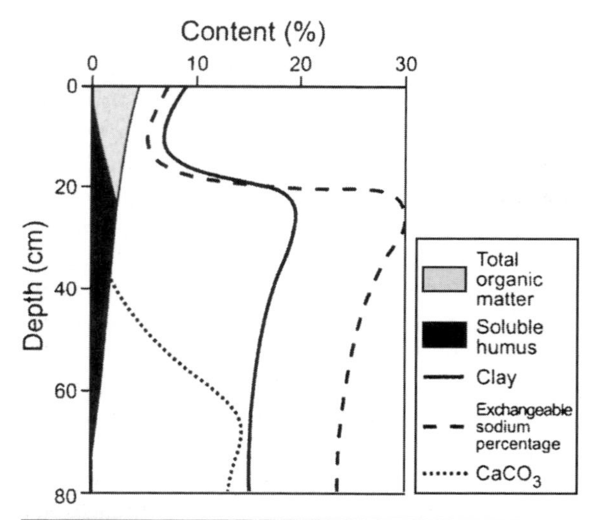

Fig. 12.54 Various physical and chemical properties of a Solonetz soil profile. After Szabolcs (1974).

tough and nearly impermeable, even when wet. Organic matter is also dispersed, giving the soil a very dark color. Again, gypsum in the lower profile is common (Bowser *et al.* 1962).

Solonetz soils (usually Natrudalfs or Natrustalfs) have columnar structure in the dense Btn horizon, often with rounded biscuit tops covered with a pulverescent white powder which is essentially a thin eluvial zone of clean sand and silt (White 1964) (Fig. 12.52c). Columnar structure is diagnostic of Solonetz soils and its genesis is interesting (Fig. 2.11e). In the initial stages of this process, translocation pathways for soil water into the lower profile are few, along desiccation cracks, as the sodium-saturated parts of the soil, i.e., the Btn horizon, are nearly impermeable. These cracks eventually become leached and form the outsides of the B horizon columns, just as occurs in a degrading fragipan. Light-colored eluvial materials form on the column tops too, for the same reasons. Sodium is then replaced by hydrogen (see solodization, below), allowing clays and clay coatings on the larger soil particles to be translocated downward, while uncoated sand and silt grains remain. The tops of the columns get rounded with time, as percolating water strips material more rapidly from their edges than from the centers (Fig. 12.55). Arshad and Pawluk (1966), however, suggested that the E horizon character in these soils may be due to degradation of organic matter in the lower A, rather than to eluviation. White (1964) suggested another mechanism that can lead to the rounded "biscuit tops": as the soil in the columns expands upon wetting, the centers of the columns swell upward.

Thorough leaching of the soil, the third stage, is called *solodization* or *dealkalization*. Now, sodium is gradually *removed* from the profile, often replaced by calcium, which facilitates flocculation of the clays, thereby forming or re-establishing good structure and enhancing leaching (White and Papendick 1961, Bowser *et al.* 1962). The columns break down, and the cracks between the columns become increasingly stripped of clay and whitish colored, taking on a strong glossic look. The upper profile may become sandier and more acidic while the unleached lower profile retains sodium and has a pH near 9. The terms *Solodized Solonetz* and *Degraded Alkali* are often applied to soils that were once saturated with sodium, but have since been partially leached (Mogen *et al.* 1959, White 1961). If enough calcium is retained within the profile, the soil may resemble a Planosol (White 1961).

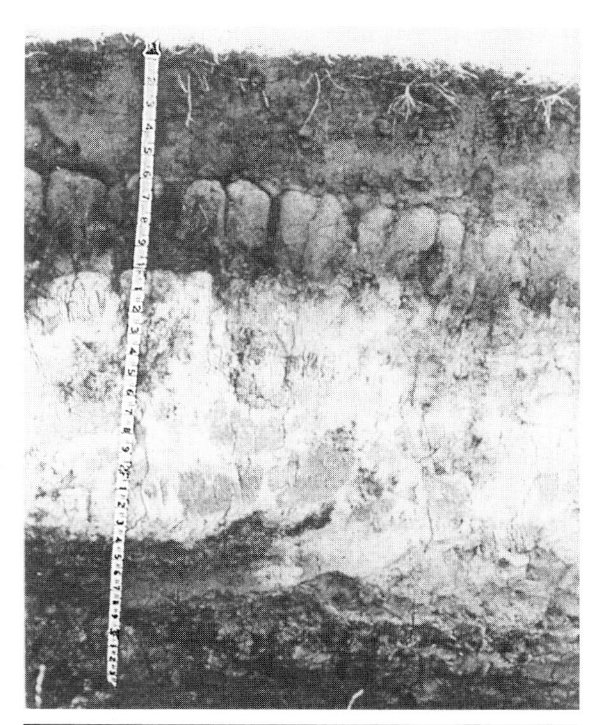

Fig. 12.55 A natric soil (Natrudoll) from Foster County, North Dakota, illustrating the columnar, "biscuit top" structure, and a calcic horizon at depth. Natural Resources Conservation Service file photo.

With continued solodization, more Na is removed. Hydrolysis becomes a dominant process and the sodium on the exchange sites is replaced by hydrogen. The columnar structure completely degrades, resulting in a soil called a *Solod*, *Soloth* or *Solidi* (White 1964) (Fig. 12.52c). Solods are not common. In wetter landscape positions they have an E horizon over a dark B horizon rich in illuvial humus (Anderson *et al.* 1979). Solodization (leaching) is inhibited in extremely dry climates or where shallow bedrock impedes leaching (Arshad and Pawluk 1966). In older classification systems these soils were called Planosols; many classify as Albaqualfs (White 1964). On uplands, the resultant soil resembles an Ustoll; the E horizon is often absent. White (1971) showed that grasses, when growing on solodized soils, do not selectively biocycle sodium; if they did, it would represent a mechanism by which solodization could be offset. Invading grasses do, however, cycle calcium, which further assists in the flocculation of clay and the formation of structure. Eventually, Solods may be so well leached of salts that they lose all outward signs of ever having been sodium-affected.

DISTRIBUTION OF SODIUM-AFFECTED SOILS

Sodium-affected soils often occur at isolated and predictable locations on the landscape. Many develop where sodium-rich bedrock is near the surface, as on summit and shoulder slopes, or where salty groundwater is near to the surface (Parakshin 1974). On many low-relief landscapes, the distribution of sodium in soils is spotty, occurring in slight depressions called *slick spots* or *panspots* (Norton and Smith 1930, Westin 1953). The dispersed clays of the exposed B horizon make these areas slick or slippery when wet. Slick spots range in diameter from 1 m or less to 15 m or more and are often devoid of vegetation, or at the most are vegetated with halophytes due to their high sodium contents and low infiltration capacities (Jordan *et al.* 1958, Munn and Boehm 1983, Hopkins *et al.* 1991, Reid *et al.* 1993). Due to minimal vegetative cover and dispersed clays (in the presence of Na), soils at slick spots are susceptible to erosion. Johnson *et al.* (1985) reported on a slick spot in South Dakota with an erosional scarp on its upslope side that had retreated 7 cm in less than 3 years.

Four factors allow this type of soil pattern to form and maintain itself (Westin 1953, Anderson 1987):

(1) An underlying source of sodium, such as shale, which also limits infiltration (Munn and Boehm 1983, Johnson *et al.* 1985).
(2) Movement of water and sodium from microhighs to microlows (spots). Good infiltration on microhighs fosters infiltration and loss of sodium; the sodium-rich water accumulates in the less permeable lows. Water movement is primarily driven by matric potential gradients, wherein moisture is preferentially pulled into clay-rich low areas. As water (and clay) continue to move into low areas, they get more clay-rich, further enhancing matric suction, and creating a positive feedback (Munn and Boehm 1983).
(3) Deflation of A horizon material from the spots, which brings any salty subsurface water into even closer contact with the surface. Conversely, lusher plant growth on non-sodic soils can trap dust, thickening the profile and, in effect, dampen the influence of groundwater.
(4) An increase in bulk density as the soil structure collapses due to sodium effects, helping to maintain slick spots at low points on the landscape.

In California, soils next to (but not *on*) slick spots had infiltration rates 4 to 97 times greater than did soils beneath slick spots (Reid *et al.* 1993). Low infiltration rates in slick spots is not so much due to lack of porosity, but to the low degree of interconnectivity of the pores. Soils not affected by sodium have blocky or prismatic structure in the B horizon.

Below, we use several examples of soil/water/landscape interactions to visualize how soils develop various patterns of sodium enrichment, e.g., Lewis and Drew (1973). For example, Munn and Boehm (1983) reported on the genesis of panspots in semi-arid Montana where soils have deep water tables (Fig. 12.56). Percolating water, carrying salts, gypsum and carbonates, encounters salty shale bedrock at shallow depths. At a high point in the subcropping bedrock, Na replaces other base cations on the exchange complex, dispersing the subsoil. Feedback processes set in, as the dispersed layer becomes drier than adjacent subsoils. Water, carrying silica and sodium, moves in response to matric potential toward the dispersed zone, enriching it in sodium. The dispersed zone grows upward and outward. Vegetation on the panspot thins, facilitating erosion. Low infiltration rates in the panspot limit how much water can enter it via

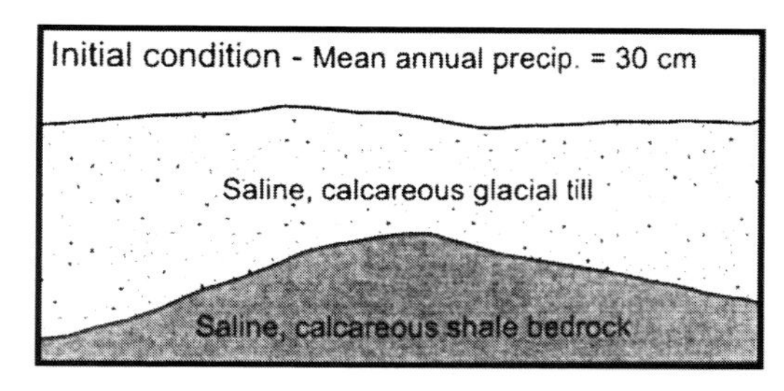

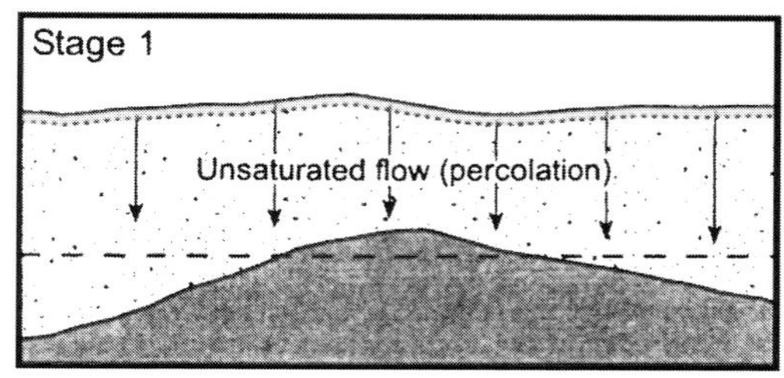

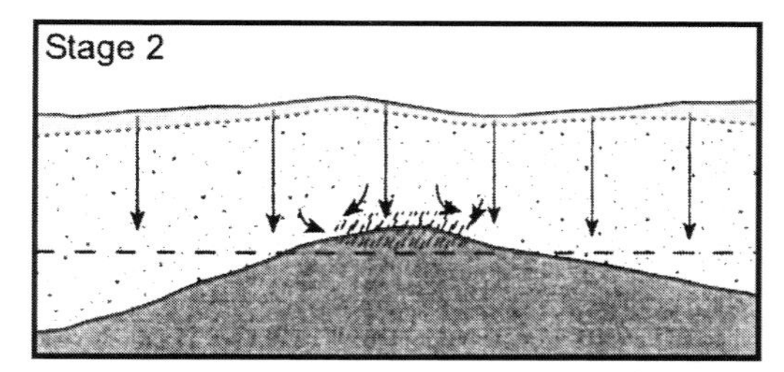

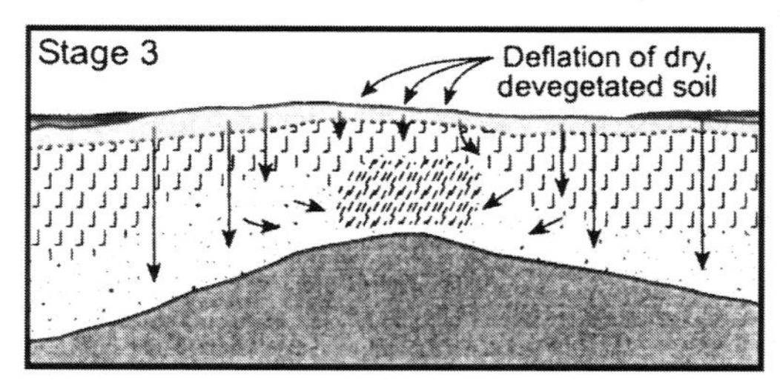

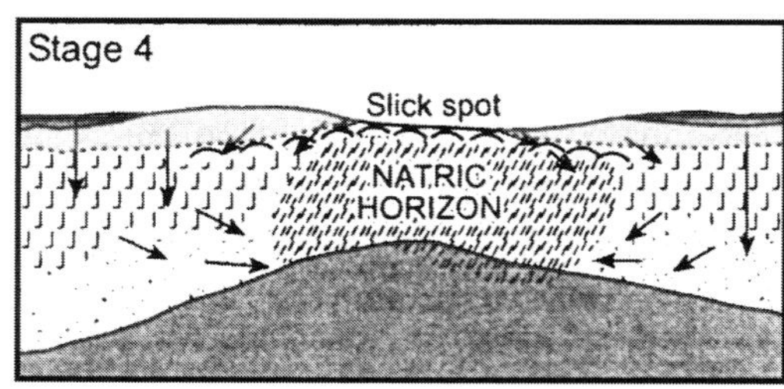

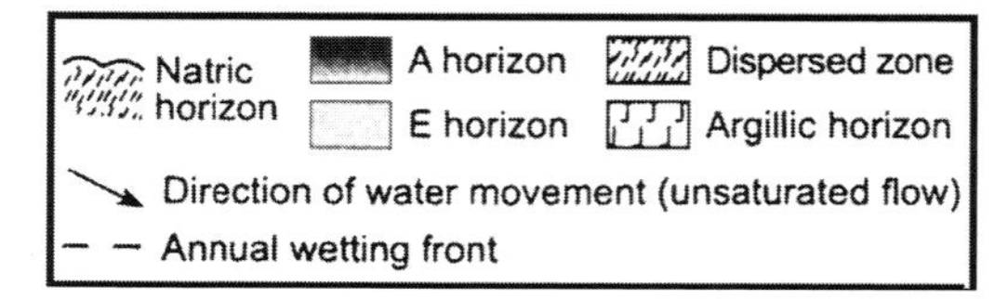

Fig. 12.56 Proposed genetic pathway for panspots in Montana. After Munn and Boehm (1983).

vertical infiltration, keeping it dry and further enhancing unsaturated, lateral flow into it (Jordan *et al.* 1958). Eroded A horizon materials from the panspot expose the Btn horizon, while concurrently, adjacent, vegetated soils trap eolian sediment and thicken. Although the panspots are sites of preferred run-on, low permeabilities prevent leaching of Na from the panspots.

In humid southern Illinois, where one would normally think that Na would be leached from soils, slick spots of Natraqualfs and Natralbolls occur on flat uplands where 0.5–2.5 m of loess overlies weathered Illinoian drift containing a paleosol (Fehrenbacher *et al.* 1963) (Fig. 12.57). These sodium-enriched soils are most common where the loess cap is 125–190 cm in thickness. The source of sodium is Na-feldspar in the loess (Indorante 1998). In thick loess areas, the soils are younger and less weathered, and therefore little Na has been weathered out of the loess. In thin loess areas where weathering should be maximal, mixing of loess with the weathered (leached) Illinoian till below dilutes the sodium-bearing mineral content. The intricate pattern of sodium-affected soils on these nearly-level landscapes points to redistribution of sodium. Wilding *et al.* (1963) suggested the following model to account for sodium-enriched soils in shallow loess above an Illinoian (Sangamon) paleosol by assuming highly variable permeabilities in the underlying, dense paleosol (Fig. 12.58). Wetting fronts carrying low amounts of sodium are diverted toward sites where the paleosol is more permeable. Precipitation of Ca and Mg carbonates occurs within the lower parts of the loess, due to the (eventual) supersaturation of the soil solution, leaving the soil solution enriched in Na. The soluble Na ions accumulate above and within the paleosol, where it is (was) more permeable. Ca and Mg continue to precipitate, aided by summer dryness, and act as nuclei for subsequent precipitation; carbonate nodules form. Thus, more and more of the exchange complex becomes dominated by sodium, and colloids in the B horizon of the surface (loess) soil become dispersed. Water movement becomes restricted to unsaturated flow, as the dispersed B horizon in the loess soil gets less and less permeable. Sodium

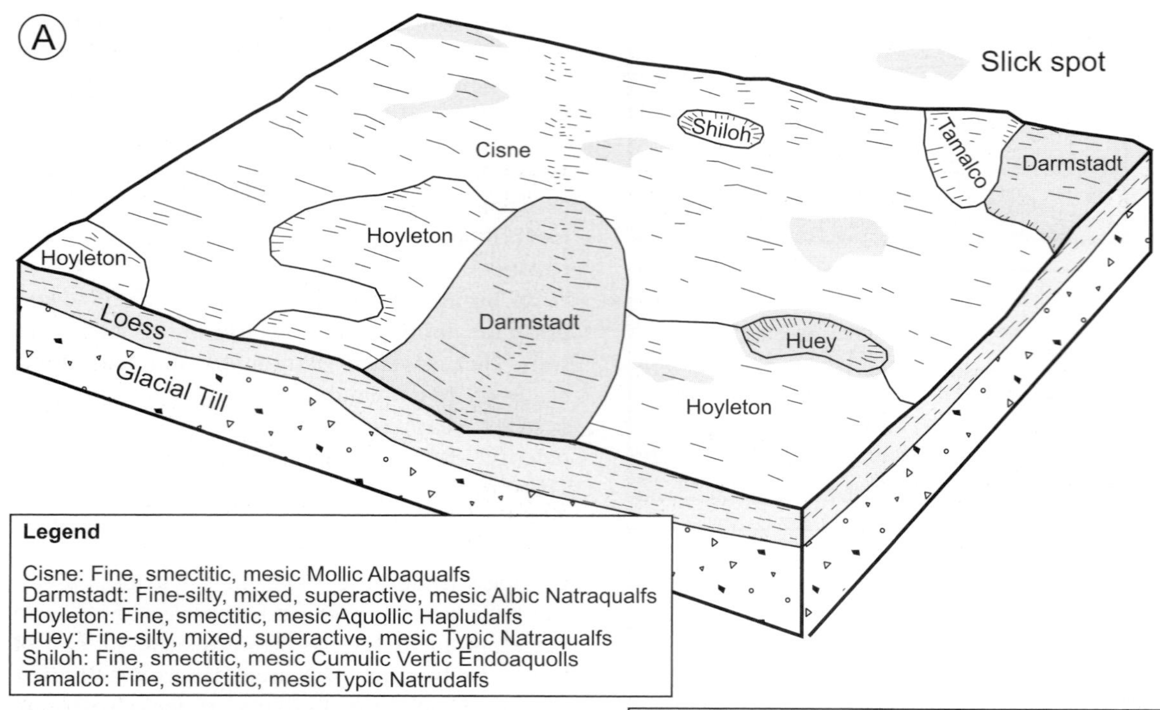

Fig. 12.57 The slick spot landscape of southern Illinois. (a) The Cisne–Hoyleton–Darmstadt soil association of southern Illinois, where thin loess overlies Illinoian till. After Bramstadt (1992). (b) The flat Illinoian till plain of southern Illinois: a Cisne landscape. Dark area in the distance is about 1 m lower than at the drill rig. Photo by L. R. Follmer. (c) Pipeline ditch in a Cisne–Newberry association in Fayette County, Illinois, showing the E horizon of the Cisne soil. Photo by L. R. Follmer.

continues to move into the drier, dispersed soil via unsaturated flow, setting up a feedback mechanism for preservation of the sodium already there, and even for continued enrichment of sodium. The sodium-enriched, slowly permeable B horizon grows laterally and upward, close to the surface. From this point on, a kind of inversion of drainage characteristics takes place – what were once the most permeable pedons are now sodium-saturated and almost impermeable. The pedons next to the Solonetzic ones receive water that has moved off the slowly permeable soils, furthering their development and leaching and providing a self-limiting mechanism to

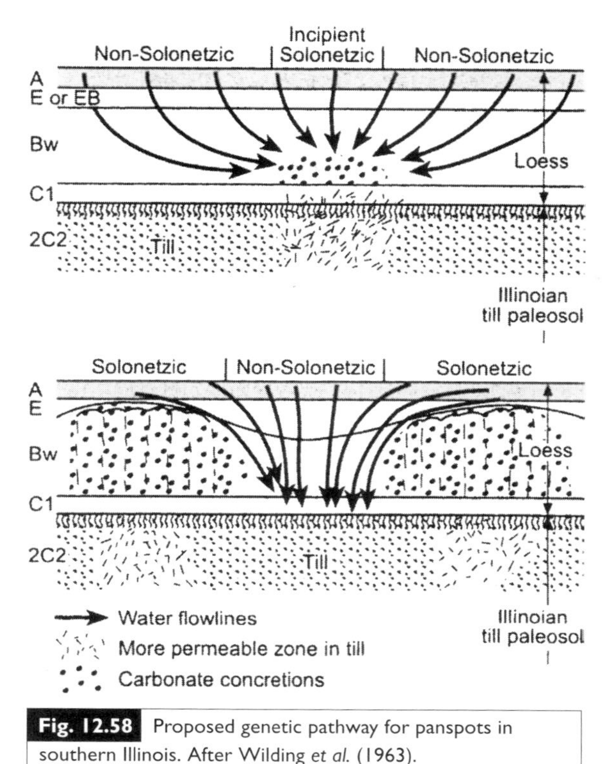

Fig. 12.58 Proposed genetic pathway for panspots in southern Illinois. After Wilding *et al.* (1963).

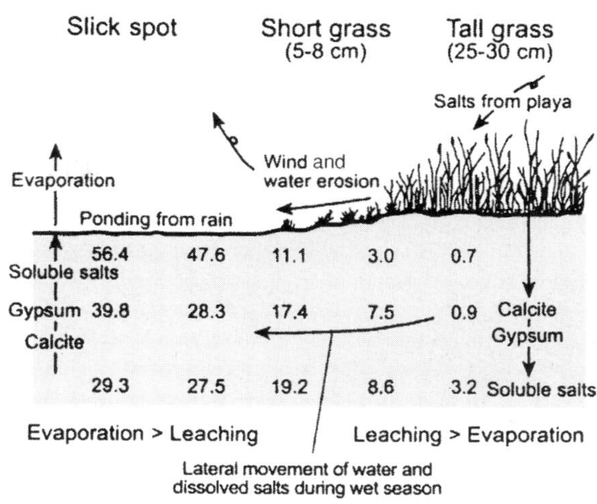

Fig. 12.59 Proposed genetic pathway for slickspots in a California grassland. After Reid *et al.* (1993).

the lateral expansion of the sodium-enriched soils. Recently, Indorante (1998) suggested an alternate genesis. On the most level sites on the Illinoian till plain, sodium can become enriched on microtopographic highs by evaporation of soil water. Once the high spots are eroded, an extremely flat landscape is formed that has a seemingly random pattern of sodium-enriched spots.

Reid *et al.* (1993) proposed a bioclimatic origin for slick spots in a California, suggesting that they begin as patches are laid bare by grazing or burrowing (Fig. 12.59). The bare areas are subjected to wind and water erosion, exposing the Bt horizon. Surface crusting and dispersion of clays follow. Infiltration capacities are lower on the bare areas, and thus these spots become drier and less leached than adjacent soils. Matric gradients facilitate water movement, as unsaturated flow, into the bare spots. The soil water, which contains salts blown in from nearby playas, moves upward, toward the dry surface, also in response to matric gradients (see also Ballantyne 1978). The least soluble substances (calcite) precipitate first, followed by gypsum and sodium (nearest the slick spot surface). In the end, solutes in soil water continue to enrich the slick spot, a feedback mechanism (Munn and Boehm 1983). Silica and base cations in the soil water can cause clays to neoform within the slick spot, further decreasing its permeability (Munn and Boehm 1983). Eventually, the vegetation on the slick spot dies, facilitating erosion, drying and continued growth of the spot at the expense of adjacent vegetated pedons. Reid *et al.* (1993) used ^{14}C ages from relict roots in the slick spot to show that this type of salinization can occur in less than 300 years. We conclude by noting that Hopkins *et al.* (1991) showed that that panspots can eventually become revegetated via normal pedogenic and geomorphic processes, effectively ending the cycle.

Near-surface processes in desert soils

Near-surface pedo-geomorphic processes in some deserts involve the formation of surficial concentrations of stones and gravel (Sharon 1962, Ollier 1966). The term *desert pavement* has been used for these surficial stony zones. The stones in most pavements almost appear to interlock (Fig. 12.60). In a strongly developed desert pavement there is little bare soil exposed. Often, the pavement is

Fig. 12.60 Desert pavement overlying a stone-poor, vesicular Av horizon. The pavement stones have a strong coating of desert varnish. Near Las Cruces, New Mexico. Photo by RJS.

only one layer in thickness. If a person walks on it, the gravel that comprises the pavement sinks a few millimeters into the soft, vesicular soil below (Nikiforoff 1937). Although desert pavements are common, we point out that many desert soils lack both a desert pavement and an Av horizon (Boettinger and Southard 1991).

Beneath most desert pavements is a porous, stone-poor A horizon that typically has more clay and silt than the underlying parent material (Springer 1958, McFadden *et al.* 1987, Williams and Zimbelman 1994). If this horizon has vesicular fabric, it is informally termed an Av horizon. Because the genesis of Av horizons and desert pavements are linked (McFadden *et al.* 1998), we will discuss them concurrently.

VESICULAR A HORIZONS

Most vesicular (Av) horizons are less than 20 cm thick, although their thickness can exceed 3 m on old, stable surfaces (Wells *et al.* 1985). Soil structure in Av horizons is usually weak due to lack of agents that could bind peds together (organic matter, clay, carbonate); Jessup (1960) referred to it as apedal. Textures range from sandy clay to silt to sandy loam (McFadden *et al.* 1998). The many pores in the Av horizon tend to be vesicular and not well interconnected. Springer (1958) demonstrated that vesicular pores and tubules will form simply by wetting of dry soil, as the wetting front effectively traps air bubbles. Yaalon (pers. comm. 1983) suggested that the pores form by expansion of air due to diurnal changes in temperature and after summer rainfall events (Evenari *et al.* 1974, McFadden *et al.* 1998). Expanding air which cannot escape forms vesicular-shaped pores. The CO_2 in the pores combines with Ca^{2+} ions in the soil to form thin $CaCO_3$ coatings on the insides of the pores, making them more stable than they would otherwise be. The vesicular nature of the horizon, therefore, can form very rapidly, perhaps in just a few months (Peterson 1980).

It is now accepted that the material in the vesicular horizon is predominantly dust that has accumulated on the surface and then been translocated in – hence the high silt + clay contents (McFadden *et al.* 1987, 1998, Anderson *et al.* 2002) (Fig. 12.61). Av and upper B horizons in the soils of Cajon Pass, California, routinely contain 40% more silt than do the C horizons (McFadden and Weldon 1987). Likewise, modern dust

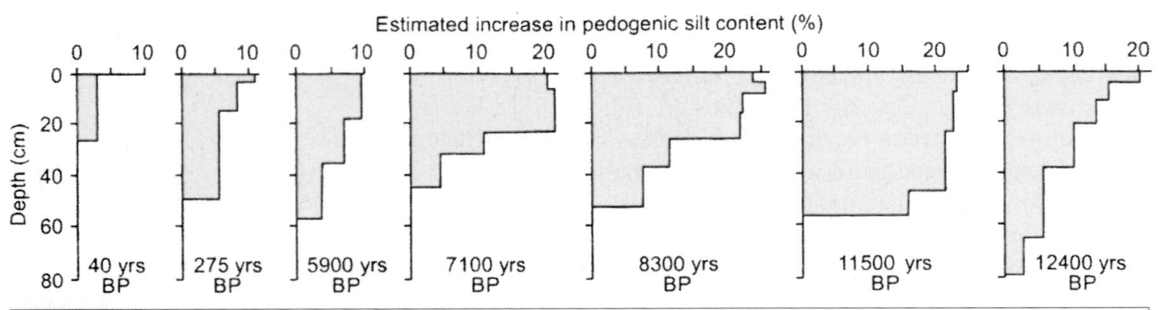

Fig. 12.61 Increases in silt content, above that of the parent material, in soils on alluvial terraces in southern California. After McFadden and Weldon (1987), who considered the minimum increase in silt content due to pedogenesis to be 3%.

contains significant amounts of salt and $CaCO_3$, which many desert parent materials lack (Reheis *et al.* 1995). The eolian origin of the Av horizon sediment is confirmed mineralogically and texturally, since Av horizons are unlike the C horizon and strongly resemble dust caught in mechanical traps (Reheis *et al.* 1995). Some of the silt and clay in Av horizons, however, may have been weathered from coarser materials in situ, although Gerson and Amit (1987) found no evidence for this process. Gerson and Amit (1987) suggested that only 1000–5000 years are required for a continuous Av horizon to form in the Negev of Israel (Fig. 12.62).

McFadden *et al.* (1998) dated the sediment in Av horizons by thermoluminescence, a technique which determines the last time that the sediment was exposed to sunlight (see Chapter 14). They found that most of the Av material was about 5000 years old, while Bw material below was last exposed to light about 13 000 years ago. Dating by ^{14}C on the organic matter within Av horizons gave comparable dates (Anderson *et al.* 2002). These data support the eolian origin of the fine material in Av horizons, and illustrate why soils older than a few thousand years also contain an Av-Bw horizon sequence.

DESERT PAVEMENT

Armored surfaces of intricate mosaics of rocks, usually one or two layers thick, overlying stone-poor soil horizons are referred to as desert pavement in the United States. In Australia they are *gibber plains* and *stony mantles*; *hamada*, *reg* and *serir* are Arabic terms for the same, while in parts of Asia the term *gobi* is used (Sharon 1962, Cooke 1970).

The rocks in pavements are often aligned such that they have a flat surface parallel to the soil surface, i.e., individual rocks usually do not overtly protrude above the general level of the pavement (Cooke 1970). The rocks are all about the same size and close-packed, often appearing pieced together like a jigsaw puzzle mosaic (Fig. 12.60). In rare instances they can be cemented. Once formed, pavements are an important stabilizing factor for both the slope and the soil; they act as a protective armor (Sharon 1962). Pavements are prominent in deserts worldwide, commonly being found on gently sloping sites where vegetation is minimal and precipitation is very scarce, i.e., in the driest deserts (Dan *et al.* 1982). The lack of vegetation on desert pavements is probably a positive feedback mechanism, as the pavement itself may inhibit vegetation from getting established, while the paucity of plants minimizes the risk of the pavement being destroyed. Cool, high-elevation areas in deserts tend not to have pavements, probably because of increased disturbances from roots and animals (Fig. 12.63). Quade's (2001) work in the Mojave Desert showed that pavements form continuously only at elevations *below 400 m*, which stay hot and dry even during glacial periods. Sites above that elevation are wetter and densely vegetated during glacial periods; pavements form here only during dry, warm interglacial climates.

Early work on the origin of desert pavement attributed its development to wind and/or water, which eroded the fines and left behind a lag concentration of rocks (Lowdermilk and Sundling 1950, Symmons and Hemming 1968, Parsons *et al.* 1992) (Fig. 12.64). This model is not widely accepted, or at least applicable, largely because any surface crust would inhibit deflationary processes, and because it does not explain the presence of the Av horizon (Cooke 1970, Williams and Zimbelman 1994). Deflation (erosion of fines by wind) may help to *maintain* an existing pavement (Jessup 1960), but probably cannot solely *create* one. Erosion of fines by water has been shown to be important in some desert pavement settings (Parsons *et al.* 1992, Williams and Zimbelman 1994) and it does appear that rainsplash and water do play *some* sort of role in the re-establishment, if not in the original formation, of pavements (Wainwright *et al.* 1999). However, these processes do little to explain the apparent interconnectiveness of the rocks or the planar surface nature of the pavement (Fig. 12.60); running water should form at least a few rills and gullies.

Others (e.g., Springer 1958, Jessup 1960, Ollier 1966) thought that desert pavements form due to the upward migration of stones through clayey soil horizons, via argilliturbation (see Chapter 10). This argument seems logical where sediments are clay-rich and the surface has gilgai,

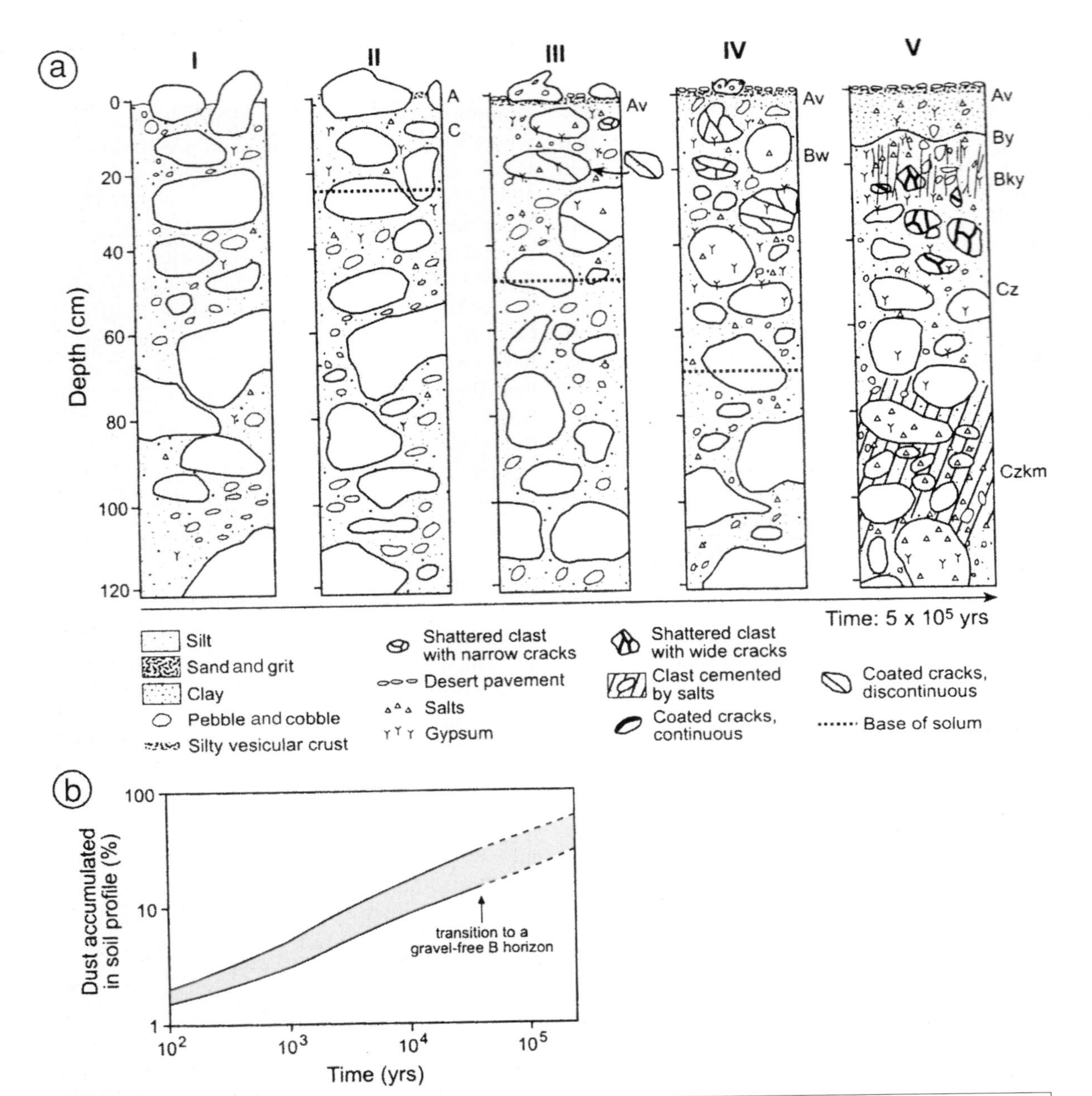

Fig. 12.62 Impact of dust on the morphology of a reg soil in the Negev Desert, Israel. After Gerson and Amit (1987) and Amit and Yaalon (1996). (a) Developmental stages associated with pedogenesis and dust inputs to gravelly parent material. Most, but not all, horizons are named. I, Recently deposited gravelly alluvium. Low salinity values exist in the soil. Typic Torrifluvent less than 1000 years old. II, Dust accumulates in the porous alluvium, forming a thin A horizon. Salt accumulations lead to the formation of a thin crust, partly biogenic in origin, and thin coatings on the undersides of clasts. Typic Torrifluvent 1000–2000 years old. III, Continued dust accumulation leads to the formation of an Av horizon. Salts (mostly gypsum) in near-surface horizons cause clast shattering. Desert pavement in early stages of formation. Gypsum begins to accumulate in the subsurface. Typic Torrifluvent 2000–7000 years old. IV, Dust continues to accumulate. Cambic Bw horizon forms below the Av horizon. High amounts of salts, especially in the C horizon. Some gypsum nodules begin to form. Salt weathering of clasts is intense in upper solum. Desert pavement is nearly continuous. Typic Haplogypsid 7000–14 000 years old. V, Soil has a strong desert pavement; Av horizon and Bky horizon underlie a continuous desert pavement. Halite dominates the C horizon, which is a petrosalic horizon composed mostly of halite and no shattered gravel. Thick, continuous salt cutans have led to intense rock shattering in the upper 60 cm. Gypsum peaks high in the profile. Gypsic Haplosalid 14 000–500 000 years old. (b) Accumulated dust in reg soils as a function of time. After Gerson and Amit (1987).

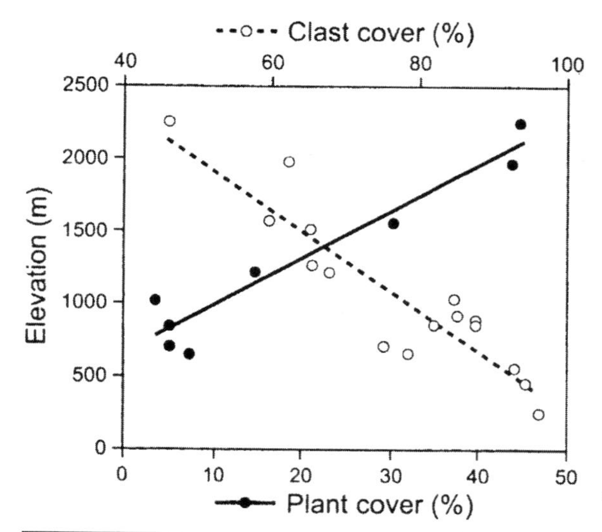

Fig. 12.63 Plant cover (trees and shrubs) vs. elevation and desert pavement coverage for some sites in the Mojave Desert, California. "Interglacial" pavements formed during oxygen isotope stage 5 when the climate was warmer and drier, and plant cover was less. After Quade (2001).

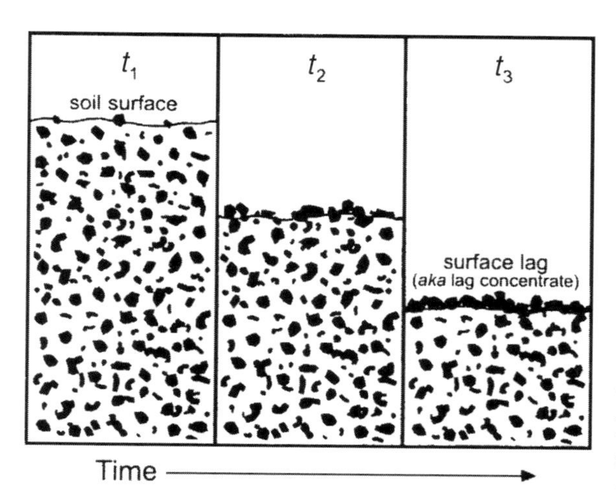

Fig. 12.64 Formation of a surface lag gravel concentration by erosion of the finer material. Erosion can occur by wind or water.

but would not be applicable to deserts where the underlying soils are silty or in hyperarid settings. Cooke (1970) extended this argument to include shrink–swell processes associated with freezing and thawing, and salt crystallization and solution.

A more recent model (McFadden *et al.* 1987), supported by data from Wells *et al.* (1995) and Haff and Werner (1996), attributed pavement genesis to two major processes: (1) accumulation of clasts on slopes and in low areas by mass movement and slopewash, in conjunction with (2) the detachment and uplifting of clasts as eolian fines infiltrate into the clast-supported matrix. The model requires that there is a source of small rocks (presumably formed by physical weathering) *and* inputs of eolian dust (Anderson *et al.* 2002). Large interstices between rocks and gravel facilitate dust accumulation. Dust that lands on desert pavements and is not incorporated below it is susceptible to deflation and water erosion. Thus, the desert pavement acts to protect the underlying, accumulated dust from further erosion, while at the same time the rocks help to trap some dust which is later incorporated into the profile. McFadden *et al.* (1987) stressed that pavements "are born and maintained at the surface" (Wells *et al.* 1995). That is, they need not evolve from rocks scattered at depth. Surface exposure dating of basalt clasts from a lava flow, now with a distinct desert pavement and Av horizon, confirmed that the rocks on the pavement have been *exposed at the surface* as long as the basalt on nearby bedrock highs, i.e., they had never been buried (Wells *et al.* 1995).

All the while, Av horizons are *evolving* as dust is trapped and incorporated into the soil (McFadden *et al.* 1987, 1998). During times of low eolian input, infiltration events lead to the formation of weak, shallow Btk and Bky horizons below the Av horizon. Small amounts of clay in the Av horizon can lead to shrink–swell processes, very small vertical cracks and weak structure (Cooke 1970). Once the Av forms the entire horizon is actually lifted with the pavement as dust moves into fractures. Additionally, if the dust contains some salts it will expand when wet, forcing rocks upward, allowing the fine materials to settle and further enhance the separation of the clasts from the subsurface. Though small, the cracks facilitate translocation of silt and fine sand, allowing the uppermost Av horizon to become more clay-rich through time (Gerson and Amit 1987). Prism faces, therefore, are silt-coated and noncalcareous, while ped interiors are reddened and carbonate-enriched. As these processes continue, the soil surface grows upward by eolian accretion, and columns/prisms (once within the Av horizon proper) are displaced lower. Coalescence of these peds forms a weak, essentially continuous Bk or By horizon. This model may apply to stone pavements in Australia that were explained incompletely by other hypotheses (e.g., Mabbutt 1965).

The thickening of the Av horizon profoundly affects the hydrology of the profile (Yair 1987, McFadden *et al.* 1998). Thicker Av horizons lead to shallower penetration of wetting fronts, as the silty material has much more surface area than did the gravelly materials that pre-dated it. This, in turn, affects the depth of carbonate accumulation within the profile, such that older soils with Av horizons accumulate carbonates at shallower depths than do younger, gravelly soils. This concept can be extended to other desert soils that have accumulated eolian materials over time: as their eolian mantles thicken, wetting fronts are able to penetrate more shallowly, rendering the soil more pedogenically arid (Yair 1987, Eppes and Harrison 1999). For this reason, B horizons in gravelly and rocky soils, other things being equal, are deeper than in soils on the same geomorphic surface that have accumulated a silty mantle. The fact that pavements develop and change predictably over time make certain of their characteristics useful as relative dating indicators (Al-Farraj and Harvey 2000) (see Chapter 14).

Because of the intimate, genetic relationship between the formation of desert pavement and the Av horizon, disturbance of the pavement can very quickly lead to erosion of the underlying material. Once its protective armor is lost, the Av horizon becomes vulnerable to wind and water erosion. Disturbance by foot traffic of all kinds, as well as off-road vehicles in deserts, can destroy the interlocking pavement and make the surface susceptible to erosion. Fortunately, some of the same processes that lead to its formation also facilitate rapid healing (Sharon 1962, Haff and Werner 1996, Wainwright *et al.* 1999). Although we have made great strides in understanding the origins of desert pavements, there is still much that we do not know (Amundson *et al.* 1997, McFadden *et al.* 1998).

ROCK VARNISH

Many rocks on relatively old geomorphic surfaces in arid and semi-arid climates contain a glossy, dark brownish-black or gray layer called *desert varnish*, *rock varnish*, *desert lacquer* or *patina*. Broecker and Liu (2001) referred to rock varnish as a thin coating of a cocktail rich in Mn, Fe and clay minerals. Varnish is typically only on the upper surfaces of rocks that have been unmoved for long periods of time. Underneath, where buried, rocks lack this varnish and take on an orange or yellow hue; some retain their original surface color (Dorn and Oberlander 1981b). Rock varnishes are usually composed of alternating, micron-scale layers of detrital clay minerals, which range from 2 to 500 μm thick overall (Potter and Rossman 1977, Perry and Adams 1978, Taylor-George *et al.* 1983). The coatings usually get thicker, darker and more continuous over time, as patches of varnish coalesce. The oldest varnishes can be essentially black. Color, content of manganese and thickness/completeness of the coatings have all been used as relative age or surface exposure indicators (Reneau 1993).

Rock varnish tends to form only in *periodically* dry regions because the microbial flora are suited to short growth periods after rain events. In humid regions, soil formation and rock dissolution are so rapid that any varnishes that might form are quickly stripped (Krumbein and Jens 1981). Similarly, rock varnish usually forms only on hard, resistant rocks; any varnish that forms on soft limestones and granites is removed as the rock disintegrates.

Rock varnish gradually gets thicker, shinier and darker over time, and slowly covers larger percentages of the rock surface (Perry and Adams 1978). Time estimates for varnish formation range from 25 (Engel and Sharp 1958) to 300 000 years (Knauss and Ku 1980); the latter places some varnishes well beyond the range of radiocarbon dating. Only young varnishes are therefore readily amenable to radiocarbon dating, but most are useful as a relative dating tool, bearing in mind that the rate of formation varies as a function of location, especially as influenced by climate (Hayden 1976). Rock varnish also has utility as a paleoenvironmental indicator (Dorn 1988, 1990, Broecker and Liu 2001).

The formation of rock varnish is only partially understood, but most researchers agree that the clastic constituents of the varnish, e.g., clay minerals, are externally derived and transported to the rock by wind (Perry and Adams 1978, Dorn and Oberlander 1982). The Mn : Fe ratio of varnishes usually exceeds 1 : 1, forcing the question regarding how Mn gets enhanced relative to Fe (Jones 1991). A possible abiotic answer to this question may center on Eh and pH conditions. Mn is considerably more soluble than Fe under slightly acidic surface conditions (Jones 1991). If dew or other water wets the rocks it can release Mn and Fe into solution. Then, as the water evaporates, clay minerals may buffer the solution and precipitate the Mn and Fe out as varnish. Under this model, wind-blown dust combined with iron and manganese compounds released via weathering may contribute to the formation of a type of varnish that has a distinctly abiotic origin (Smith and Whalley 1988).

Recent research has, however, documented that manganese-fixing bacillococci bacteria and microcolonial fungi play the dominant, active role in rock varnish formation (Dorn and Oberlander 1981b, Krumbein and Jens 1981, Staley *et al.* 1982, Taylor-George *et al.* 1983, Palmer *et al.* 1986). Dorn and Oberlander (1981a) reviewed the evidence supporting this claim. These microbes concentrate manganese and some iron, which may then combine with particulates from the atmosphere to form manganese-rich varnish (Dorn and Oberlander 1981b) (Fig. 12.65). This model does not, however, fully explain the rare forms of Fe-rich and Mn-poor varnish (Jones 1991).

The source of the Mn and Fe in rock varnish is the rocks themselves, running water, dew and dust (Allen 1978, Krumbein and Jens 1981). The Mn and Fe compounds, especially oxyhydroxides, are effective at fixing clay minerals, brought in by wind, to the host rock. As rocks wet and dry, microbial activity, clay adsorption and Mn oxidation are enhanced. Higher concentrations of clays and Mn results in more fixation of Mn by the clays, resulting a type of positive feedback mechanism (Dorn and Oberlander 1981b).

Taylor-George *et al.* (1983) proposed a model for the formation of rock varnish in deserts. Microcolinial fungi establish themselves on rocks and help to trap eolian silt and clay on the rock surface. Under favorable conditions, the fungi grow into clusters. Dead fungi decompose, possibly by bacterial action or UV oxidation. Occasionally, dissolution and reprecipitation of the inorganic materials associated with these fungi occur; it is aided by rain and dew. Then Mn is either concentrated by the microorganisms directly, or inorganically during dissolution/reprecipitation. Patches of varnish enlarge and coalesce as fungi growing at the edges get incorporated. With time, layer upon layer of fungi and precipitated solutes develop, making the varnish thicker and more extensive.

High UV radiation levels, extremely dry conditions, high diurnal temperature and humidity changes, with virtually no chance for displacement upward or downward in/on the rock, make life on a rock surface in a desert a difficult existence! And yet, bacteria and fungi live and die there, leaving behind layer upon layer of coatings (Krumbein and Jens 1981). There is some evidence, however, that the clay and particulate matter in the varnish help shield the microbes from the intense heat and radiation of desert regions. Krumbein and Jens (1981) noted that the "solution front community" of microbes that exists at the rock–varnish interface is clearly protected from the harsh desert environment by the overlying rock varnish layers, much like a protective shield. Mn oxide is a strong absorber of UV light and may thus provide some shielding to these microorganisms (Taylor-George *et al.* 1983). Thus, varnish can be viewed as an ecological protective mechanism, formed to protect the microbial communities below from intense UV radiation and desiccation.

BIOCRUSTS

Many Aridisols that lack a desert pavement have instead a thin but hard surface crust. Although only a few millimeters thick, the crust can dramatically reduce infiltration rates and increase

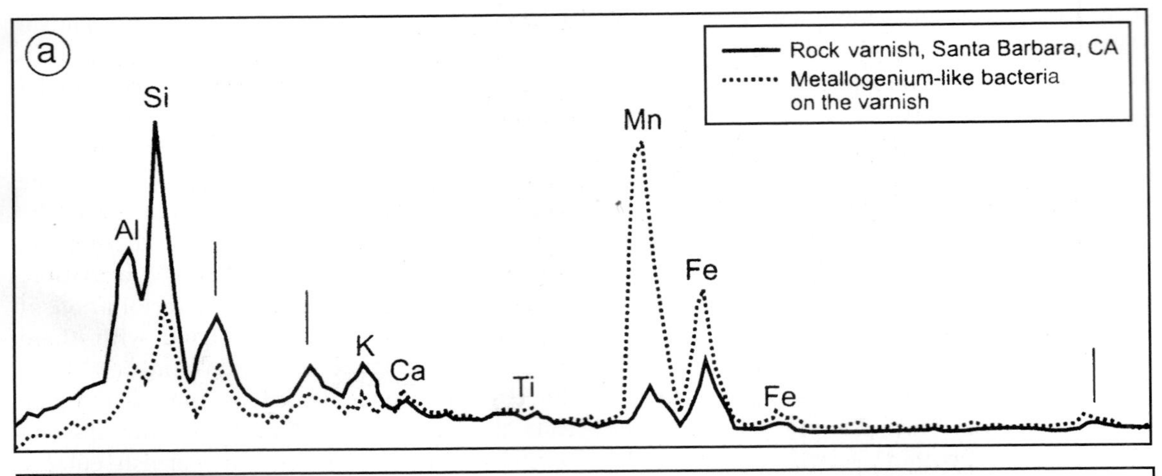

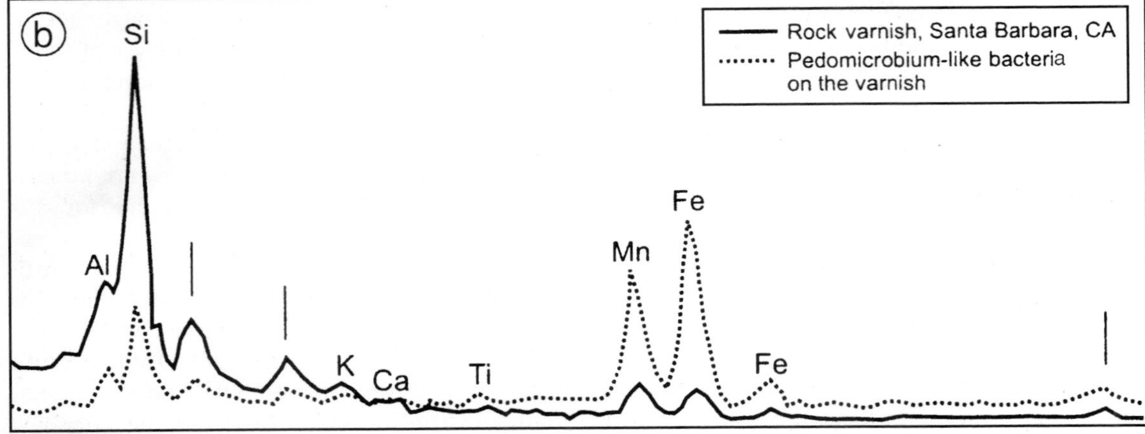

Fig. 12.65 Energy dispersive X-ray analysis spectra of rock varnish and bacteria present on that varnish; (a) metallogenium-like bacteria, (b) pedomicrobium-like bacteria. Peaks show relative abundances of various elements. Peaks designated by vertical lines are emissions from the gold and palladium coating on the samples. After Dorn and Oberlander (1981b).

runoff after even small storms, in part because they are slightly water-repellent (Jungerius and van der Meulen 1988, Yair 1990). The crusts are common on bare surfaces that are occasionally wet or even flooded, such as playas and interdunes.

The crusts have long been assumed to be rainfall-induced, because they form after rain events on bare surfaces. Raindrop impact is responsible for the close-packed structure, but the source of the binding agent has been more elusive. Small amounts of salts may provide some cohesive force (Sharon 1962). However, most crusts are now assumed to be biologic in origin (Fletcher and Martin 1948, Macgregor and Johnson 1971, Dor and Danin 1996, Danin *et al.* 1998). Algae, molds and cyanobacteria grow in the upper few millimeters of the soil after rain events. Their mycelia and filaments bind the soil particles together, forming biocrusts enriched in C, N, clay and silt (Fletcher and Martin 1948).

Podzolization

In *podzolization* organic carbon, Fe^{3+} and Al^{3+}, in some combination, are translocated from the upper profile to an illuvial (sometimes, *spodic*) horizon (Petersen 1976, DeConinck 1980, Buurman and van Reeuwijk 1984, Courchesne and Hendershot 1997, Lundström *et al.* 2000a) (Table 12.1). It is prominent in cool, humid climates where there is an excess of precipitation over

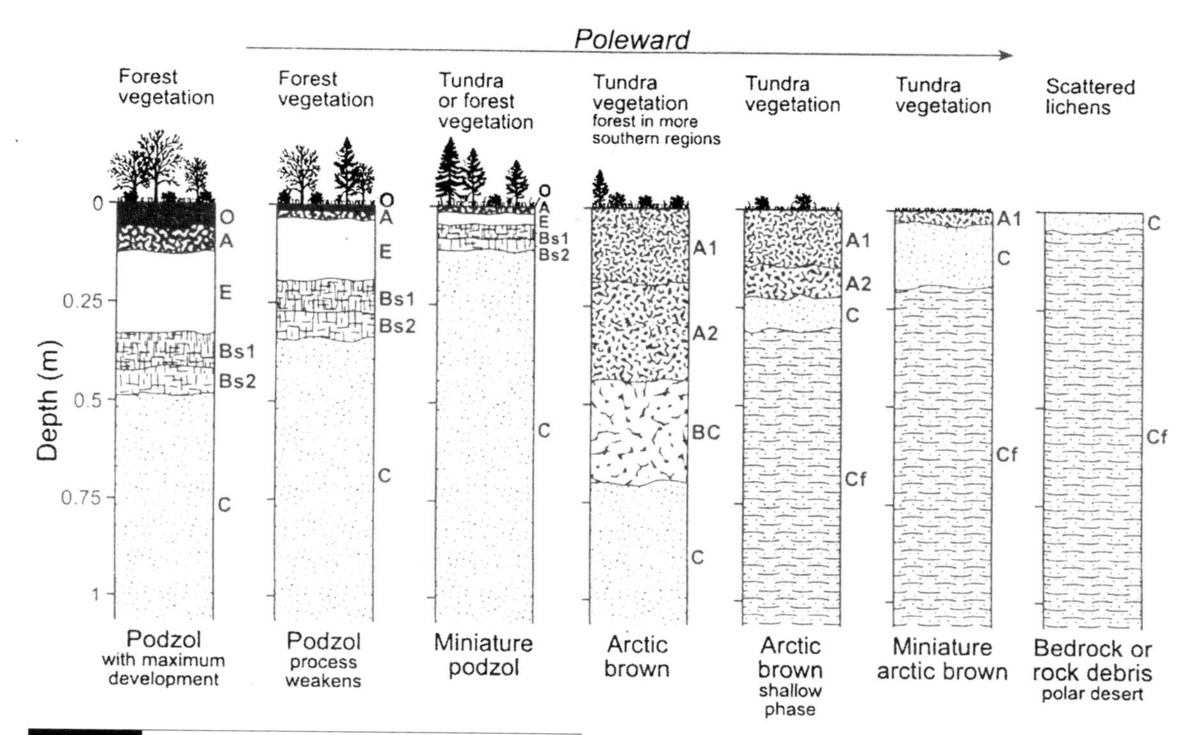

Fig. 12.66 Idealized soil morphology sequence from the boreal forest of central Canada, where podzolization is strong, to the polar deserts. After Tedrow (1977).

evapotranspiration, such that water frequently moves completely through the profile (McKeague *et al.* 1983). Podzolization is best expressed under vegetation that produces acidic litter, such as coniferous forest and heather, and on coarse-textured parent materials. It is especially strong in the coniferous forests of the high midlatitudes, and weakens in intensity to the north and south (Fig. 12.66). Cool temperatures keep evapotranspiration rates low and inhibit decomposition of the acidic litter.

In some literature, podzolization is taken to include not only translocation of metals and organics, but also associated processes of (1) chemical weathering by organic and carbonic acids and (2) clay destruction and/or translocation (Fridland 1958, Alekseyev 1983). Under this broader definition, the focus is on mineral decomposition by chemical and biochemical agents in the upper profile. Some of these weathered minerals are lost from the profile, but the great majority are simply redistributed, in suspension, to the lower profile (Guillet *et al.* 1975) (Fig. 12.67). The narrower, more traditional definition involves mainly the translocation of organic carbon, Fe^{3+} and/or Al^{3+} from an eluvial to and illuvial zone, usually in solution. Nonetheless, in both "types" of podzolization pH values are low, and weathering and translocation are so intense in the upper profile that many minerals are completely weathered, only to be translocated completely (as ions) from the profile or to form amorphous materials in the B horizon. Degradation of primary minerals, usually facilitated by abundant organic acids, is intense (Guillet *et al.* 1975). In either scenario, an acidic E horizon is formed; in the more narrow definition that E horizon need only be depleted of Fe, Al and humus, while in the broader definition it may also have lost clay. The "true" podzol (Spodosol) soil is generally so coarse-textured that it has few primary minerals and silicate clays. In finer-textured soils that are also undergoing podzolization in the broader sense, translocation of clay is strong. These soils are podzolized but in a slightly different manner. In this book, we focus on podzolization in the narrower sense – the translocation of metal cations and organic matter in coarse-textured soils.

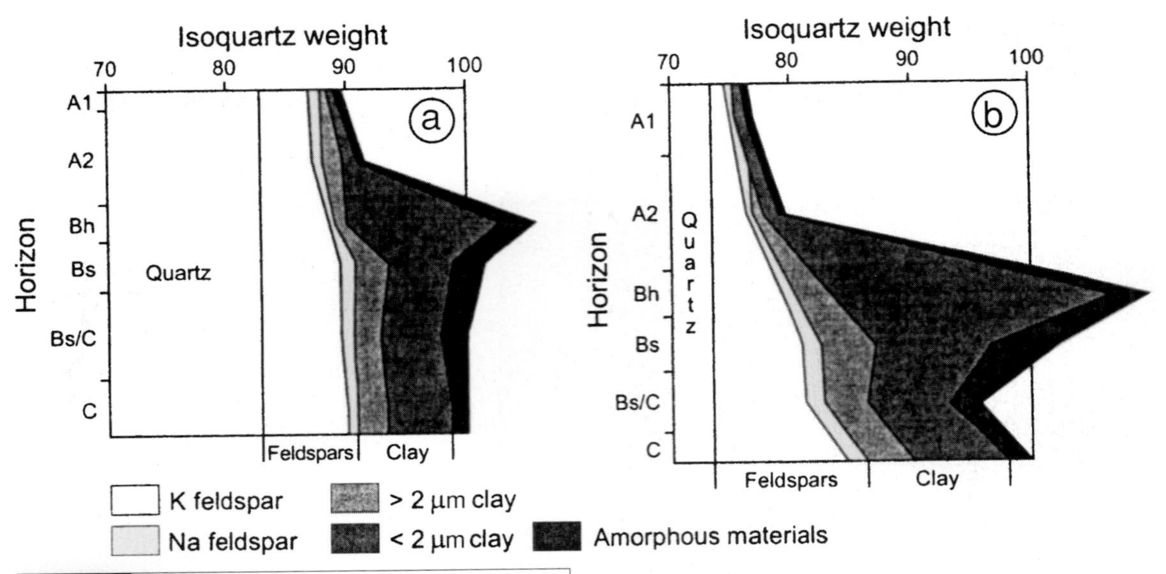

Fig. 12.67 Weathering and mineral balance, with depth, in two soils undergoing podzolization in Eastern France. Soil (a) is finer-textured and shows strong redistribution of clays and amorphous materials, whereas soil (b), which is coarser-textured, shows a greater amount of dissolution (subtractive pedogenesis) in the upper profile, along with redistribution of plasma components. Amounts of the various soil components are based on the isoquartz method of mass balance reconstruction (see Chapter 14). After Guillet *et al.* (1975).

The podzolic soil profile

Soils undergoing podzolization have distinct horizons and horizon boundaries, usually due to low biotic activity, especially with regard to faunalturbators (Moore 1974). Where pedoturbation is common, horizon boundaries get blurred, broken and distorted (Schaetzl 1986b). The typical podzolized soil profile has a thick O horizon, due to the low rates of decomposition more than to high rates of litter production. The thickest O horizons are found under coniferous forests which retain their needles for more than a year. The low decomposition rates of the litter are due to the cool climate, short growing season, acidic character and high contents of slowly degradable compounds such as lignins and waxes. The O horizon varies in its degree of decomposition, although Oi and Oe horizons are common, sometimes referred to as *mor* horizons (see Chapter 3). Worms are almost non-existent and burrowing mammals are rare. Most decomposition, slow as it is, is performed by fungi. Mycelial mats are common in the decomposing O horizon.

The A horizon is typically thin in podzolic soils, especially in coarser-textured soils in which the amount of macroorganisms is low. Bioturbation of organics into the mineral soil, normally important to A horizon formation, is minimal; thus, the primary way that the A horizon can form is through translocation of large organic materials into the mineral soil by percolating water. *Small* organic molecules move *through* the A horizon and participate in chemical reactions in the middle and lower profile.

The E horizon in podzolic soils is a primary morphological indicator of podzolization. Its coarse texture is partially due to continued eluviation and chemical weathering. The E horizon is typically lighter-colored (high Munsell values) than other horizons because (1) coatings of iron oxides and organic matter have been stripped from the quartz grains that dominate the mineralogy and (2) dark ferromagnesian minerals have been chemically weathered. In many profiles the E immediately underlies the O horizon. The E is typically the most acidic horizon, having been leached of most of its base cations (Lundström *et al.* 2000b). Well-developed E horizons qualify as *albic* horizons (Soil Survey Staff 1999).

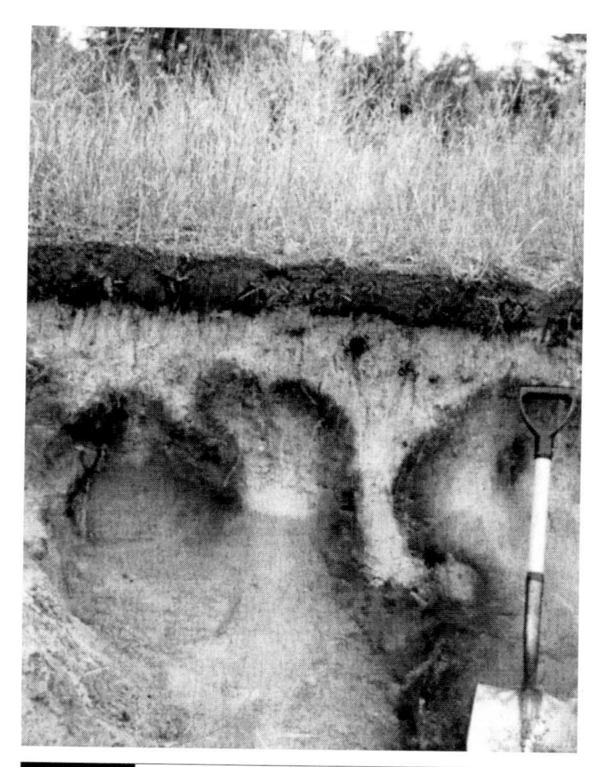

Fig. 12.68 Tonguing in a Typic Haplorthod in Michigan. This soil has been plowed in the past, as reflected in the smooth lower boundary of the Ap horizon. Photo by RJS.

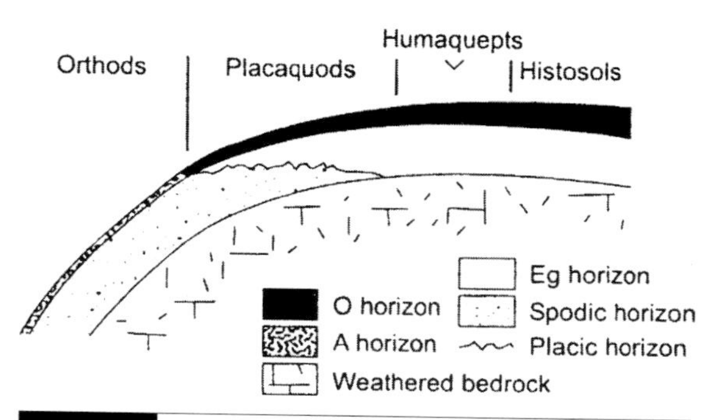

Fig. 12.69 A toposequence of soils on the British uplands, illustrating the relationship of the placic horizon to landscape position. After Clayden *et al.* (1990).

The reddish-brown to black B horizon is usually a Bs, Bh or Bhs, indicating translocated sesquioxides, humus or both (Mokma and Evans 2000). In soils with a high water table (Aquods), the B horizon may be low in Fe compounds, having been lost to the groundwater in ferrous forms. The boundary between the E and B horizons is typically wavy and sometimes even tongued, reflecting the manner in which water moves through coarse-textured soils, or following root pathways (Fig. 12.68). Schaetzl (1986b, 1990) determined that at least some of these tongues are due to preferential infiltration beneath microtopographic pits in the forest floor. If the B horizon meets certain chemical and morphological criteria, it qualifies as a *spodic* horizon (Soil Survey Staff 1999). Soils with spodic horizons are central to the order Spodosols.

Well-developed spodic horizons can become cemented, either as ortstein (Bsm, Bhsm) or as a *placic* horizon. Cementation in a placic horizon is by iron, organic matter and/or manganese, in the form of a thin (2–10 mm), black to dark reddish pan with sharp upper but diffuse lower boundary (Table 3.4). The pan is often wavy and it may bifurcate (Clayden *et al.* 1990). Research on placic horizon genesis suggests that this type of horizon may form as reduced iron gets mobilized in the surface horizons, translocated downward in the profile, and oxidized and precipitated in the B horizon where it can adsorb soluble organic matter but not necessarily form organo-metallic complexes (Hseu *et al.* 1999, Soil Survey Staff 1999). The iron is usually present as ferrihydrite and poorly crystalline goethite. Many placic horizons form at landscape positions where conditions change from reducing to oxidizing (Fig. 12.69). Wet, reducing conditions necessary for placic horizon genesis are usually caused by (1) high rainfall and cool temperatures, typical of perhumid climates and often associated with Histosols, and (2) some sort of slowly permeable subsurface layer.

When the cementation of the spodic horizon is of greater thickness and extent, the material is termed ortstein (Wang *et al.* 1978, Mokma 1997). Ortstein is commonly so well cemented that it restricts root penetration, but when sandy it still maintains good permeability (Lambert and Hole 1971). In Aquods, ortstein is generally planar and associated with a water table. In better-drained soils, ortstein is more spatially variable, coming and going in a seemingly random pattern (Mokma *et al.* 1994). It is often more vertically oriented than planar, perhaps reflecting tongues of percolating water. The cementing agent in ortstein is generally assumed to be an amorphous, Al-dominated material (Lee *et al.* 1988), although Fe and Si may also play a role.

The podzolization process

There are two primary theories regarding podzolization; both explain how Fe and Al become depleted from the A and E horizons and concomitantly enriched in the B. They also explain why some B (spodic) horizons are also rich in humus, while others have very little. In short, the mobilization, translocation and immobilization of the oxidized, metal ions and organic compounds are central to podzolization theory. By association, the theory must also explain the intense weathering that is also typical of Spodosols. All theories on podzolization assume that the parent material has been preconditioned or exists in such a state that it is conducive to the process; it has minimal clay and is acid. It is also assumed that the vegetation produces litter that is rich in low-molecular-weight organic acids and fulvic acids, and that the climate is cool and humid (Fig. 12.1).

Intense weathering by organic acids is a central part of the podzolization process (van Hees *et al.* 2000). As minerals weather, cations are released to the soil solution. The solubility of the main types of cations in acid soils is (Pedro *et al.* 1978):

$$\text{bases} > Fe^{3+} > Al^{3+} > Si^{4+}.$$

Thus, bases must first be depleted from the profile before podzolization, per se (which by definition involves the translocation of Fe and Al) can begin. Likewise, if the soil pH is not low enough, the Al and Fe cations will react to form relatively immobile compounds as $Al(OH)^{2+}$ and Fe_2O_3.

PROTO-IMOGOLITE THEORY

The *proto-imogolite theory*[1] was prompted by the observation that Al and Fe can exist in humus-poor Spodosols as amorphous, inorganic compounds such as imogolite and allophane (Farmer *et al.* 1980, Anderson *et al.* 1982, Farmer 1982, Childs *et al.* 1983, Gustafsson *et al.* 1995). Lundström *et al.* (2000a) called this general group of compounds imogolite type materials, or ITM. In this theory, Al-Si hydroxy sols (ITM) form in the soil solution from weathering products released from O and E horizons, and percolate until immobilized. Most ITM precipitate in the B horizon due to its higher (>5) pH values. Al is transported as a positively charged hydroxy–aluminum–silicate complex (Anderson *et al.* 1982). The Al and Fe in these materials are assumed to have been dissolved/weathered by non-complexing organic and inorganic acids, as well as by readily biodegradable small, complexing organic acids (Lundström *et al.* 2000b). The next step in this process involves negatively charged, colloidal organic matter, which migrates out of the upper profile, into the B horizon, and precipitates onto the positively charged ITM that are already there. This process is supported by thin-section data that show dark organs surrounding allans (allophane-rich cutans) (Freeland and Evans 1993) or Al- and Fe-rich cutans overprinted onto Si-rich cutans in B horizons (Jakobsen 1989). Dissolution/weathering processes continue to act on ITM in the B horizon. Because ITM are more easily weathered than crystalline Fe oxides, they (ITM) will continue to eluviate, leaving behind an Fe-rich B horizon. The lower B horizon becomes enriched in Al by the continued dissolution and remobilization of ITM, especially those that are rich in Al.

In a hybrid model of sorts, Ugolini and Dahlgren (1987) proposed that ITM are formed as organo-metallic complexes migrate into the B horizon and interact with an Al-rich residue or proto-imogolite formed there by CO_2 weathering (Lundström *et al.* 2000b).

CHELATE-COMPLEX THEORY

The most accepted and oldest theory regarding podzolization is the *chelate-complex* or *fulvate complex* theory. In this theory, organic acids form chelate complexes with Fe and Al cations, rendering these normally insoluble cations soluble and allowing them to be translocated from eluvial to illuvial zones (DeConinck 1980, Buurman and van Reeuwijk 1984). Normally, Fe^{3+} and Al^{3+} are not

[1] This is not its "formal" name. It is used here simply for the sake of discussion.

soluble in soils, but when *chelated* they are readily translocated in percolating water. In short, the process is driven by *organic* acids, as opposed to the dominantly *inorganic* pathway outlined above.

Many organic acids and phenolic compounds, e.g., oxalic, malic, succinic, vanillic, cinnamic, formic, benzoic, acetic, *p*-hydroxybenzoic, *p*-coumaric, have been identified in soils and leachate from litter (Vance *et al.* 1986, Krzyszowska *et al.* 1996). For the sake of discussion, these acids are placed into three groups based on their molecular weights: low-molecular-weight acids (LMW) <1000 Daltons (Da), fulvic acids (FA) 1000–3000 Da, and humic acids >3000 Da (Bravard and Righi 1991, Lundström *et al.* 2000b). Soluble, higher-molecular-weight acids, as well as other, low-molecular-weight acids, e.g., protocatechuic, are readily produced during litter decay and carried in soil water (Schnitzer and Desjardins 1969). They are also produced as root and fungal exudates. It has been long known, from experiments on aqueous extracts from plant litter, that organic acids can dissolve ferric and aluminum oxides (Bloomfield 1953, Schnitzer and Kodama 1976, Kodama *et al.* 1983, Lundström *et al.* 1995). Fungal hyphae are also now known to be effective at weathering of minerals, especially in the more acidic E horizon, thereby releasing Fe and Al to the soil solution (van Breeman *et al.* 2000). LMW, fulvic and humic acids are effective at forming chelate complexes with certain cations (Fig. 9.9). LMW acids are particularly important as chelates, because of their high complexation ability (van Hees *et al.* 2000). Thus, a predominantly organically driven mechanism, central to the chelate-complex theory, exists that is also an efficient weathering mechanism and can chemically complex with the otherwise-insoluble weathering by-products (Al^{3+} and Fe^{3+} cations) and render them soluble within the normal pH range of slightly acid to acid soils. Pedro *et al.* (1978) called this process *cheluviation* (Table 12.1).

Organo-metallic (chelate) complexes are readily translocated within acidic soil solutions (Riise *et al.* 2000). As they move they continue to chelate more metal cations and these chelate complexes remain soluble until a certain level of saturation, as indicated by the carbon : metal ratio, is achieved (McKeague *et al.* 1971, Petersen 1976). At this point, the chelate complex reaches its zero point charge, is rendered immobile and precipitates on ped faces, roots and mineral surfaces. An increase in pH in the lower profile also facilitates immobilization (Gustafsson *et al.* 1995), as does microbial decomposition of the organo-metallic complex (Lundström *et al.* 1995). In sum, these complexes can precipitate for a number of possible reasons: as the microbes break down the organic molecules, as the chelates become saturated with metal cations, or at a water table, lithologic discontinuity or aquitard (DeConinck 1980, Buurman and van Reeuwijk 1984). In the lower B horizon, alumina is released by microbial breakdown of organo-metallic complexes and can combine with silica to form ITM. Iron so released will form ferric oxyhydroxides (Buurman and van Reeuwijk 1984). With time, organo-metallic coatings in the B horizon tend to thicken. Because they are not crystalline, they tend to shrink and crack upon drying; cracked grain coatings are diagnostic of illuvial organo-metallic compounds (Stanley and Ciolkosz 1981) (Fig. 2.17a).

The net effect of podzolization is not only to weather primary minerals but to translocate some of their by-products to greater depths. In the initial stages of podzolization, the release of Fe and Al from primary minerals is so rapid that the chelate complexes are quickly saturated and the Fe and Al is translocated only a few millimeters (Fig. 12.70). Eventually, a zone of depleted Fe and Al forms (the E horizon), such that any new organic molecules that infiltrate can penetrate this eluvial zone without getting saturated with metal cations. They are then free to move to the Bs horizon and pick up metal cations that had previously been translocated there but had been released from their chelate complexes due to microbial degradation. Thus, over time, the E horizon grows downward, while metal cations are continually stripped from the top of the B horizon, remobilized and redeposited lower. The B horizon continues to gain organic matter, Fe and Al, as grain coatings (Fig. 2.17a), but the majority of the illuvial sesquioxides and humus reside near the top of the B horizon (Fig. 12.71). The E horizon may be chemically present but not

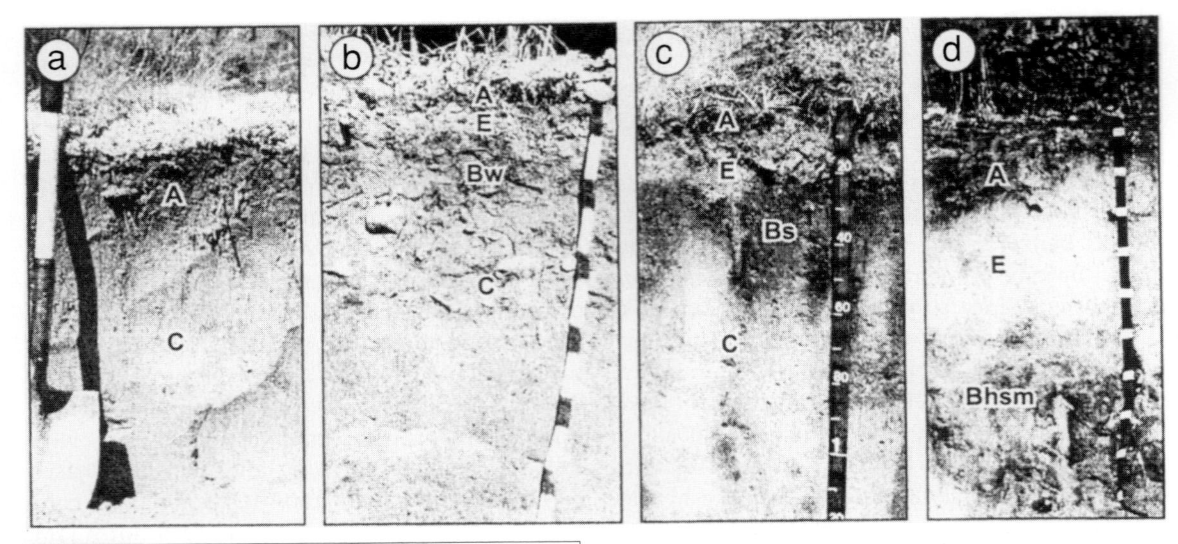

Fig. 12.70 A sequence of soils from the Great Lakes region, showing increasing amounts of podzolization. Tick marks on the tape are at 10 cm increments. Photos by RJS. (a) A Typic Udipsamment that has an A–C profile, with slight reddening in the upper solum. (b) A Spodic Udipsamment, which has formed a thin E horizon. (c) A Typic Haplorthod with the traditional Oe–A–E–Bs–C horizonation. (d) A Typic Haplorthod with ortstein. The E horizon has continued to thicken while the B horizon has developed cementation.

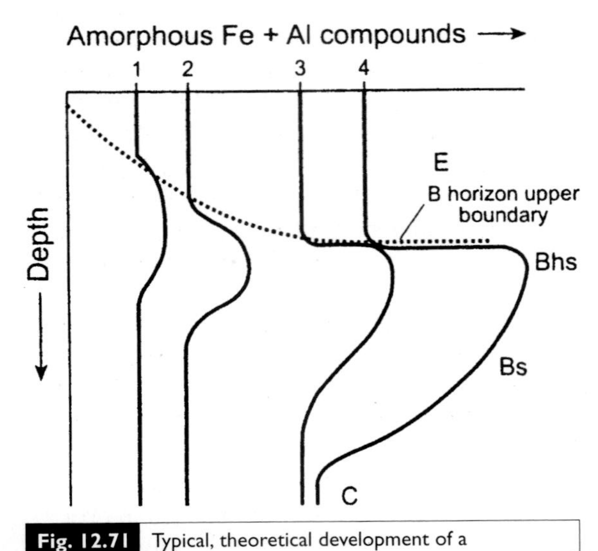

Fig. 12.71 Typical, theoretical development of a well-drained soil undergoing podzolization. After Franzmeier and Whiteside (1963).

visible until it deepens past the depth of mixing. In sandy, upland soils, mixing is minimal; E horizons can be visible within a few millimeters of the surface (Fig. 12.70).

For many pedogenic processes we can only view and examine the chemical and morphological results of the processes, i.e., the profile itself. For podzolization, however, there is another option; it is possible to monitor the *process itself* (Schnitzer and Desjardins 1969). By extracting *in situ* soil solutions, usually by placing a porous plate at the base of various horizons and removing the soil solution under suction, it is possible to examine the types and amounts of dissolved substances leaving that horizon in the soil water (Singer *et al.* 1978). Riise *et al.* (2000) and van Hees *et al.* (2000) simply extracted soil water samples by centrifugation of moist samples removed from each horizon. Another option is to implant cation-exchange resin in porous bags within or at the base of horizons (Ranger *et al.* 1991, Barrett and Schaetzl 1998). Chelates and metals that come into contact with these bags are retained and can be chemically extracted in the laboratory. These types of data not only provide an indication of the strength of the process at various sites (Ugolini *et al.* 1987), but also at different times of the year (Schaetzl 1990). Typically, one finds that dissolved organic carbon (DOC) values are high in soil solutions exiting the O horizon, but are arrested below; little DOC leaves the

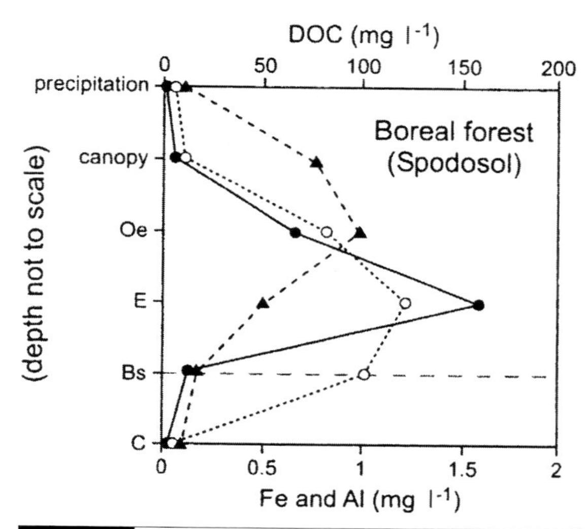

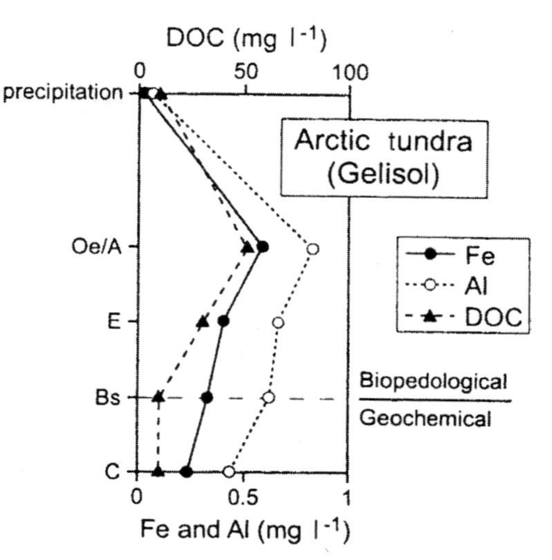

Fig. 12.72 Soil solution data for a Spodosol and Gelisol in northern Alaska. Data represent dissolved organic carbon (DOC), Fe and Al contents from soil solutions *exiting* each horizon. After Ugolini *et al.* (1987).

profile (Fig. 12.72). Similarly, higher amounts of Fe and Al in soil solutions leaving the E (but not the B) horizon confirms that active podzolization is occurring.

Ugolini *et al.* (1977) used soil solution data to document the existence of two connected yet discrete pedogenic compartments in Spodosols. The upper, biopedological compartment, contains the O, A, E and upper B horizons. Ionic movement here is governed by soluble organics that acidify the soil solution and depress bicarbonate concentrations. It is here that van Breeman *et al.* (2000) documented evidence for intense weathering by fungal hyphae. Podzolization per se is limited to the upper compartment. Most of the organic acids in the soil solution are captured and neutralized in the upper B horizon, leading to a rise in pH. In the geochemical compartment below, higher pH values form a weathering environment dominated by the dissociation of carbonic acid (H_2CO_3). Here, the bicarbonate ion (HCO_3^-) is seen as the main agent of ionic transport.

Because podzolization is expressed chemically as coatings (or the lack thereof) on skeletal grains, it is of interest to know the types of Fe and Al compounds, and their amounts, in such coatings. Fortunately, soil chemists have developed a number of chemical extractants that can be used to generate such data (McKeague and Day 1966, McKeague 1967, McKeague *et al.* 1971, Higashi *et al.* 1981, Parfitt and Childs 1988). It should be noted that the relationships between extractants and the various forms of metal cations that they extract appear to be better established for Fe than Al. Some forms of extractable Al are ill-defined and problematic, i.e., we are not sure that the extractant is really removing the exact form of Al that the literature suggests. Nonetheless, soil samples are shaken in a solution that attacks the coatings on the mineral particles; the extractant is then analyzed for Fe, Al and/or Si content.

Three major types of extractants are commonly used. Sodium citrate–dithionite extracts "free" (amorphous and crystalline) forms of Fe and Al. Any Fe and Al not tied up in the lattice of primary minerals, i.e., that which has been weathered and released to the soil solution, regardless of its fate since that time, is captured by this extractant. Sodium pyrophosphate primarily extracts organically bound forms of Fe and Al, providing a good indicator of the abundance of organo-metallic complexes. Acid ammonium oxalate extracts amorphous Fe, in both organic (as organo-metallic complexes) and inorganic (as ITM) forms. The distributions of Fe and Al, as extracted by these different chemicals, provides a great deal of information

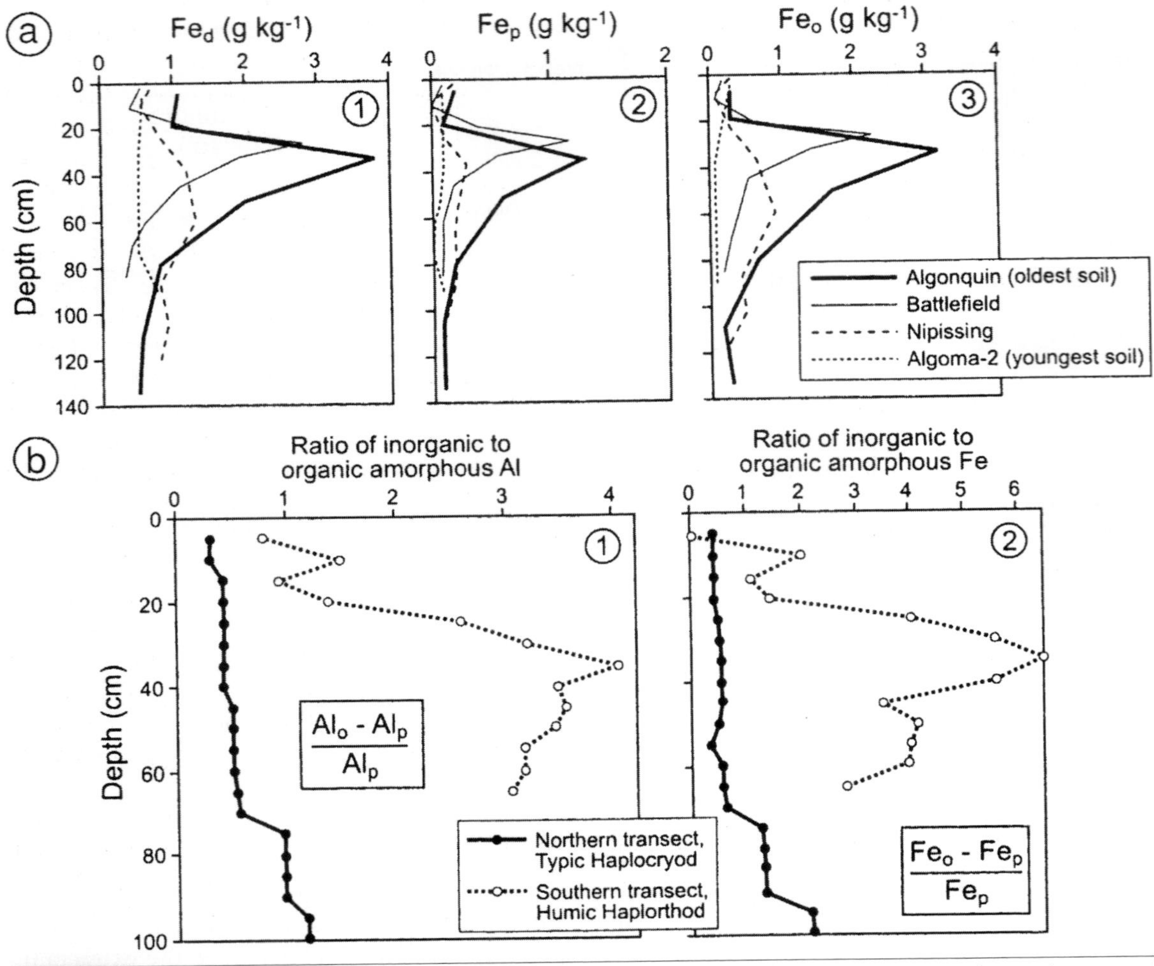

Fig. 12.73 Typical Spodosol depth functions. (a) Fe and Al depth functions for four sandy, upland soils in Michigan, as indicated by three different chemical extractants. The two youngest soils are Entisols and the two oldest are Spodosols. Note that the *x*-axis scales are not uniform among the three sub-figures. After Barrett and Schaetzl (1992). 1, Sodium citrate-dithionite; 2, sodium pyrophosphate; 3, acid ammonium oxalate. (b) Depth functions of inorganic to organic, amorphous Al and Fe in two Canadian Podzols. Organic, amorphous compounds were interpreted as pyrophosphate values. Inorganic, amorphous compounds were interpreted as oxalate minus pyrophosphate values. The profile from the northern transect, with the high inorganic/organic ratio, contained ITM. After Wang *et al.* (1986).

about the podzolization process (Fig. 12.73). Ratios and differences of extractant data are also useful for teasing out the various forms of metal compounds in soils (Barrett 1997). Relative crystallinity of Fe oxides is indicated by the iron activity ratio: Fe_o/Fe_p (McKeague and Day 1966). The amount of inorganic, amorphous material, often viewed as an indicator of ITM, is given by the ratio shown in Fig. 12.73. Inorganic, amorphous Al ($Al_o - Al_p$) is also taken to represent poorly crystalline aluminosilicates like ITM (Jersak *et al.* 1995), as is $[(Al_o - Al_p) / Si_o]$ (Jakobsen 1991, Gustafsson *et al.* 1995). When the latter ratio exceeds 2.0 allophane-like materials are probably present. Values for $Fe_o - Fe_p$ and $Al_o - Al_p$ provide some indication of the relative amounts of inorganically vs. organically bound metals in the soil.

PROS, CONS AND CONTEMPLATIONS

The arguments favoring one or the other approach to podzolization are interesting. One point made by ITM proponents is that insufficient iron and aluminum could be released by microbial decomposition of chelate complexes to account for their high concentrations in Bs horizons, because the breakdown process is too slow. The point is: there are too many free (not bound to organic molecules) sesquioxides in Bs horizons if their only mode of translocation were via organic chelates. The counter argument points out that the mean residence time (MRT) (see Chapter 14) ages of carbon in Bhs and Bs horizons is very low, sometimes less than 300 years and almost always <1000–2000 years (Guillet and Robin 1972, Schaetzl 1992a), implying that turnover of organo-metallic complexes is quite rapid. Many waves of chelate complexes could have migrated into the B horizon and degraded to free, inorganic forms over a few thousand year time-span if the turnover rate of the organic molecules is only a few hundred to a thousand years. Many metal cations are also transported by short-lived, simple organic acids which turn over even more rapidly than do fulvic acids.

In the proto-imogolite theory, migration of ITM and iron oxides is assumed to be the first stage of podzolization; later, organic materials are deposited within the Bs horizon to form a Bhs horizon. If this sequence were to hold, soil profiles would exist that had E horizons depleted in Al and Fe, and Bs horizons enriched in ITM and iron compounds, but little or no illuvial organic matter. Such profiles are rare, and where they do occur (Freeland and Evans 1993), they might be better explained by rapid turnover of organo-metallic complexes in the B horizon (Buurman and van Reeuwijk 1984). Also, if organic matter were to migrate after ITM, grain coatings in the B horizons would universally reflect this. Skeletal grains would have cutans with inner layers of ITM and outer skins of organic materials. Such coatings have been documented (Freeland and Evans 1993), but are rare.

Farmer *et al.* (1980) and Anderson *et al.* (1982) considered the presence of ITM in B horizons strong evidence that Al is transported as an Al–orthosilicate complex. The mere presence of a compound in a B horizon, however, does not imply that it was illuviated as such (Buurman and van Reeuwijk 1984). And while many spodic Bs horizons contain ITM (Farmer 1982, Wang and Kodama 1986, Kodama and Wang 1989, Arocena and Pawluk 1991), many do not (Mokma and Buurman 1982). Thus, podzolization may involve ITM, but it does not necessarily *have to*. Buurman and van Reeuwijk (1984) argued that the *absence* of allophane from E horizons can, conversely, be used to support the chelate-complex theory. Al–organic complexes are more stable than is orthosilicate; thus, any Al that is released from primary minerals is more likely to form chelate complexes than allophane. For the same reason, allophane will not be found in B horizons that are high in organics. However, in B horizons low in humus and in the lower B horizon, allophane and ITM are likely to be found (Wang *et al.* 1986); their presence there supports, rather than negates the chelate-complex theory. Lastly, the proto-imogolite theory, based on studies of the solid phase of the soil, provides no argument to explain why soil solution studies (Dahlgren and Ugolini 1989) find that most of the aluminum in E horizons is organically bound.

In all likelihood, some components of both theories are applicable to most Spodosols, but perhaps a hybrid theory is closest to reality (e.g., Wang *et al.* 1986, Jakobsen 1991, Ugolini and Sletten 1991). The main arguments that persist today among pedologists are not so much concerned with ITM vs. organo-metallic complexes, as whether the chelation is accomplished more by high-molecular-weight fulvic acids or by low-molecular-weight organic acids, and the nature of the arresting mechanism for these chelate complexes (Lundström *et al.* 2000a, van Breeman *et al.* 2000). Many researchers invoke saturation of the chelate complex by metal cations as the mechanism causing them to precipitate in the B horizon. Others point to biodegradation of the organic ligand of the less-complex LMW acids, which releases ionic forms of Al and Fe to the soil solution (Jakobsen 1989, Gustafsson *et al.* 1995, Lundström *et al.* 1995, van Hees *et al.* 2000). These ions are then free to form ITM and ferrihydrite, while fulvic acids, released from the O horizon, are free to percolate into and precipitate on the free Al–Fe–Si material in the B horizon. Thus, a type of hybrid podzolization theory, involving both ITM and chelates, is emerging.

Podzolization is often a subsidiary or background process in many soils. For example, it may occur after or concomitant with lessivage (DeConinck and Herbillon 1969, Guillet *et al.* 1975). Podzolization is inhibited in fine-textured soils, often because the organic compounds that are the major driver of the process are arrested by clay minerals. Many podzolic soils also have fragipans.

Factors affecting podzolization

Vegetation is an important, active factor in the podzolization process, because it provides essential organic compounds. Podzolization is best expressed under coniferous vegetation and ericaceous shrubs, whose litter decays slowly to compounds rich in fulvic acids and low-molecular-weight compounds. Two trees with very acidic litter, the northern hemlock and kauri, are particularly effective in this regard (Bloomfield 1953). Soil beneath these trees can be much more intensively developed than are soils farther from the tree's influence.

When the vegetation changes, podzolization processes change. Barrett and Schaetzl (1998) coined the term *depodzolization* for situations in which a vegetation (or climate) change causes soils that had been undergoing podzolization to alter their pedogenic pathway (Miles 1985, Nielsen *et al.* 1987, Almendinger 1990). Podzolic morphology slowly degrades or goes static under the new pedogenic regime. The effects of depodzolization are often chemically manifested long before there is a morphological change (Nørnberg *et al.* 1993). Hole (1975) examined the effects of logging of hemlock trees from Spodosols and concluded that the half-life of the spodic Bs horizon was about a century; in 100 years these soils lose half their accumulated organic matter.

Podzolization is found only where soils are throughly wetted during the year. It is also a dominant process in poorly drained and somewhat poorly drained soils in thermic soil temperature regimes, such as the southeastern United States (Brandon *et al.* 1977). The process tends to increase in intensity as leaching increases (Jauhiainen 1973b). Translocation of chelate complexes and inorganic compounds cannot be accomplished without at least an occasional deep wetting event. High amounts of precipitation also facilitate weathering. Podzolization does occur in the humid subtropics and tropics, but is restricted to sandy parent materials and/or sites with a high water table (Brandon *et al.* 1977, Bravard and Righi 1989, 1990). In the high and midlatitudes, podzolization is stronger on cooler sites, probably because the breakdown of organics in the O horizon is slower (Stanley and Ciolkosz 1981, Hunckler and Schaetzl 1997). Thicker, more acidic O horizons can release small amounts of organic acids at each and every infiltration event; in warmer climates the O horizon is quickly mineralized.

Schaetzl and Isard (1990, 1991, 1996) studied the effects of climate on podzolization. They noted that sandy soils in northern Michigan and Wisconsin, where podzolization is a dominant process, were best developed in areas where the climate is coolest and snowfall thickest, despite the fact that the soils there are a few thousand years younger (Fig. 12.74). Their work verified that snowmelt infiltration is important to the podzolization process in this region, by showing that the areas with the best-developed Spodosols had the deepest snowpacks (Schaetzl 2002). Not only is there more total infiltration in thick snowpack areas, there are also more *large* (>13 mm) infiltration events and more *continuous* infiltration events (Fig. 12.74). They suggested that large, continuous infiltration events are more effective at translocation than are smaller events, because they wet the entire profile and "keep things moving." Deep snowpacks also tend to coincide with lack of soil frost (Isard and Schaetzl 1995, 1998), facilitating uninterrupted infiltration of snowmelt water. Soil solution data taken during snowmelt and summer illustrate that translocation of sesquioxides is accentuated during snowmelt, perhaps because this type of infiltration is slow and continuous and because it first must pass through fresh litter (Schaetzl and Isard 1990). Their work also suggested that podzolization is best expressed where mean summer soil temperatures are <17 °C.

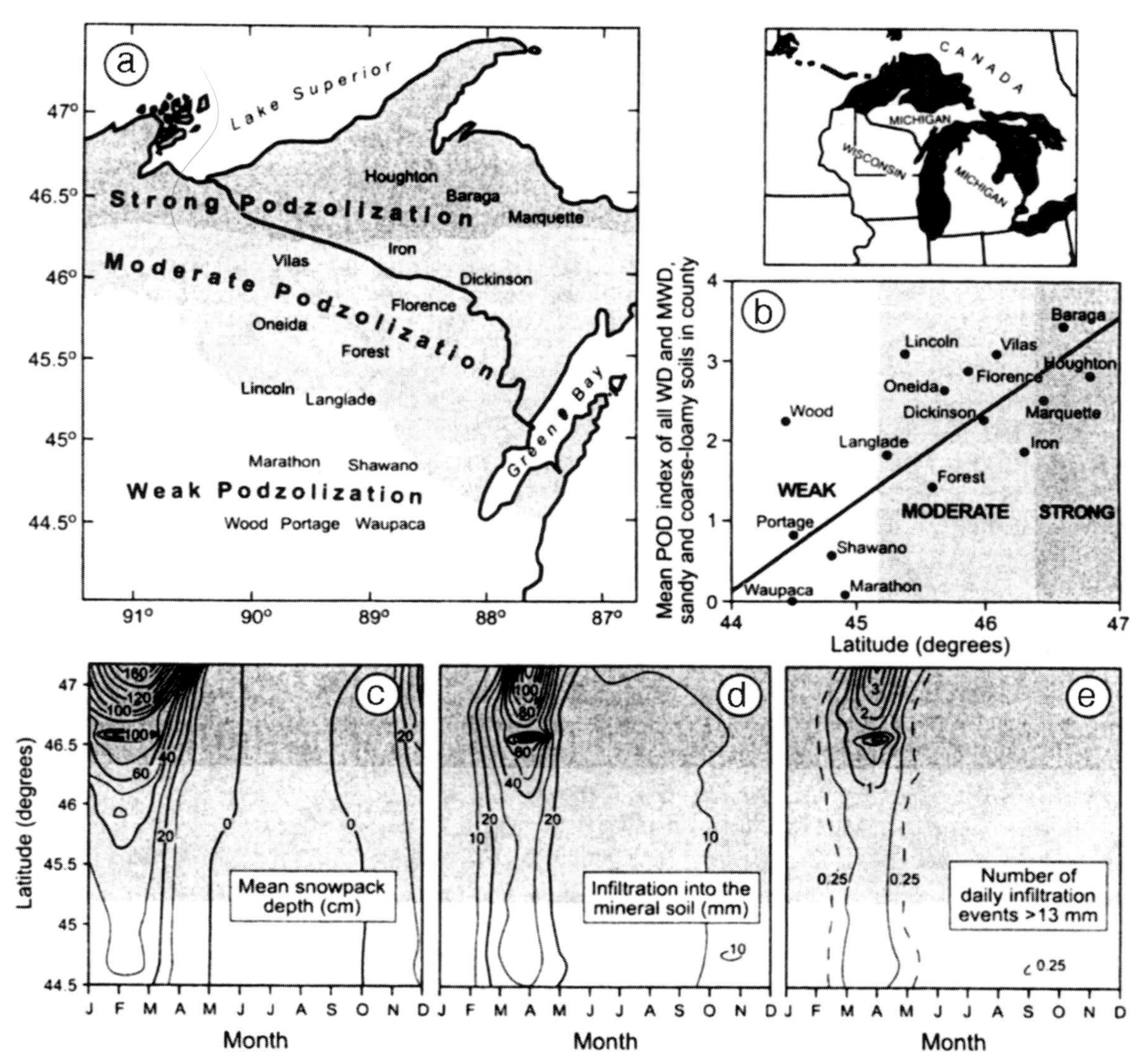

Fig. 12.74 Effects of climate on podzolization in the Great Lakes region. After Schaetzl and Isard (1996), with some additional data. (a) Strength of podzolization in the western Great Lakes region. (b) Strength of podzolization, as indicated by the POD index (see Chapter 14), along a north–south transect through this region. The best-fit linear regression line (shown) has the following equation:

$$\text{POD} = 1.13\,(\text{latitude}) - 49.42,\ r^2 = 0.58$$

(c) Time–space diagram of the mean monthly snowpack depths across the region, for the period 1951–92. (d) Time–space diagram of the mean monthly infiltration for the period 1951–92. (e) Time–space diagram of the number of "large" infiltration events per month, for the period 1951–92.

Podzolization is favored in coarse-textured parent materials with low surface areas, as they can be readily stripped of their particle coatings (Barrett and Schaetzl 1992, Barrett 2001). Neutral or acidic parent materials are also advantageous, for they are more quickly acidified and leached of bases, both of which enhance silicate weathering. Podzolization on carbonate-rich parent materials is delayed until the material is leached and the pH is lowered to $\leq$6.0. Soils on carbonaceous parent materials often have a podzolic sequum over an argillic sequum, indicating that base depletion occurs first, followed by lessivage and then, finally, podzolization (Guillet *et al.* 1975, Schaetzl 1996). Coarse-textured parent materials also tend to be lower in weatherable minerals, allowing for more rapid and intensive acidification because these minerals, upon weathering, will release bases to the soil solution. Coarse-textured soils also foster rapid infiltration, translocation and deep leaching. In them, the Fe and Al chelate complexes are relatively free to move, but in soils with clay, they may bind to the clay, slowing their translocation (Duchaufour and Souchier 1978). This last example would be typical of loamy Udepts with a reddened Bw horizon (a "color" B) but little indication of Fe and Al translocation.

PODZOLIZATION AND RUBIFICATION VS. BRUNIFICATION

Brunification (or braunification) involves the release of iron from primary minerals followed by the formation of goethite and other iron-rich minerals, giving the soil brownish or reddish-brown colors (Table 12.1). If this were the entire story, however, we would have just described something akin to rubification. The difference between rubification and brunification is that, in the latter, the iron oxides are quickly tied up with organic matter (Cointepas 1967). Brunification is more likely to occur in midlatitude soils which have higher contents of organic matter, while rubification is more typical of tropical climates where organic matter contents are low (Schwertmann *et al.* 1982). Humus plays little role in rubification, which is why the Fe oxides released by weathering in tropical upland soils can readily precipitate to form hematite or ferrihydrite.

Brunification is typical of young soils that have not been fully leached and which are not highly weathered. Many soils with cambic (Bw) horizons (Inceptisols, Cambisols) may be considered brunified. Organic matter turnover is high, and much of it is converted into forms that are not conducive to podzolization. Cation exchange capacity values are higher in brunified soils, making base leaching more difficult. Eventually, as the brunified soils become more weathered and acidic, brunification yields to lessivage.

Brunification differs from podzolization because of its effects on iron mobility. Brunification usually occurs in soils that are more base- and clay-rich than does podzolization. Thus, there exist more clay minerals to complex with iron. In podzolization, Fe released from primary minerals is quickly complexed with organic matter because there is little clay to compete with, while in brunification clay and organics are competing for the iron cations. In brunification, iron oxides, clay and humus bind *together* in the upper solum, forming essentially *immobile* complexes. When biological degradation frees the iron from the clay–humus complexes, it quickly precipitates to form brown goethite, which is stable within most soils.

An additional reason that soils become brunified instead of podzolized centers on pH; brunification operates under less acidic conditions due to more base-rich litter in the O horizon. Aluminum is mobile in slightly acidic soils but Fe is not. Instead, it forms stable Fe–humus–clay complexes. Thus, Fe-poor E horizons do not form under brunification; since Al is colorless, Al-poor E horizons may develop but they have brownish colors.

The primary advantage that coarse-textured materials have in podzolization is that they release so few bases or weathering products, and at such slow rates, that they do not *overwhelm* the system's ability to translocate them, allowing the soil to acidify and "clean" E horizons to develop (Alexander *et al.* 1994). Most coarse-textured materials are rich in quartz, which is not only slowly weatherable but also does not release bases upon weathering. In soils with large amounts of weatherable minerals which can release abundant bases to the soil system, E horizons may not even form, even if all other factors favor the development of a Spodosol profile.

In an interesting study of podzolization vs. brunification, Duchaufour and Souchier (1978) found that these processes are affected as much by iron content of the parent material as clay content (Fig. 12.75). Clay inhibits podzolization by immobilizing the iron and by releasing iron to the soil solution as it weathers, overwhelming

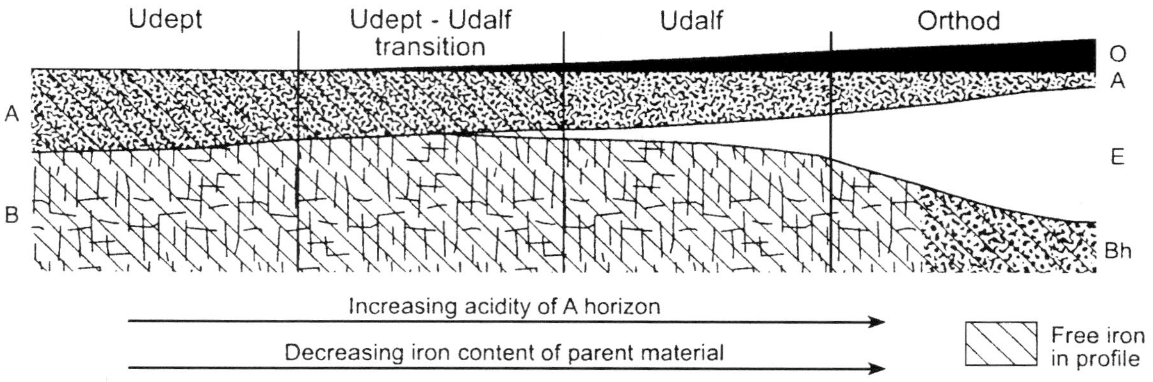

Fig. 12.75 Effects of iron and clay contents on podzolization, as illustrated by soil morphologies along a lithosequence of soils showing increasing podzolization. After Duchaufour and Souchier (1978).

it with iron – E horizons do not form. Finer-textured soils also have more biological activity, fueling proisotropic pedogenic pathways and slowing the development of anisotropic, podzolic horizons. Toutain and Vedy (1975) showed that soils with high Fe contents had very different litter, generally with lower amounts of low-molecular-weight organic compounds, which are important in podzolization. The iron released by weathering in brunified soils forms thin coatings around clay particles, or bridges between clay and organic molecules. The clay–iron–humus clusters thus formed favor the development of crumb structure, which is lacking in sandy Spodosols which do not have adequate amounts of clay. In sum, under similar climate and vegetation, podzolization tends to be favored in iron-, base- and clay-poor parent materials; *brunification* is favored where iron, base and clay contents are higher (Duchaufour and Souchier 1978).

Rates of podzolization

Because it involves only the stripping of coatings and the translocation of soluble materials a short distance, the morphological expression of podzolization can be manifested rapidly. In the most extreme example, Paton *et al.* (1976) found that thin E horizons had formed in mined dune sand materials in as little as 4.5 years! This material had been preconditioned to the process because it was an amalgamation of previously mined A and E horizon materials; thus, there were few coatings that had to be removed to form the E horizon. Their study, however, illustrates an important point: podzolization rates are largely dependent upon parent material. In base-rich, fine-textured materials, podzolic morphologies may take many millennia to form, despite conducive climate, vegetation, etc. (Ragg and Ball 1964).

Chronosequence studies provide data on the time necessary for Spodosols to form. Generally, in cool, humid areas where Spodosols are common on uplands, 1000 to 5000 years are typically necessary for their development (Franzmeier and Whiteside 1963, Ellis 1980b, Protz *et al.* 1984, Barrett and Schaetzl 1992, Petäja-Ronkainen *et al.* 1992). In some wet, cool areas, podzolic profiles can form in less than 1000 years (Chandler 1942, Jauhiainen 1973a, Singleton and Lavkulich 1987). Spodosols forming in the presence of a high water table develop at varying rates, dependent upon water table fluctuations and chemistry. In such soils, Fe is typically removed in ferrous form in groundwater, while additions of humus from the same source can be substantial. In essence, the B horizons in these Aquods develop as much from additions from the groundwater as they do from illuviation. For that reason, their B horizons tend to be more planar, while B horizons in well-drained Spodosols tend to be more undulating (Mokma *et al.* 1994).

Sulfidization and sulfuricization

These two related process bundles center on the reactions and movement of sulfur-bearing

Landscapes: the pygmy forest Podzol ecosystem

Along the Mendocino coast of northern California lies a staircase of marine terraces, formed as the land slowly rose out of the Pacific Ocean (see Figure). Known as the pygmy forest ecological staircase, the soils on the terraces represent a long chronosequence of podzolization and ecosystem succession; it is a very special place (Jenny 1960, Jenny *et al.* 1969).

On each terrace are beach deposits and, in some places, dunes. The lowest terrace became subaerial about 100 000 years ago, while the older terraces may exceed 1 million years in age. In all cases, the dunes are younger than the terraces, and the dune soils are less weathered than are the terrace soils, as indicated by the more majestic and productive forest ecosystems on the dunes. The beach deposits are shallow to bedrock, generating a fluctuating water table during the rainy winter season.

The soils here follow a unique development sequence. On the first terrace, base-rich Mollisols support a grassland. Acid, wet Alfisols and Ultisols occur on the much older, second terrace. A forest of Bishop pine, hemlock and redwood is found here. Things get interesting on the third and fourth terraces, where a globally unique pygmy forest of stunted Bishop and Bolander pines exists on a highly acid Spodosol. The Blacklock soil, a Typic Epiaquod, has a 36-cm E horizon with a pH of 2.8! Feldspar : quartz ratios, which are about 0.2 in the parent material, are zero in the Blacklock E horizon. The Bsm horizon below forms a hardpan that restricts rooting and perches water. The trees that grow here have so few nutrients (their roots cannot penetrate any deeper than the E horizon) that they are permanently stunted. Most of the twisted and stunted "pygmy" trees are only 1.5 to 3 m tall. Large parts of the surface have no trees at all, only patches of lichen.

This chronosequence is an example of extreme long-term weathering and pedogenesis with no significant rejuvenation or pedoturbation. The bedrock below has been so compartmentalized from the soil that fresh influxes of nutrients are rare. Nonetheless, even where bedrock is not limiting, podzolization taken to extremes will often result in loss of nutrients, and this will be reflected in the vegetation.

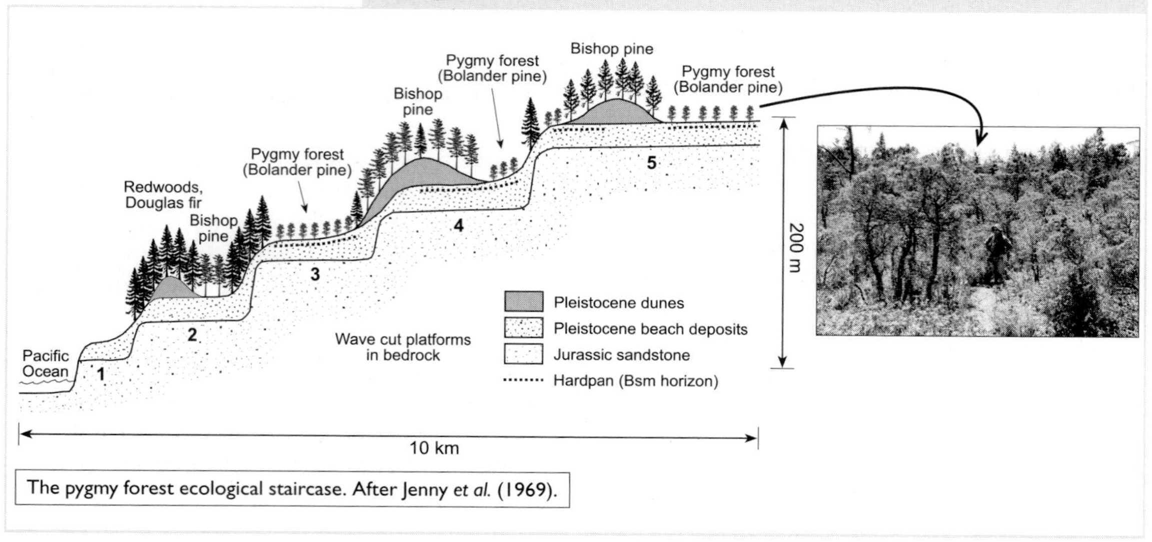

The pygmy forest ecological staircase. After Jenny *et al.* (1969).

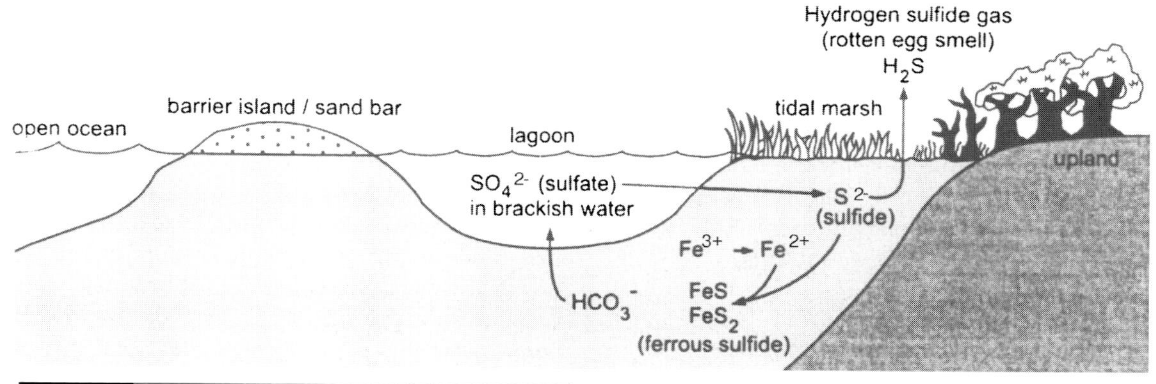

Fig. 12.76 Chemical reactions involved in the process of sulfidization. After Fanning and Fanning (1989).

minerals within soils, sometimes leading to the formation of acid sulfate soils (Kittrick *et al.* 1982, Fanning and Fanning 1989) (Table 12.1). Sulfur is not common in most soils, but in the reducing environments of tidal marshes sulfur from sea water is a ready source. Sulfides are also found in areas recently mined for high-sulfur coal, and in soils formed on pyrite-bearing sandstones and shales (Fitzpatrick *et al.* 1996).

Sulfidization refers to the accumulation of soil materials rich in sulfides, usually by the biomineralization of sulfate-bearing water (Fig. 12.76). It is best exemplified in anaerobic, humus-rich tidal marshes. Here, sulfate-bearing water comes into contact with tidal marsh soils and sediments, where sulfur-reducing bacteria change sulfate sulfur to sulfide. Ferrous iron, already present in the sediments, combines with the sulfides and crystallizes to produce secondary ferrous sulfide (FeS) or pyrite (FeS_2). Organic matter is required as an energy source for the reduction of sulfide, explaining why sulfur contents tend to increase as the organic matter contents of tidal marsh soils increase (Darmody *et al.* 1977). If enough sulfidic materials are present, the soils may classify as Sulfaquents or Sulfihemists (Soil Survey Staff 1999).

In *sulfuricization* sulfuric acid is released as sulfide-bearing minerals (formed via sulfidization) are exposed to oxidizing conditions, become oxidized and produce jarosite ($KFe_3(SO_4)_2(OH)_6$) and/or sulfuric acid (H_2SO_4). The acids create extremely low (<3.5) pH values and intensively weather soil minerals (Fig. 12.77). The jarosite may undergo slow hydrolysis, leading to further production of sulfuric acid (Soil Survey Staff 1999). The iron from the ferrous sulfide or pyrite forms a reddish precipitate, commonly ferrihydrite, which later may crystallize as maghemite, goethite or hematite. Jarosite has a straw-yellow color and frequently lines pores in these soils. If there is ample calcium in the soil system, gypsum

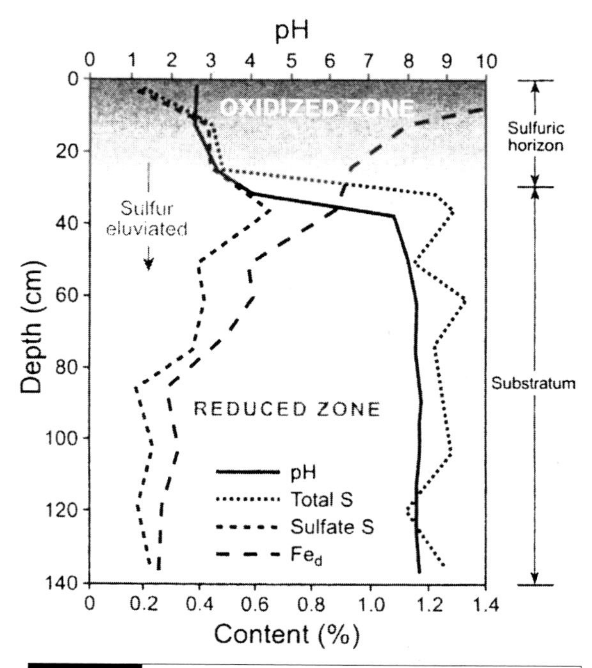

Fig. 12.77 Depth plot of pH, sulfate S and total sulfur in an acid-sulfate soil developed on sulfidic sediments on a scalped land surface near the Washington DC beltway. The loss of sulfur from the oxidized upper solum is presumably due to leaching/eluviation. After Wagner *et al.* (1982).

($CaSO_4 \cdot 2H_2O$) can also form. Sulfuric horizons and acid sulfate soils (Sulfaquepts and Sulfaquents) require the presence of jarosite.

Fitzpatrick *et al.* (1996) described a third phase of sulfur-related pedogenesis. In *ferritization*, the dissolved ferrous sulfates produced by sulfuricization move in solution to oxidizing zones. Here, they undergo oxidation and recrystallization to form ferrihydrite and ferric sulfates.

Surface additions and losses

Most of the previous discussion centered around *internal* pedogenic processes – translocations and transformations that occur within the soil profile. Losses from, and additions to, the soil profile are, however, also important. Leaching is one type of profile loss. However, many forms of whole-profile losses and additions, i.e., mass fluxes, involve mineral materials that are added to the soil surface or removed from it. These processes affect the thickness of the profile and also the mineralogy, because many of the mass flux additions and losses are clastic mineral materials.

Upbuilding refers to allochthonous surficial *additions* of mineral and organic materials to the top of the soil, e.g., via eolian additions, additions of organic material from plants, slopewash. When this happens, the soil upbuilds or gets thicker. According to Johnson (1985), upbuilding can be either *developmental* or *retardant* (see Chapter 11). In *developmental upbuilding* the surface additions are slow enough that pedogenesis can keep pace and effectively incorporate them into the profile (McDonald and Busacca 1990, Almond and Tonkin 1999). This concept is similar to cumulization (see below). Developmental upbuilding commonly occurs due to *slow* additions of loess, dust or alluvium. In *retardant upbuilding*, thick, often rapid, additions of material outpace the soil's ability to pedoegeneticize them and the soil, momentarily at least, gets buried (Schaetzl and Sorenson 1987). In this case, the soil per se gets no thicker, only buried by new sediment. In both cases, fresh mineral matter has entered the pedogenic system, and with it ions that can affect soil chemistry, flora, etc. Soil processes get altered or even stopped, as may happen if an influx of base-rich alluvium flocculates many of the clay colloids and causes lessivage to cease.

Sediment is *removed* from the soil surface through erosion and mass wasting. Subsurface removals by throughflow, leaching and biomechanical processes are also important. Oxidation and mineralization of organic matter are also considered removals. A major pedogenic impact of removals is that, by thinning the solum, they bring the weathering front at the base of the profile closer to the surface. This may increase the rate of primary mineral weathering and foster the more rapid release of base cations. It can also foster an increase or a change in biocycling, because roots can now reach bases and cations that they previously could not.

Cumulization

How a soil and or a surface reacts to upbuilding depends on whether the aggradation occurs at a constant or variable rate, whether periods of non-deposition or erosion are interspersed, and how the rate of aggradation compares to the rate of soil formation (McDonald and Busacca 1990). Rapid rates of influx will bury the soil, while slower rates may lead to *cumulization* (Fig. 12.78). Cumulization includes eolian, hydrologic or human-induced additions of mineral particles to the surface of a soil, usually as at the base of a slope or on a floodplain. Slow, gradual additions are implied, as the surface aggrades and the profile thickens (Riecken and Poetsch 1960). Intermediate and rapid rates of upbuilding may lead to soil profiles that are distinct yet overlapping e.g., compound soils (Morrison 1967), polymorphic soils (Simonson 1978), complex soils (Bos and Sevink 1975), welded soils (Ruhe and Olson 1980) and superimposed soils (Busacca 1989) (Fig. 12.79).

Cumulization and upbuilding are subtly different. Soils may *upbuild* their surfaces but not become cumulic. Cumulic soils have undergone developmental upbuilding to the point that some horizon that is *overthickened*, beyond its "normal" morphological expression. Especially key is A horizon thickness, as cumulic soils usually have overthickened A horizons (Riecken and Poetsch

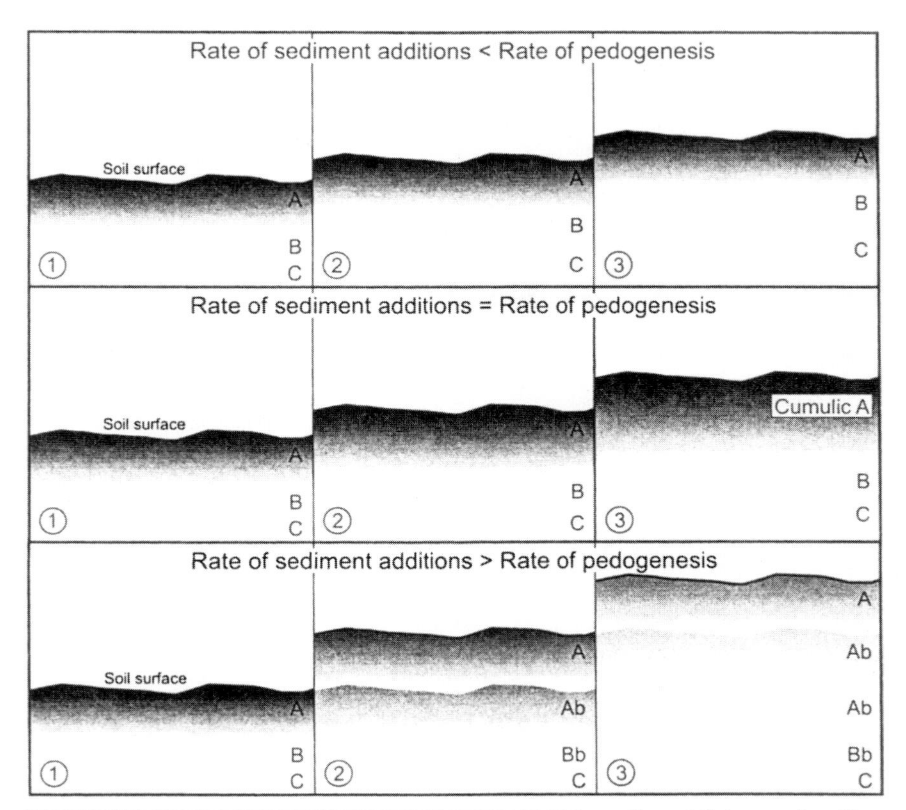

Fig. 12.78 Diagrammatic model of three different soil burial situations over time. The ultimate fate of the soil profile is dependent upon an interplay between the rate of burial vs. the rate of pedogenesis, i.e., assimilation of the surface additions into the profile.

1960). All soils thicken (upbuild) as materials are added to their surface. However, the new material results in *developmental upbuilding*, rather than cumulization, if it is assimilated into the profile with little long-lasting morphologic change. If the surficial additions occur at a rate faster than they can be assimilated into the profile, the soil grows by *retardant upbuilding* and may become buried (Schaetzl and Sorenson 1987). There are a number of related processes. Hole and Nielsen (1970) coined the term non-cumulative soil genesis to refer to the literal collapse of soils as soluble materials, e.g., carbonates, are leached from it. By definition, non-cumulative soil genesis processes do not allow for surface additions or erosion. Acknowledging that many surfaces receive periodic additions of sediment as dust, Almond and Tonkin (1999) defined two types of pedogenesis (Fig. 12.80):

(1) Pedogenesis on stable surfaces, where traditional pedogenic processes are operative and little influenced by additions of material, is called topdown pedogenesis. In topdown pedogenesis most processes impact the soil from the surface and filter their way to the subsurface over time.

(2) Upbuilding pedogenesis occurs in soils that accommodate significant surface additions contemporaneously with topdown pedogenesis. In upbuilding pedogenesis, it is assumed that each increment of soil below an aggrading surface has experienced processes characteristic of all the horizons above it.

One can argue that cumulization is more a geologic/sedimentologic process than a pedogenic one. But because it impacts soils directly and because the soil predictably reacts to the new sediment, cumulization is often a focus of soil

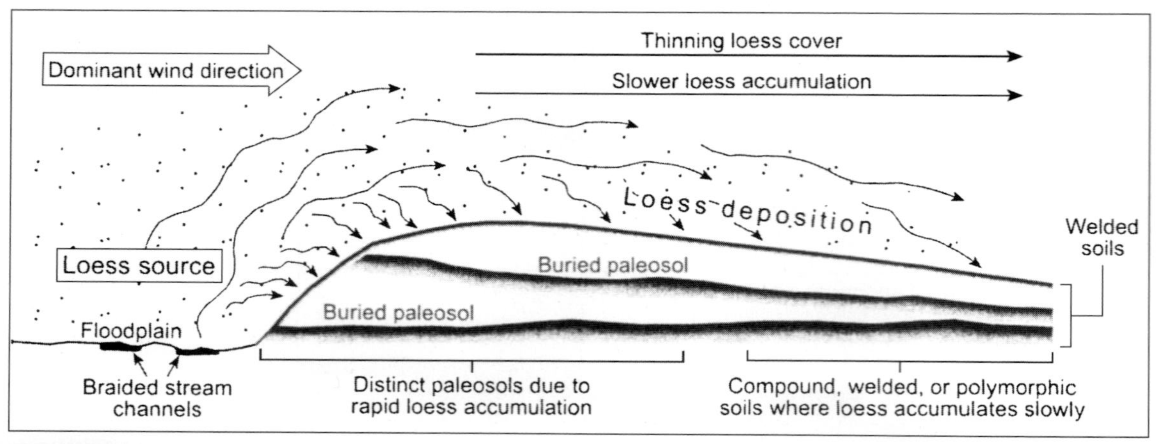

Fig. 12.79 Diagrammatic model of soil burial by loess, showing how a buried soil far from a loess source can bifurcate and become distinct soils at sites nearer the source. Based on McDonald and Busacca (1990).

geomorphology. To quote McDonald and Busacca (1990: 449):

> Soil formation on an aggrading surface is a competition between pedologic and sedimentologic processes, a competition that can lead to very complex soils and soil-stratigraphic relationships. Complexity results from the compression or dilution of soil profile features with variations in aggradation rate and from the sometimes extreme overlap of features from different episodes of soil development. Although the complexity can be great, . . . the opportunities for insight into soil and geomorphic processes can be equally great.

How does cumulization work, pedogenically? As sediment is slowly added to the surface, the A horizon thickens, resulting initially in a cumulic soil with an overthickened, dynamically changing A horizon. A form of dynamic pedogenic equilibrium keeps the A horizon to a thickness limit of about 20–50 cm, as the lower part of the A gradually transforms to a B or E horizon, which also grows upward with the upbuilding surface (Wang and Follmer 1998). The B horizon "upgrowth" occurs because the lower part of the A is now so deep that its losses of organic matter outpace any gains by roots and in-mixing of surface litter. Eventually, many characteristics of the horizons below fade completely away or are blurred by the acquisition of new characteristics, e.g., the former A acquires B horizon morphology (Hole and Nielsen 1970). Almond and Tonkin (1999) observed that the lower profiles of soils that have undergone slow upbuilding may can be more weathered than normal, because these zones were once nearer the surface where weathering is more intense. With this in mind, three possible scenarios can occur on an aggrading surface (Fig. 12.78):

(1) *Slow upward migration of the profile into the new sediment, without cumulization.* If additions to the soil surface are slower than the rate at which pedogenesis can assimilate them, they can be effectively incorporated into the soil profile. The soil grows upward slowly, and does not become cumulic. The A horizon can respond faster to the changing (upbuilding) conditions than can the B horizon. A horizon characteristics can develop into the new sediment more rapidly, and likewise they can be obliterated from the old, lower A horizon more rapidly, than can B horizon morphology. Thus, the A horizon may not become cumulic under this scenario, but the B may overthicken. Overthickened B horizons are commonly observed in buried soils that were (at first) very slowly buried, presumably due to the process outlined above (Follmer 1982).

(2) *Cumulization.* If sedimentation rates generally match the rate of pedogenic assimilation/overprinting of the new material, the soil becomes overthickened (cumulic). Grass roots quickly spread into the new, thin layer

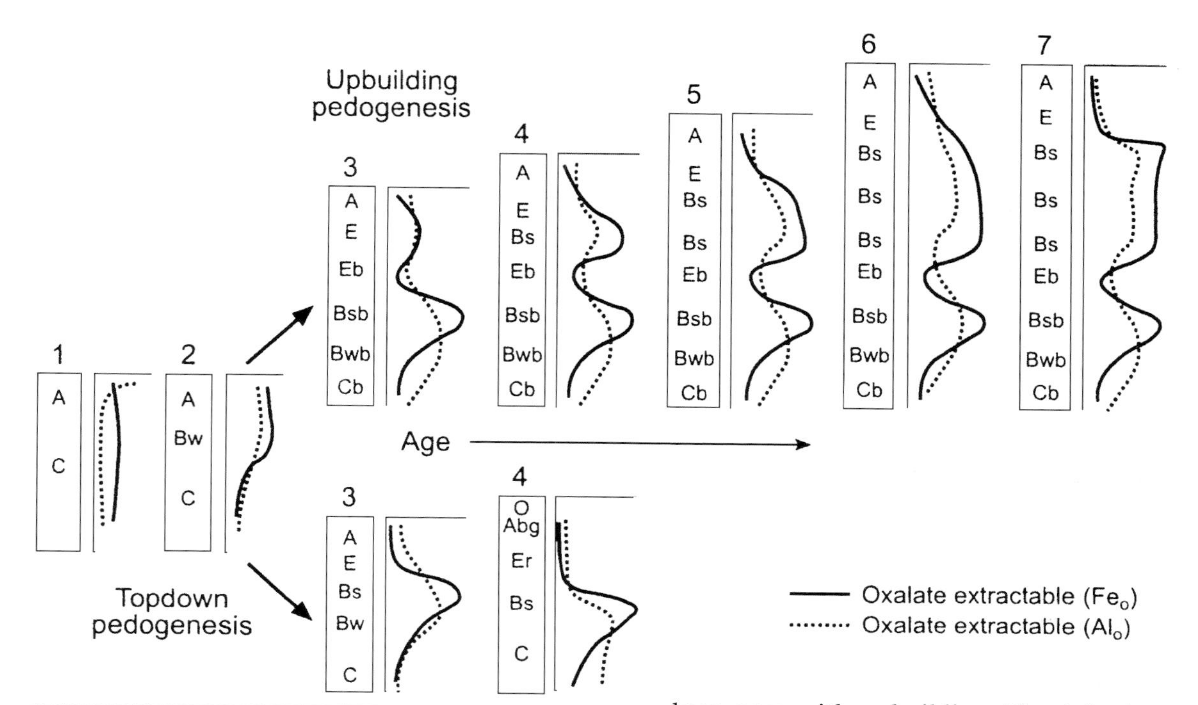

Fig. 12.80 A 60 000-year-old chronosequence showing two pedogenic pathways (topdown pedogenesis and upbuilding pedogenesis) for upland soils in New Zealand that have developed in coarse-textured parent materials. The lower pathway consists of a chronosequence of soils on stable surfaces. The chronosequence in the upper part of the figure shows how the horizonation and Fe and Al depth functions for these soils are affected by small incremental additions of loess. After Almond and Tonkin (1999).

of sediment and as they decay in it, melanize the new sediment and incorporate it into the A horizon. Equally important, bioturbation is fast and extensive enough in grasslands to mix the new sediment into the profile, even as melanization is occurring on it. The base of the A horizon gradually grows upward but not as rapidly as does its upper boundary. The lower part of the A horizon transforms to a B or E horizon. B horizons cannot grow upward at this faster pace, and thus rather than acquiring an overthickened B, the soil develops a thick, weak and perhaps intermittent B horizon below its cumulic A horizon. A Cumulic Hapludoll is an example.

(3) *Burial*. If sediment additions outpace soil-development processes, pedogenesis cannot keep pace with upbuilding. The A horizon thickens, but its upper part (in the new sediment) is not as dark as the lower, since melanization in the new material cannot keep pace with sedimentation. For a short period of time, the soil appears cumulic. Soon, however, the B and lower part of the A are rendered relict. At this point, the soil is considered effectively buried and the newly forming A horizon in the fresh sediment is thinner and lighter. Strong B horizons cannot form, as the pace of sedimentation exceeds that of B horizon formation (Smith *et al.* 1950). Evidence of rapid sedimentation can be found as a distinct boundary between the old, now buried, soil surface and the new deposits (Follmer 1982). Occasionally, in accretionary sediments like alluvium, one can find buried surfaces, where the sedimentation system stalled and the surface stabilized for a short period of time. Buried paleosols may exist here.

Cumulization is common on uplands and lowlands in certain grasslands and deserts, due to slow additions of loess or dust and by alluvial sedimentation (McDonald and Busacca 1990, Feng *et al.* 1994b, Wang and Follmer 1998). In humid

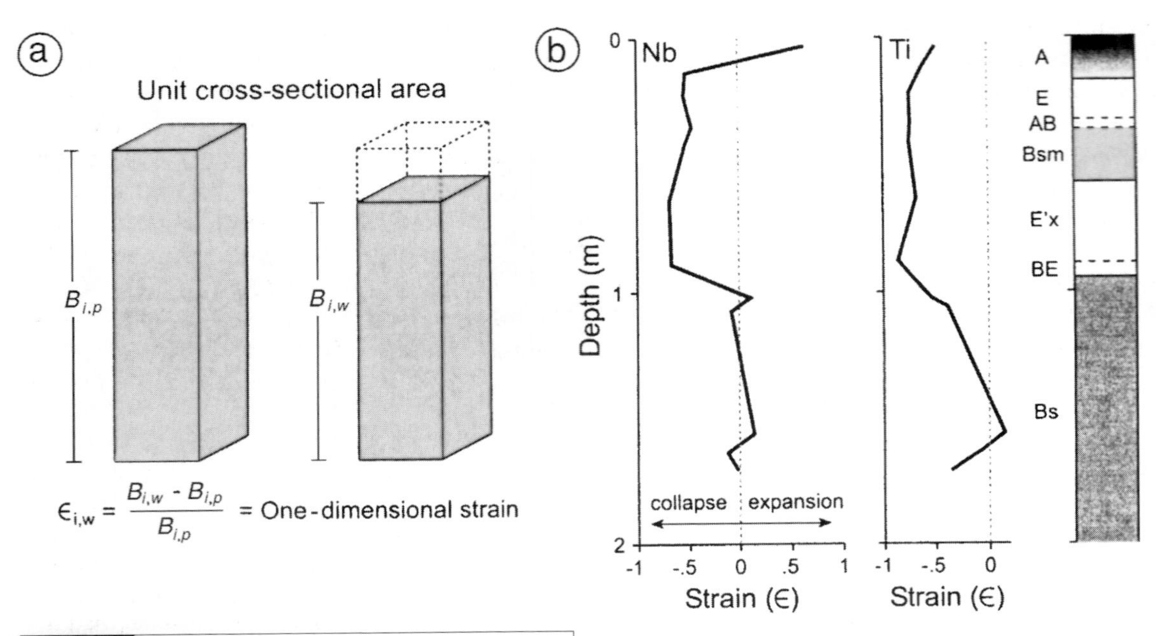

Fig. 12.81 Strain and mass balance concepts in soils. After Brimhall and Dietrich (1987). (a) Diagrammatic representation of conservation of mass and strain (collapse) for a soil profile at two periods in its development. The mass of element *i* in the soil before weathering is equal to the mass in the weathered product. (b) Depth functions of net strain, calculated using contents of immobile Nb and Ti found in rutile. The soil is a Spodosol from the pygmy forest of Mendocino County, CA (Jenny *et al.* 1969).

areas, cumulization is confined to lowlands, where sediments wash down from upslope. In Mollisols, where the major pedogenic process (melanization) is rapid and where grasses above are adept at capturing sediment, cumulization can occur slowly and continually.

Mass balance analysis, strain and self-weight collapse

In mass balance analysis and soil reconstructions, equations utilizing the amounts of stable constituents are used to quantify the gains and losses of materials during pedogenesis (Marshall and Haseman 1942, Chadwick *et al.* 1990, Brimhall *et al.* 1991a, b, Stolt *et al.* 1993). For example, mass balance analysis might indicate that an E horizon has lost clay, relative to what it inherited from the parent material, while a Bt horizon may have gained clay (Stolt and Baker 1994, 2000). Some rock-to-soil processes are isovolumetric, e.g., the formation of saprolite from bedrock. In such cases, mass balance analysis usually reveals a loss of mass but not of volume (Stolt and Baker 2000). However, when parent material is converted to soil, there is usually also a loss of volume, as the material collapses. Collapse is facilitated by solutional losses and pedoturbation.

The term *strain* is used for the deformational, volumetric changes that soils undergo. It is usually expressed as a change in profile thickness, i.e., a one-dimensional change in the vertical divided by the original length (Fig. 12.81). Strain is, therefore, a measure of volume change due to mass flux, weathering, root growth and decay, etc. (Stolt and Baker 2000). Where strain is positive, i.e., the soil has gained volume, it is termed *dilation*; where negative it is called *collapse* (Brimhall and Dietrich 1987, Brimhall *et al.* 1988). Strain is usually calculated by comparing volumes of parent material and soil. It can also be calculated for soil at any two time periods, provided that the data can be ascertained (Fig. 12.81). Strain is estimated by the change in concentration of an immobile, resistant mineral or element in the soil. Collapse, for example, is indicated when there is an increase in the concentration of the immobile element, caused by the loss of mobile constituents that is not exactly compensated by an inversely proportional decrease in bulk density or increase in porosity. Many soils undergo slow collapse in eluvial horizons and dilation in illuvial horizons (Fig. 12.81).

Another useful, related term is the *unit volume factor*, which is the number of unit volumes of parent material necessary to form a given unit volume of a soil horizon (Smeck and Wilding 1980). Thus, not only do soils evolve because of additions, removals and transformations (Simonson 1959), but their fabrics and hence their volumes do as well. Volume change and deformation are essential parts of pedogenesis, and reflect the ongoing evolution of not just the soil plasma, but also the skeleton.

The notion of volume *deformation* of soils and sediments, as a part of their long-term genesis, is an important part of various quantitative pedology studies (Brimhall and Dietrich 1987, Brimhall *et al.* 1988, 1991a). These studies and others have highlighted slow, sustained collapse in soils, leading to the concentration of certain immobile elements. George Brimhall's work has been particularly important to the understanding of how metal ores become concentrated in some saprolites. In these situations, the collapse occurs over long periods of time, usually by loss of soluble materials from the soil profile, thereby enriching it in insoluble elements such as Ni.

An emergingly important pedogenic process is self-weight collapse. Bryant (1989) first proposed this type of soil densification process – one that does not fit easily within the four-fold, process-systems framework of Simonson (1959). Bryant invoked self-weight collapse to explain the high bulk densities of Btx and Bx horizons (fragipans). It is, essentially, a new type of pedogenic process that borrows part of its theory from the disciplines of geology and civil engineering. Self-weight collapse implies that a soil or surficial sediment collapses in on itself, increasing its overall density. No external force or weight, such as the heel of a glacier, is required. The process is envisioned as occurring when a saturated, almost slurry-like deposit undergoes desiccation and compacts or collapses in on itself primarily through the rearrangement of its fabric or skeleton. When rewetted, pores fill with water but because the saturated moisture content of the already-collapsed soil in no way approaches that of the pre-collapse soil, it cannot collapse to the extent that it did initially. Small, additional increments of collapse can occur with continued, further desiccation. Eventually, after the driest stage is reached, repeated cycles of wetting and drying cause little additional fabric rearrangement and the soil is considered fully *ripened*. Pedogenesis can now act on it to form a fragipan, if conditions permit (see above). According to Bryant (1989), sediments likely to undergo self-weight collapse have: (1) loose granular contact structure, (2) a metastable condition and (3) bonding agents that reduce upon wetting. Loamy sediments with clay contents between 5% and 30%, and most loesses, generally meet these criteria.

Part III

Soil geomorphology

Chapter 13

Soil geomorphology and hydrology

Geomorphology is the study of landforms and the evolution of the Earth's surface. Because soils are so strongly linked to the landforms upon which they develop, a discipline has emerged that deals with those relationships: *soil geomorphology*. Birkeland (1999) defined soil geomorphology as the study of soils and their use in evaluating landform evolution, age and stability, surface processes and past climates. Wysocki *et al.* (2000) more broadly defined it as the scientific study of the origin, distribution and evolution of soils, landscapes and surficial deposits and the processes that create and alter them. Perhaps the definition we like best is that soil geomorphology is an assessment of the genetic relationships between soils and landforms (Gerrard 1992).

Soil geomorphology was first studied as a project area by the US National Cooperative Soil Survey (NCSS) program in the 1930s, as interests had developed among geographers, geologists and soil scientists on the relationships between soils and landforms (Effland and Effland 1992). Seeing the merit in this type of approach, the NCSS adopted soil geomorphology as a paradigm for studying and classifying soils. Early soil geomorphology studies were grounded in the work of Carl Sauer, a geographer at the University of California (Effland and Effland 1992). Many focussed on soil erosion. Later, under the leadership of Charles Kellogg, assisted by Guy Smith, the NCSS program embarked upon a research mission to understand soil–landform relationships in the major climatic areas of the United States, in support of its soil mapping program. Smith and the NCSS established sites where soil geomorphology was to be studied in detail: subhumid Iowa, an arid desert site in New Mexico, a humid Pacific Northwest site in Oregon and one in North Carolina.

The NCSS's soil geomorphology work led to a resurgence within the university community, where it had profound impacts on theories of soil and landscape genesis, and greatly influenced the way soils were classified (Effland and Effland 1992). Much of this work culminated in a book by Peter Birkeland in 1974. Several books (Knuepfer and McFadden 1990, Daniels and Hammer 1992, Gerrard 1992, Birkeland 1999) and book chapters and monographs have also been devoted to the topic (Daniels *et al.* 1971a, Hall 1983, Olson 1997). Of particular importance to the development of soil geomorphology was the establishment of a soil geomorphology committee by the Soil Science Society of America, and the many field trips run under its auspices.

Soils and landforms develop together, and soil geomorphology is designed to examine and elucidate the nature of that genetic "dance." This development is a two-way street. Soils are affected by landforms, and through their developmental accessions and features, they in turn influence geomorphic evolution.

What are the main components of soil geomorphology? What do soil geomorphologists usually study within this vast field? Although wide-reaching and inclusive as a field, soil

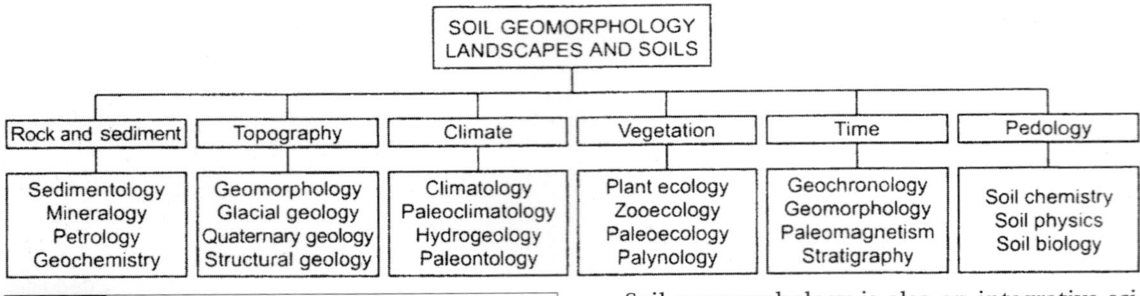

Fig. 13.1 Areas of study that contribute important concepts and background information to the field of soil geomorphology. After Ruhe (1975b).

geomorphology usually embraces at least one of the following topics:

(1) Soils as indicators of environmental/climate change.
(2) Soils as indicators of landscape and geomorphic stability.
(3) Developmental/genetic studies of soils (chronosequences).
(4) Soil–rainfall–runoff relationships, especially as they pertain to slope processes.
(5) Soils as indicators of past or ongoing sedimentological and depositional processes.
(6) Soils as indicators of Quaternary stratigraphy and parent materials.

To this list we feel compelled to add an overarching, geographic component. Many soil geomorphology studies have a strong geographic component, which more often than not assists in the interpretation of soil patterns on the landscape, whether that be the modern landscape or the paleolandscape.

Soil geomorphology is a field-based science. Knowledge of landforms in the field is essential to obtaining a representative soil sample, for most applications. No amount of laboratory work, "number crunching" or library research can make up for poor site selection or sampling technique. Once the samples have been taken, the outcome of the research is essentially "set." Knowing *where* to sample on the landscape, i.e., obtaining a representative sample, is as important as knowing *how* to sample. Both are imperative to the successful outcome of a soils-based research project.

Soil geomorphology is also an integrative science. To do it well, one must have a solid knowledge of many related fields (Fig. 13.1). Jungerius (1985) pointed out that too often soil geomorphology is essentially abiotic, largely due to the influence of William Morris Davis' cycle of erosion on the early geomorphology community. This cycle invokes little in the way of biological processes or controls on landscape evolution. Darwin's (1881) equally logical and perhaps more quantitative work had little influence on early soil scientists and geomorphologists (Johnson 2000). If Darwin's work had been more influential, soil geomorphology would today have a much stronger biophysical component (Jungerius 1985). An aim of this book is to begin to rectify the overemphasis on abiotic processes in soil geomorphology.

The geomorphic surface

Soils form on land surfaces. The surfaces may be flat or sloping, face north or south, be of all one age (having formed at the same time in the past) or time transgressive. Parent materials may change within the surface or they may be uniform across it. But the *surface* defines the soil in space and time (Ruhe 1969). The concept of the geomorphic surface captures that time–space notion; hence it must be accurately dated. The geomorphic surface must be datable; it must also be mappable. Ideally, its relationships to adjoining surfaces with respect to age, lithology, development, etc. are either known or can be discerned. Geomorphic surfaces are formed by one or more surficial processes (Daniels *et al.* 1971a). Many geomorphic surfaces are also named, particularly older, buried surfaces such as the Sangamon surface (see Chapter 15).

When exposed to the atmosphere, i.e., not buried, a surface is referred to as being *subaerial*

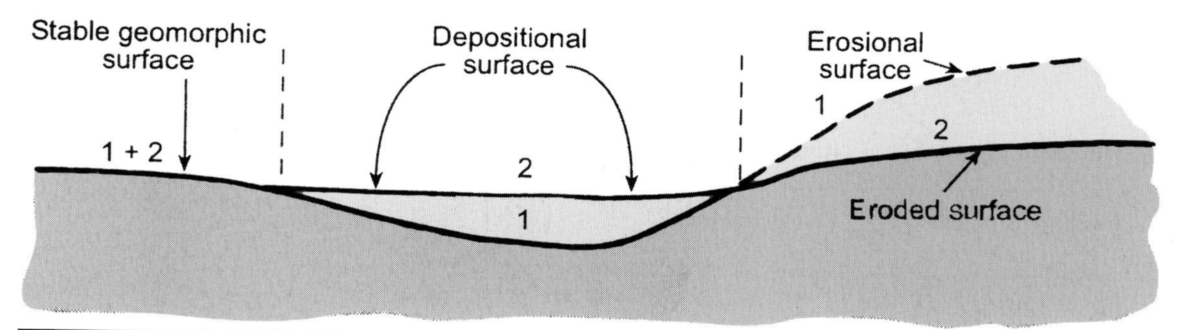

Fig. 13.2 The three main types of geomorphic surfaces. The figure shows the position of each surface at time$_1$ and time$_2$. After Follmer (1982).

or *epigene* (Watchman and Twidale 2002). All subaerially exposed surfaces eventually become either buried or eroded, but while they are exposed, soils develop on them. Some surfaces get inundated by water and become *subaqueous*.

Geomorphic surfaces can be erosional, constructive or a combination of both (Ruhe 1956) (Fig. 13.2). Erosion (or erosional) surfaces are those that have been formed by destructional processes, usually involving running water or wind. Constructional surfaces are aggraded by deposits such as eolian sand, glacial drift or loess. If the constructional event was short-lived, such as a volcanic eruption, the age of the geomorphic surface is the same as that of the parent material. Across the geomorphic surface each component slope is the same age, and thus it may be possible to precisely determine the age of the sediment and the surface (see Chapter 14). Similarly, large-scale erosion of a surface may create a new surface, of similar but younger age than the sediment below. Most geomorphic surfaces, however, are *time-transgressive*, meaning that the age of the surface *changes* across it, often in a systematic manner. For example, loess deposition in the Midwestern United States formed a number of time-transgressive, constructional surfaces. Loess deposition ceased first at sites farthest from the source areas, while sites nearest the source continued to receive loess until much later (see Chapters 8 and 15). The loess blanket is thus time-transgressive, in that the surface itself (and the loess that underlies it) is not all the same age. Obviously, knowing this, and knowing the span of ages on the surface, is imperative to understanding soil development *on* that surface, since soils that form on unstable surfaces soon get eroded or buried while soils that form on stable surfaces continue to develop. Using soils to determine the length of time that a geomorphic surface has been stable or if it is currently unstable (eroding or aggrading) is a focus of soil geomorphology (see Chapter 14).

Many surfaces are stable in a geologic sense but are slowly undergoing downwasting by natural processes, collectively referred to as *geologic erosion* or *natural erosion*. All slopes and surfaces are subject to these slow, natural downwasting processes. Often, on stable slopes, the rate of weathering or soil formation equals (after an initial, iterative period) the rate of geologic erosion. For example, Pavich (1989) described saprolite-mantled, upland surfaces in the Appalachian Piedmont that exceed 1 million years in age. These surfaces, despite their great age and apparent stability, are like all surfaces, not *truly* stable in that they experience geologic erosion at rates of about 20 m per million years. Normally, geologic erosion here keeps pace with bedrock weathering and erosion. If geologic erosion outpaces weathering and soil formation for a period of time, a positive-feedback mechanism can set up. The bedrock–saprolite contact would then grow closer to the surface, allowing weathering to accelerate, thereby deepening the bedrock–saprolite contact. During periods when geologic erosion is exceedingly slow, the "weathering front" will deepen faster than erosion can strip the weathering by-products, and weathering will slow as well because the weathering front is farther from the surface and because weathering by-products will accumulate, slowing chemical weathering reactions (see Chapter 9). Thus, over long periods of time, geologic erosion tends to keep pace with those processes that produce

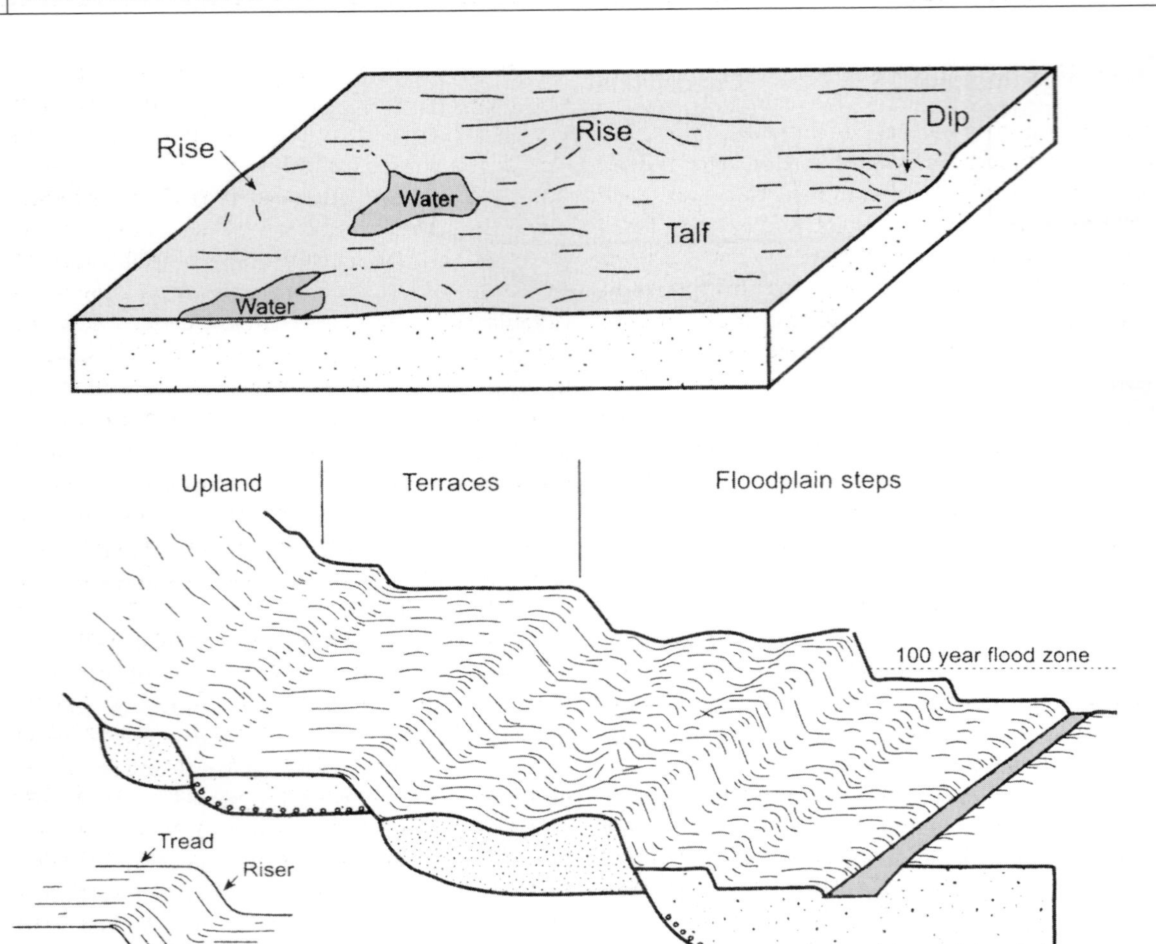

Fig. 13.3 Geomorphic components of monotonously flat landscapes and stepped landforms. After Schoeneberger and Wysocki (2001).

sediment/soil, by a complex set of feedback mechanisms, and the slope may lower in elevation but not change appreciably in other ways (Ahnert 1987). Cultivation or deforestation may result in *accelerated* or *anthropogenic erosion*, which may outstrip the system's ability to produce saprolite and soil horizons, thereby resulting in a thinner soil mantle.

Surface morphometry

Surfaces can be defined genetically (as to origin) and morphometrically (as to shape and geometry). Although soil geomorphologists must always be aware of the probable *geomorphic origins* of a slope, an involved discussion of the processes responsible for them is beyond the scope of this book (Holmes 1955). We can, however, provide terms that can be used to *describe* a landform or landscape, regardless of origin. Various slope elements were derived long ago for describing rolling or hilly landscapes (Ruhe 1975b). Elements such as summit, shoulder and footslope are typically used to describe locations on a sequence of soils in hilly terrain. On flat landscapes, which are quite common worldwide, typically the only relief is in areas of slight rises or depressions (Fig. 13.3). Schoeneberger and Wysocki (2001) describe these areas as rises and dips, respectively. Flat, broad summit areas are called *talfs* ("flat" spelled backwards). Often these flat surfaces occur on stepped or "tread-and-riser" landscapes (Fig. 13.3).

The catena concept

The geomorphic surface concept is applied primarily to broad surfaces of known age. Within each larger surface are many smaller ones, each with its own history and soils. Although soil patterns and complexity on large geomorphic surfaces may be extremely difficult to unravel, studying a pedon in isolation holds little promise for helping to explain soil–landscape evolution. Thus, we seek out a middle ground by examining small segments of the landscape, and then extrapolate what we learn onto larger areas; this is fundamentally how soil survey work is done. On these small segments, much of the variation in the soil cover is a function of *relief*. The *catena*, best expressed on landscapes of moderate relief, is an embodiment of Jenny's (1941b) relief factor (see Chapter 11). Relief is essentially a passive factor, but it has the very important function of providing potential and kinetic energy to the soil system through its impact on water movement. It "conditions" the redistribution of matter and energy within the soil landscape system. Redistribution is a three-dimensional process (Huggett 1975), although for sake of simplicity we begin with the two-dimensional catena.

Studying soils along a slope is one of the simplest yet most elegant ways to discern spatial interrelationships between soils and topography (Sommer and Schlichting 1997). A *catena* is a transect of soils from the top to the base of a hill, perpendicular (or nearly so) to the contour lines. Its name comes from the Latin *catena*, chain. Soils in a catena are often viewed as links in a chain; imagine two ends of the chain held firmly, such that the rest of it hangs freely between. This concept is not unlike that of soils from hilltop to hilltop, with a valley between. Catenas include information on soils, surficial stratigraphy and hillslope hydrology and shape. Applying the catena concept involves putting all these pieces together into a functioning soil–geomorphic system.

Based on his work in Africa, Milne (1936a) originally defined a catena as the sequence of soils between the crest of a hill and the floor of the adjacent swamp. Soil profiles change along this sequence in accordance with conditions of drainage and geomorphic history. Milne's original catena concept did not exclude catenas where the soils had developed on different parent materials, i.e., lithology did not have to be homogeneous along the catena. In 1942, Bushnell extended and at the same time constrained the application of the term to include all the various possible topographic, denudational and hydrologic situations, but *on a given parent material*. He limited catenas to a single parent material because to not do so would "spoil . . . its simplicity of connotation" (Bushnell 1942: 467). In essence, Bushnell wanted a catena to differ only with regard to drainage; today we call this a *toposequence*. The toposequence concept primarily involves morphologic connotations such as changes in soil color due to changing wetness conditions (Hall 1983). Soils along a catena, on the other hand, differ in morphology because of drainage condition variation *and* fluxes of sediment along the catena. Ruhe (1960) stressed that one cannot understand the evolution of a catena of soils without incorporating a knowledge of the geomorphic history of the landscape itself, and disagreed with the notion that catenas should be restricted to one parent material. In this book we will not restrict catenas to a uniform parent material.

Soils change predictably along catenas. To enhance this predictability, however, the two-dimensional catena must be examined in light of the three-dimensional landscape, or its curvature as seen from above. The plan curvature of a slope largely controls the directions of water and sediment transport on the slope. Catenas on nose slopes, head slopes or side slopes, for example, are quite different, as nose slopes are water- and sediment-diffusing slopes, head slopes are water- and sediment-gathering slopes, and side slopes are intermediate (Fig. 13.4). To determine if a slope is diffusing or gathering, examine the water flowlines, which run generally perpendicular to the contour lines. In nose or "spur" slope positions, flowlines will diverge downslope. In coves or head slopes they will converge. In theory, catenas should be defined such that water will flow downslope directly along them, perpendicular to the contour lines.

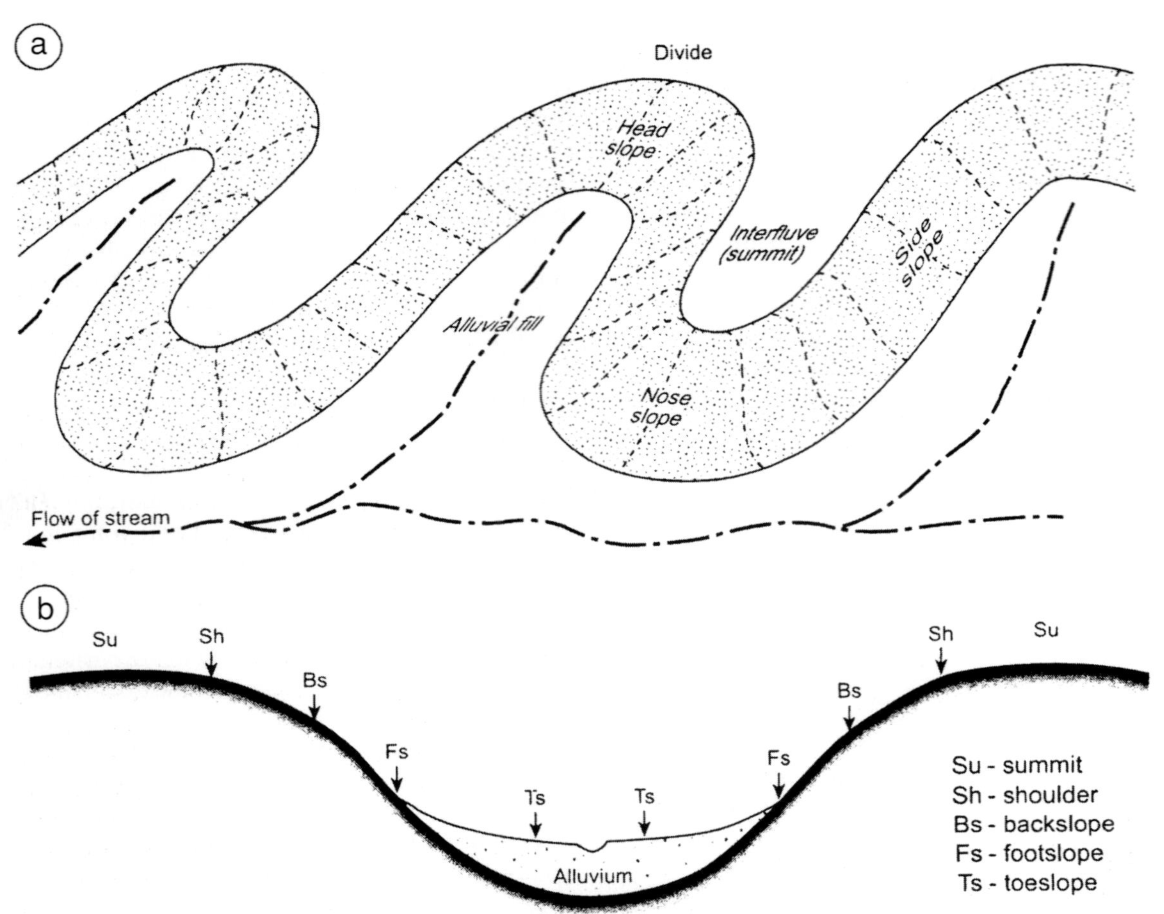

Fig. 13.4 Geomorphic components of slopes in a landscape with an open drainage system. After Ruhe (1975b).
(a) Three-dimensional components of slopes.
(b) Two-dimensional components, i.e., the five elements of "fully developed" slopes.

Reasons for soil variation within catenas

The slope is a highly explanatory framework within which to study soil development (Kleiss 1970, Furley 1971, Sommer and Schlichting 1997). Early literature even went so far as to state that, with time, the influence of relief could dominate "mature soils" to the point that parent material effects get masked (Norton and Smith 1930). To a degree, this is correct; with time the effects of topography and geomorphology become increasingly important to soil development.

Soils on one part of the landscape impact soils nearby, especially those downslope. Materials, solutions and suspensions move through and over landscapes, creating genetic linkages among pedons on slopes (Hall 1983). Translocations and transformations, espoused by Simonson (1959) as a useful method of explaining intra-profile (i.e., horizon-related) differences, also apply to catenas. Inter-profile additions and removals create many of the pedogenic differences along the catena itself. As Dan and Yaalon (1964: 757) put it, "the morphology of each (catena) member is determined by its position in the landscape and related to its adjoining members." In short, the catena is an excellent way of illustrating the geographic relationships of soils on landscapes.

Soils vary along catenas for two main reasons: (1) the slope affects *fluxes* of water and matter (generally but not always in the downslope direction) and (2) water table effects. Fluxes are of two main types: debris flux (sediment and organics) and moisture flux (Malo *et al.* 1974). In

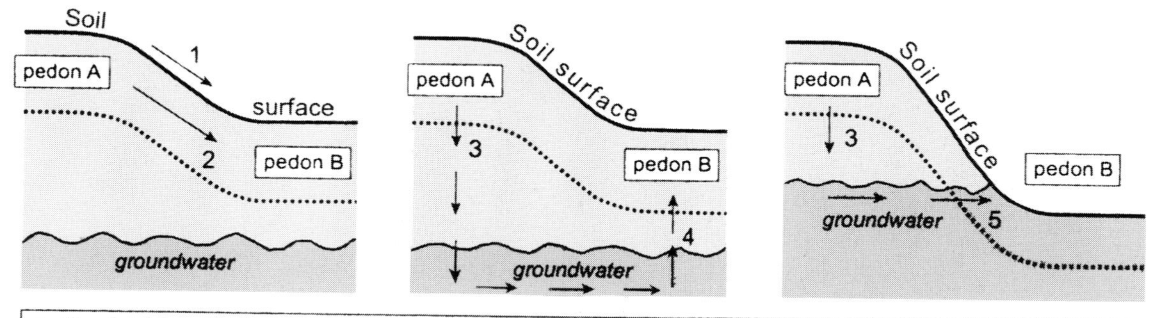

Fig. 13.5 Idealized scheme showing the various, direct and indirect, flux-related linkages among soils on a catena. Fluxes driven by water are emphasized. After Sommer and Schlichting (1997).

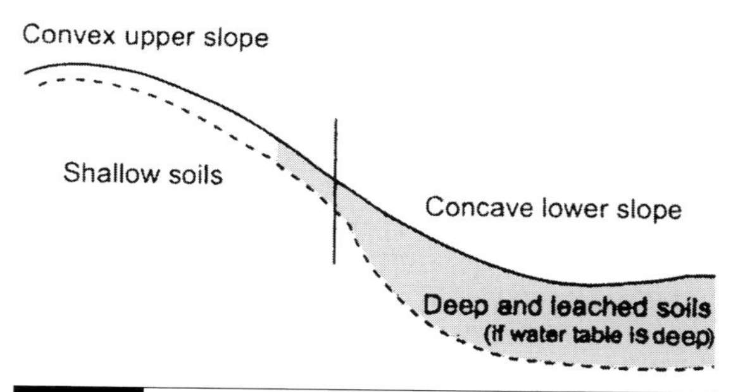

Fig. 13.6 Conceptual model illustrating how soils differ along a typical catena. After King *et al.* (1983).

closed catenas, e.g., a bolson or a kettle, all sediment and water stay at the base of the slope; debris lost from the upper slopes will accumulate at the base (Walker and Ruhe 1968). In open drainage systems, the debris can potentially be removed from the hillslope system by rivers and streams. Debris flux is so important that on many slopes soil properties reflect more strongly the hillslope sedimentation system than the pedologic system (Kleiss 1970). Sommer and Schlichting (1997) reviewed the importance of fluxes to soil variation within catenas and developed an idealized scheme to represent all the possible types of flux. Their scheme (Fig. 13.5) includes fluxes caused by overland flow, lateral subsurface flow, vertical infiltration/seepage, capillary rise and return flow and shows the linkages between these fluxes.

Debris flux involves erosional and depositional components. When transport is primarily by gravity-driven processes it has been referred to as *colluviation* (Goswami *et al.* 1996). When overland flow and rainsplash are the primary processes, the term *slopewash* is used. On summits, water tends to infiltrate or slowly run off. Because they are the steepest slopes on the landscape, upper backslope and shoulder slope areas have the most potential for runoff, and hence are commonly the most eroded, exhibit the thinnest soil profiles and are the most likely areas for rock outcrops or free faces (Gregorich and Anderson 1985) (Fig 13.6). Slope gradient, however, is not the only factor that can focus erosion on a particular site. Cover density and type, sediment texture, soil infiltration capacity and biotic (especially burrowing animal) activity all influence slope erodability (Yair and Shachak 1982). Because finer sediment is more easily eroded, coarser materials are commonly left behind on shoulder and upper backslope positions (Walker and Ruhe 1968). The mid backslope position is one of transportation. Farther downslope, debris flux slows and deposition of sediment predominates. Cumulization can occur on toeslopes (Fig. 13.6; see Chapter 12). The base of the slope is also prone to receiving inputs of sediment from outside of the slope system, e.g., overbank deposits from streams.

During transport, the material in "flux" along the hillslope tends to become more or less sorted. In general, finer material moves farther along the slope while coarser sediment is left behind (Kleiss 1970, Malo *et al.* 1974). This trend is best expressed in closed basins (Walker and Ruhe 1968), because in open basins the fine sediment on toeslopes

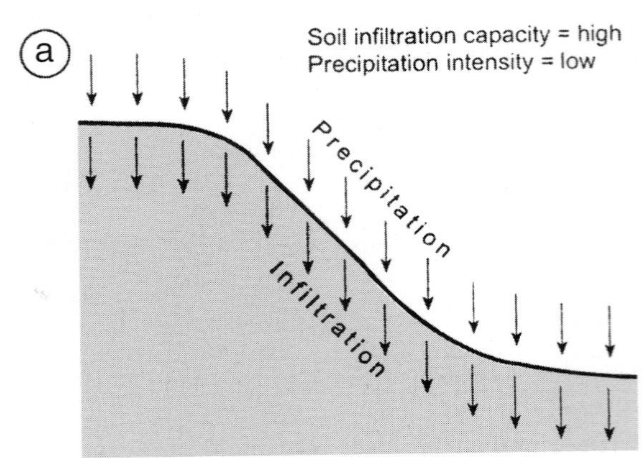

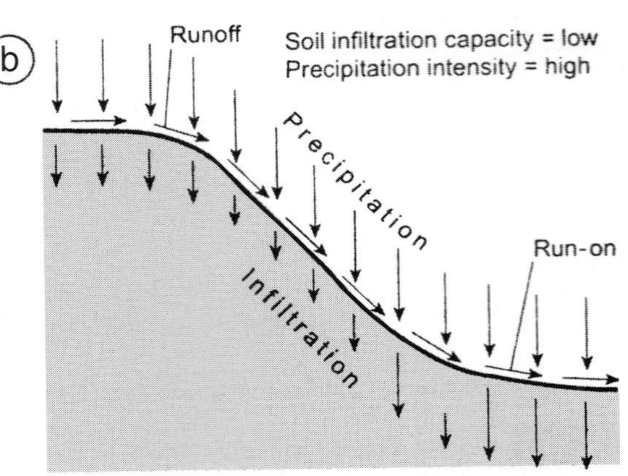

Fig. 13.7 Two end member scenarios for moisture flux along a catena of soils. (a) Precipitation rate < infiltration rate. Little runoff develops and soils are similarly leached along the catena. (b) Precipitation rate > infiltration rate. Runoff develops on steep slope segments, causing these soils to be drier, thinner and more eroded. Run-on areas at the base of the slope are deeply leached, unless leaching is inhibited by a high water table.

may be removed by streams or sediment may be added by the fluvial system.

Precipitation is usually evenly distributed along slopes. On a slope, the amount of water infiltrating vs. running off depends on a number of factors, such as infiltration capacity, slope steepness and plan curvature, intensity/frequency of rainfall events and cover type. If soil infiltration capacities exceed rates of precipitation input, most of the precipitation will enter the soil and catenary differences due to debris and moisture flux will be small (Fig. 13.7). Runoff occurs when precipitation inputs exceed the infiltration capacity of the soil (on all or part of the slope). Then, slope steepness and curvature affect the rate at which the water runs off that slope, and where it goes (Huggett 1975, 1976b). Runoff is often the driving force behind debris flux. Water falling (or melting out) on steep backslopes is most likely to run off. Lower slope elements receive run-on water and are usually wetter and potentially more leached (Fig. 13.7). High water tables in these lower slope positions may inhibit certain pedogenic processes, although if they do not these soils may be some of the best developed on the catena, due to large amounts of water and the kinetic energy it contains (Runge 1973). If all soils are equally permeable and water table effects are minimal, those on the backslope will be the least leached, those on the toeslope will be the most leached, and summit positions will be intermediate.

Catenas and soil hydrology

Water that falls onto a site can impact soil development along any of several pathways. In order to facilitate this discussion, we must first define a few terms (Table 13.1). Standard rain gauges report their data as P (gross precipitation). Some P is intercepted (I) by plants as surface detention on leaves and stems (Grah and Wilson 1944). This water then can take one of two pathways: it can be evaporated or move from the plant to the soil surface. Water lost to the atmosphere by evaporation from plant surfaces, termed canopy interception loss (C), is a significant proportion of the hydrology of forests, exceeding 15–20% of gross precipitation (Voigt 1960, Helvey and Patric 1965, Mahendrappa and Kingston 1982, Alcock and Morton 1985, Freedman and Prager 1986). The precipitation that remains after C is called net precipitation (N):

$$N = P - C.$$

Net precipitation can reach the soil surface by dripping off plants and directly from precipitation (that which is not intercepted); this input is called throughfall (T). Net precipitation can also reach the surface by running down stems as stemflow (S). Thus,

$$N = T + S$$

Table 13.1 Terms applicable to soil and forest hydrology

Term	Abbreviation used in this book	Definition/description
Gross precipitation	P	All the incident precipitation that falls from the sky
Canopy interception	I	Water intercepted by plants and not permitted to fall directly to the soil surface
Canopy interception loss	C	Water lost to the atmosphere by passive evaporation from plant surfaces
Net precipitation	N	Water remaining after gross precipitation has been reduced by canopy interception loss
Throughfall	T	Water that falls through the vegetative canopy, onto the soil surface
Stemflow	S	Water that runs down stems, onto the soil surface
Runoff	R	Water that impacts the soil surface and runs off, downslope
Litter interception loss	L	Water that impacts the O horizon and (1) is lost to the atmosphere via evaporation and (2) taken up by plants and transpired to the atmosphere
Infiltration or pedogenic precipitation	I or PP	Water that enters the mineral soil and is capable of participating in pedogenic processes within

and

$$P = C + T + S.$$

Water potentially available for pedogenesis includes throughfall (T) and stemflow (S); they are the main water inputs to the soil surface. However, not all of this water infiltrates into the mineral soil. Depending on slope gradient, surface cover and the infiltration capacity, some will run off (R). Some will be absorbed by the litter (O horizon) and evaporated. Other water within the litter is removed by plant roots and transpired, resulting in a cumulative litter interception loss (L). Litter interception losses are not inconsequential; they are usually between 2% and 4% of P (Blow 1955, Rowe 1955, Helvey 1964, Swank *et al.* 1972). Water lost as L does little to directly impact pedogenesis, although it does facilitate decomposition of the O horizon and foster a biotic community there. The remaining water enters the mineral soil as infiltration (I) (or pedogenic precipitation) and fuels pedogenesis. Thus,

$$I = T + S - L$$

and

$$I = P - C - L.$$

Slope description and catenas

Slope description must be a part of the soil geomorphologist's "bag of tricks." Slopes can be described in a number of ways. The nature and geometry of slopes can be used to not only define and describe the slope itself, but can be used as a predictor of soil character on that slope. Terms or characteristics that are frequently used

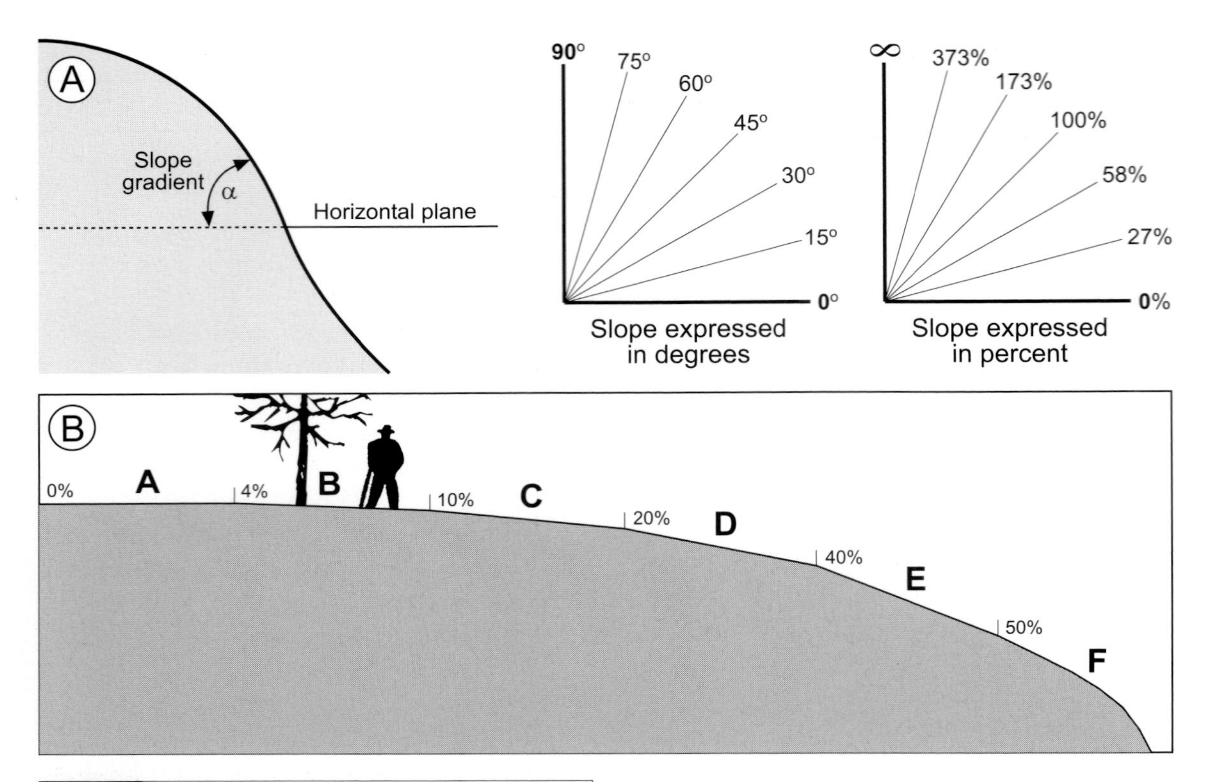

Fig. 13.8 The concept of slope gradient. (a) How slope is measured on the landscape and the two ways in which it can be expressed (degrees and percent). (b) Limits and definitions of slope classes used by the Natural Resources Conservation Service. Map units with complex slopes require the use of more than one term, e.g., "steep and moderately steep." The overlap of ranges allows them to be specifically set for each survey area. After the Soil Survey Division Staff (1993).

to describe a slope or part of a slope include its (1) gradient or steepness, (2) length, (3) aspect, (4) curvature/shape, (5) elevation and (6) position on the larger slope complex (Aandahl 1948).

Slope gradient

Gradient refers to the steepness or inclination of a slope from a horizontal plane. It is commonly referred to simply as "slope." If expressed in degrees, slopes can range from 0 to 90°, with a 90° slope being vertical. However, most soil scientists express slope in percent; a 45° (or 100%) slope has one unit of rise for every unit of run (Fig. 13.8). Because slope gradient is a good predictor of soil patterns and because gradient affects use and management of the soil, soil map units usually contain information on slope.

Perhaps no other slope attribute is as closely related to soil development as is gradient, since it affects the rates of movement of water and sediment on the slope (Norton and Smith 1930, Acton 1965, Furley 1971). Slope gradient changes along most catenas, both laterally and along flowlines (Ahnert 1970). Thus, for many catenas the gradient that is of most geomorphic interest is the flowline with the maximum gradient; this is also commonly the shortest flowline. Slope gradient is a proxy for potential energy that the slope possesses. Steep slopes impose a great deal of potential energy onto the water and debris that exist on slopes. Water impending on the soil surface has two main options: it can run off or infiltrate. If it infiltrates soil development is generally promoted whereas runoff does not promote soil development and in fact may regress it if it causes erosion. For these reasons, steeper slopes tend to have thinner soil profiles and less developed soils (Carter and Ciolkosz 1991) (Fig. 13.9). Vreeken (1973) found strong statistical relationships between slope gradient and various soil

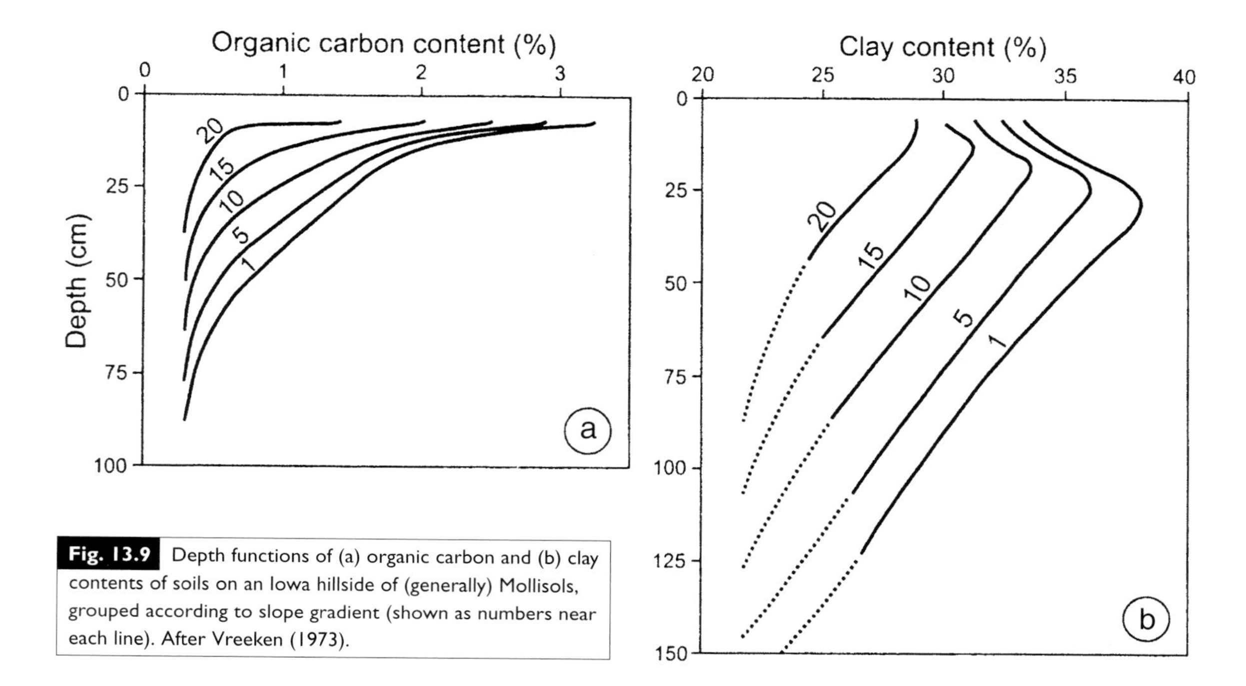

Fig. 13.9 Depth functions of (a) organic carbon and (b) clay contents of soils on an Iowa hillside of (generally) Mollisols, grouped according to slope gradient (shown as numbers near each line). After Vreeken (1973).

developmental properties, all of which could be explained by assuming that less water moves into and through the soil as slope gradient increases (see also Daniels *et al.* 1971a).

Many catenas have two or more inflection points along the flowline; the first occurs on the shoulder position and the second at the toeslope (Fig. 13.4). Inflection points may mark the location of lithologic discontinuities, from one rock type to another, and at the very least mark natural boundaries between soil bodies. Soil mappers, therefore, use inflection points to guide their estimation of soil map unit boundaries, across which drainage class and parent material often change (see Chapter 7).

Slope length

Slope length is directly correlated to erosion potential, and therefore correlates with soil development (Musgrave 1935, Gard and Van Doren 1949). On longer slopes, more and faster runoff can be expected. Thus, at the base of long slopes more slopewash or colluvium might accumulate.

Although theoretically a simple concept, slope length is difficult to quantify in a meaningful fashion (Aandahl 1948). Because slopes change gradient and curvature, one never really knows where to begin and end measuring slope length. On complex slopes, should the entire length be used, or only a subset? If the summit is broad and gently sloping, should slope length information begin on the shoulder (edge)? On highly permeable slopes the length may not be meaningful, at least from a soil erosion perspective. Therefore, the length of each slope *element* (see below) is often determined. Slope length for a slope element can be expressed absolutely, in a unit of measurement, or relatively, in terms of a percentage of the total slope length.

Slope aspect

Aspect indicates the compass direction toward which the slope faces, looking downslope. The incidence angle of the sun varies from one slope aspect to another, due to differences in local topography and the subsequent partial shading of the landscape. Equatorial latitudes are virtually unaffected by this shading phenomenon because solar radiation is received from angles both north and south of (and always very close to) celestial zenith during the entire year. In areas of high relief located between 30° and 60° latitude, however, aspect becomes a major factor in the amount of solar radiation received. Generally, in the northern hemisphere, north- and northeast-facing slopes are cooler and moister than those of other aspects, while the driest slopes face south and southwest. Differences due to aspect tend to increase as slope gradient increases. This topic is discussed more throughly in Chapter 11.

Slope curvature or shape

Slope curvature refers to the change in aspect along the slope face, and is normally best demonstrated by the manner in which contour lines (lines of equal elevation) bend or curve. When slope curvature is minimal, contours are linear in plan view, water runs directly up and down the slope, and the two-dimensional catena is able to capture and explain most of the variation on that slope. Where slopes are more complex, however, they must be considered more holistically, as a three-dimensional system.

On head slopes or in "coves" with concave contour/plan curvature (Fig. 13.4), flowlines of water and sediment will converge on lower parts of the slope (Aandahl 1948); these are *convergent slopes*. Less precipitation will be required to "wet up" the soils here, or to initiate throughflow and runoff. Debris and water will accumulate there (Huggett 1976b) and soil profiles tend to be thicker or even cumulic (King *et al.* 1983). On nose slopes, knobs and spurs, the opposite situation holds. Flowlines diverge and the bases of the slopes are not as subject to deposition and run-on; these are *divergent slopes*. Here, soil profiles are thinner (King *et al.* 1983). The formation of gullies, the deposition of sediment and to a certain extent the depth to the water table are affected by flowlines.

In a landmark paper, Huggett (1975) made the point that soil development must be examined within the three-dimensional soil system, rather than just up and down the slope. He argued that, instead of a catena, the "valley basin" concept should be employed to understand soil–topography relationships, for it includes flowlines. In this regard, he equated the basic organizational unit of the soil system to that of the fluvial system: the single-order drainage basin. The valley basin is bounded by the soil surface on top and the weathering front at the bottom, and functions as an open system. Fluxes of sediment, plasma and water move through the system, accumulating in hollows and disseminating away from nose slopes. King *et al.* (1983) supported this concept, finding that soils on a rolling landscape were much more strongly correlated to depressions and knolls, i.e., three-dimensional slope parameters, than to two-dimensional slope positions like shoulder, backslope, etc. (Nizeyimana and Bicki 1992). The reason for this is that any one pedon rarely occupies only one distinct slope position, as might be indicated along a transect. It may be influenced by slopes in other directions. For example, a site on a backslope might be sediment and water "shedding" if that backslope is also a nose slope, while a nearby backslope with a much thicker and better-developed soil might be located within a sediment-receiving site, such as a cove. Thus, no one slope parameter is adequate to explain the complexity of soil–landscape relationships, yet all seem to influence soil development.

Elevation

When describing a slope, some mention of its elevation is normally given. Elevation affects the temperature of a site, in that the normal environmental lapse rate is 6.4 °C decrease in mean air temperature per 1000 m increase in elevation. Such applications, however, are normally considered only in terrain with high relief (see Chapter 11).

Elevation also comes into play locally, when examined in the context of *local relief*. Cold air drains into low areas on calm nights, making them colder than surrounding high points or slopes. Although cold air drainage is of most interest to agriculturalists and fruit growers, it also impacts soils. Basins of interior drainage, e.g., glacial kettles in the Great Lakes region, often have a different vegetation assemblage in them, which is reflected in the soils. Our experience has shown that soils on the adjoining uplands are warmer and soils there have lower amounts of organic matter. Some deep kettles in the Great Lakes region experience frost in all months of the year, and have only stunted trees and shrubs in them, while upland areas have mature forest,

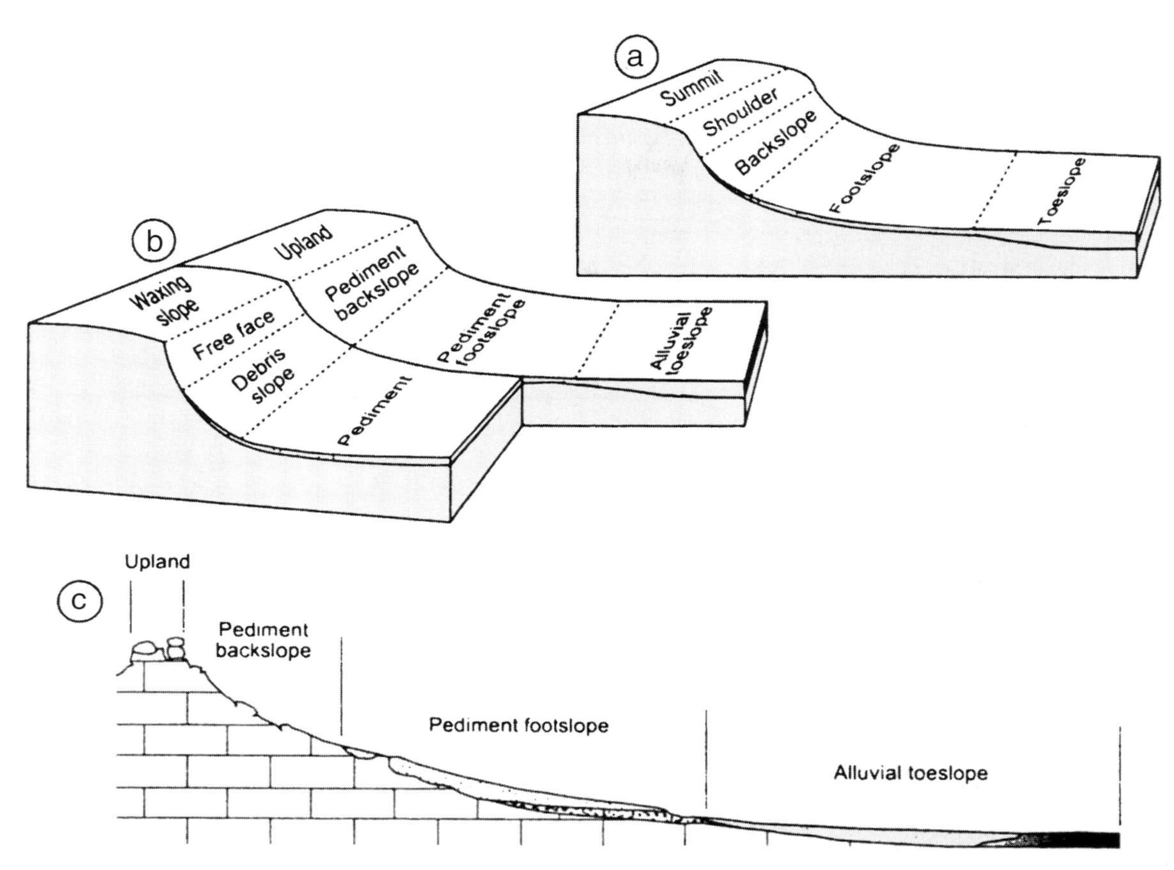

Fig. 13.10 Slope elements, based on different models. (a) After Ruhe (1960 and 1975b); (b) after Wood (1942) and Ruhe (1960); (c) after Milne (1936b).

again illustrating that even a few meters of local relief (elevation) can have pronounced impacts.

Slope elements

Soil properties vary consistently and predictably as a function of landscape position (e.g., Malo *et al.* 1974). By examining slopes as an amalgamation of slope elements, or pieces, two advantages are achieved: (1) scale becomes immaterial, as all slopes, regardless of size, have some or all of the elements, and (2) questions of genesis of the slope itself are avoided. Hillslope elements, which are separated more or less by inflections in slope gradient, also correlate to changes in surficial processes and pedogenic pathways. With respect to water, elements near the top of the slope are recharge or runoff sites, while elements nearer the base of the slope may be recharge or discharge sites, depending on local weather and subsurface geology (Khan and Fenton 1994). Because soil geomorphology is very much concerned with the *codevelopment* of slopes and soils, we discuss the matter in more detail below.

Figure 13.10 illustrates the various schemes that have been specifically developed to categorize the various hillslope elements or components (Schoeneberger and Wysocki 2001). In order to facilitate communication, various terms have been coined to describe these elements (Wood 1942, Dalrymple *et al.* 1968). In the simplest view, all hillslopes can be seen as having an erosional component (near the top), a transportational component and a depositional component (at the base). Soils respond to these general slope development processes, in that some may thicken or even get buried, others may erode, while others may maintain a balance between pedogenic and slope development. Thus, along a given slope or catena, the actual age of the soil may change, as some areas may be undergoing long-term,

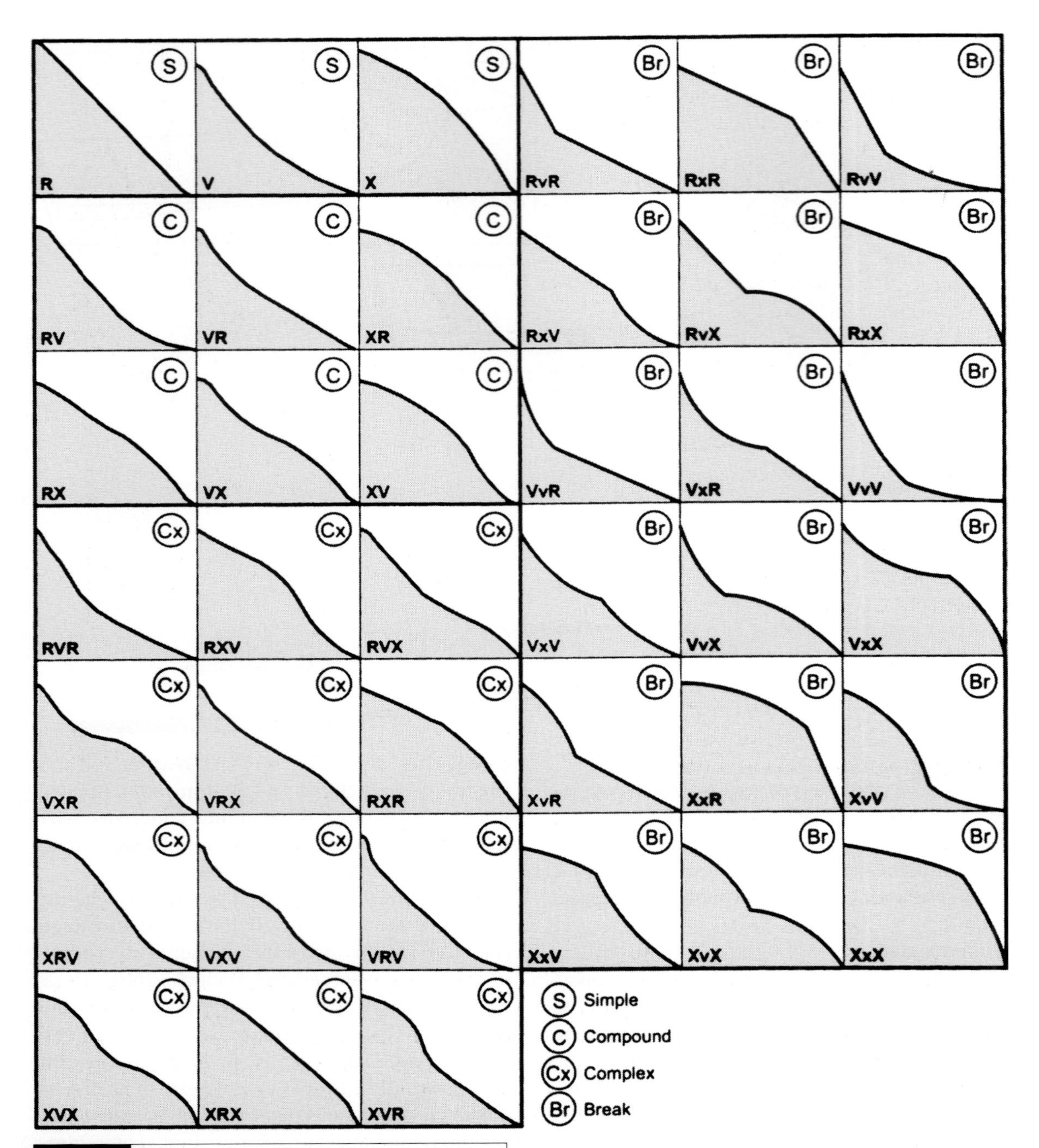

Fig. 13.11 Examples of simple (S), compound (C), complex (Cx) hillslope profiles, and those with slope breaks (Br), as viewed in cross-section. After Ahnert (1970).

gradual erosion while others are getting slowly buried by the sediment eroded from upslope. Slopes like these are time transgressive. The introduction of differing soil ages and $time_{zero}$ considerations complicates the catena concept, but is important to keep in mind. Taken further, a slope can be thought of as having three main components: a rounded upper edge or concave downward waxing slope, a constant slope of varying length (also called a pivotal point or junction point) and a concave upward waning slope at the base where sediment and debris accumulate (Wood 1942, Ahnert 1970, Ruhe 1975a). Some slopes also have a flat, horizontal component at their base, even if this segment is very small. Thus, the simplest pattern is: convex, straight, concave. Compound and complex slopes vary from this pattern but still retain the convex–straight–concave pattern, in a downslope direction, over at least some of the slope (Savigear 1965) (Fig. 13.11). Profile breaks have considerable diagnostic value in geomorphology, since each break indicates a change in slope process, lithology and/or depositional/erosional geomorphic history. Most "fully developed" slopes are described using Ruhe's (1960) system of five slope elements: summit or crest, shoulder, backslope, footslope and toeslope (Ruhe 1960) (Fig. 13.4). These elements do not need to occur consecutively down the slope, and all five need not be present.

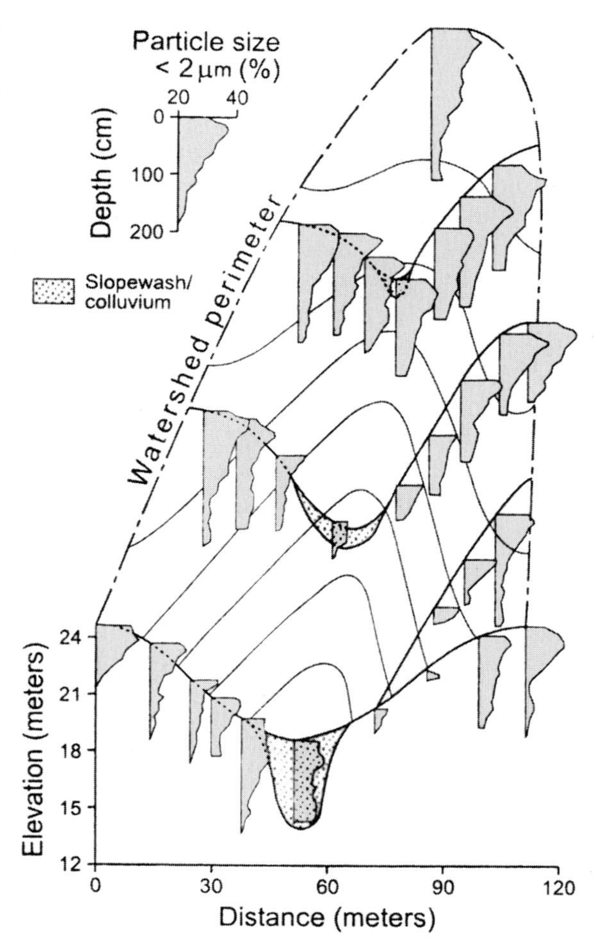

Fig. 13.12 Depth functions of clay content at various locations within a first-order watershed in Iowa. After Vreeken (1973).

SUMMIT POSITIONS

Summit (crest) positions are generally quite stable, with minimal erosion or accretion. According to King (1957), this part of the landscape is dominated more by chemical weathering than by physical weathering and erosion. Exceptions occur where the summit is narrow, where precipitation totals are high or where the soil is slowly permeable. Any of these situations facilitate runoff which is antagonistic to pedogenesis. Wide summits may be the oldest and most stable of the five slope elements, since water is unlikely to run off. Most of the water that falls on summits will infiltrate vertically, leading to wetter soils with more organic matter, thicker A horizons and better horizonation than slope elements that are immediately downslope (Fig. 13.9). Clay maxima tend to be deeper here than on steeper slope elements, because more water infiltrates (Fig. 13.12). However, if the soil is slowly permeable or has a slowly permeable horizon at depth, such as a fragipan or a thick Bt horizon, water may perch on the soil surface or within the profile, making the "upland" soils much wetter than would otherwise be expected (see "Landscapes" below). Similar situations can occur on upland summits in "bedrock country," where flat bedrock benches extend for some distance. Water may perch above the bedrock, making the upland soils quite wet. In such a setting, the soil–bedrock interface becomes a focus of illuviation (Schaetzl 1992a). On

Landscapes: the Illinoian till plain of southern Illinois

The flat landscapes of southern Illinois are underlain by Illinoian-age till (Fig. 12.57). In the > 100 millennia of generally humid conditions since that glaciation, the landscape developed a strong soil with a thick, clay-rich Bt horizon: the Sangamon soil, now a paleosol (see Chapter 15). Much of the inherited relief was lost from the landscape. The Wisconsin ice sheet later advanced to positions nearby, but never covered this landscape. Meltwater from it, however, did flow in the nearby Mississippi River valley, which served as a source of calcareous loess. About a meter of loess was deposited on parts of this low relief landscape. Between the loess and the paleosol is a layer of silty erosional sediment (pedisediment) that formed as the loess was beginning to be deposited. In actuality, the loess is composed of two units: a thin layer of Roxana (early Wisconsin) loess overlain by Peoria (late Wisconsin) loess. The Sangamon paleosol, the pedisediment and the overlying loess "welded" together to form one soil, with the paleosol becoming the B horizon of the surface (Cisne series) soil. The paleosol/Bt horizon acts as an aquitard, perching water in the loess above, trapping illuvial clay at its upper boundary and growing upward by accretion into the overlying loess. Countering this process is clay destruction by ferrolysis, which occurs in the eluvial zone of the surface soil when water seasonally perches in the upper solum. The presence of an acidic paleosol at depth also helps to acidify the Cisne soils.

Despite being on the top of the landscape, Cisne soils are poorly drained, with gleying in the upper B horizon (Fig. 12.57). This landscape illustrates that summit landscape positions need not be the best drained. Here, the soils with the lowest water tables (relative to the solum) are on the steepest slopes, either on small ridges or knolls that rise above the flat uplands or on the sides of small drainageways. Sodium-affected soils occur in depressions (Huey) and on preferred upland sites (Darmstadt and Tamalco). The Huey soils are the famous "slick spot" soils of southern Illinois (Fig. 12.57).

gently sloping summits, water may flow laterally at the top of an aquitard, eventually reaching the shoulder, where it may continue as subsurface flow or emerge as a seep. Because water often behaves similarly across the summit, soils are commonly quite uniform across it. Exceptions occur on sharp-crested and undulating summits.

SHOULDER AND FREE FACE POSITIONS

Slope convexity is the operative concept associated with shoulders, where runoff and erosion are maximal (Walker and Ruhe 1968). Steep shoulder slopes are called *free faces*. On free faces, runoff dominates to the point that erosion outstrips pedogenesis, and thin or non-existent soils are the result. Bare bedrock is common (Roy *et al.* 1967). Detritus may accumulate in rock crevices and shallow pockets within the free face, as well as at the base, in what is referred to as a debris slope (Fig. 13.10).

Shoulders are usually the youngest and least stable of the surfaces on a catena (Furley 1971, Malo *et al.* 1974). Soils there tend to be comparatively thin, low in organic matter and relatively dry (Aandahl 1948) (Figs. 13.9, 13.13, 13.14). Surface instability on shoulder slopes is the norm. If steep, mass movements may be commonplace; soils and regolith slip, slide and flow downhill. Much of this instability is initiated by lateral flow of water in the subsurface. Subsurface flow is often concentrated in preferred flowlines, leading to gullying and seeping at a few spots, rather than uniformly across the slope. Where

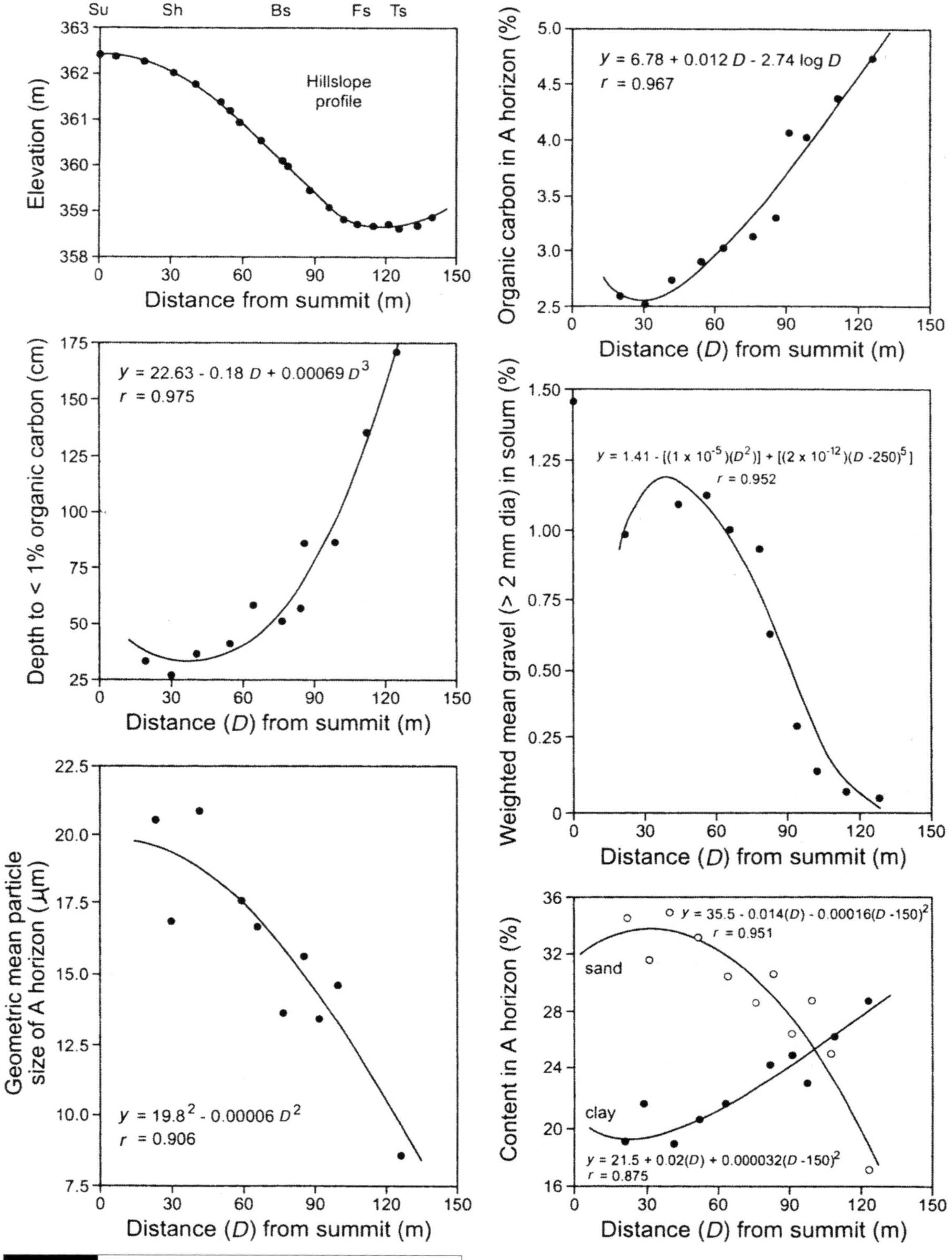

Fig. 13.13 Various scatterplots and curves illustrating the strong relationship between soil properties and slope position. After Malo *et al.* (1974).

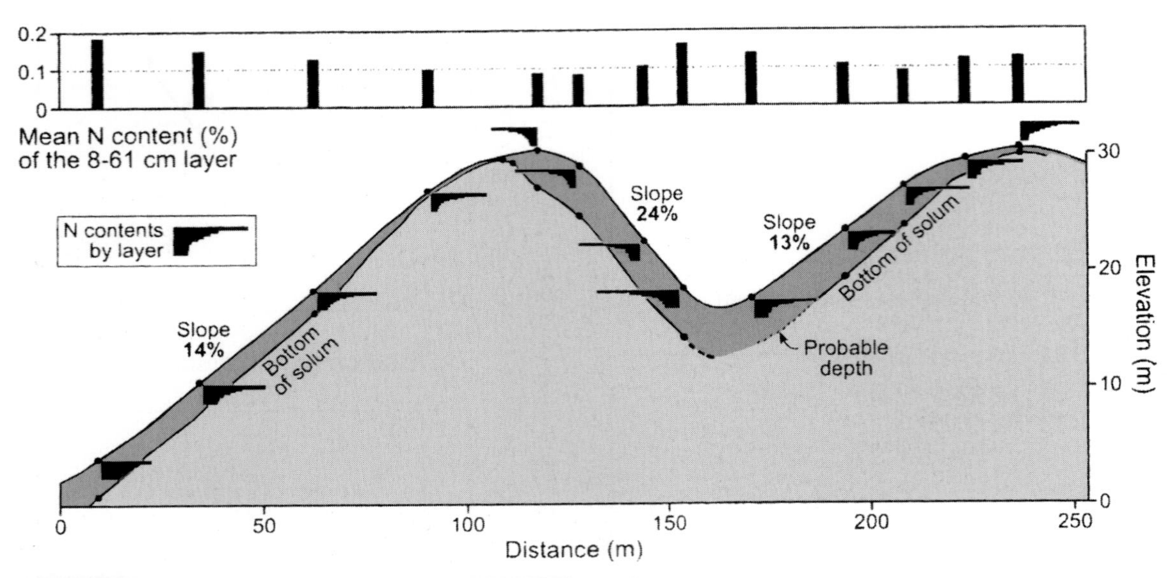

Fig. 13.14 Nitrogen contents and solum thicknesses (to calcareous loess) along three catenas in northwest Iowa. After Aandahl (1948).

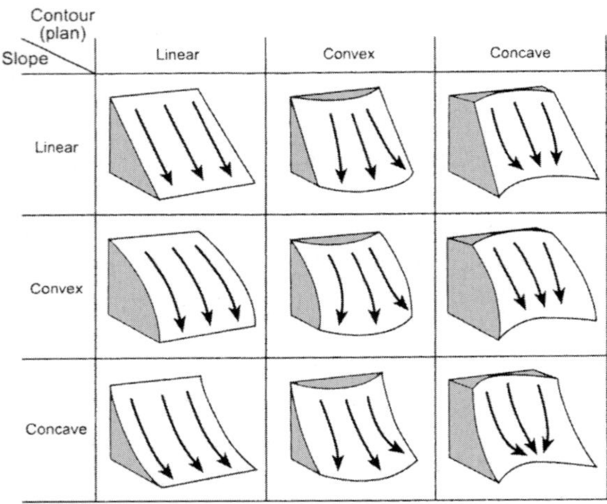

Fig. 13.15 The nine basic geometric forms of hillslopes, with flowlines illustrating how water and debris (theoretically) moves on them. After Ruhe (1975b) and Huggett (1975).

water emerges at the surface, it may deposit soluble materials such as carbonates, iron and salts. Nonetheless, for much of the year the shoulder is the driest position on the landscape. It also experiences the greatest water table fluctuation (Khan and Fenton 1994). Sites farther downslope are more uniformly wet with high water tables, while flat summit positions are occasionally wet due to lesser amounts of runoff. Because erosion preferentially strips the finer material from the shoulders, soils here may also be coarser-textured than elsewhere.

BACKSLOPE POSITIONS

Backslopes are transportational slopes that lie at the pivotal point between upslope areas dominated by erosion, and lower slopes which accumulate sediment (Furley 1971). Where backslopes are short, soils can change markedly from those just upslope to those downslope (King *et al.* 1983). Debris and water move over or through backslopes, sometimes on top of subsurface aquitards (Gile 1958, Young 1969, Huggett 1976b, Schlichting and Schweikle 1980). The pathways along which the water flows will depend on the *curvature* of the slope (Fig. 13.15). Slope length also determines how much material moves through and along the backslope, as does the stratification of materials within it, including sediments below the soil profile. Mass movements such a creep, slump and solifluction can also occur here, sometimes producing hummocky topography (Hall 1983).

FOOTSLOPE POSITIONS

Footslopes, the most concave parts of the slope, are sediment- and water-receiving positions. Material is carried in solution and in suspension, as throughflow and overland flow. The sediment

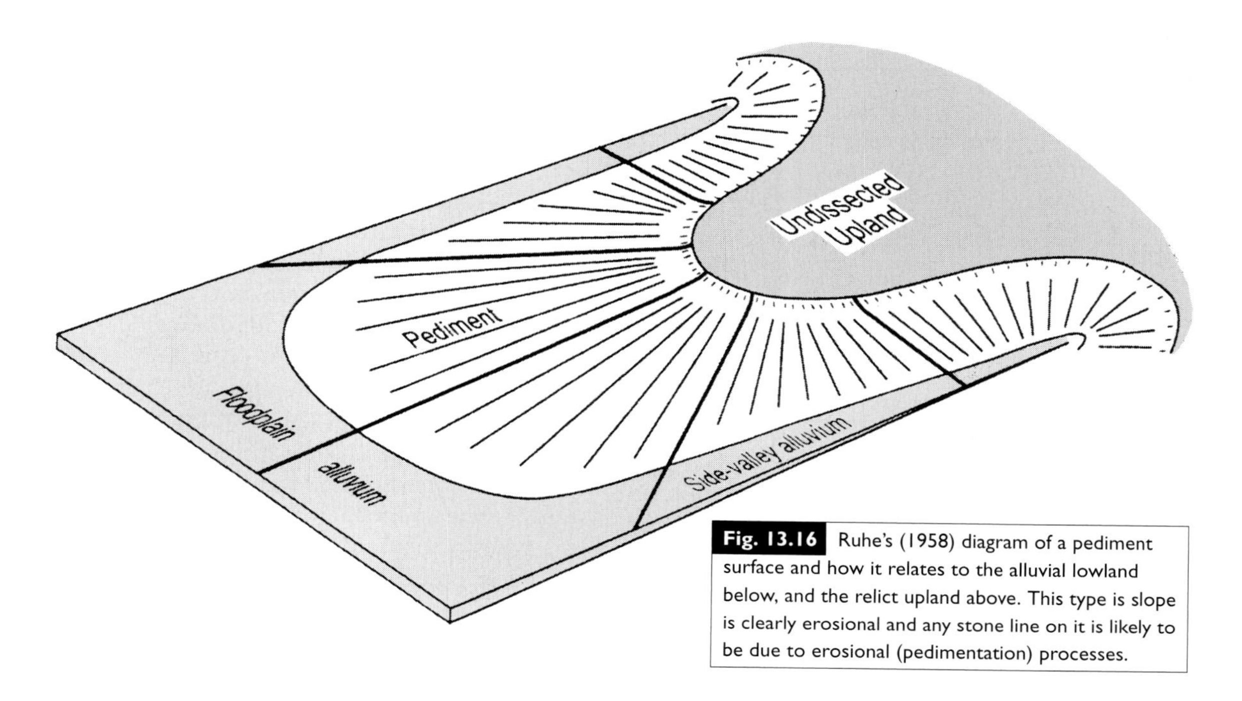

Fig. 13.16 Ruhe's (1958) diagram of a pediment surface and how it relates to the alluvial lowland below, and the relict upland above. This type is slope is clearly erosional and any stone line on it is likely to be due to erosional (pedimentation) processes.

that is most likely to be deposited here is from the upper profile (A horizon) of soils upslope, leading to overthickened A horizons and sola (Fig. 13.14). The wetness of the lower slope positions also accentuates net primary productivity of the plants there, which in turn provide more litter to these sites. Lastly, cool, wet conditions at the base of the slope may inhibit decomposition of these organics. All of these factors combine to make soils in the foot and toeslope positions high in organic matter (Kleiss 1970) (Fig. 13.13).

Dissolved material and clastic sediment can be forced to the surface from groundwater that wells up at spring sapping sites. Gullies can form there and work their way upslope. As they do, the backslope will become increasingly the focus of erosion and small fans will form on the footslope. Footslope surfaces are therefore commonly constructional, except for localized erosion at gully sites. Burial of soils can occur. Indications of wetness, such as mottles, Fe–Mn concretions and even gleying, are common though not widespread.

Perhaps no other slope position is more influenced by slope curvature than is the footslope. Water that follows flowlines down the slope is dramatically affected by them, and the influence of topography on flowlines is greatly diminished as waters slow at the footslope (Fig. 13.15).

In desert areas and erosional landscapes where bedrock is a dominant part of the landscape, the footslope position is marked by a transportational slope called a *pediment* (Hallberg *et al.* 1978b). Pediments are broad, concave-upward surfaces extending away from the backslope or debris slope, down to a lower part of the landscape where sediment accumulates in a playa or an alluvial plain (Fig. 13.16).

TOESLOPE POSITIONS

Your toes are at the end of your foot; similarly, toeslopes are the outward extension of footslopes. Toeslopes, aka *alluvial toeslopes*, are constructional sites, with sediment accumulating not only from above but also from streams that flood and deposit overbank alluvium. The latter type of sediment accumulation was especially common in the United States after European settlers cleared and cultivated the forest and prairies. Sediment washed onto toeslopes and plugged river channels, forcing the rivers to flood more frequently and each time they did they deposited sediment on toeslopes. In parts of Wisconsin, this "post-settlement alluvium" approaches a meter

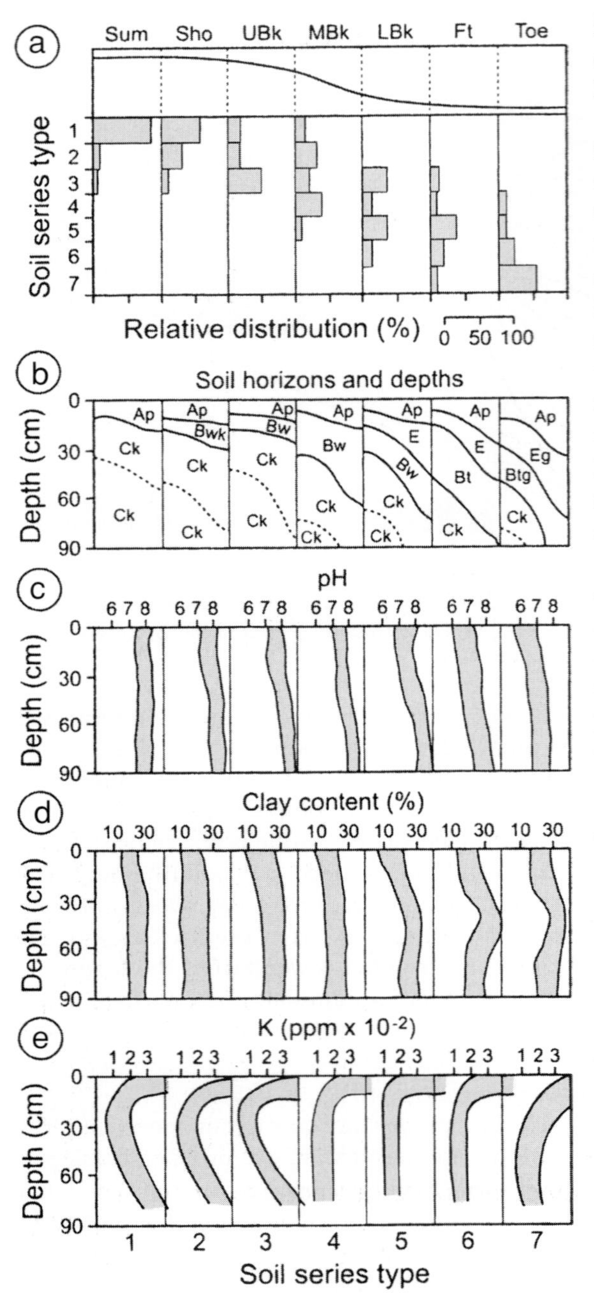

Fig. 13.17 Soil characteristics along a catena of the Weyburn Association, Saskatchewan, Canada. After King *et al.* (1983). (a) Relative distribution of seven soil series along the catena, classified according to slope position. (b) Soil horizonation for each of the seven soil series, portrayed roughly according to slope position. (c) Depth distributions of pH for each of the seven soil series. (d) Depth distributions of clay for each of the seven soil series. (e) Depth distributions of K for each of the seven soil series.

in thickness (Lecce 1997, Faulkner 1998). Thus, cumulization and burial are dominant processes here.

Sediment on toe and footslopes tends to be finer-textured and more uniform than material upslope because slopewash processes transport the finer material farther (Nizeyimana and Bicki 1992) (Fig. 13.13). Likewise, slope-derived sediment gets thicker farther out onto the toeslope (Walker and Ruhe 1968). Accumulation of sediment at the base of slopes is especially important in basins of closed drainage. Here, there is no mechanism to remove sediment (except wind), and accumulations may be thick (Walker and Ruhe 1968). Nonetheless, even in open drainage systems, some sediment is likely to accumulate at the bases of slopes (Vreeken 1973).

Soil development on toeslopes reflects wetness and sediment accumulation (Fig. 13.17). A horizons tend to be thicker than anywhere else on the slope (Gregorich and Anderson 1985). Indicators of wetness will generally be stronger and more prominent here than all other slope positions except for perhaps wide, flat summits underlain by aquitards.

THE NINE-UNIT LANDSURFACE MODEL

Recognizing the utility but also the simplicity of the catena concept, and that many pedogenic processes operate not only in the vertical dimension only but also parallel to the slope, Conacher and Dalrymple (1977) developed their nine-unit landsurface model (Fig. 13.18). Largely based on work in Australia and England, Conacher and Dalrymple's model stressed that catenas can be composed of up to nine interrelated *landsurface units*. Each units is affected by interactions among water- and gravity-based processes, by translocation and redeposition of soil materials by overland flow, throughflow and streamflow, and by creep and mass movements. The model is, above

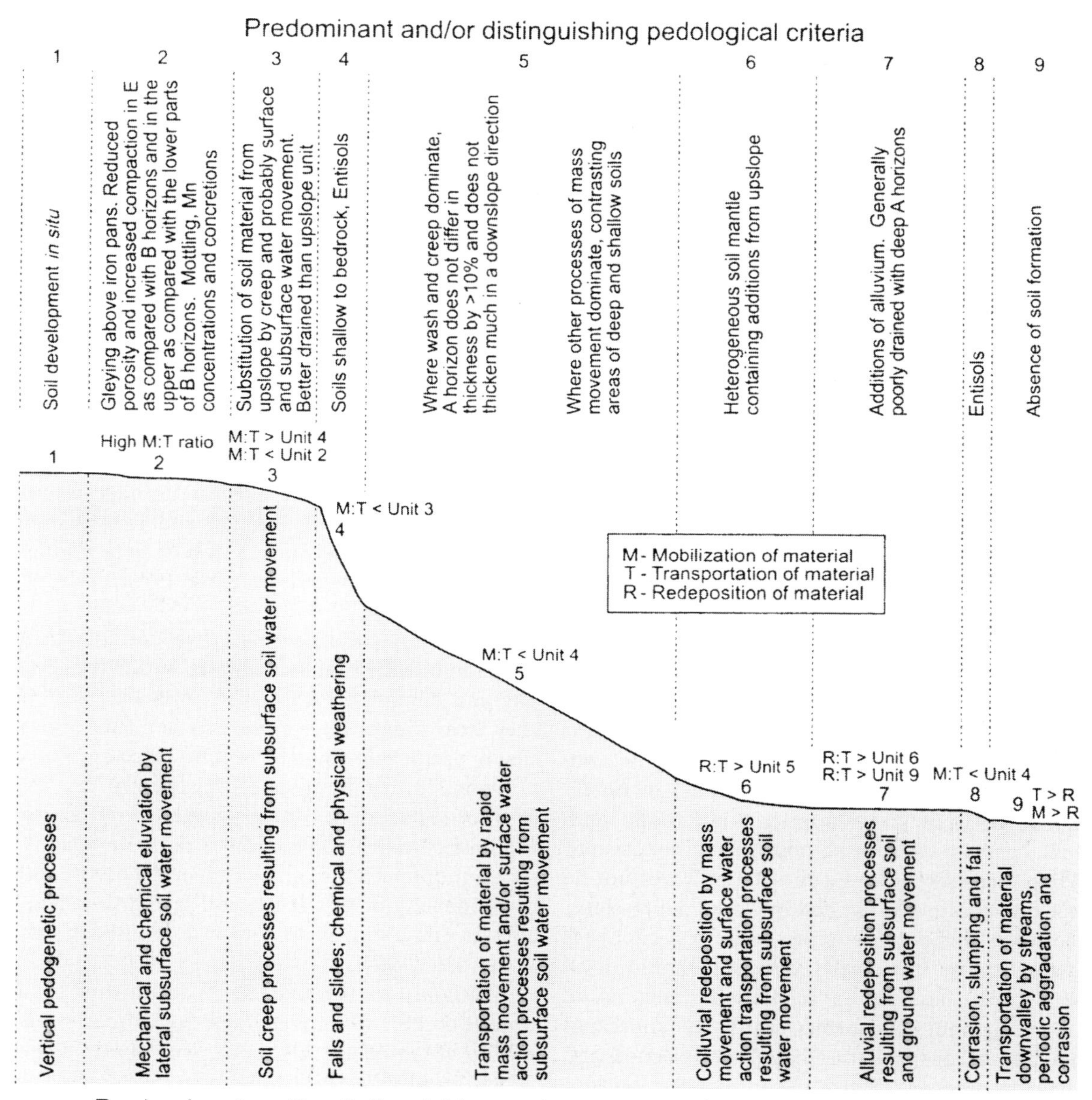

Fig. 13.18 The nine-unit landsurface model of Conacher and Dalrymple (1977), slightly modified from the original.

all, a soil/water/gravity model. It is essentially an extension of the five slope units in Ruhe's (1975b) slope model, but including processes of alluvial erosion and deposition at the base of the slope. It is a "universal" soil-geomorphic model that has wide application in older, more incised and developed landscapes, and re-emphasizes the importance of slope processes (creep, mass movement, throughflow, alluviation) on the development of catenary soil patterns.

Conacher and Dalrymple (1977) emphasized Milne's (1935, 1936a, b) original intention that the catena concept should include soil differences that result from variations in drainage, as well as from differential transport of eroded materials and chemical elements, but that through the years the latter concepts have been lost. They also stressed that the catena concept should include soils formed in redeposited materials. As a result, the landsurface model employs more geomorphology and slope processes than do most "pedogenic catenas." It is also more areally inclusive than the traditional catena. For example, the lower two units are associated with fluvial processes and sediments; unit 8 is essentially a riverine cut bank and unit 9 is the stream bed itself (Fig. 13.18). Runoff and downslope transport are given great weight. The model contains a stable summit with well-drained soils, and an adjoining, wetter unit with shallow water tables due to the pedogenic development of a subsurface aquitard. A free face, dominated by slope processes of mass movement, is also a central component. Downslope of the free face (unit 4) are slope positions that are primarily sediment-receiving. In short, their model is a response to the rather "sterile" catena concept that had dominated the mid twentieth century – one that focussed strongly on pedogenic processes and drainage, while deemphasizing hillslope processes.

Because the model was developed on older, "mature" landscapes, many of the landsurface units may not be evident on low relief and younger landscapes, and especially on the constructional, Pleistocene-age landscapes of northern Europe and North America. Some of the landsurface units will not be observed on landscapes where bedrock is deep. Units 8 and 9 will not be present in catenas that end in closed depressions. However, this is to be expected for a model that is essentially "universal"; parts of it should fit all landscapes, but few landscapes will utilize all of its component landsurface units. As landscapes age and as local relief and stream dissection and incision proceed, more and more of the landsurface units will "emerge."

The water table

Because of the strong relationship between topography and soil water, the term *hydrosequence* is used to refer to series of soils with differing wetness, usually along a catena (Zobeck and Ritchie 1984, Cremeens and Mokma 1986). Understanding the water regime of a soil is vital to interpreting its development, as well as to proper management. Drainage, an important component of that water regime, refers to the rapidity and completeness of removal of water added to a soil, as well as the frequency and duration of periods when the soil is not saturated. Drainage has internal and external drivers. Internally, drainage is impacted by permeability and water table relations. Externally, it is largely a function of slope configuration (Simonson and Boersma 1972, Crabtree and Burt 1983).

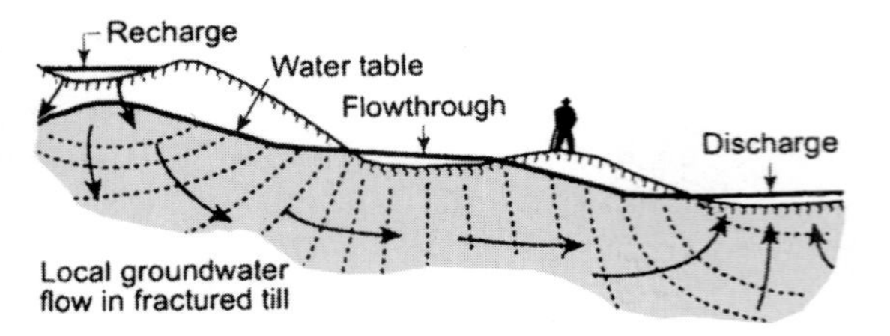

Fig. 13.19 Generalized and simplified groundwater flow net for a prairie pothole landscape underlain by fractured till. Equipotential lines represent lines of equal head. After Richardson *et al.* (1992).

Whether a soil is normally "wet" or "dry" is a function of many factors, among which topography and climate are foremost. Soils are "wet" if they have a high water table, or when they are so slowly permeable that they retain large amounts of water. In dry climates most soils are dry, except for some along stream courses or on playas. In moist climates, topography is more important, for even upland soils can be dry, provided the soil is permeable, or wet if the soil is slowly permeable or underlain by an aquitard (Mausbach and Richardson 1994).

Wetlands and hydric soils occur due to a unique combination of geologic and climatic conditions (Mausbach and Richardson 1994). High rainfall and cool conditions generally favor the formation of wetlands. Flat topography minimizes runoff, favoring the development of broad areas of wet soils, while rolling topography promotes the development of wetlands only within localized depressions and valleys. Government agencies use geologic, hydrologic and biotic criteria to determine if wet soils meet the definitions of hydric soils, or if parts of the landscape can be classified as a wetland.

In humid areas, most wetland depressions are groundwater *discharge* areas where groundwater emerges to become surface water (Figs. 13.19,

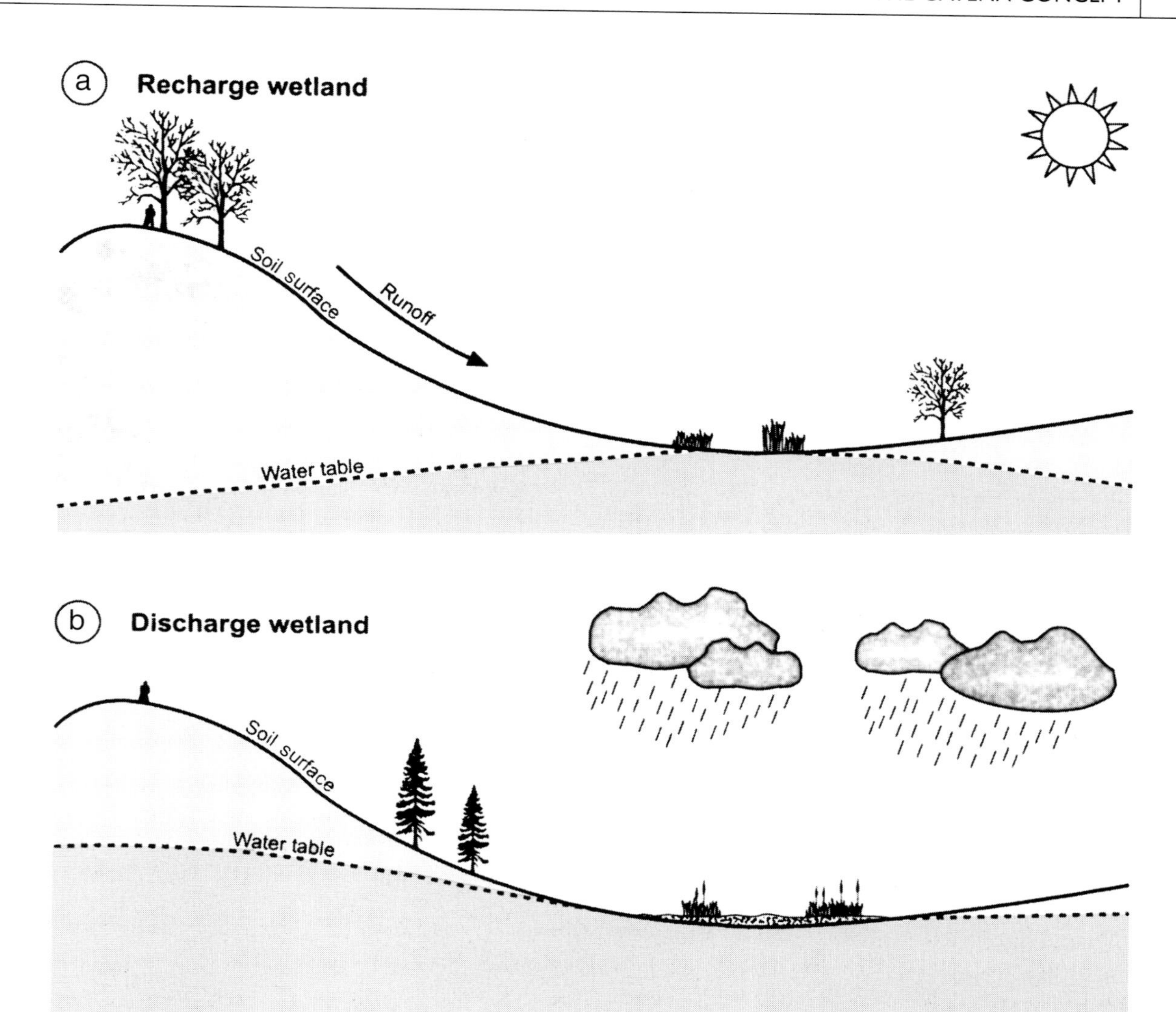

Fig. 13.20 Simplified diagram of the relationships among landscape and water table for a recharge wetland (a) and a discharge wetland (b).

13.20). The water table is mounded under topographic highs, and is lowest (in elevation) but closest to the surface beneath depressions. Water is thus moved from recharge (upland) areas to the discharge wetlands, keeping them wet even during climatically dry intervals. Capillary flow of water and dissolved substances from the shallow water table can, in such settings, lead to the formation of saline, sodic or carbonaceous soils in these low spots (Richardson and Bigler 1984, Miller *et al.* 1985, Mausbach and Richardson 1994) (Fig. 13.20). In cool, humid climates, such depressions may develop into Histosols while upland mineral soils are leached.

Conversely, depressions on uplands, where the water table is deep, may be sites of focussed infiltration; soils there may exhibit increased leaching, more organic matter production and stronger soil development (Runge 1973, Miller *et al.* 1985) (Fig. 13.21). Depressional soils such as these may be the most leached soils on the landscape. Many of these areas are *recharge* sites where surface water infiltrates and becomes groundwater. In both cases, topography facilitates the additional influx of water, and the depth to the water table determines the fate of the additional, site-focussed infiltration.

Landscapes: pocosins of North Carolina

In parts of humid, eastern North Carolina, flat upland soils are often very wet. Precipitation exceeds evapotranspiration requiring that the excess water must either infiltrate or run off (Daniels *et al.* 1977). Runoff is minimal and lateral subsurface flow is slow because the landscape is so flat. Under such conditions, as much as 2 m of organic soils develop on the wettest, flattest uplands (Daniels *et al.* 1999). These organic accumulations are, locally, called *pocosins* ("swamp-on-a-hill") (see Figure below). The accumulation of organic materials in the pocosins is aided by slow decomposition under waterlogged conditions, due to the low relief and the great distances between streams. Under the thermic soil temperature regime of North Carolina, where evapotranspiration levels are high, the development of organic soils on uplands is remarkable.

Pocosin peat blankets the broad, flat interstream divide parts of the landscape; they are not in depressions as are so many other Histosols worldwide. Rather, pocosins develop into large, low domes of organic material. The sapric organic materials generally are all younger than 10 000 years old, suggesting that they started to develop as sea levels rose at the end of the last glaciation (Whitehead 1972, Daniels *et al.* 1977). Sediments beneath the pocosins may range from sand to clay; sandy beach ridges occasionally poke through the pocosins. The pocosins are not unlike the blanket peats found in parts of northern Europe, which are formed in part because of the cool climate. During dry periods, occasional wildfires will burn parts of the pocosin, leaving behind a depression which fills with water. Some of North Carolina's largest lakes have originated in this way, and are today rimmed by the remains of pocosin Histosols.

Similar peat accumulations occur in southeastern Wisconsin, for slightly different reasons. Here, the glacial stratigraphy has created flowing springs and seeps on flat but otherwise well-drained plains. Where wet enough, mounds of peat have formed above these seep areas. The peat and muck mounds can be over 1 ha in area and 4 m in thickness (Ciolkosz 1965).

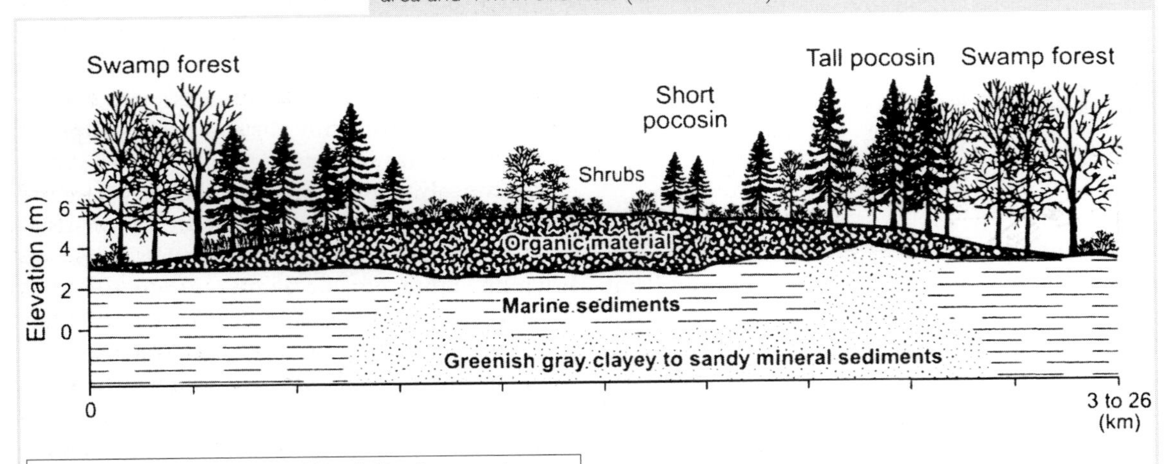

Cross-section through a typical North Carolina pocosin, illustrating the relationships among soils, topography and vegetation.

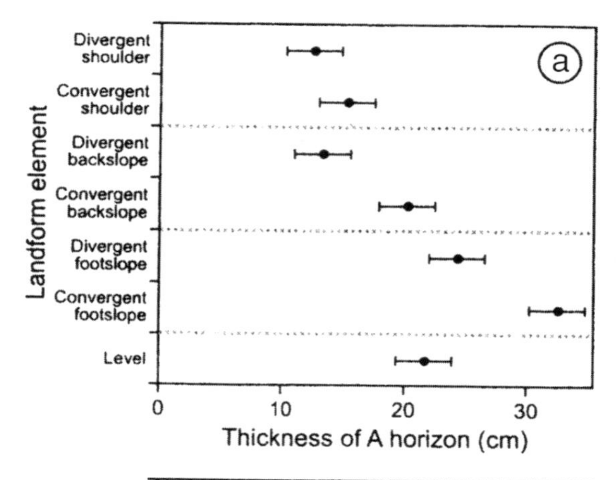

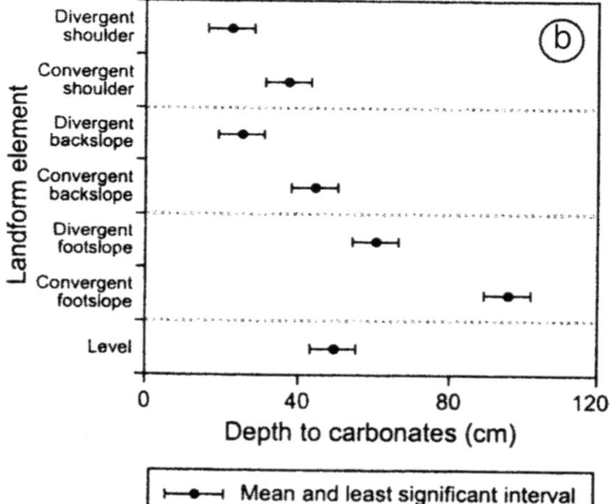

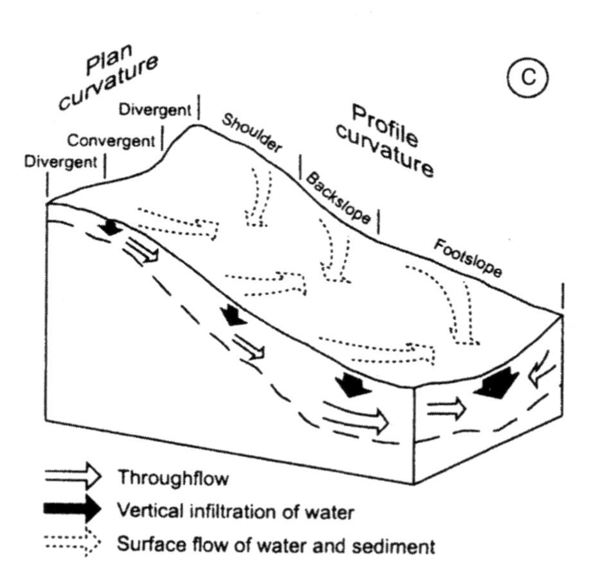

Fig. 13.21 Diagram illustrating water flow pathways on a slope and how those pathways impact soil development. After Pennock *et al.* (1987). (a) Thickness of the A horizon on various types of slope positions. (b) Depth to carbonates (a measure of leaching effectiveness) on various slope positions. (c) Block diagram illustrating the various slope positions and how water moves along and within the slope.

One of the most important components of the soil water regime is the depth to the zone of saturation (water table), how this depth changes through time and the oxygen status of the groundwater. Horizons are considered saturated when the soil water pressure is zero or positive (Soil Survey Staff 1999). Soil water held under negative pressure will generally not "flow" as saturated flow out of a soil pore (see Chapter 5). In saturated situations, "free" water will flow out of the soil into an auger hole or pit and stabilize after a period of time (Vepraskas 1999). The eventual elevation of the water that fills the hole is, for most non-clayey soils, equivalent to the water table. Water that is wicked upward into the soil, a few centimeters or more above the water table, is referred to as the *capillary fringe* (Gillham 1984). Soluble materials in the groundwater can be moved upward in solution, into the capillary fringe and deposited in the soil.

Depth to the water table is a function of several factors: (1) landscape position, which affects runoff vs. run-on tendencies (most important), (2) precipitation (temporal patterns and amounts), (3) evapotranspiration and evaporation, and (4) permeability in the surface and subsurface, especially with regard to slowly or rapidly permeable layers in the subsoil (Gile 1958, Fritton and Olson 1972, Khan and Fenton 1994). In highly permeable soils, the water table can quickly respond to precipitation (Hyde and Ford 1989), while in less permeable soils more water will run off and if there is a water table response it may have a significant lag. Upslope areas tend to be recharge areas where meteoric water percolates into the saturated zone. In lowland areas where the water table intercepts the surface, soil and surface water is discharged as water flows up, out of the soil (Figs. 13.19, 13.20). Discharge can also occur on sideslopes where a perched zone of

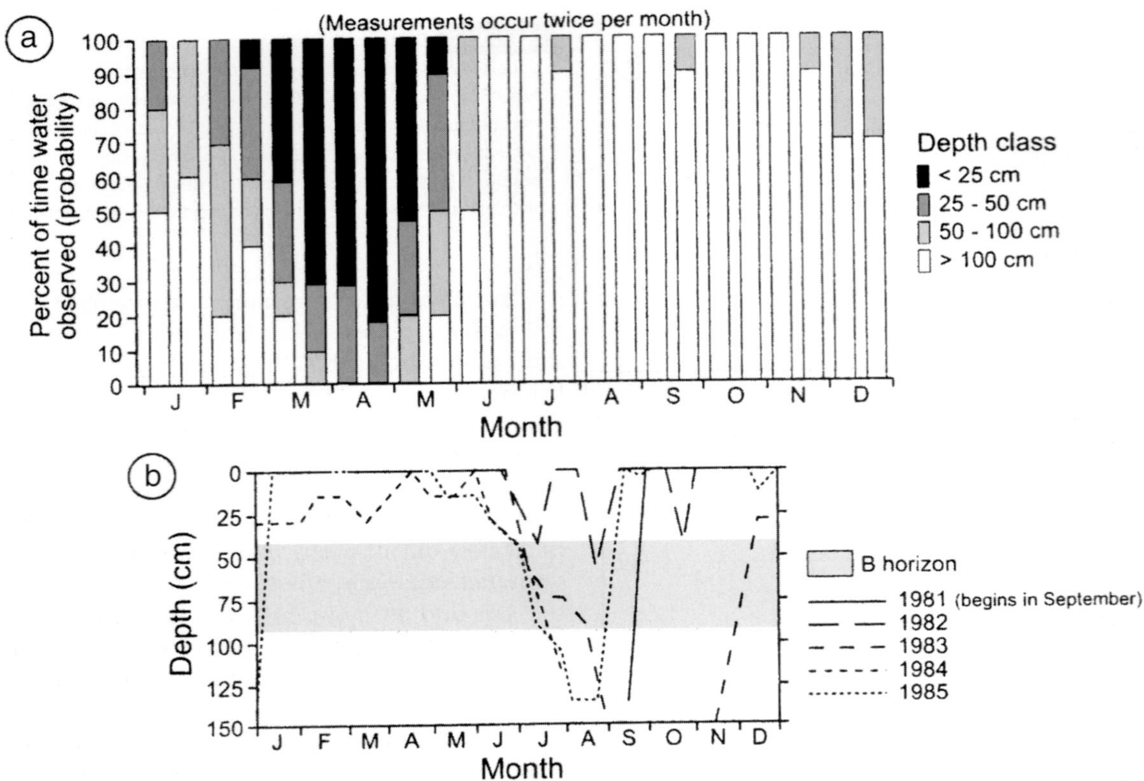

Fig. 13.22 Two methods of displaying information about water table variability in a soil, over a period of time. (a) Water table probability diagram for an Aeric Haplaquept in central Ohio. After Zobeck and Ritchie (1984). (b) Depths to the water table for a Typic Argiaquoll in southern Michigan, over a 5-year span of time. After Cremeens and Mokma (1986).

saturation intercepts the surface. Discharge areas can be determined by direct observation, but also by sampling the soil for soluble products deposited from the groundwater. Khan and Fenton (1994), for example, noted that calcite : dolomite ratios were higher in discharge areas near the bases of slopes, due to deposition of secondary carbonates from shallow groundwater (see also Knuteson *et al.* 1989).

Water tables fluctuate from year to year, and these fluctuations are usually greatest on higher landscape positions (Khan and Fenton 1994). Intra-annually, water tables are usually highest in spring or after a cool and wet season, and lowest after a warm, dry period (Khan and Fenton 1994). Thus, measuring the depth to the water table at any one time may give only a partial picture of the soil water regime. A better way to depict the water table level and variability within a soil is to produce data or diagrams of its change in depth over time, preferably including data on oxygen status. Many people refer to this as "depth and duration" data, although knowing if the water table is apparent or perched is also useful (Fig. 13.22).

Measuring water table depths, while potentially difficult, is done by coring into the soil and allowing water to flow into this shallow well called a *piezometer* (Guthrie and Hajek 1979, Mokma and Cremeens 1991). Usually the well is lined with porous tubing to prevent the walls from caving in but still allowing water to pass through (Zobeck and Ritchie 1984). An alternative method of water table monitoring involves neutron probe technology (Schaetzl *et al.* 1989a, Khan and Fenton 1994).

Soils are saturated below the water table. Such soils are said to exhibit *aquic conditions*, i.e., they undergo continuous or periodic saturation and reduction of Fe (Soil Survey Staff 1999, Vepraskas

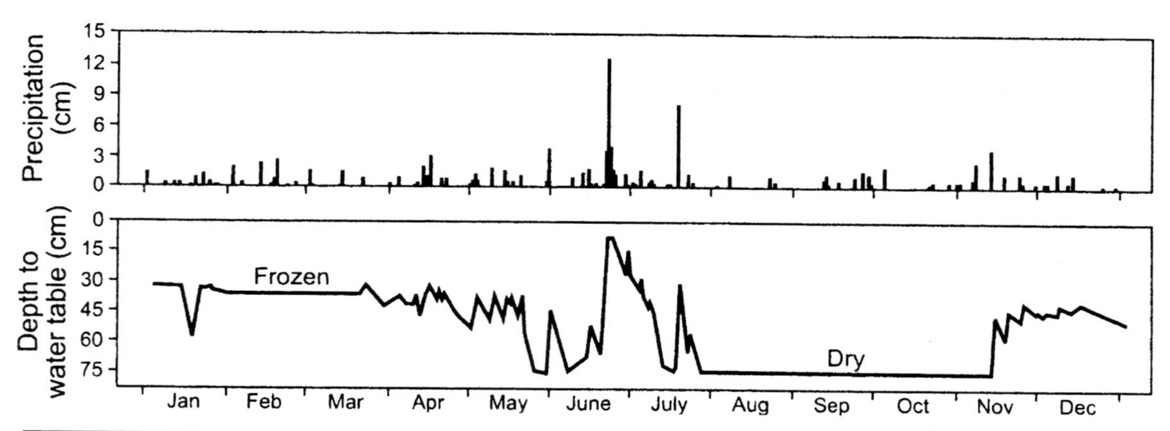

Fig. 13.23 A yearly cycle of perched water in an Aquic Fragiudalf, as affected by precipitation. Each incidence of water table presence, i.e., when the chart does not show "dry," indicates a perched water table above the fragipan. After Palkovics *et al.* (1975).

1999). The presence of these conditions is indicated by patches of red, gray, blue-gray, brown and other colors, called mottles or redoximorphic features (Vepraskas 1999) (see below). The Soil Survey Staff (1999) defines three types of aquic conditions:

(1) *Endosaturation.* In endosaturation, the soil is saturated in all layers from the upper boundary of the water table to a depth of ≥200 cm. Generally, endosaturation implies saturation below a regional or at least local water table, and the saturated zone continues to some depth. The top of this zone is referred to as an *apparent water table*, defined as the level at which water stands in a freshly dug, unlined borehole, given adequate time for adjustment (Fritton and Olson 1972, Mackintosh and van der Hulst 1978, Hyde and Ford 1989). An example of a soil with periodic endosaturation is a Mollic Endoaqualf. Endosaturation is a slow-to-change feature. Apparent water tables respond slowly to recharge or drawdown (Jacobs *et al.* 2002). Conversely, perched zones of saturation (see below) are more ephemeral features and respond more quickly but are also "lost" more quickly because they do not actually retain large amounts of water.

(2) *Episaturation.* In episaturation, the soil is saturated in one or more layers, but it *also* has one or more *unsaturated layers* below, within 200 cm. The water table is *perched* on top of a relatively impermeable layer, below which the soil is unsaturated. Episaturation is more correctly described as a *perched zone of saturation*, rather than a perched water table. In many parts of Europe this condition is referred to as the pseudogley model. An example of a soil with periodic episaturation is a Mollic Epiaqualf.

Perched water can occur on top of soil (Bx or Bt) horizons or on inherited sedimentary layers like a clay lens in till or dense basal till (Fletcher and McDermott 1957, Gile 1958, Simonson and Boersma 1972, Palkovics *et al.* 1975, McDaniel and Falen 1994, McDaniel *et al.* 2001). Water may perch on dense C horizons, while the solum, with good structure, is more permeable and allows water to move more freely in lateral directions where the surface is sloping (Harlan and Franzmeier 1974, King and Franzmeier 1981). Perched water tends to develop for short periods of time after precipitation or snowmelt events, illustrating the quick response time that episaturation requires to establish itself, or to be eliminated (McDaniel *et al.* 2001) (Fig. 13.23). On sloping surfaces, episaturated water may flow along the top of an aquitard as throughflow (Gile 1958, Evans and Franzmeier 1986). Episaturation is conducive to the formation of mottles. However, because the perched water may not be short-lived, the soil horizons involved may be saturated with oxygenated water and low (≤2) chroma mottles may not form (Evans and Franzmeier 1986). This is especially likely if the horizons are saturated at a time when the soil temperature is below 5 °C, e.g., after snowmelt.

Clothier *et al.* (1978) described a situation that is analogous to, but slightly different from, perched water. Infiltrating water will tend to "hang" at a lithologic discontinuity where finer-textured materials overlie coarse sediments (see Chapter 12). Mottles may then form in the overlying, fine material even though it may never be technically saturated.

(3) *Anthric saturation.* This special, human-induced aquic condition occurs in soils that are cultivated and irrigated, especially by flood irrigation. Examples of anthric saturation would be rice paddies, cranberry bogs and treatment wetlands.

Natural drainage classes

In order to better assist land managers with use and management decisions, the NCSS developed the concept of natural soil drainage classes. These classes refer to the frequency and duration of wet periods, and generally correlate to water table depths, as inferred from morphologic indicators such as mottles and gleying (Table 13.2, Fig. 13.24). Drainage classes are defined for soils under relatively undisturbed conditions similar to those under which the soil developed (Soil Survey Division Staff 1993). Alteration of the water regime by humans, either through drainage or irrigation, is not usually a consideration. The use of natural drainage classes in describing soils is tailored more to land and soil managers than to pedologists, as the classes are rather loosely defined and difficult to quantify.

Natural drainage classes usually vary with topography (Mackintosh and van der Hulst 1978). Exceptions occur where perched water may cause a soil to be in a wetter drainage class than would otherwise be indicated by topography, or where very fine-textured parent materials force water to permeate so slowly that increased wetness results. But, in general, water tables get nearer the surface in lower landscape positions, and natural drainage classes reflect this. Natural drainage classes are loosely associated with Soil Taxonomy, allowing the user to estimate the soils's natural drainage class from its subgroup classification (Soil Survey Division Staff 1993).

Morphologic expressions of wetness in soils

Because it is expensive and time-consuming to monitor water table depths, only rarely do we have actual data on water tables. Additionally, short-term water table data – the type that usually exist – may not be representative of long-term conditions (Zobeck and Ritchie 1984, Hyde and Ford 1989). Studies that report on water table fluctuations over a significant period of time, e.g., Pickering and Veneman (1984), Zobeck and Ritchie (1984), Hyde and Ford (1989), Mokma and Cremeens (1991) and Khan and Fenton (1994), are the exception rather than the rule. Usually we have but one observation – the one we are seeing (or not) at the moment! Therefore, our best recompense is to use *morphologic indicators* as proxy evidence about the water table (Boersma *et al.* 1972). These color-based *redoximorphic features* correlate to redox processes in soils, developing when Fe and Mn oxides are chemically reduced, oxidized and translocated. Gray, olive or pale colors (chroma ≤ 2) suggest that most of the iron in the soil is reduced, indicating wetness or saturation (although low chromas can also be caused by organic matter). Ferrous iron is essentially colorless and thus a soil dominated by it tends to take on the gray-white color of quartz, which is often the dominant soil mineral. Red or brown hues suggest that most of the iron-bearing minerals are oxidized as Fe^{3+} (Table 2.2). Oxidized iron occurs in soils where oxygen is not limiting, usually because the soils are unsaturated, thereby allowing oxygen to diffuse in from the atmosphere (Pickering and Veneman 1984). Mixtures of colors, often referred to as mottles, suggest alternating conditions of wetness and dryness. Our focus here is on the morphological manisfestations of oxidation–reduction processes in soils.

Before we enter into a discussion of how soil morphology can be used to infer soil hydrology, a few notes of caution must be mentioned. Some pedo-morphologic features may be relict, in which case they provide information on a *former* water regime (Vepraskas and Wilding 1983a, b). In soil horizons low in organic matter or those

Table 13.2	Descriptions of the USDA–NRCS designated natural drainage classes of soil		
Natural drainage class	**Characteristics[a]**	**Normal water table depth (cm)**	**Typical locations of mottles, gleying, and redoximorphic features**
Excessively drained	Water is removed from the soil very rapidly. Soil is commonly very coarse textured or rocky.	>100	None in profile
Somewhat excessively drained	Similar to excessively drained soils, but the water table may not be as deep and the soil may be slightly finer-textured.	>100	None in profile
Well drained	Water is removed from the soil readily but not rapidly. Water is available to plants throughout most of the growing season in humid regions. Wetness does not inhibit growth of roots for most or all of the growing season.	≈ 100	Mottles in C or BC horizon
Moderately well drained	Water is removed from the soil somewhat slowly. Soil is wet for only a short time within the rooting zone during the growing season, but long enough that most mesophytic crops are affected. These soils commonly have a slowly pervious layer within the upper 1 m, periodically receive high rainfall, or both.	75–100	Mottles in lower or middle B horizon, and in C horizon
Somewhat poorly drained ("imperfectly drained" in older and some non-US publications)	The soil is wet at a shallow depth for significant periods during the growing season. Wetness restricts the growth of mesophytic crops unless artificial drainage is provided. The soils commonly have (1) a slowly pervious layer, (2) a high water table, (3) additional water from seepage and/or (4) nearly continuous rainfall.	30–75	Mottles in upper B horizon; C and lower B horizons are often gleyed
Poorly drained	Water is removed so slowly that the soil is wet at shallow depths, sometimes for long periods. Water table is persistently shallow, such that most mesophytic crops cannot be grown unless the soil is artificially drained. The shallow water table is commonly the result of a slowly pervious layer, nearly continuous rainfall, or a combination of these.	<30	Mottles throughout profile; soil is gleyed in the upper B and lower horizons
Very poorly drained	Similar to poorly drained soils except that the soils are commonly level or depressed and frequently ponded. Thick O horizons are typical.	At surface or <15	Entire profile has mottles and can be gleyed

[a] Conditions in this table apply to soils that have not been artificially drained.

Source: Soil Survey Division Staff (1993).

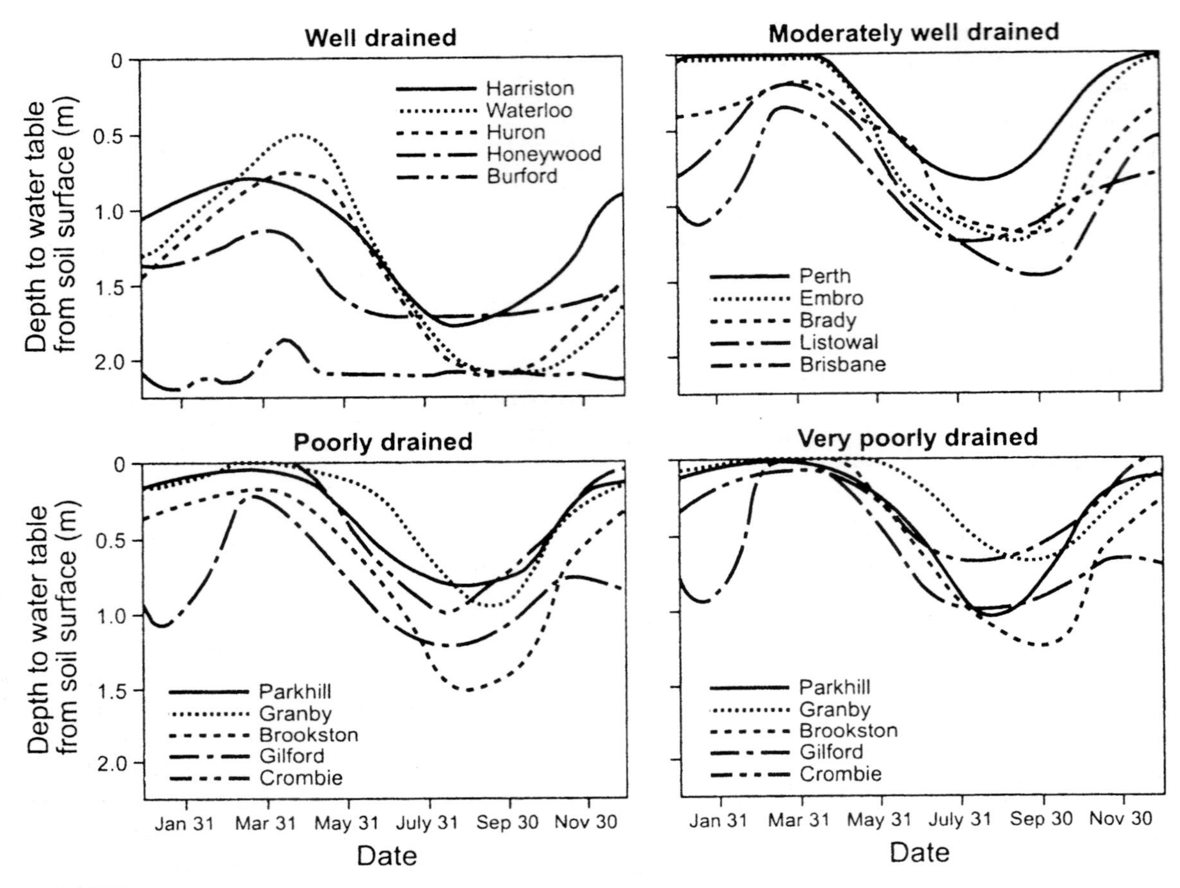

Fig. 13.24 Water table depths in four groups of soils, each group being a different drainage class. Observe the annual water table cycle, as well as the variability within and between drainage classes. After Mackintosh and van der Hulst (1978).

that are cold, reduction in saturated soils may not occur due to low biological activity (Couto *et al.* 1985). Soils formed from red parent materials may appear drier than they actually are, due to the inherited color (Sprecher and Mokma 1989). Mottle patterns also may not be readily observable in sandy Spodosols, because colors imparted by eluviation–illuviation processes overwhelm them (Hyde and Ford 1989, Mokma and Sprecher 1994). Nonetheless, many people continue to accurately judge soil hydrology and water table depths based on mophologic indicators, especially soil color (Evans and Franzmeier 1986, Mausbach and Richardson 1994, Hayes and Vepraskas 2000).

Small, gray, reduced areas in soils have recently come to be referred to as *redox depletions*, reflective of their lower iron contents (Vepraskas 1999). Redox depletions are localized areas of decreased pigmentation that are grayer, lighter or less red than the adjacent matrix (Schoeneberger *et al.* 1998). In older literature, small redox depletions were referred to as gray mottles or low-chroma (≤ 2) mottles. We still use the term mottles to indicate variegated patterns of soil colors due to redox depletions and concentrations, while acknowledging that "redoximorphic features" is now the preferred term (Schoeneberger *et al.* 1998, Vepraskas 1999). If a soil horizon has $>50\%$ low-chroma colors it is referred to as *gleyed*. *Hues* of 2.5Y or 5Y are also good indicators of saturated conditions (Simonson and Boersma 1972), though not necessarily of the presence of ferrous iron (Daniels *et al.* 1961).

In contrast to redox depletions, areas of concentrations of oxidized iron and manganese are referred to as *redox or iron concentrations* (red mottles or high-chroma mottles in older literature). Reduced Fe moves with the soil solution until it comes to an oxidizing location, where it gets reoxidized and rendered immobile, usually converting to ferrihydrite, lepidocrocite or ferric hydroxide. The color of oxidized iron concentrations can be related, in a very general sense, to the types of oxide minerals present (Table 2.3). Over time, certain microsites may become enriched in iron in this manner, relative to surrounding areas with lower chromas and grayer hues (Richardson and Hole 1979).

Redox concentrations can take on several forms within the soil: nodules, concentrations, masses and pore/ped linings (Schwertmann and Fanning 1976, Vepraskas 1999). Pore and ped linings are most readily observed, since masses and concretions are commonly within peds. Fe concentrations tend to be reddish or brown, while Fe–Mn concentrations tend to be black or very dark red. Concentrations of Fe–Mn usually occur as small masses or concretions that range in size from small flecks to pea-size particles; they can be distinguished from Fe masses/concretions chemically. Concretions of Mn will effervesce upon exposure to weak H_2O_2, while pure iron masses will not. Concentrations of Fe tend to be primarily goethite while the Mn mineral is birnesite (Schwertmann and Fanning 1976). They are indicators of saturated conditions and poor drainage (Simonson and Boersma 1972), being most abundant near the depth of maximum length of saturation where wide Eh fluctuations also occur. Wetter soils will have more and larger concretions, and they will be located throughout more of the profile, than will drier soils (Simonson and Boersma 1972).

Soil color patterns vs. soil hydrology

Patterns of soil color, mottles and redox features provide a great deal of information about soil hydrology (Harlan and Franzmeier 1974). Indeed, knowing how to interpret these patterns is vital to the correct interpretation of hydric soils. The most common pedogenic features used to infer wetness are (1) low (≤2) chroma mottles and matrix colors, (2) gray ped and channel coatings (albans or albic neoskeletans), (3) sesquioxide nodules or concretions and (4) high-chroma mottles in ped interiors.

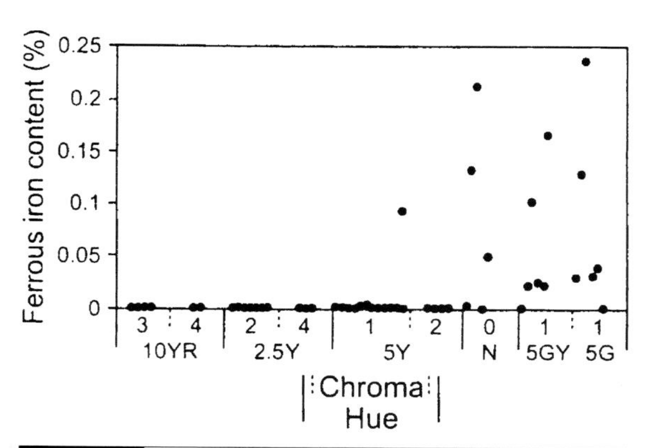

Fig. 13.25 Relationship between Munsell hue and content of Fe^{2+} in sediments from Iowa. After Daniels *et al.* (1961).

FULLY OXIDIZING CONDITIONS

In porous soils that have a deep water table, matrix colors indicative of oxidizing conditions prevail. The entire soil matrix is red, brown or yellow, with hues of 10YR (brown), 7.5YR (brownish-red), 5YR (red) or nearly so, depending in large part on the type and concentration of iron minerals (Table 2.3). These soils are well drained or excessively drained (Table 13.2). If they have a high water table for brief periods of time, it probably coincides with a cold season.

FULLY REDUCING CONDITIONS

Wet, organic-rich soils with high water tables (endosaturation) are usually gleyed with a gray, olive-gray or blue-gray soil matrix in which chromas of ≤2 dominate (Franzmeier *et al.* 1983). Matrix hues are typically 2.5Y or 5Y, or even GY. These soil materials are dominated by ferrous iron compounds (Fig. 13.25). In fully reducing conditions, wetness prevails almost to the surface, and any colors indicative of oxidation are in the upper profile and are masked by organs. The soil may or may not have a thick O horizon, depending on whether water ponds at the surface. At great depth, below the solum, these soils may show browner hues despite their saturated status because the demand for oxygen there is low (Franzmeier *et al.* 1983).

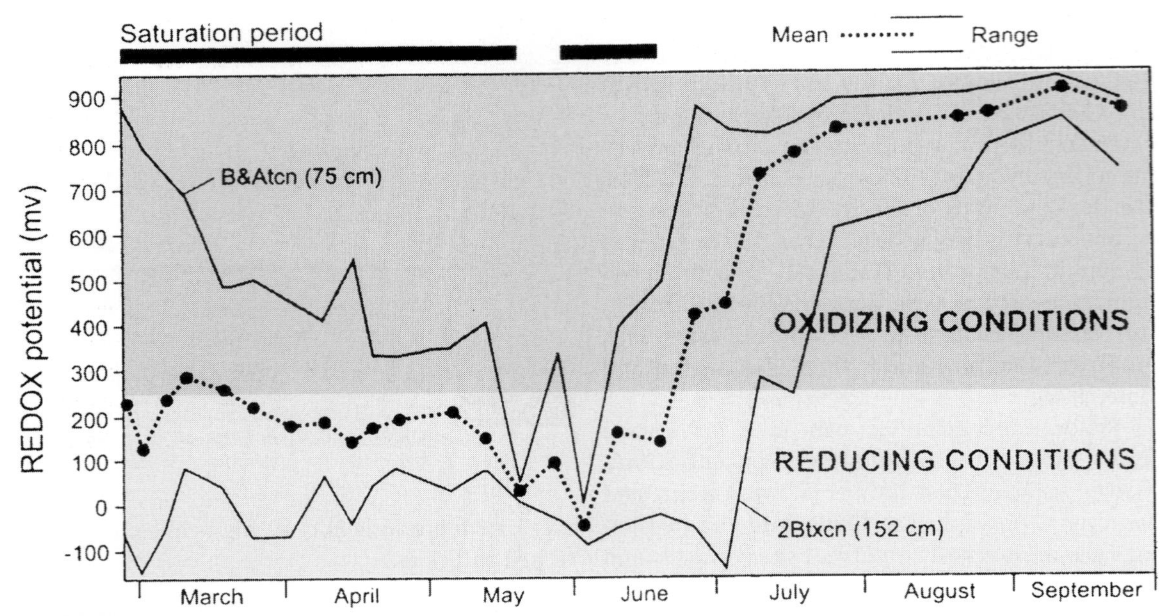

Fig. 13.26 Redox potentials during part of 1979 for two horizons of an Aeric Glossaqualf in southern Texas. After Vepraskas and Wilding (1983b).

FLUCTUATING WATER TABLE

Horizons with a fluctuating water table cross the reducing–oxidizing "threshold" periodically (Fig. 13.26). Redox depletions and concentrations occur in different parts of the horizon(s). For the profile, a typical sequence has gleyed horizons at depth, forming under reducing conditions, that change to mottled horizons nearer the surface and, depending on water table depth, to red-brown horizons near the surface.

It should now be apparent that, in soils where redoximorphic features are common, information about the location of the water table can be gleaned from the patterns of the redox depletions and concentrations vs. the soil matrix. The matrix colors are the colors within the centers of peds; they reflect the long-term oxidative status of the soil. Gray ped interiors result from long-term, reducing conditions (Veneman *et al.* 1976), but during brief periods when the water table drops, iron can move from the ped interiors to the exteriors, where it oxidizes (Veneman and Bodine 1982). The oxidized iron on the ped faces and channels gives them a brown or red color, which may persist through the ensuing period of saturation. Similarly, peds with brown/red interiors may have bleached, gleyed or reduced faces in horizons that are normally oxidizing but which undergo brief periods of saturation. During saturation, oxygen is first consumed by microbes and roots in the voids between peds. Thus, the soil solution becomes reducing and iron and manganese on the faces of peds is reduced. Water carrying reduced Fe may penetrate the ped, wherein it encounters some oxygen. The Fe in the soil solution oxidizes, creating a red "halo" within the ped interior, while the ped face is gray or has gray mottles (Richardson and Hole 1979). This situation is typical of soil horizons that perch water after snowmelt or heavy rain.

Redoximorphic features

Morphologic features associated with redox processes are grouped under the general name "redoximorphic features" (Table 13.3). Coatings (concentrations) of oxidized iron form under reducing conditions, where water with ferrous iron diffuses out of a ped or the soil matrix and oxidizes on contact with the surface (Fig. 13.27b). The pore or ped face must be aerated and oxygenated, which can occur if it has a direct pathway to the surface which is not saturated with water. The reddish brown coating is called a *ferran*. Ferrans that coat channels, such as those formed by roots, are *channel ferrans* (Veneman *et al.* 1976, Veneman and Bodine 1982). *Ferro-mangans*

Table 13.3 Location and genesis of some redoximorphic features[a] in soils

Feature name(s)	Location within soil	Probable genesis
Ferrans, neoferrans, quasiferrans, mangans	On and adjacent to ped faces, channel exteriors	Reduced iron moves with soil solution, out of pore interior onto ped or channel surface, where it encounters an oxidizing environment. Fe oxidizes into Fe^{3+} compound (ferran) on ped surface.
Ferriargillans	Ped faces, channel exteriors	Similar process, but Fe^{3+} compounds coat ped faces that originally contain argillans.
Albans, neoalbans, gray mottles	Ped faces, channel exteriors	Iron and manganese compounds on ped or channel surfaces are reduced and removed by leaching, which may occur after heavy rain. Matrix may be oxidizing.
Albic neoskeletans, silt coatings, skeletans, grainy gray ped coatings	On and adjacent to ped faces, channel exteriors	Similar process, but Mn, Fe^{2+} and clay are eluviated from the surface.
Neopedferrans	Inside peds	As ped faces and channels get reduced upon wetting, forming albans reduced iron compounds move into ped interiors and oxidize into neopedferrans, which form a ring inside the albans.

[a] Exclusive of matrix features indicative of redox processes.

and *ferriargillans* are also possible in this type of microenvironment, although if both Fe and Mn are in solution, Fe tends to precipitate first. In this case, pore linings may consist of clearly separated Mn oxides in the macropore and Fe oxides in the ped matrix, indicating that Mn and Fe ions were diffusing from ped interiors toward the ped face. Modifiers such as "quasi" and "neo" are sometimes added to indicate the degree of development or the location of the ferran or mangan (Table 13.3).

Albans or *neoalbans* are whitish or gray ped coats that form where iron and manganese compounds on ped or channel surfaces have been *reduced* and *removed* by leaching, i.e., they are a form of iron depletion (Veneman *et al.* 1976) (Fig. 13.27a). They form in horizons that are only periodically saturated, and may be best expressed in finer-textured soils (Vepraskas and Wilding 1983a). Their outer boundary may be quite sharp.

The loss of Fe and Mn from ped exteriors may cause clay present there to disperse and become translocated to lower horizons, giving the Fe depletion a whitish appearance and sandy feel. Redox cycles can also lead to loss of clay from ped exteriors by ferrolysis (see Chapter 12). These eluvial features are called skeletans, neoskeletans, albic neoskeletans, albans and grainy gray ped coatings (Arnold and Riecken 1964, Vepraskas and Wilding 1983a, Ransom *et al.* 1987a) (Table 13.3). They are common in degrading argillic horizons. Albic neoskeletans should not be confused with silans (silt coats), which can form under fully oxidizing conditions. Formation of albic neoskeletans occurs during saturation, but is facilitated after heavy rain events, as the areas between peds get saturated by infiltrating water and, due to high summer temperatures and biological activity, the areas near pores and channels become anaerobic. Roots supply the soluble organic matter needed for Fe reduction and subsequent solubilization, while the rapidly infiltrating water has the ability to translocate the Fe and clay (Vepraskas and Wilding 1983a). Skeletans are so closely tied to reducing conditions that their abundance is related to the length of time the soil is reduced (Vepraskas and Wilding 1983a).

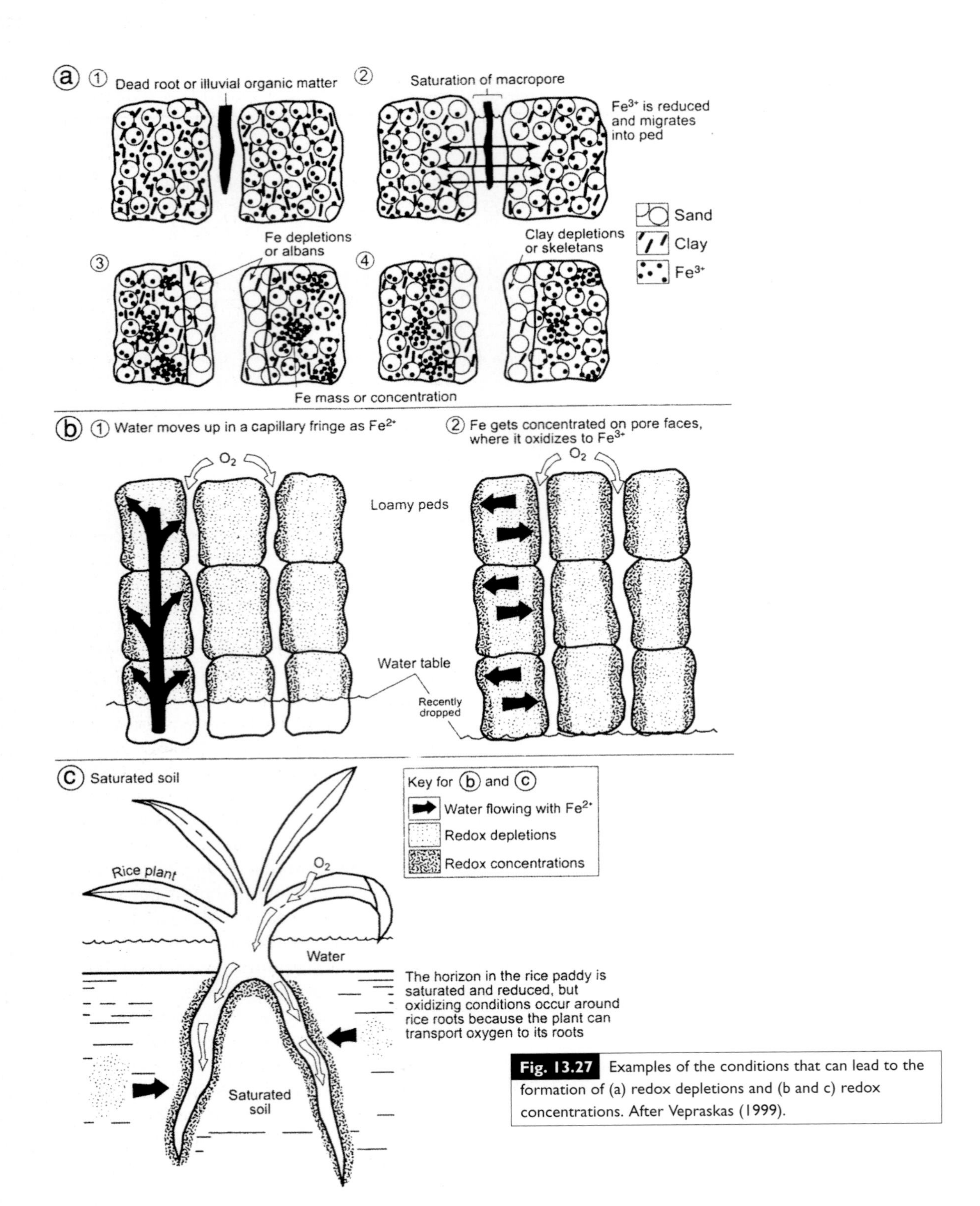

Fig. 13.27 Examples of the conditions that can lead to the formation of (a) redox depletions and (b and c) redox concentrations. After Vepraskas (1999).

The ped matrix itself may be oxidizing and stay oxidized, and may accumulate Fe as it diffuses inward. In extreme cases, the amount of oxidized iron inside the ped can increase to the point where it becomes nodule-like, a feature that Veneman and Bodine (1982) referred to as a sesquioxidic nodule. Often, these concretions or nodules occur in the centers of the peds, the locus of inwardly moving iron compounds (Richardson and Hole 1979). Because Fe depletions form slowly (Vepraskas 1999), thick albans are reflective of stable pores and ped faces which have been used by many different generations of roots.

Sometimes, gray argillans can be observed in lower horizons, reflecting clay and reduced Fe that have been translocated from the skeletans. Here, it diffuses into and is deposited onto peds, where it forms, via oxidation by "trapped" air within the peds, *neopedferrans* or Fe–Mn masses within the ped (Pickering and Veneman 1984) (Fig. 13.27a). Thus, the albans or Fe depletions grow from the ped surface inward, as a result of eluviation of Fe and Mn (Vepraskas and Wilding 1983a).

In soils that are at least periodically saturated and which have Fe and Mn contents above some minimum threshold, Fe–Mn nodules can form in zones of water table fluctuation (Richardson and Hole 1979, Vepraskas and Wilding 1983a). Nodule size and abundance generally increase with increasing duration of saturation. The Fe in these nodules or concretions generally has a low degree of crystallinity.

Landscapes: the edge effect

On many low-relief, slightly dissected landscapes, shoulder slopes are dry microsites (Daniels *et al.* 1971a). This is especially applicable to stepped landscapes and flat landscapes that drop down onto, for various reasons, a steep escarpment. Daniels and Gamble (1967) referred to this type of site as the landscape "edge." If the escarpment drops off to a drier landscape, it is called a dry edge or red edge (see Figure). Escarpments that fall off into a wet depression are wet edges. Based on their work on Paleudults in North Carolina, Daniels and Gamble (1967) described how soils on convex edges differed from other soils. They pointed out that soils change the most, pedon to pedon, at edges. Much of this rapid change has to do with the way the water table is influenced by the edge (Daniels *et al.* 1971b). At a dry edge water tables are generally deeper than farther inland, and the amount of water table variation throughout the year is greater. Dry edge soils are redder, especially the B horizons, where water tables are deeper (Daniels *et al.* 1971b). Farther from the edge, soils are grayer, with mottles and redoximorphic features higher in the profile. Changes in soil color can occur over as little as a few meters. The mottled, gray soils far from the edge are lower in Fe, presumably because they have lost reduced iron in solution. Soils at a dry edge also have more clay and more contrast between E and Bt horizon clay contents. Translocation of clay is clearly a more dominant process in edge soils, perhaps because the dry edge experiences more wet–dry cycles per unit time than do other parts of the landscape. Pedons away from the edge do not show strong E-Bt contrast; rather, they have thin E horizons, a glossic-like transitional horizon and a shallow, sandier B horizon.

At wet edges, soils predictably change in opposite ways to dry edge soils. Wet soils immediately off (downslope from) the wet edge are gray and low in iron, and have lost much of their clay to weathering and translocation (Daniels and Gamble 1967). Sandy textures are common. Strong fragipans have developed at some wet edges in North Carolina.

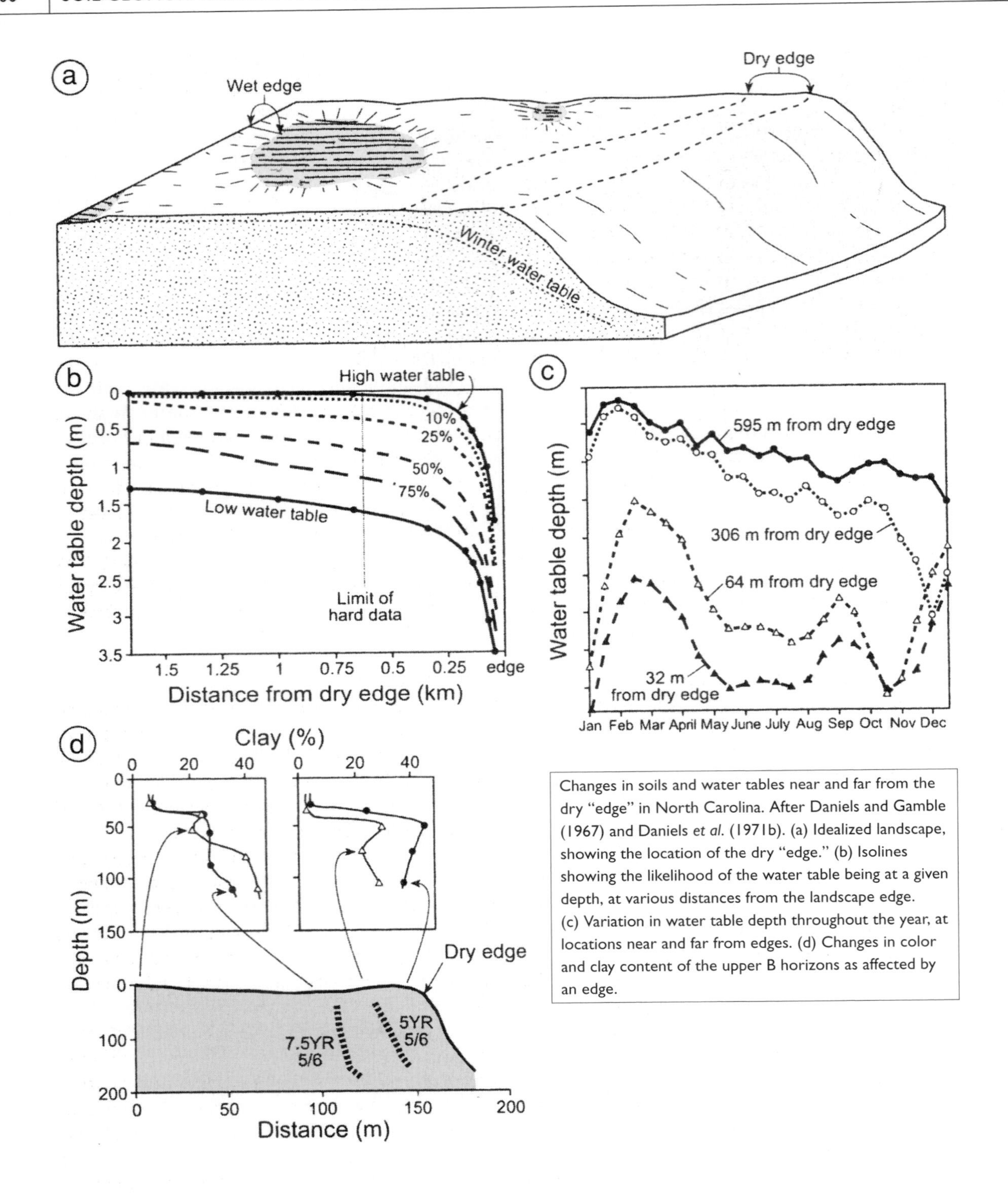

Changes in soils and water tables near and far from the dry "edge" in North Carolina. After Daniels and Gamble (1967) and Daniels *et al.* (1971b). (a) Idealized landscape, showing the location of the dry "edge." (b) Isolines showing the likelihood of the water table being at a given depth, at various distances from the landscape edge. (c) Variation in water table depth throughout the year, at locations near and far from edges. (d) Changes in color and clay content of the upper B horizons as affected by an edge.

Fig. 13.28 Pit-and-mound microtopography, formed by tree uprooting, on a once-forested but now pastured slope in northern Michigan. Photo by RJS.

Quantification of wetness

Redox patterns can provide a great deal of information about soil hydrology (Franzmeier *et al.* 1983, Mokma and Cremeens 1991). Interpreting redox patterns is also critical to soil classification, for on-site waste disposal or when delineating jurisdictional wetlands. Soils with aquic soil moisture regimes must be saturated and virtually free of dissolved oxygen for at least a few days each year (Soil Survey Staff 1999). Although color patterns related to wetness do appear to change after drainage by, for example, ditches (Hayes and Vepraskas 2000), most soils retain their color patterns for some time. Sandy soils, with lower surface areas, change most quickly in response to altered hydrological status (Jacobs *et al.* 2002).

In Indiana, Franzmeier *et al.* (1983) were able to quantify the following soil morphology–soil water relationships: (1) gleyed horizons, with matrix or argillan colors that are dominantly gray (chroma $\leq$2), are saturated most of the year, (2) horizons with a dominantly brown matrix but which also have gray mottles are saturated a few months of the year when they occur above a gleyed horizon, but are saturated most of the time if they lie below a gleyed horizon, (3) horizons with a chroma of 5–6 but which lack mottles with chromas $\leq$3 are seldom saturated. They reiterate the point that soils with a matrix chroma $\geq$5 are never saturated during the growing season unless they are close to gleyed horizons. Jacobs *et al.* (2002) were able to quantify water table relationships in some Georgia Ultisols, although they cautioned that the relationships are not valid in soils that have reticulately mottled plinthite. Soil horizons with a gray ($\leq$2 chroma) matrix are saturated >50% of the time. Low-chroma Fe depletions can form in horizons that are saturated as little as 18% of the time. Likewise, high-chroma Fe concentrations (red mottles) have formed in horizons saturated about 25% of the time, but they point out that these types of features can develop in the capillary fringe, which technically is not saturated.

Microrelief

Often overlooked, small-scale forms of topography are very important to soil development. Features with vertical or horizontal dimensions of less than a few meters could be considered microtopographic in scale. Mesotopographic features are 10 to a few tens of meters in size.

The origins of microtopographic highs and lows, swells and swales, humps and hollows, are myriad (Fig. 13.28). Tree uprooting, perhaps the most common microtopography-former in forested regions, creates pit-and-mound, or cradle knoll, microtopography (Lyford and MacLean 1966, Hamann 1984, Schaetzl *et al.* 1989b, 1990). Argilliturbation forms gilgai microrelief (see Chapter 10). Frost heave is responsible for various forms of microtopography in areas of permafrost (see Chapter 10). Ridge-and-swale microtopography is common on floodplains. Many glacial landscapes inherit hummocks of various sizes (Gracanin 1971, Attig and Clayton 1993, Johnson *et al.* 1995). Animal mounds, e.g., those of termites and ants, form a vast array of different microtopographic forms.

Fig. 13.29 Pit-and-mound microtopography, formed by tree uprooting, on a somewhat poorly drained soil in northern Michigan. The large amount of water that enters the system with the melting snow has brought the water table up so high that only the mounds are "dry." Photo by RJS.

The pedogenic impact of microtopography often depends on macroclimate and water table variables. Below microtopographic low sites in leaching regimes where the water table is high, profile differentiation may be hindered due to the high water table (Fig. 13.29). Soils on adjacent microtopographic highs, however, may be better developed. Conversely, field data indicate that, in humid climate soils where a high water table is *not* a consideration, soils in pits and small depressions are almost always better developed (Låg 1951, Denny and Goodlett 1956, Veneman *et al.* 1984, Miller *et al.* 1985, Schaetzl *et al.* 1990). Pits are more leached, and have thicker O horizons, better horizonation and many other attributes associated with progressive pedogenesis (Schaetzl 1990) because more water runs onto and through them (Table 13.4, Fig. 13.30). This application is in support of the Energy Model of Runge (1973) (see Chapter 11). As mentioned above, pits on sites with a high water table will not benefit from the extra potential energy associated with their lower site, and may be less developed than mounds. Contributing to soil development in freely draining pits are conditions that would tend to favor more runoff from surrounding sites, such as (1) a thick mat of broadleaf litter (Oi and Oe horizons), (2) slowly permeable soils due to clayey textures or frost and (3) steep slopes upslope from the microtopography. Finally, snowpacks may be thicker in pits, leading to more meltwater inputs (Schaetzl 1990) (Fig. 13.31). Another factor that can contribute to better-developed soils beneath pits is the thicker litter layers (O horizons), since forest litter tends to collect there and the cooler, moister conditions that prevail in pits inhibit decomposition (Armson and Fessenden 1973, Shubayeva and Karpachevskiy 1983, Schaetzl 1986b, Mueller *et al.* 1999) (Table 13.5). Litter is a source of organic acids and can promote soil development (see Chapter 12). The overthickened O horizons in pits also protect the soil from drying events and the extra moisture may facilitate weathering. Microtopography also affects soil temperature, which again impacts pedogenic processes and biotic communities that inhabit these microsites (Troedsson and Lyford 1973) (Table 13.4).

Microtopography is also important to pedogenesis in dry climates (Sharma *et al.* 1998). Small variations in microclimate may lead to significant differences in soil moisture, vegetation and soil development in grasslands and deserts, especially with respect to pedogenic properties that can be changed readily by small amounts of water. For example, White (1964) pointed out the differences between sodium-affected soils in depressions vs mounds. On the Coast Prairie of Texas, calcic horizons are restricted to microhighs while subtle depressions lack calcic horizons, due to capillary

Table 13.4 Comparison of soil characteristics between microtopographic locations formed by tree uprooting

Characteristic[a]	Mound	Level or undisturbed	Pit	References
Soil development or profile differentiation	Low		High	Moore 1974, Schaetzl 1990, Veneman *et al.* 1984
	Low	High		Denny and Goodlett 1956, Goodlett 1954
Winter temperature	Cool		Warm	Federer 1973, Schaetzl 1990
Spring temperature	Cool		Warm	Beatty 1984, Beatty and Stone 1986, Federer 1973
Summer temperature	Warm		Cool	Beatty 1984, Beatty and Stone 1986, Federer 1973
H_2O content	Low		High	Beatty 1984, Beatty and Sholes 1988, Beatty and Stone 1986, Lyford and MacLean 1966, Schaetzl 1990, Shubayeva and Karpachevskiy 1983
Saturated infiltration capacity	High	Low		Goodlett 1954, Lutz 1940
Pore volume	High	Low		Lutz 1940
Organic matter	Low		High	Beatty 1984, Beatty and Sholes 1988, Beatty and Stone 1986, Schaetzl 1990, Stone 1975
pH	High	Low		Lutz 1940
	Low		High	Beatty and Sholes 1988, Shubayeva and Karpachevskiy 1983
Cation exchange capacity	Low		High	Beatty 1984
Available nitrogen	Low		High	Beatty 1984
Calcium	High		Low	Stone 1975
	Low		High	Beatty 1984, Beatty and Stone 1986
Magnesium	High		Low	Stone 1975
Heavy mineral content	High	Low		Lutz 1940
Leaf litter accumulation	Low		High	Beatty and Sholes 1988, Beatty and Stone 1986, Hart *et al.* 1962, Schaetzl 1990, Stone 1975
O horizon thickness	Thin		Thick	Beatty 1984, Goodlett 1954, Hart *et al.* 1962, Lyford and MacLean 1966, Moore 1974, Schaetzl 1990, Shubayeva and Karpachevskiy 1983, Veneman *et al.* 1984
A horizon thickness	Thin		Thick	Beatty 1984, Beatty and Sholes 1988, Beatty and Stone 1986

(cont.)

Table 13.4 (*cont.*)

Characteristic[a]	Mound	Level or undisturbed	Pit	References
Texture	Coarse	Fine		Lutz 1940
Frost action	High		Low	Beatty 1984, Denny and Goodlett 1956, Goodlett 1954, Hart *et al.* 1962, Lutz 1940, Schaetzl 1990
Snow depth in midwinter	Thin		Thick	Beatty 1984, Beatty and Stone 1986, Federer 1973, Schaetzl 1990
Snow depth at snowmelt	Thin		Thick	Beatty 1984, Schaetzl 1990
Likelihood of being snow-free during the snowmelt period	High		Low	Beatty 1984, Schaetzl 1990

[a]"High," "Low," "Thick," "Thin," etc. indicate direction, not absolute magnitude, of variability.

Source: Schaetzl *et al.* (1990).

Fig. 13.30 Soil development in and below a small depression formed by tree uprooting. Spodosols are dominant in the sandy parent materials of this part of northern Michigan. The E horizon gets thicker as it approaches the pit and develops deep tongues immediately below the pit proper. Note also how the Bhs horizon is better developed in the pit. Photo by W. Kreznor.

rise of carbonate-rich water into the microhighs (Sobecki and Wilding 1982). Sobecki and Wilding (1983) proposed that carbonates are leached from microlows and redistributed to microhighs via lateral flow on top of a slowly permeable (3C) horizon (Fig. 13.32). Thus, moisture flow is driven by a hydraulic gradient that sets up between the dry knolls and the wetter depressions.

Many of the best and deepest Vertisols are in shallow depressions, which act as settling basins for clays eroded nearby, and where water can pond during the wet season (see Chapter 10). Ponding of water on these landscapes for extended periods of time further assists in smectite neoformation, and virtually assures complete wetting of the profile, accentuating the wet–dry seasonality of the site.

In sum, microtopography variously impacts all landscapes. It accelerates or decelerates soil development by redirecting dissolved and

Table 13.5 Properties of mounds, pits and undisturbed sites formed by uprooting in a beech-maple forest in New York state

Microsite	Soil moisture (%)	Summer soil temperature (°C at 10 cm)	Spring and fall soil temperature (°C at 10 cm)	Organic matter in upper profile (%)	Thickness of O horizon (cm)	Thickness of A horizon (cm)	Thickness of snowpack (cm)
Mound	20–35	10–14	3–4	5.7	1.2	2.7	35
Undisturbed	30–45	9–13	4–5	10.0	3.5	5.2	41
Pit	40–60	8–11	5–6	17.8	5.7	9.4	47

Source: Beatty and Stone (1986).

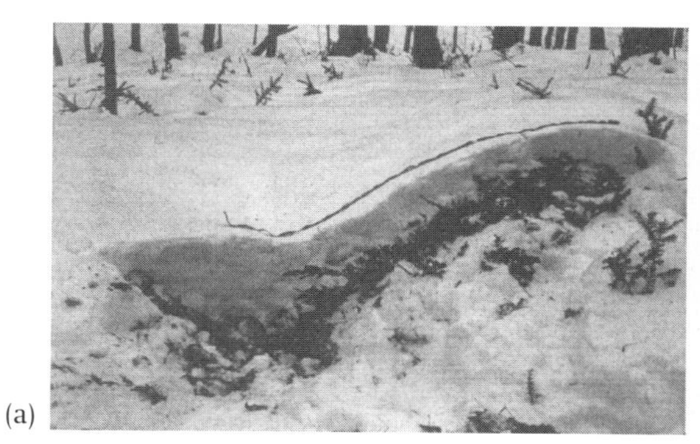

(a)

(b)

Fig. 13.31 Relationship between soil microtopography and snowpack thickness. The microtopography in both figures is due to tree uprooting. The snowpack is thicker in pits than on mounds, as shown in (a) and persists longer in pits, as shown in (b). Photos by RJS.

Fig. 13.32 Cross-section showing soil horizonation in a microtopographic high on the Texas Coast Prairie. After Sobecki and Wilding (1982).

suspended subtances in water toward or away from certain sites (Beatty and Stone 1986). Microtopography can also impact soils indirectly through its interaction with vegetation establishment patterns and productivity (Beatty 1984, Beatty and Sholes 1988). It is therefore vital to maintaining spatial heterogeneity in soil landscapes, which in turn is important to plant and animal biodiversity.

Examples of catenas

Wisconsin till plain, Iowa

In north–central Iowa and southern Minnesota, Mollisols have developed in calcareous, loam glacial till and other surficial sediments. Although the landscape was deglaciated about 14 000 years ago (Ruhe and Scholtes 1959), the soils have developed their characteristics primarily over the past 3000 years since grasslands have invaded the area (Van Zant 1979, Steinwand and Fenton 1995). The landscape is hummocky, with many basins of interior drainage. Much of this landscape is mapped within the Clarion–Nicollet–Webster catena (CNW) which occupies over 31 000 km^2 on the Des Moines glacial lobe (Steinwand and Fenton 1995) (Fig. 13.33). Soil variation is due to drainage (water table relations), carbonate status and parent material texture. Upland Hapludolls (Clarion) have developed in oxidized, leached, loam till (Fig. 13.34). Oxidizing conditions occur in upland soils, but in lowlands and at depth, soils are reduced.

In swales, postglacial sediments have accumulated above the till (Ruhe 1969). This sediment, which exhibits a fining upward sequence, typical of fluvially deposited materials, is thickest in the centers of swales and basins (Burras and Scholtes 1987). The lowermost material is sandy, and may have been deposited during the waning stages of ice retreat (Steinwand and Fenton 1995). Collectively, it may be best described as co-alluvium or slopewash. Many rolling glacial landscapes have similar sediments in swales, dating back to a period of landscape instability (Walker and Ruhe 1968, Pennock and Vreeken 1986). Finer sediments that lack coarse fragments overlie the sandy slope alluvium and form the parent material in the swales. Stone lines, in this case indicative of an erosional contact, are located at the discontinuity between the two materials (Burras and Scholtes 1987). Sediments get finer toward the swales, suggestive of sorting during downslope transport; finer sediments were transported into the swales while coarser sands remained on the slopes. The finer sediments are assumed to reflect a late Holocene period of landscape instability (Burras and Scholtes (1987).

Nicollet (Aquic Hapludolls) and Webster (Halpaquolls) soils have developed in various thicknesses of slopewash material (Fig. 13.33). In the wettest parts of the landscape, the Harps and Canisteo soils with Bkg horizons have formed where calcareous groundwater is discharged (Khan and Fenton 1994). The cumulic Okoboji soils (Cumulic Vertic Endoaquolls) form in wet areas where the silty alluvium/slopewash has been leached of carbonates.

Negev Desert, Israel

In this desert, like all deserts, soil development is slow and soil moisture determines, in large part, the pathways of pedogenesis. Annual precipitation in the Negev is less than 250 mm. Inputs of Saharan dust are present, but are quickly redistributed across the landscape. Much of it gets washed into lower slope positions, while the upper portions of slopes are rocky with thin soils (Kadmon *et al.* 1989). Thus, most catenas have a rocky upper portion and a lower portion which has thick accumulations of silty and sandy colluvium. In high-relief bedrock terrain, free faces in limestone and chalk occur on shoulder slopes (Fig. 13.35). Soils on these surfaces are thin and stony for two reasons: (1) the high Ca^{2+} contents keep pH values high and inhibit weathering and (2) the slopes are eroding and unstable. On similar landscapes in a Mediterranean climate one might find terra rossa soils (Dan *et al.* 1972). In bedrock pockets, however, soils may be more leached than almost anywhere else, since water will run off the scattered rock surfaces and become concentrated in these sites.

Despite the low precipitation amounts, lateral transport of sediment (dominated by loess and dust) downslope is a dominant factor in

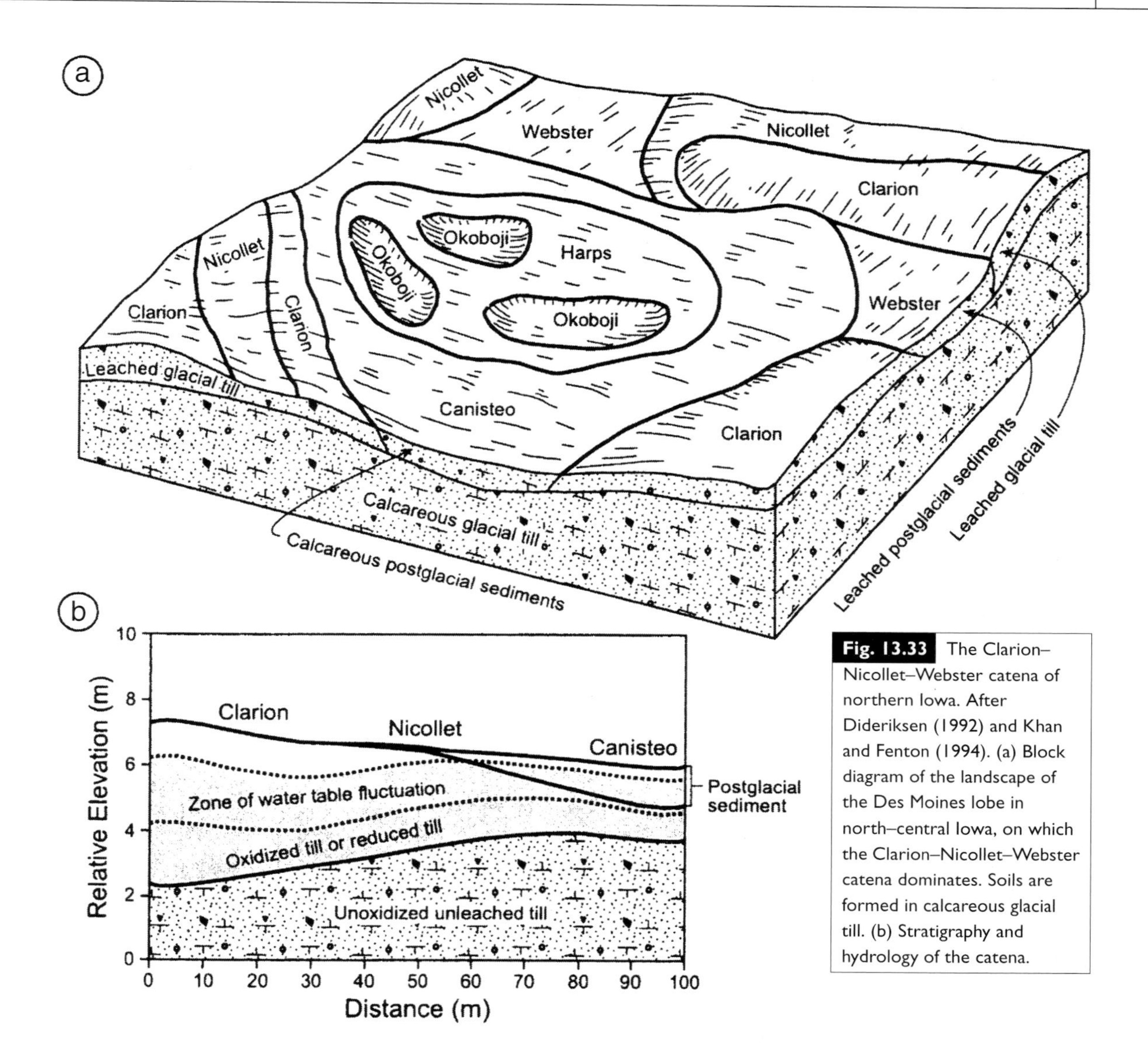

Fig. 13.33 The Clarion–Nicollet–Webster catena of northern Iowa. After Dideriksen (1992) and Khan and Fenton (1994). (a) Block diagram of the landscape of the Des Moines lobe in north–central Iowa, on which the Clarion–Nicollet–Webster catena dominates. Soils are formed in calcareous glacial till. (b) Stratigraphy and hydrology of the catena.

determining the soil and vegetation associations here (Yair and Danin 1980) (Fig. 13.36). Leaching intensity here is highly dependent on the amount of soil pore space and run-on, both of which are largely a function of surface bedrock exposures. Soils in the footslope and toeslope positions are developed in deep silts and sands that have washed down from upslope; upslope areas are rocky with thin soils. Water infiltrates more shallowly in the colluvium/alluvium here than it does on bedrock uplands, because the entire surface is permeable, there is more pore space than in rocky soils and there is essentially no runoff that impinges from upslope. Much of what infiltrates is evaporated later from the surface, leaving salts behind and making the soils drier than they might otherwise appear. Soils in this deep sediment have therefore developed calcic horizons overlying calcic + gypsic horizons (Fig. 13.37). Salts and carbonates are translocated more deeply and soils remain wetter for longer periods of time on the rocky upper slopes than on the alluvial flats at the bases of the slopes, because on the uplands more of the limited

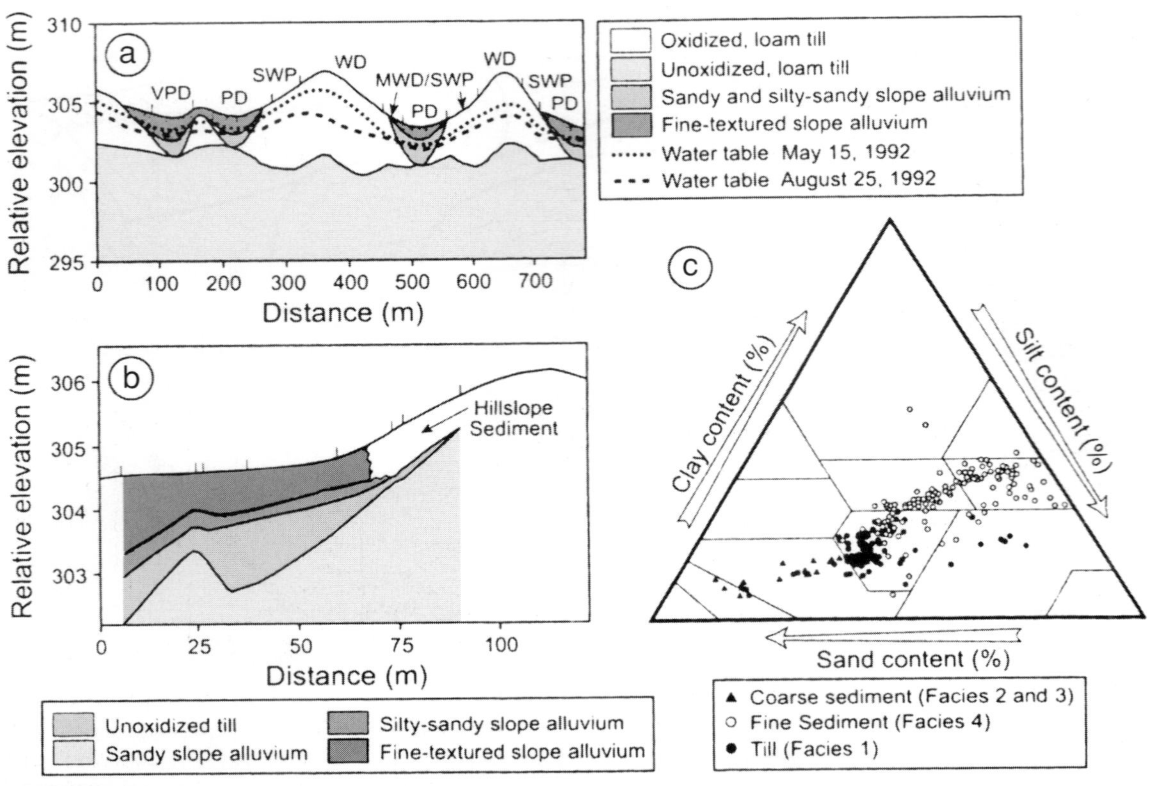

Fig. 13.34 The hummocky landscape of the Des Moines lobe, Iowa. Cored with till, the landscape has accumulated various thicknesses of slopewash or slope alluvium since deglaciation. After Steinwand and Fenton (1995) and Burras and Scholtes (1987). (a) Cross-section showing the relationship of the sediments to the landscape, water table positions and the drainage classes of the soil landscape. (b) Detailed cross-section of a hillslope showing the facies relationships along a catena. (c) Textures of the sediments from the various sedimentary facies.

infiltration is directed deeply into a few small fissures in the rock (Fig. 13.36). Vegetation reflects this pattern; it is more dense and lush on the rocky, upper slopes than on the silty toeslopes (Yair and Danin 1980). The most deeply leached areas of the slope are at the base of the rocky upper segment, where runoff is maximal, and yet the amount of land surface that can accept water is still low because rocks are exposed at the surface. Thus, depth to gypsic and calcic horizons is greatest at the base of the rocky footslope and gets progressively shallower farther downslope in the colluvial materials (Yair and Berkowicz 1989) (Fig. 13.37). Likewise, soils in colluvium have calcic horizons only near the rocky footslope. Farther out on the colluvium they have calcic *and* gypsic horizons, while those farthest out have only gypsic horizons, reflecting the increasing aridity with distance from the rocky, runoff-generating slopes (Wieder *et al.* 1985).

This catena illustrates that across the desert landscape the degree of soil aridity, soil development and leaching is highly variable, being primarily a function of location and substrate. Water availability is greatest where the ratio of hard bedrock to soil is high (Yair and Berkowicz 1989).

Front Range, Colorado

Soils are often linked geochemically along a catena, as materials in upslope positions are translocated to lower positions, where they either

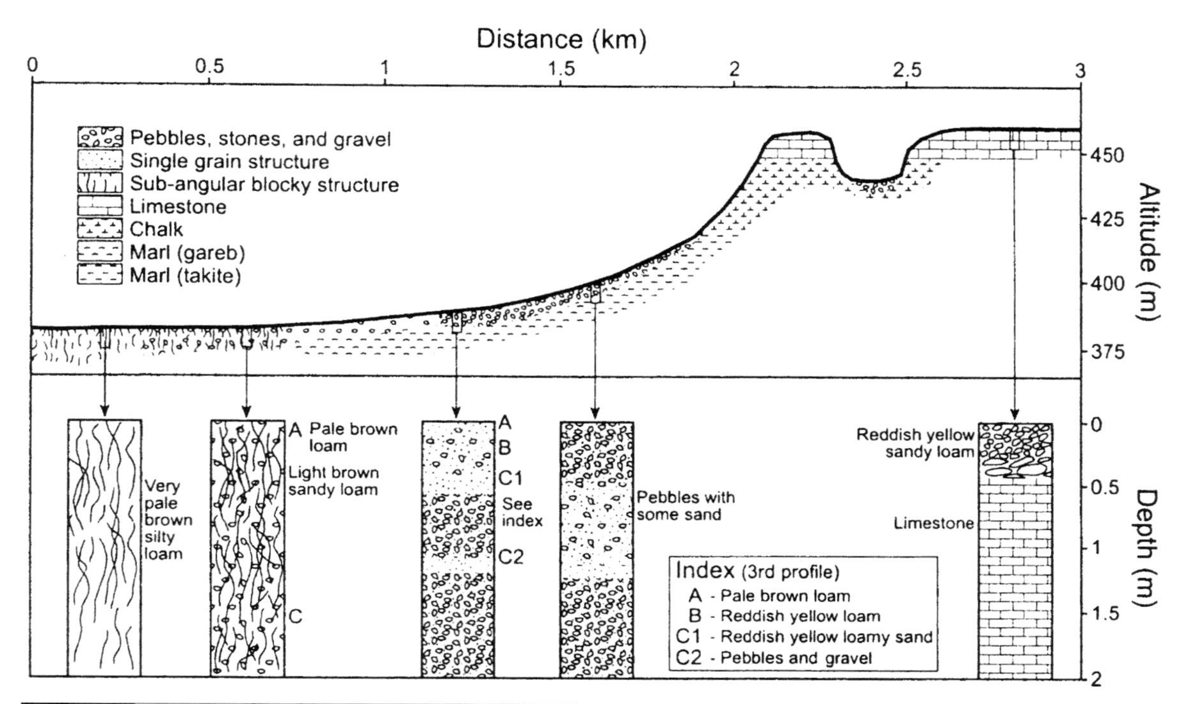

Fig. 13.35 Catena cross-section on carbonate sediments in the southern Negev Desert, Israel. After Dan and Yaalon (1964).

precipitate or make their way into a stream, lake or underground aquifer (Glazovskaya 1968). Many soil attributes may be due to *lateral* losses or gains of soluble material, which is directly due to the effects of slope (Huggett 1976b, Hillel and Talpaz 1977, Knuteson *et al.* 1989, Reuter *et al.* 1998, Sommer *et al.* 2000).

Litaor's (1992) study of soils along a catena in the Colorado Front Range focussed on movement of aluminum in solution. The site is an alpine meadow with about 3–4 m of glacial till above biotite gneiss. Litaor sampled the Typic Cryumbrepts and the solution that moved through soil macropores at summit, backslope and toeslope positions (Fig. 13.38). This catena is a good location to examine lateral transport of soluble materials because (1) there is little mass movement, (2) surface runoff is minimal because of thick vegetation and numerous cracks and fissures in the surface formed by cryoturbation and (3) subsurface water flow, mainly fed by spring snowmelt and summer thunderstorms, is accentuated by frozen subsoil. Much of this water flows laterally within the subsurface, taking ions and some clay with it. Lastly, vertical translocation of clay is insignificant, and silt is translocated vertically only in summit positions; vertical translocation of aluminum, to Bw horizons, does occur in summer. Thus, most of the variation along the catena is due to lateral transport of soluble materials.

The catenary distribution of some materials, such as organic carbon, may be due to slowed decomposition rates in the wet soils at the base of the slope (Fig. 13.38). Nonetheless the soil solution contained more dissolved organic carbon in downslope locations, suggesting that subsurface lateral flow of carbon is an important, ongoing process. Increases in Al–organic complexes at downslope locations point to lateral flow of soil solution containing these soluble products, probably being most active at snowmelt when the subsoil is frozen. This study illustrates the importance of slope on subsurface translocation processes, which can occur whenever water inputs exceed subsurface vertical infiltration capacities.

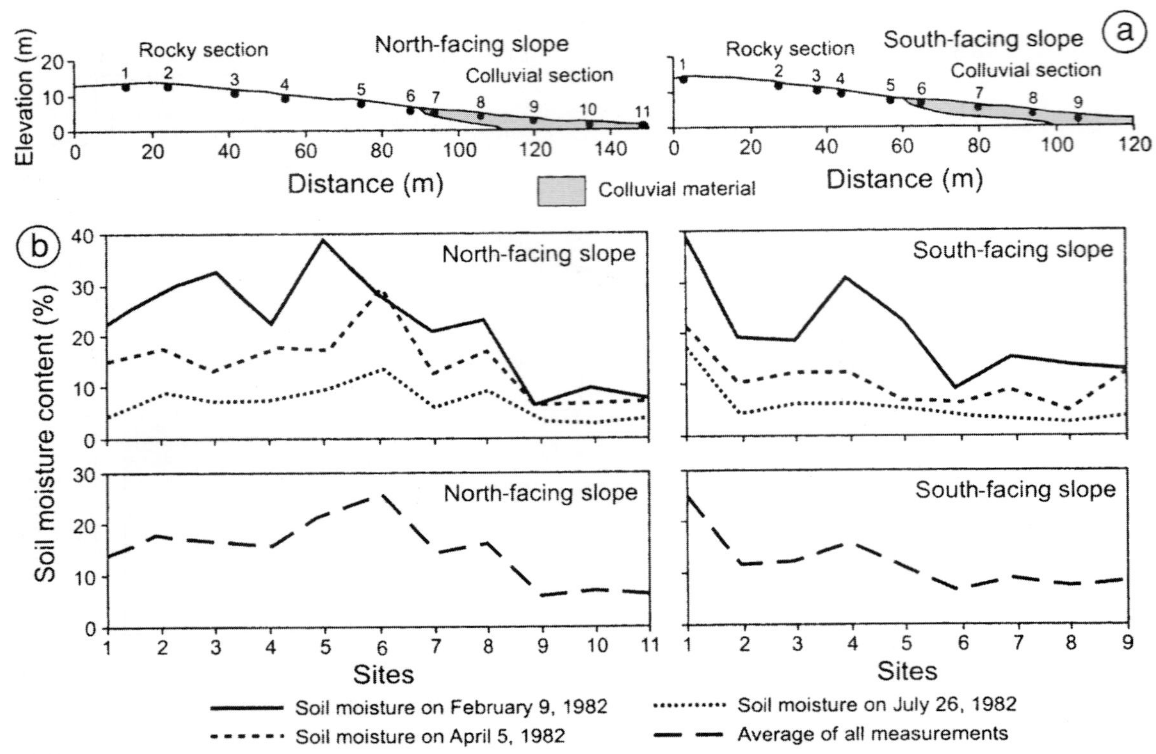

Fig. 13.36 Typical soil catena in the Negev Desert. After Kadmon *et al.* (1989). (a) Catena cross-section for north- and south-facing slopes. (b) Spatial and temporal variation in soil moisture along the same catenas.

Coastal Plain, Israel

Normally, catenas on sand dunes are texturally uniform and soils on them vary as a function of depth to the water table. Wet, sandy soils in swales between dunes are subject to different processes, e.g., oxidation–reduction, but otherwise the soils are often relatively similar. Dan *et al.* (1968) described a very different situation for a catena on a dune, less than 1 km from the Mediterranean Sea (Fig. 13.39). The xeric climate is humid enough (the moisture surplus is 150–200 mm in winter) that soils in the inter-dune swales are saturated in winter while soils on the dune crest are leached. The swale retains so much runoff in winter that it supports marsh vegetation. In the dry summers all the soils experience a severe soil moisture deficit.

The textures of the soils on this catena range from sand to clay, with the clay having been brought in as eolian dust from surrounding deserts (Yaalon and Ganor 1975). Although carbonate-rich dust blankets the landscape evenly, the plants debris and litter upon which it is deposited are preferentially washed and blown into swales, and clay accumulates there. Thus, soils in the swale are very clay-rich, but underlain by dune sand. Because there is not enough leaching in the swale to remove bases, the pH has remained high, allowing smectites to form. Haploxererts dominate the swale; cracking occurs in summer. On the dune crest some of the eolian dust has been translocated into the soil, forming a Bt horizon (Fig. 13.39). Sandy eluvial zones and low pH values in the upland soils attest to the strong leaching environment. All the soils are free of soluble salts and carbonates. Kaolinite, typical of acid soils, is a prominent clay mineral in the soils on the slope. Soils on the steepest slopes of the dune are the least developed, since this is a runoff-generating site. Even here, however, there is enough infiltration to have formed a weak Bt horizon. During winter, throughflow moves along the top of the Bt horizons, translocating still more clay and bases to the swale area.

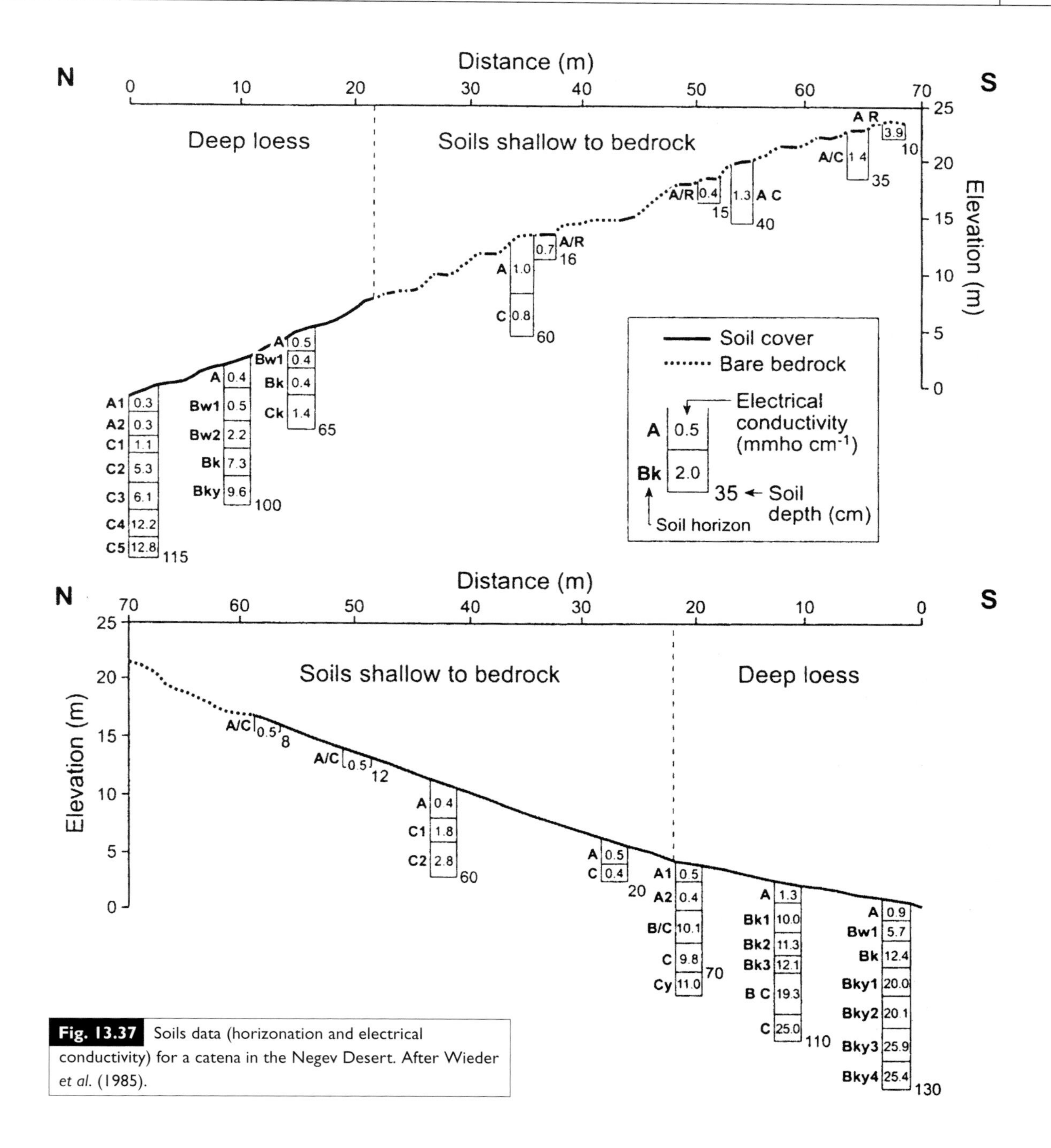

Fig. 13.37 Soils data (horizonation and electrical conductivity) for a catena in the Negev Desert. After Wieder *et al.* (1985).

Footslope soils display columnar structure in the Bt horizon not associated with natric conditions, the genesis of which is explained by Dan *et al.* (1968) as follows: the Bt horizon breaks into prisms in the dry summer, as clays slightly shrink and contract. At the onset of the winter rains, clay, sand and other coarse materials are translocated into the gaps between the prisms, preserving their gross structure. Rewetted, the prisms expand, but the primary avenue for expansion is the centers of the prisms, which are forced upward, creating the round tops.

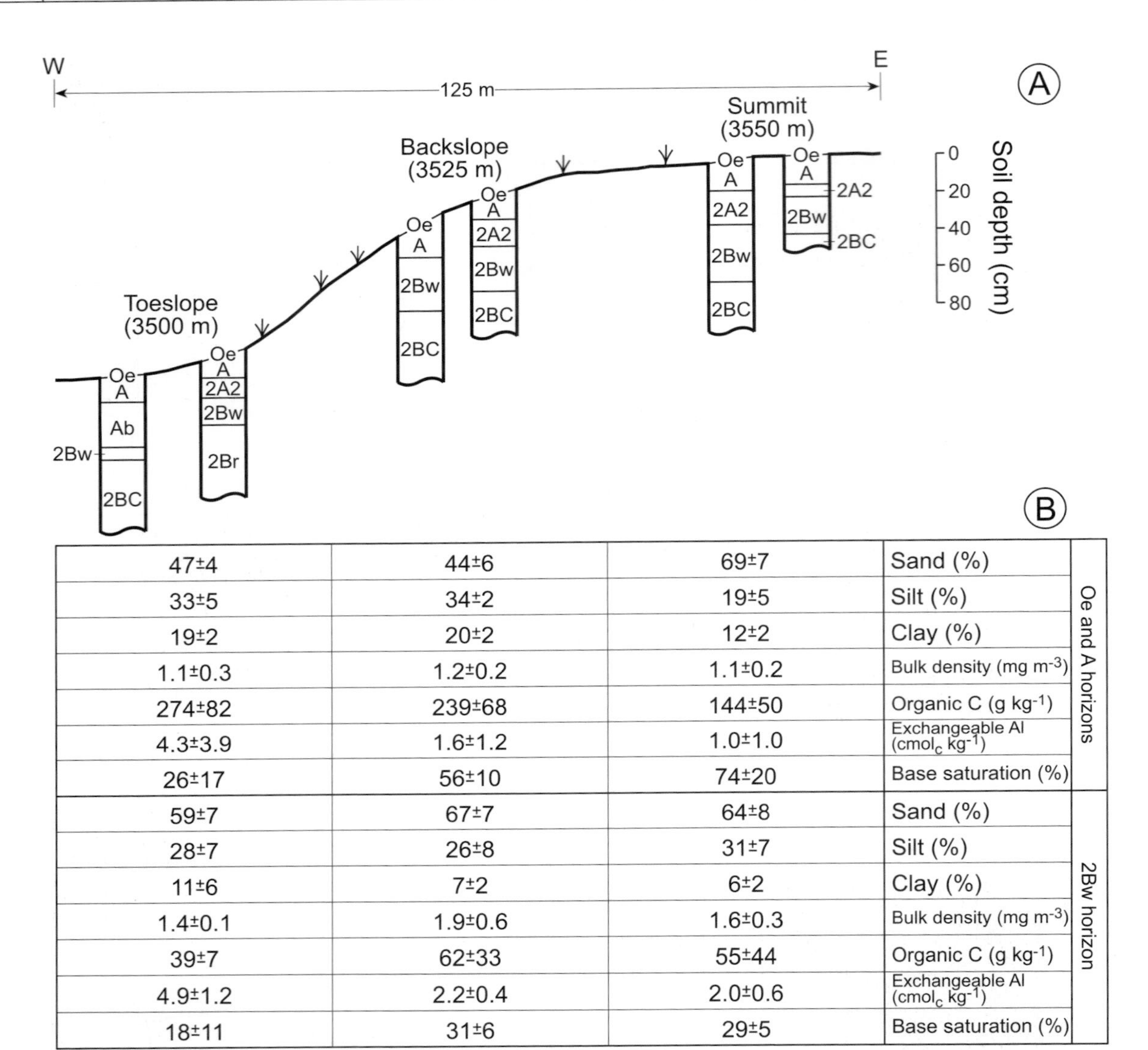

47±4	44±6	69±7	Sand (%)	Oe and A horizons
33±5	34±2	19±5	Silt (%)	
19±2	20±2	12±2	Clay (%)	
1.1±0.3	1.2±0.2	1.1±0.2	Bulk density (mg m^{-3})	
274±82	239±68	144±50	Organic C (g kg^{-1})	
4.3±3.9	1.6±1.2	1.0±1.0	Exchangeable Al (cmol$_c$ kg^{-1})	
26±17	56±10	74±20	Base saturation (%)	
59±7	67±7	64±8	Sand (%)	2Bw horizon
28±7	26±8	31±7	Silt (%)	
11±6	7±2	6±2	Clay (%)	
1.4±0.1	1.9±0.6	1.6±0.3	Bulk density (mg m^{-3})	
39±7	62±33	55±44	Organic C (g kg^{-1})	
4.9±1.2	2.2±0.4	2.0±0.6	Exchangeable Al (cmol$_c$ kg^{-1})	
18±11	31±6	29±5	Base saturation (%)	

Fig. 13.38 Data for soils along an alpine catena in the Front Range, Colorado. After Litaor (1992). (a) Slope morphology and soil horizonation. (b) Changes in soil components along the catena.

This catena illustrates the importance of dust and the local intensity of leaching processes in xeric climates. Although there is only a moisture surplus of 150–200 mm, it is concentrated in a few months when the vegetation is dormant and leaching can be maximized.

Rio Negro watershed, Amazon rain forest

Perhaps the most challenging and comprehensive study of soil geomorphology in the humid tropics is that of Dubroeucq and Volkoff (1998). Their work illustrates the complexity and the dynamism of these old landscapes (Fig. 13.40). Thick sandy deposits occur on the plains. Dubroeucq and Volkoff (1998) divided the landscape into geomorphic types and examined typical soil associations on each. On landscapes with low, rounded hills, they described a generally continuous mantle of Oxisols and Ultisols over saprolite. The saprolite is red beneath uplands but is often white beneath lowlands (Fig. 13.41). Colors of soil horizons also change from red to yellow along the summit–footslope transects, and textures get continuously sandier toward the bases of the slopes. Dubroeucq and Volkoff

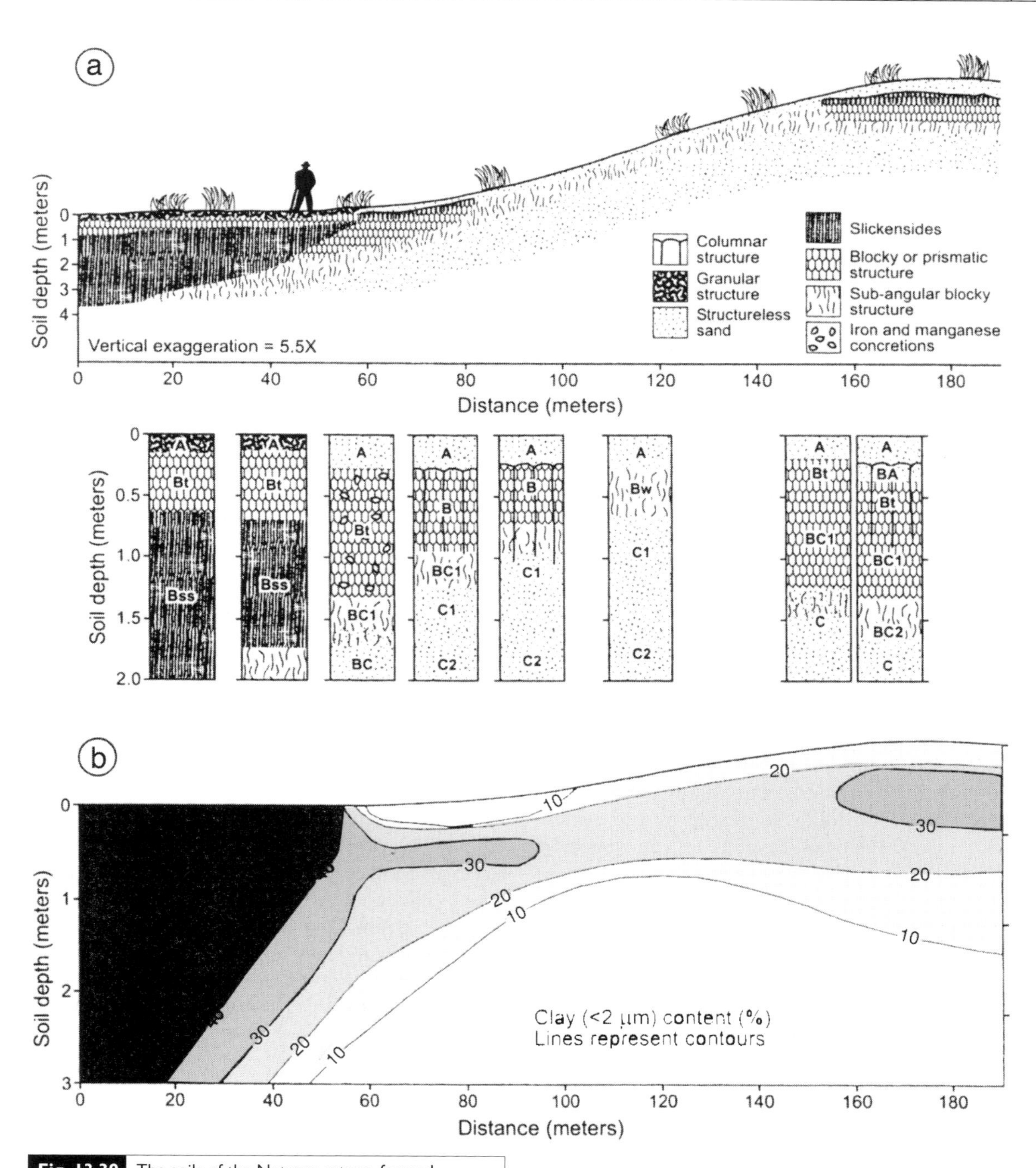

Fig. 13.39 The soils of the Netanya catena, formed on a coastal dune in Israel. After Dan *et al.* (1968). (a) Cross-section of the catena, showing soil morphological changes along the slope. (b) Distribution of clay within the catena.

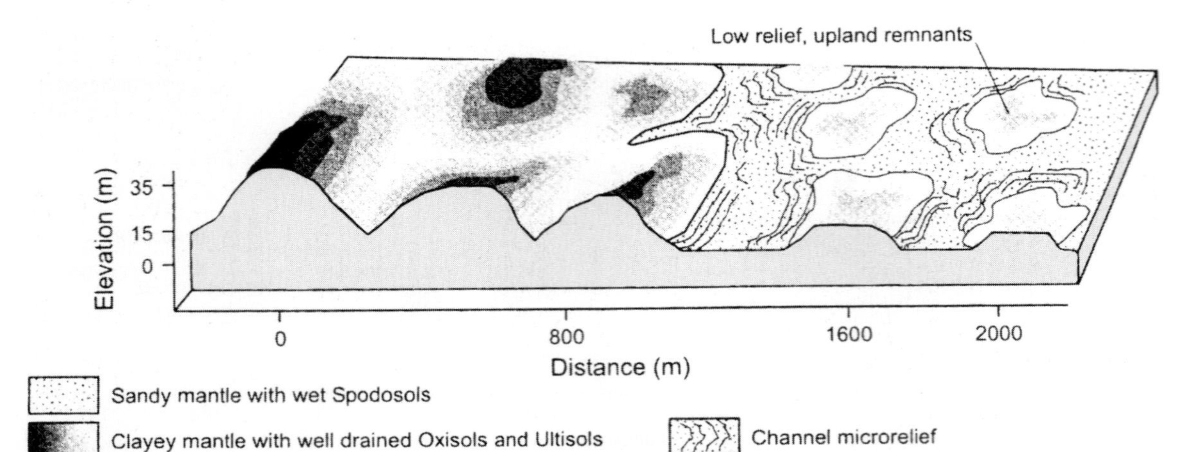

Fig. 13.40 The soil landscape of a part of the Rio Negro basin, southern Venezuela, showing the relationship between landforms, sediments and soil types in this hot, humid and generally low-relief landscape. After Dubroeucq and Volkoff (1998).

(1998) hypothesized that the source of the sand is the upland soils themselves – transported to the footslopes and toeslopes by surficial processes. In the pure sand sediment at the bases of the hills, wet Spodosols have developed. And on the very wettest sites, Histosols (Tropofibrists) have developed above sands, but Aquults have developed where saprolite is nearer to the surface (Fig. 13.42). Histosols develop as the quartz in the E horizons gets fragmented, holding up water and making the soil even wetter (Dubroeucq and Volkoff 1988). One possible endpoint of landscape evolution in this hot, humid environment may be a flat, wet landscape dominated by wet, generally sandy Spodosols and Ultisols, underlain by white saprolite rich in kaolinite.

This example illustrates the point that, while parts of the tropical landscape are old and deeply weathered, many other parts continue to be geomorphically "rejuvenated" by exposure to less weathered parent materials (by erosion of weathered topsoil) or additions of fresh materials from above (colluvium, slopewash, alluvium or ash). In rejuvenated areas, any of a number of soils might be found: Andisols in ash, Entisols or Inceptisols in alluvium, or Alfisols or Ultisols on eroding sideslopes.

This type of situation was described by Lepsch *et al.* (1977a, b) for the Occidental Plateau in Brazil (Fig. 13.43). The uplands here are held up not by a laterite layer but by sandstone. Oxisols are found on this old, stable surface. Colluvium and alluvium are common on the dissected sideslopes; in these materials various Alfisols, Inceptisols and Ultisols have developed. Where this material is sandy and water tables are high, Spodosols and Histosols can form (Richards 1941, Andriesse 1969, Tan *et al.* 1970, Schwartz *et al.* 1986). Lateral water movement on these slopes has, presumably, initiated removal of some iron, facilitating lessivage and hence Bt horizon development. Mollisols formed on sideslopes where erosion exposed calcareous sandstone, providing bases to the soil system. The bases facilitate the formation of smectite clays which in turn retain high amounts of organic matter.

Soil geomorphology case studies, models and paradigms

This section is devoted to a select few landmark studies that illustrate concepts in, and applications of, soil geomorphology. Landmark studies, like other forms of breakthroughs in science, are noteworthy because they advance the discipline farther in a short time than many other,

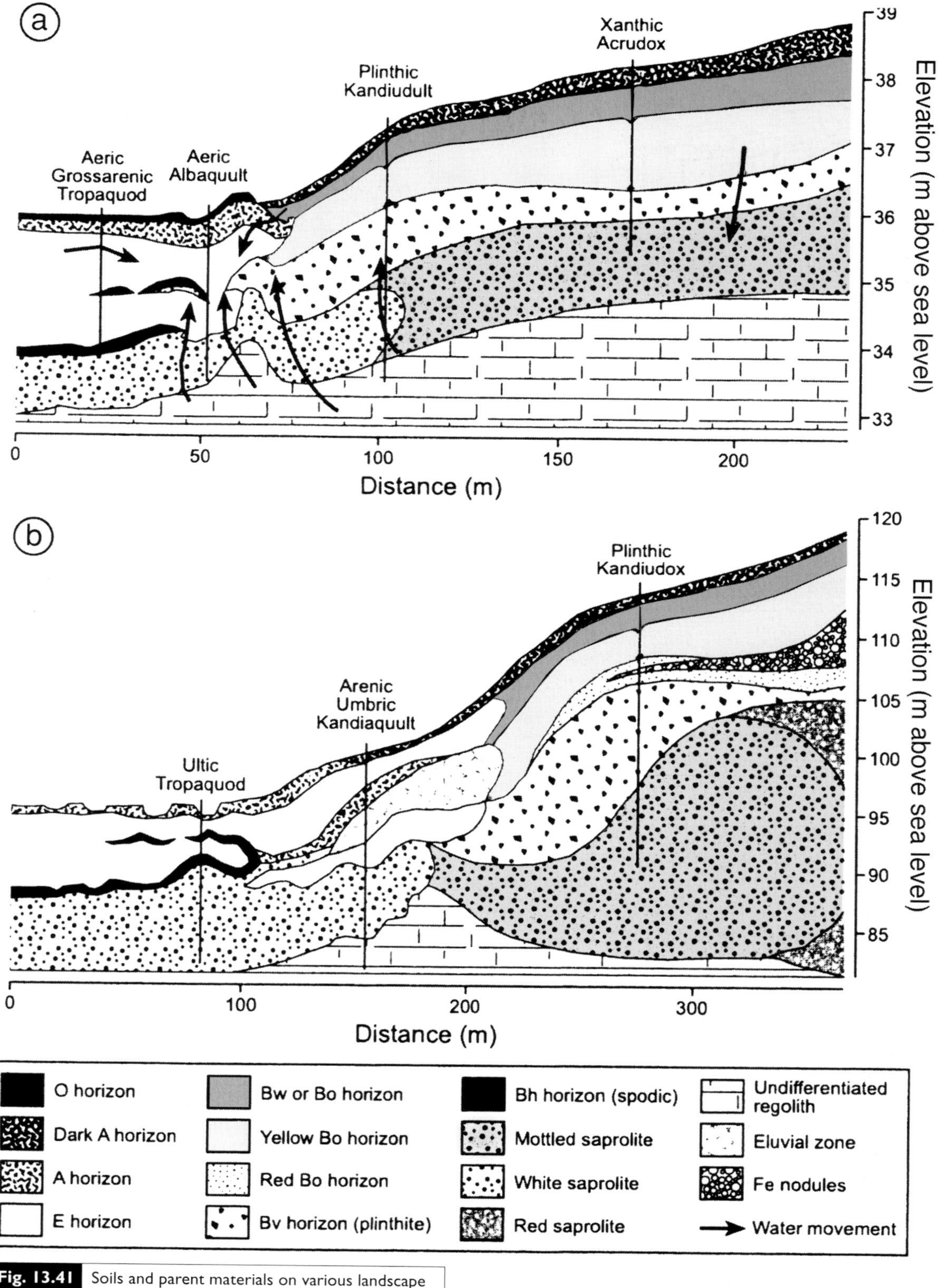

Fig. 13.41 Soils and parent materials on various landscape positions in the Rio Negro basin, Amazonia. After Dubroeucq and Volkoff (1998).

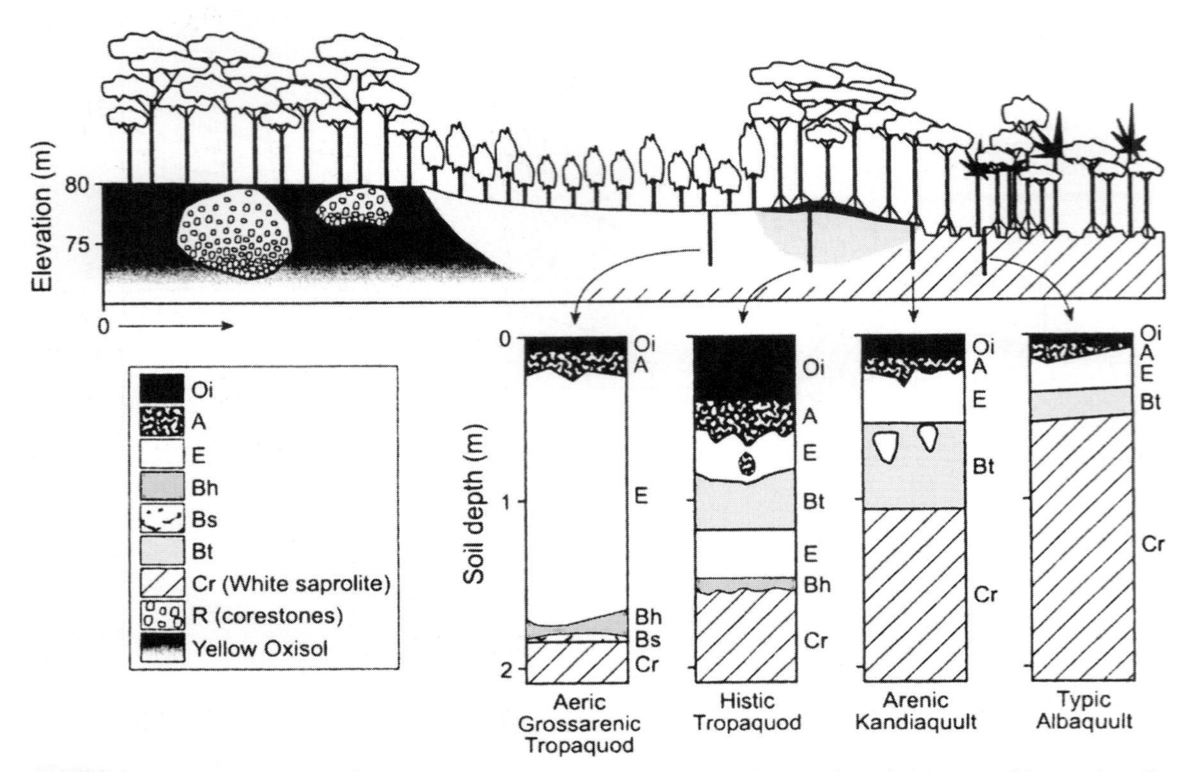

Fig. 13.42 Cross-section of the soils, vegetation and topography in a generally wet area in the Rio Negro basin, South America. After Dubroeucq and Volkoff (1998).

empirically driven studies might do, collectively, over a long period of time. And even though we present these studies independently, it should become clear that the breakthroughs of one person were seldom due to that person working in isolation. Rather, they were due, in part, to their knowledge of the work of others. Connectedness is very important in science, and soil geomorphology is no exception.

Robert Ruhe's work in Iowa

We begin in southwestern Iowa where the work of Robert Ruhe (Fig. 13.44) and many others "ushered in an era of landscape evolution and soil formation research and established the importance of paleosol studies" (Olson 1989: 133–134). Ruhe was a geologist who spent much of his career studying the stratigraphy and soil geomorphology of Iowa and nearby areas. A meticulous researcher and a tireless field man, his work refocussed and energized many in soil geomorphology, soil stratigraphy and paleopedology (Effland and Effland 1992). His two books, *Quaternary Landscapes in Iowa* (1969) and *Geomorphology* (1975b) are classics. Ruhe furthered our understanding of Quaternary processes and Quaternary stratigraphy, and stressed the importance of paleosols and comparative soil development to the understanding of landscape evolution.

Southern Iowa: stratigraphy and constructional surfaces

Iowa has several different landform regions, most of which can be delineated geomorphically and stratigraphically (Fig. 13.45). Northeast Iowa is driftless in that it does not have any glacial drift; bedrock controls the topography. The Woodfordian advance of the Wisconsin glacier formed the Des Moines lobe landscapes in north-central Iowa (Fig. 13.45). Southern Iowa was not glaciated during the Wisconsin advance; the uppermost tills here are Middle Pleistocene (Pre-Illinoian) in age. Early studies named these tills after the

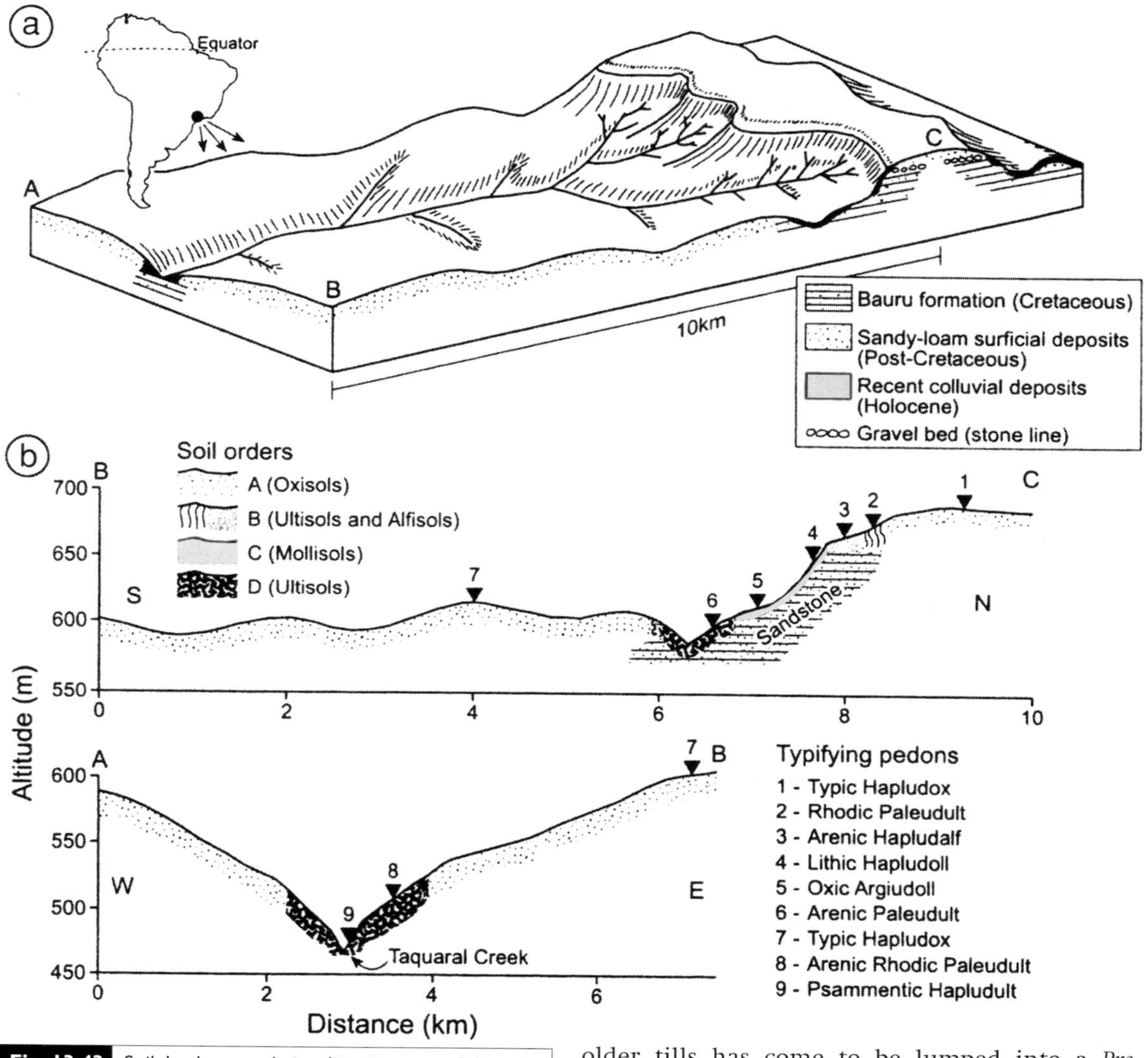

Fig. 13.43 Soil–landscape relationships in a part of the Occidental Plateau, near São Paulo, Brazil. After Lepsch *et al.* (1977a, b). (a) Block diagram of the area, showing geology and topography. (b) North–south and east–west cross-sections showing geomorphic surfaces and their relationships to soils.

states of Nebraska and Kansas (Chamberlain 1895, Shimek 1909), but as more information was generated it was discovered that there were more than two tills and these were not always correlated to the type Nebraskan and Kansan till sites. For example, it is now known that several tills predate the type Nebraskan till (Hallberg *et al.* 1978a, Hallberg 1986). Thus, the entire series of older tills has come to be lumped into a *Pre-Illinoian* category until better stratigraphic information becomes available (Guccione 1983, Richmond and Fullerton 1986, Aber 1991, 1999, Rovey and Kean 1996). Most of Iowa is floored with these pre-Illinoian tills, but only in the Southern Iowa Drift Plain are they the uppermost tills (Fig. 13.46). Ruhe (1969) named this landscape the Kansan Drift region for the uppermost drift. There is some indication that the type Kansan till may be between 780 000 and 620 000 years old (Colgan 1999). One or more deeply weathered paleosols have developed in these tills.

Above the tills and their paleosols in southern Iowa are various Late Pleistocene loess deposits

Fig. 13.44 Robert V. Ruhe (1919–92), on a 1980 field trip. Photo by L. Follmer, via John Tandarich.

(Figs. 13.46, 13.47). The lower Loveland loess is Illinoian in age (Leighton and Willman 1950, Colgan 1999). Both the Mississippi and the Missouri River valleys were sources for the Loveland loess (Shimek 1909, Ruhe 1956, Follmer 1982), which is >7 m thick near the source areas but thins to <2 m as far as 70 km inland (Ruhe 1956). In south–*central* Iowa, therefore, the Loveland loess is quite thin; much of it has been incorporated into the underlying paleosols developed in drift. During the Wisconsin glaciation, thick Peoria loess sequences were deposited across the entire Southern Iowa Drift Plain (Fig. 13.46).

Where stream incision has been deep enough to expose the lowermost Pre-Illinoian till, a buried paleosol can be seen at the top of this till.[1] This soil, formed in "Nebraskan till," could be as old as Late Pliocene (Rovey and Kean 1996); it was later buried by "Kansan" drift.[2] Extremely well developed and highly weathered, this soil was initially named the Afton soil, for the interglaciation between the so-called Nebraskan and the Kansan glaciations (Table 11.6). Early researchers thought that the high clay contents of this (and similar) paleosols were due to long-term weathering, and coined the term *gumbotil* for them (Kay 1916, Kay and Pierce 1920, Kay and Apfel 1929). They did not initially recognize it as a buried soil but thought it was simply a strongly weathered zone in till, possibly with a soil at its top (Alden and Leighton 1917). Kay (1916) defined gumbo or gumbotil as a gray to dark-colored, thoroughly leached, non-laminated, deoxidized clay, very sticky and breaking with a starch-like fracture when wet, but very hard when dry. Simonson (1941) and Scholtes *et al.* (1951) later pointed out that gumbotil was the B horizon of a buried soil. Pedostratigraphic work by Ruhe (1956, 1969, Ruhe *et al.* 1967) on the Kansan surface, and Frye in Illinois (Frye *et al.* 1960a, b), showed that gumbotil was restricted to swales, and that its morphology changed to a red-brown, oxidized B horizon on uplands. Thus, gumbotil was actually just part of a paleosol and the large amounts of clay in it had originated due to long-term chemical weathering and slopewash, accumulating to great thicknesses only in depressions on the paleosurface. These clay-rich soils have since been termed *accretion gleys* (Frye *et al.* 1960a, b, Willman 1979). The gray, clayey accretion gleys on the buried Nebraskan surface could not have formed without a long period of subaerial exposure. The long period necessary for their formation was available because the ice did not advance back over the landscape for tens of thousands of years, with the "Kansan" advance.

Gumbotil-like paleosols are also present on the low-relief *Kansan* surface of southern Iowa (Woida and Thompson 1993). These soils are

[1] For ease of communication, we will use the older chronostratigraphic names in this discussion. We encourage the use of new terminology in research, but for the sake of discussion in this text, the older terms create less confusion.

[2] The stratigraphic "story" of this region is, in actuality, much more complex than is explained here, based on information gained from recent advances in tephrostratigraphy and till stratigraphy (see Hallberg 1986). For example, there may not be a single surficial "Kansan" till across all of southern Iowa and northern Missouri. However, the point of this discussion is not to elucidate the precise stratigraphic column as we know it, for that is due to change as more knowledge emerges. Rather, we focus on the logic behind the early discoveries, the methodologies that led to them, and how these "lessons" are useful in soil geomorphology studies elsewhere.

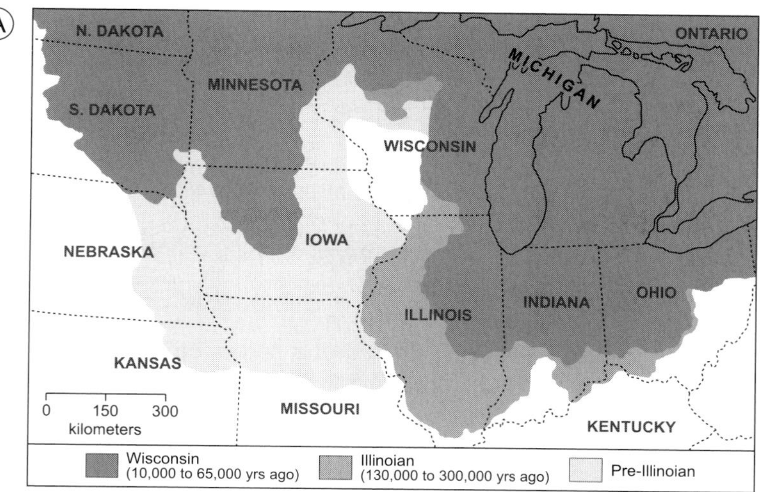

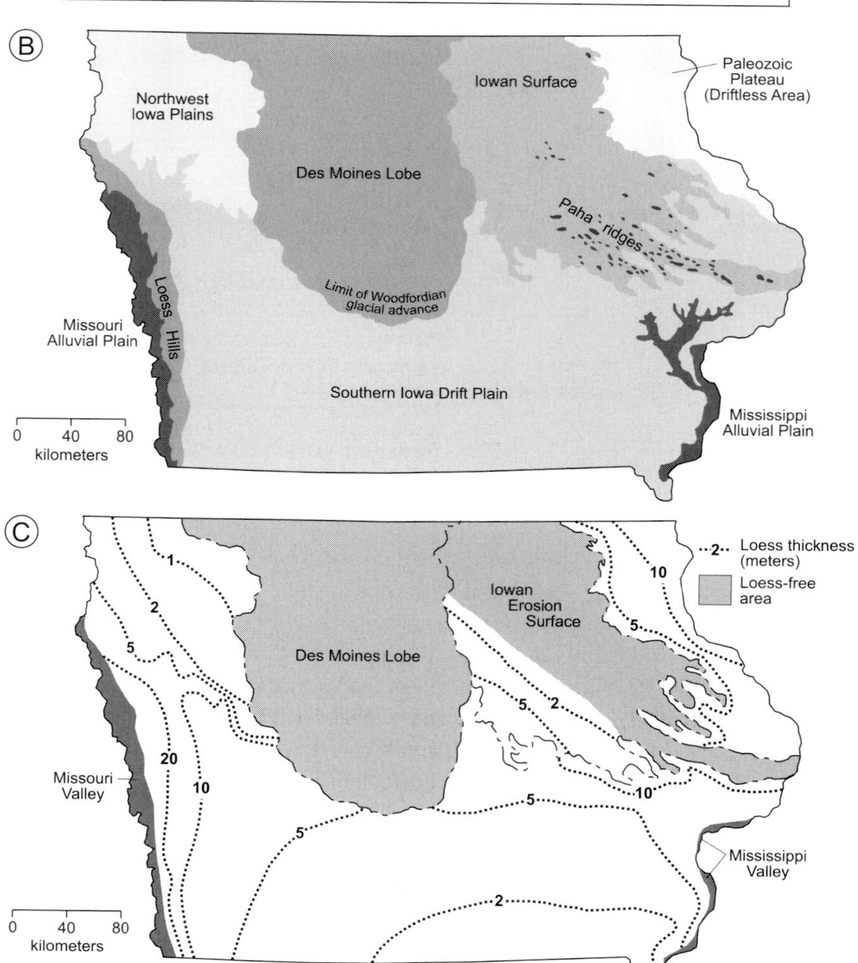

Fig. 13.45 Background information about Iowa's geomorphology. After Prior (1991). (a) Limits of the major Quaternary glaciations in the upper Midwest. (b) Landform regions of Iowa. (c) Loess thickness and distribution.

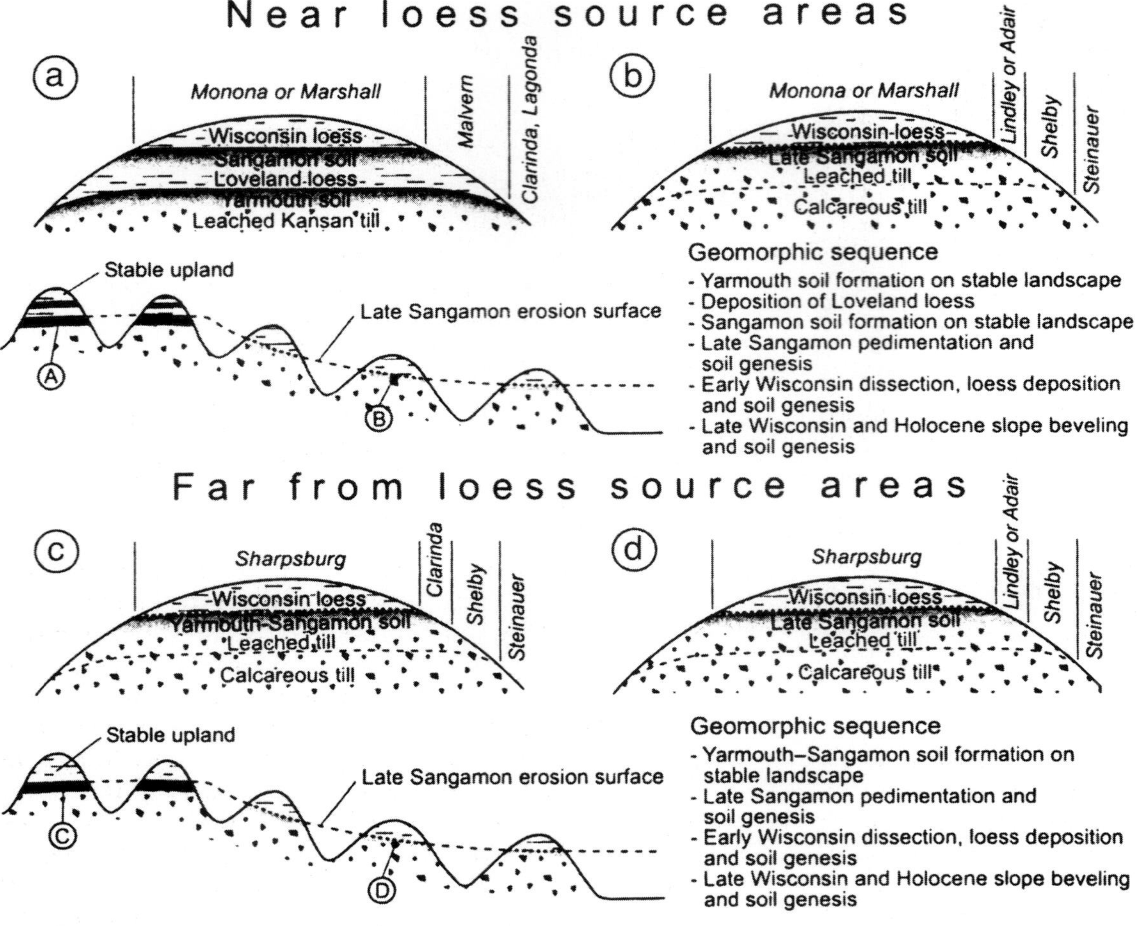

Fig. 13.46 Major types of stratigraphic sequences and associated surface soil series in southwestern Iowa. After Ruhe *et al.* (1967). (a) Interfluve sites near the Missouri and Mississippi Rivers that have discernible deposits of Loveland (Illinoian) loess. (b) Sites near the Missouri and Mississippi Rivers that have discernible deposits of Loveland (Illinoian) loess, and are located on the Late Sangamon erosion surface. (c) Interfluve sites far from the Missouri and Mississippi Rivers, i.e., south–central Iowa, that lack discernible deposits of Loveland (Illinoian) loess. (d) Sites far from the Missouri and Mississippi Rivers, i.e., south–central Iowa, that lack discernible deposits of Loveland (Illinoian) loess, and are located on the Late Sangamon erosion surface.

Yarmouth–Sangamon paleosols, for they formed more or less continuously from the Yarmouth interglaciation (between the Kansan and the Illinoian), through the Illinoian glaciation (which did not cover this part of Iowa) to the Sangamon interglaciation. They are perhaps the thickest paleosols in North America, highly weathered and extremely polygenetic. During the latter part of the Illinoian glaciation, the soils that were developing on the Kansan surface accumulated Loveland (Illinoian) loess – thickest near the river bluffs. However, this loess was not thick enough in most places to completely bury the soil below, in the till. Thus, a polygenetic, *welded paleosol* formed. Soil welding is the process where a surface soil develops downward and pedogenically "connects" with the solum of a buried soil (Ruhe and Olson 1980). Thus, in the thin loess areas of south–central Iowa, Ruhe and his colleagues believed that a thick, highly weathered, welded soil had developed in a thin covering of Loveland loess over Kansan till. In thicker loess areas near the Missouri and Mississippi rivers (Fig. 13.45), the Yarmouth paleosol was not welded with the

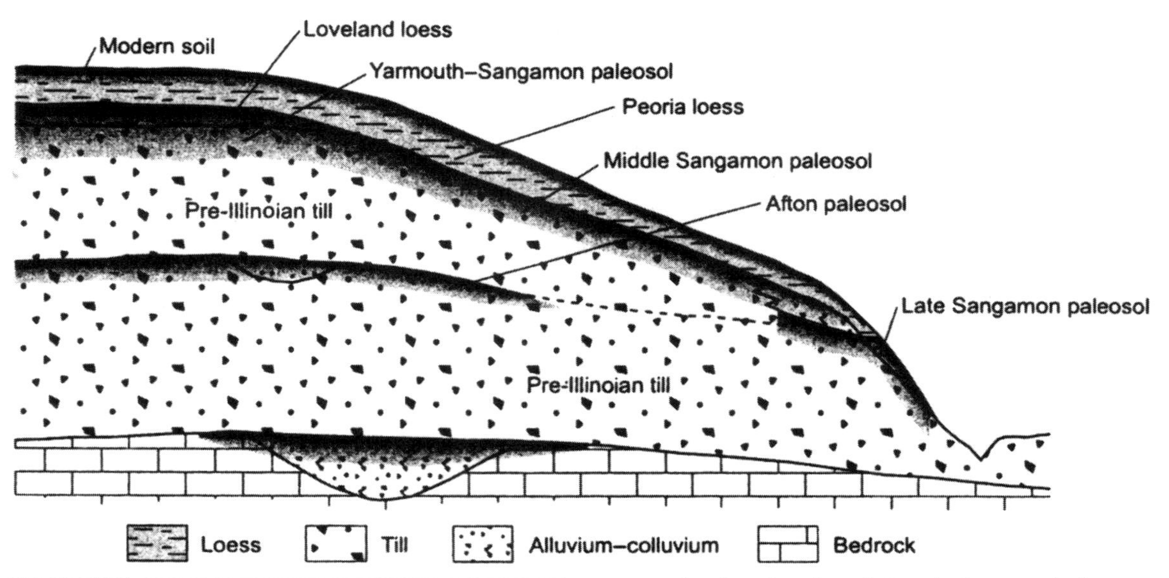

Fig. 13.47 Idealized Quaternary stratigraphy of north–central Missouri and south–central Iowa, with some vertical exaggeration. After Guccione (1983).

surface soil in the Loveland loess because the loess was too thick, and thus two distinct paleosols can be seen: the lower Yarmouth paleosol developed in till and the Sangamon paleosol above, developed in Loveland loess. And below it all is the Afton paleosol formed in Nebraskan till.

The Sangamon paleosol, formed in Loveland loess, is strongly developed, as indicated by weathering data and depth functions of, particularly, clay content (Fig. 13.48). In swales, the soil is gray and usually is an accretion gley. On flat uplands it is often an *in situ*, gleyed profile. This paleosol should not be confused with one we will soon discuss: the Late Sangamon soil, or its erosion surface.

Last, Peoria loess was deposited across all of southern Iowa, in association with the melting Woodfordian (Wisconsin) glacier. Ruhe (1956) placed the age of the base of this loess at 16 500 to 29 000 years; the wide age range illustrates that the surface was covered by loess in a time-transgressive manner (see Chapter 8). The age of the base of the loess decreases as one gets farther from the source areas (i.e., the rivers). An early Wisconsin loess, known as either the Roxana silt or the Pisgah Formation, was also deposited between the Loveland and Peoria loesses (Johnson and Follmer 1989, Leigh and Knox 1993, Leigh 1994, Grimley 2000), but it is not a major player in the Iowa story. The thin Roxana silt was incorporated into the Yarmouth–Sangamon paleosol as it aggraded upward via developmental upbuilding (Woida and Thompson 1993). Where it is distinguishable, a paleosol has formed in it; this soil is sometimes informally called the Basal Loess Paleosol. The Peoria loess was deposited rapidly enough and eventually became thick enough to bury all of the pre-existing surfaces, and of course it, too, was thickest near the main river source areas.

The Peoria loess was deposited on a low-relief, swell-and-swale surface with deeply weathered soils (Ruhe 1956). Low areas had thick, gray, accretion gley paleosols while uplands had thick, red or brown paleosols with clay-rich Bt horizons. All these surfaces contained either a Yarmouth and a Sangamon soil, or a polygenetic Yarmouth–Sangamon soil, depending on whether Loveland loess deposits were thick enough to separate the two sola (Fig. 13.46a, c).

Southern Iowa: erosional surfaces and landscape evolution

One of Ruhe's major contributions was the identification and genetic interpretation of stepped/beveled erosion surfaces on the southern Iowa landscape. Through careful examination of cores,

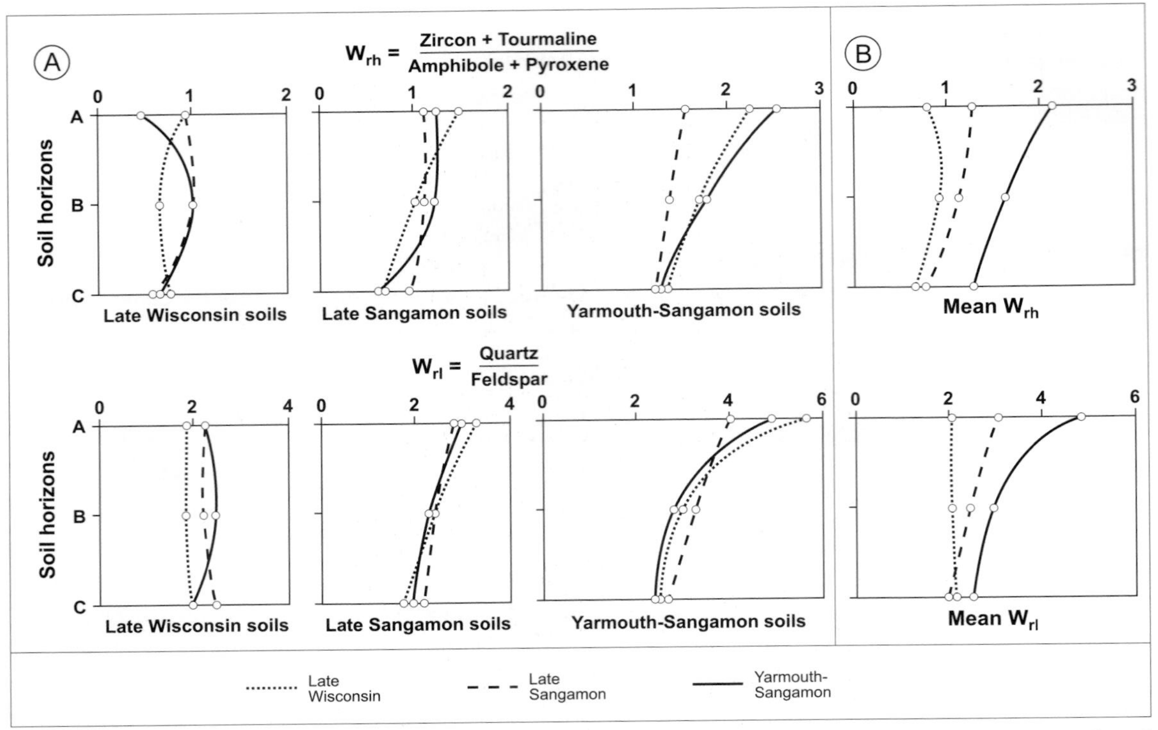

Fig. 13.48 Depth plots of weathering ratios for soils formed in Kansan till in southern Iowa. The soils have formed on surfaces of three different ages: Yarmouth–Sangamon, Late Sangamon and Late Wisconsin. After Ruhe (1956). (a) Counts of heavy and light minerals, with depth, for three soils on each surface. (b) Mean values of the same data, summarized for each surface.

roadcuts and railroad cuts, he pieced together the geomorphic history of what he called the Late Sangamon erosion surface and acsertained the chronology of these constructional and erosional events, some of which were beyond the range of radiocarbon dating. Ruhe's use of Quaternary stratigraphy coupled with paleopedology was a highly useful marriage, and has found application elsewhere.

It became clear to Ruhe that the swell-and-swale topography of this surface was not simple. Slopes were "stepped," that is, they were not long and continuous from uplands to lowlands, but rather they had discrete steps, risers and slope breaks that represented contacts between different geomorphic surfaces, some of which were erosional while others were constructional (Figs. 13.47, 13.49, 13.50).

Soil development on the various geomorphic surfaces, particularly the classic study of the Turkey Creek watershed (Ruhe *et al.* 1967), provided many details about landscape evolution in southwestern Iowa (Fig. 13.49). The flat uplands at the interfluves of this watershed are remnants of the Kansan drift plain, and have Wisconsin (Peoria) loess over a Yarmouth–Sangamon paleosol with accretion gleys in the swales (Fig. 13.46c). Near the major rivers, Loveland loess caps the uplands and a Sangamon soil has developed within it (Fig. 13.46a). The old, constructional surface has a relief of about 2–3 m, between swales that are about 200 m apart (Ruhe 1956) (Fig. 13.50). Turkey Creek is far enough from the Missouri River that Loveland loess is not recognizable as a C horizon between the Peoria loess and the Yarmouth–Sangamon paleosol (Fig. 13.49). The Peoria loess here *is* thick enough, however, that the modern soil is not welded to the Yarmouth–Sangamon paleosol below.

At some time prior to the Wisconsin glaciation, a gently sloping erosion surface formed on

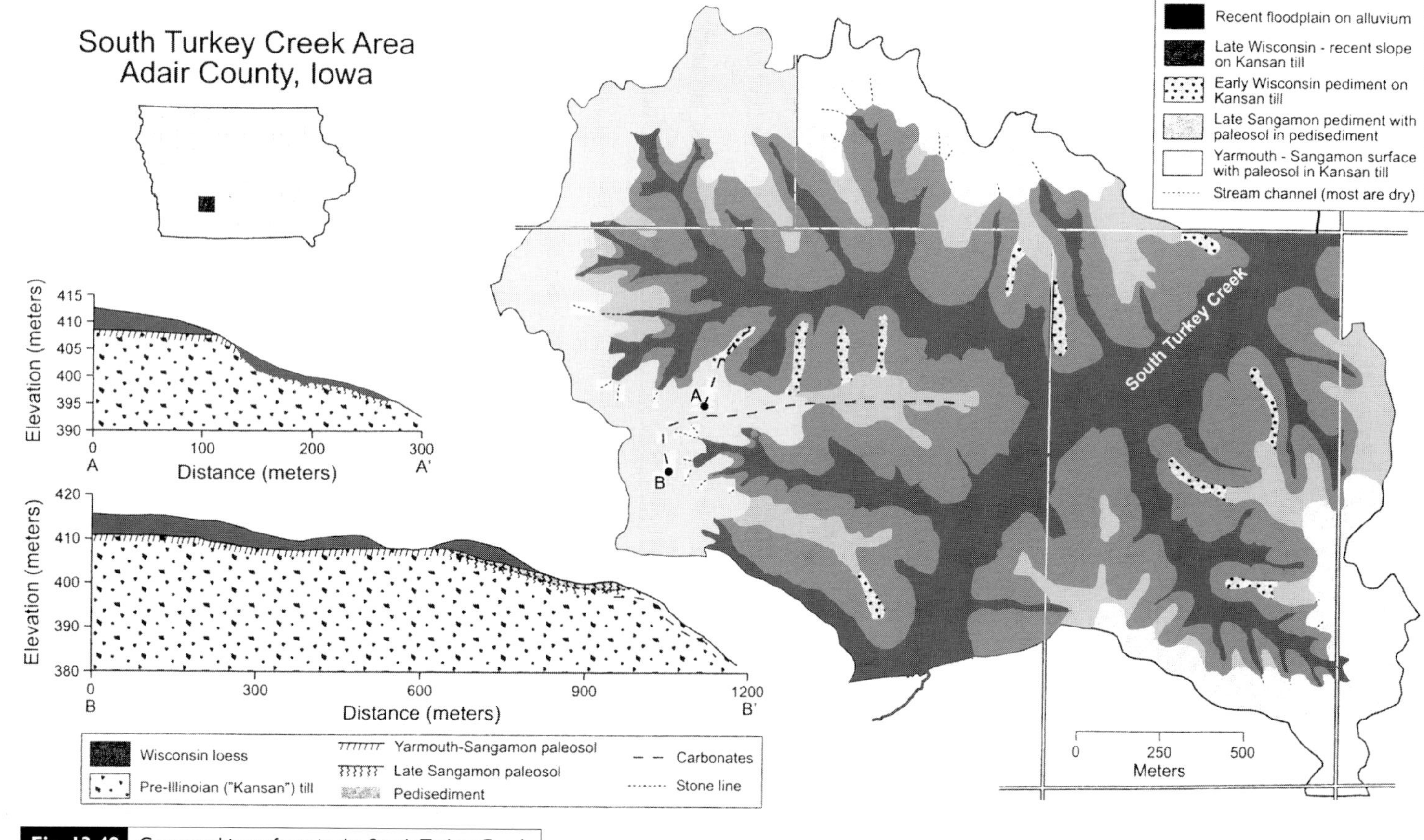

Fig. 13.49 Geomorphic surfaces in the South Turkey Creek watershed, Adair County, southwestern Iowa. After Ruhe *et al.* (1967).

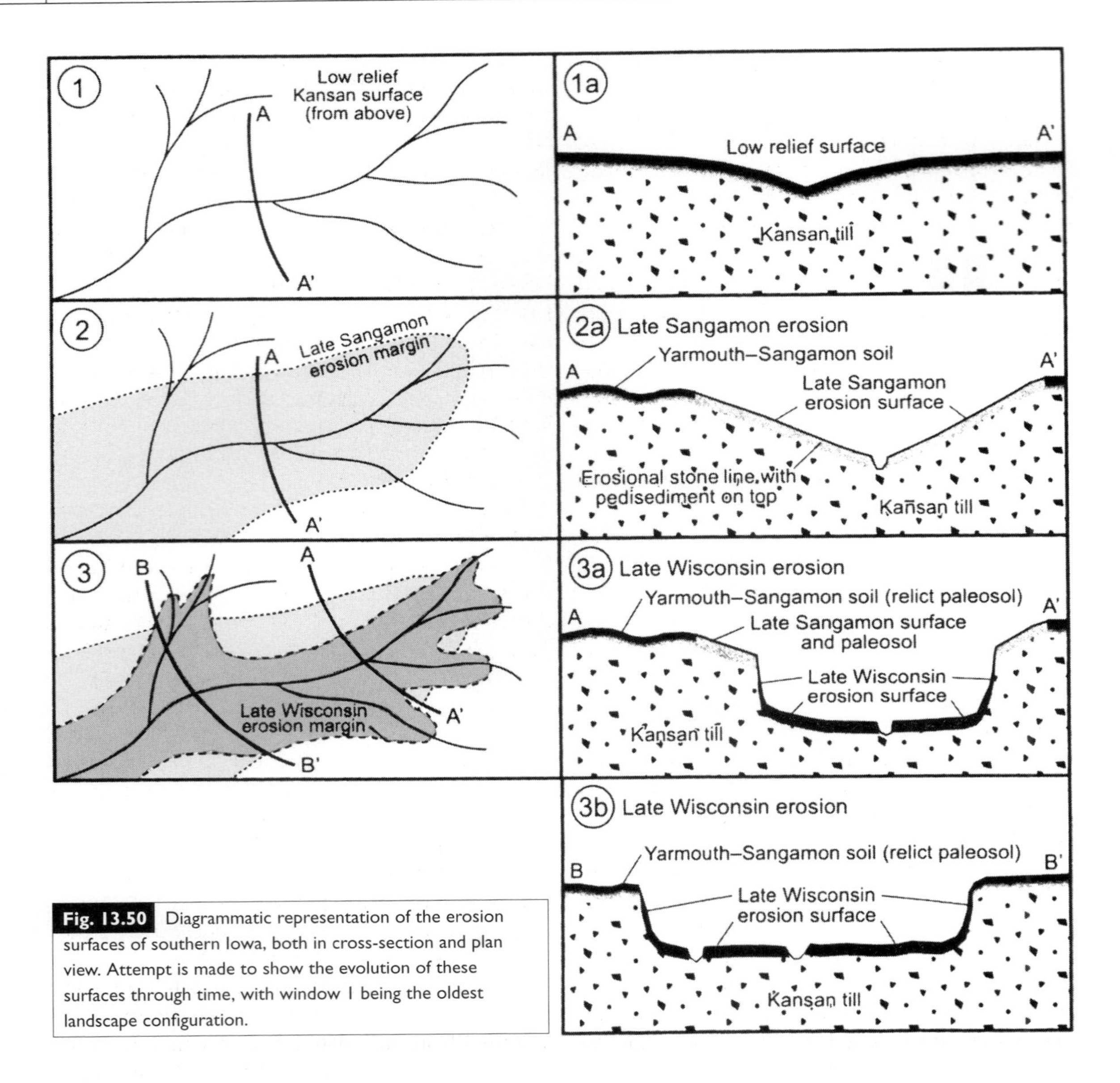

Fig. 13.50 Diagrammatic representation of the erosion surfaces of southern Iowa, both in cross-section and plan view. Attempt is made to show the evolution of these surfaces through time, with window 1 being the oldest landscape configuration.

parts of this landscape, completely removing the Yarmouth–Sangamon paleosol there (Fig. 13.50). Ruhe surmised that the age of this surface was Late Sangamon because the soils on it were not as strongly developed as the Sangamon soils developed in Loveland loess or Illinoian till in nearby areas (see Chapter 13). That is, he used relative soil development data to guide his interpretations of surface age. Why this erosion episode occurred is unknown, but it probably had something to do with the rapid landscape and climate changes that were occurring in the Midwest at this time. That such an erosion event *did* occur seems more likely than to envision long periods of stability. There is evidence that the Yarmouth–Sangamon soil on the stable uplands of southern Iowa also withstood episodic periods of erosion (Woida and Thompson 1993).

The Late Sangamon surface is incised into Kansan till; the evidence for this lies in a prominent stone line that mantles it (Ruhe 1956). The stones from the till have become concentrated as a lag on the erosion surface. Along with, and generally above, them is sediment that was being transported downslope as slopewash and colluvium. All this material was eventually buried by Wisconsin loess. Ruhe termed the process of that produced these surfaces *pedimentation*; it was essentially a type of slope backwasting in which slope gradients are lowered by erosion of the upper slope elements and deposition on lower slope elements. Ruhe (1956) also called the sandy/gravelly/stony material that mantles the erosion surfaces *pedisediment*. After the erosion event, a Late Sangamon paleosol developed in the pedisediment, stone line and underlying till. The degree of development of the reddish paleosol on the Late Sangamon erosion surface implies that the surface was cut, then stabilized, allowing the soil to form. Soil formation proceeded for some time before it was buried by Wisconsin loess, about 14 000 years ago. Estimates indicate that the Late Sangamon surface may have been stable and soils may have been developing in it for as long as 40 000 years.

Another erosion event occurred during the Early Wisconsin glacial period (Figs. 13.49, 13.50). This erosion surface, unlike the Late Sangamon surface, has no paleosol developed on it, suggesting that it was continually undergoing erosion until it became buried by the Wisconsin loess. This surface is called the Early Wisconsin pediment (Fig. 13.49). It also contains a stone line and pedisediment. The Early Wisconsin erosion event was, geomorphically, different than that of the Late Sangamon. The Late Sangamon erosion surface is relatively gently sloping and widespread, suggesting that it may have been driven mostly by slope processes such as solifluction, creep and other mass movements. It is not confined to areas near stream valleys (Fig. 13.50). The Early Wisconsin pediment and erosion event, however, are quite different. Incision is deeper and more confined to stream valleys, suggesting that it was driven more by running water. During the Holocene, streams widened their valley bottoms by lateral corrasion, developing floodplains that are in places erosional and in places filled with Holocene alluvium; and agriculturally driven erosion caused valleys to fill with sediment.

Early Wisconsin incision in south-central Iowa reached farther into the uplands along stream courses, leading to the development of two types of landscape profiles (Fig. 13.50). Some landscapes have all three surfaces: the low-relief Yarmouth–Sangamon (or Sangamon) upland surface, the Late Sangamon surface and the Early Wisconsin pediment. These landscape are found between modern stream courses, on nose slopes that escaped some of the Early Wisconsin erosion (Fig. 13.49). Farther up modern stream courses, the Early Wisconsin erosion event completely removed the Late Sangamon surface. Here, the upland surface with Yarmouth–Sangamon soils is beveled and directly connected to the Early Wisconsin pediment (Fig. 13.50).

Loess thickness is an important part of the Turkey Creek stratigraphy and evolution. On the flat Yarmouth–Sangamon landscape the Wisconsin loess is about 4.5 m thick, while on the Wisconsin erosion surface the loess is only about 2.3 m thick. Thus, Ruhe *et al.* (1967) argued that the erosion surface was cut during the early to middle period of Wisconsin loess deposition.

One of the most important applications of the "Iowa story" centers around how the stratigraphy and paleopedology can assist modern soil mappers. Ruhe's work put the soil units that were being mapped into a formal stratigraphic context, and stimulated more detailed and exhaustive work on stratigraphy and paleopedology in the Midwest. It gave mappers a clear conceptual model to use in the field (Fig. 13.46).

Perhaps the most academically important contributions of Ruhe's work in southern Iowa stemmed from his 1956 paper, in which he compared soil development in three soils, all developed in Kansan till. Yarmouth–Sangamon and Late Sangamon soils, formed in till, are widespread and easily located. The latest cycle of erosion exposed some till areas for soil development as well by stripping off the loess cover, stabilizing no earlier than 6800 years ago (Ruhe 1956). Ruhe estimated that the Late Sangamon soils were surficially exposed for at least 13 000 years, although a longer period of formation seems likely. And of course, the Yarmouth–Sangamon soils could have been forming for hundreds of thousands of years.

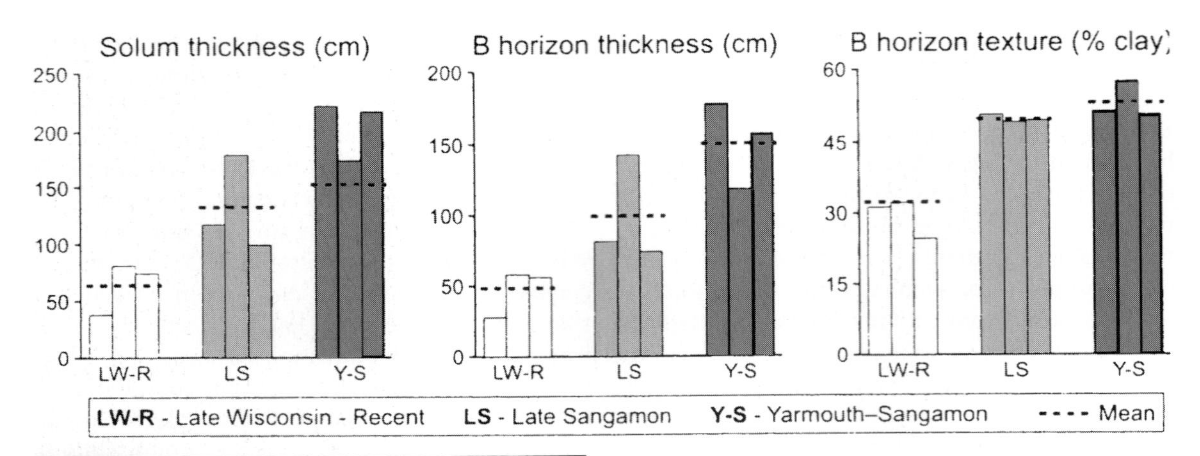

Fig. 13.51 Comparative data for soils on three different geomorphic surfaces in southern Iowa. After Ruhe (1956).

Because the chemistry of buried soils is so easily altered by solutions percolating into them from above and vice versa (see Chapter 15), an inherently valid way of comparing the development of these three soils is to evaluate their mineralogy. In theory, the ratio of resistant/weatherable minerals should increase as the soil develops, and should be higher in near-surface horizons (see Chapter 14). Data on the abundance of heavy and light minerals of varying resistances to weathering exhibited these predicted trends (Fig. 13.48). The most weathered soils are on the Yarmouth–Sangamon surface, while the least weathered soils are Late Wisconsin in age (Fig. 13.51). Solum and B horizon thickness, as well as B horizon texture, all change consistently along this development sequence. Ruhe's work paved the way for many other studies of relative soil development (see Chapter 14).

The Iowan erosion surface

One of the most controversial soil geomorphic problems in the Midwestern United States is the Iowan erosion surface (Olson 1989). Once thought to be a landscape formed during a separate ice advance, i.e., the Iowan, it is now known to have formed by long-term erosion of a loess-mantled, glaciated landscape. Most of the surface is of low relief and covered with an erosional stone line and pedisediment. Many of the stones are large and concentrated in farm fields; this is "stone country" (Fig. 13.52)! In short, the Iowan surface is a regional window through which we can view the great reduction of topography that was imposed during the early and mid Wisconsin glacial period.

The Iowan erosion surface covers almost 23 000 km^2 in northeastern Iowa (Fig. 13.45). On its northwestern edge it abuts Wisconsin-aged deposits of the Des Moines lobe, which shows little sign of Iowan-like erosion, suggesting that erosion of the Iowan surface was completed by about 14 000 years ago. Although the Iowan surface has low relief, it lacks many of the short slopes and kettles typical of recent, constructional landscapes like the Des Moines lobe (Fig. 13.53). Drainage networks are well integrated, indicating an erosional origin. Drift on the Iowan surface is Pre-Illinoian in age, much like the Kansan drift in southern Iowa (Fig. 13.45). Scattered remnants of the old, Pre-Illinoian paleosurface do exist on the very tops of the interstream divides (Alden and Leighton 1917). Here, the entire pre-erosional stratigraphy is preserved. Many of the sediments eroded off the uplands of the Iowan surface were deposited in alluvial lowlands; stones and large boulders, left behind, dot the uplands (Fig. 13.52). The Iowan surface contains only thin loess deposits, as any loess that would have been deposited here would have been quickly eroded from these unstable slopes. Only toward the very end of the erosional event could loess accumulate, and it is thin.

Erosion of the soils and upper stratigraphy of the Iowan surface was probably a late-Wisconsin event, circa 29 000 to 18 000 years ago (Ruhe

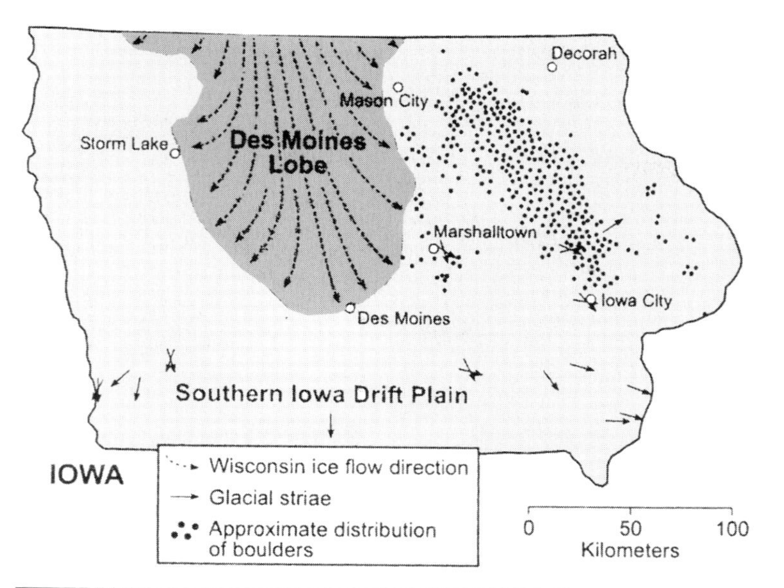

Fig. 13.52 Distribution of large granite boulders in Iowa. Inset: several large boulders in a field on the Iowan erosion surface, Chickasaw County. Image source unknown; map after Alden and Leighton (1917).

1969). At this time, cold, windy periglacial conditions existed here, giving rise to slope instability and erosion; ice wedge casts and polygons, unequivocal indicators of a periglacial climate and permafrost, are common. Faunal remains buried in lowland sediments indicate a cold, almost tundra-like climate, slightly wetter than today (Prior 1991). Walters (1994) placed the timing of periglacial conditions here at about 21 000 to 16 500 years ago. Such a climate would have facilitated extensive freeze–thaw action, widespread mass movements and slopewash. Existing soils would have been stripped from the landscape, with the exception of paleosols buried beneath relict hills. Stone lines, ubiquitous across the Iowan surface and associated with the erosion event, are slightly disrupted and slump down

(a)

(b)

Fig. 13.53 Landscape views of (a) the Iowan surface in Black Hawk County, Iowa, and (b) the dissected Kansan (Pre-Illinoian) drift plain in Jones County, Iowa. From Plate V of Alden and Leighton (1917).

Table 13.6 Frequency distribution of slopes on typical Iowa landscapes: Iowan Erosion Surface and Southern Iowa Drift Plain

Slope group (%)	Frequency on Iowan surface (%)	Frequency on Kansan (Southern Iowa) surface (%)
0–3	50.7	8.6
3–6	22.9	10.7
6–16	26.3	27.1
16–40	0	53.7

Source: Hallberg *et al.* (1978b).

Fig. 13.54 The Iowan erosion surface in northeastern Iowa. Photo by RJS.

when they cross the ice wedge casts, implying that the periglacial conditions coincided with slope degradational processes (Walters 1994). The erosion event that formed the Iowan surface only slightly impacted the Kansan drift in southern Iowa, but eroded most of this drift in the classic "Iowan" area; the boundary between the two is usually abrupt and irregular.

The net effect of this long period of slow-but-steady erosion was to decrease the relief and lessen the slopes, but to preserve the moderately integrated nature of the drainage network that was present on the pre-existing landscape (Table 13.6, Figs. 13.53, 13.54). Because of this landscape morphology, many *early* researchers saw the Iowan landscape as one that was clearly older than the kettled Des Moines lobe drift plain with its deranged drainage pattern, but one that was younger than the high-relief, dissected drift plains of southern Iowa. The drainage pattern on the Iowan surface was, therefore, thought to have been inherited by deposition of thin Iowan drift onto a dissected Kansan surface. Based on geomorphology alone, these were valid first assumptions. However, there are a number of topographic/geomorphic features associated with the Iowan landscape that suggest it is *not* a separate drift sheet: (1) it lacks an end moraine, (2) it extends farther on interstream divides than in valleys and (3) the landscape is topographically lower than the Kansan drift plain (Hallberg *et al.* 1978b).

The events that led to the current interpretation of the genesis of the Iowan erosion surface illustrate the value of pedo-stratigraphy to studies of landscape evolution. Arguments were once made for this region having a cover of Iowan drift, the age of which would have been between Illinoian and Wisconsin (Alden and Leighton 1917, Kay and Apfel 1929, Leverett 1939, Kay and Graham 1943), or perhaps early Wisconsin (Leighton 1933, Kay and Graham 1943, Ruhe *et al.*

1957, Ruhe and Scholtes 1959). Because this area has such low relief, it yields few exposures of subsurface sediments and stratigraphy, which made stratigraphic interpretations difficult (Hallberg *et al.* 1978b).

Stratigraphy on the Iowan surface, so critical to the correct interpretation of its genesis, was only worked out after extensive drilling and coring operations, largely under Ruhe's supervision. Eventually, stratigraphic and paleopedologic studies led to the conclusion that there was no "Iowan drift" and that, in fact, the region represented an erosional area where entire stratigraphic units were *missing*. Ruhe and his co-workers drilled several cores across the southern boundary of the Iowan surface, onto the Southern Iowa Drift Plain. The stratigraphy on the drift plain immediately south of the Iowan surface consisted of till containing a Yarmouth–Sangamon paleosol, covered with over 10 m of Wisconsin loess (Fig. 13.55a). Nearer the Iowan surface, drill core data showed increasing amounts of truncation of this paleosol, with loess overlying it. Eventually, on the Iowan surface, the paleosol is missing, and loess overlies either leached or calcareous till. Because the till can be traced under the paleosol, it must be Kansan, and thus there is no Iowan drift. Wisconsin loess lies directly above eroded Kansan till on the Iowan erosion surface. Ruhe's work had shown that the Iowan surface was cut into Kansan till, below the level of the Yarmouth–Sangamon paleosol (Ruhe 1969). A prominent stone line occurs, stratigraphically, between the loess and the underlying till, marking the contact and pointing a smoking gun at an erosion event as the reason for the missing stratigraphy. The sandy zone in the thick Wisconsin loess (Fig. 13.55) probably formed as sand blew off the eroding Iowan surface, onto the adjoining, higher and stable landscapes. Ventifacts on the Iowan erosion surface confirm this assumption (Walters 1994). After erosion ceased, the sand also stopped being deposited; however, loess was still coming into the area, burying the sandy zone (Fig. 13.55a). Thus, the erosion event is neatly placed between 29 000 years ago (a date on organics in the basal soil that would later be buried by loess) and 18 300 years ago (a date on organic matter from the base of the loess that immediately overlies the erosion surface). The last increment of loess covers the entire area (Fig. 13.55a), while the earliest increment (below the sand zone) equates with the hiatus represented by the stone line in the till. Thus, Wisconsin loess deposition started before the erosion event and ended after it.

Another geomorphic/stratigraphic assemblage used to show the origin of the Iowan surface occurs at its many *paha*. Paha are loess-capped hills – relicts that preserve the underlying stratigraphy (Figs. 13.56, 13.57). They are essentially southern Iowa drift plain outliers on the Iowan surface. At 4-Mile Creek paha, the complete stratigraphic assemblage is preserved, even the Afton paleosol below the Kansan till (Fig. 13.55). Paha retain the full complement of loess, in contrast to the lower-lying erosion surface, which does not (Ruhe 1969). It was the discovery and correct interpretation of the buried soils and stratigraphy that allowed Ruhe to correctly interpret the soil geomorphic history of the Iowan erosion surface.

Ruhe and his associates made good use of radiocarbon dating and stratigraphy, usually via drill cores – tools that previous researchers had limited access to. Data derived from these two new methods changed everything. His work emphasized the importance of slope stability to soil formation and showed that many geomorphic surfaces are erosional, illustrating how soil and stratigraphic data could help identify such erosion. Nonetheless, there remain unexplained questions about, in particular, the Iowan surface. Why does it, in places, have such abrupt boundaries? Why are some paha sandy? And could the Iowan erosion have been assisted by catastrophic outbursts from ice sitting at or near the Des Moines lobe margin?

Butler's K cycle concept

Soils develop best on stable surfaces. Soils on unstable surfaces are either being eroded or buried. If the surface and the soil developed on it gets eroded, a new soil forms on the surface once it stabilizes. If the unstable surface gets buried, a buried paleosol is preserved as evidence of the past period of stability (see Chapter 15). Examples

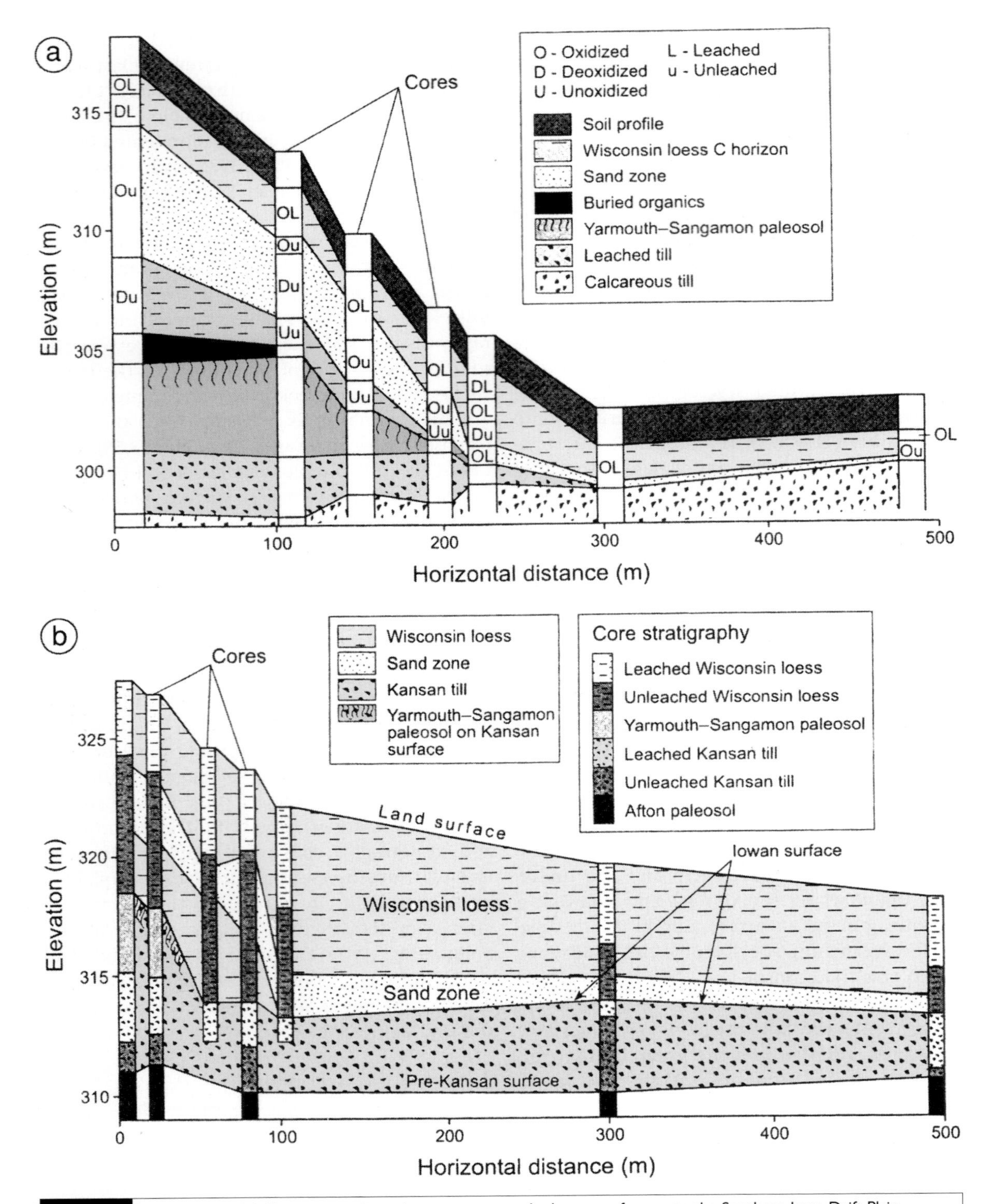

Fig. 13.55 Stratigraphy of drill cores at various sites near the Iowan erosion surface. After Ruhe et al. (1967) and Ruhe (1969). (a) Transect of drill cores across the southern edge of the Iowan surface, onto the Southern Iowa Drift Plain. (b) Transect of drill cores from the Iowan surface onto the 4-Mile Creek paha.

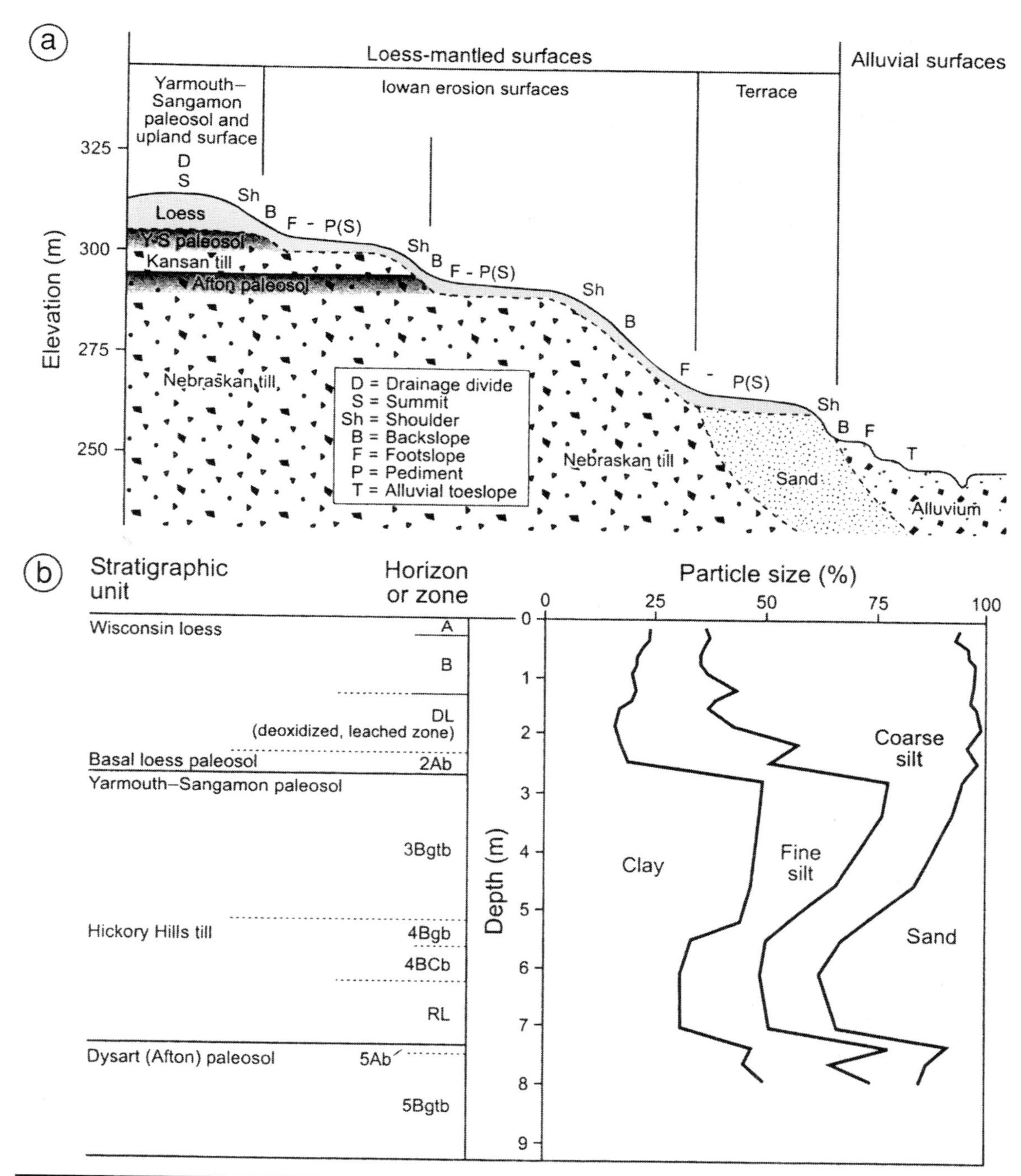

Fig. 13.56 Stratigraphy and particle size data for the sediments within Hayward's Paha. After Hallberg *et al.* (1978b). (a) Schematic representation of a section from Hayward's Paha down into the adjacent stream valley. Stepped erosion surfaces, typical of much of Iowa, are evident. (b) Detailed stratigraphy, paleopedology and particle size data from a core through the paha.

Fig. 13.57 A paha near Mount Vernon, Iowa. Photo by T. Kemmis.

abound of stacked sequences of paleosols, either buried by episodic increments of loess, dune sand, alluvium or glacial drift (e.g., Thorp *et al.* 1951) (see Chapter 15). De Villiers (1949) pointed out that this sort of landscape "periodicity" is, in fact, the norm over the past 2 million years. This premise – that all landscapes undergo periods of stability and instability – led Bruce Butler, an Australian soil scientist, to develop his K cycle model (Butler 1959). Fundamental to the model and to the notion of landscape periodicity is the *soil cycle*, which includes an alternating phase of instability and stability. During the instability phase, the older surface is either buried or eroded. During the stability phase, soil development proceeds on a new surface.

An implicit assumption of Butler's model is that there are times when certain processes overtake soil development, leading to soil burial. Similarly, at times soils get eroded because sediment is removed from the surface faster than the soils can develop on it.

Butler's model, built strongly on alternating cycles of stability and instability, contrasts with Nikiforoff's (1949) earlier model, which postulated that, on many gently rolling landscapes, the rate of surface erosion and deposition was in balance with pedogenesis. On uplands, the rate of sediment removal equaled the rate at which the B horizon changed into A horizon material, as the soil grew downward. The result was that A and B horizons maintained a standard thickness as the profile sank into the landscape at the same rate as the surface was eroded. Likewise, on lowlands, deposition took place at a slow enough rate that the soil could grow upward at an equal pace. A horizon material changed into B material just fast enough that the soils never became cumulic. On the landscape as a whole, there existed an equilibrium between slope erosion or deposition and soil development. Nikiforoff's model, now seemingly trivial, de-emphasized slope processes and made soil development into a suite of processes that seemed much more important than perhaps they are. Butler's (1959) K cycle model showed how slope processes can, at times, overwhelm pedogenesis and thus put more emphasis on the geomorphic, i.e., slope, component of soil development.

A soil cycle begins with soil development on a stable geomorphic surface (Butler preferred the term *groundsurface*) and concludes when that soil is either buried or eroded. One cycle leads immediately to the next, although a cycle need not go through to completion. For example, a soil could begin developing on a surface that is only ephemerally stable, only to be eroded or buried. This is a "partial" cycle because it did not experience *full* development. The soil cycle is common and widespread. It is most useful when used as a type of *time unit*, called the *K cycle* (K for Greek *khronos*, time) referring to relative time, not time in an absolute or geochronometric sense. The K cycle is defined as the interval of time covering

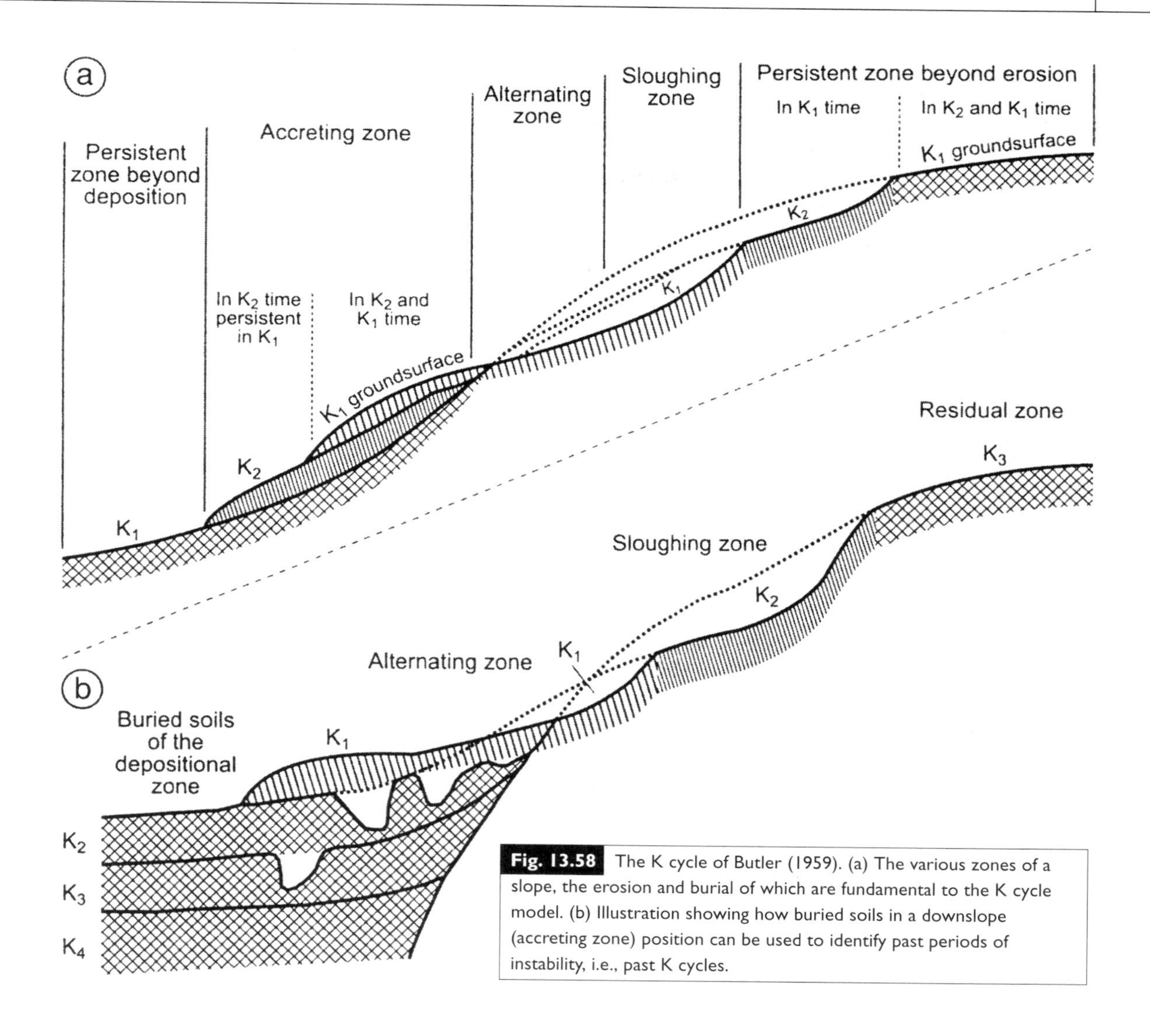

Fig. 13.58 The K cycle of Butler (1959). (a) The various zones of a slope, the erosion and burial of which are fundamental to the K cycle model. (b) Illustration showing how buried soils in a downslope (accreting zone) position can be used to identify past periods of instability, i.e., past K cycles.

the formation (whether by erosion or deposition) of the new geomorphic surface, the period in which soils develop on that surface, and ends when the surface becomes buried or eroded. Each cycle can be designated as K_1, K_2, etc., with larger numbers implying older times or earlier cycles. Using this terminology allows the investigator to communicate about periods of past stability or instability, by referring to them as K cycle 1, K cycle 3, etc. The present cycle of soil formation is not the K_1 cycle; the term K_1 is reserved for the most recent cycle back from the present. Further information could be presented about the paleo-cycles by using subscripts for stable or unstable phases of the K cycle. Thus, one could envision cycles, proceeding back from the present: K_{1s}, K_{1u}, K_{2s}, K_{2u}, K_{3s}, etc.

The unstable phase, when soils are eroded or buried, can be envisioned as the part of the cycle in which sites get conditioned for pedogenesis by exposing new parent material This phase can be spatially continuous, e.g., as when loess buries large areas. Conversely, erosion may be concentrated on certain slope types or aspects, or on surfaces underlain by coarse-textured soils (Pereira *et al.* 1978). The instability can be so spatially discontinuous that some parts of the landscape may be undergoing erosion while others are being buried (Fig. 13.58). This could be easily envisioned in a periglacial landscape such as the

Iowan surface, where solifluction is removing material from upper slopes and burying lower slope segments and soils. Butler (1959, 1982) described the various zones/sites on the land surface as *sloughing zones* (those that erode and lose sediment), *accreting zones* (those that become buried by new sediment) and *persistent or residual zones* (those that are stable and the soils on them continue to develop). Recognizing that surface stability/instability is both time and space transgressive, Butler also referred to some areas as *alternating zones* (Fig. 13.58).

We provide one example to illustrate the K cycle concept; others are available in the literature (Pereira *et al.* 1978, Dijkerman and Miedema 1988). At the Farm Creek stratigraphic section in Illinois, a stacked sequence of paleosols is exposed; each formed in loess that had accrued due to a glacial cycle, and each represents the stable part of the cycle. Loess deposition occurs here due to widespread landscape instability. (The Farm Creek section is explored in more detail in Chapter 15.)

Butler's (1959) K cycle concept is useful for linking soil development processes with surface stability. It brought to the fore the notion that surfaces and soils have linked histories. Both Butler's (1959) and Nikiforoff's (1949) models provide mechanisms by which pedogenesis can proceed on a landscape despite periods of surface erosion and deposition. A more recent model (Paton *et al.* 1995) provides another attempt to incorporate slope processes into pedogenesis.

The work of Paton *et al.* in Australia

The geomorphology specialty group of the Association of American Geographers occasionally awards the G. K. Gilbert Award for Excellence in Geomorphic Research to the author(s) of a landmark geomorphology publication. In 1995 a book by Tom Paton, Geoff Humphreys and Peter Mitchell, entitled *Soils: A New Global View*, won that award. These three Australian soil scientists and geomorphologists authored a book that outlines a view of pedogenesis that is heavily dependent upon landscape, slope processes and pedoturbation. Although it has come under a bit of scrutiny, as all models do (and should) (Beatty 2000, Johnson 2000), it brings together many excellent points.

The model of Paton *et al.* (1995) is perhaps best viewed in an historical context. Nikiforoff's (1949) earlier model de-emphasized slope processes by rationalizing a way that soil development could keep pace with slope processes, to form a "normal soil." Because buried soils are so common, Butler (1959) knew that Nikiforoff's ideas and model could not be universally true. Rather than most soils and surfaces being in equilibrium, Butler saw landscapes as being subject to periodic stability and instability. However, like Nikiforoff, Butler saw slope processes as *opposing* pedogenic processes, or at least separate types of processes. Both Nikiforoff and Butler treated erosion and deposition processes differently, but in each model *pedogenesis* was restricted to vertically operating processes on a stable substrate. This is where the model of Paton *et al.* (1995) differs. Paton and his colleagues proposed that the genesis of texture-contrast soils is due to *interacting* slope and pedogenic processes. Unlike the models of Nikiforoff (1949) and Butler (1959), slope processes are fully *integrated into* pedogenic processes in their model, and help explain the distribution of soils on slopes.

The model of Paton *et al.* (1995) is ostensibly a global model, although it was developed based primarily on observations made in Australia and is perhaps best applied to old, tropical landscapes. It should be viewed in this context. In Australia, long-term weathering and bioturbation are dominant processes. Landscapes are old. Ant and termite mounds are common, and these mounds eventually deteriorate into stone-poor biomantles. If the underlying parent material has coarse fragments, a stone line or stone zone will form below this biomantle (Fig. 13.59; see Chapter 10). In many soils the biomantle and stone line rest on a more clay-rich substrate that grades into saprolite.

One of the main purposes of the Paton *et al.* model is to explain the formation of texture-contrast soils (those with a clay-poor eluvial zone over a clay-rich Bt horizon). The genesis of such soils has traditionally been viewed as involving processes operating in the vertical dimension, i.e., lessivage (see Chapter 12). While lessivage is

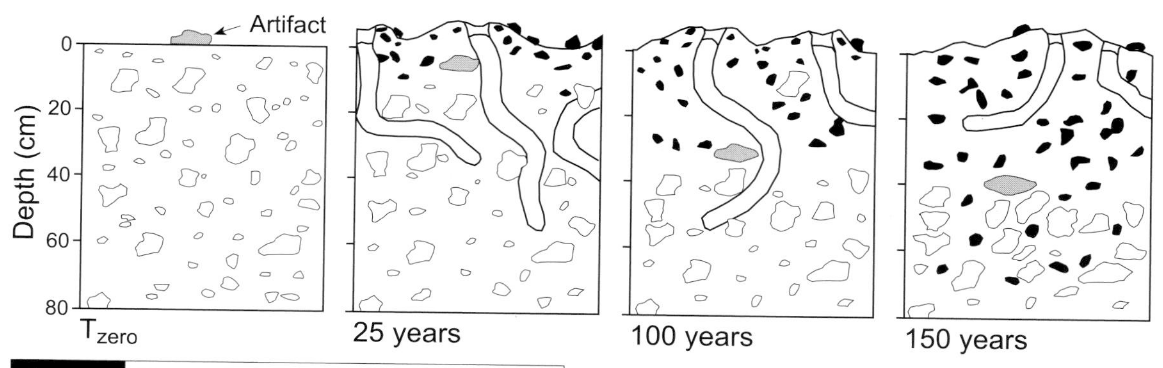

Fig. 13.59 Diagram showing how a stone zone can be formed by intense burrowing by animals – in this case mammals. Any coarse fragments in the soil that are too large for the fauna to bring to the surface settle to the depth of burrowing, forming a stone zone or stone line at that depth, and a biomantle above. After Johnson *et al.* (1987).

certainly operative in many soils, it is not as important, Paton *et al.* (1995) argued, to the formation of texture-contrast soils as are slope-related processes. Watson (1961) argued long ago that vertical soil processes in the humid tropics may not be as effective as they are in the midlatitudes, because the intense tropical rainfall on the less vegetated slopes leads to more runoff. In the tropics the focus is on water running *parallel* to the surface. Watson (1961) noted that tropical soils, therefore, are not derived so much from the rock below as they are from the rock upslope. Paton *et al.* (1995) saw similar relationships.

We begin by defining *epimorphic processes* (weathering, leaching, new mineral formation and inheritance) and distinguishing them from surface processes (including pedoturbation, rainwash, eolian processes and soil creep). Far too often, pedogenic theory ignores the latter set of processes. Paton *et al.* (1995) argued that texture-contrast soils form first as epimorphism produces soil material from rock, via saprolite. Next, biota mine the saprolite and transport some parts of it to the surface, as topsoil or a biomantle (Watson 1962). Then, surficial processes, e.g., slopewash, creep, wind transport and bioturbation, winnow and translocate this material downslope. Eventually, an ever-so-slowly downslope-moving blanket of soil material, essentially a type of altered biomantle topsoil, develops above saprolite. If the saprolite has clasts that are difficultly weatherable, a stone line will form at the contact between the biomantle and the saprolite, or at least above the more dense horizons below. Figure 13.60 illustrates the variety of morphologic and stratigraphic evidence that can be used to show that the upper portion of the soil, usually including the coarser-textured horizons, is in fact creeping downslope (Aleva 1987). Laterite (plinthite) is more typical of flat uplands, and it may contribute coarse clasts to the stone line, but stone lines are not normally found beneath thick ironstone layers, duricrusts or plinthite on flat uplands. Stone lines may be intermittent, a continuous layer about one clast thick, or they may be a zone of gravel up to several meters in thickness. They are usually not consolidated (Brückner 1955). The upper contact of the stone line is usually fairly abrupt. Together, these data (and others) indicate that stone lines are due to a combination of weathering and biomechanical processes, and that the slope and soils are linked together by a variety of dynamic processes, all leading to the slow denudation of uplands.

Material in the upper profile, typically coarser-textured than the material below, is a product of repeated winnowing by biota, slopewash, wind, and other surficial vectors, rather than eluvial processes winnowing out clay. Brückner (1955) noted the sandy character of biomantles on slopes in the Gold Coast, Africa. The effect of the surficial winnowing penetrates to the maximum depth of bioturbation. Below that depth, the subsoil is denser and more clay-rich, corresponding to a traditional Bt or BC horizon. This near-surface morphology is continually maintained as

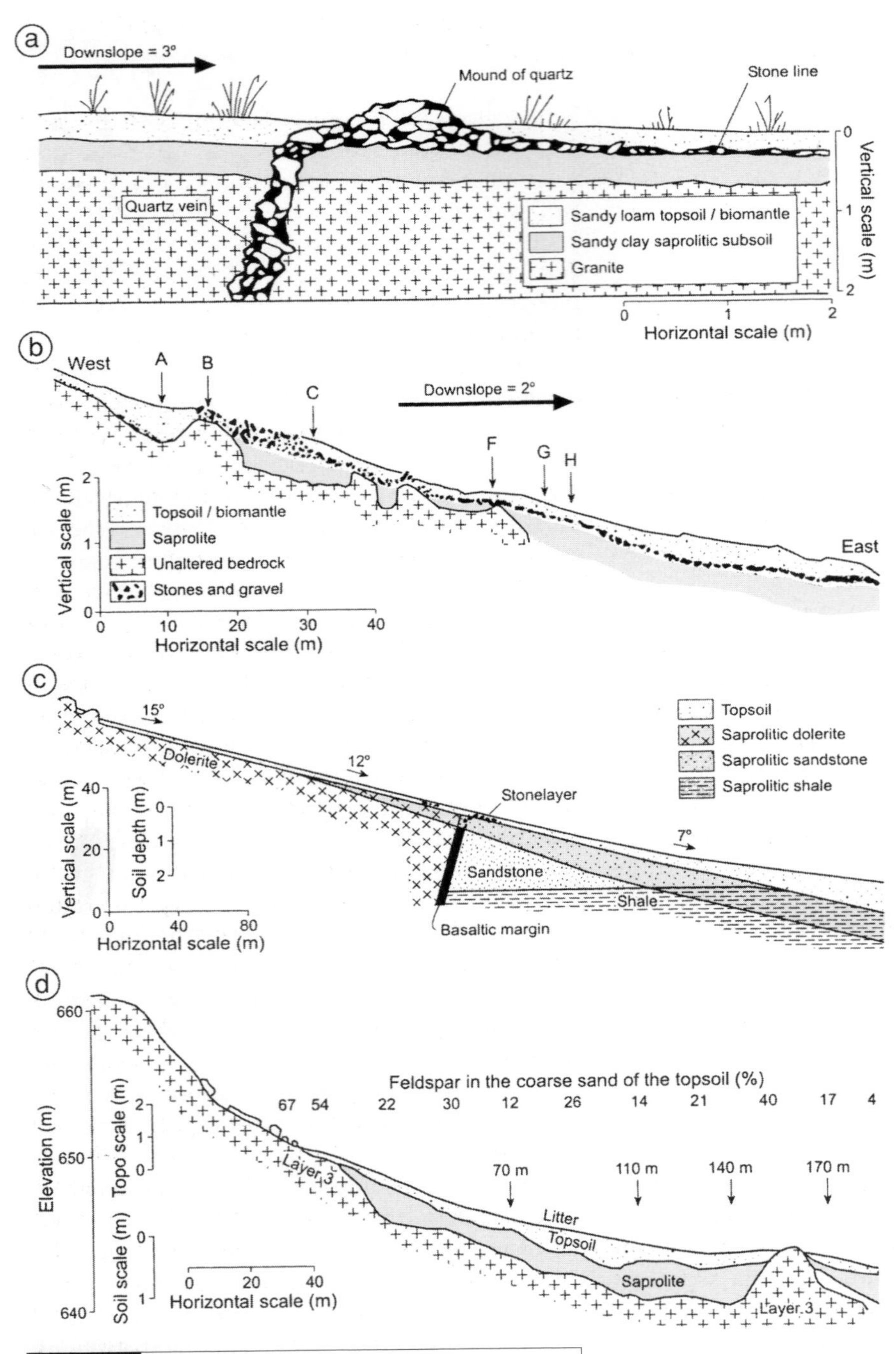

Fig. 13.60 Cross-sections through soil, saprolite and bedrock at several sites in Australia, each illustrating a mobile, downwardly moving, coarse-textured surface layer above a zone dominated by epimorphism (weathering, leaching, new mineral formation, and inheritance). (a) After Paton (1978); (b) after Bishop *et al.* (1980); (c) after Paton *et al.* (1995); (d) after Hart (1988).

biota mine the subsoil. With time, the surface is lowered and lowlands infill, but the soil profiles do not change markedly. Rather, they evolve with the surface.

The biomantle material gets thicker, coarser and more weathered in a downslope direction, as this set of processes runs to completion (Fig. 13.60). With time, it also becomes coarser, as fine material is blown or washed out of the surficial system. Weatherable minerals are, over time, depleted, leaving behind a quartz-rich, sandy layer that is an E horizon. As a result, the topsoil becomes better sorted and coarser-textured over time, moving slowly downhill over the subjacent material, not en masse but grain by grain, leading to an ever more marked differentiation, i.e., *texture contrast*, between topsoil and subsoil (Paton *et al.* 1995: 78).

Throughout their book Paton *et al.* (1995) provide evidence that vertical clay movement via eluviation has been overemphasized, implying instead that subsoil clay is due to inheritance from lithospheric material. Their work questioned the evidence used in support of lessivage, and its importance in comparison to other genetic mechanisms. Nonetheless, there are many examples of tropical soils with illuvial clay in Bt horizons in which the mechanism must have been vertically driven lessivage. Nye (1955b: 65) stated that it seemed likely that clay had been "washed vertically down these profiles," when referring to some tropical soils.

Paton *et al.* (1995) argued that many traditional pedogenic paradigms are flawed, and proposed to replace them with one involving more geomorphology, with surficial and biological processes acting upon an inherently mobile mantle of sediment (Schaetzl 2000). Their model helps explain the ubiquitous texture-contrast soils found across Australia, Africa and South America, which are typically composed of a bioturbated topsoil with a basal stonelayer over a saprolitic subsoil (Nye 1954, Watson 1961, McFarlane and Pollard 1989, Courchesne 2000). Percolating water, stressed in the northern hemisphere as an overriding vector in soil genesis processes, is de-emphasized. Old soil landscapes with their texture-contrast soils (or fabric-contrast soils) overlying landscapes floored by bedrock and saprolite are seen as the end product of landscape evolution.

Johnson's dynamic denudation model

Donald Johnson, a soil geographer, also developed a model of soil–landscape evolution that incorporates many of the ideas put forth by Paton *et al.* (1995); again, connections are important to the development of new pedogenic theory. Some of Johnson's work (Johnson 1993a, b, 1994) preceded the publication of *Soils: A New Global View* and so it is not so much a follow-up but a subtly different way of viewing soil-slope development. Johnson's dynamic denudation model incorporates geomorphologic *and* pedologic processes; it is highly integrative. What Johnson has done is to take the best of the literature and weave it together into a model that has wide utility.

Johnson's dynamic denudation model is centered on *biomechanical soil processes*, i.e., processes that change the soil physically/mechanically. They are driven primarily by fauna. Johnson (1993a) coined the term and reviewed how biomechanical processes have had a long history in the literature, beginning with Darwin (1881) and Shaler (1890). Strictly defined, biomechanical processes are those processes, impacts and effects that are *physically imparted* by biota to the landscape (see Chapter 11). Together with *biochemical* processes, they constitute the impact of all biotic processes, captured in the *o* soil-forming factor (Jenny 1941b). Biomechanical processes include disturbance processes such as bioturbation, loosening activities of hooves and feet, and anchoring by roots. Too often, however, one tends to think of the *o* factor as the biochemical part only, such as acidification by organic acids, base cycling by plants, and the chemical fertility of plant and animal residues. Though it was one of the earliest of the "soil processes" to be discovered and discussed (Johnson 1993a), the idea that biomechanical processes are highly influential in soil and slope development had to wait a century to be recognized and ranked equally with other forms of pedogenic processes.

Why were *biochemical* processes stressed for so long? Part of the answer lies in the dominance of the plant component of the *o* factor in the state-factor model, particularly the amount and

Table 13.7 The five fundamental elements of dynamic denudation theory

Major element(s)	Components
Process theories in geomorphology and pedology	Triple planation processes, etchplanation, chemical denudation and leaching processes, biomantle, soil evolution and soil thickness processes, mass transport and soil creep processes
Soil horizon conventions	O-A-E-B-C horizon designations M-S-W horizon designations
Hydrologic principle	Throughflow, interflow and eluviation–illuviation processes
Key pedogeomorphic agents	Biological (biota) Chemical (biota, air, water) Physical (biota, gravity, water)
General soil definition	Soil: rock material or sediment at the surface of planets and similar bodies altered by biological, chemical or physical agents, or a combination of them

Source: Johnson (1993a, b, 2002).

quality of organic matter produced by *plants* relative to biochemical and nutrient cycling. Also, within the "organisms" factor, animals were never given equal status with plants (Jenny 1941b) (see Chapter 11). And although animals can affect soils biomechanically, they were most often viewed as impacting soils biochemically, through their excreta.

The dynamic denudation approach is a staircase built on the work of others (Table 13.7). The model (Fig. 13.61) incorporates three pedogeomorphic agents, gravity, water and biota, to explain "why three tiered, stone line bearing soils and their equivalents are so common" on relatively stable but actively evolving surfaces of "every continent but Antarctica" (Johnson 1993b: 68). In this respect, it is similar to the model of Paton *et al.* (1995) and differs little from the early work of Watson (1961) and Nye (1954). Dynamic denudation works best on landscapes that have not undergone recent rejuvenation, such as those recently glaciated, on mountains and high hills with unstable slopes or in areas with additions of volcanic ejecta. Thus, it is most applicable to stable interiors of continents like Africa, Australia and South America. In the model, soils are assumed to be bioturbated such that a stone line forms at the lower limit of bioturbation. Like the model of Paton *et al.* (1995), the material above the stone line is seen as slowly being transported downslope by a variety of slow but persistent processes. Weathering proceeds at depth, producing saprolite from hard bedrock (Fig. 13.61). If rock is not present at shallow depths, however, the model still has explanatory power.

Aleva's (1983, 1987) triple planation model is a part of the dynamic denudation model. Triple planation refers to three "surfaces" so often found in tropical soils (Büdel 1957). In the triple planation model, P_1 is a weathering front surface, at the rock–saprolite interface, that gradually migrates downward through time (Fig. 13.61). Many of the materials released by weathering are removed laterally by groundwater, along the P_1 surface. The P_2 surface is largely a wash surface, on which sediment and soil materials are transported laterally by wind, water and biota, including tree uprooting (Brückner 1955, Schaetzl and Follmer 1990, Norman *et al.* 1995). Finer-textured materials tend to move farther, such that the bottoms of slopes have finer-textured soils and thicker sediment accumulations (Watson 1961) (Fig. 13.61). The P_3 surface is often associated with a stone line or coarse-textured zone at the base of an actively moving biomantle. Underneath is a denser zone that sometimes functions as an aquitard – either a Bt or C horizon. The P_3 surface is a zone of lateral throughflow, within which soluble materials leached from the overlying soil can leave the soil system. It is often convoluted, rather than planar as shown in Fig. 13.61 (Johnson 1993b).

Over time, weathering (epimorphism) proceeds on the underlying bedrock, and as it does the rock-turned-saprolite collapses due to solutional losses. Weathering is concentrated on the rock divides and outcrops, while the broad, flatly concave valleys, so typical of large areas of humid tropical landscapes, are areas of transportation of weathering by-products. The evidence for downslope transport is unequivocal, as dikes and quartz veins that transgress the saprolite appear to curve in a manner that could only happen if the weathering rock was collapsing (Aleva 1987, Bremer and Späth 1989) (Fig. 13.61a). Because the stones in the stone line are often resistant, economically important minerals, sampling the stone line may be an inexpensive way to establish the presence or absence of certain minerals in the rocks upslope (Aleva 1987).

Horizon nomenclature within the constraints of the model

This brief digression into the history of soil horizonation will be of value to understanding the dynamic denudation model. The traditional A–B–C horizon nomenclature system, developed by Dokuchaev in the late nineteenth century has been the standard throughout the midlatitudes (Johnson 1994). The A–B–C system is, according to Johnson, primarily a descriptive one which has become linked to various genetic processes. For example, the A horizon became associated with certain processes such as the accumulation of organic matter and eluviation, while the B horizon concept evolved into one of illuviation and weathering. By 1951, the US Department of Agriculture encoded the A–B–C nomenclature as "policy" by publishing it in their Soil Survey Manual (Soil Survey Staff 1951). The A and B horizons came to be known as the solum while everything below about 8 or 10 feet (2.4–3.0 m) was considered C horizon material or "not soil" (Marbut 1935). The logic behind this was linked to practical mapping concerns; limiting the solum to about 2–3 m in thickness allowed soil mappers to focus on near-surface soil properties and saved many hours of augering and excavation into materials that geologists cared more about than did soil scientists. By the 1980s, the basic A–B–C horizon scheme had expanded to O–A–E–B–C (Guthrie and Witty 1982, Soil Survey Division Staff 1993). Fitting soils on old landscapes to the A–B–C system is difficult because they are much thicker and their horizon boundaries are more blurred, and because slope processes are clearly an important part of their genesis (Johnson 1994). Their subsurface stone lines are also difficult to fit into the A–B–C horizon system. These landscapes had not been disrupted by ice sheets or buried by loess (Buol 1973, Johnson 1994).

The layered nature of the soils, so typical of tropical landscapes, is distinctly different from the layering of A–B–C midlatitude soils. The upper layers of tropical soils are sandy, porous and apparently heavily influenced by soil biota. Nye (1954, 1955a, b, c) distinguished two such layers, which overall were also influenced by creep processes, and hence both layers were named "Cr" (De Villiers 1965). The upper, allochthonous, "creep horizon" was called Crw, for the dominant activities of worms, while he called the lower one Crt for termite activity (Fig. 13.62). The Crw horizon was only a few centimeters thick and contained almost no coarse sand, which worms do not ingest and hence do not bring to the surface (Nye 1955b). The Crt horizon does contain coarse sand, which ants and termites can move. Strongly influenced by Nye's work, Watson (1961) later combined the Crw and Crt layers into a master horizon, M, for mineral soil (Fig. 13.62). Aleva (1983) also chose to differentiate the creep layer from a layer above it which was more strongly influenced by biotic activity and biomechanical processes. In 1987, he went on record as saying that all necessary downslope movement of the M horizon can occur by gravity movement – "there is no need to involve boring animals, such as termites" (Aleva 1987: 200). Both Nye and Watson routinely observed that a gravelly/stony layer lay below the "creep" zone or biomantle.[3] Nye (1954) called this stone layer CrG, while Watson

[3] They did not use this term, because it had not yet been coined, but it does fit their intent.

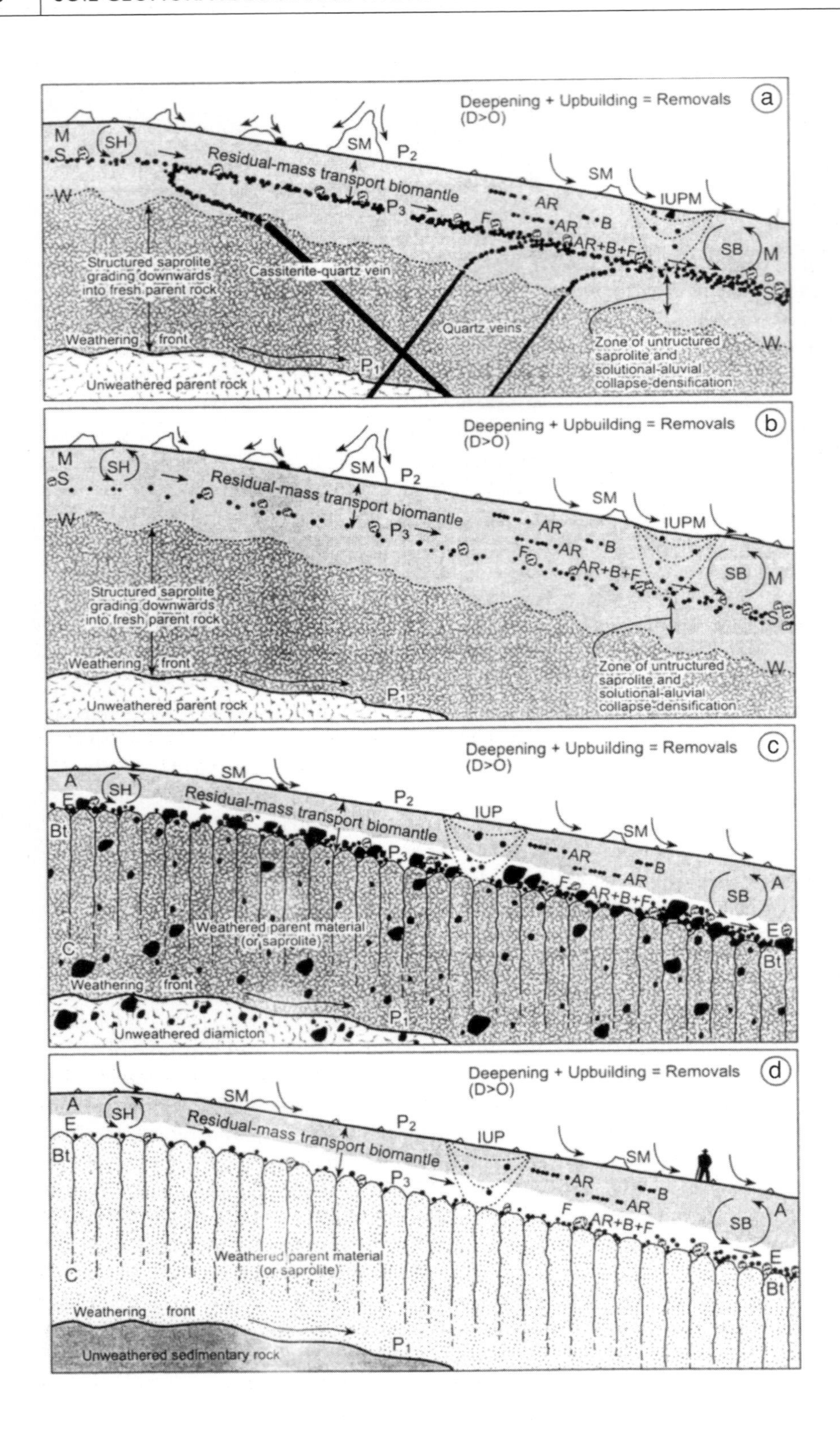
Deepening + Upbuilding = Removals (D>O)
a
M
S
SH
SM
P2
Residual-mass transport biomantle
SM
AR
IUPM
W
P3
F
AR
B
AR+B+F
SB
M
S
Structured saprolite grading downwards into fresh parent rock
Cassiterite-quartz vein
Quartz veins
Weathering front
Zone of untructured saprolite and solutional-aluvial collapse-densification
W
P1
Unweathered parent rock
Deepening + Upbuilding = Removals (D>O)
b
Weathered parent material (or saprolite)
Unweathered diamicton
c
A
E
Bt
C
IUP
d
Unweathered sedimentary rock

Fig. 13.61 Examples of dynamic denudation processes on various types of substrates. After Johnson (1993b).
(a) Dynamic denudation in landscapes underlain by dike- and vein-bearing intrusive igneous and metamorphic rock. The dashed arrow indicates the first planation (P_1) level, P_2 is the second planation (wash) surface (curved arrows) and P_3 is the third planation (throughflow) level (downslope-directed straight arrows). With regard to horizons, M is the mineral horizon, S is the stone horizon (stone line) and W is the weathered horizon (saprolite). The base of the S horizon is the top of the W horizon. AR refers to human artifacts and manuports. B indicates bioclasts (mainly avian, reptilian and mammalian gastroliths). F symbolizes ferricrete and other metallic concretions. IUPM is an infilling pit and associated mound, formed by an uprooted tree, which contain artifacts, bioclasts, and other clasts mixed upward from the stone–line. SB refers to subsurface bioturbation and attendant jostling and "loosening" by soil fauna, such that large clasts tend to sink. SM indicates surface mounds formed by invertebrates ants, termites, gophers, mole-rats, etc. Note that the residual-mass transport biomantle, which extends from the soil surface to the base of the S horizon, gets thicker downslope. A Bt horizon may or may not be present below the stone line. Note also that upper saprolite above the wavy broken line is unstructured and more dense than is either the unstructured (isovolumetric) saprolite below or the S horizon above, and thus functions as an aquitard. The S horizon, underlain by an aquitard or aquicude, functions as an aquifer for lateral throughflow. The two surface clasts (above the M of mass) were dug up from the stone line by large vertebrates. (b) Dynamic denudation in landscapes underlain by crystalline rocks that lacks veins and dikes. Symbols and horizons are as in (a). (c) Dynamic denudation in landscapes underlain by stony and gravelly sediments. Symbols are as in (a), except for the following: A, an A horizon with abundant organic matter; E, an eluvial horizon that coincides with the stone line, Bt: a horizon of illuvial clay that functions as an aquitard – the vertical structures with round tops symbolize subsoil pedogenic structure; C, weathered parent material or saprolite; and IUP, a completely infilled tree uprooting pit that contains artifacts, bioclasts, and other clasts previously ripped up from the stone line. (d) Dynamic denudation in landscapes underlain by non-stony sedimentary rock. Symbols are as in (a). In non-stony terrains residual clasts are exclusively artifacts, bioclasts and metallic nodules.

(1961, 1962) labeled it S for stones (Fig. 13.62). The bottommost layer, autochthonous saprolite, Nye (1954) believed to be sedentary and thus he labeled it S. Watson (1961) preferred W for weathered rock. The W horizon grades downward into structured saprolite.

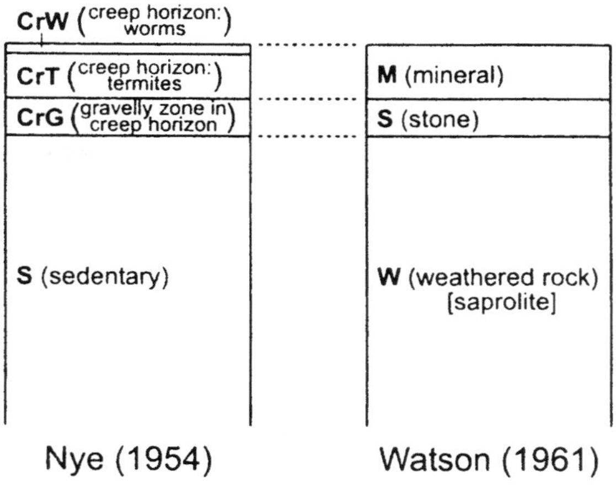

CrW (creep horizon: worms)
CrT (creep horizon: termites)
CrG (gravelly zone in creep horizon)
S (sedentary)
Nye (1954)
M (mineral)
S (stone)
W (weathered rock) [saprolite]
Watson (1961)

Fig. 13.62 Two early horizon schemes for tropical soils, and how they compare. After Nye (1954) and Watson (1961), but patterned after a figure in Johnson (1994).

To summarize, the M, S and W horizons of old, tropical landscapes correspond to layers of mineral soil (M), stones or a stone line (S) and weathered rock (W) (Williams 1968, Johnson 1994). In most cases, the W horizon corresponds to the BC, C or upper Cr horizons, the M horizon corresponds to the A horizon or the upper solum, and the S horizon is equivalent to an eluvial zone, possibly the E (Johnson 1994). Watson's (1961) M-S-W terminology, so useful for many upland tropical sites, was nonetheless never adopted by midlatitude soil scientists.

M (A) horizons are maintained on a dynamic slope by an interplay of processes such as wash, erosion, bioturbation and mass transport. Fine particles are brought to the surface by termites, ants, worms and other biota as a biomantle. The maximum grain size in the upper, M, horizon is the same as observed in termite mounds, and in the Gold Coast the thickness of the M horizon on summits and upper slope locations rarely exceeds the normal depth of termite burrowing (Brückner 1955). Some, although not a lot, of organic matter is added to the M horizon, forming a dark A horizon. That the soil has a distinct A horizon amidst all this dynamism implies that organic matter accumulation is progressing more rapidly, at least in the uppermost few centimeters, than are mixing and disturbance processes (Johnson and Watson-Stegner 1987).

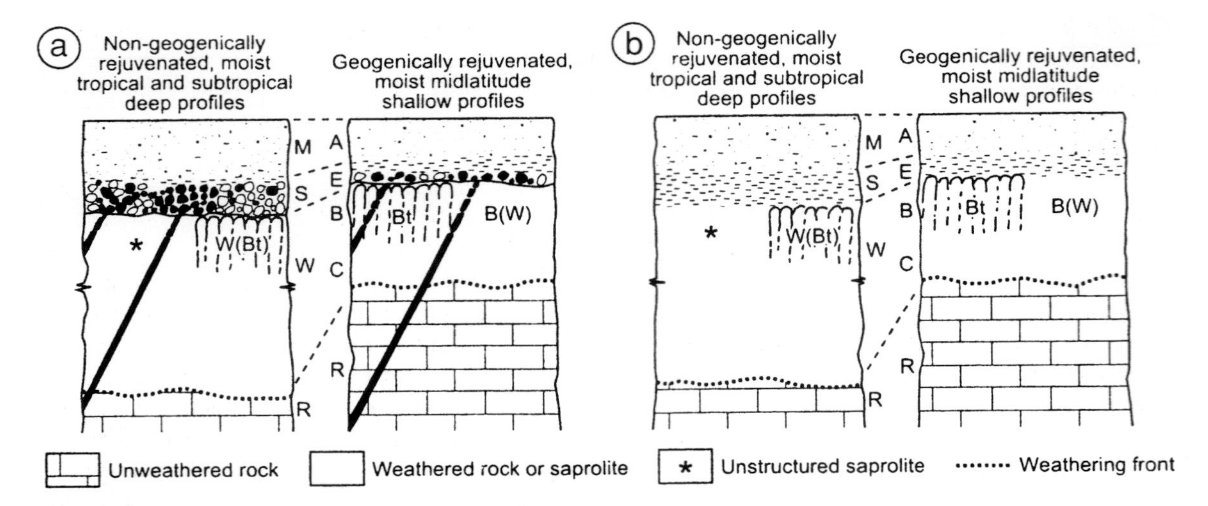

Fig. 13.63 Comparison of master horizons in the two main horizonation schemes: tropical soils vs. midlatitude soils. Designating the E horizon assumes that lateral throughflow has at least contributed some of its eluvial character. After Johnson (1994). (a) Morphologies associated with saprolite that contains dikes and veins of resistant rock. (b) Morphologies associated with parent materials that lack coarse fragments.

The coarse-textured, eluvial S (E) horizon lies at the base of the biomantle and corresponds to Aleva's (1983) P_3 planation surface. Most of the dissolved material that exits the soil–slope system leaves through the S (E) layer (Johnson 1993b). The stones in the stone line are also creeping downslope, when on sloping surfaces (Figs. 13.60, 13.63). These stones are relicts from the saprolite below, commonly with resistant lithologies of quartz, cassiterite, tourmaline, topaz, anatase, corundum, rutile and greisen (Aleva 1983, Dijkerman and Miedema 1988). Clasts are fed into the S zone from dikes and veins that are weathering out from below, and bioturbation keeps most of the coarse fragments below the M (A) horizon. Watson (1961) argued that the S horizon may be essentially stationary on gentle slopes. He also did not equate the S horizon to the E, as did Johnson (1994), noting instead that any vertically eluviated E horizon in tropical soils is probably obliterated by faunalturbation. At locations where the saprolite does not contain hard relicts, the stone line may be absent. However, Fe-rich, laterite-like nodules can and do form at the saprolite–soil contact, especially when infiltrating water perches there (Dijkerman and Miedema 1988, Faure and Volkoff 1998). Brückner (1955) referred to these features as limonite concretions. Nye (1955a, b) noted that the concretions will form only when soils wet and dry, and that they can weather if carried by creep to the wetter, lower slopes. Archeological artifacts, if present, will also be found at the stone line or above it, depending on how long they have been subjected to the biomechanical processes that operate near the surface (Brückner 1955). Bioclasts, e.g., gizzard stones and gastroliths, can also accumulate at the stone line. After large birds ingest stones (to aid in digestion), they pass them out as excreta or cough them up. Bioclasts are fairly frequent components of soils and stone lines, but are only noted in soils where coarse fragments would otherwise be absent, such as on loess uplands (Cox 1998).

The W (BC or C) horizons reflect a complex interplay between weathering and pedogenic/translocation processes (Johnson 1994). Commonly, a Bt horizon exists in the upper part of the W zone, especially if the degree of slope downwasting is slow. Illuvial clay within the Bt horizon is presumably entering the horizon laterally, along the stone line, as well as vertically from above (Nye 1955a). If the slope is downwasting rapidly, any Bt horizon that might form is stripped and degraded, as it comes into contact with the biomantle and the S (E) zone, explaining perhaps why some three-tiered soils have a Bt horizon while others lack one. Under this scenario, Bt horizons would be best developed on slope elements with the lowest gradient. Certainly, the presence or absence of a Bt horizon, on surfaces like those in Figure 13.61, also has a great deal to do with climate. In dry climates Bt horizons will not form, even on stable slopes.

With this background in mind, we now present two contrasting examples of stone line genesis, as a means to examine the utility of the dynamic denudation model, but also to illustrate the fact that many physical features in soils can have more than one possible origin.

Pedisedimentation and erosion surfaces

As surfaces erode, certain indicators remain behind as clues to the types of processes involved. Interpreting this information, usually of a stratigraphic nature, is vital to a correct interpretation of landscape evolution. These *erosion surfaces* must be interpreted and identified with respect to the rock and sediment beneath and above them, as well as their variation in space (Ruhe 1975b).

One way that landscapes are denuded is by pedimentation, which forms pediments – broad, concave-upward erosional surfaces, usually *cut into* the base of a larger slope (Ruhe 1958, Mammerickx 1964) (Fig. 13.4). When a slope is cut into another, there often exists a knickpoint (or a break in slope) at its upper end, where it grades into a higher, older surface (Fig. 13.50). The morphology of the pediment, in this case, would support an erosional origin for stone lines that may exist within or on top of it (Ruhe 1958). Pediments have a linear or slightly upwardly concave profile because descending tributaries coalesce on the lower parts of the pediment, allowing for more sediment deposition there (Fig. 13.16).

Rocks and stones, many of which have been slightly rounded during erosion, are common to pediments. A layer with this type of origin, where coarse materials and rocks have been left behind, is a *lag concentrate*. When seen in vertical section, the erosion pavement or lag appears as a stone line or stone zone. Similar terms are *carpetolith*, *carpedolite* and *carpedolith*, literally translated as "carpet of stones" (Parizek and Woodruff 1957, Aleva 1983).

Sediment is also transported across pediments, making them as much a slope of transportation as a simple erosion surface. Indeed, every erosion surface must also be a surface of transport, since the eroded material must be removed (Ruhe 1969, 1975b). *Pedisediment* refers to sediment that is, episodically, in transport across a pediment; similar terms include co-alluvium, pedoluvium and the German *Decklehm*. Some of it is always present on the erosion surface. During the waning stages of the pedimentation process some of the pedisediment that remains coats the pediment (Ruhe and Scholtes 1956, Ruhe and Daniels 1958). Because much of the pedisediment has been carried by water, it has some degree of sorting, often becoming more clay-rich downslope (Ruhe 1975b).

Theories of stone line development

Stone lines were first defined by Sharpe (1938) as lines of angular to subangular fragments which parallel the slope at a depth of several feet. Bremer and Späth (1989) listed several possible mechanisms that can lead to their formation: (1) colluviation, (2) bioturbation, (3) soil creep, (4) chemical formation as concretions and nodules and (5) denudation. They go on to point out that, as a first step in determining the origin of the stone line, one should first ascertain whether the stones are autochthonous (formed in place) or allochthonous (transported in). Perhaps the earliest theory of stone line genesis is that the stones

> have become detached from dikes or other resistant layers and have been drawn along at the base of the creeping soil, which have worked their way downward in the soil profile because of the more rapid movement of the surface layers, or which have been concentrated toward the base owing to the more rapid disintegration near the soil surface. (Sharpe 1938: 24)

Sharpe was essentially describing a situation like the texture-contrast soil model of Paton *et al.* (1995), in which bioturbation is responsible for lowering stones, ultimately forming a stone line at the maximum depth of burrowing. Darwin

(1881: 228–229) proposed the same mechanism for the lowering of stones and coins, and the burial of ancient Roman foundations and walls:

> worms have played a considerable part in the burial and concealment of several Roman and other old buildings in England; but no doubt the washing down of soil from the neighbouring higher lands, and the deposition of dust, have together aided largely the work of concealment. Dust would be apt to accumulate wherever old broken-down walls projected a little above the then existing surface and thus afforded some shelter. The floors of old rooms, halls and passages have generally sunk, partly from the settling of the ground, but chiefly from having been undermined by worms . . . The walls themselves, whenever their foundations do not lie at a great depth, have been penetrated and undermined by worms, and have subsequently subsided.

and again (Darwin 1881: 308–309):

> Archaeologists ought to be grateful to worms, as they protect and preserve for an indefinitely long period every object, not liable to decay, which is dropped on the surface of the land, by burying it beneath their castings.

Biogenic stone lines such as these can develop in any soil, as long as the soil contains coarse fragments and numerous soil biota are available to bring small particles to the surface. Biogenic stone line formation has been advocated by many, especially those working in the tropics (Tricart 1957, Ollier 1959, Webster 1965a, Williams 1968, Humphreys and Adamson 2000). Fine sediment mining is best done by soil invertebrates such as termites, ants and worms (see Chapter 10). Because it takes some time for a biomantle of this type to form, some degree of slope and surface stability is assumed.

The biomantle model (Johnson 1990, 1993a, b, 2002, Johnson and Balek 1991) does not involve erosion; stone lines form contemporaneously with soils. In this model, stone lines are biogenically induced, subsurface deposits covered by a biomantle (Johnson and Balek 1991). Johnson and Balek (1991) refer to this as the *dynamic denudation–soil evolution–biomantle model*, since the sediment above the stone line is, in essence, a stone-poor biomantle (Johnson 1993a, b, 1994). A type of hybrid model, discussed by Brückner (1955), involves erosion of gravelly soils to form a surface gravel lag (stone line), which is then buried by a biomantle.

A second model of stone line formation involves pedimentation, in which the stone line is a part of the pedisediment or a lag deposit. Stone lines of this type occur on contemporary or paleo-erosion surfaces. Flat or tabular stones in these types of stone lines usually lie roughly parallel to the surface (Wallace and Handy 1961). The stone line is formed during the same erosion event that cuts the erosion surface, and thus the stones overlie, at least momentarily, a truncated soil profile. Any pedisediment above the stone line is also contemporaneous with the erosion event. Ruhe called this process of stone line formation "pedimentation–pedisedimentation," viewing pedimentation (i.e., erosional) processes as those that produced the surface lag, and pedisedimentation processes as those that buried it. In landscapes where bedrock is near the surface, erosional stone lines may rest on bedrock. In short, the stone line is seen as the morphological expression of an erosion pavement cut by running water, and subsequently covered with surficial sediment (Ruhe 1975b: 130). Stratified deposits above stone lines argue strongly for burial of the stone line by slopewash or colluvium, rather than by a biomantle. This type of stone line marks a lithologic discontinuity and represents a break in sedimentation. It also represents a temporal discontinuity within the stratigraphic section (Parizek and Woodruff 1957), and marks many erosion surfaces cut into Quaternary drift in the Midwestern United States (Ruhe 1969).

Most mid-twentieth-century studies ascribe stone line formation to erosion–pedimentation processes (Parizek and Woodruff 1957, Fairbridge and Finkl 1984). Even the Soil Survey Manual (Soil Survey Staff 1951), produced under Charles Kellogg, presupposes that stone lines have an erosional genesis. The reason for this bias is primarily historical, and worth exploring (Johnson 2002). Curtis Marbut's successor at the US Department of Agriculture Division of the NCSS, Charles Kellogg, interpreted subsurface stone lines he saw in the Belgian Congo as erosional lags that had later become buried (Johnson and Balek 1991).

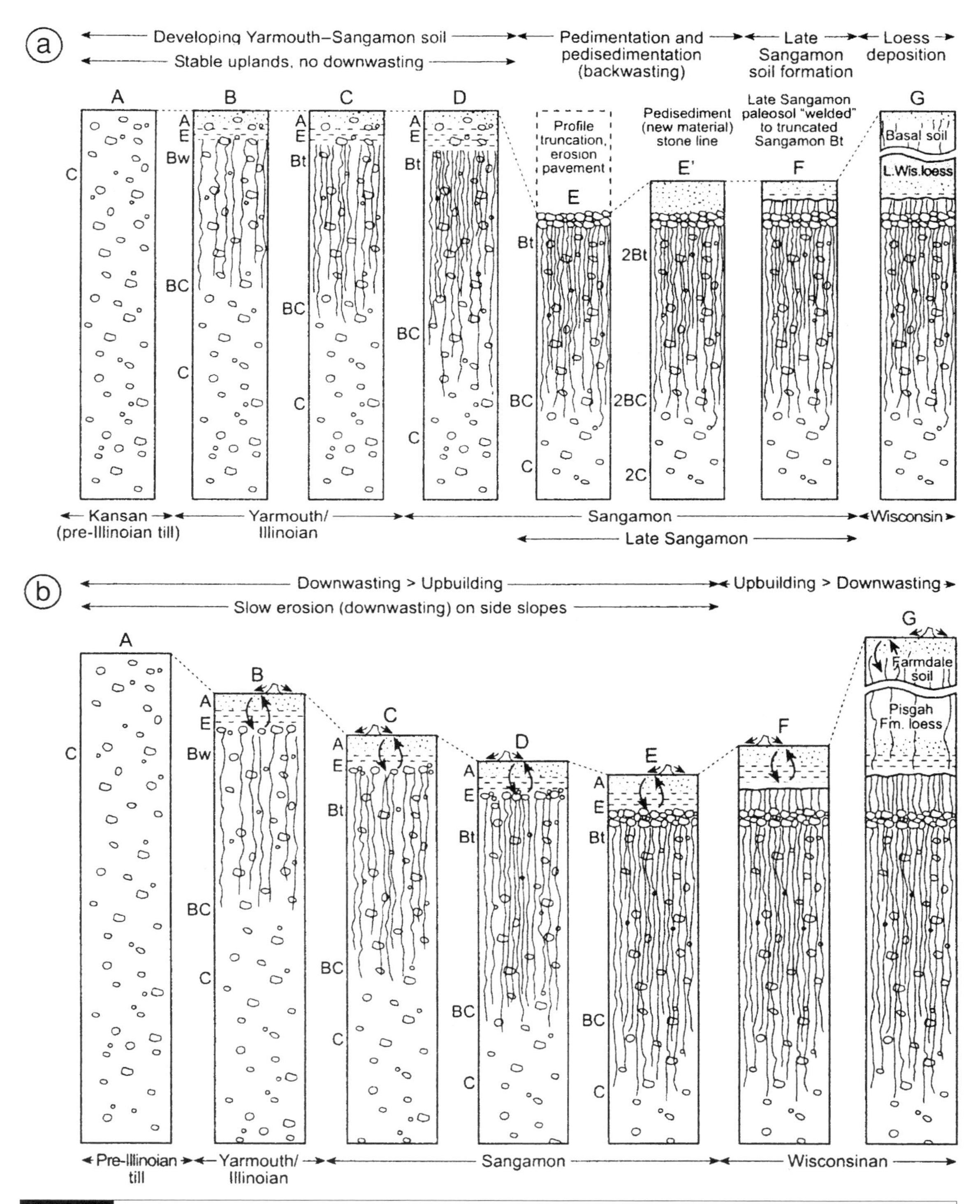

Fig. 13.64 Soil chronographs (windows in time, showing soil morphology as it changes) for the southern Iowa Drift Plain. After Johnson and Balek (1991). (a) Model of landscape evolution and stone line formation, using the erosion–pedimentation–pedisedimentation model of Ruhe and Kellogg. (b) Model of landscape evolution and stone line formation, using the biomechanical model of Darwin and Johnson.

Kellogg later arranged for a detailed study of this area by Robert Ruhe, whose findings agreed with Kellogg's erosional interpretation of stone line genesis, despite his knowledge of other work in the region that described stone lines formed by bioturbation and soil creep (Ruhe 1958). Kellogg's and Ruhe's stone line interpretations have had far-reaching consequences; stone lines have since been interpreted by many as geologic–erosional entities (Johnson 2000).

The popularity of the erosion–pedimentation model is due to its main proponent, Robert Ruhe; his impact on theories of stone line genesis was great. Ruhe's strong stance on the erosion–pedimentation origin of stone lines in Africa was carried to Iowa. His position on the stone line issue persisted despite (1) Darwin's (1881) work that had documented how soil biota could lower stones and artifacts by bioturbation (Johnson 2002, Van Nest 2002) and (2) his knowledge that non-erosional landscapes of the Des Moines lobe (in Iowa) also had occasional stone lines (Wallace and Handy 1961). He (Ruhe 1975b: 129) rejected the biogenic theory of stone line formation for sites in the United States because "mass movement processes cause mixing of particles of all sizes." Thus, mass movement alone could not explain the presence of a stone-poor sediment (i.e., the biomantle) above the stone line. In this regard, he was correct, but he (as did most others at this period in time) ignored bioturbation. Thus, almost by happenstance, the erosional–pedimentation theory of stone line formation became the preferred view, while another, proposed decades earlier by Darwin (1881), languished.

Which perspective/model has the most applicability? Certainly both are plausible models that have great utility; the two models are *not* mutually exclusive. Stone lines of both types exist in nature. Biogenic stone lines tend to be small and are often intermittent, while erosional lag stone lines are "big" and often associated with sandy pedisediment. Either model could be applied to the Southern Iowa Drift Plain (see above), with essentially the same resultant soil morphology (Fig. 13.64). The main difference in assumptions between the two is that Ruhe's erosion–pedimentation model assumes a period of backwasting, during which slope gradients were lowered (Fig. 13.64a). The pedimentation model places the formation of the stone lines at some time in the *past* (Fairbridge and Finkl 1984). In the biomechanical, dynamic denudation model (Fig. 13.64B), bioturbation, soil formation and downwasting are all assumed to occur *simultaneously*, but because soil formation outpaces bioturbation and downwasting, the profile has A–B–C horizonation. Soil formation keeps pace with the slow downwasting process, and bioturbation leads to the formation of biomantle above a subsurface stone line.

Chapter 14

Soil development and surface exposure dating

Soils and soil data can help determine the age of geomorphic surfaces (e.g., Mahaney 1984, Amit *et al.* 1996). In this regard, a soil can be no older than the *deposit within* which it formed, or the last period of stability of the geomorphic *surface upon* which it formed. That is, if a surface has undergone several periods of instability and stability, the current soil on it probably had its time$_{zero}$ no earlier than the onset of the last period of stability. Other soils on, or related to, that surface would have been either buried or eroded by earlier periods of instability (Butler 1959).

To determine the actual length of time that the soil may have been forming, one must first establish the age of the surface. This is usually done using geochronometric techniques such as radiocarbon dating or tephrochronology. Given enough pedogenic data, information regarding the amount of time required for certain soil properties to develop, or to be lost, begins to emerge. Armed with this knowledge, a soil geomorphologist can then use soil properties to provide reasonable estimates of the ages of other surfaces, for which strict geochronometric control is lacking. Thus, the first step in a soil geomorphic study is to ascertain what is known about the geologic past, what the main temporal markers are, and the duration and starting points of the soil-forming intervals, i.e., when surfaces were stable.

Colman *et al.* (1987) pointed out the need for consistency in terminology among those using dating techniques in soils and geomorphology, and we agree. For example, a *date* is a specific point in time, e.g., 5330 BP (before present), while an *age* is an interval of time measured back from the present. The terms *ka* and *Ma* are to be used for ages (thousands and millions of years ago, respectively).

Stratigraphic terminology, principles and geomorphic surfaces

In 1983, the North American Commission on Stratigraphic Nomenclature published its North American Stratigraphic Code. The code defined and formalized stratigraphic nomenclature, including various units categorized by physical attributes (lithostratigraphic, biostratigraphic and allostratigraphic units, etc.) and age (chronostratigraphic, geochronologic and geochronometric units). These units provide a basis for the systematic ordering of the time and space relations of rock/sediment bodies and establish a time framework for the discussion of geologic history (North American Commission on Stratigraphic Nomenclature 1983 : 849). Lithostratigraphic units are based/defined on lithology, chronostratigraphic units are based/defined on time or age, etc.

Isochronous means of equal duration, while *synchronous* means simultaneously, or at the same time. The term *diachronous* is used for units whose bounding surfaces are not synchronous. Diachronous is a synonym for *time-transgressive*. Many geomorphic surfaces are time-transgressive, meaning that not all parts are of the same age. A good example is a surface being buried by loess. Burial occurs first and most rapidly near the loess source and only later do sites farther away get buried. The buried surface is time-transgressive since its burial was not synchronous. Many subaerially exposed, i.e., not buried, geomorphic surfaces are undergoing erosion on some parts and burial (by slopewash or alluvium) on others; they are also time-transgressive.

In this book we are most concerned with two types of time-defined stratigraphic units. A *chronostratigraphic unit* is a body of rock or sediment that serves as the material reference for all rocks formed during the same span of time; its boundaries are synchronous. A *geochronometric unit* is a division of time, based on the rock record. It corresponds to the time-span of an established chronostratigraphic unit; its beginning and end coincide with the base and top of the referent chronostratigraphic unit. Types of geochronometric units include eons, eras, periods, epochs and ages (Table 12.3).

Of special importance to us is the *pedostratigraphic unit*. It is defined as a body of rock or sediment that consists of one (or more) pedologic horizon(s) that is overlain by one or more formally defined stratigraphic units. Thus, the pedostratigraphic unit must be buried; surface soils do not count! Soil horizons, recognized by their morphology or micromorphology, are used to distinguish pedostratigraphic units from other stratigraphic units. As in surface soils, it is often much more difficult to identify the base of a pedostratigraphic unit than it is to identify the top. The boundaries of pedostratigraphic units are almost always time transgressive.

The fundamental pedostratigraphic unit is the *geosol*, which is not dissimilar to a paleosol except that it (geosol) formally fits within the guidelines of the North American Stratigraphic Code. The term geosol refers to the entire soil; it is not formally divisible into horizons. A geosol is named for a location where it was first formalized, studied or described, e.g., the Sangamon and Farmdale Geosols.

Although the Code formalizes units of known age, or at least helps correlate among them, geologists have been correlating and comparing stratigraphic units for centuries, and have developed a few "laws" to assist in this endeavor. They may seem obvious today, but these laws were important breakthroughs at a time when geology as a science was young. Nicolaus Steno (1638–86), a Danish geologist and priest, developed a very simple rule – the *principle of superposition*. It states that, in an undeformed sequence of sedimentary rocks or other similar sediments, each bed is older than the one above and younger than the one below (Fig. 14.1). In other words, older rocks are deeper in the geologic column. Steno also recognized the *principle of original horizontality*, which states that layers of sediment are generally deposited in a horizontal position. Thus, rock layers that are flat have not been disturbed and maintain their original horizontality. Those not currently flat-lying must have been disturbed at a subsequent time. Other "laws" that are useful in interpreting the relative age of the rock column include the *principle of cross-cutting relationships* – when a fault or igneous intrusion, e.g., a dike, cuts across another body of rock, it must be younger than the host rock. A similar relative age relationship holds for rock inclusions. Material that is included within a host, such as fragments of basalt within a mud flow deposit, must be older than the host.

For geomorphic surfaces, one of the first determinations must be whether it is erosional or depositional. Depositional surfaces are often similar in age to, or slightly younger than, the sediments below them (Daniels *et al.* 1971a). Erosional surfaces are more complex; their age is usually determined by examining the slope or material they grade toward and merge with, since the erosional surface is the same age as the depositional surface it grades to. Erosional surfaces are invariably younger than the material underlying them. Such a surface is younger than the youngest material it cuts or bevels, but older than valleys cut into it or deposits that lie within those inset valleys.

Many of these slope–age relationships are captured in the *principle of ascendancy and descendancy*: an erosion surface is younger than the youngest deposit or surface that it cuts across or truncates (Hallberg *et al.* 1978a, Watchman and Twidale 2002). This principle is especially useful in fluvial settings, where younger materials are inset within older sediments. Corollaries to this principle, as outlined by Ruhe (1975b), state that an erosion surface (1) is the same age or older than other deposits lying on it, (2) is older than

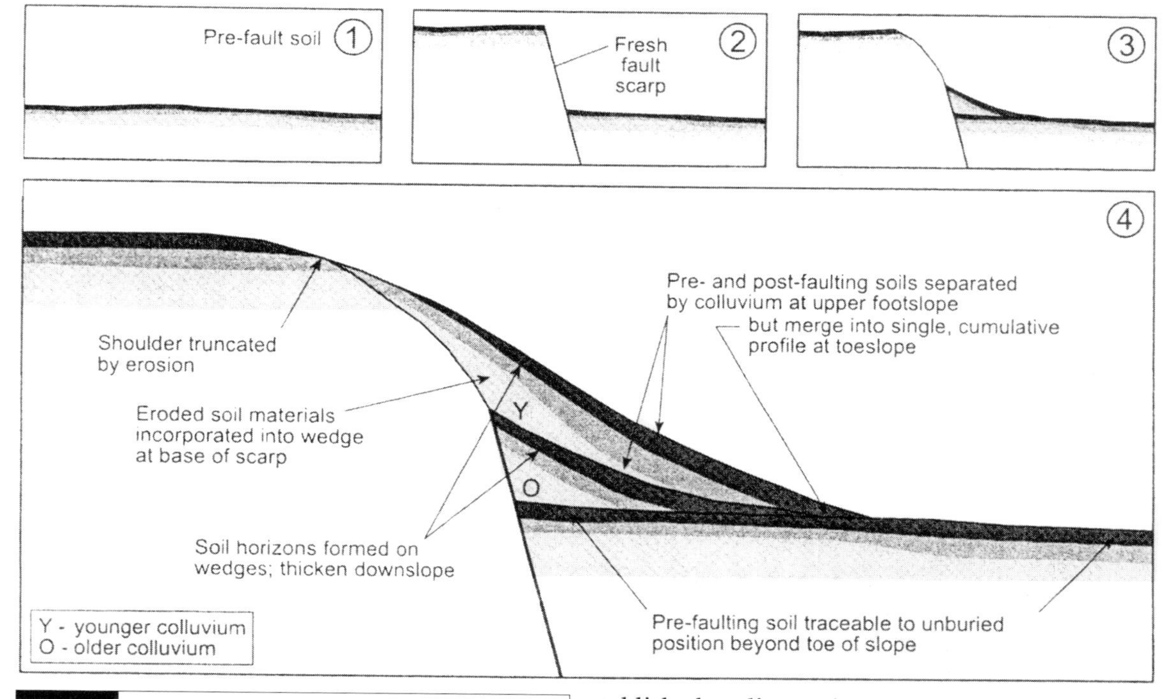

Fig. 14.1 Examples of soil development and buried soils (surfaces), and how they can be used to infer past periods of surface stability or instability. The fresh fault scarp in (2) has no soil development, as it is the youngest surface. The buried soils in (4) illustrate at least two subsequent periods of stability in this landscape. Even within geomorphic surfaces, varying degrees of stability are expressed, as indicated by the shallow soils on the shoulder slope. After McCalpin and Berry (1996).

valleys cut below it and (3) is younger than erosion remnants above it. Likewise, a hillslope is the same age as the alluvial fill to which it descends, but is younger than a higher surface to which it ascends. Many of the above-mentioned stratigraphic principles are illustrated in Figs. 14.1 and 14.2, and discussed by Daniels *et al.* (1971a).

Numerical dating

Land surfaces and surficial sediments can be dated in two basic ways, using numerical or relative dating techniques (Watchman and Twidale 2002). In numerical, aka *absolute*, dating, an established radiometric method is used to read the "natural chronometer" built into sedimentary system (Colman *et al.* 1987). This method allows for quantification of absolute differences in age among sediments or units. Numerical dating of geomorphic *surfaces* is more challenging. Often we must first determine whether we are dating the *sediment* or the *surface*. It is usually easier to date the *sediment* or something within it that is age-contemporaneous, e.g., a piece of wood that lies within alluvium can be dated with ^{14}C. If we do this, however, all we can then say about the age of the overlying *surface* is that it is no older than the numerical age obtained on the sediment or wood. If the surface has experienced a period of erosion, its age is younger than any numerical dates we could obtain from the sediment. Numerical dating of geomorphic *surfaces* can be done only if datable materials exist on that surface, which are contemporaneous and associated with it, i.e., they are strictly surficial phenomena. Some of the best types of material in this regard are archeological, e.g., hearths, storage pits, post molds, etc., since human occupation is always associated with a surface and not the sediment below. Numerical dating of both sediment and surfaces is further limited by the availability

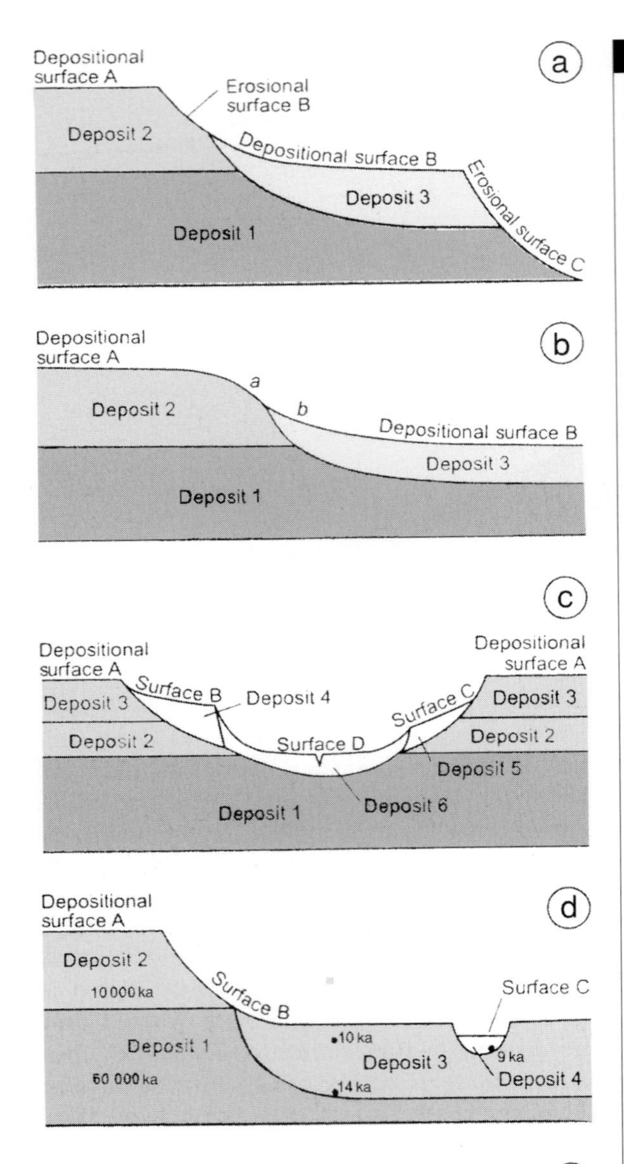

Fig. 14.2 Hypothetical examples of relative age and stratigraphic relationships in surficial deposits. After Daniels *et al.* (1971a) and Hall (1983). (a) Deposit 1 is older than deposit 2 because of the principle of superposition. Deposit 3 is younger than 1 and 2. Erosional surface B is younger than deposit 2 because it truncates that deposit. Surface B is also younger than surface A. Erosional surface C is the youngest surface in this figure. (b) The basal of deposit 3 is an erosion surface cut prior to or at the time of its emplacement. Surface A is a depositional surface and thus is younger than deposit 1 but the same age as deposit 2. Surface B is younger than surface A because it bevels the deposits underlying surface A and surface B cuts surface A. Portion *a* of surface B is the erosional element and portion *b* is the depositional element. This diagram illustrates that many surfaces have both erosional and depositional components. Without datable material no numerical ages can be assigned to the various surfaces and deposits. (c) Deposit 4 is emplaced on a surface that bevels deposits 2 and 3 and cuts surface A; therefore deposit 4 is younger than deposit 3 and surface A. Because surface B cuts surface A, it is younger. Surface C is younger than surface B because it is at a lower level. Surface D is younger than surfaces B and C because it cuts them. Deposit 6 is younger than 1 and 2 because it bevels them. (d) In this figure, numerical ages can be assigned to deposits and surfaces. In this example, it is known that the ages of deposits 1 and 2 are 60 Ma and 10 Ma, respectively. The erosional portion of surface B is therefore less than 10 Ma. A radiocarbon date of 14 ka, from the base of deposit 3, establishes that the deposit is younger than that age. A 10 ka radiocarbon date from material just below surface B implies that the depositional phase associated with deposit 3 took less than 4000 years. The dating of material in deposit 4 which fills a surface cut into and below surface B establishes that surface B erosion and deposition occurred between 10 and 9 ka. Surface C is less than 9 ka old but from the information given a more precise date cannot be established. (e) Relative ages of surfaces A–C are the same as discussed in C. Deposit 5 (loess) covers the entire landscape as a relatively uniform blanket. It has a tendency to homogenize the landscape. Although the surficial expression of each of the buried surfaces is still present, surface D has a single (synchronous) age.

of datable materials, and many numerical dating methods are expensive. Thus, many researchers are forced to turn to relative dating techniques (Hall 1983).

Relative dating

Before the advent of numerical dating, sediments and rocks were dated only by relative means. In relative dating, rocks and sediments are placed in their proper *sequence of formation*, i.e., which came first, second, etc., or which is younger or older. Relative dating methods cannot normally

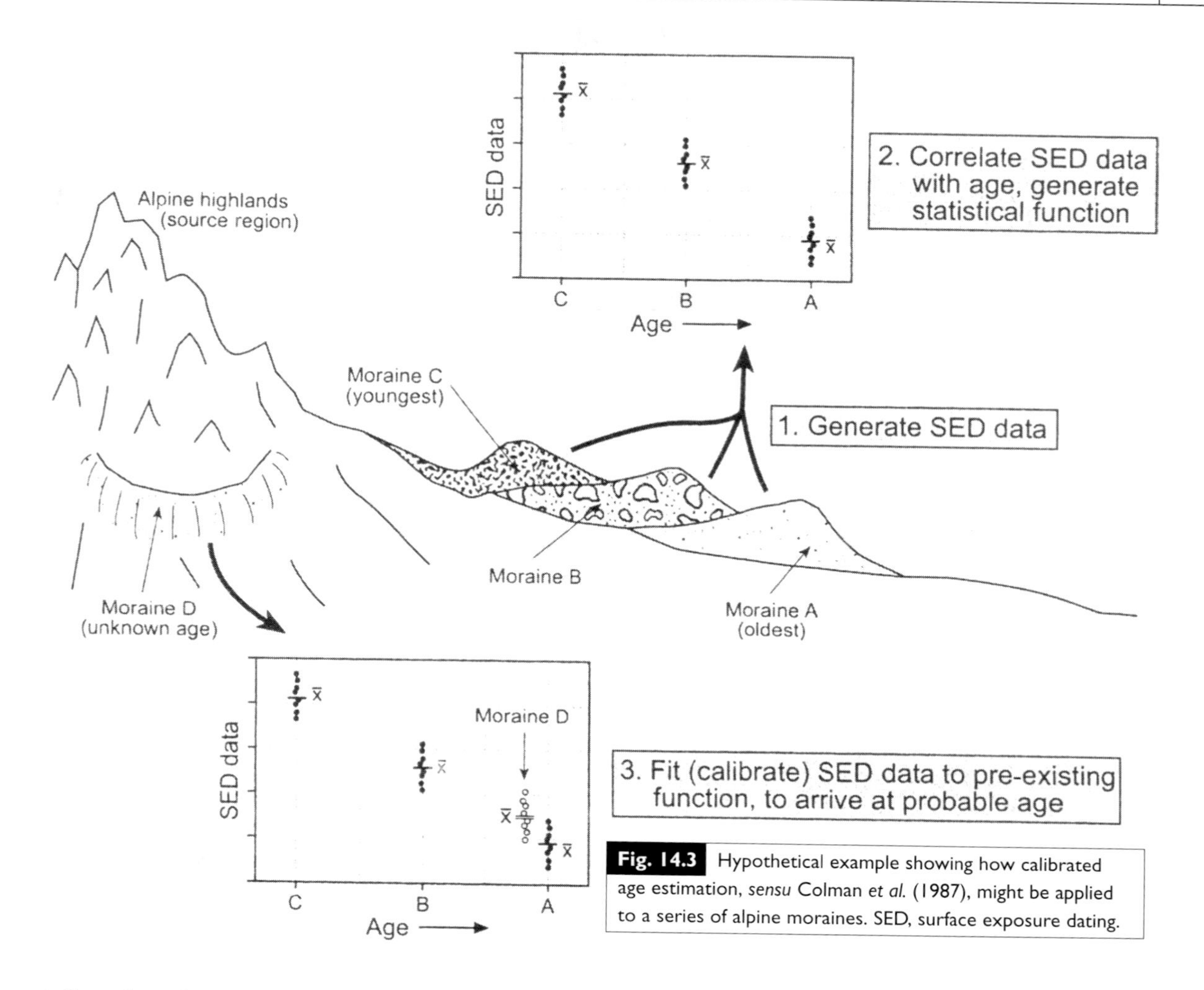

Fig. 14.3 Hypothetical example showing how calibrated age estimation, *sensu* Colman *et al.* (1987), might be applied to a series of alpine moraines. SED, surface exposure dating.

tell us how long ago a sediment was deposited or how long ago a geomorphic surface became stable, only that it followed or preceded the deposition of another sediment (principle of superposition), or perhaps the erosion of a geomorphic surface. Often, however, this degree of age assessment is adequate. Theoretically, relative dating has no temporal limits (Watchman and Twidale 2002). However, the older the surface the more likely that the necessary relative age data are either not preserved or not exposed.

With some skill and a little luck, however, relative dating can be taken one step further – these methods can be used to determine the general *magnitude* of the age differences between two or more deposits. For example, we may use soils to determine that not only is surface A older than surface B, but that it is at least *three times older*. We may be able to discriminate among three alpine moraines, for example, and determine whether they resulted from three distinct advances (MIS stages 2, 4 and 6) or from one advance with three short pulses/surges (all MIS stage 2).

Taken still further, relative dating can include a group of dating techniques called *calibrated age methods* (Colman *et al.* 1987). Calibrated age assessment requires that the numerical ages have been determined for some of the surfaces that have been dated using relative dating techniques. This widely used group of methods, therefore, involves informal "calibration" of data from soils or rocks between surfaces of known and unknown age (Fig. 14.3). For example, weathering rind data from rocks might be generated for moraines of known age to arrive at a statistical function, which can then be extrapolated (within

Table 14.1 A partial list of the major surface exposure (relative) dating methods

Surface exposure dating method	Common methods and/or assumptions	Maximum timespan utility (years)
Geology- and Biology-based Methods	Assume that the rocks or minerals used were not pre-weathered and that climatic variation over time has been minimal, or at least comparable among sites.	
Rock weathering, general	Degree of clast weathering is determined by striking the rock with a hammer and estimating the degree and ease with which it breaks apart. Rocks are then placed in a category of weatheredness.	10^4
Clast sound velocity	Compressional (P) waves are made to impact clasts; the speed with which the waves move through the rock is proportional to their degree of weathering.	$>10^5$
Rock angularity	Rates the degree of angularity of rock edges, assuming they get more rounded with time.	10^4
Etching of sand-sized hornblende minerals	Degree of etching of hornblende (or other) minerals in fine sand fraction is determined under a microscope and placed into categories. Also possible with some feldspars.	10^5
Rock surface oxidation, weathering rind formation and obsidian hydration	Correlates the depth (mean, max or mode), color, hardness or mineralogy of a weathering rind on the outsides of rocks to age/exposure. Obsidian hydration dating has the advantage in that obsidian can be numerically dated by the K–Ar method.	10^5 (10^7 for obsidian)
Pitting of rock surfaces	Pits get deeper, wider and more common on older rocks.	10^5
Vein height	Uses the height of veins on weathered rock, assuming that the veins stand out higher as the softer matrix weathers away over time.	10^4
Lichenometry	Assumes that the size of lichen thalli on surface-exposed rocks, usually in cold climates, increases with time. Lichen growth rate is, however, affected by other factors such as climate and substrate.	4.5×10^3
Desert (rock) varnish	Assumes that rock varnish thickens, darkens and accumulates more Mn with time. Indices of surface exposure correlate best to Mn content in varnish, varnish color and areal coverage.	10^4
Cation ratios of rock varnish	Assumes that the ratio of (K + Ca) : Ti in rock varnish decreases with time as the more stable Ti is relatively enriched.	10^6
Soil-based Methods[a]	Assumes that as many other state factors as possible are held constant, and that soil development has been, for the most part, progressive, over time.	

(cont.)

Table 14.1 (*cont.*)

Surface exposure dating method	Common methods and/or assumptions	Maximum timespan utility (years)
Horizon thickness or depth to a particular horizon type	Depth (thickness) is determined and compared among sites, assuming that thickness increases with time	10^5
Solum thickness, sometimes equivalent to depth of leaching or oxidation	Solum thickness is most accurate as a relative age indicator in young soils developing on calcareous materials, wherein it is equivalent to depth of leaching. Again, it is assumed that solum thickness increases with time.	10^5
Content of a mobile constituent (clay, $CaCO_3$, Fe)	The content (on a whole soil-weighted, or horizon-weighted basis) is determined. Holding parent material constant, among the surfaces being compared, is essential. Some constituents increase with time; others decrease.	10^5
Soil chemistry	Examples include CEC, electrical conductivity, pH or contents of ions such as Fe, Al and Ca. Measures changes in soil properties that reflect clay mineralogy evolution, leaching and base cycling, which are indirectly a function of time.	10^4
Horizon color, especially rubification	Expected color change is based on pedogenic theory.	10^4
Clay mineralogy	Changes in clay mineral abundance and type are used to infer degree of weathering. Ratios of the abundance of one clay mineral to another are commonly used.	10^5
Soil morphology and micromorphology	Expected morphologic change, e.g., structure, cutans, texture, silt coatings, based on pedogenic theory.	5×10^4
Sand mineralogy (light or heavy fraction)	It is assumed that resistant/weatherable mineral ratios increase over time.	10^5
Formation of desert pavement	Based on knowledge of its formative mechanisms and morphologies on surfaces of known age.	10^5

[a]Most quantitative criteria can be evaluated on a mass-based basis, weighted by horizon or solum thickness, or on a volumetric basis (weighted by horizon thickness and bulk density). See text for details.

limits) to nearby moraines of unknown age. Thus, calibrated age assessment can provide semi-quantitative estimates of the magnitude of age differences among sites, surfaces or materials.

Relative dating techniques have varying degrees of accuracy and "longevity." Some, such as lichenometry, cannot be used on surfaces older than a few millennia. Others, like obsidian hydration dating, can provide age estimates for rocks and surfaces as old as 1 million years (Table 14.1). Additionally, the types and quality of relative age data vary. Interval- and ratio-level data that can be generated for some methods (e.g., solum thickness in cm, clay content in percent, and weathering rind thickness in mm) contrast with other methods that produce only grouped ordinal data (e.g., weathered rock categories: "high," "medium" and "low").

Because each relative dating method has its own strengths and weaknesses, using a variety

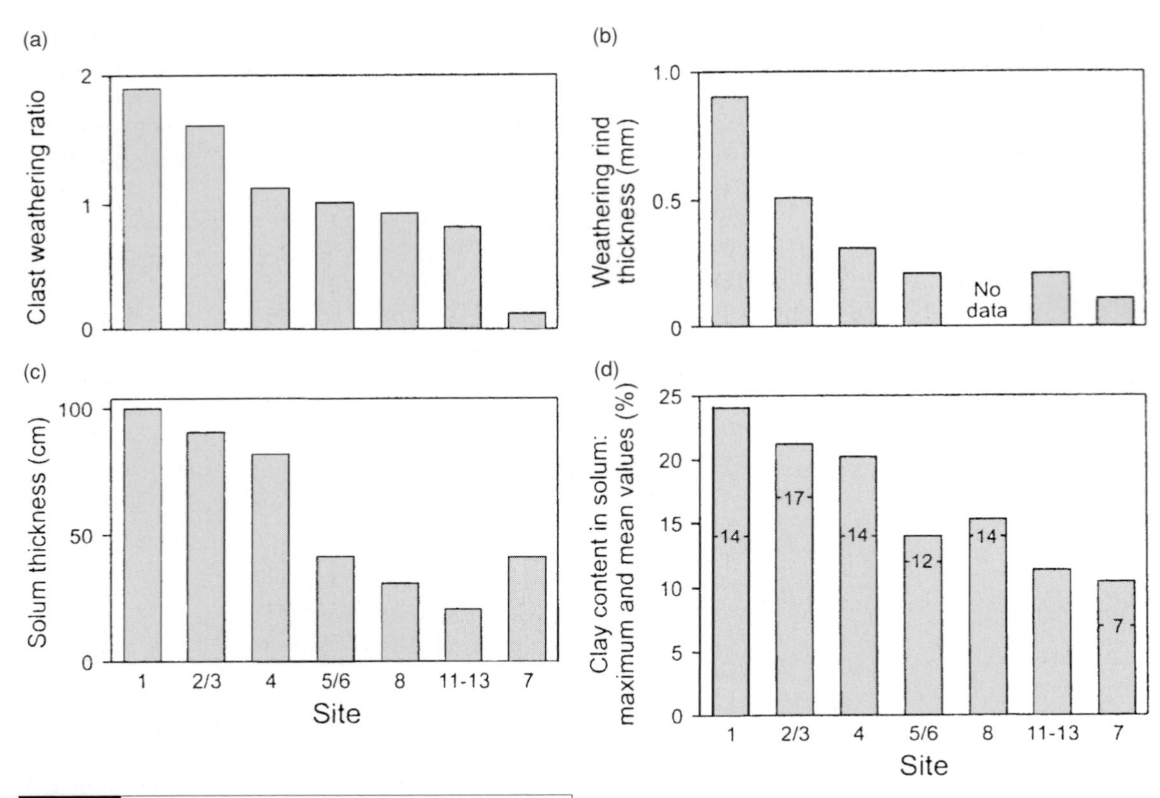

Fig. 14.4 Comparison of four relative dating methods for surfaces of different age, Tobacco Root Range, Montana. After Hall and Michaud (1988). (a) Weathering ratio. Higher ratios imply that a higher percentage of the rocks are considered weathered, as indicated by a blow from a rock hammer. For the actual equation, see Hall and Michaud (1988). (b) Mean weathering rind thicknesses. (c) Mean solum thicknesses. (d) Clay contents of soil horizons (average value is shown as a number within the bar; maximum value is shown as the height of the bar itself).

of methods is recommended (Hall and Michaud 1988) (Table 14.1, Fig. 14.4). It should also be apparent that data from any relative dating study can be "extended" only a small distance spatially. In many instances, the data may be applicable only within the region from which they originated.

Principles of surface exposure dating (SED)

Soil geomorphologists are interested in relative *and* numerical dating. With these two methods, one can date the *sediment* within which a soil has developed, or the *geomorphic surface* it is developing on – *surface exposure dating* (SED) (Colman *et al.* 1987). The age of the surface can never be greater than the age of the sediment, and can be much less. For thin deposits and rapidly formed geomorphic surfaces like moraines, the two ages (surface exposure and sediment) are so close in time that they are essentially the same. For erosional surfaces and for some thick deposits, however, the two ages may be widely disparate (Knuepfer 1988).

Knowing the exposure age of a surface has many applications within soil geomorphology. First, such knowledge should assist in the interpretation of the origin of the surface. Second, knowing the age of a surface permits estimates of the rate at which various processes operate, particularly pedogenic processes. Lastly, one might use surface exposure data, when developed within a known climatic context, to add to paleogeographic interpretations and chronologies of climatic change (Watchman and Twidale 2002). Bear in mind that most surfaces exhibit a wide range of ages, i.e., they are time-transgressive.

The Iowan erosion surface in Iowa is a good example of surface exposure vs. sediment age (see Chapter 13). It was formed by widespread erosion between about 29 and 18 ka (Ruhe 1969). $\text{Time}_{\text{zero}}$ of the soils on the Iowan surface date to the period when erosion stopped, probably due to climatic amelioration as the Wisconsin glacier began to retreat (MIS 2). The till that these soils formed in, however, dates back to the so-called Kansan glaciation, MIS 13 or older. Thus, soil development on the Iowan surface reflects the age of the *surface*, not the *sediment*. On a different landscape, surface age could equal sediment age. For this to happen soil formation would have to have started immediately after the sediment (parent material) was deposited; there could have been no intervening period of instability, e.g., a lava flow. Small inputs of (younger) dust can be added to a surface without noticeably changing it or its soils. This type of sediment post-dates the establishment of the surface but it does not really create a new surface.

When we date the surface, we determine the time when the surface became subaerially exposed and geomorphically stable, based on data from materials on and within the surface that change, predictably and quantifiably, through time (Birkeland 1982). Occasionally, we may wish to date the exposure interval of a *buried* surface or soil; this would be equivalent to the time from initial surface stabilization to the time of burial.

Each SED method has its own resolution precision, accuracy and time limit (Table 14.1). In general, accuracy and resolution are sacrificed in methods that can have long time limits, and vice versa. Soil-based methods can provide high-resolution SED information for young surfaces but may not be useful on older ones in which the soils are more comparable. Instead, slow-to-change parameters like rock weathering rinds may be a better SED tool. Thus, an SED method that may be *possible* may not necessarily be *useful*, given the resolution required or the age range of the surfaces.

All SED methods involve development and application of *post-depositional modifications* (PDMs) to geomorphic surfaces, such as changes in landform morphology and physical and chemical changes in the rocks and soils on those surfaces (Kiernan 1990). A primary assumption is that PDMs are correlated to surface age or exposure. However, complicating factors can and do influence PDMs, regardless of the dating application. Climate is a particularly important one; often its effects can overwhelm that of time (Locke 1979). Every researcher using relative or numerical dating tools must be aware of the potential complications introduced to the data by agents and factors besides age. For these reasons, as many relative dating techniques as possible should be used to assess the age of surfaces and sediments (Birkeland 1973, Miller 1979, Kiernan 1990). More is almost always better, especially when the methods are in general agreement (Burke and Birkeland 1979, McFadden *et al.* 1989) (Fig. 14.5). Combining relative *and* numerical age estimations is optimal, although not always possible.

SED methods based on geomorphology, geology and biology

There are a variety of SED–PDM techniques. We focus on those that are of most use in soil geomorphological studies (Table 14.1) and point the reader to reviews of other techniques for further information, e.g., Kiernan (1990) and Dorn and Phillips (1991).

Geomorphology and stratigraphy

Leaning on the principles of cross-cutting relationships and superposition, Dorn and Phillips (1991) observed that relative position is a highly useful SED method. Constructional landforms, such as moraines and alluvial fills, often overlap, providing unequivocal information about relative ages (Gibbons *et al.* 1984) (Fig. 14.2). Geomorphologists also use the form or shape of a landform as a key element in estimating its age (Coates 1984). Many landforms originate with sharp edges and slope breaks, but become more rounded with time (Sharp and Birman 1963, Miller 1979, Nash 1984, Nelson and Shroba 1998). Moraines become breached by water gaps (Nelson 1954). Slope inclination and surface roughness can be used to estimate surface exposure (Colman and Pierce 1986) (Fig. 14.6).

The formation of desert pavement is a PDM that follows a somewhat predictable sequence (see Chapter 12). Despite the numerous forces

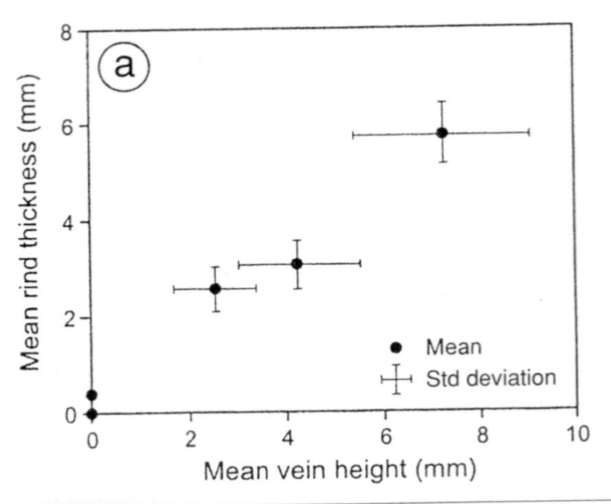

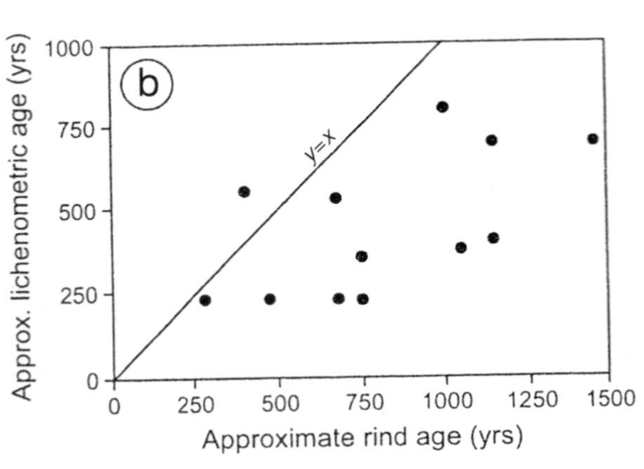

Fig. 14.5 Comparisons of two relative dating methods for the same surface. After Birkeland (1982). (a) Weathering rind thickness vs. heights of quartz veins on weathered rocks. (b) Age estimates based on lichenometry and weathering rinds.

that can destroy a pavement and reset the geomorphic clock to zero (Quade 2001), its use as a relative age indicator is well established (McFadden *et al.* 1987, 1989, 1998, Amit *et al.* 1993). Under optimal conditions, desert pavement can form in a few thousand years. On older surfaces, rocks in the pavement tend to be better packed and the vesicular (Av) horizon below is usually thicker. Over time, pavement clasts get increasingly shattered, making their angularity and size, not to mention any desert varnish on them, an SED tool. With this knowledge in mind, Al-Farraj and Harvey (2000) developed a pavement development index, scaled from 0 to 4, for alluvial fan surfaces in the Middle East. These data correlate well with relative data on soil formation, since both develop concurrently. Quade's (2001) work, however, cautioned that relative dating using desert pavement has limitations. He found that pavements on high deserts are no older than the Holocene, for these areas were cool and moist during the Pleistocene and therefore did not develop pavements. Only in the lowest, hottest deserts do pavements date back more than about 10 ka. Thus, PDMs on high-altitude surfaces in some deserts appear to reflect climate more than age.

Rock weathering and weathering rinds

A time-tested group of SED methods centers on the degree of weathering of exposed rocks on stable geomorphic surfaces (Table 14.1, Fig. 14.7). Rocks that lie on a surface or are shallowly buried beneath it undergo slow, predictable changes; most studies of rock weathering have shown a clear relationship between weathering and time (Colman 1981). Use of rock weathering as an SED or PDM tool is most useful in deposits that have many large rocks, preferably with many of them exposed at the surface (Shiraiwa and Watanabe 1991). The age–weathering correlations are most commonly applied, and therefore presumably most dependable, in dry or semi-arid regions (Colman and Pierce 1981). The PDMs that rocks undergo are usually associated with the development of weathering rinds or coatings of lichens. If the development of these features can be correlated to exposure age, then it can be used as an SED tool. Analysis of rock-related features has an advantage over soils as an SED tool because many more rocks can be sampled than can pedons. These tools usually assume that the lithology of the rock samples are similar (Kiernan 1990).

Other, less common but potentially highly useful methods include the acquisition of desert varnish (Krinsley *et al.* 1990), silica and carbonate coatings (Unger-Hamilton 1984, Curtiss *et al.* 1985) or coats of weathering by-products such as clay. For desert varnish, two time-related, relative dating methods are available to researchers: (1) the amount of Mn that has accumulated in the varnish and (2) the proportional amount of rock surfaces that are covered by varnish (McFadden *et al.* 1989, Reneau 1993) (see Chapter 12). Dorn

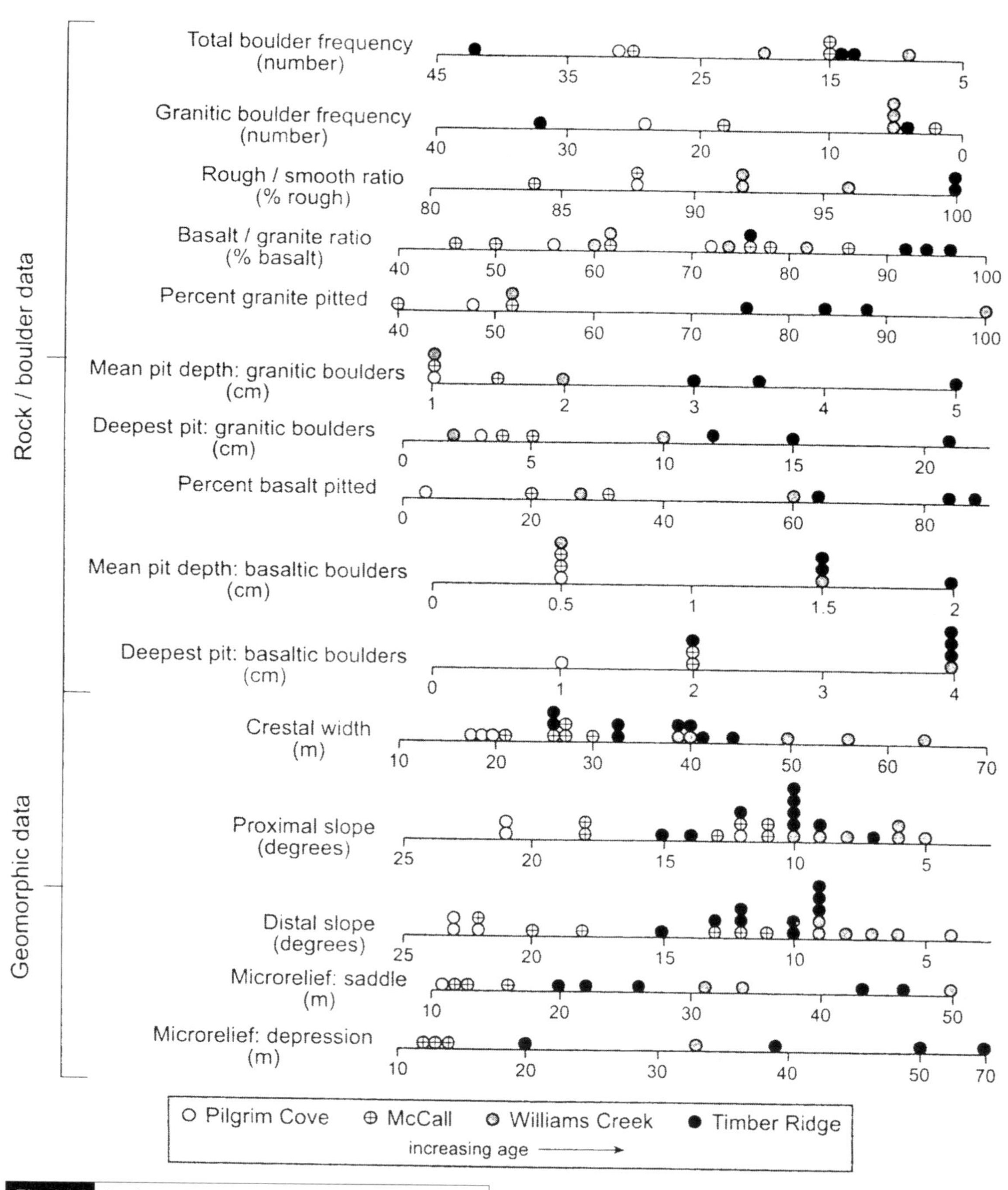

Fig. 14.6 Rock weathering and moraine morphology data, as used in an SED study on some morainic landforms in Idaho. After Colman and Pierce (1986).

Table 14.2 Weathering classes for surface-exposed boulders, as an SED tool

Weathering class	Characteristics and diagnostic weathering features
1	Completely fresh and unweathered
2	Surface staining
3	Surface is rough due to crystal relief but crystals not recoverable by hand
4	Crystals removable with fingernail
5	Crystals removable by rubbing
6	Micro-pitted (<1 cm), with exfoliation shells or weathering relief >1 cm
7	Macro-pitted (>1 cm) or inclusions protruding
8	Surface disintegrated
9	Deeply weathered or completely disintegrated

Source: Dyke (1979).

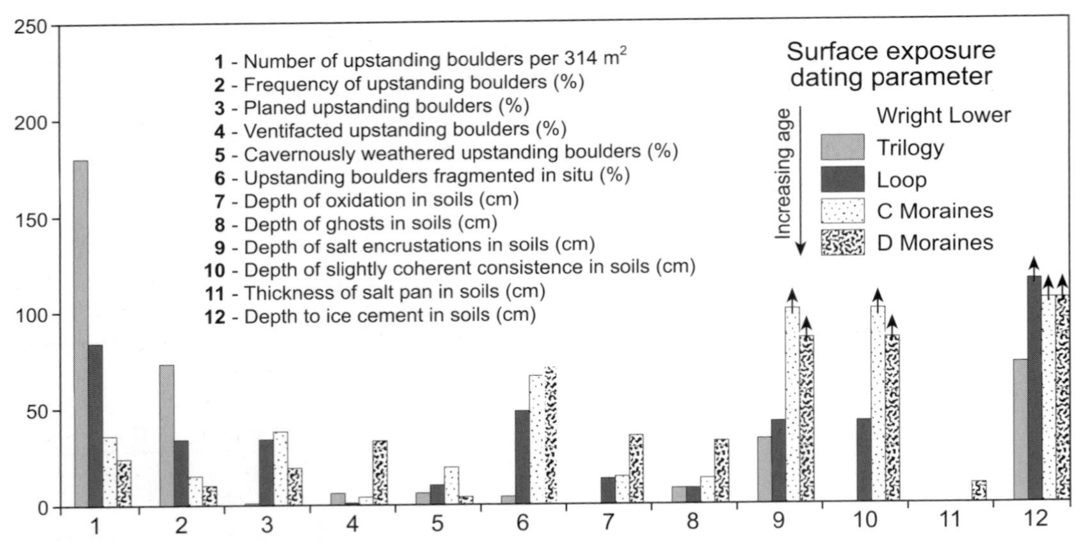

Fig. 14.7 Rock weathering and soil developmental properties on a series of moraine surfaces of increasing age in Antarctica. After Bockheim (1979b).

and Oberlander (1981a), however, cautioned that varnish forms irregularly in time and space, rendering it a potentially *supportive* SED tool, but one that should not be used as the *sole criterion* for age assessment. For example, problems can arise due to saltating sand which can remove some of the accumulated varnish by abrasion.

The mere presence or absence of rocks on a surface, i.e., frequency, decreases with time, as they weather. Thus, rock density alone, called the *surface boulder frequency* method, can be used as an SED tool (Sharp 1969, Scott 1977, Burke and Birkeland 1979, Miller 1979, Nelson *et al.* 1979).

A method used to assess degree of weathering involves striking surface-exposed rocks with a hammer and rating the ease with which they break apart. Usually, data of this sort are, at best, ordinal, forming a series of "highly weathered," "slightly weathered" and "unweathered" classes (Sharp and Birman 1963, Hall and Michaud 1988). Along these lines, several ordinal weathering scales have been devised (Miller 1973, Brookes 1982) (Table 14.2). For example, Rahn (1971) divided tombstones into three classes: unweathered, slightly weathered and extensively weathered. Angularity of rocks has been used, on the assumption that rocks get more rounded through time (Birkeland 1973, 1982). Others have measured the degree to which the rock surfaces are pitted or stained along fractures, or split along

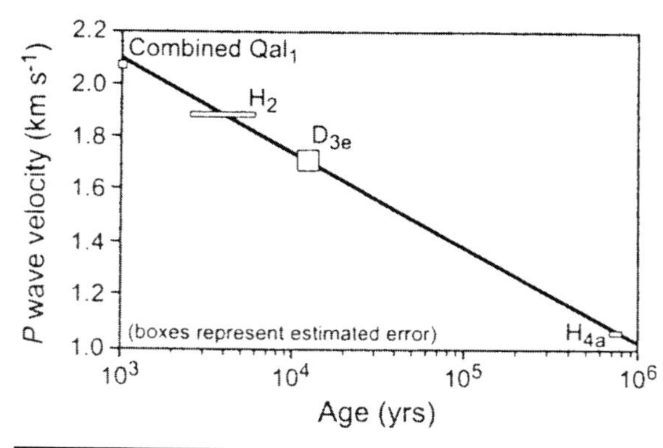

Fig. 14.8 Curve of *P*-wave velocity vs. age for rocks on geomorphic surfaces in the San Gabriel Valley, California. After Crook (1986).

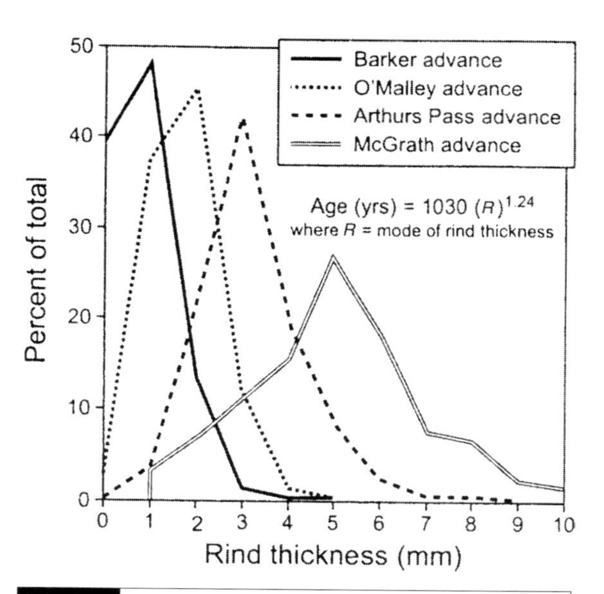

Fig. 14.9 Changes in weathering rind thickness on moraines of increasingly greater age, Southern Alps, New Zealand. After Chinn (1981).

cracks (Berry 1994). Dyke (1979) determined degree of weathering by examining how many rocks had lost evidence of glacial striae. The height of weathering-resistant veins or posts (usually quartz) increases through time as well, as the surrounding matrix weathers away (Birkeland 1982, Rodbell 1993). Quartz vein heights (and rind thicknesses) should be always viewed as *minimal relative ages* in SED studies, because parts of them could have broken off at some time in the past. Additionally, fires can cause rinds and veins to spall and fall off. Although some of these methods permit a quantitative assessment of weathering, many allow for only a subjective class-based ranking.

Several additional, more quantifiable methods for estimating rock weathering assume the integrity/hardness of the rock itself is a surrogate for its degree of weathering. For example, Crook (1986) studied compressional (*P*) waves sent through rocks; the velocity with which the waves propagate through the rock is a function of its degree of weathering. Fractures and chemical changes in the rocks, caused by weathering, force the waves to move more slowly (Fig. 14.8). This technique, termed the clast-sound velocity (CSV) method, has advantages over others: it is non-destructive, making it repeatable, and the data produced are highly precise. Another roughly equivalent method involves the Schmidt hammer – a device used to test the *hardness of a surface* by determining the distance of rebound from a controlled impact (McCarroll 1991). Initially designed to test concrete, the method has applications in rock weathering (Monroe 1966, Day and Goudie 1977, Matthews and Shakesby 1984, White *et al.* 1998), as well as on various types of soil pans, e.g., duripan, gypcrete and caliche.

As rocks weather, almost always from the outside in, their outer parts oxidize and develop a discolored layer, called a rind, which thickens with time (Fig. 14.9). Weathering rind *thickness* is one of the best and widely used SED methods (Chinn 1981, Gellatly 1984, Shiraiwa and Watanabe 1991), although one could also use rind hardness, color, specific gravity or mineralogy. Weathering rind studies routinely show that weathering rates are rapid at first and increase at slower rates with time, because weathering byproducts accumulate in the rind and because as the rind thickens, the unweathered core gets increasingly "protected." (Brookes 1982) (Fig. 14.10). Weathering rind thicknesses are, however, a function of more than age; climate, lithology and the effects of slow burial or exposure all affect weathering rind thicknesses (Chinn 1981, Knuepfer 1988). Use of weathering rinds as an SED tool is especially common in alpine settings, on moraines and rock glaciers, and has been applied frequently to debris flows.

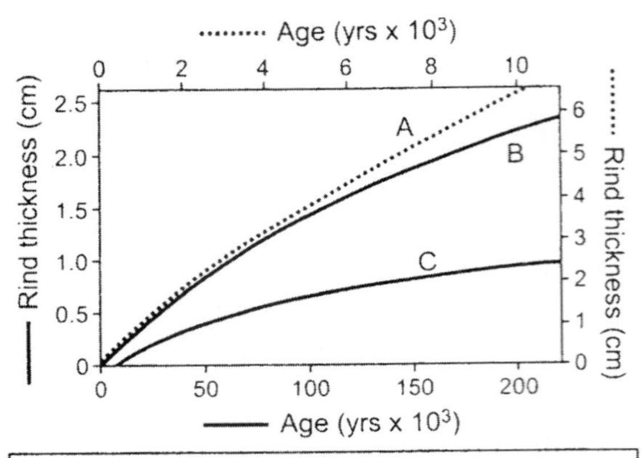

Fig. 14.10 Weathering rind thickness vs. time for three representative SED studies. After Colman's (1981) compilation.

Assessing the degree of rock weathering on a geomorphic surface is usually accomplished by examining rind thicknesses on ≈30–50 rocks. Mean, maximal or modal rind thicknesses are then calculated and used to characterize surface exposure, on the assumption that rind thickness is related to surface age (Birkeland 1973, Chinn 1981) (Fig. 14.11). Most studies identify this relationship as a power function (Fig. 14.9).

In any event, measurement accuracy is essential, since errors of less than a millimeter can dramatically affect the age estimate. Thus, the rind/no rind interface must be able to be precisely defined; this is often not possible in coarse-textured rocks. The type of rind developed, and the longevity of that rind, vary with rock type. Coarse-grained rocks, like granite, develop rinds rapidly but they quickly spall from the rock, making them of little use (Brookes 1982). Fine-textured, extrusive igneous rocks are best, as rinds form slower and persist longer than on granites. A very fine-textured rock, obsidian, is of special use in this regard. Rind development on glass-like obsidian occurs extremely slowly; most obsidian rinds are so thin that they must be measured under a microscope. *Obsidian hydration dating*, a special form of SED using weathering rinds on obsidian clasts, is therefore the most quantitative of all rind analyses. Like most extrusive volcanic materials, obsidian can be dated using K–Ar methods. It absorbs water and hydrates slowly through time, developing a whitish rind with a sharp inner "front" (Friedman and Long 1976). To use the method, a K–Ar numerical age is first obtained on some nearby obsidian, perhaps at an outcrop; hydration rinds on it are used as a numerical age datum (Pierce *et al.* 1976). Then, rocks must be located that have been eroded from the outcrop and effectively been "zeroed" (rendered rindless) by an agent of transportation, e.g., a river or glacier or by humans during tool-making. In a sediment where the obsidian fragments have not all been "zeroed," rinds on older fractures and surfaces must be sampled separately from those developed on the surfaces formed by the most recent transportation event (Adams *et al.* 1992). The hydration rinds that began forming immediately after they were redeposited can be compared to rinds on the outcrop, to arrive at a semi-quantitative, comparative age (Fig. 14.12). The value of obsidian hydration as an SED tool derives not only from the fact that it is numerically dateable by K–Ar, but also because (1) the rate of hydration is well known (Friedman and Long 1976) and (2) rinds appear to develop at similar rates in both shallowly buried and subaerial rocks. Obsidian hydration is especially useful in archeology, since many human artifacts are made of obsidian, and because the method can be extended back over 250 ka (Michels 1967, Meighan *et al.* 1968, Friedman and Trembour 1978).

Most rind thickness studies use rinds on *surface* clasts; if subsurface clasts are used, this should be indicated. As would be expected, weathering rind data for subsurface clasts is more variable, as the weathering intensity within the soil is highly spatially variable. Buried clasts tend to weather more slowly than do surface-exposed clasts, depending upon climate and soil conditions. In dry climates or in salty soils the reverse may be true.

Hornblende etching

The weathering characteristics of individual, silt- or sand-sized mineral grains within sediment is a valuable SED tool (Hall and Michaud 1988). Hornblende is an especially useful mineral in this regard (Locke 1979, Hall and Horn 1993, Mikesell *et al.* 2004), as is apatite (Lång 2000). Hornblende develops cockscomb-like terminations and etch

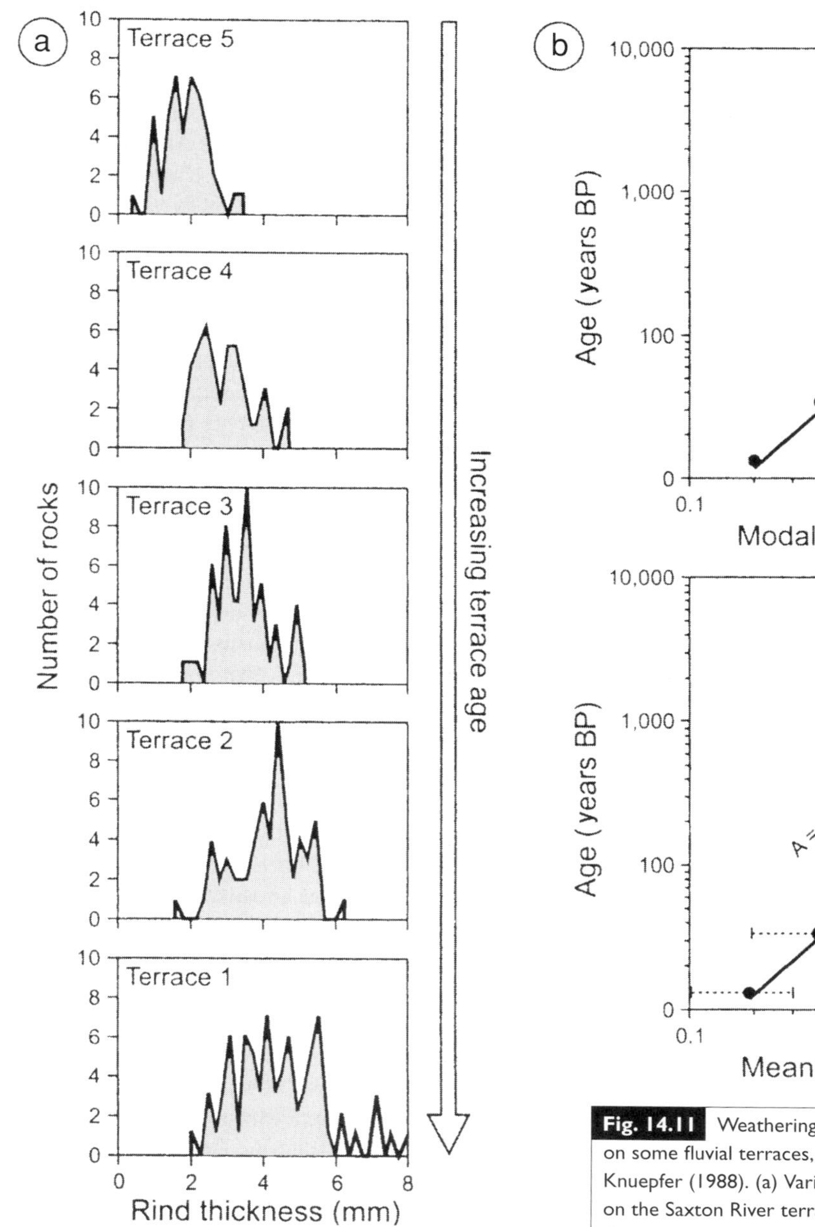

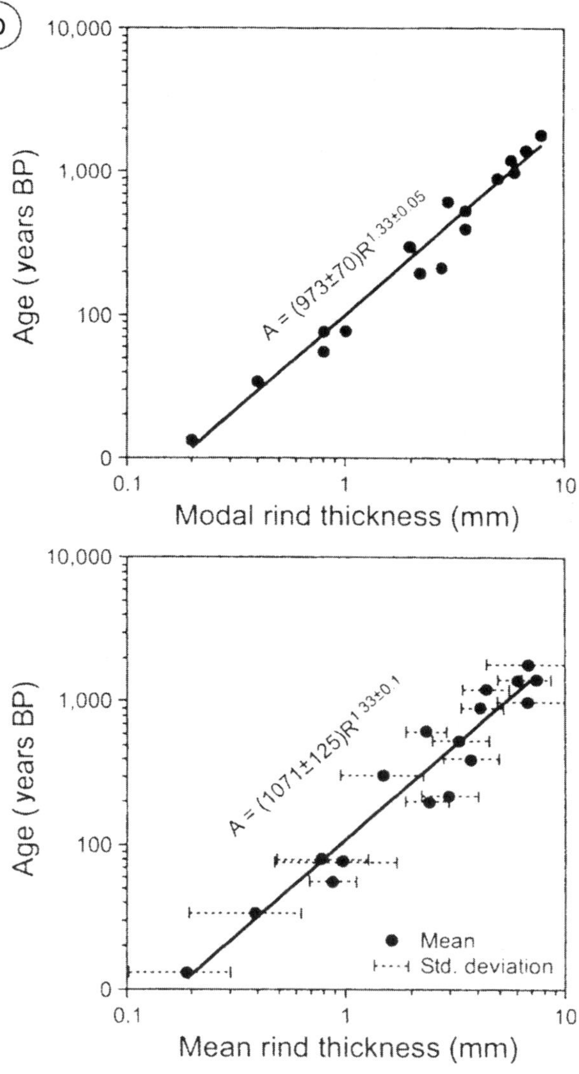

Fig. 14.11 Weathering rind thickness data for surface rocks on some fluvial terraces, South Island, New Zealand. After Knuepfer (1988). (a) Variability of rind thickness for the rocks on the Saxton River terraces. (b) Calibration curves for rind thickness vs. terrace surface age. Most of the ages are on wood from the underlying alluvium. These plots include data from a number of rivers in the region. Bars (right) indicate one standard deviation from the mean.

pits in such a predictable manner that the depth of their etch pits can be quantified and correlated to weathering intensity, which is often a function of age (Fig. 14.13). Like many age-related functions, hornblende etching is initially rapid, but the rate of etching decreases with time. And like many other weathering-related SED methods, it varies in intensity with depth in the soil, and as a function of climate. Strong etching is also associated with increased amounts of effective precipitation (Locke 1979). When applying this method and sampling for sand- and silt-sized hornblende grains, one must always be aware that they can be easily moved within soil by pedoturbation processes (see Chapter 10). Similar etching/weathering data could be generated for

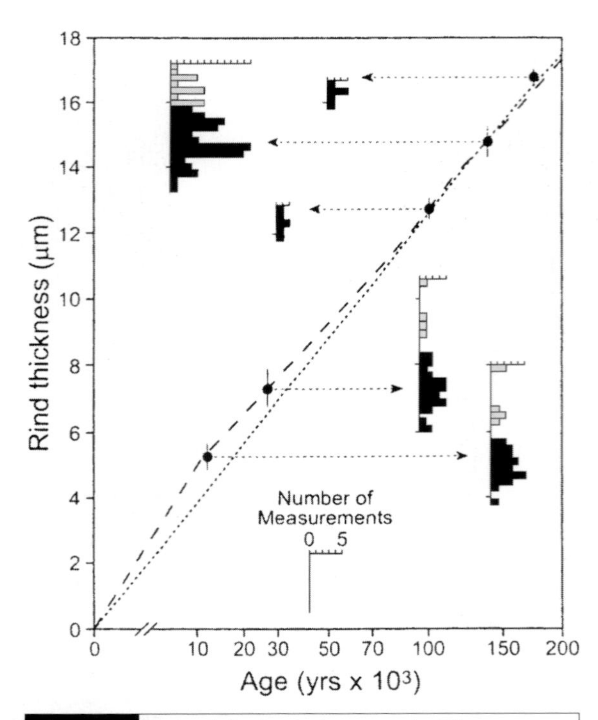

Fig. 14.12 Rate of obsidian hydration in the West Yellowstone basin, Montana. The rate is determined based on hydration of obsidian within lava flows dated at 179 and 114.5 ka. The dotted lines provides the mean rate of hydration based on these two flows. Moraine dates are from obsidian hydration on pebbles within the till. After Pierce *et al.* (1976).

minerals such as garnet, feldspars and amphiboles, but little work has been done in this area (Read 1998).

Lichenometry

In 1950, Roland Beschel established that the maximum diameter of the largest thallus of an epipetric lichen on a rock surface is proportional to time since colonization, or surface exposure, thereby providing a unique SED tool (Figs. 6.2, 14.14). A *thallus* (plural, thalli) is the body of a lichen, which usually exists in some sort of (usually circular) form on a hard substrate, like a rock. *Lichenometry* is a branch of relative dating that uses lichen sizes as an SED tool. It assumes that lichens begin colonizing rocks shortly after the rocks are subaerially exposed, and expand in area predictably over time. Lichenometry applications involve measuring lichen diameters on surfaces of different, but known, ages and using these data to calculate a lichen growth curve (Benedict 1967, Lock *et al.* 1979). This curve can then be used to estimate the ages of other surfaces for which numerical age control has not been established.

Lichenometry is especially useful in cold, harsh alpine regions, where rocky surfaces are common, lichen growth is slow and vascular plant competition is low (Webber and Andrews 1973). The absence of trees in the alpine tundra eliminates dendrochronology (see below) as an SED tool. The utility of lichenometry is further accentuated by the fact that soil development, another possible SED tool, is often minimally useful in these cold, dry environments. Lichenometry is most useful on young (late Holocene) surfaces that are <1000 years old (Birkeland 1973, Porter 1981); it is unlikely that surfaces older than about 4500 years can be dated with this method. The method has high precision on surfaces of very young age, as indicated by the fact that lichens on gravestones are commonly used to provide age control for the early stages of lichen growth curves (Beschel 1958, Carrara and Andrews 1973).

Crustose lichens, the type used in lichenometry, grow slowly in cold climates, and thus their thalli do not exceed the diameters of the rocks they are growing on for thousands of years. Beschel (1950) described an initial period of rapid growth, called a *great period*, that lasts about 20–100 years (Fig. 14.14). Growth rates for the great period are about 15–50 mm per century for most sites (Webber and Andrews 1973). The great period is followed by a longer period of much slower growth (the linear phase) (Porter 1981, McCarthy and Smith 1995). Linear phase growth rates are 2–4 mm per century, or less. High precipitation, longer growing seasons and warmer temperatures tend to favor more rapid growth rates overall (Ten Brink 1973).

An important and controversial component of lichenometry, and indeed any other SED method, centers on sampling strategy. How many lichens should be sampled per surface? Should lichen *sizes* or simply lichen *coverage* (relative amount of the rock surfaces that have lichens) be examined? Lichens of all sizes, shapes and ages occur on rocks; the small (young) ones probably reveal little about surface age, as they began growing long

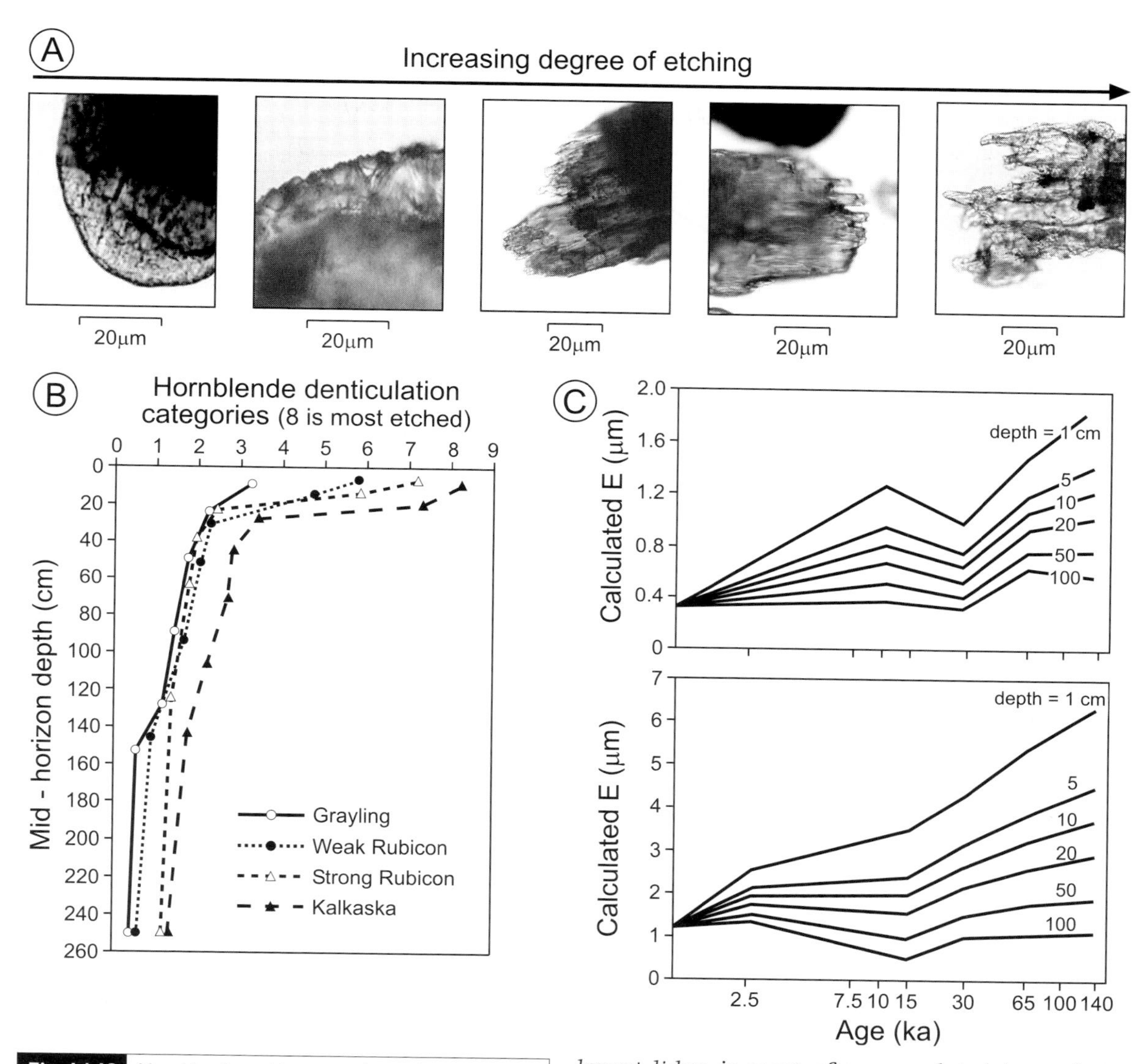

Fig. 14.13 Hornblende etching as a relative dating tool. (a) Photomicrographs of fine sand-sized hornblende grains exhibiting increasing degrees of etching on the margins of the grain. Images by M. Velbel and L. Mikesell. (b) Mean etching category vs. depth within four soils on an outwash plain in Michigan, illustrating the point that factors other than age, e.g., depth and pedogenesis, also affect etching. After Mikesell *et al.* (2004). (c) Etching vs. age and depth. Note that the *x*-axis is logarithmic, indicating that the increase in etching rates slow with time. After Hall and Horn (1993).

after the surface stabilized. Thus, in theory, we want to find the *one lichen* that began growing when the surface was first stabilized and which has continued to grow up to the present day. This theory explains why the size of the *single largest lichen* is most often correlated to surface age (Webber and Andrews 1973, Calkin and Ellis 1980, Orombelli and Porter 1983). However, the largest lichen could be an anomaly. For this reason, some researchers do not sample abnormally large thalli (Calkin and Ellis 1980, Rapp and Nyberg 1981). To quote Matthews (1975: 104):

> There is no reason why the single largest lichens should be preferred to the use of means (*averages*) of more than one largest lichen . . . Indeed there are grounds for preferring means in that a single largest lichen is more likely to be an anomaly.

Therefore, many researchers believe it more statistically valid to use the mean or modal size of

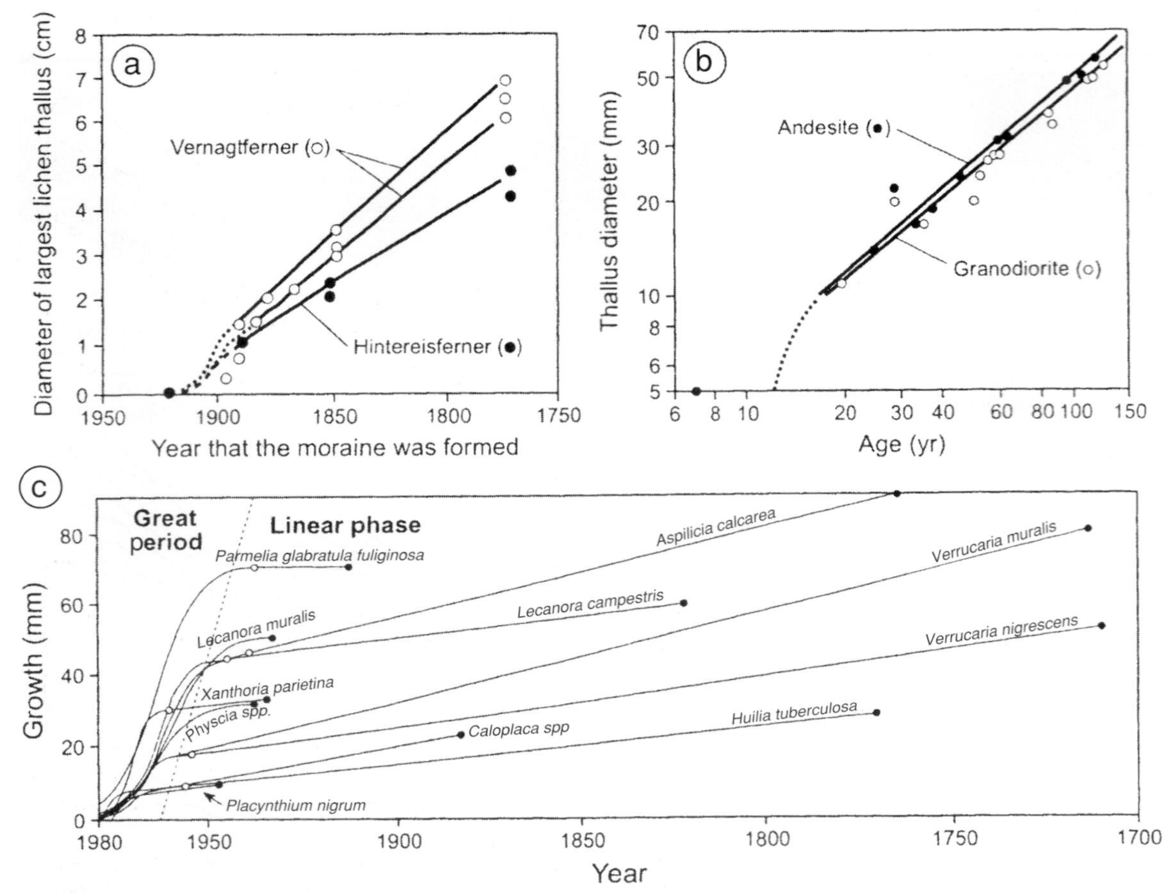

Fig. 14.14 Lichen growth curves. (a) Roland Beschel's (1950) original lichen growth curves, for moraines in the Austrian Alps. (b) Growth curves for lichens on andesite and granodiorite boulders on moraines on Mt. Ranier, Washington. After Porter (1981). (c) Growth curves for eleven different lichen species, from tombstones in southern England. After Winchester (1984).

a *sample* of large lichens, rather than relying on data from one large thallus (Matthews 1974, Innes 1985). Maizels and Petch (1985) have shown the strong relationship between these two types of data, perhaps making the distinction less critical (Fig. 14.15). Nonetheless, by sampling a larger number of lichens, the variation in lichen size on each surface decreases, as does the likelihood of an erroneous result due to one anomalously large lichen (Fig. 14.16). However, as more lichens are sampled, it becomes increasingly likely that the thalli being sampled began growing long *after* the surface was stabilized or subaerially exposed.

A related question is how large an area should be canvassed for lichens. As the area sampled gets larger, more large lichens are found (Fig. 14.17). Probably the best rule of thumb to keep in mind with regard to SED sampling strategies is to remain *consistent*, both with the literature and within your own study.

All relative dating methods have assumptions that are commonly (and sometimes, necessarily) violated; lichenometry is particularly enigmatic in this regard (Jochimsen 1973). We have discussed the problems associated with using one large lichen vs. a sample of lichens, but also consider:

(1) *The species of lichen utilized*. Lichen species grow at different rates, necessitating that sampled

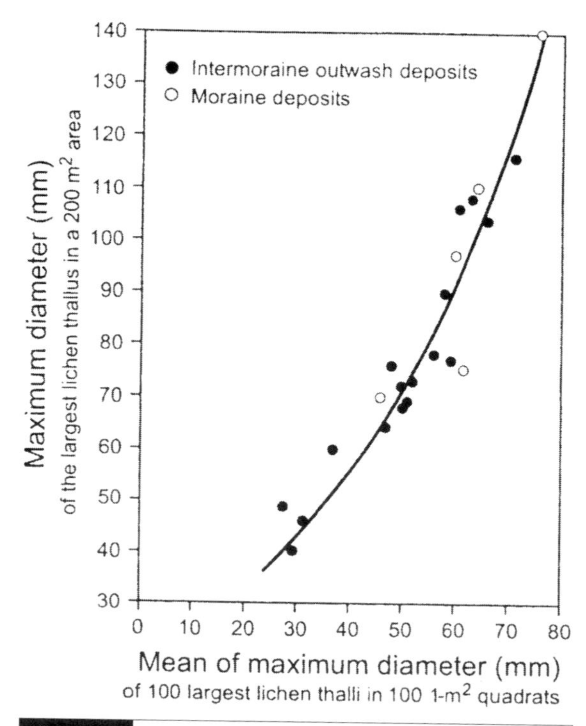

Fig. 14.15 Relationship between the maximum and mean diameters of lichen thalli in a lichenometry study in Norway. After Maizels and Petch (1985).

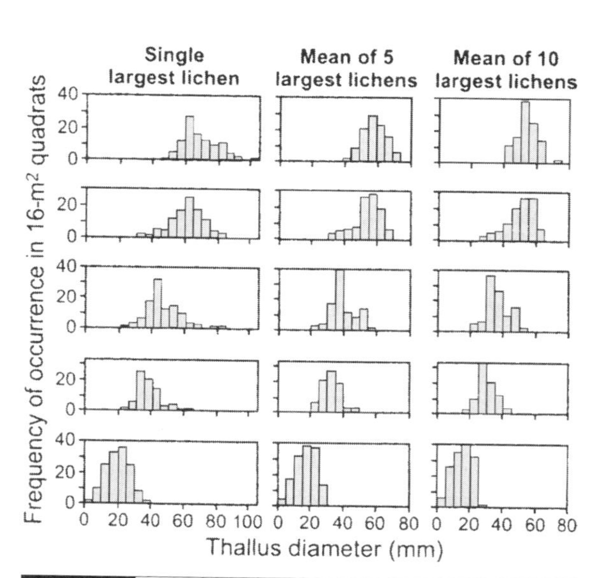

Fig. 14.16 Histogram distributions of *Rhizocarpon* lichen thalli on moraines of different age in Norway. Note that the variation in sizes decreases when sample sizes increase from 1 to 10. After Innes (1984).

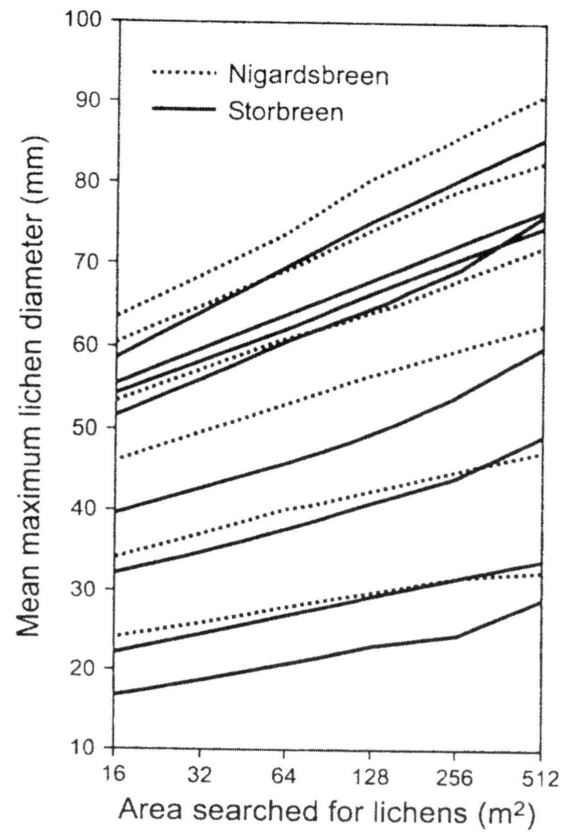

Fig. 14.17 Increase in the mean maximum lichen diameter with increasing size of the area sampled. After Innes (1984).

thalli be from the same species. It is often difficult to identify each thallus to species. Winchester (1984) suggested that the accuracy of the method is, in fact, enhanced if several lichen species are used and the data agglomerated (Fig. 14.14c).

(2) *Thallus shape*. Thalli vary in shape as well as size. Only rarely are they perfectly circular. What should be measured? Maximum or minimum length? Area? In combinational thalli (where lichens overlap) interaction may have affected growth rates.

(3) *Microclimate effects*. Climate/microclimate is a particularly important factor in affecting growth rates (Beschel 1958, Webber and Andrews 1973), although in some studies this

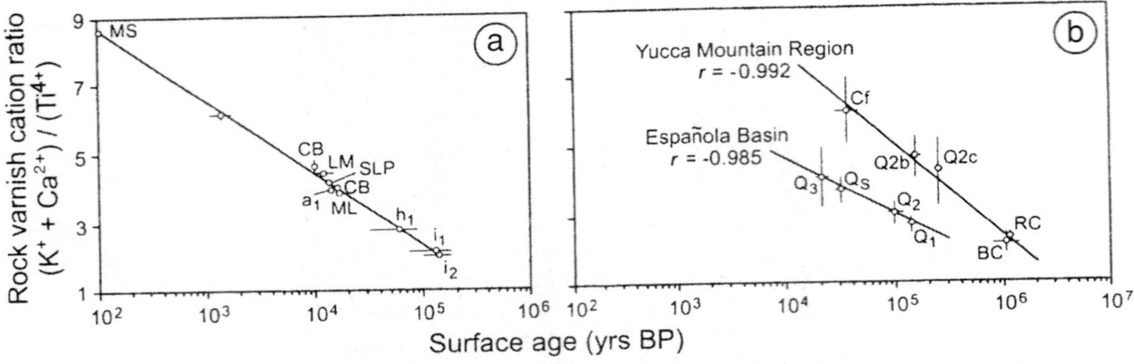

Fig. 14.18 Examples of rock varnish cation ratios as a relative dating tool, i.e., cation leaching curves. (a) Cation ratio ages of rock varnish on K–Ar dated lavas in the Mojave Desert, California. After Dorn *et al.* (1986). (b) Cation ratio ages of rock varnish on K–Ar and U-series dated lavas in the Mojave Desert, California. After Harrington and Whitney (1987).

factor can be controlled, or taken in account. Lichens growing on the top vs. sides of rocks vary in growth rates due to microclimatic effects (Birkeland 1973). Similarly, quartz veins that stand in relief cast shadows that can affect growth rates and restrict the spread of the thalli (Birkeland 1982). Also, on microclimatically cold sites or sites where snow patches persist well into summer, lichens can be much smaller than at more favorable sites. Lastly, in sites near cities, air pollution can affect lichen growth rates.

(4) *Substrate effects*. In lichenometry, one always tries to keep the substrate, i.e., the rock type, constant, as growth rates vary as a function of lithology (Innes 1984). Some rocks are not conducive to lichen growth, as they break into small fragments (Carrara and Andrews 1973).

Porter's (1981) work illustrated how a widespread lichen kill can seriously limit the technique. Nonetheless, despite all the potential pitfalls, lichenometry has been repeatedly shown to be a reliable SED tool for surfaces younger than about 4500 years. It is especially reliable when used in conjunction with other methods, such as weathering rind thickness.

Cation ratio dating of rock varnish

First proposed by Dorn and Oberlander (1981b), cation ratio dating is a calibrated relative dating method that revolves around desert varnish. In theory, the ratio of soluble to insoluble cations in the rock varnish decreases with time because they are replaced or depleted relative to less mobile cations (Dorn *et al.* 1986, Dorn 1989). Although a number of ratios could be used to reflect this theory, the most popular is:

$$[(Ca + K)/Ti)]_{\omega}.$$

This ratio, then, is a reflection of the amount of time that the varnish has been exposed to cation "leaching." It assumes that the initial varnish ratio was constant for rocks within the region of interest. Within thick varnishes, the uppermost layers should be most leached (Fig. 14.18).

Cation ratio dates are generally assumed to provide minimum ages, and obviously are applicable only to dry regions (Harry 1995). The method has some potential for dating human artifacts and petroglyphs in desert areas (Nobbs and Dorn 1988, Loendorf 1991).

A number of researchers have questioned the accuracy and validity of cation ratio dating (Reneau and Raymond 1991, Harry 1995). We refer the reader to some of the more pertinent literature (Dragovich 1988, Krinsley *et al.* 1990, Bierman and Gillespie 1994, Watchman 2000). As of this writing, cation ratio dating is viewed as a controversial but potentially useful SED technique whose precision may be unacceptable to some, especially within the archeological field (Harry 1995). It has had, therefore, few unequivocally successful applications to date.

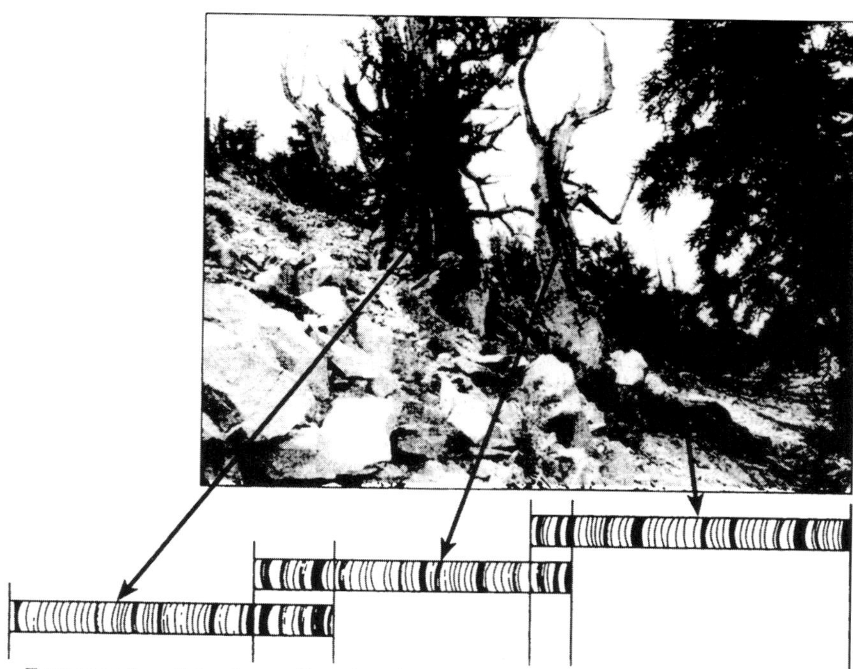

Fig. 14.19 The principle of cross-dating, using tree rings. The tree pictured is a bristlecone pine, a species that has been cross-dated to produce a tree ring record exceeding 8600 years in length. From a US Forests Service interpretive guide, Inyo National Forest. Original artwork by J. R. Janish; image by RJS.

Dendrogeomorphology

Dendrogeomorphology is the study of geomorphic processes through the use of tree ring analysis (Shroder 1978, 1980). The technique is primarily useful on surfaces younger than a few decades or centuries. By counting the annual rings of the oldest trees growing on a surface, one can establish the *minimum age* of that surface (Bryan and Hupp 1984, Hupp and Carey 1990).

Dendrochronology (tree ring dating) is a related and useful tool for determining the paleoenvironmental conditions that existed at a site. Commonly, the width of a tree ring is correlated to a particular climatic variable, usually temperature of, or moisture available in, the growing season. In cold areas, where temperature is the limiting factor to growth, e.g., at alpine treeline, the width of the rings is usually best correlated to temperature variables. In dry areas, tree ring width might be responding best to precipitation variables.

Dendrochronologists use the technique of cross-dating, illustrated in Fig. 14.19, to take the tree ring, and hence the climate, record back much farther than the current living trees on a site. By applying cross-dating techniques in the very old bristlecone pine (*Pinus longaeva*) forests of California, dendrochronologists have extended the tree ring record back more than 8600 years. However, because most sites have trees that are less than a few hundred years old, dendrochronology usually sacrifices longevity for this high (annual) level of surface exposure (and paleoclimatic proxy) accuracy.

SED methods based on soil development

Despite Daniels and Hammer's (1992: 196) warning that "probably very little if anything" can be used from soil profile data to place a surface within a relative age sequence, we feel that there are indeed many soil properties that can be used in SED applications (Markewich *et al.* 1989, Nelson and Shroba 1998). The caution that Daniels and Hammer give is that, for a number of surfaces of different age, it is difficult to hold other factors constant. Additionally, soils do not develop and soil properties are not attained and lost *linearly* over time, further complicating matters. However, because many chronosequence studies have successfully related soil development to time, we assert that soil properties can and should continue to be used as SED tools (Table 14.1).

In order to use soil-based data in an SED application, one must understand pedological theory and a bit of the paleoenvironmental history of the site. Knowledge of soil development and what properties are acquired in a predictable manner for the site are essential. For example, one would not use carbonate content of the B horizon as an SED tool for soils in which carbonates do not accumulate, or in which they accumulate sporadically. Knowledge of paleoenvironments is essential because soil processes may have shifted in the past; a soil that is accumulating carbonates today may have only begun doing so after a climatic change (see Chapter 15). The climatic change may have occurred long after the surface was exposed. Thus, we submit that soil development has great potential for SED, but is perhaps limited in its application by an incomplete knowledge of pedogenesis and paleoclimate, at least for some areas. Finally, pedogenesis is not always "predictable." Soils regress and change due to thresholds and accessions (see Chapter 12). This complication does not, however, make soil properties universally inappropriate.

Use of soils as a relative dating tool assumes that soil development is, to some degree, related to surface age or the state-factor time (see Chapter 11). To use this approach, soil development must be at least semi-quantitatively evaluated. Qualitative terms such as "infantile," "juvenile," "immature," "virile," "mature" and "senile," as well as "somewhat developed," "moderately developed," "well developed," and "poorly developed" have application and are widely used, but in an ideal SED situation soil development should be quantifiable.

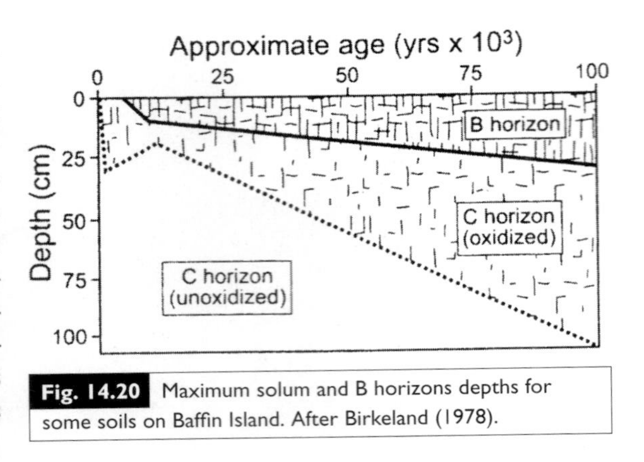

Fig. 14.20 Maximum solum and B horizons depths for some soils on Baffin Island. After Birkeland (1978).

Individual properties or attributes

Soil properties that change predictably through time are most applicable in SED studies (Tables 14.1, 14.2). The list of potential soil properties is a long one, e.g., horizon or profile thickness, accumulation or loss of mobile constituents from specific horizons and attainment and degree of development of certain morphologic features. One of the better SED tools is mineralogy, both clay mineralogy and sand or silt mineralogy (Soller and Owens 1991). As an example, iron content and types were shown to be a good indicator of soil age and development in Italy (Ardiuno *et al.* 1986b).

Solum or profile thickness is a useful parameter for young soils, especially those in which it is equivalent to depth of leaching, i.e., humid climate soils developing on calcareous parent materials (Allen and Whiteside 1954, Wang *et al.* 1986b). Thickness of individual horizons is also useful (Fig. 14.20). So are the attainment or loss of certain pedogenic features, e.g., a fragipan, plinthite or a petrocalcic horizon. For soils that develop illuvial and eluvial zones, various measures of illuviation, as related to clay, Fe, Al, organic carbon and $CaCO_3$ content are appropriate. Argillic horizons, for example, form relatively slowly. Their presence indicates that the geomorphic surface has been stable for a significant amount of time, usually at least 3–5 ka (Bilzi and Ciolkosz 1977, Ciolkosz *et al.* 1989, Cremeens 1995). For example, on many landscapes, Holocene age soils lack argillic horizons while those of Pleistocene age have

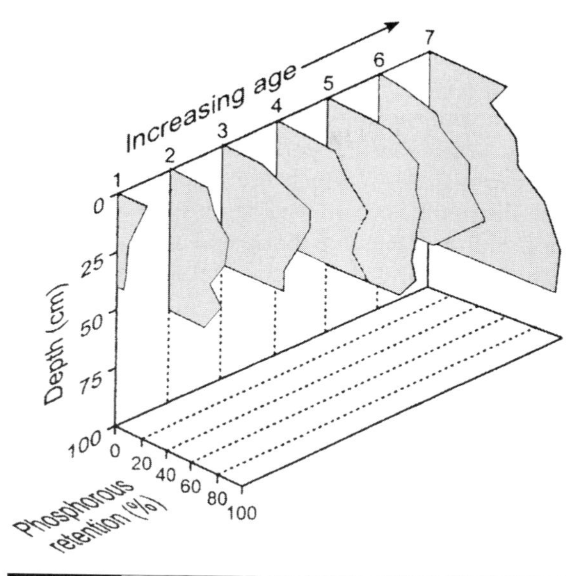

Fig. 14.21 Phosphorous retention values for soils on a series of Holocene moraines in New Zealand. After Gellatly (1985).

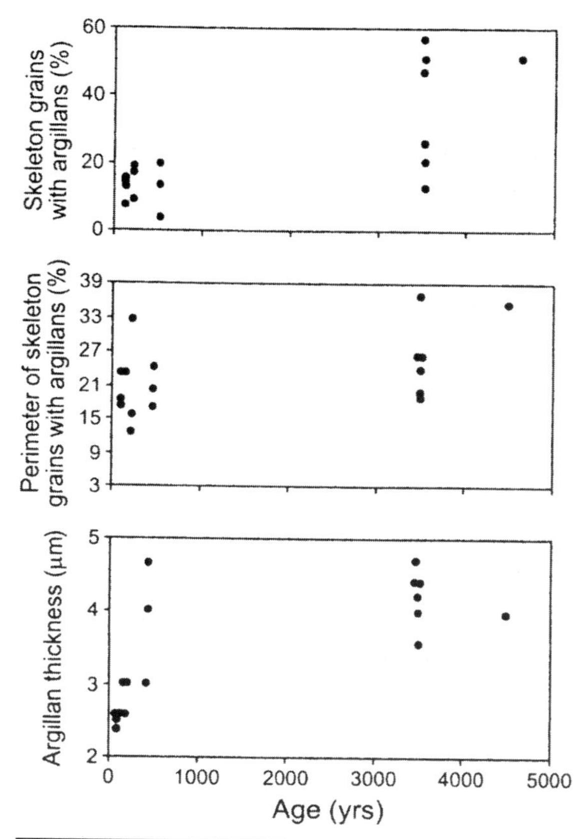

Fig. 14.22 Comparisons of various properties of argillans, viewed in thin section, with soil age. The soils are Ustepts, Ustalfs and Ustolls. After Holliday (1988).

well-developed argillic horizons. A particularly unique pedogenic property used in relative dating studies is phosphorous retention (Fig. 14.21). A soil's ability to retain more phosphorous from the soil solution increases as they get older and more weathered (Smeck 1973). Micromorphological features are also useful as soil development indicators (Berry 1994) (Fig. 14.22).

To be particularly useful in SED studies, soil properties must change systematically and, preferably, unidirectionally through time. Conversely, soil properties that are short-lived or reversible or that vary in rate of development are less useful or even problematic in SED studies. Yaalon (1971) provided much of the theory which is used to judge whether a soil property will be useful or not. In the context of determining which soil attributes were most useful for examining *buried* soils, his focus was on their relative *persistence*. Properties that persist in a soil after burial would be useful for ascertaining what the soil was like at the time of burial. Those that do not persist and are quickly altered after burial, such as pH, would tell us little about the original soil. His three-fold categorization has great utility for SED; soil properties that tend to persist are often the most useful as SED tools, especially on older surfaces (Fig. 14.23). Yaalon's categories are:

(1) *Reversible, self-regulating processes and properties.* These properties rapidly attain a state of dynamic equilibrium and are therefore also subject to rapid alteration, or even reversal, when environmental conditions change. These properties might be useful for determining relative ages on young surfaces, however, because they change rapidly over short time-spans. Examples include pH, base saturation, organic matter, contents of soluble salts and gypsum, mottles, gilgais, slickensides and most soil structures.
(2) *Processes and properties that slowly achieve a state of equilibrium.* These are slow to develop and relatively resistant to change, e.g., spodic,

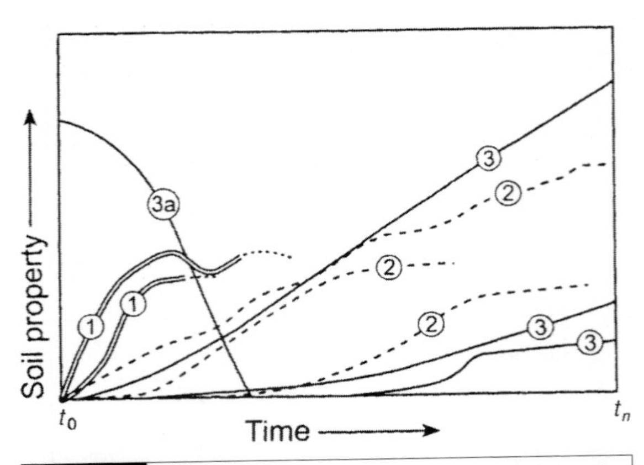

Fig. 14.23 Schematic representation of the three types of soil properties and how they theoretically change over time. After Yaalon (1971). (a) Reversible, self-regulating processes such as pH, salt content and soil structure. These properties quickly achieve or approach a steady-state or equilibrium condition. (2) Processes and properties that more slowly achieve a state of equilibrium or steady-state condition. (3) Irreversible, self-terminating processes and properties. Many of these properties, e.g., chlorite minerals, are exhausted from the soil after a period of time (3a). Others increase with little short-term likelihood of achieving a steady-state condition.

natric, argillic, gyspic and calcic horizons, histic epipedons, fragipans and some types of redox features.

(3) *Irreversible, self-terminating processes and properties.* Once formed, these pedogenic features or properties are very difficult to obliterate. Many form extremely slowly, such as oxic, petrocalcic, placic and petrogypsic horizons, plinthite, laterite, clay mineralogy, and certain strongly developed argillic and natric horizons. Embodied within the group is the concept of loss of certain properties, such as weatherable minerals. This is an irreversible process, for once these minerals are gone they do not neoform. These properties all are potentially useful on the oldest of surfaces.

Yaalon's groupings provide a sound theoretical basis for deciding which soil properties to choose for an SED study (Table 14.1).

Weighted soil properties

Many soil properties, if they are ratio- or interval-scale, i.e., not ordinal, and evaluated on a horizon basis, can be mathematically weighted by multiplying them by the *horizon thickness*. If a master horizon has subhorizons, each can be weighted in turn and summed to arrive at a weighting for the entire master horizon. For example, imagine two Alfisols in which the amount of illuvial clay in the Bt horizon was being used as an SED tool, and assume that both had formed in the same loamy sand parent material. Soil A has 25% clay in a 10-cm thick Bt horizon, while the other (B) has 20% clay in a 20-cm thick Bt horizon. Prior to weighting the data, it might be assumed that soil A was better developed, due to more illuvial clay. In actuality, soil B is better developed, because it has accumulated more overall clay, as reflected in the weighted clay contents (250 vs. 400).

Weighted soil properties can be examined in a number of different ways (Hall and Shroba 1995). One can determine *horizon-weighted* values, i.e., property value × horizon thickness, as discussed above. Similarly, one can sum the horizon-weighted values to arrive at a *profile-weighted* value for various soil properties. Thicker sola would be advantaged in such a scheme. In order to correct for overly thick profiles, profile-weighted data can be divided by overall profile thickness, thereby producing data that are not as biased by profile thickness (Goodman *et al.* 2001). Birkeland and Burke (1988) called this the profile-weighted mean (PWM) value. Another type of quantification is the accumulation index (AI). To calculate the AI, the value of a soil property in each horizon is compared to that of the C horizon. For most properties that are assumed to increase with time (e.g., clay, iron, organic matter content) the C horizon value is subtracted from the values of the other horizons. Negative numbers are ignored. For soil properties that are assumed to decrease with time, such as feldspar content, horizon values should be subtracted from that of the C horizon. The difference obtained for each horizon is then multiplied by horizon thickness and summed for the profile (Table 14.3). The AI value can be calculated in this way or adjusted by dividing it by profile thickness.

The next logical step in weighting soil properties is to multiply them by *bulk density* to

Table 14.3 Quantitative data for soils on moraines of three different ages, Wyoming

Soil property[a]	Soils on Pinedale moraines (youngest)	Soils on an Early Wisconsin moraine	Soils on Bull Lake moraines (oldest)
Solum thickness (cm)	38	58	>125
B horizon thickness (cm)	28	45	>78
Profile development index (Harden 1982) (index – cm)	25	28	39
Profile maximum clay content (%)	10	11	19
Profile-weighted mean clay content (%)	6	6	15
Clay accumulation index	122	324	1452
Profile maximum $CaCO_3$ content (%)	1	1	12
Profile-weighted mean $CaCO_3$ content (%)	1	0.5	7
Accumulation index for $CaCO_3$ content	31	58	809

[a] Mean values for a number of soils, unless otherwise indicated.

Source: Hall and Shroba (1995).

arrive at a *volumetrically weighted* estimate. Many soil properties, e.g., clay, quartz and Fe content, are frequently determined on a mass basis. That is, their values are based on the weight of a soil constituent in a given weight of soil; such properties are commonly expressed on a percentage $[(w/w) \times 100]$ basis. However, in a dense horizon there will actually be more of this constituent than in a "light" or void-filled horizon, because there is more actual soil material, i.e., more mass, in the former. Multiplying the horizon-weighted value by the bulk density compensates for this artifact and provides a weighted, *volumetric estimate* of the soil constituent. This method also has the advantage of compensating for coarse fragment content, since bulk density data are so adjusted (see Chapter 2). Volumetrically determined soil data can also be summed over each horizon or the entire profile by multiplying the data by the horizon or profile thickness. In fact, this is perhaps the best way to provide soil data on a profile-to-profile or horizon-to-horizon comparative basis. The resulting mass–volumetric data are one of the best estimates that can be made of actual *mass* of a soil constituent. These data types are frequently referred to as *mass data*, e.g., clay mass is the mass of clay per unit area (usually per m^2 or cm^2) integrated through a predetermined profile thickness or soil column (Markewich and Pavich 1991, Liebens and Schaetzl 1997) (Fig. 14.24).

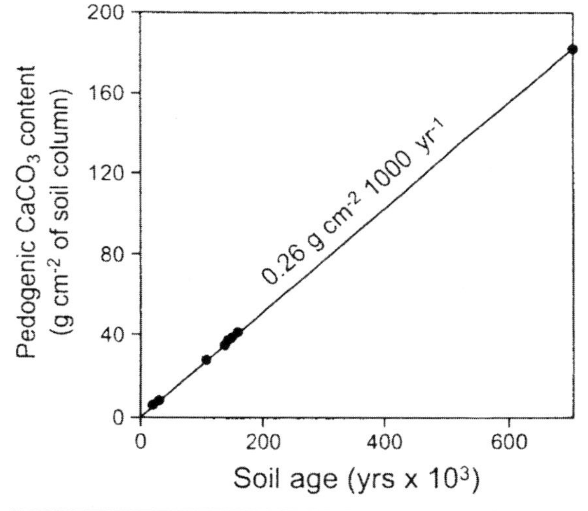

Fig. 14.24 Accumulation of, in this case, secondary carbonates, in soils of a chronosequence. The data are reported as mass per unit (1 cm^2) column of soil, which is a highly useful means of expressing soil data. After Harden *et al.* (1985).

There are many acceptable ways to "manipulate" soil data; each has its advantages and

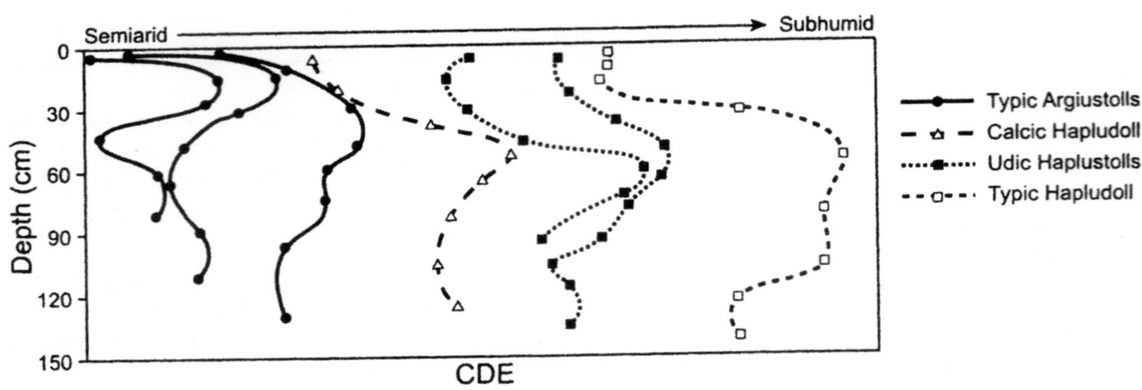

Fig. 14.25 Color development equivalent curves (CDEs) for a transect of soils across the ustic–udic boundary in the northern Great Plains. The CDE is the same as the Buntley–Westin (B–W) color index. After Buntley and Westin (1965).

disadvantages. However, soil data are only as good as the sampling technique used to acquire them, reinforcing the point that representative site selection and unbiased sampling are as important to soil analysis and interpretation as mathematical manipulation of data.

Indices of soil development

Researchers have long known that it is difficult to compare a variety of soils data in a meaningful way. Rather than compare a number of different data sets and lines of evidence between soils of differing development, soil scientists have devised a number of schemes and mathematical indices, designed to incorporate various data into *one* value. Such indices provide an easier and yet more conceptually integrative way of comparing soils (Goodman *et al.* 2001). No index is universal and each has limited utility for the types of soils for which it was developed.

Some of the early indices were primarily morphologic, facilitating comparisons on an *ordinal scale*, e.g., 1, 2, 3, 4 (Gile *et al.* 1966). Later, these indices became highly quantitative, providing *interval scale* data on soil development, allowing researchers to not only rank order soils from strongest to weakest developed, but to assign place-holders along that scale. For example, an index which increases in value as soil development increases, could assign one soil a value of 1, another a value of 3.5 and a third soil might be 6.7. Such data provide a quantitative development "ladder." While the same could be done for individual soil properties, their validity is usually increased if the data being ranked are indexed, rather than raw or even weighted.

Color indices

Some of the simpler yet highly useful indices of soil development center on only one morphological aspect: color. As discussed in Chapters 2 and 12, color is often highly correlated with various soil properties (Fernandez *et al.* 1988, Mokma 1993). Indices that incorporate hue, value and/or chroma provide an integration of soil color that can be even more insightful.

One of the earliest color indices was the Buntley–Westin (B–W) index (Buntley and Westin 1965). Developed for Mollisols, the index converts hue to a single number from 0 to 7, generally with each Munsell page separated from another by a whole number, e.g., 10YR = 3, 7.5YR = 4, 5YR = 5. Redder hues are assigned higher numbers because they often signify increased weathering and development. The number for hue is then multiplied by chroma to arrive at a color development equivalent, or CDE. This B–W index was not so much designed to be used as an SED tool as it was a discriminatory tool, to separate soils based on depth distributions of their horizon colors (Fig. 14.25). Nonetheless, it has been applied as an SED tool in the Arctic (Bockheim 1979a).

Because many soils and regoliths redden with age (Ardiuno *et al.* 1984, McFadden and Hendricks 1985, Markewich and Pavich 1991, Howard *et al.*

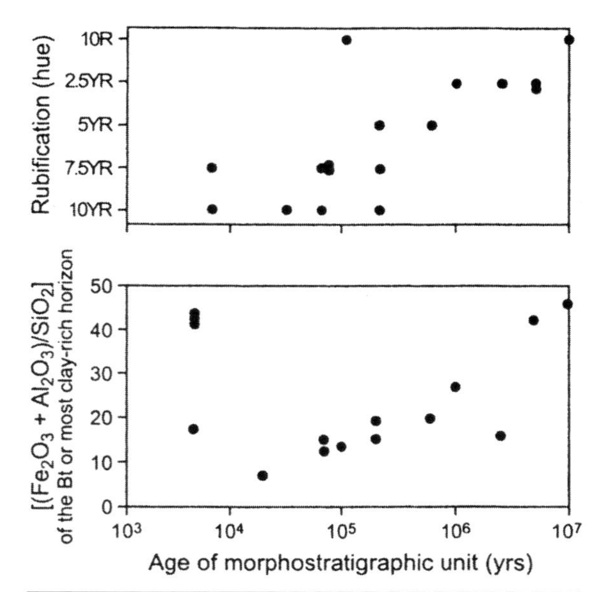

Fig. 14.26 Two scatterplots for some soils on the Coastal Plain of the southeastern United States. Rubification was defined as the Munsell hue of the Bt horizon or the reddest B horizon. Both rubification and sesquioxide content correlate well with age, despite the wide variety of soils included in the data set. After Markewich *et al.* (1989).

1993), quantification of *rubification* is important as an SED index (Fig. 14.26). Rubification is generally ascribed to increased amounts of Fe minerals, especially hematite, but because determining Fe_2O_3 content in the laboratory is a fairly time-consuming process, indices were developed to estimate its content based solely on color. Early attempts (Soileau and McCracken 1967) found no consistent relationship between hue and free iron oxide content. The first successful attempt, therefore, incorporated hue, value and chroma into one index value. It was developed by a geologist for use in saprolite (Hurst 1977). Like the B–W index, the Hurst index converts hue to a single numerical value (5R = 5, 10R = 10, 5YR = 15, 10YR = 20). These numbers are then multiplied by the Munsell value/chroma quotient to arrive at the Hurst index, which decreases as iron content and redness increase. Although developed for saprolite, it has been successfully applied to soils (Shiraiwa and Watanabe 1991, Leigh 1996, Liebens and Schaetzl 1997) (Fig. 12.29) and mine wastes (Shum and Lavkulich 1999). Alexander's (1985) suggested changes to the Hurst index formulation have also been utilized (Ajmone Marsan *et al.* 1988).

Torrent *et al.* (1980, 1983) modified the Hurst index and called their index the redness rating (Fig. 12.29). The redness rating behaves oppositely to the Hurst index; it increases as iron oxide and hematite content of soils increases. Recently, modifications to the Hurst index and redness rating have been developed for purposes of discriminating among various soil groups (Gobin *et al.* 2000).

Thompson and Bell (1996) developed a profile darkness index (PDI), which recognized that darker colors in Mollisols imply not only more organic carbon, but also wetter conditions:

$$\text{PDI} = \sum \frac{\text{A horizon thickness}}{(V \cdot C) + 1}$$

where the PDI is calculated for each A subhorizon and then summed; V and C are Munsell value and chroma. As A horizons get thicker and/or darker, the PDI will increase. In a catena of Mollisols in Minnesota, PDI consistently increased downslope and was highly correlated with A horizon organic carbon content (Thompson and Bell 1996).

Field/morphology indices

There is another family of soil development indices that is based on field morphology. These indices have the advantage of not requiring laboratory data for their application, and they typically take into consideration more morphological indicators than simply color. Most assume a uniform parent material at $\text{time}_{\text{zero}}$.

Perhaps the simplest and most elegant index is the one proposed by Follmer (1998a) (Table 14.4). Originally developed for buried soils (paleosols) in which gross morphology is preserved but potentially little else, the 1–10 ordinal scheme should be applicable to surface soils as well. It will perhaps be applied most in situations where soils of wide ranges of development are being compared, and for buried soils whose chemical properties have been altered since burial.

The Bilzi–Ciolkosz (B–C) index (Bilzi and Ciolkosz 1977) was developed for leached soils of humid regions, but can potentially be applied worldwide. Although not developed as an SED

Table 14.4 An ordinal, 1–10 scale for ranking the degree of soil and paleosol development

Morphological expression	Key character	Horizons[a]	Defining characteristics	Degree of mineral alteration
0 None	Geologic	D or R	Unaltered sediment	None
1 "Not soil" (protosoil)	"Geologic"	C or Cr D R	Evidence of a landsurface-altered horizon (C) over unaltered material	Detectable
2 Weak ("band")	Weak solum	A C D/R	Evidence of an A/C profile	Slight
3 Weak	Weak B	A Bw C D/R	Evidence of an A/Bw/C profile	Weak
4 Moderate	Weak E or Bt	A E Bt C D R	First evidence of Bt or E/Bt horizons	Weak
5 Strong	"Normal" E and Bt	A E Bt BCt C D/R	"normal" Bt or E/Bt horizons	Weak
6 Very strong	Thick E and Bt	A E Bt Bt Ct D/R	Thick Bt or E/Bt horizons	Moderate
7 Very strong	Maximum Bt	A E Btt Bt Ct D/R	Occurrence of a Bt horizon with >50% clay (Btt)	Moderate
8 Strong	Thick horizons	AE Bt Bto Ct Cr R	Occurrence of a Bto horizon	Strong
9 Moderate	"?"	EA Bto Bo Ct Cr R	Transitional	Very strong
10 Weak	Poor horizonation	EA Bo Bo Ct Cr R	Occurrence of a Bo horizon	Complete

[a]The use of the D and Btt horizon designations is informal.

Source: Follmer (1998a).

tool, it nonetheless has SED applicability. The index is used to compare the morphologies of adjacent horizons to each other and/or to the C horizon, hence its alternative name, the relative horizon distinctness (RHD) index. To operationalize the index, points are arbitrarily given for differences between horizons. One point is assigned for each "unit difference" in color, texture, structure, consistence, mottles, horizon boundary and argillans (the index was first applied to Alfisols). In short, most major morphologic criteria are determined for each horizon in the field, and the *magnitude* of the combined differences is determined. When used to compare the morphologies of adjacent horizons to each other, the method is useful for evaluating profile anisotropy, which for many soils is an indicator of development and age (Duchaufour and Souchier 1978, Meixner and Singer 1981, Ajmone Marsan *et al.* 1988). When comparing soil horizons to the C horizon, points are assigned based on the differences between them and the C horizon, implying that the B–C index value for the C horizon is always zero (Fig. 14.27). Used in this way, the index provides information on roughly how "far" soil development has proceeded beyond conditions at $time_{zero}$, i.e., the C horizon state. The larger the rating scale for a particular horizon the greater its pedological development, which again is often equivalent to or associated with age. The method

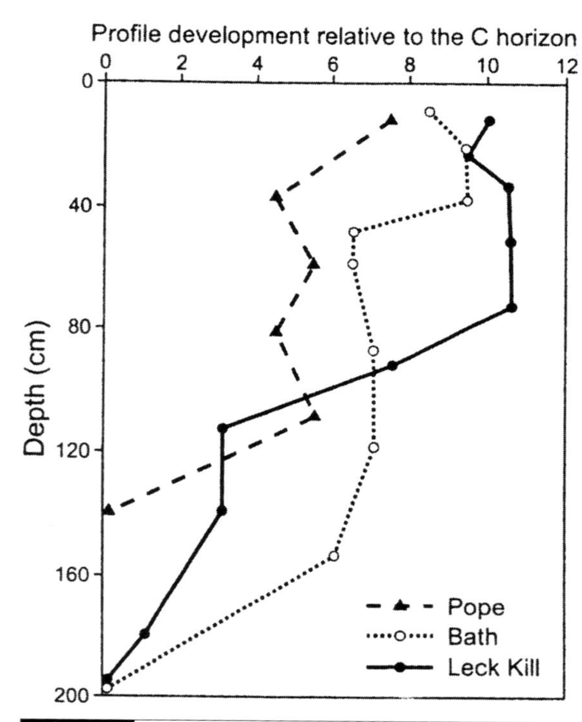

Fig. 14.27 Relative profile development, using the Bilzi-Ciolkosz (B–C) index (Bilzi and Ciolkosz 1977), for some soils in Pennsylvania. Larger values indicate greater differentiation of that horizon, with respect to its parent material. Pope soils are presumably younger than are the other two, and hence have lower B–C index values.

is not applicable to soils in which the C horizon is in a different parent material than is the profile, for to use the index in that situation would be equivalent to comparing "apples and oranges."

Perhaps the most utilized field morphology index is that of Jennifer Harden (1982). The Harden index, commonly called the profile development index (PDI), is a modification of the B–C index. In both schemes, points are assigned to each horizon as particular properties are developed or increase in magnitude. In the B–C index, the points for each horizon are compared to that of the C horizon; in the Harden index the horizon values are compared to the assumed parent material. If the parent material is not available in the soil pit, it must be sought out from a nearby location. The index examines any or all of the following soil properties: argillans, texture plus wet consistence, rubification, structure, dry consistence, moist consistence, color value and pH. First applied to soils in central California, the index was modified by Harden and Taylor (1983) to make it more applicable to arid climate soils by adding two properties typical of desert soils: color paling and color lightening. The index design is open-ended such that other researchers can add properties to this list or delete those that are not applicable (Knuepfer 1988).

In the B–C index, the values for each horizon are calculated and plotted graphically (Fig. 14.27). The PDI is mathematically more involved (Fig. 14.28). Like the B–C index, PDI values can be obtained on a per-horizon basis (Fig. 14.29). To normalize these so that one PDI value is produced for the entire profile, the value for each property for each horizon is divided by the maximum value, yielding a rating between 0 and 1. All the properties' normalized values are then summed, the total is divided by the number of properties used, and the latter is multiplied by horizon thickness. A final sum of these values yields a PDI for the profile. Modifications to the PDI include adjusting it so that all profiles are of the same thickness; for thinner profiles the C horizon thickness is increased so that the data for all profiles is based on an equivalent thickness of soil plus regolith. Birkeland (1999: 21) discussed the advantages and disadvantages of this modified PDI approach. Higher PDI values indicate increased soil development and the values usually increase logarithmically with time (Figs. 14.28, 14.30). Although the index is more complicated than others, it compensates for that by having a wide range of applicability. It remains, even today, the most-used index of soil development, e.g., Amit *et al.* (1996), Vidic (1998), Evans (1999), Treadwell-Steitz and McFadden (2000), Al-Farraj and Harvey (2000) and Dahms (2002).

A simplified version of the PDI, applicable only to soils that lack lithologic discontinuities, is the index of profile anisotropy (IPA) (Walker and Green 1976). The IPA assumes that profiles are isotropic at $time_{zero}$, at which time the IPA also equals zero. As the soil develops, anisotropy, taken as a surrogate for development, increases (Walker *et al.* 1996). The IPA is defined as

$$\text{IPA} = \sum D(100/M)$$

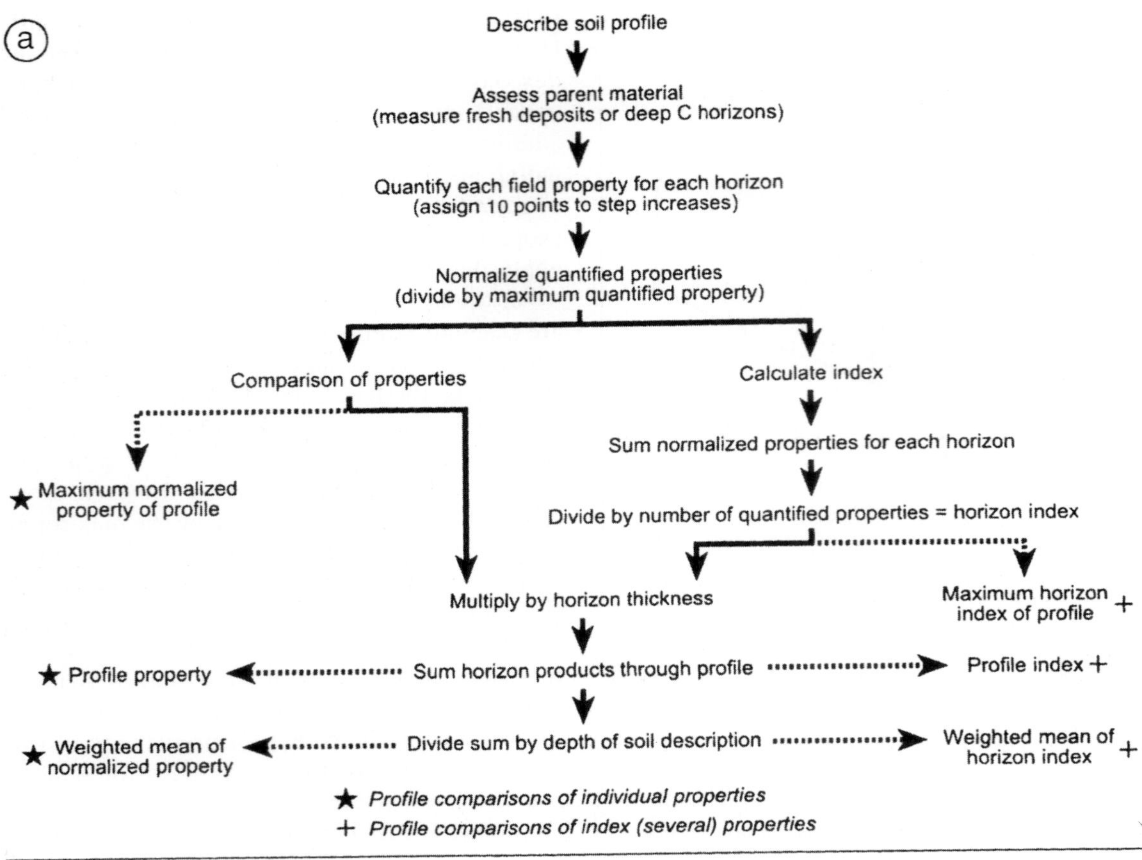

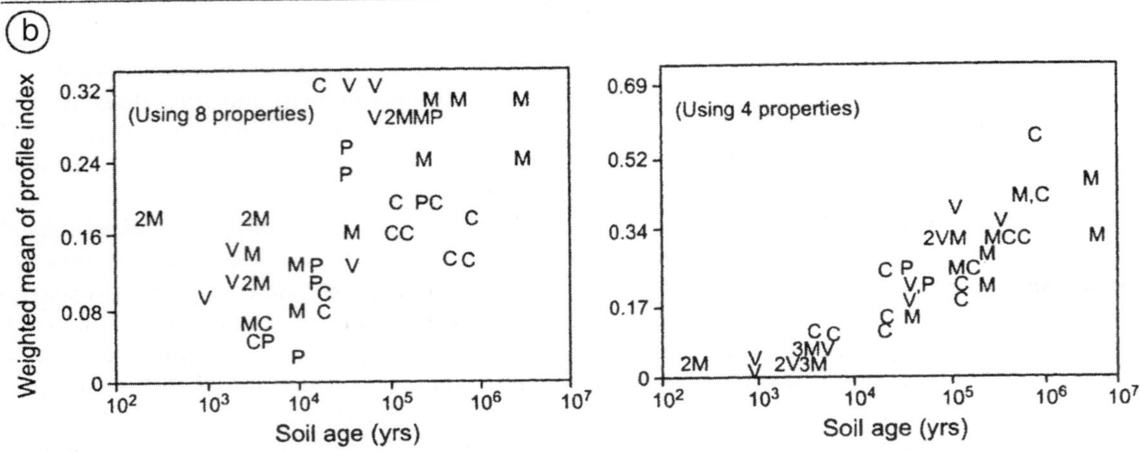

Fig. 14.28 The profile development index (PDI). After Harden (1982) and Harden and Taylor (1983). (a) Flowchart to assist in the derivation of the PDI from morphological data. (b) PDI data for soils of four chronosequences, each in a different climate. V, Ventura, California (xeric soil moisture regime, near the coast); P, Pennsylvania (udic soil moisture regime); C, Las Cruces, New Mexico (aridic soil moisture regime); M, Merced, California (xeric soil moisture regime, inland).

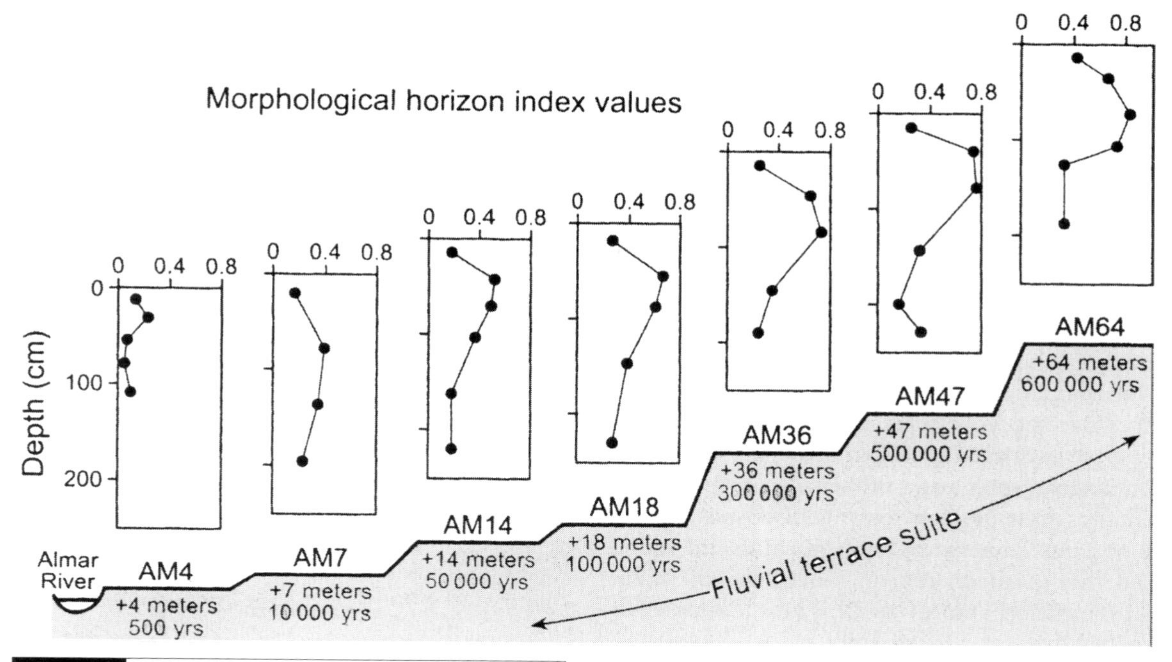

Fig. 14.29 Distribution of PDI values (Harden 1982) by horizon, for soils along a chronosequence of fluvial terraces in western Spain. After Dorronsoro and Alonso (1994).

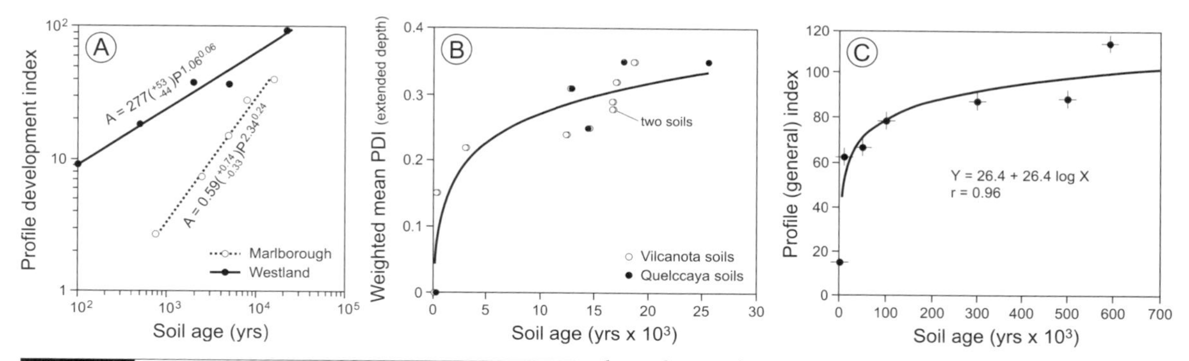

Fig. 14.30 Examples of application of the profile development index (Harden 1982). Note that the PDI increases logarithmically with time; data plot as a straight line on a logarithmic axis. (a) Stream terraces, South Island, New Zealand. After Knuepfer (1988). (b) Alpine moraines, southern Peru. After Goodman *et al.* (2001). (c) Stream terraces, Spain. After Alonso *et al.* (1994).

where D is the mean deviation of a horizon from the overall weighted mean value for the profile (M). Horizon deviation values are summed for each horizon to arrive at the IPA. The investigator has the option to choose which soil property or properties to use when calculating the IPA.

Birkeland (1999) suggested a modified formulation:

$$\mathrm{mIPA} = \frac{\sum[(t \cdot D)/\mathrm{PM}]}{T}$$

where D is the numerical deviation of a soil property from that of the parent material (PM). Deviations are determined for each horizon and then multiplied by horizon thickness (t). The sum of these values is then divided by the profile thickness (T).

Recognizing that the PDI requires knowledge of parent material, Langley-Turnbaugh and Evans (1994) developed a modified index which does not require parent material data. It weights structure more heavily than does the PDI, and minimizes the quantitative influence of horizon thickness. Its primary use is in distinguishing soil from not-soil sediment, and therefore has great utility in paleopedology.

Schaetzl and Mokma (1988) developed a field-based index specifically for Spodosols and soils developing toward that morphology. It is not unlike the IPA and mIPA in that it assumes that, as they develop, soils become more anisotropic with respect to color. The POD index uses information on horizon color and number. The index assumes that, as soils develop toward Spodosols, (1) their E horizons increase in Munsell value and get less red (hue) and (2) their B horizons get thicker and develop more subhorizons, and attain redder hues and lower color values (Goldin and Edmonds 1996). Most soils with POD indices ≤ 2 are Entisols, whereas those with POD indices ≥ 6 are within Typic groups of Spodosols, e.g., Typic Haplorthods (Fig. 14.31). Entic subgroups of Spodosols commonly have POD values between 2 and 6. Thus, POD index value increases as soils show increased evidence of podzolization (see Chapter 12). Its formula is

$$\text{POD index} = \sum \Delta V \cdot 2^{\Delta H}$$

where ΔV is the Munsell color value difference between the E and B subhorizons, and ΔH is the difference in the number of Munsell pages between the horizons (Fig. 14.31). The data are summed over all the B subhorizons.

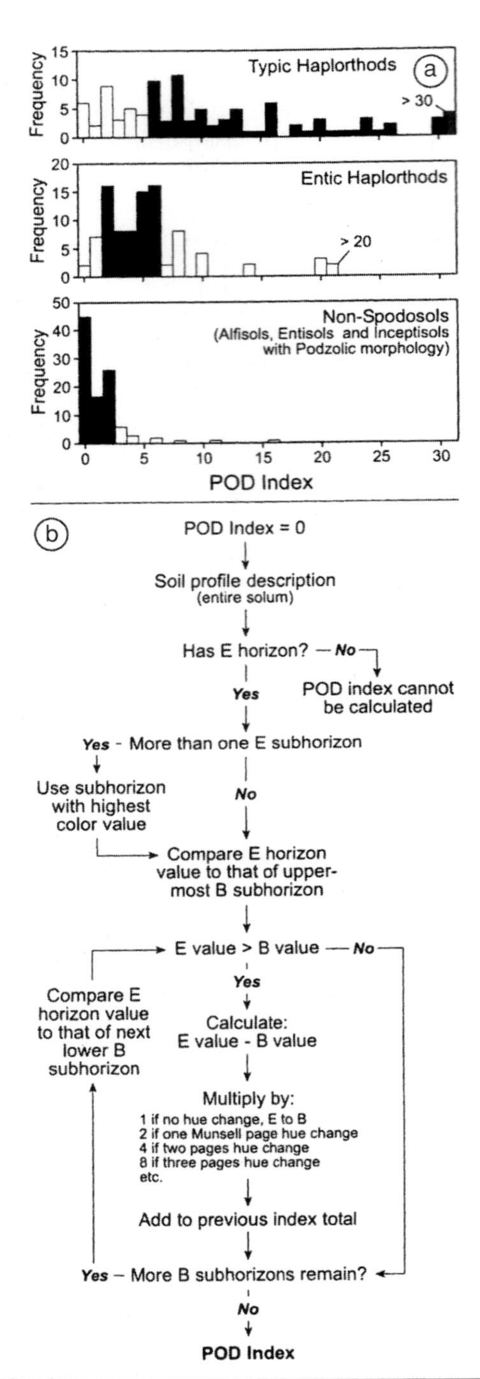

Fig. 14.31 The POD index, a numerical index of soil development for soils trending toward Spodosol morphologies. After Schaetzl and Mokma (1988). (a) POD index values vs. classification for a large sample of soil series in the northeastern United States. (b) Flowchart to assist in the derivation of the POD index from morphological data.

Laboratory indices

Soil development indices that incorporate laboratory data, such as mineralogy, clay content, pH and contents of ions, have advantages over field-based indices. Having more data is, generally, better, and thus lab + field indices tend to be even more indicative of soil development. However, laboratory data come at a cost since any index that requires laboratory data cannot be generated in the field and is therefore more expensive with regard to time and money. In lab-based indices it is critical that soils being compared have formed in similar and uniform parent materials, and that there has been little (or at least comparable amounts of) additions of materials to the soils since $time_{zero}$, e.g., as dust. Textural data are some of the most commonly utilized types of laboratory data for soil development indices.

As tropical soils get better developed (to a point), they tend to get more clay-rich. In very old soils, however, clay begins to be destroyed, and in some the potential for argillan formation decreases as pedoturbation continues to mix the soil. Van Wambeke (1962) felt that *clay : silt ratios* were a good indicator of soil age and weathering status in the humid tropics. He rationalized a developmental index by assuming that the fine silt content is a good indicator of the soil's remaining weatherable mineral storehouse, and as these minerals are weathered and lost from the silt fraction they become a part of the clay fraction.

Martini (1970) developed a weathering index for red tropical soils that takes into account that (1) soils become more clay-rich through time, (2) most of the CEC is tied up in clay and the little organic matter that there is in most of these soils and (3) as the soils weather, clay mineral suites change from amorphous materials to 2 : 1 clays to 1 : 1 clays and finally to oxide clays, each of which has lower CEC. Thus, CEC values decrease as soils become more weathered. Contrary to this trend, however, is the tendency for older, more weathered soils to have more clay. Martini (1970) suggested that the lowering of CEC through time was the more important factor and therefore developed his weathering index (Iw) of tropical soils as:

$$\mathrm{Iw} = \frac{\mathrm{CEC(meq/100\ g\ clay)}}{\mathrm{percent\ clay}}.$$

Iw appears to reflect tropical weathering in soils quite well; older soils have lower values. Generally, Oxisols have Iw values below 1.0.

An index of desilication, formalized by Singh *et al.* (1998) but in use for decades, is the molar ratio of silica to resistant oxides:

$$\mathrm{molar\ ratio} = SiO_2/R_2O_3$$

where $R_2O_3 = Fe_2O_3 + Al_2O_3 + TiO_2$ (see Chapter 12). Given the potential mobility of titanium, substituting ZrO_2 for TiO_2 might be a useful modification. These values can be calculated by horizon, weighted by horizon or summed for the entire profile. This index is of most value in humid tropical soils (see Chapter 12).

Another textural index that has been used effectively is Levine and Ciolkosz' (1983) clay accumulation index (CAI). The index is essentially a measure of the difference in clay content between the Bt and C horizons, weighted by horizon thickness:

$$\mathrm{CAI} = \sum[(\mathrm{Clay_{Bt}} - \mathrm{Clay_C}) \cdot T]$$

where $\mathrm{Clay_{Bt}}$ is the clay content (in %) of the Bt and other clay-enriched B horizons, $\mathrm{Clay_C}$ is the clay content of the parent material and T is the thickness (in cm) of the B horizon used in the equation. The values are summed for as many Bt or Bw subhorizons as apply. In essence, this equation represents a measure of illuvial clay. It works well in soils where lessivage is a primary pedogenic process, i.e., Alfisols and Ultisols. The index has been applied in only a few studies (Singh *et al.* 1998), but has great potential. Modification of the index to include bulk density data, making it a volumetrically based index, might be advantageous.

A fruitful avenue for future research involves extending the CAI to other illuvial materials, e.g., Fe in Spodosols, Ca in Calcids or gypsum in Gypsids, e.g., Machette's (1985) cS index. It was developed for use on soils that are accumulating secondary carbonates. It is defined as the weight of $CaCO_3$ in a 1-cm^2 vertical column through the soil, down to the parent material. It is the difference between the total carbonate content in a horizon (cT) and the amount present in the parent material (cP) (both expressed on a mass, not volumetric, basis):

$$\mathrm{cS} = \mathrm{cT} - \mathrm{cP}.$$

The value for cP is obtained from the parent material, while cT values are obtained for each horizon within the profile. The differences are then determined on each horizon as follows:

$$\mathrm{cS} = \sum \mathrm{cT} \cdot \mathrm{cT_{bd}} \cdot \mathrm{cT_t} - \mathrm{cP} \cdot \mathrm{cP_{bd}} \cdot \mathrm{cP_t}$$

where the subscript bd is the bulk density of the horizon and t is horizon thickness. In short,

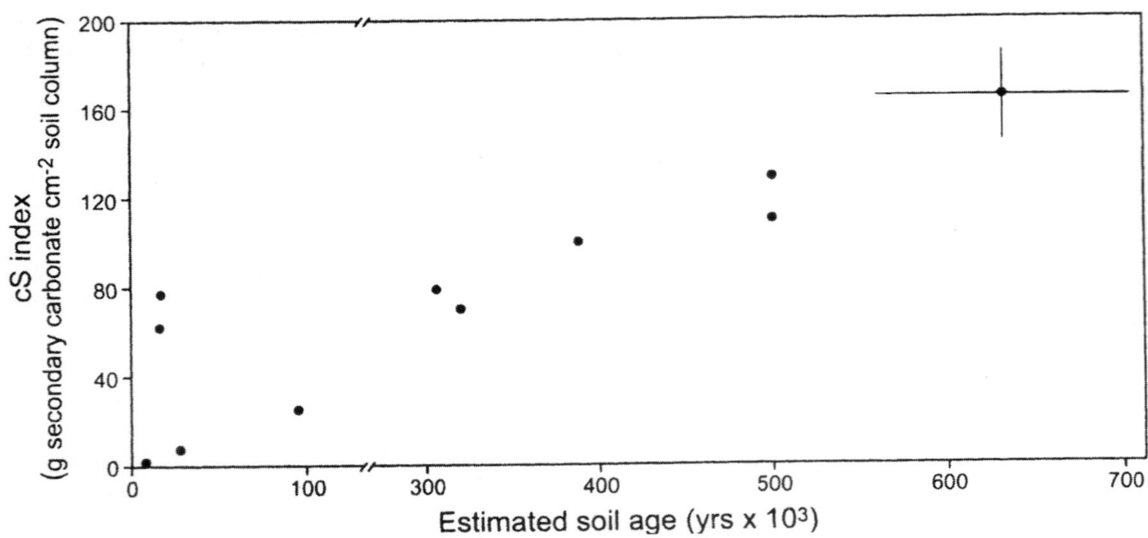

Fig. 14.32 Scatterplot showing how the ages of calcic soils in New Mexico compare to their cS indexes. Note the break in the *x*-axis between 100 and 300 ka. After Machette (1985).

for each horizon a *volumetric* estimate of the carbonate present is determined by multiplying its mass content by its bulk density and thickness. This value is then subtracted from an estimate of the volumetric carbonate that horizon had at $time_{zero}$, obtained from the parent material. Often, cP and parent material bulk density must be estimated. Machette (1985) reviewed the strengths and weaknesses of this index, which nonetheless performs well (Fig. 14.32). The cS index formulation, if viewed generically, provides the blueprint for indexing any soil based on amount of illuvial materials it contains.

Duchaufour and Souchier (1978) developed a podzolization index based on laboratory data. Knowing that soils undergoing podzolization lose aluminum from their eluvial zones and gain it in their B horizons, they developed an index to reflect this trend. The K_{Al} index is best envisioned graphically (Fig. 14.33). To generate it, the amount of total aluminum in the A and C horizons is determined and plotted as a depth function. A line connecting these two points is drawn, as is a horizontal line at the depth of maximum aluminum content. The ratio of two subsets of the horizontal line, as illustrated in Fig. 14.33, is the K_{Al} index, which represents the ratio of translocated to inherited aluminum in the maximally developed B horizon. While the original index uses total aluminum, other podzolization-related products can also be used to generate this index, e.g., total Fe, Fe_d, Fe_p, Fe_o, Al + Fe, etc. (see Chapter 12).

Resistant/weatherable mineral ratios

As early as 1956, Ruhe had successfully applied ratios of resistant/weatherable (R/W) minerals in soils as a relative dating, and a comparative soil development, tool. Over time, contents of weatherable minerals decrease in soils while resistant minerals, because they are less affected by weathering, increase proportionately (Howard *et al.* 1993) (Fig. 14.34). In theory, the R/W mineral ratio increases as soils develop and weather (Dorronsoro and Alonso 1994). In well-developed soils, these ratios also decrease with depth, indicating the degree to which the parent material has been altered by weathering (Brophy 1959). There are a number of generally accepted mineral weathering sequences, from resistant to weatherable, which vary as a function of pH (Fig. 9.1).

To apply the method, the content of resistant, primary minerals within a sand or silt fraction is divided by the amount of weatherable minerals in the same fraction (Soller and Owens 1991). The mineral grains are typically isolated in the laboratory and the identification and counting are

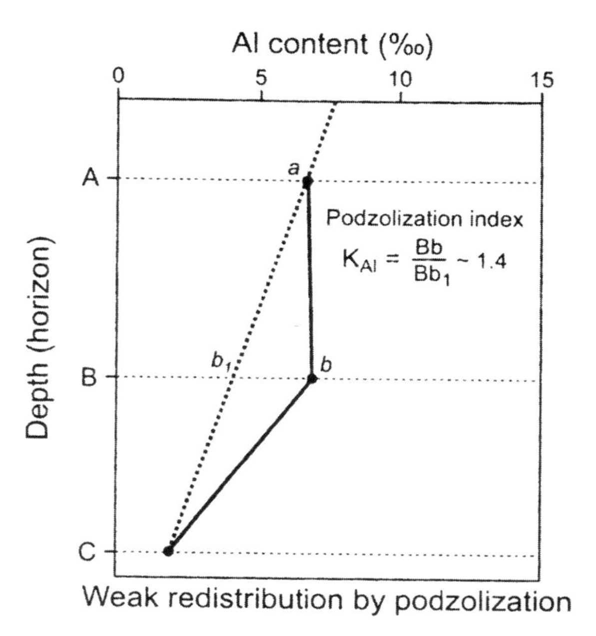

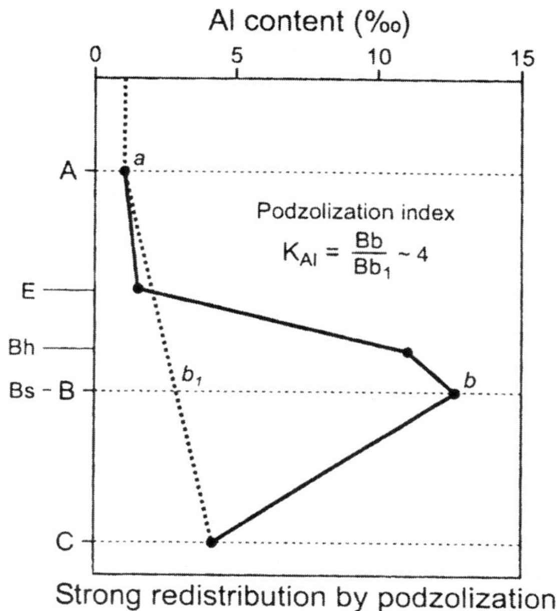

Fig. 14.33 Graphical representation of the K_{Al} index of Duchaufour and Souchier (1978).

done using a petrographic microscope. A minimum of 300 grains are usually counted for each sample. If this is done for each horizon in a number of soils, data on R/W mineral ratios within the soil (surface) as well as between soils (surfaces) can be compared. Fine sand is often the grain size of choice for this type of analysis, for several reasons: (1) in larger size fractions the weatherable minerals are almost completely depleted, especially in older soils, (2) while the smaller sand and silt size fractions may retain more of the relatively rare resistant minerals (Chapman and Horn 1968), this fraction is more difficult to "work with" on the petrographic microscope and (3) because they are skeleton grains, they are assumed to have been immobile during soil formation, pedoturbation notwithstanding.

Resistant minerals are often referred to as *index minerals*. Since they are immobile and resistant to weathering, they provide the index against which other, mobile and weatherable minerals, are compared (Table 14.5). The choice of resistant minerals is a function of their abundance (rare minerals are generally avoided, to save time) and resistance to weathering, which has generally been well established by geochemists and mineralogists based on thermodynamics. Some minerals, e.g., certain feldspars, are easier to identify under the petrographic microscope and may be preferred for that reason. Dyes and stains that are selective to certain minerals can assist in the identification of feldspars and other minerals (Reeder and McAllister 1957, Norman 1974, Houghton 1980).

Ratios of resistant/weatherable minerals can entirely utilize light minerals, heavy minerals or a combination of the two (Brophy 1959) (Fig. 14.35). The relative degree of stability of various heavy minerals in soils is known (Fig. 9.1). Before each is used, the specific conditions under which the sequence was determined should be ascertained. For example, was it a lab- or field-based sequence? What were the pH values of the soils in the study? Howard *et al.* (1993) reported the following general stability suite for heavy minerals in soils:

tourmaline > zircon > rutile > silimanite
≥ kyanite > hornblende/amphiboles
> augite/pyroxenes.

Although augite is last on this list, it is still fairly resistant to weathering, in comparison to

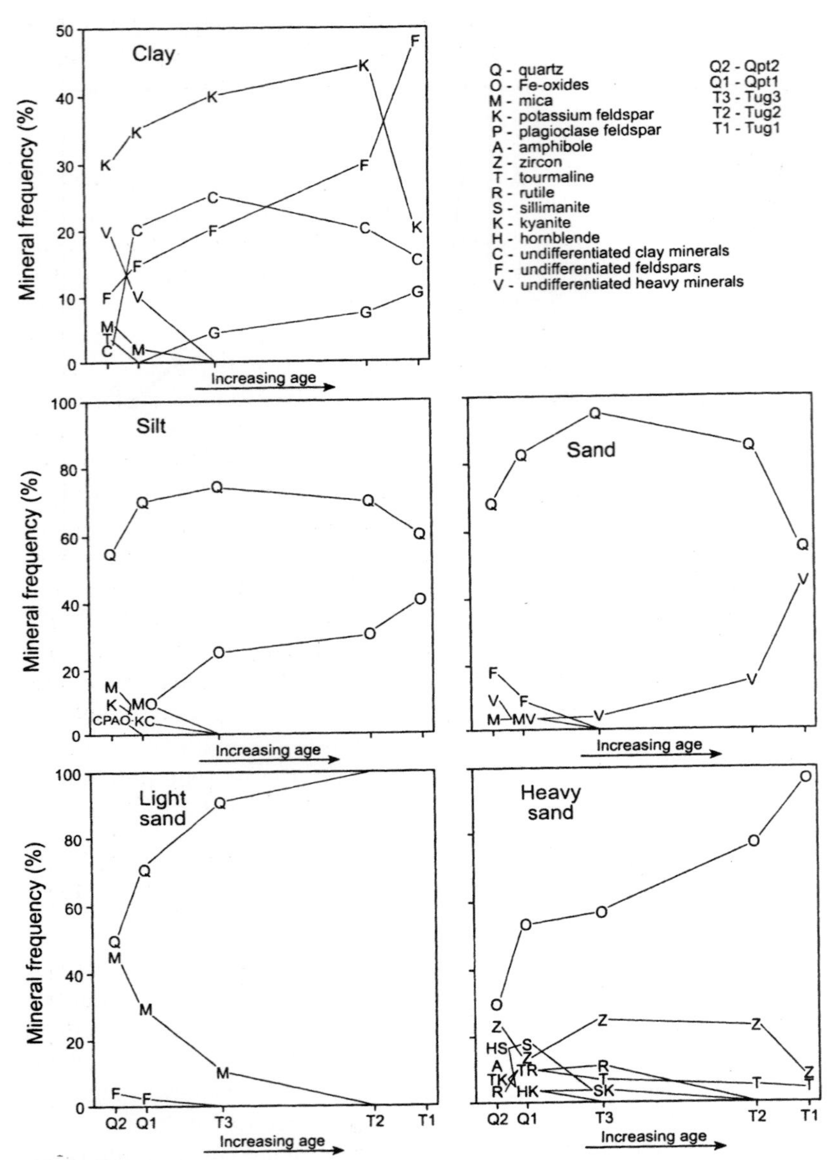

Fig. 14.34 Changes in mineral assemblage frequencies in a chronosequence of soils (Alfisols and Ultisols) in Virginia. After Howard *et al.* (1993).

many other, light minerals like feldspars. Compare the above list to one determined in the laboratory, under acid (pH <5.6) condition, by Nickel (1973):

zircon > epidote > amphiboles > garnet > apatite.

R/W mineral ratios have a number of applications in pedology and paleopedology. In addition to their standard use in chronosequences (Bockheim *et al.* 1996), they are particularly useful in discerning the weathered nature of buried soils, in which standard pedogenic tests of development, especially those associated with chemistry, do not work (Brophy 1959). A consideration, and sometimes a problem, associated with this analysis centers on the relative paucity of some types of resistant minerals (Chittleborough *et al.*

Table 14.5 Resistant/weatherable mineral ratios that have been used to assess weathering and soil development

Quartz / feldspar
(Zircon + tourmaline / amphibole + pyroxene)
(Zircon + tourmaline) / hornblende
Garnet / hornblende
(Zircon + tourmaline) / garnet
Tourmaline / (kyanite + staurolite)[a]
Tourmaline / biotite
Tourmaline / zoisite

[a]Kyanite and staurolite are only of moderate resistance. Use of other minerals with less resistance to weathering is recommended.

1984). For example, zircon and tourmaline can be rare in soils. Extreme care must be taken to insure that large, *representative* samples are taken from the profile. These must then be carefully split into subsamples to assure that they are fully representative of the horizon or profile. Statistically valid conclusions can be drawn only on representative samples, and then only if a minimum amount of the mineral is obtained.

Pedogenic mass balance

A related but more quantitative technique, pioneered by Haseman and Marshall (1945), centers on creating a balance sheet of elemental or mineralogical gains and losses for the soil profile (Bourne and Whiteside 1962). This type of study is called pedogenic mass balance analysis. Over time, the soil profile loses material as minerals weather and the soluble by-products are removed in solution. The focus in pedogenic mass balance is on the minerals and elements that are lost vis-à-vis the minerals or elements that are gained. The gains and losses are calculated on a relative *and* a total basis. Losses or gains can be examined over the entire profile, for the eluvial zone only or by horizon. It cannot be stressed enough that this type of study can be performed only on soils developing in uniform parent materials; determining that the soil has no lithologic

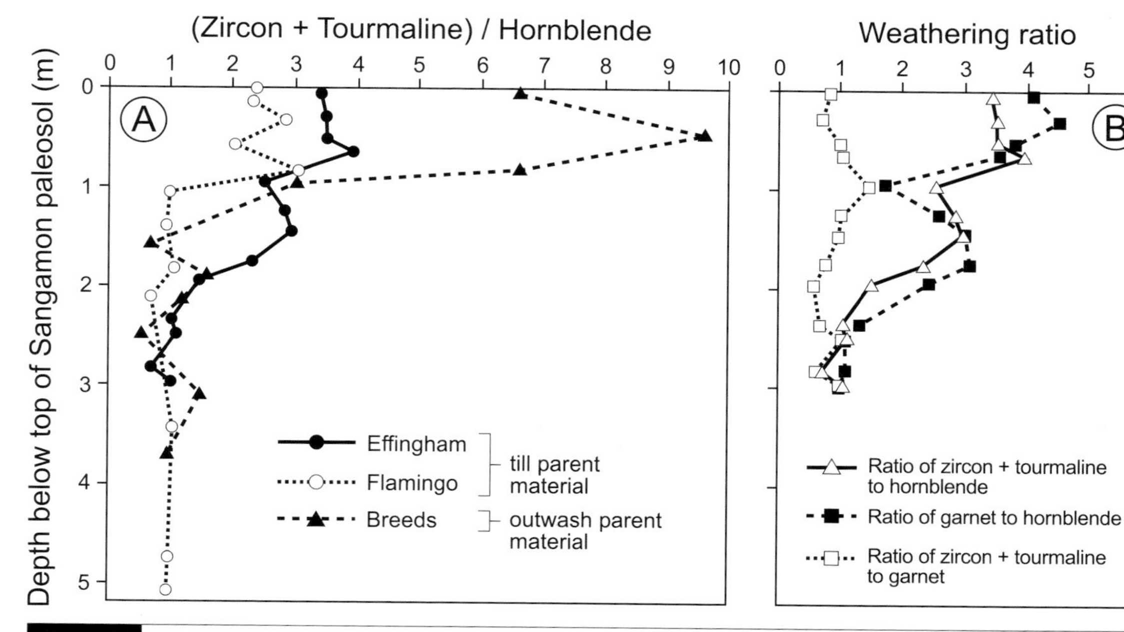

Fig. 14.35 Weathering profiles, based on resistant/weatherable heavy mineral suites in the fine sand fraction, in the buried Sangamon paleosol in Illinois. After Brophy (1959). (a) Variation in weathering ratios as a function of parent material. (b) Variation in weathering ratios in a Sangamon paleosol formed in till, as a function of minerals chosen for the ratios.

discontinuities is imperative to the success of a mass balance study.

Applying pedogenic mass balance principles is much like using R/W mineral ratios in that it assumes that certain minerals, usually zircon, rutile, anatase, tourmaline, ilmenite and monzanite are resistant to weathering and thereby provide a certain measure of chemical stability. Quartz is occasionally used in this context as well and has an advantage over the less-common heavy minerals (Sudom and St. Arnaud 1971, Guillet *et al.* 1975); quartz data are not as dramatically affected by sampling technique as are rare minerals like zircon and tourmaline. Souchier (1971) outlined a mass balance formula that relies on quartz as an immobile and resistant mineral. The *isoquartz method* compares the distributions of other minerals within the profile to quartz, as a way of determining how they have been weathered and either lost from the profile or redistributed within it (Fig. 12.68). Like all mass balance formulae, it assumes a homogeneous parent material at $time_{zero}$. The isoquartz formula for a given horizon is:

$$\Delta X_i = h_i d_i \cdot (Q_i/Q_o) \cdot (X_i - X_o)$$

in which ΔX_i is the gain (positive) or loss (negative) of a component X from a horizon, h_i is horizon thickness, d_i is the bulk density, X_i is the isoquartz content of a specified component of the horizon, X_o is the isoquartz content of the same component in the parent material, and Q_i/Q_o is the ratio of the quartz contents of the given horizon and the parent material.

Because they use only sand-sized minerals, these methods assume that the grains have been pedogenically immobile. This is not true in many profiles where pedoturbation has been ongoing. Nonetheless, depth distributions of stable and slowly weatherable, sand-sized minerals are one of the best ways to assess parent material uniformity – a critical first step in any mass balance study (Evans and Adams 1975a, Chittleborough *et al.* 1984) (Fig. 14.36; see Chapter 8). Over time, resistant minerals increase in abundance as other, less resistant ones, weather and their byproducts are removed from the profile (Fig. 14.37).

There are often two options for determining the amounts of certain minerals in soils: (1) point counts under a microscope (as discussed above) or (2) complete chemical dissolution of a mass of soil followed by elemental analysis using, for example, atomic absorption, mass spectrophotometry, inductively coupled plasma or ion chromatography (Busacca and Singer 1989). Fortunately, the most useful resistant minerals are almost the only sources of certain ions; zircon is the main source of zirconium, while anatase, rutile and tourmaline are the main sources of titanium. Yttrium, which occurs in the resistant mineral xenotime, has also been used (Murad 1978, Chittleborough *et al.* 1984).

Complete dissolution is fast and can provide elemental data on Ti and Zr contents for a number of horizons in a short period of time, which is why index *elements* have largely replaced index *minerals* in mass balance studies (Rabenhorst and Wilding 1986a, Santos *et al.* 1986). One can use Ti and Zr as immobile elements, while (depending on the location) Na, K, Ca, Mg, Fe and Al can be used as mobile elements. The choice of elements is dependent upon the expected pedogenic processes, e.g., in areas of podzolization, Fe is mobile while in dry climates it is not. The use of elements has an additional advantage: the silt fraction, which has a great content of these rare minerals, can be included. However, it is difficult to work with silt grains under the microscope.

Data on resistant minerals and elements provide a standard against which mobile elements and minerals can be compared (Beavers *et al.* 1963, Evans and Adams 1975b). Data for Ti and Zr are assumed to reflect the state of the soil at $time_{zero}$, since they have presumably not been lost by weathering or translocation. Evidence is mounting that titanium is not as stable in soils as initially thought (Sudom and St. Arnaud 1971, Brinkman 1977b, Smeck and Wilding 1980, Busacca and Singer 1989, Cornu *et al.* 1999). Thus, we favor Zr as a stable index element, and zirconium as a stable index mineral, in mass balance studies.

The main difference between mass balance calculations and R/W mineral ratios is that in mass balance analysis the *actual amount* of loss or gain, relative to the parent material, of a certain mineral or element is quantitatively determined.

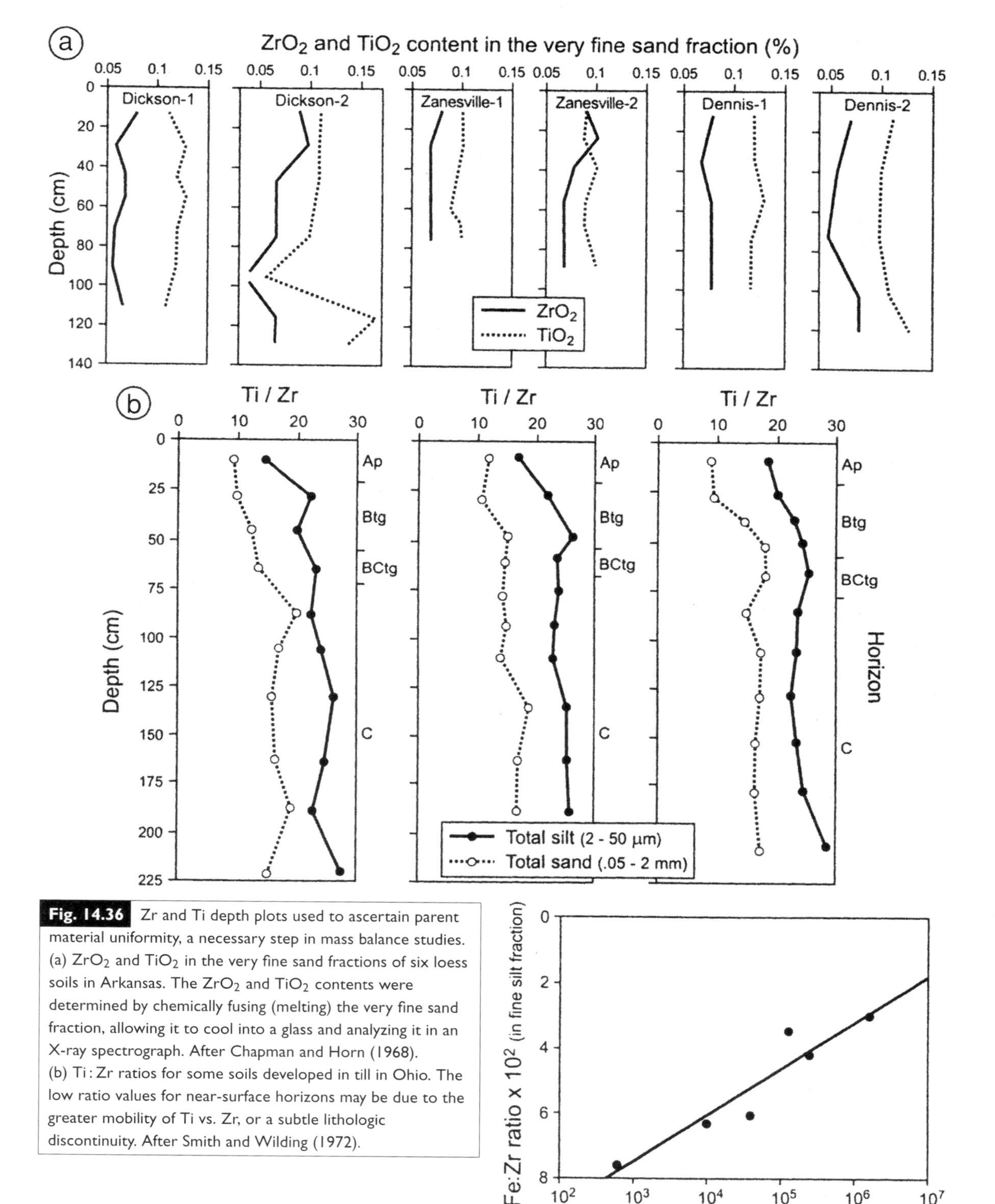

Fig. 14.36 Zr and Ti depth plots used to ascertain parent material uniformity, a necessary step in mass balance studies. (a) ZrO_2 and TiO_2 in the very fine sand fractions of six loess soils in Arkansas. The ZrO_2 and TiO_2 contents were determined by chemically fusing (melting) the very fine sand fraction, allowing it to cool into a glass and analyzing it in an X-ray spectrograph. After Chapman and Horn (1968).
(b) Ti : Zr ratios for some soils developed in till in Ohio. The low ratio values for near-surface horizons may be due to the greater mobility of Ti vs. Zr, or a subtle lithologic discontinuity. After Smith and Wilding (1972).

Fig. 14.37 Changes in content of (Zr : Fe) in soils of varying age in the Sacramento Valley, California. After Busacca and Singer (1989).

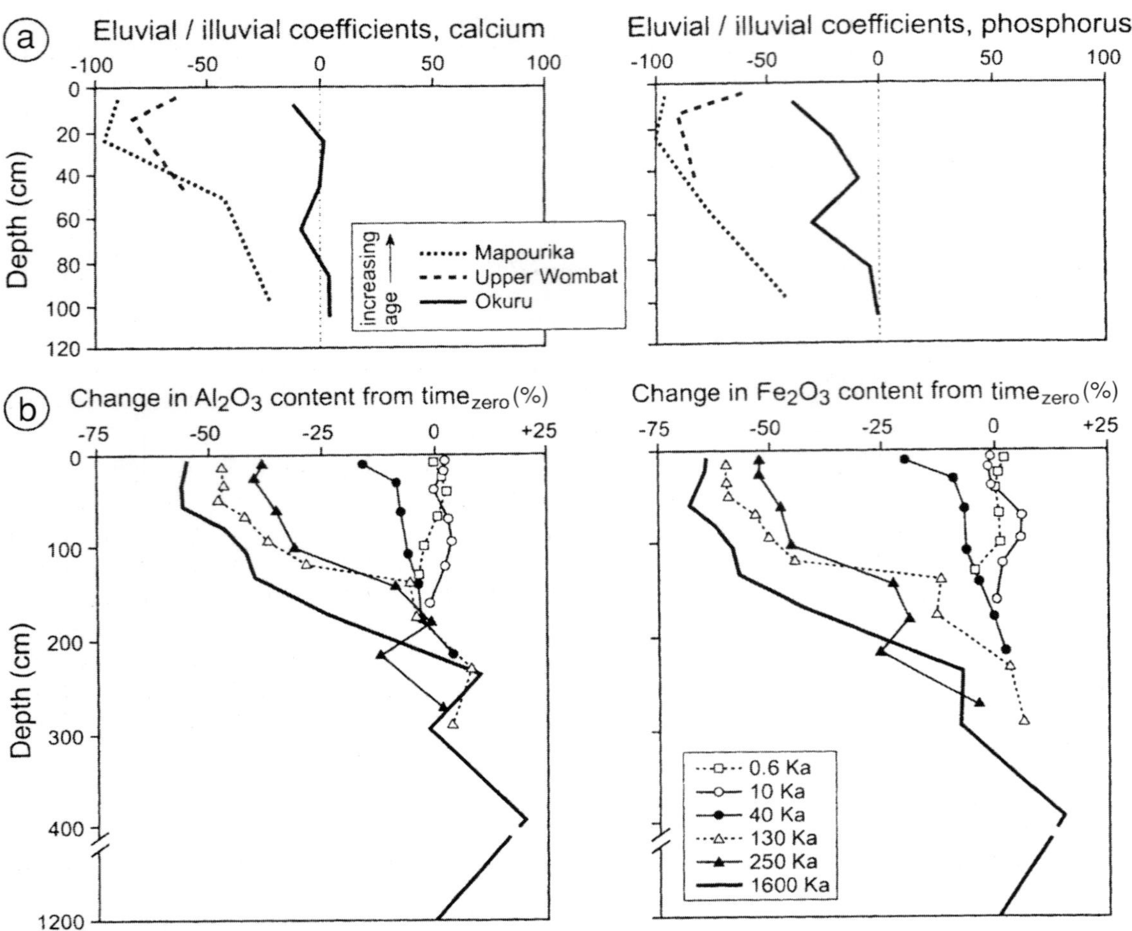

Fig. 14.38 Examples of mass balance studies in which relative gains or losses of elements are assessed for soils, by comparing them to amounts of a resistant and immobile element. (a) Eluvial/illuvial coefficient (EIC) values for soils on a series of Quaternary stream terraces in New Zealand. Negative EIC values indicate a loss of the element relative to the parent material. After Knuepfer (1988). (b) Percentage gains and losses, relative to the parent material, for soils on a series of stream terraces in the Sacramento Valley, California. After Busacca and Singer (1989).

For example, Muir and Logan (1982) determined the eluvial/illuvial coefficient, or EIC:

$$\mathrm{EIC} = \{[(S_\mathrm{h}/R_\mathrm{h})/(S_\mathrm{p}/R_\mathrm{p})] - 1\} \cdot 100$$

where S_h and S_p are the concentrations of element S (not sulfur) in the horizon and parent material, respectively, and R_h and R_p are the concentrations of a resistant element such as Zr or Ti in the horizon and parent material, respectively (Knuepfer 1988). If unaltered parent material is not available for sampling, this method can still be employed; the EIC values will change but their depth trends will remain unaltered. A positive EIC value means that the element has been enriched relative to the parent material (Fig. 14.38). This method is also useful for determining the relative solubility and mobilities of elements, and hence the intensities of various pedogenic processes (Fig. 14.39).

To determine the percentage of a weatherable element remaining in a horizon, use the following equation (Busacca and Singer 1989):

$$\%\ \text{of element remaining} = [(S_\mathrm{h} \cdot R_\mathrm{p})/(S_\mathrm{p} \cdot R_\mathrm{h})] \cdot 100.$$

To incorporate bulk density in these equations and, hence, determine losses and gains on a

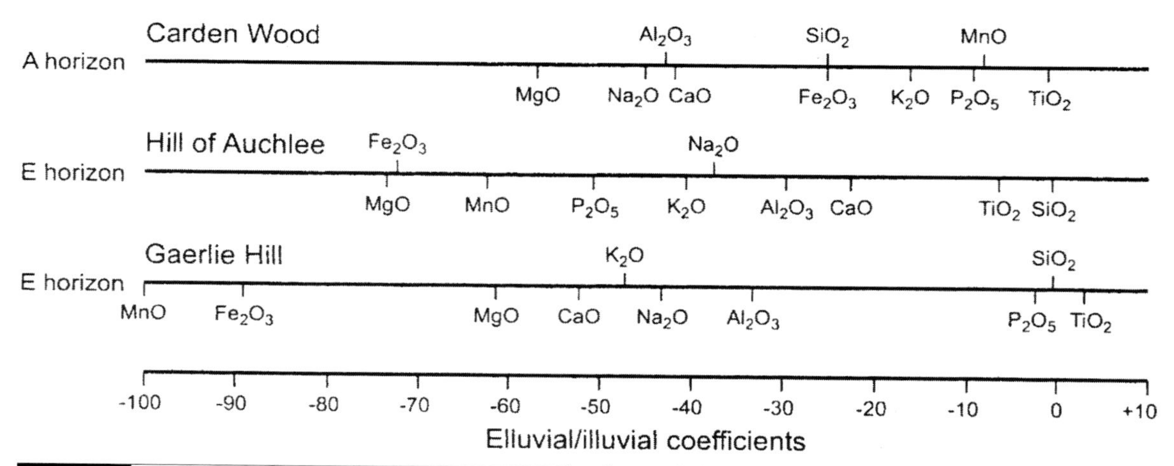

Fig. 14.39 Eluvial/illuvial coefficients for the A and E horizons of three soils undergoing podzolization in Scotland. The low values for Fe and Mn for the latter two soils are due to wet, reducing conditions in the E horizon. After Muir and Logan (1982).

volumetric basis, we suggest the equation of Bain *et al.* (1993):

$$W = t_w \cdot d_w \cdot (X_w - X_c)$$

where W is the amount of an element lost or gained from a horizon, and t_w and d_w are the thickness and bulk density of the horizon. Weathering losses (X_w) are calculated as:

$$X_w = x_w \cdot (R_p / R_h)$$

where x_w is the proportion of element x in the horizon; R_p and R_h are defined as above. When W is negative, there has been a net loss of element x from the horizon.

Chronosequences

A *chronosequence* is a series of soils of known age, as originally defined within the functional–factorial model (Jenny 1941b). In a chronosequence, time (soil age) is allowed to vary while, assumedly, all other soil-forming factors are held constant (Stevens and Walker 1970, Yaalon 1975, Huggett 1998b). While the latter condition is never fully realized, in most chronosequences the impact of the time factor on soil development so outweighs the other state factors that the effect of time on soil development can be generally determined. Bockheim's (1980b) review of chronofunctions illustrated the value, from a process standpoint, that such studies can contribute to soil geomorphology and pedology (Fig. 14.40). Huggett's (1998b) recent study reaffirmed it.

If soil data from a chronosequence are plotted against age, with age as the independent variable, the resultant statistical equation is called a *chronofunction* (Fig. 11.1):

$$S = f_t(\text{time})_{cl,o,r,p} \ldots$$

Bain *et al.* (1993: 276) referred to chronofunctions as "rate-equations of soil formation."

Theoretical considerations

Soils form on geomorphic surfaces when they are stable and when environmental conditions are suitable. This period of time, its duration and characteristics are often referred to as a *soil-forming interval* or pedogenic interval (Morrison 1967, Vreeken 1984a) (see Chapter 15). For surface soils, the soil-forming interval began at some time in the past and continues today, while the soil-forming interval for buried soils ceased upon burial. Many soil-forming intervals have gradational beginnings and endings. A major goal of chronofunction studies is to determine the pedogenic outcomes of the soil-forming interval: What type of soil formed? What degree of development did it obtain? What rate of formation was occurring during that time period?

The theoretical underpinnings of chronosequences involve the ergodic hypothesis, aka the

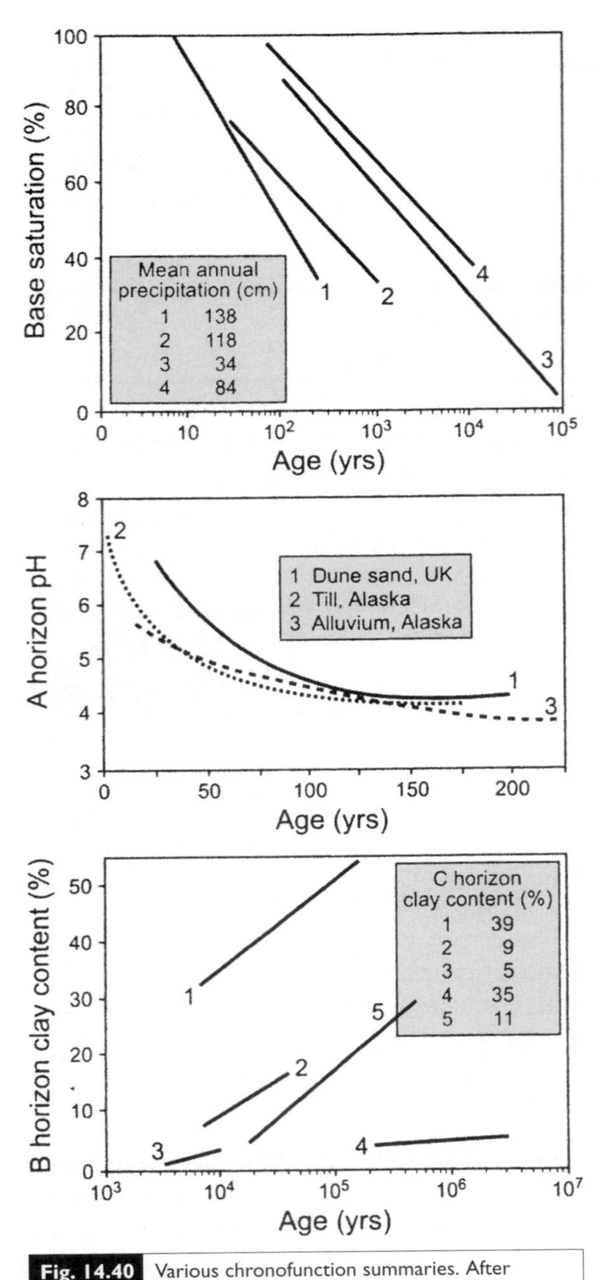

Fig. 14.40 Various chronofunction summaries. After Bockheim (1980b).

comparative geographical approach, in which space is substituted for time (Huggett 1998b). For example, since we cannot remain in one *place* and examine soil development over long periods of *time*, we substitute space for time by examining a number of soils *at the same moment in time but at different places*. For example, a series of raised beach ridges, all of different ages but otherwise similar, are allowed to substitute for time and thereby provide the experimental construct for the chronosequence.

Chronosequence assumptions are that (1) the soil sequence represents successive stages of one or several pedogenic processes and (2) the soils all pass through stages characterized by some preceding member of the sequence (Vreeken 1984a, Huggett 1998b). Both assumptions involve some sort of progressive pedogenic development which, although commonly observed, is not always the case. In fact, Huggett (1998b: 155) attributed the popularity and widespread applicability of chronofunctions to the fact that many researchers support the notion of a developmental view of pedogenesis. However, a progressive/developmental viewpoint is counter to notions of regressive pedogenesis (Johnson and Watson-Stegner 1987). Because most chronosequences report progressive soil development or steady-state conditions (Gile *et al.* 1966, Reheis 1987b, Holliday 1988), it can be assumed that the progressive pedogenic pathway in many soils is at least as strong as the regressive one (see Chapter 11). On the other hand, Hall (1999b) explained chronofunctions that did not have good age–time trends as indicative of soil regression, cryoturbation and erosion, due to changes in external climatic forcings and pedogenic pathways. In perhaps one of the longest chronosequences, on alluvial terraces in Virginia, Howard *et al.* (1993: 201) made the point that

> Not all soil properties show unidirectional development, nor is a steady state of pedon development observed even after approximately 10^7 yr of chemical weathering. Soil development... is episodic. The transition from one phase to the next is marked by a change in rate, and sometimes a reversal in the direction, of development of one or more soil properties.

Obviously, the progressive–regressive–steady-state argument is not over, and chronosequence data will continue to provide valuable fodder for it.

Chronosequences have inherent problems that must be considered if they are to be interpreted correctly. First, rarely can all soil-forming factors except time be held constant over the duration of the chronosequence. Climate is almost certain to have changed, and often vegetation evolves in conjunction with climate and soils. Topography also evolves and changes over time, as attested to by innumerable geomorphologic studies. Most chronofunctions have a limited range of time within they can be applied. Pedogenic thresholds and accessions create problems, as they dramatically change the rate and direction of pedogenesis. Huggett (1998b) also pointed out that not all pedogenic events are recorded in the soil's morphology or chemistry, rendering the chronosequence only a *partial* record of the past. All geomorphic surfaces are spatially variable, prompting questions as to which soil on a surface is most representative (Sondheim and Standish 1983, Harrison *et al.* 1990, Vidic 1998, Eppes and Harrison 1999). Barrett and Schaetzl (1993) sampled a number of soils on each surface and only used data from the modal profile in their chronofunction. Certainly, chaos and deterministic uncertainty also contribute to the unpredictability of chronosequences; soil development may not be unidirectional, but multidirectional, displaying evolutionary divergence (Huggett 1998b) (see Chapter 11).

Vreeken (1975) defined different kinds of chronosequences, based on the time of initiation and/or termination (Fig. 14.41). Soils in a chronosequence may all begin developing at the same time but cease development at different times, or they may begin development at different times and all cease development at the same time (or they may still all be developing today). Additionally, the beginning and ending times for the soils may be highly variable among the group, and may be time-transgressive. In all cases, the length of development (or time period of development) is different among the soils – that is what makes it a chronosequence. Soil development can be ended by erosion or burial, but burial is the option we discuss here, because if the soil is eroded it cannot be part of a chronosequence!

The simplest and most common type of chronosequence is post-incisive (Huggett 1998b) (Fig. 14.41a). Soils in a *post-incisive chronosequence* all began developing at different times in the geologic past, i.e., they each had a different

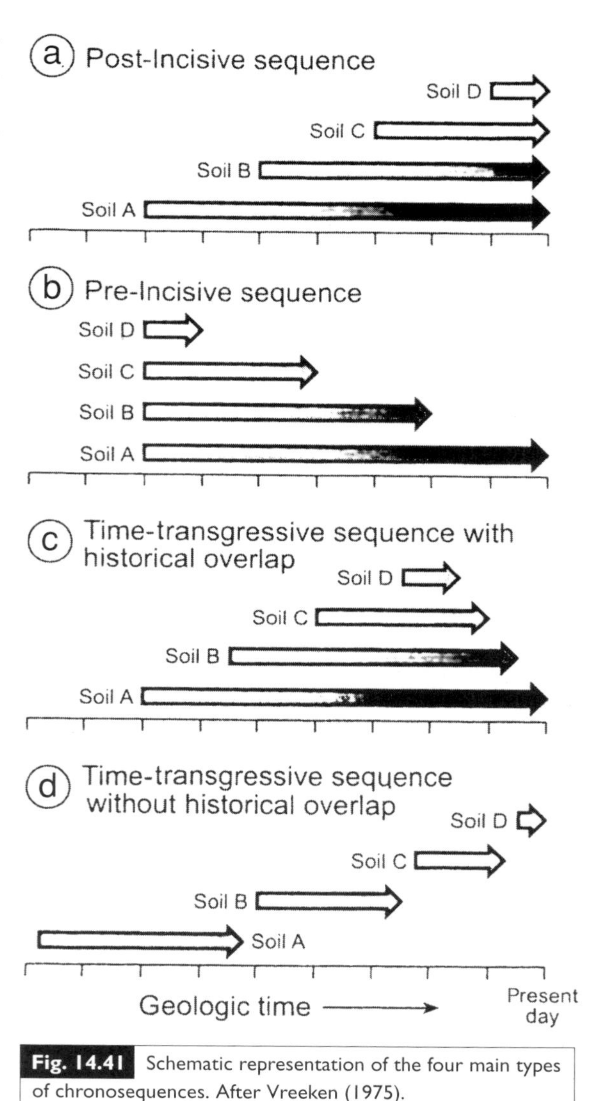

Fig. 14.41 Schematic representation of the four main types of chronosequences. After Vreeken (1975).

$\text{time}_{\text{zero}}$. These soils may still be developing now, or burial may have forced pedogenesis to stop, but in all cases their endpoints are the same. Soils in *pre-incisive chronosequences* started forming at the same time but their development was ended, usually by burial, at different times in the past (Khokhlova *et al.* 2001) (Fig. 14.41b). If neither the starting nor the ending times of soil development are coincident, the chronosequence falls into Vreeken's third category: *time-transgressive with historical overlap* (Fig. 14.41c). This type of chronosequence often forms when landscapes get progressively buried, but the burial is space- and time-transgressive. Soils in these three types of chronosequences always have some degree of historical overlap, i.e., there exists some time in the past when at least two of them were concurrently undergoing pedogenesis. However, many surfaces are exposed to soil development and later buried, and the times during which these surfaces are undergoing pedogenesis do not overlap. Vreeken (1975) called this situation *time-transgressive without historical overlap* (Fig. 14.41d). This type of chronosequence usually occurs in a stacked series of buried soils, as in a till or loess column with intercalated paleosols (Karlstrom and Osborn 1992). Interpretations drawn from this type of chronosequence, in which no two soils were ever forming at the same time in the past, are difficult (Stevens and Walker 1970).

Most chronosequences are post-incisive, e.g., James (1988), Barrett and Schaetzl (1992), Scalenghe *et al.* (2000). A series of moraines or stream terraces of different age provide a possible post-incisive chronosequence, for all the soils on these surfaces are still forming but have different $\text{time}_{\text{zeroes}}$. The analogy in biology would be a sequoia forest with numerous individual trees that started growing at different points in the past. By examining their morphologies today, we could learn about the growth rates of the species and how its morphology changes over time. Pre-incisive chronosequences, which Vreeken (1975) favored on theoretical grounds but which are fairly rare, are equivalent to a stand of sequoias that all started growing after a disturbance event in the past, but parts of which were cut at different times since (Gardiner and Walsh 1966). In soils, the event that stops the pedogenic clock is usually burial, and like the sequoias that run the risk of decomposition after they are cut, buried soils are variously altered after burial (Mausbach *et al.* 1982, Olson and Nettleton 1998). Another problem associated with pre-incisive chronosequences centers on burial itself – one can never determine the degree to which burial is time-transgressive.

Chronofunctions are developed primarily to ascertain rates of soil development (Reheis *et al.*, 1989). Another application centers on the steady-state condition that many soils assumedly develop, or develop toward. What *is* this state, this theoretical endpoint of soil development? And how long does it take to get there? If the rate of soil development slows and appears to approach an asymptote, then one can assume that the soil system is either approaching steady state or may already be there. The chronofunction can then be used to determine (1) *whether* a steady state is achieved, (2) *what* the steady-state value is, and (3) *how long* it takes the system to reach it. Certainly not all soils achieve steady-state conditions (Bockheim 1980b, Dorronsoro and Alonso 1994). In these cases, pedogenic theory such as deterministic uncertainty, chaos theory or soil evolution principles may help explain the non-linear and perhaps multidimensional aspects of the soil's development. Often, certain soil *properties* achieve more or less steady-state conditions, while others continue to change. Soils or soil properties that *do* achieve steady-state conditions are not useful as a measure of soil age for those time periods after they have reached that state (Catt 1990). Those properties that take a long time to achieve steady-state conditions are most useful for dating old soils, while rapidly adjusting properties such as organic matter content are most useful in "short" chronofunctions.

Many soil properties are assumed to develop in a way that is analogous to the classic, S-shaped growth curve (Crocker and Major 1955, Yaalon 1975, Sondheim *et al.* 1981, Birkeland 1999). In this model, growth (development) is slow initially and then increases rapidly, only to slow as it approaches a steady state (Fig. 14.42). However, the many studies which point to logarithmic change in soils suggest that the S-shaped, sigmoidal growth curve does not fit most soil properties. Even soils that do not show rapid pedogenic gains in their early stages of development, as would have been predicted by a logarithmic curve, do not necessarily support the sigmoidal growth curve. Rather, their growth or development could simply be delayed and later, after a threshold has been passed, development proceeds along a logarithmic pathway (James 1988, Schaetzl 1994) (Fig. 14.42b).

Statistical considerations

Once the data for a chronosequence have been attained, a chronofunction can be developed. To

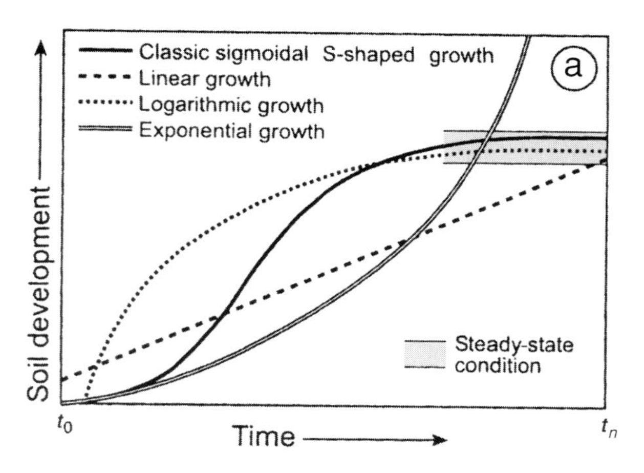

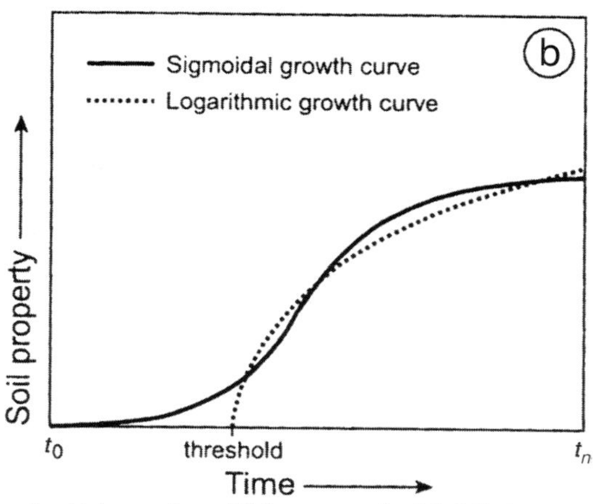

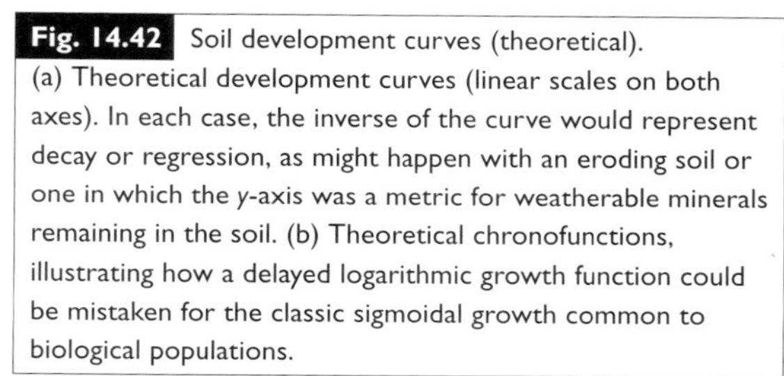
Fig. 14.42 Soil development curves (theoretical). (a) Theoretical development curves (linear scales on both axes). In each case, the inverse of the curve would represent decay or regression, as might happen with an eroding soil or one in which the *y*-axis was a metric for weatherable minerals remaining in the soil. (b) Theoretical chronofunctions, illustrating how a delayed logarithmic growth function could be mistaken for the classic sigmoidal growth common to biological populations.

create a chronofunction, one must utilize numerical or semi-quantitative estimates of surface ages and correlate these ages to a soil property or properties. Surface exposure dating is often used to provide the age estimates. Chronofunctions not only lean on SED in their development, they can also provide information for future SED studies, regarding the likely minimum or maximum age of a surface (Vincent *et al.* 1994). Such data, however, should not be used as the *primary* age determinant for geomorphic surfaces.

Once established, chronofunctions can be used to develop and enhance pedogenic theory, which is then often reapplied to surfaces of known age. In a type of circular logic, chronofunctions therefore provide much of the aforementioned "pedological theory" that is used in SED; they tell us how long it takes for soils to develop property *X* or to lose property *Y* or to thicken to depth *Z*. For example, a chronosequence of soils in northern Michigan established that at least 4000 years are required to form a spodic (Bs) horizon in this region (Barrett and Schaetzl 1992). Future SED studies can now "lean on" this finding and use it as a key baseline datum in this region. This example highlights the differences between chronofunctions and SED studies involving soils. A SED study can only be as accurate as the "library" of chronofunction data allow. In short, chronofunctions provide the *theory* and *age estimation* for pedologic features that are then *applied* in SED. We caution, however, against using chronofunction data circularly, i.e., using time–soil relationships to infer age of surfaces and then using soils information on those surfaces to generate additional chronofunctions, etc. (Vidic 1998). Be aware of the limits of the data and do not overextend their applicability.

Although strictly defining soil development, or a part of it, as a statistical function has been questioned (Yaalon 1975, James 1988, Harrison *et al.* 1990), the advantages far outweigh the shortcomings. Not only do chronofunctions allow us to better understand the soil system today and in the past, many can also provide a measure of prediction. How we analyze and provide order and explanation to the array of these soil data is not only challenging but will dramatically affect the interpretations we make.

Bockheim (1980b), Schaetzl *et al.* (1994) and Huggett (1998b) provided summaries of the many types of dependent soil data that have been applied in chronofunctions. The dependent, i.e., soils, data are usually regressed against surface age in a chronofunction, using least-squares methods (Dorronsoro and Alonso 1994). Early attempts at the creation of chronofunctions involved simply hand-fitting a line to a scatter of points (Crocker 1952, Wilson 1960). This technique is still used with some success (Schaetzl

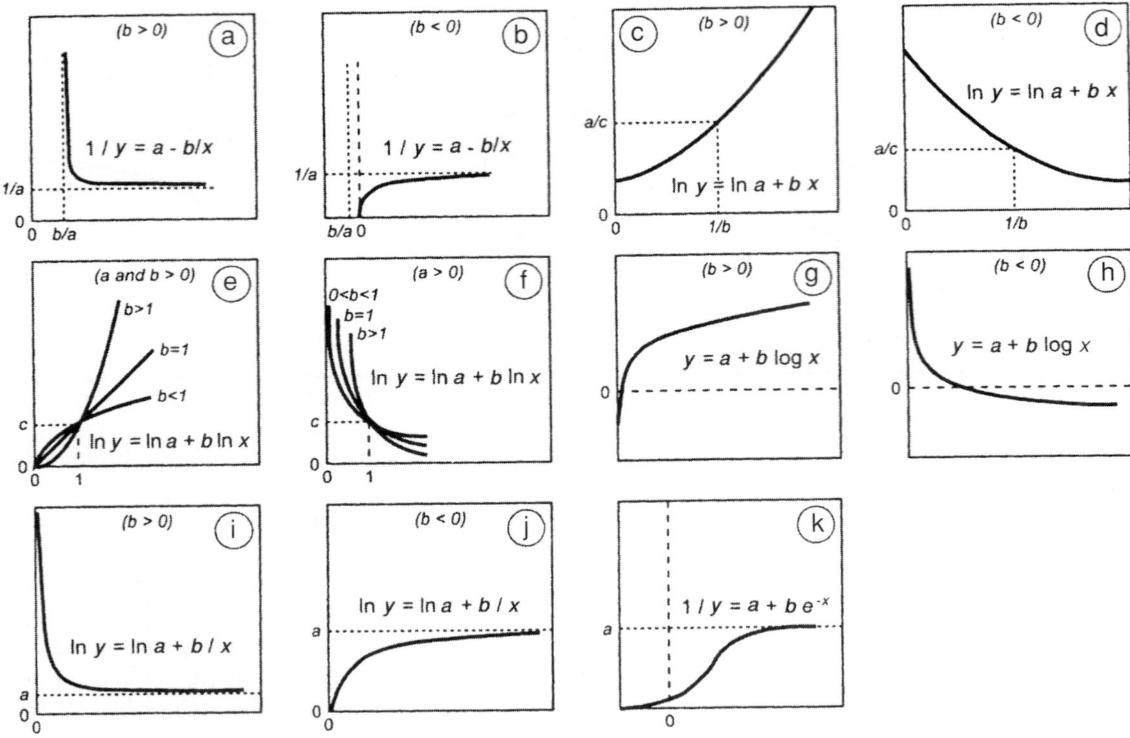

Fig. 14.43 The major types of statistical functions that may be fitted to chronofunction data. All are linearizable; the linear form of the curves is provided within each subfigure. (a) and (b) are hyperbolic functions. (c) and (d) are exponential functions. (e) and (f) are power functions. (g) and (h) are logarithmic functions. (i) and (j) are other functions. See Schaetzl *et al.* (1994) for the general form of each of these functions.

and Mokma 1988, Amit *et al.* 1993, Vincent *et al.* 1994). Raw soil data are typically used as the dependent variable, although factor analysis and principle components analysis are attractive alternatives (Sondheim *et al.* 1981, Scalenghe *et al.* 2000). In chronosequences where parent material cannot be held constant, the use of *ratios* as the dependent variable holds great promise (Mellor 1985).

An inherent dilemma in chronofunctions is that the data may fit any of a number of different statistical models. The model that is chosen should be *based on the theoretical understanding of pedogenesis* that is occurring or has occurred in the soils (Schaetzl *et al.* 1994). Because soil systems function at different rates and along different pedogenic pathways, chronofunctions can be fit to any of at least four statistical models (Levine and Ciolkosz 1983, Mellor 1985) (Fig. 14.43):

$$Y = a + bt \quad \text{(linear model)}$$

$$Y = a + b(\log t) \quad \text{(single logarithmic model)}$$

$$\log Y = a + bt \quad \text{(exponential model)}$$

$$\log Y = a + b(\log t) \quad \text{(power function model).}$$

In these equations Y is the soil property being examined, t is time, a is the y-axis intercept and b is the slope of the regression line. There are other models (Schaetzl *et al.* 1994), including polynomial models ($Y = a + bt + ct^2$), but the four above are the most common. Of these, the first two are the most popular (Little and Ward 1981, Muhs 1982, Dorronsoro and Alonso 1994).

Once a model is chosen and a regression line is calculated, the slope of the line can be used to infer rates of pedogenesis over specific timespans of the chronofunction (Harden *et al.* 1991a). If the best-fit model is linear, it is justifiable to infer that pedogenesis has been proceeding at a generally constant rate for the period of study, and that it may continue to do so in the near

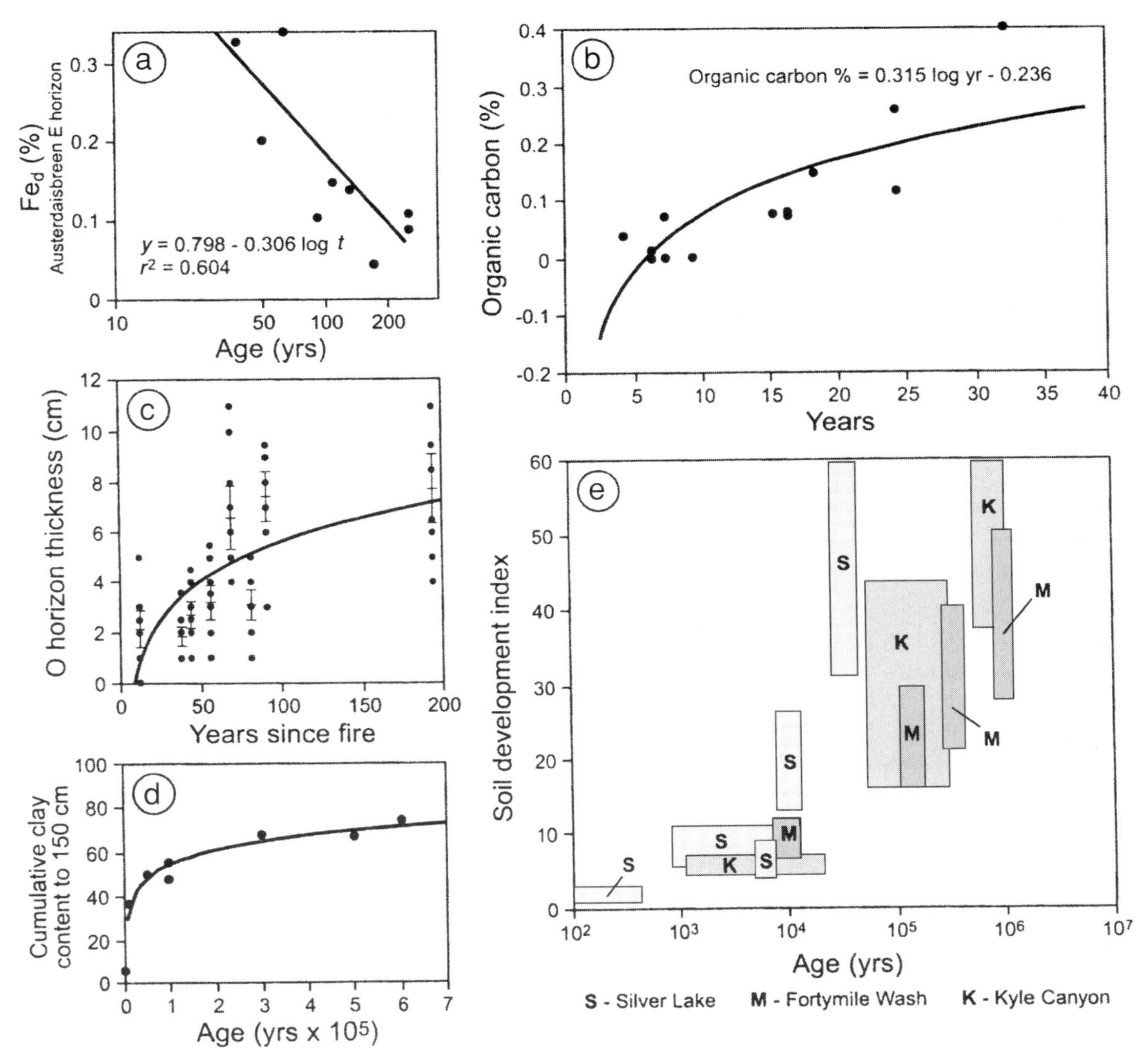

Fig. 14.44 Examples of positive and negative logarithmic chronofunctions. (a) After Mellor (1985); (b) after James (1988); (c) after Schaetzl (1994); (d) after Dorronsoro and Alonso (1994); (e) after Harden *et al.* (1991a).

future. Although linear models are frequently employed in chronofunction research (Mellor 1986, Koutaniemi *et al.* 1988, Merritts *et al.* 1991), logarithmic models are also common (Fig. 14.44). Their application often suggests that the soil system is or will approach a steady state, although Dorronsoro and Alonso (1994) disagreed. Some (Muhs 1982, Busacca 1987) feel that unless the slope of the chronofunction goes entirely to zero, a steady state has not been achieved for the soil system. We argue that, because of measurement error and statistical uncertainty, if the slope of the chronofunction is *near* zero, steady-state conditions may hold or develop.

The interpretation of logarithmic models is an important part of chronofunction research. Log statistical functions plot as curvilinear lines when both *x*- and *y*-axes are linearly scaled. If the same chronofunction equation, however, is plotted on log-linear axes, the regression is straight, with a constant slope (Fig. 14.45). When viewed as a straight line, the soil property may not appear to be approaching equilibrium. One might then (erroneously) interpret the chronofunction as indicating that the soil system is *not*

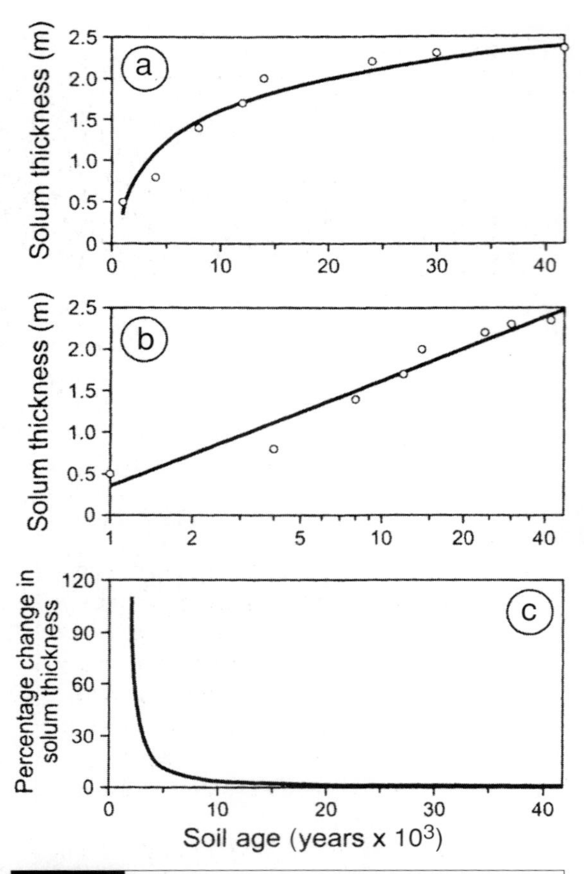

Fig. 14.45 Theoretical chronofunction showing a soil property that increases logarithmically through time. The same chronofunction is plotted in (a) and (b), but on different axes. If plotted on linear axes (a), one might (correctly) conclude that the soil is approaching a steady state. Plot (c) verifies this conclusion by showing that the percentage change with time is rapidly approaching zero. If the same regression equation is plotted on log-linear axes (b), however, one might (erroneously) conclude that the soil property is not approaching any sort of equilibrium or steady state. This figure points out how the conclusions drawn from chronofunctions can be impacted by the graphical method of presentation.

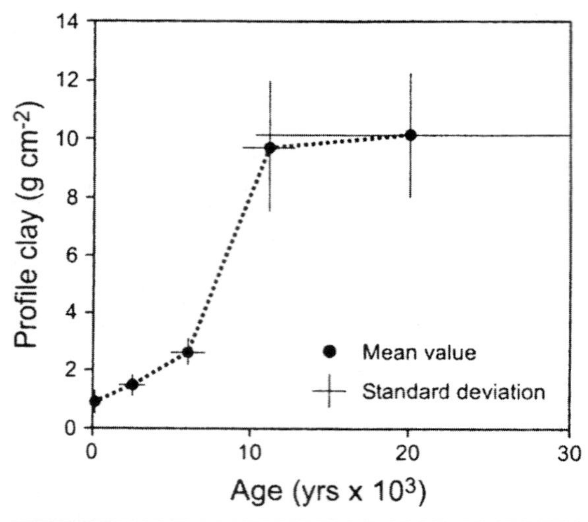

Fig. 14.46 Profile clay content in a chronofunction of desert soils in the southwestern United States. The chronofunction has a step function form. After Harden *et al.* (1991a).

approaching a steady state (Levine and Ciolkosz 1983, Howard *et al.* 1993). A review of rates of soil development, gleaned from published chronofunctions (Bockheim 1980b: 81), observed that chronofunction "trends . . . cast some doubt as to whether soils reach a steady state." This conclusion is flawed because Bockheim evaluated time, i.e., the *x*-axis, logarithmically. *Time is linear*, necessitating that one *must* interpret the slope of the line based on *linear axes*. Indeed, most chronofunction equations can be "linearized," but this fact does not imply that the soil has not reached a steady state; it is simply a statistical artifact. Chronofunctions, therefore, must *always* be interpreted as if the data were presented on *linearly scaled* axes (Fig. 14.45).

Chronofunction formulation is almost always affected by the paucity of data. Because of this, use of exponential or polynomial functions is not encouraged (Bockheim 1980b), although they have been used with some success (Bockheim 1990, Harden *et al.* 1991a). In two-phase regression, similar to a step function, a scatter of points is fitted to two separate linear chronofunctions (Fig. 14.46). This has a great deal of as-yet untapped potential in pedology, for (as noted in Chapter 11) many soils change pedogenic pathway (Bacon and Watts 1971). For example, the slope of the regression equation might be significantly different in the time periods before and after the development of a pedogenic accession. Two-phase regression might, therefore, allow for the unbiased determination of the existence of a pedogenic threshold. Again, one must be careful

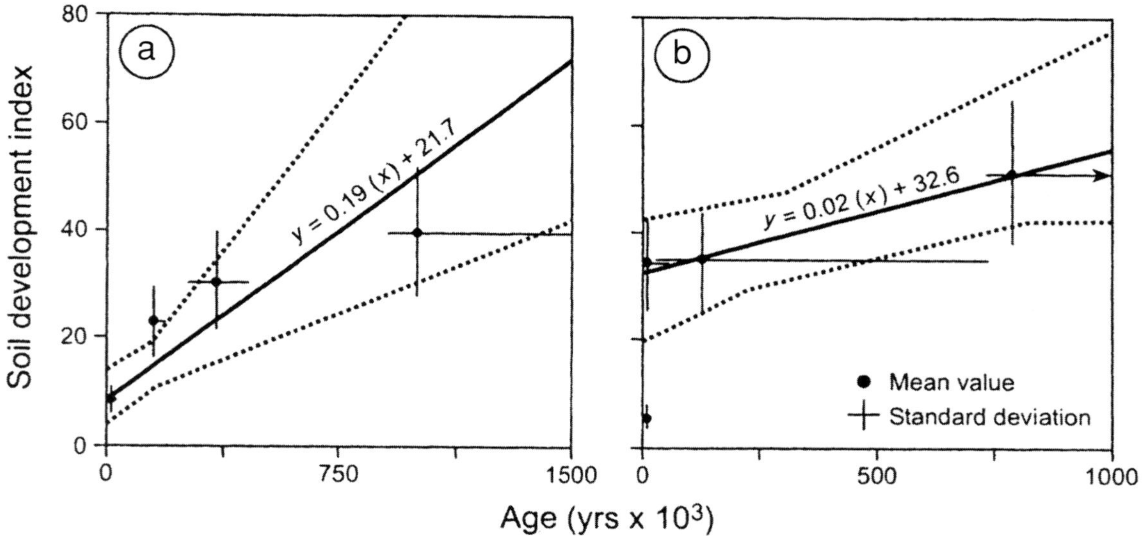

Fig. 14.47 Chronofunctions for soils at (a) Fortymile Wash and (b) Kyle Canyon, Nevada, showing the uncertainty in the statistical function, using maximum likelihood estimation. After Harden *et al.* (1991a).

because the paucity of data that plagues most chronofunctions does not lend itself to two-phase regression.

Chronofunction data have a great deal of statistical uncertainty, especially with regard to age but also for dependent data, due to sampling constraints (Harden *et al.* 1991a). To address this problem, Switzer *et al.* (1988) developed a Monte Carlo approach of refitting the regression line to various data combinations. Their method is iterative, developing many different chronofunctions, given the range of possible data from the one data set. The standard deviation of the various possible chronofunctions represents the uncertainty in the rate of soil development (Fig. 14.47).

For many chronosequences, rates of change near $time_{zero}$ are of unique concern. For example, remediation and recovery of disturbed soils, like minesoils or urban soils, is a branch of pedology that focusses on incipient pedogenic processes (Leisman 1957). However, the processes and pathways in the early phases of pedogenesis are often different than similar processes in later stages. In order to address the need for information near $time_{zero}$, aka the *boundary condition*, it is tempting to extend the chronofunction regression line beyond the range of the data, either back to zero or forward in time, as a potential predictive tool. Almost all researchers warn against the latter practice (James 1988, Yaalon 1992, Schaetzl *et al.* 1994), especially for chronofunctions where the confidence limits on the regression equation are broad (confidence limits are actually widest at the ends of the regression).

Obtaining information about boundary conditions from chronofunctions can be done in several ways. One involves the use of the origin (0,0) in the chronofunction. In chronofunctions where it can be assumed that the value of Y (i.e., the soil property) is zero at $time_{zero}$, insertion of the 0,0 point may be warranted. Examples might be solum or horizon thicknesses, horizon-weighted or solum-weighted data, various pedogenic indices, and pedogenically *acquired properties* such as organic carbon or illuvial clay. If this option is not feasible or optimal, Schaetzl *et al.* (1994) suggest one of three *additional options* for chronofunctions in which no data are available for $time_{zero}$ (shown graphically in Fig. 14.48):

(1) Statistically force the regression line through the origin. This option should only be used for soils in which it can be assumed that the value of Y was zero at $time_{zero}$.

(2) Infer the state of the soil system at $time_{zero}$ from deep C horizon data and use that value

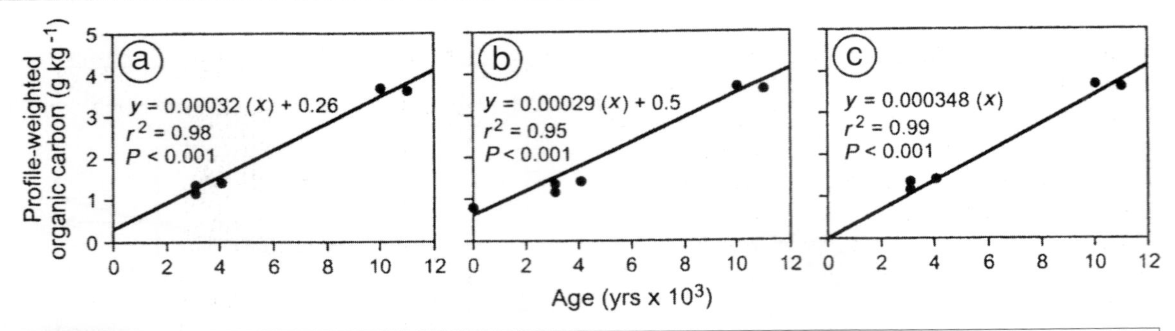

Fig. 14.48 An illustration of how treatment of data near time$_{zero}$ can affect the chronofunction. After Schaetzl *et al.* (1994). (a) The original chronofunction, with its five data points. (b) The same five data points, but with a sixth added, reflecting the condition of the soil at time$_{zero}$. The data for the sixth point came from analysis of the contemporary C horizon. Its use was based on the assumption that the profile-weighted organic carbon content of the soil system at time$_{zero}$ was the same as that of the C horizon. A second option here is to include the point (0, 0) into the data set, assuming the soil had no organic carbon at time$_{zero}$. (c) The same five data points, but with the chronofunction statistically forced to go through the origin. The assumption here is that the soil system had no organic carbon at time$_{zero}$.

in the chronofunction. This is generally an acceptable option, although in older soils it is difficult to obtain samples of unaltered parent material.

(3) Retain the chronofunction in its original form, even if it does not pass through the origin and the researcher knows that at time$_{zero}$ $Y = 0$.

Theoretically, chronofunction research is only in its infancy. As more chronofunction data accrue and as pedogenic theory advances, researchers will make better use of these types of data. It is vital, however, that statistical theory used to generate the chronofunction also match, as best as possible, pedogenic theory.

Numerical dating techniques applicable to soils

There are a number of dating techniques that, when applied to soils or sediments, provide a numerical estimate of surface or sediment age. In this section, we discuss the major methods of numerical, or geochronometric, dating that are especially applicable to soil geomorphology. For a review of other methods, see Watchman and Twidale (2002). Almost all geochronometric techniques used have 1950 as their zero or datum year, i.e., "years BP" refers to years before 1950.

Paleomagnetism

The intensity and orientation of the Earth's magnetic field, as preserved in the orientation of ferromagnetic minerals (particularly magnetite) in rocks and sediments, is called *paleomagnetism*. When initially deposited in a loosely packed body of sediment or as they grow from a melt, these minerals acquire *remanent magnetism*, i.e., they get magnetically aligned (Barendregt 1984). The minerals align themselves to the Earth's magnetic field at the time of deposition, and retain this orientation until disturbed. This orientation is a record of the Earth's magnetic field at the time of deposition. The best types of unconsolidated sediments for this method contain grains of silt and fine sand size that can be strongly magnetized and are free from secondary mineralization, weathering or pedoturbation. Igneous rocks and volcanic deposits are particularly good at preserving paleomagnetic information, although other fine-grained sediments such as loess and marine and lacustrine sediments are also applicable (Barendregt 1981).

Paleomagnetic studies of rocks and ocean sediments have shown that the orientation of the Earth's magnetic field has changed dramatically over geologic time. Our current polarity is considered "normal", i.e., the north-seeking end of the compass needle points toward the north magnetic pole. Periods of *normal polarity* have, however, alternated with periods of *reversed polarity*, when the north-seeking end of the compass needle pointed to the south magnetic pole. Essentially, at many times in the geologic past, the magnetic field of the Earth has done a complete flip-flop; south becomes north. The cause of these magnetic reversals is not clearly understood, but because the transitional (changeover) periods are usually quite short (<10 ka), the long periods of normal or reversed polarity are useful stratigraphic markers for worldwide correlation and dating. Within these long *epochs* of reversed or normal polarity are short (100–1000 years) periods in which the magnetic pole moves sharply toward the equator for 100–1000 years and then returns to a more stable, polar position (Mankinen and Dalrymple 1979). These geomagnetic or polarity *excursions* or *events*, of which there have been several, are also important magnetostratigraphic markers (Harrison and Ramirez 1975, Barendregt 1984). Some of these excursions are considered questionable, however, and may actually reflect problems with the sediment or the analytical procedures rather than representing real magnetic changes (Barendregt 1981). Finally, within the polarity record are still shorter periods of polarity reversal (10–100 ka), called polarity subchronozones.

Geophysicists who study *magnetostratigraphy* have been able to determine the timing of these paleomagnetic changes from the rock record (Cox *et al.* 1963). There have been about 171 paleomagnetic reversals in the past 76 million years. Our current epoch of normal polarity, called the Brunhes epoch, began 774 ka (Sarna-Wojcicki *et al.* 2000). Previous to that, the Matuyama reversed polarity event spanned about 2480-774 ka. The Gauss normal epoch preceded the Matuyama. Each of these events contained within them short-lived geomagnetic excursions (Fig. 14.49). Given that these events are preserved in stratigraphic sequences of hard rock, most soil geomorphologic studies are not concerned with paleomagnetic reversals prior to the Pleistocene. One exception is Early Pleistocene deposits and surfaces, with their reversed polarity.

Scientists use paleomagnetic data in Pleistocene stratigraphic studies for correlations as well as for relative and numerical dating of deposits (Barendregt 1981). Although paleomagnetism seldom allows for great precision in dating the *entire* stratigraphic column, it does provide two baselines at ≈770 and ≈2500 ka (Harden *et al.* 1985, Busacca 1989, Jacobs and Knox 1994). The former is especially important in thick loess sections, which preserve paleomagnetism fairly well (Nabel 1993), or Early Pleistocene lacustrine sediments. The plasma portions of soils, e.g., illuvial clay, sesquioxides and carbonates, acquire paleomagnetic signatures well, making paleosols (if they are suspected of being at least 700 ka old) excellent targets for this method as well (Cioppa *et al.* 1995).

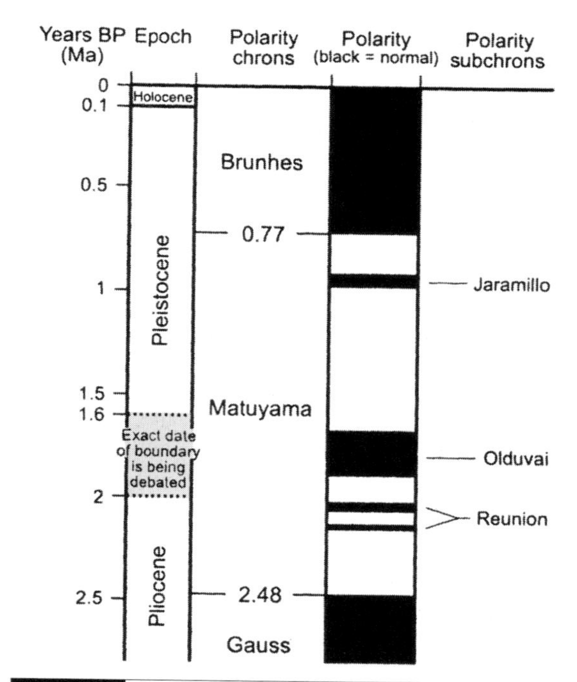

Fig. 14.49 The paleomagnetic time scale for the past 2.8 Ma. After Harland *et al.* (1982).

Tephrochronology

The dating technique of *tephrochronology* (*tephra* are volcanic materials, usually ash and pyroclastics, transported aerially) uses volcanic materials of known age to date associated sediments, and to provide marker beds within the stratigraphic column. Sediments above ash marker beds are younger than the ash, thereby establishing a maximum age for soil development within the

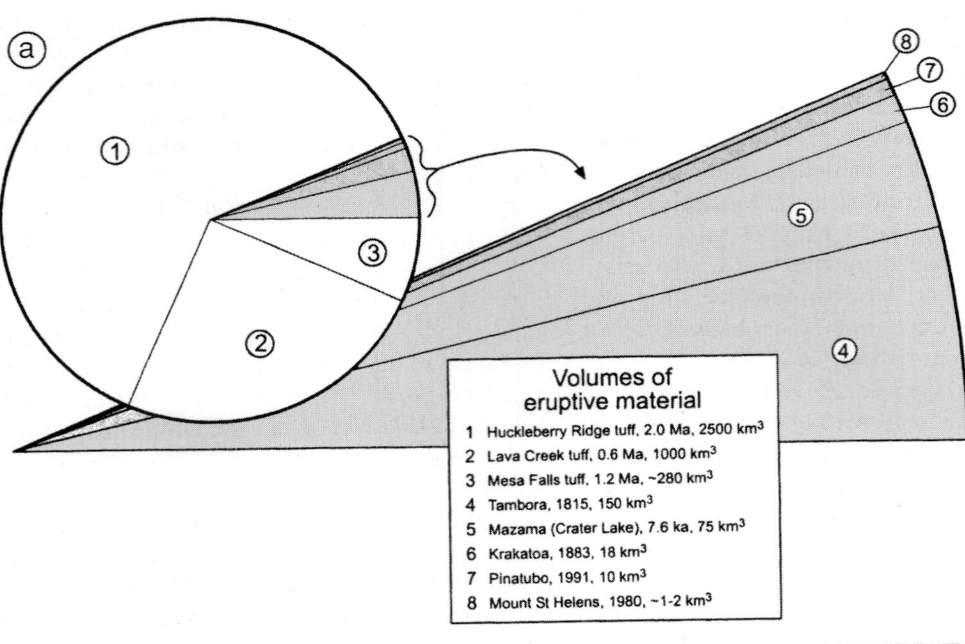
a
1
2
3
4
5
6
7
8
Volumes of eruptive material
1 Huckleberry Ridge tuff, 2.0 Ma, 2500 km3
2 Lava Creek tuff, 0.6 Ma, 1000 km3
3 Mesa Falls tuff, 1.2 Ma, ~280 km3
4 Tambora, 1815, 150 km3
5 Mazama (Crater Lake), 7.6 ka, 75 km3
6 Krakatoa, 1883, 18 km3
7 Pinatubo, 1991, 10 km3
8 Mount St Helens, 1980, ~1-2 km3

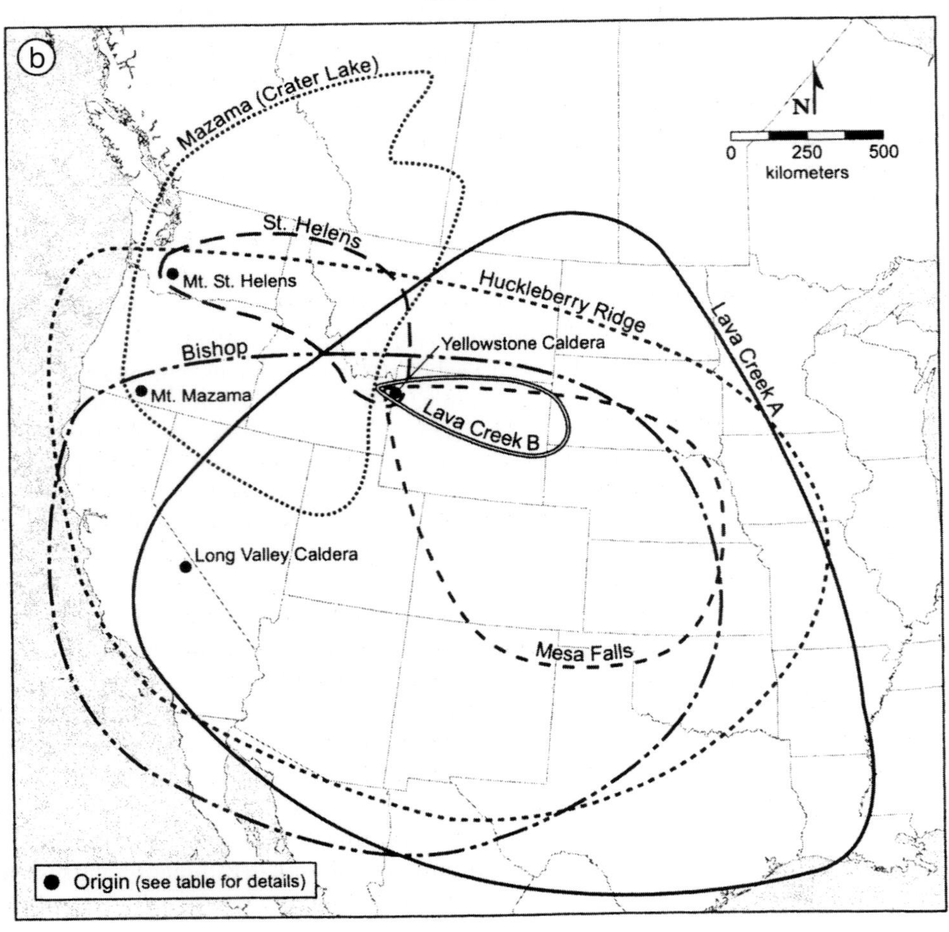
b
Mazama (Crater Lake)
N
0
250
500
kilometers
St. Helens
Mt. St. Helens
Huckleberry Ridge
Lava Creek A
Bishop
Yellowstone Caldera
Mt. Mazama
Lava Creek B
Long Valley Caldera
Mesa Falls
Origin (see table for details)

overlying sediments (Nettleton and Chadwick 1991, Kemp *et al.* 1998). Likewise, sediments (including paleosols) below tephra beds provide clues about climate and environmental conditions before the ashfall (Ward and Carter 1998). Tephrochronology has two strong advantages as a numerical dating technique: (1) ash layers often span large areas, allowing many different sediments to be dated and providing a single chronostratigraphic unit that cuts across (potentially) many landscapes and (2) on a geologic timescale ash is deposited in an instant.

Ashes begin as relatively uniform layers across the landscape but soon are modified by wind and water. When viewed in exposures, many ashes appear to have been redeposited into low-lying parts of the paleolandscape; ash beds thicken in a downslope direction and may be absent on uplands (Carter *et al.* 1990).

Glass shards are the primary constituent of volcanic ash, rendering them amenable to dating by K–Ar, fission track dating or thermoluminescence methods (Dalrymple *et al.* 1965, Westgate and Briggs 1980). Young ashfall layers can be dated by historical means or by radiocarbon dating of organic materials within the layers or immediately above or below them.

Ash is an isochronous horizon, or marker bed, that can not only be definitively dated but also is recognizable based on physical and mineralogical characteristics (Davis 1985, Westgate *et al.* 1987). Therefore, the ash marker bed does not always need to be dated by numerical means after the initial ash dating is completed. Rather, it can often be identified based on shard size and shape, composition, mineralogy, hydration, weathering or other intrinsic characteristics, and its age correlated to equivalent beds that have been previously dated. Thus, ashfall units are chronostratigraphic layers that can often be utilized across wide areas (Izett 1981, McDonald and Busacca 1988). Many are tens of meters in thickness near the source, and are identifiable hundreds of kilometers away. Eventually, the layers get so thin that they are not identifiable, although even in cases where the ashfall layer is not distinct, glass shards that are identifiable to a particular eruption can sometimes be detected, mixed within soils and sediments. For example, ash of known age can be identified in soils, using micromorphological techniques. Kemp *et al.* (1998) identified argillans coating ash shards and tephra pockets, thereby providing evidence of contemporary lessivage as well as quantitative information on the amount of lessivage that has occurred since the ashfall event.

Fig. 14.50 The major ashfall marker beds of central North America. (a) Volumes of some of some Pleistocene and recent volcanic eruptions. After Smith and Braile (1994). (b) Distribution of the major ashfall beds (tephrostratigraphic units) in the western United States. From many sources, but primarily after Izett and Wilcox (1982) and Smith and Braile (1994).

←

Tephrochronology is a highly useful dating tool in volcanically active areas of the world, e.g., New Zealand (Vucetich 1968) and Iceland. It can be used in a macro-sense to date sediments and stratigraphic layers and in a micro-sense to date pedogenic processes. To the former point, hundreds of distinct ash beds have also been identified in the United States (Izett 1981, Bacon 1983, Ward *et al.* 1993). Tephrochronology has great utility in western North America, where most of the major ashfalls are from the Cascade and Rocky Mountains; these ash beds are seldom discernible east of the Mississippi River (Fig. 14.50). Most of the major ashfalls and tuff beds of Quaternary age are named (Table 14.6); many more Pliocene and older beds are known but not included here (Smith *et al.* 1999). Most come from three major regions: the Yellowstone Park area, the Long Valley area of California and the northern Cascades (Izett *et al.* 1970, Porter 1978). Ashes from these different centers are readily distinguishable, but different-aged eruptions from the same center are more difficult to differentiate. In older literature, and still in informal discussions, the name "Pearlette Ash" is mentioned (Swineford and Frye 1946). Pearlette Ash refers to any of a number of ash beds on the Great Plains, which have since been recorrelated and renamed (Izett 1981) (Table 14.6).

Radiometric dating

A large family of numerical dating techniques that has gained increasing application centers

Table 14.6 Approximate ages and origins of major ashfall beds in central North America

Tephrochronological unit[a]	Age (years BP)	Area of origin
White River ash	1 250[b]	Mount Bona, Alaska
Mazama (Crater Lake) ash	7 630[b]	Mount Mazama, Oregon
Mount St. Helens J ash	8 000–12 000[b]	Mount St. Helens, Oregon
Glacier Peak B ash	11 250[b]	Glacier Peak, Washington
Glacier Peak G, M ash	12 750[b]	Glacier Peak, Washington
Mount St. Helens S ash	12 120–13 650[b]	Mount St. Helens, Oregon
Mount St. Helens M tephra	18 560–20 350[b]	Mount St. Helens, Oregon
Lava Creek A and B ash[c] (Pearlette O)	580 000–850 000 (620 000 most likely)	Yellowstone caldera, Wyoming–Idaho
Bishop ash (and overlying ash flow Bishop tuff)	759 000	Long Valley caldera, California
Mesa Falls ash (Pearlette S)	1 270 000–1 290 000	Yellowstone caldera, Wyoming–Idaho
Huckleberry Ridge ash (Pearlette B)	1 800 000–2 180 000 (1 970 000 most likely)	Yellowstone caldera, Wyoming–Idaho

[a] Most ashfall units consists of a number of smaller units that cluster together. For simplicity, we refer to them here as one unit.
[b] In ^{14}C years BP. All others dates have been determined by other means.
[c] Lava Creek A is the older of the two ashes.

Sources: Westgate and Briggs (1980), Izett *et al.* (1981), Davis (1985), Porter *et al.* (1983), Ward *et al.* (1993), Williams (1994), Ward and Carter (1998) and Sarna-Wojcicki *et al.* (2000).

around the use of various isotopes, some stable and some unstable (radioactive). The isotopes produced are of great use as SED tools (Nishiizumi *et al.* 1986). Related radiometric methods, for example, K–Ar and U–Th, utilize the build-up of daughter products from primordial radionuclides; because they are most useful in geology and not soil geomorphology, we do not dwell on them here.

Radiocarbon

Carbon-containing compounds are widely distributed and occur in many forms that cycle among their major reservoirs, including the atmosphere as CO_2, the biosphere as biomass, and the pedosphere/lithosphere as humus, peat, limestone, marl and many other substances. Radiocarbon dating is used to determine the ages of these carbon-bearing materials.

THEORY AND MEASUREMENT

Radiocarbon theory is based on the variability that exists with respect to carbon isotopes. About 98.9% of all carbon atoms are ^{12}C, with an atomic number of 12. Another 1.1% are ^{13}C, while a very small proportion are ^{14}C, also called *radiocarbon*. Both ^{12}C and ^{13}C are stable, but ^{14}C is unstable and radioactive; it decays at a known rate, relative to ^{12}C, producing beta particles as it decays (Taylor 1987). Radiocarbon, with two extra neutrons in its nucleus, is produced when neutrons produced from cosmic rays impact nitrogen ($^{14}N_2$) gas, which comprises over 78% of the atmosphere (Vogel 1969).

The radiocarbon "cycle" is well understood (Fig. 14.51). Radiocarbon in the atmosphere is quickly converted to $^{14}CO_2$ and mixed throughout. Much of it is dissolved in the oceans as bicarbonate, but some is also taken up by plants during photosynthesis and incorporated into their biomass. The amount of radiocarbon in plants quickly equilibrates to that in the atmosphere. (Actually, for most plants, the amount of radiocarbon in this biomass is 3–4% below that of the atmosphere, but this can be taken into account.) As long as the plant is alive, it is absorbing

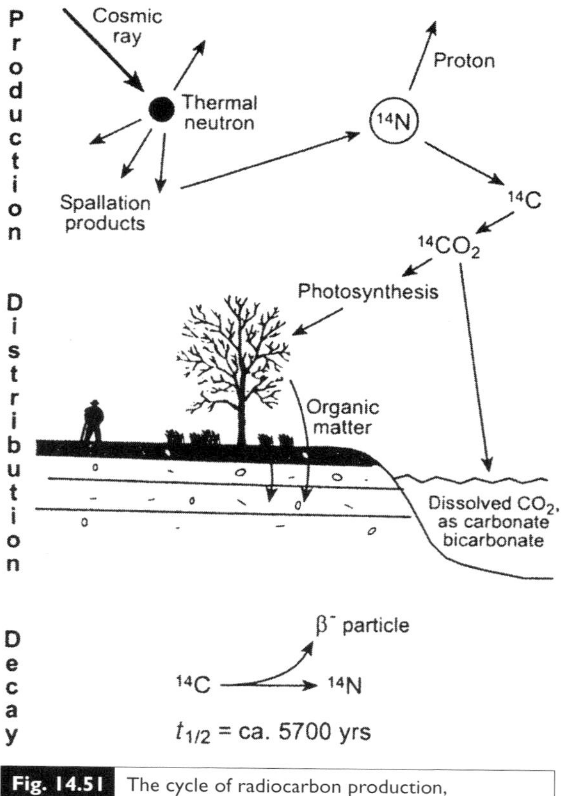

Fig. 14.51 The cycle of radiocarbon production, distribution and decay. After Taylor (1987).

radiocarbon from the atmosphere at a rate that maintains an equilibrium state. Green plants form the base of the food chain, such that higher organisms are ultimately formed from plant material and thus have radiocarbon in their bodies in an amount that is in approximate equilibrium with the atmosphere (Burleigh 1974). However, after plant or animal metabolism ceases, i.e., the organism dies, uptake of radiocarbon stops and the store of radiocarbon in the dead bone, log, leaf or tooth is determined only by radioactive decay, a function of the time since death. The ^{14}C concentration in the biomass diminishes logarithmically, having a half-life of 5730 ± 40 years (Linick *et al.* 1989). Thus, in 5730 years, half of the radiocarbon has decayed to ^{14}N, and after another 5730 years another half is gone (leaving only ¼ of the original store of radiocarbon), etc. After about 10 half-lives, the amount of residual ^{14}C is so small that it cannot be determined with current technology, meaning that samples older than about 50 ka are not datable by the ^{14}C method.

A radiocarbon date is an estimate of the time since the death of the plant or animal. The radiocarbon age is based on a measurement of a sample's residual ^{14}C content, a comparison to modern standards, and a knowledge of the half-life decay constant (Taylor 1987). Determining the amount of residual ^{14}C can be done in either of two ways. Conventionally, the sample is isolated in a chamber and the number of beta particles (electrons) emitted over a period of time, usually at least 24 hs, is determined (Fig. 14.51). Usually about 2–200 g of carbonaceous sample is required, depending on its age and type, to obtain a ^{14}C date using this method. Old and/or small samples will not yield enough beta particles to accurately determine an age. For example, in a gram of "young" carbonaceous material only 14 beta particles are produced each minute. In an "old" sample there might only be this many each hour. Eventually, the cost of laboratory time and the increased size of the error term becomes prohibitive.

A second technique is also available for determining the radiocarbon content of samples (Linick *et al.* 1989). In this method, *accelerator mass spectrometry* or AMS dating, the total amount of radiocarbon is determined by directly counting the numbers of ^{14}C atoms in a sample with a mass spectrometer. AMS dating can be used on samples as small as 2–30 mg (depending on carbon content). It is faster, more accurate and requires almost 1000 times less sample than the conventional decay-counting method. With AMS the required sample size is so small that one has the option to date only the tiniest, most pristine fragments of carbon from a possible sample locale, reducing the potential for contamination (but increasing the likelihood that the sample is not representative). It is also possible to obtain datable samples from sediments that typically have only finely comminuted fragments of carbonaceous material, further widening the possible applications (Pohl 1995). AMS dating has extended the radiocarbon dating method to 50 or even 60 ka.

Because the amount of ^{14}C detected (in either method) is based on statistical probabilities, one must always consider the *statistical uncertainty* in the reported age. The amount of radiocarbon in a sample can be determined only within a certain range, yielding age estimates that are usually reported with an envelope of error, expressed as a standard deviation, or sigma (σ), e.g., 7560 $\pm$ 340 BP. The radiocarbon date is only a *range of years* within which the *true age* lies, according to a certain probability. If the reported range is one standard deviation or σ, the true age has a 68% probability of falling within that range. The most probable date, however, remains the figure obtained by the measurement. It is advisable to report dates to two standard deviations, which widens the range but increases the likelihood of that range containing the actual age to 95%.

ASSUMPTIONS

The radiocarbon method has several assumptions. First, we assume that the amount of radiocarbon in each carbon reservoir has remained constant over the time-span of the method. Detailed ^{14}C analysis of the wood in tree rings, from long-lived trees like bristlecone pine, redwood, sequoia and European oak, which can be very accurately dated, have shown that this assumption does not hold (Willis *et al.* 1960, Ferguson and Graybill 1983). The reason for the variations in the amount of radiocarbon in the reservoirs is due to (1) variation in production, which is commonly traced back to changes in the intensity of the Earth's magnetic field and (2) for very young or near-surface samples, large-scale burning of fossil fuels and open-air testing of nuclear devices (Stuiver and Quay 1980). Thus, radiocarbon ages vary somewhat from the standard astronomical calendar, necessitating that they be reported in radiocarbon years BP, not calendar years.

In an effort to fully compare radiocarbon dates with astronomical time, there is a growing tendency to convert, i.e., calibrate, traditional radiocarbon dates to calendar years. The parameters used for this calibration come from the tree ring record, which provides detailed information to about 10 ka. Use of information from varves (carbonate-rich lake sediment which accumulates in annual layers and thus can be dated to the exact calendar year of formation) and sea coral have extended the radiocarbon-to-calendar age calibration curve to 24 ka (Stuiver *et al.* 1986, 1998). In general, the divergence between radiocarbon dates and tree rings dates is not significant after about 3500 BP, but before this time the difference gets progressively larger (Fig. 14.52). It can be as much as 700 years by 4500 BP. It is often suggested that only *calibrated* ^{14}C dates be reported on everything except soil carbon mean residence time (MRT) dates (see below). Most radiocarbon laboratories report both uncalibrated and calibrated ^{14}C dates.

A second basic assumption applied to ^{14}C dating is that all plants accumulate radiocarbon in equal proportions to each other; this has been shown to be incorrect (Olsson and Osadebe 1974). However, the small variation among plant species is usually not a large source of error.

Another assumption centers on the lack of sample contamination, either by "old" carbon such as coal, black shale or any carbonaceous material that no longer retains radiocarbon, or by "young" carbon with high amounts of ^{14}C. The most common contaminants in soils are roots, especially fine root hairs, humic substances and carbonates from "hard" water and those deposited by pedogenesis (Olsson 1974). Hard water contamination should be seriously evaluated when dating material from soils with high water tables. Contamination by young carbon has the greater potential to skew the date, especially if the date is old (Fig. 14.53a). Even so, contamination is not a large problem, statistically, for samples that are less than a few thousand years old (Vogel 1969). Fortunately, many radiocarbon laboratories are able to pretreat and leach samples of possible secondary carbon contaminants, especially if using the AMS method. Nonetheless, care must be taken during sampling and subsequent storage to avoid (1) allowing the sample to touch skin or hair, (2) the growth of mold or other organisms on the sample (always keep the samples refrigerated or frozen) and (3) samples with large numbers of roots growing in and among them. Samples are best stored in aluminum foil, as plastic is also a carbon source.

APPLICATIONS

In principle, any material containing carbon which at some time has been in exchange with

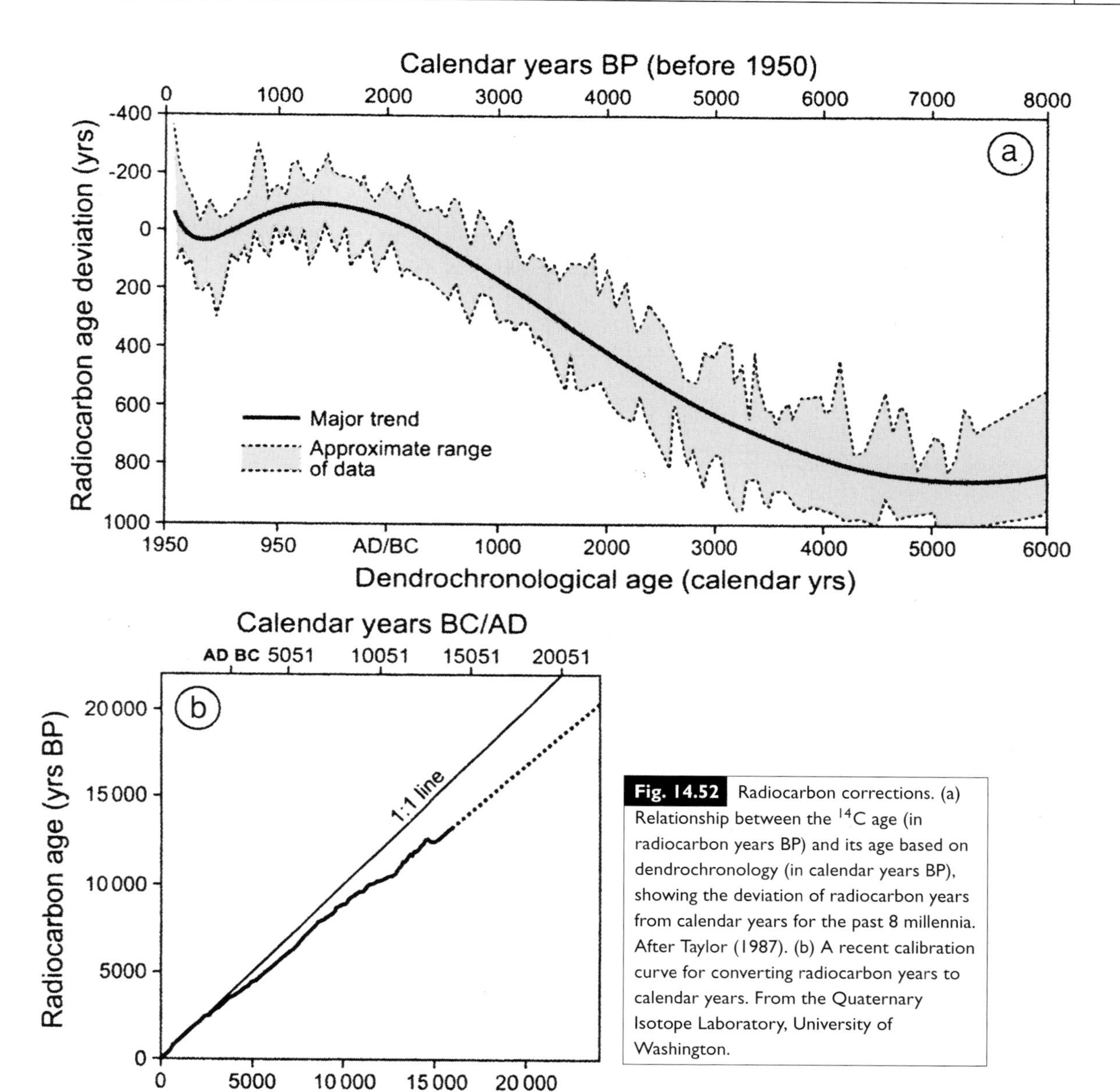

Fig. 14.52 Radiocarbon corrections. (a) Relationship between the ^{14}C age (in radiocarbon years BP) and its age based on dendrochronology (in calendar years BP), showing the deviation of radiocarbon years from calendar years for the past 8 millennia. After Taylor (1987). (b) A recent calibration curve for converting radiocarbon years to calendar years. From the Quaternary Isotope Laboratory, University of Washington.

atmospheric CO_2 can be radiocarbon dated (Vogel 1969). By far the most common type of materials used in ^{14}C applications are organic forms such as wood and charcoal, although inorganic C such as cave speleothems and soil carbonate are also datable (Pohl 1995). Wood and charcoal fragments are easy to recognize in sediments and because of their high-molecular-weight components, they allow for rigorous pretreatment procedures to extract possible contaminants, rendering them the "safest" materials to date (Geyh *et al.* 1971, Taylor 1987). Charcoal is particularly attractive because it is inert and readily preserved. However, it does have the potential to absorb humic substances from soils, which are a source of contamination. Dating charcoal, like any other substance, should not be regarded as reliable until evidence is found, usually via field observations, that its radiocarbon age relates to the date of deposition (Watchman and Twidale 2002). Non-woody plant

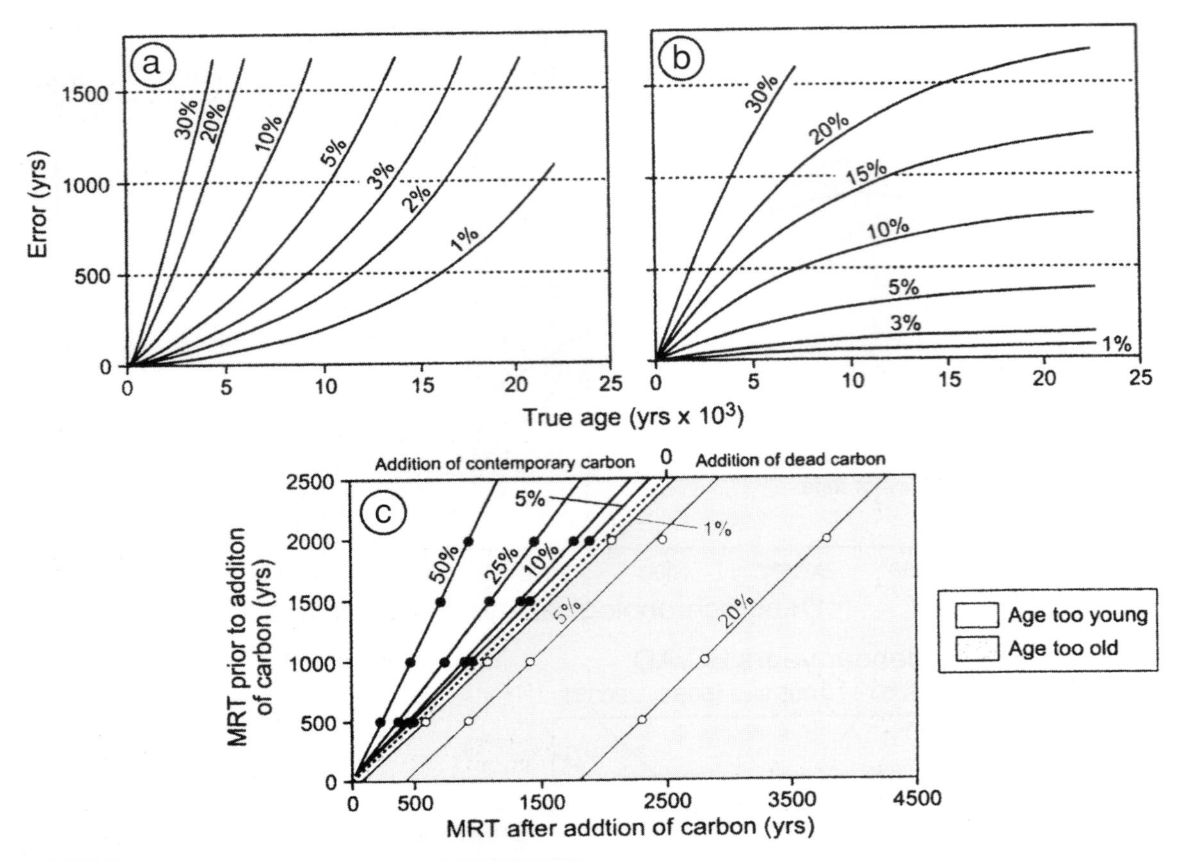

Fig. 14.53 Potential errors due to contamination of ^{14}C samples. (a) Effects of the introduction of "old" carbon (lower ^{14}C activity) to the sample. Values on the *y*-axis indicate the difference, in years, between the true and reported sample ages. After Olsson (1974). (b) Effects of the introduction of "young" carbon (higher ^{14}C activity) to the sample. Values on the *y*-axis indicate the difference, in years, between the true and reported sample ages. After Olsson (1974). (c) Effects of "old" and "young" carbon on mean residence time (MRT) dates on soil organic matter. Because MRT dates are a statistical average of many different organic fractions, the effects of "contamination" are different. After Campbell *et al.* (1967).

parts such as leaves, seeds, reed, roots and their derivatives (papyrus, paper) are also candidates for dating.

In carbonate-rich shells, the ^{14}C determinations are based on the inorganic fraction, as $CaCO_3$. The carbon source for the animal that made the shell was the ^{14}C in the water at the time the shell was constructed. Thin shells from freshwater sources, such as snails, are easily contaminated by other carbon sources (Goodfriend and Hood 1983, Wilson and Farrington 1989). With proper pretreatment, however, ^{14}C dates on shell material, especially thick shells, can be reliable. Dating the shells of land snails, which are common in loess deposits, is important in establishing loess chronologies (Fig. 14.54). Bone is also commonly dated, especially within archeological contexts. Because bone is highly susceptible to contamination, however, proper pretreatment is essential (Taylor 1982).

In soil geomorphology, ^{14}C dates usually provide a minimum- or maximum-limiting age for a *sediment* or *geomorphic surface* (Nelson *et al.* 1979, Pohl 1995). Maximum-limiting dates imply that the sediment or surface can be *no older than* the ^{14}C (or other type of) date. Conversely, minimum-limiting dates indicate that a surface or sediment can be *no younger than* that date. The technique is therefore used to provide numerical age control on stratigraphy, providing a

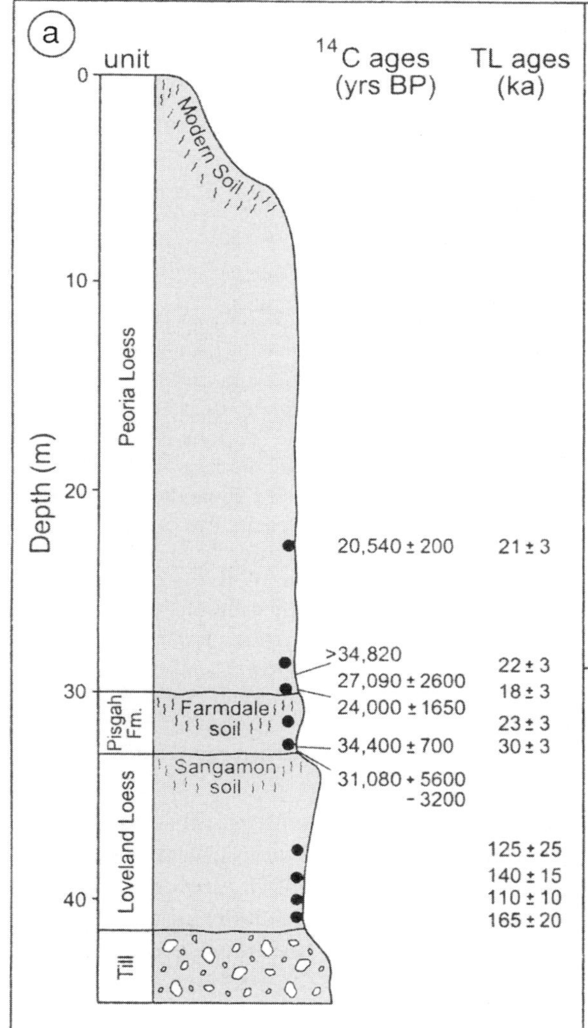

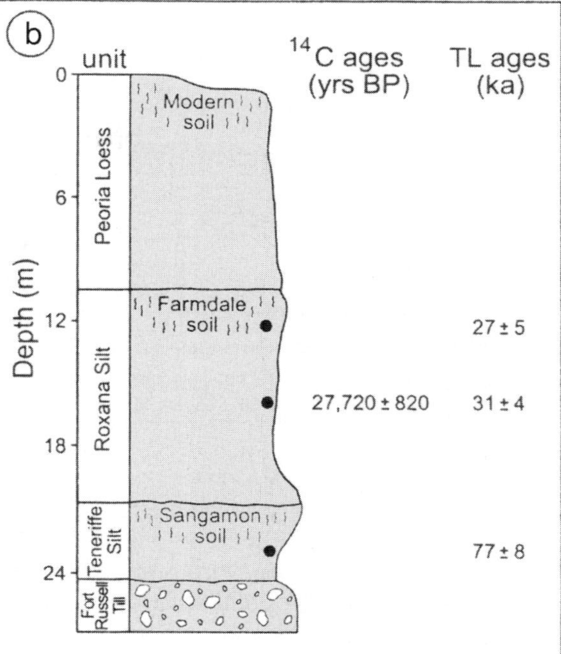

Fig. 14.54 Typical loess and paleopedological stratigraphy for the midcontinent United States. The major loess units and paleosols are shown, along with ^{14}C ages (on shells) and thermoluminescence (TL) ages (on the loess itself). After Forman *et al.* (1992). (a) Loess stratigraphy and associated radiocarbon and luminescence ages estimates for the Loveland loess type locality in Pottawattamie County, Iowa. (b) Loess stratigraphy and associated radiocarbon and luminescence ages estimates for the Pleasant Grove School section in Madison County, Illinois.

temporal framework for the timing of erosional and depositional events. For example, envision wood within an alluvial deposit dated at 11 300 BP[1] (Fig. 14.55) This date suggests that the alluvium is *no older than* 11 300 years, for on approximately that date the tree died. If it was deposited within the alluvium that same year, then the alluvium is 11 300 years old. More likely is the situation in which the wood was intermittently carried to the site and deposited there later. Thus, the ^{14}C date provides a *maximum-limiting date* for the alluvium and the geomorphic surface that overlies it; neither the alluvium nor the geomorphic surface above can be older than 11 300 years. Likewise, a geomorphic surface or sediment that lies stratigraphically below this alluvium can be no younger than 11 300 years. This wood therefore provides a *minimum-limiting date* for that surface or sediment. Another example of a minimum-limiting date would be an alluvial terrace that contains fragments of an *in situ* tree stump that dates at 1560 BP. In this case, the terrace surface can be *no younger than* 1560 BP. That is, the surface had to be at least that old because a tree was growing on it at that time.

[1] In this book we will typically omit the error terms on radiocarbon dates, for the sake of brevity.

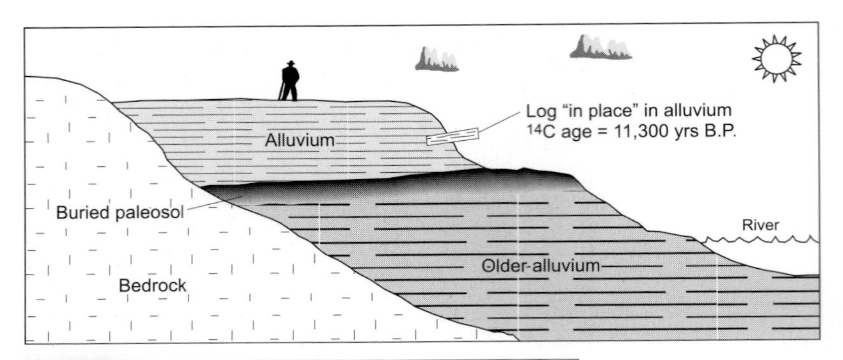

Fig. 14.55 Hypothetical valley cross-section with two fluvial terraces. The alluvium in the upper terrace contains a log that has been dated at 11 300 BP. See text for discussion.

DATING SOIL ORGANIC MATTER

Despite being somewhat problematic, dating the organic matter fraction of soils, both modern and buried, is frequently done. Interpreting the results must, however, be done with care (Perrin *et al.* 1964, Scharpenseel and Schiffmann 1977, Gilet-Blein *et al.* 1980). Nonetheless, if the assumptions and limitations are understood beforehand, dating soil organic matter can be applied with success.

Soils accumulate organic forms of carbon from plants and animals, as well as inorganic forms of carbon such as $CaCO_3$. We will first discuss soil organic matter. Soils are a storehouse of organic materials in various stages of decomposition and of innumerable ages. Early studies of the age of soil carbon focussed on dating extracted humic acids from soils (Felgenhauser *et al.* 1959). In 1964, Paul *et al.* reported on a new way to study the age of soil carbon – dating the entire suite of soil humic compounds and arriving at a mean age. This *mean residence time* (MRT) method drew attention to the dynamics of soil organic matter, and is still in use.

Organic matter begins to accumulate as humus and as macroscopic forms almost from $time_{zero}$ and this accumulation continues to the present, or until the soil gets deeply buried (Schaetzl and Sorenson 1987). Largely due to pedoturbation, the radiocarbon age of the entire mass of soil organic matter is therefore a *weighted mean* of the ages of all the various organic components. Geyh *et al.* (1983: 409) stated the inherent problems this can cause:

> "Soil dating" is a questionable attempt to date only a small part of the total humic matter of a soil horizon and to interpret the result as representative of the whole sample. The discrepancy between ^{14}C soil dates and true ages results from the complexity of soil genesis, which is a continuous process of accumulation and decomposition of organic substances. Penetration of rootlets, bioturbation, and percolation of soluble humic substances . . . cause rejuvenation, and the admixture of allochthonous plant residues may cause apparent aging. As a result, the organic matter in a soil is a mixture of an unknown number of compounds of unknown chemical composition, concentration, and age.

MRT dates use the entire soil organic matter fraction, usually obtained from alkaline extracts (Campbell *et al.* 1967). Also called *apparent mean residence time* (AMRT) dates, they reflect the *weighted mean age* of the many organic components within the soil. The MRT concept takes into account that younger carbon is continually added to the soil while older carbon is continually lost through decay and mineralization (Anderson *et al.* 2002). The degree to which each component is represented is a function of its *residence time* within the soil and the amount and type of carbon cycling within the soil (Geyh *et al.* 1971, Stout and Goh 1980) (Fig. 14.53, 14.56). Organic materials within the soil that are slow to decompose, i.e, have long residence times, will be overrepresented. Thus, the residence time of the organics governs what proportion of the sample they comprise. MRT dates *should never be calibrated* to calendar years, since the date is an average of many organic fractions, each of which would have its own dendrochronological correction (Geyh *et al.* 1971).

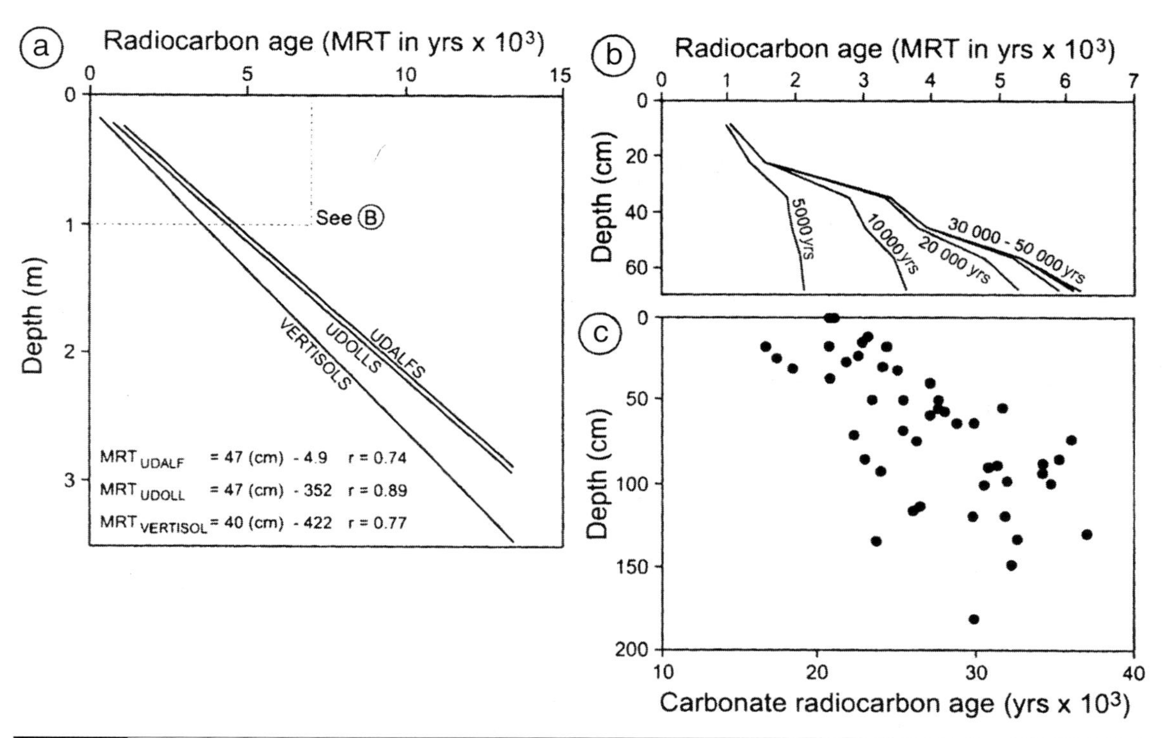

Fig. 14.56 Ages obtained by ^{14}C for soil organic matter and carbonate, as a function of depth. (a) MRT of soil organic matter vs. depth for soils of three different orders. After Scharpenseel (1972a). (b) Theoretical evolution of MRT ages of soil organic matter, with depth, for a forest soil at different stages in its development. Numbers along the lines indicate the true age of the surface soil. After Wang *et al.* (1996). (c) Published ages of soil carbonate with depth for a number of soils in Texas. The younger ages nearer the surface could reflect the true age of the carbonate or may be due to the drier Holocene climate, which caused carbonates to be deposited at shallower depths than they were during the Pleistocene. After Rightmire (1967) and Amundson *et al.* (1994).

MRT dates provide a *minimum age* for surface soils, i.e., the soil can be no younger than the MRT age (Polach and Costin 1971, Gilet-Blein *et al.* 1980, Cherkinsky and Brovkin 1991, Wang *et al.* 1996a). Visualize this by realizing that the oldest carbon in the soil can be no older than $time_{zero}$, unless it had been brought in from outside the soil system. If we were to date the one micro-fragment of humus that dates to $time_{zero}$, we could obtain the exact age of the soil, i.e., the date of $time_{zero}$. But since we, in fact, date a *mix* of carbon compounds, even if they did contain that one old fragment, the mean age is still statistically weighted toward a younger date. The mean date obtained on the whole organic matter fraction of the soil, therefore, must be younger than the age of the soil (Fig. 14.57a). Although there are innumerable age fractions in the soil, all of which are a part of the MRT determination, only four are shown in Fig. 14.57. The soil in Fig. 14.57a must be at least 5000 years old, because it contains humus of that age (we are assuming none was inherited in the parent material). If we assume the soil is 5000 years old, then the MRT age of 825 years is only 16.5% of the actual age. In the upper solum, much of the organic matter is composed of recently added organic materials (Geyh *et al.* 1971). Note that, in Figure 14.57b, the proportions of older carbon increase, producing an older MRT age. In this case, the MRT age is closer to the actual age (52.5%), but it is still a *minimum* age estimate. Figure 14.57b illustrates the point that when sampling surface soils for MRT purposes, the deepest carbon-rich horizon should be sampled to obtain the oldest possible date and hence, the one closest to the $time_{zero}$.

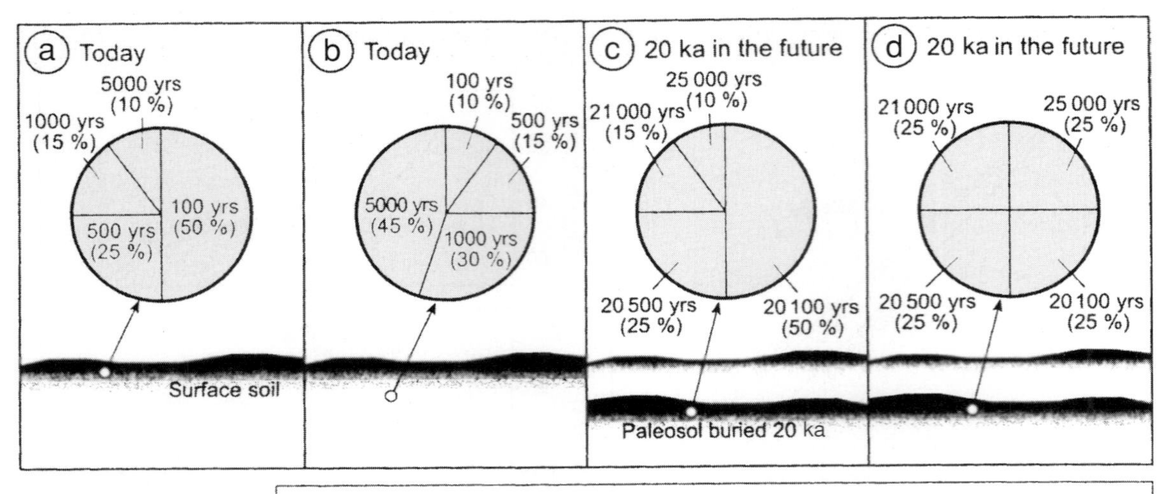

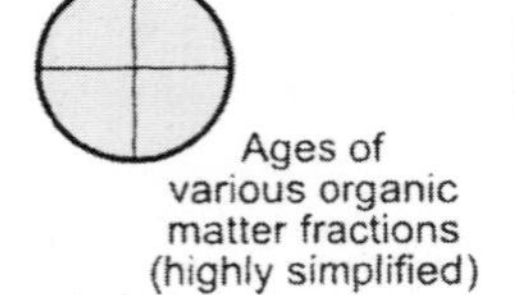

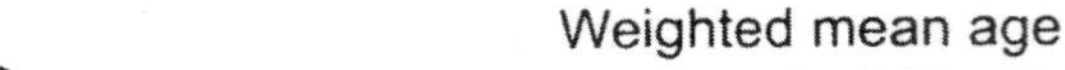

(a) MRT = (0.5 x 100) + (0.25 x 500) + (0.15 x 1000) + (0.1 x 5000) = 825 yrs
(b) MRT = (0.1 x 100) + (0.15 x 500) + (0.3 x 1000) + (0.45 x 5000) = 2635 yrs
(c) MRT = (0.5 x 20100) + (0.25 x 20500) + (0.15 x 21000) + (0.1 x 25000) = 20825 yrs
(d) MRT = (0.25 x 20100) + (0.25 x 20500) + (0.25 x 21000) + (0.25 x 25000) = 21650 yrs

Fig. 14.57 Simplified diagram showing how MRT dates on soil organic matter from a surface soil reflect, roughly, the ages of the various organic fractions in the soil. Although there are innumerable age fractions in the soil, all of which are a part of the MRT determination, only four are shown here for simplicity.

MRT ages in soils increase with depth and time (Catt 1990, Wang *et al.* 1996a) (Fig. 14.56). Dates on soil organic matter are usually youngest in A horizons (Herrera and Tamers 1970, Ganzhara 1974, Stout and Goh 1980) (Fig. 14.58). (Ballagh and Runge (1970) reported on an exception.) Although this depth–age relationship holds for soils in which organic matter is not generally mobile within the profile, e.g., Mollisols and Aridisols (Table 14.7), in soils where soluble organic matter is readily translocated as part of their genesis (Spodosols and some Andisols), MRT ages in the B horizon may be similar to those of the A (Tamm and Holmen 1967). Likewise, soils that are strongly pedoturbated may show irregular depth–age trends.

The effect of contamination (both young and old) on MRT dates is different than it is on whole fragments of carbonaceous material that are assumed to have one age, because MRT dates are a statistical mean (Fig. 14.56b). Additions of "dead" carbon, therefore, have a more dramatic effect on MRT dates.

When dating the organic fraction of surface soils, one must always consider the impact of "bomb carbon" from the open-air testing of nuclear devices (Broecker and Olson 1960, Wang *et al.* 1996a). Since about 1955, thermonuclear tests have added considerably to the amount of ^{14}C in the atmosphere. The effect of this testing has been to almost double the amount of ^{14}C activity in terrestrial carbon-bearing materials, making MRT dates on modern surface soils much younger than they normally would have been (Taylor 1987). Work by Campbell *et al.* (1967), however, indicated that bomb carbon may not be a major influence on MRT dates because of the rapid turnover of fresh carbon residues in soils.

The most common application of MRT dates occurs in paleopedology, where the goal could be either (1) determining $time_{zero}$ of soil formation for the buried soil or (2) ascertaining the time since burial (Kusumgar *et al.* 1980, Matthews and Dresser 1983, Nesje *et al.* 1989) (see Chapter 15). The MRT age on a buried soil reflects the MRT age plus the length of burial. In Fig. 14.57c, we hypothetically bury the same soil for 20 ka

Table 14.7 Age of humus fractions in Chernozems (Ustolls) on the Russian Plain

Soil	Depth (cm)	Organic matter content (%)	^{14}C age of humic acids bound to Ca (years BP)	^{14}C age of humin (years BP)
Kursk	10–20	7.7	1680 ± 80	1110 ± 70
Chernozem	50–60	4.7	2970 ± 110	1230 ± 180
	70–80	2.3	4020 ± 90	2970 ± 90
	120–130	0.7	6100 ± 200	—
Voronezh	20–30	8.1	660 ± 80	1590 ± 100
Chernozem	40–50	5.6	2200 ± 100	—
Tambov	60–70	3.1	—	3000 ± 160
Chernozem	0–22	7.8	1330 ± 80	2120 ± 130
	30–40	7.5	2800 ± 120	2380 ± 150
	70–80	5.4	2460 ± 120	2970 ± 130

Source: After Ganzhara (1974).

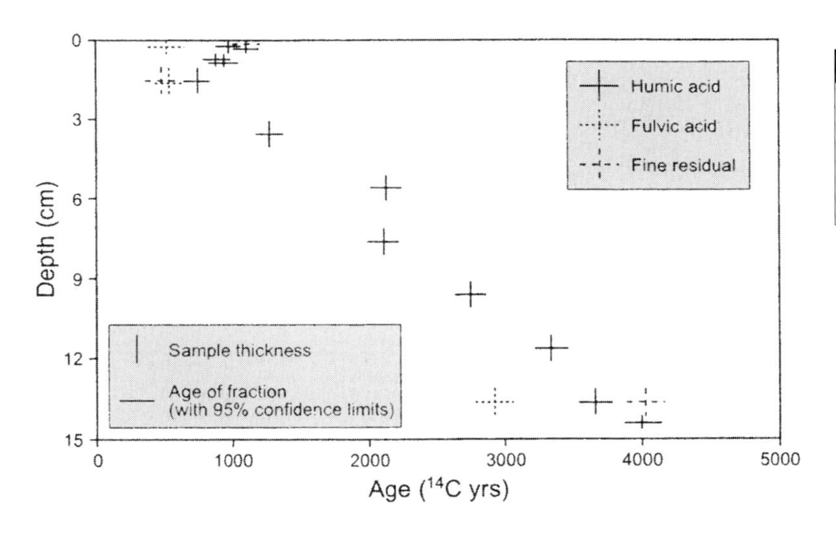

Fig. 14.58 Depth plot of radiocarbon ages on various soil organic matter fractions, for a soil buried beneath a Neoglacial moraine in Norway. After Matthews and Dresser (1983).

and re-evaluate its MRT age. Now, because all the organic fractions are 20 ka older, the analysis yields an MRT age of 20 825 years BP. This MRT age is a *maximum estimate* of the time since burial, i.e, the soil could not have been buried longer than the ^{14}C date obtained (Haas *et al.* 1986, Nesje *et al.* 1989). Recall that the soil in Fig. 14.57c is actually 25 000 years old (it formed for 5 ka before being buried for 20 ka). The MRT age on the buried soil is (20 825/25 000) 83.3% of the actual age, which is a better estimate of its time$_{zero}$ than when it was a surface soil. Thus, MRT dates on buried soils have advantages over those on surface soils. Add to this the fact that, after burial, the younger carbon fractions will decay more rapidly than will the older fractions, yielding a more even distribution as shown in Fig. 14.57d. Thus, Fig. 14.57d may be a closer reflection of reality, and yields an MRT age of 21 650 years BP, or 86.6% of actual age. Note, however, that this latter age is farther from the actual length of time that the soil has been buried. Thus, to obtain the length of time that the soil has been buried, sample the part of the soil with the youngest carbon, i.e., the uppermost part of the A horizon (Haas *et al.* 1986, Nesje *et al.* 1989). Matthews and Dresser (1983) used an application of this theory. By regressing ^{14}C ages with depth in a buried soil they obtained the expected age of the organic matter at the very surface of the buried soil, and assumed that this age was the best maximum estimate of the time of burial (Fig. 14.58). To obtain the overall age of the soil from its $time_{zero}$, or essentially the minimum age for the deposit, sample the lowermost part of the A horizon. One problem with sampling deeper horizons is that they may not contain adequate carbon for a date.

Several other factors must be considered when dating the carbon in buried soils. All buried soils are modified by the burial process and after burial (see Chapter 15). Thus, one potential problem centers on erosion prior to burial. In such circumstances, MRT dates will be slightly older, possibly resulting in an overestimate of the time since burial (Matthews 1980). Similar problems can result if older carbon is mixed into the soil prior to burial (Geyh *et al.* 1983). From a dating perspective, post-burial alteration of the organic carbon storehouse in the buried soil is so common as to be expected in all but the very driest environments. Translocation of dissolved organic substances from overlying sediments can seriously impact the MRT age of the buried soil; this is so problematic in cool, humid environments as to make dating of all but the most deeply buried soils impractical (Matthews 1980, Geyh *et al.* 1983). Oxidation and mineralization processes, after burial, reduce the amount of carbon in the buried soil and increase the error in the estimate of time since burial because the younger organic fractions are disproportionately lost. Haas *et al.* (1986) also cautioned that MRT dates on buried soils that have been exposed for several years, e.g., in a trench wall, are less reliable than those taken from freshly exposed faces.

Because of the inherent problems in using MRT dates, efforts have continually been made to isolate and date a *fraction* of the soil organic matter that is (ideally) the oldest and most biologically inert, so as to better estimate $time_{zero}$ of soil formation (Scharpenseel 1972a, b, Martin and Johnson 1995). The soil organic matter suite is fractionated through chemical or physical means, on the assumption that one of those fractions has a greater age than the others and hence (in surface soils) can provide a better approximation of the soil's age (Table 14.7). Most laboratories break the organic matter into the NaOH-soluble fraction (humic acid, aka humate) and the NaOH-insoluble fraction (residue, or humin). Others have extracted and dated the lipid and fulvic acid fractions (Gilet-Blein *et al.* 1980). The humic acid fraction is probably more mobile and usually yields older ages, possibly because humin is an earlier intermediary than is humic acid in the humification process (Polach and Costin 1971, Dalsgaard and Odgaard 2001). Whichever fraction yields the oldest ages (preferred if dating surface soils) remains, nonetheless, debatable (Gilet-Blein *et al.* 1980, Haas *et al.* 1986, Martin and Johnson 1995).

DATING SOIL CARBONATE

In theory, it is possible to date pedogenic $CaCO_3$. The results of ^{14}C dating of soil carbonates have generally been consistent with other dating methods (Fig. 14.59). Like MRT dates on soil organic matter, ^{14}C dates on carbonate are not highly precise but often can provide stratigraphic age restrictions, or bracketing, that may be helpful in certain contexts (Kusumgar *et al.* 1980).

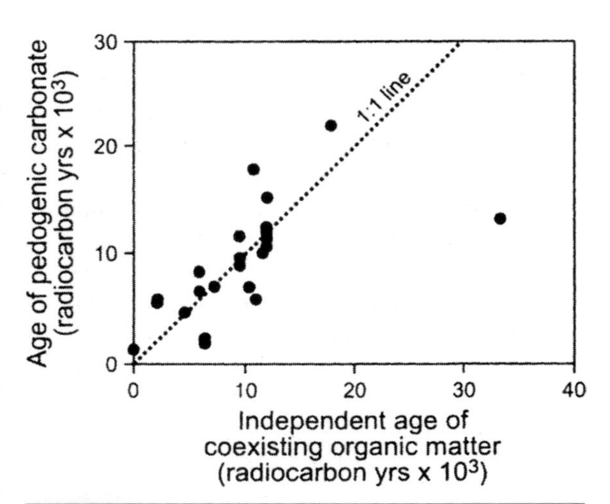

Fig. 14.59 Radiocarbon ages of pedogenic carbonate as compared to other, established ages for the same sediment or surface. See the original source (Amundson *et al.* 1994) for the references to the ^{14}C dates.

Pedogenic carbonate is datable because $CaCO_3$ incorporates ^{14}C into its structure, as $Ca^{14}CO_3$ (Leamy 1974). As pedogenic carbonate forms, it acquires ^{14}C from the soil air and from dissolved gases in soil water. Therefore, pedogenic carbonate forms, more or less, in isotopic equilibrium with soil CO_2 (Cerling 1984, Quade *et al.* 1989), which is derived mostly from biological processes: roots, respiring fauna and microbes that are consuming organic matter (Dörr and Münnich 1986). The first two of these are generally in isotopic equilibrium with the atmosphere, but depending upon the age of the organic matter being decomposed, soil micobes could be releasing older ^{14}C to the soil air. Therefore, the radiocarbon content of soil CO_2 is assumed to be only *generally* in equilibrium with the atmosphere. A ^{14}C date on soil carbonate should, therefore, provide a minimum age for the soil. The potential problem lies in the effects of contamination from pre-existing, older caliche or even limestone that gets dissolved and reprecipitates in the "new" caliche (Williams and Polach 1969, Amundson *et al.* 1989b). This is especially problematic in highly polygenetic soils that have undergone humid and dry climatic intervals. During humid periods soil carbonate is dissolved and translocated to greater depths, where it can reprecipitate. During dry periods, carbonate-rich soils are eroded and the dust derived from them, rich with "old" carbonate, impacts broad areas (Chen and Polach 1986). Still another source of contamination is "dead" carbon from groundwater.

An important source of carbon in soil CO_2 derives from the respiration/mineralization of soil organic matter. In young soils with low organic matter contents, little of the soil CO_2 content is derived from this source. Knowing the degree of contamination from this source is very difficult without the use of stable isotopes, like $\delta^{13}C$ (see Chapter 15). The mass of soil organic matter always has a ^{14}C activity that is less than that of the atmosphere because much of it is in various stages of decay from plants that died long ago. However, this carbon source is always younger than the surface age, since essentially none of it formed prior to time$_{zero}$. Since some soil CO_2 has a modern age while other CO_2, derived from the decay of old organic matter, can be thousands of years old, pedogenic carbonate *in very young soils* may overestimate the soil age, i.e., they are maximum ages. Soils with high respiration rates due to high organic matter contents also may have low $^{14}CO_2$ contents and thereby yield disproportionately *old* ^{14}C carbonate ages (Amundson *et al.* 1994).

Despite these complications, ^{14}C ages on soil carbonate from all but the youngest soils are usually *minimum ages* because (1) most of the carbonate comes from soil air that is in equilibrium with the atmosphere and (2) the date on the carbonate is essentially an MRT date (Callen *et al.* 1983).

Dating of soil carbonate has the advantage that one is able to utilize carbonate nodules. Nodules of carbonate grow in roughly concentric rings; inner rings are older than outer ones. Dates obtained from ^{14}C on carbonate nodules confirm this and thus provide a means of establishing the length of time necessary for these types of pedogenic features to form (Chen and Polach 1986). Related to this, pedogenic carbonates have a built-in advantage over MRT dates: they often occur in laminated sequences (Bkm horizons) and mixing of old and young carbonate is usually minimal. Inclusion of detrital carbonates, such as limestone gravel or pebbles, within a carbonate mass would obviously cause errors in the age (Callen *et al.* 1983); the likelihood of this is minimal in laminar Bkm horizons but can be quite high in some dryland soils.

Like MRT dates on organic matter, ^{14}C dates on soil carbonate increase with depth and time (Gile and Grossman 1968). This occurs either because (1) carbonates continue to build upward in the profile (younger carbonates form nearer the surface), (2) of contamination by near-surface carbonates by younger C and/or (3) drier conditions in the Holocene have led to carbonate deposition at shallower depths while Pleistocene carbonate occurs deeper. The last line of reasoning should remind us that desert soils are generally polygenetic; most date back to the wetter, cooler Pleistocene.

Dating pedogenic carbonate and silica is also possible using the uranium-series method (Ku *et al.* 1979, Schwarcz and Gascoyne 1984, Ludwig and Paces 2002). Care must be taken to date only

Table 14.8 Major cosmogenic isotopes used in surface exposure dating, and their half-lives

Isotope[a]	Useful dating range (years × 10^3)	Half-life (years)
^{41}Ca	30–50	10 000
^{36}Cl	1000	300 000
^{26}Al	2500	710 000
^{10}Be	5000–6000	1 500 000
^{21}Ne	Unlimited	Stable
^{3}He	Unlimited	Stable

[a]Only those isotopes that are most applicable to soil geomorphology and SED. For a more complete list, see Lal (1988).

pedogenic carbonates that are free of old, detrital limestone fragments (Bischoff *et al.* 1981, Rabenhorst *et al.* 1984, Radtke *et al.* 1988). U-series dates of soil carbonate are minimum ages (Slate *et al.* 1991).

Cosmogenic isotopes

Cosmogenic isotopes are produced as solar radiation interacts with atoms (Lal 1988, Cerling and Craig 1994). This process occurs in situ, e.g., within rocks or soils, and in the atmosphere. Within the family of cosmogenic isotope dating methods, one group uses the decay of meteoric, cosmogenic isotopes, such as ^{14}C and ^{10}Be, that are *produced in the atmosphere* and then incorporated into terrestrial reservoirs. These nuclides impact the surface as "fallout," after which they may or may not be leached, taken up by plants, eroded or remain in place. The other group of cosmogenic isotope dating methods utilizes isotopes that form directly in rocks due to exposure to cosmic rays (Table 14.8). Some of these isotopes are stable such as ^{21}Ne and ^{3}He, while some decay over time, e.g., ^{36}Cl. To determine the exposure time of rocks on the surface one must know or estimate (1) the rate of nuclide production, (2) its half-life, (3) the burial/erosion history of the surface and (4) the amount of inherited nuclides in the rock or reservoir. Burial/erosion history and the amount of nuclide inheritance are particularly important and are often difficult to determine (Anderson *et al.* 1996, Hancock *et al.* 1999). Many parent materials, e.g., alluvium, loess, colluvium and residuum, have a history of nuclide inheritance from previous sites which must be taken into consideration before an accurate age can be determined (Phillips *et al.* 1998).

The applications of cosmogenic dating are many-faceted (Cerling and Craig 1994). They have been used to date moraines, fluvial terraces and glacial depositional landforms (Brown *et al.* 1991, Gosse *et al.* 1995, Zreda and Phillips 1995) (Fig. 14.60). Isotopic methods have the advantage of not only providing information about surface exposure, but also about erosional or burial histories of surfaces (Lal 1988, 1991).

FALLOUT FROM THE ATMOSPHERE

The primary fallout nuclides are ^{14}C and ^{10}Be. In the case of ^{14}C, the isotope does not so much fall out as it is mixed within the atmosphere and taken up by plants. Beryllium-10 is a cosmogenic nuclide that actually does fall onto surfaces, where it accumulates in soils; it has been widely used as an SED tool (McHargue and Damon 1991, Morris 1991). Many researchers refer to this form of ^{10}Be as the "garden variety" beryllium, as opposed to that produced *in situ* in rocks (Nishiizumi *et al.* 1986, Monaghan and Elmore 1994). It is produced in the upper atmosphere as cosmic radiation interacts with O and N atoms. Some also is recycled back to the atmosphere, from the oceans, as hygroscopic nuclei. Beryllium-10 is delivered to land surfaces primarily in precipitation, where it is either fixed in soils or carried away in overland flow; some is locked in ice sheets. Deposition of ^{10}Be onto the surface is primarily dependent upon precipitation (since most of it is brought down with rain and snow) and latitude. The theoretical global average rate of ^{10}Be fallout is 2.15 to 4.0 × 10^{-2} atoms cm^{-2} s^{-1} (O'Brien 1979, Pavich *et al.* 1984), or about 1.7 to 1.8 × 10^{24} atoms yr^{-1} (McHargue and Damon 1991).

Of the ^{10}Be that impacts the surface, that which is not removed by runoff is free to infiltrate. Many researchers have assumed that ^{10}Be is quickly adsorbed onto clay minerals because it is found in only very low concentrations in surface waters. In soils, ^{10}Be behaves like other cations

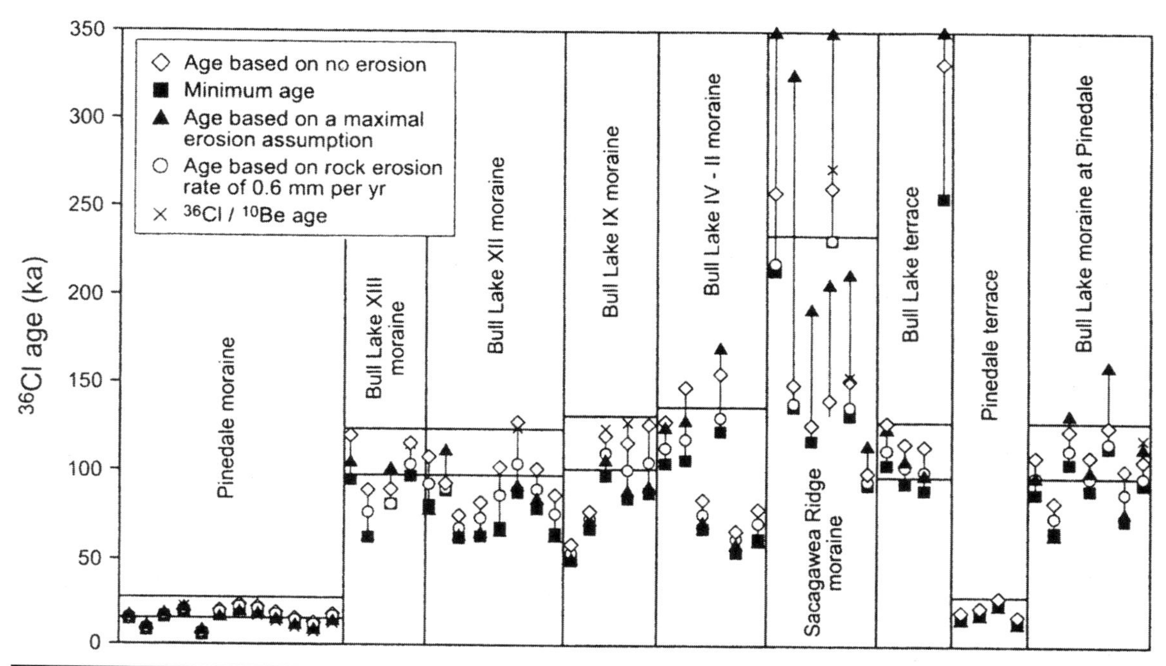

Fig. 14.60 Ages of moraines and glacially derived terraces in the Wind River Range, Wyoming, based on ^{36}Cl accumulation in surface boulders, and a few assumptions about erosion rates. After Phillips *et al.* (1997).

and is subject to many of the same translocation and mixing processes. Strong correlations between clay and ^{10}Be contents (Pavich *et al.* 1986) support the assumption that it adsorbs onto, and moves with, clay and humus. Monagahan *et al.* (1983) suggested that it moves with organic ligands (see Chapter 12). The assumption of ^{10}Be retentivity in soils is critical to its application in SED studies (You *et al.* 1988, Morris 1991). Therefore, many have concluded that soils are potentially long-term dosimeters for beryllium (Fig. 14.61).

Assuming some degree of surface stability, one need only know the rate of ^{10}Be influx and the decay constant to estimate surface exposure. In soils where cations readily leave the profile, accumulation of ^{10}Be may not be as effective an SED method as in more clay-rich, slowly permeable soils (Pavich *et al.* 1984). In all soils, however, surficial erosion and deep leaching removes a substantial amount of the adsorbed ^{10}Be, making these minimum dates (Pavich *et al.* 1985). Ages of marine terraces determined with "garden-variety" ^{10}Be, by Monaghan *et al.* (1983), were 16–380 times too young. Even in clayey soils with low organic matter contents, where ^{10}Be mobility should, theoretically, be low, the mean residence time of the isotope, probably on the order of 10^4 to 10^5 years, puts an upper limit on the method (Pavich *et al.* 1985, 1986). Also bear in mind that ^{10}Be is also being produced within the soil as cosmic rays impact rocks within (see below), again making any

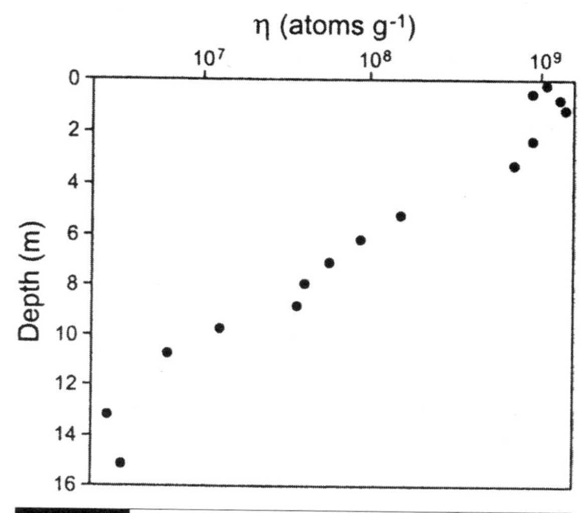

Fig. 14.61 Concentrations of ^{10}Be of with depth, in a soil–saprolite sequence in Virginia. After Pavich *et al.* (1985).

surface exposure estimate based on accumulated ^{10}Be a minimum one. However, *in situ* production rates of ^{10}Be are two to three orders of magnitude lower than are fallout rates (McHargue and Damon 1991, Monaghan *et al.* 1992).

"Garden-variety" ^{10}Be has its most common application in the determination of local- to regional-scale erosion rates (Valette-Silver *et al.* 1986, Brown 1987, Brown *et al.* 1988). It can also be used to determine rock-to-soil conversion rates (Pavich 1989, Monaghan *et al.* 1992, Monaghan and Elmore 1994).

FORMATION IN ROCKS AT THE SURFACE

Nuclides accumulate *in situ* within rocks at the surface, as cosmic rays interact with atoms in minerals by high-energy spallation, neutron-capture reactions and muon-induced nuclear disintegrations (Lal 1988, 1991). Each type of nuclide preferentially accumulates in certain "target" minerals; for many, the favored target is quartz (Nishiizumi *et al.* 1986). The rate of accumulation is dependent upon a number of site factors, such as altitude, latitude, rock chemistry and density, geometry of the exposed rock, depth of burial or shielding (if any), as well as the cosmic ray flux (Dorn and Phillips 1991). Because the cosmic ray flux has varied with time, data on the accumulation of nuclides primarily reflects long-term mean rates. Most cosmic rays can penetrate to about 50–60 cm in rock, meaning that below that depth the rock will accumulate almost no nuclides (Nishiizumi *et al.* 1993). In soils, because of porosity, a depth of 3 m is equivalent. Rocks buried more shallowly in regolith, therefore, receive only partial doses of radiation.

For stable isotopes, like ^{3}He and ^{21}Ne, accumulation rates are linear, making them useful dating tools on stable surfaces with no prior exposure (Cerling 1990) (Fig. 14.62). The stable isotopes are also particularly useful for studies of surfaces that are eroding (Lal 1988). Use of ^{3}He is advantageous because it is stable and has the highest production rate of any cosmogenic nuclide (Kurz *et al.* 1990).

For other nuclides, net accumulation is more complicated, reflecting gains vs. loss by radioactive decay. Half-lives of the radionuclides, and hence effective time-spans of the method, vary

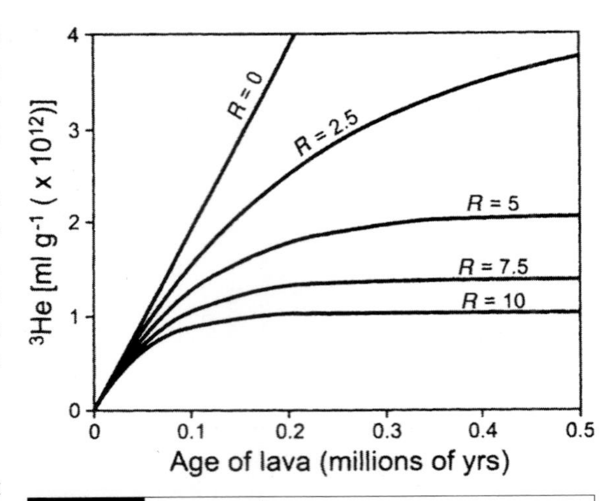

Fig. 14.62 Build-up of ^{3}He in surface-exposed rock, as a function of time and erosion rate. *R* represents the erosion rate in meters per million years. After Craig and Poreda (1986).

(Table 14.8). Saturation of nuclides is eventually reached after about four times the half-life, effectively determining the timespan of the method (Fig. 14.63). Thus, by using a number of different isotopes with widely varying half-lives more information on surface exposure and erosion history can be obtained (Nishiizumi *et al.* 1986).

EXPOSURE HISTORIES

In rocks on surfaces that have been stable for a long period of time, determining the surface exposure time is relatively straightforward (Fig. 14.64). However, few surfaces have such simple histories (Liu *et al.* 1996, Phillips *et al.* 1998), and therein lies the utility of cosmogenic isotopes. They can be used to unravel the exposure history of a geomorphic surface, even if it has had periods of stability, erosion and burial (Nishiizumi *et al.* 1993). The simplest scenario involves rocks at the surface today which have had no prior exposure history, usually because they have always been deeply buried and then immediately exposed by some geomorphic process, e.g., an impact structure, or because they solidified from lava (Phillips *et al.* 1986). The exposure of the rock to cosmogenic radiation therefore coincides with the formation of a subaerial geomorphic surface. This situation is typical of most moraines and volcanic materials. A similar situation (Fig. 14.64b)

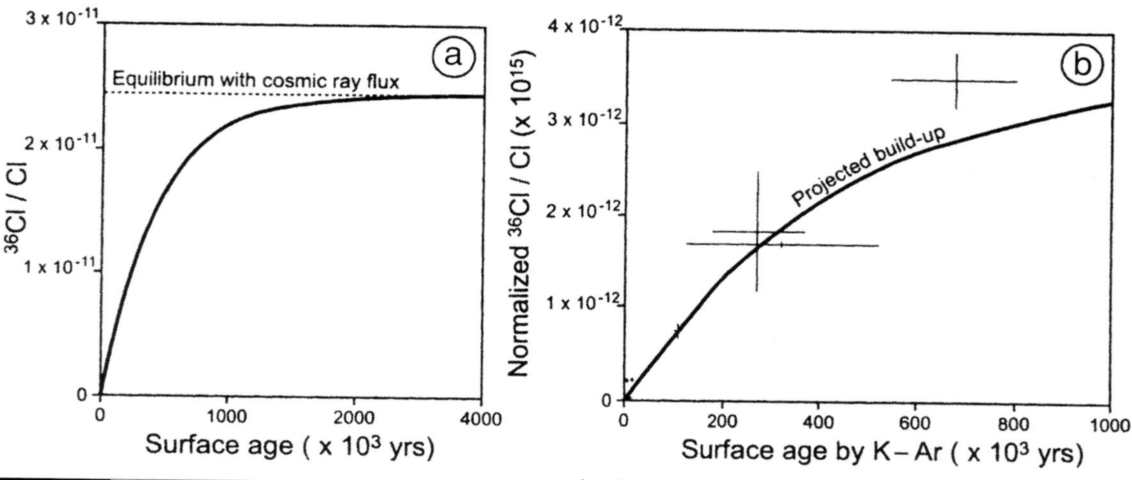

Fig. 14.63 Accumulations of ^{36}Cl with time. (a) Calculated ^{36}Cl accumulation in a lava exposed at 3500 m elevation on Mauna Kea, Hawaii. After Dorn and Phillips (1991). (b) $^{36}Cl/Cl$ ratios for rocks from lava flows in the western United States. After Phillips *et al.* (1986).

involves rapid erosion of overburden to create a new surface that lacks an exposure history. In both cases, isotopes begin to accumulate in the rock as soon as it is exposed. The signal given by radionuclides within the rock then will reflect the age of the surface. Interpreting the age of the surface is easiest when using stable isotopes, for the accumulation rate is linear.

Complicating factors that can affect rocks dated by cosmogenic isotopes include burial or erosion of the surface, as well as pre-exposure of the rocks from another time or place (Phillips *et al.* 1998) (Fig. 14.62, 14.64). Also of concern, especially for ^{10}Be applications, is the additive effect of fallout from the atmosphere, which Lal (1988) called a "hindrance" to the study of *in situ* isotopes. In some cases, the "garden-variety" isotopes can overwhelm those produced *in situ*.

For surfaces of known age, one way to determine the effect and magnitude of surface erosion or burial, or even if either of these two processes has been operative, is to compare the amounts of nuclides in the rock to the amounts expected. An example of this approach, by Craig and Poreda (1986), used cosmogenic nuclides to calculate erosion rates of lavas. The age of the lavas was known by other means, and because the content of ^{3}He in the lavas was below what would have been expected for lavas of this age, an erosion rate could be generated (Fig. 14.62).

Most researchers, however, do not know the surface age; in this case one has to compare the amounts of two or more nuclides in surficial rocks. An alternative way to determine exposure or burial time is to compare cosmogenic nuclide data from nearby, similar rocks that have never been buried to those that have been exposed and later buried. Wells *et al.* (1995) compared ^{3}He contents in desert pavement clasts to uneroded basalt nearby. Similar values for both areas indicated that the rocks on the pavement had been exposed at the surface continually and had never been buried.

Efforts have been made to "tease out" the effect of nuclide inheritance (initial nuclides brought to the site in the sediment) from that of accumulation after a geomorphic surface has been formed. Anderson *et al.* (1996) showed that inheritance can be corrected by examining depth profiles of stable nuclides. Phillips *et al.* (1998) extended their work by illustrating different depth profiles of stable isotopes with respect to conditions of burial and bioturbation. Similar depth plot inferences can be used for radioisotopes if they are adjusted for decay. The theory is as follows. Beneath (erosional or constructional) surfaces that form quickly, nuclide contents will initially be uniform with depth; eventually (assuming surface stability), nuclide contents will decrease exponentially with depth because they accumulate fastest in near-surface

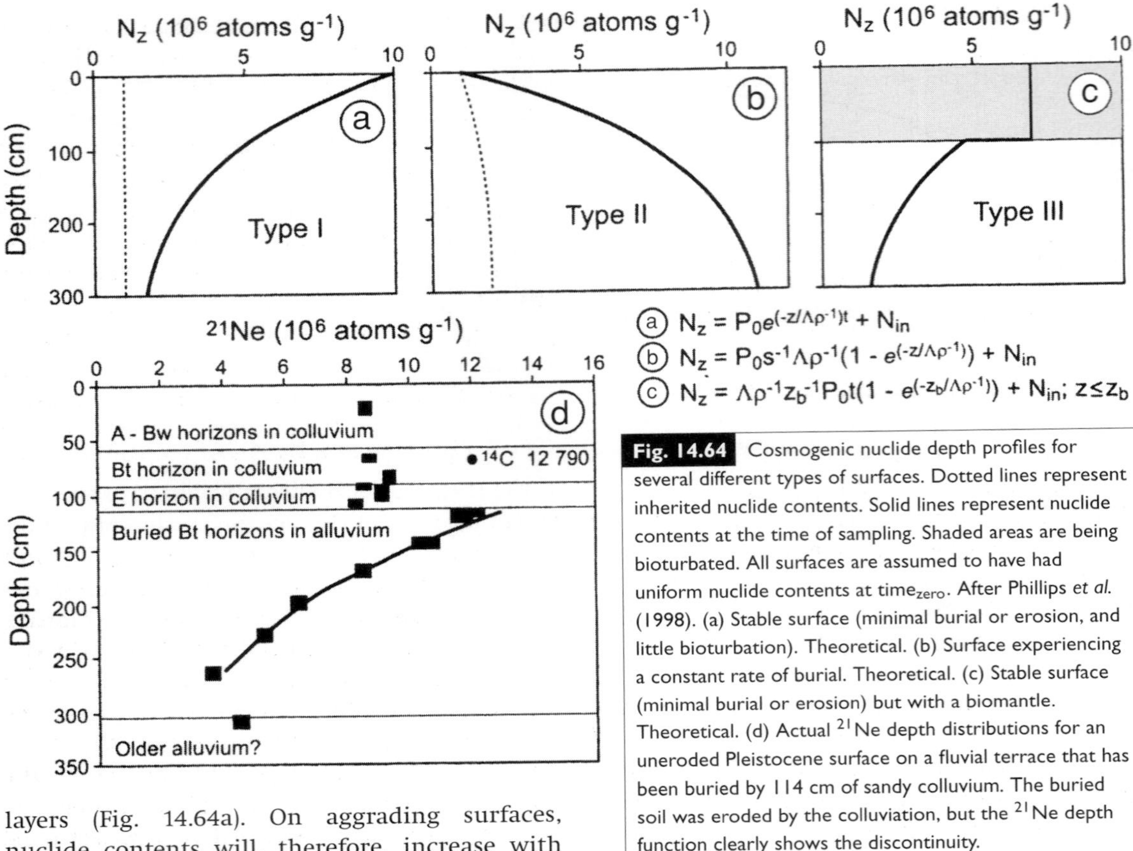

ⓐ $N_z = P_0 e^{(-z/\Lambda\rho^{-1})t} + N_{in}$

ⓑ $N_z = P_0 s^{-1} \Lambda\rho^{-1}(1 - e^{(-z/\Lambda\rho^{-1})}) + N_{in}$

ⓒ $N_z = \Lambda\rho^{-1} z_b^{-1} P_0 t(1 - e^{(-z_b/\Lambda\rho^{-1})}) + N_{in};\ z \leq z_b$

Fig. 14.64 Cosmogenic nuclide depth profiles for several different types of surfaces. Dotted lines represent inherited nuclide contents. Solid lines represent nuclide contents at the time of sampling. Shaded areas are being bioturbated. All surfaces are assumed to have had uniform nuclide contents at $time_{zero}$. After Phillips *et al.* (1998). (a) Stable surface (minimal burial or erosion, and little bioturbation). Theoretical. (b) Surface experiencing a constant rate of burial. Theoretical. (c) Stable surface (minimal burial or erosion) but with a biomantle. Theoretical. (d) Actual ^{21}Ne depth distributions for an uneroded Pleistocene surface on a fluvial terrace that has been buried by 114 cm of sandy colluvium. The buried soil was eroded by the colluviation, but the ^{21}Ne depth function clearly shows the discontinuity.

layers (Fig. 14.64a). On aggrading surfaces, nuclide contents will, therefore, increase with depth (Fig. 14.64b). Strongly bioturbated layers will show uniform nuclide concentrations (Fig. 14.64c).

Interpretations involving rocks with multiple exposures and burials is complicated, requiring the use of several isotopes with different half-lives (Lal 1991, Bierman and Turner 1995). In the case of a rock with multiple-exposure histories, it is not possible to determine the length of each exposure event; only cumulative exposure can be determined.

Luminescence dating

Luminescence dating is one of a suite of methods used to date *sediment*. Luminescence is the light emitted from mineral grains when heated or exposed to light. Thermoluminescence (TL) is the dating method which utilizes heat to produce luminescence while optically stimulated luminescence (OSL) utilizes light. These similarities between OSL and TL justify discussing them together. A third and less used method, IRSL, refers to luminescence that is stimulated when a mineral is exposed to infrared (IR) radiation. In essence, OSL and TL dates provide ages for the *burial* of sediment that *had been* exposed to sunlight, usually during a prior transportation event (Huntley *et al.* 1985, Aitken 1994). Both TL and OSL are routinely used for sediments like loess and eolian sand (Dreimanis *et al.* 1978, Pye and Johnson 1988, Wintle 1993).

Luminescence dating is possible because most common silicate minerals (mainly quartz and feldspars) contain crystal lattice defects/imperfections which become potential sites of electron storage. The source of the electrons, which get trapped in these crystal "gaps," is the decay of isotopes in the surrounding (within a few decimeters) sediment. The electrons given off by the decay of isotopes get caught in the crystal defects and stored there, making the minerals

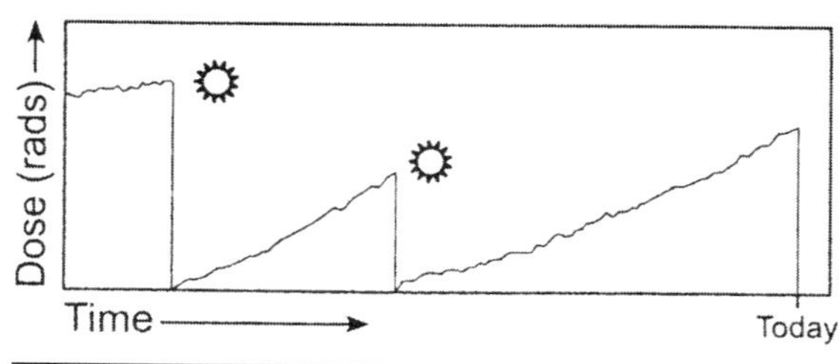

Fig. 14.65 Hypothetical optically stimulated luminescence (OSL) or thermoluminescence (TL) dose rate vs. time. Exposure to sunlight "zeroes" the clock and resets the dose.

essentially long-term dosimeters. The amount of stored electrons is called the *equivalent dose*. Exposure to light releases the stored electrons as luminescence, and "zeroes" the clock (Fig. 14.65). Thus, a key constraint to the method is that the sample to be dated must have no residual luminescence signal from the period before the event being dated, i.e., it must have been "zeroed." Only exposure to direct sunlight (or in some cases, heat) can do that. OSL has an advantage over TL in that less light, or a more brief exposure to light, is required to zero the sample.

The primary factor that impacts the number of trapped electrons is the number emitted from the surrounding sediment, or the *dose rate*. With this information, the length of time since the sample last was exposed to light can be calculated as:

$$\text{Age} = \frac{\text{equivalent dose (in rads)}}{\text{dose rate (in rads yr}^{-1})}.$$

A sample for TL or OSL dating is removed from sediment below the soil profile (pedogenesis can impact the signal) in the dark, lest the accumulated electrons be released from the sand and silt grains. Usually, a pit is dug and an opaque tarpaulin is laid on top. Within the dark pit, the face is further scraped back, a sample is removed and placed in a light-restrictive container that can be sealed, such as a metal can. In order to calculate the dose rate, a sample of the surrounding sediment must also be collected. This sample, which is used to determine the rate of electron emission from the surrounding sediment, need not be collected in the dark. In the laboratory, the main sample is exposed to light (in OSL) or heat (in TL) and the number of photons emitted from it determined. The more photons emitted, the more electrons had been trapped in the minerals and the longer the sample had presumably been buried.

Luminescence dating was slow to catch on with Quaternary scientists (Dreimanis *et al.* 1978). Many viewed it as a relative dating tool, able to rank order sediments by age but little more. As technological advances were made during the 1970s, 1980s and 1990s, the accuracy and precision of the methods improved markedly. Luminescence dating is now critical to studies of Quaternary sediments and soils because it provides an SED and sediment dating tool with acceptable accuracy that extends well beyond the range of radiocarbon (Dreimanis *et al.* 1978).

Applications for OSL and TL dating are myriad, and growing (G. W. Berger 1984, 1988, Rees-Jones and Tite 1997). Whereas eolian sediments (loess, dune sand) are particularly ideal for OSL and TL dating (Wintle and Catt 1985, Duller and Augustinus 1997, Arbogast 2000, Murray and Clemmensen 2001, Arbogast *et al.* 2002), other applicable sediment types include colluvium, glacial outwash and fluvial sediment, providing that the water below which they were deposited was clear enough to allow for solar resetting of the inherited luminescence signal (Forman 1989). Fluvial and beach sands are popular applications of OSL dating (Ollerhead *et al.* 1994, Bateman and Catt 1996, Forman 1999, Wallinga *et al.* 2001). Because most pedogenic carbonates are brought in as dust, they are also datable by OSL (Singhvi and Krbetschek 1996). Luminescence dating can provide excellent age constraints on buried soils, providing that the sediment above and below the soil can be dated (Berger and Mahaney 1990, Hutt *et al.* 1993). Because heat, in addition to light, can also "reset the clock," TL can be used determine the last time a sample was heated. Therefore, it is especially applicable to dating of pottery, hearths, baked flints (Zimmerman 1971, Wintle 1996) and lavas.

Like any dating method, TL and OSL have some problems that must be taken into consideration. Samples that have not been sufficiently zeroed before being buried will report ages that are too old, while in older samples saturation of the

Table 14.9 Characteristics of numerical dating methods discussed in the text

Method	Useful range (years)	Materials best applied to
^{14}C (radiocarbon)	50–60 ka	Charcoal, wood, shell carbonate (corals, speleothems), soil organic matter, groundwater, cloth, etc. (anything with organic carbon in it)
Optically stimulated luminescence (OSL)	0–200 ka on quartz 0–1 Ma on feldspar	Silicate minerals, usually of silt or sand size
Thermoluminescence (TL)	0–300 ka	Silicate minerals, usually of silt or sand size
Tephrochronology	0–several Ma	Volcanic ash and tephra
Paleomagnetism	Millions of years	Fine sediments, volcanic materials
U–Th (U-series)	10–350 ka	Carbonates, limestone, travertine, volcanics
Cosmogenic isotopes	Varies	Quartz or other minerals

electron traps can cause ages to be anomalously young. Anomalous fading is the loss of electrons from the lattice traps on a time scale that is short compared to the lifetime predicted on the basis of trap size (Wintle 1973, Huntley and Lamothe 2001, Strickertsson *et al.* 2001). Anomalous fading is primarily a problem with feldspar, where it will cause an age estimate that is too young. One should also be aware that sediments that have been affected by pedogenesis are unreliable for luminescence dating. Samples should not be taken from the soil profile per se because dose rates are affected by soil materials. Illuvial materials like carbonate, Fe or silica typically have higher contents of radioisotopes. In such a case, for example, the laboratory would determine an anomalously high (current) *dose rate*, which would be quite different from the *long-term* dose rate. Thus, samples from the profile or even the upper C horizon may yield anomalously young *ages*. Finally, variations in dose rate can be caused by soil or ground water, which greatly attenuates radiation and possibly leads to anomalously old ages.

The age limitations of TL and OSL dating are a function of the dose rate and the capacity of the sample to store electrons. Generally, K-feldspars have more capacity than quartz, rendering their effective age range nearly three times greater (Table 14.9). Most analysts assume that TL and OSL ages less than 200 ka are reliable, and ages up to 800 ka have been reported. The precision on OSL and TL dates is generally thought to be ±10% of the actual age.

Chapter 15

Soils, paleosols and paleoenvironmental reconstruction

A palimpsest (Latin *palimpsestus*, scraped again) is a parchment that has been used one or more times after earlier writings have been erased. Recently erased passages are not that difficult to make out, as erasures are rarely complete. Previous, older passages on the palimpsest are much more difficult, but not impossible, to read and interpret. It simply takes skill and patience. In this chapter we provide information on the interpretation of pedo-palimpsests, i.e., soils. Each and every soil (except for the very youngest ones) is a palimpsest that has information written on it; interpreting it is an exciting challenge and an important application of soil geomorphology (Catt 1990). The information on pedo-palimpsests is usually indicative of past landscape change, whether that change refers to *climate*, *vegetation* or *geomorphology*.

Soil morphology and chemistry are all influenced by the various soil-forming factors. In this chapter we discuss how a careful "reading" of soils as palimpsests, coupled with a knowledge of how soil development is related to *contemporary* soil-forming processes, can often (for older soils) provide a wealth of information about conditions associated with periods in the *geologic past*. This chapter must follow chapters on soil genesis, weathering, pedoturbation, parent materials and geomorphology, for the type of paleopedologic, paleoclimatic and paleogeomorphic interpretations we discuss require a thorough knowledge of all these facets of soil geomorphology. Our approach involves, at a minimum, these two questions.

- *What soil properties should we examine to better understand the evolution of a soil?*
- *What can and do they tell us about past evolutionary changes?*

In other words, if we were dropped out of the sky into a soil pit, with the intent of determining all that we could about the evolution of the landscape and its soils, what should we do and what should we be cautious/mindful of? Included in the plethora of possible answers to this question are data related to soil mineralogy, horizonation, sedimentology, various isotopes and biogenic materials.

An additional focus of this chapter is *paleopedology* – the science of ancient soils, or paleosols – because buried soils can provide important information about the environment at a time when a geomorphic surface, now buried, was subaerially exposed.

One must always be mindful of the fact that soils respond to more than external forcings and inputs (see Chapter 11). Many soils react to intrinsic thresholds and develop accessions (Muhs 1984, Johnson and Watson-Stegner 1987, Phillips 2001c). They often change pedogenic pathway after they reach and pass a pedogenic threshold or develop an accession; this change may have little or nothing to do with external climatic or geomorphic forcings. Feedbacks are also an important part of the soil geomorphic story. Soil development impacts landscape change and evolution, e.g., when soils develop Bkm horizons and act as caprocks, or when well-developed Btx horizons become aquicludes and force water to move laterally across them, producing surface seepage and springs at downslope areas, thereby initiating gully formation. Indeed, understanding soil history is a complex but exciting challenge.

Paleosols and paleopedology

One of the most common ways that soil palimpsests are utilized to provide a window into the past is via *paleosols*, aka fossil soils. Ruhe (1965) defined a paleosol as a soil that formed on a landscape of the geologic past. It is defined based on pedogenic information and characteristics. When used in formal *stratigraphic* applications, the term *geosol* is suggested in the North American Stratigraphic Code (Morrison 1967, 1998, Catt 1998). When capitalized, a Geosol is the fundamental pedostratigraphic unit. Such geosols are always accompanied by a geographic name, e.g., the Farmdale Geosol or the Sangamon Geosol. The term is *defined on buried soils* that have formed in a consistent stratigraphic position, and like all stratigraphic units it has a type locality. It also carries with it a specific lithostratigraphic connotation, but because soils are diachronous units that begin forming and get buried at different times at different parts of the landscape (Hall 1999a), they should not be assigned strict chronostratigraphic status. When considered in its less formal, lower-case form, the term geosol is considered equivalent to paleosol, which is defined more on pedogenic than on stratigraphic characteristics, and need not be buried. There is, however, an important difference between geosol and paleosol: geosol communicates that the observation of a soil has some stratigraphic and chronologic value, whereas the term paleosol is more ambiguous. Some researchers use the informal term *pedoderm*, which by definition can include buried or non-buried paleosols.

Paleopedology is the study of paleosols, their genesis, morphology, stratigraphy and classification. Much of the early work in this field was done in the Midwestern (Olson 1989) and southwestern United States (Holliday 1998). Although there is a growing interest in pre-Quaternary paleosols (Retallack 1990, Nesbitt 1992, Dahms 1998), our focus here is on those formed in the Quaternary.

Why study paleosols? First, when soil scientists try to understand the response of various substrata to actual climatic and topographic conditions, they are often confronted with anomalous data caused by paleosols that are either at the surface or shallowly buried. For them, knowledge of these paleosols is necessary to understand the surficial pedogenic system. Probably the main utility of paleopedology, however, is the information it can provide about paleoenvironments (Kemp 1999). It should be kept in mind, however, that the information provided by paleosols is *indirect* and their interpretation is *difficult*; they are only rarely complete physically and/or monogenetic.

Like most surface soils, paleosols and catenas of paleosols are palimpsests (and usually good as chronological marker beds) from a time when the surface was stable and soil development was outpacing erosional or burial processes (Nesje *et al.* 1989). Maher (1998: 26) referred to them as "natural archives of paleoclimatic information" and stressed that the integrity, accuracy and retrieval of this information depends on the degree of development and preservation of pedogenic properties, and whether or not these properties can be interpreted via modern analogues. Paleosols change across space, just as surface soils do, prompting many investigators to stress that the best way to interpret (and, indeed, to identify) paleosols is to examine them on a landscape basis, as we do for surface soils (Follmer 1982). Paleosols exist in catenas (Valentine and Dalrymple 1976). Likewise, since paleosols are stratigraphic entities, their spatial variation should be fully known if they are to be used accurately as stratigraphic markers, just like other rock-stratigraphic units (Ruhe 1965).

Because paleosols are soils and therefore can provide a wealth of paleoenvironmental information, there have been attempts to classify them, as we do for surface soils (Retallack 1990, Buurman 1993, Mack *et al.* 1993, Nettleton *et al.* 1998, 2000). Most of these classifications, however, have been fraught with non-acceptance, controversy or indifference.

Classes of paleosols

We begine with Ruhe's (1956, 1965, 1970) early definitions of the three main types of paleosols (recognizing that Catt (1998) has recently updated them).

Relict paleosols are soils that have formed on pre-existing landscapes, under previous and presumably different paleoenvironmental conditions, and remain at the surface today (Ruellan 1971). They have never been buried. Most relict paleosols are polygenetic; many have endured major climatic shifts during the Pleistocene (Woida and Thompson 1993, Heinz 2002). Evidence for polygenesis in soils is discussed in later sections of this chapter, but we agree with the statement of Chadwick *et al.* (1995: 2) in reference to polygenetic soils:

> Optimal conditions for *interpreting* paleoclimates from polygenetic soils occur when precipitation and/or temperature changes are great enough to produce new soil properties without obliterating existing properties. (Emphasis ours.)

How long a soil must have remained at the surface to be considered a relict paleosol is a topic of considerable debate. Nettleton *et al.* (1989) argued that a relict soil is one that has its $time_{zero}$ in the Pleistocene, has remained at the surface since that time and exhibits horizons or features that formed in pre-Holocene environments that were different from today. By this definition, soils formed in the Holocene cannot be relict paleosols. Catt (1990) made the point that, should a surface soil become buried by a debris flow, the soil beneath it would unquestionably be defined as a buried soil and a paleosol. However, this raises the issue of what to call the unburied soil that is pedostratigraphically continuous with the buried soil. The portion of the soil that becomes a paleosol due to burial is perhaps best considered a special kind or variant of the surface soil, rather than the other way around.

If one argues that *all soils* are polygenetic (Johnson and Watson-Stegner 1987, Johnson 2002), then either (1) all surface soils, even the youngest ones, must be considered relict paleosols or (2) the term relict paleosol must be abandoned, because it would have no discriminating power. In this book, we recognize the legitimacy of this debate but take the position that, just as there are many degrees of polygenesis, there are soils that have been influenced in different ways by paleoenvironments. Thus, we acknowledge that there are excellent examples in the world of relict paleosols, as originally defined. They usually occur on surfaces that date back to the Middle Pleistocene or earlier (Daniels *et al.* 1978a, Machette 1985, Nettleton *et al.* 1989). Many might also argue that many Holocene-aged soils with clear evidence of polygenesis, e.g., the Mollic Hapludalfs of the Midwestern United States (White and Riecken 1955), might also be considered minimal relict paleosols. However, there are also many soils in a "gray zone" that might or might not fit with the definition. Ruhe did not concern himself with these soils, and for lack of space, neither do we. In short, the term relict paleosol has value and should be retained.

A term that has been slow to catch on, *Vetusol*, refers to a soil, surface or buried, that has evolved under the same or similar processes from its $time_{zero}$ to the present (Cremaschi 1987, Busacca and Cremaschi 1998), but exhibits little or no impact of polygenesis. A Vetusol has developed "under the influence of a single set of major pedogenic processes throughout their history" and has been little affected by Pleistocene environmental change (Busacca and Cremaschi 1998: 96).

Buried paleosols have formed on landscapes of the past, but have since been buried by younger sediment such as loess, till or alluvium (Catt 1998, Johnson 1998b). For this reason, they are the easiest kind of paleosol to recognize (Ruhe 1965, Ruellan 1971, Nettleton *et al.* 1989). The Soil Survey Staff (1999) defined buried soils as having a surface mantle material of new, pedogenically unaltered material that is generally $\geq$50 cm thick. Bos and Sevink (1975) suggested that even thicker accumulations are necessary, whereas Schaetzl and Sorenson (1987) felt that the depth of burial should be irrelevant to the definition. Rather, they suggested that any soil buried by material that has not yet been pedogenically incorporated into the profile is, effectively, buried. Thus, any sediment can be viewed as burying a soil, but the soil will *stay* buried only if the sediment is sufficiently thick such that pedogenesis will not or cannot incorporate it into the buried profile (Johnson 1985). Then, because any amount of sediment can bury a soil (at least for a while), Schaetzl and Sorenson (1987) also defined a subtype of buried paleosol – an *isolated paleosol* – that is buried so deeply that it is essentially cut off from surficial pedogenic processes (Fig. 15.1). Isolated paleosols are affected only by diagenesis, not pedogenesis (Ruellan 1971). Diagenesis refers to deep-seated processes associated with compaction, heat and circulation of fluids. Use of paleosols to interpret paleoenvironmental conditions requires that one be able to identify what, if any, post-burial modification that the paleosol has undergone (see below). Isolated paleosols, by definition, are not currently being modified by surficial processes.

Buried paleosols have been widely used and recognized by Quaternary scientists for over a century. Many of those originally studied were associated with soil formation during warm interglacial periods, and therefore provided key information about glacial stratigraphy (Leverett 1898, Shimek 1909, Leighton 1926, Leighton and MacClintock 1930).

When the sediment that overlies a buried paleosol is removed, the soil is said to be exhumed (Ruhe 1965). *Exhumed paleosols* are once again subject to the entire suite of surficial pedogenic and geomorphic processes (Bushue *et al.* 1974). Most have been variously truncated, or sediment from the current surface has been mixed into their upper sola (Nettleton *et al.* 2000). Thus, the B horizon is often the main criterion used to recognize exhumed paleosols. Many have been so changed that they can only be recognized as an exhumed paleosol by tracing it laterally to the point where it becomes buried.

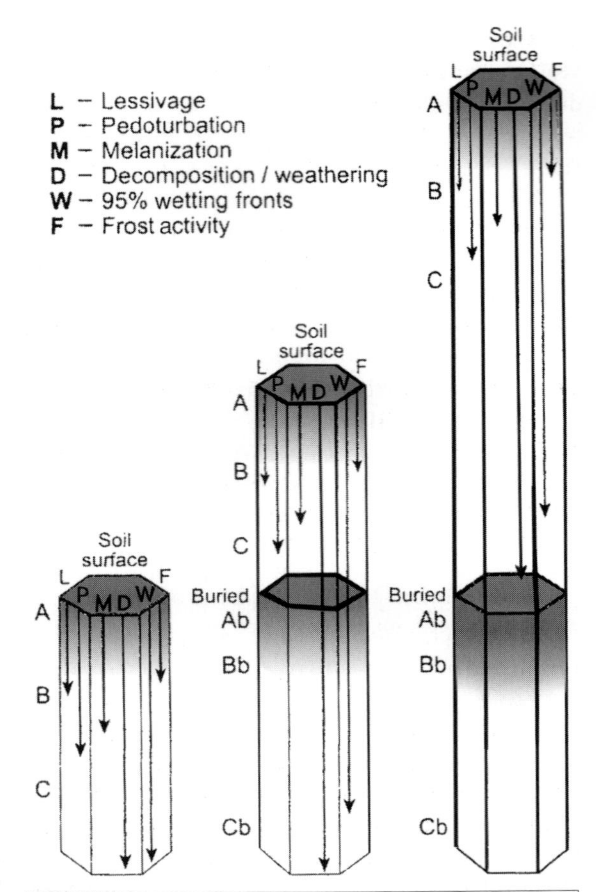

Fig. 15.1 Idealized model of paleosols buried at different depths, showing how paleosols buried at varying depths are impacted by surficial processes. The degree to which a buried soil is impacted by surficial processes determines whether it is considered to be pedogenically "isolated." After Schaetzl and Sorenson (1987).

Recognition of paleosols

Recognizing a buried soil as such, or as a non-pedogenic layer, is a highly important conclusion, for it will affect every interpretation thereafter. If it is a soil, it represents a former stable surface and thus could have been occupied by humans – information that is of importance to paleopedologists, archeologists (Mandel and Bettis 2001), geomorphologists and paleoenvironmental scientists. Recognition of buried paleosols ranges from very obvious to extremely difficult (Ruhe 1965, Ruellan 1971). Two main problems arise when trying to recognize a paleosol: (1) in relict paleosols it is difficult to differentiate relict features from those associated with contemporary pedogenesis, and (2) post-burial modifications (diagenesis) sometimes make recognition of buried paleosols problematic (Simonson 1954a, Catt 1990). Use of multiple criteria, usually assessed in the field and followed up with laboratory data, can assist in the identification and recognition of paleosols, and their differentiation from unmodified sediment.

Burial of a geomorphic surface and its associated soils is a time- and space-transgressive process, and represents a period of time when the pedo-geomorphic system was unstable. During burial, vegetative cover may have been changing, fires, floods or drought may have been rampant, or a glacier may have been advancing onto the landscape. Any of these, and several others not mentioned here, could have caused widespread erosion of all or parts of the surface, prior to burial. Thus, buried soils exist in any number of states of erosion, or lack thereof. Buried soils that have an abrupt upper boundary or that lack O and A horizons are likely to have had *some* erosion prior to burial (Catt 1990). On the other hand, an abrupt upper boundary in a buried paleosol may be due only to rapid burial, such as by eolian sand or volcanic ash. Nonetheless, many buried soils have been *interpreted* as having been truncated during burial simply because the post-burial loss (mineralization) of organic matter has made it appear as though it lacks an A horizon (Busacca 1989). In such cases, the micromorphology of the humus-poor A horizon, specifically its granular structure, can be used to indicate the existence of an Ab horizon.

While many soils are eroded during these periods of geomorphic instability, others are rapidly buried and neatly "preserved" or may even develop progressively during burial. When buried within sedimentation systems that were slow, such as eolian systems (loess), alluvial systems (alluvium) or volcanic systems (ash), the soil may grow upwards into the burying sediment and become overthickened, before finally getting overwhelmed and effectively buried (see Chapter 12). These types of buried paleosols usually have thick Ab horizons that gradually merge into unaltered sediment above. They also typically have "mixed zones" at the top of the buried soil, in which materials from the upper paleosolum and overlying, burying sediment are intermingled (Ransom *et al.* 1987b).

Although there are many buried paleosols that are unmistakable and easily recognizable, many are weak and difficult to discern. Thus, a set of criteria should be established and used when describing paleosols. When recognizing, describing or interpreting a paleosol, it is usually best to treat it as any other soil, using standard pedologic techniques and assumptions while looking for pedogenic features (Woida and Thompson 1993). A key point of discrimination for the recognition of buried (and surface) soils is that, unlike many other sedimentary units, soils have distinct ebbs and flows with respect to depth functions (see "Landscapes," below). Clay contents may decrease and then increase with depth, as do many other pedogenic attributes. Organic matter and phytolith contents are commonly maximal at the top of the profile and decrease with depth, sometimes with a secondary peak in the upper B horizon. All these sorts of trends are uniquely pedogenic; geologic strata frequently have much more abrupt upper or lower boundaries. Retallack (1990) listed three main features that differentiate paleosols from sediment: (1) root traces, which sometimes become more prominent *after* burial (although there are problems with using *only* traces; see Berry and Staub (1998)), (2) soil horizons (see above) and (3) pedogenic structure and other pedogenically acquired features such as cutans and nodules. Indeed, cutans are one of the most unique type of pedogenic signatures, and they can be observed in the field or in thin section. Micromorphology, therefore, has been shown to be a vital technique for the differentiation of pedogenic and sedimentary fabrics (Dalrymple 1958, Botha and Fedoroff 1995). It is also one of the better ways to determine previous pedogenic pathways (Fedoroff and Goldberg 1982).

Mineralogical data, e.g., clay mineralogy and weathering data on primary minerals, provide some of the best criteria for the identification of buried paleosols. They are useful because of their slow-to-change nature, as well as the fact that immobile skeletal grains can be used. Mineralogical methods also have the advantage of providing excellent data on the relative degree of paleosol development, prior to burial (Ruhe 1956) (Figs. 13.48, 13.51). Clay and primary mineral "signatures" are also unlikely to change markedly after burial, as they frequently are more stable when buried than when at the surface (Droste and Tharin 1958, Mausbach *et al.* 1982, Markewich *et al.* 1998, Nettleton *et al.* 2000). For this reason, weathering ratios of sand or silt-sized grains are some of the most useful indicators of paleosol development (Ruhe 1956, 1965). Likewise, contents and depth functions of certain Fe and Al species are also useful for paleosol identification and for determining the degree of development within certain constraints (Evans 1982). Use of mineralogical measures for comparison of development between buried and surface soils, however, generally necessitates that they both be formed in similar materials.

Post-burial modification of paleosols

Post-burial modification of buried paleosols occurs along two main pathways: pedogenesis (impacts from pedogenic processes occurring at the surface and "trickling" down into the shallowly buried soil) and diagenesis. Taken together, these processes are *pedodiagenesis*. The longer a soil is buried the more likely pedodiagenesis has occurred and, to a point, the greater impact it will have had on the buried soil. Processes associated with pedodiagenesis include additions of bases (from overlying calcareous material by the percolating soil solution), loss of organic matter by oxidation/mineralization, changes in redox patterns, additions or losses of Fe and Mn (depending on redox conditions), soil color changes, pedoturbation and blurring of horizon boundaries (Simonson 1941, Stevenson 1969, Ruhe *et al.* 1974, Valentine and Dalrymple 1976, Mausbach *et al.* 1982, Ransom *et al.* 1987b). To this list we add those processes associated with *during-burial modification*, primarily erosion.

Diagenesis of a buried soil involes formation of new minerals (sulfides, carbonates, clay minerals, etc.), redistribution and recrystallization of existing minerals, termination of much of the pre-existing microbial activity and compaction (Catt 1990). Associated with compaction are subsidiary processes such as loss of interstitial pore water and transformation of some minerals, e.g., ferrihydrite to goethite to hematite. Peds are "welded" and pedality is lost or muted. The entire buried profile is squeezed – a process that is especially pronounced in Histosols. Despite compaction, many buried paleosols remain more permeable than the surrounding sediment, readily conducting fluids. When the fluid flow ceases, precipitation of secondary minerals may occur. Deep burial environments are usually mildly alkaline and reducing, favoring carbonate and silica precipitation (Catt 1990).

Paleopedological research requires a knowledge of which soil properties are most prone to post-burial modification and which are persistent. To this end, Yaalon (1971) placed soil properties into a series of categories based on the degree of persistence after burial (see Chapter 14). Properties that persist in a soil after burial are most useful for ascertaining what the soil was like when it was buried. Those that do not persist and are quickly altered after burial, such as pH, potentially provide little information about the original soil. Yaalon's three-fold categorization includes:

(1) Reversible, self-regulating processes and properties, such as pH, organic matter, contents of soluble salts and gypsum, mottles, gilgai, slickensides and most soil structures. These properties usually change after burial, rendering them minimally useful in determining the characteristics of the pre-burial soil (Valentine and Dalrymple 1976, Karlstrom and Osborn 1992).

(2) Processes and properties that slowly achieve a state of equilibrium, such as spodic, natric, argillic, albic, gyspic and calcic horizons, histic epipedons, fragipans and some types of redox features.

(3) Irreversible, self-terminating processes and properties, such as oxic, petrocalcic, placic and petrogypsic horizons, plinthite, laterite, clay mineralogy, and strongly developed argillic and natric horizons. Like weatherable minerals that have been lost from the profile, many of these pedogenic properties, once acquired, are persistent.

Catt (1990), however, noted that organic matter content, a property that is routinely altered by oxidation after burial, may be preserved if the soil is rapidly buried in a stagnant, anaerobic environment. Thus, whether a soil property is preserved or not may depend not only on what it *is* but also on the physicochemical conditions at the location where it is buried.

Under most conditions, it can be broadly assumed that morphological properties, e.g., horizonation, structure, root casts, krotovinas, are the most resistant to modification, whereas chemical properties (unless manifested as morphological features) are least persistent (Busacca 1989). For example, many paleosols buried by calcareous loess are totally saturated with bases, although in highly clay-rich paleosols the Bt horizon may be a sufficient aquiclude that it limits how much alteration can occur (Mausbach *et al.* 1982). Although post-burial enrichment in bases does erase most of the chemical signatures that the paleosol may have had, it also is advantageous because it effectively stops lessivage into or out of the buried soil, allowing the investigator the opportunity to use clay depth plots to interpret pedogenic development. Clay and primary mineral signatures are therefore unlikely to change markedly after burial. For this reason, clay mineralogy and weathering ratios of sand- or silt-sized grains are some of the most useful indicators of paleosol development (Ruhe 1956, 1965, Droste and Tharin 1958, Mausbach *et al.* 1982).

Shallowly buried paleosols are not only prone to modification by post-burial processes, but have dramatic effects on surface soil genesis as well. Particularly noteworthy are the cases where an acid paleosol is overlain by a thin layer of burying sediment. The acidity of the underlying soil enhances the rate of pedogenesis in the surface soil, especially leaching (Kleiss 1973, James *et al.* 1995). For example, Ruhe's (1969) work on the Sangamon paleosol in Iowa, buried by Peoria loess, shows that the soil is better developed, i.e., has a much stronger Bt horizon, at sites that are progressively farther from the loess source (Fig. 15.2). The amount of clay in the Sangamon Bt horizon increases mathematically as:

$$\text{clay}\,(\%) = 19.7 + 1.23\ (\text{distance in miles}).$$

The increased development of the Sangamon paleosol at sites farther from the river is coincident with three distance-related trends: (1) the Yarmouth paleosol below it becomes increasingly closer to the surface, i.e., the loess thins, (2) the loess gets finer-textured and (3) the period of time within which the Sangamon soil has had to form, i.e., its soil-forming interval, increases. Kleiss (1973) confirmed a similar trend in *surface* soils developed in Peoria loess in Illinois, first observed by Guy Smith; he called this the Illinois soil development sequence.

In many parts of the Midwestern United States, strongly developed paleosols buried by only a meter or two of loess have led to the formation of fragipans in the surface soil, presumably due to perching of silica-rich water on top of the dense, clay-rich paleosol (Harlan and Franzmeier 1977, Ruhe and Olson 1980, Ransom *et al.* 1987b). In southern Illinois, variable thicknesses of loess above a shallowly buried paleosol has led to a spotty pattern of Na-rich slick spots (Wilding *et al.* 1963, Indorante 1998). Water flowing laterally along the top of a buried paleosol can form seeps and springs where it crops out on side slopes.

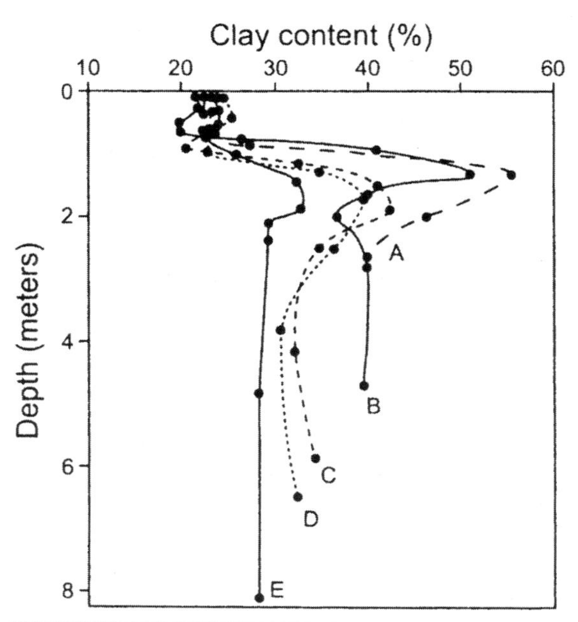

Fig. 15.2 Clay content of the Bt horizons of the Sangamon paleosol, formed in Loveland loess, in southeastern Iowa, along an east–west traverse away from the Missouri River. Soil E is closest to the River; A is farthest. After Ruhe (1969).

Shallowly buried paleosols may become overprinted by the downward-developing profile of the surface soil. When these two soil profiles pedogenically "join" they are said to be *welded* (Ruhe and Olson 1980) or superimposed. In a sense, a welded paleosol is intermediate between a buried and relict paleosol (Fig. 15.3). It is affected by contemporary pedogenesis but retains a cover of sediment that did not exist when it was initially forming. Soil welding complicates recognition and interpretation of buried soils, and makes assessments of their pedostratigraphic relationships more difficult (Kemp *et al.* 1998) (Fig. 15.4).

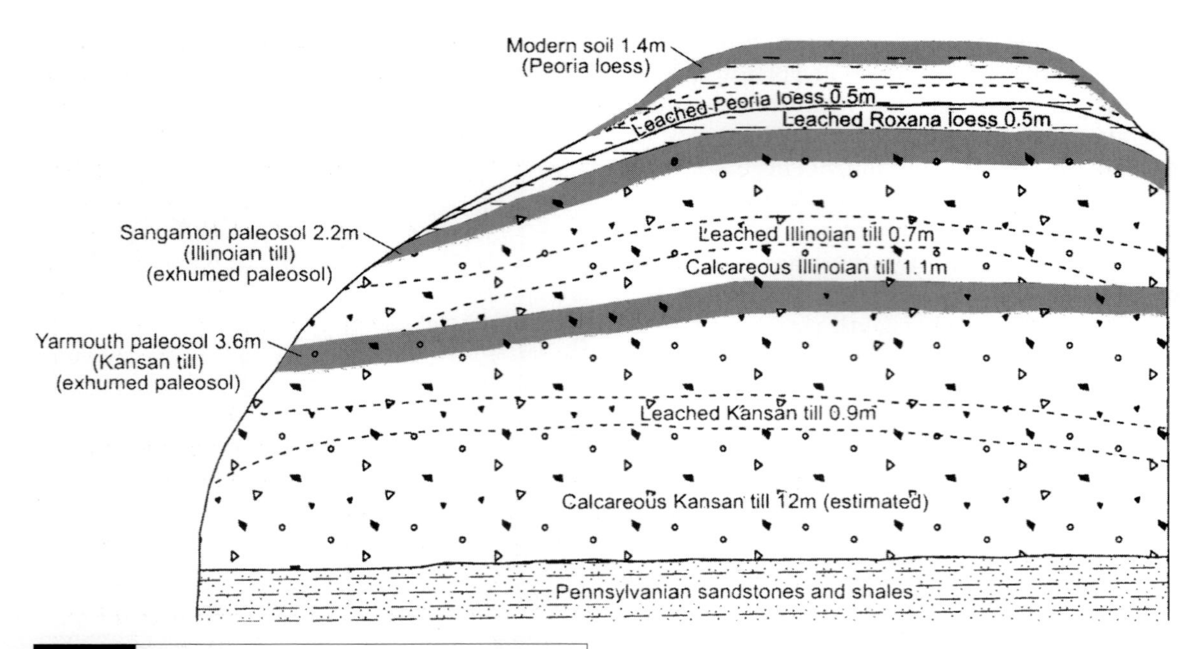

Fig. 15.3 Quaternary stratigraphy in a part of western Illinois, illustrating the concepts of buried, exhumed and welded paleosols. After Bushue *et al.* (1974).

Buried paleosols and Quaternary stratigraphy

Paleosols provide definite and distinct stratigraphic markers, representing a period of time when a geomorphic surface was both stable and subaerial (Ruhe *et al.* 1971). Paleosols represent some of the best records of metastability episodes in landscape history when little or no erosion or sediment deposition took place at a given site. Indeed, much of the early interest in buried paleosols stemmed from their use as stratigraphic marker beds, rather than as indicators of paleoenvironmental conditions (Leverett 1898, Kay and Apfel 1929, Leighton and MacClintock 1930, Frye and Leonard 1949). Stratigraphically, a buried paleosol can be used to (1) discriminate between deposits of different ages, (2) correlate deposits across landscapes and (3) provide information on the length of depositional hiatuses, i.e., length of the soil-forming interval, or erosional episodes within the stratigraphic column (Ely *et al.* 1996, Olson *et al.* 1997). The most common use of buried paleosols in this context occurs in areas of constructional topography like the Midwestern United States and the loess plains of China, Russia

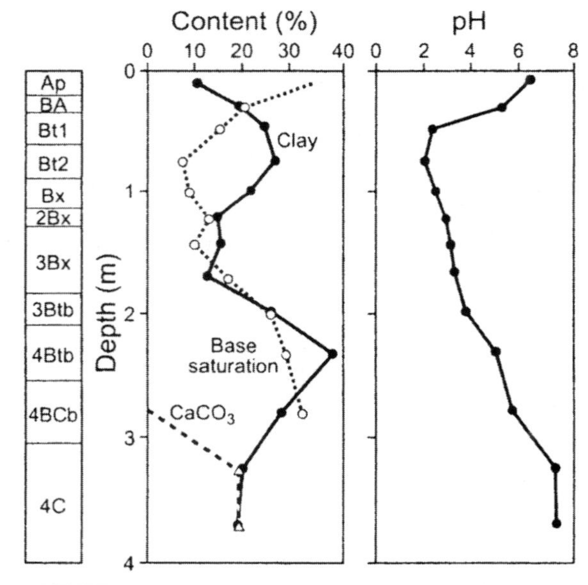

Fig. 15.4 Depth functions for a surface soil welded to a Sangamon paleosol in southern Indiana. Note the Bt horizon "clay bulge" for both the surface soil and the paleosol. Note also that the paleosol has retained some signature of its pre-burial, acidic status (low base saturation) but that pH, a more ephemeral property, does not provide any paleopedological discriminating information. After Ruhe and Olson (1980).

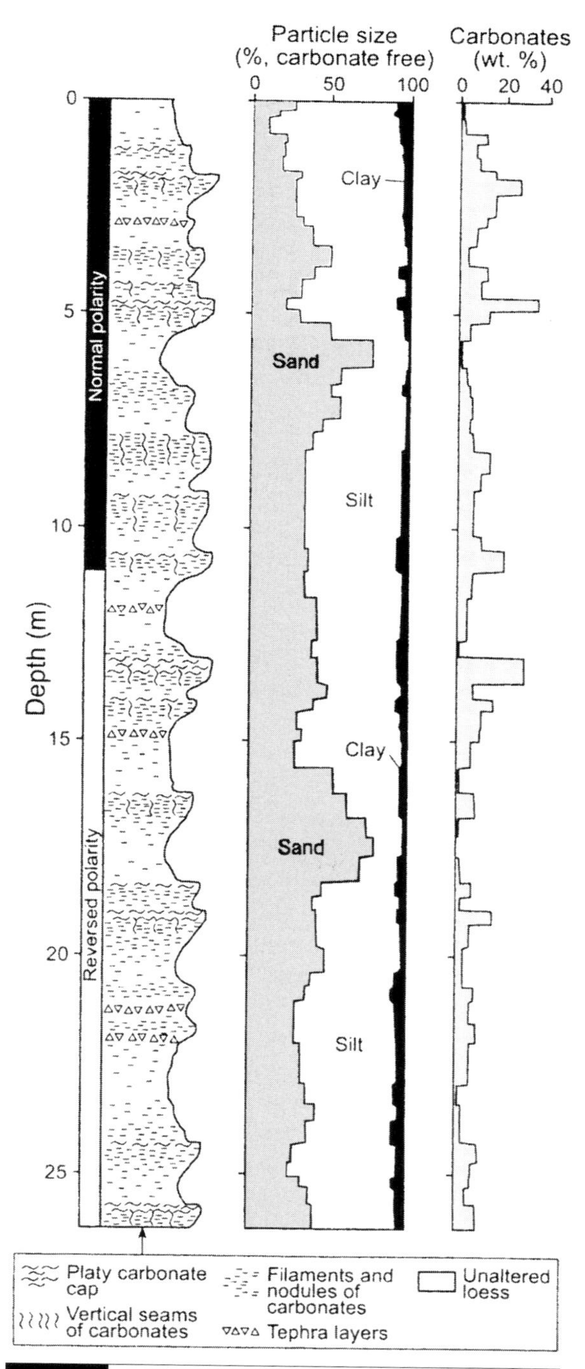

Fig. 15.5 Stratigraphic and paleopedologic information for a deep loess sequence in the Palouse Hills of eastern Washington. After Busacca (1989).

and Europe where sediments that were deposited during periods of instability are intercalated with paleosols (Wascher *et al.* 1948). For example, in the Midwestern United States, the Brady paleosol separates the lower Peoria loess from the Bignell loess above (Feng *et al.* 1994a).

The loess–paleosol record, which in places can be traced back to the Pliocene, can be evaluated in a number of ways and provides many different kinds of paleoenvironmental proxy data. Paleoenvironmental records are variously preserved in loess–paleosol sequences as: magnetic susceptibility and other magnetic properties, paleomagnetism, Fe^{2+} : Fe^{3+} ratios, pollen and macrofossil remains, carbonate contents, clay mineralogy and grain size, among several others (Farnham *et al.* 1964, Feng and Johnson 1995, Botha and Fedoroff 1995) (Fig. 15.5).

The thickness of the China loess column and the number of buried paleosols within it is truly amazing and worthy of note (Liu 1985, Pye 1987, Kemp 2001). The loess forms a huge plateau southwest of Beijing. Although slow eolian deposition started here almost 7 Ma, loess deposition did not begin in earnest until about 2.6–2.4 Ma. The loess source region was the Takla Makan and Gobi Deserts to the northwest and north (see Chapter 8). Loess was blown in from these deserts by the winds of the strong, cold, winter monsoon. Deposition occurred during glacial periods, when the cold landscapes to the northwest destabilized; soil formation occurred as surfaces stabilized during the warmer interglacial periods, which were dominated by the wet summer monsoon (Hovan *et al.* 1989, Zhongli *et al.* 1993, Zhang *et al.* 1994). For this reason, the loess carries with it a strong paleoenvironmental signal and its study has provided a wealth of information on paleoclimate (Verosub *et al.* 1993) (Fig. 15.6).

The most extensively studied and, one could argue, impressive paleosol in the United States is the Sangamon soil (Fig. 15.7). It is widely developed and exposed throughout the central United States, where it is developed on Illinoian drift and its correlative loess, Loveland loess (Ruhe 1974b, Follmer 1978, 1979a, 1982, Schaetzl 1986c, Jacobs 1998, Hall and Anderson 2000) (Fig. 14.54). In places where these deposits are absent, it occurs directly on bedrock or residuum. In almost every instance, the buried Sangamon soil is better developed than its counterpart on the surface, with respect to illuvial clay in the Bt horizon, clay mineralogy, redness (if on a well-drained part of the paleolandscape), weathered character and solum thickness (Ruhe *et al.* 1974). The Sangamon soil is time transgressive. Within the margin of the Illinoian glaciation and its loesses, the $time_{zero}$ of the Sangamon soil is placed between 125 and 75 ka (MIS 5); throughout this region is was effectively buried between 55 and 25 ka (Hall and Anderson 2000). The Sangamon soil is commonly leached to 3 m or more in coarse-textured materials like outwash, but is usually about 2 m thick when developed in till. Where formed on sloping surfaces, it typically has a stone line, ascribed either to pedimentation processes immediately preceding burial or to bioturbation (Fig. 15.7b; see Chapter 13).

Known formally as the Sangamon Geosol, it has many known soil types (paleosols) that range in color, thickness and other features. Depending

Landscapes: The Farm Creek stratigraphic section, central Illinois

The Farm Creek section near Peoria, Illinois has been a classic Pleistocene exposure for over a century (Leighton 1926, Willman and Frye 1970, Follmer *et al.* 1979). It is the type section for the Farmdale soil, a paleosol that formed during a brief interstadial (MIS 3, about 28 to 24 ka) during the Wisconsin glaciation (Frye and Willman 1960, Baker *et al.* 1989). On upland sites, the Farmdale paleosol is an Entisol or Inceptisol, formed in the middle Wisconsin-age Roxana silt. At many places, the Roxana silt is so thin that the Farmdale soil is welded to the Sangamon soil below. It was later buried by Morton loess, which was being deposited in front of the advancing Wisconsin ice margin (see Figure). Eventually, the Wisconsin ice sheet overrode the Farm Creek area and deposited the Delavan till, and as it retreated, an outwash (the Henry Formation). Finally, the Richland loess (a Peoria loess correlative) was deposited on top of the Henry Formation, after the ice margin had retreated from the area. Beyond the Wisconsin glacial margin, the Farmdale soil is buried only by loess. The Farmdale paleosol defines the Farmdalian substage (Johnson *et al.* 1997), an interstadial period in the Wisconsin stage during which the Woodfordian ice margin was advancing, the climate was getting progressively cooler, and there was no loess deposition. Hence, soils were able to form on stable geomorphic surfaces, and the climate was warm enough for soil formation.

Illinoian till forms the base of the Farm Creek section. The Sangamon paleosol, with its well-developed Bt horizon, has formed in this till (Follmer 1979a). The Bt horizon is the best preserved part of the Sangamon. It has retained strong blocky structure and prominent argillans, despite deep burial and great age. At Farm Creek, the Sangamon soil is a poorly drained Albaqualf or Albaquult. It is clayey and dark-colored, with iron stains from many millennia of soil water. Like many paleosols buried by loess, there is a significant mixed zone between the Sangamon soil and the Roxana silt, making its upper boundary difficult to distinguish. And also like many strongly developed acid paleosols, there often exists a leached zone above, in the Roxana loess. This leached zone may have formed by percolating water, due to mixing of acid sediment from below or simply because the acid soil promoted strong leaching during the brief period when the loess above was thin and subaerially exposed.

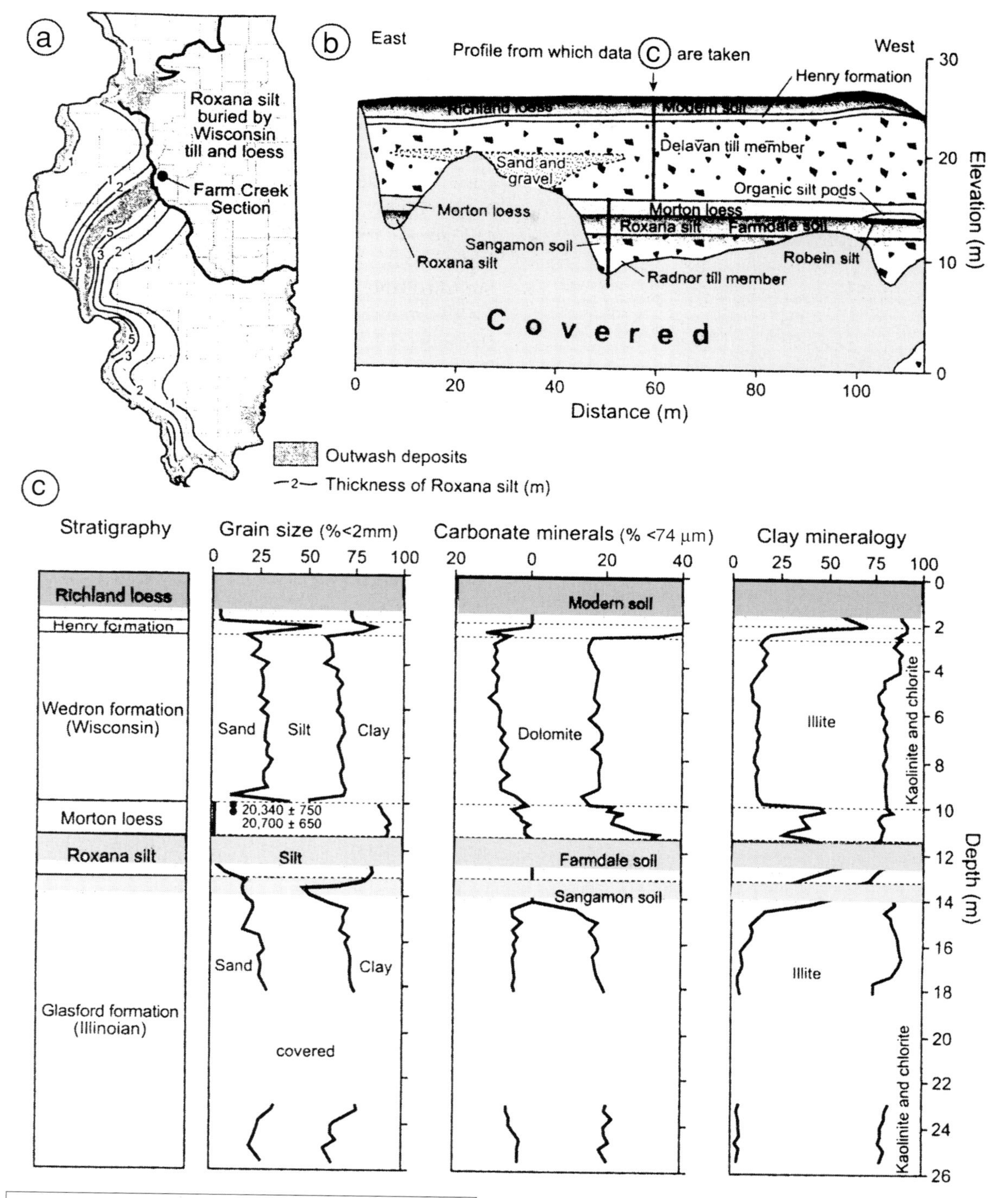

Stratigraphic and quantitative pedologic data for the Farm Creek Section, Peoria, Illinois. Inset map: Thickness of the Roxana silt and extent of the Woodfordian glaciation in Illinois. After Follmer *et al.* (1979).

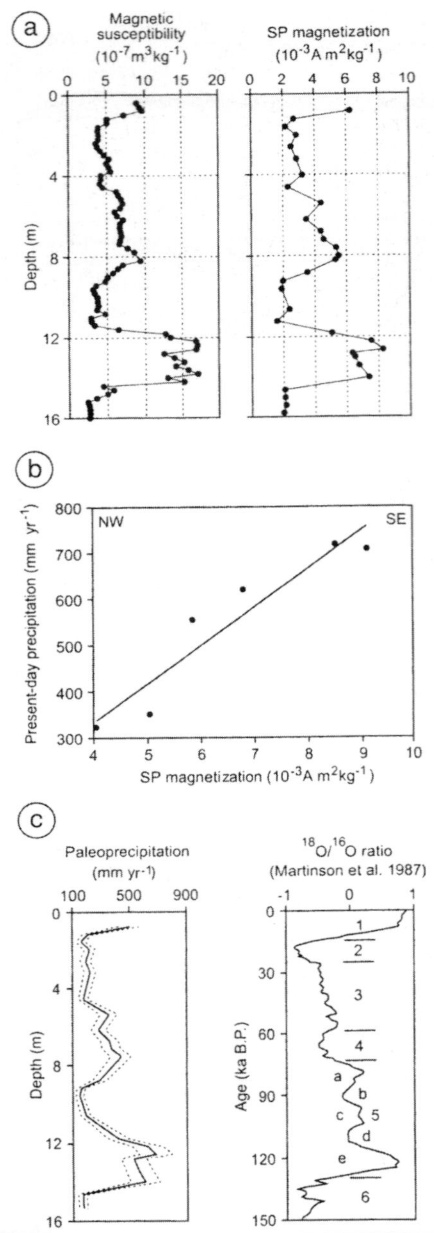

Fig. 15.6 Three graphs illustrating the types of information that can be gleaned from paleomagnetism and magnetic susceptibility. (a) Depth plot of superparamagnetic (SP) magnetization and magnetic susceptibility, for a section of the Chinese loess plateau. (b) Relation between contemporary annual precipitation totals and SP magnetism at six sites on the Chinese loess plateau. (c) Reconstructed paleoprecipitation during the last 140 ka for an area in China, using loess data and the relationship in (a), compared to the marine oxygen isotope record. After Liu *et al.* (1995) and Martinson *et al.* (1987).

upon landscape position, it can be a deep, reddish, clay-rich Ultisol on uplands or a thick, gleyed, grayish-black Aquoll or Aquult in swales. Many of these paleosols reflect the length of time they had to develop within the warm Sangamon interglacial (Fig. 15.8). Where developed on deposits older than Illinoian age, the Sangamon soil is the highly polygenetic Yarmouth–Sangamon soil (see Chapter 13). In parts of the Midwestern United States, however, the soil-forming interval included only the Late Sangamon.

The morphological variation in the Sangamon Geosol is reflective of the variety of environments that occurred when it was forming, as well as the geomorphology of the surface it developed on (Follmer 1979a). On paleosummits the soil has developed in situ in Illinoian till or Loveland loess. It commonly has a strong Aquoll- or Aqualf-like profile in many places in Illinois, while on the drier, western end of its range within the central United States the Sangamon takes on redder colors, resembling a highly weathered Ustoll; it crops out as a relict paleosol throughout much of Oklahoma, where it is mapped as a red Ustoll. As with all buried paleosols, recognition of its full variation is essential if it is to be properly correlated as a stratigraphic marker bed.

The soil-forming interval

Soils form on stable geomorphic surfaces, and continue to form and evolve with them, as long as they remain relatively stable. Eventually, all surfaces will become (or have become) buried or eroded. The period of time during which the soil formed, before interruption, is called its *soil-forming interval*. For a buried paleosol this interval is usually bracketed by the ages of adjacent depositional units. For surface soils, the soil-forming interval starts at the cessation of an erosion or depositional episode, i.e., when the surface last stabilized, and continues to the present. For soils on surfaces that have long been stable, soil-forming intervals might be associated with a period or periods of particularly strong pedogenesis – punctuated by periods of slowed or regressive development (Morrison 1967, Chadwick and Davis 1990, Hall and Anderson 2000).

The presence of a buried paleosol implies that there existed a stable period on that landscape.

(a)

(b)

Fig. 15.7 Exposures of the Sangamon paleosol in the central United States. Images by D. L. Johnson. (a) Formed in Loveland loess, buried by Peoria loess, Iowa. (b) Formed in Illinoian till, buried by Peoria loess, Iowa. A stone line is evident.

The fact that the soil is now buried also implies that the pedogeomorphic processes have since changed. Thus, as suggested by Butler (1959), buried soils provide excellent proxy information about slope and landscape stability (Nesje *et al.* 1989, Anderton and Loope 1995, Olson *et al.* 1997), which when dated can be connected/related to the drivers of that instability, e.g., climatic change, human activity.

Soil properties can reveal a great deal about paleoenvironmental conditions during the soil-forming interval, as well as the length of that interval. Often, however, the relative importance of *length* vs. *intensity* of the soil-forming interval is unclear. For example, the Sangamon soil was long assumed to have attained its strong morphology due to an intense (warm) soil-forming climate during MIS 5, particularly the hottest initial part, stage 5e. However, the counter-argument is that pedogenesis was not abnormally strong or intense during MIS 5, but rather, the great *length* of the soil-forming interval facilitated the strong development (Ruhe 1965, Boardman 1985, Hall and Anderson 2000). Length or intensity? We will revisit this topic below, under magnetic

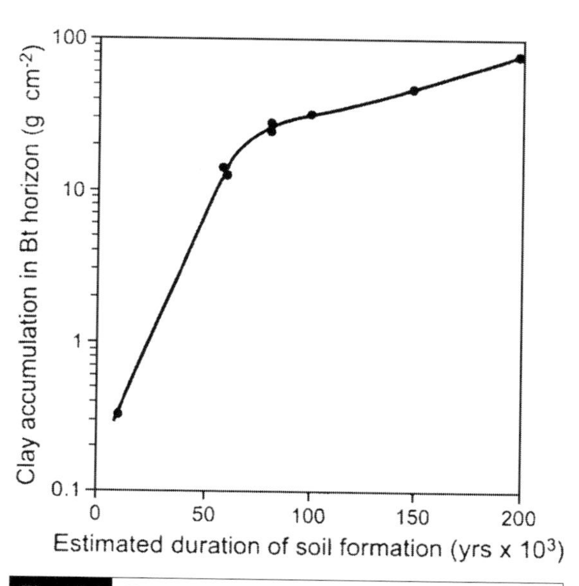

Fig. 15.8 Generalized diagram showing clay accumulation in modern (Holocene age) soils vs. the major paleosols of the Midwestern United States. This diagram makes the point that the larger clay accumulations in the major Pleistocene paleosols are due to longer soil-forming intervals, not a more favorable climate. After Boardman (1985).

susceptibility. Soils, paleosols and soil-forming intervals are the essence of paleopedology, and provide a suitable lead-in to the next section, which focusses on pedogenic features that are useful for interpreting the duration and intensity of the soil-forming interval.

Dating of buried paleosols

Studies of buried paleosols usually are concerned with one or more of the following questions: (1) when was its $time_{zero}$, (2) how long was the soil-forming interval and (3) when was it effectively buried? Much of these types of data can be obtained using various dating techniques (see Chapter 14). Some of the best material that can be used, in a radiocarbon context, for estimating the date of burial of a soil is plant and animal remains, e.g., charcoal, seeds, wood and shells, that were on the soil surface immediately prior to burial (Catt 1990). However, since such material is not always available, one usually has to turn to a mean residence time (MRT) (see Chapter 14) date on soil organic matter as an acceptable but less desirable estimate of time since burial (Scharpenseel and Schiffmann 1977). We remind the reader that, when dating soil organic matter in buried paleosols, the length of time that the soil has been buried (if it is not highly polygenetic) is best accomplished by sampling the uppermost part of the A horizon, because it will have the youngest MRT age (Haas *et al.* 1986, Nesje *et al.* 1989). And even then, the date obtained will be a maximum age of the time since burial. Put another way, the soil could not have been buried any longer than the MRT date obtained from the A horizon of the buried paleosol.

To establish the date of $time_{zero}$ of the buried soil from soil organic matter, it is advisable to sample the lowermost part of the A horizon, or the oldest carbon possible. Sampling deeper horizons might (but often does not) result in older ages but often they do not contain adequate carbon for a date. The age obtained will be a minimum estimate of the $time_{zero}$ of the buried soil, i.e, the soil could not have begun forming any later than that date. For soils that contain some ancient or "dead" organic materials such as coal, caution must be used, since MRT dates on these soils could significantly pre-date $time_{zero}$ (Catt 1990).

Estimating the length of the soil-forming interval based on radiocarbon ages from the youngest and oldest organic carbon in the soil is theoretically possible but usually not advisable, given the error terms and uncertainties of the method. If a more reliable age estimate of the time of burial were to be obtained, e.g., on charcoal that was on the very top of the buried soil, then one could compare that age to an MRT date from the lowermost A horizon to arrive at a minimum estimate of the soil-forming interval of the buried paleosol. A potentially useful method to estimate the length of the soil-forming interval involves garden-variety ^{10}Be, which accumulates in surface soils as a function of exposure time (Pavich *et al.* 1985). Other surface exposure dating (SED) techniques, such as mineral weathering indices, might also provide information on the relative length of soil-forming intervals of buried paleosols vs. others in which the length of that interval is known with more certainty (see Chapter 14).

For soils that are shallowly buried, contamination of the soil organic matter fraction with modern carbon from the surface soil is possible and even likely. Likewise, roots may penetrate the buried soil in search of nutrients, and proliferate there.

Environmental pedo-signatures

Landforms and soils provide "clues" about paleoenvironments. Most geomorphology texts discuss how certain *landforms* provide indications of paleoprocesses. Examples include stone circles and frost wedges as indicators of permafrost conditions and periglacial activity (Kozarski 1974, Washburn 1980, Ballantyne and Matthews 1982, Mader and Ciolkosz 1997), debris flows as indicators of slope instability caused largely by high-magnitude rainfall events (Liebens 1999) and innumerable others. Our focus is on *soils* and *pedogenic features/attributes* as paleoenvironmental indicators (Gao and Chen 1983, Chadwick *et al.* 1995, Jiamao *et al.* 1996). The best features to use, in this regard, are those that are not easily obliterated or erased.

Many soil properties are associated with a particular type of environmental situation or condition, and thus carry a unique environmental signature. Their development is strictly or loosely associated with processes that are driven by a certain set of vectors which characterize an environmental setting. "Passive" environmental factors such as sedimentology of the site, including the presence or absence of lithologic discontinuities and the texture and mineralogy of the parent material, as well as water table effects and slope/aspect, all set the "environmental stage" for the more active vectors. Pedogenic processes are driven by these active vectors, e.g., precipitation, temperature, vegetation, snowfall and aerosol inputs. Together, these two sets of vectors form unique pedo-signatures. Every soil *has* these pedo-signatures, although they may be weak and difficult to ascertain. For example, because A horizons are common to all soils, one might assume that the A horizon can provide little or no information about the soil vectors or processes that have affected it. In a general sense, this is correct; the mere presence of an A horizon tells little about paleoenvironmental conditions. However, a more careful examination might reveal information that is highly useful in this regard, e.g., thickness and organic matter content, isotopic composition of the organic matter, types and density of krotovinas, etc. Thus, the pedo-signature information *is there*, and as our knowledge of pedogenesis increases we will accentuate our abilities to decipher and understand it.

Pedo-signatures from paleosols or surface soils, when coupled with geomorphic data, can provide a variety of paleoenvironmental information on, for example, paleotemperature, paleoprecipitation, wind direction and speed, and paleoecology (Ruhe 1974a, b). These signatures have varying degrees of persistence within soils; those that are highly persistent have the most value in the interpretation of paleosols and paleoclimates (Yaalon 1971) (see Chapter 14). Examples abound of buried paleosols with characteristics that imply that they formed in a different paleoenvironment than that of the surface soil (e.g., Bryson *et al.* 1965, Follmer 1978, Reider *et al.* 1988). In such cases, there has been an environmental change since the paleosol's soil-forming interval; the task of the paleoenvironmental scientist is to determine what that change(s) have been.

Our goal is to provide examples of soil properties that *have* a usable signature or provide a discernable environmental, and paleoenvironmental, signal. The main soil properties that apply in this context are mineralogy and morphology (Dormaar and Lutwick 1983, Karlstrom and Osborn 1992). Soil chemistry is more ephemeral and even if it does change under a different environmental regime, the signals it leaves behind usually reside in the soil mineralogical package (Barshad 1966).

Horizonation/morphology

The *horizonation* of soils is usually the first thing that one observes, and it generally sets the tone for the overall pedogenic interpretation. For example, soils with carbonate-rich horizons provide definite evidence of a climatically dry period of formation in which carbonate-rich dust was impinging on the soil surface (Bryan and Albritton 1943). When such soils are present in areas that, today, are so humid that carbonates would get leached, this morphology alone is a clear indicator of a dry-to-wet climate change (Reider 1983). Similarly, soils with accumulations of salt-, gypsum- or sesquioxide-rich horizons all clearly indicate a distinct type of pedogenesis. Ruhe and Scholtes (1956) used the morphology of a buried Alfisol in Iowa to infer that it formed in a forested paleoenvironment. Ruhe *et al.* (1974) suggested that the buried Sangamon soil in southern Indiana was morphologically like the forested soils that are forming, today, several hundred kilometers to the south. Lastly, van Ryswyk and Okazaki (1979) observed that soil profiles in the alpine zone of the Cascade Range had spodic-like morphologies. This type of soil genesis usually occurs under forest, suggesting that treeline was once higher. In sum, the ability to deduce bioclimatic signatures from profile morphologies, which comes from studies of modern soil genesis, is a focus of paleoenvironmental reconstruction.

Before an assessment of pedogenic pathways can be undertaken, the parent material(s) that the pedogenic processes have been acting upon must be discerned and evaluated (Fig. 15.9;

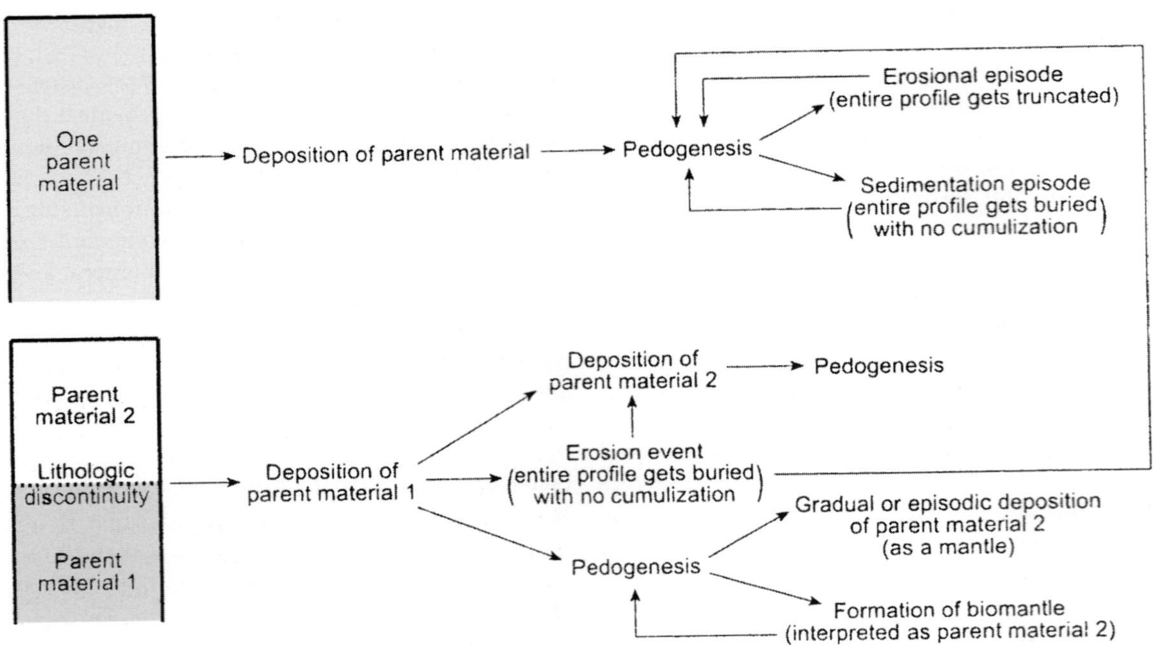

Fig. 15.9 Paleosedimentological pathways associated with soils that have developed in one vs. two parent materials. Knowledge of the lithologic nature of the soil parent material enables one to ascertain the sedimentological history of a soil, which then has utility in determining the paleoenvironmental history of that site/soil.

see Chapter 8). In short, knowing the sedimentological and geomorphic history of the site is often as important as knowing the paleoclimate or paleovegetation (Amba *et al.* 1990). Soils provide many clues about their geomorphic history in their parent materials (Evans 1978, Meixner and Singer 1981, Schaetzl 1996, 1998). Thus, another useful aspect of soil morphology that can be used to unravel paleoenvironmental conditions centers on parent material and its stratification/sedimentology. This type of information, always of great use in determining the nature and rate of past surficial geomorphic and sedimentological processes, can also provide information on paleoenvironments (Benedict 1970, Reider 1983, Rindfleisch and Schaetzl 2001). For example, a period of dust accumulation is often used to infer dry conditions while lack of dust could be associated with more humid conditions when surfaces are better vegetated.

Identification of a geomorphic surface buried within the near-surface stratigraphic column has implications for climatic change, landscape stability and archeology, among many others (Holliday 1988, Cremeens and Hart 1995, Curry and Pavich 1996). Thus, by examining the parent material closely, one can ascertain whether periods of erosion and deposition occurred, what agents were responsible for those depositional or erosion episodes, and if a biomantle exists (Johnson 1990, Humphreys 1994, Nooren *et al.* 1995).

For soils developed in a single parent material, only one period of deposition needs to be inferred. Another option for soils with one parent material, however, involves a period of deposition, then a large erosion episode (in which the entire, pre-existing soil is eroded), followed by a period of stability (Fig. 15.9). In any event, at some point in the past a surface became stable and subaerial, the material beneath that surface was relatively uniform and probably developed in association with a single geomorphic event, and nothing significant has been added to the top of that surface since $\text{time}_{\text{zero}}$.

Soils that have developed in two or more parent materials, as indicated by one or more lithologic discontinuities, probably have more complicated histories (Amba *et al.* 1990). At some time in the past, there were at least two periods of deposition (unless the upper material is determined to be a biomantle; see Chapters 9 and 14). These two periods may have been separated by an erosion event. A period of stability and pedogenesis may have intervened between deposition events, or it may not have, i.e., the two depositional events followed closely on the heels of each other. If one envisions no erosion event, it follows that pedogenesis probably began acting on the lower material, followed by additions of a second material at a later time, but this second material was pedogenically incorporated into the profile, i.e, there is no buried paleosol. Dixon (1991) used sedimentologic data of this kind for some alpine soils to infer a period of eolian influx (during the Holocene) onto a soil developed in till. Dixon's hypothesis was strengthened because the upper material was less weathered than the material below, suggesting that it was younger. His study is a good example of how a lithologic discontinuity can be used to understand changes in pedogenic/depositional systems over the lifespan of a soil or geomorphic surface. Too often, a detailed analysis of parent material and subtle lithologic breaks is an ignored component of soil geomorphic research. Soil geomorphologists *must* do this work, for frequently geologists "look" only below the profile, and many pedologists are not armed with enough geomorphic and sedimentologic training to properly interpret near-surface discontinuities.

Soil classification can often provide information on soil-forming climate, and some of this information can be used loosely to infer paleoenvironmental conditions (Arkley 1967) (Table 15.1). Be aware, however, that several of the major taxonomic groups, e.g., Entisols, Inceptisols, Andisols, Fluvents, Anthrepts, etc. can form in a variety of climates. In the United States, all soils are classified to soil temperature and moisture regime, but these reflect very short-term conditions in the soil and are not intended to correspond to past conditions (Soil Survey Staff 1999). Many taxonomic groups are influenced most by local circumstances, such as a wet site that leads to the formation of a Histosol, or the swale that is shallow to saline shale that leads to the formation of a Natrustoll, rather than macroclimate. Still others provide virtually no pedoclimatic signature because were designed to reflect incipient development. Other taxonomic classes or materials are so ephemeral that they can be used only to understand recent conditions. Still other groups or horizons are reflective of parent material, e.g., Sulfaquents, Vitrands, while others reflect human impact (Plaggepts).

Organic matter content

Soils store great amounts of organic carbon – globally nearly three times that of the aboveground biomass (Eswaran *et al.* 1993). The amount of organic matter in a soil, its C:N ratio and its distribution through the profile are useful indicators of paleoclimate and paleovegetation. Generally, organic matter contents increase as the climate gets cooler and moister (Jenny 1941b). Obviously, this relationship does not hold for the entire range of climate types on Earth, since in the coldest climates the decrease in organic matter production outstrips the decrease in mineralization rate; these cold soils do not have inordinately high organic matter contents.

Many studies have examined the rates of organic matter production vs. decomposition vs. storage in soils across the many global ecosystems and taxonomic categories. The purpose of these studies is usually to determine an inventory of soil carbon reservoirs, and in so doing examine the soil carbon cycle. Obviously, Histosols contain the most organic matter of all soils, because the decomposition rates of the raw organics is inhibited by cold conditions, as well as a high water table that keeps the soil system anaerobic. Many wet mineral soils have high carbon contents for the same reasons. The high contents of organic carbon in these soils reflects local conditions, however, and not regional climate, making the data of limited utility. Most studies of upland mineral soils (excluding Entisols) tend to show that cool grassland Udolls and high alpine Gelisols and Cryepts soils store the most organic carbon, while hot Aridisols have the least (Burke *et al.* 1989). Thus, with a few exceptions, soil organic carbon contents appear to be related to climatic parameters (soil temperature and moisture), as influenced by parent material, especially clay content (Nichols 1984).

Table 15.1 Taxonomic groups, horizons and materials that provide a climatic pedo-signature

Climatic signal/type	Taxonomic classes indicative of that climate[a,b]	Horizon types[a] or materials indicative of that climate
Cold climate with permafrost (tundra)	Gelisols, Cryids	Gelic materials
Cold desert	Gelisols, Cryids	Gelic materials, salic horizon, desert pavement and varnish
Hot desert	Calcids, Gypsids, Durids, Salids	Calcic, gypsic and salic horizons (and petro-versions thereof), duripan, desert pavement and varnish
Subhumid and semiarid grassland and savanna	Mollisols, Vertisols	Mollic epipedon (possibly a calcic or natric horizon), gilgais, slickensides
Cool, humid (conifer or mixed conifer–hardwood forest)	Spodosols	Spodic materials, umbric and spodic horizons (possibly a fragipan)
Temperate humid (deciduous forest)	Udalfs, Udolls, Udults	Argillic and kandic horizons, plinthite (possibly a fragipan)
Humic tropics	Oxisols	Oxic horizons, plinthite
Mediterranean climate	Xeralfs, Xerolls, Xerults	Duripans

[a] Based on Soil Taxonomy (Soil Survey Staff 1999). This table is our interpretation and is not meant to imply endorsement by the NRCS.
[b] Classes with lack of development, i.e., Entisols, are common on many landscapes and provide little climatic information. They are not included here.

The distribution of organic carbon with depth is also an indicator of pedogenesis and climate (Fig. 15.10). Soils in dry climates tend to retain humus in near-surface horizons for two main reasons: (1) the high base saturation in these soils facilitates Ca–humus bonds, which tend to render the humus immobile, and (2) the generally low precipitation values limit how deeply humus can be translocated. In some soils, the depth distribution of organic carbon often shows a second peak in the B horizon. Many Spodosols, especially Humods, and Andisols display such trends due to the soluble nature of their organic acids. Fluvents are defined as having irregular decreases in content of organic carbon with depth, due to burial of A horizons by recent alluvium. In forest soils such as Alfisols, Ultisols and Oxisols, organic matter peaks at the surface. In sum, both the total amount of organic carbon and its distribution with depth have a gross climatic signature, although it does not always persist after burial (Holliday 1988, Dahms 1998).

Clay mineralogy

Clay mineralogy is a slow-to-change and sometimes almost irreversible pedogenic property that has, in most soils, strong ties to parent material and bioclimate (Folkoff and Meentemeyer 1985). Parent material sets the starting point for the clay mineral suite in a soil, while the ions contained within that parent material determine the limits on what can eventually neoform. In young soils, the clay mineral suite has been so recently inherited that it is generally of little value as a paleoenvironmental indicator. However, as time progresses and chemical weathering has a greater impact on soils, their clay mineral suites become increasingly in equilibrium with the chemistry and leaching regime of the soil, i.e., the *soil climate* (Fig. 15.11). Thus, clay mineralogy is usually reflective of climate, as impacted and modified by the biota (van der Merwe and Weber 1963, Arkley 1967). We include the caveat "as affected by the biota and soil" because, for many soils, the soil climate is different than the macroclimate. The soil

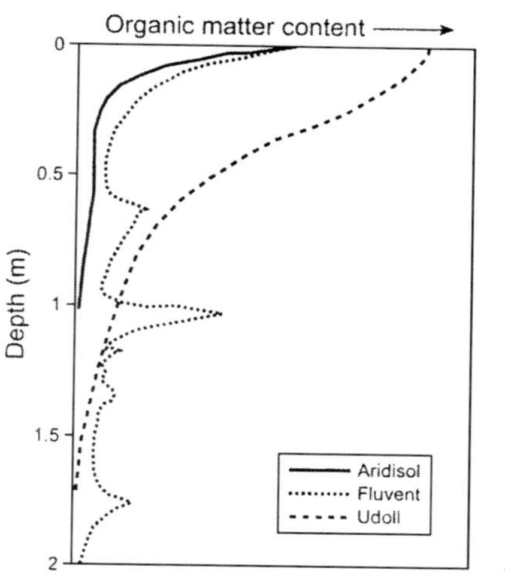

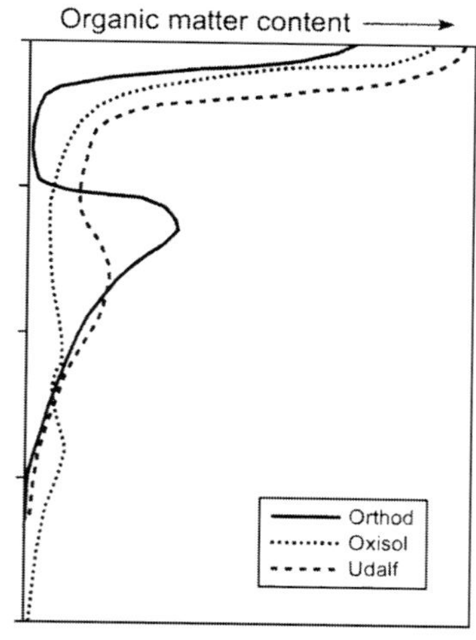

Fig. 15.10 Theoretical depth distributions of organic matter or humus in a few selected, representative mineral soil types.

climate of an upland soil might be wetter than expected because of an aquiclude that perches water. Soils in aquic soil moisture regimes are always wetter than macroclimatic data might indicate. Wet, clayey soils often show redox features even though they are technically not frequently saturated, because water percolates so slowly through them. Depletion of bases may be minimal in a poorly drained soil while well-drained counterparts on uplands are throughly leached and acidified. Conversely, some soils are drier than the macroclimate would suggest because they are coarse-textured or on a convex part of the landscape. Also, vegetation impacts clay mineralogy through base cycling and acidification, or lack thereof.

With time, clay minerals, especially those in the A horizon, develop a quasi-equilibrium with the prevailing climate (Barshad 1966, Folkoff and Meentemeyer 1985). There is a scale factor at work here as well; on local scales clay mineralogy may continue to reflect the influence of parent material or topography, but after longer periods of time the impact of climate will come to dominate the clay mineral suite across larger and larger segments of the landscape. The clay mineral suites of soils that have had several thousand or more years to "acclimate" confirm this bioclimate–clay mineralogy relationship.

The primary clay mineral–environment relationship is centered around leaching and weathering regimes, driven primarily by precipitation. Specifically, Folkoff and Meentemeyer (1985) argued that the two main climatic factors that relate most to clay mineralogy are the magnitude of seasonal leaching and (if present) drying (Chadwick *et al.* 1995). Along the leaching–weathering sequence, various types and suites of clay minerals exist, according to their stabilities. Increased leaching tends to decrease silica activity and increase weathering, resulting in the formation of more stable, weathering-resistant minerals and an increase in aluminum activities. Where leaching is minimal, weathering by-products and cations, especially Si, remain in the soil solution and can neoform into new clay minerals that are stable in, and therefore reflective of, that environment. Temperature impacts all this primarily through its impact on the *rate* of the weathering reactions, not the *direction* of the weathering pathways.

Both theoretical/geochemical models and field observations indicate a strong linkage between clay mineralogy and environmental conditions related to leaching and weathering (Beaven and Dumbleton 1966, Rai and Lindsay 1975) (Fig. 15.11; see Chapter 4). Smectites tend to form and remain stable in non-leaching environments: dry climates and areas of shallow groundwater (Beaven and Dumbleton 1966, Birkeland 1969, Ojanuga 1979). The 2 : 1 smectite minerals are associated with abundant Ca and Mg – bases that would normally be leached from other soils. They are especially common in soils with distinct wet

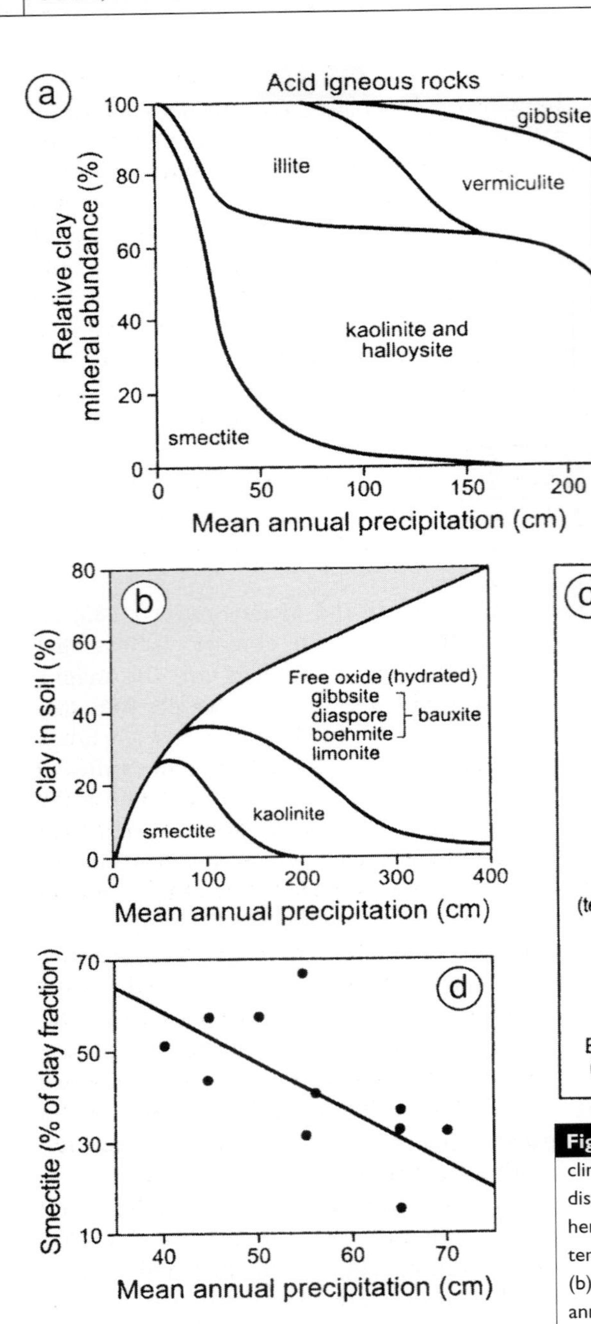

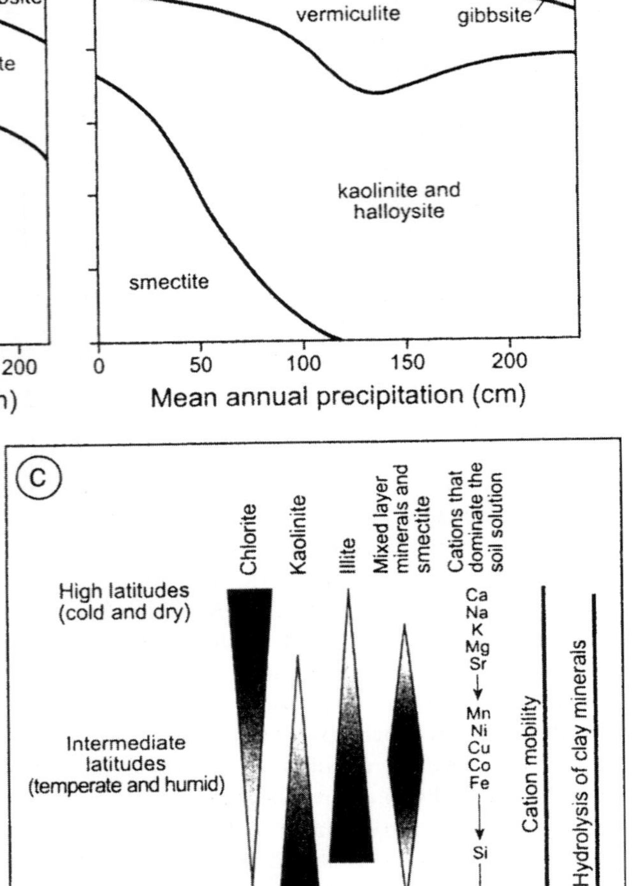

Fig. 15.11 Relationships between clay mineral suites in soils and climatic parameters. (a) Effect of precipitation on the frequency distribution of clay minerals in residual soils in California. The climate here is xeric, i.e., with a winter rainfall maximum, with mean annual temperatures between 10 and 15 °C. After Barshad (1966). (b) Generalized relationship between clay mineralogy and mean annual precipitation in a continuously wet climate (Hawaii). After Sherman (1952). (c) Generalized relationships between clay mineral suites and the latitudinal process drivers that influence them. After Yemane *et al.* (1996). (d) Relationship between smectite content in soils near Galilee (Israel) and mean annual precipitation. After Singer (1966).

and dry seasons (Singer and Navrot 1977) (see Chapter 10). Many soils on base-rich parent materials or with calcic horizons are rich in smectite. The smectite example illustrates the point that it is the *soil climate*, not the *regional atmospheric climate*, that directly affects clay mineralogy. And soil climate affects and drives clay mineralogy by impacting soil solution chemistry.

Other linkages between climate and clay mineralogy are also useful for paleoenvironmental reconstruction. Allophane forms in humid or at least semi-humid climates; in climates with a dry season where desiccation prevents allophane formation, halloysite forms instead. Vermiculite is an alteration product of biotite, forming where annual precipitation exceeds about 100 cm (Barshad 1966). In dry climates where base saturation values are high, illite weathers to smectite. In intense leaching environments, silica is depleted from the soil system and oxide clays come to dominate. Oxide minerals, e.g., goethite, gibbsite and hematite, are associated with strong weathering regimes with ample precipitation and strong desilication (van der Merwe and Weber 1963). The association of hematite-rich, deep, red soils with hot, dry conditions was discussed in Chapter 12 (Fig. 12.26). In slightly drier, but still intensively leached environments, bases and silica are leached from the soil system, leading to the formation of 1 : 1 clays such as kaolinite and halloysite (van der Merwe and Weber 1963, Ojanuga 1979). Palygorskite and sepiolite are associated with aridic climates, alkaline pH values and relatively high concentrations of Mg and Si (Bachmann and Machette 1977).

There are, however, some parent material limitations on clay mineral formation. For example, granitic rocks give rise to illite, regardless of climate (van der Merwe and Weber 1963, Singer 1980). Likewise, in areas where the parent material is slowly permeable, leaching is inhibited and the clay mineralogy may be reflective of a drier climatic regime; the same relationship holds on wet sites due to topography and a high water table (Kantor and Schwertmann 1974).

In a classic study, Folkoff and Meentemeyer (1987) identified the dominant clay minerals in 99 representative pedons in the United States (Fig. 15.12). They studied well-developed soils which had had time to develop and adjust to climatic imputs. Folkoff and Meentemeyer developed five statistical models, one each for mica, vermiculite, smectite, kaolinite and chlorite, based on correlations among the clay mineral suites and various soil (texture, drainage class, pH, etc.), parent material (acid rocks, basic rocks, till, loess, alluvium, carbonate rocks, etc.) and climatic variables (annual temperature, precipitation, water surplus and deficit variables, leaching indexes, etc.). For each clay mineral model, a climate variable was the first one selected by the regression equation, illustrating the strong link between clay mineralogy and climate. With the exception of the mica model, a climatic variable was also the second variable selected. Their models suggested the following:

Mica model Mica, particularly biotite, dominates in cold, arid climates, on calcareous parent materials. Its abundance is negatively correlated with annual precipitation, which accounted for 41% of the explained variation. This model suggests that as the climate gets colder and more arid, weathering is increasingly inhibited, allowing more weathering-susceptible mica to remain in the soil.

Vermiculite model Water surplus, an index of leaching intensity, was the variable most strongly correlated to vermiculite content. Leaching removes the interlayer K^+ cations from biotite, forming vermiculite. Seasonality and intensity of the dry season may also play a role, probably because when the soil dries, the unit structure of mica pulls away from the interlayer K^+, increasing its susceptibility to leaching (Scott and Smith 1968).

Smectite model Smectite formation is favored in low leaching environments with intense, long dry seasons. Its abundance is most strongly (but negatively) correlated to water surplus. Unlike mica, smectite formation does require a minimal level of leaching, tending to make it most common in dry climates with a short wet season and some types of warm and humid, continental climates. These are also areas where Vertisols are common (see Chapter 10).

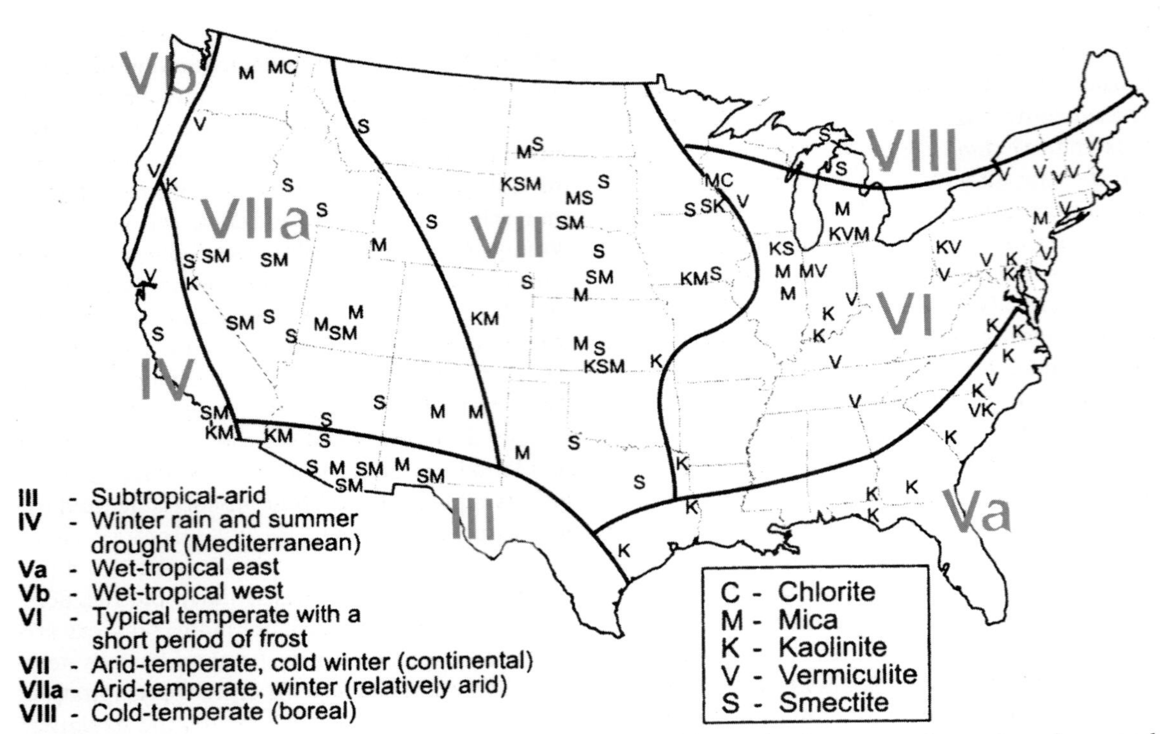

Fig. 15.12 Dominant clay minerals in the A horizons of 99 soils in the contiguous United States. The map is divided into climatic regions to show the strong relationship between climate and clay mineralogy. After Folkoff and Meentemeyer (1985, 1987).

Kaolinite/halloysite model These 1 : 1 minerals are optimally formed in warm, humid climates with a prolonged leaching season. The first variable to enter this model was mean annual temperature, while the second and third variables were positively correlated to leaching and water availability.

Chlorite model Chlorite is generally an unstable clay mineral; its relationship to climate is the weakest of the five discussed here. Its presence is limited to areas with weakly developed soils derived from chlorite-bearing parent materials, commonly in cold climates (Gao and Chen 1983, Yemane *et al.* 1996).

The linkages between clay mineralogy and environment of formation are complex and continue to be the focus of much study. It is not our intent to summarize this literature in great detail, but to simply point out that strong relationships exist between clay mineralogy and environment, particularly leaching regime and parent material, and that such knowledge can be highly useful in interpreting soil evolution.

However, to assess the *paleo*environmental significance of clay mineralogy, one must know not only the environmental conditions under which clays form (above), but also their resistance to alteration as the environment changes or after the soil is buried (Birkeland 1969). The clay mineral suite in older *surface* soils commonly reflects only the most long-lasting and intense paleoclimatic interval. In general, the low-silica minerals (kaolinite, gibbsite) formed in humid conditions will persist if there is a change to a drier or cooler environment. Conversely, the more siliceous minerals, perhaps inherited from an earlier cool or dry period, are likely to weather to less siliceous minerals in a wetter and warmer climate (Catt 1990). Thus, clay mineralogy is a one-way palimpsest; they reflect the most intense climate that the polygenetic soil has undergone. Nonetheless, if this caveat is taken into consideration, clay minerals are a good paleoenvironmental indicator for both surface and buried soils. The only drawbacks we note are (1) they do not provide a great deal of *detail* about the formative pedo-environment

(Fernández Sanjurjo *et al.* 2001) and (2) they are mobile and can be translocated into or out of paleosols. Clay minerals in *buried* soils have an advantage, however, over other paleoenvironmental proxies in that they are slow to change and undergo little post-burial alteration (Yaalon 1971), but unfortunately they tend to alter to other forms primarily along a unidirectional weathering sequence (see Chapter 9). Singer (1980) reviewed the advantages and potential problems in using clay mineralogy of buried soils to indicate paleoclimatic conditions. He cautioned that, because clay mineral suites change with depth in the profile and can reflect paleotopography as well as paleoclimate, they must be viewed carefully.

Cutans

The advantage of cutans as paleoenvironmental palimpsests lies in the fact that their signature is often not erased, but simply written over, building up as thick cutans. Also, multiple paleoenvironmental "signatures" are also less likely to get mixed when they are preserved as cutans, because the original layering is preserved (Pustovoytov 2002). In addition, cutans are frequently the consequence of illuvial processes that have a distinct climatic signature, e.g., lessivage, calcification, gypsification, podzolization (Jakobsen 1989). Argillans carry with them not only the climatic signature associated with a humid (but not perhumid) leaching regime with a short dry season, but their mineralogy can be used to further fine-tune paleoenvironmental interpretations. Thus, cutans provide a potentially powerful environmental pedo-signature (Chadwick *et al.* 1995). In some cases, however, cutans can be ephemeral features. Pedoturbation, for example, can destroy argillans (Gile and Grossman 1968, Nettleton *et al.* 1969), voiding their utility as a pedo-palimpsest.

Like many paleoenvironmental indicators, cutans have the most potential to be palimpsests in ecotonal areas (Busacca 1989). An ecotone is a transition zone, in this instance from one bioclimatic regime to another, e.g., grassland to forest. Subtle climatic shifts are recorded at the ecotone (as it shifts position) while the central part of the environmental region may not undergo noticeable change, and thus the soils within will not register a record of it. With this in mind, Reheis (1987a) studied soils in an area of Montana that experienced dramatic climatic shifts over the past 2 Ma, associated with advancing and retreating glaciers. During warm interglacials, soils accumulated calcans in the upper profile, while during the cool glacial periods argillans were overprinted onto the calcans. In the oldest soils, nine distinct cutan layers (five calcans alternating with four argillan layers) point to the highly dynamic nature of the climate in this region. A similar though less complicated polygenetic soil is described by Reider (1983) for an arid lowland to humid upland climosequence in Wyoming. Here, carbonates engulf argillans in the lower B horizon. The current climate is humid, leading one to suspect that the carbonate-forming episode dates to the mid-Holocene warm period. Dates obtained with ^{14}C on soil carbonates (5230 and 4735 BP) confirmed this assumption.

These examples highlight the fact that perhaps the most usable and well-preserved cutans in humid climates are argillans, and in dry climates calcans are of great utility (Kleber 2000). They can attain great thicknesses and are readily preserved. As discussed in Chapter 12, the depth to carbonates is a rough indicator of paleoprecipitation. Even upon a change to a more humid climate, some of the calcans may remain, especially on the undersides of clasts. The presence of pedogenic carbonates in buried paleosols provides a reliable indicator of a subhumid or drier environment of formation (Feng *et al.* 1994) and the amount of secondary carbonates might be related to the length or intensity of the soil-forming interval. The paleoenvironmental signatures contained *within* calcans are multifaceted (Courty *et al.* 1994). Chadwick *et al.* (1989b) demonstrated that the crystal form of calcite in carbonate coatings can reveal a climatic signature. In desert areas the micrite of the calcans takes the form of equant or parallel prismatic crystals while in moister areas the crystals are randomly oriented, euhedral, prismatic and fibrous. Pustovoytov (2002) suggested that the microlaminae of calcans provide clear paleoenvironmental information. Laminae of "pure" $CaCO_3$ with better calcite crystals indicate the driest climates with low biological activity whereas laminae with admixtures of organic matter and other minerals, containing poorly formed calcite crystals, are indicative of more humid paleoclimates (see also Pustovoytov and Targulian 1996).

Magnetic susceptibility

A useful paleoenvironmental proxy that has been widely applied in stacked loess paleosols is magnetic susceptibility (MS). Proportional to the amount of strongly magnetic minerals in the sediment or soil, MS is easily measured in the laboratory and the field, enabling a great number of samples to be analyzed in a few hours (Mullins 1977). High MS values in a soil or sediment are generally due to large amounts of magnetic minerals (ferrimagnets), particularly magnetite and maghemite but also titanomagnetites. The size of the ferrimagnetic minerals also affects the intensity of the MS signal (Mullins 1977).

Higher numbers of ferrimagnets in a soil or sediment are usually ascribed to either (1) dilution of surrounding sediments with materials low in magnetite and maghemite or (2) *in situ* alteration of the material, creating more ferrimagnets (Reynolds and King 1995). In the former case, it is assumed that magnetic enhancement occurs in soils buried by loess because dilution of the magnetic fallout occurs during periods of loess deposition (Kukla *et al.* 1988). The second explanation ascribes the increased MS to pedogenesis (Singer and Fine 1989, Zhou *et al.* 1990, Fine *et al.* 1993, Heller *et al.* 1993, Verosub *et al.* 1993). Indeed, increased amounts of ferrimagnets in a sediment column are most commonly associated with paleosols. Because intensity of the MS signal is often correlated to soil development, it is a highly useful tool for recognizing even the weakest paleosols within a stratigraphic column, as well as correlating between columns where paleosols are the primary type of stratigraphic marker bed (Jiamao *et al.* 1996). It is also a useful SED tool, as has been shown by Grimley (1998) for a loess column with intercalated paleosols of varying strength in Illinois (Fig. 15.13).

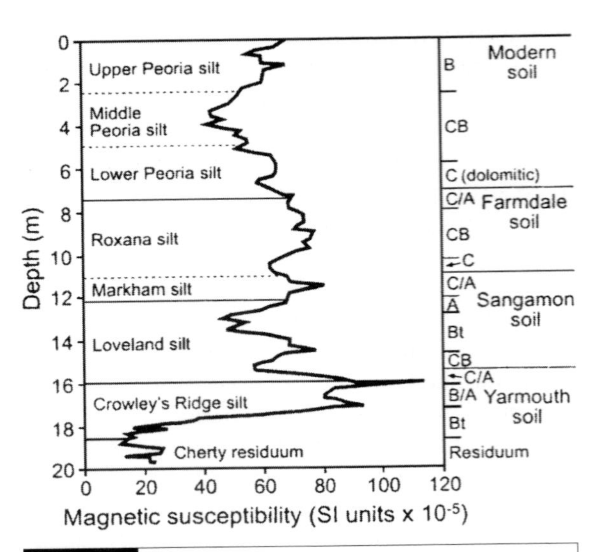

Fig. 15.13 Magnetic susceptibility and paleopedology of a loess–paleosol sequence from southern Illinois. Low susceptibility values for the Bt horizon on the Sangamon soil are ascribed to weathering of ferrimagnets under acidic or reducing conditions. After Grimley (1998).

Paleosols, especially those formed in loess, usually have higher MS values than their intervening parent material. However, in certain settings, where chemical reduction, weathering or chelation of iron was a dominant process, e.g., Spodosols or poorly drained soils, a buried or surface soil may actually exhibit *lower* MS values than the overlying C horizon (Hayward and Lowell 1993, Grimley 1998). A buried Spodosol, for example, may have high quantities of iron, but because it exists in chelated forms the soil will have low MS. This example makes the point that there is often little direct correlation between iron content and MS values (Fine *et al.* 1995, Maher 1998). Think of the relationship between iron content and MS in this way: Fe contents are rarely a limiting factor for magnetic enhancement; instead, *opportunities for ferrimagnetic conversion* and *preservation* may be the key constraint. It follows, then, that we need to know what pedogenesis actually *does* to create ferrimagnets, and in what kinds of soil environments MS is particularly enhanced, if we are to use it as a paleoenvironmental indicator (Verosub *et al.* 1993).

The formation of ferrimagnets by pedogenesis has been the topic of considerable study. Ferrimagnet formation is perhaps most favored in well-drained, organic-rich, mildly acidic soils forming in weatherable, Fe-bearing (but *not* Fe-rich) materials, in a warm climate (Grimley 1998, Maher 1998). The first mineral *formed*, by precipitation from the soil solution, is commonly magnetite or maghemite. Maher (1998) listed several reasons why MS is increased by pedogenesis, the most important being "fermentation," in which iron-reducing bacteria in soils reduce ferric iron. It is then (later) oxidized to ultrafine-grained magnetite, which then may or may not be subsequently oxidized to maghemite (Grimley 1998). More than 30 species of Fe-reducing bacteria have already been identified in soils (Maher 1998). However, if the iron is released from silicates is already in reduced (ferrous) form, reduction by bacteria is not necessary.

Magnetite is susceptible to reduction and removal from the profile. For that reason, the *persistence* of the MS signal is dependent on a mechanism to convert magnetite to a more stable form – maghemite. Maghemite formation is favored by oxidizing conditions or excessive heating of the soil, as with fires. Therefore, the optimal conditions for the formation of ferrimagnets involves a warm, wet–dry climate – magnetite is formed by weathering during periods of intermittent soil wetness and is oxidized to maghemite in the dry season (Maher 1998). In soils where ferrimagnet production and preservation are favored, there appears to be a correlation between the maximum value of MS and annual rainfall, at least for situations in which the annual precipitation exceeds about 140 cm (Vidic and Verosub 1999). Ferrimagnet depletion (or non-formation) occurs in permanently wet soils, acid Spodosols, hyperarid soils and those forming in the humid tropics where rainfall exceeds 2 m yr^{-1}.

In unaltered sediments, the amount of ferrimagnets is dependent upon the source sedimentology and, therefore, the source area (Grimley 1998). For this reason, any comparisons of MS among various soils or paleosols must strive for parent material uniformity; fortunately, this is common in sequences of stacked paleosols in loess. Thus, it should come as no surprise that the MS signal in the 150-m thick Chinese loess column, with its many intercalated paleosols, has been fruitful for MS study. The MS signals of the soils in the sequence are not only highly correlated to pedogenesis, but also have a strong climatic signal (Zhou *et al.* 1990, Verosub *et al.* 1993, Chlachula *et al.* 1998). Generally, paleosols in the loess–paleosol sequence correspond to interglacial periods in the marine record, indicating a strong climatic coupling of marine and continental climatic processes (Liu 1985, Hovan *et al.* 1989). The MS signal here correlates so well to pedogenesis because many properties have come together to create a situation where highly contrasting amounts of ferrimagnets are formed (or not) in the column as the glacial–interglacial cycles switch on and off. The loess parent material is uniform and rich in Fe-bearing primary minerals, so that the signal can be ascribed to pedogenesis and not depositional anomalies. Most of the Chinese paleosols have also formed under well-drained conditions, which has favored preservation of pedogenically formed ferrimagnets. Vegetation has covaried with climate (Maher 1998). The climate of formation was wet–dry (monsoonal), which led to near-neutral pH values in the loess soils. Thus, soils within the loess column record intensity of pedogenesis, which here is a reflection of paleorainfall (Fig. 15.14b). The amount of paleorainfall, in turn, correlates to the intensity of the summer monsoon, which changed in concert with the glacial cycles recorded in the marine isotope record. And lastly, the well-drained nature of the soils facilitated the preservation of the pedogenically formed ferrimagnets.

The paleoclimatic information we have gleaned from the >200-m thick Chinese loess record, which spans 2.4 Ma, is truly amazing. Its use has been particularly important to our understanding of the regional climate–geomorphic shifts, but also because it has reinforced the accuracy of the marine oxygen isotope record as a climatic proxy (Fig. 15.14). The record shows not only changes in paleoclimate through time (expressed vertically in the loess column) but also through space (Maher *et al.* 1994; see also Busacca 1989). Data gleaned from this record include not only susceptibility and morphological–paleopedological information from the many paleosols within, but also information on the paleofloristic composition of past landscapes based on isotopic data from carbonates contained within the buried soils (Ding and Yang 2000). Study and correlation of the China loess and marine isotope records have shown that cool,

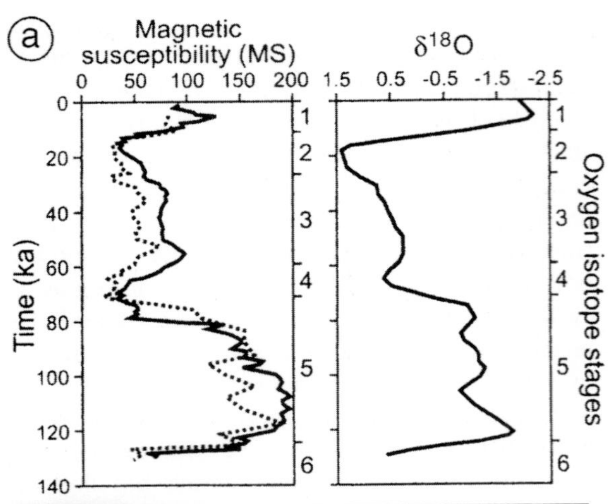

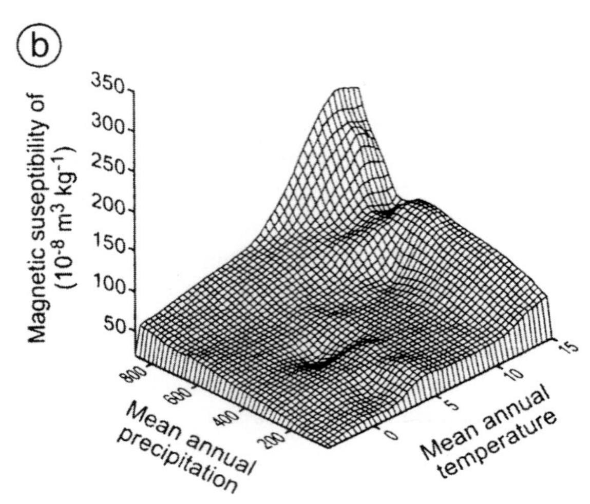

Fig. 15.14 Illustration of the utility of magnetic susceptibility in buried soils and sediments as a paleoenvironmental indicator. (a) Comparison of the magnetic susceptibility of the Chinese loess at Xifeng and Luochuan (after Kukla *et al.* 1988) and the marine oxygen isotope record (Prell *et al.* 1986). (b) Magnetic susceptibility of modern soils in the loess plateau of China, as it relates to contemporary values of precipitation and temperature. After Jiamao *et al.* (1996).

glacial periods in the Late Pleistocene loess were times of loess deposition and worldwide dust transport, while during interglacials the landscape stabilized and soil formation occurred (Hovan *et al.* 1989). These soils would later be buried and preserve within them a record of pedogenesis and regional geomorphic stability.

The role of duration vs. intensity of pedogenesis applies to studies of magnetic susceptibility. Are high values of MS due to a *longer* period of pedogenesis (and hence, a longer period of surface stability) or are they due to a more *intense* period of pedogenesis (perhaps a warmer and wetter soil-forming interval)? For magnetic susceptibility, or any other parameter, to reflect short-term changes in paleoclimate, it helps if the pedogenic regime favors rapid development, with soils reaching a steady state with regard to climate in a few thousand years or less. This may or may not be the case over parts of the Chinese loess plateau, where pedogenic properties reflect *intensity* of pedogenic drivers, in this case climate. Conversely, a steady, continually developing soil property, one that requires many thousands of years to reach a steady state, will reflect *duration* of soil development more than intensity. Thus, clay mineralogy would not be a good paleorainfall proxy within the loess column, but it might reflect length of surface exposure for individual paleosols better than MS does. This rule of thumb about intensity vs. duration is, however, countermanded by studies that show that magnetic susceptibility in soils continues to increase with time (Singer *et al.* 1992). Maher (1998) dodged this bullet by suggesting that, as soils weather and thicken, *cumulative*, i.e., profile-weighted, susceptibility might increase but the *maximum* susceptibility of any one horizon might attain an equilibrium value with climate.

Carbon-13 in soil carbonate and organic matter

Like ^{14}C, ^{13}C is a carbon isotope that can be used to great advantage in soil geomorphology. And also like ^{14}C, ^{13}C accumulates in biomass and soil carbonate. Unlike ^{14}C, ^{13}C is a stable isotope.

Natural isotopic fractionations in plant tissue and soil carbonate are so small that they are commonly reported in parts per thousand (‰). The isotopic composition of the sample being measured is expressed as delta ^{13}C ($\delta^{13}C$), which represents the *parts per thousand* difference (per mille) between the sample's $^{13}C/^{12}C$ ratio and the same ratio for the international PDB standard

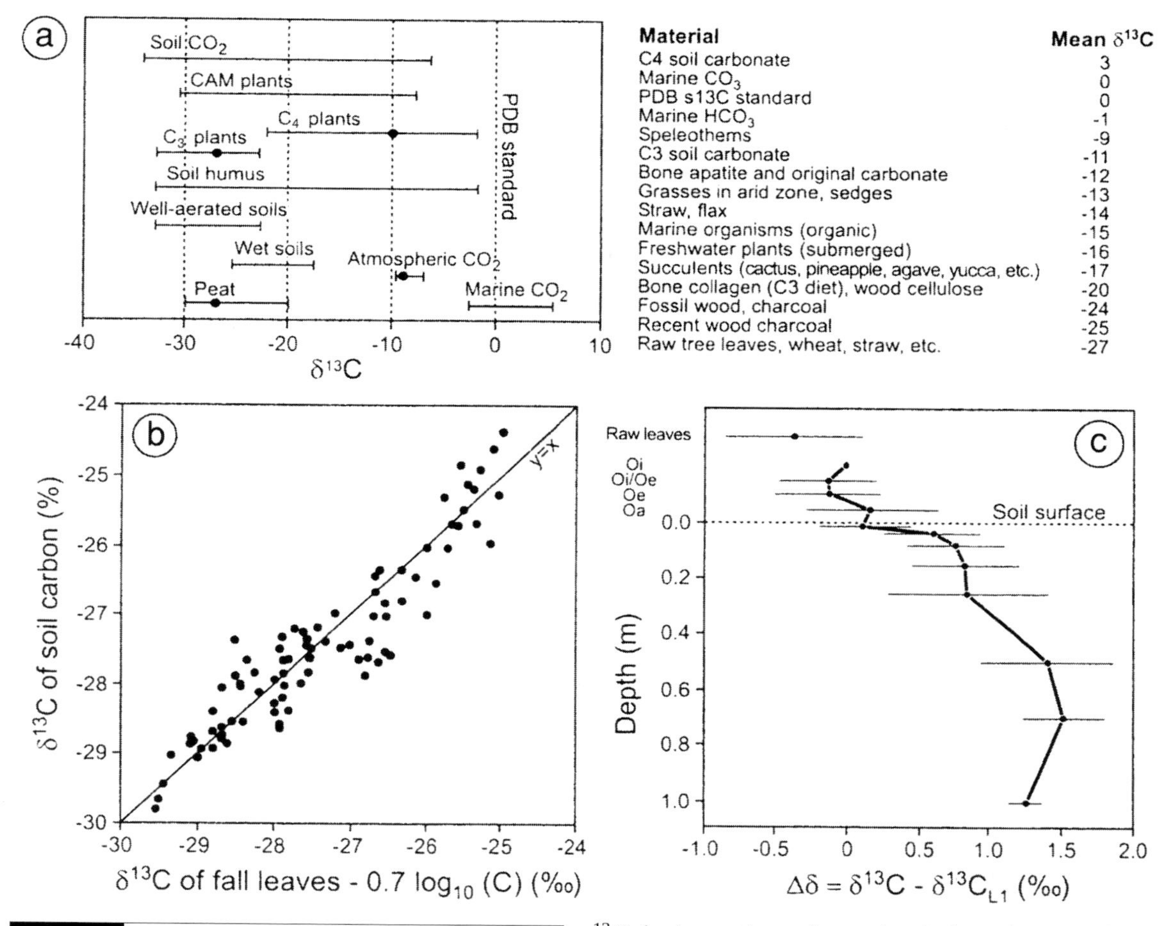

Material	Mean $\delta^{13}C$
C4 soil carbonate	3
Marine CO_3	0
PDB s13C standard	0
Marine HCO_3	-1
Speleothems	-9
C3 soil carbonate	-11
Bone apatite and original carbonate	-12
Grasses in arid zone, sedges	-13
Straw, flax	-14
Marine organisms (organic)	-15
Freshwater plants (submerged)	-16
Succulents (cactus, pineapple, agave, yucca, etc.)	-17
Bone collagen (C3 diet), wood cellulose	-20
Fossil wood, charcoal	-24
Recent wood charcoal	-25
Raw tree leaves, wheat, straw, etc.	-27

Fig. 15.15 Values of $\delta^{13}C$ and soils. (a) Values of $\delta^{13}C$ for some important carbon compounds and reservoirs. After Pfeiffer and Janssen (1994). (b) Correlations between $\delta^{13}C$ contents of recently fallen leaves and organic matter for the same soil. After Balesdent *et al.* (1993). (c) Mean isotopic ($\delta^{13}C$) enrichment or depletion with depth in some Alfisols, Inceptisols and Spodosols in France, as compared to that of the O horizon. After Balesdent *et al.* (1993).

carbonate (PDB refers to a Cretaceous belemnite formation at Peedee, South Carolina.):

$$\delta^{13}C\ (‰) = [^{13}C/^{12}C\ (\text{sample}) - {}^{13}C/^{12}C\ (\text{standard})]/[^{13}C/^{12}C\ (\text{standard})] \cdot 1000.$$

For most plants, the atmosphere and pedogenic carbonate, the $\delta^{13}C$ value is negative (Fig. 15.15a), indicating that they contain less of the ^{13}C isotope than does the belemnite standard (Troughton 1972).

The $\delta^{13}C$ ratio of humus and carbonates in soils and paleosols contains a paleobiotic signature. The signature for soil organic matter is easier to interpret than it is for carbonate. The $\delta^{13}C$ value in soil humus is reflective of the types of plants that contributed to it (Tieszen *et al.* 1997). Humus derived from forest litter, for example, has a much more negative $\delta^{13}C$ value than does humus from a shortgrass prairie (Fig. 15.15a). Values of $\delta^{13}C$ in soils can also be used to examine carbon dynamics and turnover (Balesdent *et al.* 1987, Martin *et al.* 1990).

Plants are grouped into one of three metabolism pathways, based on how they utilize CO_2. Plants require CO_2 for photosynthesis, and take it in through through small pores in their leaves called stomata. When plants open their stomata, however, they risk losing water vapor through those same vents. Put simply, they can't have it both ways, and therefore plants have adapted mechanisms to utilize CO_2 efficiently, given their particular environmental constraints. For example, under hot and dry environmental conditions the stomata close during the daytime to reduce the loss of water vapor, but this also results in a greatly diminished intake of CO_2.

There are three pathways of carbon fixation: C3, C4 and the CAM (crassulacean acid metabolism) (Smith and Epstein 1971, Ehleringer and Monson 1993). Most humid climate plants use the Calvin (C3) cycle to fix carbon dioxide; it is the "default" pathway for all trees, most shrubs and herbs and many grasses. The stereotypical photosynthetic plant is called a C3 plant because the first stable compound formed from CO_2 is a three-carbon compound at the beginning of the Calvin cycle. Values of $\delta^{13}C$ for C3 plants range from −20‰ to −35‰ with a mean value of −27‰. In the C4 pathway the photosynthetic cycle is generally restricted to cells that are interior to the plant, reducing the need to have open stomata on their exteriors. The C4 (aka Hatch–Slack) pathway is also more CO_2 efficient than the C3 pathway. Plants that use this pathway are known as C4 plants because the initial carboxylation reaction in photosynthesis produces a four-carbon compound. C4 plants fix CO_2 so efficiently that they do not need to have their stomata open as much as plants operating by other pathways, enabling them to be more water-efficient. This pathway is mostly found among tropical grasses and some sedges and herbs growing in warm, sunny environments; it allows for more efficient carbon fixation in dry, warm environments. C3 plants outcompete C4 plants in moist, colder and less sunny environments. C4 plants generally grow better than C3 plants in warm or dry, arid climates, whereas C3 plants grow better than C4 plants in cool, moist climates. Although fewer than 1% of plant species use the C4 photosynthetic pathway, they are important in temperate prairies, tropical savannas and arid grasslands. Values of $\delta^{13}C$ for C4 plants range from −9‰ to −17‰ with a mean value of −13‰.

A related pathway, CAM, is found in plants that live in very dry, desert-like conditions, e.g., cacti and succulents, as well as some tropical succulents. It is the least common pathway. The name points to the fact that this pathway occurs mainly in succulent plants of the Crassulaceae and Cactaceae families. CAM plants have adapted to the dry conditions by opening their stomata only at night, whereupon they store CO_2 in their tissue. During the day when the stomata are closed (avoiding unnecessary loss of water vapor), the CO_2 is removed from storage and enters into photosynthetic reactions, which are fueled by light energy from the sun.

The basic assumption used to determine the paleovegetation of a paleosol or surface soil, from soil organic matter contained within, is that the storehouse of organic matter in the soil or paleosol *isotopically reflects* the vegetation that existed when the soil was forming (Guillet *et al.* 1988). Most plants in humid and/or cool climates, and most shade-tolerant plants, follow the C3 pathway while most C4 plants are in warm, semiarid or subhumid climates (Ehleringer and Monson 1993) (Fig. 15.15a). A rule of thumb for grassland soils is that the warmer the climate is, the more C4 plants are likely to be in that grassland. The $\delta^{13}C$ values derived from soil organic matter, along a climatic gradient, follow this rule (Koch 1998). For example, Quade and Cerling (1990) noted that, along an altitudinal transect in a desert climate, the plants changed from CAM and C4 plants to C3 plants at higher altitudes. The $\delta^{13}C$ values of pedogenic carbonates within the soils varied as well: about zero in the creosote bush–desert holly zone at the base of the mountain, about −7 in the pinyon–juniper shrubland upslope, and −9 in the ponderosa pine forests still farther upslope. It should also be noted that, within soils, there is a tendency for the $\delta^{13}C$ value to become slightly less negative (1–2‰) with depth (Fig. 15.15c).

Knowing the types of plants that typify the C3 and C4 pathways is also useful in conjunction with ^{14}C analysis (Guillet *et al.* 1988). Most ^{14}C laboratories routinely report the $\delta^{13}C$ value of samples, allowing the investigator the opportunity to know both sample age *and* likely floristic composition of the site at a particular period in the past. We caution that bulk soil organic carbon can only yield long-term mean $\delta^{13}C$ values, which may be difficult to interpret in polygenetic soils.

Like cutans, the $\delta^{13}C$ method is particularly advantageous in areas where environmental shifts force a change in biota that have one pathway to biota that have another pathway. For example, glacial–interglacial climatic shifts on Great Plains grasslands may have forced a biotic change from C3-dominated forest to C4 grasses. The organic matter in the soils of such places may contain an isotopic signal of that paleobotanical/paleoclimatic change. This method is particularly useful in ecotonal areas, where small paleoenvironmental changes are easily registered, e.g., where grassland has invaded forest or vice versa (Steuter *et al.* 1990, Ambrose and Sikes 1991). Paleosols formed under one botanical association often get buried when the climate changes, because periods of climatic change are also periods of geomorphic instability. The end result might be a buried soil with carbon isotope ratios indicative of the pre-burial paleoenvironment, and the modern soil (or a soil higher in the stratigraphic column) that is reflective of the environment at a later time. Such paleosols would carry carbon isotopes containing information about the paleovegetation that occupied the site while they formed, and by proxy, information about paleoclimate (Khokhlova *et al.* 2001).

Applications of $\delta^{13}C$ values of organic matter in surface soils are also interesting. Because the isotopic composition of soil organic matter remains independent of the vegetative cover for some time after a vegetative change, the method can be used to track the extent of recent, or even ongoing, floristic shifts. Steuter *et al.* (1990) did precisely this, on a landscape where C3 trees were presumably invading a C4 grassland. They were able to show that, in sites that had been invaded by forest in the recent past, the isotopic composition of roots was significantly more negative than that of the soil organic matter. On sites that had been invaded less recently, the isotopic difference was less.

Because pedogenic *carbonate carbon* is derived largely from soil CO_2, which in turn derives primarily from decay of soil organic matter and respiration from plant roots, its $\delta^{13}C$ value can also provide a useful paleoenvironmental signature – one which is especially applicable in dry climates. The $\delta^{13}C$ values of pedogenic carbonate are usually correlated to the type and density of the overlying flora (Amundson *et al.* 1988, Koch 1998, Ding and Yang 2000). They are also, however, correlated to soil respiration rates, which are low in deserts and cold climates, in which case much of the soil CO_2 could have come from the atmosphere and is therefore less reflective of the flora (Dörr and Münnich 1986). Therefore, the $\delta^{13}C$ values of soil carbonate can range from that of the atmosphere to that of the existing, more isotopically negative, vegetation (Emrich *et al.* 1970) (Figs. 15.15a, 15.16). In humid and subhumid areas, where C3 and C4 plants dominate, soil respiration is high and thus little atmospheric CO_2 mixes with that of the soil, meaning that the isotopic composition of soil carbonate (if any actually forms) is more reflective of the flora (Cerling 1984). In arid climates, however, where soil respiration rates are low, the $\delta^{13}C$ values of soil carbonate may be related to the density of vegetation and therefore, rates of soil respiration (Amundson *et al.* 1988, Wang *et al.* 1996b). Along these lines, Cerling *et al.* (1989) pointed out that the relationship between the isotopic composition of flora and carbonates is less useful in desert soils (<35 mm annual precipitation) because they have respiration rates that are so low that diffusion of atmospheric CO_2 into the soil

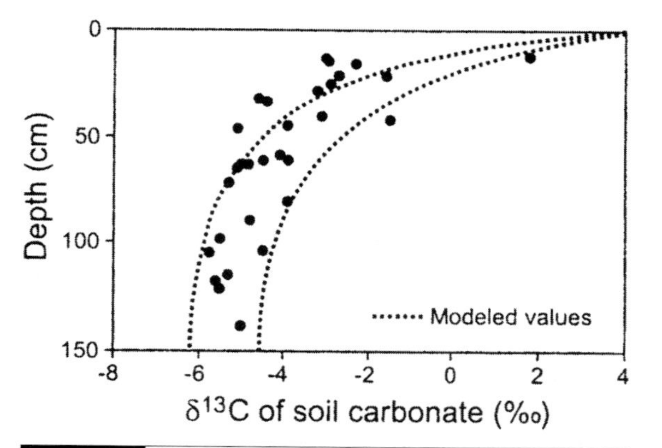

Fig. 15.16 Variation in $\delta^{13}C$ values for soil carbonate with depth, in some desert soils in southern California. Predicted values, based on a diffusion model, are also shown. After Wang *et al.* (1996b).

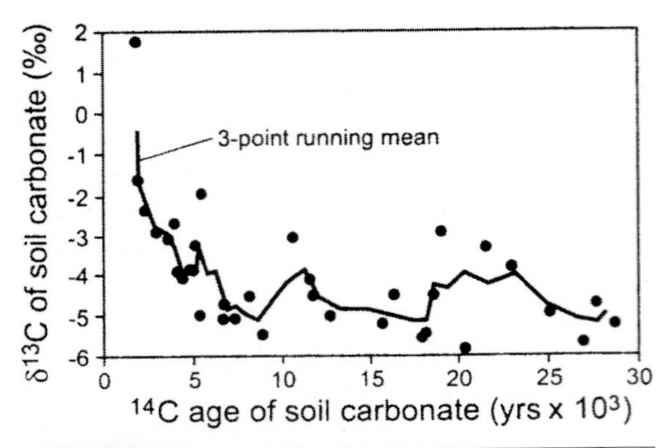

Fig. 15.17 Relationship between the $\delta^{13}C$ values and ^{14}C ages of soil carbonates for some desert soils in southern California. Note the shift to less negative values shortly after the beginning of the Holocene (last 10 ka). After Wang *et al.* (1996b).

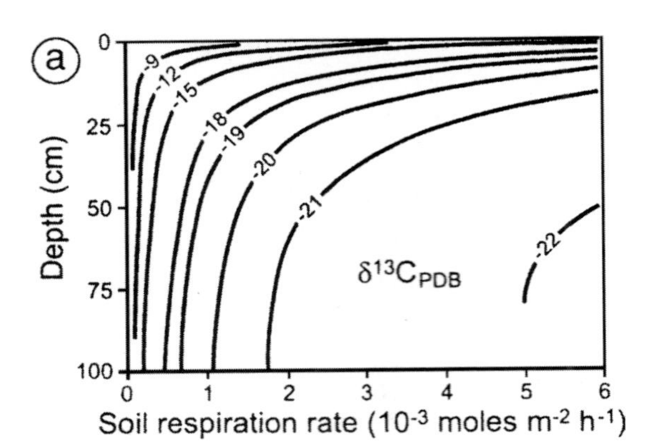

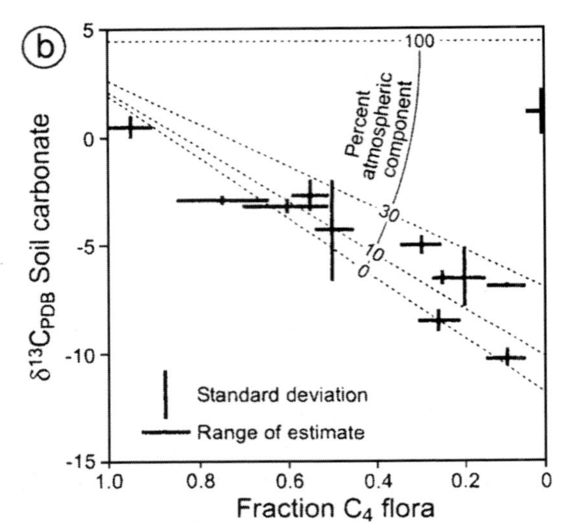

Fig. 15.18 Isotopic composition of soil carbonate and paleoenvironments. After Cerling (1984). (a) Calculated steady state isotopic composition of soil CO_2 for different soil respiration rates, assuming that the $\delta^{13}C$ values for the atmosphere and soil air are −6‰ and −27‰, respectively. (b) Isotopic composition of soil carbonate compared to the fraction of C4 plants in the local flora.

raises the isotopic values above what would be expected from the flora alone.

This circumstance, obviously, presents a potential problem in the use of soil carbonate isotopes[1] for paleoenvironmental reconstruction. Data from Wang *et al.* (1996b) show clearly that, despite the potential problems in the method, the isotopic composition of soil carbonates can be used to provide information on paleoenvironments (Fig. 15.17). They documented a change in isotopic composition of soil carbonates after about 9 ka, which suggests either an increase in C4 and CAM plants or a decrease in plant density. Either way, the trend is toward increasing climatic dryness. Cerling (1984) developed equations that can be used to determine the fraction of C4 plants in the local flora, using the isotopic composition of soil carbonate (Fig. 15.18). He assumed that the $\delta^{13}C$ values for the atmosphere and soil air (the source of which is plant root respiration and decaying litter) were −6‰ and −27‰, respectively. Using these assumptions, he was able to calculate the steady-state isotopic composition of soil carbonate under various C4–C3 floral "mixes."

For reasons too complex to discuss here, the isotopic composition of pedogenic carbonate precipitated in equilibrium with soil CO_2 should be enriched by 14‰ to 17‰ relative to CO_2 respired from plant roots and soil microbes (Fig. 15.19). That is, isotopic values of *soil organic matter* are

[1] This discussion has focussed on the $\delta^{13}C$ values of soil organic matter and carbonate as paleoenvironmental proxies. The reader should also be aware of a third type of isotope ($\delta^{18}O$) the values of which in carbonates are primarily controlled by the $\delta^{18}O$ values of soil water. On a general level, therefore, there is a relationship between the $\delta^{18}O$ values of precipitation and that of soil carbonate (Cerling 1984), making this isotope a potentially usable indicator of paleoclimate. However, because the oxygen isotopes of desert soil carbonates are still fairly poorly understood (Barnes and Allison 1983, Wang *et al.* 1996b), we chose not to examine them in detail.

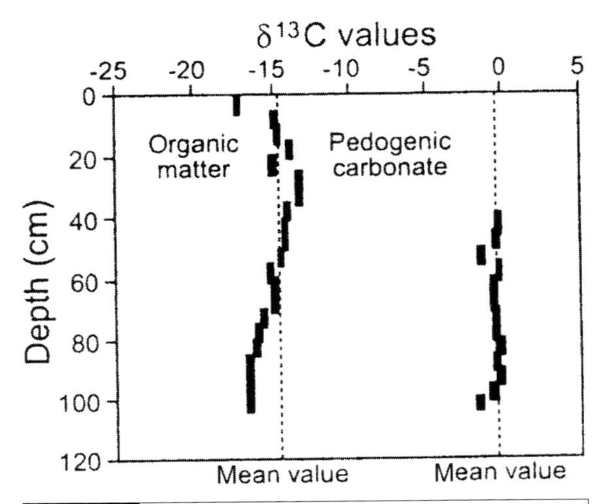

Fig. 15.19 Depth plots of the isotopic composition of soil organic matter and pedogenic carbonate from an Argiustoll in Kansas. After Cerling *et al.* (1989).

preserved with little or no isotopic fractionation, while *pedogenic carbonate* is enriched in ^{13}C by 14‰ to 17‰. This means that the $\delta^{13}C$ value for soil *organic matter* is a direct reflection to the flora that produced it, while *carbonate* values are shifted by 14‰ to 17‰ (Cerling 1984, Quade *et al.* 1989). This shift will increase as the environment gets drier and, to a lesser extent, colder, largely due to the increased effects of atmospheric CO_2. Under conditions of moisture stress, therefore, $\delta^{13}C$ values of soil carbonate of −8‰ and +4‰ indicate nearly pure C3 and C4 ecosystems, respectively (Ding and Yang 2000).

Our discussion has focussed on secondary pedogenic carbonate, which should not be confused with carbonate inherited from parent material. The $\delta^{13}C$ values of soil carbonate can actually help distinguish between the two. Carbonate derived from dissolution of limestone has near-zero to slightly positive $\delta^{13}C$ values, whereas purely pedogenic carbonate, where the main C source is respiration from plant roots, has $\delta^{13}C$ values that are much more negative (Magaritz and Amiel 1980, Rabenhorst *et al.* 1984, Quade and Cerling 1990) (Fig. 15.15a). Marine carbonate has a much higher $\delta^{13}C$ value than does pedogenic carbonate (Fig. 15.15a). Thus, in soils where some of the pedogenic carbonate has resulted from the dissolution of limestone, it should be evident because the caliche will have an abnormally high $\delta^{13}C$ value. However, it is also likely that, once the marine carbonate dissolves, it equilibrates, partially or wholly, with soil-respired CO_2. Pedogenic carbonate formed by reprecipitation of dissolved marine carbonates may therefore have the respiration isotopic signature and not the marine carbonate signature. As always, interpretations must be performed based on knowledge of diffusion coefficients for $^{12}CO_2$ and $^{13}CO_2$ (Cerling *et al.* 1989).

Use and interpretation of isotopic values for soil carbonate has a distinct advantage over organic matter, since carbonates can persist in an unaltered state despite burial or slight changes in climate, whereas organic matter is quickly oxidized and lost (Amundson *et al.* 1989b). Carbonate is readily preserved in dry paleosols, giving the $\delta^{13}C$ value of carbonates from paleosols particular application to paleoenvironmental study (Cerling and Hay 1986, Amundson *et al.* 1989b). However, there are several potential complications. In soils with shallow Bk horizons, much of the soil CO_2 is derived from the atmosphere, which has a less negative $\delta^{13}C$ value than CO_2 derived from the decay of organic matter, regardless of the type of plant (Fig. 15.15a). Soil carbonate at depth will have little of this influence, with most of its carbon having been derived from plant sources. There is also the problem of "overprinting" or the imposition of later climatic information contained in soil carbonate onto information from much older carbonate-generating climatic cycles (Cerling 1984). Polygenesis is common in desert environments; it remains to be seen whether isotopic analysis of soil carbonate will ultimately help to resolve these climatic and floristic changes, or whether the method will be defeated by them.

Opal phytoliths

Phytoliths are microscopic mineral, usually silica, deposits formed in the epidermal cells of plants, especially grasses. They are formed as the plants uptake monosilicic acid through their roots (Fig. 15.20). The silica precipitates as water is transpired from the plant. Most of the silica precipitates in the leaves; some phytoliths are formed in roots. Phytoliths range in size from 2 to 1000 μm, but most are between 20 and 200 μm in length. In transmitted light, phytoliths are usually translucent, yellowish-brown in color. Opal (silica-rich)

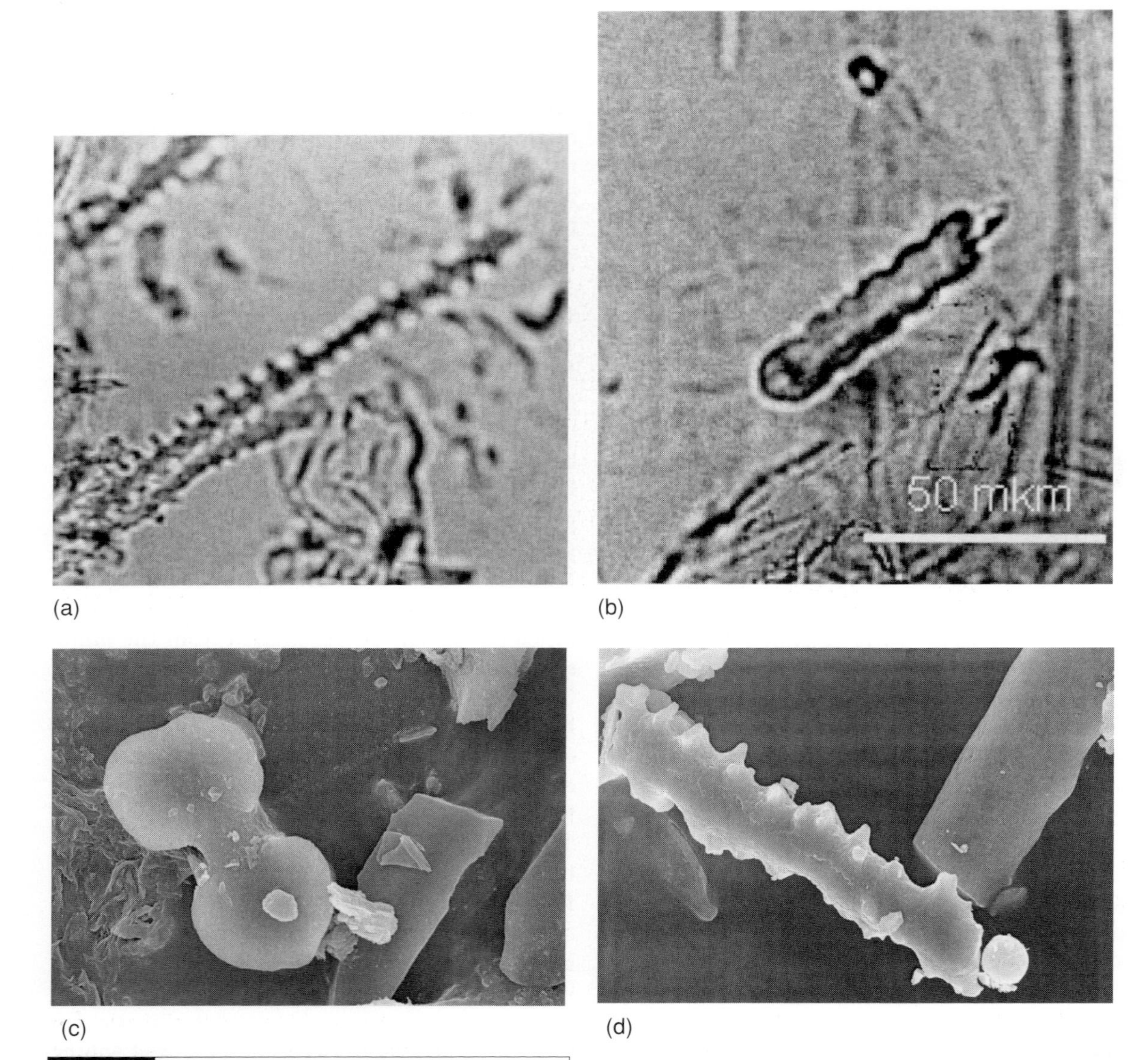

Fig. 15.20 Examples of opal phytoliths, viewed under the light microscope ((a) and (b)) and the scanning electron microscope ((c) and (d)). Images by M. Blinnikov. (a) Strongly indented *Agropyron spicatum*-type phytolith. (b) *Stipa*-type bilobates. (c) *Calamagrostis rubescens*. (d) *Festuca idahoensis*.

phytoliths have been the subject of most study, although calcium phytoliths also occur, especially in cacti (Jones and Handreck 1967, Franceschi and Horner 1980). Phytoliths are liberated from plants by normal decay processes and enter the soil with plant litter. Because they are fairly resistant to decomposition, they accumulate in soils (Smithson 1958, Baker 1959, Fredlund *et al.* 1985, Zhao and Pearsall 1998), mostly in A horizons where roots are most dense (Beavers and Stephen 1958) and within the coarse silt fraction.

Like pollen grains (below), opal phytoliths can be identified to plant genus or even species, providing for paleobotanical and paleoenvironmental proxy data (Fredlund and Tieszen 1997). Examinination of the phytolith content of a soil provides a signature of the flora that once inhabited it. As with pollen, phytolith study requires the development of a regional phytolith (specimen) database before it can be successfully

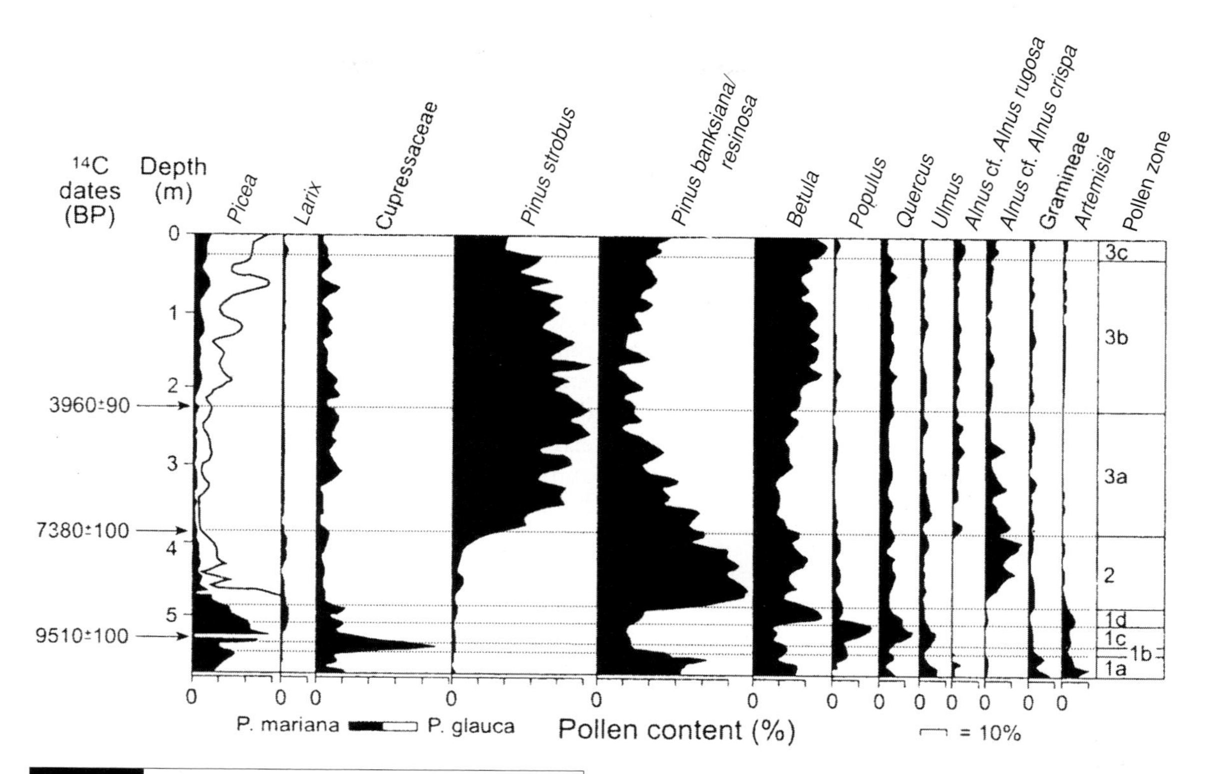

Fig. 15.21 Pollen percentage diagram for Nina Lake in central Ontario, just north of Lake Huron. Because of space limitations, not all the pollen spectra originally reported by Liu (1990) are shown here.

implemented. Phytoliths have an advantage over pollen because they do not rely on lake and swamp sediments – some landscapes have few such wetland deposits (Lewis 1981, Piperno and Becker 1996). Unlike pollen grains, which accumulate in wet sites and reflect the regional "pollen rain," phytoliths are more site specific, coming directly from the plants that inhabited a few square meters area. Phytoliths are potentially datable by radiocarbon, although they require a significant amount of pretreatment. However, they rarely occupy more than about 3% of a horizon by volume, making isolation of individual phytoliths a laborious and painstaking task (Wilding 1967).

The combination of fairly widespread production coupled with resistance to decomposition gives phytoliths good potential as paleoenvironmental indicators (Rovner 1971, 1988). In the first instance, their abundance can be an indication of the long-term dominance of grasses vs. forest on a site, or of changes in grass vs. forest dominance (Beavers and Stephen 1958, Witty and Knox 1964, Reider *et al.* 1988, Fuller and Anderson 1993). Additionally, phytoliths provide information on the types of grasses that have been at the site, as they can be taxonomically identified to various grass genera. Mulholland and Rapp (1992) reported that the following families are consistent accumulators of opal phytoliths: Poaceae or Gramineae (grasses), Cyperaceae (sedges), Ulmaceae (elm trees), Fabaceae or Leguminosae (beans and peas), Cucurbitaceae (squash) and Asteraceae or Compositae (sunflowers).

Content of opal phytoliths can also be used for the identification of buried paleosols, and to help locate the very top of a buried soil (since they are most concentrated in the A horizon) (Beavers and Stephen 1958, Dormaar and Lutwick 1969). They can also be used to determine if a horizon rich in organic matter is a buried A horizon or simply a Bh horizon (Anderson *et al.* 1979). An Ab horizon, especially if it had been formed under grass, would have high numbers of phytoliths, with decreases in phytolith occurrence below.

Alternatively, a Bh horizon would contain fewer phytoliths than the horizons above it, because root numbers (and hence, phytolith production) would have been higher in overlying horizons, and because phytoliths are not normally translocated in soils.

Pollen and macrofossils

Pollen and macrofossils have many characteristics in common with opal phytoliths. All are produced by plants and are somewhat resistant to decomposition in sediments or soils, thereby providing a record of the paleovegetation. Pollen is particularly resistant to decomposition in acidic environments like peat bogs. In this book, we provide only a succinct treatment of pollen and its field of study, *palynology* (Bryant and Holloway 1985, Faegri and Iverson 1989). The reason for this centers not only on the great breadth of the field but also on the fact that most pollen is extracted from organic soils in bogs, making the analysis perhaps more sedimentological than pedological. Only occasionally has pollen been extracted from upland mineral soils, and these studies have shown mixed results (Dimbleby 1957, Havinga 1974, Schaetzl and Johnson 1983). Preservation is poor in mineral soils, often because of abrasion as well as neutral or high pH values. Stratification is soon lost due to pedoturbation and translocation processes (Ray 1959, Walch *et al.* 1970, Kelso 1974). Only in soils with pH values <5.5 is there any significant preservation of pollen, and even then one must be wary of preferential preservation which will distort the palynological signal. Faunal macrofossils in mineral soils are, however, somewhat useful as paleoenvironmental indicators, e.g., diatoms and sponge spicules generally suggest severe waterlogging. Examples of macrofossils might include seeds, fragments of stems and roots, or leaf buds or scales.

The primary way that palynological studies contribute to soil genesis/geomorphology is by providing paleoecological information for segments of the geologic past. In a typical palynological study, a core of sediment is removed from stratified, acidic, lacustrine sediments or peat. Radiocarbon dates on *in situ* organic materials (charcoal is especially useful because it is less easily contaminated) from the core provide a chronostratigraphic framework for the pollen and/or macrofossils extracted from it. The palynologist operates on the assumption that the pollen and macrofossils for each layer within the core are representative of the regional or local vegetation at that moment in the past, or statistically determines the relationship between modern pollen rain and modern plant communities and then uses those data to back-forecast pollen–flora relationships (Liu and Lam 1985, Overpeck *et al.* 1985). Obviously, this assumption is violated somewhat by differential preservation and because some plants, particularly conifer trees, produce pollen in amounts that far exceed their abundance in the local flora. Presumably, these problems can be eliminated statistically to some extent.

The final step in the process is the development of a depth plot of pollen – a pollen diagram. Examination of the pollen diagram often provides a reasonably clear picture of the temporal changes in vegetation that have occurred at a site (Maher 1972). These vegetation changes can then be correlated to climate, paleoindian activities, natural plant succession, soil development, or any of a number of other factors (Baker *et al.* 1989). Figure 15.21 provides an example of a typical pollen diagram from Ontario. The diagram, with its pollen assemblage biozones, illustrates the climate-induced changes in Holocene vegetation rather nicely – from a spruce-dominated boreal forest assemblage to a jack pine–red pine community to the current white pine–white birch forest. Similar examples abound in the literature (e.g., Watts and Wright 1966, Davis 1969, Webb 1974, Davis *et al.* 1975, Birks 1976, Bartlein *et al.* 1984).

Chapter 16

Conclusions and Perspectives

With the exception of ice-covered surfaces, areas of bare bedrock and perhaps a few others, soils of one kind or another exist on every part of the Earth's land surface. The wide variety and range of expression that soils have is truly remarkable, especially when one considers that this plethora of colors, mineral assemblages and horizon types and sequences is due to the interaction of only a few (5) soil-forming factors. In mathematics, 5 is a small number but 5^5 is much, much larger. So it is in soils, as those five soil-forming factors can team together in myriad ways to form a world of soils that is complex, spatially diverse and many-faceted. Unraveling that world, or at least some of the better-understood parts of it, is a goal of this book.

Almost all of our food and sustenance comes from the soil. The oxygen that we breathe come from plants, most of which are rooted in the soil. Soil is one of the best natural filters we have for contaminated surface waters. Many Christians may know that the name "Adam" comes from the Hebrew *adama*, which means earth, or soil. Adam's name, in Genesis, is meant to capture humankind's intimate link with the soil, to which we are tied through our entire life and to which we return upon our death (Hillel 1991). Clearly, soils have been a part of humankind for as long as we have existed and will be here long after we are gone. The importance of soils cannot be overestimated.

And despite this, soils are usually treated with indifference, disdain or lack of respect (Hillel 1991). We call it "dirt." Our clothes get "soiled." Paving over cropland is considered a form of "progress." How many people do you know that really *care* about soils? That see them as *essential* to our existence? That consider these natural systems *fascinating*? That like to smell and taste and examine them closely? We hope that, after reading this book, you would add yourself to that list! We hope that this book has made you a land and soil lover (Hole 1997). We hope we have in some small way enlightened you, as the giants of the field have done for us, in even larger ways. We mention especially Vasili Dokuchaev, Curtis Marbut, Guy Smith, Hans Jenny, Roy Simonson, James Thorp, Kirk Bryan, John Frye, Gerry Richmond, Ray Daniels, Bob Ruhe, Roger Morrison, Lee Gile, Francis Hole, Dan Yaalon, Don Johnson and Pete Birkeland, among many, many others.

With few exceptions, most soils have horizons. The near omnipresence of genetic soil horizons immediately below the Earth's land surface implies that soils represent a degree of stability and equifinality. To be sure, some types of horizons form rather quickly, but in most cases the typical "soil" that one finds, with its distinct O, A and B horizons, should send the signal that pedogenesis has been ongoing and that this soil's profile is forming on a relatively stable location, from a geomorphic standpoint. The characteristics of the soil, especially its colors and the relationship of these characteristics to nearby soils, should provide additional evidence that this soil has evolved on, within and concurrently with its landscape. That is a major point of this book: soils form on landscapes and it is best to study the two systems together. That is, indeed, the essence of soil geomorphology, which forms the core of this book. Soil genesis is more easily explained if we know about the landscape upon which the soil is forming. And each landscape has its own unique geography, pointing to the undeniable conclusion that knowing the *geography* of soils is vital to understanding the *essence* of the physical landscape.

But where does the discipline of soil geomorphology and the study of soil genesis go from here? What are the challenges and wherein lie the greatest potentials for advancing our understanding of the soil geomorphic system? First, we emphasize dating. Dating of the $time_{zero}$ for soils is essential to understanding many aspects of its genesis, including the rates of many pedogenic processes. Knowing $time_{zero}$ with some degree of accuracy is also a highly useful piece of information from a geomorphic standpoint. Much progress has been made in the arena of dating, but vast potentials remain yet to be realized. There is much more to be done.

Biologists commonly lament, and rightfully so, the rapid rate at which the human species, by its actions, is wiping out other species across the globe. Extinctions occur daily. Few dispute the fact that this is highly regrettable. But what of soils? Unique soil types are destroyed as well (Amundson et al. 2003), and information from *them* is lost forever. Our position here is not necessarily one of soil preservation, as one might do for buffalo, red-legged frogs or Peregrine falcons. Instead we need to realize that there are many, many soils across the globe that have never been studied, some of which we do not even know *exist*! Consider this a call for soil scientists, soil geomorphologists and pedologists everywhere to beat the corners of the globe for unique and interesting soils to study. Often, the study of unique, almost freak-like soils can lead to major advances in pedogenic theory, e.g., Jenny et al. (1969). Travel, travel, travel, to the ends of the earth, especially travel to and through the tropics, and glean what you can! There is much to learn, opportunities to be had, and no time to lose.

The definition of soil should be broadened, especially with regard to soil depth, but also regarding the definition of a soil horizon. Are stone lines soil horizons, since most are pedogenically formed? Many "C horizons" have been highly altered by pedogenic processes. Stone lines at great depth in tropical soils are almost always pedogenic features. We should rethink the definition of *soil* and broaden it, so as to bring more practitioners "into the fold." More geologists and geomorphologists would and could "use" soils and contribute to soil knowledge if they realized (or were permitted to acknowledge) that pedogenic processes operate to great depth in many landscapes.

Allow soil research and knowledge to drive its taxonomy, not vice versa. Taxonomic systems should work *for* soil users, researchers and theorists. The pendulum is swinging in this direction already and this trend should continue. Taxonomy is necessary and it is good. But it is the *tail* on the horse, not the horse.

Basic research in soil genesis and geomorphology should continue to be encouraged. Applied soils research is laudable and necessary, but the trend toward more and more soils-based research becoming clearly "applied" in not necessarily healthy, especially if taken too far. The long-term applications of basic research can rarely be perceived at the time it is done, particularly in a discipline like soils where the system is still only minimally understood and most scholars acknowledge its extreme complexity.

Lastly, we encourage all scholars of soils and geomorphology to be firmly educated in the literature, resolute field scientists, but above all free thinkers. "Thinking outside the box" has become such a common platitude that we tend to ignore it or downplay it as something we either all do anyway, or as a useless dead end. We disagree. This book is dedicated to two soil scientists and free thinkers, Francis Hole and Donald Johnson, both of whom were thinking outside the box before it was popular. Many heroes of science invariably operated at least partly outside of the box. Unfortunately, they are often (now) dead heroes, because it often takes a generation or more after they die before they are heroized. We exhort all soil geomorphologists to be free thinkers, to think of new and exciting ways to better study, explore and understand the soil system. The rewards will be many, for you and for science.

Soils are important. We have made that point. Soils are interesting, exciting and complex. Surely you know that by now. Soils are genetically intertwined with the landscapes on which they have formed. That is our mantra. The soil system is a highly complex one, and we are optimistic that our treatise has explained some of these complexities to you, the reader. Thank you for your trust and attention.

References

Aandahl, A. R. 1948. The characterization of slope positions and their influence on the total nitrogen content of a few virgin soils of western Iowa. *Soil Sci. Soc. Am. Proc.* **13**: 449–454.

Aber, J. S. 1991. The glaciation of northeastern Kansas. *Boreas* **20**: 297–314.

1999. Pre-Illinoian glacial geomorphology and dynamics in the central United States, west of the Mississippi. *Geol. Soc. Am. Spec.* Paper **337**: 113–119.

Abrahams, A. D., Parsons, A. J., Cooke, R. U., *et al.* 1984. Stone movement on hillslopes in the Mojave Desert, California: a 16-year record. *Earth Surf. Proc. Landforms* **9**: 365–370.

Abrahams, P. W. and J. A. Parsons. 1996. Geophagy in the tropics: a literature review. *Geog. J.* **162**: 63–72.

1997. Geophagy in the tropics: an appraisal of three geophagical materials. *Env. Geochem. Health* **19**: 19–22.

Abtahi, A. 1980. Soil genesis as affected by topography and time in highly calcareous parent materials under semiarid conditions in Iran. *Soil Sci. Soc. Am. J.* **44**: 329–336.

Acquaye, D. K., Dowuona, G. N., Mermut, A. R., *et al.* 1992. Micromorphology and mineralogy of cracking soils from the Accra Plains of Ghana. *Soil Sci. Soc. Am. J.* **56**: 193–201.

Acton, D. F. 1965. The relationship of pattern and gradient of slopes to soil type. *Can. J. Soil Sci.* **45**: 96–101.

Acworth, R. I. 1987. The development of crystalline basement aquifers in a tropical environment. *Q. J. Eng. Geol.* **20**: 265–272.

Adams, K. D., Locke, W. W., and R. Rossi. 1992. Obsidian-hydration dating of fluvially reworked sediments in the West Yellowstone region, Montana. *Quat. Res.* **38**: 180–195.

Adams, W. A. 1973. The effect of organic matter on the bulk and tree densities of some uncultivated podzolic soils. *J. Soil Sci.* **24**: 10–17.

Afanasiev, J. N. 1927. *The Classification Problem in Russian Soil Science*, Russian Pedological Investigation no. 5. Moscow, USSR Academy of Science.

Afanas'yeva, Ye. A. 1966. Thick Chernozems under grass and tree coenoses. *Sov. Soil Sci.* 615–625.

Aghamiri, R. and D. W. Schwartzman. 2002. Weathering rates of bedrock by lichens: a mini watershed study. *Chem. Geol.* **188**: 249–259.

Ahlbrandt, T. S. and S. G. Fryberger. 1980. *Eolian Deposits in the Nebraska Sand Hills*, Prof. Paper no. 1120. Washington, DC, US Geological Survey.

Ahmad, N. 1983. Vertisols. In L. P. Wilding, N. E. Smeck and G. F. Hall (eds.) *Pedogenesis and Soil Taxonomy*. New York, Elsevier, pp. 91–123.

Ahmad, N. and R. L. Jones. 1969. Genesis, chemical properties and mineralogy of Caribbean grumusols. *Soil Sci.* **107**: 166–174.

Ahmad, N., Jones, R. L., and A. H. Beavers. 1966. Genesis, mineralogy and related properties of West Indian soils. I. Bauxite soils of Jamaica. *Soil Sci. Soc. Am. Proc.* **30**: 719–722.

Ahmadjian, V. 1993. *The Lichen Symbiosis*. New York, John Wiley.

Ahn, J. H. and D. R. Peacor. 1987. Kaolinitization of biotite: TEM data and implications for an alteration mechanism. *Am. Mineral.* **72**: 353–356.

Ahnert, F. 1970. An approach towards a descriptive classification of slopes. *Z. Geomorph.* **9**: 71–84.

1987. Process-response models for denudation at different spatial scales. *Catena (Suppl.)* **10**: 31–50.

1996. *Introduction to Geomorphology*. London, Edward Arnold.

Aitken, M. J. 1994. Optical dating: a non-specialist review. *Quat. Geochron.* **13**: 503–508.

Ajmone Marsan, F. and J. Torrent. 1989. Fragipan bonding by silica and iron oxides in a soil from northwestern Italy. *Soil Sci. Soc. Am. J.* **53**: 1140–1145.

Ajmone Marsan, F., Barberis, E., and E. Ardiuno. 1988. A soil chronosequence in northwestern Italy: morphological, physical and chemical characteristics. *Geoderma* **42**: 51–64.

Akamigbo, F. 1984. The role of the nasute termites in the genesis and fertility of Nigerian soils. *Pédologie* **36**: 179–189.

Alban, D. H. 1974. *Soil Variation and Sampling Intensity under Red Pine and Aspen in Minnesota*, Research Paper no. NC-106. St Paul, MN, US Forest Service.

Albert, D. A. and B. V. Barnes. 1987. Effects of clearcutting on the vegetation and soil of a sugar maple-dominated ecosystem, western upper Michigan. *For. Ecol. Mgt* **18**: 283–298.

Alcock, M. R. and A. J. Morton. 1985. Nutrient content of throughfall and stem-flow in woodland recently established on heathland. *J. Ecol.* **73**: 625–632.

Alden, W. C. and M. M. Leighton. 1917. The Iowan drift, a review of the evidences of the Iowan stage of glaciation. *Iowa Geol. Survey Ann. Rept* **26**: 49–212.

Aleinikoff, J. N., Muhs, D. R., Sauer, R. R., *et al.* 1999. Late Quaternary loess in northeastern Colorado. II. Pb isotopic evidence for the variability of loess sources. *Geol. Soc. Am. Bull.* **111**: 1876–1883.

Alekseyev, V. Ye. 1983. Mineralogical analysis for the determination of podzolization, lessivage, and argillation. *Sov. Soil Sci.* **15**: 21–28.

Aleva, G. J. J. 1965. The buried bauxite deposits of Onverdacht, Suriname, South America. *Geol. Mijnbouw* **44**: 45–58.

1983. On weathering and denudation of humid tropical interfluves and their triple planation surfaces. *Geol. Mijnbouw* **62**: 383–388.

1987. Occurrence of stone-lines in tin-bearing areas in Belitung, Indonesia, and Rondônia, Brazil. *Intl J. Trop. Ecol. Geog.* **11**: 197–203.

Alexander, E. B. 1980. Bulk densities of California soils in relation to other soil properties. *Soil Sci. Soc. Am. J.* **44**: 689–692.

1982. Volume estimates of coarse fragments in soils: a combination of visual and weighing techniques. *J. Soil Water Cons.* **37**: 62–63.

1985. Estimating relative ages from iron-oxide/total-iron ratios of soils in the western Po Valley, Italy: a discussion. *Geoderma* **35**: 257–259.

1986. Stones: an earth scientist's conception. *Soil Surv. Horiz.* **27**: 15–17.

1990. Influences of inorganic substrata on chemical properties of Lithic Cryofolists in southeast Alaska. *Soil Surv. Horiz.* **31**: 92–99.

1991. Soil temperatures in forest and muskeg on Douglas Island, southeast Alaska. *Soil Surv. Horiz.* **32**: 108–116.

Alexander, L. T. and J. G. Cady. 1962. *Genesis and Hardening of Laterite in Soils*, US Department of Agriculture. Tech. Bull. no. 1282.

Alexander, J. B., Beavers, A. H., and P. R. Johnson. 1962. Zirconium content of coarse silt in loess and till of Wisconsin age in northern Illinois. *Soil Sci. Soc. Am. Proc.* **26**: 189–191.

Alexander, L. T., Cady, J. G., Whittig, L. D., *et al.* 1956. Mineralogical and chemical changes in the hardening of laterite. *Proc. 6th Intl Congr. Soil Sci. Paris* **11**: 67–72.

Alexander, E. B., Mallory, J. I., and W. L. Colwell. 1993. Soil–elevation relationships on a volcanic plateau in the southern Cascade Range, Northern California, USA. *Catena* **20**: 113–128.

Alexander, E. B., Ping, C. L., and P. Krosse. 1994. Podzolisation in ultramafic materials in southeast Alaska. *Soil Sci.* **157**: 46–52.

Alexandrovskaya, E. I. and A. L. Alexandrovskaya. 2000. History of the cultural layer in Moscow and accumulation of anthropogenic substances in it. *Catena* **41**: 249–259.

Al-Farraj, A. and A. M. Harvey. 2000. Desert pavement characteristics on wadi terrace and aluvial fan surfaces: Wadi Al-Bih, UAE and Oman. *Geomorphology* **35**: 279–297.

Allan, R. J. and F. D. Hole. 1968. Clay accumulation in some Hapludalfs as related to calcareous till and incorporated loess on drumlins in Wisconsin. *Soil Sci. Soc. Am. Proc.* **32**: 403–408.

Allen, B. L. and B. F. Hajek. 1989. Mineral occurrence in soil environments. In: J. B. Dixon and S. B. Weed (eds.) *Minerals in Soil Environments*, 2nd edn, Madison, WI, Soil Science Society of America, pp. 331–378.

Allen, B. L. and E. P. Whiteside. 1954. The characteristics of some soils on tills of Cary and Mankato age in Michigan. *Soil Sci. Soc. Am. Proc.* **18**: 203–206.

Allen, C. C. 1978. Desert varnish of the Sonoran Desert: optical and electron probe microanalysis. *J. Geol.* **86**: 743–752.

Allen, J. R. L. 1970. A quantitative model of grain size and sedimentary structures in lateral deposits. *Geol. J.* **7**: 129–146.

Allison, R. J. and G. E. Bristow. 1999. The effects of fire on rock weathering: some further considerations of laboratory experimental simulation. *Earth Surf. Proc. Landforms* **24**: 707–713.

Almendinger, J. C. 1990. The decline of soil organic matter, total-N, and available water capacity following the late-Holocene establishment of jack pine on sandy Mollisols, north-central Minnesota. *Soil Sci.* **150**: 680–694.

Almond, P. C. and P. J. Tonkin. 1999. Pedogenesis by upbuilding in an extreme leaching and weathering environment, and slow loess accretion, south Westland, New Zealand. *Geoderma* **92**: 1–36.

Aloni, K. and J. Soyer. 1987. Cycle des matériaux de construction des termitières d'humivores en savane au Shaba méridional (Zaïre). *Rev. Zool. Africa* **101**: 329–358.

Aloni, K., Malaisse, F., and I. Kapinga. 1983. Rôles des termites dans la décomposition du bois et la transfert de terre dans une forét claire zambézienne. In P. Lebrun, H. André, A. de Medts, *et al.* (eds.) *New Trends in Soil Biology*. Louvain-la-Neuve, Dieu-Brichart, pp. 600–602.

Alonso, P., Sierra, C., Ortega, E., *et al.* 1994. Soil development indices of soils developed on fluvial terraces (Peñaranda de Bracamonte, Salamanca, Spain). *Catena* **23**: 295–308.

Alonso-Zarza, A. M., Silva, P. G., Goy, J. L., *et al.* 1998. Fan-surface dynamics and biogenic calcrete development: interactions during ultimate phases of fan evolution in the semiarid SE Spain (Murcia). *Geomorphology* **24**: 147–167.

Amba, E. A., Smeck, N. E., Hall, G. F., *et al.* 1990. Geomorphic and pedogenic processes operative in soils of a hillslope in the unglaciated region of Ohio. *Ohio J. Sci.* **90**: 4–12.

Ambrose, G. J. and R. B. Flint. 1981. A regressive Miocene lake system and silicified strandlines, in northern Australia: implications for regional stratigraphy and silicrete genesis. *J. Geol. Soc. Austral.* **28**: 81–94.

Ambrose, S. H. and N. E. Sikes. 1991. Soil carbon isotope evidence for Holocene habitat change in the Kenya Rift Valley. *Science* **253**: 1402–1405.

Amit, R. and R. Gerson. 1986. The evolution of Holocene reg (gravelly) soils in deserts – an example from the Dead Sea region. **13**: 59–79.

Amit, R. and J. B. J. Harrison. 1995. Biogenic calcic horizon development under extremely arid conditions, Nizzana Sand Dunes, Israel. *Adv. GeoEcol.* **28**: 65–88.

Amit, R. and D. H. Yaalon. 1996. The micromorphology of gypsum and halite in reg soils: the Negev Desert, Israel. *Earth Surf. Proc. Landforms* **21**: 1127–1143.

Amit, R., Gerson, R., and D. H. Yaalon. 1993. Stages and rate of the gravel shattering process by salts in desert Reg soils. *Geoderma* **57**: 295–324.

Amit, R., Harrison, J. B. J., and Y. Enzel. 1995. Use of soils and colluvial deposits in analyzing tectonic events: the southern Arava Rift, Israel. *Geomorphology* **12**: 91–107.

Amit, R., Harrison, J. B. J., Enzel, Y., *et al.* 1996. Soils as a tool for estimating ages of Quaternary fault scarps in a hyperarid environment: the southern Arava Valley, the Dead Sea Rift, Israel. *Catena* **28**: 21–45.

Amoozegar, A. and A. W. Warrick. 1986. Hydraulic conductivity of saturated soils. In A. Klute (ed.) *Methods of Soil Analysis*, Part 1, *Physical and Mineralogical Methods*, Agronomy Monograph no. 9, 2nd edn. Madison, WI, American Soil Association and Soil Science Society of America, pp. 735–770.

Amos, D. F. and E. P. Whiteside. 1975. Mapping accuracy of a contemporary soil survey in an urbanizing area. *Soil Sci. Soc. Am. Proc.* **39**: 937–942.

Amundson, R. 1998. Discussion of the paper by J. D. Phillips. *Geoderma* **86**: 25–27.

Amundson, R. G., Chadwick, O. A., Sowers, J. M., *et al.* 1988. Relationship between climate and vegetation and the stable carbon isotope chemistry of soils in the eastern Mojave Desert, Nevada. *Quat. Res.* **29**: 245–254.

1989a. Soil evolution along an altitudinal transect in the eastern Mojave Desert of Nevada, USA *Geoderma* **43**: 349–371.

1989b. The stable isotope chemistry of pedogenic carbonates at Kyle Canyon, Nevada. *Soil Sci. Soc. Am. J.* **53**: 201–210.

Amundson, R. G., Graham, R. C., and E. Franco-Vizcaino. 1997. Orientation of carbonate laminations in gravelly soils along a winter/summer precipitation gradient in Baja California, Mexico. *Soil Sci.* **162**: 940–952.

Amundson, R. Guo, Y., and P. Gong. 2003. Soil diversity and land use in the United States. *Ecosystems* **6**: 470–482.

Amundson, R. G., Wang, Y., Chadwick, O. A., *et al.* 1994. Factors and processes governing the ^{14}C content of carbonate in desert soils. *Earth Planet. Sci. Lett.* **125**: 385–405.

Anderson, D. W. 1979. Processes of humus formation and transformation in soils of the Canadian Great Plains. *J. Soil Sci.* **30**: 77–84.

Anderson, D. W. 1987. Pedogenesis in the grassland and adjacent forests of the Great Plains. *Adv. Soil Sci.* **7**: 53–93.

Anderson, D. W., De Jong, E., and D. S. McDonald. 1979. The pedogenetic origin and characteristics of organic matter of Solod soils. *Can. J. Soil Sci.* **59**: 357–362.

Anderson, D. W., Paul, E. A., and R. J. St. Arnaud. 1974. Extraction and characterization of humus with reference to clay-associated humus. *Can. J. Soil Sci.* **54**: 317–323.

Anderson, H. A., Berrow, M. L., Farmer, V. C., *et al.* 1982. A reassessment of Podzol formation processes. *J. Soil Sci.* **33**: 125–136.

Anderson, J. U., Bailey, O. F., and D. Rai. 1975. Effects of parent materials on the genesis of Borolls and Boralfs in south-central New Mexico mountains. *Soil Sci. Soc. Am. Proc.* **39**: 901–904.

Anderson, J. U., Fadul, K. E., and G. A. O'Connor. 1973. Factors affecting the coefficient of linear extensibility in Vertisols. *Soil Sci. Soc. Am. Proc.* **37**: 296–299.

Anderson, K., Wells, S., and R. Graham. 2002. Pedogenesis of vesicular horizons, Cima Volcanic Field, Mojave Desert, California. *Soil Sci. Soc. Am. J.* **66**: 878–887.

Anderson, M. G. and T. P. Burt. 1978. The role of topography in controlling throughflow generation. *Earth Surf. Proc.* **3**: 331–334.

Anderson, R. S., Repka, J. L., and G. S. Dick. 1996. Explicit treatment of inheritance in dating depositional surfaces using *in situ* ^{10}Be and ^{26}Al. *Geology* **24**: 47–51.

Anderson, S. P. 1988. The upfreezing process: experiments with a single clast. *Geol. Soc. Am. Bull.* **100**: 609–621.

Anderton, J. B. and W. L. Loope. 1995. Buried soils in a perched dunefield as indicators of Late Holocene lake-level change in the Lake Superior basin. *Quat. Res.* **44**: 190–199.

Andrews, J. T. 1987. The Late Wisconsin glaciation and deglaciation of the Laurentide ice sheet. In W. F. Ruddiman and H. E. Wright, Jr. (eds.) *North America and Adjacent Oceans during the Last Deglaciation. Geol. Soc. Am.* Vol. K-3. pp. 13–37. Boulder, CO.

Andriesse, J. P. 1969. A study of the environment and characteristics of tropical podzols in Sarawak (East-Malaysia). *Geoderma* **2**: 201–226.

Anduaga, S. and G. Halffter. 1991. Beetles associated with rodent burrows (Coleoptera: Scarabaedae, Scararaeinae). *Folia Entomol. Mexicana* **81**: 185–197.

Antoine, P. P. 1970. Mineralogical, chemical and physical studies on the genesis and morphology of a Rockwood sand loam (Typic Fragiboralf). Ph.D. dissertation, University of Minnesota.

Applegarth, M. T. and D. E. Dahms. 2001. Soil catenas of calcareous tills, Whiskey Basin, Wyoming, USA. *Catena* **42**: 17–38.

Aran, D., Gury, M., and E. Jeanroy. 2001. Organo-metallic complexes in an Andisol: a comparative study with a Cambisol and a Podzol. *Geoderma* **99**: 65–79.

Arbogast, A. F. 1996. Stratigraphic evidence for Late-Holocene aeolian sand mobilization and soil formation in south-central Kansas, USA. *J. Arid Envs.* **34**: 403–414.

Arbogast, A. F. 2000. Estimating the time since final stabilization of a perched dune field along Lake Superior. *Prof. Geog.* **52**: 594–606.

Arbogast, A. F. and D. R. Muhs. 2000. Geochemical and mineralogical evidence from eolian sediments for northwesterly mid-Holocene paleowinds, central Kansas, USA. *Quat. Intl* **67**: 107–118.

Arbogast, A. F., Scull, P., Schaetzl, R. J., *et al.* 1997. Concurrent stabilization of some interior dune fields in Michigan. *Phys. Geog.* **18**: 63–79.

Arbogast, A. F., Wintle, A. G., and S. C. Packman. 2002. Widespread middle Holocene dune formation in the eastern Upper Peninsula of Michigan and the relationship to climate and outlet-controlled lake level. *Geology* **30**: 55–58.

Arduino, E., Barberis, E., Ajmone Marsan, F., 1986. Iron oxides and clay minerals within profiles as indicators of soil age in northern Italy. *Geoderma* **37**: 45–55.

Ardiuno, E., Barberis, E., Carraro, F., *et al.* 1984. Estimating relative ages from iron-oxide/total-iron ratios of soils in the western Po Valley, Italy. *Geoderma* **33**: 39–52.

Aristarain, L. F. 1970. Chemical analyses of caliche profiles from the High Plains, New Mexico. *J. Geol.* **78**: 201–212.

Arkley, R. J. 1963. Calculation of carbonate and water movement in soil from climatic data. *Soil Sci.* **96**: 239–248.

1967. Climates of some Great Soil Groups of the western United States. *Soil Sci.* **103**: 389–400.

Armson, K. A. and R. J. Fessenden. 1973. Forest windthrows and their influence on soil morphology. *Soil Sci. Soc. Am. Proc.* **37**: 781–783.

Arnalds, O. and J. Kimble. 2001. Andisols of deserts in Iceland. *Soil Sci. Soc. Am. J.* **65**: 1778–1786.

Arnalds, O., Hallmark, C. T., and L. P. Wilding. 1995. Andisols from four different regions of Iceland. *Soil Sci. Soc. Am. J.* **59**: 161–169.

Arno, S. F. 1979. *Forest Regions of Montana.* US Department of Agriculture Forest Service Research Paper INT-218. Ogden, UT.

Arnold, R. W. 1968. Pedological significance of lithologic discontinuities. *Trans. 9th Intl Congr. Soil Sci.* (Adelaide) **4**: 595–603.

Arnold, R. W. 1983. Concepts of soils and pedology. In L. P. Wilding, N. E. Smeck and G. F. Hall (eds.) *Pedogenesis and Soil Taxonomy.* New York, Elsevier, pp. 1–21.

Arnold, R. W. 1994. Soil geography and factor functionality: interacting concepts. In *Factors of Soil Formation: A Fiftieth Anniversary Retrospective*, Special Publication no. 33. Madison, WI, Soil Science Society of America, pp. 99–109.

Arnold, R. W. and F. F. Riecken. 1964. Grainy gray ped coatings in Brunizem soils. *Proc. Iowa Acad. Sci.* **71**: 350–360.

Arocena, J. M. and S. Pawluk. 1991. The nature and origin of nodules in Podzolic soils from Alberta. *Can. J. Soil Sci.* **71**: 411–426.

Arshad, M. A. and S. Pawluk. 1966. Characteristics of some solonetzic soils in the Glacial Lake Edmonton basin of Alberta. I. Physical and chemical. J. Soil Sci. **17**: 36–47.

Arshad, M. A., Schnitzer, M., and C. M. Preston. 1988. Characterization of humic acids from the termite mounds and surrounding soils, Kenya. *Geoderma* **42**: 213–225.

Asadu, C. L. A. and F. O. R. Akamigbo. 1987. The use of abrupt changes in selected soil properties to assess lithological discontinuities in soils of eastern Nigeria. *Pédologie* **37**: 43–56.

Asady, G. H. and E. P. Whiteside. 1982. Composition of a Conover–Brookston map unit in southeastern Michigan. *Soil Sci. Soc. Am. J.* **46**: 1043–1047.

Asamoa, G. K. and R. Protz. 1972. Influence of discontinuities in particle size on the genesis of two soils of the Honeywood catena. *Can. J. Soil Sci.* **52**: 497–511.

Ascaso, C. and J. Galvan. 1976. Studies on the pedogenetic action of lichen acids. *Pedobiologia* **16**: 321–331.

Ashley, G. M., Shaw, J., and N. D. Smith (eds.) 1985. *Glacial Sedimentary Environments*, Society for Sedimentory Geology Short Course no. 16. Publ. by SEPM, Tulsa, OK.

Assallay, A. M., Rogers, C. D. F., Smalley, I. J., *et al.* 1998. Silt: 2–62 um, 9–4. *Earth Sci. Revs.* **45**: 61–88.

Atalay, I. 1997. Red Mediterranean soils in some karstic regions of Taurus mountains, Turkey. *Catena* **28**: 247–260.

Atkinson, H. J. and J. R. Wright. 1957. Chelation and the vertical movement of soil constituents. *Soil Sci.* **84**: 1–11.

Atkinson, R. J. C. 1957. Worms and weathering. *Antiquity* **31**: 219–233.

Attig, J. W. and L. Clayton. 1993. Stratigraphy and origin of an area of hummocky glacial topography, northern Wisconsin, USA. *Quat. Intl* **18**: 61–67.

Avakyan, Z. A., Karavaiko, G. I., Melnikova, *et al.* 1981. Role of microscopic fungi in weathering of rocks and minerals from a pegmatite deposit. *Mikrobiologiya* **50**: 156–162.

Bachmann, G. O. and M. N. Machette. 1977. *Calcic Soils and Calcretes in the Southwestern United States.* US Geol. Surv. Open File Rept. no. 77–794. Denver, CO.

Bacon, C. R. 1983. Eruptive history of Mount Mazama and Crater Lake caldera Cascade Range, USA. *J. Volcan. Geotherm. Res.* **18**: 57–115.

Bacon, D. W. and D. G. Watts. 1971. Estimating the transition between two intersecting straight lines. *Biometrika* **58**: 525–534.

Bagine, R. K. N. 1984. Soil translocation by termites of the genus *Odontotermes* (Holgren) (Isoptera: Macrotermitinae) in an arid area of northern Kenya. *Oecologia* **64**: 263–266.

Bailey, H. P. 1979. Semi-arid climates: their definition and distribution. In A. E. Hall, G. H. Cannell and H. W. Lawton (eds.) *Agriculture in Semi-arid Environments.* Berlin, Springer-Verlag, pp. 73–97.

Bailey, L. W., Odell, R. T., and W. R. Boggess. 1964. Properties of selected soils developed near the forest–prairie border in east-central Illinois. *Soil Sci. Soc. Am. Proc.* **28**: 257–263.

Bailey, S. W. 1980. Structures of layer silicates. In G. W. Brindley and G. Brown (eds.) *Crystal Structures of Clay Minerals and their X-Ray Identification, Monograph no. 5.* London, Mineralogical Society, pp. 1–123.

Bain, D. C., Mellor, A., Robertson-Rintoul, M. S. E., *et al.* 1993. Variations in weathering processes and rates with time in a chronosequence of soils from Glen Feshie, Scotland. *Geoderma* **57**: 275–293.

Baker, D. G. 1971. *Snow Cover and Winter Soil Temperatures at St. Paul, Minnesota*, US Dept. of Interior. Bull. no. 37. St. Paul, MN, Water Resources Research Center, University of Minnesota.

Baker, G. 1959. Opal phytoliths in some Victorian soils and "red rain" residues. *Austral. J. Bot.* **7**: 64–87.

Baker, R. G., Schwert, D. P., Bettis, E. A. III, *et al.* 1991. Mid-Wisconsinan stratigraphy and paleoenvironments at the St. Charles site in south-central Iowa. *Geol. Soc. Am. Bull.* **103**: 210–220.

Baker, R. G., Sullivan, A. E., Hallberg, G. R., *et al.* 1989. Vegetational changes in western Illinois during the onset of Late Wisconsinan glaciation. *Ecology* **70**: 1363–1376.

Bakker, L., Lowe, D. J., and A. G. Jongmans. 1996. A micromorphological study of pedogenic processes in an evolutionary soil sequence formed on Late Quaternary rhyolitic tephra deposits, North Island, New Zealand. *Quat. Intl* **34–36**: 249–261.

Baldwin, M., Kellogg, C. E., and J. Thorp. 1938. Soil classification. In *Soils and Men: Yearbook of Agriculture.* US Department of Agriculture, Washington, DC, US Government Printing Office, pp. 979–1001.

Balek, C. L. 2002. Buried artifacts in stable upland sites and the role of bioturbation: a review. *Geoarchaeology* **17**: 41–51.

Balesdent, J., Girardin, C., and A. Mariotti. 1993. Site-related ^{13}C of tree leaves and soil organic matter in a temperate forest. *Ecology* **74**: 1713–1721.

Balesdent, J., Mariotti, A., and B. Guillet. 1987. Natural ^{13}C abundance as a tracer for soil organic matter dynamics studies. *Soil Biol. Biochem.* **19**: 25–30.

Ball, D. F. and W. M. Williams. 1968. Variability of soil chemical properties in two uncultivated Brown Earths. *J. Soil Sci.* **19**: 379–391.

Ballagh, T. M. and E. C. A. Runge. 1970. Clay-rich horizons over limestone: Illuvial or residual? *Soil Sci. Soc. Am. Proc.* **34**: 534–536.

Ballantyne, A. K. 1978. Saline soils in Saskatchewan due to wind deposition. *Can. J. Soil Sci.* **58**: 107–108.

Ballantyne, C. K. and J. A. Matthews. 1982. The development of sorted circles on recently deglaciated terrain, Jotunheimen, Norway. *Arctic Alpine Res.* **14**: 341–354.

Barendregt, R. W. 1981. Dating methods of Pleistocene deposits and their problems. VI. Paleomagnetism. *Geosci. Can.* **8**: 56–64.

1984. Using paleomagnetic remanence and magnetic susceptibility data for the differentiation, relative correlation and absolute dating of Quaternary sediments. In W. C. Mahaney (ed.) *Quaternary Dating Methods*. New York, Elsevier, pp. 101–122.

Barnes, C. J. and G. B. Allison. 1983. The distribution of deuterium and ^{18}O in dry soils. I. *Theory. J. Hydrol.* **60**: 141–156.

Barnes, G. W. 1879. The hillocks or mound formations of San Diego, California. *Am. Nat.* **13**: 565–571.

Barnhisel, R. I. and P. M. Bertsch. 1989. Chlorites and hydroxy-interlayered vermiculite and smectite. In J. B. Dixon and S. B. Weed (eds.) *Minerals in Soil Environments*, 2nd edn. Madison, WI, Soil Science Society of America, pp. 729–788.

Barnhisel, R. I. and C. I. Rich. 1967. Clay mineral formation in different rock types of a weathering boulder conglomerate. *Soil Sci. Soc. Am. Proc.* **31**: 627–631.

Barnhisel, R. I., Bailey, H. H., and S. Matondang. 1971. Loess distribution in central and eastern Kentucky. *Soil Sci. Soc. Am. Proc.* **35**: 483–487.

Barrett, L. R. 1997. Podzolization under forest and stump prairie vegetation in northern Michigan. *Geoderma* **78**: 37–58.

2001. A strand plain soil development sequence in Northern Michigan, USA. *Catena* **44**: 163–186.

Barrett, L. R. and R. J. Schaetzl. 1992. An examination of podzolization near Lake Michigan using chronofunctions. *Can. J. Soil Sci.* **72**: 527–541.

1993. Soil development and spatial variability on geomorphic surfaces of different age. *Phys. Geog.* **14**: 39–55.

1998. Regressive pedogenesis following a century of deforestation: evidence for depodzolization. *Soil Sci.* **163**: 482–497.

Barshad, I. 1966. The effect of a variation in precipitation on the nature of clay mineral formation in soils from basic and acid igneous rocks. *Proc. Intl Clay Conf.* (Jerusalem) **1**: 167–173.

Barshad, I. and F. M. Kishk. 1969. Chemical composition of soil vermiculite clays as related to their genesis. *Contrib. Mineral. Petrol.* **24**: 136–155.

Barshad, I., Halevy, E., Gold, H. A., *et al.* 1956. Clay minerals in some limestone soils from Israel. *Soil Sci.* **81**: 423–437.

Bartelli, L. J. 1973. Soil development in loess in the southern Mississippi Valley. *Soil Sci.* **115**: 254–260.

Bartelli, L. J. and R. T. Odell. 1960a. Field studies of a clay-enriched horizon in the lowest part of the solum of some Brunizem and grey-brown podzolic soils in Illinois. *Soil Sci. Soc. Am. Proc.* **24**: 388–390.

1960b. Laboratory studies and genesis of a clay-enriched horizon in the lowest part of the solum of some Brunizem and grey-brown podzolic soils in Illinois. *Soil Sci. Soc. Am. Proc.* **24**: 390–395.

Bartlein, P. J., Webb, T. III, and E. C. Fleri. 1984. Holocene climatic change in the northern Midwest: Pollen-derived estimates. *Quat. Res.* **22**: 361–374.

Barton, R. N. 1987. Vertical distribution of artefacts and some post-depositional factors affecting site deformation. In P. Rowley-Conwy, M. Zvelebil and H. P. Blankholm (eds.) *Mesolithic in Northwest Europe: Recent Trends*. Sheffield, UK, Dept of Archaeology and Prehistory, University of Sheffield pp. 55–62.

Bateman, M. D. and J. A. Catt. 1996. An absolute chronology for the raised beach and associated deposits at Sewerby, East Yorkshire, England. *J. Quat. Sci.* **11**: 389–395.

Baxter, F. P. and F. D. Hole. 1967. Ant (*Formica cinerea*) pedoturbation in a prairie soil. *Soil Sci. Soc. Am. Proc.* **31**: 425–428.

Beatty, S. W. 1984. Influence of microtopography and canopy species on spatial patterns of forest understory plants. *Ecology* **65**: 1406–1419.

2000. On the rocks: shaken, not stirred. *Ann. Assoc. Am. Geog.* **90**: 785–787.

Beatty, S. W. and O. D. V. Sholes. 1988. Leaf litter effect on plant species composition of deciduous forest treefall pits. *Can. J. For. Res.* **18**: 553–559.

Beatty, S. W. and E. L. Stone. 1986. The variety of soil microsites created by tree falls. *Can. J. For. Res.* **16**: 539–548.

Beaven, P. J. and M. J. Dumbleton. 1966. Clay minerals and geomorphology in four Caribbean islands. *Clay Mins.* **6**: 371–382.

Beavers, A. H. and I. Stephen. 1958. Some features of the distribution of plant-opal in Illinois soils. *Soil Sci.* **86**: 1–5.

Beavers, A. H., Fehrenbacher, J. B., Johnson, P. R., *et al.* 1963. CaO–ZrO^2 molar ratios as an index of weathering. *Soil Sci. Soc. Am. Proc.* **27**: 408–412.

Bech, J., Rustullet, J., Garrigó Tobías, F. J., *et al.* 1997. The iron content of some Mediterranean soils from northeast Spain and its pedogenic significance. *Catena* **28**: 211–229.

Beckel, D. K. B. 1957. Studies on seasonal changes in the temperature gradient of the active layer of soil at Fort Churchill, Manitoba. *Arctic* **10**: 151–183.

Beckett, P. H. T. and R. Webster. 1971. Soil variability: a review. *Soils and Fert.* **34**: 1–15.

Beckmann, G. G. 1984. The place of "genesis" in the classification of soils. *Austral. J. Soil Res.* **22**: 1–14.

Beckmann, G. G., Thompson, C. H., and B. R. Richards. 1984. Relationships of soil layers below gilgai in black earths. In J. W. McGarity, E. H. Hoult and H. B. So (eds.) *The Properties and Utilization of Cracking Clay Soils.* Reviews in Rural Science no. 5. Armidale, NSW, University of New England, pp. 64–72.

Beckwith, R. S. and R. Reeve. 1964. Studies on soluble silica in soils. II. The release of monosilicic acid from soils. *Austral. J. Soil Res.* **2**: 33–45.

Beer, J., Shen, C. D., Heller, F., *et al.* 1993. Be and magnetic susceptibility in Chinese loess. *Geophys. Res. Lett.* **20**: 57–60.

Beinroth, F. H. 1982. Some highly weathered soils of Puerto Rico. I. Morphology, formation and classification. *Geoderma* **27**: 1–73.

Beinroth, F. H., Eswaran, H., Uehara, G., *et al.* 2000. Oxisols. In M. E. Sumner (ed.) *Handbook of Soil Science.* Boca Raton, FL, CRC Press, pp. E-373–E-392.

Beinroth, F. H., Uehara, G., and H. Ikawa. 1974. Geomorphic relationships of Oxisols and Ultisols on Kauai, Hawaii. *Soil Sci. Soc. Am. Proc.* **38**: 128–131.

Bell, J. C., Cunningham, R. L., and M. W. Havens. 1992. Calibration and validation of a soil-landscape model for predicting soil drainage class. *Soil Sci. Soc. Am. J.* **56**: 1860–1866.

Benac, C. and G. Durn. 1997. Terra rossa in the Kvarner area: geomorphological conditions of formation. *Acta Geog. Croatica* **32**: 7–17.

Benedict, J. B. 1967. Recent glacial history of an alpine area in the Colorado Front Range, USA. I. Establishing a lichen-growth curve. *J. Glac.* **6**: 817–832.

1970. Downslope soil movement in a Colorado alpine region: rates, processes and climatic significance. *Arctic Alpine Res.* **2**: 165–226.

1976. Frost creep and gelifluction features: a review. *Quat. Res.* **6**: 55–76.

Bennema, J. 1974. Organic carbon profiles in Oxisols. *Pédologie* **24**: 119–146.

Bennett, D. R. and T. Entz. 1989. Moisture-retention parameters for coarse-textured soils in southern Alberta. *Can. J. Soil Sci.* **69**: 263–272.

Benson, L. V., Currey, D. R., Dorn, R. I., *et al.* 1990. Chronology of expansion and contraction of four Great Basin lake systems during the past 35 000 years. *Palaeogeog. Palaeoclimat. Palaeoecol.* **78**: 241–286.

Berg, A. W. 1990. Formation of Mima mounds: a seismic hypothesis. *Geology* **18**: 281–284.

Berg, R. C. 1984. The origin and early genesis of clay bands in youthful sandy soils along Lake Michigan, USA *Geoderma* **32**: 45–62.

Berger, A. 1988. Milankovitch theory and climate. *Rev. Geophys.* **26**: 624–657.

Berger, G. W. 1984. Thermoluminescence dating studies of glacial silts from Ontario. *Can. J. Earth Sci.* **21**: 1393–1399.

Berger G. W. 1988. TL dating studies of tephra, loess and lacustrine sediments. *Quat. Sci. Rev.* **7**: 295–303.

Berger, G. W. and W. Mahaney. 1990. Test of thermoluminescence dating of buried soils from Mt. Kenya, Kenya. *Sed. Geol.* **66**: 45–56.

Berger, I. A. and R. U. Cooke. 1997. The origin and distribution of salts on alluvial fans in the Atacama Desert, Northern Chile. *Earth Surf. Proc. Landforms* **22**: 581–600.

Bernier, N. 1998. Earthworm feeding activity and the development of the humus profile. *Biol. Fert. Soils* **26**: 215–223.

Bernoux, M., Arrouays, D., Cerri, C., *et al.* 1998. Bulk densities of Brazilian Amazon soils related to other soil properties. *Soil Sci. Soc. Am. J.* **62**: 743–749.

Berry, E. C. and J. K. Radke. 1995. Biological processes: relationships between earthworms and soil temperatures. *J. Minnesota Acad. Sci.* **59**: 6–8.

Berry, L. and B. P. Ruxton. 1959. Notes on weathering zones and soils on granitic rocks in two tropical regions. *J. Soil Sci.* **10**: 54–63.

Berry, M. E. 1987. Morphological and chemical characteristics of soil catenas on Pinedale and Bull Lake moraine slopes in the Salmon River Mountains, Idaho. *Quat. Res.* **28**: 210–225.

1994. Soil-geomorphic analysis of Late-Pleistocene glacial sequences in the McGee, Pine, and Bishop Creek Drainages, East-Central Sierra Nevada, California. *Quat. Res.* **41**: 160–175.

Berry, M. E. and J. R. Staub. 1998. Root traces and the identification of paleosols. *Quat. Intl* **51/52**: 9–10.

Beschel, R. E. 1950. Flechten als Altersmasstab rezenter Moränen. *Gletscherkd. Glazialgeol.* **1**: 152–161.

1958. Lichenometrical studies in West Greenland. *Arctic* **11**: 254.

Beshay, N. F. and A. S. Sallam. 1995. Evaluation of some methods for establishing uniformity of profile parent materials. *Arid Soil Res. Rehabil.* **9**: 63–72.

Bettis, E. A. III. 1988. Pedogenesis in late prehistoric Indian mounds, upper Mississippi Valley. *Phys. Geog.* 263–279.

1998. Subsolum weathering profile characteristics as indicators of the relative rank of stratigraphic breaks in till sequences. *Quat. Intl* **51/52**: 72–73.

Beyer, L., Knicker, H., Blume, H.-P., *et al.* 1997. Soil organic matter of suggested spodic horizons in relic ornithogenic soils of coastal continental Antarctica (Casey Station, Wilkes Land) in comparison with that of spodic soil horizons in Germany. *Soil Sci.* **162**: 518–527.

Beyer, L., Pingpank, K., Wriedt, G., *et al.* 2000. Soil formation in coastal Antarctica (Wilkes Land). *Geoderma* **95**: 283–304.

Bhattacharjee, J. C., Landey, R. J., and A. R. Kalbande. 1977. A new approach in the study of Vertisol morphology. *J. Ind. Soc. Soil Sci.* **25**: 221–232.

Bhattacharyya, T., Pal, D. K., and M. Velayutham. 1999. A mathematical equation to calculate linear distance of cyclic horizons in Vertisols. *Soil Surv. Horiz.* **40**: 127–133.

Bidwell, O. W. and F. D. Hole. 1965. Man as a factor of soil formation. *Soil Sci.* **99**: 65–72.

Bierman, P. R. and A. R. Gillespie. 1994. Evidence suggesting that methods of rock-varnish cation-ratio dating are neither comparable nor consistently reliable. *Quat. Res.* **42**: 82–90.

Bierman, P. and J. Turner. 1995. ^{10}Be and ^{26}Al evidence for exceptionally low rates of Australian bedrock erosion and the likely existence of pre-Pleistocene landscapes. *Quat. Res.* **44**: 378–382.

Bigham, J. M., Golden, D. C., Buol, S. W., *et al.* 1978. Iron oxide mineralogy of well-drained Ultisols and Oxisols. II. Influence on color, surface area, and phosphate retention. *Soil Sci. Soc. Am. J.* **42**: 825–830.

Bigham, J. M., Smeck, N. E., Norton, L. D., *et al.* 1991. Lithology and general stratigraphy of Quaternary sediments in a section of the Teays River valley of southern Ohio. In W. N. Melhorn and J. P. Kempton (eds.) *Geology and Hydrogeology of the Teays–Mahomet Bedrock Valley System*, Special Paper no. 258. Boulder, CO, Geological Society of America, pp. 19–27.

Billings, W. D. and K. M. Peterson. 1980. Vegetational change and ice-wedge polygons through the thaw-lake cycle in Arctic Alaska. *Arctic Alpine Res.* **12**: 413–432.

Bilzi, A. F. and E. J. Ciolkosz. 1977. A field morphology rating scale for evaluating pedological development. *Soil Sci.* **124**: 45–48.

Birkeland, P. W. 1969. Quaternary paleoclimatic implications of soil clay mineral distribution in a Sierra Nevada–Great Basin transect. *J. Geol.* **77**: 289–302.

1973. Use of relative age-dating methods in a stratigraphic study of rock glacier deposits, Mt. Sopris, Colorado. *Arctic Alpine Res.* **5**: 401–416.

1974. *Pedology, Weathering and Geomorphological Research.* New York, Oxford University Press.

1978. Soil development as an indication of relative age of Quaternary deposits, Baffin Island, N.W.T., Canada. *Arctic Alpine Res.* **10**: 733–747.

1982. Subdivision of Holocene glacial deposits, Ben Ohau Range, New Zealand, using relative dating methods. *Geol. Soc. Am. Bull.* **93**: 433–449.

1999. *Soils and Geomorphology*, 3rd edn. New York, Oxford University Press.

Birkeland, P. W. and R. M. Burke. 1988. Soil catena chrono sequences on eastern Sierra Nevada moraines, California, USA. *Arctic Alpine Res.* **20**: 473–484.

Birkeland, P. W., Berry, M. E., and D. K. Swanson. 1991. Use of soil catena field data for estimating relative ages of moraines. *Geology* **19**: 281–283.

Birks, H. J. B. 1976. Late-Wisconsinan vegetation history at Wolf Creek, central Minnesota. *Ecol. Monogr.* **46**: 395–429.

Biscaye, P. E., Chesselet, R., and J. M. Prospero. 1974. Rb-Sr, $^{87}Sr/^{86}Sr$ isotope system as an index of provenance of continental dusts in the open Atlantic Ocean. *J. Rech. Atmos.* **8**: 819–829.
Bischoff, J. L., Shlemon, R. J., Ku, T. L., *et al.* 1981. Uranium-series and soil-geomorphic dating of the Calico archaeological site, California. *Geology* **9**: 576–582.
Bisdom, E. B. A., Nauta, R., and B. Volbert. 1983. STEM-EDXRA and SEM-EDXRA investigations of iron-coated organic matter in thin sections with transmitted, secondary and backscattered electrons. *Geoderma* **30**: 77–92.
Bishop, P. M., Mitchell, P. B., and T. R. Paton. 1980. The formation of duplex soils on hillslopes in the Sydney Basin, Australia. *Geoderma* **23**: 175–189.
Bjelland, T. and I. H. Thorseth. 2002. Comparative studies of the lichen–rock interface of four lichens in Vingen, western Norway. *Chem. Geol.* **192**: 81–98.
Black, R. F. 1969. Climatically significant fossil periglacial phenomena in northcentral United States. *Biul. Peryglac.* **20**: 227–238.
Black, R. F. 1976. Periglacial features indicative of permafrost: ice and soil wedges. *Quat. Res.* **6**: 3–26.
Blackwelder, E. 1926. Fire as an agent in rock weathering. *J. Geol.* **35**: 134–140.
Blake, G. R. and K. H. Hartage. 1986. Bulk density. In A. Klute (ed.) *Methods of Soil Analysis*, 2nd edn., Agronomy Monograph no. 9. Madison, WI, American Society Agron., pp. 363–375.
Blanchart, E. 1992. Restoration by earthworms (Megascolecidae) of the macroaggregate structure of a destructed savanna soil under field conditions. *Soil Biol. Biochem.* **24**: 1587–1594.
Bland, W. and D. Rolls. 1998. *Weathering: An Introduction to the Scientific Principles*. London, Edward Arnold.
Blank, H. R. and E. W. Tynes. 1965. Formation of caliche *in situ*. *Geol. Soc. Am. Bull.* **76**: 1387–1392.
Blank, R. R. and M. A. Fosberg. 1991. Duripans of Idaho, USA: *in situ* alteration of eolian dust (loess) to an opal-A/X-ray amorphous phase. *Geoderma* **48**: 131–149.
Blank, R. R., Young, J. A., and T. Lugaski. 1996. Pedogenesis on talus slopes, the Buckskin range, Nevada, USA. *Geoderma* **71**: 121–142.
Blokhuis, W. A. 1982. Morphology and genesis of Vertisols. *Proc. 12th Intl Congr. Soil Sci.* 23–47.
Bloomfield, C. 1953. A study of podzolization. II. The mobilization of iron and aluminum by the leaves and bark of *Agathis australis* (Kauri). *J. Soil Sci.* **4**: 17–23.
Blow, F. E. 1955. Quantity and hydrologic characteristics of litter upon upland oak forests in eastern Tennessee. *J. For.* **53**: 190–195.
Blum, W. E. and R. Ganssen. 1972. Bodenbildende prozesse der Erde, ihre erscheinungsformen und diagnostischen Merkmale in tabellarischer Darstellung. *Die Erde* **103**: 7–20.
Blume, H.-P., Beyer, L., Bölter, M., *et al.* 1997. Pedogenic zonation in soils of the southern circum-polar region. *Adv. GeoEcol.* **30**: 69–90.
Blümel, W. D. 1982. Calcretes in Namibia and SE Spain: relations to substratum, soil formation and geomorphic factors. *Catena (Suppl.)* **1**: 67–82.
Boardman, J. 1985. Comparison of soils in midwestern United States and western Europe with the interglacial record. *Quat. Res.* **23**: 62–75.
Bocek, B. 1986. Rodent ecology and burrowing behavior: predicted effects on archaeological site formation. *Am. Antiq.* **51**: 589–603.
Bochter, R. and W. Zech. 1985. Organic compounds in cryofolists developed on limestone under subalpine coniferous forest, Bavaria. *Geoderma* **36**: 145–157.
Bockheim, J. G. 1979a. Properties and relative age of soil of southwestern Cumberland peninsula, Baffin Island, NWT, Canada. *Arctic Alpine Res.* **11**: 289–306.
1979b. Relative age and origin of soils in eastern Wright Valley, Antarctica. *Soil Sci.* **128**: 142–152.
1980a. Properties and classification of some desert soils in coarse-textured glacial drift in the Arctic and Antarctic. *Geoderma* **24**: 45–69.
1980b. Solution and use of chronofunctions in studying soil development. *Geoderma* **24**: 71–85.
1982. Properties of a chronosequence of ultraxerous soils in the Trans-Antarctic Mountains. *Geoderma* **28**: 239–255.
1990. Soil development rates in the Transantarctic Mountains. *Geoderma* **47**: 59–77.
1997. Properties and classification of cold desert soils from Antarctica. *Soil Sci. Soc. Am. J.* **61**: 224–231.
Bockheim, J. G. and A. N. Gennadiyev. 2000. The role of soil-forming processes in the definition of taxa in Soil Taxonomy and the World Soil Reference Base. *Geoderma* **95**: 53–72.
Bockheim, J. G. and C. Tarnocai. 2000. Gelisols. In M. E. Sumner (ed.) *Handbook of Soil Science*. Boca Raton, FL, CRC Press, pp. E-256–E-269.

Bockheim, J. G. and F. C. Ugolini. 1990. A review of pedogenic zonation in well-drained soils of the southern circumpolar region. *Quat. Res.* **34**: 47–66.

Bockheim, J. G., Everett, L. R., Hinkel, K. M., *et al.* 1999. Soil organic carbon storage and distribution in Arctic Tundra, Barrow, Alaska. *Soil Sci. Soc. Am. J.* **63**: 934–940.

Bockheim, J. G., Marshall, J. G., and H. M. Kelsey. 1996. Soil-forming processes and rates on uplifted marine terraces in southwestern Oregon, USA. *Geoderma* **73**: 39–62.

Bockheim, J. G., Tarnocai, C., Kimble, J. M., *et al.* 1997. The concept of gelic materials in the new Gelisol order for permafrost-affected soils. *Soil Sci.* **162**: 927–939.

Bocock, K. L., Jeffers, J. N. R., Lindley, D. K., *et al.* 1977. Estimating woodland soil temperature from air temperature and other climatic variables. *Agric. Metr.* **18**: 351–372.

Boero, V. and U. Schwertmann. 1989. Iron oxide mineralogy of terra rossa and its genetic implications. *Geoderma* **44**: 319–327.

Boersma, L., Simonson, G. H., and D. G. Watts. 1972. Soil morphology and water table relations. I. Annual water table fluctuations. *Soil Sci. Soc. Am. Proc.* **36**: 644–648.

Boettinger, J. L. and R. J. Southard. 1991. Silica and carbonate sources for Aridisols on a granitic pediment, western Mojave Desert. *Soil Sci. Soc. Am. J.* **55**: 1057–1067.

Bold, H. D. and M. J. Wynne. 1979. *Introduction to the Algae*. Englewood Cliffs, NJ, Prentice-Hall.

Boll, J., van Rijn, R. P. G., Weiler, K. W., *et al.* 1996. Using ground penetrating radar to detect layers in a sandy field soil. *Geoderma* **70**: 117–132.

Bond, W. J. 1986. Illuvial band formation in a laboratory column of sand. *Soil Sci. Soc. Am. J.* **50**: 265–267.

Borchardt, G. 1989. Smectites. In J. B. Dixon and S. B. Weed (eds.) *Minerals in Soil Environments*, 2nd edn. Madison, WI, Soil Science Society of America, pp. 675–727.

Borchardt, G. A., Hole, F. D., and M. L. Jackson. 1968. Genesis of layer silicates in representative soils in a glacial landscape of southeastern Wisconsin. *Soil Sci. Soc. Am. Proc.* **32**: 399–403.

Borgaard, O. K. 1983. Effect of surface area and mineralogy of iron oxides on their surface charge and anion adsorption properties. *Clays Clay Min.* **31**: 230–232.

Bormann, F. H., Siccama, T. G., Likens, G. E., *et al.* 1970. The Hubbard Brook ecosystem study: composition and dynamics of the tree stratum. *Ecol. Monogr.* **40**: 373–388.

Borst, G. 1968. The occurrence of crotovinas in some southern Californian soils. *Trans. 9th Intl Congr. Soil Sci.* (Adelaide) **2**: 19–27.

Bos, R. H. G. and J. Sevink. 1975. Introduction of gradational and pedomorphic features in descriptions of soils. *J. Soil Sci.* **26**: 223–233.

Botha, G. A. and N. Fedoroff. 1995. Palaeosols in Late Quaternary colluvium, northern KwaZnhr–Natal, South Africa. *J. Afr. Earth Sci.* **21**: 291–311.

Bouabid, R., Nater, E. A., and P. Barak. 1992. Measurement of pore size distribution in a lamellar Bt horizon using epifluorescence microscopy and image analysis. *Geoderma* **53**: 309–328.

Bouchard, M. and M. J. Pavich. 1989. Characteristics and significance of pre-Wisconsinan saprolites in the northern Appalachians. *Z. Geomorphol.* **72**: 125–137.

Boulaine, J. 1975. *Géographie des soils*. Paris, Presses Universités de France.

Bouma, J. and V. van Schuylenborgh. 1969. On soil genesis in a temperate humid climate. VII. The formation of a glossaqualf in a silt–loam terrace deposit. *Neth. J. Agric. Sci.* **17**: 261–271.

Bourne, W. C. and E. P. Whiteside. 1962. A study of the morphology and pedogenesis of a medial Chernozem developed in loess. *Soil Sci. Soc. Am. Proc.* **26**: 484–490.

Bowen, D. Q. 1978. *Quaternary Geology: A Stratigraphic Framework for Multidisciplinary Work*. New York, Pergamon Press.

Bowen, L. H. and S. B. Weed. 1981. Mössbauer spectroscopic analysis of iron oxides in soil. In J. G. Stevens and G. K. Shenoy (eds.) *Mössbauer Spectroscopy and its Applications.* Washington, DC, American Chemical Society, pp. 247–261.

Bowler, J. M. 1973. Clay dunes: their occurrence, formation and environmental significance. *Earth Sci. Rev.* **9**: 315–338.

Bowser, W. E., Milne, R. A., and R. R. Cairns. 1962. Characteristics of the major soil groups in an area dominated by solonetzic soils. *Can. J. Soil Sci.* **42**: 165–179.

Boyer, P. 1973. Actions de certains termites constructeurs sur l'évolution des soils tropicaux. *Ann. Sci. Nat. Zool.* (Paris) **15**: 329–498.

1975. Etude particulières de Bellioositermes et de leur action sur sols tropicaux. *Ann. Sci. Nat. Zool.* (Paris) **17**: 273–446.

Boynton, D. and W. Reuther. 1938. A way of sampling soil gas in dense subsoil and some of its advantages and limitations. *Soil Sci. Soc. Am Proc.* **3**: 37–42.

Bradbury, J. P., Dean, W. E., and R. Y Anderson. 1993. Holocene climatic and limnologic history of the north-central United States as recorded in the varved sediments of Elk Lake, Minnesota: a synthesis. In J. P. Bradbury, and W. E. Dean (eds.) *Elk Lake, Minnesota: Evidence for Rapid Climatic Change in the North-Central United States*, Special Paper no. 276. 309–328.

Bradley, W. C. 1963. Large-scale exfoliation in massive sandstones of the Colorado Plateau. *Geol. Soc. Am. Bull.* **74**: 519–528.

Brady, N. C. 1974. *The Nature and Properties of Soils*. Macmillian, New York.

Brady, N. C. and R. R. Weil. 1999. *The Nature and Properties of Soils*, 12th edn. Upper Saddle River, NJ, Prentice-Hall.

2001. *The Nature and Properties of Soils*, 13th edn. Englewood Cliffs, NJ, Prentice-Hall.

Brakensiek, D. L. and W. J. Rawls. 1994. Soil containing rock fragments: effects on infiltration. *Catena* **23**: 99–110.

Bramstedt, M. W. 1992. *Soil Survey of Jasper County, Illinois*. Washington, DC, US Department of Agriculture Soil Conservation Service, US Government Printing Office.

Branch, L. C. 1993. Inter- and intra-group spacing in the plains vizcacha (*Lagostomus maximus*). *J. Mammal.* **74**: 890–900.

Brandon, C. E., Buol, S. W., Gamble, E. E., *et al.* 1977. Spodic horizon brittleness in Leon (Aeric Haplaquod) soils. *Soil Sci. Soc. Am. J.* **41**: 951–954.

Bravard, S. and D. Righi. 1989. Geochemical differences in an Oxisol–Spodosol toposequence of Amazonia, Brazil. *Geoderma* **44**: 29–42.

1990. Podzols in Amazonia. *Catena* **17**: 461–475.

1991. Characterization of fulvic and humic acids from an Oxisol–Spodosol toposequence of Amazonia, Brazil. *Geoderma* **48**: 151–162.

Bray, J. R. and E. Gorham. 1964. Litter production in forests of the world. *Adv. Ecol. Res.* **2**: 101–157.

Breckenfield, D. J. 1999. *Soil Survey of Tohono O'odham Nation, Arizona: Parts of Maricopa, Pima, and Pinal* Counties. Washington, DC, US Department of Agriculture Natural Resources Conservation Service, US Government Printing Office.

Bremer, H. and H. Späth. 1989. Geomorphological observations concerning stone-lines. *Intl J. Trop. Ecol. Geog.* **11**: 185–195.

Brener, A. G. F. and J. F. Silva. 1995. Leaf-cutting ant nests and soil fertility in a well-drained savanna in western Venezuela. *Biotropica* **27**: 250–254.

Brewer, R. 1976. *Fabric and Mineral Analysis of Soils*. Huntington, NY, Robert E. Krieger.

Bridges, E. M. and P. A. Bull. 1983. The role of silica in the formation of compact and indurated horizons in the soils of South Wales. In P. Bullock and C. P. Murphy (eds.) *Soil Micromorphology*. New York, Academic Press, pp. 605–613.

Briese, D. T. 1982. The effects of ants on the soil of a semi-arid salt bushland habitat. *Insectes Soc.* **29**: 375–386.

Brimecombe, M. J., DeLeij, F. A., and J. M. Lynch. 2001. The effect of root exudates on rhizosphere microbial populations. In R. Pinton, Z. Varanini and P. Nannipieri (eds.) *The Rhizosphere: Biochemistry and Organic Substances at the Soil–Plant Interface*. New York, Marcel Dekker, pp. 95–140.

Brimhall, G. H. and W. E. Dietrich. 1987. Constitutive mass balance relations between chemical composition, volume, density, porosity, and strain in metasomatic hydrochemical systems: results of weathering and pedogenesis. *Geochim. Cosmochim. Acta* **51**: 567–587.

Brimhall, G. H., Chadwick, O. A., Lewis, C. J., *et al.* 1991a. Deformational mass transport and invasive processes in soil evolution. *Science* **255**: 695–702.

Brimhall, G. H., Lewis, C. J., Ague, J. J., *et al.* 1988. Metal enrichment in bauxites by deposition of chemically mature aeolian dust. *Nature* **333**: 819.

Brimhall, G. H., Lewis, C. J., Ford, *et al.* 1991b. Quantitative geochemical approach to pedogenesis: importance of parent material reduction, volumetric expansion, and eolian influx in lateritization. *Geoderma* **51**: 51–91.

Brindley, G. W. 1980. Quantitative x-ray mineral analysis of clays. In G. W. Brindley and G. Brown (eds.) *Crystal Structures of Clay Minerals and their X-Ray Identification*, Monograph no. 5. London, Mineralogical Society, pp. 411–438.

Brinkman, R. 1970. Ferrolysis, a hydromorphic soil forming process. *Geoderma* **3**: 199–206.

1977a. Problem hydromorphic soils in north-east Thailand. II. Physical and chemical aspects, mineralogy and genesis. *Neth. J. Agric. Sci.* **25**: 170–181.

1977b. Surface-water gley soils in Bangladesh: genesis. *Geoderma* **17**: 111–144.

Britton, M. E. 1957. Vegetation of the arctic tundra. *Proc. Biol. Colloq.* (Corvallis) 26–61.

Broecker, W. S. and E. A. Olson. 1960. Radiocarbon from nuclear tests. II. *Science* **132**: 712–721.

Broecker, W. S. and T. Liu. 2001. Rock varnish: recorder of desert wetness? *Geol. Soc. Am. Today* **11**: 4–10.

Broersma, K. and L. M. Lavkulich. 1980. Organic matter distribution with particle-size in surface horizons of some sombric soils in Vancouver Island. *Can. J. Soil Sci.* **60**: 583–586.

Bronger, A. 1991. Argillic horizons in modern loess soils in an ustic soil moisture regime: comparative studies in forest–steppe and steppe areas from Eastern Europe and the United States. *Adv. Soil Sci.* **15**: 41–90.

Bronswijk, J. J. B. 1991. Relations between vertical soil movements and water-content changes in cracking clays. *Soil Sci. Soc. Am. J.* **55**: 1220–1226.

Brook, G. A., Folkoff, M. E., and E. G. Box. 1983. A world model of soil carbon dioxide. *Earth Surf. Proc. Landforms* **8**: 79–88.

Brookes, I. A. 1982. Dating methods of Pleistocene deposits and their problems. VIII. Weathering. *Geosci. Canada.* **9**: 188–199.

Brophy, J. A. 1959. Heavy mineral ratios of Sangamon weathering profiles in Illinois. *Illinois Geol. Surv. Circ.* **273**: 1–22

Brown, C. N. 1956. The origin of caliche in the northeast Llano Estacado, Texas. *J. Geol.* **46**: 1–15.

Brown, E. T., Edmond, J. M., Raisbeck, G. M., *et al.* 1991. Examination of surface exposure ages of Antarctic moraines using *in situ* produced ^{10}Be and ^{26}Al. *Geochim. Cosmochim. Acta* **55**: 2269–2283.

Brown, G. and G. W. Brindley. 1980. X-ray diffraction procedures for clay mineral identification. In G. W. Brindley and G. Brown (eds.) *Crystal Structures of Clay Minerals and their X-Ray Identification*, Monograph no. 5. London, Mineralogical Society, pp. 305–359.

Brown, J. 1969. Soils of the Okpilak River region, Alaska. In T. L. Péwé (ed.) *The Periglacial Environment Past and Present.* Montreal, McGill–Queens' University Press, pp. 93–128.

Brown, J. L. 1979. Etude systématique de la variabilité d'un sol Podzolique, le long d'une tranchée dans une erabliére à bouleau jaune. *Can. J. Soil Sci.* **59**: 131–146.

Brown, L. 1987. ^{10}Be as a tracer of erosion and sediment transport. *Chem. Geol.* **65**: 189–196.

Brown, L., Pavich, M. J., Hickman, R. E., *et al.* 1988. Erosion of the eastern United States observed with ^{10}Be. *Earth Surf. Proc. Landforms* **13**: 441–457.

Bruce, J. G. 1996. Morphological characteristics and interpretation of some polygenetic soils in loess in southern South Island, New Zealand. *Quat. Intl* **34–36**: 205–211.

Bruce, R. R. and R. J. Luxmoore. 1986. Water retention: field methods. In A. Klute (ed) *Methods of Soil Analysis*, Part 1, *Physical and Mineralogical Methods*, Agronomy Monograph no. 9, 2nd edn. Madison, WI, American Soil Association and Soil Science Society of America, pp. 663–686.

Brückner, W. 1955. The mantle rock ("laterite") of the Gold Coast and its origin. *Geol. Rundschau* **43**: 307–327.

Bryan, B. A. and C. R. Hupp. 1984. Dendrogeomorphic evidence for channel changes in an east Tennessee coal area stream. *Trans. Am. Geophys. Union* **65**: 891.

Bryan, D. 1994. Factors controlling the occurrence and distribution of hematite and goethite in soils and saprolites derived from schists and gneisses in western North Carolina. M.S. thesis, Michigan State University.

Bryan, K. 1946. Cryopedology: the study of frozen ground and intensive frost-action with suggestions on nomenclature. *Am. J. Sci.* **244**: 622–642.

Bryan, K. and C. C. Albritton, Jr. 1943. Soil phenomena as evidence of climatic changes. *Am. J. Sci.* **241**: 469–490.

Bryant, R. B. 1989. Physical processes of fragipan formation. In N. E. Smeck and E. J. Ciolkosz (eds.) *Fragipans Their Occurrence, Classification, and Genesis.* Special Publication no. 24. Madison, WI, Soil Science Society of America, pp. 141–150.

Bryant, V. M., Jr. and R. G. Holloway (eds.) 1985. *Pollen Records of Late Quaternary North American Sediments.* American Association of Stratigraphic Palynologists Foundation. Dallas, TX.

Bryson, R. A., Irving, W. N., and J. A. Larsen. 1965. Radiocarbon and soil evidence of former forest in the southern Canadian tundra. *Science* **147**: 46–48.

Bucher, E. H. and R. B. Zuccardi. 1967. Significación de los hormigueros de *Atta vollenweideri* Foreal como alteradores del suelo en la Provinces de Tucuman. *Acta Zool. Lilloana* **23**: 83–96.

Büdel, J. 1957. Die "Doppelten Einebnungsflächen" in den feuchten Tropen. *Z. Geomorph.* **2**: 201–228.

Büdel, J. 1982. *Climatic Geomorphology*. Princeton, NJ, Princeton University Press.

Buhmann, C. and P. L. C. Grubb. 1991. A kaolin–smectite interstratification sequence from a red and black complex. *Clay Mins.* **26**: 343–358.

Bullock, P. and C. P. Murphy (eds.). Berkhamsted, UK, A. B. Academic Publications. 1983. *Soil Micromorphology*, vol. 2, *Soil Genesis*.

Bullock, P., Federoff, N., Jongerius, A., *et al.* 1985. *Handbook for Soil Thin Section Description*. Wolverhampton, UK, Waine Research Publications.

Bullock, P., Milford, M. H., and M. G. Cline. 1974. Degradation of argillic horizons in Udalf soils in New York State. *Soil Sci. Soc. Am. Proc.* **38**: 621–628.

Bulmer, C. E. and L. M. Lavkulich. 1994. Pedogenic and geochemical processes of ultramific soils along a climatic gradient in southwestern British Columbia. *Can. J. Soil Sci.* **74**: 165–177.

Buntley, G. J., Daniels, R. B., Gamble, E. E., *et al.* 1973. Soil genesis, morphology and classification. In P. A. Sanchez (ed.) *A Review of Soils Research on Tropical Latin America*, Technical Bulletin no. 219. North Carolina Agricultural Experimental Station, pp. 1–37. Raleigh, NC.

1977. Fragipan horizons in soils of the Memphis–Loring–Grenada sequence in west Tennessee. *Soil Sci. Soc. Am. J.* **41**: 400–407.

Buntley, G. J. and R. I. Papendick. 1960. Worm-worked soils of eastern South Dakota: their morphology and classification. *Soil Sci. Soc. Am. Proc.* **24**: 128–132.

Buntley, G. J. and F. C. Westin. 1965. A comparative study of developmental color in a Chestnut–Chernozem–Brunizem soil climosequence. *Soil Sci. Soc. Am. Proc.* **29**: 579–582.

Buol, S. W. and M. G. Cook. 1998. Red and lateritic soils of the world: concept, potential, constraints and challenges. In J. Sehgal, W. E. Blum and K. S. Gajbhiye (eds.) *Red and Lateritic Soils: Managing Red and Lateritic Soils for Sustainable Agriculture*, vol. 1. New Delhi, Oxford and IBH Pub. Co., pp. 49–56.

Buol, S. W. and H. Eswaran. 2000. Oxisols. *Adv. Agron.* **68**: 151–195.

Buol, S. W. and F. D. Hole. 1961. Clay skin genesis in Wisconsin soils. *Soil Sci. Soc. Am. Proc.* **25**: 377–379.

Buol, S. W. and M. S. Yesilsoy. 1964. A genesis study of a Mohave sandy loam profile. *Soil Sci. Soc. Am. Proc.* **28**: 254–256.

Buol, S. W., Hole, F. D., McCracken, R. J., *et al.* 1997. *Soil Genesis and Classification*, 4th edn. Ames, IA, Iowa State University Press.

Buondonno, C., Ermice, A., Buondonno, A., *et al.* 1998. Human-influenced soils from an iron and steel works in Naples, Italy. *Soil Sci. Soc. Am. J.* **62**: 694–700.

Burges, A. 1968. The role of soil micro-flora in the decomposition and synthesis of soil organic matter. *Trans. 9th Intl Congr. Soil Sci.* (Adelaide) **2**: 29–35.

Burghardt, W. 1994. Soils in urban and industrial environments. *Z. Pflanzen.* Bodenkd. **157**: 205–214.

Burke, I. W., Yonker, C. M., Parton, W. J., *et al.* 1989. Texture, climate, and cultivation effects on soil organic matter content in US grassland soils. *Soil Sci. Soc. Am. J.* **53**: 800–805.

Burke, R. M. and P. W. Birkeland. 1979. Re-evaluation of multiparameter relative dating techniques and their application to the glacial sequence along the eastern escarpment of the Sierra Nevada, California. *Quat. Res.* **11**: 21–51.

Burleigh, R. 1974. Radiocarbon dating: some practical considerations for the archaeologist. *J. Archaeol. Sci.* **1**: 69–87.

Burns, S. F. and P. J. Tonkin. 1982. Soil-geomorphic models and the spatial distribution and development of alpine soils. In C. Thorn (ed.) *Space and Time in Geomorphology*. London, Allen and Unwin, pp. 25–43.

Burras, C. L. and W. H. Scholtes. 1987. Basin properties and postglacial erosion rates of minor moraines in Iowa. *Soil Sci. Soc. Am. J.* **51**: 1541–1547.

Busacca, A. J. 1987. Pedogenesis of a chronosequence in the Sacramento Valley, California, USA. I. Application of a soil development index. *Geoderma* **41**: 123–148.

1989. Long Quaternary record in eastern Washington, USA, interpreted from multiple buried paleosols in loess. *Geoderma* **45**: 105–122.

Busacca, A. J. and M. Cremaschi. 1998. The role of time versus climate in the formation of deep soils of the Apennine fringe of the Po Valley, Italy. *Quat. Intl* **51/52**: 95–107.

Busacca, A. J. and M. J. Singer. 1989. Pedogenesis of a chronosequence in the Sacramento Valley, California, USA. II. Elemental chemistry of silt fractions. *Geoderma* **44**: 43–75.

Bushnell, T. M. 1942. Some aspects of the soil catena concept. *Soil Sci. Soc. Am. Proc.* **7**: 466–476.

Bushue, L. J., Fehrenbacher, J. B., and B. W. Ray. 1974. Exhumed paleosols and associated modern till soils in western Illinois. *Soil Sci. Soc. Am. Proc.* **34**: 665–669.

Busson, G. and J. P. Perthuisot. 1977. Intérêt de la Sabkha el Melah (Sud Tunisien) pour l'interpretation des séries évaporitiques anciennes. *Sed. Geol.* **19**: 139–164.

Butler, B. E. 1956. Parna: an eolian clay. *Austral J. Sci.* **18**: 145–151.

1959. *Periodic Phenomena in Landscapes as a Basis for Soil Studies*, Soil Publication no. 14. Melbourne, Australia, CSIRO.

1982. A new system for soil studies. *J. Soil Sci.* **33**: 581–595.

Buurman, P. 1993. Classification of paleosols: a comment. *INQUA/ISSS Paleopedology Comm. Newslett.* **9**: 3–7.

Buurman, P. and L. P. van Reeuwijk. 1984. Proto-imogolite and the process of podzol formation: a critical note. *J. Soil Sci.* **35**: 447–452.

Byers, H. G., Kellogg, C. E., Anderson, M. S., *et al.* 1938. Formation of soil. In *Soils and Men: Yearbook of Agriculture*. Washington, DC, US Department of Agriculture, US Government Printing Office, pp. 948–978.

Cabrera-Martinez, F., Harris, W. G., Carlisle, V. W., *et al.* 1989. Evidence for clay translocation in coastal plain soils with sandy/loamy boundaries. *Soil Sci. Soc. Am. J.* **53**: 1108–1114.

Cady, J. G. 1950. Rock weathering and soil formation in the North Carolina Piedmont region. *Soil Sci. Soc. Am. Proc.* **15**: 337–342.

Cady, J. G. and R. B. Daniels. 1968. Genesis of some very old soils: the Paleudults. *Trans. 9th Intl Congr. Soil Sci.* (Adelaide) **4**: 103–112.

Cady, J. G., Wilding, L. P., and L. R. Drees. 1986. Petrographic microscope techniques. In A. Klute (ed.) *Methods of Soil Analysis*, Part 1, *Physical and Mineralogical Methods*, Agronomy Monograph no. 9. Am. Madison, WI, American Society for Agronomy, pp. 185–218.

Cailleux, A. 1942. Les actions éoliennes periglaciaires en Europe. *Mém. Soc. Geol.* France **46**: 1–166.

Caldwell, R. E. and J. Pourzad. 1974. Characterization of selected Paleudults in west Florida. *Soil Crop Sci. Soc. Florida Proc.* **33**: 143–147.

Calhoun, F. G., Smeck, N. E., Slater, B. L., *et al.* 2001. Predicting bulk density of Ohio soils from morphology, genetic principles, and laboratory characterization data. *Soil Sci. Soc. Am. J.* **65**: 811–819.

Calkin, P. E. and J. M. Ellis. 1980. A lichenometric dating curve and its application to Holocene glacier studies in the Central Brooks Range, Alaska, USA. *Arctic Alpine Res.* **12**: 245–264.

Callen, R. A. 1983. Late Tertiary "grey billy" and the age and origin of surficial silicifications (silcrete) in South Australia. *J. Geol. Soc. Austral.* **30**: 393–410.

Callen, R. A., Wasson, R. J., and R. Gillespie. 1983. Reliability of radiocarbon dating of pedogenic carbonate in the Australian arid zone. *Sed. Geol.* **35**: 1–14.

Calvert, C. S., Buol, S. W., and S. B. Weed. 1980a. Mineralogical characteristics and transformations of a vertical rock–saprolite–soil sequence in the North Carolina Piedmont. I. Profile morphology, chemical composition, and mineralogy. *Soil Sci. Soc. Am. J.* **44**: 1096–1103.

1980b. Mineralogical characteristics and transformations of a vertical rock–saprolite–soil sequence in the North Carolina Piedmont. II. Feldspar alteration products: their transformations through the profile. *Soil Sci. Soc. Am. J.* **44**: 1104–1112.

Calvet, F., Pomar, L., and M. Esteban. 1975. Las Rizocreciones del Pleistoceno de Mallorca. *Inst. Invest. Geol., Univ. Barcelona* **30**: 36–60.

Campbell, C. A., Paul, E. A., Rennie, D. A., *et al.* 1967. Factors affecting the accuracy of the carbon-dating method of soil humus studies. *Soil Sci.* **104**: 81–85.

Campbell, G. S., and G. W. Gee. 1986. Water potential: miscellaneous methods. In A. Klute (ed.) *Methods of Soil Analysis*, Part 1, *Physical and Mineralogical Methods*, Agronomy Monograph no. 9, 2nd edn. Madison, WI, American Soil Association and Soil Science Society of America, pp. 619–633.

Campbell, I. B. and G. C. C. Claridge. 1990. Classification of cold desert soils. In J. M. Kimble and W. D. Nettleton (eds.) *Characterization, Classification, and Utilization of Aridisols, Proc. 4th Intl Soil Correlation Meeting (ISCOM IV)*. Lincoln, NE, US Department of Agriculture–Soil Conservation Service, pp. 37–43.

Campbell, J. B. 1977. Variation of selected properties across a soil boundary. *Soil Sci. Soc. Am. J.* **41**: 578–582.

1978. Spatial variation of sand content and pH within single contiguous delineations of two mapping units. *Soil Sci. Soc. Am. J.* **42**: 460–464.

1979. Spatial variability of soils. *Ann. Assoc. Am. Geog.* **69**: 544–556.

Campbell, J. B. and W. J. Edmonds. 1984. The missing geographic dimension to Soil Taxonomy. *Ann. Assoc. Am. Geog.* **74**: 83–97.

Cantlon, J. E. 1953. Vegetation and microclimates on north and south slopes of Cushetunk Mountains, New Jersey. *Ecol. Monogr.* **23**: 241–270.

Cárcamo, H. A., Abe, T. A., Prescott, C. E., *et al.* 2000. Influence of millipedes on litter decomposition,

N mineralization, and microbial communities in a coastal forest in British Columbia, Canada. *Can. J. For. Res.* **30**: 817–826.

Carlson, D. C. and E. M. White. 1987. Effects of prairie dogs on mound soils. *Soil Sci. Soc. Am. J.* **51**: 389–393.

1988. Variations in surface-color, texture, pH, and phosphorous content across prairie dog mounds. *Soil Sci. Soc. Am. J.* **52**: 1758–1761.

Carpenter, C. C. 1953. A study of hibernacula and hibernating associations of snakes and amphibians in Michigan. *Ecology* **34**: 445–453.

Carrara, P. E. and J. T. Andrews. 1973. Problems and application of lichenometry to geomorphic studies, San Juan Mountains, Colorado. *Arctic Alpine Res.* **5**: 373–384.

Carson, M. A. 1984. The meandering-braided river threshold: a reappraisal. *J. Hydrol.* **73**: 315–334.

Carter, B. J. and E. J. Ciolkosz. 1980. Soil temperature regimes of the central Appalachians. *Soil Sci. Soc. Am. J.* **44**: 1052–1058.

1991. Slope gradient and aspect effects on soils developed from sandstone in Pennsylvania. *Geoderma* **49**: 199–213.

Carter, B. J. and W. P. Inskeep. 1988. Accumulation of pedogenic gypsum in western Oklahoma soils. *Soil Sci. Soc. Am. J.* **52**: 1107–1113.

Carter, B. J., Ward, P. A. III, and J. T. Shannon. 1990. Soil and geomorphic evolution within the Rolling Red Plains using Pleistocene volcanic ash deposits. *Geomorphology* **3**: 471–488.

Carter, G. F. and R. L. Pendleton. 1956. The humid soil: process and time. *Geog. Rev.* **46**: 488–507.

Cary, J. W., Campbell, G. S., and R. I. Papendick. 1978. Is the soil frozen or not? An algorithm using weather records. *Water Resources Res.* **14**: 1117–1122.

Cassell, D. K. and A. Klute. 1986. Water potential: tensiometry. In A. Klute (ed.) *Methods of Soil Analysis*, Part 1, *Physical and Mineralogical Methods*, Agronomy Monograph no. 9, 2nd edn. Madison, WI, American Soil Association and Soil Science Society of America, pp. 563–596.

Cates, K. J. and F. W. Madison. 1993. *Soil-Attentuation-Potential Map of Trempealeau County*, Wisconsin, Soil map no. 14. Madison, WI, Wisconsin Geology and Natural History Survey.

Catt, J. A. 1990. Paleopedology manual. *Quat. Intl* **6**: 1–95.

1998. Report from working group on definitions used in paleopedology. *Quat. Intl* **51/52**: 84.

Cerling, T. E. 1984. The stable isotope composition of modern soil carbonate and its relationship to climate. *Earth Planet. Sci. Lett.* **71**: 229–240.

1990. Dating geomorphologic surfaces using cosmogenic ^{3}He. *Quat. Res.* **33**: 148–156.

Cerling, T. E. and H. Craig. 1994. Geomorphology and *in-situ* cosmogenic isotopes. *Ann. Rev. Earth Planet. Sci.* **22**: 273–317.

Cerling, T. E. and R. L. Hay. 1986. An isotopic study of paleosol carbonates from Olduvai Gorge. *Quat. Res.* **25**: 63–78.

Cerling, T. E., Quade, J., Wang, Y., *et al.* 1989. Carbon isotopes in soils and palaeosols as ecology and palaeoecology indicators. *Nature* **341**: 138–139.

Černohouz, J. and I. Šolc. 1966. Use of sandstone wanes and weathered basalt crust in absolute chronology. *Nature* **212**: 806–807.

Chadwick, O. A. and J. O. Davis. 1990. Soil-forming intervals caused by eolian pulses in the Lahontan Basin, northwestern Nevada. *Geology* **18**: 243–246.

Chadwick, O. A., Brimhall, G. H., and D. M. Hendricks. 1990. From a black box to a gray box: a mass balance interpretation of pedogenesis. *Geomorphology* **3**: 369–390.

Chadwick, O. A., Hall, R. D., and F. M. Phillips. 1997. Chronology of Pleistocene glacial advances in the central Rocky Mountains. *Geol. Soc. Am. Bull.* **109**: 1443–1452.

Chadwick, O. A., Hendricks, D. M., and W. D. Nettleton. 1987. Silica in duric soils. I. A depositional model. *Soil Sci. Soc. Am. J.* **51**: 975–982.

1989a. Silicification of Holocene soils in northern Monitor Valley, Nevada. *Soil Sci. Soc. Am. J.* **53**: 158–164.

Chadwick, O. A., Nettleton, W. D., and G. J. Staidl. 1995. Soil polygenesis as a function of Quaternary climate change, northern Great Basin, USA. *Geoderma* **68**: 1–26.

Chadwick, O., Sowers, J., and R. Amundson. 1989b. Morphology of calcite crystals in clast coatings from four soils in the Mojave Desert region. *Soil Sci. Soc. Am. J.* **53**: 211–219.

Chamberlain, E. J. and A. J. Gow. 1979. Effect of freezing and thawing on the permeability and structure of soils. *Engin. Geol.* **13**: 73–92.

Chamberlain, E. J., Iskandar, I., and S. E. Hunsicker. 1990. Effect of freeze–thaw cycles on the permeability and macrostructure of soils. In *Frozen Soil Impacts on Agricultural, Range, and Forest Lands.* Hanover, NH, US Army Cold Regions Research and Engineering Laboratory, pp. 144–155.

Chamberlain, T. C. 1895. The classification of American glacial deposits. *J. Geol.* **3**: 270–277.

Chandler, R. F., Jr. 1942. The time required for Podzol profile formation as evidenced by the Mendenhall glacial deposits near Juneau, Alaska. *Soil Sci. Soc. Am. Proc.* **7**: 454–459.

Chang, C. W. 1950. Effects of long-time cropping on soil properties. *Soil Sci.* **69**: 359–368.

Chapman, S. L. and M. E. Horn. 1968. Parent material uniformity and origin of silty soils in northwest Arkansas based on zirconium–titanium contents. *Soil Sci. Soc. Am. Proc.* **32**: 265–271.

Chappel, J. 1983. Thresholds and lags in geomorphologic changes. *Austral. Geog.* **15**: 357–366.

Chartres, C. J. 1987. The composition and formation of grainy void cutans in some soils with textural contrast in southeastern Australia. *Geoderma* **39**: 209–233.

Chen, J., Blume, H. P., and L. Beyer. 2000. Weathering of rocks induced by lichen colonization: a review. *Catena* **39**: 121–146.

Chen, Y. and H. Polach. 1986. Validity of ^{14}C ages of carbonates in sediments. *Radiocarbon* **28**: 464–472.

Chenery, E. M. 1951. Some aspects of the aluminium cycle. *J. Soil Sci.* **2**: 97–109.

Cheng, W., Coleman, D. C., Carroll, C. R., *et al.* 1993. *In situ* measurement of root respiration and soluble C concentrations in the rhizosphere. *Soil Biol. Biochem.* **25**: 1189–1196.

Cherkinsky, A. E. and V. A. Brovkin. 1991. A model of humus formation in soils based on radiocarbon data of natural ecosystems. *Radiocarbon* **33**: 186–187.

Chertov, O. G. 1966. Description of humus-profile types for podzolic soils of Leningrad Oblast. *Sov. Soil Sci.* 266–275.

Chester, R., Baxter, G. G., Behairy, A. K. A., *et al.* 1977. Soil-sized eolian dusts from the lower troposphere of the eastern Mediterranean Sea. *Marine Geol.* **24**: 201–217.

Chesworth, W. 1973. The parent rock effect in the genesis of soil. *Geoderma* **10**: 215–225.

Childs, C. W., Parfitt, R. L., and R. Lee. 1983. Movement of aluminum as an inorganic complex in some podzolised soils, New England. *Geoderma* **29**: 139–155.

Chinn, T. J. H. 1981. Use of rock weathering-rind thickness for Holocene absolute age-dating in New Zealand. *Arctic Alpine Res.* **13**: 33–45.

Chittleborough, D. J. 1992. Formation and pedology of duplex soils. *Austral. J. Exp. Agric.* **32**: 815–825.

Chittleborough, D. J., Oades, J. M., and P. H. Walker. 1984. Textural differentiation in chronosequences from eastern Australia. III. Evidence from elemental chemistry. *Geoderma* **32**: 227–248.

Chlachula, J., Evans, M. E., and N. W. Rutter. 1998. A magnetic investigation of a Late Quaternary loess/palaeosol record in Siberia. *Geophys. J. Intl* **132**: 128–132.

Christopherson, R. W. 2000. *Geosystems: An Introduction to Physical Geography*. Upper Saddle River, NJ, Prentice-Hall.

Ciolkosz, E. J. 1965. Peat mounds in southeastern Wisconsin. *Soil Surv. Horiz.* **6**: 15–17.

Ciolkosz, E. J., Cronce, R. C., Cunningham, R. L., *et al.* 1985. Characteristics, genesis and classification of Pennsylvania minesoils. *Soil Sci.* **139**: 232–238.

Ciolkosz, E. J., Peterson, G. W., Cunningham, R. L., *et al.* 1979. Soils developed from colluvium in the Ridge and Valley area of Pennsylvania. *Soil Sci.* **128**: 153–162.

Ciolkosz, E. J., Waltman, W. J., Simpson, T. W., *et al.* 1989. Distribution and genesis of soils of the northeastern United States. *Geomorphology* **2**: 285–302.

Ciolkosz, E. J., Waltman, W. J., and N. C. Thurman. 1995. Fragipans in Pennsylvania soils. *Soil Surv. Horiz.* **36**: 5–20.

Cioppa, M. T., Karlstrom, E. T., Irving, E., *et al.* 1995. Paleomagnetism of tills and associated paleosols in southwestern Alberta and northern Montana: evidence for Late Pliocene–Early Pleistocene glaciations. *Can. J. Earth Sci.* **32**: 555–564.

Cipra, J. E., Bidwell, O. W., Whitney, D. A., *et al.* 1972. Variations with distance in selected fertility measurements of pedons of Western Kansas Ustoll. *Soil Sci. Soc. Am. Proc.* **36**: 111–115.

Claridge, G. G. C. and I. B. Campbell. 1968. Soils of the Shackleton Glacier region, Queen Maud Range, Antarctica. *N. Z. J. Sci.* **11**: 171–218.

Claridge, G. G. C. and I. B. Campbell. 1982. A comparison between hot and cold desert soils and soil processes. *Catena (Suppl.)* **1**: 1–28.

Clark, M. M. 1972. Intensity of shaking estimated from displaced stones. *US Geol. Surv. Prof. Paper* **787**: 175–182.

Clayden, B., Daly, B. K., Lee, R., *et al.* 1990. The nature, occurrence and genesis of placic horizons. In Kimble, J. M. and R. D. Yeck (eds.) *Characterization, Classification, and Utilization of Spodosols, Proc. 5th Intl Soil Correlation Mtg (ISCOM)*. Lincoln, NE, US Department of Agriculture–Soil Conservation Service, pp. 88–104.

Clayton, L. and J. W. Attig. 1989. *Glacial Lake Wisconsin*, Memoir no. 173. Madison, WI, Geological Society of America.

Cleaves, E. T., Godfrey, A. E., and O. P. Bricker. 1970. Geochemical balance of a small watershed and its geomorphic implications. *Geol. Soc. Am. Bull.* **81**: 3015–3032.

Clements, F. E. 1936. Nature and structure of the climax. *J. Ecol.* **24**: 252–284.

Cline, A. J. and D. D. Johnson. 1963. Threads of genesis in the Seventh Approximation. *Soil Sci. Soc. Am. Proc.* **27**: 220–222.

Cline, M. G. 1949a. Basic principles of soil classification. *Soil Sci.* **67**: 81–91.

1949b. Profile studies of normal soils of New York. I. Soil profile sequences involving Brown Forest, Gray-Brown Podzolic, and Brown Podzolic soils. *Soil Sci.* **68**: 259–272.

1955. *Soil Survey of the Territory of Hawaii*. Washington, DC, Soil Conservation Service, US Department of Agriculture, US Government Printing Office.

1963. Logic of the new system of soil classification. *Soil Sci.* **96**: 17–22.

1975. Origin of the term Latosol. *Soil Sci. Soc. Am. Proc.* **39**: 162.

1977. Historical highlights in soil genesis, morphology, and classification. *Soil Sci. Soc. Am. J.* **41**: 250–254.

Clothier, B. E., Pollok, J. A., and D. R. Scotter. 1978. Mottling in soil profiles containing a coarse-textured horizon. *Soil Sci. Soc. Am. J.* **42**: 761–763.

Coates, D. R. 1984. Landforms and landscapes as measures of relative time. In W. C. Mahaney (ed.) *Quaternary Dating Methods*. New York, Elsevier, pp. 247–267.

Coates, D. R. and J. D. Vitek (eds.) 1980. *Thresholds in Geomorphology*. London, Allen and Unwin.

Coester, M., Pfisterer, U., and H. K. Siem. 1997. Soil development and heavy metal contents of a street sweepings dump. *Z. Pflanzen. Bodenk.* **160**: 89–92.

Coile, T. S. 1953. Moisture content of a small stone in soil. *Soil Sci.* **75**: 203–207.

Cointepas, J. P. 1967. Les sols rouges et bruns meditérranéens de Tunisie. *Trans. Conf. Mediterranean Soils* (Madrid). 187–194.

Colgan, P. M. 1999. Early middle Pleistocene glacial sediments (780 000–620 000 BP) near Kansas City, northeastern Kansas and northwestern Missouri, USA. *Boreas* **28**: 477–489.

Collins, J. F. and T. O'Dubhain. 1980. A micromorphological study of silt concentrations in some Irish podzols. *Geoderma* **24**: 215–224.

Colman, S. M. 1981. Rock-weathering rates as functions of time. *Quat. Res.* **15**: 250–264.

Colman, S. M. and K. L. Pierce. 1981. Weathering rinds on basaltic and andesitic stones as a Quaternary age indicator, western United States. *US Geol. Surv. Prof. Paper* **1210**: 1–56.

Colman, S. M. and K. L. Pierce. 1986. Glacial sequence near McCall, Idaho: weathering rinds, soil development, morphology, and other relative-age criteria. *Quat. Res.* **25**: 25–42.

Colman, S. M., Pierce, K. L., and P. W. Birkeland. 1987. Suggested terminology for Quaternary dating methods. *Quat. Res.* **28**: 314–319.

Conacher, A. J. and J. B. Dalrymple. 1977. The nine unit landscape model: an approach to pedogeomorphic research. *Geoderma* **18**: 1–154.

Conkin, J. E. and B. M. Conkin. 1961. Fossil land snails from the loess at Vicksburg, Mississippi. *Trans. Kentuclces Acad. Sci.* **22**: 10–15.

Cooke, H. B. S. 1973. Pleistocene chronology: long or short? *Quat. Res.* **3**: 206–220.

Cooke, R. U. 1970. Stone pavements in deserts. *Ann. Assoc. Am. Geog.* **60**: 560–577.

Cooke, R. U., Warren, A. and A. S. Goudie. 1993. *Desert Geomorphology*. London, UCL Press.

Cooper, A. W. 1960. An example of the role of microclimate in soil genesis. *Soil Sci.* **90**: 109–120.

Cooper, W. S. 1913. The climax forest of Isle Royale, Lake Superior, and its development. *I. Bot. Gaz.* **55**: 1–44.

Cornu, S., Lucas, Y., Lebon, E., *et al.* 1999. Evidence of titanium mobility in soil profiles, Manaus, central Amazonia. *Geoderma* **91**: 281–295.

Corte, A. E. 1963. Particle sorting by repeated freezing and thawing. *Science* **142**: 499–501.

Cortes, A. and D. P. Franzmeier. 1972. Climosequence of ash-derived soils in the central Cordillera of Columbia. *Soil Sci. Soc. Am. Proc.* **36**: 653–659.

Corti, G., Ugolini, F. C., and A. Agnelli. 1998. Classing the soil skeleton (greater than two millimeters): proposed approach and procedure. *Soil Sci. Soc. Am. J.* **62**: 1620–1629.

Costin, A. B. 1955a. A note on gilgaies and frost soils. *J. Soil Sci.* **6**: 32–34.

Costin, A. B. 1955b. Alpine soils in Australia with reference to conditions in Europe and New Zealand. *J. Soil Sci.* **6**: 35–50.

Coulombe, C. E., Wilding, L. P., and J. B. Dixon. 1996. Overview of Vertisols: characteristics and impacts on society. *Adv. Agron.* **57**: 289–375.

Courchesne, F. 2000. Breaking the barrier of conceptual locks. *Ann. Assoc. Am. Geog.* **90**: 782–785.

Courchesne, F. and G. R. Gobran. 1997. Mineralogical variations of bulk and rhizosphere soils from a Norway Spruce stand. *Soil Sci. Soc. Am. J.* **61**: 1245–1249.

Courchesne, F. and W. H. Hendershot. 1997. La Genèse des Podzols. *Geog. Phys. Quat.* **51**: 235–250.

Courty, M. A., Goldberg, P., and R. Macphail. 1989. *Soils and Micromorphology in Archaeology*. New York, Cambridge University Press.

Courty, M.-A., Marlin, C., Dever, L., *et al.* 1994. Morphology, geochemistry and origin of calcitic pendants from the High Arctic (Spitsbergen). *Geoderma* **61**: 71–102.

Couto, W., Sanzonowicz, C., and A. de O. Barcellos. 1985. Factors affecting oxidation–reduction processes in an Oxisol with a seasonal water table. *Soil Sci. Soc. Am. J.* **49**: 1245–1248.

Coventry, R. J., Holt, J. A., and D. F. Sinclair. 1988. Nutrient cycling by mound-building termites in low-fertility soils of semi-arid tropical Australia. *Austral. J. Soil Res.* **26**: 375–390.

Covey, C. 1984. The Earth's orbit and the ice ages. *Sci. Am.* **250(2)**: 58–66.

Cowan, J. A., Humphreys, G. S., Mitchell, P. B., *et al.* 1985. An assessment of pedoturbation by two species of mound-building ants, *Camponotus intrepidus* (Kirby) and *Iridomyrmex purpureus* (F. Smith). *Austral. J. Soil Res.* **22**: 95–107.

Cox, A., Doell, R. R., and G. B. Dalrymple. 1963. Geomagnetic polarity epochs and Pleistocene geochronometry. *Nature* **198**: 1049–1051.

Cox, G. W. 1984. The distribution and origin of Mima mound grasslands in San Diego County, California. *Ecology* **65**: 1397–1405.

1990. Soil mining by pocket gophers along topographic gradients in a Mima moundfield. *Ecology* **71**: 837–843.

Cox, G. W. and D. W. Allen. 1987a. Soil translocation by pocket gophers in a Mima moundfield. *Oecologia* **72**: 207–210.

1987b. Sorted stone nets and circles of the Columbia Plateau: a hypothesis. *Northwest Sci.* **61**: 179–185.

Cox, G. W. and C. G. Gakahu. 1986. A latitudinal test of the fossorial rodent hypothesis of Mima mound origin. *Z. Geomorph.* **30**: 485–501.

1987. Biogeographical relationships of rhizomyid mole rats with Mima mound terrain in the Kenya highlands. *Pedobiologia* **30**: 263–275.

Cox, G. W. and J. Hunt. 1990. Form of Mima mounds in relation to occupancy by pocket gophers. *J. Mammal.* **71**: 90–94.

Cox, G. W. and W. T. Lawrence. 1983. Cemented horizon in subarctic Alaskan sand dunes. *Am. J. Sci.* **283**: 369–373.

Cox, G. W. and V. G. Roig. 1986. Agentinian Mima mounds occupied by *Ctenomyid* rodents. *J. Mamm.* **67**: 428–432.

Cox, G. W. and V. B. Scheffer. 1991. Pocket gophers and mima terrain in North America. *Natural Areas J.* **11**: 193–198.

Cox, G. W., Gakahu, C. G., and D. W. Allen. 1987a. Small-stone content of Mima mounds of the Columbia Plateau and Rocky Mountain regions: implications for mound origin. *Gt Basin Natural.* **47**: 609–619.

Cox, G. W., Gakahu, C. G., and J. M. Waithaka. 1989. The form and small stone content of large earth mounds constructed by mole rats and termites in Kenya. *Pedobiologia* **33**: 307–314.

Cox, G. W., Lovegrove, B. G., and W. R. Siegfried. 1987b. The small stone content of Mima-like mounds in the South African cape region: implications for mound origin. *Catena* **14**: 165–176.

Cox, T. 1998. Origin of stone concentrations in loess-derived interfluve soils. *Quat. Intl* **51/52**: 74–75.

Crabtree, R. W. and T. P. Burt. 1983. Spatial variation in solutional denudation and soil moisture over a hillslope hollow. *Earth Surf. Proc. Landforms* **8**: 151–160.

Craig, H. and R. J. Poreda. 1986. Cosmogenic ^{3}He in terrestrial rocks: the summit lavas of Maui. *Proc. Natl Acad. Sci.* USA **83**: 1970–1974.

Crawford, T. W., Jr., Whittig, L. D., Begg, E. L., *et al.* 1983. Eolian influence on development and weathering of some soils of Point Reyes peninsula, California. *Soil Sci. Soc. Am. J.* **47**: 1179–1185.

Cremaschi, M. 1987. *Paleosols and Vetusols in the Central Po Plains (Northern Italy): A Study in Quaternary Geology and Soil Development*, Studi e ricerche sul territorio no. 28. Milan, Edizioni Unicopli.

Cremeens, D. L. 1995. Pedogenesis of Cotiga Mound, a 2100-year-old Woodland mound in southwest West Virginia. *Soil Sci. Soc. Am. J.* **59**: 1377–1388.

2000. Pedology of the regolith–bedrock boundary: an example from the Appalachian plateau of northern West Virginia. *Southeast. Geol.* **39**: 329–339.

2003. Geoarchaeology of soils on stable geomorphic surfaces: mature soil model for the glaciated northeast. In D. L. Cremeens and J. P. Hart (eds.) *Geoarchaeology of Landscapes in the Glaciated Northeast*, Bulletin no. 497. New York State Museum, Albany, NY, pp. 49–60.

Cremeens, D. L. and J. P. Hart. 1995. On chronostratigraphy, pedostratigraphy, and archaeological context. *Soil Sci. Soc. Am. Spec. Publ.* **44**: 15–33.

Cremeens, D. L. and J. C. Lothrop. 2001. Geomorphology of upland regolith in the unglaciated Appalachian Plateau. In L. P. Sullivan and S. C. Prezzano (eds.) *Archaeology of Appalachian Highlands*, Knoxville, TN, University of Tennessee Press, pp. 31–48.

Cremeens, D. L. and D. L. Mokma. 1986. Argillic horizon expression and classification in the soils of two Michigan hydrosequences. *Soil Sci. Soc. Am. J.* **50**: 1002–1007.

Cremeens, D. L., Norton, L. D., Darmody, R. G., *et al.* 1988. Etch-pit measurements on scanning electron micrographs of weathered grain surfaces. *Soil Sci. Soc. Am. J.* **52**: 883–885.

Crocker, R. L. 1952. Soil genesis and the pedogenic factors. *Q. Rev. Biol.* **27**: 139–168.

1960. The plant factor in soil formation. *Proc. 9th Pacific Sci. Congr.* **18**: 84–90.

Crocker, R. L. and J. Major. 1955. Soil development in relation to vegetation and surface age at Glacier Bay, Alaska. *J. Ecol.* **43**: 427–448.

Crompton, E. 1962. Soil formation. *Outlook Agric.* **3**: 209–218.

Crook, R. Jr. 1986. Relative dating of Quaternary deposits based on P-wave velocities in weathered granitic clasts. *Quat. Res.* **25**: 281–292.

Crosson, L. S. and R. Protz. 1974. Quantitative comparison of two closely related soil mapping units. *Can. J. Soil Sci.* **54**: 7–14.

Crownover, S. H., Collins, M. E., and D. A. Lietzke. 1994. Parent materials and stratigraphy of a doline in the valley and ridge province. *Soil Sci. Soc. Am. J.* **58**: 1738–1746.

Cullity, B. D. 1978. *Elements of X-Ray Diffraction*, 2nd edn. New York, Addison-Wesley.

Culver, J. R. and F. Gray. 1968a. Morphology and genesis of some grayish claypan soils of Oklahoma. I. Morphology, chemical and physical measurements. *Soil Sci. Soc. Am. Proc.* **32**: 845–851.

1968b. Morphology and genesis of some grayish claypan soils of Oklahoma. II. Mineralogy and genesis. *Soil Sci. Soc. Am. Proc.* **32**: 851–857.

Curi, N. and D. P. Franzmeier. 1984. Toposequence of Oxisols from the Central Plateau of Brazil. *Soil Sci. Soc. Am. J.* **48**: 341–346.

Curry, B. B. and M. J. Pavich. 1996. Absence of glaciation in Illinois during Marine Isotope Stages 3 through 5. *Quat. Res.* **46**: 19–26.

Curtis, J. T. 1959. *The Vegetation of Wisconsin*. Madison, WI, University of Wisconsin Press.

Curtis, R. O. and B. W. Post. 1964. Estimating bulk density from organic-matter content in some Vermont forest soils. *Soil Sci. Soc. Am. Proc.* **28**: 285–286.

Curtiss, B., Adams, J. B., and M. K. Ghiorso. 1985. Origin, development and chemistry of silica-alumina rock coatings from the semi-arid regions of the island of Hawaii. *Geochim. Cosmochim. Acta* **49**: 49–56.

Cutler, E. J. B. 1981. The texture profile forms of New Zealand soils. *Austral. J. Soil Res.* **19**: 97–102.

Dahlgren, R. A. and F. C. Ugolini. 1989. Aluminum fractionation of soil solutions from unperturbed and tephra-treated spodosols, Cascade Range, Washington, USA. *Soil Sci. Soc. Am. J.* **53**: 559–566.

1991. Distribution and characterization of short-range-order minerals in Spodosols from the Washington Cascades. *Geoderma* **48**: 391–413.

Dahlgren, R. A., Boettinger, J. L., Huntington, G. L., *et al.* 1997. Soil development along an elevational transect in the western Sierra Nevada, California. *Geoderma* **78**: 207–236.

Dahms, D. E. 1993. Mineralogical evidence for eolian contributions to soils of Late Quaternary moraines, Wind River Mountains, Wyoming, USA. *Geoderma* **59**: 175–196.

1998. Reconstructing paleoenvironments from ancient soils: a critical review. *Quat. Intl* **51/52**: 58–60.

2002. Glacial stratigraphy of Stough Creek Basin, Wind River Range, Wyoming. *Geomorphology* **42**: 59–83.

Dalrymple, G. B., Cox, A., and R. R. Doell. 1965. Potassium–argon age and paleomagnetism of the Bishop tuff, California. *Geol. Soc. Am. Bull.* **76**: 665–674.

Dalrymple, J. B. 1958. The application of soil micromorphology to fossil soils and other deposits from archaeological sites. *J. Soil Sci.* **9**: 199–209.

Dalsgaard, K. and B. V. Odgaard. 2001. Dating sequences of buried horizons of podzols developed in wind-blown sand at Ulfborg, Western Jutland. *Quat. Intl* **78**: 53–60.

Dalsgaard, K., Baastrup, E., and B. T. Bunting. 1981. The influence of topography on the development of Alfisols on calcareous clayey till in Denmark. *Catena* **8**: 111–136.

Dan, J. and D. H. Yaalon. 1964. The application of the catena concept in studies of pedogenesis in Mediterranean and desert fringe regions. *Trans. 8th Intl Congr. Soil Sci.* **83**: 751–758.

1982. Automorphic saline soils in Israel. *Catena (Suppl.)* **1**: 103–115.

Dan, J., Yaalon, D. H., and H. Koyumdjisky. 1968. Catenary soil relationships in Israel. I. The Netanya catena on coastal dunes of the Sharon. *Geoderma* **2**: 95–120.

1972. Catenary soil relationships in Israel. II. The Bet Gurvin catena on chalk and nari limestone crust in the Shefela. *Isr. J. Earth Sci.* **21**: 99–114.

Dan, J., Yaalon, D. H., Moshe, R., *et al.* 1982. Evolution of reg soils in southern Israel and Sinai. *Geoderma* **28**: 173–202.

Daniels, R. B. and S. W. Buol. 1992. Water table dynamics and significance to soil genesis. In J. M. Kimble (ed.) *Characterization, Classification, and Utilization of Wet Soils, Proc. 8th Intl Soil Correlation Mtg (ISCOM)*. Lincoln, NE, US Department of Agriculture–Soil Conservation Service, pp. 66–74.

Daniels, R. B. and E. E. Gamble. 1967. The edge effect in some Ultisols in the North Carolina coastal plain. *Geoderma* **1**: 117–124.

Daniels, R. B. and R. D. Hammer. 1992. *Soil Geomorphology*. New York, John Wiley.

Daniels, R. B., Buol, S. W., Kleiss, H. J., *et al.* 1999. *Soil Systems in North Carolina*, Technical Bulletin no. 314. Raleigh, NC, North Coroline State University.

Daniels, R. B., Gamble, E. E., and L. J. Bartelli. 1968. Eluvial bodies in B horizons of some Ultisols. *Soil Sci.* **106**: 200–206.

Daniels, R. B., Gamble, E. E., and J. G. Cady. 1971a. The relation between geomorphology and soil morphology and genesis. *Adv. Agron.* **23**: 51–88.

Daniels, R. B., Gamble, E. E., and L. A. Nelson. 1971b. Relations between soil morphology and water-table levels on a dissected North Carolina coastal plain surface. *Soil Sci. Soc. Am. Proc.* **35**: 781–784.

Daniels, R. B., Gamble, E. E., and W. H. Wheeler. 1978a. Age of soil landscapes in the coastal plain of North Carolina. *Soil Sci. Soc. Am. J.* **42**: 98–105.

Daniels, R. B., Gamble, E. E., Wheeler, W. H., *et al.* 1977. The stratigraphy and geomorphology of the Hofmann Forest pocosin. *Soil Sci. Soc. Am. J.* **41**: 1175–1180.

Daniels, R. B., Handy, R. L., and G. H. Simonson. 1960. Dark-colored bands in the thick loess of western Iowa. *J. Geol.* **68**: 450–458.

Daniels, R. B., Perkins, H. F., Hajek, B. F., *et al.* 1978b. Morphology of discontinuous phase plinthite and criteria for its field identification in the southeastern United States. *Soil Sci. Soc. Am. J.* **42**: 944–949.

Daniels, R. B., Simonson, G. H., and R. L. Handy. 1961. Ferrous iron content and color of sediments. *Soil Sci.* **91**: 378–382.

Daniels, W. L., Everett, C. J., and L. W. Zelazny. 1987. Virgin hardwood forest soils of the southern Appalachian Mountains. I. Soil morphology and geomorphology. *Soil Sci. Soc. Am. J.* **51**: 722–729.

Danin, A., Dor, I., Sandler, A., *et al.* 1998. Desert crust morphology and its relations to microbiotic succession at Mt. Sedom, Israel. *J. Arid Envs.* **38**: 161–174.

Danin, A., Gerson, R., and J. Garty. 1983. Weathering patterns on hard limestone and dolomite by endolithic lichens and cyanobacteria: supporting evidence for eolian contribution to terra rossa soil. *Soil Sci.* **136**: 213–217.

Dansgard, W., White, J. W. C., and S. J. Johnsen. 1989. The abrupt termination of the Younger Dryas climate event. *Nature* **339**: 532–534.

Dare-Edwards, A. J. 1984. Aeolian clay deposits of southeastern Australia: parna or loessic clay? *Trans. Inst. Brit. Geog.* **9**: 337–344.

Darmody, R. G. and J. E. Foss. 1979. Soil–landscape relationships of the tidal marshes of Maryland. *Soil Sci. Soc. Am. J.* **43**: 534–541.

Darmody, R. G., Fanning, D. S., Drummond, W. J., Jr., *et al.* 1977. Determination of the total sulfur in tidal marsh soils by X-ray spectroscopy. *Soil Sci. Soc. Am. J.* **41**: 761–765.

Darwin, C. 1881. *On the Formation of Vegetable Mould Through the Action of Worms*. London, John Murray.

Darwish, T. M. and R. A. Zurayk. 1997. Distribution and nature of Red Mediterranean soils in Lebanon along an altitudinal sequence. *Catena* **28**: 191–202.

Dasog, G. S., Acton, D. F., and A. R. Mermut. 1987. Genesis and classification of clay soils with vertic properties in Saskatchewan. *Soil Sci. Soc. Am. J.* **51**: 1243–1250.

Davey, B. G., Russell, J. D., and M. J. Wilson. 1975. Iron oxide and clay minerals and their relation to colours of red and yellow podzolic soils near Sydney, Australia. *Geoderma* **14**: 125–138.

Davis, J. O. 1985. Correlation of Late Quaternary tephra layers in a long pluvial sequence near Summer Lake, Oregon. *Quat. Res.* **23**: 38–53.

Davis, M. B. 1969. Climatic changes in southern Connecticut recorded by pollen deposition at Rogers Lake. *Ecology* **50**: 409–422.

Davis, R. B., Bradstreet, T. E., Stuckenrath, R., *et al.* 1975. Vegetation and associated environments during the past 14000 years near Moulton Pond, Maine. *Quat. Res.* **5**: 435–465.

Day, M. J. and A. S. Goudie. 1977. Field assessment of rock hardness using the Schmidt test hammer. *Br. Geom. Res. Group Tech. Bull.* **18**: 19–29.

De Kimpe, C. R. 1970. Chemical, physical and mineralogical properties of a Podzol soil with fragipan derived from glacial till in the province of Quebec. *Can. J. Soil Sci.* **50**: 317–330.

De Kimpe, C. R. and J. L. Morel. 2000. Urban soil management: a growing concern. *Soil Sci.* **165**: 31–40.

De Kimpe, C. R., Baril, R. W., and R. Rivard. 1972. Characterization of a toposequence with fragipan: the Leeds–Ste. Marie–Brompton series of soils, province of Quebec. *Can. J. Soil Sci.* **52**: 135–150.

De Ploey, J. 1964. Nappes de gravats et couvertures argilo-sableuses au Bas-Congo: leur genèse et l'action des termites. In A. Bouillon (ed). *Etudes sur les termites africains.* Léopoldville, Congo, Editions de l'Université de Léopoldville, pp. 399–414.

De Ploey, J., Savat, J., and J. Moeyersons. 1976. The differential impact of some soil loss factors on flow, runoff creep and rainwash. *Earth Surf. Proc.* **1**: 151–161.

De Villiers, J. M. 1965. Present soil-forming factors and processes in tropical and subtropical regions. *Soil Sci.* **99**: 50–57.

De Vos, J. H. and K. J. Virgo. 1969. Soil structure in Vertisols of the Blue Nile Clay Plains, Sudan. *J. Soil Sci.* **20**: 189–206.

de Vries, W. and A. Breeuwsma. 1984. Causes of soil acidification. *Neth. J. Agric. Sci.* **32**: 159–161.

Dean, W. R. J. and S. J. Milton. 1991. Disturbances in semi-arid shrubland and arid grassland in the Karoo, South Africa: mammal diggings as germination sites. *Afr. J. Ecol.* **29**: 11–16.

DeConinck, F. 1980. Major mechanisms in formation of spodic horizons. *Geoderma* **24**: 101–128.

DeConinck, F. and A. Herbillon. 1969. Evolution minéralogique et chémique des fractions argileuses dans des Alfisols et des Spodosols de la Campine (Belgique). *Pédologie* **19**: 159–272.

DeConinck, F., Herbillon, A. J., Tavernier, R., *et al.* 1968. Weathering of clay minerals and formation of amorphous material during the degradation of a Bt horizon and podzolization in Belgium. *Trans. 9th Intl Congr. Soil Sci.* (Adelaide) **4**: 353–365.

Dekker, L. W. and C. J. Ritsema. 1996. Uneven moisture patterns in water repellent soils. *Geoderma* **70**: 87–99.

Del Villar, E. H. 1944. The tirs of Morocco. *Soil Sci.* **57**: 313–339.

Denny, C. S. and J. C. Goodlett. 1956. Microrelief resulting from fallen trees. *US Geol. Surv. Prof. Paper* **288**: 59–66.

1968. Tree-throw origin of patterned ground on beaches of the ancient Champlain Sea near Plattsburgh, New York. *US Geol. Surv. Prof. Paper* **600B**: 157–164.

Dexter, A. R. 1978. Tunnelling of soil by earthworms. *Soil Biol. Biochem.* **10**: 447–449.

Diamond, D. D. and F. E. Smeins. 1985. Composition, classification and species response patterns of remnant tallgrass prairies in Texas. *Am. Midl. Nat.* **113**: 294–308.

Dideriksen, R. O. 1992. *Soil Survey of Wright County, Iowa.* Washington, DC, US Department of Agriculture Natural Resources Conservation Service, US Government Printing Office.

Dijkerman, J. C. 1974. Pedology as a science: the role of data, models and theories in the study of natural soil systems. *Geoderma* **11**: 73–93.

Dijkerman, J. C. and R. Miedema. 1988. An Ustult–Aquult–Tropept catena in Sierra Leone, West Africa. I. Characteristics, genesis and classification. *Geoderma* **42**: 1–27.

Dimbleby, G. W. 1957. Pollen analysis of terrestrial soils. *New Phytol.* **56**: 12–28.

Dimmock, G. M., Bettenay, E., and M. J. Mulcahy. 1974. Salt content of lateritic soil profiles in the Darling Range, Western Australia. Austral. *J. Soil Res.* **12**: 63–69.

Ding, Z. L. and S. L. Yang. 2000. C_3/C_4 vegetation evolution over the last 7.0 Myr in the Chinese Loess Plateau: evidence from pedogenic carbonate ^{13}C. *Palaeogeog. Palaeoclimat. Palaeoecol.* **160**: 291–299.

Ding, Z. L., Liu, T. S., Rutter, N. W., *et al.* 1995. Ice-volume forcing of East Asia winter monsoon variations in the past 800000 years. *Quat. Res.* **44**: 149–159.

Ding, Z. L., Rutter, N., Han, J. T., *et al.* 1992. A coupled environmental system formed at about 2.5 Ma in

East Asia. *Palaeogeog. Palaeoclimat. Palaeoecol.* **94**: 223–242.

Ding, Z. L., Rutter, N., and L. Tungsheng. 1993. Pedostratigraphy of Chinese loess deposits and climatic cycles in the last 2.5 Myr. *Catena* **20**: 73–91.

Ding, Z. L., Sun, J., Rutter, N. W., *et al.* 1999. Changes in sand content of loess deposits along a north–south transect of the Chinese Loess Plateau and the implications for desert variations. *Quat. Res.* **52**: 56–62.

Dixit, S. P. 1978. Measurement of the mobility of soil colloids. *J. Soil Sci.* **29**: 557–566.

Dixon, J. B. 1989. Kaolin and serpentine group minerals. In J. B. Dixon and S. B. Weed (eds.) *Minerals in Soil Environments*, 2nd edn. Madison, WI, Soil Science Society of America, pp. 467–525.

Dixon, J. B. and S. B. Weed (eds.) 1989. *Minerals in Soil Environments*, 2nd edn. Madison, WI, Soil Science Society of America.

Dixon, J. C. 1991. Alpine and subalpine soil properties as paleoenvironmental indicators. *Phys. Geog.* **12**: 370–384.

1994. Duricrusts. In A. D. Abrahams and A. J. Parsons (eds.) *Geomorphology of Desert Environments.* London, Chapman and Hall, pp. 64–81.

Dixon, J. C. and R. W. Young. 1981. Character and origin of deep arenaceons weathering mantles on the Bega Batholith, Southeastern Australia. *Catena* **8**: 97–109.

Dokuchaev, V. V. 1883. *Russkii Chernozem*. Moscow.

Dokuchaev, V. V. 1893. *The Russian Steppes/Study of the Soil in Russia in the Past and Present.* St. Petersburg, Department of Agricultural Ministry of Crown Domains for the World's Columbian Exposition at Chicago.

Dokuchaev, V. V. 1899. *On the Theory of Natural Zones*. St. Petersburg.

Donahue, R. L., Miler, R. W., and J. C. Shickluna. 1983. *Soils: An Introduction to Soils and Plant Growth.* Englewood Cliffs, NJ, Prentice-Hall.

Donald, R. G., Anderson, D. W., and J. W. B. Stewart. 1993. The distribution of selected soil properties in relation to landscape morphology in forested Gray Luvisol soils. *Can. J. Soil Sci.* **73**: 165–172.

Donaldson, N. C. 1980. *Soil Survey of Whitman County, Washington.* Washington, DC, US Department of Agriculture Soil Conservation Service, US Government Printing Office.

Doner, H. E. and W. C. Lynn. 1989. Carbonate, halide, sulfate, and sulfide minerals. In J. B. Dixon and S. B. Weed (eds.) *Minerals in Soil Environments*, 2nd edn. Madison, WI, Soil Science Society of America, pp. 279–330.

Dong, H., Peacor, D. R., and S. F. Murphy. 1998. TEM study of progressive alteration of igneous biotite to kaolinite throughout a weathered soil profile. *Geochim. Cosmochim. Acta* **62**: 1881–1887.

Doornkamp, J. C. and H. A. M. Ibrahim. 1990. Salt weathering. *Progr. Phys. Geog.* **14**: 335–348.

Dor, I. and A. Danin. 1996. Cyanobacterial desert crusts in the Dead Sea Valley, Israel. *Algol. Stud.* **83**: 197–206.

Dormaar, J. F. and L. E. Lutwick. 1969. Infrared spectra of humic acids and opal phytoliths as indicators of paleosols. *Can. J. Soil Sci.* **49**: 29–37.

1983. Extractable Fe and Al as an indicator for buried soil horizons. *Catena* **10**: 167–173.

Dorn, R. I. 1988. A rock varnish interpretation of alluvial-fan development in Death Valley, California. *Natl Geog. Res.* **4**: 56–73.

1989. Cation-ratio dating: a geographic perspective. *Progr. Phys. Geog.* **13**: 559–596.

1990. Quaternary alkalinity fluctuations recorded in rock varnish microlaminations on western USA volcanics. *Palaeogeog. Palaeoclimat. Palaeoecol.* **76**: 291–310.

Dorn, R. I. and T. M. Oberlander. 1981a. Microbial origin of rock varnish. *Science* **213**: 1245–1247.

1981b. Rock varnish origin, characteristics, and usage. *Z. f. Geomorphol.* **25**: 420–436.

1982. Rock varnish. Progr. Phys. *Geog.* **74**: 308–322.

Dorn, R. I. and F. M. Phillips. 1991. Surface exposure dating: review and critical evaluation. *Phys. Geog.* **12**: 303–333.

Dorn, R. I., Bamforth, D. B., Cahill, T. A., *et al.* 1986. Cation-ratio and accelerator radiocarbon dating of rock varnish on Mojave artifacts and landforms. *Science* **231**: 830–833.

Dörr, H. and K. O. Münnich. 1986. Annual variations of the ^{14}C content of soil CO_2. *Radiocarbon* **28**: 338–406.

Dorronsoro, C. and P. Alonso. 1994. Chronosequence in Almar River fluvial-terrace soil. *Soil Sci. Soc. Am. J.* **58**: 910–925.

Douglas, C. L., Jr., Fehrenbacher, J. B., and B. W. Ray. 1967. The lower boundary of selected Mollisols. *Soil Sci. Soc. Am. Proc.* **31**: 795–800.

Douglas, L. A. 1989. Vermiculites. In J. B. Dixon and S. B. Weed (eds.) *Minerals in Soil Environments*, 2nd edn. Madison, WI, Soil Science Society of America, pp. 635–674.

Douglas, L. A. and M. L. Thompson (eds.) 1985. *Soil*

Micromorphology and Soil Classification, Special Publication no. 15. Madison, WI, Soil Science Society of America.

Downes, R. G. 1954. Cyclic salt as a factor in soil genesis. *Austral. J. Soil Res.* **5**: 448–464.

Dragovich, D. 1988. A preliminary electron probe study of microchemical variations in desert varnish in western South Wales, Australia. *Earth Surf. Proc. Landforms* **13**: 259–270.

Dredge, L. A. 1992. Breakup of limestone bedrock by frost shattering and chemical-weathering, eastern Canadian Arctic. *Arctic Alpine Res.* **24**: 314–323.

Drees, L. R. and L. P. Wilding. 1973. Elemental variability within a sampling unit. *Soil Sci. Soc. Am. Proc.* **37**: 82–87.

Dreimanis, A. 1989. Tills: their genetic terminology and classification. In R. P. Goldthwait and C. L. Matsch (eds.) *Genetic Classification of Glacigenic Deposits.* International Quaternary Association pp. 17–83.

Dreimanis, A., Hütt, G., Raukas, A., *et al.* 1978. Dating methods of Pleistocene deposits and their problems. I. Thermoluminescence dating. *Geosci. Can.* **5**: 55–60.

Drever, J. I. 1994. The effect of land plants on weathering rates of silicate minerals. *Geochim. Cosmochim. Acta* **58**: 2325–2332.

Droste, J. B. and J. C. Tharin. 1958. Alteration of clay minerals in Illinoian till by weathering. *Geol. Soc. Am. Bull.* **69**: 61–68.

Drury, W. H. and I. C. T. Nisbet. 1971. Interrelationships between developmental models in geomorphology, plant ecology, and animal ecology. *Gen. Syst.* **16**: 57–68.

Du Preez, J. W. 1949. Laterite: a general discussion with a description of Nigerian occurrences. *Bull. Agric. Congo Belge* **40**: 53–66.

Duan, L., Hao, J. M., Xie, S. D., *et al.* 2002. Determining weathering rates of soils in China. *Geoderma* **110**: 205–225.

Dubroeucq, D. and B. Volkoff. 1988. Evolution des couvertures pédologiques sableuses à podzols géants d'Amazonie (Bassin du Haut rio Negro). *Cahiers ORSTOM (Série pédol.)* **24**: 191–214.

1998. From Oxisols to Spodosols and Histosols: evolution of the soil mantles in the Rio Negro basin (Amazonia). *Catena* **32**: 245–280.

Dubroeucq, D., Geissert, D., and P. Quantin. 1998. Weathering and soil-forming processes under semi-arid conditions in two Mexican volcanic ash soils. *Geoderma* **86**: 99–122.

Duchaufour, P. 1976. Dynamics of organic matter in soils of temperate regions: its action on pedogenesis. *Geoderma* **15**: 31–40.

1982. *Pedology*. Winchester, MA, Allen and Unwin.

Duchaufour, P. H. and B. Souchier. 1978. Roles of iron and clay in genesis of acid soils under a humid, temperate climate. *Geoderma* **20**: 15–26.

Ducloux, J., Meunier, A., and B. Velde. 1976. Smectite, chlorite and a regular interstratified chlorite–vermiculite in soils developed on a small serpentinite body Massif Central, France. *Clay Mins.* **11**: 121–135.

Dudal, R. 1965. *Dark Clay Soils of Tropical and Sub-tropical Regions.* Agricultural Development Paper no. 83. Rome, Food and Agriculture Organization.

Dudal, R. and H. Eswaran. 1988. Distribution, properties and classification of Vertisols. In L. P. Wilding and R. Puentes (eds.) *Vertisols: Their Distribution, Properties, Classification and Management.* College Station, TX, Texas A&M University, pp. 1–22.

Dudas, M. J. and S. Pawluk. 1969. Naturally occurring organo-clay complexes of Orthic Black Chernozems. *Geoderma* **3**: 5–17.

Duller, G. A. T. and P. Augustinus. 1997. Luminescence studies of dunes from north-eastern Tasmania. *Quat. Sci. Rev.* **16**: 357–365.

Durand, J. H. 1963. Les croûtes calcaires et gypseuses en Algérie: formation et âge. *Bull. Géol. Soc. France* **7**: 959–968.

Durn, G., Ottner, F., and D. Slovenec. 1999. Mineralogical and geochemical indicators of the polygenetic nature of terra rossa in Istria, Croatia. *Geoderma* **91**: 125–150.

Dutta, P. K., Zhou, Z., and P. R. dos Santos. 1993. A theoretical study of mineralogical maturation of eolian sand. *Geol. Soc. Am. Spec. Paper* **284**: 203–209.

Dyke, A. S. 1979. Glacial and sea-level history, southwestern Cumberland Peninsula, Baffin Island, Canada. *Arctic Alpine Res.* **11**: 179–202.

Eaqub, M. and H.-P. Blume. 1982. Genesis of a so-called ferrolysed soil of Bangladesh. *Z. Pflanzen. Bodenkd.* **145**: 470–482.

Eaton, J. S., Likens, G. E., and F. H. Bormann. 1973. Throughfall and stemflow chemistry in a northern hardwood forest. *J. Ecol.* **61**: 495–508.

Eden, D. N., Qizhong, W., Hunt, J. L., *et al.* 1994. Mineralogical and geochemical trends across the Loess Plateau, North China. *Catena* **21**: 73–90.

Edmonds, W. J., Campbell, J. B., and M. Lemtner. 1985. Taxonomic variation within three soil mapping units in Virginia. *Soil Sci. Soc. Am. J.* **49**: 394–401.

Edwards, C. 1999. *Earthworms*, PA-1637. Washington, DC, US Department of Agriculture–NRCS Soil Quality Institute, pp. H1–H8.

Edwards, W. M., Norton, L. D., and C. E. Redmond. 1988. Characterizing macropores that affect infiltration into nontilled soil. *Soil Sci. Soc. Am. J.* **52**: 483–487.

Effland, A. B. W. and W. R. Effland. 1992. Soil geomorphology studies in the US soil survey program. *Agric. Hist.* **66**: 189–212.

Effland, W. R. and R. V. Pouyat. 1997. The genesis, classification, and mapping of soils in urban areas. *Urban Ecosyst.* **1**: 217–228.

Ehleringer, J. R. and R. K. Monson. 1993. Evolutionary and ecological aspects of photosynthetic pathway variation. *Ann. Rev. Ecol. Syst.* **24**: 411–439.

Ehrlich, W. A., Rice, H. M., and J. H. Ellis. 1955. Influence of the composition of parent materials on soil formation in Manitoba. *Can. J. Agric. Sci.* **35**: 407–421.

Eidt, R. C. 1977. Detection and examination of Anthrosols by phosphate analysis. *Science* **197**: 1327–1333.

El Abedine, Z., Robinson, G. H., and A. Commissaris. 1971. Approximate age of the Vertisols of Gezira, central clay plain, Sudan. *Soil Sci.* **111**: 200–207.

Eldredge, N. and S. J. Gould. 1972. Punctuated equilibria: an alternative to phyletic gradualism. In T. J. M. Schopf (ed.) *Models of Paleobiology*. San Francisco, CA, Freeman, Cooper and Co., pp. 82–115.

Elghamry, W. and M. Elashkar. 1962. Simplified textural classification triangles. *Soil Sci. Soc. Am. Proc.* **26**: 612–613.

Elliott, J. A. and E. DeJong. 1992. Quantifying denitrification on a field scale in hummocky terrain. *Can. J. Soil Sci.* **72**: 21–29.

Ellis, S. 1979. The identification of some Norwegian mountain soil types. *Norsk. Geogr. Tidsskr.* **33**: 205–211.

1980a. An investigation of weathering in some Arctic–Alpine soils on the northeast flank of Oksskolten, north Norway. *J. Soil Sci.* **31**: 371–385.

1980b. Physical and chemical characteristics of a podzolic soil formed in neoglacial till, Okstindan, northern Norway. *Arctic Alpine Res.* **12**: 65–72.

Elvidge, C. D. and R. M. Iverson. 1983. Regeneration of desert pavement varnish. In R. H. Webb and H. G. Wilshire (eds.) *Environmental Effects of Off-Road Vehicles: Impacts and Management in Arid Regions*. New York, Springer-Verlag, pp. 225–243.

Ely, L. L., Enzel, Y., Baker, V. R., *et al.* 1996. Changes in the magnitude and frequency of late Holocene monsoon floods on the Narmada River, central India. *Geol. Soc. Am. Bull.* **108**: 1134–1148.

Emiliani, C. 1955. Pleistocene temperatures. *J. Geol.* **63**: 538–578.

Emiliani, C. 1966. Palaeotemperature analysis of Caribbean cores and a generalized temperature curve for the last 425 000 years. *J. Geol.* **74**: 109–126.

Emrich, K. D., Ehalt, H., and J. C. Vogel. 1970. Carbon isotope fractionation during the precipitation of calcium carbonate. *Earth Planet. Sci. Lett.* **8**: 363–371.

Engel, C. G. and R. P. Sharp. 1958. Chemical data on desert varnish. Geol. *Soc. Am. Bull.* **69**: 487–518.

Engel, R. J., Witty, J. E., and H. Eswaran. 1997. The classification, distribution, and extent of soils with a xeric moisture regime in the United States. *Catena* **28**: 203–209.

Eppes, M. C. and J. B. J. Harrison. 1999. Spatial variability of soils developing on basalt flows in the Potrillo volcanic field, southern New Mexico: prelude to a chronosequence study. *Earth Surf. Proc. Landforms* **24**: 1009–1024.

Erlandson, J. M. 1984. A case study in faunalturbation: delineating the effects of the burrowing pocket gopher on the distribution of archaeological materials. *Am. Antiq.* **49**: 785–790.

Eschner, A. R. and J. H. Patric. 1982. Debris avalanches in eastern upland forests. *J. For.* **80**: 343–347.

Escolar, R. P. and M. A. L. López. 1968. Nature of aggregation in two tropical soils of Puerto Rico. *Univ. Puerto Rico J. Agric.* **52**: 227–232.

Eswaran, H. and W. C. Bin. 1978a. A study of a deep weathering profile on granite in peninsular Malaysia. I. Physico-chemical and micromorphological properties. *Soil Sci. Soc. Am. J.* **42**: 144–149.

1978b. A study of a deep weathering profile on granite in peninsular Malaysia. II. Mineralogy of the clay, silt, and sand fractions. *Soil Sci. Soc. Am. J.* **42**: 149–153.

Eswaran, H. and C. Sys. 1970. An evaluation of the free iron in tropical basaltic soils. *Pédologie* **20**: 62–85.

Eswaran, H. and G. Zi-Tong. 1991. Properties, genesis, classification, and dsitribution of soils with gypsum. In W. D. Nettleton (ed.) *Occurrence, Characteristics, and Genesis of Carbonate, Gypsum, and Silica Accumulations in Soils. Soil Sci. Soc. Am. Special Publication* **26**: 89–119.

Eswaran, H., van den Berg, E., and P. Reich. 1993. Organic carbon in soils of the world. *Soil Sci. Soc. Am. J.* **57**: 192–194.

Eswaran, H., van Wambeke, A., and F. H. Beinroth. 1979. A study of some highly weathered soils of Puerto Rico: micromorphological properties. *Pédologie* **29**: 139–162.

Etienne, S. and J. Dupont. 2002. Fungal weathering of basaltic rocks in a cold oceanic environment (Iceland): comparison between experimental and field observations. *Earth Surf. Proc. Landforms* **27**: 737–748.

Evans, C. V. and W. A. Bothner. 1993. Genesis of altered Conway Granite (grus) in New Hampshire, USA. *Geoderma* **58**: 201–218.

Evans, C. V. and D. P. Franzmeier. 1986. Saturation, aeration, and color patterns in a toposequence of soils in north-central Indiana. *Soil Sci. Soc. Am. J.* **50**: 975–980.

Evans, D. J. A. 1999. A soil chronosequence from neoglacial moraines in western Norway. *Geog. Annal.* **81A**: 47–62.

Evans, L. J. 1978. Quantification of pedological processes. In W. C. Mahaney (ed.) *Quaternary Soils*. Norwich, UK, Geo Abstracts, pp. 361–378.

1982. Dating methods of Pleistocene deposits and their problems. VII. Paleosols. *Geosci. Can.* **9**: 155–160.

Evans, L. J. and Adams, W. A. 1975a. Quantitative pedological studies on soils derived from Silurian mudstones. IV. Uniformity of the parent material and evaluation of internal standards. *J. Soil Sci.* **26**: 319–326.

1975b. Quantitative pedological studies on soils derived from Silurian mudstones. V. Redistribution and loss of mobilized constituents. *J. Soil Sci.* **26**: 327–335.

Evenari, J., Yaalon, D. H., and Y. Gutterman. 1974. Note on soils with vesicular structures in deserts. *Z. Geomorphol.* **18**: 162–172.

Everett, K. R. 1979. Evolution of the soil landscape in the sand region of the Arctic coastal plain as exemplified at Atkasook, Alaska. *Arctic* **32**: 207–223.

1980. Distribution and variability of soils near Atkasook, Alaska. *Arctic Alpine Res.* **12**: 433–446.

Expert Committee on Soil Science, Agriculture Canada Research Branch. 1987. *The Canadian System of Soil Classification*, 2nd edn, Publication no. 1646. Ottawa, Agriculture Canada.

Fadl, A. E. 1971. A mineralogical characterization of some Vertisols in the Gezira and the Kenana clay plains of the Sudan. *J. Soil Sci.* **22**: 129–135.

Faegri, K. and J. Iverson. 1989. *Textbook of Pollen Analysis*. New York, John Wiley.

Fairbridge, R. W. and C. W. Finkl, Jr. 1984. Tropical stone lines and podzolized sand plains as paleoclimatic indicators for weathered cratons. *Quat. Sci. Rev.* **3**: 41–72.

Fanning, D. S. and M. C. B. Fanning. 1989. *Soil Morphology, Genesis, and Classification*. New York, John Wiley.

Fanning, D. S., Keramidas, V. Z., and M. A. El-Desoky. 1989. Micas. In J. B. Dixon and S. B. Weed (eds.) *Minerals in Soil Environments*, 2nd edn. Madison, WI, Soil Science Society of America, pp. 551–634.

Farmer, V. C. 1982. Significance of the presence of allophane and imogolite in Podzol Bs horizons for podzolization mechanisms: a review. *Soil Sci. Plant Nutr.* **28**: 571–578.

Farmer, V. C., Russell, J. D., and M. L. Berrow. 1980. Imogolite and proto-imogolite allophane in Spodic horizons: evidence for a mobile aluminum silicate complex in Podzol formation. *J. Soil Sci.* **31**: 673–684.

Farnham, R. S., McAndrews, J. H., and H. E. Wright, Jr. 1964. A Late-Wisconsin buried soil near Aitkin, Minnesota, and its paleobotanical setting. *Am. J. Sci.* **262**: 393–412.

Faulkner, D. J. 1998. Spatially variable historical alluviation and channel incision in west-central Wisconsin. *Ann. Assoc. Am. Geog.* **88**: 666–685.

Faure, P. and B. Volkoff. 1998. Some factors affecting regional differentiation of the soils in the Republic of Benin (West Africa). *Catena* **32**: 281–306.

Favre, F., Boivin, P., and M. C. S. Wopereis. 1997. Water movement and soil swelling in a dry, cracked Vertisol. *Geoderma* **78**: 113–123.

Feddema, J. and T. C. Meierding. 1987. Marble weathering and air pollution in Philadelphia. *Atmos. Env.* **21**: 143–157.

Federer, C. A. 1973. *Annual Cycles of Soil and Water Temperatures at Hubbard Brook*. US Forest Service Experimental Station Research Note no. NE-167. Newton Square, PA.

Fedoroff, N. and P. Goldberg. 1982. Comparative micromorphology of two Late Pleistocene paleosols (in the Paris basin). *Catena* **9**: 227–251.

Fedorova, N. N. and A. Yarilova. 1972. Morphology and genesis of prolonged seasonally frozen soils in Western Siberia. *Geoderma* **7**: 1–13.

Fehrenbacher, J. B., Olson, K. R., and I. J. Jansen. 1986. Loess thickness in Illinois. *Soil Sci.* **141**: 423–431.

Fehrenbacher, J. B., White, J. L., Ulrich, H. P., *et al.* 1965. Loess distribution in southeastern Illinois and southwestern Indiana. *Soil Sci. Soc. Am. Proc.* **29**: 566–572.

Fehrenbacher, J. B., Wilding, L. P., Odell, R. T., *et al.* 1963. Characteristics of solonetzic soils in Illinois. *Soil Sci. Soc. Am. Proc.* **27**: 421–431.

Felgenhauser, F., Fink, J., and H. de Vries. 1959. Studien zur absoluten und relativen Chronologie der fossil Böden in Österreich. *Archaeol. Austr.* **25**: 35–73.

Feng, Z.-D. and W. C. Johnson. 1995. Factors affecting the magnetic susceptibility of a loess-soil sequence, Barton County, Kansas, USA. *Catena* **24**: 25–37.

Feng, Z.-D., Johnson, W. C., Lu, Y.-C., *et al.* 1994a. Climatic signals from loess-soil sequences in the central Great Plains, USA. Palaeogeog. *Palaeoclimat. Palaeoecol.* **110**: 345–358.

Fenton, T. E. 1983. Mollisols. In L. P. Wilding, N. E. Smeck and G. F. Hall (eds.) *Pedogenesis and Soil Taxonomy.* New York, Elsevier. pp. 125–163.

Ferguson, C. W. and D. A. Graybill. 1983. Dendrochronology of bristlecone pine: a progress report. *Radiocarbon* **25**: 287–288.

Fernández Sanjurjo, M. J., Corti, G., and F. C. Ugolini. 2001. Chemical and mineralogical changes in a polygenetic soil of Galicia, NW Spain. *Catena* **43**: 251–265.

Fernandez, R. N., Schulze, D. G., Coffin, D. L., *et al.* 1988. Color, organic matter, and pesticide adsorption relationships in a soil-landscape. *Soil Sci. Soc. Am. J.* **52**: 1023–1026.

Ferrians, O. J., Jr., Kachadoorian, R., and G. W. Greene. 1969. *Permafrost and Related Engineering Problems in Alaska*, US Geological Survey Professional Paper no. 678. Washington, DC, US Government Printing Office.

Fine, P., Singer, M. J., and K. L. Verosub. 1992. Use of magnetic-susceptibility measurements in assessing soil uniformity in chronosequence studies. *Soil Sci. Soc. Am. J.* **56**: 1195–1199.

1993. New evidence for the origin of ferrimagnetic minerals in loess from China. *Soil. Sci. Soc. Am. J.* **57**: 1537–1542.

Fine, P., Verosub, K. L., and M. J. Singer. 1995. Pedogenic and lithogenic contributions to the magnetic susceptibility record of the Chinese loess/palaeosol sequence. *Geophys. J. Intl* **122**: 97–107.

Finlayson, B. 1981. Field measurements of soil creep. *Earth Surf. Proc. Landforms* **6**: 35–48.

Finney, H. R., Holowaychuk, N., and M. R. Heddleson. 1962. The influence of microclimate on the morphology of certain soils of the Allegheny Plateau of Ohio. *Soil Sci. Soc. Am. Proc.* **26**: 287–292.

Fiskell, J. G. A. and V. W. Carlisle. 1963. Weathering of some Florida soils. *Soil Crop Soc. Florida Proc.* **23**: 32–44.

FitzPatrick, E. A. 1956. An indurated soil horizon formed by permafrost. *J. Soil Sci.* **7**: 248–254.

1975. Particle size distribution and stone orientation patterns in some soils of north east Scotland. In A. M. D. Gemmell (ed.) *Quaternary Studies in North East Scotland*. Aberdeen, UK, Quaternary Research Association, pp. 49–60.

1993. *Soil Microscopy and Micromorphology*. New York, John Wiley.

Fitzpatrick, R. W., Fritsch, E., and P. G. Self. 1996. Interpretation of soil features produced by ancient and modern processes in degraded landscapes. V. Development of saline sulfidic features in non-tidal seepage areas. *Geoderma* **69**: 1–29.

Flach, K. W., Cady, J. G., and W. D. Nettleton. 1968. Pedogenic alteration of highly weathered parent materials. *Trans. 9th Intl Congr. Soil Sci.* (Adelaide) **4**: 343–351.

Flach, K. W., Holzhey, C. S., DeConinck, F., *et al.* 1980. Genesis and classification of Andepts and Spodosols. In B. K. G. Theng (ed.) *Soils with Variable Charge.* Palmerston North, New Zealand, New Zealand, Society for Soil Science, pp. 411–426.

Fletcher, J. E. and W. P. Martin. 1948. Some effects of algae and molds in the rain-crust of desert soils. *Ecology* **29**: 95–100.

Fletcher, P. W. and R. E. McDermott. 1957. Moisture depletion by forest cover on a seasonally saturated Ozark ridge soil. *Soil Sci. Soc. Am Proc.* **21**: 547–550.

Folkoff, M. E. and V. Meentemeyer. 1985. Climatic control of the assemblages of secondary clay minerals in the A-horizon of United States soils. *Earth Surf. Proc. Landforms* **10**: 621–633.

1987. Climatic control on the geography of clay minerals genesis. *Ann. Assoc. Am. Geog.* **77**: 635–650.

Follmer, L. R. 1978. The Sangamon soil in its type area: a review. In W. C. Mahaney (ed.) *Quaternary Soils.* Norwich, UK, Geo Abstracts, pp. 125–165.

1979a. A historical review of the Sangamon soil. In *Wisconsinan, Sangamonian, and Illinoian Stratigraphy in Central Illinois*, Illinois State Geological Survey Guidebook no. 13. Urbana-Champaign, IL, Midwest Friends of the Pleistocene Field Conference pp. 79–91.

1979b. Explanation of pedologic terms and concepts used in the discussion of soils for this guidebook. In *Wisconsinan, Sangamonian, and Illinoian Stratigraphy in central Illinois*, Illinois State Geological Survey Guidebook no. 13. Champaign-Urbana, IL, Midwest Friends of the Pleistocene Field Conference Guidebook, pp. 129–134.

1982. The geomorphology of the Sangamon surface: its spatial and temporal attributes. In C. Thorn (ed.) *Space and Time in Geomorphology*. Boston, MA, Allen and Unwin, pp. 117–146.

1984. Soil: an uncertain medium for waste disposal. *Proc. 7th Ann. Madison Waste Conf.* 296–311.

1998a. A scale for judging degree of soil and paleosol development. *Quat. Intl* **51/52**: 12–13.

1998b. Preface. *Quat. Intl* **51/52**: 1–3.

Follmer, L. R., McKay, E. D., Lineback, J. A., *et al.* 1979. *Wisconsinan, Sangamonian, and Illinoian stratigraphy in central Illinois*, Illinois State Geological Survey Guidebook no. 13. Midwest Friends of the Pleistocene Field Conference Guidebook.

Fölster, H., Kalk, E., and N. Moshrefi. 1971. Complex pedogenesis of ferrallitic savanna soils in South Sudan. *Geoderma* **6**: 135–149.

Fookes, G. 1997. *Tropical Residual Soils*, a Geological Society Engineering Group Working Party Revised Report. London, Geological Society.

Forman, S. and G. H. Miller. 1984. Pedogenic processes and time-dependent soil morphologies on raised beaches, Bröggerhalvöya, Spitzbergen, Norway. *Arctic Alpine Res.* **16**: 381–394.

Forman, S. L. 1989. Application and limitations of thermoluminescence to date Quaternary sediments. *Quat. Intl* **1**: 47–59.

1999. Infrared and red stimulated luminescence dating of Late Quaternary near-shore sediments from Spitsbergen, Svalbard. *Arctic Alpine Res.* **31**: 34–49.

Forman, S. L., Bettis, E. A. III, Kemmis, T. J., *et al.* 1992. Chronologic evidence for multiple periods of loess deposition during the Late Pleistocene in the Missouri and Mississippi River Valley, United States: implications for the activity of the Laurentide Ice Sheet. *Palaeogeog. Palaeoclimat. Palaeoecol.* **93**: 71–83.

Forman, S. L., Nelson, A. R., and J. P. McCalpin. 1991. Thermoluminescence dating of fault-scarp-derived colluvium: deciphering the timing of paleoearthquakes on the Weber Segment of the Wasatch fault zone, north central Utah. *J. Geophys. Res. B* **96**: 595–605.

Fosberg, M. A. 1965. Characteristics and genesis of patterned ground in Wisconsin time in a chestnut soil zone of southern Idaho. *Soil Sci.* **99**: 30–37.

Foss, J. E. and R. H. Rust. 1968. Soil genesis study of a lithologic discontinuity in glacial drift in western Wisconsin. *Soil Sci. Soc. Am. Proc.* **32**: 393–398.

Foss, J. E., Fanning, D. S., Miller, F. P., *et al.* 1978. Loess deposits of the eastern shore of Maryland. *Soil Sci. Soc. Am. J.* **42**: 329–334.

Foster, R. C. 1988. Microenvironments of soil microorganisms. *Biol. Fert. Soils* **6**: 189–203.

Foster, R. J. 1971. *Physical Geology*. Columbus, OH, Merrill.

Fox, B. J., Fox, M. D., and G. A. McKay. 1979. Litter accumulation after fire in a eucalypt forest. *Austral. J. Bot.* **27**: 157–165.

Fox, C. A., Trowbridge, R., and C. Tarnocai. 1987. Classification, macromorphology and chemical characteristics of Folisols from British Columbia. *Can. J. Soil Sci.* **67**: 765–778.

Frakes, L. A. and S. Jianzhong. 1994. A carbon isotope record of the upper Chines loess sequence: estimates of plant types during stadials and interstadials. *Palaeogeogat. Palaeoclimat. Palaeoecol.* **108**: 183–189.

Franceschi, V. R. and H. T. Horner, Jr. 1980. Calcium oxalate crystals in plants. *Bot. Rev.* **46**: 361–427.

Franzmeier, D. P. and E. P. Whiteside. 1963. A chronosequence of Podzols in northern Michigan. II. Physical and chemical properties. *Michigan. State Univ. Agr. Exp. St. Q. Bull.* **46**: 21–36.

Franzmeier, D. P., Bryant, R. B., and G. C. Steinhardt. 1985. Characteristics of Wisconsinan glacial tills in Indiana and their influence on argillic horizon development. *Soil Sci. Soc. Am. J.* **49**: 1481–1486.

Franzmeier, D. P., Pedersen, E. J., Longwell, T. J., *et al.* 1969. Properties of some soils in the Cumberland Plateau as related to slope aspect and position. *Soil Sci. Soc. Am. Proc.* **33**: 755–761.

Franzmeier, D. P., Yahner, J. E., Steinhardt, G. C., *et al.* 1983. Color patterns and water table levels in some Indiana soils. *Soil Sci. Soc. Am. J.* **47**: 1196–1202.

Frazee, C. J., Fehrenbacher, J. B., and W. C. Krumbein. 1970. Loess distribution from a source. *Soil Sci. Soc. Am. Proc.* **34**: 296–301.

Frazier, B. E. and G. B. Lee. 1971. Characteristics and classification of three Wisconsin Histosols. *Soil Sci. Soc. Am. Proc.* **35**: 776–780.

Frazier, C. S. and R. C. Graham. 2000. Pedogenic transformation of fractured granitic bedrock, southern California. *Soil Sci. Soc. Am. J.* **64**: 2057–2069.

Fredlund G. G. and L. L. Tieszen. 1997. Phytolith and carbon isotope evidence for Late Quaternary vegetation and climate change in the southern Black Hills, South Dakota. *Quat. Res.* **47**: 206–217.

Fredlund, G. G., Johnson, W. C., and W. Dort, Jr. 1985. A preliminary analysis of opal phytoliths from the Eustis Ash Pit, Frontier County, Nebraska. *TER-QUA Symp. Ser.* **1**: 147–162.

Freedman, B. and U. Prager. 1986. Ambient bulk deposition, throughfall, and stemflow in a variety of forest stands in Nova Scotia. *Can. J. For. Res.* **16**: 854–860.

Freeland, J. A. and C. V. Evans. 1993. Genesis and profile development of Success soils, northern New Hampshire. *Soil Sci. Soc. Am. J.* **57**: 183–191.

Frei, E. and M. G. Cline. 1949. Profile studies of normal soils of New York. II. Micromorphological studies of the Gray-Brown Podzolic-Brown Podzolic soil sequence. *Soil Sci.* **68**: 333–344.

French, H. M. 1971. Slope asymmetry of the Beaufort Plain, Northwest Banks Island, NWT, Canada. *Can. J. Earth Sci.* **8**: 717–731.

1993. Cold-climate processes and landforms. In H. M. French and O. Slaymaker (eds.) *Canada's Cold Environments.* Montreal, McGill–Queen's University Press, pp. 143–167.

French, M. H. 1945. Geophagia in animals. *E. Afr. Med. J.* **22**: 103–110.

Frenot, Y., Van Vliet-Lanoë, B., and J.-C. Gloaguen. 1995. Particle translocation and initial soil development on a glacier foreland, Kerguelen Islands, Subantarctic. *Arctic Alpine Res.* **27**: 107–115.

Fridland, V. M. 1958. Podzolization and illimerization (clay migration). *Sov. Soil Sci.* 24–32.

1965. Makeup of the soil cover. *Sov. Soil Sci.* **4**: 343–354.

1974. Structure of the soil mantle. *Geoderma* **12**: 35–41.

Friedman, G. M. 1961. Distinction between dune, beach, and river sands from their textural characteristics. *J. Sed. Petrol.* **31**: 514–529.

Friedman, I. and W. Long. 1976. Hydration rate of obsidian. *Science* **191**: 347–352.

Friedman, I. and F. W. Trembour. 1978. Obsidian: the dating stone. *Am. Sci.* **66**: 44–52.

Fritton, D. D. and G. W. Olson. 1972. Depth to the apparent water table in 17 New York soils from 1963 to 1970. *NY Food Life Sci. Bull.* **13**.

Frolking, T. A., Jackson, M. L., and J. C. Knox. 1983. Origin of red clay over dolomite in the loess-covered Wisconsin Driftless uplands. *Soil Sci. Soc. Am. J.* **47**: 817–820.

Frye, J. C. and A. B. Leonard. 1949. Pleistocene stratigraphic sequence in northeastern Kansas. *Am. J. Sci.* **247**: 883–899.

Frye, J. C. and H. B. Willman. 1960. Classification of the Wisconsin Stage in the Lake Michigan glacial lobe. *Illinois Geol. Surv. Circ.* **285**.

Frye, J. C., Shaffer, P. R., Willman, H. B., *et al.* 1960a. Accretion gley and the gumbotil dilemma. *Am. J. Sci.* **258**: 185–190.

Frye, J. C., Willman, H. B., and H. D. Glass. 1960b. Gumbotil, accretion-gley, and the weathering profile. *Illinois Geol. Survey Circ.* **295**.

Fuller, L. G. and D. W. Anderson. 1993. Changes in soil properties following forest invasion of black soils of the aspen parkland. *Can. J. Soil Sci.* **73**: 613–627.

Fuller, L. G., Wang, D., and D. W. Anderson. 1999. Evidence for solum recarbonation following forest invasion of a grassland soil. *Can. J. Soil Sci.* **79**: 443–448.

Furian, S., Barbiero, L., Boulet, R., *et al.* 2002. Distribution and dynamics of gibbsite and kaolinite in an oxisol of Serra do Mar, southeastern Brazil. *Geoderma* **106**: 83–100.

Furley, P. A. 1971. Relationships between slope form and soil properties developed over chalk parent materials. *Inst. Br. Geog. Spec. Pub.* **3**: 141–163.

Furman, T., Thompson, P., and B. Hatchl. 1998. Primary mineral weathering in the central Appalachians: a mass balance approach. *Geochim. Cosmochim. Acta* **62**: 2889–2904.

Gaikawad, S. T. and F. D. Hole. 1965. Characteristics and genesis of a gravelly Brunizemic regosol. *Soil Sci. Soc. Am. Proc.* **29**: 725–728.

Galloway, W. E. and D. K. Hobday. 1983. *Terrigenous Clastic Depositional Systems.* New York, Springer-Verlag.

Gamble, E. E., Daniels, R. B., and R. J. McCracken. 1969. A2 horizons of coastal plain soils pedogenic or geologic origin. *Southeast. Geol.* **11**: 137–152.

Ganzhara, N. F. 1974. Humus formation in Chernozem soils. *Sov. Soil Sci.* **6**: 403–407. (Translated from *Pochvovedeniye* **7**: 39–43.)

Gao, Y.-X. and H.-Z. Chen. 1983. Salient characteristics of soil-forming processes in Xizang (Tibet). *Soil Sci.* **135**: 11–17.

Gard, L. E. and C. A. Van Doren. 1949. Soil losses as affected by cover, rainfall and slope. *Soil Sci. Soc. Am. Proc.* **14**: 374–378.

Gardiner, M. J. and T. Walsh. 1966. Comparison of soil material buried since Neolithic times with those of the present day. *Proc. Roy. Irish Acad.* **65C**: 29–35.

Gardner, D. R. and E. P. Whiteside. 1952. Zonal soils in the transition region between the Podzol and Gray-Brown Podzolic regions in Michigan. *Soil Sci. Soc. Am. Proc.* **16**: 137–141.

Gardner, L. R. 1972. Origin of the Mormon Mesa caliche, Clark County, Nevada. *Geol. Soc. Am. Bull.* **83**: 143–156.

Gardner, W. H. 1986. Water content. In A. Klute (ed.) *Methods of Soil Analysis*, Part 1, *Physical and Mineralogical Methods*, Agronomy Monograph no. 9, 2nd edn. Madison, WI, American Soil Association and Soil Science Society of America, pp. 493–544.

Gaston, L. A., Mansell, R. S., and H. M. Selim. 1992. Predicting removal of major soil cations and anions during acid infiltration: model evaluation. *Soil Sci. Soc. Am. J.* **56**: 944–950.

Gates, F. C. 1942. The bogs of northern lower Michigan. *Ecol. Monogr.* **12**: 216–254.

Geis, J. W., Boggess, W. R., and J. D. Alexander. 1970. Early effects of forest vegetation and topographic position on dark-colored, prairie-derived soils. *Soil Sci. Soc. Am. Proc.* **34**: 105–111.

Gellatly, A. F. 1984. The use of rock weathering-rind thickness to redate moraines in Mount Cook National Park, New Zealand. *Arctic Alpine Res.* **16**: 225–232.

1985. Phosphate retention: relative dating of Holocene soil development. *Catena* **12**: 227–240.

Gerakis, A. and B. Baer. 1999. A computer program for soil textural classification. *Soil Sci. Soc. Am. J.* **63**: 807–808.

Gerasimov, I. P. 1973. Chernozems, buried soils and loesses of the Russian plaIn their age and genesis. *Soil Sci.* **116**: 202–210.

Gerrard, A. J. 1992. *Soil Geomorphology: An Integration of Pedology and Geomorphology*. New York, Chapman and Hall.

Gerson, R. and R. Amit. 1987. Rates and modes of dust accretion and deposition in an arid region: the Negev, Israel. In L. E. Frostick and I. Reid (eds.) *Desert Sediments: Ancient and Modern*. Boston, MA, Blackwell Scientific Publications, pp. 157–169.

Gessler, P. E., Chadwick, O. A., Chamran, F., *et al.* 2000. Modeling soil-landscape and ecosystem properties using terrain attributes. *Soil Sci. Soc. Am. J.* **64**: 2046–2056.

Geyh, M. A., Benzler, J.-H., and G. Roeschmann. 1971. Problems of dating Pleistocene and Holocene soils by radiometric means. In D. H. Yaalon (ed.) *Paleopedology Origin, Nature and Dating of Paleosols*. Jerusalem, Israel University Press, pp. 63–75.

Geyh, M. A., Roeschmann, G., Wijmstra, T. A., *et al.* 1983. The unreliability of ^{14}C dates obtained from buried sandy Podzols. *Radiocarbon* **25**: 409–416.

Gibbons, A. B., Megeath, J. D., and K. L. Pierce. 1984. Probability of moraine survival in a succession of glacial advances. *Geology* **12**: 327–330.

Gibbs, J. A. and H. F. Perkins. 1966. Properties and genesis of the Hayesville and Cecil series of Georgia. *Soil Sci. Soc. Am. Proc.* **30**: 256–260.

Gile, L. H. 1958. Fragipan and water-table relationships of some Brown Podzolic and Low Humic-Gley soils. *Soil Sci. Soc. Am. Proc.* **22**: 560–565.

1970. Soils of the Rio Grande Valley border in southern New Mexico. *Soil Sci. Soc. Am. Proc.* **34**: 465–472.

1975a. Causes of soil boundaries in an arid region. I. Age and parent materials. *Soil Sci. Soc. Am. Proc.* **39**: 316–323.

1975b. Causes of soil boundaries in an arid region. II. Dissection, moisture, and faunal activity. *Soil Sci. Soc. Am. Proc.* **39**: 324–330.

1979. Holocene soils in eolian sediments of Bailey County, Texas. *Soil Sci. Soc. Am. J.* **43**: 994–1003.

Gile, L. H. and R. R. Grossman. 1968. Morphology of the argillic horizon in desert soils of southern New Mexico. *Soil Sci.* **106**: 6–15.

1979. *The Desert Project*, Washington, DC, US Department of Agriculture, US Soil Conservation Service, US Government Printing Office.

Gile, L. H. and J. W. Hawley. 1968. Age and comparative development of desert soils at the Gardner Spring Radiocarbon Site, New Mexico. *Soil Sci. Soc. Am. Proc.* **32**: 709–716.

1972. The prediction of soil occurrence in certain desert regions of the southwestern United States. *Soil Sci. Soc. Am. Proc.* **36**: 119–123.

Gile, L. H., Hawley, J. W., and R. B. Grossman. 1981. *Soils and Geomorphology in the Basin and Range Area of Southern New Mexico: A Guidebook to the Desert Project*, Memoir no. 39. Socorro, NM, New Mexico Bureau of Mines and Mineral Research.

Gile, L. H., Peterson, F. F., and R. B. Grossman. 1965. The K horizon: a master soil horizon of carbonate accumulation. *Soil Sci.* **99**: 74–82.

1966. Morphological and genetic sequences of carbonate accumulation in desert soils. *Soil Sci.* **101**: 347–360.

Gilet-Blein, N., Marien, G., and J. Evin. 1980. Unreliability of ^{14}C dates from organic matter of soils. *Radiocarbon* **22**: 919–929.

Gilichinsky, D. A., Barry, R. G., Bykhovets, S. S., *et al.* 1998. A century of temperature observations of soil climate: methods of analysis and long-term trends. In A. G. Lewkowicz and M. Allard (eds.) *Permafrost. Proc. 7th Intl Conf.* Yellowknife, Canada, Université Laval, Centre d'Etudes Nordiques, pp. 313–317.

Gilkes, R. J., Scholz, G., and G. M., Dimmock. 1973. Lateritic deep weathering of granite. *J. Soil Sci.* **24**: 523–536.

Gillham, R. A. 1984. The capillary fringe and its effect on water table response. *J. Hydrol.* **67**: 307–324.

Gillman, L. R., Jeffries, M. K., and G. N. Richards. 1972. Non-soil constituents of termite (*Coptotermes acinaciformes*) mounds. *Austral. J. Biol. Sci.* **25**: 1005–1013.

Giovannini, G. and P. Sequi. 1976. Iron and aluminum as cementing substances of soil aggregates: changes in stability of soil aggregates following extraction of iron and aluminum by acetylacetone in a non polar solvent. *J. Soil Sci.* **27**: 148–153.

Glaccum, R. A. and J. M. Prospero. 1980. Saharan aerosols over the tropical North Atlantic: mineralogy. *Marine Geol.* **37**: 295–321.

Glazovskaya, M. A. 1968. Geochemical landscapes and types of geochemical soil sequences. *Trans. 9th Intl Congr. Soil Sci.* (Adelaide) **4**: 303–312.

Glazovskaya, M. A. and E. I. Parfenova. 1974. Biogeochemical factors in the formation of terra rossa in the southern Crimea. *Geoderma* **12**: 57–82.

Glinka, K. D. 1914. *Die Typen der Bodenbildung, irhe Klassifikation und Geographische Verbreitung.* Berlin, Borntraeger. (See Marbat (1927).)

Gobin, A., Campling, P., Deckers, J., *et al.* 2000. Quantifying soil morphology in tropical environments: methods and application in soil classification. *Soil Sci. Soc. Am. J.* **64**: 1423–1433.

Godfrey, C. L. and F. F. Riecken. 1954. Distribution of phosphorous in some genetically related loess-derived soils. *Soil Sci. Soc. Am. Proc.* **18**: 80–84.

Gokceoglu, C., Ulusay, R., and H. Sonmez. 2000. Factors affecting the durability of selected weak and clay-bearing rocks from Turkey, with particular emphasis on the influence of the number of drying and wetting cycles. *Engin. Geol.* **57**: 215–237.

Goldich, S. S. 1938. A study of rock weathering. *J. Geol.* **46**: 17–58.

Goldin, A. and J. Edmonds. 1996. A numerical evaluation of some Florida Spodosols. *Phys. Geog.* **17**: 242–252.

Goldthwait, R. P. 1976. Frost sorted patterned ground: a review. *Quat. Res.* **6**: 27–35.

Goodfriend, G. A. and D. G. Hood. 1983. Carbon isotope analysis of land snail shells: implications for carbon sources and radiocarbon dating. *Radiocarbon* **27**: 33–42.

Goodlett, J. C. 1954. Vegetation adjacent to the border of the Wisconsin drift in Potter County, Pennsylvania. *Harvard For. Bull.* **25**.

Goodman, A. Y., Rodbell, D. T., Seltzer, G. O., *et al.* 2001. Subdivision of glacial deposits in southeastern Peru based on pedogenic development and radiometric ages. *Quat. Res.* **56**: 31–50.

Gordon, A. and C. B. Lipman. 1926. Why are serpentine and other magnesian soils infertile. *Soil Sci.* **22**: 291–302.

Gordon, M., Jr., Tracey, J. I., Jr., and M. W. Ellis. 1958. *Geology of the Arkansas Bauxite Region.* Professional Paper no. 299. Washington, DC, US Geological Survey, US Government Printing Office.

Goryachkin, S. V., Karavaeva, N. A., Targulian, V. O., *et al.* 1999. Arctic soils: spatial distribution, zonality and transformation due to global change. *Permafrost Periglac. Proc.* **10**: 235–250.

Goss, D. W., Smith, S. J., and B. A. Stewart. 1973. Movement of added clay through calcareous materials. *Geoderma* **9**: 97–103.

Gosse, J. C., Klein, J., Evenson, E. B., *et al.* 1995. Beryllium-10 dating of the duration and retreat of the last Pinedale glacial sequence. *Science* **268**: 1329–1333.

Goswami, A., Das, A. L., Sah, K. D., *et al.* 1996. Pedological studies in Great Nicobar Island. *Geog. Rev. Ind.* **58**: 162–168.

Gosz, J. R., Likens, G. E., and F. H. Bormann. 1976. Organic matter and nutrient dynamics of the forest and forest floor in the Hubbard Brook Forest. *Oecologia* **22**: 305–320.

Goudie, A. S. 1996. Organic agency in calcrete development. *J. Arid. Envs.* **32**: 103–110.

Goudie, A. S. and H. Viles. 1997. *Salt Weathering Hazard.* New York, John Wiley.

Gracanin, Z. 1971. Age and development of the hummocky meadow (buckelwiese) in the Lechtaler Alps (Austria). In D. H. Yaalon (ed.) *Paleopedology: Origin, Nature and Dating of Paleosols.* Jerusalem, Israel University Press, pp. 117–127.

Grah, R. F. and C. C. Wilson. 1944. Some components of rainfall interception. *J. For.* **42**: 890–898.

Graham, R. C. and E. Franco-Vizcaino. 1992. Soils on igneous and metavolcanic rocks in the Sonoran

Desert of Baja California, Mexico. *Geoderma* **54**: 1–21.

Graham, R. C. and A. R. Southard. 1983. Genesis of a Vertisol and an associated Mollisol in northern Utah. *Soil Sci. Soc. Am. J.* **47**: 552–559.

Graham, R. C., Daniels, R. B., and S. W. Buol. 1990a. Soil-geomorphic relations on the Blue Ridge Front. I. Regolith types and slope processes. *Soil Sci. Soc. Am. J.* **54**: 1362–1367.

Graham, R. C., Diallo, M. M., and L. J. Lund. 1990b. Soils and mineral weathering on phyllite colluvium and serpentine in northwestern California. *Soil Sci. Soc. Am. J.* **54**: 1682–1690.

Graham, R. C., Weed, S. B., Bowen, L. H., *et al.* 1989. Weathering of iron-bearing minerals in soils and saprolite on the North Carolina Blue Ridge front. II. Clay mineralogy. *Clay Mins.* **37**: 29–40.

Graham, S. A. 1941. Climax forests of the upper peninsula of Michigan. *Ecology* **22**: 355–362.

Green, R. D. and G. P. Askew. 1965. Observations on the biological development of macropores in soils of Romney Marsh. *J. Soil Sci.* **16**: 342–349.

Green, R. E., Ahuja, L. R., and S. K. Chong. 1986. Hydraulic conductivity, diffusivity, and sorptivity of unsaturated soils: field methods. In A. Klute (ed.) *Methods of Soil Analysis*, Part 1, *Physical and Mineralogical Methods*, Agronomy Monograph no. 9 2nd edn. Madison, WI, American Soil Association and Soil Science Society of America, pp. 771–798.

Greene, H. 1945. Classification and use of tropical soils. *Soil Sci. Soc. Am. Proc.* **10**: 392–396.

Greenslade, P. J. N. 1974. Some relations of the meat ant, *Iridomyrmex purpureus* (Hymenoptera: Formicidae) with soil in South Australia. *Soil Biol. Biochem.* **6**: 7–14.

Gregorich, E. G. and D. W. Anderson. 1985. Effects of cultivation and erosion on soils of four toposequences in the Canadian prairies. *Geoderma* **36**: 343–354.

Grieve, I. C. 2000. Effects of human disturbance and cryoturbation on soil iron and organic matter distributions and on carbon storage at high elevations in the Cairngorm Mountains, Scotland. *Geoderma* **95**: 1–14.

Griffey, N. J. and S. Ellis. 1979. Three *in situ* paleosols buried beneath neoglacial moraine ridges, Okstindan and Jotunheimen, Norway. *Arctic Alpine Res.* **11**: 203–214.

Grimley, D. A. 1998. Pedogenic influences on magnetic susceptibility patterns in loess-paleosol sequences of southwestern Illinois. *Quat. Intl* **51/52**: 51.

2000. Glacial and nonglacial sediment contributions to Wisconsin Episode loess in the central United States. *Geol. Soc. Am. Bull.* **112**: 1475–1495.

Groffman, P. M. and J. M. Tiedje. 1989. Denitrification in north temperate forest soils: relationships between denitrification and environmental factors, at the landscape scale. *Soil. Biol. Biochem* **21**: 621–626.

Grossman, R. B. and F. J. Carlisle. 1969. Fragipan soils of the eastern United States. *Adv. Agron.* **21**: 237–279.

Grossman, R. B. and M. G. Cline. 1957. Fragipan horizons in New York soils. II. Relationships between rigidity and particle size distribution. *Soil Sci. Soc. Am. Proc.* **21**: 322–325.

Grossman, R. B., Odell, R. T., and A. H. Beavers. 1964. Surfaces of peds from B horizons of Illinois soils. *Soil Sci. Soc. Am. Proc.* **28**: 792–798.

Grube, S. 2001. Soil modification by the harvester termite *Hodotermes mossambicus* (Isoptera; Hodotermitidae) in a semiarid savanna grassland of Namibia. *Sociobiology* **37**: 757–767.

Gubin, S. V. 1994. Relic features in recent tundra soil profiles and tundra soils classification. In J. M. Kimble and R. J. Ahrens (eds.) *Classification Correlation, and Management of Permafrost-Affected Soils, Proc. Intl Soil Correlation Mtg (ISCOM)*. Lincoln, NE, US Department of Agriculture–Soil Conservation Service, pp. 63–65.

Guccione, M. 1983. Quaternary sediments and their weathering history in northcentral Missouri. *Boreas* **12**: 217–226.

Guillet, B. and A. M. Robin. 1972. Interprétation de datations par le ^{14}C d'horizons Bh de deux podzols humo-ferrugineux, l'un formé sous callune, l'autre sous chênaie-hêtraie. *C. R. Acad. Sci.* (Paris) **274D**: 2859–2861.

Guillet, B., Faivre, P., Mariotti, A., *et al.* 1988. The ^{14}C dates and $^{13}C/^{12}C$ ratios of soil organic matter as a means of studying the past vegetation in intertropical regions: examples from Colombia (South America). *Palaeogeog. Palaeoclimat. Palaeoecol.* **65**: 51–58.

Guillet, B., Rouiller, J., and B. Souchier. 1975. Podzolization and clay migration in Spodosols of eastern France. *Geoderma* **14**: 223–245.

Gunatilaka, A. and S. Mwango 1987. Continental sabkha pans, and associated nebkhas in southern Kuwait, Arabian Gulf. In L. E. Frostick and I. Reid (eds.) *Desert Sediments: Ancient and Modern*. Boston, MA, Blackwell Scientific Publications, pp. 187–203.

Gundlach, H. F., Campbell, J. E., Huffman, T. J., *et al.* 1982. *Soil Survey of Shawano County, Wisconsin.* Washington, DC, US Department of Agriculture Soil Conservation Service, US Government. Printing Office.

Gunn, R. H. 1967. A soil catena on denuded laterite profiles in Queensland. *Austral. J. Soil Res.* **5**: 117–132.

Gunn, R. H. and D. P. Richardson. 1979. The nature and possible origins of soluble salts in deeply weathered landscapes of eastern Australia. *Austral. J. Soil Res.* **17**: 197–215.

Gunnarsson, T., Sundin, P., and A. Tunlid. 1988. Importance of leaf litter fragmentation for bacterial growth. *Oikos* **52**: 303–308.

Gupta, S. R., Rajvanshi, R., and J. S. Singh. 1981. The role of the termite *Odontotermes gurdaspurensis* (Isoptera: Termitidae) in plant decomposition in a tropical grassland. *Pedobiologia* **22**: 254–261.

Gustafsson, J. P., Bhattacharya, P., Bain, D. C., *et al.* 1995. Podzolization mechanisms and the synthesis of imogolite in northern Scandinavia. *Geoderma* **66**: 167–184.

Guthrie, R. L. and B. F. Hajek. 1979. Morphology and water regime of a Dothan soil. *Soil Sci. Soc. Am. J.* **43**: 142–144.

Guthrie, R. L. and J. E. Witty. 1982. New designations for soil horizons and layers and the new soil survey manual. *Soil Sci. Soc. Am. J.* **46**: 443–444.

Haantjens, H. A. 1965. Morphology and origin of patterned ground in a humid tropical lowland area, New Guinea. *Austral. J. Soil Res.* **3**: 111–129.

Haas, H., Holliday, V., and R. Stuckenrath. 1986. Dating of Holocene stratigraphy with soluble and insoluble organic fractions at the Lubbock Lake archaeological site, Texas: an ideal case study. *Radiocarbon* **28**: 473–485.

Habecker, M. A., McSweeney, K., and F. W. Madison. 1990. Identification and genesis of fragipans in Ochrepts of North Central Wisconsin. *Soil Sci. Soc. Am. J.* **54**: 139–146.

Haberman, G. M. and F. D. Hole. 1980. Soilscape analysis in terms of pedogeomorphic fabric: an exploratory study. *Soil Sci. Soc. Am. J.* **44**: 336–340.

Haff, P. K. and B. T. Werner. 1996. Dynamical processes on desert pavements and the healing of surficial disturbances. *Quat. Res.* **45**: 38–46.

Haile-Mariam, S. and D. L. Mokma. 1990. Soils with carbonate-rich zones in east central Michigan. *Soil Surv. Horiz.* **31**: 23–29.

Hairston, A. B. and D. F. Grigal. 1994. Topographic variation in soil water and nitrogen for two forested land forms in Minnesota, USA. *Geoderma* **64**: 125–138.

Hale, M. G., More, L. D., and G. J. Griffin. 1978. Root exudates and exudation. In Y. R. Dommergues and S. V. Krupa (eds.) *Interactions Between Non-Pathogenic Soil Microorganisms and Plants.* New York, Elsevier, pp. 163–204.

Hall, G. F. 1983. Pedology and geomorphology. In L. P. Wilding, N. E. Smeck, and G. F. Hall (eds.) *Pedogenesis and Soil Taxonomy*, vol. 1, *Concepts and Interactions.* Amsterdam, Elsevier, pp. 117–140.

Hall, K. and M. F. Andre. 2001. New insights into rock weathering from high-frequency rock temperature data: an Antarctic study of weathering by thermal stress. *Geomorphology* **41**: 23–35.

Hall, K. and A. Hall. 1996. Weathering by wetting and drying: some experimental results. *Earth Surf. Proc. Landforms* **21**: 365–376.

Hall, K., Thorn, C. E., Matsuoka, N., *et al.* 2002. Weathering in cold regions: some thoughts and perspectives. *Progr. Phys. Geog.* **26**: 577–603.

Hall, R. D. 1999a. A comparison of surface soils and buried soils: factors of soil development. *Soil Sci.* **164**: 264–287.

1999b. Effects of climate change on soils in glacial deposits, Wind River Basin, Wyoming. *Quat. Res.* **51**: 248–261.

Hall, R. D. and A. K. Anderson. 2000. Comparative soil development of Quaternary paleosols of the central United States. *Palaeogeog. Palaeoclimat. Palaeoecol.* **158**: 109–145.

Hall, R. D. and L. L. Horn. 1993. Rates of hornblende etching in soils in glacial deposits of the northern Rocky Mountains (Wyoming–Montana, USA): influence of climate and characteristics of the parent material. *Chem. Geol.* **105**: 17–29.

Hall, R. D. and D. Michaud, 1888. The use of hornblende etching, clast weathering, and soils to date alpine glacial and periglacial deposits: a study from southwestern Montana. *Geol. Soc. Am. Bull.* **100**: 458–467.

Hall, R. D. and R. R. Shroba. 1995. Soil evidence for a glaciation intermediate between the Bull Lake and Pinedale glaciations at Fremont Lake, Wind River Range, Wyoming, USA. *Arctic Alpine Res.* **27**: 89–98.

Hallberg, G. R. 1984. The US system of Soil Taxonomy: from the outside looking. In R. B. Grossman, H. Eswaran and R. H. Rust (eds.) *Soil Taxonomy: Achievements and Challenges.* Madison, WI, Soil Science Society of America Press, pp. 45–59.

1986. Pre-Wisconsin glacial stratigraphy of the Central Plains region in Iowa, Nebraska, Kansas, and Missouri. *Quat. Sci. Rev.* **5**: 11–15.

Hallberg, G. R., Fenton, T. E., and G. A. Miller. 1978a. Part 5: Standard weathering zone terminology for the description of Quaternary sediments in Iowa. In G. R. Hallberg (ed.) *Standard Procedures for Evaluation of Quaternary Materials in Iowa*, Technical Information Series 8. Iowa City, IA, Iowa Geological Survey, pp. 75–109.

Hallberg, G. R., Fenton, T. E., Miller, G. A., *et al.* 1978b. *The Iowan Erosion Surface: An Old Story, an Important lesson, and Some New Wrinkles*, Guidebook, 42nd Annual Tri-State Geological Field Conference, Geology of East-Central Iowa. Iowa City, IA, Iowa Geological Survey, pp. 2-1–2-94.

Hallet, B. and S. Prestrud. 1986. Dynamics of periglacial sorted circles in western Spitsbergen. *Quat. Res.* **26**: 81–99.

Hallet, B., Anderson, S. P., Stubbs, C. W., *et al.* 1988. Surface soil displacements in sorted circles, western Spitzbergen. *Proc. 5th Intl Permafrost Conf.* (Trondheim) **1**: 770–775.

Hallsworth, E. G. and G. G. Beckmann. 1969. Gilgai in the Quaternary. *Soil Sci.* **107**: 409–420.

Hallsworth, E. G., Robertson, G. K., and F. R. Gibbons. 1955. Studies in pedogenesis in New South Wales. VII. The "gilgai" soils. J. Soil Sci. **6**: 1–31.

Halsey, D. P., Mitchell, D. J., and S. J. Dews. 1998. Influence of climatically induced cycles in physical weathering. *Q. J. Engin. Geol.* **31**: 359–367.

Hamann, C. 1984. Windwurfals ursache der bodenbuckelung an sudrand des Tennengebirges, ein beitrag zur genese der Buckelwiesen. *Berl. Geogr. Abh.* **36**: 69–76.

Hanawalt, R. B. and R. H. Whittaker. 1976. Altitudinally coordinated patterns of soils and vegetation in the San Jacinto Mountains, California. *Soil Sci.* **121**: 114–124.

Hancock, G. S., Anderson, R. S., Chadwick, O. A., *et al.* 1999. Dating fluvial terraces with ^{10}Be and ^{26}Al profiles: application to the Wind River, Wyoming. *Geomorphology* **27**: 41–60.

Hannah, P. R. and R. Zahner. 1970. Nonpedogenetic texture bands in outwash sands of Michigan: their origin, and influence on tree growth. *Soil Sci. Soc. Am. Proc.* **34**: 134–136.

Hanson, C. T. and R. L. Blevins. 1979. Soil water in coarse fragments. *Soil Sci. Soc. Am. J.* **43**: 819–820.

Harden, D. R., Biggar, N. E., and M. L. Gillam. 1985. Quaternary deposits and soils in and around Spanish Valley, Utah. *Geol. Soc. Am. Spec. Paper* **203**: 43–64.

Harden, J. W. 1982. A quantitative index of soil development from field descriptions: examples from a chronosequence in central California. *Geoderma* **28**: 1–28.

1987. Soils developed on granitic alluvium near Merced, California. *US Geol. Surv. Bull.* **1590-A**.

1988. Measurements of water penetration and volume percentage water-holding capacity for undisturbed, coarse-textured soils in southwestern California. *Soil Sci.* **146**: 374–383.

Harden, J. W. and E. M. Taylor. 1983. A quantitative comparison of soil development in four climatic regimes. *Quat. Res.* **20**: 342–359.

Harden, J. W., Taylor, E. M., Hill, C., 1991a. Rates of soil development from four soil chronosequences in the southern Great Basin. *Quat. Res.* **35**: 383–399.

Harden, J. W., Taylor, E. M., McFadden, L. D., *et al.* 1991b. Calcic, gypsic, and siliceous soil chronosequences in arid and semiarid environments. In W. D. Nettleton (ed.) *Occurrence, Characteristics, and Genesis of Carbonate, Gypsum, and Silica Accumulations in Soils. Soil Sci. Soc. Am. Special Publ.* **26**: 1–16.

Hardy, F. 1933. Cultivation properties of tropical red soils. *Emp. J. Exp. Agric.* **1**: 103–112.

Hardy, M. 1993. Influence of geogenesis and pedogenesis on clay mineral distribution in northern Vietnam soils. *Soil Sci.* **156**: 336–345.

Harlan, P. W. and D. P. Franzmeier. 1974. Soil-water regimes in Brookston and Crosby soils. *Soil Sci. Soc. Am. Proc.* **38**: 638–643.

1977. Soil formation on loess in southwestern Indiana. I. Loess stratigraphy and soil morphology. *Soil Sci. Soc. Am. J.* **41**: 93–98.

Harlan, P. W., Franzmeier, D. P., and C. B. Roth. 1977. Soil formation on loess in southwestern Indiana. II. Distribution of clay and free oxides and fragipan formation. *Soil Sci. Soc. Am. J.* **41**: 99–103.

Harland, W. B., Cox, A. V., Llewellyn, P. G., *et al.* 1982. *A Geologic Time Scale*. New York, Cambridge University Press.

Harper, W. G. 1957. Morphology and genesis of Calcisols. *Soil Sci. Soc. Am. Proc.* **21**: 420–424.

Harrington, C. D. and J. W. Whitney. 1987. Scanning electron microscope method for rock-varnish dating. *Geology* **15**: 967–970.

Harris, C. 1982. The distribution and altitudinal zonation of periglacial landforms, Okstindan, Norway. *Z. Geomorphol.* **26**: 283–304.

Harris, W. G., Zelazny, L. W., Baker, J. C., *et al.* 1985. Biotite kaolinization in Virginia Piedmont soils. II. Zonation in single grains. *Soil Sci. Soc. Am. J.* **49**: 1297–1302.

Harrison, C. G. A. and E. Ramirez. 1975. Areal coverage of spurious reversals of the earth's magnetic field. *J. Geomag. Geoelec.* **27**: 139–151.

Harrison, J. B. J., McFadden, L. D., and R. J. Weldon III. 1990. Spatial soil variability in the Cajon Pass chronosequence: implications for the use of soils as a geochronological tool. *Geomorphology* **3**: 399–416.

Harry, D. G. and J. S. Gozdzik. 1988. Ice wedges: growth, thaw transformation, and palaeoenvironmental significance. *J. Quat. Sci.* **3**: 39–55.

Harry, K. G. 1995. Cation-ratio dating of varnished artifacts: testing the assumptions. *Am. Antiq.* **60**: 118–130.

Hart, D. M. 1988. A fabric contrast soil on Dolerite in the Sydney Basin, Australia. *Catena* **15**: 27–37.

Hart, G. and H. W. Lull. 1963. Some relationships among air, snow, and soil temperatures and soil frost. *US For. Serv. Res. Note* **NE-3**.

Hart, G., Leonard, R. E., and R. S. Pierce. 1962. Leaf fall, humus depth, and soil frost in a northern hardwood forest. *US For. Serv. Res. Note* **NE-131**.

Hartgrove, N. T., Ammons, J. T., Khiel, A. R., *et al.* 1993. Genesis of soils on two stream terrace levels on the Tennessee River. *Soil Surv. Horiz.* **34**: 78–88.

Hartshorn, J. H. 1958. Flowtill in southeastern Massachusetts. *Geol. Soc. Am. Bull.* **69**: 477–481.

Haseman, J. F. and C. E. Marshall. 1945. The use of heavy minerals in studies of the origin and development of soils. Missouri Agric. *Exp. Stn Res. Bull.* **387**: 1–75.

Havinga, A. J. 1974. Problems in the interpretation of pollen diagrams of mineral soils. *Geol. Mijnbouw* **53**: 449–453.

Hawker, H. W. 1927. A study of the soils of Hildago County, Texas, and the stages of their soil lime accumulation. *Soil Sci.* **23**: 475–485.

Hay, R. L. 1960. Rate of clay formation and mineral alteration in a 4000-year-old volcanic ash soil on St. Vincent, B. W. I. *Am. J. Sci.* **258**: 354–368.

Hayden, J. D. 1976. Pre-altithermal archaeology in the Sierra Pinacate, Sonora, Mexico. *Am. Antiq.* **41**: 274–289.

Hayes, W. A., Jr. and M. J. Vepraskas. 2000. Morphologic changes in soils produced when hydrology is altered by ditching. *Soil Sci. Soc. Am. J.* **64**: 1893–1904.

Haynes, C. V. 1982. Great Sand Sea and Selima Sand Sheet, eastern Sahara: geochronology of desertification. *Science* **217**: 629–633.

Hayward, R. K. and T. V. Lowell. 1993. Variations in loess accumulation rates in the mid-continent, United States, as reflected by magnetic susceptibility. *Geology* **21**: 821–824.

Heath, G. W. 1965. The part played by animals in soil formation. In E. G. Hallsworth and D. V. Crawford (eds.) *Experimental Pedology*. London, Butterworth, pp. 236–243.

Heiberg, S. O. and R. F. Chandler. 1941. A revised nomenclature of forest humus layers for the northeastern United States. *Soil Sci.* **52**: 87–99.

Heimsath, A. M., Chappell, J., Spooner, N. A., *et al.* 2002. Creeping soil. *Geology* **30**: 111–114.

Heinonen, R. 1960. Das volumgewicht als Kennzeichen der "normalen" Bodenstruktur. *J. Sci. Soc. Finland* **32**: 81–87.

Heinselman, M. L. 1970. Landscape evolution, peatland types, and the environment in the Lake Agassiz Peatlands Natural Area, Minnesota. *Ecol. Monogr.* **40**: 235–261.

Heinz, V. 2002. Relict soils as paleoclimatic indicators: examples from the Austrian Alps and the Central Andes. *Proc. 17th World Congr. Soil Sci.* (Bangkok) **4**: 1520.

Heller, F., Shen, C. D., Beer, J., *et al.* 1993. Quantitative estimates of pedogenic ferromagnetic mineral formation in Chinese loess and paleoclimatic implications. *Earth Planet. Sci. Lett.* **114**: 385–390.

Heller, J. L. 1963. The nomenclature of soils or what's in a name? *Soil Sci. Soc. Am. Proc.* **27**: 216–220.

Helmisaari, H. S. 1995. Nutrient cycling in *Pinus sylvestris* stands in eastern Finland. *Plant Soil* **169**: 327–336.

Helvey, J. D. 1964. Rainfall interception by hardwood forest litter in the southern Appalachians. *US For. Serv. Res. Paper* **SE-8**.

Helvey, J. D. and J. H. Patric. 1965. Canopy and litter interception of rainfall by hardwoods of eastern United States. *Water Resources Res.* **1**: 193–206.

Henry, J. L., Bullock, P. R., Hogg, T. J., *et al.* 1985. Groundwater discharge from glacial and bedrock aquifers as a soil salinization factor in Saskatchewan. *Can. J. Soil Sci.* **65**: 749–768.

Herrera, R. and M. A. Tamers. 1970. Radiocarbon dating of tropical soil associations in Venezuela. *Symp. Age of Parent Materials and Soils* (Amsterdam) 109–115.

Hesse, P. R. 1955. A chemical and physical study of the

soils of termite mounds in East Africa. *J. Ecol.* **43**: 449–461.

Higashi, T., DeConinck, F., and F. Gelaude. 1981. Characterization of some spodic horizons of the Campine (Belgium) with dithionite-citrate, pyrophosphate, and sodium hydroxide-tetraborate. *Geoderma* **25**: 131–142.

Hilgard, E. W. 1906. *Soils*. New York, Macmillan.

Hill, D. E. and B. L. Sawhney. 1971. Electron microprobe analysis of soils. *Soil Sci.* **112**: 32–38.

Hillel, D. 1991. *Out of the Earth*. University of California Press, Berkeley. 321 pp.

Hillel, D. and H. Talpaz. 1977. Simulation of soil water dynamics in layered soils. *Soil Sci.* 123:54–62.

Hiller, D. A. 2000. Properties of Urbic Anthrosols from an abandoned shunting yard in the Ruhr area, Germany. *Catena* **39**: 245–266.

Hintikka, V. 1972. Wind-induced root movements in trees. *Commun. Inst. For. Fenn.* **76**: 1–56.

Hirschfield, E. and R. S. Hirschfield. 1937. Soil problems in Brigalow and Belah country. *Qld. Agric. J.* **47**: 586.

Hoffland, E., Giesler, R., Jongmans, T., *et al.* 2002. Increasing feldspar tunneling by fungi across a north Sweden podzol chronosequence. *Ecosystems* **5**: 11–22.

Höfle, C., Ping, C.-L., and J. M. Kimble. 1998. Properties of permafrost soils on the Northern Seward Peninsula, Northwest Alaska. *Soil Sci. Soc. Am. J.* **62**: 1629–1639.

Hole, F. D. 1953. Suggested terminology for describing soils as three-dimensional bodies. *Soil Sci. Soc. Am. Proc.* **17**: 131–135.

1961. A classification of pedoturbations and some other processes and factors of soil formation in relation to isotropism and anisotropism. *Soil Sci.* **91**: 375–377.

1975. Some relationships between forest vegetation and Podzol B horizons in soils of Menominee tribal lands, Wisconsin, USA. *Sov. Soil Sci.* **7**: 714–723.

1976. *Soils of Wisconsin*. Madison, WI, University of Wisconsin Press.

1978. An approach to landscape analysis with emphasis on soils. *Geoderma* **21**: 1–13.

1980. *Soil Guide for Wisconsin Land Lookers*, Bulletin no. 88, Soil Series no. 63. Madison, WI, Wisconsin Geological and Natural History Survey.

1981. Effects of animals on soil. *Geoderma* **25**: 75–112.

1988. Terra vibrata: some observations on the dynamics of soil landscapes. *Phys. Geog.* **9**: 175–185.

1997. The Earth beneath our feet: Explorations in community. *Soil Survey Hor.* **38**: 40–53.

Hole, F. D. and G. A. Nielsen. 1970. Soil genesis under prairie. *Proc. Symp. Prairie and Prairie Restoration*, Knox College, Galesburg, IL, pp. 28–34.

Hole, F. D. and J. B. Campbell. 1985. *Soil Landscape Analysis*. Totowa, NJ, Rowman and Allanheld.

Holliday, V. T. 1988. Genesis of a late-Holocene soil chronosequence at the Lubbock Lake archaeological site, Texas. *Ann. Assc. Am. Geog.* **78**: 594–610.

Holliday, V. T. 1994. The "state factor" approach in geoarcheology. *Soil Sci. Soc. Am. Spec. Publ.* **33**: 65–86.

1998. Origins of soil-stratigraphic and soil-geomorphic research in the southwestern United States. *Quat. Intl* **51/52**: 20–21.

2001. Stratigraphy and geochronology of upper Quaternary eolian sand on the southern High Plains of Texas and New Mexico, USA. *Geol. Soc. Am. Bull.* **112**: 88–108.

Holmer, B. 1998. Flaking by insolation drying and salt weathering on the Swedish west coast. *Z. Geomorphol.* **42**: 39–55.

Holmes, C. D. 1955. Geomorphic development in humid and arid regions: a synthesis. *Am. J. Sci.* **253**: 377–390.

Holmes, W. H. 1893. Vestiges of early man in Minnesota. *Am. Geol.* **11**: 219–240.

Holt, J. A., Abe, T., and N. Kirtibutr. 1998. Microbial biomass and some chemical properties of Macrotermes carbonarius mounds near Korat, Thailand. *Sociobiology* **31**: 1–8.

Holt, J. A., Coventry, R. J., and D. F. Sinclair. 1980. Some aspects of the biology and pedological significance of mound building termites in a red and yellow earth landscape near Charters Towers, North Queensland. *Austral. J. Soil Res.* **18**: 97–109.

Holtmeier, F.-K. and G. Broll. 1992. The influence of tree islands and microtopography on pedoecological conditions in the forest-alpine tundra ecotone on Niwot Ridge, Colorado Front Range, USA. *Arctic Alpine Res.* **24**: 216–228.

Hoosbeek, M. R. and R. B. Bryant. 1992. Towards the quantitative modeling of pedogenesis: a review. *Geoderma* **55**: 183–210.

1994. Developing and adapting soil process submodels for use in the pedodynamic orthod model. *Soil Sci. Soc. Am. Spec. Publ.* **39**: 111–128.

Hoover, M. D. and H. A. Lunt. 1952. A key for the classification of forest humus types. *Soil Sci. Soc. Am. Proc.* **16**: 368–370.

Hoover, M. T. and E. J. Ciolkosz. 1988. Colluvial soil parent material relationships in the ridge and valley physiographic province of Pennsylvania. *Soil Sci.* **145**: 163–172.

Hopkins, D. G., Sweeney, M. D., and J. L. Richardson. 1991. Dispersive erosion and Entisol-panspot genesis in sodium-affected landscapes. *Soil Sci. Soc. Am. J.* **55**: 171–177.

Host, G. E. and K. S. Pregitzer. 1992. Geomorphic influences on ground-flora and overstory composition in upland forests of northwestern lower Michigan. *Can. J. For. Res.* **22**: 1547–1555.

Host, G. E., Pregitzer, K. S., Ramm, C. W., *et al.* 1988. Variation in overstory biomass among glacial landforms and ecological land units in northwestern Lower Michigan. *Can. J. For. Res.* **18**: 659–668.

Houghton, H. F. 1980. Refined techniques for staining plagioclase and alkalai feldspars in thin section. *J. Sed. Petrol.* **55**: 629–631.

Hovan, S. A., Rea, D. K., Pisias, N. G., *et al.* 1989. A direct link between the China loess and marine $\delta^{18}O$ records: aeolina flux to the north Pacific. *Nature* **340**: 296–298.

Howard, J. L., Amos, D. F., and W. L. Daniels. 1993. Alluvial soil choronosequence in the Inner Coastal Plain, Central Virginia. *Quat. Res.* **39**: 201–213.

Hseu, Z.-Y., Chen, Z.-S., and Z.-D. Wu. 1999. Characterization of placic horizons in two subalpine forest Inceptisols. *Soil Sci. Soc. Am. J.* **63**: 941–947.

Hsu, P. H. 1989. Aluminum oxides and hydroxides. In J. B. Dixon and S. B. Weed (eds.) *Minerals in Soil Environments*, 2nd edn. Madison, WI, Soil Science Society of America, pp. 199–278.

Huayu, L. U. and A. N. Zhisheng. 1998. Paleoclimatic significance of grain size of loess-palaeosol deposit in Chinese Loess Plateau. *Sci. China* **41D**: 626–631.

Huggett, R. J. 1975. Soil landscape systems: a model of soil genesis. *Geoderma* **13**: 1–22.

1976a. Conceptual models in pedogenesis: a discussion. *Geoderma* **16**: 261–262.

1976b. Lateral translocations of soil plasma through a small valley basin in the Northaw Great Wood, Hertfordshire. *Earth Surf. Proc. Landforms* **1**: 99–109.

1991. *Climate, Earth Processes, and Earth History*. Berlin, Springer-Verlag.

1998a. Discussion of the paper by J. D. Phillips. *Geoderma* **86**: 23–25.

1998b. Soil chronosequences, soil development, and soil evolution: a critical review. *Catena* **32**: 155–172.

Hugie, V. K. and H. B. Passey. 1963. Cicadas and their effect upon soil genesis in certain soils in southern Idaho, northern Utah, and northeastern Nevada. *Soil Sci. Soc. Am. Proc.* **27**: 78–82.

Humphreys, G. S. 1981. The rate of ant mounding and earthworm casting near Sydney, New South Wales. *Search* **12**: 129–131.

1994. Bioturbation, biofabrics and the biomantle: an example from the Sydney Basin. In A. J. Ringrose-Voase and G. S. Humphreys (eds.) *Soil Micromorphology: Studies in Management and Genesis.* London, Elsevier, pp. 421–436.

Humphreys, G. S. and D. A. Adamson. 2000. Inadequate pedogeomorphic evidence for dry and cold climatic conditions in southeastern Brazil. *Z. Geomorphol.* **44**: 529–531.

Humphreys, G. S. and P. B. Mitchell. 1983. A preliminary assessment of the role of bioturbation and rainwash on sandstone hillslopes in the Sydney Basin. In R. W. Young and G. C. Nanson (eds.) *Aspects of Australian Sandstone Landscapes.* Wollongong, NSW, Dept of Geography, University of Wollongong, pp. 66–80.

Hunckler, R. V. and R. J. Schaetzl. 1997. Spodosol development as affected by geomorphic aspect, Baraga County, Michigan. *Soil Sci. Soc. Am. J.* **61**: 1105–1115.

Hunt, C. B. 1972. *Geology of Soils*. San Francisco, CA, W. H. Freeman.

Hunter, C. R., Frazier, B. E., and A. J. Busacca. 1987. Lytell series: a nonvolcanic Andisol. *Soil Sci. Soc. Am. J.* **51**: 376–383.

Hunter, J. M. 1961. Morphology of a bauxite summit in Ghana. *Geog. J.* **77**: 469–476.

1973. Geophagy in Africa and in the United States: a culture–nutrition hypothesis. *Geog. Rev.* **63**: 170–195.

1984a. Geophagy in Central America. *Geog. Rev.* **74**: 157–169.

1984b. Insect geophagy in Sierra Leone. *J. Cult. Geog.* **4**: 2–13.

1993. Macroterme geophagy and pregnancy clays in southern Africa. *J. Cult. Geog.* **14**: 69–92.

Hunter, J. M., Horst, O. H., and R. N. Thomas. 1989. Religious geophagy as a cottage industry: the Holy Clay Tablet of Esquipulas, Guatemala. *Natl Geog. Res.* **5**: 281–295.

Huntley, D. J. and M. Lamothe. 2001. Ubiquity of anomalous fading in K-feldspars, and the measurement and correction for it in optical dating. *Can. J. Earth Sci.* **38**: 1093–1106.

Huntley, D. J., Godfrey-Smith, D. I., and M. L. W. Thewalt. 1985. Optical dating of sediments. *Nature* **313**: 105–107.

Hupp, C. R. and W. P. Carey. 1990. Dendrogeomorphic approach to estimating slope retreat, Maxey Flats, Kentucky. *Geology* **18**: 658–661.

Hurst, V. J. 1977. Visual estimation of iron in saprolite. *Geol. Soc. Am. Bull.* **88**: 174–176.

Hussain, M. S., Amadi, T. H., and M. S. Sulaiman. 1984. Characteristics of soils of a toposequence in northeastern Iraq. *Geoderma* **33**: 63–82.

Hutcheson, T. B., Jr. and H. H. Bailey. 1964. Fragipan soils: certain genetic implications. *Soil Sci. Soc. Am. Proc.* **28**: 684–685.

Hutchins, R. B., Blevins, R. L., Hill, J. D., *et al.* 1976. The influence of soils and microclimate on vegetation of forested slopes in eastern Kentucky. *Soil Sci.* **121**: 234–241.

Hutt, G., Jungner, H., Kujansuu, R., *et al.* 1993. OSL and TL dating of buried podsols and overlying sands in Ostrobothnia, western Finland. *J. Quat. Sci.* **8**: 125–132.

Hutton, C. E. 1947. Studies of loess-derived soils in southwestern Iowa. *Soil Sci. Soc. Am. Proc.* **12**: 424–431.

Hyde, A. G. and R. D. Ford. 1989. Water table fluctuations in representative Immokalee and Zolfo soils of Florida. *Soil Sci. Soc. Am. J.* **53**: 1475–1478.

Ibanez, J. J., Ballexta, R. J., and A. G. Alvarez. 1990. Soil landscapes and drainage basins in Mediterranean mountain areas. *Catena* **17**: 573–583.

Indorante, S. J. 1998. Introspection of natric soil genesis on the loess-covered till plain in south central Illinois. *Quat. Intl* **51/52**: 41–42.

Indorante, S. J. and I. J. Jansen. 1984. Perceiving and defining soils on disturbed land. *Soil Sci. Soc. Am. J.* **48**: 1334–1337.

Ingham, E. R. 1999a. Soil fungi. *USDA-NRCS Soil Qual. Inst. PA* **1637**: D1–D4.

1999b. Soil nematodes. *USDA-NRCS Soil Qual. Inst. PA* **1637**: F1–F4.

1999c. Soil protozoa. *USDA-NRCS Soil Qual. Inst. PA* **1637**: E1–E4.

1999d. The soil food web. *USDA-NRCS Soil Qual. Inst. PA* **1637**: A1–A8.

Inglis, D. R. 1965. Particle sorting and stone migration by freezing and thawing. *Science* **148**: 1616–1617.

Innes, J. L. 1984. The optimal sample size in lichenometric studies. *Arctic Alpine Res.* **16**: 233–244.

1985. An examination of some factors affecting the largest lichens on a substrate. *Arctic Alpine Res.* **17**: 99–106.

Innis, R. P. and D. J. Pluth. 1970. Thin section preparation using an epoxy impregnation for petrographic and electron microprobe analysis. *Soil Sci. Soc. Am. Proc.* **34**: 483–485.

Isard, S. A. and R. J. Schaetzl. 1995. Estimating soil temperatures and frost in the lake effect snowbelt region, Michigan, USA. *Cold Regions Sci. Tech.* **23**: 317–332.

1998. Effects of winter weather conditions on soil freezing in southern Michigan. *Phys. Geog.* **19**: 71–94.

Isbell, R. F. 1991. Australian Vertisols. In J. M. Kimble (ed.) *Characterization, Classification, and Utilization of Cold Aridisols and Vertisols, Proc. 6th Intl Soil Correlation Mtg.* Lincoln, NE, US Department of Agriculture Soil Conservation Service, pp. 73–80.

Isherwood, D. and A. Street. 1976. Biotite-induced grusification of the Boulder Creek Granodiorite, Boulder County, Colorado. *Geol. Soc. Am. Bull.* **87**: 366–370.

Izett, G. A. 1981. Volcanic ash beds: recorders of upper Cenozoic silicic pyroclastic volcanism in the western United States. *J. Geophys. Res.* **88**: 10200–10222.

Izett, G. A. and R. E. Wilcox. 1982. Map showing localities and inferred distributions of the Huckleberry Ridge, Mesa Falls, and Lava Creek ash beds (Pearlette family ash beds) of Pliocene and Pleistocene age in the western United States and southern Canada. *US Geol. Survey Misc. Invest. Map* **1325**.

Izett, G. A., Wilcox, R. E., Powers, H. A., *et al.* 1970. The Bishop ash bed, a Pleistocene marker bed in the western United States. *Quat. Res.* **1**: 121–132.

Jackson, M. L. 1965. Clay transformations in soil genesis during the Quaternary. *Soil Sci.* **99**: 15–22.

Jackson, M. L. and T. A. Frolking. 1982. Mechanism of terra rosa red coloration. *Clay Res.* **1**: 1–5.

Jackson, M. L., Levelt, T. W. M., Syers, J. K., *et al.* 1971. Geomorphological relationships of tropospherically derived quartz in the soils of the Hawaiian Islands. *Soil Sci. Soc. Am. Proc.* **35**: 515–525.

Jackson, M. L., Tyler, S. A., Willis, A. L., *et al.* 1948. Weathering sequence of clay size minerals in soils and sediments. *J. Phys. Coll. Chem.* **52**: 1237–1260.

Jacobs, P. M. 1998. Influence of parent material grain size on genesis of the Sangamon Geosol in south-central Indiana. *Quat. Intl* **51/52**: 127–132.

Jacobs, P. M. and J. C. Knox. 1994. Provenance and pedology of a long-term Pleistocene depositional sequence in Wisconsin's Driftless Area. *Catena* **22**: 49–68.

Jacobs, P. M., West, L. T., and J. N. Shaw. 2002. Redoximorphic features as indicators of seasonal saturation, Lowndes County, Georgia. *Soil Sci. Soc. Am. J.* **66**: 315–323.

Jacobson, G. L., Jr. and H. J. B. Birks. 1980. Soil development on recent end moraines of the Klutlan Glacier, Yukon Territory, Canada. *Quat. Res.* **14**: 87–100.

Jakobsen, B. H. 1989. Evidence for translocations into the B horizon of a subarctic Podzol in Greenland. *Geoderma* **45**: 3–17.

1991. Aspects of soil geography in South Greenland. *Folia Geog. Danica* **14**: 155–164.

James, H. R., Ransom, M. D., and R. J. Miles. 1995. Fragipan genesis in polygenetic soils on the Springfield Plateau of Missouri. *Soil Sci. Soc. Am. J.* **59**: 151–160.

James, L. A. 1988. Rates of organic carbon accumulation in young mineral soils near Burroughs Glacier, Glacier Bay, Alaska. *Phys. Geog.* **9**: 50–70.

Janeau, J. L., and C. Valentin. 1987. Relations entre les termitières *Trinervitermes* sp. et la surface du sol: réorganisations, ruissellement et érosion. *Rev. Ecol. Biol. Sol* **24**: 637–647.

Jarvis, N. L., Ellis, R., Jr., and O. W. Bidwell. 1959. A chemical and mineralogical characterization of selected Brunizem, Reddish Prairie, Grumusol, and Planosol soils developed in Pre-Pleistocene materials. *Soil Sci. Soc. Am. Proc.* **23**: 234–239.

Jauhiainen, E. 1973a. Age and degree of podzolization of sand soils on the coastal plain of northwest Finland. *Comment. Biol.* **68**: 5–32.

1973b. Effect of climate on podzolization in southwest and eastern Finland. *Comment. Phys.-Math.* **43**: 213–242.

Jayawardane, N. S. and E. L. Greacen. 1987. The nature of swelling in soils. *Austral. J. Soil Res.* **25**: 107–113.

Jenkins, D. A. and R. P. Bower. 1974. The significance of the atmospheric contribution to the trace element content of soils. *Trans. 10th Intl Congr. Soil Sci.* (Moscow) **6**: 466–474.

Jenny, H. 1935. The clay content of the soil as related to climatic factors, particularly temperature. *Soil Sci.* **40**: 111–128.

1941a. Calcium in the soil. III. Pedologic relations. *Soil Sci. Soc. Am. Proc.* **6**: 27–37.

1941b. *Factors of Soil Formation*. New York, McGraw-Hill.

1946. Arrangement of soil series and types according to functions of soil-forming factors. *Soil Sci.* **61**: 375–391.

1958. Role of the plant factor in the pedogenic functions. *Ecology* **39**: 5–16.

1960. Podsols and pygmies: special need for preservation. *Sierra Club Bull.* 8–9.

1961. Derivation of state factor equations of soils and ecosystems. *Soil Sci. Soc. Am. Proc.* **25**: 385–388.

Jenny, H. and C. D. Leonard. 1934. Functional relationships between soil properties and rainfall. *Soil Sci.* **38**: 363–381.

Jenny, H., Arkley, R. J., and A. M. Schultz. 1969. The Pygmy forest-Podsol ecosystem and its dune associates of the Mendocino Coast. *Madrono* **20**: 60–74.

Jensen, H. I. 1911. The nature and origin of gilgai country. *Proc. Roy. Soc.* NSW **45**: 337–358.

Jensen, M. E. 1984. Soil moisture regimes on some rangelands of southeastern Idaho. *Soil Sci. Soc. Am. J.* **48**: 1328–1330.

Jersak, J., Amundson, R., and G. Brimhall, Jr. 1995. A mass balance analysis of podzolization: examples from the northeastern United States. *Geoderma* **66**: 15–42.

Jessup, R. W. 1960. The stony tableland soils of the southeastern portion of the Australian arid zone and their evolutionary history. *J. Soil Sci.* **11**: 188–196.

Jha, P. P. and M. G. Cline. 1963. Morphology and genesis of a Sol Brun Acide with fragipan in uniform silty material. *Soil Sci. Soc. Am. Proc.* **27**: 339–344.

Jiamao, H., Houyuan, L., Naiqin, W., *et al.* 1996. The magnetic susceptibility of modern soils in China and its use for paleoclimate reconstruction. *Stud. Geoph. Geodet.* **40**: 262–275.

Jochimsen, M. 1973. Does the size of lichen thalli really constitute a valid measure for dating glacial deposits? *Arctic Alpine Res.* **5**: 417–424.

Joffe, J. S. 1936. *Pedology*. New Brunswick, NJ, Rutgers University Press.

1949. *Pedology*, 2nd edn. Somerville, NJ, Somerset Press.

Johnson, D. L. 1985. Soil thickness processes. *Catena (Suppl.)* **6**: 29–40.

1989. Subsurface stone lines, stone zones, artifact-manuport layers, and biomantles produced by bioturbation via pocket gophers (*Thomomys bottae*). *Am. Antiq.* **54**: 370–389.

1990. Biomantle evolution and the redistribution of earth materials and artifacts. *Soil Sci.* **149**: 84–102.

1993a. Biomechanical processes and the Gaia paradigm in a unified pedo-geomorphic and pedo-archaeologic framework: dynamic denudation. In J. E. Foss, M. E. Timpson and M. W. Morris (eds.) *Proc. 1st Intl Conf. Pedo-Archaeology*, University of Tennessee Agricultural Experimental Station Special Publication no. 93-03, pp. 41–67. Knoxville, TN.

1993b. Dynamic denudation evolution of tropical, subtropical and temperate landscapes with three tiered soils: toward a general theory of landscape evolution. *Quat. Intl* **17**: 67–78.

1994. Reassessment of early and modern soil horizon designation frameworks and associated pedogenetic processes: are midlatitude A E B-C horizons equivalent to tropical M S W horizons? *Soil Sci. (Trends Agric. Sci.)* **2**: 77–91.

1998a. A universal definition of soil. *Quat. Intl* **51/52**: 6–7.

1998b. Paleosols are buried soils. *Quat. Intl* **51/52**: 7.

2000. Soils and soil-geomorphology theories and models: the Macquarie connection. *Ann. Ass. Am. Geog.* **90**: 775–782.

2002. Darwin would be proud: bioturbation, dynamic denudation, and the power of theory in science. *Geoarchaeology* **17**: 7–40.

Johnson, D. L. and C. L. Balek. 1991. The genesis of Quaternary landscapes with stone lines. *Phys. Geog.* **12**: 385–395.

Johnson, D. L. and N. C. Hester. 1972. Origin of stone pavements on Pleistocene marine terraces in California. *Proc. Ass. Am. Geog.* **4**: 50–53.

Johnson, D. L. and F. D. Hole. 1994. Soil formation theory: a summary of its principal impacts on Geography, Geomorphology, Soil-Geomorphology, Quaternary Geology and Paleopedology. *Soil Sci. Soc. Am. Spec. Publ.* **33**: 111–126.

Johnson, D. L. and D. Watson-Stegner. 1987. Evolution model of pedogenesis. *Soil Sci.* **143**: 349–366.

Johnson, D. L., Johnson, D. N., and D. M. Moore. 2001. Deep and actively forming illuvial clay in the regolith and on bedrock. *Proc. 12th Intl Clay Conf.*, Bahia Blanca, Argentina, pp. 205–210.

Johnson, D. L., Keller, E. A., and T. K. Rockwell. 1990. Dynamic pedogenesis: new views on some key soil concepts, and a model for interpreting Quaternary soils. *Quat. Res.* **33**: 306–319.

Johnson, D. L., Watson-Stegner, D., Johnson, D. N., *et al.* 1987. Proisotropic and proanisotropic processes of pedoturbation. *Soil Sci.* **143**: 278–292.

Johnson, M. D., Mickelson, D. M., Clayton, L., *et al.* 1995. Composition and genesis of glacial hummocks, western Wisconsin, USA. *Boreas* **24**: 97–116.

Johnson, W. F., Mausbach, M. J., Gamble, E. E., *et al.* 1985. Natric horizons on some erosional landscapes in northwestern South Dakota. *Soil Sci. Soc. Am. J.* **49**: 947–952.

Johnson, W. H. 1990. Ice-wedge casts and relict patterned ground in central Illinois and their environmental significance. *Quat. Res.* **33**: 51–72.

Johnson, W. H. and L. R. Follmer. 1989. Source and origin of Roxana silt and Middle Wisconsinan midcontinent glacial activity. *Quat. Res.* **31**: 319–331.

Johnson, W. H., Hansel, A. K., Bettis, E. A. III, *et al.* 1997. Late Quaternary temporal and event classifications, Great Lakes Region, North America. *Quat. Res.* **47**: 1–12.

Johnson, W. M. 1963. The pedon and the polypedon. *Soil Sci. Soc. Am. Proc.* **27**: 212–215.

Johnson, W. M., Cady, J. G., and M. S. James. 1962. Characteristics of some Brown Grumusols of Arizona. *Soil Sci. Soc. Am. Proc.* **26**: 389–393.

Johnsson, H. and L.-C. Lundin. 1991. Surface runoff and soil water percolation as affected by snow and soil frost. *J. Hydrol.* **122**: 141–159.

Johnston, C. A., Lee, G. B., and F. W. Madison. 1984. The stratigraphy and composition of a lakeside wetland. *Soil Sci. Soc. Am. J.* **48**: 347–354.

Jones, C. E. 1991. Characteristics and origin of rock varnish from the hyperarid coastal deserts of Northern Peru. *Quat. Res.* **35**: 116–129.

Jones, L. H. P. and K. A. Handreck. 1967. Silica in soils, plants and animals. *Adv. Agron.* **19**: 107–149.

Jones, M. J. 1985. The weathered zone aquifers of the basement complex areas of Africa. *Q. J. Engin. Geol.* **18**: 35–46.

Jones, P. E. 1970. The occurrence of *Chtonius ischnocheles* (Hermann) (Chelonethi: Chtoniidae) in two types of hazel coppice leaf litter. *Bull. Br. Arachnol. Soc.* **1**: 77–79.

Jones, R. L. and A. H. Beavers. 1963. Some mineralogical and chemical properties of plant opal. *Soil Sci.* **96**: 375–379.

Jones, R. L., Hay, W. W., and A. H. Beavers. 1963. Microfossils in Wisconsinan loess and till from western Illinois and eastern Iowa. *Science* **140**: 1222–1224.

Jones, R. L., Ray, B. W., Fehrenbacher, J. B., *et al.* 1967. Mineralogical and chemical characteristics of soils in loess overlying shale in northwestern Illinois. *Soil Sci. Soc. Am. Proc.* **31**: 800–804.

Jones, T. A. 1959. Soil classification: a destructive criticism. *J. Soil Sci.* **10**: 196–200.

Jordan, J. V., Lewis, G. C., and M. A. Fosberg. 1958. Tracing moisture movement in slick-spot soils with radiosulfur. I. *Trans. Am. Geophys. Union* **39**: 446–450.

Joseph, K. T. 1968. A toposequence on limestone parent materials in North Kedah, Malaya. *J. Trop. Geog.* **27**: 19–22.

Jouquet, P., Mamou, L., Lepage, M., *et al.* 2002. Effect of termites on clay minerals in tropical soils: fungus-growing termites as weathering agents. *Eur. J. Soil Sci.* **53**: 521–527.

Jouzel, J., Lorius, C., Petit, J. R., *et al.* 1987. Vostok ice core: a continuous isotope temperature record over the last climatic cycle (160 000 years). *Nature* **329**: 403–408.

Junge, C. E., and R. T. Werby. 1958. The concentration of chloride, potassium, calcium, and sulfate in rain water over the United States. *Journal of Meteorology* **15**: 417–425.

Jungerius, P. D. 1985. Soils and geomorphology. *Catena (Suppl.)* **6**: 1–18.

Jungerius, P. D. and F. van der Meulen. 1988. Erosion processes in a dune landscape along the Dutch coast. *Catena* **15**: 217–288.

Jury, W. A. and B. Bellantouni. 1976. Heat and water movement under surface rocks in a field soil. II. Moisture effects. *Soil Sci. Soc. Am. J.* **40**: 509–513.

Jury, W. A., Gardner, W. R., and W. H. Gardner. 1991. *Soil Physics*, 5th edn. New York, John Wiley.

Kabrick, J. M., Clayton, M. K., and K. McSweeney. 1997. Spatial patterns of carbon and texture on drumlins in northeastern Wisconsin. *Soil Sci. Soc. Am. J.* **61**: 541–548.

Kadmon, R., Yair, A., and A. Danin. 1989. Relationship between soil properties, soil moisture, and vegetation along loess-covered hillslopes, northern Negev, Israel. *Catena (Suppl.)* 14:43–57.

Kahle, C. F. 1977. Origin of subaerial Holocene calcareous crusts: role of algae, fungi and sparmicritisation. *Sedimentology* **24**: 413–435.

Kahle, P. 2000. Heavy metals in garden soils from the urban area of Rostock. *J. Plant Nutr. Soil Sci.* **163**: 191–196.

Kämpf, N. and U. Schwertmann. 1982. Goethite and hematite in a climo-sequence in southern Brazil and their application in classification of kaolinitic soils. *Geoderma* **29**: 27–39.

Kamprath, E. J. 1973. Phosphorous. In P. A. Sanchez (ed.) *A Review of Soils Research on Tropical Latin America*, Technical Bulletin no. 219. North Carolina Agricultural Experimental Statim, pp. 138–161. Raleigh, NC.

Kantor, W. and U. Schwertmann. 1974. Mineralogy and genesis of clays in red-black soil toposequences on basic igneous rocks in Kenya. *J. Soil Sci.* **25**: 67–78.

Karathanasis, A. D. 1987a. Mineral solubility relationships in Fragiudalfs of western Kentucky. *Soil Sci. Soc. Am. J.* **51**: 474–481.

1987b. Thermodynamic evaluation of amorphous aluminosilicate binding agents in fragipans of western Kentucky. *Soil Sci. Soc. Am. J.* **51**: 819–824.

Karathanasis, A. D. and P. A. Golrick. 1991. Soil formation on loess/sandstone toposequences in west-central Kentucky. I. Morphology and physiochemical properties. *Soil Sci.* **152**: 14–24.

Karathanasis, A. D. and B. R. Macneal. 1994. Evaluation of parent material uniformity criteria in loess-influenced soils of west-central Kentucky. *Geoderma* **64**: 73–92.

Karlstrom, E. T. and G. Osborn. 1992. Genesis of buried paleosols and soils in Holocene and late Pleistocene tills, Bugaboo Glacier area, British Columbia, Canada. *Arctic Alpine Res.* **24**: 108–123.

Kaufman, I. R. and B. R. James. 1991. Anthropic epipedons in oyster shell middens of Maryland. *Soil Sci. Soc. Am. J.* **55**: 1191–1193.

Kay, B. D. and D. A. Angers. 2000. Soil structure. In M. E. Sumner (ed.) *Handbook of Soil Science.* Boca Raton, FL, CRC Press, pp. A-229–A-276.

Kay, G. F. 1916. Gumbotil, a new term in Pleistocene geology. *Science* **44**: 637–638.

Kay, G. F. and E. T. Apfel. 1929. The Pre-Illinoian Pleistocene geology of Iowa. *Iowa Geol. Surv. Ann. Rept* **34**: 1–304.

Kay, G. F. and J. B. Graham. 1943. The Illinoian and Post-Illinoian Pleistocene geology of Iowa. *Iowa Geol. Survey Ann. Rept* **38**: 1–262.

Kay, G. F. and J. N. Pierce. 1920. The origin of gumbotil. *J. Geol.* **28**: 89–125.

Keen, K. L. and L. C. K. Shane. 1990. A continuous record of Holocene eolian activity and vegetation change at Lake Ann, east-central Minnesota. *Geol. Soc. Am. Bull.* **102**: 1646–1657.

Kellogg, C. E. 1930. Preliminary study of the profiles of the principal soil types of Wisconsin. *Wisc. Geol. Nat. Hist. Surv. Bull.* **77A**.

1934. Morphology and genesis of the Solonetz soils of western North Dakota. *Soil Sci.* **38**: 483–500.

1936. *Development and Significance of the Great Soil Groups of the United States*, Miscellaneous Publication no. 229. Washington, DC, US Department of Agriculture, US Government Printing Office.

1941. Climate and soil. In *Climate and Men: Yearbook of Agriculture*. Washington, DC, US Department of Agriculture, US Government Printing Office, pp. 276–277.

1949. Preliminary suggestions for the classification and nomenclature of great soil groups in tropical and equatorial regions. *CAB Soil Sci. Tech. Commun.* **46**: 76–85.

1950. Tropical soils. *Trans. 4th Intl Congr. Soil Sci.* (Amsterdam) **1**: 266–276.

1974. Soil genesis, classification, and cartography: 1924–1974. *Geoderma* **12**: 347–362.

Kellogg, C. E. and I. J. Nygard. 1951. *The Principal Soil Groups of Alaska*, Agriculture Monograph no. 7. Washington, DC, US Department of Agriculture, US Government Printing Office.

Kelso, G. K. 1974. Pollen percolation rates in Euroamerican-era cultural deposits in the northeastern United States. *J. Archaeol. Sci.* **21**: 481–488.

Kemp, R. A. 1999. Micromorphology of loess-paleosol sequences: a record of paleoenvironmental change. *Catena* **35**: 181–198.

2001. Pedogenic modification of loess: significance for palaeoclimatic reconstructions. *Earth Sci. Rev.* **54**: 145–156.

Kemp, R. A. and P. D. McIntosh. 1989. Genesis of a texturally banded soil in Southland, New Zealand. *Geoderma* **45**: 65–81.

Kemp, R. A., McDaniel, P. A., and A. J. Busacca. 1998. Genesis and relationship of macromorphology and micromorphology to contemporary hydrological conditions of a welded Argixeroll from the Palouse in Idaho. *Geoderma* **83**: 309–329.

Kevan, D. K. M. 1968. Soil fauna and humus formation. *Trans. 9th Intl Congr. Soil Sci.* (Adelaide) **2**: 1–10.

Khakural, B. R., Lemme, G. D., and D. L. Mokma. 1993. Till thickness and argillic horizon development in some Michigan Hapludalfs. *Soil Surv. Horiz.* **34**: 6–13.

Khan, F. A. and T. E. Fenton. 1994. Saturated zones and soil morphology in a Mollisol catena of central Iowa. *Soil Sci. Soc. Am. J.* **58**: 1457–1464.

Khangarot, A. S., Wilding, L. P., and G. F. Hall. 1971. Composition and weathering of loess mantled Wisconsin- and Illinoian-age terraces in central Ohio. *Soil Sci. Soc. Am. Proc.* **35**: 621–626.

Khokhlova, O. S., Kovalevskaya, I. S., and S. A. Oleynik. 2001. Records of climatic changes in the carbonate profiles of Russian Chernozems. *Catena* **43**: 203–215.

Kieffer, S. W. 1981. Blast dynamics at Mount St. Helens on 18 May 1980. *Nature* **291**: 568–570.

Kiernan, K. 1990. Weathering as an indicator of the age of Quaternary glacial deposits in Tasmania. *Austral. Geog.* **21**: 1–17.

Kimble, J. M., Ping, C. L., Sumner, M. E., *et al.* 2000. *Andisols*. In M. E. Sumner (ed.) *Handbook of Soil Science*. Boca Raton, FL, CRC Press, pp. E-209–E-224.

King, D., Bourennane, H., Isambert, M., *et al.* 1999. Relationship of the presence of a non-calcareous clay-loam horizon to DEM attributes in a gently sloping area. *Geoderma* **89**: 95–111.

King, G. J., Acton, D. F., and R. J. St. Arnaud. 1983. Soil-landscape analysis in relation to soil distribution and mapping at a site within the Weyburn association. *Can. J. Soil Sci.* **63**: 657–670.

King, J. J. and D. P. Franzmeier. 1981. Estimation of saturated hydraulic conductivity from soil morphological and genetic information. *Soil Sci. Soc. Am. J.* **45**: 1153–1156.

King, L. C. 1957. The uniformitarian nature of hillslopes. *Trans. Edinburgh Geol. Soc.* **17**: 81–102.

King, T. J. 1977. The plant ecology of ant-hills in calcareous grasslands. *J. Ecol.* **65**: 235–316.

Kirkby, A. and M. J. Kirkby. 1974. Surface wash at the semi-arid break in slope. *Z. Geomorphol. (Suppl.)* **21**: 151–176.

Kirkham, V. R. D., Johnson, M. M., and D. Holm. 1931. Origin of Palouse Hills topography. *Science* **73**: 207–209.

Kittrick, J. A. 1971. Montmorillonite equilibria and the weathering environment. *Soil Sci. Soc. Am. J.* **35**: 815–820.

1973. Mica-derived vermiculites as unstable intermediates. *Clay Mins.* **21**: 479–488.

Kittrick, J. A., Fanning, D. S., and L. R. Hossner (eds.) 1982. *Acid Sulphate Weathering*, Soil Science Society of America Special Publication no. 10. Madison, WI, Soil Science Society of America.

Kladivko, E. J. and H. J. Timmenga. 1990. Earthworms and agricultural management. In J. E. Box and L. C. Hammond (eds.) *Rhizosphere Dynamics*. American Association for Advancement Science Selected Symposium, Boulder, Westview Press, pp. 192–216.

Klappa, C. F. 1979. Calcified filaments in Quaternary calcretes: organo-mineral interactions in the subaerial vadose environment. *J. Sed. Petrol.* **49**: 955–968.

Kleber, A. 1997. Cover-beds as soil parent materials in midlatitude regions. *Catena* **30**: 197–213.

2000. Compound soil horizons with mixed calcic and argillic properties: examples from the northern Great Basin, USA. *Catena* **41**: 111–131.

Kleiss, H. J. 1970. Hillslope sedimentation and soil formation in northeastern Iowa. *Soil Sci. Soc. Am. Proc.* **34**: 287–290.

1973. Loess distribution along the Illinois soil-development sequence. *Soil Sci.* **115**: 194–198.

Kline, J. R. 1973. Mathematical simulation of soil–plant relationships and soil genesis. *Soil Sci.* **115**: 240–249.

Klingbiel, A. A., Horvath, E. H., Moore, D. G., *et al.* 1987. Use of slope, aspect and elevation maps derived from digital elevation model data in making soil surveys. *Soil Sci. Soc. Am. Spec. Publ.* **20**: 77–90.

Klute, A. 1986. Water retention: laboratory methods. In A. Klute (ed.) *Methods of Soil Analysis*, Part 1, *Physical and Mineralogical Methods*, Agronomy Monograph no. 9, 2nd edn. Madison, WI, American Soil Association and Soil Science Society of America, pp. 635–662.

Knapp, B. D. 1993. *Soil Survey of Presque Isle County, Michigan.* Washington, DC, US Department of Agriculture Soil Conservation Service, US Government Printing Office.

Knauss, K. G. and T. Ku. 1980. Desert varnish: potential for age dating via uranium-series isotopes. *J. Geol.* **88**: 95–100.

Knight, M. J. 1980. Structural analysis and mechanical origins of gilgai at Boorook, Victoria, Australia. *Geoderma* **23**: 245–283.

Knollenberg, W. G., Merritt, R. W., and D. L. Lawson. 1985. Consumption of leaf litter by *Lumbricus terrestris* (Oligochaeta) on a Michigan woodland floodplain. *Am. Midl. Nat.* **113**: 1–6.

Knox, E. G. 1957. Fragipan horizons in New York soils. III. The basis of rigidity. *Soil Sci. Soc. Am. Proc.* **21**: 326–330.

Knuepfer, P. L. K. 1988. Estimating ages of late Quaternary stream terraces from analysis of weathering rinds and soils. *Geol. Soc. Am. Bull.* **100**: 1224–1236.

Knuepfer, P. L. K. and L. D. McFadden (eds.) 1990. Soils and landscape evolution. *Geomorphology* **3**: issues (3) and (4).

Knuteson, J. A., Richardson, J. L., Patterson, D. D., *et al.* 1989. Pedogenic carbonates in a Calciaquoll associated with a recharge wetland. *Soil Sci. Soc. Am. J.* **53**: 495–499.

Koch, P. L. 1998. Isotopic reconstruction of past continental environments. *Ann. Rev. Earth Planet. Sci.* **26**: 573–613.

Kocurek, G. and N. Lancaster. 1999. Aeolian system sediment state: theory and Mojave Desert Kelso dune field example. *Sedimentology* **46**: 505–515.

Kodama, H. and C. Wang. 1989. Distribution and characterization of noncrystalline inorganic compounds in Spodosols and Spodosol-like soils. *Soil Sci. Soc. Am. J.* **53**: 526–534.

Kodama, H., Schnitzer, M., and M. Jaakkimainen. 1983. Chlorite and biotite weathering by fulvic acid solutions in closed and open systems. *Can. J. Soil Sci.* **63**: 619–629.

Kohnke, H., Stuff, R. G., and P. A. Miller. 1968. Quantitative relations between climate and soil formation. *Z. Pflanzen. Bodenkd.* **119**: 24–33.

Konrad, J.-M. 1989. Physical processes during freeze–thaw cycles in clayey silts. *Cold Regions Sci. Tech.* **16**: 291–303.

Koppi, A. J. and D. J. Williams. 1980. Weathering and development of two contrasting soils formed from grandodiorite in south-east Queensland. *Austral. J. Soil Res.* **18**: 257–271.

Koshel'kov, S. P. 1961. Formation and subdivision of forest floor in southern taiga coniferous forests. *Sov. Soil Sci.* 1065–1073.

Koutaniemi, L., Koponen, R., and K. Rajanen. 1988. Podzolization as studied from terraces of various ages in two river valleys, northern Finland. *Silvia Fennica* **22**: 113–133.

Kouwenhoven, J. K. and R. Terpstra. 1979. Sorting action of tines and tine-like tools in the field. *J. Agric. Engin. Res.* **24**: 95–113.

Kozarski, S. 1974. Evidences of Late-Wurm permafrost occurrence in north-west Poland. *Quaest. Geog.* **1**: 65–86.

Krause, H. H., Rieger, S., and S. A. Wilde. 1959. Soils and forest growth on different aspects in the Tanana watershed of interior Alaska. *Ecology* **40**: 492–495.

Krinsley, D. H., Dorn, R. I., and S. W. Anderson. 1990. Factors that interfere with the age determination of rock varnish. *Phys. Geog.* **11**: 97–119.

Krishnamani, R. and W. C. Mahaney. 2000. Geophagy among primates: adaptive significance and ecological consequences. *Anim. Behav.* **59**: 899–915.

Kroonenberg, S. B. and P. J. Melitz. 1983. Summit levels, bedrock control and the etchplain concept in the basement of Suriname. *Geol. Mijnbouw* **62**: 389–399.

Krumbein, W. E. and C. Giele. 1979. Calcification in a coccoid-cyanbacterium associated with the formation of desert stromatolites. *Sedimentology* **26**: 593–604.

Krumbein, W. E. and K. Jens. 1981. Biogenic rock varnishes of the Negev Desert (Israel): an ecological study of iron and manganese transformation by cyanobacteria and fungi. *Oecologia* **50**: 25–38.

Krzyszowska, A. J., Blaylock, M. J., Vance, G. F., *et al.* 1996. Ion chromatographic analysis of low molecular weight organic acids in spodosol forest floor solutions. *Soil Sci. Soc. Am. J.* **60**: 1565–1571.

Ku, T. L., Bull, W. G., Freeman, S. T., *et al.* 1979. $^{230}Th/^{234}U$ dating of pedogenic carbonates in gravelly desert soils of Vidal Valley, southeastern California. *Geol. Soc. Am. Bull.* **90**: 1063–1073.

Kubiëna, W. L. 1938. *Micropedology*. Ames, IA, Collegiate Press.

1970. *Micromorphological Features of Soil Geography*. New Brunswick, NJ, Rutgers University Press.

Kukla, G. 1987. Loess stratigraphy in central China. *Quat. Sci. Rev.* **6**: 191–219.

Kukla, G. and Z. An. 1989. Loess stratigraphy in central China. *Palaeogeog. Palaeoclimat. Palaeoecol.* **72**: 203–225.

Kukla, G., Heller, F., Ming, L. X., *et al.* 1988. Pleistocene climates in China dated by magnetic susceptibility. *Geology* **16**: 811–814.

Künelt, W. 1961. *Soil Biology with Special Reference to the Animal Kingdom*. London, Faber and Faber.

Kunze, G. W. and J. B. Dixon. 1986. Pretreatment for mineralogical analysis. In A. Klute (ed), *Methods of Soil Analysis*, Part 1, *Physical and Mineralogical Methods*, Agronomy Monograph no. 9, 2nd edn. Madison, WI, American Soil Association and Soil Science Society of America, pp. 91–100.

Kunze, G. W., Oakes, H., and M. E. Bloodworth. 1963. Grumusols of the Coast Prairie of Texas. *Soil Sci. Soc. Am. Proc.* **27**: 412–421.

Kurz, M. D., Colodner, D., Trull, T. W., *et al.* 1990. Cosmic ray exposure dating with *in situ* produced cosmogenic ^{3}He: results from young Hawaiian lava flows. *Earth Planet. Sci. Lett.* **97**: 177–189.

Kusumgar, S., Agrawal, D. P., and R. V. Krishnamurthy. 1980. Studies of the loess deposits of the Kashmir valley and ^{14}C dating. *Radiocarbon* **22**: 757–762.

Kutiel, P. 1992. Slope aspect effect on soil and vegetation in a Mediterranean ecosystem. *Isr. J. Bot.* **41**: 243–250.

Kuzila, M. S. 1995. Identification of multiple loess units within modern soils of Clay County, Nebraska. *Geoderma* **65**: 45–57.

La Roi, G. H. and R. J. Hnatiuk. 1980. The *Pinus contorta* forests of Banff and Jasper national parks: a study in comparative synecology and syntaxonomy. *Ecol. Monogr.* **50**: 1–29.

Låg, J. 1951. Illustration of the influence of topography on depth of A2 layer in Podzol profiles. *Soil Sci.* **71**: 125–127.

Lagacherie, P., Andrieux, P., and R. Bouzigues. 1996. Fuzziness and uncertainty of soil boundaries: from reality to coding in GIS. In P. A. Burrough and A. U. Frank (eds.) *Geographic Objects with Indeterminate Boundaries*. London, Taylor and Francis, pp. 275–286.

Laker, M. C., Hewitt, P. H., Nel, A., *et al.* 1982. Effects of the termite *Trinervitermes trinervoides* Sjöstedt on the organic carbon and nitrogen contents and particle size distribution of the soils. *Rev. Ecol. Biol. Sol* **19**: 27–39.

Lal, D. 1988. *In situ*-produced cosmogenic isotopes in terrestrial rocks. *Ann. Rev. Earth Planet. Sci.* **16**: 355–388.

1991. Cosmic ray labeling of erosion surfaces: *in situ* nuclide production rates and erosion models. *Earth Planet. Sci. Lett.* **104**: 424–439.

Lambert, J. L. and F. D. Hole. 1971. Hydraulic properties of an ortstein horizon. *Soil Sci. Soc. Am. Proc.* **35**: 785–787.

Lammers, D. A. and M. G. Johnson. 1991. Soil mapping concepts for environmental assessment. *Soil Sci. Soc. Am. Spec. Publ.* **28**: 149–160.

Landeweert, R., Hoffland, E., Finlay, R. D., *et al.* 2001. Linking plants to rocks: ectomycorrhizal fungi mobilize nutrients from minerals. *Trends Ecol. Evol.* **16**: 248–254.

Lång, L.-O. 2000. Heavy mineral weathering under acidic soil conditions. *Appl. Geochem.* **15**: 415–423.

Lang, R. 1915. Versuch einer exakten Klassifikation der Böden in klimatischer und geologischer Hinsicht. *Intl Mitteil. Bodenkd.* **5**: 312–346.

Langbein, W. B. and S. A. Schumm. 1958. Yield of sediment in relation to mean annual precipitation. *Trans. Am. Geophys. Union* **39**: 1076–1084.

Langley-Turnbaugh, S. J. and C. V. Evans. 1994. A determinitive soil development index for pedo-stratigraphic studies. *Geoderma* **61**: 39–59.

Langmaid, K. K. 1964. Some effects of earthworm invasion in virgin podzols. *Can. J. Soil Sci.* **44**: 34–37.

Langohr, R. and R. Vermeire. 1982. Well-drained soils with a "degraded" Bt horizon in loess deposits in Belgium: relationship with paleoperiglacial processes. *Biul. Peryglac.* **29**: 203–212.

Langohr, R., Scoppa, C. O., and A. Van Wambeke. 1976. The use of a comparative particle size distribution index for the numerical classification of soil parent materials: application to Mollisols of the Argentine Pampa. *Geoderma* **15**: 305–312.

Lark, R. M. and P. H. T. Beckett. 1995. A regular pattern in the relative areas of soil profile classes and possible applications in reconnaissance soil survey. *Geoderma* **68**: 27–37.

Latham, M. 1980. Ferrallitization in an oceanian tropical environment. In K. T. Joseph (ed.) *Proc. Conf. on Classification and Management of Tropical Soils*. Kuala Lumpur, Malaysian Society for Soil Science, pp. 20–26.

Lattman, L. H. 1973. Calcium carbonate cementation of alluvial fans in southern Nevada. *Geol. Soc. Am. Bull.* **84**: 3013–3028.

Lattman, L. H. and S. K. Lauffenburger. 1974. Proposed role of gypsum in the formation of caliche. *Z. Geomorphol. (Suppl.)* **20**: 140–149.

Laundré, J. W. 1989. Estimating soil bulk density with expanding polyurethane foam. *Soil Sci.* **147**: 223–224.

Laurent, T. E., Graham, R. C., and K. R. Tice. 1994. Soils of the red fir forest–barrens mosaic, Siskiyou Mountains crest, California. *Soil Sci. Soc. Am. J.* **58**: 1747–1752.

Lawrey, J. D. 1977. Elemental partitioning in *Pinus resinosa* leaf litter and associated fungi. *Mycologia* **69**: 1121–1128.

Leahy, A. 1963. The Canadian system of soil classification and the Seventh Approximation. *Soil Sci. Soc. Am. Proc.* **27**: 224–225.

Leamy, M. L. 1974. The use of pedogenic carbonate to determine the absolute age of soils and to assess rates of soil formation. *Trans. 10th Intl Congr. Soil Sci.* (Moscow) **6**: 331–338.

Lecce, S. A. 1997. Spatial patterns of historical overbank sedimentation and floodplain evolution, Blue River, Wisconsin. *Geomorphology* **18**: 265–277.

Lee, B. D., Graham, R. C., Laurent, T. E., *et al.* 2001. Spatial distributions of soil chemical conditions in a serpentinitic wetland and surrounding landscape. *Soil Sci. Soc. Am. J.* **65**: 1183–1196.

Lee, F. Y., Yuan, T. L., and V. W. Carlisle. 1988. Nature of cementing materials in ortstein horizons of selected Florida Spodosols. I. Constituents of cementing materials. *Soil Sci. Soc. Am. J.* **52**: 1411–1418.

Lee, K. E. 1967. Microrelief features in a humid tropical lowland area, New Guinea, and their relation to earthworm activity. *Austral. J. Soil Res.* **5**: 263–274.

Lee, K. E. 1985. *Earthworms: Their Ecology and Relationships with Soils and Land Use*. Orlando, FL, Academic Press.

Lee, K. E. and J. H. A. Butler. 1977. Termites, soil organic matter decomposition and nutrient cycling. *Ecol. Bull.* **25**: 544–548.

Lee, K. E. and T. G. Wood. 1968. Preliminary studies of the role of *Nasutitermes exitiosus* (Hill) in the cycling of organic matter in a Yellow Podzolic soil under dry sclerophyll forest in South Australia. *Trans. 9th Intl Congr. Soil Sci.* (Adelaide) **2**: 11–18.

1971a. Physical and chemical effects on soils of some Australian termites and their pedological significance. *Pedobiologia* **11**: 376–409.

1971b. *Termites and Soils*. New York, Academic Press.

Lee, R. and A. Baumgartner. 1966. The topography and insolation climate of a mountainous forest area. *For. Sci.* **12**: 258–267.

Leigh, D. L. 1994. Roxana silt of the Upper Mississippi Valley: lithology, source, and paleoenvironment. *Geol. Soc. Am. Bull.* **106**: 430–442.

1996. Soil chronosequence of Brasstown Creek, Blue Ridge Mountains, USA. *Catena* **26**: 99–114.

Leigh, D. L. and J. C. Knox. 1993. AMS radiocarbon age of the Upper Mississippi Valley Roxana Silt. *Quat. Res.* **39**: 282–289.

Leighton, M. M. 1926. A notable type Pleistocene section: the Farm Creek exposure near Peoria, Illlinois. *J. Geol.* **34**: 167–174.

1933. The naming of the subdivisions of the Wisconsin glacial age. *Science* **77**: 168.

1958. Principles and viewpoints in formulating the stratigraphic classifications of the Pleistocene. *J. Geol.* **66**: 700–709.

Leighton, M. M. and P. MacClintock. 1930. Weathered zones of drift sheets of Illinois. *J. Geol.* **38**: 28–53.

Leighton, M. M. and H. B. Willman. 1950. Loess formations of the Mississippi Valley. *J. Geol.* **58**: 599–623.

Leisman, G. A. 1957. A vegetation and soil chronosequence on the Mesabi Iron Range spoil banks, Minnesota. *Ecol. Monogr.* **27**: 221–245.

Leonardi, G. and M. Miglavacca. 1999. Soil phosphorus analysis as an integrative tool for recognizing buried ancient ploughsoils. *J. Archaeol. Sci.* **26**: 343–352.

Leopold, L. B. and J. P. Miller. 1954. *A Postglacial Chronology for some Alluvial Valleys in Wyoming*, Water Supply Paper no. 1261. Washington, DC, US Geological Survey, US Government Printing Office.

Lepage, M. 1972. Recherches écologiques sur une savane sahélienne du Ferlo Septentrional, Sénégal: données préliminaires sur l'écologie des termites. *Terre et la Vie* **26**: 384–409.

1973. Recherches écologiques sur une savane sahélienne du Sénégal Septentrional: termites – répartition, biomasse et récolte de nourriture. *Ann. Univ. Adidjan* **6**: 139–145.

1984. Distribution, density and evolution of *Macrotermes bellicosus* nests (Isoptera: Macrotermitinae) in the north-east of Ivory Coast. *J. Anim. Ecol.* **53**: 107–118.

Lepsch, I. F., Buol, S. W., and R. B. Daniels. 1977a. Soil-landscape relationships in the Occidental Plateau of Sao Paulo state, Brazil. I. Geomorphic surfaces and soil mapping units. *Soil Sci. Soc. Am. J.* **41**: 104–109.

1977b. Soil-landscape relationships in the Occidental Plateau of Sao Paulo state, Brazil. II. Soil morphology, genesis, and classification. *Soil Sci. Soc. Am. J.* **41**: 109–115.

Lev, A. and R. H. King. 1999. Spatial variation of soil development in a high arctic soil landscape: Truelove Lowland, Devon Island, Nunavut, Canada. *Permafrost Periglacial Proc.* **10**: 289–307.

Levan, M. A. and E. L. Stone. 1983. Soil modification by colonies of black meadow ants in a New York old field. *Soil Sci. Soc. Am. J.* **47**: 1192–1195.

Leverett, F. 1898. The weathered zone (Yarmouth) between the Illinoian and Kansan till sheets. *J. Geol.* **6**: 238–243.

1939. The place of the Iowan drift. *J. Geol.* **47**: 398–407.

Lévieux, J. 1976. Deux aspects de l'action des fourmis (Hymenoptera, Formicidae) sur le sol d'une savane préforestière de Côte-d'Ivoire. *Bull. Ecol.* **7**: 283–295.

Levine, E. R. and E. J. Ciolkosz. 1983. Soil development in till of various ages in northeastern Pennsylvania. *Quat. Res.* **19**: 85–99.

1986. A computer simulation model for soil genesis applications. *Soil Sci. Soc. Am. J.* **50**: 661–667.

Levine, S. J., Hendricks, D. M., and J. F. Schreiber, Jr. 1989. Effect of bedrock porosity on soils formed from dolomitic limestone residuum and eolian deposition. *Soil Sci. Soc. Am. J.* **53**: 856–862.

Lewis, D. T. and J. V. Drew. 1973. Slick spots in southeastern Nebraska: patterns and genesis. *Soil Sci. Soc. Am. Proc.* **37**: 600–606.

Lewis, R. O. 1981. Use of opal phytoliths in paleoenvironmental reconstruction. *J. Ethnobiol.* **1**: 175–181.

Liebens, J. 1999. Characteristics of soils on debris aprons in the Southern Blue Ridge, North Carolina. *Phys. Geog.* **20**: 27–52.

Liebens, J. and R. J. Schaetzl. 1997. Relative-age relationships of debris flow deposits in the Southern Blue Ridge, North Carolina. *Geomorphology* **21**: 53–67.

Liegh, D. S. 1998. A 12 000-year record of natural levee sedimentation along the Broad River near Columbia, South Carolina. *Southeast. Geog.* **28**: 95–111.

Likens, G. E. 2001. Biogeochemistry, the watershed approach: some uses and limitations. Mar. *Freshw. Res.* **52**: 5–12.

Likens, G. E., Driscoll, C. T., Buso, D. C., *et al.* 1998. The biogeochemistry of calcium at Hubbard Brook. *Biogeochemistry* **41**: 89–173.

Lin, H. S., McInnes, K. J., Wilding, L. P., *et al.* 1999. Effects of soil morphology on hydraulic properties. I. Quantification of soil morphology. *Soil Sci. Soc. Am. J.* **63**: 948–954.

Lindbo, D. L. and P. L. M. Veneman. 1993. Micromorphology of selected Massachusetts fragipan soils. *Soil Sci. Soc. Am. J.* **57**: 437–442.

Lindbo, D. L., Rhoton, F. E., Hudnall, W. H., *et al.* 2000. Fragipan degradation and nodule formation in Glossic Fragiudalfs of the Lower Mississippi Valley. *Soil Sci. Soc. Am. J.* **64**: 1713–1722.

Lindsay, W. L., Vlek, P. L. G., and S. H. Chien. 1989. Phosphate minerals. In J. B. Dixon and S. B. Weed (eds.) *Minerals in Soil Environments,* 2nd edn. Madison, WI, Soil Science Society of America, pp. 1089–1130.

Linick, T. W., Damon, P. E., Donahue, D. J., *et al.* 1989. Accelerator mass spectrometry: the new revolution in radiocarbon dating. *Quat. Intl* **1**: 1–6.

Litaor, M. I. 1987. The influence of eolian dust on the genesis of alpine soils in the Front Range, Colorado. *Soil Sci. Soc. Am. J.* **51**: 142–147.

1992. Aluminum mobility along a geochemical catena in an alpine watershed, Front Range, Colorado. *Catena* **19**: 1–16.

Litaor, M. I., Manicelli, R., and J. C. Halfpenny. 1996. The influence of pocket gophers on the status of nutrients in alpine soils. *Geoderma* **70**: 37–48.

Little, I. P. and W. T. Ward. 1981. Chemical and mineralogical trends in a chronosequence developed on alluvium in eastern Victoria. Australia. *Geoderma* **25**: 173–188.

Liu, B., Phillips, F. M., Pohl, M. M., *et al.* 1996. An alluvial surface chronology based on cosmogenic

^{36}Cl dating, Ajo Mountains (Organ Pipe Cactus National Monument), southern Arizona. *Quat. Res.* **45**: 30–37.

Liu, K.-B. 1990. Holocene paleoecology of the boreal forest and Great Lakes–St. Lawrence forest in northern Ontario. *Ecol. Monogr.* **60**: 179–212.

Liu, K.-B. and N. S. N. Lam. 1985. Paleovegetational reconstruction based on modern and fossil pollen data: an application of discriminant analysis. *Ann. Ass. Am. Geog.* **75**: 115–130.

Liu, T. S. 1985. *Loess and the Environment*. Beijing, China Ocean Press.

Liu, X., Rolph, T., Bloemendal, J., *et al.* 1995. Quantitative estimates of palaeoprecipitation at Xifeng, in the Loess Plateau of China. *Palaeogeog. Palaeoclimat. Palaeoecol.* **113**: 243–248.

Liu, Y., Steenhuis, T. S., Parlange, J.-Y., *et al.* 1991. Hysteretic finger phenomena in dry and wetted sands. In *Referential Flow*. St. Joseph, MI, American Society of Agricultural Engineers, pp. 160–172.

Livens, P. J. 1949. Characteristics of some soils of the Belgian Congo. *Commonw. Bur. Soil Sci. Tech. Pub.* **46**: 29–35.

Lobry de Bruyn, A. L. and A. J. Conacher. 1990. The role of termites and ants in soil modification: a review. *Austral. J Soil Res.* **28**: 55–93.

Lock, W. W., Andrews, P. J., and J. T. Webber. 1979. A manual for lichenometry. *Br. Geomorphal. Res. Group Tech. Bull.* **26**: 1–25.

Locke, W. W. III. 1979. Etching of hornblende grains in Arctic soils: an indicator of relative age and paleoclimate. *Quat. Res.* **11**: 197–212.

Loendorf, L. L. 1991. Cation-ratio varnish dating and petroglyph chronology in southeastern Colorado. *Antiquity* **65**: 246–255.

Löffler, E. and C. Margules. 1980. Wombats detected from space. *Remote Sens. Env.* **9**: 47–56.

Lokaby, B. G. and J. C. Adams. 1985. Pedoturbation of a forest soil by fire ants. *Soil Sci. Soc. Am. J.* **49**: 220–223.

Losche, C. K., McCracken, R. J., and C. B. Davey. 1970. Soils of steephy sloping landscapes in the southern Appalachian Mountains. *Soil Sci. Soc. Am. Proc.* **34**: 473–478.

Lotspeich, F. B. and H. W. Smith. 1953. Soils of the Palouse loess. I. The Palouse catena. *Soil Sci.* **76**: 467–480.

Lousier, J. D. and D. Parkinson. 1978. Chemical element dynamics in decomposing leaf litter. *Can. J. Bot.* **56**: 2795–2812.

Lovegrove, B. G. and W. R. Siegfried. 1986. Distribution and formation of Mima-like earth mounds in the western Cape Province of South Africa. *S. Afr. J. Sci.* **82**: 432–436.

Lowdermilk, W. C. and H. L. Sundling. 1950. Erosion pavement, its formation and significance. *Trans. Am. Geophys. Union* **31**: 96–100.

Lu, H. and D. Sun. 2000. Pathways of dust input to the Chinese Loess Plateau during the last glacial and interglacial periods. *Catena* **40**: 251–261.

Ludwig, K. R. and J. B. Paces. 2002. Uranium-series dating of pedogenic silica and carbonate, Crater Flat, Nevada. Geochim. *Cosmochim. Acta* **66**: 487–506.

Lundqvist, J. and K. Bengtsson. 1970. The red snow: a meteorological and pollen analytic study of longtransported material from snowfalls in Sweden. *Geol. Foreningens i Stockholm Forhandl.* **92**: 288–301.

Lundström, U. S., van Breemen, and N. Bain. 2000b. The podzolization process: a review. *Geoderma* **94**: 91–107.

Lundström, U. S., van Breemen, N., Bain, D. C., *et al.* 2000a. Advances in understanding the podzolization process resulting from a multidisciplinary study of three coniferous forest soils in the Nordic Countries. *Geoderma* **94**: 335–353.

Lundström, U. S., van Breemen, N., and A. G. Jongmans. 1995. Evidence for microbial decomposition of organic acids during podzolization. *Eur. J. Soil Sci.* **46**: 489–496.

Lutz, H. J. 1940. *Disturbance of Forest Soil Resulting from the Uprooting of Trees*, Bulletin no. 45. New Haven, CT, Yale University School of Forestry.

1960. Movement of rocks by uprooting of forest trees. *Am. J. Sci.* **258**: 752–756.

Lyford, W. H. 1938. Horizon variations of three New Hampshire Podzol profiles. *Soil Sci. Soc. Am. Proc.* **2**: 242–246.

1963. *Importance of Ants to Brown Podzolic Soil Genesis in New England*, Forest Paper no. 7. Cambridge, MA, Harvard University.

1974. Narrow soils and intricate soil patterns in southern New England. *Geoderma* **11**: 195–208.

Lyford, W. H. and D. W. MacLean. 1966. *Mound and Pit Microrelief in Relation to Soil Disturbance and Tree Distribution in New Brunswick, Canada*, Forest Paper no. 15. Cambridge, MA, Harvard University.

Lyford, W. H. and T. Troedsson. 1973. Fragipan horizons in soils on moraines near Garpenburg, Sweden. *Stud. Fort. Suecica* **108**: 1–21.

Lynn, W. and D. Williams. 1992. The making of a Vertisol. *Soil Surv. Horiz.* **33**: 45–50.

Mabbutt, J. A. 1965. Stone distribution in a stony tableland soil. *Austral. J. Soil Res.* **3**: 131–142.

Macedo, J. and R. B. Bryant. 1987. Morphology, mineralogy, and genesis of a hydrosequence of Oxisols in Brazil. *Soil Sci. Soc. Am. J.* **51**: 690–698.

Macgregor, A. N. and D. E. Johnson. 1971. Capacity of desert algal crusts to fix atmospheric nitrogen. *Soil Sci. Soc. Am. Proc.* **35**: 843–844.

Machette, M. N. 1985. Calcic soils of the southwestern United States. *Geol. Soc. Am. Spec. Paper* **203**: 1–21.

Mack, G. H., James, W. C., and H. C. Monger. 1993. Classification of paleosols. *Geol. Soc. Am. Bull.* **105**: 129–136.

Mackay, A. D., Syers, J. K., Springett, J. A., *et al.* 1982. Plant availability of phosphorous in superphosphate and a phosphate rock as influenced by earthworms. *Soil Biol. Biochem.* **14**: 281–287.

Mackay, J. R. 1984. The frost heave of stones in the active layer above permafrost with downward and upward freezing. *Arctic Alpine Res.* **16**: 439–446.

Mackay, J. R. and J. V. Matthews. 1983. Pleistocene ice and sand wedges, Hooper Island, Northwest Territories. *Can. J. Earth Sci.* **20**: 1087–1097.

MacKinney, A. L. 1929. Effects of forest litter on soil temperature and soil freezing in autumn and winter. *Ecology* **10**: 312–321.

Mackintosh, E. E. and J. van der Hulst. 1978. Soil drainage classes and soil water table relations in medium and coarse textured soils in southern Ontario. *Can. J. Soil Sci.* **58**: 287–301.

Maclean, A. H. and S. Pawluk. 1975. Soil genesis in relation to groundwater and soil moisture regimes near Vegreville, Alberta. *J. Soil Sci.* **26**: 278–293.

MacLeod, D. A. 1980. The origin of the red Mediterranean soils in Epirus, Greece. *J. Soil Sci.* **31**: 125–136.

MacNamara, E. E. 1969. Soils and geomorphic surfaces in Antarctica. Biul. *Peryglac.* **20**: 299–320.

Macyk, T. M., S. Pawluk, and J. D. Lindsay. 1978. Relief and microclimate as related to soil properties. *Can. J. Soil Sci.* **58**: 421–438.

Mader, D. L. 1963. Soil variability: a serious problem in soil-site studies in the Northeast. *Soil Sci. Soc. Am. Proc.* **27**: 707–709.

Mader, W. F. and E. J. Ciolkosz. 1997. The effects of periglacial processes on the genesis of soils on an unglaciated northern Appalachian Plateau landscape. *Soil Surv. Horiz.* **38**: 19–30.

Magaritz, M. and A. J. Amiel. 1980. Calcium carbonate in a calcareous soil from the Jordan Valley, Israel: its origin as revealed by the stable carbon isotope method. *Soil Sci. Soc. Am. J.* **44**: 1059–1062.

Mahaney, W. C. (ed.) 1984. *Quaternary Dating Methods.* New York, Elsevier.

Mahendrappa, M. K. and D. G. O. Kingston. 1982. Prediction of throughfall quantities under different forest stands. *Can. J. For. Res.* **12**: 474–481.

Maher, B. A. 1998. Magnetic properties of modern soils and Quaternary loessic paleosols: paleoclimatic implications. *Palaeogeog. Palaeoclimat. Palaeoecol.* **137**: 25–54.

Maher, B. A. and R. Thompson. 1995. Paleorainfall reconstructions from pedogenic magnetic susceptibility variations in Chinese loess and paleosols. *Quat. Res.* **44**: 383–391.

Maher, B. A., Thompson, R., and L. P. Zhou. 1994. Spatial and temporal reconstructions of changes in the Asian palaeomonsoon: a new mineral magnetic approach. *Earth Planet. Sci. Lett.* **125**: 461–471.

Maher, L. J., Jr. 1972. Absolute pollen diagram of Redrock Lake, Boulder County, Colorado. *Quat. Res.* **2**: 531–553.

Maizels, J. K. and J. R. Petch. 1985. Age determination of intermontane areas, Austerdalen, Southern Norway. *Boreas* **14**: 51–65.

Maldague, M. E. 1964. Importance des populations de termites dans les sols equatoriaux. *Trans. 8th Intl Congr. Soil Sci.* (Bucharest) **3**: 743–751.

Malde, H. E. 1964. Patterned ground on the western Snake River plain, Idaho and its possible cold-climate origin. *Geol. Soc. Am. Bull.* **75**: 191–207.

Malo, D. D., Worcester, B. K., Cassel, D. K., *et al.* 1974. Soil–landscape relationships in a closed drainage system. *Soil Sci. Soc. Am. Proc.* **38**: 813–818.

Malterer, T. J., Verry, E. S., and J. Erjavec. 1992. Fiber content and degree of decomposition in peats: review of national methods. *Soil Sci. Soc. Am. J.* **56**: 1200–1211.

Mammerickx, J. 1964. Quantitative observations on pediments in the Mojave and Sonoran Deserts (southwestern United States). *Am. J. Sci.* **262**: 417–435.

Mandel, R. D. and E. A. Bettis III. 2001. Use and analysis of soils by archaeologists and geoscientists. In P. Goldberg, V. T. Holliday, and C. R. Ferring (eds.) *Earth Sciences and Archaeology.* New York, Plenum Press, pp. 173–204.

Mandel, R. D. and C. J. Sorenson. 1982. The role of the Western Harvester ant (*Pogonomyrmex occidentalis*) in soil formation. *Soil Sci. Soc. Am. J.* **46**: 785–788.

Manikowska, B. 1982. Upfreezing of stones in boulder clay of central and north Poland. *Biul. Peryglac.* **29**: 87–115.

Mankinen, E. A. and G. B. Dalrymple. 1979. Revised geomagnetic polarity timescale for the interval 0–5 My BP. *J. Geophys. Res.* **B84**: 615–626.

Manning, G., Fuller, L. G., Eilers, R. G., *et al.* 2001. Topographic influence on the variability of soil properties within an undulating Manitoba landscape. *Can. J. Soil Sci.* **81**: 439–447.

Manrique, L. A. and C. A. Jones. 1991. Bulk density of soils in relation to soil physical and chemical properties. *Soil Sci. Soc. Am. J.* **55**: 476–481.

Mansberg, L. and T. R. Wentworth. 1984. Vegetation and soils of a serpentine barren in western North Carolina. *Bull. Torrey Bot. Club* **111**: 273–286.

Marbut, C. F. 1923. Soils of the Great Plains. *Ann. Ass. Am. Geog.* **13**: 41–66.

1927. *The Great Soil Groups of the World and Their Development*. Ann Arbor, MI, Edwards Bros. (Trans. from Glinka (1914).)

1928. A scheme for soil clasification. IV. *Proc. 1st Intl Congr. Soil Sci.* **4**: 1–31.

1935. Soils of the United States. III. In *Atlas of American Agriculture*. Washington, DC, US Department of Agriculture, pp. 1–98.

Marbut, C. F., Bennett, H. H., and J. E. Lapham. 1913. *Soils of the United States*, US Bureau of Soils Bulletin no. 96. Washington, DC, US Government Printing Office.

Marion, G. W., Schlesinger, W. H., and P. J. Fonteyn. 1985. CALDEP: a regional model for soil $CaCO_3$ (caliche) deposition in Southwestern Deserts. *Soil Sci.* **139**: 468–481.

Markewich, H. W. and M. J. Pavich. 1991. Soil chronosequence studies in temperate to subtropical, low-latitude, low-relief terrain with data from the eastern United States. *Geoderma* **51**: 213–239.

Markewich, H. W., Pavich, M. J., Mausbach, M. J., *et al.* 1989. A guide for using soil and weathering profile data in chronosequence studies of the coastal plain of the eastern United States. *US Geol. Survey Bull.* **1589**.

Markewich, H. W., Wysocki, D. A., Pavich, M. J., *et al.* 1998. Paleopedological plus TL, ^{10}Be and ^{14}C dating as tools in stratigraphic and paleoclimatic investigations, Mississippi River Valley, USA. *Quat. Intl* **51/52**: 143–167.

Marron, D. C. and J. H. Popenoe. 1986. A soil catena on schist in northwestern California. *Geoderma* **37**: 307–324.

Marshall, C. E. and J. F. Haseman. 1942. The quantitative evaluation of soil formation and development by heavy mineral studies: a Grundy silt loam study. *Soil Sci. Soc. Am. Proc.* **7**: 448–453.

Martel, Y. A. and E. A. Paul. 1974. Effects of cultivation on the organic matter of grassland soils as determined by fractionation and radiocarbon dating. *Can. J. Soil Sci.* **54**: 419–426.

Martin, A., Mariotti, A., Balesdent, J., *et al.* 1990. Estimates of the organic matter turnover rate in a savanna soil by ^{13}C natural abundance. *Soil Biol. Biochem.* **22**: 517–523.

Martin, C. W. and W. C. Johnson. 1995. Variation in radiocarbon ages of soil organic matter fractions from Late Quaternary buried soils. *Quat. Res.* **43**: 232–237.

Martin, J. P. and K. Haider. 1971. Microbial activity in relation to soil humus formation. *Soil Sci.* **111**: 54–63.

Martini, J. A. 1970. Allocation of cation exchange capacity to soil fractions in seven surface soils from Panama and the application of a cation exchange factor as a weathering index. *Soil Sci.* **109**: 324–331.

Martini, J. A. and M. Macias. 1974. A study of six "Latosols" from Costa Rica to elucidate the problems of classification, productivity and management of tropical soils. *Soil Sci. Soc. Am. Proc.* **38**: 644–652.

Martinson, D. G., Pisias, N. G., Hays, J. D., *et al.* 1987. Age dating and the orbital theory of the Ice Ages: development of a high resolution 0–300 000 yr chronostratigraphy. *Quat. Res.* **27**: 1–29.

Martius, C. 1994. Diversity and ecology of termites in Amazonian forests. *Pedobiologia* **38**: 407–428.

Marye, W. B. 1955. The great Maryland barrens. *Maryland Hist. Magazine* **50**: 11–23, 120–142, 234–253.

Mason, J. A. 2001. Transport direction of Peoria loess in Nebraska and implications for loess source areas on the central Great Plains. *Quat. Res.* **56**: 79–86.

Mason, J. A. and P. M. Jacobs. 1998. Chemical and particle-size evidence for addition of fine dust to soils of the midwestern United States. *Geology* **26**: 1135–1138.

Mason, J. A. and E. A. Nater. 1994. Soil morphology–Peoria loess grain size relationships, southeastern Minnesota. *Soil Sci. Soc. Am. J.* **58**: 432–439.

Masson, P. H. 1949. Circular soil structures in northeast California. *Calif. Div. Mines Bull.* **151**: 61–71.

Materechera, S. A., Dexter, A. R., and A. M. Alston. 1992. Formation of aggregates by plant roots in homogenised soils. *Plant Soil* **142**: 69–79.

Matsumoto, T. 1976. The role of termites in an equatorial rain forest ecosystem of West Malaysia. I. Population density, biomass, carbon, nitrogen and calorific content and respiration rate. *Oecologia* **22**: 153–178.

Matthews, J. A. 1974. Families of lichenometric dating curves from the Storbreen gletschervorfeld, Jotunheimen, Norway. *Norsk Geog. Tidsskr.* **28**: 215–235.

1975. Experiments on the reproducibility and reliability of lichenometric dates, Storbreen gletschervorfeld, Jotunheimen, Norway. Norsk Geog. *Tidsskr.* **29**: 97–109.

Matthews, J. A. 1980. Some problems and implications of ^{14}C dates from a Podzol buried beneath an end moraine at Haugabreen, southern Norway. *Geog. Ann.* **62A**: 185–208.

Matthews, J. A. and P. Q. Dresser. 1983. Intensive ^{14}C dating of a buried palaeosol horizon. *Geol. Foreningens i Stockholm Forhandl.* **105**: 59–63.

Matthews, J. A. and R. A. Shakesby. 1984. The status of the Little Ice-Age in Southern-Norway: relative-age dating of neoglacial moraines with Schmidt hammer and lichenometry. *Boreas* **13**: 333–346.

Mattson, S. and H. Lönnemark. 1939. The pedography of hydrologic podsol series. I. Loss on ignition, pH and amphoteric reactions. *Ann. Agric. Coll. Sweden* **7**: 185–227.

Mausbach, M. J. and J. L. Richardson. 1994. Biogeochemical processes in hydric soil formation. *Curr. Topics Wetland Biogeochem.* **1**: 68–127.

Mausbach, M. J., Wingard, R. C., and E. E. Gamble. 1982. Modification of buried soils by postburial pedogenesis, southern Indiana. *Soil Sci. Soc. Am. J.* **46**: 364–369.

Mayer, L., McFadden, L. D., and J. W. Harden. 1988. Distribution of calcium carbonate in desert soils: a model. *Geology* **16**: 303–306.

McBrearty, S. 1990. Consider the humble termite: termites as agents of post-depositional disturbance at African archaeological sites. *J. Archaeol. Sci.* **17**: 111–143.

McBurnett, S. L. and D. P. Franzmeier. 1997. Pedogenesis and cementation in calcareous till in Indiana. *Soil Sci. Soc. Am. J.* **61**: 1098–1104.

McCahon, T. J. and L. C. Munn. 1991. Soils developed in Late Pleistocene till, Medicine Bow Mountains, Wyoming. *Soil Sci.* **152**: 377–388.

McCaleb, S. B. 1954. Profile studies of normal soils of New York. IV. Mineralogical properties of the Gray-Brown Podzolic-Brown Podzolic soil sequence. *Soil Sci.* **77**: 319–333.

McCalpin, J. P. and M. E. Berry. 1996. Soil catenas to estimate ages of movements on normal fault scarps, with an example from the Wasatch fault zone, Utah, USA. *Catena* **27**: 265–286.

McCarroll, D. 1991. The Schmidt hammer, weathering and rock surface-roughness. *Earth Surf. Proc. Landforms* **16**: 477–480.

McCarthy, D. P. and D. J. Smith. 1995. Growth curves for calcium-tolerant lichens in the Canadian Rocky Mountains. *Arctic Alpine Res.* **27**: 290–297.

McClelland, J. E., Mogen, C. A., Johnson, W. M., *et al.* 1959. Chernozems and associated soils of eastern North Dakota: some properties and topographic relationships. *Soil Sci. Soc. Am. Proc.* **23**: 51–56.

McComb, A. L. and W. E. Lomis. 1944. Subclimax prairie. *Bull. Torrey Bot. Club* **71**: 46–76.

McDaniel, P. A. and A. L. Falen. 1994. Temporal and spatial patterns of episaturation in a Fragixeralf landscape. *Soil Sci. Soc. Am. J.* **58**: 1451–1457.

McDaniel, P. A., Fosberg, M. A., and A. L. Falen. 1993. Expression of andic and spodic properties in tephra-influenced soils of northern Idaho. *Geoderma* **58**: 79–94.

McDaniel, P. A., Gabehart, R. W., Falen, A. L., *et al.* 2001. Perched water tables on Argixeroll and Fragixeralf hillslopes. *Soil Sci. Soc. Am. J.* **65**: 805–810.

McDole, R. E. and M. S. Fosberg. 1974a. Soil temperatures in selected Southeastern Idaho soils. I. Evaluation of sampling techniques and classification of soils. *Soil Sci. Soc. Am. Proc.* **38**: 480–486.

1974b. Soil temperatures in selected Southeastern Idaho soils. II. Relation to soil and site characteristics. *Soil Sci. Soc. Am. Proc.* **38**: 486–491.

McDonald, E. V. and A. J. Busacca. 1988. Record of pre-late Wisconsin floods in the Channeled Scabland interpreted from loess deposits. *Geology* **16**: 728–731.

1990. Interaction between aggrading geomorphic surfaces and the formation of a Late Pleistocene paleosol in the Palouse loess of eastern Washington state. *Geomorphology* **3**: 449–470.

1992. Late Quaternary stratigraphy of loess in the Channeled Scabland and Palouse Regions of Washington State. *Quat. Res.* **38**: 141–156.

McDonald, E. V., Pierson, F. B., Flerchinger, G. N., *et al.* 1996. Application of a soil–water balance model to evaluate the influence of Holocene climate change on calcic soils, Mojave Desert, California, USA. *Geoderma* **74**: 167–192.

McDonnell, M. J. and S. T. A. Pickett. 1990. The study of ecosystem structure and function along the urban–rural gradients: an unexploited opportunity. *Ecology* **71**: 1232–1237.

McFadden, J. P., MacDonald, N. W., Witter, J. A., *et al.* 1994. Fine-textured soil bands and oak forest productivity in northwestern lower Michigan, USA. *Can. J. For. Res.* **24**: 928–933.

McFadden, L. D. and D. M. Hendricks. 1985. Changes in the content and composition of pedogenic iron oxyhydroxides in a chronosequence of soils in southern California. *Quat. Res.* **23**: 189–204.

McFadden, L. D. and R. J. Weldon, Jr. 1987. Rates and processes of soil development on Quaternary terraces in Cajon Pass, California. *Geol. Soc. Am. Bull.* **98**: 280–293.

McFadden, L. D., Amundson, R. G., and O. A. Chadwick. 1991. Numerical modeling, chemical, and isotopic studies of carbonate accumulation in soils of arid regions. *Soil Sci. Soc. Am. Special Publ.* **26**: 17–35.

McFadden, L. D., McDonald, E. V., Wells, S. G., *et al.* 1998. The vesicular layer and carbonate collars of desert soils and pavements: formation, age and relation to climate change. *Geomorphology* **24**: 101–145.

McFadden, L. D., Ritter, J. B., and S. G. Wells. 1989. Use of multiparameter relative-age methods for age estimation and correlation of alluvial fan surfaces on a desert piedmont, eastern Mojave Desert, California. *Quat. Res.* **32**: 276–290.

McFadden, L. D., Wells, S. G., and M. J., Jercinovich. 1987. Influence of eolian and pedogenic processes on the origin and evolution of desert pavements. *Geology* **15**: 504–508.

McFarlane, M. J. and D. J. Bowden. 1992. Mobilization of aluminum in the weathering profiles of the African surface in Malawi. *Earth Surf. Proc. Landforms* **17**: 789–805.

McFarlane, M. J. and S. Pollard. 1989. Some aspects of stone-lines and dissolution fronts associated with regolith and dambo profiles in parts of Malawi and Zimbabwe. *Intl J. Trop. Ecol. Geog.* **11**: 23–35.

McHargue, L. P. and P. E. Damon. 1991. The global beryllium 10 cycle. Rev. *Geophys.* **29**: 141–158.

McKay, E. D. 1979. Wisconsinan loess stratigraphy of Illinois. In *Wisconsinan, Sangamonian, and Illinoian Stratigraphy in Central Illinois*, Illinois State Geol. Survey Guidebook no. 13. Midwest Friends of the Pleistocene Field Conference Guidebook. pp. 95–108. Urbana-Champaign, IL.

McKeague, J. A. 1967. An evaluation of 0.1 M pyrophosphate and pyrophosphate-dithionite in comparison with oxalate as extractants of the accumulation products in podzols and some other soils. *Can. J. Soil Sci.* **47**: 95–99.

McKeague, J. A. and J. H. Day. 1966. Dithionite- and oxalate-extractable Fe and Al as aids in differentiating various classes of soils. *Can. J. Soil Sci.* **46**: 13–22.

McKeague, J. A. and R. J. St. Arnaud. 1969. Pedotranslocation: eluviation–illuviation in soils during the Quaternary. *Soil. Sci.* **107**: 428–434.

McKeague, J. A., Bourbeau, G. A., and D. B. Cann. 1967. Properties and genesis of a bisequa soil from Cape Breton Island. *Can. J. Soil Sci.* **47**: 101–110.

McKeague, J. A., Brydon, J. E., and N. M. Miles. 1971. Differentiation of forms of extractable iron and aluminum in soils. *Soil Sci. Soc. Am. Proc.* **35**: 33–38.

McKeague, J. A., DeConinck, F. and D. P. Franzmeier. 1983. Spodosols. In L. P. Wilding, N. E. Smeck and G. F. Hall (eds.) *Pedogenesis and Soil Taxonomy.* New York, Elsevier, pp. 217–252.

McKenzie, R. M. 1989. Manganese oxides and hydroxides. In J. B. Dixon and S. B. Weed (eds.) *Minerals in Soil Environments*, 2nd edn. Madison, WI, Soil Science Society of America, pp. 439–465.

McManus, D. A. 1991. Suggestions for authors whose manuscripts include quantitative clay mineral analysis by X-ray diffraction. *Marine Geol.* **98**: 1–5.

McNeil, M. 1964. Lateritic soils. *Sci. Am.* **211**: 96–102.

Meighan, C. W., Foote, L. J., and P. V. Aiello. 1968. Obsidian dating in west Mexican archeology. *Science* **160**: 1069–1075.

Meixner, R. E. and M. J. Singer. 1981. Use of a field morphology rating system to evaluate soil formation and discontinuities. *Soil Sci.* **131**: 114–123.

Melillo, J. M., Aber, J. D., Linkins, A. E., *et al.* 1989. Carbon and nitrogen dynamics along the decay continuum: plant litter to soil organic matter. *Plant Soil* **115**: 189–198.

Mellor, A. 1985. Soil chronosequences on Neoglacial moraine ridges, Jostedalsbreen and Jotunheimen, southern Norway: a quantitative pedogenic approach. In K. S. Richards, R. R. Arnett and S. Ellis (eds.) *Geomorphology and Soils.* London, Allen and Unwin, pp. 289–308.

1986. A micromorphological examination of two alpine soil chronosequences, southern Norway. *Geoderma* **39**: 41–57.

1987. A pedogenic investigation of some soil chronosequences on neoglacial moraine ridges, southern Norway: examination of soil chemical data using principal components analysis. *Catena* **14**: 369–381.

Melton, F. A. 1935. Vegetation and soil mounds. *Geog. Rev.* **25**: 430–433.

Meng, X., Derbyshire, E., and R. A. Kemp. 1997. Origin of the magnetic susceptibility signal in Chinese loess. *Quat. Sci. Rev.* **16**: 833–839.

Merritts, D. J., Chadwick, O. A., and D. M. Hendricks. 1991. Rates and processes of soil evolution on uplifted marine terraces, northern California. *Geoderma* **51**: 241–275.

Merritts, D. J., Chadwick, O. A., Hendricks, D. M., *et al.* 1992. The mass balance of soil evolution on late Quaternary marine terraces, northern California. *Geol. Soc. Am. Bull.* **104**: 1456–1470.

Mew, G. and C. W. Ross. 1994. Soil variation on steep greywacke slopes near Reefton, Western South Island. *J. Roy. Soc.* New Zealand **24**: 231–242.

Meyers, N. L. and K. McSweeney. 1995. Influence of treethrow on soil properties in Northern Wisconsin. *Soil Sci. Soc. Am. J.* **59**: 871–876.

Miall, A. D. 1985. Architectural-element analysis: a new method of facies analysis applied to fluvial deposits. *Earth Sci. Rev.* **22**: 261–308.

Michaelson, G. J., Ping, C. L., and J. M. Kimble. 1996. Carbon content and distribution in tundra soils in arctic Alaska. *Arctic Alpine Res.* **28**: 414–424.

Michels, J. W. 1967. Archaeology and dating by obsidian hydration. *Science* **158**: 211–214.

Middleton, L. T. and M. J. Kraus. 1980. Simple technique for thin-section preparation of unconsolidated materials. *J. Sed. Petrol.* **50**: 622–623.

Mikesell, L. R., Schaetzl, R. J., and M. A. Velbel. 2004. Hornblende etching and quartz/feldspar ratios as weathering and soil development indicators in some Michigan soils. *Quat. Res.* **62**: 162–171.

Miles, J. 1985. The pedogenic effects of different species and vegetation types and the implications of succession. *J. Soil Sci.* **36**: 571–584.

Miles, R. J. and D. P. Franzmeier. 1981. A lithochronosphere of soils formed in dune sand. *Soil Sci. Am. Proc.* **45**: 326–367.

Milfred, C. J. and R. W. Kiefer. 1976. Analysis of soil variability with repetitive aerial photography. *Soil Sci. Soc. Am. J.* **40**: 553–557.

Miller, C. D. 1979. A statistical method for relative-age dating of moraines in the Sawatch Range, Colorado. *Geol. Soc. Am. Bull.* **90**: 1153–1164.

Miller, G. and J. M. Greenwade. 2001. *Soil Survey of McLennan County, Texas.* Washington, DC, US Department of Agriculture Natural Resources Conservation Service, US Government Printing Office.

Miller, G. H. 1973. Late Quaternary glacial and climatic history of northern Cumberland Peninsula, Baltin Island, N.W.T., Canada. *Quat. Res.* **3**: 561–583.

Miller, J. J., Acton, D. F., and R. J. St. Arnaud. 1985. The effect of groundwater on soil formation in a morainal landscape in Saskatchewan. *Can. J. Soil Sci.* **65**: 293–307.

Miller, L. L., Hinkel, K. M., Nelson, F. E., *et al.* 1998. Spatial and temporal patterns of soil moisture and thaw depth at Barrow, Alaska, USA. In A. G. Lewkowicz and M. Allard (eds.) *Permafrost, 7th Intl Conf. Proc.* Yellowknife, Canada, Université. Laval, Centre D'Etudes Nordiques, pp. 731–737.

Miller, M. B., Cooper, T. H., and R. H. Rust. 1993. Differentiation of an eluvial fragipan from dense glacial till in northern Minnesota. *Soil Sci. Soc. Am. J.* **57**: 787–796.

Miller, N. G. and R. P. Futyma. 1987. Paleohydrological implications of Holocene peatland development in northern Michigan. *Quat. Res.* **27**: 297–311.

Millot, G., Nahon, D., Paquet, H., *et al.* 1977. L'Epigenie calcaire des roches silicatées dans les encroutements carbonates en pays subaride Antiatlas. *Maroc. Soc. Geol. Bull.* **30**: 129–152.

Mills, H. H. 1981. Some observations on slope deposits in the vicinity of Grandfather Mountain, North Carolina, USA. *Southeast. Geol.* **22**: 209–222.

Milne, G. 1935. Some suggested units for classification and mapping, particularly for East African soils. *Soil Res.*, Berlin **4**: 183–198.

1936a. *A Provisonal Soil Map of East Africa*, African Agricultural Research Station, Amani Memoirs. Tanganyika Territory.

1936b. Normal erosion as a factor in soil profile development. *Nature* **138**: 548.

Milnes, A. R. and C. R. Twidale. 1983. An overview of silicification in Cainozoic landscapes of arid central and southern Australia. *Austral. J. Soil Res.* **21**: 387–410.

Ming, D. W. and F. A. Mumpton. 1989. Zeolites in soils. In J. B. Dixon and S. B. Weed (eds.) *Minerals in Soil Environments,* 2nd edn. Madison, WI, Soil Science Society of American, pp. 873–911.

Mitchell, M. J. 1980. *Soil Survey of Winnebago County, Wisconsin.* Washington, DC, US Department of Agriculture Soil Conservation Service, US Government Printing Office.

Mitchell, P. B. and G. S. Humphreys. 1987. Litter dams and microterraces formed on hillslopes subject to rainwash in the Sydney Basin, Australia. *Geoderma* **39**: 331–357.

Mizota, C., Izuhara, H., and M. Noto. 1992. Eolian influence on oxygen isotope abundance and clay minerals in soils of Hokkaido, northern Japan. *Geoderma* **52**: 161–171.

Moeyersons, J. 1975. An experimental study of pluvial processes on granite gruss. *Catena* **2**: 289–308.

1978. The behaviour of stones and stone implements, buried in consolidating and creeping Kalahari sands. *Earth Surf. Proc. Landforms* **3**: 115–128.

1989. The concentration of stones into a stone-line, as a result from subsurface movements in fine and loose soils in the tropics. *Geo-Eco-Trop.* **11**: 11–22.

Mogen, C. A., McClelland, J. E., Allen, J. S., *et al.* 1959. Chestnut, Chernozem, and associated soils of western North Dakota. *Soil Sci. Soc. Am. Proc.* **23**: 56–60.

Mohanty, B. P., Kanwar, R. S., and C. J. Everts. 1994. Comparison of saturated hydraulic conductivity measurement methods for a glacial till soil. *Soil Sci. Soc. Am. J.* **58**: 672–677.

Mokma, D. L. 1993. Color and amorphous materials in Spodosols from Michigan. *Soil Sci. Soc. Am. J.* **57**: 125–128.

1997. Ortstein in selected soils from Washington and Michigan. *Soil Surv. Horiz.* **38**: 71–75.

Mokma, D. L. and P. Buurman. 1982. *Podzols and Podzolization in Temperate Regions*. Wageningen, the Netherlands, International Soil Museum.

Mokma, D. L. and D. L. Cremeens. 1991. Relationships of saturation and B horizon colour patterns in soils of three hydrosequences in south-central Michigan, USA. *Soil Use Mgt.* **7**: 56–61.

Mokma, D. L. and C. V. Evans. 2000. Spodosols. In M. E. Sumner (ed.) *Handbook of Soil Science*. Boca Raton, FL, CRC Press, pp. E-307-E-321.

Mokma, D. L. and S. W. Sprecher. 1994. Water table depths and color patterns in Spodosols of two hydrosequences in northern Michigan, USA. *Catena* **22**: 275–286.

1995. How frigid is frigid? *Soil Surv. Horiz.* **36**: 71–76.

Mokma, D. L., Doolittle, J. A., and L. A. Tornes. 1994. Continuity of ortstein in sandy Spodosols, Michigan. *Soil Surv. Horiz.* **35**: 6–10.

Moldenke, A. R. 1999. Soil arthropods. *USDA-NRCS Soil Qual. Inst. PA* **1637**: G1–G8.

Monaghan, M. C. and D. Elmore. 1994. Garden-variety Be-10 in soils on hill slopes. *Nucl. Instrum. Meth.* **B92**: 357–361.

Monaghan, M. C., Krishnaswami, S., and J. H. Thomas. 1983. ^{10}Be concentrations and the long-term fate of particle-reactive nuclides in five soil profiles from California. *Earth Planet. Sci. Lett.* **65**: 51–60.

Monaghan, M. C., McKean, J., Dietrich, W., *et al.* 1992. ^{10}Be chronometry of bedrock-to-soil conversion rates. *Earth Planet. Sci. Lett.* **111**: 483–492.

Monger, H. C. 2002. Pedogenic carbonate: links between biotic and abiotic $CaCO_3$. *Proc. 17th World Congr. Soil Sci. (Bangkok)* **2**: 796.

Monger, H. C., Daugherty, L. A., and L. H. Gile. 1991a. A microscopic examination of pedogenic calcite in an Aridisol of southern New Mexico. *Soil Sci. Soc. Am. Spec. Publ.* **26**: 37–60.

Monger, H. C., Daugherty, L. A., Lindemann, W. C., *et al.* 1991b. Microbial precipitation of pedogenic calcite. *Geology* **19**: 997–1000.

Monroe, W. H. 1966. Formation of tropical karst topography by limestone solution and precipitation. *Caribb. J. Sci.* **6**: 1–7.

Moore, D. M. and R. C. Reynolds, Jr. 1989. *X-Ray Diffraction and the Identification and Analysis of Clay Minerals*. Oxford, UK, Oxford University Press.

Moore, I. D., Burch, G. J., and D. H. Mackenzie. 1988. Topographic effects on the distribution of surface soil water and the location of ephemeral gullies. *Trans. Am. Soc. of Agricultural Engin.* **31**: 1383–1395.

Moore, I. D., Gessler, P. E., Nielsen, G. A., *et al.* 1993. Soil attribute prediction using terrain analysis. *Soil Sci. Soc. Am. J.* **57**: 443–452.

Moore, T. R. 1974. Pedogenesis in a subarctic environment: Cambrian Lake, Quebec. *Arctic Alpine Res.* **6**: 281–291.

Moorhead, D. L., Fisher, F. M., and W. G. Whitford. 1988. Cover of spring annuals on nitrogen-rich kangaroo rat mounds in a Chihuahuan Desert grassland. *Am. Midl. Nat.* **120**: 443–447.

Moresi, M. and G. Mongelli. 1988. The relation between the terra rossa and the carbonate-free residue of the underlying limestones and dolostones in Apulia, Italy. *Clay Mins.* **23**: 439–446.

Morris, J. D. 1991. Applications of cosmogenic ^{10}Be to problems in the Earth Sciences. *Ann. Rev. Earth Planet. Sci.* **19**: 313–350.

Morris, W. J. 1985. A convenient method of acid etching. *J. Sed. Petrol.* **55**: 600.

Morrison, R. B. 1967. Principles of Quaternary soil stratigraphy. *Proc. 7th INQUA Congr.* **9**: 1–69.

1998. How can the treatment of pedostratigraphic units in the North American Stratigraphic Code be improved? *Quat. Intl* **51/52**: 30–33.

Moser, M. and F. Hohensinn. 1983. Geotechnical aspects of soil slips in alpine regions. *Engin. Geol.* **19**: 185–211.

Moss, R. P. 1965. Slope development and soil morphology in a part of south west Nigeria. *J. Soil Sci.* **16**: 192–209.

Mossin, L., Jensen, B. T., and P. Nørnberg. 2001. Altered podzolization resulting from replacing heather with Sitka spruce. *Soil Sci. Soc. Am. J.* **65**: 1455–1462.

Mount, H. R. 1998. Global change remote soil temperature network. *Soil Surv. Horiz.* **39**: 92.

Muckenhirn, R. J., Whiteside, E. P., Templin, E. H., *et al.* 1949. Soil classification and the genetic factors of soil formation. *Soil Sci.* **67**: 93–105.

Mueller, G., Broll, G., and C. Tarnocai. 1999. Biological activity as influenced by microtopography in a Cryosolic soil, Baffin Island, Canada. *Permafrost Periglac. Proc.* **10**: 279–288.

Muggler, C. C. and P. Buurman. 2000. Erosion, sedimentation and pedogenesis in a polygenetic oxisol sequence in Minas Gerais, Brazil. *Catena* **41**: 3–17.

Muhs, D. R. 1982. A soil chronosequence on Quaternary marine terraces, San Clemente Island, California. *Geoderma* **28**: 257–283.

1984. Intrinsic thresholds in soil systems. *Phys. Geog.* **5**: 99–110.

Muhs, D. R. and E. A. Bettis III. 2000. Geochemical variations in Peoria loess of western Iowa indicate paleowinds of midcontinental North America during last glaciation. *Quat. Res.* **53**: 49–61.

Muhs, D. R. and V. T. Holliday. 2001. Origin of late Quaternary dune fields on the southern High Plains of Texas and New Mexico. *Geol. Soc. Am. Bull.* **113**: 75–87.

Muhs, D. R. and S. A. Wolfe. 1999. Sand dunes of the northern Great Plains of Canada and the United States. *Geol. Surv. Can. Bull.* **534**: 183–197.

Muhs, D. R. and M. Zárate. 2001. Late Quaternary eolian records of the Americas and their paleoclimatic significance. In V. Markgraf (ed.) *Interhemispheric Climate Linkages*. San Diego, CA, Academic Press, pp. 183–216.

Muhs, D. R., Aleinikoff, J. N., Stafford, T. W., Jr., *et al.* 1999. Late Quaternary loess in northeastern Colorado. I. Age and paleoclimatic significance. *Geol. Soc. Am. Bull.* **111**: 1861–1875.

Muhs, D. R., Bush, C. A., Stewart, K. C., *et al.* 1990. Geochemical evidence of Saharan dust parent material for soils developed on Quaternary limestones of Caribbean and western Atlantic islands. *Quat. Res.* **33**: 157–177.

Muhs, D. R., Crittenden, R. C., Rosholt, J. N., *et al.* 1987. Genesis of marine terrace soils, Barbados, West Indies: evidence from mineralogy and geochemistry. *Earth Surf. Proc. Landforms* **12**: 605–618.

Muhs, D. R., Stafford, T. W., Jr., Been, J., *et al.* 1997a. Holocene eolian activity in the Minot dune field, North Dakota. *Can. J. Earth Sci.* **34**: 1442–1459.

Muhs, D. R., Stafford, T. W., Jr., Swinehart, J. B., *et al.* 1997b. Late Holocene eolian activity in the mineralogically mature Nebraska Sand Hills. *Quat. Res.* **48**: 162–176.

Muir, J. W. and J. Logan. 1982. Eluvial/illuvial coefficients of major elements and the corresponding losses and gains in three soil profiles. *J. Soil Sci.* **33**: 295–308.

Mulholland, S. C. and G. Rapp, Jr. 1992. Phytolith systematics: an introduction. In G. Rapp, Jr. and S. C. Mulholland (eds.) *Phytolith Systematics*. New York, Plenum Press, pp. 1–13.

Muller, P. E. 1879. Studier over skovjord, som bidrag til skovdyrkningens theori. *Tidsskr. Skovbr.* **3**: 1–124.

Mullins, C. E. 1977. Magnetic susceptibility of the soil and its significance in soil science: a review. *J. Soil Sci.* **28**: 223–246.

Munday, T. J., Reilly, N. S., Glover, M., *et al.* 2000. Petrophysical characterisation of parna using ground and downhole geophysics at Marinna, central New South Wales. *Explor. Geophys.* **31**: 260–266.

Munn, L. C. 1993. Effects of prairie dogs on physical and chemical properties of soils. *US Fish and Wildlife Serv. Biol. Rept* **13**: 11–17.

Munn, L. C. and M. M. Boehm. 1983. Soil genesis in a Natrargid–Haplargid complex in northern Montana. *Soil Sci. Soc. Am. J.* **47**: 1186–1192.

Murad, E. 1978. Yttrium and zirconium as geochemical guide elements in soil and stream sediment sequences. *J. Soil Sci.* **29**: 219–223.

Murray, A. S. and L. B. Clemmensen. 2001. Luminescence dating of Holocene aeolian sand movement, Thy, Denmark. *Quat. Sci. Rev.* **20**: 751–754.

Murray, D. F. 1967. Gravel mounds at Rocky Flats, Colorado. *Mountain Geol.* **4**: 99–107.

Musgrave, G. W. 1935. Some relationships between slope length, surface runoff and the silt load of surface runoff. *Trans. Am. Geophys. Union* **11**: 472–478.

Myrcha, A. and A. Tatur. 1991. Ecological role of the current and abandoned rookeries in the land environment of the maritime Antarctic. *Polish Polar Res.* **12**: 3–24.

Nabel, P. 1993. The Brunhes–Matuyama boundary in Pleistocene sediments of Buenos Aires province, Argentina. *Quat. Intl* **17**: 79–85.

Nahon, D. 1991. *Introduction to the Petrology of Soils and Chemical Weathering*. New York, John Wiley.

Naiman, Z., Quade, J., and P. J. Patchett. 2000. Isotopic evidence for eolian recycling of pedogenic carbonate and variations in carbonate dust sources throughout the southwest United States. *Geochim. Cosmochim. Acta* **64**: 3099–3109.

Nash, D. B. 1984. Morphologic dating of fluvial terrace scarps and fault scarps near West Yellowstone, Montana. *Geol. Soc. Am. Bull.* **95**: 1413–1424.

Nash, M. H., Daugherty, L. A., Buchanan, B. A., *et al.* 1994. The effect of groundwater table variation on characterization and classification of gypsiferous soils in the Tularosa Basin, New Mexico. *Soil Surv. Horiz.* **35**: 102–110.

Neary, D. G., Swift, L. W., Jr., Manning, D. M., *et al.* 1986. Debris avalanching in the southern Appalachians: an influence on forest soil formation. *Soil Sci. Soc. Am. J.* **50**: 465–471.

Nel, J. J. C. and E. M. Malan. 1974. The distribution of the mounds of *Trinervitermes trinervoides* in the central Orange Free State. *J. Entomol. Soc. S. Afr.* **37**: 251–256.

Nelson, A. R. and R. R. Shroba. 1998. Soil relative dating of moraine and outwash-terrace sequences in the northern part of the upper Arkansas Valley, Central Colorado, USA. *Arctic Alpine Res.* **30**: 349–361.

Nelson, A. R., Millington, A. C., Andrews, J. T., *et al.* 1979. Radiocarbon-dated upper Pleistocene glacial sequence, Fraser Valley, Colorado Front Range. *Geology* **7**: 410–414.

Nelson, L. A., Kunze, G. W., and C. L. Godfrey. 1960. Chemical and mineralogical properties of a San Saba clay, a Grumusol. *Soil Sci.* **89**: 122–131.

Nelson, R. L. 1954. Glacial geology of the Frying Pan River drainage, Colorado. *J. Geol.* **62**: 325–343.

Nesbitt, H. W. 1992. Diagenesis and metasomatism of weathering profiles, with emphasis on Precambrian paleosols. In I. P. Martini and W. Chesworth (eds.) *Weathering, Soils and Paleosols*. Amsterdam, Elsevier, pp. 127–152.

Nesje, A., Kvamme, M., and N. Rye. 1989. Neoglacial gelifluction in the Jostedalsbreen region, western Norway: evidence from dated buried palaeopodzols. *Earth Surf. Proc. Landforms* **14**: 259–270.

Netterberg, F. 1969a. Ages of calcretes in southern Africa. *S. Afr. Archaeol. Bull.* **24**: 88–92.

1969b. The interpretation of some basic calcrete types. *S. Afr. Archaeol. Bull.* **24**: 117–122.

Nettleton, W. D. (ed.) 1991. *Occurrence, Characteristics, and Genesis of Carbonate, Gypsum, and Silica Accumulations in Soils*, Special Publication no. 26. Madison, WI, Soil Science Society of America.

Nettleton, W. D. and O. A. Chadwick. 1991. Soil–landscape relationships in the Wind River Basin, Wyoming. *Mountain Geol.* **28**: 3–11.

Nettleton, W. D., Brasher, B. R., Benham, E. C., *et al.* 1998. A classification system for buried paleosols. *Quat. Intl* **51/52**: 175–183.

Nettleton, W. D., Daniels, R. B., and R. J. McCracken. 1968. Two North Carolina coastal plain catenas. I. Morphology and fragipan development. *Soil Sci. Soc. Am. Proc.* **32**: 577–582.

Nettleton, W. D., Flach, K. W., and B. R. Brasher. 1969. Argillic horizons without clay skins. *Soil Sci. Soc. Am. Proc.* **33**: 121–125.

Nettleton, W. D., Gamble, E. E., Allen, B. L., *et al.* 1989. Relict soils of the subtropical regions of the United States. *Catena (Suppl.)* **16**: 59–93.

Nettleton, W. D., Goldin, A., and R. Engel. 1986. Differentiation of Spodosols and Andepts in a western Washington soil climosequence. *Soil Sci. Soc. Am. J.* **50**: 987–992.

Nettleton, W. D., Olson, C. G., and D. A. Wysocki. 2000. Paleosol classification: problems and solutions. *Catena* **41**: 61–92.

Nettleton, W. D., Witty, J. E., Nelson, R. E., *et al.* 1975. Genesis of argillic horizons in soils of desert areas of the southwestern United States. *Soil Sci. Soc. Am. Proc.* **39**: 919–926.

Neustruev, S. S. 1927. *Genesis of Soils*, Russian Pedological Investigations vol. III. Leningrad, Academy of Sciences of the USSR.

Neustruyev, S. S. 1915. Soil combinations in plains and mountainous countries. *Pochvovedeniye* **17**: 62–73.

Newcomb, R. C. 1952. Origin of Mima mounds, Thurston County region, Washington. *J. Geol.* **60**: 461–472.

Newman, A. L. 1983. Vertisols in Texas: some comments. *Soil Surv. Horiz.* **24**: 8–20.

Nichols, J. D. 1984. Relation of organic carbon to soil properties and climate in the southern Great Plains. *Soil Sci. Soc. Am. J.* **48**: 1382–1384.

Nichols, J. D., Brown, P. L., and W. J. Grant (eds.) 1984. *Erosion and Productivity of Soils Containing Rock Fragments*, Special Publication no. 13. Madison, WI, Soil Science Society of America.

Nickel, E. 1973. Experimental dissolution of light and heavy minerals in comparison with weathering and intrastratal solution. *Contrib. Sedimentol.* **1**: 1–68.

Nielsen, G. A. and F. D. Hole. 1964. Earthworms and the development of coprogenous A1 horizons in forest soils of Wisconsin. *Soil Sci. Soc. Am. Proc.* **28**: 426–430.

Nielsen, K. E., Dalsgaard, K., and P. Nornberg. 1987. Effects on soils of an oak invasion of a Calluna heath, Denmark. I. Morphology and chemistry. *Geoderma* **41**: 79–95.

Nikiforoff, C. C. 1937. General trends of the desert type of soil formation. *Soil Sci.* **43**: 105–131.

1949. Weathering and soil evolution. *Soil Sci.* **67**: 219–230.

Nikiforoff, C. C. and M. Drosdoff. 1943. Genesis of a claypan soil. II. *Soil Sci.* **56**: 43–62.

Nimlos, T. J. and M. Tomer. 1982. Mollisols beneath conifer forests in southwestern Montana. *Soil Sci.* **134**: 371–375.

Nishiizumi, K., Kohl, C. P., Arnold, J. R., *et al.* 1993. Role of *in situ* cosmogenic nuclides ^{10}Be and ^{26}Al in the study of diverse geomorphic processes. *Earth Surf. Proc. Landforms* **18**: 407–425.

Nishiizumi, K., Lal, D., Klein, J., *et al.* 1986. Production of ^{10}Be and ^{26}Al by cosmic rays in terrestrial quartz in situ and implications for erosion rates. *Nature* **319**: 134–136.

Nizeyimana, E. and T. J. Bicki. 1992. Soil and soil-landscape relationships in the north central region of Rwanda, east-central Africa. *Soil Sci.* **153**: 225–236.

Nobbs, M. F. and R. I. Dorn. 1988. Age determinations for rock varnish formation within petroglyphs: cation-ratio dating of 24 motifs from the Olary region, South Australia. *Rock Art Res.* **5**: 108–146.

Nooren, C. A. M., van Breemen, N., Stoorvogel, J. J., *et al.* 1995. The role of earthworms in the formation of sandy surface soils in a tropical forest in Ivory Coast. *Geoderma* **65**: 135–148.

Norfleet, M. L. and A. D. Karathanasis. 1996. Some physical and chemical factors contributing to fragipan strength in Kentucky soils. *Geoderma* **71**: 289–301.

Norman, M. B. II. 1974. Improved techniques for selective staining of feldspar and other minerals using amaranth. *US Geol. Soc. J. Res.* **2**: 73–79.

Norman, S. A., Schaetzl, R. J., and T. W. Small. 1995. Effects of slope angle on mass movement by tree uprooting. *Geomorphology* **14**: 19–27.

Nørnberg, P., Sloth, L., and K. E. Nielsen. 1993. Rapid changes in sandy soils caused by vegetation changes. Can. *J. Soil Sci.* **73**: 459–468.

North American Commission on Stratigraphic Nomeclature. 1983. North American Stratigraphic Code. *Am. Ass. Petrol. Geol. Bull.* **67**: 841–875.

Norton, E. A. and R. S. Smith. 1930. The influence of topography on soil profile character. *J. Am. Soc. Agron.* **22**: 251–262.

Norton, L. D. and G. F. Hall. 1985. Differentiation of lithologically similar soil parent materials. *Soil Sci. Soc. Am. J.* **49**: 409–414.

Norton, L. D., Bigham, J. M., Hall, G. F., *et al.* 1983. Etched thin sections for coupled optical and electron microscopy and micro-analysis. *Geoderma* **30**: 55–64.

Norton, L. D., Hall, G. F., Smeck, N. E., *et al.* 1984. Fragipan bonding in a Late-Wisconsinan loess-derived spoil in east-central Ohio. *Soil Sci. Soc. Am. J.* **48**: 1360–1366.

Noy-Meir, I. 1974. Multivariate analysis of the semiarid vegetation in south-eastern Australia. II. Vegetation catenae and environmental gradients. *Austral. J. Bot.* **22**: 115–140.

Nullet, D., Ikawa, H., and P. Kilham. 1990. Local differences in soil temperature and soil moisture regimes on a mountain slope, Hawaii. *Geoderma* **47**: 171–184.

Nutting, W. L., Haverty, M. I., and J. P. LaFage. 1987. Physical and chemical alteration of soil by two subterranean termite species in Sonoran Desert grassland. *J. Arid Envs.* **12**: 233–239.

Nye, P. H. 1954. Some soil-forming processes in the humid tropics. I. A field study of a catena in the west African forest. *J. Soil Sci.* **5**: 7–21.

1955a. Some soil-forming processes in the humid tropics. II. The development of the upper-slope member of the catena. *J. Soil Sci.* **6**: 51–62.

1955b. Some soil-forming processes of the humid tropics. III. Laboratory studies on the development of a typical catena over granitic gneiss. *J. Soil Sci.* **6**: 63–72.

1955c. Some soil-forming processes in the humid tropics. IV. The action of the soil fauna. *J. Soil Sci.* **6**: 73–83.

O'Brien, K. 1979. Secular variations in the production of cosmogenic isotopes in the Earth's atmosphere. *J. Geophys. Res.* **84**: 423–431.

O'Connell, A. M. 1987. Litter dynamics in Karri (*Eucalyptus diversicolor*) forests of south-western *Austral. J. Ecol.* **75**: 781–796.

Oades, J. M. 1989. An introduction to organic matter in mineral soils. In J. B. Dixon and S. B. Weed (eds.) *Minerals in Soil Environments*, 2nd edn. Madison, WI, Soil Science Society of America, pp. 89–159.

Oakes, H. and J. Thorp. 1950. Dark-clay soils of warm regions variously called Rendzina, Black Cotton Soils, Regur, and Tirs. *Soil Sci. Soc. Am. Proc.* **15**: 348–354.

Oertel, A. C. 1961. Pedogenesis of some Red-Brown Earths based on trace element profiles. *J. Soil Sci.* **12**: 242–258.

1968. Some observations incompatible with clay illuviation. *Trans. 9th Intl Congr. Soil Sci.* (Adelaide) **4**: 482–488.

Oertel, A. C. and J. B. Giles. 1966. Quantitative study of a layered soil. *Austral. J. Soil Res.* **4**: 19–28.

Offer, Z. Y. and D. Goossens. 2001. Ten years of aeolian dust dynamics in a desert region (Negev desert, Israel): analysis of airborne dust concentration, dust accumulation and the high-magnitude dust events. *J. Arid Envs.* **47**: 211–249.

Oganesyan, A. S. and N. G. Susekova. 1995. Parent materials of Wrangel Island. *Eurasian Soil Sci.* **27**: 20–35.

Ogg, C. M. and J. C. Baker. 1999. Pedogenesis and origin of deeply weathered soils formed in alluvial fans of the Virginia Blue Ridge. *Soil Sci. Soc. Am. J.* **63**: 601–606.

Ohnuki, Y., Terazono, R., Ikuzawa, H., *et al.* 1997. Distribution of colluvia and saprolites and their physical properties in a zero-order basin in Okinawa, southwestern Japan. *Geoderma* **80**: 75–93.

Ojanuga, A. G. 1979. Clay mineralogy of soils in the Nigerian tropical savanna region. *Soil Sci. Soc. Am. J.* **43**: 1237–1242.

Oliver, C. D. 1978. Subsurface geologic formations and site variation in Upper Sand Hills of South Carolina. *J. For.* **76**: 352–354.

Oliver, S. A., Oliver, H. R., Wallace, J. S., *et al.* 1987. Soil heat flux and temperature variation with vegetation, soil type, and climate. *Agric. For. Meteorol.* **39**: 257–269.

Ollerhead, J., Huntley, D. J., and G. W. Berger. 1994. Luminescence dating of sediments from Buctouche Spit, New Brunswick. *Can. J. Earth Sci.* **31**: 523–531.

Ollier, C. D. 1959. A two-cycle theory of tropical pedology. *J. Soil Sci.* **10**: 137–148.

1966. Desert gilgai. *Nature* **212**: 581–583.

Ollier, C. D. and J. E. Ash. 1983. Fire and rock breakdown. *Z. Geomorphol.* **27**: 363–374.

Ollier, C. D. and C. Pain. 1996. *Regolith, Soils and Landforms*. New York, John Wiley.

Ollier, C. D., Drover, D. P., and M. Godelier. 1971. Soil knowledge amongst the Baruya of Wonenara, New Guinea. *Oceania* **42**: 33–41.

Olson, C. G. 1989. Soil geomorphic research and the importance of paleosol stratigraphy to Quaternary investigations, midwestern USA. *Catena (Suppl.)* **16**: 129–142.

1997. Systematic soil-geomorphic investigations: contributions of R. V. Ruhe to pedologic interpretation. *Adv. Geoecol.* **29**: 415–438.

Olson, C. G. and W. D. Nettleton. 1998. Paleosols and the effects of alteration. *Quat. Intl* **51/52**: 184–194.

Olson, C. G., Nettleton, W. D., Porter, D. A., *et al.* 1997. Middle Holocene aeolian activity on the High Plains of west-central Kansas. *The Holocene* **7**: 255–261.

Olson, C. G., Ruhe, R. V., and M. J. Mausbach. 1980. The terra rossa–limestone contact phenomena in karst, southern Indiana. *Soil Sci. Soc. Am. J.* **44**: 1075–1079.

Olsson, I. U. 1974. Some problems in connection with the evaluation of ^{14}C dates. *Geol. Foreningens i Stockholm Förhandl.* **96**: 311–320.

Olsson, I. U. and F. A. N. Osadebe. 1974. Carbon isotope variations and fractionation corrections in ^{14}C dating. *Boreas* **3**: 139–146.

Olsson, M. and P.-A. Melkerud. 1989. Chemical and mineralogical changes during genesis of a Podzol from till in Southern Sweden. *Geoderma* **45**: 267–287.

Orombelli, G. and S. C. Poter. 1983. Lichen growth curves for the southern flank of the Mont Blanc Massif, Western Italian Alps. *Arctic Alpine Res.* **15**: 193–200.

Orton, G. J. 1996. Volcanic environments. In H. G. Reading (ed.) *Sedimentary Environments Processes, Facies and Stratigraphy.* Cambridge, MA, Blackwell Science, pp. 485–567.

Ovalles, F. A. and M. E. Collins. 1986. Soil–landscape relationships and soil variability in north-central Florida. *Soil Sci. Soc. Am. J.* **50**: 401–408.

Overpeck, J. T., Webb, T. III, and I. C. Prentice. 1985. Quantitative interpretation of fossil pollen spectra: dissimilarity coefficients and the method of modern analogs. *Quat. Res.* **23**: 87–108.

Pachepsky, Y. 1998. Discussion of the paper by J. D. Phillips. *Geoderma* **86**: 31–32.

Paetzold, R. F. 1990. Soil climate definitions used in Soil Taxonomy. In J. M. Kimble and W. D. Nettleton (eds.) *Characterization, Classification, and*

Utilization of Aridisols, Proc. 4th Intl Soil Correlation Mtg. Lincoln, NE, US Department of Agriculture–Natural Resources Conservation Service, pp. 151–165.

Palkovics, W. E., Petersen, G. W., and R. P. Matelski. 1975. Perched water table fluctuation compared to streamflow. *Soil Sci. Soc. Am. Proc.* **39**: 343–348.

Palmer, F. E., Staley, J. T., Murray, R. G. E., *et al.* 1986. Identification of manganese-oxidizing bacteria from desert varnish. *Geomicrobiol. J.* **4**: 343–360.

Paradise, T. R. 2002. Sandstone weathering and aspect in Petra, Jordan. *Z. Geomorphol.* **46**: 1–17.

Parakshin, Y. P. 1974. "Hill" solonetzes of the low-hill region of Kazakhstan. *Sov. Soil Sci.* **6**: 408–411. (Translated from *Pochvovedeniye* **8**: 137–140.)

Parfitt, R. L. and C. W. Childs. 1988. Estimation of forms of Fe and Al: a review, and analysis of contrasting soils by dissolution and Mössbauer methods. *Austral. J. Soil Res.* **26**: 121–144.

Parfitt, R. L. and M. Saigusa. 1985. Allophane and humus-aluminum in Spodosols and Andepts formed from the same volcanic ash beds in New Zealand. *Soil Sci.* **139**: 149–155.

Parfitt, R. L., Saigusa, M., and D. N. Eden. 1984. Soil development processes in an Aqualf–Ochrept sequence from loess with admixtures of tephra, New Zealand. *J. Soil Sci.* **35**: 625–640.

Pariente, S. 2001. Soluble salts dynamics in the soil under different climatic conditions. *Catena* **43**: 307–321.

Parisio, S. 1981. The genesis and morphology of a serpentine soil in Staten Island, New York. *Proc. Staten Island Inst. Arts Sci.* **31**: 2–17.

Parizek, E. J. and J. F. Woodruff. 1957. Description and origin of stone layers in soils of the southeastern states. *J. Geol.* **65**: 24–34.

Parker, G. R., McFee, W. W., and J. M. Kelly. 1978. Metal distribution in forested ecosystems in urban and rural northwestern Indiana. *J. Env. Qual.* **7**: 337–342.

Parsons, A. J., Abrahams, A. D., and J. R. Simanton. 1992. Microtopography and soil-surface materials on semi-arid piedmont hillslopes, southern Arizona. *J. Arid Envs.* **22**: 107–115.

Parsons, R. B. and C. A. Balster. 1966. Morphology and genesis of six "Red Hill" soils in the Oregon coast range. *Soil Sci. Soc. Am. Proc.* **30**: 90–93.

Parsons, R. B. and R. C. Herriman. 1975. A lithosequence in the mountains of southwestern Oregon. *Soil Sci. Soc. Am. Proc.* **39**: 943–947.

Parsons, R. B., Simonson, G. H., and C. A. Balster. 1968. Pedogenic and geomorphic relationships of associated Aqualfs, Albolls, and Xerolls in Western Oregon. *Soil Sci. Soc. Am. Proc.* **32**: 556–563.

Paton, T. R. 1974. Origin and terminology for gilgai in Australia. *Geoderma* **11**: 221–242.

1978. *The Formation of Soil Material.* Boston, MA, Allen and Unwin.

Paton, T. R., Humphreys, G. S., and P. B. Mitchell. 1995. *Soils: A New Global View.* New Haven, CT, Yale University Press.

Paton, T. R., Mitchell, P. B., Adamson, D., *et al.* 1976. Speed of podzolization. *Nature* **260**: 601–602.

Patrick, W. H., Jr. and R. E. Henderson. 1981. Reduction and reoxidation cycles of manganese and iron in flooded soil and in water solution. *Soil Sci. Soc. Am. J.* **45**: 855–859.

Patrick, W. H., Jr. and I. C. Mahapatra. 1968. Transformation and availability of nitrogen and phosphorous in waterlogged soils. *Adv. Agron.* **20**: 323–359.

Patro, B. C. and B. K. Sahu. 1974. Factor analysis of sphericity and roundness data of clastic quartz grains: environmental significance. *Sed. Geol.* **11**: 59–78.

Patton, P. C. and S. A. Schumm. 1975. Gully erosion, northern Colorado: a threshold phenomenon. *Geology* **3**: 88–90.

Paul, E. A., Campbell, C. A., Rennie, D. A., *et al.* 1964. Investigations of the dynamics of soil humus utilizing carbon dating techniques. *Trans. 8th Intl Congr. Soil Sci.* (Bucharest) **3**: 201–208.

Paul, E. A. and F. E. Clark (eds.) 1996. *Soil Microbiology and Biochemistry*, 2nd edn. San Diego, CA, Academic Press.

Pavich, M. J. 1989. Regolith residence time and the concept of surface age of the Piedmont "Peneplain". *Geomorphology* **2**: 181–196.

Pavich, M. J., Brown, L., Harden, J., *et al.* 1986. ^{10}Be distribution in soils from Merced River terraces, California. *Geochim. Cosmochim. Acta* **50**: 1727–1735.

Pavich, M. J., Brown, L., Klein, J., *et al.* 1984. ^{10}Be accumulation in a soil chronosequence. *Earth Planet. Sci. Lett.* **68**: 198–204.

Pavich, M. J., Brown, L., Valette-Silver, J. N., *et al.* 1985. ^{10}Be analysis of a Quaternary weathering profile in the Virginia Piedmont. *Geology* **13**: 39–41.

Pavich, M. J., Leo, G. W., Obermeier, S. F., *et al.* 1989. *Investigations of the Characteristics, Origin, and Residence Time of the Upland Residual Mantle of the Piedmont of Fairfax County, Virginia*, Professional Paper no. 1352. Washington, DC, US Geological Survey.

Pavlik, H. F. and F. D. Hole. 1977. Soilscape analysis of slightly contrasting terrains in southeastern Wisconsin. *Soil Sci. Soc. Am. J.* **41**: 407–413.

Pawluk, S. and M. J. Dudas. 1982. Floralpedoturbations in black chernozemic soils of the Lake Edmonton Plain. *Can. J. Soil Sci.* **62**: 617–629.

Payton, R. W. 1992. Fragipan formation in argillic brown earths (Fragiudalfs) of the Milfield Plain, north-east England. I. Evidence for a periglacial stage of development. *J. Soil Sci.* **43**: 621–644.

1993a. Fragipan formation in argillic brown earths (Fragiudalfs) of the Milfield Plain, north-east England. II. Post-Devensian developmental processes and the origin of fragipan consistence. *J. Soil Sci.* **44**: 703–723.

1993b. Fragipan formation in argillic brown earths (Fragiudalfs) of the Milfield Plain, north-east England. III. Micromorphological, SEM and EDXRA studies of fragipan degradation and the development of glossic features. *J. Soil Sci.* **44**: 725–729.

Peacock, E. and D. W. Fant. 2002. Biomantle formation and artifact translocation in upland sandy soils: an example from the Holly Springs National Forest, north-central Mississippi, USA. *Geoarchaeology* **17**: 91–114.

Pecher, K. 1994. Hydrochemical analysis of spatial and temporal variations of solute composition in surface and subsurface waters of a high Arctic catchment. *Catena* **21**: 305–327.

Pedro, G. 1966. Caractérisation géochimique des différents processus zonaux résultant de l'altération des roches superficielles. *C.R. Acad. Sci.* (Paris) **262**: 1828–1831.

Pedro, G. 1983. Structuring of some basic pedological processes. *Geoderma* **31**: 289–299.

Pedro, G., Jamagne, M., and J. C. Begon. 1978. Two routes in genesis of strongly differentiated acid soils under humid, cool-temperate conditions. *Geoderma* **20**: 173–189.

Peel, R. F. 1974. Insolation weathering: some measurements of diurnal temperature changes in exposed rocks in the Tibesti region, central Sahara. *Z. Geomorphol. (Suppl.)* **21**: 19–28.

Peet, G. B. 1971. Litter accumulation in jarrah and karri forest. *Austral. For.* **35**: 258–262.

Peltier, L. C. 1950. The geographic cycle in periglacial regions as it is related to climatic geomorphology. *Anna. Ass. Am. Geog.* **40**: 214–236.

Pennock, D. J. and W. J. Vreeken. 1986. Soil-geomorphic evolution of a Boroll catena in southwestern Alberta. *Soil Sci. Soc. Am. J.* **50**: 1520–1526.

Pennock, D. J., Zebarth, B. J., and E. DeJong. 1987. Landform classification and soil distribution in hummocky terrain, Saskatchewan, Canada. *Geoderma* **40**: 297–315.

Pereira, M., Vazquez, F. M., and F. G. Ojea. 1978. Pedological and geomorphological cycles in a catena of Galacia (NW Spain). *Catena* **5**: 375–387.

Pérez, F. L. 1984. Striated soil in an Andean Paramo of Venezuela: its origin and orientation. *Arctic Alpine Res.* **16**: 277–289.

1986. The effect of compaction on soil-disturbance by needle ice growth. *Acta Geocriogen.* **4**: 111–119.

1987a. Downslope stone transport by needle ice in a high Andean Area (Venezuela). *Rev. Géomorphol. Dynam.* **36**: 33–51.

1987b. Needle-ice activity and the distribution of stem-rosette species in a Venezuelan Paramo. *Arctic Alpine Res.* **19**: 135–153.

1987c. Soil surface roughness and needle ice-induced particle movement in a Venezuelan paramo. *Carib. J. Sci.* **23**: 454–460.

1991. Particle sorting due to off-road vehicle traffic in a high Andean paramo. *Catena* **18**: 239–254.

Perkins, S. M., Filippelli, G. M., and C. J. Souch. 2000. Airborne trace metal contamination of wetland sediments at Indiana Dunes National Lakeshore. *Water Air Soil Poll.* **122**: 231–260.

Perrin, R. M. S., Willis, E. H. and D. A. H. Hodge. 1964. Dating of humus Podzols by residual radiocarbon activity. *Nature* **202**: 165–166.

Perry, R. S. and J. B. Adams. 1978. Desert varnish: evidence for cyclic deposition of managnese. *Nature* **276**: 489–491.

Petäja-Ronkainen, A., Peuraniemi, V., and R. Aario. 1992. On podzolization in glaciofluvial material in northern Finland. *Ann. Acad. Sci. Fenn. Geol.-Geogr. Ser.* **A3**: 156.

Petersen, L. 1976. *Podzols and Podzolization*, Copenhagen, Royal Veterinary and Agricultural University.

Peterson, F. F. 1980. Holocene desert soil formation under sodium salt influence in a playa-margin environment. *Quat. Res.* **13**: 172–186.

1981. Landforms of the Basin & Range Province. *Nevada Agric. Exp. Station Tech. Bull.* **28**. 52 pp.

Pettijohn, F. J. 1941. Persistence of heavy minerals and geologic age. *J. Geol.* **49**: 610–625.

Péwé, T. L. 1966. Paloeclimatic significance of fossil ice wedges. *Biul. Peryglac.* **5**: 65–73.

Péwé, T. L. (ed.) 1969. *The Periglacial Environment Past and Present*. Montreal, McGill–Queens' University Press.

Pfeiffer, E.-M. and H. Janssen. 1994. Characterization of organic carbon, using the ^{13}C-value of a permafrost site in the Kolyma-Indigirka-lowland, northeast Siberia. In J. M. Kimble and R. J. Ahrens (eds.) *Classification, Correlation, and Management of Permafrost-Affected Soils.* Lincoln, NE, US Department of Agriculture–Soil Conservation Service, pp. 90–98.

Phillips, F. M. 1994. Environmental tracers for water movement in desert soils of the American southwest. *Soil Sci. Soc. Am. J.* **58**: 15–24.

Phillips, F. M., Leavy, B. D., Jannik, N. O., *et al.* 1986. The accumulation of cosmogenic chlorine-36 in rocks: a method for surface exposure dating. *Science* **231**: 41–43.

Phillips, F. M., Zreda, M. G., Gosse, J. C., *et al.* 1997. Cosmogenic ^{36}Cl and ^{10}Be ages of Quaternary glacial and fluvial deposits of the Wind River Range, Wyoming. *Geol. Soc. Am. Bull.* **109**: 1453–1463.

Phillips, J. D. 1989. An evaluation of the state factor model of soil ecosystems. *Ecol. Model.* **45**: 165–177.

1993a. Chaotic evolution of some coastal plain soils. *Phys. Geog.* **14**: 566–580.

1993b. Progressive and regressive pedogenesis and complex soil evolution. *Quat. Res.* **40**: 169–176.

1993c. Stability implications of the state factor model of soils as a nonlinear dynamical system. *Geoderma* **58**: 1–15.

1999. *Earth Surface Systems.* Oxford, UK, Blackwell Scientific Publications.

2001a. Divergent evolution and the spatial structure of soil landscape variability. *Catena* **43**: 101–113.

2001b. Inherited vs. acquired complexity in east Texas weathering profiles. *Geomorphology* **40**: 1–14.

2001c. The relative importance of intrinsic and extrinsic factors in pedodiversity. *Ann. Ass. Am. Geog.* **91**: 609–621.

Phillips, J. D., Golden, H., Capiella, K., *et al.* 1999. Soil redistribution and pedologic transformations in coastal plain croplands. *Earth Surf. Proc. Landforms* **24**: 23–39.

Phillips, J. D., Perry, D., Garbee, A. R., *et al.* 1996. Deterministic uncertainty and complex pedogenesis in some Pleistocene dune soils. *Geoderma* **73**: 147–164.

Phillips, S. E. and P. G. Self. 1987. Morphology, crystallography and origin of needle-fibre calcite in Quaternary pedogenic calcretes of South Australia. *Austral. J. Soil Res.* **25**: 429–444.

Phillips, S. E., Milnes, A. R., and R. C. Foster. 1987. Calcified filaments: an example of biological influences in the formation of calcrete in South Australia. *Austral. J. Soil Res.* **25**: 405–428.

Phillips, W. M., McDonald, E. V., Reneau, S. L., *et al.* 1998. Dating soils and alluvium with cosmogenic ^{21}Ne depth profiles: case studies from the Pajarito Plateau, New Mexico, USA. *Earth Planet. Sci. Lett.* **160**: 209–223.

Pickering, E. W. and P. L. M. Veneman. 1984. Moisture regimes and morphological characteristics in a hyrosequence in central Massachusetts. *Soil Sci. Soc. Am. J.* **48**: 113–118.

Pierce, K. L., Obradovich, J. D., and I. Friedman. 1976. Obsidian hydration dating and correlation of Bull Lake and Pinedale glaciations near West Yellowstone, Montana. *Geol. Soc. Am. Bull.* **87**: 703–710.

Pierce, R. S., Lull, H. W., and H. C. Storey. 1958. Influence of land use and forest condition on soil freezing and snow depth. *For. Sci.* **4**: 246–263.

Ping, C. L. 1987. Soil temperature profiles of two Alaskan soils. *Soil Sci. Soc. Am. J.* **51**: 1010–1018.

1997. Characteristics of permafrost soils along a latitudinal transect in arctic Alaska. *Agroborealis* **29**: 35–36.

Ping, C. L., Sumner, M. E., and L. P. Wilding. 2000. Andisols. In M. E. Sumner (ed.) *Handbook of Soil Science.* Boca Raton, FL, CRC Press, pp. E-209–E-224.

Pinton, R. , Varanini, Z. , and P. Nannipieri (eds.) 2001 *The Rhizosphere: Biochemistry and Organic Substances at the Soil–Plant Interface.* New York, Marcel Dekker.

Piperno, D. R. and P. Becker. 1996. Vegetational history of a site in the central Amazon Basin derived from phytolith and charcoal records from natural soils. *Quat. Res.* **45**: 202–209.

Pissart, A. 1969. Le mécanisme périglaciaire dressant les pierres dans le sol: résultats d'expériences. *C.R. Acad. Sci.* (Paris) **268**: 3015–3017.

Pitty, A. F. 1968. Particle size of the Saharan dust which fell in Britain in July, 1968. *Nature* **220**: 364–365.

Plaster, R. W. and W. C. Sherwood. 1971. Bedrock weathering and residual soil formation in central Virginia. *Geol. Soc. Am. Bull.* **82**: 2813–2826.

Plice, M. J. 1942. Factors affecting soil color (progress report). *Proc. Oklahoma Acad. Sci.* **23**: 49–51.

Poesen, J. 1987. Transport of rock fragments by rill flow: a field study. *Catena (Suppl.)* **8**: 35–54.

Poesen, J. and H. Lavee. 1994. Rock fragments in top soils: significance and processes. *Catena* **23**: 1–28.

Poesen, J., Ingelmo-Sanchez, F., and H. Mücher. 1990. The hydrological response of soil surfaces to rainfall as affected by cover and position of rock fragments in the top layer. *Earth Surf. Proc. Landforms* **15**: 653–671.

Pohl, M. M. 1995. Radiocarbon ages on organics from piedmont alluvium, Ajo Mountains, Arizona. *Phys. Geog.* **16**: 339–353.

Pokras, E. M. and A. C. Mix. 1985. Eolian evidence for spatial variability of late Quaternary climates in tropical Africa. *Quat. Res.* **24**: 137–149.

Polach, H. A. and A. B. Costin. 1971. Validity of soil organic matter radiocarbon dating: buried soils in Snowy Mountains, Southeastern Australia as example. In D. H. Yaalon (ed.) *Paleopedology Origin, Nature and Dating of Paleosols*. Jerusalem, Israel University Press, pp. 89–96.

Pomeroy, D. E. 1976. Some effects of mound-building termites on soils in Uganda. *J. Soil Sci.* **27**: 377–394.

Ponnamperuma, F. N., Tianco, E. M., and T. A. Loy. 1967. Redox equilibria in flooded soils. I. The iron hydroxide systems. *Soil Sci.* **103**: 374–382.

Ponomarenko, S. V. 1988. Probable mechanism of redistribution of coarse fractions in the soil profile. *Sov. Soil Sci.* **20**: 35–41.

Ponomareva, V. V. 1974. Genesis of the humus profile of Chernozem. *Sov. Soil Sci.* **6**: 393–402. (Translated from *Pochvovedeniye* **7**: 27–37.)

Pope, G. A., Dorn, R. I., and J. C. Dixon. 1995. A new conceptual model for understanding geographical variations in weathering. *Ann. Ass. Am. Geog.* **85**: 38–64.

Pope, G. A., Meierding, T. C., and T. R. Paradise. 2002. Geomorphology's role in the study of weathering of cultural stone. *Geomorphology* **47**: 211–225.

Porter, S. C. 1978. Glacier Peak tephra in the North Cascade Range, Washington: stratigraphy, distribution, and relationship to late-glacial events. *Quat. Res.* **10**: 30–41.

Porter, S. C. 1981. Lichenometric studies in the Cascade Range of Washington: eastablishment of *Rhizocarpon geographicum* growth curves at Mount Ranier. *Arctic Alpine Res.* **13**: 11–23.

Porter, S. C., Pierce, K. L., and T. D. Hamilton. 1983. Late Wisconsin mountain glaciation in the western United States. In H. E. Wright and S. C. Porter (eds.) *Late-Quaternary Environments of the United States*, vol. 1, *The Late Pleistocene*. Minneapolis, MN, University of Minnesota Press, pp. 71–111.

Post, F. A. and F. R. Dreibelbis. 1942. Some influence of frost penetration and microclimate on the water relationships of woodland, pasture, and cultivated soils. *Soil Sci. Soc. Am. Proc.* **7**: 95–104.

Potter, R. M. and G. R. Rossman. 1977. Desert varnish: the importance of clay minerals. *Science* **196**: 1446–1448.

Prasolov, I. J. 1965. Soil regions of European Russia. *Sov. Soil Sci.* **4**: 343–354.

Prell, W. L., Imbrie, J., Martinson, D. G., *et al.* 1986. Graphic correlation of the oxygen isotope stratigraphy application to the Late Quaternary. *Paleoceanography* **1**: 137–162.

Prevost, D. J. and B. A. Lindsay. 1999. *Soil Survey of Hualapai-Havasupai Area, Arizona: Parts of Coconino, Mojave, and Yavapai Counties*. Washington, DC, US Department of Agriculture Natural Resources Conservation Service, US Government Printing Office.

Price, A. G. and B. O. Bauer. 1984. Small-scale heterogeneity and soil-moisture variability in the unsaturated zone. *J. Hydrol.* **70**: 277–293.

Price, T. W., Blevins, R. L., Barnhisel, R. I., *et al.* 1975. Lithologic discontinuities in loessial soils of southwestern Kentucky. *Soil Sci. Soc. Am. Proc.* **39**: 94–98.

Price, W. A. 1949. Pocket gophers as architects of Mima (pimple) mounds of the western United States. *Texas J. Sci.* **1**: 1–17.

Prior, J. C. 1991. *Landforms of Iowa*. Iowa City, IA, University of Iowa Press.

Pritchett, W. L. 1979. *Properties and Management of Forest Soils*. New York, John Wiley.

Proctor, J. 1971. The plant ecology of serpentine. II. Plant response to serpentine soils. *J. Ecol.* **59**: 397–410.

Proctor, J. and S. R. J. Woodell. 1975. The ecology of serpentine soils. *Adv. Ecol. Res.* **9**: 256–365.

Prodi, F. and G. Fea. 1979. A case of transport and deposition of Saharan dust over the Italian peninsula and southern Europe. *J. Geophys. Res.* **84**: 6951–6960.

Protz, R., Arnold, R. W., and E. W. Presant. 1968. The approximation of the true modal profile with the use of the high speed computer and landscape control. *Trans. 9th Intl Congr. Soil Sci.* (Adelaide) **4**: 193–204.

Protz, R., Ross, G. J., Martini, I. P., *et al.* 1984. Rate of podzolic soil formation near Hudson Bay, Ontario. *Can. J. Soil Sci.* **64**: 31–49.

Pullan, R. A. 1979. Termite hills in Africa: their characteristics and evolution. *Catena* **6**: 267–291.

Pustovoytov, K. E. 2002. Pedogenic carbonate cutans on clasts in soils as a record of history of grassland ecosystems. *Palaeogeog. Paleaeoclimat. Palaeoecol.* **177**: 199–214.

Pustovoytov, K. E. and V. O. Targulian. 1996. Illuviation coatings on rock fragments as a source of pedogenic information. *Euras. Soil Sci.* **3**: 335–347.

Putman, B. R., Jansen, I. J., and L. R. Follmer. 1989. Loessial soils: their relationship to width of the source valley in Illinois. *Soil Sci.* **146**: 241–247.

Pye, K. 1987. *Aeolian Dust and Dust Deposits*. London, Academic Press.

Pye, K. 1988. Bauxites gathering dust. *Nature* **333**: 800.

Pye, K. and R. Johnson. 1988. Stratigraphy, geochemistry, and thermoluminescence ages of lower Mississippi Valley loess. *Earth Surf. Proc. Landforms* **13**: 103–124.

Pye, K. and H. Tsoar. 1987. The mechanics and geological implications of dust transport and deposition in deserts with particular reference to loess formation and dune sand diagenesis in the northern Negev, Israel. In L. E. Frostick and I. Reid (eds.) *Desert Sediments: Ancient and Modern*. Boston, MA, Blackwell Scientific Publications, pp. 139–156.

1990. *Aeolian Sand and Sand Dunes*. London, Unwin Hyman.

Quade, J. 2001. Desert pavements and associated rock varnish in the Mojave Desert: how old can they be? *Geology* **29**: 855–858.

Quade, J. and T. E. Cerling. 1990. Stable isotope evidence for a pedogenic origin of carbonates in Trench 14 near Yucca Mountain, Nevada. *Science* **250**: 1549–1552.

Quade, J., Cerling, T. E., and J. R. Bowman. 1989. Systematic variations in the carbon and oxygen isotopic composition of pedogenic carbonate along elevation transects in the southern Great Basin, United States. *Geol. Soc. Am. Bull.* **101**: 464–475.

Quade, J., Chivas, A. R., and M. T. McCulloch. 1995. Strontium and carbon isotope tracers and the origins of soil carbonate in South Australia and Victoria. *Palaeogeog. Palaeoclimat. Palaeoecol.* **113**: 103–117.

Raad, A. T. and R. Protz. 1971. A new method for the identification of sediment stratification in soils of the Blue Springs Basin, Ontario. *Geoderma* **6**: 23–41.

Rabenhorst, M. C. and J. E. Foss. 1981. Soil and geologic mapping over mafic and ultramafic parent materials in Maryland. *Soil Sci. Soc. Am. J.* **45**: 1156–1160.

Rabenhorst, M. C. and D. Swanson. 2000. Histosols. In M. E. Sumner (ed.) *Handbook of Soil Science*. Boca Raton, FL, CRC Press, pp. E-183–E-209.

Rabenhorst, M. C. and L. P. Wilding. 1984. Rapid method to obtain carbonate-free residues from limestone and petrocalcic materials. *Soil Sci. Soc. Am. J.* **48**: 216–219.

1986a. Pedogenesis on the Edwards Plateau, Texas. I. Nature and continuity of parent materials. *Soil Sci. Soc. Am. J.* **50**: 678–687.

1986b. Pedogenesis on the Edwards Plateau, Texas. III. New model for the formation of petrocalcic horizons. *Soil Sci. Soc. Am. J.* **50**: 693–699.

Rabenhorst, M. C., Foss, J. E., and D. S. Fanning. 1982. Genesis of Maryland soils formed from serpentinite. *Soil Sci. Soc. Am. J.* **46**: 607–616.

Rabenhorst, M. C., West, L. T., and L. P. Wilding. 1991. Genesis of calcic and petrocalcic horizons in soils over carbonate rocks. *Soil Sci. Soc. Am. Spec. Publ.* **26**: 61–74.

Rabenhorst, M. C., Wilding, L. P., and C. L. Girdner. 1984a. Airborne dusts in the Edwards Plateau region of Texas. *Soil Sci. Soc. Am. J.* **48**: 621–627.

Rabenhorst, M. C., Wilding, L. P., and L. T. West. 1984b. Identification of pedogenic carbonates using stable carbon isotope and microfabric analyses. *Soil Sci. Soc. Am. J.* **48**: 125–132.

Radke, J. K. and E. C. Berry. 1998. Soil water and solute movement and bulk density changes in repacked soil columns as a result of freezing and thawing under field conditions. *Soil Sci.* **163**: 611–624.

Radtke, U., Bruckner, H., Mangini, A., *et al.* 1988. Problems encountered with absolute dating (U-series, ESR) of Spanish calcretes. *Quat. Sci. Rev.* **7**: 439–445.

Ragg, J. M. and D. F. Ball. 1964. Soils of the ultra-basic rocks of the Island of Rhum. *J. Soil Sci.* **15**: 124–133.

Rahn, P. H. 1971. The weathering of tombstones and its relation to the topography of New England. *J. Geol. Educ.* **19**: 112–118.

Rai, D. and W. L. Lindsay. 1975. A thermodynamic model for predicting the formation, stability and weathering of common soil minerals. *Soil Sci. Soc. Am. Proc.* **32**: 443–444.

Raison, R. J., Woods, P. V., and P. K. Khanna. 1986. Decomposition and accumulation of litter after fire in sub-alpine eucalypt forests. *Austral. J. Ecol.* **11**: 9–19.

Ranger, J., Dambrine, E., Robert, M., *et al.* 1991. Study of current soil-forming processes using bags of vermiculite and resins placed within soil horizons. *Geoderma* **48**: 335–350.

Ranney, R. W. and M. T. Beatty. 1969. Clay translocation and albic tongue formation in two

Glossoboralfs in west-central Wisconsin. *Soil Sci. Soc. Am. Proc.* **33**: 768–775.

Ransom, M. D. and N. E. Smeck. 1986. Water table characteristics and water chemistry of seasonally wet soils of southwestern Ohio. *Soil Sci. Soc. Am. J.* **50**: 1281–1290.

Ransom, M. D., Smeck, N. E., and J. M. Bigham. 1987a. Micromorphology of seasonally wet soils on the Illinoian till plain, USA. *Geoderma* **40**: 83–99.

1987b. Stratigraphy and genesis of polygenetic soils on the Illinoian till plain of southwestern Ohio. *Soil Sci. Soc. Am. J.* **51**: 135–141.

Rapp, A. 1984. Are terra rossa soils in Europe eolian deposits from Africa? *Geol. Foreningens i Stockholm Forhandl.* **105**: 161–168.

Rapp, A. and R. Nyberg. 1981. Alpine debris flows in northern Scandinavia. *Geog. Ann.* **63A**: 183–196.

Raukas, A., Mickelson, D. M., and A. Dreimanis. 1978. Methods of till investigation in Europe and North America. *J. Sed. Petrol.* **48**: 285–294.

Raw, F. 1962. Studies of earthworm populations in orchards. I. Leaf burial in apple orchards. *Ann. Appl. Biol.* **50**: 389–404.

Rawling, J. E. 3rd. 2000. A review of lamellae. *Geomorphology* **35**: 1–9.

Rawlins, S. L. and G. S. Campbell. 1986. Water potential, thermocouple psychrometry. In A. Klute (ed.) *Methods of Soil Analysis*, Part 1, *Physical and Mineralogical Methods*, Agronomy Monograph no. 9, 2nd edn. Madison, WI, American Soil Association and Soil Science Society of America, pp. 587–618.

Rawls, W. J. 1983. Estimating bulk density from particle size analysis and organic matter content. *Soil Sci.* **135**: 123–125.

Ray, A. 1959. The effect of earthworms in soil pollen distribution. *J. Oxford Univ. For. Soc.* **7**: 16–21.

Rea, D. L. 1990. Aspects of atmospheric circulation: the Late Pleistocene (0-950 000 yr) record of eolian deposition in the Pacific Ocean. *Palaeogeog. Palaeoclimat. Palaeoecol.* **78**: 217–227.

Read, G. 1998. The establishment of a buried soil chronosequence with a feldspar weathering index. *Quat. Intl* **51/52**: 39.

Rebertus, R. A. and S. W. Buol. 1985. Iron distribution in a developmental sequence of soils from mica gneiss and schist. *Soil Sci. Soc. Am. J.* **49**: 713–720.

Rech, J. A., Reeves, R. W., and D. M. Hendricks. 2001. The influence of slope aspect on soil weathering processes in the Springerville volcanic field, Arizona. *Catena* **43**: 49–62.

Redmond, C. E. and J. E. McClelland. 1959. The occurrence and distribution of lime in calcium carbonate Solonchak and associated soils of eastern North Dakota. *Soil Sci. Soc. Am. Proc.* **23**: 61–65.

Reeder, S. W. and A. L. McAllister. 1957. A staining method for the quantitative determination of feldspars in rocks and sands from soils. *Can. J. Soil Sci.* **37**: 57–59.

Rees-Jones, J. and M. S. Tite. 1997. Optical dating results for British archaeological sediments. *Archaeometry* **39**: 177–188.

Reeves, C. C., Jr. 1970. Origin, classification, and geologic history of caliche on the southern High Plains, Texas and eastern New Mexico. *J. Geol.* **78**: 352–362.

Reeve, R. C. 1986. Water potential, piezometry. In A. Klute (ed.) *Methods of Soil Analysis*, Part 1, *Physical and Mineralogical Methods*, Agronomy Monograph no. 9, 2nd edn. Madison, WI, American Soil Association and Soil Science Society of America, pp. 545–561.

Reheis, M. C. 1987a. Climatic implications of alternating clay and carbonate formation in semiarid soils of south-central Montana. *Quat. Res.* **27**: 270–282.

1987b. Gypsic soils on the Kane alluvial fans, Big Horn County, Wyoming. *US Geol. Survey Bull.* **1590-C**.

Reheis, M. C. and R. Kihl. 1995. Dust deposition in southern Nevada and California, 1984–1989: relations to climate, source area, and source lithology. *J. Geophys. Res.* **100**: 8893–8918.

Reheis, M. C., Goodmacher, J. C., Harden, J. W., *et al.* 1995. Quaternary soils and dust deposition in southern Nevada and California. *Geol. Soc. Am. Bull.* **107**: 1003–1022.

Reheis, M. C., Harden, J. W., McFadden, L. D., *et al.* 1989. Development rates of late Quaternary soils, Silver Lake Playa, California. *Soil Sci. Soc. Am. J.* **53**: 1127–1140.

Reheis, M. C., Sowers, J. M., Taylor, E. M., *et al.* 1992. Morphology and genesis of carbonate soils on the Kyle Canyon Fan, Nevada, USA. *Geoderma* **52**: 303–342.

Reichman, O. J. and S. C. Smith. 1991. Burrows and burrowing behavior by mammals. *Curr. Mammal.* **2**: 197–244.

Reid, D. A., Graham, R. C., Southard, R. J., *et al.* 1993. Slickspot soil genesis in the Carrizo Plain, California. *Soil Sci. Soc. Am. J.* **57**: 162–168.

Reider, R. G. 1983. A soil catena in the Medicine Bow Mountains, Wyoming, USA., with reference to paleoenvironmental influences. *Arctic Alpine Res.* **15**: 181–192.

Reider, R. G., Huckleberry, G. A., and G. C. Frison. 1988. Soil evidence for postglacial forest–grassland fluctuation in the Absaroka Mountains of northwestern Wyoming, USA. *Arctic Alpine Res.* **20**: 188–198.

Reimer, A. and C. F. Shaykewich. 1980. Estimation of Manitoba soil temperatures from atmospheric meteorological measurements. *Can. J. Soil Sci.* **60**: 299–309.

Reinhart, K. G. 1961. The problem of stones in soil-moisture measurement. *Soil Sci. Soc. Am. Proc.* **25**: 268–270.

Reneau, S. L. 1993. Manganese accumulation in rock varnish on a desert piedmont, Mojave Desert, California, and application to evaluating varnish development. *Quat. Res.* **40**: 309–317.

Reneau, S. L. and R. Raymond, Jr. 1991. Cation-ratio dating of rock varnish: why does it work? *Geology* **19**: 937–940.

Retallack, G. J. 1990. *Soils of the Past*. Boston, MA, Unwin Hyman.

Retzer, J. L. 1956. Alpine soils of the Rocky Mountains. *J. Soil Sci.* **7**: 22–32.

1965. Present soil-forming factors and processes in Arctic and alpine regions. *Soil Sci.* **99**: 38–44.

Reuter, G. 1965. Tonminerals in Pseudogleyboden. *Wiss. Z. Friedrich Schiller Univ. Jena, Math. Naturwiss. Reihe.* **14**: 75–78.

Reuter, R. J., McDaniel, P. A., Hammel, J. E., *et al.* 1998. Solute transport in seasonal perched water tables in loess-derived soilscapes. *Soil Sci. Soc. Am. J.* **62**: 977–983.

Reynolds, R. L. and J. W. King. 1995. Magnetic records of climate change. *Rev. Geophys. (Suppl.)* **33**: 101–110.

Rice, T. J., Jr., Buol, S. W., and S. B. Weed. 1985. Soil-saprolite profiles derived from mafic rocks in the North Carolina piedmont. I. Chemical, morphological, and mineralogical characteristics and transformations. *Soil Sci. Soc. Am. J.* **49**: 171–178.

Richards, P. L. and L. R. Kump. 1997. Application of the geographical information systems approach to watershed mass balance studies. *Hydrol. Proc.* **11**: 671–694.

Richards, P. W. 1941. Lowland tropical podzols and their vegetation. *Nature* **148**: 129–131.

1952. *The Tropical Rainforest*. London, Cambridge University Press.

Richardson, J. L. and R. J. Bigler. 1984. Principal component analysis of prairie pothole soils in North Dakota. *Soil Sci. Soc. Am. J.* **48**: 1350–1355.

Richardson, J. L. and W. J. Edmonds. 1987. Linear regression estimations of Jenny's relative effectiveness of state factors equation. *Soil Sci.* **144**: 203–208.

Richardson, J. L. and F. D. Hole. 1979. Mottling and iron distribution in a Glossoboralf-Haplaquoll hydrosequence on a glacial moraine in northwestern Wisconsin. *Soil Sci. Soc. Am. J.* **43**: 552–558.

Richardson, J. L., Wilding, L. P., and R. B. Daniels. 1992. Recharge and discharge of groundwater in aquic conditions illustrated with flownet analysis. *Geoderma* **53**: 65–78.

Richmond, G. M. and D. S. Fullerton. 1986. Summation of Quaternary glaciations in the United States of America. *Quat. Sci. Rev.* **5**: 183–196.

Richmond, G. R. 1986. Stratigraphy and chronology of glaciations in Yellowstone National Park. *Quat. Sci. Rev.* **5**: 83–98.

Richter, D. D. and L. I. Babbar. 1991. Soil diversity in the tropics. *Adv. Ecol. Res.* **21**: 315–389.

Riecken, F. F. 1945. Selection of criteria for the classification of certain soil types of Iowa into Great Soil Groups. *Soil Sci. Soc. Am. Proc.* **10**: 319–325.

Riecken, F. F. and E. Poetsch. 1960. Genesis and classification considerations of some prairie-formed soil profiles from local alluvium in Adair County, Iowa. *Proc. Iowa Acad. Sci.* **67**: 268–276.

Riestenberg, M. M. and S. Sovonick-Dunford. 1983. The role of woody vegetation in stabilizing slopes in the Cincinnati area, Ohio. *Geol. Soc. Am. Bull.* **94**: 506–518.

Rightmire, C. T. 1967. A radiocarbon study of the age and origin of caliche deposits. M.A. thesis, University of Texas, Austin.

Riise, G., Van Hees, P., Lundstrom, U., and L. T. Strand. 2000. Mobility of different size fractions of organic carbon, Al, Fe, Mn and Si in Podzols. *Geoderma* **94**: 237–247.

Rindfleisch, P. R. and R. J. Schaetzl. 2001. Soils and geomorphic evidence for a high lake stand in a Michigan drumlin field. *Phys. Geog.* **22**: 483–501.

Ritchie, A. M. 1953. The erosional origin of Mima mounds of southwest Washington. *J. Geol.* **61**: 41–50.

Robert, M. 1973. The experimental transformation of mica toward smectite: relative importance of total charge and tetrahedral charge. *Clay Mins.* **21**: 168–174.

Roberts, B. A. 1980. Some chemical and physical properties of serpentine soils from western Newfoundland. *Can. J. Soil Sci.* **60**: 231–240.

Roberts, E. G. 1961. Soil settlement caused by tree roots. *J. For.* **59**: 25–26.

Roberts, F. J. and B. A. Carbon. 1972. Water repellence in sandy soils of south-western Australia. *Austral. J. Soil Res.* **10**: 35–42.

Robinson, G. H. and C. I. Rich. 1960. Characteristics of the multiple yellowish-red bands common to certain soils of the southeastern United States. *Soil Sci. Soc. Am. Proc.* **24**: 226–230.

Rockie, W. A. 1934. Snowdrifts and the Palouse topography. *Geog. Rev.* **24**: 380–385.

Rodbell, D. T. 1993. Subdivision of Late Pleistocene moraines in the Cordillera Blanca, Peru, based on rock-weathering features, soils, and radiocarbon dates. *Quat. Res.* **39**: 133–143.

Rogers, L. E. 1972. The ecological effects of the western harvester ant (*Pogonomyrmex occidentalis*) in the shortgrass plain ecosystem. *USA/BP Grassland Biome Tech. Rept* **206**.

Rolfsen, P. 1980. Disturbance of archaeological layers by processes in the soils. *Norw. Arch. Rev.* **13**: 110–118.

Romanovskij, N. N. 1973. Regularities in formation of frost-fissures and development of frost-fissure polygons. *Biul. Peryglac.* **23**: 237–277.

Romell, L. G. and S. O. Heiberg. 1931. Types of humus layer in the forests of northeastern US. *Ecology* **12**: 567–608.

Roose, E. 1980. *Dynamique actuelle de sols ferrallitiques et ferrugineux tropicaux d'Afrique Occidentale: étude expérimentale des transferts hydrologiques et biologiques de matières sous végétations naturelles ou cultivées.* Bordeaux, France, ORSTOM.

Roth, K., Schulin, R., Fluhler, H., and W. Attinger. 1990. Calibration of time domain reflectometry for water content measurement using a composite dielectric approach. *Water Resources Res.* **26**: 2267–2273.

Rousseau, D.-D. and G. Kukla. 1994. Late Pleistocene climate record in the Eustis loess section, Nebraska, based on land snail assemblages and magnetic susceptibility. *Quat. Res.* **42**: 176–187.

Rousseau, D.-D., Wu, N., and Z. Guo. 2000. The terrestrial molluses as new indices of the Asian paleomonsoons in the Chinese loess Plateau. *Global Planet. Change* **26**: 199–206.

Rovey, C. W. II and W. F. Kean 1996. Pre-Illinoian glacial stratigraphy in north-central Missouri. *Quat. Res.* **45**: 17–29.

Rovner, I. 1971. Potential of opal phytoliths for use in paleoecological reconstruction. *Quat. Res.* **1**: 343–359.

1988. Macro- and micro-ecological reconstruction using plant opal phytolith data from archaeological sediments. *Geoarcheology* **3**: 155–163.

Rowe, P. B. 1955. Effects of the forest floor on disposition of rainfall in pine stands. *J. For.* **53**: 342–348.

Roy, B. B. and N. K. Barde. 1962. Some characteristics of the Black Soils of India. *Soil Sci.* **93**: 142–147.

Roy, B. B., Ghose, B., and S. Pandey. 1967. Landscape–soil relationship in Chohtan Block in Barmer District in western Rajasthan. *J. Ind. Soc. Soil Sci.* **15**: 53–59.

Rubilin, Y. V. 1962. Prairie soils of North America. *Sov. Soil Sci.* 610–617.

Rudberg, S. 1972. Periglacial zonation: a discussion. *Göttinger Geog. Abh.* **60**: 221–233.

Ruddiman, W. F. and H. E. Wright, Jr. 1987. Introduction. In W. F. Ruddiman and H. E. Wright, Jr. (eds.) *North America and Adjacent Oceans during the Last Deglaciation.* Geological Society of America, Boulder, CO, pp. 1–12.

Ruellan, A. 1971. The history of soils: some problems of definition and interpretation. In D. H. Yaalon (ed.) *Paleopedology Origin, Nature and Dating of Paleosols.* Jerusalem, Israel University Press, pp. 3–13.

Ruhe, R. V. 1954. Relations of the properties of Wisconsin loess to topography in western Iowa. *Am. J. Sci.* **252**: 663–672.

1956. Geomorphic surfaces and the nature of soils. *Soil Sci.* **82**: 441–455.

1958. Stone lines in soils. *Soil Sci.* **84**: 223–231.

1960. Elements of the soil landscape. *Trans. 7th Intl Congr. Soil Sci.* (Madison, WI) **4**: 165–170.

1965. Quaternary paleopedology. In H. E. Wright and D. G. Frey (eds.) *Quaternary of the United States.* Princeton, NJ, Princeton University Press, pp. 755–764.

1967. Geomorphic surfaces and surficial deposits in southern New Mexico. *New Mexico Bur. Mines Mineral Resources Mem.* **18**.

1969. *Quaternary Landscapes in Iowa.* Ames, IA, Iowa State University Press.

1970. Soils, paleosols, and environment. In: W. Dort and J. K. Jones (eds.) *Pleistocene and Recent Environments of the Central Great Plains.* Lawrence, KS, University of Kansas Press, pp. 37–52.

1973. Background of model for loess-derived soils in the upper Mississippi Valley. *Soil Sci.* **115**: 250–253.

1974a. Holocene environments and soil geomorphology in Midwestern United States. *Quat. Res.* **4**: 487–495.

1974b. Sangamon paleosols and Quaternary environments in midwestern United States. *Geog. Monogr.* **5**: 153–167.

1975a. Climatic geomorphology and fully developed slopes. *Catena* **2**: 309–320.

1975b. *Geomorphology*. Boston, MA, Houghton Mifflin.

Ruhe, R. V. and R. B. Daniels. 1958. Soils, paleosols, and soil-horizon nomenclature. *Soil Sci. Soc. Am. Proc.* **22**: 66–69.

Ruhe, R. V. and C. G. Olson. 1980. Soil welding. *Soil Sci.* **130**: 132–139.

Ruhe, R. V. and W. H. Scholtes. 1956. Ages and development of soil landscapes in relation to climatic and vegetational changes in Iowa. *Soil Sci. Soc. Am. Proc.* **20**: 264–273.

1959. Important elements in the classification of the Wisconsinan glacial stage: a discussion. *J. Geol.* **67**: 585–593.

Ruhe, R. V., Daniels, R. B., and J. G. Cady. 1967. *Landscape Evolution and Soil Formation in Southwestern Iowa*, Technical Bulletin no. 1349. Washington, DC, US Department of Agriculture. Soil Conservation Service, US Government Printing Office.

Ruhe, R. V., Hall, R. D., and A. P. Canepa. 1974. Sangamon paleosols of southwestern Indiana, USA. *Geoderma* **12**: 191–200.

Ruhe, R. V., Miller, G. A., and W. J. Vreeken. 1971. Paleosols, loess sedimentation and soil stratigraphy. In D. H. Yaalon (ed.) *Paleopedology: Origin, Nature and Dating of Paleosols.* Jerusalem, Israel University Press, pp. 41–60.

Ruhe, R. V., Rubin, M., and W. H. Scholtes. 1957. Late Pleistocene radiocarbon chronology in Iowa. *Am. J. Sci.* **255**: 671–689.

Runge, E. C. A. 1973. Soil development sequences and energy models. *Soil Sci.* **115**: 183–193.

Rutledge, E. M., Holowaychuk, N., Hall, G. F., *et al.* 1975. Loess in Ohio in relation to several possible source areas. I. Physical and chemical properties. *Soil Sci. Soc. Am. Proc.* **39**: 1125–1132.

Ruzyla, K. and D. I. Jezek. 1987. Staining method for recognition of pore space in thin and polished sections. *J. Sed. Petrol* **57**: 777–778.

Sadleir, S. B. and R. J. Gilkes. 1976. Development of bauxite in relation to parent material near Jarrahdale, Western Australia. *J. Geol. Soc. Austral.* **23**: 333–334.

Salem, M. Z. and F. D. Hole. 1968. Ant (*Formica exsectoides*) pedoturbation in a forest soil. *Soil Sci. Soc. Am. Proc.* **32**: 563–567.

Salik, J., Herrera, R., and C. F. Jordan. 1983. Termitaria: nutrient patchiness in nutrient deficient rain forests. *Biotropica* **15**: 1–7.

Salleh, K. O. 1994. Colluvium thickness and its relationships to vegetation cover density and slope gradient: an observation for part of Murcia Province, SE Spain. *Geog. Fis. Dinam. Quat.* **17**: 187–195.

Salomons, W., Goudie, A., and W. G. Mook. 1978. Isotopic composition of calcrete deposits from Europe, Africa and India. *Earth Surf. Proc. Landforms* **3**: 43–57.

Salter, P. J. and J. B. Williams. 1965. The influence of texture on the moisture characteristics of soils. II. Available-water capacity and moisture release characteristics. *J. Soil Sci.* **16**: 310–317.

Sanborn, P. and S. Pawluk. 1983. Process studies of a chernozemic pedon, Alberta (Canada). *Geoderma* **31**: 205–237.

Sanchez, P. A. 1976. *Properties and Management of Soils in the Tropics*. New York, John Wiley.

Sancho, C. and A. Meléndez. 1992. Génesis y significado ambiental de los caliches Pleistocenos de la región del Cinca (Depresión del Ebro). *Rev. Soc. Geol. España* **5**: 81–93.

Sandor, J. A., Gersber, P. L., and J. W. Hawley. 1986a. Soils at prehistoric terracing sites in New Mexico. I. Site placement, soil morphology, and classification. *Soil Sci. Soc. Am. J.* **50**: 166–173.

1986b. Soils at prehistoric terracing sites in New Mexico. II. Organic matter and bulk density changes. *Soil Sci. Soc. Am. J.* **50**: 173–177.

1986c. Soils at prehistoric terracing sites in New Mexico. III. Phosphorous, selected micronutrients, and pH. *Soil Sci. Soc. Am. J.* **50**: 177–180.

Sandoval, F. M. and L. Shoesmith. 1961. Genetic soil relationships in a saline glacio-lacustrine area. *Soil Sci. Soc. Am. Proc.* **25**: 316–320.

Santos, M. C. D., St. Arnaud, R. J., and D. W. Anderson. 1986. Quantitative evaluation of pedogenic changes in Boralfs (Gray Luvisols) of East Central Saskatchewan. *Soil Sci. Soc. Am. J.* **50**: 1013–1019.

Sanz, M. E. and V. P. Wright. 1994. Modelo alternativo para el desarrollo de calcretas: un ejemplo del Plio-Cuaternario de la Cuenca de Madrid. *Geogaceta* **16**: 116–119.

Sarna-Wojcicki, A. M., Pringle, M. S., and J. Wijbrans. 2000. New Ar-40/Ar-39 age of the Bishop Tuff from multiple sites and sediment rate calibration for the Matuyama-Brunhes boundary. *J. Geophys. Res. Solid Earth* **105**: 21431–21443.

Savigear, R. A. G. 1965. A technique of morphological mapping. *Ann. Ass. Am. Geog.* **55**: 514–538.

Sawhney, B. L. 1986. Electron microprobe analysis. In A. Klute, (ed.) *Methods of Soil Analysis*, Part 1, *Physical and Mineralogical Methods*, Agronomy Monograph no. **9**, 2nd edn. Madison, WI, American Soil Association and Soil Science Society of America, pp. 271–290.

Scalenghe, R., Zanini, E., and D. R. Nielsen. 2000. Modeling soil development in a post-incisive chronosequence. *Soil Sci.* **165**: 455–462.

Schaetzl, R. J. 1986a. A soilscape analysis of contrasting glacial terrains in Wisconsin. *Ann. Ass. Am. Geog.* **76**: 414–425.

1986b. Complete soil profile inversion by tree uprooting. *Phys. Geog.* **7**: 181–189.

1986c. The Sangamon paleosol in Brown County, Kansas. *Kansas Acad. Sci. Trans.* **89**: 152–161.

1990. Effects of treethrow microtopography on the characteristics and genesis of Spodosols, Michigan, USA. *Catena* **17**: 111–126.

1991a. A lithosequence of soils in extremely gravelly, dolomitic parent materials, Bois Blanc Island, Lake Huron. *Geoderma* **48**: 305–320.

1991b. Factors affecting the formation of dark, thick epipdeons beneath forest vegetation, Michigan, USA. *J. Soil Sci.* **42**: 501–512.

1992a. Beta Spodic horizons in podzolic soils (Lithic Haplorthods and Haplohumods). *Pédologie* **42**: 271–287.

1992b. Texture, mineralogy and lamellae development in sandy soils in Michigan. *Soil Sci. Soc. Am. J.* **56**: 1538–1545.

1994. Changes in O horizon mass, thickness and carbon content following fire in northern hardwood stands. *Vegetatio* **115**: 41–50.

1996. Spodosol–Alfisol intergrades: bisequal soils in NE Michigan, USA. *Geoderma* **74**: 23–47.

1998. Lithologic discontinuities in some soils on drumlins: theory, detection, and application. *Soil Sci.* **163**: 570–590.

2000. Shock the world (and then some). *Ann. Ass. Am. Geog.* **90**: 772–774.

2001. Morphologic evidence of lamellae forming directly from thin, clayey bedding planes in a dune. *Geoderma* **99**: 51–63.

2002. A Spodosol–Entisol transition in northern Michigan: climate or vegetation? *Soil Sci. Soc. Am. J.* **66**: 1272–1284.

Schaetzl, R. J. and L. R. Follmer. 1990. Longevity of treethrow microtopography: implications for mass wasting. *Geomorphology* **3**: 113–123.

Schaetzl, R. J. and S. A. Isard. 1990. Comparing "warm season" and "snowmelt" pedogenesis in spodosols. In J. M. Kimble and R. D. Yeck (eds.) *Characterization, Classification, and Utilization of Spodosols, Proc. 5th Int. Soil Correlation Mtg.* Lincoln, NE, US Department of Agriculture Soil Conservation Service, pp. 303–318.

1991. The distribution of Spodosol soils in southern Michigan: a climatic interpretation. Ann. Ass. *Am. Geog.* **81**: 425–442.

1996. Regional-scale relationships between climate and strength of podzolization in the Great Lakes region, North America. *Catena* **28**: 47–69.

Schaetzl, R. J. and W. C. Johnson. 1983. Pollen and spore stratigraphy of a Mollic Hapludalf (Degraded Chernozem) in northeastern Kansas. *Prof. Geog.* **35**: 183–191.

Schaetzl, R. J. and D. L. Mokma. 1998. A numerical index of Podzol and Podzolic soil development. *Phys. Geog.* **9**: 232–246.

Schaetzl, R. J. and C. J. Sorenson. 1987. The concept of "buried" vs. "isolated" paleosols: examples from northeastern Kansas. *Soil Sci.* **143**: 426–435.

Schaetzl, R. J. and D. M. Tomczak. 2002. Wintertime soil temperatures in the fine-textured soils of the Saginaw Valley, Michigan. *Great Lakes Geog.* **8**: 87–99.

Schaetzl, R. J., Barrett, L. R., and J. A. Winkler. 1994. Choosing models for soil chronofunctions and fitting them to data. *Eur. J. Soil Sci.* **45**: 219–232.

Schaetzl, R. J., Burns, S. F., Small, T. W., *et al.* 1990. Tree uprooting: review of types and patterns of soil disturbance. *Phys. Geog.* **11**: 277–291.

Schaetzl, R. J., Drzyzga, S. A., Weisenborn, B. N., *et al.* 2002. Measurement correlation, and mapping of Glacial Lake Algonquin shorelines in northern Michigan. *Ann. Ass. Am. Geog.* **92**: 399–415.

Schaetzl, R. J., Frederick, W. E., and L. Tornes. 1996. Secondary carbonates in three fine and fine-loamy Alfisols in Michigan. *Soil Sci. Soc. Am. J.* **60**: 1862–1870.

Schaetzl, R. J., Johnson, D. L., Burns, S. F., and T. W. Small. 1989. Tree uprooting: review of terminology, process, and environmental implications. *Can. J. For. Res.* **19**: 1–11.

Schaetzl, R. J., Krist, F., Rindfleisch, P., *et al.* 2000. Postglacial landscape evolution of northeastern lower Michigan, interpreted from soils and sediments. *Ann. Ass. Am. Geog.* **90**: 443–466.

Scharpenseel, H. W. 1972a. Messung der natürlichen C-14 konzentration in der organischen substanz von rezenten Böden. Z. Pflanzen. *Bodenkd.* **133**: 241–263.

1972b. Messung der naturlichen C-14 konzentration in der organischen substanz von rezenten boden. *Z. Pflanzen. Bodenkd.* **133**: 924–927.

1972c. Natural radiocarbon measurement of soil and organic matter fractions and on soil profiles of different pedogenesis. *Proc. 8th Intl Conf.* Radiocarbon Dating (Wellington, NZ) p. E1.

Scharpenseel, H. W. and H. Schiffmann. 1977. Radiocarbon dating of soils, a review. *Z. Pflanzen. Bodenkd.* **140**: 159–174.

Schatz, A. 1963. Soil microorganisms and soil chelation: the pedogenic action of lichens and lichen acids. *Agric. Food Chem.* **11**: 112–118.

Scheffer, V. B. 1947. The mystery of the mima mounds. *Sci. Monthly* **65**: 283–294.

Schemenauer, R. S. and P. Cereceda. 1992. The quality of fog water collected for domestic and agricultural use in Chile. *J. Appl. Meteorol.* **31**: 275–290.

Schenck, C. A. 1924. Der Waldbau des Urwalds. *Allg. Forst Jagd-Zeit.* **100**: 377–388.

Schiffman, P. M. 1994. Promotion of exotic weed establishment by endangered giant kangaroo rats (*Dipodomys ingens*) in a California grassland. *Biodivers. Conserv.* **3**: 524–537.

Schlesinger, W. H. 1985. The formation of caliche in soils of the Mojave Desert, California. *Geochim. Cosmochim. Acta* **49**: 57–66.

Schlesinger, D. R. and B. L. Howes. 2000. Organic phosphorus and elemental ratios as indicators of prehistoric human occupation. *J. Archaeol. Sci.* **27**: 479–492.

Schlichting, E. and V. Schweikle. 1980. Interpedon translocations and soil classification. *Soil Sci.* **130**: 200–204.

Schlichting, E. and U. Schwertmann (eds). 1973. *Pseudogley and Gley: Genesis and Use of Hydromorphic Soils. Trans. Comm. V and VI Int. Soc. Soil Sci.*, Weinheim, Germany, Chemie Verlag. 71–80.

Schmidlin, T. W., Peterson, F. F., and R. O. Gifford. 1983. Soil temperature regimes in Nevada. *Soil Sci. Soc. Am. J.* **47**: 977–982.

Schnitzer, M. and J. G. Desjardins. 1969. Chemical characteristics of a natural soil leachate from a Humic Podzol. *Can. J. Soil Sci.* **49**: 151–158.

Schnitzer, M. and H. Kodama. 1976. The dissolution of micas by fulvic acid. *Geoderma* **15**: 381–391.

Schoeneberger, P. J. and D. A. Wysocki. 2001. *A Geomorphic Description System* (version 3.0). Lincoln, NE, National Soil Survey Center. US Department of Agriculture.

Schoeneberger, P. J., Wysocki, D. A., Benham, E. C., *et al.* 1998. *Field Book for Describing and Sampling Soils.* Lincoln, NE, US Department of Agriculture, National Soil Survey Center.

Scholten, J. J. and W. Andriesse. 1986. Morphology, genesis and classification of three soils over limestone, Jamaica. *Geoderma* **39**: 1–40.

Scholtes, W. H., Ruhe, R. V., and F. F. Riecken. 1951. Use of the morphology of buried soil profiles in the Pleistocene of Iowa. *Proc. Iowa Acad. Sci.* **58**: 295–306.

Schulmann, O. P. and A. V. Tiunov. 1999. Leaf litter fragmentation by the earthworm *Lumbricus terrestris* L. *Pedobiologia* **43**: 453–458.

Schulze, D. G. 1984. The influence of aluminum on iron oxides. VI. Unit cell dimensions of Al substituted goethites and estimation of Al from them. *Soil Sci. Soc. Am. J.* **43**: 793–799.

Schumm, S. A. 1967. Rates of surficial rock creep on hillslopes in western Colorado. *Science* **155**: 560–562.

1979. Geomorphic thresholds: the concept and its applications. *Trans. Inst. Brit. Geographers* NS **4**: 485–515.

Schumm, S. A. and R. S. Parker. 1973. Implications of complex response of drainage systems for Quaternary alluvial stratigraphy. *Nature* **243**: 99–100.

Schwarcz, H. and M. Gascoyne. 1984. Uranium-series dating of Quaternary deposits. In W. C. Mahaney (ed.) *Quaternary Dating Methods.* New York, Elsevier, pp. 33–51.

Schwartz, D. 1996. Archéologie préhistorique et processus de formation des stone-lines en Afrique central (Congo-Brazzaville et zones peripheriques). *Geo.-Eco.-Trop.* **20**: 15–38.

Schwartz, D., Guillet, B., Villemin, G., *et al.* 1986. Les alios humiques des Podzols tropicaux du Congo:

constituants, micro- et ultra-structure. *Pédologie* **36**: 179–198.

Schwenk, S. 1977. Krusten und verkrustungen in Sudtunesien. *Stuttgart. Geog. Stud.* **91**: 83–103.

Schwertmann, U. 1984. The influence of aluminium oxides on rion oxides. IX. Dissolution of Al-goethites in 6M HCl. *Clay Mins.* **19**: 9–19.

Schwertmann, U. and D. S. Fanning. 1976. Iron-manganese concretions in hydrosequences of soils in loess in Bavaria. *Soil Sci. Soc. Am. J.* **40**: 731–738.

Schwertmann, U. and R. M. Taylor. 1989. Iron oxides. In J. B. Dixon and S. B. Weed (eds.) *Minerals in Soil Environments*, 2nd edn. Madison, WI, *Soil Science Society of America*, pp. 379–438.

Schwertmann, U., Murad, E., and D. G. Schulze. 1982. Is there Holocene reddening (hematite formation) in soils of axeric temperate areas? *Geoderma* **27**: 209–223.

Scott, A. D. and S. J. Smith. 1968. Mechanism for soil potassium release by drying. *Soil Sci. Soc. Am. Proc.* **32**: 443–444.

Scott, W. E. 1977. Quaternary glaciation and volcanism, Metolius River area, Oregon *Geol. Soc. Am. Bull.* **88**: 113–124.

Scott, W. E., McCoy, W. D., Shroba, R. R., *et al.* 1983. Reinterpretation of the exposed record of the last two cycles of Lake Bonneville, western United States. *Quat. Res.* **20**: 261–285.

Scull, P., Franklin, J., Chadwick, O. A., *et al.* 2003. Predictive soil mapping: a review. *Progr. Phys. Geog.* **27**: 171–197.

Scurfield, G., Segnit, E. R., and C. A. Anderson. 1974. Silica in woody stems. *Austral. J. Bot.* **22**: 211–229.

Seelig, B. D. and A. R. Gulsvig. 1988. *Soil Survey of Kidder County, North Dakota*. Washington, DC, US Department of Agriculture Soil Conservation Service, US Government Printing Office.

Seelig, B. D., Richardson, J. L., and C. J. Heidt. 1990. Sodic soil variability and classification in coarse-loamy till of central North Dakota. *Soil Surv. Horiz.* **31**: 33–43.

Sehgal, J. H. and J. C. Bhattacharjee. 1990. Typic Vertisols of India and Iraq: their characteristics and classification. *Pédologie* **38**: 67–95.

Sehgal, J. L. and G. Stoops. 1972. Pedogenic calcite accumulation in arid and semi-arid regions of the Indo-Gangetic alluvial plain of Erstwhile Punjab (India): their morphology and origin. *Geoderma* **8**: 59–72.

Seibert, D. R., Weaver, J. B., Bush, R. D., *et al.* 1983. *Soil Survey of Greene and Washington Counties, Pennsylvania*. Washington, DC, US Department of Agriculture Soil Conservation Service, US Government Printing Office.

Seki, T. 1932. Volcanic ash loams of Japan proper (their classification, distribution, and characteristics). *Trans. 2nd Intl Congr. Soil Sci.* (Moscow) **5**: 141–143.

Selby, M. J. 1980. A rock mass strength classification for geomorphic purposes: with tests from Antarctica and New Zealand. *Z. Geomorphol.* **24**: 31–51.

Setz, E. Z. F., Enzweiler, J., Solferini, V. N., *et al.* 1999. Geophagy in the golden-faced saki monkey (*Pithecia pithecia chrysocephala*) in the Central Amazon. *J. Zool.* **247**: 91–103.

Severson, R. C. and H. F. Arneman. 1973. Soil characteristics of the forest-prairie ecotone in northwestern Minnesota. *Soil Sci. Soc. Am. Proc.* **37**: 593–599.

Shackleton, N. J. 1968. Depth of pelagic Foraminifera and isotope changes in Pleistocene oceans. *Nature* **218**: 79–80.

1977. The oxygen isotope record of the Late Pleistocene. *Phil. Trans. Roy. Soc. B* **280**: 169–182.

Shackleton, N. J. and N. D. Opdyke. 1973. Oxygen isotope and palaeomagnetic stratigraphy of equatorial Pacific core V28-238: oxygen isotope temperatures and ice volumes on a 10^5-year and 10^6-year scale. *Quat. Res.* **3**: 39–55.

1976. Oxygen-isotope and paleomagnetic stratigraphy of Pacific core V28-239, Late Pliocene to Latest Pleistocene. *Mem. Geol. Soc. Am.* **145**: 449–464.

Shafer, W. M. 1979. Variability of minesoils and natural soils in southeastern Montana. *Soil Sci. Soc. Am. J.* **43**: 1207–1212.

Shaler, N. S. 1890. The origin and nature of soils. *US Geol. Surv. Ann. Rept* **12(1)**: 213–345.

Sharma, B. D., Mukhopadhyay, S. S., and P. S. Sidhu. 1998. Microtopographic controls on soil formation in the Punjab region, India. *Geoderma* **81**: 357–367.

Sharon, D. 1962. On the nature of hamadas in Israel. *Z. Geomorphol.* **6**: 129–147.

Sharp, R. P. 1942. Periglacial involutions in north-eastern Illinois. *J. Geol.* **50**: 113–133.

1949. Pleistocene ventifacts east of the Big Horn Mountains, Wyoming. *J. Geol.* **57**: 175–195.

1969. Semi-quantitative differentiation of glacial moraines near Convict Lake, Sierra Nevada, California. *J. Geol.* **77**: 68–91.

Sharp, R. P. and J. H. Birman. 1963. Additions to the classical sequence of Pleistocene glaciations, Sierra Nevada, California. *Geol. Soc. Am. Bull.* **74**: 1079–1086.

Sharpe, C. F. S. 1938. *Landslides and Related Phenomena.* New York, Columbia University Press.

Sharpe, D. M., Cromack, K., Jr., Johnson, W. C., *et al.* 1980. A regional approach to litter dynamics in southern Appalachian forests. *Can. J. For. Res.* **10**: 395–404.

Sharratt, B. S., Baker, D. G., Wall, D. B., *et al.* 1992. Snow depth required for near steady-state soil temperatures. *Agric. For. Meteorology.* **57**: 243–251.

Sharratt, B. S., Saxton, K. E., and J. K. Radke. 1995. Freezing and thawing of agricultural soils: implications for soil, water and air quality. *J. Minnesota Acad. Sci.* **59**: 1–5.

Shaw, C. F. 1927. Report of committee on soil terminology. *Am. Soil Surv. Ass. Bull.* **8**: 66–98.

Shaw, J. 1994. Magnetic dating of Chinese loess. *Quat. Newslett.* **73**: 60–61.

Sherman, G. D. 1952. The genesis and morphology of the alumina-rich laterite clays. In: *Problems of Clay and Laterite Genesis. Am. Inst. Mining and Metall. Engineers Sympos.*, St. Louis. pp. 154–161.

Sherman, G. D. and L. T. Alexander. 1959. Characteristics and genesis of Low Humic Latosols. *Soil Sci. Soc. Am. Proc.* **23**: 168–170.

Sherman, G. D., Cady, J. G., Ikawa, H., *et al.* 1967. Genesis of the bauxitic Haliii soils. *Hawaii Agric. Exp. Stn Tech. Bull.* **56**.

Shields, J. A., Paul, E. A., St. Arnaud, R. J., *et al.* 1968. Spectrophotometric measurement of soil color and its relationship to moisture and organic matter. *Can. J. Soil Sci.* **48**: 271–280.

Shimek, B. 1909. Aftonian sands and gravels in western Iowa. *Geol. Soc. Am. Bull.* **20**: 339–408.

Shipp, R. F. and R. P. Matelski. 1965. Bulk-density and coarse-fragment determinations on some Pennsylvania soils. *Soil Sci.* **99**: 392–397.

Shiraiwa, T. and T. Watanabe. 1991. Late Quaternary glacial fluctuations in the Langtang Valley, Nepal Himalaya, reconstructed by relative dating methods. *Arctic Alpine Res.* **23**: 404–416.

Shoji, S. and T. Ono. 1978. Physical and chemical properties and clay mineralogy of Andosols from Kitakami, Japan. *Soil Sci.* **126**: 297–312.

Shoji, S., Ito, T., Saigusa, M., *et al.* 1985. Properties of nonallophanic Andosols from Japan. *Soil Sci.* **140**: 264–277.

Shoji, S., Nanzyo, M., and R. A. Dahlgren. 1993. *Volcanic Ash Soils: Genesis, Properties, and Utilization.* Amsterdam, Elsevier.

Shoji, S., Takahashi, T., Saigusa, M., *et al.* 1988. Properties of Spodosols and Andisols showing climosequential and biosequential relations in southern Hokkoda, northeastern Japan. *Soil Sci.* **145**: 135–150.

Short, J. R., Fanning, D. S., Foss, J. E., *et al.* 1986a. Soils of the Mall in Washington, DC. II. Genesis, classification and mapping. *Soil Sci. Soc. Am. J.* **50**: 705–710.

Short, J. R., Fanning, D. S., McIntosh, M. S., *et al.* 1986b. Soils of the Mall in Washington, DC. I. Statistical summary of properties. *Soil Sci. Soc. Am. J.* **50**: 699–705.

Short, N. M. 1961. Geochemical variations in four residual soils. *J. Geol.* **69**: 534–571.

Shrader, W. D. 1950. Differences in the clay contents of surface soils developed under prairie as compared to forest vegetation in the central United States. *Soil Sci. Soc. Am. Proc.* **15**: 333–337.

Shroder, J. F., Jr. 1980. Dendrogeomorphology: review and new techniques in tree-ring dating. *Progr. Phys. Geog.* **4**: 161–188.

1978. Dendrogeomorphological analysis of mass movement on Table Cliffs Plateau, Utah. *Quat. Res.* **9**: 168–185.

Shubayeva, V. I. and L. O. Karpachevskiy. 1983. Soil-windfall complexes and pedogenesis in the Siberian stone pine forests of the maritime territory. *Sov. Soil Sci.* 50–57.

Shum, M. and L. M. Lavkulich. 1999. Use of sample color to estimate oxidized Fe content in mine waste rock. *Env. Geol.* **37**: 281–289.

Siever, R. 1962. Silica solubility, 0–200 C, and the diagenesis of siliceous sediments. *J. Geol.* **70**: 127–150.

Simonson, G. H. and L. Boersma. 1972. Soil morphology and water table relations II. Correlation between annual water table fluctuations and profile features. *Soil Sci. Soc. Am. Proc.* **36**: 649–653.

Simonson, R. W. 1941. Studies of buried soils formed from till in Iowa. *Soil Sci. Soc. Am. Proc.* **6**: 373–381.

1952. Lessons from the first half century of soil survey I. Classification of soils. *Soil Sci.* **74**: 249–257.

1954a. Identification and interpretation of buried soils. *Am. J. Sci.* **252**: 705–732.

1954b. The regur soils of India and their utilization. *Soil Sci. Soc. Am. Proc.* **18**: 199–203.

1959. Outline of a generalized theory of soil genesis. *Soil Sci. Soc. Am. Proc.* **23**: 152–156.

1978. A multiple-process model of soil genesis. In W. C. Mahaney (ed). *Quaternary Soils*. Norwich, UK, Geo Abstracts, pp. 1–25.

1979. Origin of the name "Ando Soils." *Geoderma* **22**: 333–335.

1986. Historical aspects of soil survey and soil classification. I. 1899–1910. *Soil Surv. Horiz.* **27**: 3–11.

1995. Airborne dust and its significance to soils. *Geoderma* **65**: 1–43.

Simonson, R. W. and D. R. Gardner. 1960. Concept and function of the pedon. *Trans. 8th Intl Congr. Soil Sci.* **4**: 129–131.

Simonson, R. W. and C. E. Hutton. 1942. Distribution curves for loess. *Am. J. Sci.* **252**: 99–105.

Simonson, R. W. and S. Rieger. 1967. Soils of the Andept suborder in Alaska. *Soil Sci. Soc. Am. Proc.* **31**: 692–699.

Simonson, R. W., Riecken, F. F., and G. D. Smith. 1952. Great soil groups in Iowa. In: *Understanding Iowa Soils*. Dubuque, IA, Wm. C. Brown Co., pp. 23–25.

Sinai, G., Zaslavsky, D., and P. Golany. 1981. The effect of soil surface curvature on moisture and yield: Beer Sheba observations. *Soil Sci.* **132**: 367–375.

Singer, A. 1966. The mineralogy of the clay fractions from basaltic soils in the Galilee (Israel). *J. Soil Sci.* **17**: 136–147.

1980. The paleoclimatic interpretation of clay minerals in soils and weathering profiles. *Earth Sci. Rev.* **15**: 303–326.

Singer, M. J. and P. Fine. 1989. Pedogenic factors affecting magnetic susceptibility in northern California soils. *Soil Sci. Soc. Am. J.* **53**: 1119–1127.

Singer, A. and J. Navrot. 1977. Clay formation from basic volcanic rocks in a humid Mediterranean climate. *Soil Sci. Soc. Am. J.* **41**: 645–650.

Singer, M. J., Fine, P., Verosub, K. L., *et al.* 1992. Time dependence of magnetic susceptibility of soil chronosequences on the California coast. *Quat. Res.* **37**: 323–332.

Singer, M., Ugolini, F. C., and J. Zachara. 1978. *In situ* study of podzolization on tephra and bedrock. *Soil Sci. Soc. Am. J.* **42**: 105–111.

Singh, L. P., Parkash, B., and A. K. Singhvi. 1998. Evolution of the Lower Gangetic Plain landforms and soils in West Bengal, India. Catena **33**: 75–104.

Singhvi, A. K. and M. R. Krbetschek. 1996. Luminescence dating: a review and a perspective for arid zone sediments. *Ann. Arid Zone* **35**: 249–279.

Singleton, G. A. and L. M. Lavkulich. 1987. A soil chronosequence on beach sands, Vancouver Island, British Columbia. *Can. J. Soil Sci.* **67**: 795–810.

Sivarajasingham, S., Alexander, L. T., Cady, J. G., *et al.* 1962. Laterite. *Adv. Agron.* **14**: 1–60.

Sjöberg, A. 1976. Phosphate analysis of Anthropic soils. *J. Field Archaeol.* **3**: 447–454.

Slate, J. L., Bull, W. B., Ku, T.-L., *et al.* 1991. Soil-carbonate genesis in the Pinnacate Volcanic Field, northwestern Sonora, Mexico. *Quat. Res.* **35**: 400–416.

Small, T. W. 1973. Morphological properties of driftless area soils relative to slope aspect and position. *Prof. Geog.* **24**: 321–326.

1997. The Goodlett–Denny mound: a glimpse at 45 years of Pennsylvania treethrow mound evolution with implications for mass wasting. *Geomorphology* **18**: 305–313.

Small, T. W., Schaetzl, R. J., and J. M. Brixie. 1990. Redistribution and mixing of soil gravels by tree uprooting. *Prof. Geog.* **42**: 445–457.

Smalley, I. J. and J. E. Davin. 1982. *Fragipan Horizons in Soils: A Bibliographic Study and Review of Some of the Hard Layers in Loess and Other Material.* Bibliographical Report no. 30. Lower Hutt, NZ, New Zealand Soil Bureau.

Smart, P. and N. K. Tovey. 1982. *Electron Microscopy of Soils and Sediments: Examples*. New York, Oxford University Press.

Smeck, N. E. 1973. Phosphorous: an indicator of pedogenetic weathering processes. *Soil Sci.* **115**: 199–206.

Smeck, N. E. and E. J. Ciolkosz (eds.) 1989. *Fragipans: Their Occurrence, Classification, and Genesis*. Madison, WI, *Soil Science Society of America.*

Smeck, N. E. and L. P. Wilding. 1980. Quantitative evaluation of pedon formation in calcareous glacial deposits in Ohio. *Geoderma* **24**: 1–16.

Smeck, N. E., Ritchie, A., Wilding, L. P., *et al.* 1981. Clay accumulation in sola of poorly drained soils of western Ohio. *Soil Sci. Soc. Am. J.* **45**: 95–102.

Smeck, N. E., Runge, E. C. A., and E. E. MacKintosh. 1983. Dynamics and genetic modelling of soil systems. In L. P. Wilding *et al.* (eds). *Pedogenesis and Soil Taxonomy.* New York, Elsevier, pp. 51–81.

Smeck, N. E., Thompson, M. L., Norton, L. D., *et al.* 1989. Weathering discontinuities: a key to fragipan formation. *Soil Sci. Soc. Am. Spec. Publ.* **24**: 99–112.

Smeck, N. E., Wilding, L. P., and N. Holowaychuk. 1968. Genesis of argillic horizons in Celina and Morley soils of western Ohio. *Soil Sci. Soc. Am. Proc.* **32**: 550–556.

Smith, B. J. and W. B. Whalley. 1988. A note on the characteristics and possible origins of desert varnishes from southeast Morocco. *Earth Surf. Proc. Landforms* **13**: 251–258.

Smith, B. N. and S. Epstein. 1971. Two categories of $^{13}C/^{12}C$ ratios of higher plants. *Plant Physiol.* **47**: 380–384.

Smith, C. A. S., Ping, C. L., Fox, C. A., *et al.* 1999. Weathering characteristics of some soils formed from White River Tephra, Yukon Territory, Canada. *Can. J. Soil Sci.* **79**: 603–613.

Smith, G. D. 1942. Illinois loess: variations in its properties and distribution. *Univ. Illinois Agric. Exp. Stn Bull.* **490**.

1979. Conversations in taxonomy. *NZ Soil News* **27**: 43–47.

1983. Historical development of Soil Taxonomy: background. In L. P. Wilding, N. E. Smeck and G. F. Hall (eds). *Pedogenesis and Soil Taxonomy.* New York, Elsevier, pp. 23–49.

1986. *The Guy Smith Interviews: Rationale for Concepts in Soil Taxonomy*, Soil Management Support Services Technical Monograph no. 11. Ithaca, NY, Cornell University.

Smith, G. D., Allaway, W. H., and F. F. Riecken. 1950. Prairie soils of the upper Mississippi valley. *Adv. Agron.* **2**: 157–205.

Smith, G. D., Newhall, D. F., and L. H. Robbins. 1964. *Soil-Temperature Regimes, their Characteristics and Predictability*. Washington, DC, US Department of Agriculture, Soil Conservation Service, US Government Printing Offfice.

Smith, H. and L. P. Wilding. 1972. Genesis of argillic horizons in Ochraqualfs derived from fine textured till deposits of northwestern Ohio and southeastern Michigan. *Soil Sci. Soc. Am. Proc.* **36**: 808–815.

Smith, H. T. U. 1962. Periglacial frost features and releate phenomena in the United States. *Biul. Perygl.* **11**: 325–342.

Smith, J. 1956. Some moving soils in Spitzbergen. *J. Soil Sci.* **7**: 10–21.

Smith, M. E. and T. J. Price. 1994. Aztec-period agricultural terraces in Morelos, Mexico: evidence for household-level agricultural intensification. *J. Field Archaeol.* **21**: 169–179.

Smith, R. B. and L. W. Braile. 1994. The Yellowstone hotspot. *J. Vulcanal. Geotherm. Res.* **61**: 121–187.

Smithson, F. 1958. Grass opal in British soils. *J. Soil Sci.* **9**: 148–154.

Snead, R. E. 1972. *Atlas of World Physical Features.* New York, John Wiley.

Sobecki, T. M. and L. P. Wilding. 1982. Calcic horizon distribution and soil classification in selected soils of the Texas Coast Prairie. *Soil Sci. Soc. Am. J.* **46**: 1222–1227.

1983. Formation of calcic and argillic horizons in selected soils of the Texas coast prairie. *Soil Sci. Soc. Am. J.* **47**: 707–715.

Sohet, K., Herbauts, J., and W. Gruber. 1988. Changes caused by Norway spruce in an ochreous brown earth, assessed by the isoquartz method. *J. Soil Sci.* **39**: 549–561.

Soil Classification Working Group. 1998. *The Canadian System of Soil Classification*, 3rd edn. Ottawa, ON, Research Branch, Agriculture and Agri-Food Canada.

Soil Science Society of America. 1997. *Glossary of Soil Science Terms*. Madison, WI, Soil Science Society of America.

Soil Survey Division Staff. 1993. *Soil Survey Manual*, US Department of Agriculture Handbook no. 18. Washington, DC, US Government Printing Office.

Soil Survey Staff. 1951. *Soil Survey Manual*, US Department of Agriculture Handbook no. 18. Washington, DC, US Government Printing Office.

1960. *Soil Classification: A Comprehensive System – 7th Approximation*. Washington, DC, US Government Printing Office.

1975. *Soil Taxonomy*, US Department of Agriculture Handbook no. 436. Washington, DC, US Government Printing Office.

1992. *Keys to Soil Taxonomy*, 5th edn, Soil Management Support Services Technical Monograph no. 19. Blacksburg, VA, Pocahontas Press.

1999. *Soil Taxonomy*, US Department of Agriculture Handbook no. 436. Washington, DC, US Government Printing Office.

Soileau, J. M. and R. J. McCracken. 1967. Free iron coloration in certain well-drained coastal plain soils in relation to their other properties and classification. *Soil Sci. Soc. Am. Proc.* **31**: 248–255.

Sokolov, I. A. and D. E. Konyushkov. 1998. Soils and the soil mantle of the northern circumpolar region. *Euras. Soil Sci.* **31**: 1179–1193.

Soller, D. R. and J. P. Owens. 1991. The use of mineralogic techniques as relative age indicators for weathering profiles on the Atlantic Coastal Plain, USA. *Geoderma* **51**: 111–131.

Sommer, M. and E. Schlichting. 1997. Archetypes of catenas in respect to matter: a concept for structuring and grouping catenas. *Geoderma* **76**: 1–33.

Sommer, M., Halm, D., Weller, U., *et al.* 2000. Lateral podzolization in a granite landscape. *Soil Sci. Soc. Am. J.* **64**: 1434–1442.

Sondheim, M. W. and J. T. Standish. 1983. Numerical analysis of a chronosequence including an assessment of variability. *Can. J. Soil Sci.* **63**: 501–517.

Sondheim, M. W., Singleton, G. A., and L. M. Lavkulich. 1981. Numerical analysis of a chronosequence, including the development of a chronofunction. *Soil Sci. Soc. Am. J.* **45**: 558–563.

Sorenson, C. J. 1977. Reconstructed Holocene bioclimates. *Ann. Ass. Am. Geog.* **67**: 215–222.

Souchier, B. 1971. Evolution des soils sur roches cristallines à l'étage montagnard (Vosges). *Mem. Serv. Carte Geol.* Alsace Lorraine **33**.

Souirji, A. 1991. Classification of aridic soils, past and present. In J. M. Kimble (ed.) *Characterization, Classification and Utilization of Cold Aridisols and Vertisols, Proc. 6th Intl Soil Correlation Mtg.* Lincoln, NE, US Department of Agriculture Soil Conservation Service, National Soil Survey Center, pp. 175–184.

Southard, R. J. 2000. Aridisols. In M. E. Sumner (ed.) *Handbook of Soil Science.* Boca Raton, FL, CRC Press, pp. E-321–E-338.

Southard, R. J., Boettinger, J. L., and O. A. Chadwick. 1990. Identification, genesis, and classification of duripans. In J. M. Kimble and W. D. Nettleton (eds.) *Characterization, Classification, and Utilization of Aridisols, Proc. 4th Intl Soil Correlation Mtg.* Lincoln, NE, US Department of Agriculture Soil Conservation Service, pp. 45–60.

Southard, R. J. and R. C. Graham. 1992. Cesium-137 distribution in a California Pelloxerert: evidence of pedoturbation. *Soil Sci. Soc. Am. J.* **56**: 202–207.

Spain, A. V. and J. G. McIvor. 1988. The nature of herbaceous vegetation associated with termitaria in north-eastern Australia. *J. Ecol.* **76**: 181–191.

Spain, A. V., Okello-Oloya, T., and A. J. Brown. 1983. Abundance, above-ground masses and basal areas of termite mounds at six locations in tropical north-eastern Australia. *Rev. Ecol. Biol. Sol* **20**: 547–566.

Spence, D. H. N. 1957. Studies on the vegetation of Shetland. I. The serpentine debris vegetation in Unst. *J. Ecol.* **45**: 917–945.

Sprecher, S. W. and D. L. Mokma. 1989. Refining the color waiver for Aqualfs and Aquepts. *Soil Surv. Horiz.* **30**: 89–91.

Springer, M. E. 1958. Desert pavement and vesicular layer of some soils of the desert of the Lahontan Basin, Nevada. *Soil Sci. Soc. Am. Proc.* **22**: 63–66.

St. Arnaud, R. J. and E. P. Whiteside. 1964. Morphology and genesis of a chernozemic to podzolic sequence of soil profiles in Saskatchewan. *Can. J. Soil Sci.* **44**: 88–99.

Stace, H. C. T. 1956. Chemical characteristics of terra rossas and rendzinas of South Australia. *J. Soil Sci.* **7**: 280–293.

Staley, J. T., Palmer, F., and J. B. Adams. 1982. Microcolonial fungi: common inhabitants of desert rocks. *Science* **213**: 1093–1094.

Stanley, S. R. and E. J. Ciolkosz. 1981. Classification and genesis of Spodosols in the Central Appalachians. *Soil Sci. Soc. Am. J.* **45**: 912–917.

Stark, N. M. and C. F. Jordan. 1978. Nutrient retention by the root mat of an Amazonian rain forest. *Ecology* **59**: 434–437.

Steele, F., Daniels, R. B., Gamble, E. E., *et al.* 1969. Fragipan horizons and Be masses in the middle coastal plain of north central North Carolina. *Soil Sci. Soc. Am. Proc.* **33**: 752–755.

Stein, J. K. 1983. Earthworm activity: a source of potential disturbance of archaeological sediments. *Am. Antiq.* **48**: 277–289.

Steinhardt, G. C. and D. P. Franzmeier. 1979. Chemical and mineralogical properties of the fragipans of the Cincinnati catena. *Soil Sci. Soc. Am. J.* **43**: 1008–1013.

Steinhardt, G. C., Franzmeier, D. P., and L. D. Norton. 1982. Silica associated with fragipan and non-fragipan horizons. *Soil Sci. Soc. Am. J.* **46**: 656–657.

Steinwand, A. L. and T. E. Fenton. 1995. Landscape evolution and shallow groundwater hydrology of a till landscape in central Iowa. *Soil Sci. Soc. Am. J.* **59**: 1370–1377.

Stepanov, I. N. 1962. Snow cover and soil formation in high mountains. *Sov. Soil Sci.* 270–276.

Stephen, I. 1952. A study of rock weathering with reference to the soils of the Malvern Hills. I. Weathering of biotite and granite. *J. Soil Sci.* **3**: 20–33.

Stephen, I., Bellis, E., and A. Muir. 1956. Gilgai phenonmena in tropical black clays of Kenya. *J. Soil Sci.* 1–9.

Stephens, C. G. 1947. Functional synthesis in pedogenesis. *Trans. Roy. Soc. S. Austral.* **71**: 168–181.

1964. Silcretes of central Australia. *Nature* **203**: 1407.

Steuter, A. A., Jasch, B., Ihnen, J., *et al.* 1990. Woodland/grassland boundary changes in the middle Niobrara Valley of Nebraska identified by ^{13}C values of soil organic matter. *Am. Midl. Nat.* **124**: 301–308.

Stevens, P. R. and T. W. Walker. 1970. The chronosequence concept and soil formation. *Q. Rev. Biol.* **45**: 333:350.

Stevenson, F. J. 1969. Pedohumus accumulation and diagenesis during the Quaternary. *Soil Sci.* **107**: 470–479.

Stewart, V. I., Adams, W. A., and H. H. Abdullah. 1970. Quantitative pedological studies on soils derived from Silurian mudstones. II. The relationship between stone content and the apparent density of the fine earth. *J. Soil Sci.* **21**: 2484–255.

Stockdill, S. M. J. 1982. Effects of introduced earthworms on the productivity of New Zealand pastures. *Pedobiology* **24**: 29–35.

Stolt, M. H. and J. C. Baker. 1994. Strategies for studying saprolite and saprolite genesis. *Soil Sci. Soc. Am. Spec. Publ.* 34: 1–19.

2000. Quantitative comparison of soil and saprolite genesis: examples from the Virginia Blue Ridge and Piedmont. *Southeast. Geol.* **39**: 129–150.

Stolt, M. H., Baker, J. C., and T. W. Simpson. 1991. Micromorphology of the soil–saprolite transition zone in Hapludults of Virginia. *Soil Sci. Soc. Am. J.* **55**: 1067–1075.

1992. Characterization and genesis of saprolite derived from gneissic rocks of Virginia. *Soil Sci. Soc. Am. J.* **56**: 531–539.

1993. Soil–landscape relationships in Virginia. II. Reconstruction analysis and soil genesis. *Soil Sci. Soc. Am. J.* **57**: 422–428.

Stone, E. L. 1975. Windthrow influences on spatial heterogeneity in a forest soil. Eidgen. Anstalt Forstl. *Versuch. Swes. Mitt.* **51**: 77–87.

1977. Abrasion of tree roots by rock during wind stress. *For Sci.* **23**: 333–336.

1993. Soil burrowing and mixing by a crayfish. *Soil Sci. Soc. Am. J.* **57**: 1096–1099.

Stottlemyer, R. 1987. Evaluation of anthropogenic atmospheric inputs on United States national park ecosystems. *Env. Mgt* **11**: 91–97.

Stout, J. D. and K. M. Goh. 1980. The use of radiocarbon to measure the effects of earthworms on soil development. *Radiocarbon* **22**: 892–896.

Strahler, A. H. and A. N. Strahler. 1992. *Modern Physical Geography*. New York, John Wiley.

Strain, M. R. and C. V. Evans. 1994. Map unit development for sand- and gravel-pit soils in New Hampshire. *Soil Sci. Soc. Am. J.* **58**: 147–155.

Strakhov, N. M. 1967. *Principles of Lithogenesis*. Edinburgh, UK, Oliver and Boyd.

Strickertsson, K., Murray, A. S., and H. Lykke-Andersen. 2001. Optically stimulated luminescence dates for Late Pleistocene sediments from Stensnæs, Northern Jutland, Denmark. *Quat. Sci. Rev.* **20**: 755–759.

Stroganova, M. N. and M. G. Agarkova. 1993. Urban soils: experimental study and classification (exemplified by the soils of southwestern Moscow). *Euras. Soil Sci.* **25**: 59–69.

Stuart, D. M. and R. M. Dixon. 1973. Water movement and caliche formation in layered arid and semiarid soils. *Soil Sci. Soc. Am. Proc.* **37**: 323–324.

Stuiver, M. and P. D. Quay. 1980. Changes in atmospheric carbon-14 attributed to a variable sun. *Science* **207**: 11–19.

Stuiver, M., Kromer, B., Becker, B., *et al.* 1986. Radiocarbon age calibration back to 13300 years BP and the ^{14}C age matching of the German oak and US bristlecone pine chronologies. *Radiocarbon* **28**: 969–979.

Stuiver, M., Reimer, P. J., Bard, E., *et al.* 1998. INTCAL98 radiocarbon age calibration, 24000-0 cal BP. *Radiocarbon* **40**: 1041–1083.

Stützer, A. 1999. Podzolisation as a soil forming process in the alpine belt of Rondane, Norway. *Geoderma* **91**: 237–248.

Sudd, J. H. 1969. The excavation of soil by ants. *Z. Tierpsychal.* **26**: 257–276.

Sudom, M. D. and R. J. St. Arnaud. 1971. Use of quartz, zirconium and titanium as indices in pedological studies. *Can. J. Soil Sci.* **51**: 385–396.

Summerfield, M. A. 1982. Distribution, nature and probable genesis of silicrete in arid and semi-arid southern Africa. *Catena (Suppl.)* **1**: 37–65.

1983. Petrology and diagenesis of silicrete from the Kalahari Basin and Cape coastal zone, southern Africa. *J. Sed. Petrol.* **53**: 895–909.

Sunico, A., Bouza, P., and H. DelValle. 1996. Erosion of subsurface horizons in northeastern Patagonia, Argentina. *Arid Soil Res. Rehab.* **10**: 359–378.

Suprychev, V. A. 1963. New calcareous formations in the soil-forming parent material in the Sivash region. *Sov. Soil Sci.* 383–386.

Sutton, J. C. and B. R. Sheppard. 1976. Aggregation of

sand dune soil by endomycorrhizal fungi. *Can. J. Bot.* **54**: 326–333.

Suzuki, Y., Matsubara, T., and M. Hoshino. 2003. Breakdown of mineral grains by earthworms and beetle larvae. *Geoderma* **112**: 131–142.

Swank, W. T., Goebel, N. B., and J. D. Helvey. 1972. Interception loss in loblolly pine stands of the South Carolina piedmont. *J. Soil Water Conserv.* **27**: 160–164.

Swanson, D. K. 1996. Soil geomorphology on bedrock and colluvial terrain with permafrost in central Alaska, USA. *Geoderma* **71**: 157–172.

Swineford, A. and J. C. Frye. 1946. Petrographic comparison of Pliocene and Pleistocene volcanic ash from western Kansas. *Kansas St. Geol. Surv. Bull.* **64**: 1–32.

Switzer, P., Harden, J. W., and R. K. Mark. 1988. A statistical method for estimating rates of soil development and ages of geologic deposits: a design for soil-chronosequence studies. *Math. Geol.* **20**: 49–61.

Symmons, P. M. and C. F. Hemming. 1968. A note on wind-stable stone-mantles in the southern Sahara. *Geog. J.* **134**: 60–64.

Szabolcs, I. 1974. *Salt Affected soils in Europe.* The Hague, Martinus Nijhoff.

Taber, S. 1943. Perennially frozen ground in Alaska: its origin and history. *Geol. Soc. Am. Bull.* **54**: 1433–1458.

Talbot, M. 1953. Ants of an old field community on the Edwin S. George Reserve, Livingston County, Michigan. *Contrib. Lab. Vertebr. Biol. Univ. Michigan* **69**: 1–9.

Tamm, C. O. and H. Holmen. 1967. Some remarks on soil organic matter turn-over in Swedish Podzol profiles. *Meddel. Norske Skogforsoks.* **85**: 69–88.

Tan, K. H. 1965. The Andosols of Indonesia. *Soil Sci.* **99**: 375–378.

(ed.) 1984. *Andosols.* New York, Van Nostrand Reinhold.

Tan, K. H. and J. Van Schuylenborgh. 1961. On the classification and genesis of soils developed over acid volcanic materials under humid tropical conditions. II. *Neth. J. Agric. Sci.* **9**: 41–54.

Tan, K. H., Perkins, H. F., and R. A. McCreeery. 1970. The characteristics, classification and genesis of some tropical Spodosols. *Soil Sci. Soc. Am. Proc.* **34**: 775–779.

Tanaka, A., Sakuma, T., Okagawa, N., *et al.* 1989. *Agro-Ecological Condition of the Oxisol-Ultisol Area of the Amazon River System.* Sapporo, Faculty of Agriculture, Hokkaido University.

Tandarich, J. P. and S. W. Sprecher. 1994. The intellectual background for the Factors of Soil Formation. *Soil Sci. Soc. Am. Spec. Publ.* **33**: 1–13.

Tandarich, J. P., Darmody, R. G., and L. R. Follmer. 1988. The development of pedological thought: some people involved. *Phys. Geog.* **9**: 162–174.

1994. The pedo-weathering profile: a paradigm for whole-regolith pedology from the glaciated midcontinental United States of America. *Soil Sci. Soc. Am. Spec. Publ.* **34**: 97–117.

Tandarich, J. P., Darmody, R. G., Follmer, L. R., *et al.* 2002. The historical development of soil and weathering profile concepts. *Soil Sci. Soc. Am. J.* **66**: 1–14.

Tarnocai, C. 1994. Genesis of permafrost-affected soils. In J. M. Kimble and R. J. Ahrens (eds.) *Classification, Correlation, and Management of Permafrost-Affected Soils.* Lincoln, NE, US Department of Agriculture Soil Conservation Service, pp. 143–154.

Tatur, A. 1989. Ornithogenic soils of the maritime Antarctica. *Pol. Polar Res.* **10**: 481–532.

Tavernier, R. and A. Louis. 1971. La dégradation des soils limoneaux sous monoculture de hêtres de la forêt de Soignes (Belgique). *Ann. Int. St. Cerc. Pedolog.* **38**: 165–191.

Tavernier, R. and G. D. Smith. 1957. The concept of Braunerde (Brown Forest soils) in Europe and the United States. *Adv. Agron.* **9**: 217–289.

Taylor, R. E. 1982. Problems in the radiocarbon dating of bone. In L. A. Currie (ed.) *Nuclear and Chemical Dating Techniques: Interpreting the Environmental Record.* Washington, DC, American Chemistry Society, pp. 453–473.

Taylor, R. E. 1987. *Radiocarbon Dating: An Archaeological Perspective.* New York, Academic Press.

Taylor-George, S., Palmer, F., Staley, J. T., *et al.* 1983. Fungi and bacteria involved in desert varnish formation. *Microb. Ecol.* **9**: 227–245.

Tedrow, J. C. F. 1962. Morphological evidence of frost action in Arctic soils. *Biul. Peryglac.* **11**: 343–352.

1968. Pedogenic gradients of the Polar regions. *J. Soil Sci.* **19**: 197–204.

1977. *Soils of the Polar Landscapes.* New Brunswick, NJ, Rutgers University Press.

Tedrow, J. C. F. and J. Brown. 1962. Soils of the Northern Brooks Range, Alaska: weakening of the soil-forming potential at high Arctic altitudes. *Soil Sci.* **93**: 254–261.

Templin, E. H., Mowery, I. C., and G. W. Kinze. 1956. Houston Black Clay, the type Grumusol. I. Field morphology and geography. *Soil Sci. Soc. Am. Proc.* **20**: 88–90.

Ten Brink, N. W. 1973. Lichen growth rates in West Greenland. *Arctic Alpine Res.* **5**: 323–331.

Thiry, M. and A. R. Milnes. 1991. Pedogenic and groundwater silcretes at Stuart Creek Opal Field, South Australia. *J. Sed. Petrol.* **61**: 111–127.

Thomas, D. J. 1996. *Soil Survey of Macon County, North Carolina*. Washington, DC, US Department of Agriculture Natural Resources Conservation Service, US Government Printing Office.

Thomas, M. F. 1974. *Tropical Geomorphology: A Study of Weathering and Landform Development in Warm Climates*. New York, John Wiley.

Thompson, J. A. and J. C. Bell. 1996. Color index for identifying hydric conditions for seasonally saturated Mollisols in Minnesota. *Soil Sci. Soc. Am. J.* **60**: 1979–1988.

Thompson, R. and B. A. Maher. 1995. Age models, sediment fluxes and palaeoclimatic reconstructions for the Chinese loess and palaeosol sequences. *Geophys. J. Intl* **123**: 611–622.

Thorn, C. E. 1976. A model of stony earth circle development, Schefferville, Quebec. *Proc. Ass. Am. Geog.* **8**: 19–23.

1979. Bedrock freeze–thaw weathering regime in an alpine environment, Colorado Front Range. *Earth Surf. Proc. Landforms* **4**: 211–228.

Thorn, C. E. and R. G. Darmody. 1980. Contemporary eclian sediments in the alpine zone, Colorado Front Range. *Phys. Geog.* **1**: 162–171.

Thorn, C. E., Darmody, R. G., Dixon, J. C., *et al.* 2001. The chemical weathering regime of Karkevagge, arctic–alpine Sweden. *Geomorphology* **41**: 37–52.

Thornthwaite, C. W. 1931. The climates of North America according to a new classification. *Geog. Rev.* **21**: 633–655.

Thorp, J. 1948. How soils develop under grass. In *Yearbook of Agriculture*. Washington, DC, US Department of Agriculture, pp. 55–66.

1949. Effects of certain animals that live in soils. *Sci. Monogr.* **68**: 180–191.

Thorp, J. and G. D. Smith. 1949. Higher categories of soil classification. *Soil Sci.* **67**: 117–126.

Thorp, J., Cady, J. G., and E. E. Gamble. 1959. Genesis of Miami silt loam. *Soil Sci. Soc. Am. Proc.* **23**: 156–161.

Thorp, J., Johnson, W. M., and E. C. Reed. 1951. Some post-Pliocene buried soils of central United States. *J. Soil Sci.* **2**: 1–19.

Tiessen, H. and J. W. B. Stewart. 1983. Particle-size fractions and their use in studies of soil organic matter. II. Cultivation effects on organic matter composition in size fractions. *Soil Sci. Soc. Am. J.* **47**: 509–514.

Tieszen, L. L., Reed, B. C., Bliss, N. B., *et al.* 1997. NDVI, C-3 and C-4 production and distributions in Great Plains grassland land cover classes. *Ecol. Appl.* **7**: 59–78.

Tisdall, J. M., Smith, S. E., and P. Rengasamy. 1997. Aggregation of soil by fungal hyphae. *Austral. J. Soil Res.* **35**: 55–60.

Todhunter, P. E. 2001. A hydroclimatological analysis of the Red River of the north snowmelt flood catastrophe of 1997. *J. Am. Water Resources Ass.* **37**: 1263–1278.

Topp, G. C. and J. L. Davis. 1985. Time-domain reflectometry (TDR) and its application to irrigation scheduling. *Adv. Irrig.* **3**: 107–127.

Topp, G. C., Davis, J. L., and A. P. Anan. 1980. Electromagnetic determination of soil water content: measurement in coaxial transmission lines. *Water Resources Res.* **16**: 574–582.

Torrent, J. and W. D. Nettleton. 1978. Feedback processes in soil genesis. *Geoderma* **20**: 281–287.

Torrent, J., Nettleton, W. D., and G. Borst. 1980a. Genesis of a Typic Durixeralf of Southern California. *Soil Sci. Soc. Am. J.* **44**: 575–582.

Torrent, J., Schwertmann, U., Fechter, H., *et al.* 1983. Quantitative relationships between soil color and hematite content. *Soil Sci.* **136**: 354–358.

Torrent, J., Schwertmann, U., and D. G. Schulze. 1980b. Iron oxide mineralogy of some soils of two river terrace sequences in Spain. *Geoderma* **23**: 191–208.

Toutain, F. and J. C. Vedy. 1975. Influence de la végétation forestière sur l'humification et la pédogenèse en mileau acide et en climat tempéré. *Rev. Ecol. Biol. Sol* **12**: 375–382.

Treadwell-Steitz, C. and L. D. McFadden 2000. Influence of parent material and grain size on carbonate coatings in gravelly soils, Palo Duro Wash, New Mexico. *Geoderma* **94**: 1–22.

Tremocoldi, W. A., Steinhardt, G. C., and D. P. Franzmeier. 1994. Clay mineralogy and chemistry of argillic horizons, fragipans, and paleosol B horizons of soils on a loess-thinning transect in southwestern Indiana, USA. *Geoderma* **63**: 77–93.

Tricart, J. 1957. Observations sur le rôle ameublisseur des termites. *Rev. Geomorphol.* **8**: 170–172, 179.

Trifonova, T. A. 1999. Formation of the soil mantle in mountains: the geosystem aspect. *Euras. Soil Sci.* **32**: 150–156.

Troedsson, T. and W. H. Lyford. 1973. Biological disturbance and small-scale spatial variations in a forested soil near Garpenburg, Sweden. *Stud. For. Suec.* **109**. 23 pp.

Troughton, J. H. 1972. Carbon isotope fractionation by plants. *Proc. 8th Intl Conf. Radiocarbon Dating* (Wellington, NZ) E39–E57.

Tsai, C.-C. and Z.-S. Chen. 2000. Lithologic discontinuities in Ultisols along a toposequence in Taiwan. *Soil Sci.* **165**: 587–596.

Tuhkanen, S. 1986. Delimitation of climatic–phytogeographical regions at the high-latitude area. *Nordia* **20**:105–112.

Twidale, C. R. and A. R. Milnes. 1983. Aspects of the distribution and disintegration of siliceous duricrusts in arid Australia. *Geol. Mijnbouw* **62**: 373–382.

Ugolini, F. C. 1966. Soils of the Mesters Vig district, northeast Greenland. I. The arctic brown and related soils. *Meddel. Grønland* **176**: 1–22.

1986a. Pedogenic zonation in the well-drained soils of the arctic regions. *Quat. Res.* **26**:100–120.

1986b. Processes and rates of weathering in cold and polar desert environments. In S. M. Colman and D. P. Dethier (eds.) *Rates of Chemical Weathering of Rocks and Minerals*. New York, Academic Press, pp. 193–235.

Ugolini, F. C. and C. Bull. 1965. Soil development and glacial events in Antarctica. *Quaternaria* **7**: 251–269.

Ugolini, F. C. and R. Dahlgren. 1987. The mechanism of podzolization as revealed by soil solution studies. In D. Righi and A. Chauvel (eds.) *Podzols et Podzolisation*. Paris, Association Française pour l'Etude du Sol and Institut National de la Recherche Agronomique, pp. 195–203.

Ugolini, F. C. and R. S. Sletten. 1991. The role of proton donors in pedogenesis as revealed by soil solution studies. *Soil Sci.* **151**: 59–75.

Ugolini, F. C., Corti, G., Agnelli, A., *et al.* 1996. Mineralogical, physical, and chemical properties of rock fragments in soil. *Soil Sci.* **161**: 521–542.

Ugolini, F. C., Minden, R., Dawson, H., *et al.* 1977. An example of soil processes in the *Abies amabilis* zone of central Cascades, Washington. *Soil Sci.* **124**: 291–302.

Ugolini, F. C., Stoner, M. G., and D. J. Marrett. 1987. Arctic pedogenesis. I. Evidence for contemporary podzolization. *Soil Sci.* **144**: 90–100.

Ulery, A. L. and R. C. Graham. 1993. Forest fire effects on soil color and texture. *Soil Sci. Soc. Am. J.* **57**: 135–140.

Ulrich, H. P. 1947. Morphology and genesis of the soils of Adak Island, Aleutian Islands. *Soil Sci. Soc. Am. Proc.* **11**: 438–441.

Ulrich, R. 1950. Some chemical changes accompanying profile formation of nearly level soils developed from Peorian loess in southwestern Iowa. *Soil Sci. Am. Proc.* **15**: 324–329.

Unger-Hamilton, R. E. 1984. The formation of use-wear polish on flint: beyond the "deposit versus abrasion" controversy. *J. Archaeol. Sci.* **11**: 91–98.

Vacca, A., Adamo, P., Pigna, M., *et al.* 2003. Genesis of tephra-derived soils from the Roccamonfina Volcana, South Central Italy. *Soil Sci. Soc. Am. J.* **67**: 198–207.

Vadivelu, S. and O. Challa. 1985. Depth of slickenside occurrence in Vertisols. *J. Ind. Soc. Soil Sci.* **33**: 452–454.

Valentine, K. W. G. and J. B. Dalrymple. 1976. Quaternary buried paleosols: a critical review. *Quat. Res.* **6**: 209–222.

Valette-Silver, J. N., Brown, L., Pavich, M., *et al.* 1986. Detection of erosion events using ^{10}Be profiles: example of the impact of agriculture on soil erosion in the Chesapeake Bay area (USA). *Earth Planet. Sci. Lett.* **80**: 82–90.

van Breeman, N. 1982. Genesis, morphology and classification of acid sulfate soils in coastal plains. In J. A. Kittrick, D. S. Fanning, and L. R. Hossner (eds.) *Acid Sulfate Weathering*. Madison, WI, Soil Science Society of America.

1987. Effects of redox processes on soil acidity. *Neth. J. Agric. Sci.* **35**: 271–279.

van Breeman, N. and R. Protz. 1988. Rates of calcium carbonate removal from soils. *Can. J. Soil Sci.* **68**: 449–454.

van Breeman, N., Lundström, U. S., and A. G. Jongmans. 2000. Do plants drive podzolization via rock-eating mycorrhizal fungi? *Geoderma* **94**: 163–171.

van der Merwe, C. R. 1949. A few notes with regard to misconceptions concerning soils of the tropics and sub-tropics. *Commonw. Bureau Soil Sci. Tech. Pub.* **46**: 128–130.

van der Merwe, C. R. and H. W. Weber. 1963. The clay minerals of South African soils developed from granite under different climatic conditions. *S. Afr. J. Agric. Sci.* **6**: 411–454.

van Hees, P. A. W., Lundstrom, U. S., and R. Giesler. 2000. Low molecular weight organic acids and their Al-complexes in soil solution: composition, distribution and seasonal variation in three podzolized soils. *Geoderma* **94**: 173–200.

Van Nest, J. 2002. The good earthworm: how natural processes preserve upland Archaic archaeological sites of western Illinois, USA. *Geoarchaeology.* **17**: 53–90.

Van Reeuwijk, L. P. and J. M. de Villiers. 1985. The origin of textural lamellae in Quaternary coastal sands of Natal. *S. Afr. J. Plant Soil* **2**: 38–44.

van Ryswyk, A. L. and R. Okazaki. 1979. Genesis and classification of modal subalpine and alpine soil pedons of south-central British Columbia, Canada. *Arctic Alpine Res.* **11**: 53–67.

Van Steijn, H., Coutard, J. P., Filippo, H. C., 1988. Simulation expérimentale de laves de ruissellement. *Bull. Assoc. Géog. Franç.* **1**: 33–40.

Van Vliet, B. and R. Langhor. 1981. Correlation between fragipans and permafrost with special reference to silty Weichselian deposits in Belgium and northern France. *Catena* **8**: 137–154.

Van Vliet-Lanoë, B. 1985. Frost effects in soils. In J. Boardman (ed). *Soils and Quaternary Landscape Evolution.* Chichester, UK, John Wiley, pp. 117–158.

1988. The significance of cryoturbation phenomena in environmental reconstruction. *J. Quat. Sci.* **3**: 85–96.

Van Wambeke, A. 1966. Soil bodies and soil classification. *Soils Fert.* **29**: 507–510.

1962. Criteria for classifying tropical soils by age. *J. Soil Sci.* **13**: 124–132.

van Wesemael, B., Poesen, J., and T. de Figueiredo. 1995. Effects of rock fragments on physical degradation of cultivated soils by rainfall. *Soil Till. Res.* **33**: 229–250.

Van Zant, K. 1979. Late glacial and postglacial pollen and plant macrofossils from Lake West Okoboji, Northwestern Iowa. *Quat. Res.* **12**: 358–380.

Vance, G. F., Mokma, D. L., and S. A. Boyd. 1986. Phenolic compounds in soils of hydrosequences and developmental sequences of Spodosols. *Soil Sci. Soc. Am. J.* **50**: 992–996.

Vaniman, D. T., Chipera, S. J., and D. L. Bish. 1994. Pedogenesis of siliceous calcretes at Yucca Mountain, Nevada. *Geoderma* **63**: 1–17.

Vasenev, I. I. and V. O. Targul'yan. 1995. A model for the development of sod-podzolic soils by windthrow. *Euras. Soil Sci.* **27**: 1–16.

Vaughn, D. M. 1997. A major debris flow along the Wasatch front in northern Utah, USA. *Phys. Geog.* **18**: 246–262.

Veenenbos, J. S. and A. M. Ghaith. 1964. Some characteristics of the desert soils of the UAR. *Trans. 8th Intl Congr. Soil Sci.* **110**: 148–173.

Velbel, M. A. 1985. Geochemical mass balances and weathering rates in forested watersheds of the southern Blue Ridge. *Am. J. Sci.* **285**: 904–930.

1990. Mechanisms of saprolitization, isovolumetric weathering, and pseudomorphous replacement during rock weathering: a review. *Chem. Geol.* **84**: 17–18.

Veneman, P. L. M. and S. M. Bodine. 1982. Chemical and morphological characteristics in a New England drainage-toposequence. *Soil Sci. Soc. Am. J.* **46**: 359–363.

Veneman, P. L. M., Jacke, P. V. and S. M. Bodine. 1984. Soil formation as affected by pit and mound microrelief in Massachusetts, USA. *Geoderma* **33**: 89–99.

Veneman, P. L. M., Vepraskas, M. J., and J. Bouma. 1976. The physical significance of soil mottling in a Wisconsin toposequence. *Geoderma* **15**: 103–118.

Vepraskas, M. J. 1999. *Redoximorphic Features for Identifying Aquic Conditions*, N. Carolina Agricultural Research Service Technical Bulletin no. 301. Raleigh, NC, North Carolina State University.

Vepraskas, M. J. and L. P. Wilding. 1983a. Albic neoskeletans in argillic horizons as indices of seasonal saturation and iron reduction. *Soil Sci. Soc. Am. J.* **47**: 1202–1208.

1983b. Aquic moisture regimes in soils with and without low chroma colors. *Soil Sci. Soc. Am. J.* **47**: 280–285.

Verheye, W. 1974. Soils and soil evolution on limestones in the Mediterranean environment. *Trans. 10th Intl Congr. Soil Sci.* (Moscow) **6**: 387–393.

Vermeer, D. E. 1966. Geophagy among the Tiv of Nigeria. *Ann. Ass. Am. Geog.* **56**: 197–204.

Verosub, K. L., Fine, P., Singer, M. L., *et al.* 1993. Pedogenesis and paleoclimate: interpretation of the magnetic susceptibility record of Chinese loess–paleosol sequences. *Geology* **21**: 1011–1014.

Verrecchia, E. P. 1987. Le contexte morpho-dynamique des croûtes calcaires: apport des analysis séquentielles à l'escale microscopique. *Z. Geomorphol.* **31**: 179–193.

1990. Litho-diagenetic implications of the calcium oxalate-carbonate biogeochemical cycle in semi-arid calcretes, Nazareth, Israel. *Geomicrobiol. J.* **8**: 87–99.

1994. L'origine biologique et superficielle des croûtes zonaires. *Bull. Soc. Géol. France* **165**: 583–592.

Vidic, N. J. 1998. Soil-age relationships and correlations: comparison of chronosequences in the Ljubljana Basin, Slovenia and USA. *Catena* **34**: 113–129.

Vidic, N. J. and F. Lobnik. 1997. Rates of soil development of the chronosequence in the Ljubljana Basin, Slovenia. *Geoderma* **76**: 35–64.

Vidic, N. J. and K. L. Verosub. 1999. Magnetic properties of soils of the Ljubljana Basin chronosequence, Slovenia. *Chin. Sci. Bull.* **44**: 75–80.

Vilborg, L. 1955. The uplift of stones by frost. *Geog. Annal.* **37**: 164–169.

Vincent, K. R., Bull, W. B., and O. A. Chadwick. 1994. Construction of a soil chronosequence using the thickness of pedogenic carbonate coatings. *J. Geol. Ed.* **42**: 316–324.

Vincent, K. R. and O. A. Chadwick. 1994. Synthesizing bulk density for soils with abundant rock fragments. *Soil Sci. Soc. Am. J.* **58**: 455–464.

Visser, S. A. and M. Caillier. 1988. Observations on the dispersion and aggregation of clays by humic substances. I. Dispersive effects of humic acids. *Geoderma* **42**: 281–295.

Vogel, J. C. 1969. The radiocarbon time-scale. *S. Afr. Archaeol. Bull.* **24**: 83–87.

Voigt, G. K. 1960. Distribution of rainfall under forest stands. *For. Sci.* **6**: 2–10.

Volobuyev, V. R. 1962. Use of a graphical method for studying the humus composition of the major soil groups of the USSR. *Sov. Soil Sci.* 1–3.

Vreeken, W. J. 1973. Soil variability in small loess watersheds: clay and organic carbon content. *Catena* **1**: 181–196.

1975. Principal kinds of chronosequences and their significance in soil history. *J. Soil Sci.* **26**: 378–394.

1984a. Relative dating of soils and paleosols. In W. C. Mahaney (ed.) *Quaternary Dating Methods*. New York, Elsevier, pp. 269–281.

1984b. Soil landscape chronograms for pedochronological analysis. *Geoderma* **34**: 149–164.

Vucetich, C. G. 1968. Soil–age relationships for New Zealand based on tephro-chronology. *Trans. 9th Intl Congr. Soil Sci.* (Adelaide) **4**: 121–130.

Wada, K. 1985. The distinctive properties of Andosols. *Adv. Soil Sci.* **2**: 173–229.

Wada, K. and S. Aomine. 1973. Soil development on volcanic materials during the Quaternary. *Soil Sci.* **116**: 170–177.

Waegemans, G. and S. Henry. 1954. La couleur des Latosols en relation avec leurs des fer. *Trans. 5th Intl Congr. Soil Sci.* **1/2**: 384–389.

Wagner, D. P., Fanning, D. S., Foss, J. E., *et al.* 1982. Morphological and mineralogical features related to sulfate oxidation under natural and disturbed land surfaces in Maryland. *Soil Sci. Soc. Am. Spec. Publ.* **10**: 109–125.

Wainwright, J., Parsons, A. J., and A. D. Abrahams. 1999. Field and computer simulation experiments on the formation of desert pavement. *Earth Surf. Proc. Landforms* **24**: 1025–1037.

Walch, K. M., Rowley, J. R., and N. J. Norton. 1970. Displacement of pollen grains by earthworms. *Pollen Spores* **12**: 39–44.

Walder, J. and B. Hallet. 1985. A theoretical model of the fracture of a rock during freezing. *Geol. Soc. Am. Bull.* **96**: 336–346.

Walker, M. D., Everett, K. R., Walker, D. A., *et al.* 1996. Soil development as an indicator of relative pingo age, northern Alaska, USA. *Arctic Alpine Res.* **28**: 352–362.

Walker, P. H.. 1966. Postglacial environments in relation to landscape and soils in the cary drift, Iowa. *Iowa. Agric. Exp. St. Res. Bull.* **549**: 838–875.

Walker, P. H. and P. Green. 1976. Soil trends in two valley fill sequences. *Austral. J. Soil Res.* **14**: 291–303.

Walker, P. H. and R. V. Ruhe. 1968. Hillslope models and soil formation. II. Closed systems. Trans. *9th Intl Congr. Soil Sci.* **58**: 561–568.

Walker, P. H., Hall, G. F., and R. Protz. 1968. Relation between landform parameters and soil properties. *Soil Sci. Soc. Am. Proc.* **32**: 101–104.

Walker, R. B., Walker, H. M., and P. R. Ashworth. 1955. Calcium–magnesium nutrition with special reference to serpentine soils. *Plant Physiol.* **30**: 214–221.

Wallace, R. W. and R. L. Handy. 1961. Stone lines on Cary till. *Proc. Iowa Acad. Sci.* **68**: 372–379.

Wallinga, J., Murray, A. S., Duller, G. A. T., *et al.* 2001. Testing optically stimulated luminescence dating of sand-sized quartz and feldspar from fluvial deposits. *Earth Planet. Sci. Lett.* **193**: 617–630.

Waloff, N. and R. E. Blackith. 1962. The growth and distribution of the mounds of *Lasius flavus* (F.) (Hymenoptera: Formicidae) in Silwood Park, Berkshire. *J. Anim. Ecol.* **31**: 421–437.

Walters, J. C. 1994. Ice-wedge casts and relict polygonal ground in north-east Iowa, USA. *Permafrost Periglacial Proc.* **5**: 269–282.

Waltman, W. J., Cunningham, R. L., and E. J. Ciolkosz. 1990. Stratigraphy and parent material relationships of red substratum soils on the Allegheny Plateau. *Soil Sci. Soc. Am. J.* **54**: 1049–1057.

Wang, C. and H. Kodama. 1986. Pedogenic imogolite in sandy Brunisols of eastern Ontario. *Can. J. Soil Sci.* **66**: 135–142.

Wang, C., Beke, G. J., and J. A. McKeague. 1978. Site characteristics, morphology and physical properties of selected ortstein soils from the Maritime Provinces. *Can. J. Soil Sci.* **58**: 405–420.

Wang, C., McKeague, J. A. and H. Kodama. 1986a. Pedogenic imogolite and soil environments: case study of Spodosols in Quebec, Canada. *Soil Sci. Soc. Am. J.* **50**: 711–718.

Wang, C., Nowland, J. L., and H. Kodama. 1974. Properties of two fragipan soils in Nova Scotia including scanning electron micrographs. *Can. J. Soil Sci.* **54**: 159–170.

Wang, C., Ross, G. J., and H. W. Rees. 1981. Characteristics of residual and colluvial soils developed on granite and of the associated pre-Wisconsin landforms in north-central New Brunswick. *Can. J. Earth Sci.* **18**: 487–494.

Wang, C., Stea, R. R., Ross, G. J., *et al.* 1986b. Age estimation of the Shulie Lake and Eatonville tills in Nova Scotia by pedogenic development. *Can. J. Earth Sci.* **23**: 115–119.

Wang, D., McSweeney, K., Lowery, B., *et al.* 1995. Nest structure of ant *Lasius neoniger* Emery and its implications to soil modification. *Geoderma* **66**: 259–272.

Wang, H. and L. R. Follmer. 1998. A polygenetic model for pedostratigraphic units in the Chinese Loess Plateau region. *Quat. Intl* **51/52**: 52.

Wang, Y., Amundson, R., and S. Trumbore. 1996a. Radiocarbon dating of soil organic matter. *Quat. Res.* **45**: 282–288.

Wang, Y., McDonald, E., Amundson, R., *et al.* 1996b. An isotopic study of soils in chronological sequences of alluvial deposits, Providence Mountains, California. *Geol. Soc. Am. Bull.* **108**: 379–391.

Ward, P. A. III and B. J. Carter. 1998. Paleopedologic interpretations of soils buried by Tertiary and Pleistocene-age volcanic ashes: southcentral Kansas, western Oklahoma, and northwestern Texas, USA. *Quat. Intl* **51/52**: 213–221.

Ward, P. A. III, Carter, B. J., and B. Weaver. 1993. Volcanic ashes: time markers in soil parent materials of the southern plains. *Soil Sci. Soc. Am. J.* **57**: 453–460.

Warke, P. A., Smith, B. J., and R. W. Magee. 1996. Thermal response characteristics of stone: implications for weathering of soiled surfaces in urban environments. *Earth Surf. Proc. Landforms* **21**: 295–306.

Wascher, H. L., Humbert, R. P. and J. G. Cady. 1948. Loess in the southern Mississippi valley: identification and distribution of the loess sheets. *Soil Sci. Soc. Am. Proc.* **12**: 389–399.

Wascher, N. E. and M. E. Collins. 1988. Genesis of adjacent morphologically distinct soils in Northwest Florida. *Soil Sci. Soc. Am. J.* **52**: 191–196.

Washburn, A. L. 1956. Classification of patterned ground and review of suggested origins. *Geol. Soc. Am. Bull.* **67**: 823–866.

1980. Permafrost features as evidence of climatic change. *Earth Sci. Rev.* **15**: 327–402.

Watchman, A. 2000. A review of the history of dating rock varnishes. *Earth Sci. Rev.* **49**: 261–277.

Watchman, A. L. and C. R. Twidale. 2002. Relative and "absolute" dating of land surfaces. *Earth Sci. Rev.* **58**: 1–49.

Waters, A. C. and C. W. Flagler. 1929. Origin of small mounds on the Columbia River Plateau. *Am. J. Sci.* **18**: 209–224.

Watson, A. 1979. Gypsum crusts in deserts. *J. Arid Envs.* **2**: 3–20.

1985. Structure, chemistry and origins of gypsum crusts in southern Tunisia and the central Namib Desert. *Sedimentology* **32**: 855–875.

1988. Desert gypsum crusts as palaeoenvironmental indicators: a micropetrographic study of crusts from southern Tunisia and the central Namib Desert. *J. Arid Envs.* **15**: 19–42.

Watson, J. A. L. 1972. An old mound of the spinifex termite, *Nasutitermes triodiae* (Froggatt) (Isoptera: Termitidae). *Austral. J. Entomol. Soc.* **11**: 79–80.

Watson, J. A. L. and F. J. Gay. 1970. The role of grass eating termites in the degradation of a mulga ecosystem. *Search* **1**:43.

Watson, J. P. 1961. Some observations on soil horizons and insect activity in grantite soils. *Proc. Fed. Sci. Congr.* (Salisbury, S. Rhodesia) **1**: 271–276.

1962. The soil below a termite mound. *J. Soil Sci.* **13**: 46–51.

1967. A termite mound in an iron-age burial ground in Rhodesia. *J. Ecol.* **55**: 663–669.

1968. Some contributions of archaeology to pedology in central Africa. *Geoderma* **2**: 291–296.

1969. Water movement in two termite mounds in Rhodesia. *J. Ecol.* **57**: 441–451.

1977. The use of mounds of the termite *Macrotermes falciger* (Gerstacker) as a soil amendment. *J. Soil Sci.* **28**: 664–672.

Watts, N. L. 1980. Quaternary pedogenic calcretes from the Kalahari (southern Africa): mineralogy, genesis and diagenesis. *Sedimentology* **27**: 661–686.

Watts, S. H. 1977. Major element geochemistry of silicrete from a portion of inland Australia. *Geochim. Cosmochim. Acta* **41**: 1164–1167.

Watts, W. A. and H. E. Wright, Jr. 1966. Late-Wisconsin pollen and seed analysis from the Nebraska Sandhills. *Ecology* **47**: 202–210.

Wayne, W. J. 1991. Ice-wedge casts of Wisconsinan age in eastern Nebraska. *Permafrost Periglacial Proc.* **2**: 211–223.

Weaver, R. M., Jackson, M. L., and J. K. Syers. 1971. Magnesium and silicon activities in matrix solutions of montmorillonite-containing soils in relation to clay mineral stability. *Soil Sci. Soc. Am. J.* **35**: 55–86.

Webb, T. III. 1974. A vegetational history from northern Wisconsin: evidence from modern and fossil pollen. *Am. Midl. Nat.* **92**: 12–34.

Webber, P. J. and J. T. Andrews. 1973. Lichenometry: a commentary. *Arctic Alpine Res.* **5**: 295–302.

Webster, R. 1965a. A catena of soils on the northern Rhodesia plateau. *J. Soil Sci.* **16**: 31–43.

1965b. A horizon of pea grit in gravel soils. *Nature* **206**: 696–697.

1968. Fundamental objections to the 7th approximation. *J. Soil Sci.* **19**: 354–365.

Weisenborn, B. N. 2001. A model for fragipan evolution in Michigan soils. M.A. thesis, Michigan State University

Wells, S. G., Dohrenwend, J. C., McFadden, L. D., *et al.* 1985. Late Cenozoic landscape evolution on lava flow surfaces of the Cima volcanic field, Mojave Desert, California. *Geol. Soc. Am. Bull.* **96**: 1518–1529.

Wells, S. G., McFadden, L. D., Poths, J., *et al.* 1995. Cosmogenic ^{3}He surface-exposure dating of stone pavements: implications for landscape evolution in deserts. *Geology* **23**: 613–616.

Wenner, K. A., Holowaychuk, N., and G. M. Schafer. 1961. Changes in clay content, calcium carbonate equivalent, and calcium/magnesium ratio with depth in parent materials of soils derived from calcareous till of Wisconsin age. *Soil Sci. Soc. Am. Proc.* **25**: 312–316.

Werchan, L. E. and J. L. Coker. 1983. *Soil Survey of Williamson County, Texas*. Washington, DC, US Department of Agricultural Soil Conservation Service, US Government Printing Office.

Wescott, W. A. 1993. Geomorphic thresholds and complex response of fluvial systems: some implications for sequence stratigraphy. *Am. Ass. Petrol. Geol. Bull.* **77**: 1208–1218.

West, A. J., Bickle, M. J., Collins, R., *et al.* 2002. Small-catchment perspective on Himalayan weathering fluxes. *Geology* **30**: 355–358.

West, L. T., Wilding, L. P., and C. T. Hallmark. 1988b. Calciustolls in central Texas. II. Genesis of calcic and petrocalcic horizons. *Soil Sci. Soc. Am. J.* **52**: 1731–1740.

West, L. T., Wilding, L. P., Strahnke, C. R., *et al.* 1988a. Calciustolls in central Texas. I. Parent material uniformity and hillslope effects on carbonate-enriched horizons. *Soil Sci. Soc. Am. J.* **52**:1722–1731.

West, W. F. 1970. The Bulawayo Symposium papers. II. Termite prospecting. *Chamb. Mines J.* **12**: 32–35.

Westgate, J. A. and N. D. Briggs. 1980. Dating methods of Pleistocene deposits and their problems. V. Tephrochronology and fission-track dating. *Geosci. Can.* **7**: 3–10.

Westgate, J. A., Easterbrook, D. J., Naeser, N. D., *et al.* 1987. Lake Tapps tephra: an Early Pleistocene stratigraphic marker in the Puget Lowland, Washington. *Quat. Res.* **28**: 340–355.

Westin, F. C. 1953. Solonetz soils of eastern South Dakota: their properties and genesis. *Soil Sci. Soc. Am. J.* **17**: 287–293.

Weyl, R. 1950. Schwermineralverwitterung und ihr Einfluss auf die mineralführung klastischer Sedimente. *Erdöl Kohle* **3**: 209–211.

White, A. F. and S. L. Brantley (eds.) 1995. Chemical weathering rates of silicate minerals. *Rev. Mineral.* **31**: 1–583.

White, A. F., Blum, A. E., Bullen, T. D., *et al.* 1999. The effect of temperature on experimental and natural chemical weathering rates of granitoid rocks. *Geochim. Cosmochim. Acta* **63**: 3277–3291.

White, E. M. 1955. Brunizem-Gray Brown Podzolic soil biosequences. *Soil Sci. Soc. Am. Proc.* **19**: 504–509.

1961. Calcium-solodi or planosol genesis from solodized-solonetz. *Soil Sci.* **91**: 175–177.

1964. The morphological–chemical problem in solodized soils. *Soil Sci.* **98**: 187–191.

1971. Grass cycling of calcium, magnesium, potassium, and sodium in solodization. *Soil Sci. Soc. Am. Proc.* **35**: 309–311.

1978. Medium-textured soils derived from solodized soils. *Proc. S. Dakota Acad. Sci.* **57**: 81–91.

White, E. M. and R. G. Bonestell. 1960. Some gilgaied soils in South Dakota. *Soil Sci. Soc. Am. Proc.* **24**: 305–309.

White, E. M. and R. I. Papendick. 1961. Lithosolic Solodized-Solonetz soils in southwestern South Dakota. *Soil Sci. Soc. Am. Proc.* **25**: 504–506.

White, E. M. and F. F. Riecken. 1955. Brunizem-Gray Brown Podzolic soil biosequences. *Soil Sci. Soc. Am. Proc.* **19**: 504–509.

White, K., Bryant, R., and N. Drake. 1998. Techniques for measuring rock weathering: application to a dated fan segment sequence in southern Tunisia. *Earth Surf. Proc. Landforms* **23**: 1031–1043.

White, S. E. 1976. Is frost action really only hydration shattering? A review. *Arctic Alpine Res.* **8**: 1–6.

Whitehead, D. R. 1972. Development and environmental history of the Dismal Swamp. *Ecol. Monogr.* **42**: 301–315.

Whitehead, E. M. 1925. *Science and the Modern World.* New York, Macmillan.

Whitford, W. G. and F. R. Kay. 1999. Biopedoturbation by mammals in deserts: a review. *J. Arid Envs.* **41**: 203–230.

Whitney, M. I. and J. F. Splettstoesser. 1982. Ventifacts and their formation: Darwin Mountains, Antarctica. *Catena* (Suppl.) **1**: 175–194.

Whittaker, R. H., Buol, S. W., Niering, W. A., *et al.* 1968. A soil and vegetation pattern in the Santa Catalina Mountains, Arizona. *Soil Sci.* **105**: 440–450.

Whittaker, R. H. and W. A. Niering. 1965. Vegetation of the Santa Catalina Mountains, Arizona: a gradient analysis of the south slope. *Ecology* **46**: 429–452.

Whittecar, G. R. 1985. Stratigraphy and soil development in upland alluvium and colluvium north-central Virginia piedmont. *Southeast. Geol.* **26**: 117–129.

Whittig, L. D. and W. R. Allardice. 1986. X-ray diffraction techniques. In A. Klute (ed.) *Methods of Soil Analysis*, Part 1. *Physical and Mineralogical Methods*, Agronomy Monograph no. 9, 2nd edn. American Soil Association and Soil Science Society of America, Madison, WI, pp. 331–362.

Wieder, M. and D. H. Yaalon. 1972. Micromorphology of terra rossa soils in northern Israel. *Israel J. Agric. Res.* **22**: 153–154.

Wieder, M., Yair, A., and A. Arzi. 1985. Catenary soil relationships on arid hillslopes. *Catena* (Suppl.) 6:41–57.

Wiken, E. B., Broersma, K., Lavkulich, L. M., *et al.* 1976. Biosynthetic alteration in a British Columbia soil by ants (*Formica fusca* Linné). *Soil Sci. Soc. Am. J.* **40**: 422–426.

Wilde, S. A. 1946. *Forest Soils and Forest Growth.* Waltham, MA, Chronica Botanica Co.

1950. Crypto-mull humus: its properties and growth effects (a contribution to the classification of forest humus). *Soil Sci. Soc. Am. Proc.* **15**: 360–362.

1966. A new systematic terminology of forest humus layers. *Soil Sci.* **101**: 403–407.

Wilding, L. P. 1967. Radiocarbon dating of biogenic opal. *Science* **156**: 66–67.

1994. Factors of soil formation: contributions to pedology. *Soil Sci. Soc. Am. Spec. Publ.* **33**: 15–30.

Wilding, L. P. and D. Tessier. 1988. Genesis of Vertisols: shrink–swell phenomena. In L. P. Wilding and R. Puentes (eds.) *Vertisols: Their Distribution, Properties, Classification and Management.* College Station, TX, Texas A&M Printing Center, Texas A&M University, pp. 55–81.

Wilding, L. P., Jones, R. B., and G. W. Schafer. 1965. Variation in soil morphological properties within Miami, Celina, and Crosby mapping units in west-central Ohio. *Soil Sci. Soc. Am. Proc.* **29**: 711–717.

Wilding, L. P., Odell, R. T., Fehrenbacher, J. B., *et al.* 1963. Source and distribution of sodium in Solonetzic soils in Illinois. *Soil Sci. Soc. Am. Proc.* **27**: 432–438.

Wilding, L. P., Williams, D., Miller, W., *et al.* 1991. Close interval spatial variability of Vertisols: a case study in Texas. In M. Kimble (ed.) *Characterization, Classification and Utilization of Cold Aridisols and Vertisols, Proc. 6th Intl Soil Correlation Mtg.* Lincoln, NE, US Department of Agricultural Soil Conservation Service, **11**: 232–247.

Wildman, W. E., Jackson, M. L., and L. D. Whittig. 1968. Iron-rich montmorillonite formation in soils derived from serpentinite. *Soil Sci. Soc. Am. Proc.* **32**: 787–794.

Williams, D., Cook, T., Lynn, W., *et al.* 1996. Evaluating the field morphology of Vertisols. *Soil Surv. Horiz.* **37**: 123–131.

Williams, G. E. and H. A. Polach. 1969. The evaluation of ^{14}C ages for soil carbonate from the arid zone. *Earth Planet. Sci. Lett.* **7**: 240–242.

Williams, G. P. and H. P. Guy. 1973. *Erosional and Depositional Aspects of Hurricane Camille in Virginia, 1969*, Professional Paper no. 804. Washington, DC, US Geological Survey.

Williams, M. A. 1974. Surface rock creep on sandstone slopes in the northern territory of Australia. *Austral. Geog.* **12**:419–424.

Williams, M. A. J. 1968. Termites and soil development near Brocks Creek, Northern Territory. *Austral. J. Sci.* **31**:153–154.

Williams, M. E. and E. D. Rudolph. 1974. The role of lichens and associated fungi in the chemical weathering of rock. *Mycologia* **66**: 648–660.

Williams, S. H. and J. R. Zimbelman. 1994. Desert pavement evolution: an example of the role of the sheetflood. *J. Geol.* **102**: 243–248.

Williams, S. K. 1994. Late Cenozoic tephrostratigraphy of deep sediment cores from the Bonneville basin, northwest Utah. *Geol. Soc. Am. Bull.* **105**: 1517–1530.

Willimott, S. G. and D. W. G. Shirlaw. 1960. C horizon of the soil profile. *Nature* **187**:966.

Willis, E. H., Tauber, H., and K. O. Munnich. 1960. Variations in the atmospheric radiocarbon concentration over the past 1300 years. *Radiocarbon* **2**: 1–4.

Willman, H. B. 1979. Comments on the Sangamon soil. In *Wisconsinan, Sangamonian, and Illinoian Stratigraphy in Central Illinois*, Illinois State Geological Survey Guidebook no. 13. Midwest Friends of the Pleistocene Field Conference Guidebook. pp. 92–94. Urbana-Champaign, IL.

Willman, H. B. and J. C. Frye. 1970. Pleistocene stratigraphy of Illinois. *Illinois State Geology Survey Bulletin* **94**.

Willman, H. B., Glass, H. D., and J. C. Frye. 1966. Mineralogy of glacial tills and their weathering profiles in Illinois. II. Weathering profiles. *Illinois State Geological Survey Circular* **400**.

Wilshire, H. G., Nakata, J. K., and B. Hallet. 1981. Field observations of the December 1977 wind storm, San Joaquin Valley, California. *Geol. Soc. Am. Spec. Pap.* **186**: 233–251.

Wilson, K. 1960. The time factor in the development of dune soils at South Haven Peninsula, Dorset. *J. Ecol.* **48**: 341–359.

Wilson, M. J. 1987. X-ray powder diffraction methods. In M. J. Wilson (ed.) *A Handbook of Determinative Methods in Clay Mineralogy.* London, Blackie, pp. 26–98.

Wilson, P. and O. Farrington. 1989. Radiocarbon dating of the Holocene evolution of Magilligan Foreland, Co, Londonberry. *Proc. Roy. Irish Acad. Sci.* **89B**: 1–23.

Winchester, V. 1984. A proposal for a new approach to lichenometry. *Br. Geomorphol. Res. Group Tech. Bull.* **33**: 3–20.

Winkler, E. M. and E. J. Wilhelm. 1970. Salt bursts by hydration pressures in architectural stone in urban atmosphere. *Geol. Soc. Am. Bull.* **81**: 567–572.

Wintle, A. G. 1973. Anomalous fading of thermoluminescence in mineral samples. *Nature* **245**: 143–144.

1993. Luminescence dating of aeolian sands: an overview. In K. Pye (ed.) *The Dynamics and Environmental Context of Aeolian Sedimentary Systems.* London, Geological Society, pp. 49–58.

1996. Archaeologically relevant dating techniques for the next century. *J. Archaeol. Sci.* **23**: 123–138.

Wintle, A. G. and J. A. Catt. 1985. Thermoluminescence dating of soils developed in Late Devensian loess at Pegwell Bay, Kent. *J. Soil Sci.* **36**: 293–298.

Wintle, A. G., Shackleton, N. J., and J. P. Lautridou. 1984. Thermoluminescence dating of periods of loess deposition and soil formation in Normandy. *Nature* **310**: 491–493.

Witty, J. E. and R. W. Arnold. 1970. Some folists on Whiteface Mountain, New York. *Soil Sci. Soc. Am. Proc.* **34**: 653–657.

Witty, J. E. and E. G. Knox. 1964. Grass opal in some chestnut and forested soils in north central Oregon. *Soil Sci. Soc. Am. Proc.* **28**: 685–688.

Woese, C. R. 1987. Bacterials evolution. *Microbiol. Rev.* **51**: 221–271.

Woese, C. R., Kandler, O., and M. L. Wheelis. 1990. Towards a natural system of organisms: proposal for the domains Archaea, Bacteria, Eucary. *Proc. Natl Acad. Sci. USA* **87**: 4576–4579.

Woida, K. and M. L. Thompson. 1993. Polygenesis of a Pleistocene paleosol in southern Iowa. *Geol. Soc. Am. Bull.* **105**: 1445–1461.

Wood, A. 1942. The development of hillside slopes. *Geol. Ass. Proc.* **53**: 128–138.

Wood, T. G. and W. A. Sands. 1978. The role of termites in ecosystems. In M. V. Brian (ed.) *Production Ecology of Ants and Termites*. London, Cambridge University Press. pp. 245–292.

Wood, T. G., Johnson, A. A., and J. M. Anderson 1983. Modification of soil in Nigerian savanna by soil-feeding *Cubitermes* (Isoptera, Termitidae). *Soil Biol. Biochem.* **15**: 575–579.

Wood, W. R. and D. L. Johnson. 1978. A survey of disturbance processes in archaeological site formation. In M. B. Schiffer (ed.) *Advances in Archaeological Method and Theory*, vol. 1. New York, Academic Press, pp. 315–381.

Woods, W. I. 1977. The quantitative analysis of soil phosphate. *Am. Antiq.* **42**: 248–252.

Wright, R. L. 1963. Deep weathering and erosion surfaces in the Daly River basin, Northern Territory. *J. Geol. Soc. Austral.* **10**: 151–163.

Wright, V. P., Platt, N. H., Marriot, S. B., *et al.* 1995. A classification of rhizogenic (root-formed) calcretes, with examples from the Upper Jurassic–Lower Cretaceous of Spain and Upper Cretaceous of southern France. *Sed. Geol.* **100**: 143–158.

Wysocki, D. A., Schoeneberger, P. J., and H. E. LaGarry. 2000. Geomorphology of soil landscapes. In M. E. Sumner (ed.) *Handbook of Soil Science.* Boca Raton, FL, CRC Press, pp. E-5–E-39.

Yaalon, D. H. 1969. Origin of desert loess. *8th Int. Quaternary Ass. Congr. (Paris)* **2**: 755.

1971. Soil-forming intervals in time and space. In D. H. Yaalon (ed.) *Paleopedology.* Jerusalem, Israel University Press, pp. 29–39.

1975. Conceptual models in pedogenesis: can soil-forming functions be solved? *Geoderma* **14**: 189–205.

1983. Climate, time and soil development. In L. P. Wilding, N. E. Smeck, and G. F. Hall (eds.) *Pedogenesis and Soil Taxonomy.* New York, Elsevier, pp. 233–251.

1987. Saharan dust and desert loess: effect on surrounding soils. *J. Afr. Earth Sci.* **6**: 569–571.

1992. On fortuitous results and compensating factors. *Soil Sci.* **154**: 431–434.

1997a. Comments on the source, transport and deposition scenario of Saharan dust to southern Europe. *J. Arid Envs.* **36**: 193–196.

1997b. Soils in the Mediterranean region: what makes them different? *Catena* **28**: 157–169.

Yaalon, D. H. and E. Ganor. 1975. Rates of aeolian dust accretion in the Mediterranean and desert fringe environments of Israel. 9th Intl *Congr. Sedimentol.* (Nice) 169–174.

Yaalon, D. H. and D. Kalmar. 1978. Dynamics of cracking and swelling clay soils: displacement of skeletal grains, optimum depth of slickensides, and rate of intra-pedonic turbation. *Earth Surf. Proc. Landforms* **3**: 31–42.

Yaalon, D. H. and J. Lomas. 1970. Factors controlling the supply and the chemical composition of aerosols in a near-shore and coastal environment. *Agric. Meteorol.* **7**: 443–454.

Yaalon, D. H. and B. Yaron. 1966. Framework for man-made soil changes: an outline of metapedogenesis. *Soil Sci.* **102**: 272–277.

Yair, A. 1987. Environmental effects of loess penetration into the Negev Desert. *J. Arid Envs.* **13**: 9–24.

1990. Runoff generation in a sandy area: The Nizzana sands, western Negev, Israel. *Earth Surf. Proc. Landforms* **15**: 597–609.

Yair, A. and S. M. Berkowitz. 1989. Climatic and non-climatic controls of aridity: the case of the northern Negev of Israel. *Catena* (Suppl.) **14**: 145–158.

Yair, A. and A. Danin. 1980. Spatial variations in vegetation as related to the soil moisture regime over an arid limestone hillside, Northern Negev, Israel. *Oecologia* **47**: 83–88.

Yair, A. and M. Shachak. 1982. A case study of energy, water and soil flow chains in an arid ecosystem. *Oecologia* **54**: 389–397.

Yair, A., Yaalon, D. H., and S. Singer. 1978. Thickness of calcrete (nari) on chalk in relation to relief factors, Shefela, Israel. *10th Intl Congr. Sedimentol.* (Jerusalem) **2**: 754–755.

Yassoglou, N. J. and E. P. Whiteside. 1960. Morphology and genesis of some soils containing fragipans in northern Michigan. *Soil Sci. Soc. Am. Proc.* **24**: 396–407.

Yassoglou, N., Kosmas, C., and N. Moustakas. 1997. The red soils, their origin, properties, use and management in Greece. *Catena* **28**: 261–278.

Yatsu, E. 1988. *The Nature of Weathering An Introduction.* Tokyo, Sozosha.

Yemane, K., Kahr, G., and K. Kelts. 1996. Imprints of post-glacial climates and palaeogeography in the detrital clay mineral assemblages of an Upper Permian fluviolacustrine Gondwana deposit from northern Malawi. *Palaeogeog. Palaeoclimat. Palaeoecol.* **125**: 27–49.

Yerima, B. P. K., Calhoun, F. G., Senkayi, A. L., *et al.* 1985. Occurrence of interstratified kaolinite–smectite in El Salvador Vertisols. *Soil Sci. Soc. Am. J.* **49**:462–466.

Yerima, B. P. K., Wilding, L. P., Hallmark, C. T., *et al.* 1989. Statistical relationships among selected properties of Northern Cameroon Vertisols and associated Alfisols. *Soil Sci. Soc. Am. J.* **53**: 1758–1763.

Yoshinaga, N. and S. Aomine. 1962a. Allophane in some Ando soils. *Soil Sci. Plant Nutr.* **8**: 6–13.

1962b. Imogolite in some Ando soils. *Soil Sci. Plant Nutr.* **8**: 22–29.

You, C.-F., Lee, T., Brown, L., *et al.* 1988. ^{10}Be study of rapid erosion in Taiwan. *Geochim. Cosmochim. Acta* **52**: 2687–2691.

Young, A. 1960. Soil movement by denudational processes on slopes. *Nature* **188**: 120–122.

1969. The accumulation zone on slopes. *Z. Geomorphol.* **13**: 231–233.

Young, J. A. and R. A. Evans. 1986. Erosion and deposition of fine sediments from playas. *J. Arid Envs.* **10**: 103–115.

Yuretich, R., Knapp, E., Irvine, V., *et al.* 1996. Influences upon the rates and mechanisms of chemical weathering and denudation as determined from watershed studies in Massachusetts. *Geol. Soc. Am. Bull.* **108**: 1314–1327.

Zech, W., Hempfling, R., Haumaier, L., *et al.* 1990. Humification in subalpine Rendzinas: chemical analyses, IR and ^{13}NMR spectroscopy and pyrolysis-field ionization mass spectrometry. *Geoderma* **47**: 123–138.

Zelazny, L. W. and G. N. White. 1989. The pyrophyllite-talc group. In J. B. Dixon and S. B. Weed (eds.) *Minerals in Soil Environments*, 2nd edn. Madison, WI, *Soil Science Society of America*, pp. 527–550.

Zhang, H. and S. Schrader. 1993. Earthworm effects on selected physical and chemical properties of soil aggregates. *Biol. Fert. Soils* **15**: 229–234.

Zhang, H., Thompson, M. L., and J. A. Sandor. 1988. Compositional differences in organic matter among cultivated and uncultivated Argiudolls and Hapludalfs derived from loess. *Soil Sci. Soc. Am. J.* **52**: 216–222.

Zhang, X., An, Z., Chen, T., *et al.* 1994. Late Quaternary records of the atmospheric input of eolian dust to the center of the Chinese Loess Plateau. *Quat. Res.* **41**: 35–43.

Zhao, Z. and D. M. Pearsall. 1998. Experiments for improving phytolith extraction from soils. *J. Archaeol. Sci.* **25**: 587–598.

Zhongli, D., Rutter, N., and L. Tungsheng. 1993. Pedostratigraphy of Chinese loess deposits and climatic cycles in the last 2.5 Myr. *Catena* **20**: 73–91.

Zhou, L. P., Oldfield, F., Wintle, A. G., *et al.* 1990. Partly pedogenic origin of magnetic variations in Chinese loess. *Nature* **346**: 737–739.

Zimmerman, D. W. 1971. Thermoluminescence dating using fine grains from pottery. *Archaeometry* **13**: 29–52.

Zi-Tong, G. 1983. Pedogenesis of paddy soil and its significance in soil classification. *Soil Sci.* **135**: 5–10.

Zobeck, T. M. and A. Ritchie, Jr. 1984. Analysis of long-term water table records from a hydrosequence of soils in central Ohio. *Soil Sci. Soc. Am. J.* **48**: 119–125.

Zoltai, S. C. and C. Tarnocai. 1975. Perennially frozen peatlands in the western Arctic and subarctic of Canada. *Can. J. Earth Sci.* **12**: 28–43.

Zonn, S. V. 1995. Use of elementary soil processes in genetic soil identification. *Euras. Soil Sci.* **27**: 12–20.

Zreda, M. G. and F. M. Phillips. 1995. Insights into alpine moraine development from cosmogenic ^{36}Cl buildup dating. *Geomorphology* **14**: 149–156.

Zuzel, J. R. and J. L. Pikul, Jr. 1987. Infiltration into a seasonally frozen agricultural soil. *J. Soil Water Conserv.* **42**: 447–450.

Glossary

A horizon The topmost mineral soil horizon, usually showing signs of organic matter accumulation.

ablation till General term for loose, relatively permeable material, either contained within or accumulated on the surface of a glacier deposited during the downwasting of nearly static glacial ice.

abrasion The physical weathering of a rock surface by running water, glaciers or wind laden with fine particles.

absolute dating See *numerical dating.*

absolute time Geologic time expressed in years before the present.

absorption The physical uptake of water and/or ions by a substance.

accelerated erosion Erosion in excess of natural rates, usually as a result of anthropogenic activities such as cultivation.

accretion gley A type of dark, clay-rich, sticky (when wet) soil, once termed gumbotil, that formed in swales on landscapes, by slopewash. Many are exposed today as clay-rich, gleyed, overthickened paleosols.

acid sulfate soil Soils that contain high amounts of various sulfur compounds.

acid(ic) soil Soil with a pH value <7.0.

acid(ic) rock An igneous rock that contains more than 60% silica and free quartz.

acid-insoluble residue Material left behind after a soil sample has been subjected to harsh acids.

acidic cations Cations that, on being added to water, undergo hydrolysis, resulting in an acidic solution. Hydrated acidic cations donate protons to water to form hydronium ions (H_3O^+) and thus in aqueous solutions are acids. Examples in soils are Al^{3+} and Fe^{3+}.

acidification The formation of acidic conditions in a soil or sediment.

acidity The hydrogen ion activity in the soil solution expressed as a pH value.

actinomycetes A group of organisms intermediate between bacteria and true fungi, mainly resembling the latter. A form of filamentous, branching bacteria.

active factors Of the five state factors of Jenny (climate, organisms, relief, parent material and time), the two that are not passive, i.e., climate and organisms.

active layer The top layer of soil subject to annual thawing and freezing in areas underlain by permafrost. The active layer is frozen in the winter but thawed in summer.

adsorption The process by which atoms, molecules or ions are taken up from the soil solution or soil atmosphere and retained on the surfaces of solids by chemical or physical binding. The attachment of a particle, ion or molecule to a surface.

adsorption complex Any of various organic and inorganic substances in soil that are capable of adsorbing ions and molecules, mainly clay or humus.

aeolian See *eolian.*

aeration The process by which atmospheric air enters the soil. The rate and amount of aeration depends on the size and continuity of the pore spaces and the degree of waterlogging.

aerobic (i) Having molecular oxygen as a part of the environment. (ii) Growing only in the presence of molecular oxygen, such as aerobic organisms. (iii) Occurring only in the presence of molecular oxygen (said of chemical or biochemical processes such as aerobic decomposition).

aerobic organism Organisms living or becoming active in the presence of molecular oxygen.

aeroturbation Soil and sediment mixing by gases, air and wind.

aggradation The building-up of the Earth's surface by deposition; specifically, the accumulation of material by any process in order to establish or maintain uniformity of grade or slope.

aggregate A group of primary (e.g., sand, silt, clay, gravel) soil particles that cohere to each other more strongly than to other surrounding particles and are held together in a single mass or cluster. Discrete clusters of particles formed naturally or artificially are called peds. Types of aggregates include crumbs, granules, clods, fecal pellets, fragments of fecal pellets and concretions. Clods are aggregates produced by tillage or logging.

agric horizon A mineral soil horizon in which clay, silt and humus derived from an overlying cultivated and fertilized layer have accumulated. Wormholes

Portions of this glossary come from or have been adapted from the Soil Science Society of America (1997), Peterson (1981), Hole and Campbell (1985), Thomas (1996) and Cady *et al.* (1986).

and illuvial clay, silt and humus occupy at least 5% of the horizon by volume. The illuvial clay and humus occur as horizontal lamellae or fibers, as coatings on ped surfaces or in wormholes.

agronomy The theory and practice of crop production and soil management.

ahumic soils Those with virtually no organic matter, originally defined for ultraxerous soils in Antarctica.

air dry The state of dryness at equilibrium with the water content in the surrounding atmosphere. The actual water content will depend upon the relative humidity and temperature of the surrounding atmosphere.

alban A cutan that is light-colored in thin section because the material comprising it is either dominated by quartz or because the iron minerals in it have been reduced or translocated from it.

albedo The ratio of the amount of solar radiation reflected from an object to the total amount incident upon it.

albic horizon A mineral soil horizon from which clay and free iron oxides have been removed or in which the oxides have been segregated to the extent that the color of the horizon is determined primarily by the color of the primary sand and silt particles rather than by coatings on these particles.

albic neoskeletans Whitish coatings on ped faces that have been stripped of most of their clay and Fe-bearing minerals, usually due to chemical reduction in an anoxic environment. Also called grainy gray ped coatings, gray ped and channel coatings, and albans.

alcrete A soil horizon cemented by aluminum-rich compounds.

Alfisols A soil order. Alfisols have an umbric or ochric epipedon and an argillic horizon, and hold water at <1.5 MPa tension during at least 90 days when the soil is warm enough for plants to grow. Alfisols have a mean annual soil temperature of <8 °C or a base saturation in the lower part of the argillic horizon of 35% or more when measured at pH 8.2.

alkali soil (i) A soil with a pH of 8.5 or higher or with a exchangeable sodium ratio greater than 0.15. (ii) A soil that contains sufficient sodium to interfere with the growth of most crop plants.

alkaline soil Soil with a pH value >7.0 (or >7.3 by some definitions).

alkalinity The degree or intensity of alkalinity in a soil, expressed by a pH value >7.0.

allochthonous A term which connotes that something, usually a sediment, is derived from someplace else, or is not indigenous to a site or area. For example, the allochthonous parent material of an alluvial soil, or an allochthonous community of organisms that invaded an area (e.g., an "allochthonous flora"). See also *autochthonous*.

allophane An aluminosilicate with primarily short-range structural order. Occurs as exceedingly small spherical particles especially in soils formed from volcanic ash.

alluvial Pertaining to processes or materials associated with transportation or deposition by running water.

alluvial fan Land counterpart of a delta. A constructional landform built of stratified alluvium with or without debris flow deposits that occurs on the pediment slope, downslope from their source. These form as streams carrying sediment, flowing rapidly along a steep gradient, enter a lowland with lower slopes and deposit much of their sediment at the contact area. Typical of arid and semiarid climates but not confined to them.

Alluvial soil (i) A soil developing in, or developed from, fairly young alluvium, often exhibiting essentially no horizon development or modification of the recently deposited materials. (ii) When capitalized the term refers to a great soil group of the azonal order (1938 system of soil classification) consisting of soils with little or no modification of the recent alluvium in which they are forming.

alluvial toeslope Concave-upward landscape position at the lowest part of a hillside, below the footslope position. See also *toeslope*.

alluvium Sediment deposited by running water of streams and rivers. It may occur on terraces well above present streams, on the present flood plains or deltas, or as a fan at the base of a slope.

Alpine Meadow soils A great soil group of the intrazonal order (1938 system of soil classification), comprising of dark soils of grassy meadows at altitudes above the timberline.

alpine zone Areas in mountainous terrain that are devoid of trees, due to high altitude and cold temperatures.

aluminosilicates Minerals containing Al, Si and O as their main constituents.

amorphous material Non-crystalline soil constituents that either do not fit the definition of allophane or are unable to meet allophane criteria. The term is used generally with reference to Andisols and Spodosols. Lacking crystalline structure.

amphibole A family of silicate minerals forming prism or needle-like crystals. Amphibole minerals generally contain Fe, Mg, Ca and Al in varying amounts, along with water. Hornblende is a common dark green to black variety of amphibole.

AMS dating A form of radiocarbon dating that utilizes accelerator mass spectrometry and involves the direct measurement of the ratio of different carbon isotopes. It can be applied to very small samples.

anaerobic (i) Characterized by the absence of molecular oxygen. In soils this is usually caused by excessive wetness. (ii) Growing in the absence of molecular oxygen (such as anaerobic bacteria). (iii) Occurring in the absence of molecular oxygen (as a biochemical process). See also *anoxic*.

anaerobic organism One that lives in an environment without molecular oxygen.

andesite A fine-grained volcanic rock of intermediate composition, consisting largely of plagioclase and one or more mafic minerals.

andic Soil properties related to volcanic origin of materials. The properties include organic carbon content, bulk density, phosphate retention, and iron and aluminum extractable with ammonium oxalate.

Andisols A soil order. Andisols are dominated by andic soil properties.

angle of repose The maximum angle of slope (measured from a horizontal plane) at which loose, cohesionless material will come to rest.

angular blocky structure A form of soil structure, common in clay-rich B horizons, in which the equant peds have angular edges and sharp corners.

anion An ion having a negative charge, e.g., OH^-.

anion exchange capacity The sum of exchangeable anions that a soil can adsorb. Usually expressed as centimoles, or millimoles, of charge per kilogram of soil (or of other adsorbing material such as clay).

anisotropy A condition of a soil profile of being layered (horizonated) and in that sense, regularly heterogenous, in contrast to the (possible) isotropic nature of the original initial material. In the context of place-to-place variability, "anisotropic" refers to the fact that some soil characteristics may vary at different rates as direction varies.

annelid Any red-blooded worm such as an earthworm.

anomalous fading In luminescence dating, the loss of electrons from the lattice traps on a timescale that is short compared to the lifetime predicted on the basis of trap size. It is primarily a problem with feldspar, leading to an age estimate that is too young.

anoxic See *anaerobic*.

anthric saturation Human-induced aquic condition that occurs in soils that are cultivated and irrigated, especially by flood irrigation. Examples of anthric saturation would be rice paddies, cranberry bogs and treatment wetlands.

anthropic epipedon A surface layer of mineral soil that has the same requirements as the mollic epipedon, but has >110 mg P kg^{-1} soluble in 0.05 M citric acid, or is dry >300 days (cumulative) during the period when not irrigated. Anthropic epipedons form under long continued cultivation and fertilization.

anthropogenic erosion Soil erosion that has been induced or accelerated by human actvities such as agriculture or mining.

anthrosols Soils that have been strongly impacted by human agency, such as in cities or on mine spoil.

anthroturbation Soil and sediment mixing by human activities.

apedal Condition of a soil that has no structure, i.e., no peds, but rather is massive or composed of single grains.

aquaturbation Soil and sediment mixing by water, usually on a very small scale.

aquic conditions Continuous or periodic saturation and reduction. The presence of aquic conditions is indicated by redoximorphic features and can be verified by measurement of saturation and reduction.

aquic moisture regime A reducing soil moisture regime, occurring when the soil is virtually free of dissolved oxygen because it is saturated by groundwater or by water of the capillary fringe.

aquiclude A sediment body, rock layer or soil horizon that is incapable of transmitting significant quantities of water under ordinary hydraulic gradients, i.e., it is nearly impermeable.

aquifer A saturated, permeable geologic unit of sediment or rock that can transmit significant quantities of water under hydraulic gradients.

aquitard A slowly permeable body of rock or sediment that retards but does not prevent the flow of water through it. It does not readily yield water to wells or springs but may serve as a storage unit for groundwater.

arenization Physical disintegration of a rock induced by the chemical weathering of some of its weatherable minerals.

argillan A cutan composed dominantly of oriented phyllosilicate clay minerals.

argillic horizon A soil horizon that is characterized by the illuvial accumulation of phyllosilicate clays. The argillic horizon has a certain minimum thickness depending on the thickness of the solum, a minimum quantity of clay in comparison with an overlying eluvial horizon depending on the clay content of the eluvial horizon, and usually has coatings of oriented clay on the surface of pores or peds or bridging sand grains.

argilliturbation Soil and sediment mixing by shrinking and swelling of clays, such as smectite, usually in a wet–dry climate.

aridic A soil moisture regime in which soils have no water available for plants for more than half the cumulative time that the soil temperature at 50 cm depth is >5 °C, and has no period as long as 90 consecutive days when there is water for plants while the soil temperature at 50 cm is continuously >8 °C. Typical of arid regions (deserts).

Aridisols A soil order. Aridisols have an aridic moisture regime, an ochric epipedon, and other pedogenic horizons, but lack an oxic horizon.

arkose A sedimentary rock formed by the cementation of sand-sized grains of feldspar and quartz.

arthropods A group of animals in the animal kingdom, characterized by the presence of a hard, outer skeleton (exoskeleton) and jointed body parts (appendages). Examples include spiders, scorpions, crabs, crustaceans, millipedes, mites, centipedes and insects such as termites and ants.

asepic fabric In soil micromorphology, plasmic fabrics that have dominantly anisotropic plasma with anisotropic domains that are unoriented with regard to each other; they have a flecked extinction pattern and no plasma separations.

ash (volcanic) Unconsolidated, pyroclastic material less than 2 mm in diameter.

aspect The direction toward which a slope faces with respect to the compass or to the rays of the sun.

attapulgite See *palygorskite*.

autochthonous Microorganisms and/or substances indigenous to a given site or ecosystem; the true inhabitants of an ecosystem; referring to the common microbiota of the body of soil microorganisms that tend to remain constant. Also may refer to sediment that is derived from that place, i.e., not from outside that place. See also *allochthonous*.

autotroph Organism that utilizes carbon dioxide as a source of carbon and obtains its energy from the sun or by oxidizing inorganic substances such as S, H, ammonium and nitrate salts.

available elements Elements (nutrients) in the soil solution that can readily be taken up by plant roots.

available water That part of the soil water that can be taken up by plant roots.

available water capacity The amount of water released between *in situ* field capacity and the permanent wilting point (usually estimated by water content at soil matric potential of −15 MPa). The weight percentage of water which a soil can store in a form available to plants. It is equal to the moisture content at field capacity minus that at the wilting point. It is commonly expressed as length units of water per length units of soil.

azonal soils Soils without distinct genetic horizons, in the 1938 system of soil classification.

B horizon A subsoil, mineral horizon that is formed by illuviation of materials or weathering in place. The horizon has pedogenic, not rock, structure.

backslope The hillslope position that forms the steepest, and generally linear, middle portion of the slope. In profile, backslopes are bounded by a convex shoulder above and a concave footslope below.

backswamp A floodplain landform. Extensive, marshy or swampy depressed areas of floodplains between the higher natural levees (near the channel) and valley sides or terraces (far away from the channel).

bacteria Unicellular or multicellular microscopic organisms. They occur everywhere and in very large numbers in favorable habitats such as soil and sour milk where they number many millions per gram.

badland A type of area that is generally devoid of vegetation, is intricately dissected by a fine, drainage network with a high drainage density and has short, steep slopes with narrow interfluves resulting from erosion of soft geologic materials. Most common in arid or semiarid regions.

basal till Unconsolidated material deposited and compacted beneath a glacier and having a relatively high bulk density. See also *till, ablation till, lodgement till*.

basalt A fine-grained, dark-colored igneous rock forming from lava flows or minor intrusions. It is composed of calcic plagioclase, augite and magnetite; olivine may be present. Extrusive equivalent of gabbro, it comprises oceanic crust.

base cation Any cation common to soils other than H and Al. Common base cations include Ca, Mg, K, Na and Fe.

base cycling The cycling of bases (base cations) between the soil and biosphere, as plants take them up and later release them back to the soil.

base flow Groundwater that enters a stream channel, maintaining stream flow at times when it is not raining.

base level The theoretical limit or lowest level toward which fluvial erosion of the Earth's surface constantly progresses but seldom, if ever, reaches; essentially, the level below which a stream cannot erode its bed. The general or ultimate base level for the land surface is sea level, but temporary base levels for rivers may exist locally.

base saturation percentage The extent to which the adsorption complex of a soil is occupied or saturated with exchangeable cations other than H and Al, i.e., with bases. It is expressed as a percentage of the total cation exchange capacity.

basin A synclinal geologic structure, roughly circular in its outcrop pattern, in which beds dip gently toward the center from all directions.

bauxite A rock composed of aluminum hydroxides and impurities in the form of silica, clay, silt and iron hydroxides. A residual weathering product, exploited as the primary ore for aluminum.

bed load The sediment in a river channel that moves by sliding, rolling or saltating on or very near the streambed; sediment moved mainly by tractive or gravitational forces or both, but at velocities less than the surrounding flow.

bedding plane Surface separating layers of sedimentary rocks and deposits. Each bedding plane marks the termination of one deposit and the beginning of another of different character, such as a surface separating a sandstone bed from an overlying mudstone bed. Rock tends to break or separate along bedding planes.

bedrock A general term for the solid rock that underlies the soil and other unconsolidated material, or that is exposed at the surface.

bentonite A relatively soft rock formed by chemical alteration of glassy, high silica content volcanic ash. Bentonite shows extensive swelling in water and has a high specific surface area. The principal mineral constituent is clay-size smectite.

beta horizon, beta B horizon The second (lower) horizon of the same general type in a soil, e.g., a Bt horizon that is below and disjunct from the main Bt horizon of the upper solum.

biochemical processes Processes, impacts and effects that are *chemically* imparted, by biota, to the landscape. Compare to *biomechanical processes*.

bioclast Stone or rock that has a partial biological origin. Often refers to stones coughed up or passed through the guts of birds.

biocycling Translocation of minerals and elements from the soil to plants and back again.

biofabric A type of soil fabric that owes it existence to soil fauna and/or flora.

biofunction See *biosequence*.

biological availability That portion of a chemical compound or element that can be taken up readily by living organisms.

biomantle A layer of material that has been brought to the surface and intimately mixed by biota, usually soil fauna such as ants, termites and worms. Biomantles typically have few or almost no coarse fragments.

biomass The total mass of living organisms in a given volume or mass of soil, or in a particular location.

biomechanical processes Processes, impacts and effects that are *physically* imparted, by biota, to the landscape. Compare to *biochemical processes*.

biopore A soil pore formed by biota, such as a worm burrow or a plant root.

biorelict Inherited biological feature (such as a mollusc shell or chitonic remnant of soil animal) in the mineral soil that is stable under the present soil conditions.

biosequence A group of related soils that differ, one from the other, primarily because of differences in kinds and numbers of plants and soil organisms as a soil-forming factor. When expressed as a mathematical equation, it is referred to as a biofunction.

biotite A common rock-forming mineral consisting primarily of ferromagnesian silicate minerals. A brown, trioctahedral layer silicate of the mica group with Fe^{2+} and Mg in the octahedral layer and Si and Al in a ratio of 3 : 1 in the tetrahedral layer. Its color ranges from dark brown to green in thin section. Biotite is commonly referred to as "black mica" because of the natural black color.

bioturbation The mixing of soils and sediments by organisms.

biscuit tops The name given to the rounded tops of columnar peds in soils that are typically high in sodium. The tops of the columns are often coated with a residual material that is whitish-colored and clay-poor.

bisequal soils Soils in which two sequa have formed, one above the other, in the same deposit.

bleicherde A light-colored, leached E horizon in Podzolic soils.

blocky soil structure A type of soil structure where the peds take on a block-like shape: many sided with angular or rounded corners.

blowout A hollow or depression of the land surface, which is generally saucer or trough-shaped, formed by wind erosion especially in an area of shifting sand or loose soil, or where vegetation is disturbed or destroyed.

boehmite The most common crystalline form of alumina monohydrate, AlO(OH). A constituent of bauxite, boehmite is a common Al-rich oxide clay in humid tropical soils.

bog A peat-accumulating wetland that has no significant inflows or outflows and supports acidophilic mosses, particularly *Sphagnum*.

Bog soil A great soil group of the intrazonal order and hydromorphic suborder (1938 system of soil classification). Includes muck and peat.

bolson A basin of interior drainage, common to areas with fault block mountains.

bottomland The normal floodplain of a stream, subject to flooding.

boulder A rock or mineral fragment >600 mm in diameter.

bouldery Containing appreciable quantities of boulders.

Bowen's reaction series A series of minerals formed during crystallization of a magma, in which the formation of minerals alters the composition of the remaining magma. Mafic minerals comprise a discontinuous series, in which successive minerals form at the expense of early-formed ones. The plagioclase feldspars form in a continuous series, in which the composition of plagioclase becomes progressively sodium rich, but the crystal structure of the mineral does not change.

braided stream A channel or stream with multiple channels that interweave as a result of repeated bifurcation and convergence of flow around interchannel bars, resembling (in plan view) the strands of a complex braid. Braiding is generally confined to broad, shallow streams of low sinuosity, high bedload, non-cohesive bank material, and steep gradients.

breccia A clastic rock in which angular, gravel-sized particles make up an appreciable volume of the rock.

Brown Forest soils A great soil group of the intrazonal order and calcimorphic suborder (1938 system of soil classification), formed on calcium-rich parent materials under deciduous forest, and possessing a high base status but lacking a pronounced illuvial horizon.

Brown Podzolic soils A zonal great soil group (1938 system of soil classification) similar to Podzols but lacking the distinct E horizon that is characteristic of the Podzol group.

Brown soils A great soil group (1938 system of soil classification) of the temperate to cool, arid regions, composed of soils with a brown surface and a light-colored transitional subsurface horizon over calcium carbonate accumulation.

brunification Pedogenic process bundle involving the release of iron from primary minerals, followed by the dispersion of particles of iron oxide in increasing amounts. Their progressive oxidation or hydration gives the soil mass brownish, reddish brown and red colors, respectively, often producing cambic Bw horizons.

Brunizem A synonym for Prairie soils (1938 system of soil classification).

buffer A substance that prevents a rapid change in pH when acids or alkalis are added to the soil, including clay, humus and carbonates.

bulk density Mass per unit volume of undisturbed soil, dried to constant weight at 105 °C. Usually expressed as g cm^{-3}.

buried soil, buried paleosol Soil covered by an alluvial, loessal or other surface mantle of more recent depositional material, usually to a depth greater than 50 cm. Some consider a soil "buried" even if the burying deposit is thin, as long as it is identifiable. Buried soils are referred to as geosols or paleosols.

bypass flow See *macropore flow.*

C horizon The presumed parent material of a soil. Although many C horizons exhibit some alteration from their original state, the concept implies lack of alteration by surficial processes.

C3 pathway The most common pathway of carbon fixation in plants. Most humid climate plants such as trees, most shrubs and herbs and many grasses use the Calvin (C3) cycle to fix carbon dioxide.

C4 (Hatch–Slack) pathway An alternative carbon fixation pathway. C4 plants are mostly found among tropical grasses and some sedges and herbs growing in warm, sunny environments. The C4 pathway allows for more efficient carbon fixation in dry, warm environments.

calcan A light-colored cutan composed of carbonates.

calcareous soil Soil containing sufficient free $CaCO_3$ and other carbonates to effervesce visibly or audibly when treated with weak HCl. These soils usually contain from 10 to almost 1000 g kg^{-1} $CaCO_3$ equivalent.

calcic horizon A mineral soil horizon of secondary carbonate enrichment that is >15 cm thick, has a $CaCO_3$ equivalent of >150 g kg^{-1}, and has at least 50 g kg^{-1} more calcium carbonate equivalent than the underlying C horizon.

calcification The pedogenic process of accumulation of calcium in a soil horizon, such as the calcic horizon of some Aridisols and Mollisols.

calcite Crystalline calcium carbonate, $CaCO_3$.

calcium carbonate equivalent The content of carbonate in a liming material or calcareous soil, calculated as if all of the carbonate is in the form of $CaCO_3$.

calcrete See *caliche.*

caliche A zone near the surface, more or less cemented by secondary carbonates of Ca or Mg precipitated from the soil solution. It may occur as a soft thin soil horizon, as a hard thick bed, or as a surface layer exposed by erosion. Also known as calcrete. Usually forms as illuvial carbonates are deposited in a soil horizon.

CAM (crassulacean acid metabolism) pathway The least common of the three carbon fixation pathways in plants. CAM plants, typical of desert-like conditions, include mainly plants in the Crassulaceae and Cactaceae families.

cambic horizon A non-sandy, mineral soil horizon that has soil structure rather than rock structure, contains some weatherable minerals and is characterized by the alteration or removal of mineral material as indicated by mottling or gray colors, stronger chromas or redder hues than in underlying horizons, or the removal of carbonates. Cambic horizons lack cementation or induration and have too few evidences of illuviation to meet the requirements of argillic or spodic horizons.

canopy interception loss Water that falls onto the canopy as precipitation, is intercepted and evaporates, never reaching the soil surface.

capillarity The process by which moisture moves in any direction through the fine pore spaces and as films around particles.

capillary fringe A zone in the soil just above the water table that remains saturated or almost saturated with water, due to "wicking" of water from below, upward within soil pores. Capillary fringe thickness depends upon the size distribution of pores.

capillary water The water held in the "capillary" or small pores of a soil, usually with a tension >60 cm of water. It is held by adhesion and surface tension as films around particles and in the finer pore spaces. Surface tension is the adhesive force that holds capillary water in the soil.

carbon to nitrogen ratio (C:N ratio) Ratio representing the quantity of carbon (C) in relation to the quantity of nitrogen (N) in a soil or organic material.

carbonate rock, carbonaceous rock A rock consisting primarily of a carbonate mineral such as calcite or dolomite, the chief minerals in limestone and dolostone, respectively.

carbonation A form of chemical weathering usually involving carbonic acid (H_2CO_3).

carpedolith, carpetolith, carpedolite See *stone line.*

cat clay A type of poorly drained, clayey soil, commonly formed in an estuarine environment, that becomes very acidic when drained, due to oxidation of ferrous sulfide.

catena A sequence of soils along a slope, having different characteristics due to variation in relief, elevation and drainage (depth to water table), as well as the influence of slope processes on sediment removal and delivery. Bushnell (1942) conceived of the catena as a topohydrosequence of soils developed in a single parent material, such as a glacial till.

cation An ion having a positive electrical charge. The common soil cations are Ca, K, Mg, Na, Al, Fe and H.

cation exchange The exchange between cations in solution and cations held on the negatively charged exchange sites of minerals and organic matter.

cation exchange capacity (CEC) The potential of soils for adsorbing cations, expressed in millimoles of charge per kg (mmolc kg^{-1}) of soil; the sum of exchangeable bases plus total soil acidity at a specific pH values, usually 7.0 or 8.0. In essence, soils with high CEC values have large amount of negative charges per unit mass of soil. Determined by the amount of organic matter, the proportion of clay to sand and the mineralogy of the clay fraction. See also *effective cation exchange capacity (ECEC).*

cation ratio dating Used to date desert varnish on rock surfaces, it is, in theory, the ratio of soluble (Ca and K) to insoluble Ti cations in the varnish. This ratio should decrease with time because the soluble cations are replaced or depleted relative to less mobile cations.

cellulose Carbon-rich component of plants, not easily digested by microorganisms.

cemented Having a hard, brittle consistency because the particles are held together by cementing substances such as humus, $CaCO_3$, silica or the oxides of silicon, iron and aluminum. The hardness and brittleness persist even when wet. See also *consistence.*

Cenozoic era The current geologic era, which began 66.4 million years ago and continues to the present.

chalk Soft white limestone composed of very pure calcium carbonate, which leaves little residue when treated with hydrochloric acid, sometimes consisting largely of the remains of foraminifera, echinoderms, molluscs and other marine organisms.

chamber In soil micromorphology, a relatively large circular or ovoid pore with smooth walls and an outlet through channels, fissures or planar pores. Vesicles or vughs connected by a channel or channels.

channel In soil micromorphology, a tubular-shaped pore or void.

channel neoferrans Coatings of (usually oxidized) Fe on channel walls.

channer In Scotland and Ireland, gravel; in the USA, thin, flat rock fragments up to 150 mm on the long axis, e.g., fragments of shale or limestone.

channery Having large amounts of channers.

chelate-complex theory A theory of podzolization in which the mobility of Fe and Al cations within the soil is ascribed to complexation by organic molecules (chelators), especially fulvic and low-molecular-weight organic acids.

chelates Organic chemicals with two or more functional groups that can bind with metals to form a ring structure. Soil organic matter can form chelate structures with some metals, especially transition metals, but much metal ion binding in soil organic matter probably does not involve chelation.

chelation The condition of being chelated.

chemical weathering The chemical breakdown of rocks and minerals due to the presence of water and other components in the soil solution, or changes in redox potential; for example, the transformation of orthoclase to kaolinite. Also known as decomposition.

Chernozem A zonal great soil group (1938 system of soil classification) consisting of soils with a thick, nearly black or black, organic-matter-rich A horizon high in exchangeable calcium, underlain by a lighter-colored transitional horizon above a zone of calcium carbonate accumulation; occurs in a cool subhumid climate under a vegetation of tail and midgrass prairie. Many Chernozems are equivalent to Ustolls or Udic Ustolls in Soil Taxonomy.

chert A cryptocrystalline form of quartz, microscopically granular. Occurs as nodules and as thin, continuous layers. Duller, less waxy luster than chalcedony. Occurs in limestone, dolostone and mudstones.

Chestnut soil A zonal great soil group (1938 system of soil classification) consisting of soils with a moderately thick, dark-brown A horizon over a lighter-colored horizon that is above a zone of calcium carbonate accumulation. Many Chestnut soils are equivalent to Ustolls or Aridic Ustolls in Soil Taxonomy.

chlorite A group of 2:1 layer silicate minerals that has the interlayer filled with a positively charged, metal-hydroxide octahedral sheet. There are both trioctahedral (e.g., $M = Fe^{2+}$, Mg^{2+}, Mn^{2+}, Ni^{2+}) and dioctahedral ($M = Al^{3+}$, Fe^{3+}, Cr^{3+}) chlorites.

chroma The relative purity, strength or saturation of a color; directly related to the dominance of the determining wavelength of the light and inversely related to grayness. One of the three variables of color.

chronofunction See *chronosequence.*

chronosequence A group of related soils that differ, one from the other, primarily as a result of differences in time as a soil-forming factor. When expressed as a mathematical equation, it is referred to as a chronofunction.

chronostratigraphic unit A sequence of rocks deposited during a particular interval of geological time.

clast A large fragment, such as a rock or pebble, that is significantly larger than the surrounding material.

clastic materials Pertaining to rock or sediment composed mainly of fragments derived from pre-existing rocks or minerals, i.e., not organic.

clay (i) A soil separate consisting of particles <2 μm in equivalent diameter. (ii) A soil textural class. (iii) In reference to clay mineralogy, a naturally occurring material composed primarily of fine-grained minerals, which is generally plastic at appropriate water contents and will harden when dried or fired. Although clay usually contains phyllosilicates, it may contain minerals with other structures.

clay coating See *argillan.*

clay domain A small stack of clay mineral platelets.

clay film See *argillan.*

clay flow A term often used to describe an argillan that appears wavy or distorted, suggesting that it has "flowed" along the ped face.

clay-free basis A mathematical treatment in which one of the other particle size separates in a soil, e.g., sand, is ratioed not to the *entire* fine earth fraction (sand, silt and clay) but on the clay-free fraction only (sand and silt).

clay–humus complex Complex formation between humic acids and clay substances, mechanically inseparable and chemically completely divisible only with great difficulty, whose nature is little known.

clay mineral A phyllosilicate mineral or a mineral that imparts plasticity to clay and which hardens upon drying or firing. Crystalline or amorphous mineral material <2 μm in diameter.

clay mineralogy Term used to describe the mineral structure of clay-sized particles. Also used to describe the general "mix" of clay minerals in a soil or a sample. Phyllosilicate clay minerals have layer structures composed of shared octahedral and tetrahedral sheets.

clay pan A dense, compact, slowly permeable layer in the subsoil, with significantly more clay than the horizon above. It is usually very dense and has a sharp upper boundary. Clay pans generally impede drainage, are usually plastic and sticky when wet and hard when dry.

clay skin See *argillan.*

climax community The most advanced successional community of plants capable of development under, and in dynamic equilibrium with, the prevailing environment.

climofunction See *climosequence.*

climosequence A group of related soils that differ one from another primarily as a result of differences in climate as a soil-forming factor. When expressed as a mathematical equation, it is referred to as a climofunction.

clod A compact, coherent mass of soil varying in size, usually produced by plowing, digging, etc., especially when these operations are performed on soils that are either too wet or too dry and usually formed by compression, or breaking off from a larger unit, as opposed to a building-up action as in aggregation.

close packing The condition in which a soil's matrix is tightly packed, with little large void space.

co-alluvium Material that has some history of movement as alluvial and as colluvium.

coarse fragments Unattached pieces of rock 2 mm in diameter or larger that are strongly cemented or more resistant to rupture.

coarse-textured Soil texture term referring to the dominance of sand and coarse fragments in the soil.

coastal plain A plain of unconsolidated fluvial or marine sediment which had its margin on the shore of a large body of water, particularly the sea, e.g., the coastal plain of the southeastern United States, extending from New Jersey to Texas.

cobbles Rounded or partially rounded rock or mineral fragments between 75 and 250 mm in diameter.

coefficient of linear extensibility (COLE) The percent shrinkage in one dimension of a molded soil between two water contents, e.g., between its plastic limit to air-dry state. The ratio of the difference between the moist and dry lengths of a clod to its dry length, $L_m - L_d/L_d$, when L_m is the moist length at (1/3 atmospheres) and L_d is the air-dry length. The measure correlates with the volume change of a soil upon wetting and drying.

colloid Organic and inorganic material with very fine particle sizes (about 0.1 to 0.001 μm) and therefore high surface areas, which usually exhibits exchange properties.

colluvial Pertaining to material or processes associated with transportation and/or deposition by mass movement (direct gravitational action) and local, unconcentrated runoff on side slopes and/or at the base of slopes.

colluviation The process of downslope, gravity-driven transport of sediment, resulting in a colluvial deposit downslope.

colluvium Unconsolidated, unsorted earth material transported under the influence of gravity, assisted by water, and deposited on lower slopes. Material transported by mass movement (e.g., direct gravitational action) and by local, unconcentrated runoff.

color hue The dominant spectral color. One of the three variables of color.

color value The degree of lightness or darkness of a color in relation to a neutral gray scale. On a neutral gray scale, value extends from pure black to pure white. One of the three variables of color.

columnar soil structure A shape of soil structure wherein the peds resemble long blocks, with their long dimension being the vertical one.

competence The ability of a current of water or wind to transport sediment, in terms of particle *size* rather than amount. Measured as the diameter of the largest particle capable of being transported. It depends upon velocity: a small but swift current, for example, may have greater competence than a larger but slower moving current.

composite map unit A generalized soil map unit composed of several more detailed map units.

compound structure Large peds such as prisms and columns that are themselves composed of smaller, incomplete peds.

concrete frost Ice in the soil in such quantity as to constitute a virtually solid block.

concretion A cemented concentration of a chemical compound, such as calcium carbonate, gypsum or iron oxide, that can be removed from the soil intact and that has crude internal symmetry. Usually spherical or subspherical but may be irregular in shape. It forms by precipitation of mineral matter about a nucleus such as a leaf, or a piece of shell or bone. In soil micromorphology, a concretion is a type of glaebule with a generally concentric fabric about a center which may be a point, line or plane.

conduction Energy transfer directly from atom to atom, representing the flow of energy along a temperature gradient.

conductivity A measure of the soluble salts in the soil.

conglomerate A clastic sedimentary rock composed mainly of rounded boulders of various sizes.

coniferous forest A forest consisting of predominantly cone-bearing trees with needle-shaped leaves; usually such trees are evergreen but some are deciduous, e.g., the larch forests of Siberia.

consistence The attributes of soil material as expressed in degree of cohesion and adhesion or in resistance to deformation or rupture. Terms commonly used to describe consistence are:

cemented Hard; little affected by moistening.

loose Non-coherent when dry or moist; does not hold together in a mass.

friable When moist, crushes easily under gentle pressure between thumb and forefinger and can be pressed together into a lump.

firm When moist, crushes under moderate pressure between thumb and forefinger, but resistance is distinctly noticeable.

hard When dry, moderately resistant to pressure; can be broken with difficulty between thumb and forefinger.

soft When dry, breaks into powder or individual grains under very slight pressure.

sticky When wet, adheres to other material and tends to stretch somewhat and pull apart rather than to pull free from other material.

plastic When wet, readily deformed by moderate pressure but can be pressed into a lump; will form a "wire" when rolled between thumb and forefinger.

consociation On a soil map, a delineation which is dominated by a single soil taxon (or miscellaneous area) and similar soils. Generally, at least half of the pedons in each delineation of a soil consociation are of the same soil components that provide the name for the map unit. Most of the remainder of the delineation consists of soil components so similar to the named soil that major interpretations are not significantly affected.

consolidated A term referring to compacted or cemented rocks.

constructional surface Land surface owing its origin and form to depositional processes, with little or no subsequent modification by erosion.

continuous permafrost Permafrost occurring everywhere beneath the exposed land surface throughout a geographic region.

contrasting soil A soil that does not share diagnostic criteria and does not behave or perform similar to the soil being compared.

control section The part of the soil on which classification is based. The thickness varies among different kinds of soil, but for many it is that part of the soil profile between depths of 25 and 100 or 200 cm.

convection The process by which heat, solutes or particles are transferred from one part of a fluid to another by movement of the fluid itself. In the atmosphere, convective motion is usually upward or downward.

convergent slope A slope in which the flowlines of water converge at the base, e.g., head slope.

coppice mound A small dune that forms around desert, brush-and-clump vegetation. Often partly due to human action that has disturbed nearby soil.

coprogenic material Remains of excreta and similar materials that occur in some organic soils.

corestone The generally rounded remnant of a rock, the remainder of which has weathered to saprolite or other by-products.

corrasion Erosion by friction.

cosmogenic isotope An isotope that can be used to study the age and origin of the Earth.

cove A wide, gently sloping to steep, concave colluvial area. Commonly found at the head of or along drainageways in mountainous areas.

cradle knoll The small mound of debris sloughed from the root plate (ball) of a tipped-over tree. Local soil horizons are commonly obliterated or folded over, which results in a heterogeneous mass of soil material. So named because lumberjacks used to put their babies to sleep in the pits that adjoin the mounds, i.e., cradles.

craton The stable portions of the continents that have escaped orogenic activity for the last 2 billion years. Made predominantly of granite and metamorphic rocks.

craze plane In soil micromorphology, planar voids with a highly complex conformation of the walls due to interconnection of numerous short planes.

creep Slow mass movement of soil and soil material down slopes, driven primarily by gravity but facilitated by saturation with water and by alternate freezing and thawing. Usually not perceptible except through extended observation.

crest Synonym for summit.

crotovina See *krotovina*.

crumb structure A structural condition, typical of organic-rich A horizons, in which most of the peds are crumbs. Antiquated term; crumb structure is, today, referred to as "granular."

crust Surface layer in a soil that becomes harder than the underlying horizon. In geology, the upper part of the lithosphere, divided into oceanic crust and continental crust.

cryic A soil temperature regime where the mean annual soil temperature is >0 °C but <8 °C, and there is a >5 °C difference between mean summer and mean winter soil temperatures at 50 cm, and cold summer temperatures.

cryogenic soil Soil that has formed under the influence of cold soil temperatures.

Cryosol Old term for a class of soils that are dominated by freeze–thaw activity and permafrost. If formalized, it is still used in the Canadian system of soil classification.

cryostatic pressures Pressures created in soils and sediments due to the formation of ice and the expansion that results from it.

cryoturbation Soil and sediment mixing by freeze–thaw activity, as ice crystals grow and melt.

cryptocrystalline In soil micromorphology, term used for structure in which the individual grains are too small to be identified by optical means.

crystal structure The orderly arrangement of atoms in a crystalline material.

crystalline rock A rock consisting of various minerals that have crystallized in place from magma. Usually used to refer to igneous and metamorphic rocks as a whole, so as to differentiate them from sedimentary rock.

crystalturbation Soil and sediment mixing by the growth and wastage of crystals, e.g., ice, salts.

cumulative infiltration Total volume of water infiltrated per unit area of soil surface during a specified time period.

cumulization The slow upward growth of the soil surface due to additions of sediment on top. The additions, e.g., alluvium, loess, slopewash, must occur slowly enough so that pedogensis can incorporate the sediment into the profile's horizons, and thicken the A horizon.

cutan Coating or deposit of material on the surface of peds, stones, etc. A common type is the clay cutan caused by translocation and deposition of clay particles on ped surfaces. In soil micromorphology, a cutan is considered a modification of the texture, structure or fabric at natural surfaces in soils, due to concentration of particular soil constituents or *in situ* modification of the plasma.

D horizon Soil parent material that is below the C horizon and less altered that the C horizon. D horizons are geogenic or non-pedogenic horizons (unaltered zones) of fresh sediment, excluding consolidated bedrock, characterized by original rock or sedimentary fabric, lack of tension joints and lack of alteration features of bio-oxidation origin.

Darcy's law A law in soil physics that describes the rate of flow of water through saturated porous media:

$$Q = K\,S(H + e)/e$$

where Q is the volume of water passed in unit time, S is the area of the bed through which the water is flowing (or is to flow), e is the thickness of the bed, H is the depth of water on top of the bed and "K is a coefficient dependent on the nature of the medium," and for cases "when the pressure under the filter is equal to the weight of the atmosphere."

dealkalinization See *solodization*.

debris flow Fast-moving, turbulent mass movement with a high content of both water and rock debris. Rapid debris flows rival the speed of rock slides.

decalcification The removal of calcium carbonate or calcium ions from the soil by leaching.

debris flux The movement of sediment downslope, usually referring to processes associated with a catena.

decarbonation See *decalcification*.

decay rate The rate at which a population of radioactive atoms decays into stable daughter atoms. Rate often expressed in terms of half-life of the parent isotope.

deciduous forest A forest composed of trees that shed their leaves at some season of the year. In tropical areas the trees lose their leaves during the hot season in order to conserve moisture. Deciduous trees of the cool areas shed their leaves during the autumn to protect themselves against the cold of winter.

decomposition See *chemical weathering*.

deflation The sorting out, lifting and removal of loose, dry, fine-grained soil particles by the action of wind.

deflocculation The inverse of flocculation. The state of soil particles when they act as independent entities and not as clustered groups. When soil solutions are at low ionic strength and dominated by alkali metal cations, especially at higher pH values, soil colloidal particles can be dispersed throughout the solution. See also *dispersion*.

degradation (i) The process whereby a compound is transformed into simpler compounds. (ii) In older literature, refers to the changing of a soil to a more highly leached and a more highly weathered condition; usually accompanied by morphological changes such as development of an E horizon.

degraded B horizons B horizons, usually Bt or Bx horizons, that have lost clay and various cementing agents, usually due to acidification and eluviation. Many degraded B horizons acquire a tongue-like (glossic) appearance as the E horizon engulfs them from above.

Degraded Chernozem A zonal great soil group (1938 system of soil classification) consisting of soils with a very dark brown or black A horizon underlain by a dark gray, weakly expressed E horizon and a brown B horizon. Formed in the forest–prairie transition of cool climates.

delineation An individual polygon shown by a closed boundary on a soil map that defines the area, shape and location of a map unit within a landscape.

delta A body of alluvium, nearly flat and fan-shaped, deposited at or near the mouth of a river where it enters a body of relatively quiet water, usually a sea or lake.

denitrification The biochemical reduction of nitrogen oxides (usually nitrate and nitrite) to molecular (gaseous) nitrogen or nitrogen oxides with a lower oxidation state of nitrogen by bacterial activity (denitrification) or by chemical reactions involving nitrite (chemodenitrification). The biological reduction of nitrogen to ammonia, molecular nitrogen or oxides of nitrogen, resulting in the loss of nitrogen into the atmosphere. Nitrogen oxides are used by bacteria as terminal electron acceptors in place of oxygen in anaerobic or microaerophilic respiratory metabolism.

denudation The sum of the processes that result in the wearing away or the progressive lowering of the Earth's surface by weathering, erosion, mass wasting and transportation.

depth function See *depth plot.*

depth plot A graphical way of presenting soils data. The data are shown on the *x*-axis and depth is plotted (downward) on the *y*-axis.

desert crust A hard surface layer, containing calcium carbonate, gypsum or other binding material, in a desert region; often only a few millimeters in thickness.

desert pavement A natural concentration of wind-polished, closely packed (almost interlocking) pebbles, boulders and other rock fragments, mantling the desert soil surface. It usually protects underlying, finer-grained material. Also called hammada.

desert varnish A thin, dark (brown to black), shiny film or coating of iron and manganese oxides and silica formed on the surfaces of pebbles, boulders, rock fragments and rock outcrops in arid regions. Also called rock varnish.

Desert soil A zonal great soil group (1938 system of soil classification) consisting of soils with a very thin, light-colored surface horizon, which may be vesicular and is ordinarily underlain by calcareous material. Formed in arid regions under sparse shrub vegetation.

desiccation Drying out, becoming dry.

desilication The loss of silica from a soil, primarily in solution. Typical process in hot, humid climates. Also known as desilification.

desorption The migration of adsorbed entities off adsorption sites. The inverse of adsorption.

deterministic uncertainty A perspective based on two fundamental axioms related to soil variability: (1) it can be explained with more and better measurements of the soil system, and/or (2) it may be an irresolvable outcome of complex system dynamics.

detrital grain In soil micromorphology, a mineral grain that was originally present in the parent material.

detrital sediments Sediments made of fragments or mineral grains weathered from pre-existing rocks.

developmental upbuilding See *cumulization.*

diabase An intrusive rock with a crystal structure intermediate between gabbro and basalt.

diachronous Term to describe a single rock unit or surface that exhibits varying ages in different areas.

diagenesis All the physical, chemical and biologic changes undergone by sediments from the time of their initial deposition, through their conversion to solid rock, excluding metamorphism.

diagnostic horizons Combinations of specific soil characteristics that are indicative of certain classes of soils. Those which occur at the soil surface are called epipedons, those below the surface, diagnostic subsurface horizons. Based on criteria outlined in Soil Taxonomy.

diapir An intrusion of material, usually malleable in nature, upward and into a surrounding matrix. Usually applied to rocks but a diapir-like materials are common in many Vertisols.

diatoms Algae that possess a siliceous cell wall which remains preserved after the death of the organisms. They are abundant in both fresh and salt water and in a variety of soils.

differential weathering Weathering that occurs at different rates, as the result of variations in composition and mechanical resistance of rocks, or differences in the intensity of weathering processes.

diffuse double layer A conceptual model of a heterogeneous system that consists of a solid surface (e.g., clay or oxide surface) having a net electrical charge together with an ionic swarm in solution containing ions of opposite charge, neutralizing the surface charge.

diffusion The net movement of a substance (liquid or gas) from an area of higher concentration to one of lower concentration.

dilatation See *unloading*.

dilation The condition in which a soil or material has gained volume over time.

dioctahedral An octahedral sheet or a mineral containing such a sheet that has two-thirds of the octahedral sites filled by trivalent ions such as Al^{3+} or Fe^{3+}.

diorite Intrusive igneous rock made of plagioclase feldspar and amphibole and/or pyroxene. Similar to gabbro only not so dark, and containing less iron and magnesium.

dip The angle that a structural surface such as a bedding plane or fault surface makes with the horizontal, measured perpendicular to the strike and in the vertical plane.

discharge In a stream, the volume of water passing through a channel in a given time.

discharge areas Locations where groundwater moves from beneath the surface to the surface.

disconformity A geologic term for an unconformity in which the beds above the unconformity are parallel to the beds below the unconformity.

discontinuity Any interruption in sedimentation, whatever its cause or length, usually a manifestation of non-deposition and/or accompanying erosion. See also *lithologic discontinuity*.

discontinuous permafrost Permafrost occurring in some areas beneath the exposed land surface throughout a geographic region, while other areas remain free of permafrost.

discordant Cutting across surrounding strata.

disintegration See *mechanical weathering*.

dispersion A term used in relation to solute movement which implies that individual soil particles, e.g., clay, silt, are acting independently and are not flocculated. Also, the process whereby the structure or aggregation of the soil is destroyed so that each particle is separate and behaves as a unit.

dissection Fluvial erosion of a land surface or landform by the cutting of gullies, arroyos, canyons or valleys leaving ridges, hills, mountains or flat-topped remnants separated by drainageways.

dissolution A chemical reaction in which a solid material is dispersed as ions in a liquid. Example: Halite (NaCl) undergoes dissolution when placed in water.

dissolved load The amount of dissolved material that water carries in solution.

dissolved organic carbon (DOC) Organic molecules that are so small that they will not settle out of a water-based suspension.

distributary channels Stream channels that split and fan out from the upstream point of the delta and carry the sediments that build the delta.

divergent slope A slope in which the flowlines of water diverge at the base, e.g., nose slopes.

divide The line of separation, or the summit area, or narrow tract of higher ground that constitutes the watershed boundary between two adjacent drainage basins. It divides the surface waters that flow naturally in one direction from those that flow in another direction. Also called drainage divide or interfluve.

doline, dolina A closed depression in a karst region often rounded or elliptical in shape; forms by the solution and subsidence of the limestone near the surface. Sometimes at the bottom is a sinkhole into which surface water flows and disappears underground.

dolomite, dolostone A carbonate sedimentary rock made up predominately of the mineral dolomite, $CaMg(CO_3)_2$.

drainage basin A general term for a region or area bounded by a drainage divide and occupied by a drainage system. All water entering a drainage basin either infiltrates to the water table, is evapotranspired, or leaves via the river that occupies the basin.

drainage class Under natural conditions, not artificially drained, this term refers to a group of soils defined as having a specific range in relative wetness due to a water table (apparent or perched), under conditions similar to those under which the soil developed. See Table 13.2 for definitions.

drainage divide See *divide*.

drain tile Porous concrete, ceramic, plastic, or other rigid pipe or similar buried structure used to collect and conduct free water, i.e., that water below the water table, from the soil in a field.

drift A generic term for all superficial material that has a glacial origin.

drumlin A low, smooth, elongated oval hill, mound or ridge composed of glacial drift (usually till) that may or may not have a core of bedrock or stratified drift. The longer axis is parallel to the general direction of glacier flow. Drumlins are products of streamline (laminar) flow of glaciers, which molded the subglacial floor through a combination of erosion and deposition. Drumlins usually occur in groups.

duff Another term for the litter that comprises the O horizon.

duff mull A forest humus/litter type, transitional between mull and mor, characterized by an accumulation or organic matter on the soil surface in friable Oe horizons, reflecting the dominant zoogenous decomposers. Duff is similar to mor in that it is generally an accumulation of partially to well-humified organic materials resting on the mineral soil. It is similar to mull in that it is zoologically active. Duff mulls usually have four horizons: Oi(L), Oe(F), Oa(H) and A. Also called moder (mostly in Europe).

dune A low mound, ridge, bank or hill of loose, wind-blown, granular material (generally sand), either bare or covered with vegetation, capable of further movement. Also called sand dune.

duricrust See *duripan.*

durinode A soil nodule cemented or indurated with SiO_2. Durinodes break down in concentrated KOH after treatment with HCl to remove carbonates, but do not break down on treatment with concentrated HCl alone.

duripan A subsurface soil horizon that is cemented by illuvial silica, usually opal or microcrystalline forms of silica. Less than 50% of the volume of air-dry fragments will slake in water or HCl. When exposed at the surface due to erosion it is commonly called a duricrust.

dust devil A small, dust-bearing whirlwind, common in deserts.

dynamic denudation A general model of soil and landscape evolution and archeological site formation. Developed by Donald Johnson, the model emphasizes the dynamic nature of soil and landscape evolution, and integrates surface erosion–denudation, bioturbation, soil creep, throughflow, eluviation–illuviation, leaching, weathering and saprolite production processes with several key theories of pedology and geomorphology into a unified process framework.

dysic Low level of bases in soil material.

E horizon A major horizon dominated by eluviation of clay, Fe, Al and humus, among others, usually occurring above a B horizon. E horizons are usually light-colored because the coatings on the primary minerals have been stripped, revealing the natural, light color of quartz, which often dominates the mineralogy.

earthflow A form of slow, but perceptible, mass movement, with high content of water and rock debris. Lateral boundaries are well defined and the terminus is lobed. With increasing moisture content grades into a mudflow. Slower than a debris flow, but faster than creep or solifluction.

eccentricity of the Earth's orbit A measure of the circularity of the Earth's orbit. It varies in cycles of about 100 000 and 400 000 years.

edaphic Of or pertaining to the soil. Resulting from or influenced by factors inherent in the soil or other substrate, rather than by climatic factors.

effective cation exchange capacity (ECEC) Cation exchange capacity (CEC) is the sum of exchangeable bases plus total soil acidity at a specific pH values. When acidity is expressed as salt extractable acidity, the CEC is then called the effective cation exchange capacity (ECEC) because this is considered to be the CEC of the exchanger at the native pH value. It is usually expressed in centimoles of charge per kilogram of exchanger or millimoles of charge per kilogram of exchanger.

effective precipitation That portion of the total rainfall precipitation which becomes available for plant growth. Also called net precipitation.

electrical conductivity (EC) Conductivity of electricity through water or an extract of soil. Commonly used to estimate the soluble salt content in solution.

electrolyte A molecule that separates into a cation and an anion when it is dissolved in a solvent, usually water, e.g., NaCl separates into Na^+ and Cl^- in water.

electron acceptor A compound that accepts electrons during biotic or abiotic chemical reactions and is thereby reduced.

electron donor A compound that donates or supplies electrons during metabolism and is thereby oxidized.

elementary soil areal A term, popularized in Russia by V. M. Fridland, for the simplest soil cover element. See also *pedon.*

eluvial horizon See *eluvial zone.*

eluvial zone Part of the soil profile dominated by losses of constituents in solution and suspension. The best-developed part of the eluvial zone is the E horizon.

eluviation The removal of soil material from a horizon, in colloidal suspension or in solution.

end moraine A ridge-like accumulation that is being or was produced at the outer margin of an actively flowing glacier at any given time. Commonly composed of till.

endomycorrhiza A mycorrhizal association with intracellular penetration of the host root cortical cells by the fungus as well as outward extension into the surrounding soil.

endosaturation The condition of saturation of a zone or soil horizon by groundwater (not perched water).

Entisols A mineral soil order in Soil Taxonomy that has no distinct subsurface diagnostic horizons within 1 m of the soil surface.

entrain To pick up, as in a river entraining sediment from its bed.

environmental lapse rate The normal change in air temperature with height, in air that is neither rising nor descending. Generally this rate is about 6.4 °C per 1000 m.

eolian Pertaining to earth material transported and deposited by the wind including dune sands, sand sheets, loess and parna. Also aeolian.

eon The primary division of geologic time which are, from oldest to youngest, the Hadean, Archean, Proterozoic and Phanerozoic eons.

ephemeral stream A stream, or reach of a stream, that flows only in direct response to precipitation. It receives no protracted supply from melting snow or other source, and its channel is, at all times, above the water table.

epimorphic processes Processes operating within the Earth, e.g., weathering, leaching, new mineral formation and mineral inheritance.

epipedon Combinations of specific soil characteristics that are indicative of certain classes of soils. Those which occur at the soil surface are called epipedons, those below the surface, diagnostic subsurface horizons.

episaturation The condition in which the soil is saturated with water in one or more layers within 200 cm of the surface and also has one or more unsaturated layers with an upper boundary above 200 cm depth, below the saturated layer(s). Episaturation is usually synonymous with the condition of having a perched water table. See also *endosaturation*.

epoch A division of geologic time next shorter than a period. Example: the Pleistocene epoch is in the Quaternary period.

equatorial forest A dense, luxuriant, evergreen forest of hot, wet equatorial regions containing many trees of tremendous heights, largely covered with lianas and epiphytes. Individual species of trees are infrequent but they include such valuable tropical hardwoods such as mahogany, ebony and rubber. Also known as tropical rain forest or selva.

equifinality The situation in which a number of different processes all lead to essentially the same outcome.

era A division of geologic time next smaller than the eon and larger than a period, e.g., the Paleozoic era is in the Phanerozoic eon and includes, among others, the Devonian period.

erosion (i) The wearing away of the land surface by rain or irrigation water, wind, ice or other natural or anthropogenic agents that abrade, detach and remove geologic parent material or soil from one point on the Earth's surface and deposit it elsewhere, including such processes as creep and so-called tillage erosion. (ii) The detachment and movement of soil or rock by water, wind, ice or gravity.

erosion classes A grouping of erosion conditions based on the degree of past erosion or on characteristic patterns. Applied to accelerated erosion, not to normal, natural or geological erosion. Four erosion classes are recognized for water erosion and three for wind erosion.

erosion pavement A layer of gravel or stones left on the surface of the ground after the removal of the fine particles by erosion.

erosion surface A land/geomorphic surface shaped by the erosive action of ice, wind or water; but usually the result of running water.

erratic A stone or boulder glacially transported from place of origin and left in an area of different bedrock composition.

escarpment A relatively continuous cliff or steep slope, produced by erosion or faulting, breaking the general continuity of more gently sloping land surfaces. The term is most commonly applied to cliffs produced by differential erosion and is used synonymously with "scarp."

esker A long, narrow, usually sinuous, steep-sided ridge composed of irregularly stratified sand and gravel that was deposited by a subglacial or supraglacial stream flowing between ice walls, or in an ice tunnel of a retreating glacier, left behind when the ice melted. Eskers range in length from less than 1 km to more than 160 km, and in height from 3 to 30 m.

estuary The seaward end or the widened, funnel-shaped tidal mouth of a river valley where fresh water comes into contact with sea water and where tidal effects are evident. Most have been formed due to the rise in sea level during the past 15 ka.

etch pit Hole or hollow formed in a mineral by chemical weathering.

eubacteria Prokaryotes other than archaebacteria.

euic High level of bases in soil material, specified at family level of classification.

eukaryote Cellular organism having a membrane-bound nucleus within which the genome of the cell is stored as chromosomes composed of DNA. Includes algae, fungi, protozoa, plants and animals.

eustatic change in sea level A worldwide change in sea level, such as caused by melting glaciers.

eutrophic Having concentrations of nutrients optimal, or nearly so, for plant, animal or microbial growth. Used to refer to nutrient or soil solutions, and bodies of water. The term literally means "self-feeding."

evaporite Residue of salts (including gypsum and all more soluble species) precipitated by evaporation. A mineral or rock deposited directly from a solution (commonly sea water) during evaporation, e.g., gypsum and halite are evaporite minerals.

evapotranspiration The combined loss of water by evaporation from the soil surface and from water bodies, as well as by transpiration from plants. The combined processes of evaporation and transpiration.

exchange acidity The acidity of a soil that can be neutralized by lime or a solution buffered in the range of 7 to 8.

exchange complex The suite of negatively charged sites in a soil that can adsorb cations.

exchangeable anion A negatively charged ion held on or near the surface of a solid particle by a positive surface charge and which may be easily replaced by other negatively charged ions (e.g., a Cl^- salt).

exchangeable bases Charge sites on the surfaces of soil particles that can be readily replaced with a salt solution. In most soils, Ca^{2+}, Mg^{2+}, K^+ and Na^+ predominate. Historically, these are called bases because they are cations of strong bases. Many soil chemists object to this term because these cations are not bases by any modern definition of the term.

exchangeable cation A positively charged ion held on or near the surface of a solid particle by a negatively charged surface and which may be replaced by other positively charged ions in the soil solution. Usually expressed in centimoles or millimoles of charge per kilogram.

exchangeable sodium percentage (ESP) Exchangeable sodium fraction expressed as a percentage.

exchangeable sodium ratio (ESR) The ratio of exchangeable sodium to all other exchangeable cations.

exfoliation A weathering process during which thin layers of rock peel off from the surface.

exfoliation dome A large dome-shaped form that develops in homogeneous crystalline rocks as the result of exfoliation.

exhumed paleosol A soil, geosol or paleosol, which, after formation, was buried, and now has been exhumed by removal of overburden, normally by natural erosion.

exotic river A river that is able to maintain its flow through a desert because of water received from outside the desert.

extragrade A taxonomic class at the subgroup level of Soil Taxonomy having properties that are not characteristic of any class in a higher category (any order, suborder or great group) and that do not indicate a transition to any other known kind of soil. See also *intergrade*.

extrinsic Term for processes and inputs that are external to a given system.

extrusive igneous rock Igneous rock that solidifed from lava at the Earth's surface, characterized by very small mineral crystals. Compare with *intrusive igneous rock*.

fabric The physical constitution of soil material as expressed by the spatial arrangement of the solid particles and associated voids.

facies The sum of all primary lithologic and paleontologic characteristics of sediments or sedimentary rock that are used to infer its origin and environment. The general nature of appearance of sediments or sedimentary rock produced under a given set of conditions. A distinctive group of characteristics that distinguish one group from another within a stratigraphic unit, e.g., contrasting river-channel facies and overbank-floodplain facies in alluvial valley fills.

factorial model See *state-factor model*.

facultative organism An organism that can carry out both options of a mutually exclusive process, e.g., aerobic and anaerobic metabolism.

fault A fracture or fracture zone in the earth with displacement along one side in respect to the other.

faunal passage In soil micromorphology, a tubular pore produced by a member of the soil fauna. See also *krotovina*.

faunalturbation Soil and sediment mixing by the activities of animals.

fecal pellets Rounded and subrounded aggregates of fecal material produced by soil fauna.

Fe concentrations Small areas in a soil where Fe, generally Fe^{3+}, has been concentrated. Commonly found in soils that undergo redox cycles. Also called red mottles.

Fe–Mn concretions Same as Fe concentrations but also including black spots of manganese.

feldspar Family of silicate minerals containing varying amounts of potassium, sodium and calcium along with aluminum, silicon and oxygen. Potassium feldspars contain considerable potassium. Plagioclase feldspars contain considerable sodium and calcium.

felsic rock A general term for light-colored igneous and some metamorphic crystalline rocks.

fen A peat-accumulating wetland that receives some drainage from surrounding mineral soils and usually supports marsh-like vegetation. These areas are richer in nutrients and less acidic than bogs. Soils under fens may eventually develop into thick peat (Histosols).

fermentation The metabolic process in which an organic compound serves as both an electron donor and the final electron acceptor.

ferrallitization See *laterization*.

ferran A cutan composed of iron oxides, hydroxides or oxyhydroxides.

ferri-argillan A cutan consisting of a mixture of clay minerals and iron oxides, hydroxides or oxyhydroxides.

ferric hydroxide $Fe(OH)_3$.

ferricrete See *ironstone*.

ferrihydrite $Fe_5O_7(OH) \cdot 4H_2O$. A dark reddish-brown, poorly crystalline iron oxide mineral that forms in wet soils. Occurs in concretions and placic horizons and often can be found in ditches and pipes that drain wet soils.

ferrimagnet A magnetic mineral, e.g., magnetite and maghemite.

ferrolysis A clay destroying, pedogenic process involving disintegration and solution in water, largely due to the alternate reduction and oxidation of iron.

ferromagnesian Containing iron and magnesium, applied to mafic minerals, e.g., olivine.

ferrous iron Iron in the reduced state (Fe^{2+}), usually mobile within the soil solution.

fibric material Organic soil material that contains three-quarters or more recognizable fibers (after rubbing between fingers) of undecomposed plant remains. Bulk density is usually very low and water-holding capacity very high.

field capacity The content of water, on a mass or volume basis, remaining in a soil 2 or 3 days after it has been wetted with water, and after free drainage is negligible, expressed as a percentage of oven-dry soil.

fifteen-atmosphere percentage The percentage of water contained in a soil that has been saturated, subjected to, and is in equilibrium with, an applied pressure of 15 atm. Approximately the same as fifteen-bar percentage. Commonly equated with wilting point water content.

fine earth fraction Fraction of the soil that passes through a 2 mm sieve: the sand, silt and clay fractions.

fine texture A broad group of textures consisting of or containing large quantities of the fine fractions, particularly of silt and clay. Containing >35% clay.

fine-textured soil A soil rich in clay and silt, containing little sand or gravel. Clayey soil.

firm See *consistence*.

fission track dating Numerical dating method, used in minerals. Fission tracks are damage tracks left in a mineral by spontaneous alpha emissions.

fixation The process by which available plant nutrients are rendered less available or unavailable in the soil.

flaggy Containing appreciable quantities of flagstones.

flagstone A relatively thin, flat rock fragment, from 15–38 cm on the long axis. Usually shale, limestone, slate or sandstone.

flint A variety of chert, often black because of included organic matter.

flocculation The coagulation or physical cohesion of colloidal soil particles due to the ions in solution. In most soils the clays and humic substances remain flocculated due to the presence of +2 and +3 cations.

floodplain The nearly level alluvial plain that borders a stream and is subject to inundation under flood-stage conditions unless protected artificially. It is usually a constructional landform built of sediment deposited during overflow and lateral migration of the stream.

floralfunction See *floralsequence*.

floralsequence A group of related soils that differ one from the other primarily because of differences in kinds and numbers of plants as a soil-forming factor. When expressed as a mathematical equation, it is referred to as a floralfunction.

floralturbation Soil and sediment mixing by the activities of plants, e.g., tree uprooting, root growth.

flowline Direction of water flow within or on top of a surface/soil.

flow till A supraglacial till that is modified and transported by mass flow.

footslope The colluvial, concave hillslope position that forms the inner, gently inclined surface at the base of a slope. In profile, footslopes are commonly concave and are situated between the backslope and toeslope.

forest floor All organic matter generated by forest vegetation, including litter and unincorporated humus, on the mineral soil surface.

fragipan A natural subsurface (Bx or Ex) horizon with very low organic matter, high bulk density and/or high mechanical strength relative to overlying and underlying horizons. Fragipans have hard or very hard consistence (seemingly cemented) when dry, but showing a moderate to weak brittleness when moist. They typically have redoximorphic features, are slowly or very slowly permeable to water, root restricting, and usually have roughly vertical planes which are faces of coarse or very coarse polyhedrons or prisms.

free face Slope that is nearly vertical, commonly associated with rockfalls.

free iron oxides A general term for those iron oxides that can be reduced and dissolved by a dithionite treatment. Generally includes goethite, hematite, ferrihydrite, lepidocrocite and maghemite, but not magnetite.

freely drained Term for a soil that allows water to percolate freely.

freezing front The bottom edge of frozen soil. Below the freezing front the soil temperature is assumed to be $>0\,^{\circ}C$.

friable See *consistence*.

frigid Term for a soil temperature regime that has mean annual soil temperature between 0 °C and 8 °C, with a >5 °C difference between mean summer and mean winter soil temperatures at 50 cm, and warm summer temperatures. See also *isofrigid*.

frost heave Lifting or lateral movement of soil as caused by freezing processes in association with the formation of ice lenses or ice needles.

frost wedge V-shaped body of ground ice, usually less than 4 m in depth and 2 m in width, that typically forms in areas of continuous permafrost.

frost wedge cast The morphological expression of an ice wedge after the ice has melted. Often, the ice wedge has filled with sediment, preserving the wedge shape.

fulvate-complex theory See *chelate-complex theory*.

fulvic acid The pigmented organic material that remains in solution after removal of humic acid by acidification. It is separated from the fulvic acid fraction by adsorption on a hydrophobic resin at low pH values.

fulvic acid fraction Fraction of soil organic matter that is soluble in both alkali and dilute acid.

fungi Simple plants that lack chlorophyll and are composed of cellular filamentous growth known as hyphae.

gabbro A coarse-grained, intrusive igneous rock, chemically equivalent to a basalt.

galleries Tunnels made by termites.

garden variety Colloquial term for the ^{10}Be that falls from the sky and impacts the soil surface, accumulating over time; contrasted with *in situ* ^{10}Be that accumulates directly within quartz-rich surface rocks.

gastroliths Clastic rock and gravel fragments ingested by an animal, usually a bird, in order to grind food in gastric digestion.

gelifluction Form of mass movement in periglacial environments where a permafrost layer exists. It is characterized by the movement of soil material over the permafrost layer and the formation of lobe-shaped features.

geliturbation Mixing of soils and sediments by processes associated with ice and frost.

geoarcheology The science that primarily includes the physical (geological, soils, etc.) aspects of archeology.

geographic information system (GIS) A method of overlaying spatial data of different kinds. The data are referenced to a set of geographical coordinates and encoded in a form suitable for handling by a computer.

geologic column The arrangement of rock units in chronological order.

geologic erosion Normal or natural erosion caused by natural weathering or other geological processes. Synonymous with natural erosion over a geologic time frame or large geographic area.

geology The science that deals with the study of the planet Earth – the materials of which it is made, the processes that act to change these materials from one form to another, and the history recorded by these materials.

geomorphic surface A mappable area of the Earth's surface that has a common history. The area is of similar age and is formed by a set of processes during an episode of landscape evolution. A geomorphic surface can be erosional, constructional or both. It can be planar concave or convex, or any combination of these.

geomorphology The science that studies the evolution of the Earth's surface. The science of landforms. The systematic examination of landforms and their interpretation as records of geologic history.

geophagy The deliberate ingesting of soil, often for religious or health reasons.

geosol Similar to a paleosol, but more rigorously defined, especially with respect to stratigraphic placement and nomenclature.

gibbsite $Al(OH)_3$. A mineral with a platy habit that occurs in highly weathered soils and laterite. It may be prominent in the subsoil and saprolite of soils formed on crystalline rock high in feldspar.

gilgai The microrelief of small basins and knolls or valleys and ridges on a soil surface produced by expansion and contraction during wetting and drying (usually in regions with distinct, seasonal precipitation patterns) of clayey soils that contain large amounts of smectite.

glacial drift A general term applied to all mineral material transported by a glacier and deposited directly by or from the ice, or by running water emanating from a glacier. Drift includes unstratified material (till) that forms moraines, and stratified glaciofluvial deposits that form outwash plains, eskers, kames, varves and glaciolacustrine sediments.

glacial till Unsorted and unstratified material, deposited by glacial ice, which consists of a mixture of clay, silt, sand, gravel, stones and boulders. Sometimes, till may be crudely sorted.

glacier A mass of ice, formed by the recrystallization of snow, that flows forward, or has flowed at some time in the past.

glaciofluvial deposits Material moved by glaciers and subsequently sorted and deposited by streams flowing from the melting ice. The deposits are stratified and may occur in the form of outwash plains, deltas, kames, eskers and kame terraces.

glaciolacustrine deposits Material ranging from fine clay to sand derived from glaciers and deposited in glacial lakes originating mainly from the melting of glacial ice. Many are bedded or laminated with varves.

glaebule In soil micromorphology, a three-dimensional pedogenic feature within the S-matrix of soil material that is approximately prolate to equant in shape.

glassy A texture of extrusive igneous rocks that develops as the result of rapid cooling, so that crystallization is inhibited.

glauconite An Fe-rich dioctahedral mica with tetrahedral Al (or Fe^{3+}) usually greater than 0.2 atoms per formula unit and octahedral R^{3+} correspondingly greater than 1.2 atoms. Mixtures containing an iron-rich mica as a major component can be called glauconitic.

gleization See *gleyzation.*

gleyed A soil condition resulting from prolonged soil saturation, which is manifested by the presence of bluish or greenish colors through the soil mass or in mottles (spots or streaks) among the colors. Gleying occurs under reducing conditions, by which iron is reduced predominantly Fe^{2+}.

gleyed soil Soil developed under conditions of poor drainage resulting in reduction of iron and other elements and the formation of gray colors and mottles.

gleyzation The processes involved in the gleying of soils, usually wet soils. Associated with this process is the reduction of Fe and Mn.

glossic horizon An E horizon that protrudes in a tongue-like manner into a (usually) degrading Bt or Btx horizon.

gneiss A coarse-grained, foliated metamorphic rock in which bands of granular minerals (commonly quartz and feldspars) alternate with bands of flaky or elongate minerals (e.g., micas, pyroxenes). Generally less than 50% of the minerals are aligned in a parallel orientation. Commonly formed by the metamorphism of granite.

goethite FeOOH. A yellow–brown iron oxide mineral. Goethite occurs in almost every soil type and climatic region, and is responsible for the yellowish-brown color in many soils and weathered materials.

grain cutan Cutan associated with the surfaces of skeleton grains or other discrete units such as nodules, concretions, etc.

granite Light-colored, coarse-grained, intrusive igneous rock characterized by the minerals orthoclase and quartz with lesser amounts of plagioclase feldspar and iron–magnesium minerals. Underlies large sections of the continents.

granular soil structure A shape of soil structure common to A horizons.

gravelly Containing appreciable amounts of pebbles and fragments >2 mm in diameter.

gravitational water Water which freely moves into, through or out of the soil under the influence of gravity.

graviturbation Soil and sediment mixing by mass movements, which are driven by gravity.

Gray-Brown Podzolic soil A zonal great soil group (1938 system of soil classification) consisting of soils with a thin, moderately dark A horizon and with a grayish-brown E horizon underlain by a base-rich Bt horizon. They occur on relatively young land surfaces, mostly glacial deposits, from material relatively rich in calcium, under deciduous forests in humid temperate regions.

Gray Desert soil A term used in Russia, and frequently in the United States, synonymously with Desert soil.

graywacke A variety of sandstone characterized by angular-shaped grains of quartz and feldspar, and small fragments of dark rock, all set in a matrix of finer particles.

great period In lichenometry, the initial period of rapid lichen growth that lasts about 20–100 years.

Great soil group One of the categories in the Soil Taxonomy system of soil classification. Great groups group soils according to soil moisture and temperature, base saturation status and expression of horizons.

gross precipitation The total amount of precipitation that falls from the sky.

gross primary production See *net primary production (NPP)*.

ground moraine A landscape formed on an extensive layer of till, having an uneven or undulating surface, usually formed by subglacial processes.

Ground-Water Laterite soil A great soil group of the intrazonal order and hydromorphic suborder (1938 system of soil classification), consisting of soils characterized by hardpans or concretional horizons rich in Fe and Al (and sometimes Mn) that have formed immediately above the water table.

Ground-Water Podzol soil A great soil group of the intrazonal order and hydromorphic suborder (1938 system of soil classification), consisting of soils with an organic mat on the surface over a very thin layer of acid humus material underlain by a whitish-gray leached E horizon, which may be as much as 70–100 cm in thickness. The Bsm or Bhsm horizon is brown, or very dark-brown and cemented. These soils are formed under various types of forest vegetation in cool to tropical, humid climates under conditions of poor drainage.

groundsurface The land surface.

groundwater That portion of the water below the surface of the ground at a pressure equal to or greater than atmospheric pressure.

groundwater table The upper limit of the ground water. Also called water table.

grus Weathered granite residuum.

grusification Specifically, the formation of grus from hard granite. Generally, the formation of weathered rock from unweathered rock. Also known as grusivication.

gully A shallow steep-sided valley that may occur naturally or be formed by accelerated erosion. The distinction between a gully and a rill is one of depth: a rill is of lesser depth and can be smoothed over by ordinary tillage.

gumbotil Gray to dark-colored, thoroughly leached, non-laminated, deoxidized clay, very sticky, and breaking with a starch-like fracture when wet, but very hard when dry. Antiquated term, now replaced by accretion glay. Gumbotils formed on old, stable landscapes and are often found today as buried paleosols.

gypcrete A soil horizon indurated or cemented by gypsum. Also known as a petrogypsic horizon.

gypsan A cutan composed of gypsum.

gypsic horizon A soil horizon of secondary $CaSO_4$ enrichment that is >15 cm thick and has at least 50 g kg^{-1} more gypsum than the C horizon, and in which the product of the thickness in centimeters and the amount of $CaSO_4$ is equal to or greater than 1500 g kg^{-1}.

gypsification The process whereby a soil horizon becomes enriched in illuvial gypsum.

gypsum $CaSO_4 \cdot 2H_2O$. The common name for calcium sulfate.

gyttja Peat consisting of fecal material, strongly decomposed plant remains, shells of diatoms, phytoliths and fine material particles. Usually forms in standing water.

Half-Bog soil A great soil group, of the intrazonal order and hydromorphic suborder (1938 system of soil classification) consisting of soils with dark-brown or black peaty material over gleyed and mottled soil horizons. They are formed under conditions of poor drainage under forest, sedge or grass vegetation in cool to tropical, humid climates.

half-life The amount of time that it takes for one-half of an original population of atoms of a radioactive isotope to decay.

halloysite A member of the kaolin subgroup of clay minerals. It is similar to kaolinite in structure and composition except that hydrated varieties occur that have interlayer water molecules. Halloysite usually occurs as tubular or spheroidal particles and is most common in soils formed from volcanic ash.

Halomorphic soil A suborder of the intrazonal soil order (1938 system of soil classification), consisting of saline and sodic soils formed on wet sites in arid regions and including the great soil groups Solonchak or Saline soils, Solonetz soils, and Soloth soils. In a general sense, the term means a soil containing a significant proportion of soluble salts.

halophyte A plant capable of growing in salty soil, i.e., a salt-tolerant plant.

haploidization Processes that lead to profile simplification. See also *horizonation*.

hard See *consistence*.

hardness In geology, the resistance of a mineral to scratching, determined on a comparative basis by the Mohs scale.

hardpan Colloquial term for a soil layer with physical characteristics that limit root penetration and restrict water movement.

head slope A hillslope, as seen in plan view (i.e., from above), with concave boundaries above and below, situated in a hollow between interfluves or nose slopes.

heat capacity Heat required to produce a unit increase in temperature per quantity of material.

heave In mass movement, the upward motion of material by expansion, e.g., the heaving caused by freezing water.

heavy metals Metals that have densities >5.0 Mg m^{-3}. In soils these include the elements Cd, Co, Cr, Cu, Fe, Hg, Mn, Mo, Ni, Pb and Zn. Many of these heavy or trace elements are regulated because of their potential for human, plant or animal toxicity, including cadmium (Cd), copper (Cu), chromium (Cr), mercury (Hg), nickel (Ni), lead (Pb) and zinc (Zn).

heavy soil A colloquial term for a soil with a high content of the fine separates, particularly clay. So named because these soils have a high drawbar pull and hence are difficult to cultivate, especially when wet.

hematite Fe_2O_3. A red iron oxide mineral that contributes to deep red colors in many soils.

hemic material Organic soil material at an intermediate degree of decomposition that contains one-sixth to three-quarters recognizable fibers (after rubbing) of undecomposed plant remains.

heterotroph An organism able to derive carbon and energy for growth and cell synthesis by utilizing (decomposing) organic compounds.

hibernaculum A secure area, usually a cave or a den of some sort, used by hibernating animals while in a state of torpor.

histic epipedon An organic soil horizon at or near the surface that is saturated with water at some period of the year unless artificially drained. It has a maximum thickness depending on the kind of materials in the horizon and the lower limit of organic carbon is the upper limit for the mollic epipedon.

Histosols An organic soil order. Histosols have organic soil materials in more than half of the upper 80 cm, or that are of any thickness if overlying rock or fragmental materials that have interstices filled with organic soil materials. They are composed of mucks and peats with a high concentration of organic materials in the surface soil or overlying rock.

Holocene period The period of geologic time extending from 10 000 years ago to the present.

honeycomb frost Ice in the soil in insufficient quantity to be continuous, thus giving the soil an open, porous structure permitting the ready entrance of water.

horizon A layer of soil or soil material approximately parallel to the land surface and differing from adjacent, genetically related layers in physical, chemical and biological properties or characteristics, such as structure, texture, consistency, kinds and numbers of organisms present and/or degree of acidity or alkalinity. It is assumed that these characteristics have been produced by soil-forming processes.

horizonation Processes that lead to profile complexity and/or horizonation. See *hapliodization*.

hornblende A rock-forming ferromagnesian silicate mineral of the amphibole group.

hornblende etching The use of the etched or serrated edges that develop on hornblende due to weathering as a relative dating tool.

hue A measure of the chromatic composition, or wavelength, of light that reaches the eye. One of the three variables of color. In lay terms, the "color" of something.

humic acid The ill-defined, dark-colored organic material that can be extracted from soil with dilute alkali and other reagents and that is precipitated by acidification of a dilute alkali extract of soil to pH 1 to 2. It is the main constituent of humus, composed of proteins and lignins, dark brown to black in color.

humic substances Relatively high-molecular-weight, yellow-to black-colored organic substances formed by secondary synthesis reactions in soils. The term is used in a generic sense to describe the colored material or its fractions obtained on the basis of solubility characteristics.

Humic Gley soil Soil of the intrazonal order and hydromorphic suborder (1938 system of soil classification) that includes Wisenboden and related soils, such as Half-Bog soils, which have a thin muck or peat Oi horizon and an A horizon. Developed in wet meadows and forested swamps.

humification The process whereby the carbon of organic residues is transformed and converted to humic substances through biochemical and abiotic processes. The decomposition of organic matter leading to the formation of humus. Also called maturation.

humin The fraction of the soil organic matter that cannot be extracted from soil with dilute alkali.

humus Organic compounds in soil exclusive of undecayed plant and animal tissues, their "partial decomposition" products and the soil biomass. The well-decomposed, relatively stable part of the organic matter found in soils. The principal constituents of humus are derivatives of lignins, proteins and cellulose. Humus has a high CEC. Generally synonymous with soil organic matter.

hydration The process whereby a substance takes up water. A form of chemical weathering involving the absorption of water into the molecular structure of a mineral, causing instability and decomposition.

hydraulic conductivity The rate at which water will move through soil in response to a given potential gradient.

hydraulic gradient (soil water) A vector (macroscopic) point function that is equal to the decrease in the hydraulic head per unit distance through the soil in the direction of the greatest rate of decrease. In isotropic soils, this will be in the direction of the water flux. In essence, this is the slope of the water table, measured by the difference in elevation between two points on the slope of the water table and the distance of flow between them.

hydraulic head The sum of gravitational, hydrostatic and matric water potential, expressed as head, pressure or potential. The level to which groundwater in the zone of saturation will rise. Also known as hydraulic pressure or hydraulic potential.

hydric soil One that is wet long enough to periodically produce anaerobic conditions, thereby influencing the growth of plants.

hydrogen bond An intramolecular chemical bond between a hydrogen atom of one molecule and a highly electronegative atom (e.g. O, N) of another molecule.

hydrologic cycle The various pathways of water from the time of precipitation until the water has been returned to the atmosphere by evaporation and is again ready to be precipitated.

hydrolysis A weathering process involving water, whereby hydrogen ions (H^+) or hydroxyl ions (OH^-) are exchanged for cations such as sodium, potassium, calcium and magnesium. The result is a new residual mineral. Example: the addition of water to orthoclase produces kaolinite and releases K^+ and silica into solution.

Hydromorphic soils A suborder of intrazonal soils (1938 system of soil classification), consisting of seven great soil groups, all formed under conditions of poor drainage in marshes, swamps, seepage areas or flats. In a general sense, soils developed in the presence of excess water.

hydrophilic Molecules and surfaces that have a strong affinity for water molecules.

hydrophobic Molecules and surfaces that have little or no affinity for water molecules. Hydrophobic substances have more affinity for other hydrophobic substances than for water.

hydrophobic soils Soils that are water repellent, often due to dense fungal mycelial mats or hydrophobic substances vaporized and reprecipitated during fire.

hydrosequence A sequence of related soils that differ, one from the other, primarily with regard to wetness. Similar to catena.

hydrosphere The gaseous, liquid and solid water of the Earth's upper crust, ocean and atmosphere; includes lakes, groundwater, snow, ice and water vapor.

hydrous mica See *illite*.

hydroxy–aluminum interlayers Polymers of general composition which are adsorbed on interlayer cation exchange sites. Although not exchangeable by unbuffered salt solutions, they are responsible for a considerable portion of the titratable acidity (and pH-dependent charge) in soils.

hydroxy-interlayered vermiculite (HIV) A vermiculite clay mineral with partially filled interlayers of hydroxy–aluminum groups. It is normally dioctahedral in both the interlayer and the octahedral sheet of the vermiculite layer, and is common in the coarse clay fraction of acid surface soil horizons. It has intermediate cation exchange properties between vermiculite and chlorite. Synonyms are "chlorite–vermiculite intergrade" and "vermiculite–chlorite intergrade."

hygroscopic coefficient The weight percentage of water held by, or remaining in, the soil after it has been air-dried or after it has reached equilibrium with an unspecified environment of high relative humidity, usually near saturation, or with a specified relative humidity at a specified temperature.

hygroscopic water Water adsorbed by a dry soil from an atmosphere of high relative humidity, water remaining in the soil after "air drying." Water held by the soil when it is in equilibrium with an atmosphere of a specified relative humidity at a specified temperature. Outdated term.

hyperthermic A soil temperature regime that has mean annual soil temperatures of 22 °C or more and >5 °C difference between mean summer and mean winter soil temperatures at 50 cm depth. See also *isohyperthermic*.

hyphae Filament-like, root-like structures, common to fungi.

hypoxic The situation in which there is insufficient availability of oxygen in an environment to support aerobic respiration.

hysteresis The relationship between soil-water content and soil-water matric potential, wherein the curves depend on the sequences or starting point used to observe the variables.

ice segregation Ice formed by the migration of pore water to the freezing plane, where it forms into discrete lenses, layers or seams ranging in thickness from hairline to greater than 10 m.

ice wedge cast See *frost wedge cast*.

igneous rock Rock formed from the cooling and solidification of magma and lava, and that has not been changed appreciably by weathering since its formation. Igneous rocks are generally crystalline in nature.

illite As a general term, refers to either a discrete non-expansible mica of detrital or authigenic origin or to the micaceous component of interstratified systems, as in illite–smectite. If used to refer to the mineral species, it should meet the following requirements: (1) the micaceous layers ideally are non-expansible, (2) the octahedral sheet is dioctahedral and aluminous, (3) the interlayer cation is primarily potassium and (4) the composition deviates from that of muscovite in strictly defined. More correctly referred to as hydrous mica.

illuvial horizon A soil horizon into which material carried from an overlying layer, i.e., the eluvial horizon, has been precipitated, either from solution or deposited from suspension.

illuvial materials Materials that have been moved into a horizon, usually in association with percolating water.

illuviation The process of movement of material from one horizon and its deposition in another horizon of the same soil; usually from an upper horizon to a middle or lower horizon. Movement can also take place laterally.

illuviation cutan Coating of illuvial material, often clay, on the surfaces of peds and mineral grains and lining pores.

immobilization The conversion of an element from the inorganic to the organic form in microbial or plant tissues rendering it unavailable to other organisms or plants.

imogolite A poorly crystalline aluminosilicate mineral with an ideal composition $SiO_2Al_2O_3 \cdot 2.5H_2O(^+)$. It appears as threads consisting of assemblies of a tube unit with inner and outer diameters of 1.0 and 2.0 nm, respectively. Imogolite is commonly found in association with allophane, and is similar to allophane in chemical properties. Imogolite is mostly found in soils derived from volcanic ash, and in weathered pumices and Spodosols.

imogolite type material (ITM) Illuvial material, common to the podzolization process, that resembles imogolite or allophane.

impacturbation Soil and sediment mixing that occurs as large objects, e.g., comets, asteroids, bombs, artillery shells, impact the surface and explode.

Inceptisols A mineral soil order. Inceptisols have one or more pedogenic horizons in which mineral materials other than carbonates or amorphous silica have been altered or removed but not accumulated to a significant degree. Water is available to plants more than half of the year or more than 90 consecutive days during a warm season.

inclusion An "impurity" or unnamed, different soil in an area delineated and labeled as a certain map unit. The reader of the map is not explicitly informed of the presence or identity of these soils, thereby reducing their ability to use the map as a predictive tool.

index minerals In various forms of quantitative pedology or mineralogy, a mineral that is resistant to weathering and usually difficult to translocate within the soil.

indurated Term for a very strongly cemented soil horizon.

infauna Animals that live, primarily, within the soil.

infiltration The entry of water into a porous medium, namely soil. See also percolation.

infiltration capacity The maximum rate at which water can infiltrate into a soil, over a given period of time and under a given set of conditions.

infiltration flux The volume of water entering a specified cross-sectional area of soil per unit time $[l\,t^{-1}]$. Also known as infiltration rate.

inner layer In the clay mineral–soil solution system, the inner layer refers to the clay mineral and its associated negative charges.

inorganic Term for any substance in which carbon-to-carbon bonds are absent, i.e., mineral matter.

inselberg A steep-sided residual hill composed predominantly of hard rock and rising abruptly above a plain, found mainly in tropical and subtropical areas.

insolation Solar radiation.

integrated drainage A general term for a drainage pattern in which stream systems have developed to the point where all parts of the landscape drain into some part of a stream system, the initial or original surfaces have essentially disappeared and the region drains to a common base level. Basins of interior drainage are essentially gone, having been integrated into the drainage system.

interception The stopping, interrupting or temporary holding of precipitation by mulch, a vegetative canopy, vegetation residue or any other physical barrier.

interflow That portion of rainfall that infiltrates into the soil and moves laterally through the upper soil horizons until intercepted by a stream channel or until it returns to the surface at some point downslope from its point of infiltration. Also called throughflow.

interfluve The upland or ridge between two adjacent valleys, drainage basins or drainageways. Also called divide or drainage divide.

interglacial A relatively mild (warm) period occuring between two glacial periods or advances. Longer than an interstadial.

intergrade A taxonomic class at the subgroup level of Soil Taxonomy. Intergrades have properties typical of the great group of which they are a member, but they also have properties that indicate that they are transitional to another taxonomic group of soil. See also *extragrade*.

interlayer In phyllosilicate mineral terminology, materials between structural layers of minerals, including cations, hydrated cations, organic molecules and hydroxide octahedral groups and sheets.

intermittent stream A stream, or reach of a stream, that does not flow year-round and that flows only when (1) it receives base flow solely during wet periods or (2) it receives groundwater discharge or protracted contributions from melting snow or other erratic surface and shallow subsurface sources.

interstadial A short, relatively mild period that occurs during a glacial period, in which the ice sheet melts partially back, but not entirely. A readvance of the ice is assumed to follow an interstadial. A slightly warmer phase during a glacial period.

Intrazonal soil One of the three orders in soil classification (1938 system of soil classification). Intrazonal soils have more or less well-developed soil characteristics that reflect the dominating influence of some local factor of relief, parent material or age, over the normal effect of climate and vegetation.

intrinsic From within the system.

intrusive igneous rock Igneous rock that solidifed from magma below the surface, characterized by large mineral crystals. Compare with *extrusive igneous rock*.

ion Any atom, group of atoms or compound that is electrically charged as a result of the loss of electrons (cations) or the gain of electrons (anions).

ionic radius The effective distance from the center of an ion to the edge of its electron cloud.

ionic strength A parameter that estimates the interaction between ions in solution. It is calculated as one-half the sum of the products of ionic concentration and the square of ionic charge for all the charged species in a solution.

ionic substitution The replacement of one or more ions in a crystal structure by others of similar size and electrical charge. Example: Fe^{2+} is interchangeable with Mg^{2+} in most ferromagnesian minerals.

iron oxides Group name for the oxides and hydroxides of iron. Includes the minerals goethite, hematite, lepidocrocite, ferrihydrite, maghemite and magnetite. Sometimes referred to as sesquioxides or iron hydrous oxides.

iron pan A hardpan layer within a soil profile in which iron oxide is the principal cementing agent. See also *plinthite*.

ironstone An in-place concentration of iron oxides that is at least weakly cemented.

isochronous Term referring to a body of rock or a geomorphic surface that is all of the same age.

isofrigid Term for a soil temperature regime in which the mean annual soil temperature is between 0 °C and 8 °C at 50 cm, with the summer and winter temperatures differing by <5 °C.

isohyperthermic Term for a soil temperature regime that has a mean annual soil temperature of 22 °C or more at 50 cm depth, with the summer and winter temperatures differing by <5 °C.

isolated paleosol A soil buried so deeply that it is essentially no longer influenced by surficial pedogenic processes.

isomesic Term for a soil temperature regime in which the mean annual soil temperature is between 8 °C and 15 °C, with the summer and winter temperatures differing by <5 °C.

isomorphous replacement See *isomorphous substitution*.

isomorphous substitution The replacement of one atom by another of similar size in a crystal structure without disrupting or seriously changing the structure. When a substituting cation is of a smaller valence than the cation it is replacing, there is a net negative charge on the structure.

isothermic Term for a soil temperature regime in which the mean annual soil temperature is between 15 °C and 22 °C, with the summer and winter temperatures differing by <5 °C.

isotope Atom that differs from another in atomic mass number, but not in atomic number. For example, oxygen (atomic number 8) may have an atomic mass number of 16, 17 or 18, depending on whether it has 8, 9 or 10 neutrons. It therefore has three isotopes.

isotropic A condition of a volume of soil material that is homogeneous in all directions.

isovolumetric weathering Weathering in which the rock loses material but does not collapse, i.e., it retains its orginal volume. Common when some types of rock weather to saprolite.

jarosite $KFe_3(OH)_6(SO_4)_2$. A pale yellow, potassium iron sulfate mineral.

joint A surface of fracture in a rock, without displacement parallel to or along the fracture.

joint planes In soil micromorphology, planar voids that traverse soil in a regular parallel or subparallel pattern.

kame A mound, knob, hummock or short irregular ridge composed of stratified sand and gravel deposited by a glacial meltwater as a fan or delta at the margin of a melting glacier; by a supraglacial stream in a low place or hole on the surface of the glacier; or as a ponded deposit on the surface or at the margin of stagnant ice. Often gravelly and sandy in texture. Sediment comprising kames is called ice-contact stratified drift.

kandic horizon Subsoil diagnostic horizon having a clay increase relative to overlying horizons, and is dominated by low-activity clays, i.e., <160 cmolc kg^{-1} clay.

kaolin (i) A subgroup name of aluminum silicates with a 1 : 1 layer structure. Kaolinite is the most common clay mineral in the subgroup. (ii) A soft, usually white, rock composed largely of kaolinite.

kaolinite A phyllosilicate clay mineral of the kaolin subgroup. It has a 1 : 1 layer structure composed of shared sheets of Si-O tetrahedrons and Al-(O,OH) octahedrons with very little isomorphous substitution.

karst Topography or landscape with sinkholes, caves and underground drainage that is formed in limestone, gypsum or other rocks, by dissolution. Drainage is usually by underground streams.

kettle Depression in ground surface formed by the melting of a block of glacial ice that was once buried or partially buried by drift. If filled with water: Kettle lake.

krotovina Irregular tubular streak within material, transported from another layer by filling of tunnels made by burrowing animals with material from outside the layer in which they are found. In filled: Faunal burrow. Also called crotovina.

lacustrine deposit Clastic sediments and chemical precipitates deposited in lakes.

lacustrine soil Soil formed on or from lacustrine deposits.

lag concentrate Layer of rocks and gravel that forms on the surface as finer materials are removed by erosion.

lahar A volcanic mudflow composed chiefly of pyroclastic material.

lamella (pl. lamellae) Thin, usually <3 mm in thickness, contorted and wavy, illuvial clay layers in sandy soils.

land capability class One of eight classes of land in the land capability classification of the US Natural Resource Conservation Service; distinguished according to the risk of land damage or the difficulty of land use.

landform Any physical, recognizable form or feature on the Earth's surface, having a characteristic shape, and produced by natural causes. Landforms can be of almost any size, such as a shrub–coppice dune that can be a few meters across or a seif dune which can be up to 100 km long.

landscape A section or portion of the land. Examples are high, intermediate and low mountains; low rolling hills; and floodplains. Parts of a landscape include side slopes, back slopes, toeslopes, footslopes, ridgetops, ridge noses and spurs.

landscape position A particular location on a landscape. Examples are summit of a ridge, shoulder of a ridge, ridge nose, side slope, back slope, footslope, toeslope, cove and drainageway.

lapilli Non-vesicular or slightly vesicular pyroclastics, 2 to 76 mm in at least one dimension, with an apparent specific gravity of 2.0 or more.

large-scale map A map having a scale >1 : 100000.

latent heat The energy required to change a substance to a higher state of matter (solid > liquid > gas). This same energy is released from the substance when the change of state is reversed (gas > liquid > solid).

laterite See *plinthite*.

Lateritic soil A suborder of zonal soils (1938 system of soil classification) formed in warm, temperate and tropical regions and including the following great soils groups: Yellow Podzolic, Red Podzolic, Yellowish-Brown Lateritic and Lateritic.

laterization A bundle of soil-forming processes, common in humid tropical regions, whereby Fe compounds migrate into and out of the soil, leading to Fe concentrations at preferred sites; it places a premium on Fe mobility. It is the suite of processes whereby soils develop into Latosols.

Latosol A suborder of zonal soils (1938 system of soil classification) including soils formed under forested, tropical, humid conditions and characterized by low silica : sesquioxide ratios of the clay fractions, low base exchange capacity, low activity of the clays, low content of most primary minerals and soluble constituents, a high degree of aggregate stability and usually having a red color.

latosolization The bundle of pedogenic processes that lead to the formation of oxic horizons, or soils formerly known as Latosols. In latosolization, residual sesquioxides accrue, as bases and silica are leached from the profile, under long-term weathering in a hot, humid climate; there is little assumed in-migration or translocation of Fe.

lava Molten rock that flows at the Earth's surface.

lava flow A solidified body of rock formed from the lateral, surficial outpouring of molten lava from a vent or fissure.

layer In phyllosilicate mineral terminology, it refers to a combination of sheets in a 1 : 1 or 2 : 1 assemblage.

layer charge Magnitude of charge per formula unit of a clay, which is balanced by ions of opposite charge external to the unit layer.

layer silicate minerals See *phyllosilicates*.

leachate Liquids that have percolated through a soil and that contain substances in solution or suspension.

leaching The removal of soluble materials from soil or other material by percolating water.

lee Opposite of windward. Also called leeward.

lepidocrocite FeOOH. An orange iron oxide mineral that is found in mottles and concretions in wet soils.

lessivage The mechanical migration of clay particles from the A and E horizons to the B horizons of a soil, producing Bt horizons enriched in clay.

leucinization The lightening of a soil horizon, usually due to translocation of clay, Fe and organic matter from that horizon, leaving behind mostly uncoated sand and silt grains.

lichen An organism formed by the symbiotic association of an alga and a fungus, functioning as a single plant.

lichenometry The relative dating technique that estimates the age of a surface based on the size of lichen thalli (bodies) that have developed on rocks that are exposed on that surface. Usually applied in alpine or cold regions.

light soil A colloquial term for a coarse-textured soil. Light soils have low drawbar pull and hence are easy to cultivate. Contrast to heavy soil.

lignin The component of wood or other plant matter responsible for its rigidity.

lime, agricultural A soil amendment containing calcium carbonate, magnesium carbonate and other materials, used to neutralize soil acidity and furnish calcium and magnesium for plant growth.

limestone A sedimentary rock composed mostly of the mineral calcite, $CaCO_3$.

limnic material One of the common components of some organic soils. Includes both organic and inorganic materials that were either deposited in water by precipitation or through the action of aquatic organisms, or derived from underwater and floating aquatic plants and aquatic animals.

lipids General term for all fats, oils and related fatty compounds.

lithic contact The boundary between soil and continuous, coherent, underlying material. The underlying material must be sufficiently coherent to make hand-digging with a spade impractical.

lithification The process by which an unconsolidated deposit of sediments is converted in to solid rock. Compaction, cementation and recrystallization are involved.

lithofunction See *lithosequence*.

lithologic discontinuity The contact, usually manifested as a horizontal plane in soils, between two genetically unlike sediments. Soil horizons in the material below the discontinuity are preceded by the numeral 2 to indicate the second parent material in the sequence. In the uppermost parent material the numeral 1 is customarily omitted.

lithorelict A micromorphological feature derived from the parent rock that can be recognized by its rock structure and fabric.

lithosequence A group of related soils that differ one from the other in certain properties primarily as a result of differences in the parent material as a soil-forming factor. When expressed as a mathematical equation, it is referred to as a lithofunction.

Lithosols A great soil group of azonal soils (1938 system of soil classification) characterized by an incomplete solum or no clearly expressed soil morphology and consisting largely of freshly and imperfectly weathered rock or rock fragments.

lithosphere The rigid outer shell of the Earth. It includes the crust and uppermost mantle and is on the order of 100 km in thickness.

litter The surface (Oi) horizon of the forest floor which is not in an advanced stage of decomposition, usually consisting of freshly fallen leaves, needles, twigs, stems, bark and fruit.

litter interception loss Water that falls or drips onto the O horizon (litter), is intercepted and either evaporates or is taken up by plants, never making it into the mineral soil.

litter mat The O horizon. Also called the forest floor.

littering Addition of litter or organic materials to the soil surface.

loamy A soil texture class and family. If a class, loamy refers to coarse sandy loam, sandy loam, fine sandy loam, very fine sandy loam, loam, silt loam, silt, clay loam, sandy clay loam, and silty clay loam soil textures. If a family particle-size class, loamy refers to soils with textures finer than very fine sandy loam but with $<35\%$ clay and $<35\%$ rock fragments in the upper subsoil horizons.

loess Material transported and deposited by wind and consisting of predominantly silt-sized particles. It is usually of yellowish-brown color and has a widely varying calcium carbonate content.

loose See *consistence.*

low-activity clay A clay mineral generally with a CEC <16 cmol(+) kg^{-1} clay and ECEC <12 cmol(+) kg^{-1}clay). Typical of oxic and kandic horizons, and highly weathered soils in general.

lower plastic limit See *plastic limit.*

luminescence dating A numerical dating technique that establishes the point of time at which a substance was last exposed to light (and then buried) by measuring trapped electrons in crystal defects in certain minerals. Can be used to date loess, eolian sand, some types of alluvium and coastal sediments fired clay, pottery, brick and burned stones.

lysimeter A device for measuring (a) percolation and leaching losses from a column of soil under controlled conditions or (b) gains (irrigation, precipitation, and condensation) and losses (evapotranspiration) by a column of soil.

M horizon In some systems of tropical soil nomenclature, the M horizon is the uppermost mineral soil layer, usually which has been worked extensively by soil infauna such as termites.

macroelement See *macronutrient.*

macrofauna Animals that fall within an arbitrary size range, e.g. between 2 mm and 20 mm body width. Larger than microfauna but smaller than megafauna.

macronutrient A plant nutrient found at relatively high concentrations (>500 mg kg^{-1}) in plants. Usually refers to N, P, and K, but may include Ca, Mg, Fe and S. Also called macroelement.

macropore A comparatively large (>75 μm dia.) pore in a soil. Contrast to micropore.

macropore flow The tendency for water applied to the soil surface (at rates exceeding the upper limit of unsaturated hydraulic conductivity) to move into the soil profile mainly via saturated flow through macropores, thereby bypassing micropores and rapidly transporting any solutes to the lower soil profile. Also called preferential flow and bypass flow.

mafic Term referring to a generally dark-colored igneous rock with significant amounts of one or more ferromagnesian minerals, or to a magma with significant amounts of iron and magnesium, but containing little quartz, feldspar, or muscovite mica.

maghemite Fe_2O_3. A dark reddish-brown, magnetic iron oxide mineral chemically similar to hematite, but structurally similar to magnetite. Often found in well-drained, highly weathered soils of tropical regions.

magma Molten rock, containing dissolved gases and suspended solid particles. When at the Earth's surface, magma is called lava.

magnetic declination Angle of divergence between true north and magnetic north. Measured in degrees east or west of true, or geographic, north.

magnetic pole The point on the Earth's surface where a magnetic needle points vertically downward (north magnetic pole) or vertically upward (south magnetic pole).

magnetic susceptibility The degree to which which a sediment is affected by a magnetic field.

magnetite Fe_3O_4. A black, magnetic iron oxide mineral usually inherited from igneous rocks. Often found in soils as black, magnetic sand grains.

magnetostratigraphy Use of magnetism in rocks to determine the history of events in record of changes in the Earth's magnetic field in past geologic ages, and to ascertain the age of a rock body or sediment.

mangan A cutan composed of manganese oxide or hydroxide, usually identifiable because it will effervesce upon application of H_2O_2.

manganese concretions Small concentrations of MnO_2 in soils.

manganese oxides Oxides of manganese, typically black and frequently occurring in soils as nodules and coatings on ped faces usually in association with iron oxides. Birnessite and lithiophorite are common manganese oxide minerals in soils.

map unit A polygon on a soil map, defined to correspond to a certain soil or set of soils. Limitations of map scale in relation to landscape complexity mean that map units are composed of several kinds of soil, not all of which are actually specified in the legend. That is, most map units have inclusions of other soils. Also called mapping unit.

marble A metamorphic rock composed largely of calcite. The metamorphic equivalent of limestone.

marine isotope stage (MIS) See *oxygen isotope stage.*

marl Soft and unconsolidated calcium carbonate, usually mixed with varying amounts of clay or other impurities.

marsh A wet area, periodically inundated with standing or slow-moving water, that has grassy or herbaceous vegetation and often little peat accumulation. The water may be salt, brackish or fresh. Sometimes called wet prairies.

mass balance Procedure in which the changes in elements or minerals in a soil, from a certain period in the past to the present, are calculated, either on a horizon basis or a profile basis.

massive A type of soil structure in which the soil breaks along no preferred planes, i.e., it has not pattern of repeating structural elements.

mass movement Material that has been moved downslope under the direct influence of gravity.

matric potential The amount of work that must be done per unit of a specified quantity of pure water in order to transport reversibly and isothermally an infinitesimal quantity of water from a specified source to a specified destination. An alternative but less preferred term is water suction.

matrix In soils, the fine material (generally <2 mm dia.) forming a continuous phase and enclosing coarser material and/or pores.

maturation See *humification.*

mature soil An outdated term for a soil with well-developed soil horizons produced by processes of soil formation and essentially "in equilibrium" with its present environment.

maximum-limiting date A numerical date on a surface or sediment that implies that the surface or sediment can be no older than that date.

mean residence time (MRT) A type of radiocarbon date usually applied to soils. MRT dates reflect the weighted mean age of the many organic components within the soil.

mechanical analysis Determination of the various amounts of the different soil separates in the fine earth (<2 mm dia.) fraction of a soil sample, usually by sedimentation, sieving, micrometry or combinations of these methods. Often referred to as particle size analysis.

mechanical weathering The process of weathering by which frost action, salt-crystal growth, absorption of water and other physical processes break down a rock into smaller fragments. No chemical change is involved. Also called disintegration.

Mediterranean climate Climate typical of the west sides of continents at about 30 to 40 degrees latitude. Summers are dry and hot, while winters are cool and rainy.

medium-textured Texture group consisting of very fine sandy loam loam, silt loam and silt textures.

melanization The darkening of a soil horizon, usually by additions of humus.

mesic A soil temperature regime that has mean annual soil temperatures of $8\,^{\circ}C$ or more but $<15\,^{\circ}C$, but with $>5\,^{\circ}C$ difference between mean summer and winter soil temperatures at 50 cm depth. See also *isomesic.*

mesofauna Small organisms such as nematodes, oligochaete worms, insects and smaller insect larvae, and microarthropods.

Mesozoic An era of time during the Phanerozoic eon lasting from 245 million years ago to 66.4 million years ago.

metamorphic rock Rock derived from pre-existing rocks that have been altered physically chemically and/or mineralogically as a result of natural geological processes, principally heat and pressure, originating within the Earth. The pre-existing rocks may have been igneous, sedimentary or another form of metamorphic rock. During metamorphosis the rocks are altered but not completely melted.

metamorphism The processes of recrystallization, textural and mineralogical change that take place in solid rock, under conditions beyond those normally encountered during diagenesis.

mica A group of layer-structured aluminosilicate minerals of the 2 : 1 type that is characterized by its nonexpandability and high layer charge, which is usually satisfied by potassium. The major types of mica are muscovite, biotite and phlogopite.

micrite Calcite crystals less than 4 μm in size.

microbial biomass The total mass of living microorganisms in a given volume or mass of soil, or the total weight of all microorganisms in a particular environment.

microbial population The sum of living microorganisms in a given volume or mass of soil.

microbiota Microflora and protozoa.

microclimate The climatic condition of a small area, which is usually different from the climate of the larger region (macroclimate), because it is modified by local differences in elevation, aspect or other local phenomena.

microcrystalline In soil micromorphology, term used for structure in which the individual grains can only be seen with the aid of a microscope.

microelement See *micronutrient.*

microfauna Protozoa, nematodes and arthropods and other soil animals of microscopic size.

microflora Small plants that can only be seen with a microscope, including bacteria, actinomycetes, fungi, algae and viruses.

micronutrient A plant nutrient found in relatively small amounts (<100 mg kg^{-1}) in plants. Sometimes called trace elements or trace nutrients. They are usually B, Cl, Cu, Fe, Mn, Mo, Ni, Co and Zn. Also called microelement.

micropore Any pore smaller than about 30 μm in diameter. Contrast to macropore.

microrelief The local, slight irregularities in form and height of a land surface that are superimposed upon a larger landform, including such features as low mounds, swales and shallow pits. Examples include gilgai, shrub–coppice dunes and tree-tip mounds and pits. Generally smaller than 2 m in total relief.

midden A pile of domestic refuse consisting of waste food, dung and animal bones (basically anything that the local inhabitants wanted to dispose of).

Milankovitch cycles Earth's orbital cycles that were first studied by Milutin Milankovitch, thought to have influenced climate to the extent that they cause or at least have largely affected the glacial cycles of the Pleistocene.

mineral A naturally occurring homogeneous solid, inorganically formed, with a definite chemical composition and an ordered atomic arrangement.

mineral soil A soil consisting predominantly of, and having its properties determined predominantly by, mineral matter. Contrast with organic soil.

mineralization The conversion of an element from an organic form to an inorganic state as a result of microbial activity.

mineralogically mature Having lost most of its weatherable minerals.

minimum-limiting date A numerical date on a surface or sediment that implies that the surface or sediment can be no younger than that date.

mites Very small arachnids. They occur in large numbers in many O horizons.

moder See *duff mull.*

Mohs scale The 10-point scale of mineral hardness, keyed arbitrarily to the minerals talc (1), gypsum, calcite, fluorite, apatite, orthoclase, quartz, topaz, corundum and diamond (10).

moisture content The mass of water lost per unit dry mass when the material is dried at $\approx$105 °C for $>$8 h. When expressed as a percentage, moisture content is water weight/wet weight.

moisture flux The redistribution of moisture inputs on and within a slope.

mollic epipedon A dark-colored surface horizon of mineral soil that is relatively thick, contains at least 5.8 g kg^{-1}organic carbon, is not massive and hard or very hard when dry, has a base saturation of $>$50% when measured at pH 7, has $<$110 mg P kg^{-1}soluble in 0.05 M citric acid, and is dominantly saturated with divalent cations. Typical of soils formed under grassland vegetation.

Mollisols A soil order. Mollisols have a mollic epipedon overlying mineral material with a base saturation of 50% or more when measured at pH 7.

monogenetic soil A soil that presumably has formed under only one set of soil-forming factors, i.e., in its history there is little evidence of major changes in climate, vegetation, etc.

monsoon A name for seasonal winds, first applied to the winds over the Arabian Sea that blow for 6 months from the northeast and for 6 months from the southwest. The term has been extended to similar winds in other parts of the world, e.g., the prevailing west to northwest winds of summer in Europe have been called the European monsoon.

montmorillonite $Si_4Al_{1.5}Mg_{0.5}O_{10}(OH)_2Ca_{0.25}$. An aluminum silicate (smectite) with a 2:1 layer structure composed of two silica tetrahedral sheets and a shared aluminum and magnesium octahedral sheet. Montmorillonite has a permanent negative charge that attracts interlayer cations that exist in various degrees of hydration thus causing expansion and collapse of the structure (i.e., shrink–swell). The calcium in the formula above is readily exchangeable with other cations.

mor A type of acidic forest humus characterized by an accumulation or organic matter on the soil surface in matted Oe (F) horizons, reflecting the dominant mycogenous decomposers. The boundary between the organic horizon and the underlying mineral soil is usually abrupt.

moraine An accumulation of glacial drift with an initial topographic expression of its own, built chiefly by the direct action of active glacial ice. Subsets include end, ground, lateral, recessional and terminal moraines.

mottled zone A layer that is marked with mottles, usually redoximorphic features associated with oxidation–reduction processes in wet soils.

mottles Spots or blotches of different color or shades of color interspersed with the dominant color.

muck Highly decomposed organic materials in soil.

muck soil An organic soil (Histosol) in which the plant residues have been altered beyond recognition, i.e., they are predominantly sapric (Oi) materials. Contains more mineral matter and is usually darker in color than peat.

mucky peat Organic soil material in which a significant part of the original plant parts are recognizable and a significant part are not. Intergrade between muck and peat, but more like the latter.

mudstone A fine-grained detrital sedimentary rock made up of clay- and silt-sized particles.

mull A forest humus type characterized by intimate incorporation of organic matter into the upper mineral soil (i.e., a well developed A horizon), in contrast to accumulation on/above the surface. The crumbly intimate mixture of organic and mineral material is formed mainly by earthworms.

Munsell color system A color designation system that specifies the relative degrees of the three simple variables of color: hue, value and chroma. For example: 10 YR 6/4 is a color (of soil) with a hue = 10 YR, value = 6 and chroma = 4.

muscovite A non-ferromagnesian, clear, dioctahedral layer silicate of the mica group with Al^{3+} in the octahedral layer and Si and Al in a ratio of 3:1 in the tetrahedral layer. Sometimes called white mica.

mycelium (pl. **mycelia**) A mass of interwoven filamentous hyphae, such as that of the vegetative portion of the thallus of a fungus.

mycorrhiza (pl. **mycorrhizae**) Literally "fungus root." The association, usually symbiotic, of specific fungi with the roots of higher plants.

natric horizon A mineral soil horizon that satisfies the requirements of an argillic horizon but that also has prismatic, columnar or blocky structure and at least one subhorizon having >15% saturation with exchangeable Na^+.

natural erosion Wearing away of the Earth's surface by water, ice or other natural agents under natural environmental conditions of climate, vegetation, etc., undisturbed by humans. Also called geologic erosion.

natural soil drainage class The conditions of frequency and duration of periods of saturation or partial saturation that existed during the development of the soils, as opposed to human-altered drainage. Different classes are described by such terms as excessively drained, well drained, and poorly drained (Table 13.2).

natural soil individual A real soil body, whether affected by human activity or not, that may be observed in a terrain.

nematodes Elongated, cylindrical, unsegmented worms, ranging in size from about 0.1 to over 1.0 mm in length.

neocutan A cutan with a consistent relationship with natural surfaces of soil material. It does not occur immediately at ped surfaces. Similar to hypo-coating.

neoformation In soil micromorphology, a feature or a new substance formed *in situ*, e.g., clay coatings, concretions and synthesized clay minerals.

net precipitation Water that enters the mineral soil from precipitation, not including runoff and water that is intercepted by plants and litter and later lost to evaporation.

net primary productivity (NPP) Net carbon assimilation by plants. NPP = gross primary production − respiration losses. NPP can be estimated for a given time period as $B + L + H$, where B = biomass accumulation for the period, L = biomass of material produced in the period and shed (i.e. foliage, flowers, branches) and H = biomass produced in the period and consumed by animals and insects.

neutral soil A soil in which the pH of the surface layer is approximately between 6.5 and 7.3.

neutron A particle in the nucleus of an atom, which is without electrical charge and with approximately the same mass as a proton.

nitrate reduction (biological) The process whereby nitrate is reduced by plants and microorganisms to ammonium for cell synthesis (nitrate assimilation, assimilatory nitrate reduction) or to nitrite by bacteria using nitrate as the terminal electron acceptor in anaerobic respiration (respiratory nitrate reduction, dissimilatory nitrate reduction). Sometimes used synonymously with denitrication.

nitrification Biological oxidation of ammonium to nitrite and nitrate, or a biologically induced increase in the oxidation state of nitrogen.

nitrogen fixation The transformation of elemental nitrogen to an organic form by microorganisms.

nodule (i) A cemented concentration of a chemical compound, such as calcium carbonate or iron oxides, that can be removed from the soil intact and that has no orderly internal organization. Small and irregular, nodules differ in composition from the soil or sediment that surround them. (ii) In soil micromorphology, glaebules with an undifferentiated rock and/or soil fabric.

normal polarity A section of geologic time when the compass needle points to the magnetic north pole. See also *reversed polarity*.

normal soil See *zonal soil*.

north magnetic pole The point on the Earth where the north-seeking end of a magnetic needle, free to swing in space, points directly down.

nose slope The projecting end of an interfluve, where contour lines connecting the opposing side slopes form convex curves. Overland flow of water is divergent.

not-soil body A volume of not-soil, such as open water, glacial ice, flowing hot lava, large mass of salt, bedrock, or even unconsolidated sediments, provided such materials have not been influenced by soil genesis processes to produce a soil body that differs from the initial material.

numerical dating Establishing the actual or near-actual age of a surface or sediment, in calendar years, usually by radiometric means. Also known as chronometric or absolute dating.

nutrient Element or compound essential as raw material for organism growth and development.

O horizon A soil horizon composed prediminantly of organic soil materials.

Oa horizon (H layer) A layer occurring in mor humus consisting of well-decomposed organic matter of unrecognizable origin (sapric material).

obliquity of the Earth's ecliptic Tilt of the Earth's rotational axis in relation to the plane in which the Earth circles the Sun. Cycles from about 21.5° through 24.5° and back to 21.5° every 41 000 years.

obsidian hydration dating A relative dating technique which relates the thickness of a weathering (hydration) rind on obsidian to the time that the rock has been exposed at the surface.

ochric epipedon A surface horizon of mineral soil that is too light in color, too high in chroma, too low in organic carbon or too thin to be a plaggen, mollic, umbric, anthropic or histic epipedon. Most weakly expressed epipedons are ochric.

octahedral sheet Sheet of horizontally linked, octahedral-shaped units that are a basic structural component of phyllosilicate clay minerals. Each octahedral unit consists of a central, six-coordinated metallic cation surrounded by six hydroxyl groups that, in turn, are linked to other nearby metal cations. Also see *tetrahedral sheet*.

Oe horizon (F layer) A layer of partially decomposed litter with portions of plant structures still recognizable (hemic material). Occurs below the L layer on the forest floor in forest soils. Formerly referred to as the fermentation layer.

Oi horizon (L layer (litter)) A layer of organic material having undergone little or no decomposition (fibric material). On the forest floor this layer consists of freshly fallen leaves, needles, twigs, stems, bark and fruits. The Oi horizon may be very thin or absent during the late growing season.

oligotrophic Term for environments in which the concentration of nutrients available for growth is limited; nutrient-poor habitats.

oligotroph Organism able to grow in environments with low nutrient concentrations.

olivine Common silicate mineral found in rocks formed from mafic magma. Its chemical composition varies between Mg_2SiO_4 and Fe_2SiO_4.

one-third-atmosphere percentage Old term referring to the percentage of water contained in a soil that has been saturated, subjected to, and is in equilibrium with, an applied pressure of one-third atmosphere. Approximately the same as one-third-bar percentage.

opal An amorphous variety of hydrous silica that occurs in various colors, often exhibiting "play of color," where small areas within the opal flash with bright prismatic colors.

opal phytolith A microscopic body of non-crystalline silica (opal) that is secreted by a plant. Phytoliths sometimes accumulate in soils to such a degree that they can be isolated and analyzed to ascertain information about paleovegetational composition.

order The largest taxonomic group (taxa) in Soil-Taxonomy. There are currently 12 soil orders.

organan A cutan composed of organic matter or humus. Also called organ.

organic matter Portion of the soil that includes microflora and microfauna (living and dead) and residual decomposition products of plant and animal tissue; any carbon assembly (exclusive of carbonates), large or small, dead or alive, inside soil space. It generally consists primarily of humus.

organic soil A soil in which the sum of the thicknesses of layers containing organic soil materials is generally greater than the sum of the thicknesses of mineral layers. A soil that is composed predominantly of organic matter. In Soil Taxonomy, a Histosol.

organic soil materials Materials in saturated soil that have $\geq$12% organic carbon and no clay, $\geq$18% more organic carbon and $\geq$60% clay, or a proportional amount of organic carbon, between 12% and 18%, if the clay content is between 0% and 60%.

organo-metallic complex A chemical association between an organic molecule and a metal, usually Fe or Al. Also called chelate-complex.

ornithogenic soils Soils dominated by bird feces, common in penguin rookeries.

orogeny The process of mountain building.

ortstein A cemented spodic horizon, usually indicated as Bsm or Bhsm.

outer layer In the clay mineral–soil solution system, the outer layer refers to the soil solution near to the clay mineral, and the cation swarm it contains.

outwash Stratified and sorted detritus (chiefly sand and gravel) removed or "washed out" from a glacier by meltwater streams and deposited in front of or beyond the end moraine or the margin of the glacier. The coarser material is deposited nearer to the ice. Also called glaciofluvial materials.

outwash plain A constructional plain underlain by thick sequences of glacial outwash, sometimes pitted with kettles.

oven-dry soil Soil that has been dried at $\approx$105 °C until it reaches a constant mass.

overbank deposits Sediments deposited from flood water on floodplains.

overburden Recently transported and deposited material that lies immediately superjacent to the surface horizon of a soil. Also used to designate disturbed or undisturbed material of any nature, consolidated or unconsolidated, that overlies a deposit of useful materials, ores, lignites or coals, especially those deposits mined from the surface by open cuts.

overland flow Water that flows across the surface, but not in channels.

oxbow lake A crescent-shaped, often ephemeral body of standing water situated by the side of a stream in the abandoned channel (oxbow) of a meander, formed when the stream develops a neck cutoff and the ends of the original bend silt in.

oxic horizon A mineral soil horizon that is at least 30 cm thick, characterized by the virtual absence of weatherable primary minerals or 2:1 layer silicate clays. Oxic horizons contain 1:1 layer silicate clays and highly insoluble minerals such as quartz sand, hydrated oxides of iron and aluminum, but lack water-dispersible clay. They have low CECs and small amounts of exchangeable bases.

oxidation (i) Energy-releasing process involving the loss of one or more electrons by an ion or molecule. (ii) A decomposition (chemical weathering) process by which iron or other metallic elements in a rock combine with oxygen to form residual oxide minerals. See also *reduction*.

oxidation–reduction Term refering to both a class of chemical reactions and chemical conditions in a water body, soil or sediment. In an oxidation reaction, an element loses one or more electrons and its oxidation state increases. In a reduction reaction, an element gains one or more electrons and its oxidation state decreases. Oxidation and reduction reactions occur simultaneously. The substance (element, compound or ion) that gets oxidized is called the reductant and the substance that gets reduced is called the oxidant. Also called redox processes.

oxidation state The number of electrons to be added (or subtracted) from an atom in a combined state to convert it to the elemental form.

oxides Clays that lack a sheet silicate structure and which usually have the form R_2O_3 or RO(OH).

Oxisols Mineral soils that have an oxic horizon within 2 m of the surface, or plinthite as a continuous phase within 30 cm of the surface. They lack a spodic or argillic horizon above the oxic horizon.

oxyaquic conditions Pertaining to soils that are saturated but are not reduced and do not contain redoximorphic features.

oxygen isotope stage Period in the Earth's past in which ^{18}O was either enriched or depleted from oceanic waters, due to cold–warm climate cycles and their effects on oceanic evaporation and the formation of ice sheets. Also called marine isotope stage (MIS).

packing voids (simple) In soil micromorphology, voids formed by the random packing of skeletal grains.

paha A loess-capped hill or prominence, usually elongate in plan view, that stands above an erosional surface. Named for the many paha that are erosional relicts on the Iowan erosion surface, in Iowa.

paleoclimate The climate of a period in the geologic past.

paleomagnetism The Earth's magnetism as it is recorded in rocks and sediments.

paleopedology The study of paleosols and the environments in which they formed.

paleosol A soil that formed on a landscape in the past with distinctive morphological features resulting from a soil-forming environment that no longer exists at the site. The former pedogenic process was either altered because of external environmental change or interrupted by burial. A paleosol (or component horizon) may be classed as relict if it has persisted on the land surface. An exhumed paleosol is one that formerly was buried and has been re-exposed by erosion of the covering mantle. Roughly synonymous with geosol.

Paleozoic An era of geologic time lasting from 570 to 245 million years ago.

palimpsest Something that has been written over (on top of); the notion that past landscapes or soils preserve within them information about the past that has not yet been "written over" and obliterated by modern processes.

paludification The expansion of a bog caused by the gradual rising of the water table as accumulations of peat impede drainage.

palygorskite $Si_8Mg_2Al_2O_{20}(OH)_2(OH_2)\cdot 4H_2O$. A fibrous clay mineral composed of two silica tetrahedral sheets and one aluminum and magnesium octahedral sheet that make up the 2 : 1 layer which occurs in strips. The strips that have an average width of two linked tetrahedral chains are linked at the edges forming tunnels where water molecules are held. Palygorskite is most common in soils of arid regions. Also referred to as attapulgite.

palynology The study of pollen, usually as a key to past environments.

pan A soil horizon that is strongly compacted, cemented or has a high content of clay.

panspot See *slick spot.*

papules In soil micromorphology, glaebules composed dominantly of clay minerals with continuous and/or lamellar fabric and sharp external boundaries.

paralithic contact Similar to a lithic contact except that it is softer, can be dug with difficulty with a spade.

parent A radioactive element whose decay produces stable daughter elements.

parent material The unconsolidated and more or less chemically weathered mineral or organic matter from which a soil has developed. The relatively unaltered lower material in soils is often similar to the material in which the horizons above have formed.

parna Term used, especially in southeast Australia, for silt and sand-sized aggregates of eolian clay.

particle density The weight per unit volume of soil solids only. The density of the soil particles, the dry mass of the particles being divided by the solid (not bulk) volume of the particles, in contrast with bulk density. Units are in mass $volume^{-1}$.

particle size The effective diameter of a particle measured by sedimentation, sieving, or micrometric methods.

particle size analysis See *mechanical analysis.*

particle size distribution The fractions of the various soil separates in a soil sample, often expressed as mass percentages.

patterned ground A general term for any ground surface exhibiting a discernibly ordered, more or less symmetrical, morphological pattern of ground and, where present, vegetation. It is characteristic of, but not confined to, permafrost regions or areas subjected to intense frost action.

peat Organic soil material in which the original plant parts are recognizable (fibric material). Often forming a layer many meters deep, it is only slightly decomposed due to being waterlogged.

pebbles Rounded or partially rounded rock or mineral fragments between 2 and 75 mm in diameter. Size may be further refined as fine pebbles (2–5 mm dia.), medium pebbles (5–20 mm dia.) and coarse pebbles (20–75 mm dia.).

ped The basic unit of soil structure. A natural soil aggregate, such as a block, column, granule, plate or prism (in contrast with a clod, which is formed artificially). Volumes of soil that are structureless (massive, or single-grain) are called apedal.

pedal In soil micromorphology, applied to soil materials, most of which consist of peds.

pedalfer Old term for soils in which sesquioxides increased relative to silica during soil formation, typical of formation under humid climates.

pediment A gently sloping, erosional surface developed at the foot of a receding hill or mountain slope.

The surface can be bare or it may be thinly mantled with alluvium and colluvium in transport to the adjacent valley. The pediment backslope (the receded slope) is generally concave upward and rises from the pediment footslope to the upland.

pedimentation Formation of a pediment.

pedisediment A layer of sediment, eroded from the shoulder and backslope of an erosional slope, that lies on and is, or was, being transported across a pediment. Often coarser textured than the material being eroded, because the finer material gets transported farther downslope while the coarser material remains behind as pedisediment.

pedocal An old term for soils in which calcium compounds have accumulated during soil formation, typical of dry climates.

pedogenesis Natural processes involved in the formation of soils.

pedogeneticize To form/evolve soil or soil properties, however slight, upon rock, sediment, peaty materials, or other parent material.

pedogenic Referring to soils or soil genesis.

pedogenic accession A morphologic feature that is acquired by a soil through normal pedogenic processes.

pedogenic overprinting Emplacement of pedogenic properties upon a pre-existing soil, similar to soil welding but the preexisting soil need not be a paleosol.

pedogenic pathway A set of pedogenic processes leading to a given soil morphology. Can be either progressive or regressive.

pedological features In soil micromorphology, recognizable units within a soil material which are distinguishable from the enclosing material for any reason such as origin (deposition as an entity), differences in concentration of some fraction of the plasma or differences in arrangement of the constituents (fabric).

pedology The branch of soil science that addresses soils, their properties, origins, distribution and occurrence on the landscape, as well as their evolution through time. The study of soils as naturally occurring phenomena taking into account their composition, distribution and method of formation.

pedon The smallest volume that can be called a "soil." A pedon is a three-dimensional body of soil with lateral dimensions large enough to permit the study of horizon shapes and relations. Its area ranges from 1 to 10 m^2.

pedoplasmation The bundle of processes associated with the tranformation of saprolite to soil material with pedogenic fabric. Originally coined for use in humid tropical soils.

pedostratigraphic unit In stratigraphy, a layer defined based on soil properties, typically associated with a named Geosol.

pedotubule A cemented tubular structure in soil or sediment formed around roots and other organic or inorganic materials.

pedoturbation All forms of soil mixing. Pedoturbation can be accomplished by any of a number of vectors: plants, animals, gravity, impacts from meteoroids, water, shrink–swell clays, human activities, frost action and several others.

peneplain An old term for area which has been reduced by erosion to a low, gently rolling surface resembling a plain.

peraquic Constantly wet due to a high water table.

perched water table A saturated layer of soil which is separated from any underlying saturated layers by an unsaturated layer. Also known as perched zone of saturation.

percolation The downward or lateral movement of water through soil. Especially, the downward flow of water in saturated or nearly saturated soil under hydraulic gradients of the order of 1.0 or less.

pergelic A soil temperature regime that has mean annual soil temperatures <0 °C. Permafrost is present.

peridotite An ultramafic igneous rock, the major constituent of the Earth's mantle.

periglacial Pertaining to processes, conditions, areas, climates and topographic features occurring at the immediate margins of glaciers and ice sheets, and influenced by cold temperature of the ice.

period In the geologic timescale, a unit of time less than an era and greater than an epoch, e.g., the Tertiary period was the earliest period in the Cenozoic era and included, among others, the Eocene epoch.

permafrost Permanently frozen soil or unconsolidated surficial material.

permafrost table The upper boundary of the permafrost coincident with the lower limit of seasonal thaw.

permanent charge The net negative (or positive) charge of clay particles inherent in the crystal structure of the particle, not affected by changes in pH or by ion-exchange reactions.

permanent wilting point The largest water content of a soil at which indicator plants, growing in that soil, wilt and fail to recover when placed in a humid chamber. Often estimated by the water content at -1.5 MPa soil matric potential.

permeability The ease with which air or plant roots penetrate into or pass through a soil horizon.

perudic A udic soil moisture regime in which water moves through the soil in all months when it is not frozen.

pervection The pedogenic process whereby silt is translocated, usually associated with freeze–thaw cycles.

petrocalcic horizon A continuous, indurated soil horizon that is cemented by calcium carbonate and/or magnesium carbonate. It cannot be penetrated with a spade or auger when dry, and dry fragments do not slake in water. It is impenetrable to roots.

petroferric contact A boundary between soil and a continuous layer of indurated soil in which iron is an important cement. Contains little or no organic matter.

petrogypsic horizon A continuous, strongly cemented, massive soil horizon that is cemented by calcium sulfate (gypsum). It can be chipped with a spade when dry. Dry fragments do not slake in water. It is impenetrable to roots.

pH In soils, the negative logarithm of the hydrogen ion concentration of a soil solution. It is the quantitative expression of the acidity and alkalinity of a solution and has a scale that ranges from about 0 to 14. pH 7 is neutral, <7 is acid and >7 is alkaline. The degrees of acidity or alkalinity, expressed as pH values, are:

Ultra acid	Below 3.5
Extremely acid	3.5–4.4
Very strongly acid	4.5–5.0
Strongly acid	5.1–5.5
Moderately acid	5.6–6.0
Slightly acid	6.1–6.5
Neutral	6.6–7.3
Mildly alkaline	7.4–7.8
Moderately alkaline	7.9–8.4
Strongly alkaline	8.5–9.0
Very strongly alkaline	9.1 and higher

pH-dependent charge The portion of the cation or anion exchange capacity that varies with pH.

Phanerozoic The most recent eon of geologic time beginning 570 million years ago and continuing to the present.

phase A utilitarian grouping of soils defined by soil or environmental features that are not class differentia used in Soil Taxonomy, e.g., surface texture, surficial rock fragments, rock outcrops, substratum, special soil water conditions, salinity, physiographic position, erosion, thickness, etc.

phreatophyte General plant type that lives in dry climates, but is able to survive by virtue of long, deep roots that tap shallow groundwater reserves. Typically found along stream courses. Compare with *xerophyte*.

phyllite A metamorphosed mudstone with a silky sheen, more coarse-grained than a slate and less coarse-grained than a schist.

phyllosilicate Silicate structure in which the SiO_4 tetrahedra are linked together in two-dimensional sheets and are condensed with layers of AlO or MgO octahedra in the ratio 2 : 1 or 1 : 1. Isomorphous substitution of certain elements often occurs.

physical ripening Desiccation of an initially slurried material that occurs naturally following the geomorphic processes of draining of a lacustrine deposit, deposition of a saturated colluvial deposit, wet deposition of glacial till and melting of congeliturbate materials, permafrost and solifluction materials.

physical weathering. See *mechanical weathering.*

phytolith An inorganic body derived from replacement of plant cells; it is usually opaline. See *opal phytolith.*

phytotoxic Detrimental to plant growth.

phytoturbation See *floralturbation.*

piezometer A device used to measure the depth to an unconfined aquifer. Usually, it is simply a perforated tube that is emplaced in the soil to a depth at least as deep as the water table. Eventually, the water table in the soil equilibrates with the water level in the piezometer tube.

pipe Subsurface tunnel or pipe-like cavity formed by water moving through soil.

pit-and-mound topography A form of microrelief created by numerous cradle knolls and their attendant pits. Usually formed over long periods of time in forested regions, due to periodic but ongoing tree uprooting. Thus, the microtopography is associated with forested sites or cleared sites that have not been plowed. Also called cradle-knoll topography.

placic horizon A black to dark-reddish colored mineral soil horizon that is usually thin but that may range from 1 to 25 mm in thickness. Placic horizons are commonly cemented with iron and are slowly permeable or impenetrable to water and roots.

plaggen epipedon A man-made surface horizon more than 50 cm thick that is formed by long-continued manuring and mixing.

plan curvature Curvature of a slope as seen from above, or on a topographic map.

Planosol A great soil group of the intrazonal order and hydromorphic suborder (1938 system of soil classification) consisting of soils with eluviated surface horizons underlain by B horizons more strongly eluviated, cemented or compacted than associated normal soil. The B horizons are often so strongly developed that they perch water at certain times of the year.

plasma In soil micromorphology, that part of the soil material that is capable of being or has been moved, reorganized and/or concentrated by pedogenic processes. It includes all the material, mineral or organic, of colloidal size and relatively soluble material that is not contained in the skeleton grains.

plasma aggregate In soil micromorphology, the preferential alignment of individual plasma grains into larger anisotropic domains that can be recognized in thin section.

plasma concentration In soil micromorphology, concentration of any of the fractions of the plasma in various parts of the soil material.

plasma separation In soil micromorphology, features characterized by a significant change in the arrangement of the constituents rather than a change in concentration of some fraction of the plasma. For example, aligning of plasma aggregates by stress at or near the surface of slickensides.

plastic See *consistence.*

plastic limit The minimum water mass content at which a small sample of soil material can be deformed without rupture, i.e., when it changes from a semisolid to a plastic. Synonymous with lower plastic limit.

platy soil structure A shape of soil structure in which the peds form crude plates that lie more or less horizontal to the soil surface, i.e., the soil aggregates are developed predominantly along the horizontal axes.

playa An ephemerally flooded, vegetatively barren area on a basin floor that is veneered with fine-textured sediment and acts as a temporary or as the final sink for drainage water. Commonly contains salts that accumulate as salty water evaporates.

Pleistocene The period following the Pliocene, extending from about 1.6 million years ago to the Holocene (10 000 years BP). The Pleistocene and Holocene periods comprise the Quaternary epoch of geologic time. In Europe and North America, there is evidence for several periods of intense cold during this period, when large areas of the land surface were covered by glacial ice.

pleochroism The changes in color when some transparent minerals are rotated in plane polarized light, usually under a petrographic microscope. It is expressed in terms of the nature and intensity of the color change.

plinthite A weakly cemented iron-rich, humus-poor mixture of clay with other diluents that commonly occurs as dark red redox concentrations that form platy, polygonal or reticulate patterns. Plinthite changes irreversibly to ironstone hardpans or irregular aggregates on exposure to repeated wetting and drying. Also called laterite.

Pliocene The last of the five geologic epochs of the Tertiary period, extending from the end of the Miocene epoch (about 5 million years ago) to the beginning of the Quaternary period (and Pleistocene epoch) about 1.6 million years ago.

plow layer The part of a soil profile that is mixed by plowing. Also called an Ap horizon.

plow pan A dense layer that forms immediately under the plow layer due to compression of the plow sole.

pluvial A climatic period of of heavy rainfall, generally thought of as occurring in the geologic past and usually restricted to areas that are normally climatically dry.

pluvial lake A lake formed during a pluvial (wet) climatic period.

pocosin A swamp, usually containing organic soil, and partly or completely enclosed by a sandy rim.

Podzol A great soil group of the zonal order (1938 system of soil classification) consisting of soils formed in cool-temperate to temperate, humid climates, under coniferous or mixed coniferous and deciduous forest, and characterized particularly by a highly leached, whitish-gray E horizon. Roughly synonymous with Spodosols in Soil Taxonomy.

podzolization Pedogenic processes resulting in the genesis of Podzols and Podzolic soils. Translocation of Al and humus, and sometimes Fe, are the trademarks of podzolization.

polygenesis The condition in which a soil undergoes pedogenesis involving more than one different pedogenic regime. Commonly, it is assumed that polygenesis involves pedogenesis under at least two different climate or vegetation types. Polygenesis may be brought about through extrinsic changes, e.g., climatic change, or it may occur due to intrinsic changes within the soil, e.g., the development of a slowly permeable horizon. Many or all soils are polygenetic, depending on how one defines "contrasting processes."

polygenetic soil A soil that has been formed by two or more different and somewhat contrasting processes so that all of the horizons or pedogenic properties are not genetically related.

polypedon A group of contiguous similar pedons with a minimum area of 1 m^2 and an unspecified maximum area. The limits of a polypedon are reached at a place where there is no soil or where the pedons have characteristics that differ significantly. Every polypedon can be classified into a soil series, but a series normally has a wider range than that shown by a single polypedon.

pore A discrete volume of soil atmosphere completely surrounded by soil material.

pore ice Frozen water in the interstitial pores of a porous medium.

pore-size distribution The volume fractions of the various size ranges of pores in a soil, expressed as percentages of the soil bulk volume (soil particles plus pores).

pore space The portion of soil bulk volume occupied by soil pores. Also called pore volume.

pore volume See *pore space.*

porosity The volume of pores in a soil sample (nonsolid volume) divided by the bulk volume of the sample. In effect, the fraction of material occupied by pore space.

post-depositional modification (PDM) A time-dependent change in landform morphology, and physical and chemical change in the rocks and soil on a geomorphic surface. PDM data can be used as a relative age determinant.

potassium fixation The process of converting exchangeable or water-soluble potassium to that occupying the position of K^+ in micas. This potassium is in the form of counter-ions entrapped in the ditrigonal voids in the plane of basal oxygen atoms of some phyllosilicates as a result of contraction of the interlayer space. The fixation may occur spontaneously with some minerals in aqueous suspensions or as a result of heating to remove interlayer water in others. Fixed K^+ ions are exchangeable only after expansion of the interlayer space.

pothole Shallow marsh-like pond, particularly as found in the Dakotas. Most potholes are glacial kettles.

prairie Name given to grassland ecosystems, but usually restricted to grasslands where the grasses are taller than about 0.5 m.

Prairie soils A zonal great soil group (1938 system of soil classification) consisting of soils formed in temperate to cool-temperate, humid regions under tall grass vegetation. Roughly synonymous with Udolls in Soil Taxonomy.

precession of the equinox The wobble of the Earth as it spins changes the direction in which its axis of rotation points. One "wobble" takes about 23 000 years.

preferential flow See *macropore flow.*

primary mineral A mineral that has not been altered chemically since deposition and crystallization from molten lava. It occurs or occurred originally in an igneous rock. See also *secondary mineral.*

principle of ascendancy and descendancy Stratigraphic principle that centers on surfaces and how their location and relationship to each other can be used as a relative age dating tool. For example, an erosion surface is younger than the youngest deposit or surface that it cuts across or truncates.

principle of cross-cutting relationships Stratigraphic principle that states that if a feature, e.g., a fault, igneous intrusion or a dike, cuts across another body of rock, it must be younger than the host rock.

principle of original horizontality Stratigraphic principle that states that layers of sediment are generally deposited in a horizontal position. Thus, rock layers that are flat have not been disturbed and maintain their original horizontality.

principle of superposition The principle that in an undeformed sequence of sedimentary rocks or other similar sediments, each bed is older than the one above and younger than the one below.

principle of uniformitarianism See *uniformitarianism.*

prismatic soil structure A shape of soil structure, common in the B and C horizons, in which the peds are elongated in the vertical dimension (taller than they are wide), have nearly flat tops and angular vertices on the sides of the prisms.

proanisotropic pedoturbation A form of soil mixing involving processes that form or aid in forming/maintaining horizons, subhorizons or genetic layers and/or cause an overall increase in profile order. See also *proisotropic pedoturbation.*

profile See *soil profile.*

progressive pedogenesis Soil formation, development and organization. It includes those processes and factors which singularly or collectively lead to organized and differentiated (more anisotropic) profiles. When progressive pedogenesis dominates, a soil develops more, thicker and better-expressed genetic horizons.

proisotropic pedoturbation A form of soil mixing involving processes that disrupt, blend or destroy horizons, subhorizons or genetic layers and/or impede their formation, and cause morphologically simplified profiles to evolve from more ordered ones. See also *proiansotropic pedoturbation.*

prokaryote Cell or organism lacking a membrane-bound, structurally discrete nucleus and other subcellular compartments, such as in bacteria. Rather, prokaryotes have their genetic material in the form of loose strands of DNA found in the cytoplasm.

Proterozoic The geologic eon lying between the Archean and Phanerozoic eons, beginning about 2.5 billion years ago and ending about 0.57 billion years ago.

proto-imogolite theory In podzolization, one of two widely held theories of soil genesis in which the accumulated Fe, Si and Al compounds (commonly referred to as imogolite-type materials or ITM) are assumed to be translocated within acid soil profiles as positively charged, amorphous sols. Later, negatively charged organic matter compounds illuviate into the B horizon and precipitated onto the ITM.

proton A fundamental particle of matter. Protons provides positive charges in the nucleus of atoms.

pumice A light, porous volcanic rock that forms during explosive eruptions. It resembles a sponge because it consists of a network of gas bubbles frozen amidst fragile volcanic glass and minerals.

push probe A type of soil coring device consiting of a hollow tube that is pushed into the soil. When removed, a core soil sample is revealed.

pyrite The mineral iron sulfide (FeS_2).

pyroclastic materials A general term applied to detrital volcanic materials that have been explosively or aerially ejected from a volcanic vent.

pyroxene A silicate mineral containing two metal oxides: $CaMgSi_2O_6$,$CaFeSi_2O_6$,$(Mg,Fe)SiO_3$.

quartz A light-colored or clear, glassy, resistant silicate mineral (SiO_2).

quartzite A metamorphic rock consisting largely of interlocking quartz grains: the metamorphic equivalent of sandstone or chert.

Quaternary period The period of the Cenozoic era of geologic time, extending from the end of the Tertiary period (about 1.6 million years ago) to the present and comprising two epochs, the Pleistocene and Holocene.

radiocarbon ^{14}C derived from ^{14}N as cosmic ray bombardment adds a neutron to its nucleus and the nucleus emits a proton. Radiocarbon decays back to ^{14}N by beta decay. Its half-life is 5730 ± 30 years.

radiocarbon dating Technique whereby the age of a sediment or geomorphic surface is estimated based on the radiocarbon age of material within, on top of or below it.

radionuclide A radioisotope; see *isotope.*

rain forest See *equatorial forest.*

rainshadow A dry area or desert formed when moisture-bearing winds are blocked by a mountain barrier. Also called rainshadow desert.

recarbonation The process whereby carbonates (and to a lesser extent, bases) are added to an acidic soil profile, often via base cycling.

recharge upland areas Areas where water infiltrates into the soil and percolates to the groundwater, thereby recharging it. Commonly these areas are depressions on uplands.

Red Desert soil Highly leached, red clayey soils of the humid tropics (1938 system of soil classification), usually with very deep profiles that are low in silica and high in sesquioxides.

Red-Yellow Podzolic soils A combination of the zonal great soil groups, Red Podzolic and Yellow Podzolic (1938 system of soil classification), consisting of soils formed under warm-temperate to tropical, humid climates, under deciduous or coniferous forest vegetation and usually, except for a few members of the Yellow Podzolic Group, under conditions of good drainage.

redox Adjectival abbreviation for reduction–oxidation processes, or morphologies that have resulted from those processes.

redox concentrations Zones of apparent accumulation of Fe–Mn oxides in soils. Also called red or brown mottles.

redox depletions Zones of low (≤ 2) chroma and high (≥ 4) value where Fe–Mn oxides have been stripped or where both Fe–Mn oxides and clay have been stripped. Redox depletions contrast distinctly or prominently with the matrix. Also called gray mottles.

redoximorphic feature A soil property associated with wetness that results from the reduction and oxidation of iron and manganese compounds after saturation with water and desaturation, respectively. Formed by the processes of reduction, translocation, and/or oxidation of Fe and Mn oxides. Formerly called mottles and low chroma colors. Alternate spelling: redoxymorphic.

reduced matrix Condition in which the soil matrix has a low chroma, but undergoes a change in hue or chroma within 30 min after the soil material is exposed to air. The color change is due to the oxidation of iron.

reduction The gain of one or more electrons by an ion or molecule. See also *oxidation.*

regolith The unconsolidated mantle of weathered rock and soil material on the Earth's surface. Loose earth materials above solid rock.

Regosol Any soil of the azonal order (1938 system of soil classification) without genetic horizons and developing from or on deep, unconsolidated, soft mineral deposits such as sands, loess or glacial drift.

regressive pedogenesis Regressive pedogenesis, or soil regression, reverses, stops or slows soil progression and development. It includes those processes and factors which singularly or collectively lead to simpler and less differentiated (more isotropic) profiles. When regressive pedogenesis dominates, soil horizons become thinner, blurred and/or mixed, and even eroded.

Regur An intrazonal group of dark calcareous soils (1938 system of soil classification) high in clay, mainly smectite, formed mainly from rocks low in quartz; occurring extensively on the Deccan Plateau of India. Roughly equivalent to Vertisols.

relative dating Dating of rocks, soils, stratigraphic layers and geologic events by their position in chronological order without reference to number of years before the present.

relict paleosol An old, usually polygenetic soil that has never been buried and remains at the surface today.

relief The relative difference in elevation between the upland summits and the lowlands or valleys of a given region or area. If used to refer to the relief of a localized area, the term local relief is applicable.

remanent magnetism Magnetism acquired by a rock or sediment at some time in the past.

remote sensing Refers to activities that collect information from a distance. Remote sensing employs such devices as the camera, lasers, radio frequency receivers, radar systems, sonar, seismographs, gravimeters magnetometers and scintillation counters to gather information.

Rendzina A great soil group of the intrazonal order and calcimorphic suborder (1938 system of soil classification) comprising soils that have brown or black, friable surface horizons underlain by light gray to pale yellow calcareous material developed from soft, highly calcareous parent material under grass vegetation or mixed grasses and forest in humid and semiarid climates.

reserve acidity See *residual acidity.*

residual acidity Soil acidity that is neutralized by lime or a buffered salt solution to raise the pH to a specified value (usually 7.0 or 8.0) but which cannot be replaced by an unbuffered salt solution. It can be calculated by subtraction of salt replaceable acidity from total acidity. Also called reserve acidity.

residual soil A soil formed from, or resting on, consolidated rock of the same kind as that from which it was formed, and in the same location. A soil formed in residuum.

residual mineral A mineral that persists in soil after weathering, either because it was resistant to weathering or because it was formed during the weathering process. Also called resistant mineral.

residual soil material See *residuum.*

residuum Unconsolidated, weathered or partly weathered mineral material that accumulated as consolidated rock disintegrated in place. Also called residual soil material.

resistant mineral See *residual mineral.*

retardant upbuilding Additons of material onto the soil surface that occur so rapidly that pedogenesis cannot incorporate them into the profile, essentially burying it.

reticulate mottling A network of mottles with no dominant color, most commonly found in deeper horizons of soils containing plinthite.

reversed polarity A period of geologic time when a magnetic needle would have pointed to the south pole.

rhizobia Bacteria able to live symbiotically in roots of leguminous plants, from which they receive energy and often utilize molecular nitrogen. Collective common name for the genus *Rhizobium.*

rhizosphere The zone of soil immediately adjacent to plant roots in which the kinds, numbers or activities of microorganisms differ from (usually implying that there are more) those of the bulk soil.

rhyolite A fine-grained silica-rich igneous rock, the extrusive equivalent of granite.

rill A small, intermittent watercourse with steep sides; usually only several centimeters deep.

riparian Land adjacent to a body of water or river that is at least periodically influenced by flooding.

rock A solid aggregate of one or more minerals in varying proportions.

rock cycle The sequence of events involving the formation, alteration, destruction and reformation of rocks as a result of geologic processes and which is recurrent, returning to a starting point. It represents a closed system.

rock flour Finely divided rock material ground by glacial action, found in streams issuing from melting glaciers. When dry, it may be blown away as loess. Usually silt-sized.

rock glacier A mass of ice-cemented rock rubble found on slopes of some high mountains. Movement is slow, averaging 30–40 cm yr^{-1}.

rock land Areas containing frequent rock outcrops and shallow soils. Rock outcrops usually occupy from 25% to 90% of the area.

rock outcrop In soil survey applications, a map unit that consists of exposures of bedrock other than lava flows and rock-lined pits.

rock varnish A thin, dark, shiny veneer of clay minerals and iron and manganese oxides that forms on rocks in a desert environment.

root exudate Low-molecular-weight metabolites that enter the soil from plant roots.

root plate The soil plus root mass that is torn up when a tree uproots.

root trace The path or conduit in soil material left after a root has decayed.

root zone The part of the soil that can be penetrated by plant roots. Also called rooting zone.

rough broken land Areas with very steep topography and numerous intermittent drainage channels but usually covered with vegetation.

roundness The degree to which a particle's corners and edges are rounded.

rubefaction See *rubification*.

rubification The development of red color in soil, i.e., reddening. Also called rubefaction.

runoff That portion of precipitation or irrigation on an area which does not infiltrate, but instead is discharged from the area, usually into stream channels. Surface runoff is lost without entering the soil.

salic horizon A mineral soil horizon enriched with secondary salts more soluble than gypsum. A salic horizon is 15 cm or more in thickness, contains at least 20 g kg^{-1} salt, and the product of the thickness in centimeters and amount of salt by weight is >600 g kg^{-1}.

saline–alkali soil An old term for a soil containing sufficient exchangeable sodium to interfere with the growth of most crop plants and containing appreciable quantities of soluble salts. A saline–alkali soil has a combination of harmful qualities of salts and either a high alkalinity or high content of exchangeable sodium, or both, so distributed in the profile that the growth of most crop plants is reduced.

saline seep Intermittent or continuous saline water discharge at or near the soil surface under dryland conditions, which reduces or eliminates crop growth.

saline soil A non-sodic soil containing sufficient soluble salt to adversely affect the growth of most crop plants.

salinization The pedogenic process whereby soluble salts accumulate in soils.

salt-affected soil Soil that has been adversely modified for the growth of most crop plants by the presence of soluble salts, with or without high amounts of exchangeable sodium.

saltation A type of transport involving the rolling, bouncing or jumping action of soil particles, due to the action of wind or flowing water.

salt balance The quantity of soluble salt removed from an irrigated area in the drainage water minus that delivered in the irrigation water.

salt tolerance The ability of plants to resist the adverse, non-specific effects of excessive soluble salts in the rooting medium.

sand Mineral fragments that range in diameter from 2–0.05 mm in the US Department of Agriculture texture system. Most sand grains have quartz mineralogy.

sandstone A clastic sedimentary rock in which the particles are dominantly of sand size, from 0.05 to 2 mm in diameter.

sandy Texture group consisting of sand and loamy sand textures. Family particle-size class for soils with sand or loamy sand textures and <35% rock fragments in upper subsoil horizons.

Sangamon soil A prominent soil, usually buried within the Quaternary sediments of the glaciated and nearby areas of the globe, that formed during the Sangamon interglacial period (marine isotope stage 5).

sapric material Organic soil (Oa) material that contains less than one-sixth recognizable fibers (after rubbing) of undecomposed plant remains. Bulk density is usually very low, and water-holding capacity very high.

saprolite Soft, friable, isovolumetrically weathered bedrock that retains the fabric and structure of the parent rock exhibiting extensive inter-crystal and intra-crystal weathering. In pedology, the term saprolite was formerly applied to any unconsolidated residual material underlying the soil and grading to hard bedrock below.

saprolitization The formation of saprolite from unweathered rock.

saprophyte An organism that lives on dead organic material.

saturated flow The movement of water in a soil that is completely filled with water.

scarp Abbreviation for escarpment. A form of cliff or steep slope of some extent along the margin of a plateau, mesa, terrace or structural bench. A scarp may be of any height.

schist A strongly foliated, coarsely crystalline metamorphic rock, produced during regional metamorphism, that can readily be split into slabs or flakes because more than 50% of its mineral grains are parallel to each other.

secondary mineral A mineral resulting from the decomposition of a primary mineral or from the reprecipitation/recrystallization of the decomposition products of a primary mineral. The main secondary minerals in soils are clays and oxides.

sedimentary facies An accumulation of deposits that exhibits specific characteristics and grades laterally into other sedimentary accumulations that were formed at the same time but exhibit different characteristics.

sedimentary rock Rock formed from the accumulation of sediment, which may consist of fragments and mineral grains of varying sizes from pre-existing rocks, remains or products of animals and plants, the products of chemical action, or mixtures of these. The principal sedimentary rocks are sandstones, shales, limestones and conglomerates.

segregated ice Massive ice in a soil, which is relatively free of soil particles.

seismiturbation Soil and sediment mixing by earthquakes and the surficial settling that occurs after them.

self-mulching soil A soil in which the surface layer becomes so well aggregated that it does not crust and seal under the impact of rain but instead serves as a surface mulch upon drying. Common in some Vertisols.

self-weight collapse The condition in which a soil is wetted and then, upon drying, collapses, resulting in an increase in bulk density and a loss of pore space.

sepic fabric In soil micromorphology, plasmic fabrics with recognizable anisotropic domains that have various patterns of preferred orientation; i.e., plasma separations with a striated extinction pattern are present.

sequum (pl. sequa) The B horizon together with any overlying eluvial horizons. It is essentially an eluvial and illuvial couplet of horizons.

serpentine A group of predominately green minerals that occur in masses of tiny inter-grown crystals. The two main types used in jewelry are bowenite (translucent green or blue-green) and the rarer williamsite (translucent, oily green, veined or spotted with inclusions).

serpentinite A rock rich in serpentine.

sesquan A cutan composed of a concentration of sesquioxides.

sesquioxide General term for oxides and hydroxides of iron and aluminum.

shale A mudstone that splits or fractures readily; a sedimentary rock.

shear strength The resistance of a body to shear stress.

shear stress The stress on an object operating parallel to the slope on which it lies.

sheet In phyllosilicate mineral terminology, a flat array of more than one atomic thickness and composed of one level of linked coordination polyhedra. A sheet, as in a tetrahedral sheet or octahedral sheet, is thicker than a plane and thinner than a layer.

sheet erosion The removal of a relatively uniform, thin layer of soil by rainfall and largely unchanneled surface runoff (sheet flow).

sheet flow See *sheet erosion.*

shoulder The hillslope position that forms the uppermost inclined surface near the top of a slope. It comprises the transition zone from backslope to summit. Shoulder slopes are dominantly convex in profile and erosional in origin.

shrink–swell The shrinking of soil when dry and the swelling when wet. Shrinking and swelling can damage plant roots, roads, dams, building foundations and other structures.

side slope The slope bounding a drainageway (stream channel) and lying between the drainageway and the adjacent interfluve. It is generally linear along the slope width and overland flow is parallel down the slope. Side slopes, as seen in plan view (i.e., from above), have fairly straight boundaries above and below, and are situated on the side of an interfluve.

Sierozem A zonal great soil group (1938 system of soil classification) consisting of soils with pale grayish A horizons grading into calcareous material at a depth of 1 foot or less. They form in temperate to cool, arid climates under a vegetation of desert plants, short grass, and scattered brush.

silan A cutan of silt-sized material.

silcrete A silica-enriched hardpan, usually a duripan.

silica The basic chemical constituent common to all silicate minerals and magmas. Silicon dioxide (SiO_2) as a pure crystalline substance makes up quartz and related forms such as flint and chalcedony.

silica : alumina ratio The molecules of silicon dioxide (SiO_2) per molecule of aluminum oxide (Al_2O_3) in clay minerals or in soils.

silica : sesquioxide ratio The molecules of silicon dioxide (SiO_2) per molecule of aluminum oxide (Al_2O_3) plus ferric oxide (Fe_2O_3) in clay minerals or in soils.

silicate A rock-forming mineral that contains silicon.

silica tetrahedron The basic structural unit of which all silicates are composed, consisting of a silicon atom surrounded symmetrically by four oxygen atoms. The structure, therefore, has the form of a tetrahedron with an oxygen atom at each corner.

silicic A rock or material rich in silica.

silicification The process whereby a soil horizon becomes enriched in illuvial silica.

silt Mineral particles that range in diameter from 0.02 to 0.002 mm in the international system or 0.05 to 0.002 mm in the US Department of Agriculture system.

siltstone A sedimentary rock composed primarily of silt-sized clastic particles.

single-grain structure A soil structure classification in which the soil particles occur almost completely as individual or primary particles with essentially no secondary particles or aggregates present. Usually found only in extremely coarse-textured soils; structureless.

skeletan A cutan composed of skeleton grains, usually silt-sized or larger.

skeleton grains Individual grains that are relatively stable and not readily translocated, concentrated or reorganized by soil-forming processes. They include mineral grains and resistant siliceous and organic bodies larger than colloidal size.

skew planes In soil micromorphology, planar voids that traverse the soil material in an irregular manner and are formed mostly by desiccation.

slate A compact, fine-grained metamorphic rock that has slaty cleavage. Formed by the low-grade, regional metamorphism of shale.

slaty cleavage A style of foliation common in metamorphosed mudstones, characterized by nearly flat, sheet-like planes of breakage, similar in appearance to a deck of playing cards.

slick spot A small area of surface soil that is slick when wet because of alkalinity or high exchangeable sodium. Slick spots have a puddled or crusted, very smooth, nearly impervious surface. The underlying material is dense and massive. Also called panspot.

slickenside A stress surface that is polished and striated produced by one mass sliding past another as the soil material expands in response to wetting. Slickensides are common below 50 cm in swelling clay soils subject to large changes in water content.

slope The inclination of the land surface from the horizontal. Slope percentage is the vertical slope distance divided by horizontal distance, then multiplied by 100. Thus, a slope of 20% has a drop of 20 m in 100 m of horizontal distance. Slope classes vary slightly between soil surveys, but a listing from one is:

Nearly level	0–3%
Gently sloping	1–8%
Strongly sloping	8–15%
Moderately steep	15–30%
Steep	30–50%
Very steep	50–95%

slope element One of five parts of a slope: summit, shoulder, backslope, footslope and alluvial toeslope.

slope gradient The angle of inclination of a slope, from the horizontal.

slopewash Movement of material downslope by the combined impact of rainsplash and overland flow, possibly including some small channel flow.

slump Landform or mass of material that slipped down during, or was produced by, a mass movement process (also called a slump) characterized by a landslide involving a shearing and rotary movement of a generally independent mass of rock or earth along a curved slip surface (concave upward).

small-scale map A map having a scale smaller than 1 : 1 000 000.

S-matrix In soil micromorphology, the material within the simplest peds, or composing apedal soil materials, in which the pedological features occur. It consists of the plasma, skeleton grains and voids that do not occur as pedological features other than those expressed by specific extinction (orientation) patterns. Pedological features also have an internal S-matrix.

smectite A group of 2 : 1 layer silicates with a high cation exchange capacity, about 110 cmolc kg^{-1} for soil smectites, and variable interlayer spacing. Formerly called the montmorillonite group. The group includes dioctahedral members montmorillonite, beidellite and nontronite, and trioctahedral members saponite, hectorite and sauconite.

sodic soil A non-saline soil containing sufficient exchangeable sodium to adversely affect crop production and soil structure under most conditions of soil and plant type. The sodium adsorption ratio of the saturation extract is at least 13.

sodium adsorption ratio (SAR) A relation between soluble sodium and soluble divalent cations which can be used to predict the exchangeable sodium fraction of soil equilibrated with a given solution.

soft See *consistence*.

soil The unconsolidated mineral and organic material on the surface that serves as a natural medium for the growth of land plants, or that responds to diurnal and seasonal climatic and microclimatic conditions in the absence of plants (as in parts of Antarctica) OR the unconsolidated mineral and organic matter on the surface that has been subjected to and influenced by genetic and environmental factors of parent material, climate (including moisture and temperature effects), macro- and microorganisms, topography, all acting over a period of time and producing a product – soil – that differs from the material from which it is derived in many physical chemical, biological and morphological properties and characteristics.

soil aeration The exchange of soil air with air from the atmosphere. The air in a well-aerated soil is similar to that in the atmosphere; the air in a poorly aerated soil is considerably higher in carbon dioxide and lower in oxygen.

soil amendment A substance added to soil to improve chemical or physical characteristics or as a means of treating a waste material.

soil association A group of defined and named taxonomic soil units, i.e., map units, occurring together in an individual and characteristic pattern over a geographic region, comparable to plant associations. On some soil maps a soil association is created when two or more map units occurring together in a characteristic pattern are combined because the scale of the map or the purpose for which it is being made does not require delineation of the individual soils.

soil boundary A line, or band, composed of points on the landscape at which soil characteristics are transitional between those of adjacent soil bodies, within which gradients of change are spatially less rapid than at the boundary.

soil classification The systematic arrangement of soils into groups or categories on the basis of their characteristics.

soil complex A map unit composed of soil units occurring in a pattern too complex to be individually represented at the scale of the survey.

soil continuum The soil cover or blanket on the land surface, interrupted though it is by bodies of not-soil.

soil cover The entirety of soils occurring in a region.

soil creep The slow downslope movement of soil and sediment due to heave induced by freeze–thaw and wet–dry cycles.

soil fabric In soil micromorphology, the arrangement or mutual relationship of the individual soil constituents within a feature or within the soil as a whole.

soil family In soil classification, one of the categories intermediate between the subgroup and the soil series. Families provide groupings of soils with ranges in texture, mineralogy, temperature and thickness.

soil fertility The relative ability of a soil to supply the nutrients essential to plant growth.

soil formation factors See *state factors*.

soil-forming interval The period of time during which a soil is forming on a surface, usually from the initiation of soil genesis to the present (if the soil is still exposed at the surface) or burial (if it is currently buried).

soil genesis The mode of origin of the soil with reference to the processes or soil-forming factors responsible for its development from not-soil material, such as fresh sediment or bedrock. The continuing response of a complex body of mineral and organic materials to a never-ending sequence of impacts and overprints by climatic and organic events and attendent physical, chemical and biotic processes.

soil geography A subspecialization of physical geography concerned with describing and explaining the areal distributions of soils.

soil geomorphology The science that studies the genetic relationships between soils and landforms.

soil horizon See *horizon*.

soil infauna Soil fauna that reside and live most of their lives within the soil proper.

soil landscape The soils portion of the landscape. Soilscape is an abbreviation.

soil landscape analysis The study of soil pattern, especially as it pertains to landforms.

soil map unit complexes A delineation on a soil map that consists of two or more dissimilar components occurring in a regularly repeating pattern.

soil micromorphology The study of of intact soil material to determine the arrangement of the various components of the soil matrix.

soil moisture regime The moisture status of a soil, primarily a function of climate (water supply) and topography/relief as it affects the local water table.

soil moisture tension Equal in magnitude but opposite in sign to the soil water pressure, soil moisture tension is equal to the pressure that must be applied to the soil water to bring it to a hydraulic equilibrium, through a porous permeable wall or membrane, with a pool of water of the same composition.

soil monolith A vertical section through the soil preserved with resin and mounted for display.

soil morphology The physical entirety of a soil, including for example its horizonation, color, structure, etc.

soil order The highest level of soil classification. There are presently 12 soil orders: (1) Entisols, (2) Inceptisols, (3) Spodosols, (4) Ultisols, (5) Alfisols, (6) Vertisols, (7) Oxisols, (8) Histosols, (9) Andisols, (10) Aridosols, (11) Mollisols and (12) Gelisols.

soil permeability The ease with which gases, liquids or plant roots penetrate or pass through a mass of soil. Measured in units of length $time^{-1}$.

soil phase A subdivision of a unit of soil classification having characteristics that affect the use and management of the soil but which do not vary sufficiently to differentiate it as a separate type. A variation in a property or characteristic such as surface texture, surficial rock fragments, rock outcrops, substratum, special soil water conditions, salinity, physiographic position, erosion, thickness, etc.

soil profile A vertical section of the soil through all its horizons and extending into the C horizon.

soil reaction A measure of the acidity or alkalinity of a soil, usually expressed as a pH value. See also pH.

soil salinity The amount of soluble salts in a soil. The conventional measure of soil salinity is the electrical conductivity of a saturation extract.

soilscape See *soil landscape*.

soil separates Mineral particles less than 2 mm in equivalent diameter and ranging between specified size limits. The names and sizes, in mm, of separates recognized in the United States are as follows:

Very coarse sand	2.0–1.0
Coarse sand	1.0–0.5
Medium sand	0.5–0.25
Fine sand	0.25–0.10
Very fine sand	0.10–0.05
Silt	0.05–0.002
Clay	<0.002

soil series The basic unit of soil classification, being a subdivision of a family and consisting of soils that are essentially alike in all major profile characteristics except texture of the A horizon. The soil series is the lowest category in Soil Taxonomy.

soil solution The water surrounding soil particles.

soil structure The spatial distribution and total organization of the soil system as expressed by the degree, size and type of aggregation, i.e., peds, and the nature and distribution of pores and pore spaces.

soil survey The systematic examination and mapping of soil.

soil taxonomy The systematic arrangement of soils into groups or categories on the basis of their characteristics.

Soil Taxonomy The USDA-NRCS system of soil classification used in the United States.

soil temperature regime In Soil Taxonomy, a class of soil temperature, usually measured at 50 cm depth.

soil textural triangle Three-phase scale, i.e., a ternary diagram, used to define soils into textural groups or classes.

soil texture The physical nature of the soil, according to its relative proportions of sand, clay and silt.

soil water content The water lost from the soil upon drying to constant mass at 105 °C, expressed either as the mass of water per unit mass of dry soil, or as the volume of water per unit bulk volume of soil. Same as soil moisture content.

soil water diffusivity The hydraulic conductivity divided by the differential water capacity (care being taken to be consistent with units), or the flux of water per unit gradient of water content in the absence of other force fields.

soil welding The processes that cause the horizons in a surface soil to deepen and connect with a buried soil below, such that C horizon material no longer exists between the two.

solifluction Slow, viscous downslope flow of water-saturated regolith, usually on top of frozen substrate. Characteristic of, though not confined to, regions subjected to alternate periods of freezing and thawing.

Solod A solonetzic soil that has been leached of most of its salts and sodium, resembling an Alboll or Aquoll.

solodization The process whereby sodium is gradually removed from a profile, often replaced by calcium.

Solonchak A great soil group of the intrazonal order and halomorphic suborder (1938 system of soil classification) consisting of soils with gray, thin, salty crusts on the surface, and with fine granular mulch immediately below being underlain with grayish, friable, salty horizons. Solonchaks are formed under subhumid to arid, hot or cool climates, under conditions of poor drainage, and under a sparse growth of halophytic grasses, shrubs and some trees.

Solonetz A great soil group of the intrazonal order and halomorphic suborder (1938 system of soil classification) consisting of soils with a very thin, friable, surface which is underlain by a dark, hard columnar layer usually highly alkaline. Solonetz soils are formed under subhumid to arid, hot to cool climates, under better drainage than Solonchaks, and under a native vegetation of halophytic plants.

solum (pl. sola) The upper and most weathered part of the soil profile; the A and B horizons.

sombric horizon A subsurface mineral horizon that is darker in color than the overlying horizon but that lacks the properties of a spodic horizon. Common in cool, moist soils of high altitude in tropical regions.

sorption The removal of an ion or molecule from solution by adsorption and absorption. It is often used when the exact nature of the mechanism of removal is not known.

sorting The range of particle sizes in a sedimentary deposit. A deposit with a narrow range of particle sizes is termed well-sorted.

spatial variability The variation in soil properties (1) laterally across the landscape, and/or (2) vertically downward through the soil.

specific adsorption The strong adsorption of ions or molecules on a surface. Specifically adsorbed materials are not readily removed by ion exchange.

specific surface The solid-particle surface area (of a soil or porous medium) divided by the solid-particle mass or volume.

sphericity Term relating to the overall shape of a feature irrespective of the sharpness of its edges; a measure of the degree of its conformity to a sphere. A descriptive term to describe how close a particle's shape is to a sphere.

splash erosion The detachment and airborne movement of small soil particles caused by the impact of raindrops on soils.

spodic horizon A mineral soil horizon that is characterized by the illuvial accumulation of amorphous materials composed of Al and organic carbon, with or without Fe. The spodic horizon has a certain minimum thickness, and a minimum quantity of extractable carbon plus Fe plus Al in relation to its content of clay. Diagnostic horizon for Spodosols.

Spodosols A mineral soil order. Spodosols have a spodic horizon or a placic horizon that overlies a fragipan.

springtails Very small insects of Order Collembola that live near the surface and feed on organic matter.

stable isotope An isotope that does not decay radioactively over time.

stadial A part of geologic time in which a glacier makes a brief readvance.

state-factor model A model of soil development in which the soil (S) is envisioned as forming due to the complex interplay among five state factors See also *state factors*.

state factors In the state-factor model of soil genesis, $S = f(cl, o, r, p, t, \ldots)$, the five state factors are cl (climate), o (organisms), r (relief), p (parent material) and t (time).

static pedogenesis A condition in soils that occurs for only short periods of time. In static pedogenesis, the forces of regressive and progressive pedogenesis exactly balance out, such that the soil neither develops nor regresses.

steady state A condition in soils and other systems in which the average condition of the system remains unchanged over a period of time.

stemflow That portion of precipitation or irrigation water that is intercepted by plants and then flows down their stems to the ground.

steppe Name given to the dry, cool, shortgrass grasslands of Russia and Ukraine.

sticky See *consistence*.

stoma (pl. stomata) A pore and two guard cells in the epidermis of a stem or leaf.

stone line A sheet-like concentration of coarse fragments in sediments. In cross-section, the line may be marked only by scattered fragments or it may be a discrete layer of fragments. Also called stone zone or stone layer.

stones Rock or mineral fragments between 250 and 600 mm in diameter if rounded, and 380 to 600 mm if flat.

stone zone See *stone line.*

stoniness Classes based on the relative proportion of stones at or near the soil surface. Used as a phase distinction in mapping soils.

stony A stoniness class in which there are enough stones at or near the soil surface to be a continuing nuisance during operations that mix the surface layer, but they do not make most of such operations impractical.

strain The deformational changes that a soil undergoes through time, usually expressed as a change in profile thickness.

stratification The accumulation of material in distinct layers or beds.

stratified The characteristic of being in distinct layers or beds.

stratified drift Debris washed from a glacier and laid down in well-defined layers.

stratigraphy The succession and age relation of layered rocks.

stream order An integer system applied to tributaries (stream segments) that documents their relative position within a drainage basin network as determined by the pattern of its confluences. The order of the drainage basin is determined by the highest integer. In most stream ordering systems, the smallest unbranched tributaries are designated order 1, and larger streams have successively higher orders.

stream terrace A relatively flat, former floodplain surface along a valley, with a steep bank separating it either from the floodplain, or from a lower terrace. Originally formed near the level of the stream, and representing the dissected remnants of an abandoned floodplain, stream bed or valley floor.

striations Scratches or small channels gouged by glacier action into rock. Striations occur on boulders, pebbles and bedrock. Striations along bedrock indicate direction of ice movement.

structural charge The charge (usually negative) on a mineral resulting from isomorphous substitution within the mineral layer.

subaerial A land surface not buried or covered with water, but which intersects the atmosphere.

subangular blocky structure A type of soil structure in which the peds resemble rounded cubes.

Subaqueous A surface that is at present under water.

Subarctic Brown Forest soils Soils similar to Brown Forest soils (1938 system of soil classification) except having more shallow sola and average temperatures of <5 °C at >18 inches depth.

subgroup In Soil Taxonomy, the fourth largest taxonomic grouping.

suborder In Soil Taxonomy, the second largest taxonomic grouping of soils, below order.

subsoil Technically, the B horizon; roughly, the part of the solum below the plow depth.

sulfidic material Waterlogged material or organic material that contains 7.5 g kg^{-1} or more of sulfide-sulfur.

sulfidization The accumulation of soil materials rich in sulfides, usually by the biomineralization of sulfate-bearing water. It is best exemplified in the anaerobic, humus-rich environments of tidal marshes.

sulfur cycle The sequence of transformations undergone by sulfur wherein it is used by living organisms, transformed upon death and decomposition of the organism, and ultimately converted to its original oxidation state.

sulfuric horizon A horizon composed either of mineral or organic soil material that has both pH <3.5 and jarosite mottles.

sulfuricization The process whereby sulfuric acid is produced in soils or sediments as sulfide-bearing minerals (formed via sulfidization) are exposed to oxidizing conditions, become oxidized and produce jarosite and/or sulfuric acid.

summit The highest point of any landform remanant, hill or mountain.

supraglacial Carried upon, deposited from or pertaining to the top surface of a glacier or ice sheet; said of meltwater streams, till, drift, etc.

surface area The area of the solid particles in a given quantity of soil or porous medium.

surface-charge density The excess of negative or positive charge per unit of surface area of soil or soil mineral.

surface exposure dating (SED) The technique whereby the amount of time that a geomorphic surface has been stable and subaerially exposed is determined.

surface soil The uppermost part of the soil, ordinarily moved in tillage, or its equivalent in uncultivated soils. Frequently designated as the plow layer, the surface layer, or the Ap horizon.

surface tension The elastic-like force in a body, especially notable between a liquid and the air, that tends to minimize or constrict the area of the surface.

suspended load The amount of detrital material a stream carries in suspension. Also called wash load.

swale A topographically low area.

swamp An area saturated with water throughout much of the year but with the surface of the soil usually not deeply submerged. Usually characterized by tree or shrub vegetation.

talc $Si_4Mg_3O_{10}(OH)_2$. A trioctahedral magnesium silicate mineral with a 2 : 1 type layer structure but without isomorphous substitution. May occur in soils as an inherited mineral.

talf Geomorphic descriptor for a flat area.

talus Rock fragments of any size or shape (usually coarse and angular) derived from and lying at the base of a cliff or very steep rock slope. The accumulated mass of such loose, broken rock formed chiefly by falling, rolling or sliding.

taxadjunct A soil that is correlated as a recognized, existing soil series for the purpose of expediency. They are so like the soils of the defined series in morphology, composition and behavior that little or nothing is gained by distinguishing them as a new series.

taxon (pl. taxa) In the context of soil survey, a class at any categorical level in a system of soil taxonomy. Taxa are chosen because of their natural significance and their usefulness in discriminating between soils.

taxonomic generalization Simplification of a map pattern by combining those map units on the detailed, small-scale map that are most nearly similar to one another in respect to pedologic characteristics.

taxonomic unit A subdivision within a taxonomic system. For example, the Mollisols constitute a taxonomic unit. Contrast with *genetic unit* and map unit.

tectonic Rock structures produced by movements in the Earth's crust.

tensiometer A device for measuring the soil-water matric potential *in situ*. It has a porous, permeable ceramic cup connected through a water-filled tube to a manometer, vacuum gauge, pressure transducer or other pressure measuring device.

tephra Collective term for all clastic volcanic materials that are ejected from a vent during an eruption and transported through the air, including ash (volcanic), blocks (volcanic), cinders, lapilli, scoria and pumice.

tephrochronology Dating technique that uses volcanic materials, usually tephra, of known age to date associated sediments, and to provide marker beds within the stratigraphic column.

termitarium (pl. termitaria) A mound made by a colony of termites.

terra rosa Residual, red, clayey materials and the soils formed from them. Terra rosa originates from the chemical weathering of limestone and/or the influx of eolian particles onto old, stable surfaces.

terrace (i) A step-like surface, bordering a stream or shoreline, that represents the former position of a floodplain, lake or seashore. (ii) A raised, generally horizontal strip of earth and/or rock constructed along a hill on or nearly on a contour to make land suitable for tillage and to prevent accelerated erosion.

terrain The landscape, or "lay of the land."

terrane An area of a certain structure or rock type, such as granitic terrane or limestone terrane.

Tertiary Geologic period that occurred roughly 65 to 1.6 million years ago. The ensuing period is the Quaternary.

tessera (pl **tesserae**.) Like a pedon but includes both soil and vegetation. It is usually viewed as being smaller than a pedon, but larger than a hand specimen.

tetrahedral sheet Sheet of horizontally linked, tetrahedron-shaped units that are a basic structural component of phyllosilicate clay minerals. Each octahedral unit consists of a central, four-coordinated cation surrounded by four oxygen atoms that, in turn, are linked to other nearby cations. Also see *octahedral sheet*.

texture-contrast soil A soil that has a clay-impoverished upper solum and a clay-enriched B horizon.

thallus (pl. thalli) The vegetative part of simple plants, e.g., lichens, showing no differentiation into root, stem and leaves.

thermic A soil temperature regime that has mean annual soil temperatures of $>15°C$ but $<22°C$, and >5 °C difference between mean summer and mean winter soil temperatures at 50 cm depth. See also *isothermic*.

threshold The point in a system in which a change occurs.

throughfall That portion of precipitation that falls through or drips off of a plant canopy.

throughflow See *interflow*.

tidal flats Areas of nearly flat, barren mud, periodically covered by tidal waters. Normally these materials have an excess of soluble salt.

tile drain Concrete, ceramic, plastic, etc. pipe, or related structure, placed at suitable depths and spacings in the soil or subsoil to enhance and/or accelerate drainage of groundwater from the soil profile by collecting it and leading it to an outlet.

till See *glacial till.*

till plain An extensive flat to undulating surface underlain by till. Also called ground moraine.

tilth The physical state of the soil that determines its suitability for plant growth, taking into account texture, structure, consistence and pore space. It is a subjective estimation and is judged by experience.

time-domain reflectometry (TDR) A method that uses the timing of wave reflections to determine the properties of various materials, such as the dielectric constant of soil as an indication of water content.

time transgressive Usually applied to geomorphic surfaces, the concept that parts of the surface are of different age, or that the age of the surface changes systematically across that surface.

$time_{zero}$ The moment at which soil formation begins and soil profiles and soil bodies and arrangements of them are initiated. The sudden draining of a lake, with subsequent exposure of the bottom to impact of climate and organisms, illustrates how $time_{zero}$ may be introduced into an area. Theoretically each body of soil had a $time_{zero}$. Abrupt changes in environmental conditions may superimpose upon a soil body a succession of $time_{zero}$s, such that it is polygenetic.

toeslope The outermost hillslope position that forms a gently inclined surface at the base of a slope. Toeslopes in profile are commonly gentle and linear, and are constructional surfaces forming the lower part of a slope continuum that grades to a valley or closed depression. Also called alluvial toeslope.

topofunction See *toposequence.*

topography The relative positions and elevations of the natural or man-made features of an area that describe the configuration of its surface.

toposequence A sequence of related soils that differ, one from the other, primarily because of topography as a soil-formation factor. When expressed as a mathematical equation, it is referred to as a topofunction.

topsoil The layer of soil moved in cultivation. Frequently designated as the Ap horizon.

torric A soil moisture regime defined like aridic moisture regime but used in a different category of Soil Taxonomy.

tortuosity The non-straight nature of soil pores.

total acidity The acidity that includes residual and exchangeable acidity. It is often calculated by subtraction of exchangeable bases from the cation exchange capacity determined by ammonium exchange at pH 7.0. It can be determined directly using pH buffer-salt mixtures (e.g. $BaCl_2$ plus triethanolamine, pH 8.0 or 8.2) and titrating the basicity neutralized after reaction with a soil.

toxicity Adverse biological effect due to toxins and other compounds.

toxin Unstable poison-like compound of biological origin which may cause a reduction of viability or functionality in living organisms.

trace elements Usually referred to as micronutrients. In environmental applications, those elements exclusive of the eight abundant rock-forming elements: oxygen, aluminum, silicon, iron, calcium, sodium, potassium and magnesium.

translocation Migration of material in solution or suspension from one horizon to another.

tree-tip mound The small mound of debris sloughed from the root plate (ball) of a tipped-over tree. Local soil horizons are commonly obliterated or folded over, which results in a heterogeneous mass of soil material. Also called cradle knoll.

tree-tip pit The small pit or depression resulting from the area vacated by the root plate(ball) from tree-tip. Such pits are commonly adjacent to small mounds composed of the displaced material. Subsequent infilling usually produces a heterogeneous soil matrix. Pits are commonly sites of increased infiltration and enhanced soil development.

treeline In alpine areas, the ecotone between treeless alpine tundra and the forests downslope.

trioctahedral An octahedral sheet or a mineral containing such a sheet that has all of the sites filled, usually by divalent ions such as Mg^{2+} or Fe^{2+}.

truncated Having lost all or part of the upper soil horizon or horizons by soil removal (erosion, excavation, etc.).

tuff A compacted deposit that is 50% or more volcanic ash and dust. A general term for all consolidated pyroclastic rock. Not to be confused with tufa.

tundra A generally treeless habitat that has an extremely cold climate, low biotic diversity, short growth and reproduction season, and often overlies a layer of permafrost. Found in high alpine and Arctic/Antarctic regions.

Tundra soils A zonal great soil group (1938 system of soil classification) consisting of soils with dark-brown peaty layers over grayish mottled horizons and having continually frozen substrata. Tundra soils form under cold, humid climates, with poor drainage, and native vegetation of lichens, moss, flowering plants and shrubs.

udic A soil moisture regime that is neither dry for as long as 90 cumulative days nor for as long as

60 consecutive days in the 90 days following the summer solstice.

Ultisols A mineral soil order. Ultisols have an argillic horizon with a base saturation of <35% when measured at pH 8.2. They have a mean annual soil temperature of >8 °C.

ultramafic Being very rich in Mg and Fe.

ultraxerous Extremely dry.

umbric epipedon A surface layer of mineral soil that has many of the same requirements as the mollic epipedon, but has a base saturation <50% when measured at pH 7.

unconformity A substantial break or gap in the geologic record where a unit is overlain by another that is not in stratigraphic succession. A buried erosion surface separating two rock masses.

unconsolidated Sediments that are loose and not hardened.

underfit stream A stream that appears to be too small to have eroded the valley in which it flows. It is a common result of drainage changes effected by capture, glaciers or climatic change.

uniformitarianism The geologic principle that assumes that the laws of nature are constant. As originally used it meant that the processes operating to change the Earth in the present also operated in the past and at the same rate and intensity and produced changes similar to those we see today. The meaning has evolved and today the principle of uniformitarianism acknowledges that past processes, even if the same as today, may have operated at different rates and with different intensities than those of the present.

unit structure In phyllosilicate mineral terminology, the total assembly of a layer plus interlayer material.

unloading The release of confining pressure associated with the removal of overlying material. May result in expansion of rock, accompanied by the development of joints or sheets. Also known as dilatation.

unsaturated flow The movement of water in soil in which the pores are not filled to capacity with water.

upfreezing The process whereby rocks and other particles are gradually moved upward by freeze–thaw activity.

urban land On soil maps, areas so altered or obstructed by urban works or structures that identification of individual soils is not feasible.

urbanthroturbation Human-initiated, non-agronomic soil mixing. Often refers to excavations such as mines and pits, road construction, urban land moving, cut-and-fill operations, etc.

ustic A soil moisture regime that is intermediate between the aridic and udic regimes and common in temperate subhumid or semiarid regions, or in tropical and subtropical regions with a monsoon climate. A limited amount of water is available for plants but occurs at times when the soil temperature is optimum for plant growth.

vadose water Water in the vadose zone, above the water table.

vadose zone See *zone of aeration*.

valency The number of bonds that can be formed by an atom of an element (or the numerical value of the charge on an ion of the element).

valley wall The steep slope that borders the outside edge of a river floodplain.

value See *color value*.

van der Waals forces Non-polar bonds caused by accumulations of positive or negative charges at the ends of molecules.

vapor flow The gaseous flow of water vapor in soils from a moist or warm zone of higher potential to a drier or colder zone of lower potential.

variable charge A solid surface carrying a net electrical charge which may be positive, negative or zero, depending on the activity of one or more species of potential-determining ions in the solution phase contacting the surface.

varve A sedimentary layer, lamina or sequence of laminae, deposited in a body of still water within one year. Specifically, a thin *pair* of graded glaciolacustrine layers seasonally deposited, usually by meltwater streams, in a glacial lake or other body of still water in front of a glacier. It usually consists of a lighter and darker portion due to the change in rate of decomposition in the winter vs warm season.

ventifact A stone or pebble that has been shaped, worn, faceted or polished by the abrasive action of wind-blown sand, usually under arid conditions. The flat facets meet at sharp angles.

vermiculite A highly charged layer silicate of the 2 : 1 type that is formed from mica. It is characterized by adsorption preference for potassium, ammonium and cesium over smaller exchange cations. It may be di- or trioctahedral.

vermiform Worm-like; in soil micromorphology, the term refers to the fecal material of worms or fillings in faunal passages. The material may be extraneously or locally derived.

Vertisols A mineral soil order. Vertisols have 30% or more clay, deep wide cracks when dry, and either gilgai microrelief, intersecting slickensides or wedge-shaped structural aggregates tilted at an angle from the horizon.

vesicle (i) In geology, a cavity in a lava, formed by the entrapment of a gas bubble during solidification of the lava. (ii) In soil micromorphology, a circular, ovoid or prolate soil pore with a smooth surface.

vesicular A textural term applied to an igneous rock containing abundant vesicles.

viscosity The resistance to flow in a liquid.

void In soil micromorphology, a void is the same as a pore, which is the preferred term.

volcanic ash Fine particles ejected during a volcanic eruption, usually highly silicic, dust-sized, sharp-edged and glassy.

volcanic ejecta Material physically ejected from a volcano, i.e., tephra.

volcaniclastic Pertaining to the entire spectrum of fragmental materials with a preponderance of clasts of volcanic origin. The term includes not only pyroclastic materials but also epiclastic deposits derived from volcanic source areas by normal processes of mass movement and stream erosion.

vughs In soil micromorphology, relatively large voids, usually irregular and not normally interconnected with other voids of comparable size; at the magnifications at which they are recognized they appear as discrete entities.

waterlogged Saturated (or nearly so) with water.

water suction See *matric potential.*

water table The upper surface of groundwater or that level in the ground where the water is at atmospheric pressure. The surface between the zone of saturation and the zone of aeration.

weatherable minerals Those minerals that are relatively easily weatherable in the soil/surficial environment.

weathering The breakdown and changes in rocks and sediments at or near the Earth's surface produced by biological, chemical and physical agents or combinations of them.

weathering front The lower edge of the weathering profile. Materials below the weathering front are either not weathered or weathered to a much lesser degree than are materials above.

weathering profile A vertical cut from the surface down through the weathered materials at the Earth's surface, including the soil profile and weathered materials below, but not including solid bedrock.

weathering rind A weathered layer on the outer edges of a rock.

welded soil A soil that is buried so shallowly and for a long enough period of time that pedogenesis has linked it to the surface soil above, i.e., there is no C horizon material existing between the solum of the surface soil and that of the buried soil.

wetland A transitional area between aquatic and terrestrial ecosystems that is inundated or saturated for long enough periods to produce hydric soils and support hydrophytic vegetation.

wetting front The boundary between the wetted region and the dry region of soil, during infiltration.

white mica See *muscovite.*

W horizon A surface horizon, usually in the tropics, formed as worms bring material to the surface. It is dominated by worm casts.

wilting point See *permanent wilting point.*

xeric In Soil Taxonomy, a soil moisture regime common to Mediterranean climates that have moist cool winters and warm dry summers. A limited amount of water is present but does not occur at optimum periods for plant growth. In general usage, xeric refers to dry climatic conditions.

xerophyte Plant that grows in extremely dry areas and does not have its roots in wet soil below the water table.

X-ray diffraction The diffraction of a beam of X-rays by the three-dimensional array of atoms in a crystal structure. The identity and arrangement of atoms in the structure can be determined by interpreting the angles at which X-rays are scattered by the structure and the intensities of scattered beams.

zircon $ZrSiO_4$. A hard, resistant mineral, found as an accessory mineral in many kinds of rocks.

zonal soil (i) A soil characteristic of a large area or zone. (ii) One of the three primary subdivisions (orders) in soil classification as used in the United States (1938 system of soil classification). The term is used today only in an historical context.

zone of aeration Zone immediately below the ground surface but above the water table within which pore spaces are partially filled with water and partially filled with air. Also called the vadose zone.

zone of saturation The zone below the zone of aeration in which all pore spaces are filled with water.

Index

A horizons 36, 46–48, 208, 386, 541, 608, 633, 637, 640
  and cumulization 153, 206
  definition/concept 37, 539
  formation 103, 244, 249, 254, 355, 356, 362, 458, 541
  images 46
  plow zone (Ap horizon) 254, 292, 318
  processes associated with 94, 240, 265, 321, 335, 356, 393, 398, 430, 623, 650
  structure 283, 357
  suffixes 38
  thickness 105, 151, 271, 287, 289, 315, 349, 356, 392, 442, 456–458, 479, 483, 484, 489, 503, 505, 573
  vesicular (Av) 332, 434–438, 556
abiotic processes 466
absolute age (of a soil or surface) *see* numerical age dating
accreting zones 532, 534
accretion gleys 518, 521
Accumulation Index 570, 571
acid ammonium oxalate 447, 448, 459
acid rain 230, 297, 320
acid soils/soil acidity
  changes with depth 176, 386, 442
  factors that lead to acidity 184, 382
  implications/effects of acidity 206, 370, 373, 394, 429, 439, 441, 442, 510, 642
  situations/settings with acid soils 210, 377, 387
acid sulfate soils 194, 195, 455–456
acidification 278, 305, 349, 359, 360, 363, 365, 373, 384, 480, 537, 637
acidity effect 204
acidocomplexolysis 349, 351
acids
  inorganic 444
  organic 174, 209, 210, 234, 235, 236, 350, 351, 359, 365, 373, 382, 394, 396, 441, 444, 445, 447, 449, 502, 537
Acrudox 514
Acrustox 177
actinolite 62, 202
actinomycetes 15, 44, 97, 99, 354, 419
active layer 265, 266–267, 272, 274
additions/gains (to a soil or its horizons) 47, 169, 321, 322, 325, 329, 330, 348, 353, 456, 460, 461, 470
adhesion/adhesive forces 82, 84
aeration, soil 94, 103, 259, 382
aerial photographs *see* air photos
Afton soil/paleosol 518, 520, 521, 528, 529, 530
age (interval of time measured back from the present) 547
  maximum ages/dates 412, 604, 605
  minimum ages/dates 566, 567, 604, 605
aggradation (additions to soil surface) 326, 328, 331, 456–458, 467, 521
aggregate, soil 17, 18
aggregation 58, 94, 97, 99, 101
agric horizon 118, 292
agriculture 292
ahumic soils 277
air photos 150, 158
Albaqualfs 384, 430, 628
Albaquults 514, 515, 628
albedo 87–88, 231, 343
albic horizons/materials 117, 122, 373, 379, 442, 624
albite 62, 233
Albolls 116, 426
alcrete 391
Alfisols 127, 223, 282, 388, 645
  pedogenesis 126, 244, 262, 335, 354, 579
  relationships to geomorphology 396, 454, 514
  taxonomy/classification 121, 362
algae 93, 236, 418, 420, 440
alkaline conditions 351, 624, 639
  alkalai salts 427
  alkalai soils 427
  alkalinolysis 351
  alkalization 112, 278, 349, 428
allitization 350
allophane 208, 209, 210, 373, 444, 448, 449, 639
alluvial fans 191, 193, 407, 409, 556
alluvial surfaces 169
Alluvial soils 109, 114, 123
alluviation 348
alluvium 191–193, 212, 213, 605, 617
  characteristics 191, 212, 239, 612
  classification/categorization
  implications for pedogenesis/geomorphology 152, 254, 324, 332, 387, 396, 456, 459, 483, 514, 532, 548, 621, 623, 636
almandine 202
alpine environments/soils 231, 237, 312, 313, 314, 316, 345, 512, 559, 562, 633
  Alpine Meadow soils 109, 113, 123
  alpine treeline/tundra 313, 315
  treeline 567
alternating zones 532, 534
altitude 87
aluminosilicates *see* minerals, primary silicate; minerals, secondary silicate
aluminosilicic acid 233
aluminum 209, 210, 308, 388, 393, 395, 407, 444, 449, 452, 580
  ^{36}Cl 612, 637
  aluminum chlorite 371
  exchangeable 209, 352, 359, 373, 382, 392, 440, 444
  oxides 205, 232, 390, 391, 392, 399, 445, 579, 586, 587
  toxicity 209
AMS dating 601, 602
ammonium 402
amoebae 99
amorphous materials 27, 208, 210, 376, 378, 441, 579
amphibians 104
amphiboles 61, 62, 68, 181, 521, 562, 580, 581, 583
amphibolite 179
anaerobic/anoxic conditions 17, 92, 352, 380, 382, 455, 624, 635
anatase 202, 386, 542, 584
andalusite 202
andesite 180, 182, 227
andic properties 122
andisolization 450
Andisols 207, 208, 396, 514

Andisols (*cont.*)
  pedogenesis 126, 210, 354, 362, 608, 635, 636
  taxonomy/classification 117, 121, 208–210
Ando soils 123, 208
Andosols 208
angle of incidence 87
anhydrite 232
animals, burrowing 231
anion adsorption 209
anions, soil 236
anisotropy 7, 239, 323, 327, 348, 574
anomalous fading 618
anorthite 62
anoxic/anaerobic conditions 190
Anthrepts 635
anthropedogenesis 320
anthropogenic soils 319
anthroposequence 320
anthrosols 293, 319
antigorite 65, 68
ants 97, 101, 253, 255
  anthills 245, 250, 252, 253, 283, 501
  implications for pedogenesis 47, 244, 250, 388, 534, 539
  implications for pedoturbation 243, 246, 247, 248, 249, 328, 544
apatite 61, 202, 582
Ap horizon 48
Appalachian Mountains 172, 212, 358
apparent mean residence time date 606
aquitards
  aquicludes 212, 267, 335, 363, 366, 367, 379, 382, 385, 400, 404, 620, 625, 637
  development/examples 363, 367, 373, 412, 429, 541
  impact on geomorphology/slopes 482, 486, 491
  impact on pedogenesis/pedogenic pathways 299, 366, 370, 378, 404, 443, 445, 479–480, 538
Aquods 443, 453
Aquolls 630
Aquorthels 273
Aquults 163, 514, 630
arachnids 101
aragonite 186
Archaea 93
Archaic Period 254
archeology 217, 239, 240, 254, 264, 319, 560, 566, 622, 634
Arctic landscapes/soils 277
areals, elementary soil 158–159
arenization 171, 180, 351
argillation 350, 361
argillic horizons
  Argiaquolls 490
  Argids 407
  Argiudolls 516
  Argiustolls 403, 571, 649
  Argixerolls 293
  genesis 240, 306, 370, 452, 534, 543, 568
  taxonomy/classification/concept 115, 117, 122, 128, 131, 362, 368
  utility in soil geomorphology 185, 570, 624, 625, 636
  see also cutans, argillans
argilluviation; *see also* lessivage 350, 361
Aridisols 127, 247, 293, 400–422, 439, 608, 634, 635, 647
  pedogenesis 126, 169, 362
  relationships to geomorphology 407
  taxonomy/classification 118, 121
arkose sandstone 180, 184
arthropods 43, 45, 91, 94, 253
artifacts, human/archeological 240, 242, 243, 254, 259, 260, 264, 541, 542, 546, 560, 566, 617
ash, fly 229
ash, volcanic 208, 407, 423, 623
  Bishop ash 599, 600
  Crater Lake ash 599
  as a dating tool 597, 599, 600, 618
  Glacier Peak ash 600
  Huckleberry Ridge ash 599, 600
  impacts on pedogenesis 422
  Lava Creek ash 599, 600
  Mazama (Crater Lake) ash 600
  Mesa Falls ash 599
  Mount St. Helens ash 599, 600
  Pearlette ash 599, 600
  physical and geomorphic characteristics 207, 387
  and taxonomy/classification 117, 121, 129, 132
  White River ash 600
aspect, topographic/slope 127, 153, 262, 267, 292, 295, 302–310, 312, 315, 399, 474, 475–476, 510, 633
auger, hand 150, 151
augite 61, 62, 202, 385, 581
autotrophs 97
available water 388
Av horizons *see* vesicular A horizons
Azonal soils 108, 109

B horizons 36, 48–50, 368, 392, 622
  color 280, 443, 499
  "color B" horizon 120, 452
  definition/concept 37, 240, 539
  development/degradation of 305, 373, 431, 608
  formation of 245, 453, 459
  processes associated with 209, 335, 361, 382, 427, 458, 518, 578, 580
  structure 283
  suffixes 38–39
  thickness 151
  see also subsoil
backscattered electrons 26
backswamps 191
bacteria 93, 97
  ecosystem function 91, 95
  role in pedogenesis/humus formation 44, 352, 354, 419–420, 439, 440, 455, 643
badgers 231, 247, 416
Baldwin, M. 337
ball-and-stick diagrams 55, 63, 64
barrier islands 194, 195
basal plane 63
basalt 179, 180, 182, 184, 207, 227, 278, 289, 393, 407, 437, 557, 560, 615
bases/base cations
  and clay minerals 280, 433, 639
  base cycling 182, 323, 348, 357, 358–360, 373, 384, 456, 537, 553, 637
  base saturation (cases, examples) 116, 131, 188–189, 209, 219, 222, 318, 370, 383, 387, 398–399, 511, 569, 587, 625, 626, 636
  base status 179, 180, 182, 250, 262, 332, 360, 392, 456
  gain of, impact on pedogenesis 366, 514, 624
  loss of (depletion), impact on pedogenesis 173, 174, 206,

322, 333, 350, 351, 352, 359, 360, 365, 371, 373, 382, 384, 393, 394, 430, 442, 444, 451, 452
basic soil unit 106
Basin and Range 407–408
basins of interior drainage *see* catenas, closed
bauxite 170, 174, 391–392, 395
Bayerite 56, 58
beaches 194
bedding planes 227
bedrock
  and saprolite 174, 460, 538
  impact on pedogenesis 226, 291, 367, 411, 427, 479, 536
  impact on soil taxonomy/classification/horizonation 34, 37
  impacted by weathering 32
  shallow to bedrock soils/situations 190, 201, 212, 256, 314, 454, 544
beetles 97, 101, 189
beidellite 66, 67, 72, 277
belemnite standard 645
beryl 62, 219
  ^{10}Be 612, 613
  garden variety 612, 615, 632
  in situ type 614
Beschel, R. E. 562, 564
beta horizon 367
beta particles 601
bicarbonate(s) 186, 234, 352, 400, 404, 424, 447, 600
Bilzi–Ciolkosz Index 573, 575
biochemical processes 299, 310, 323, 398, 537
bioclasts 541, 542
biocrusts 439–440
biocycling/biocycled 186, 234, 236, 327, 330, 348, 356, 357–359, 377, 393, 407, 430, 456
biodiversity 163
biofabric 242
biogeochemical cycling 348, 358
biogeomorphology 256
biological activity 641
biomantles 242–244, 538, 539
  formation/genesis 250, 257, 261, 262, 329, 363, 398, 534–537, 541, 544, 546, 631
  and lithologic discontinuities 216–217, 219, 634
biomass/biomass production 149, 257, 313, 323
biomechanical processes/agents 5, 295, 299, 310, 325, 388, 398, 456, 535, 537–538, 539, 545, 546
biopedological compartment 447
biopores, soil 17, 18, 252, 292
biosequences 310
  biofunctions 299, 310
biota, soil 93, 238, 535, 539, 541, 544, 546
  biological activity in soils 87, 93
  biotic activity in soils 50
  see also organisms, soil
biotite 66, 67, 68, 69, 79, 179, 181, 184, 202, 385, 583, 639
birds 91, 94, 104
Birkeland, P.W. 465
birnessite 60, 495
"biscuit tops" 20, 429
bisequal soils 35, 50, 373, 374
bisiallitization 350
Bk horizon 92
black alkali soils 427
Blackland Prairie 280, 293
Black soils 147
block diagrams 156, 168, 213, 281, 307, 418, 489, 507
blockfield zone 314
boehmite 56, 58, 59, 174, 178, 388, 392
bogs 190
  Bog soils 109, 113, 123
bonding agents 379
bonds, ionic 54
bone 604
boulders/bouldery 12
boundary condition; *see also* time$_{zero}$ 595
Bowen's series 226
Brady paleosol 627
Bragg's law 76–77, 78
Braunerde 109, 113
breccia/brecciation 180, 413, 415
bristlecone pine 567
brittleness 375, 376, 378, 379, 422
Brown (Light Chestnut) soils 147
Brown Forest soils 109, 113, 123
Brown Podzolic soils 109, 111, 123
Brown soils 109, 110, 123
Brunhes Epoch 597
brunification 69, 276, 278, 352, 353
Brunisolic soils 147
Brunizem soils 124
Bryant, R. B. 375
bulk density 18, 83, 209, 219, 222, 387, 511
  factors that affect bulk density 14, 19, 172, 247, 253, 350, 374, 377, 378, 461
  impact on pedogenesis 88, 90, 373, 430
  use in pedogenic quantification 570, 579, 584, 586
Bull Lake glaciation 571
Buntley–Westin index 572
buried soils/paleosols 52–53, 587, 620, 621
  burial of soils/surfaces, the process 300, 467, 615, 621, 632
    impact on pedogenesis 295, 453, 458, 589–590, 606, 631, 636
    processes associated with burial 50, 326, 544
    rates of burial 209, 324–325, 349, 457, 616, 620
    settings/situations of burial 215, 274, 477, 483, 484, 529–534, 544, 621, 644
  during- and post-burial modification 458, 569, 622, 623, 624
  examples/settings 206, 211, 332, 459, 626
  impact on surface soils/pedogenesis 375–378
  isolated paleosols 169, 622
  utility in soil geomorphology 529, 534, 549, 582, 608–610, 617, 630, 635, 641, 642, 643, 651
burrows/burrowing animals 240, 247, 249, 333, 351, 427, 433, 442, 471
Butler, B. E. 532

C3 plants/pathway 646
C4 plants/pathway 646, 648
C horizons 50–51, 167, 575, 595, 618
  definition/concept 36, 37, 51–52, 539
C:N ratio 354, 355, 358, 635
calcification 110–112, 113, 276, 277, 313, 332, 349, 367, 402–420, 421, 641
  calcareous materials 366

calcification (*cont.*)
calcic horizons 118, 122, 128, 131, 306, 329, 403, 404, 405–406, 412, 418, 422, 424, 429, 438, 502, 507, 570, 624, 636, 639, 649
Calcic Solonchak soils 428
Calcids 407, 579, 636
Calcisols 123
Calcium Carbonate Solonchak soils 123, 428
calcium, exchangeable 252, 262, 355, 358–359, 365, 366, 424, 429, 503, 553, 637
Calcomorphic soils 109
calcosiallitization 349
calcrete 49, 391, 403
moisture-limited vs. influx-limited 410
per ascensum model 403, 422
per descensum model 403, 411–415, 422
Calcium-41 612
calcium-humus bonds/complexes (Ca-humates) 189, 191, 355, 356, 358, 636
calibrated age dating/methods 551
Calvin (C3) cycle 646
CAM pathway/plants 646, 648
cambic horizons 17, 115, 120, 122, 128, 354, 452
Ca : Mg ratio 183
Canadian system of soil classification 147
canopy interception 473
canopy interception loss 472, 473
capillarity/capillary water 83, 400, 403, 407, 421, 422, 425, 487, 489, 501, 502
carbon-13 611, 644
carbon cycle/dynamics 103, 635, 645
carbon dioxide 92, 227, 234, 236, 274, 402–413, 415, 419, 600, 645, 646, 647, 648
carbonates 16, 38, 55, 60, 78, 92, 194, 199, 203, 287, 348, 355, 359, 400, 402, 407–411, 418, 423, 427, 602, 618
accumulation rates 410, 412
caliche 49, 403, 611, 649
calcite 60, 194, 197, 202, 230, 293, 418, 420, 423, 433, 641
calcium carbonate 232, 302, 349, 400, 402–413, 422, 428, 435, 571, 604
coatings 301, 412, 556
content 261, 286, 412, 627
content of ^{13}C (isotopes) 409, 644
dating of 606, 610, 612, 648
depth to 152, 406, 425, 607, 641
distribution by depth 289, 411, 626
filaments 411, 413, 420
modeling the accumulation thereof 414, 417
nodules 413, 431, 611, 626
precipitation in soils 404–405, 431, 624
secondary/pedogenic; *see also* calcification 130, 403, 416, 426, 571
affected by pedoturbation 247, 289
depth functions 285, 287, 405, 411, 438
impacts on taxonomy/classification 118, 131
formation/pedogenesis 257, 287, 333, 335, 349, 354, 402–424, 430, 434, 438
from groundwater 490
utility in soil geomorphology/dating 579, 597, 603, 610, 617–618, 633, 641, 643
carbonation 186, 234, 235, 278, 403, 405, 406, 418
carbonic acid 227, 233, 234, 359, 403, 415, 418, 441, 444, 447
carpetolith 169, 543
carpedolite 543
carpedolith 543
cassiterite 542
catenas 152, 620
closed 471, 476, 484, 486, 506, 507
examples 153, 311, 427, 482, 484, 506–514
and soil mapping/maps 157, 158
theoretical/conceptual 114, 149, 302, 469–486
cations
base *see* bases/base cations
cation exchange 233
cation exchange capacity (CEC) 219, 383, 553
impacts on pedogenesis 371, 394, 452
factors that affect CEC 71, 73, 352, 382, 503, 579
impact on classification/taxonomy 117, 120, 131, 133, 386, 387
cation exchange resin 446
exchangeable 180
general 54, 236, 566, 613
leaching 566
cation ratio dating 552, 566
caves 234
celadonite 66, 67, 70
cellulose 354
cementation/cementing agents 132, 375, 378, 388, 392, 396, 413, 422, 427, 443
Cenozoic Era 342
centipedes 97, 101
central concept 133, 141
cesium-137 290
chalcedony 181, 424
chalk 181, 201, 227, 280
channel ferrans 496
channers/channery 12
chaos *see* deterministic chaos
charcoal 215, 603, 652
chelates/chelation 210, 234–236, 314, 324, 347, 352, 365, 393, 396, 446, 449, 642
chelate-complex theory/model 444–448, 449
cheluviation 352, 445
Chernozem soils 109, 110, 123, 296, 609
Chernozemic soils 147
chert 181, 187, 188
Chestnut soils 109, 110, 123
chitin 45, 354
chlorides 60, 349, 400, 402
chlorine-36 612, 613, 615
chlorite 62, 66, 67, 70, 72–73, 79, 80, 180, 181, 640
chlorite schist 180
chloritization 352, 382
conditions that favor its formation 352
what it alters/weathers to 184
chromium 182, 183, 393
chronosequences 42, 300–301, 333, 335, 453, 454, 459, 466, 568, 571, 576, 577, 582, 587–596
chronofunctions 44, 299, 300, 301, 335, 338, 341, 570, 585, 587, 588, 590–596

post-incisive 588, 589, 590
pre-incisive 588, 589, 590
time transgressive with historical overlap 588, 589
time transgressive without historical overlap 588, 590
chrysotile 65, 68
cicadas 101, 240, 250, 255, 332, 412
ciliates 99
cinder cones 305
cinders 208
classification, soil 34, 106–146, 297, 324, 347, 390, 393, 399, 501, 620, 635
1938 soil classification system 109
class limits (taxonomic) 107, 114, 145, 146
Soil Taxonomy 123
formative elements 115, 121, 128–129, 130, 131–132, 134–139, 140
taxa/taxonomic classes 107, 144, 145, 163
clast sound velocity, as a dating tool 559
clastic particles 9, 456
clay 9, 321, 371, 424, 439, 452, 509, 579, 613
accumulation rates 410, 631
bridging 376
clay–humus complexes 168, 355, 358, 359, 452, 453
clayey soils 168, 502, 613
content
genesis/distribution 196, 282, 284, 301, 313, 453, 481, 587, 593, 594, 623, 635
utility in soil geomorphology/pedogenic quantification 219, 312, 412, 553, 554, 571, 578–579
depth functions of 311, 475, 479, 484, 513, 625
dissolution/degradation 382, 384, 441
domain 366
eating/ingestion *see* geophagy
films; *see also* cutans, argillans 49
fine clay fraction 277, 280, 282, 355, 363, 367
fine/coarse clay ratio 363
low activity 370, 392
minerals/mineralogy *see* minerals, clay
silicate clays 121, 191, 441
skins; *see also* cutans, argillans 49
tablets 75
translocation *see* lessivage
Clay Accumulation Index (CAI) 571, 579
clay-free calculations/basis 217, 219, 222, 223
clayfree sand 219, 220
clayfree silt 219, 220
claypan 384
clay/silt ratio 579
cleavage 70
climate/macroclimate 122–127
biofactor 303
climate change *see also* polygenetic soils/polygenesis 312, 589, 611, 621
how soils can help identify past climate changes 217, 466, 631, 633
impact on geomorphology/landforms 196, 524
impact on pedogenesis/soil morphology 190, 295, 328–333, 335, 349, 367, 384, 410, 415, 450, 568, 611, 641
climate regions 129
cold 237, 253, 358, 552, 639, 640
humid 312, 610
impact on pedogenesis 356, 358, 359, 363, 393, 422, 427, 440, 502, 634, 636, 646, 647
impact on weathering 209, 237, 394, 438, 640
humidity factor 303
Mediterranean/xeric 335, 399, 422, 506, 510, 636
microclimate 292, 302, 310, 316, 399, 502, 565, 566
perhumid 366, 443
impact on pedogenesis 168, 313, 337, 339, 425, 439, 444, 450, 486, 502, 635, 640
impact on weathering 236, 238
proxies 643
semi-arid and arid 253, 312, 402, 438–439, 510
clay minerals 182, 280, 394, 637, 639
dating applications 556, 560, 566
impact on pedogenesis 207, 305, 363, 406, 425, 502, 506, 543, 634, 636, 641
impact on weathering 209, 231, 236–237
soils information used to identify dry paleoclimates 636, 643, 646
tropical *see also* tropics/tropical landscapes/environments 128, 234, 236, 257, 300, 385, 392, 450, 452, 636
wet–dry 277, 284, 333, 382, 425, 637, 643
see also state-factor model/approach
climax community 337
climosequences 312, 313, 317, 425, 641
climofunctions 299, 313
clorpt model *see* state-factor model/approach
close-packing 54, 376
co-alluvium 211, 212, 215
coal 227
coarse fragments/fraction 13, 571
definition 6, 12–14
how redistributed by surficial processes 246, 262, 264, 265, 266, 271, 290, 292, 293, 535, 542
impacts on pedogenesis 47, 302, 354, 359, 406, 535
utility in interpreting soils/sediments 220, 240, 301, 542
see also gravels/gravelly
coarse-textured layers/materials 87, 441, 442, 451, 452, 471, 482, 506, 538, 637
coastal plain deposits 194
cobalt 183
cobbles/cobbly 12
cockroaches 97, 101
coefficient of linear extensibility 19, 277, 280, 282, 287, 378
coefficient of thermal expansion 231
Coffey, G. N. 296
cohesive forces 82, 84
collapse in clay minerals 72, 77, 79
of materials, landforms or landscapes 172, 232, 250, 256, 264, 318, 349, 378, 393–395, 457, 460–461, 539
collembola 43–45, 354

colloids 359, 369
colluvium 156, 172, 212, 213, 215, 216, 217, 309, 387, 396, 475, 506, 508, 514, 525, 544, 612, 617
  colluvial soils 212
  colluviation 212, 471, 543
color, soil 14–17
  chroma 15, 16
  color chips, Munsell 15
  Color Development Equivalent (CDE) 572
  color (soil development indices) 572–573
  darkening/dark colors; *see also* melanization 356, 427, 429
  factors that affect 58, 169, 209, 302, 399, 438, 469, 494, 624
  gleyed colors *see also* gleyed, gleying 381, 492, 495
  hue 15, 495, 573
  lightening/light colors; *see also* leucinization 442, 575
  Munsell soil color system 15, 16, 572
  patterns in soils 311, 501
  red/redness 178, 305, 352, 368, 381, 386, 387, 392, 452, 496, 628, 630
  use as a diagnostic pedogenic property 219, 222, 238, 289, 369, 394, 443, 494, 495, 572
  value 15, 16
  of various materials/sediments 198, 455
Colorado Plateau 185
combinations, soil 149, 158
compaction, soil 14, 104, 624
comparative particle size distribution (CPSD) index 221, 223, 224
complex landscapes 156
complex soils 156, 456
complexes, organo-metallic; *see also* chelation 30, 209–210, 443, 444–447, 452, 509
compound soils 456
concretions 24, 174, 219, 335, 390, 495, 543
conduction 90
conductivity, electrical 425
conglomerate 180
consistence, soil 20–22, 219, 222, 289
  friable 22
convection 90, 92
coordination number 55
coprogenic materials 45
coquina 181
coral 602
corestones 32, 33, 173, 174, 177, 515
corraded/corrasion 226
corundum 202, 542
cover, soil 160
coversand 196
Cox, G. W. 256
Cox horizon 52
crabhole 283
cracks in soils
  depth of 286, 290
  desiccation 269, 280–289, 335, 355, 429, 437, 510
  shrinkage 264
  thermal 275
cradle knoll microtopography; *see also* mounds 501
crayfish 283
creep, soil/slope 173, 196, 212, 216, 217, 246, 254, 292, 397, 482, 484, 485, 525, 535, 538, 539, 542, 543, 546
CrG horizons 539, 541
crickets 97
cross bedding 191
cross dating; *see also* dendrochronology 567
croûte de nappe 421
Crt horizons 539, 541
crust *see* surface crust
crustaceans 101
Crw horizons 539, 541
crystal lattice 232
cS Index 579, 580
cumulization 169, 206, 209, 310, 325, 349, 456–460, 471, 476, 484
cutans 15, 18, 24, 28, 48, 49, 356, 553, 623, 641
  albans 371, 495, 497, 498, 499
  albic neoskeletans 495, 497
  allans 444
  argillans 38, 117, 388, 497
    definition/concept 31, 49
    images 25, 27, 30, 364
    utility in paleoenvironmental reconstruction 599, 641
    utility in soil geomorphology 51, 173, 363, 367, 373, 397, 499, 569
  calcans 49, 641
  ferrans 496–497
  ferriargillans 376, 497
  ferro-mangans 496
  mangans 49, 497
  neoalbans 497
  neocutans 371
  neoferrans 497
  neopedferrans 497, 499
  organs 30, 354, 444, 495
  quasiferrans 497
  sesquans 49
  silans 49, 371, 382, 384, 497
  skeletans 49, 371, 382, 497, 498, 499
cyanobacteria 93, 97, 201, 440
cyclosilicates 61, 62, 63

Darcy's Law 86
Dark Brown (Dark Chestnut) soils 147
Dark Gray soils 147
Darwin, C. 4, 5, 242, 254, 466, 543
date (specific point in time); *see also* age 547
dating techniques 208, 341, 549
Davis, William M. 5, 108, 466
dealkalization 112, 347, 349, 429
  dealkalinization 350
debris flows 212, 216, 239, 246, 559, 621, 632
debris flux 470–472, 476, 482
decalcification 325, 349, 359
  decarbonation 189, 348, 365
decomposers 94, 97
decomposition 41–42, 44, 231, 232, 351, 352, 354, 355, 358, 387, 441, 483, 502
  decomposition and synthesis *see* dissolution and synthesis
  rates 190, 191, 509
deepening, profile; *see also* thickness, soil 325, 326, 328, 330, 332, 349
deflation 195, 196, 246, 275, 316, 330, 421, 430, 435–437, 438
deforestation 318
degraded horizons/degradation of soil horizons 48, 338, 369–373, 385, 429, 497, 542
Degraded Alkalai soils 429
Degraded Chernozem soils 109, 111, 123
dehydration 210, 392, 394
delineation 146
deltas 191, 194
dendrochronology 562, 567, 602, 603

dendrogeomorphology 567
denitrification 149, 324
densification 375
denudation, soil/landscape 104, 172, 226, 236, 238, 543
depodzolization 318, 349, 450
deposition zones 212
depth plots/functions 22, 219, 225, 623
  examples 22, 179, 218, 263, 289, 311, 315, 377, 386, 423, 448, 460, 475, 479, 484, 522, 563, 583, 585, 609, 613, 626
  utility in soil geomorphology 216, 218, 223, 388, 580, 615, 623
desalinization 112, 349, 428
desert lacquer 438
deserts
  desert pavement 196, 277, 278, 433, 434, 437, 553, 555, 615, 636
  Desert soils 109, 110, 123
  desert varnish (see "rock varnish")
  pavement development index 556
  see also climate, arid; climate, semi-and 556
desiccation 48, 276, 316, 366, 375, 376, 378, 404, 405, 461, 639
desilication 174, 178, 186, 210, 234, 350, 386, 388, 394–399, 579, 639
deterministic chaos/uncertainty 339–341, 589, 590
devegetation 212
D horizons 32, 33, 36, 37, 51–52, 167
diabase 180
diachronous units/surfaces 547
diagenesis 622, 624
diapir 271, 285
diaspore 56, 58, 59
diatoms 423, 652
dickite 65, 67
dielectric number 83
differential thermal analysis (DTA) 80
diffusion 90, 92
digitization 156
dilatation 227
dilation 173, 349, 460
dioctahedral sheet 64, 65, 68
diopside 61, 62, 202
diorite 179, 180, 182
discriminant analysis 225
disintegration 227, 229, 351, 543
dispersion/deflocculation (of clay minerals) 72, 78, 363–366, 388, 427, 428, 433, 497
dissimilar series/soils 149, 157
dissolution and synthesis 71, 233, 234, 245, 351, 362, 403, 424, 439
  dissolution, congruent 231, 234
  dissolution, incongruent 231
distributary channels 194
disturbance/disturbed lands 42, 320
ditrigonal cavity 71
ditrigonal ring 71
diversity, pedologic 163
Dokuchaev, V. V. 4–5, 36, 52, 108, 109, 158, 296, 297, 310, 337, 539
dolerite 180, 227
dolomite 60, 181, 186, 197, 201, 227, 293
dose rate; *see also* luminescence dating 617, 618
drainage, artificial 492
drainage classes, soil 146, 151–154, 156, 158, 160–163, 492
  excessively drained 493, 495
  imperfectly drained 493
  moderately well drained 153, 493, 494
  poorly drained 153, 450, 480, 493, 494, 642
  somewhat excessively drained 493
  somewhat poorly drained 153, 493
  very poorly drained 153, 493, 494
  well drained 153, 313, 366, 493, 494, 495, 642, 643
drainage index, soil 149, 162, 163
drift, glacial 210, 211, 532
drumlins 161, 162
dry edge 499
dry soil 122
d-spacing 70, 76, 77, 78
duff/duff mull 43, 44
dunes; *see also* eolian materials/sediments 191, 193–196, 200, 454, 510, 532
dunite 182
duripans 49, 119, 122, 128, 131, 256, 422, 423, 424, 636
  duricrusts 173, 391, 396, 422, 535
  Durids 636
  durinodes 422
dust, eolian 207, 295, 409, 421, 579
  as a state factor 297
  impact on pedogenesis 182, 322, 348, 394, 406, 419, 424, 430, 436, 459, 544, 617
  influx rates 411
  sources 176, 194, 205, 420, 506, 611
  utility in soil geomorphology 174, 334, 409–410, 423, 434, 437, 439, 457, 510, 633, 634
dynamic denudation 537–543, 546
dynamic equilibrium 238
dynamical instability 340
dynamical system theory 339
Dystric Brunisol soils 147
Dystrudepts 163, 170

E horizons 117, 128, 131, 363, 430, 442
  color of 578
  definition/concept 36, 37, 48, 368
  formation of 361, 362, 441, 445, 449, 452, 453, 537
  images 46, 504
  processes associated with 210, 385, 429, 445, 514, 542
  thickness 151, 305, 370, 373
earth hummocks 275
Earth's crust 54
Earth's magnetic field 596
earthworms/worms 46, 91, 97, 101, 247, 252–253, 255
  anecic worms 103
  casts 103, 240, 242, 252, 283, 357, 544
  endogeic worms 103
  epigeic worms 103
  impact on pedogenesis 43, 44, 45, 47, 94, 103, 189, 240, 242, 246, 249, 254, 332, 354, 442, 539, 544
  krotovinas 252
  middens 47, 104, 249
  see also nightcrawlers; red wigglers
earwigs 97
eccentricity of the orbit 342, 343
ECEC 386
ecotones 641
ectorhizosphere 95, 99
edge effect 499
Eh 439, 495
ejecta 208
elastic limit 231
electrical conductivity 219, 222, 511, 553
electrolytes 365, 366

elemental composition 220, 222
elevation 312, 474, 476–477
Eluviated Gleysol soils 147
eluviation 48, 145, 178, 305, 348, 352, 353–354, 429, 442, 460, 499, 535, 537, 539
  eluvial bodies 371
  eluvial/illuvial coefficient (EIC) 586, 587
  eluvial zones/horizons; *see also* E horizon 247, 373, 380, 428, 441, 461, 534, 541, 568, 580
enchytraeid worms 45
endorhizosphere 95
endosaturation 131, 491
  Endoaqualfs 29, 491
  Endoaquents 398
  Endoaquolls 506
energy dispersive spectroscopy (EDS) 28
energy dispersive X-ray analysis (EDXRA) 28, 377, 440
energy in soils 82
energy model (Runge) 323–324, 502
  capacity factors 323
  intensity factors 323
    water available for leaching 323
  organic matter production 323
  organizing vector 323, 324
  renewing/rejuvenating vector 323
  see also priority factors: energy model
energy transport in soils 90
enrichment 348
enstatite 62
Entisols 168, 314, 388, 514, 578, 628
  pedogenesis 93, 126, 262, 362
  relationships to geomorphology 185, 279, 407
  taxonomy/classification 121
entrainment (by wind or water) 191
environmental lapse rate 312, 476
eolian materials/sediments 195–207, 456, 617
  dune sand; *see also* dunes 158, 169, 191, 195, 197, 198, 212, 239, 269, 453, 467, 616, 623
  impacts on pedogenesis/geomorphology 173, 182, 187, 208, 216–217, 246, 256, 314–316, 325, 335, 378, 393, 406, 425, 435–438, 439, 456, 535, 623
  processes/influx/transport 196, 535, 627, 635
  salts 276, 424
epidote 62, 202, 582
epimorphism 535, 536, 539
epipedons 115, 151, 357
  anthropic 116, 128, 131, 319
  folistic 116
  histic 116, 267, 570, 624
  melanic 117, 122, 208
  mollic 116, 122, 132, 318, 636
  ochric 117
  plaggen 117, 132, 318
  umbric 116, 132, 636
episaturation
  Epiaqualfs 491
  Epiaquods 454
  see water table, perched
equifinality 145, 295, 370, 374
equivalent dose; *see also* luminescence dating 617
ergodic hypothesis 587
erosion 202, 244, 348, 437, 526, 544, 626
  accelerated/anthropogenic 318, 468
  concept/theory 169, 216, 224, 226, 246, 295, 325, 326, 328, 330, 335, 362, 456, 457, 588, 620, 634
  degree of 623
  effect on dating techniques 613–615
  factors that affect erosion rates 14, 104, 149, 183, 185, 239, 388, 415, 430–433, 438
  geologic/natural 467, 532
  impacts on pedogenesis/soil morphology 279, 280, 300, 327, 335, 410, 435, 474, 506, 514, 541, 542, 589, 611, 631
  impacts on weathering 172
  microscale 18
  rates 247, 614
  vs. slope characteristics/position 254, 471, 472, 475, 479, 480–483
  surfaces *see* surfaces, erosion/erosional
  truncation (of a soil profile) 325
  wind *see* deflation
estuaries 194
etch pits 28, 172, 561
etchplanation 538
Eukarya/Eukaryotes 93
Eutric Brunisol soils 147
evaporation 86, 87, 88
evapotranspiration 85, 90, 312, 323, 441, 488, 489
exfoliation 229
expansion/expansibility (of clay minerals) 282
extragrade, taxonomic 133, 142, 143, 157
extrinsic factors, soil 299, 328

fabric contrast soils 537
factor analysis 592
factors of soil formation (the concept) *see* state-factor model/approach
Factors of Soil Formation (the book) 6, 297
falls; *see also* mass wasting 291
family, soil 115, 121–143
Farm Creek stratigraphic section 534, 628
Farmdale paleosol 604, 620, 628, 640
fauna, soil 46, 96–105
  impact on pedogenesis 18, 46, 173, 189, 231, 234, 242, 325, 355, 359, 361, 398, 435, 537–538, 611
  infauna 97, 242, 243, 247, 253, 257
  mainly pedoturbation-related 245, 247–250
feces, 19, 24, 27, 45, 101, 244, 252, 354, 355, 357
feedback mechanisms 268, 274, 287, 306, 309, 329, 330, 334, 335, 339, 340, 367, 404, 430–433, 435, 439, 467, 619
feldspars 29, 48, 62, 179, 180, 183, 184, 197, 423, 535, 570, 581, 616, 618
  plagioclase feldspars; *see also* albite; anorthite 62, 181, 184, 202, 407, 431, 580
  potassium feldspars 62, 63, 202, 233, 234, 361, 580, 618
  weathering of 63, 172, 178, 552, 562
fens 190
fermentation 380
ferrallitization 350, 386, 395, 399
ferric hydrates 350
ferric hydroxides/oxyhydroxides 381, 382, 394, 445, 495
ferricrete 49, 391, 541

ferric sulfates 456
ferrihydrite 16, 57, 59–60, 187, 208, 209, 352, 381, 388, 443, 449, 452, 455, 456, 495, 624
ferrimagnets 642, 643
ferripyrophyllite 66
ferritization 352, 456
Ferro-Humic Podzol soils 147
ferrolysis 204, 299, 347, 351, 352, 367, 370, 371, 373, 382–385, 480, 497
ferrous sulfates 352
ferrous sulfide 455
ferrugination 352, 395, 398
  ferruginous crusts 174
fersialitization 395, 398
fertility *see* nutrient content/fertility
fertilizers 253, 318, 387
Fibrisol soils 147
Fick's law of diffusion 92
field capacity 84, 86
filter effect 367
fine earth fraction 6, 9, 240, 242, 244, 354
fingered flow (of soil water); *see also* preferential flow 87, 225
fingers, fingering *see* glossic tongues/tongueing
fining upward sequence 191, 208
fire 43, 231, 297, 356, 358, 359, 488, 559, 643
fission track dating 599
flagstones 12
flaking 227
F layer 38, 42
flint 187
flocculation 201, 287, 330, 364–366, 369, 370, 388, 429–430, 456
floccules 364
floodplains 128, 131, 149, 169, 191, 192, 200, 330, 332, 398, 456, 501, 525
flows; *see also* mass wasting 212, 291, 294, 480
fluorite 418
fluted landscapes 162, 285
Fluvents 192, 634, 635, 636
Folisol soils 147
Folists 45, 190–191
food web 93, 101
foraminifera 345
forest floor *see* litter
forest soils 48, 262
fossil soils; *see also* paleosols 52
fossorial rodents 256
foxes 249
fragipans 39, 119, 122, 131, 211, 373–380, 450, 491, 568, 570, 624
  Fragiorthods 25, 374, 378
  Fragiudalfs 370, 377, 378, 491
  genesis considerations 325, 370, 371, 372, 379, 461, 499, 625, 636
fragmentation (of litter) 44, 45
free face 477, 480, 486, 506
freeze–thaw processes 87, 230, 243, 246, 264, 266, 267, 275, 306, 314, 349, 351, 357, 375, 437, 504
  cycles 238, 263, 266, 267, 271
  freezing front 264, 265, 275, 276, 375
  typical areas affected 237, 527
frost in soils
  boils 275
  frozen soil 86, 133
  heave 265, 271, 275, 501
  shattering/wedging 231, 236
  in soils 87, 375, 378
  wedges 269, 270, 271, 632
fulvic acids 191, 210, 324, 355, 444, 445, 449–450, 610
functional-factorial models of soil development *see* state-factor model/approach
fungi 91, 93, 95, 97, 98–99, 101, 102, 236, 258, 418, 439, 445
  ectomycorrhizal 236
  fungus gardens 258
  hyphae 98, 100, 227, 236, 419, 420, 445, 447
  mycelia 98, 236, 442
  mycorrhizae 94, 99, 100
    arbuscular 99
    ericoid 99
  role in pedogenesis 43, 44, 45, 94, 354, 358, 419, 442
  saprophytic 98

gabbro 179, 180, 182, 227
gamma ray attenuation 83
garnet 62, 202, 562, 582, 583
gas tortuosity factor 92
gastroliths 243, 290, 541, 542
Gauss Epoch 597
Gelisols 127, 168, 263, 275, 314, 354, 447, 636
  gelic materials 122, 636
  gelifluction 267, 269
  gelifraction 231
  geliturbation 263
  pedogenesis 126, 271–276, 635
  taxonomy/classification 121
genesis, soil *see* pedogenesis
geoarcheology 298
geochemical compartment 447
geochronometric dating 596
geography
  geographic approach 3, 4, 465, 466
  geographic information systems (GIS) 149
  geographical cycle (cycle of erosion) 108, 466
  soil geography 167
geologic disorder 239
geologic timescale 342
geomagnetic excursions 597
geomorphology 202, 267, 331, 337, 465, 470, 479, 486, 528, 537, 538, 547
  soil geomorphology 167, 238, 295, 336, 396, 457, 465–546, 547, 554, 555, 587, 596, 604, 619, 635, 644, 652
geophagy 73, 74, 252, 259
geosols; *see also* paleosols 52, 548, 620
gibber plains 435
gibbsite 56–58, 59, 65
  conditions that favor its formation 171, 174, 350, 386, 391, 392, 395, 399, 639, 640
  what alters/weathers to it 178, 202
gilgai 245, 283–285, 286, 289, 435, 501, 569, 624, 636
glaciers and glacial deposits; *see also* drift 158, 346
  glacial flour 200
  glacial grinding 200
  glacial/interglacial cycles 199, 339, 342, 343, 346, 534, 643, 647
  glaciations/glacials 343, 346, 363, 641
  glaciofluvial sediment 107, 211, 345, 617, 628
  ice sheets 195, 200, 203, 210, 342, 345, 409
glaebules 24, 422
glass, volcanic 208
glauconite 66, 67, 70, 71, 72

gleying 17, 276, 318, 381–382, 400, 480, 483, 494, 495–496, 501, 521
  gleization 110, 113, 276, 352, 381
  Gleysolic soils 147
  Gleysol soils 147
  impact on horizonation/classification/taxonomy 39, 51, 152, 492, 493
Glinka, K. 5, 108, 296
global warming 274
glossic horizons 118, 131, 370, 371, 373, 379, 428
  Glossaqualfs 496
  glossic tongues/tongueing 363, 371, 429, 443, 499, 504
  Glossudalfs 364, 370, 371
gneiss 171, 172, 179, 180, 227, 509
gobi 435
goethite 16, 57, 58, 59, 187, 388, 398
  conditions that favor its formation 68, 178, 382, 386, 391, 392, 395, 399, 443, 452, 455, 495, 639
  what alters/weathers to it 184, 232, 391, 394, 624
  what it alters/weathers to gold
goniometer 76
gophers 104, 243, 247, 249
grainy gray ped coatings 371, 497
Grand Prairie 280, 293
granite 172, 179, 180, 182, 226, 227, 230, 393, 407, 438, 557, 560, 639
granodiorite 179, 184
grassland soils/landscapes; *see also* Mollisols 48, 240, 253, 324, 355, 358–359, 377, 384, 402, 459, 646
gravels/gravelly; *see also* coarse fragments/fraction 12, 13, 191, 242, 243, 290, 302, 354, 361, 393, 406, 411, 418, 434, 438, 481, 544
gravitational drainage 86
gravitational energy 323
gravitational water 84
Gray Brown Luvisol soils 147
Gray Luvisol soils 147
Gray Wooded soils 123
Gray-Brown Podzolic soils 109, 111, 123
great groups, taxonomic 115, 130, 131–132, 133
greisen 542
greywacke 184
ground penetrating radar (GPR) 83
ground squirrels 246, 250
Ground Water Laterite 393
ground wedges 269
groundsurface 532
groundwater
  discharge areas/wetlands 396, 471, 487, 489, 506
  flow patterns/rates 428, 486
  Ground Water Laterite soils 109, 113, 123
  Ground Water Podzol soils 109, 113, 123
  impact on (or impacted by) pedogenesis 173, 236, 257, 359, 381, 394, 403, 407, 421, 422, 425–427, 430, 453, 490, 538, 637
  recharge areas/wetlands 487, 489
  see also water table
Grumusol soils 123
grus 180
  grusification 180
guano 104
gullies 199, 435, 476, 480, 483
gumbotil 518
gypsum 60, 181, 427
  depth to 425
  gypcrete 49, 420, 422
  gypsic horizons 118, 122, 128, 131, 329, 420–422, 438, 507, 570, 636
  Gypsids 579, 636
  gypsification 352, 400, 420–422, 641
  impact on pedogenesis/soil morphology 38, 400, 404, 409, 411, 428, 433, 455, 569, 624
  petrogypsic horizons 119, 122, 420, 570, 624
  secondary/pedogenic 118, 131, 287, 293, 333, 389, 406, 410, 420, 430, 436
  weathering-related 202, 229

H layer 38, 42
Half Bog soils 109, 113, 124
halite 181, 232
halloysite 65, 67, 70, 79, 178, 182, 209, 210, 639, 640
Halomorphic soils 109
hamadas 435
Haplaquepts 490
Haplaquolls 506
Haplargids 416, 460
Haplocalcids 460
Haplocambids 415, 460
Haplocryalfs 317
Haplodurids 423
Haplogypsids 436
haploidization 239, 326, 327, 328, 330, 333, 348
Haplorthels 268, 317
Haplorthods 332, 364, 446, 578
Haplosalids 436
Haploxererts 510
Haploxerolls 293
Hapludalfs 29, 42, 152, 170, 216, 225, 516, 621
Hapludolls 459, 506, 516, 571
Hapludox 179, 386, 516
Hapludults 163, 213, 516
Haplumbrepts 216
Haplusterts 293
Haplustolls 293, 571
Harden Index 575–578
hardening 349
H-bonds 82
H^+ cations (protons) 233, 234, 236, 352, 359, 365, 382, 402
heat capacity, soil/water 87, 268
heat flux, soil 87, 88–91
heating/warming 229
heaving 263, 264, 265, 291
hectorite 66, 67
helium-3 612, 614, 615
hematite 57, 58, 59, 388, 398
  conditions that favor its formation 59, 178, 392, 395, 639
  impact on soil morphology/classification 16, 201, 354, 386, 399–400, 401, 573
  what alters/weathers to it 187, 391, 394, 452, 455, 624
  what it alters/weathers to 202, 232
hemicellulose 354
Hemistels 268, 273
hemlock trees 359
herbivores 101–103
heterotrophs 97
Hilgard, E. W. 296, 312
Histels 274
Histosols 42, 314, 652
  histic materials 122
  pedogenesis 126, 351, 356, 443, 624, 635

relationships to geomorphology 153, 443, 487, 488, 514, 635
taxonomy/classification 116, 121, 147, 189
Histoturbels 273
Hole, F.D. 158, 239, 357
Holocene Epoch 199, 206, 339, 344, 346, 406, 415, 488, 506, 556, 568, 611, 621
homogenization 239
horizons, soil 35, 53, 150, 333, 348, 538, 539, 624, 633
boundaries and boundary characteristics 169, 216, 219, 222, 224, 274, 287, 386, 624
definition/concept 36
diagnostic 115, 116–120, 122, 151, 347, 373, 400
differentiation from parent material 239–240, 320, 321, 323, 324, 328, 479, 546
discontinuous/interrupted 244, 262, 272
homogenization 173
importance/utility 108, 619
master 36, 322
nomenclature/terminology 4, 40
patterns and spatial variability 311, 505, 511
subsurface 151
suffixes 36
thickness 301, 526, 532, 570, 595
transitional 36, 41
various types of 36
weighted values (statistically weighted by horizon thickness) 570
hornblende 61, 62, 172, 181–182, 202, 208, 580, 581, 582, 583
etching 552, 560–562, 563
hornblende gneiss 180
humans/human agency 317, 373, 456, 525, 631, 635
Humaquepts 443
humic acids 191, 210, 324, 355, 445, 610
Humic Ferruginous Latosol soils 124
Humic Gley soils 124
Humic Gleysol soils 147
Humic Latosol soils 124
Humic Podzol soils 147
humin 610
Humisol soils 147
hummocky landscapes 508
Humo-Ferric Podzol soils 147
Humods 636
Humphreys, G. S. 534
humus 16, 20, 37, 46, 145, 258, 323, 351, 354–356, 362, 453, 600, 602, 606–607
formation 44, 45, 46
humification 44–46, 102, 103, 274, 276, 278, 305, 321, 348, 351, 354–356, 610
illuvial/pedogenic 120, 292, 429, 443, 444, 445, 449, 452, 613
Hurst Index 573
hydration 232, 314, 352, 560, 599
hydraulic conductivity 86, 263
saturated hydraulic conductivity 86
hydraulic head 86, 87
hydric soils 486, 495
hydrogen sulfide 420
Hydrol Humic Latosol soils 124
hydrologic cycle 295
hydrolysis 186, 232–234, 236, 382, 429, 455
hydromorphism 352
hydromorphic soils 109
hydrophobicity 86
hydrosequence 302, 486
hydrous mica 70
hydrous oxides 209, 375, 378
hydroxy-interlayered clays 72–73
hydroxy-interlayered smectite (HIS) 67, 70, 73, 79, 80, 184
hydroxy-interlayered vermiculite (HIV) 67, 70, 73, 79, 80, 184, 209
hydroxyl anion (OH^-) 233, 361, 382, 402
hypersthene 62, 202
hysteresis/hysteresis effect 85

ice in soils
crystals 229, 230, 271, 292
ice wedges 264, 269, 270, 272, 274
casts 270, 271, 272, 527, 528
polygons 271, 274
pseudomorphs 270
lenses 264, 265, 275, 375
segregation 264, 271
I/E Index 363
ignimbrite 227, 423
Illinoian glaciation/deposits 211, 344, 431, 480, 518, 520, 524, 528, 579, 625, 627, 628
illite 66, 67, 70, 71, 180, 184, 186, 188, 202, 639
illuviation 48–49, 50, 145, 212, 305, 348, 353–354, 368, 453, 479, 539, 641
illuvial clay 29, 364, 367, 374, 380, 427, 480, 597
translocation of 172, 292, 542
theoretical/process related 189, 212, 225, 537, 579
see also lessivage
illuvial materials 37, 49, 225, 245
illuvial zone/horizons 51, 131, 225, 441, 461, 568
lateral 353
vertical 353
ilmenite 101, 202, 584
immobile fraction/elements; *see also* skeleton, soil 219, 222, 225, 461, 584
imogolite 208, 209, 210, 444
imogolite type materials (ITM) 444, 445, 447–449
impermeable layers *see* aquiclude
Inceptisols 168, 314, 388, 396, 514, 628, 645
pedogenesis 126, 173, 262, 362, 452, 635
relationships to geomorphology 280, 397
taxonomy/classification 121
inclusions (in rocks) 548
Index of Profile Anisotropy (IPA) 575
indices of soil development *see* soil development, indices of
infertility 184
infiltration 40, 85, 94, 99, 101, 103, 129, 234, 239, 250, 262, 263, 324, 437, 451, 472
infiltrating water 323, 471
factors affecting 192, 225, 253, 335, 450, 472–474, 487, 507
impact on pedogenesis/weathering 177, 189, 306, 310, 324, 437, 479, 510, 542
infiltration rate/capacity 85–86, 387, 388, 430–433, 439, 471, 472, 473, 503, 509
initiation zones 212, 216
inner layer 365
inosilicates 61, 62, 63
insects, soil 97, 101, 242, 247, 253
insoluble components/materials/fraction 179, 187, 232, 234

interception (of precipitation by plants) 472
interfingering 118
intergrade, taxonomic 133, 141
interlayer cation(s) 67, 68, 70–71, 72, 78, 79
interlayer hydroxide polymers 73
interlayer hydroxide sheet 73
interlayer space 72, 73, 280
interlayer water 67
interstadials 345
Intrazonal soils 108, 109
intrinsic factors, soil 299, 328
invertebrates, soil 96
ions in soils 231, 232, 245, 353, 357, 359, 360–361, 509, 578
  ion-dipole bonds 82
  ionic charge 232
  ionic potential 232, 233
  ionic radius 55, 232, 361
  transport in soils 82
Iowan erosion surface 272, 518, 526–529, 530, 555
iron (cations and minerals) 352, 440, 452, 453, 553, 568, 618, 624, 627
  concentrations *see* redox concentrations
  crystallization 392
  depletions *see* redox depletions
  extractable 178
  ferric 380–381, 382, 388, 396, 399, 440, 444, 492, 499, 643
  ferrous 380, 381, 382, 393, 396, 443, 453–455, 492, 494, 495–496, 497, 499, 643
  hydroxides 187, 391
  iron activity ratio 448
  iron-manganese concretions/nodules 182, 370, 371, 380, 381, 384, 483, 494, 495, 497, 499, 512
  ironstone 177, 178, 259, 350, 391, 392, 396, 535
  loss via reduction/translocation 371
  minerals 219, 308, 350, 392, 399, 438–439, 443, 573
  oxides 172, 184, 201, 205, 225, 318, 365, 373, 390, 394, 395, 398, 399, 400, 442, 444–445, 449, 452, 492, 497, 573, 579, 580, 583, 586
  residual 392, 393
  stains/staining 173
irrigation 492
isochronous units/surfaces 547
isomorphous substitution 65–67, 68, 71, 72, 78
isopods 44, 354
isoquartz method 442, 584
isothermal 133
isotopes, cosmogenic 207, 409, 415, 600, 612–616, 618, 619, 633, 644
  isotopic signal 647
  stable 600, 611
  unstable/radioactive 600
isotropy/isotropic 240, 323, 325, 575

jarosite 16, 38, 61, 293, 455
jasper 187
Jenny, H. 5, 6, 296–300, 302, 310, 320, 323, 337
Johnson, D.L. 324, 325, 537
jointing/joints 51, 227

K Cycle 529–534
  instability phase 532
  stability phase 532
K horizons 403
K–Ar dating 412, 552, 560, 566, 599, 600
kandic horizons 117, 370, 398, 399, 636
  Kandiaquults 514, 515
  Kandihumults 398
  Kandiudalfs 396
  Kandiudox 398, 514
  Kandiudults 396, 514
  Kanhaplustults 398
Kansan Drift Plain/region 517, 518, 522
Kansan till 524, 529
kaolin group of minerals 62, 67, 78
  kaolinite 65, 67, 70, 79–202
    conditions that favor its formation 178, 371, 399, 639, 640
    impacts on pedogenesis/soils 363
    soils/settings where it is found 120, 174, 176, 177, 386, 392, 396, 399, 510, 514
    what weathers to it 171, 178, 182, 184, 188, 233, 234, 391, 393, 395
karren 235
karst 187, 201, 232, 234, 235
Kellogg, C.E. 296, 337, 390, 465, 544
kitchen middens 319
krotovina 247, 250, 252, 255, 256, 355, 412, 416, 624, 633
kyanite 202, 580, 581, 583

lacustrine sediments 107, 158, 191, 193, 194, 211, 597, 652
lag concentrate/surface gravel 169, 196, 243, 244, 262, 291, 334, 435, 437, 525, 543, 544
lag time, soil temperature 91
lagoons 195
lamellae 38, 117, 197, 304, 363, 368–370
landforms 256, 466
landscapes
  complexity 243
  evolution 170, 172
  position 127, 427
  soil landscapes 3, 107, 160
lapis 187
lapilli 208
latent heat
  capacity 87
  transport of 90
laterization 112, 113, 350, 386, 390, 391, 392–400
  laterite 49, 176, 177, 350, 387, 388–392, 542, 570, 624
  Laterite soils 109, 112, 124
  Lateritic soils 109
latitude 87
latosolization 350, 386, 392, 393–394
  Latosol soils 124, 176, 388, 390, 392, 393
  Latosolic Brown Forest soils 124
lava/lava flows 207, 208, 407, 412, 437, 562, 566, 614, 615
layer charge 65–67, 68–72
leaching 32, 122, 236, 325, 356, 359, 428, 429, 456, 538
  definition/concept 348, 359–361
  factors that affect intensity 58, 233, 234, 262, 287, 303–305, 306, 313, 316, 324, 384, 427, 429, 431, 433, 450, 472, 502, 507, 639
  impact on pedogenesis 145, 163, 168, 178, 182, 183, 210, 360, 362, 370, 382, 386, 393, 395, 427, 429

leached zone/depth of leaching 49, 50, 224, 289, 302, 312, 325, 360, 361, 366, 378, 489, 553, 568
leaching (of carbonates and bases) 37, 51, 186, 204, 225, 289, 305, 318, 338, 365, 366, 377–378, 400, 406, 410, 451, 457, 506, 628, 637
potential 163, 232, 423
regime 420, 636
unleached or cessation of leaching 32, 33, 318, 425
utility in soil geomorphology 149, 508, 553, 573, 613
leaf mold 43
Leighton, M. M. 52
lentils 283
lepidocrocite 16, 57, 58, 59, 388, 391, 394, 399, 495
lepidolite 66
lessivage 325, 350, 534, 579, 599, 641
effects of soil morphology 186, 245, 385, 398, 510, 514
factors/settings that affect the process 184, 189, 197, 247, 276, 323, 331, 332, 359, 384, 396, 399, 452, 456, 499, 625
process 329, 330, 333, 361, 370, 388, 450, 452
leucinization 352, 361, 362
levees 191
lichens 93, 96, 97, 236, 454, 556, 564, 565
great period 96, 562, 563
growth curve 562
lichenometry 552, 553, 556, 562–566
linear phase 562, 563
lignin 44, 45, 354, 442
lignocellulose index 44
limestone 181, 600
applications in geomorphology and pedogenesis 172, 185, 191, 399, 438, 611
fossiliferous 181
oolitic 181
weathering-related 186, 187, 201, 227, 232, 234, 332, 407, 419, 649
limnic sediment 194
limonite 542
Linnean system 115
lipids 610
lithiophorite 60
lithologic discontinuities 37, 215–225, 633, 634
definition/concept 51
detection techniques/principles 218, 220, 222, 223, 224, 585
examples 35, 209, 215, 225, 367
impacts on/by pedogenesis 86, 192, 239, 366, 367, 375, 378, 393, 404, 418, 445, 492
utility in soil geomorphology 150, 217, 506, 544, 583, 631
lithosequences 301, 304, 331, 453
lithochronosequences 304
lithofunctions 299, 300
Lithosols 109, 114, 124
lithotrophs 97
litter
leaf litter 37, 41, 43, 47, 87, 93, 97, 240, 354, 355, 357, 361, 370, 373, 394, 441, 444, 445, 450, 510, 645, 650
litter interception loss 473
litter layer 43, 93, 97, 190, 247, 358, 450, 458, 473, 502
littering 348, 351, 354, 356, 442, 503
see also O horizon
lixivation 349
lizardite 65, 68
L layer 38, 42
local relief 163, 476, 477, 486
loess 197–207, 306, 346, 517
as an agent of soil/sediment burial 211, 324, 330, 458, 459, 467, 533, 547, 621, 624
Bignell 627
characteristics 158, 190, 204, 205, 239, 243, 324, 461, 542, 604, 612, 616, 642
China Loess 199, 203, 346, 627, 630, 643, 644
dunes 200
genesis/environment of formation 200, 480
impacts on pedogenesis 169, 203, 424, 431, 459
Loveland 518, 520, 521, 522, 524, 604, 625, 627, 630, 631
non-glacigenic 199
Peoria 205, 480, 518–525, 528, 529, 604, 625, 627, 628, 631
Roxana 625, 628, 640
sequence/column 532, 590, 597, 605, 627, 643
thickness and location (maps) 199, 200, 518
loosening 349, 537, 541
losses (from a soil or its horizons) 47
lowering of materials within a soil 240, 250, 254
Low Humic Gley soils 124
Low Humic Latosol soils 124
low molecular weight organic acids 444, 445, 449, 450, 453
luminescence dating 435, 599, 605, 618
infrared stimulated luminescence (IRSL) dating 616
optically stimulated luminescence (OSL) dating 616, 617, 618
thermoluminescence (TL) dating 616, 618
Luvisolic soils 147, 362

macrocombinations, soil 149
macrofauna/macroorganisms, soil 43, 44–45
macrofossils 190, 627, 652
macronutrients 184, 358
macroterme 74
maghemite 16, 57, 60, 386, 388, 391, 394, 455, 642, 643
magnesium 180, 182, 184, 262, 358, 365, 366, 382, 426–427, 503, 637, 639
exchangeable 252
magnesium chlorides 424
magnetic reversals 597
magnetic susceptibility 206, 219, 220, 222, 627, 630, 642
magnetite 16, 57, 60, 183, 202, 391, 596, 642, 643
magnetostratigraphy 597
mammals 91, 101, 104, 242, 244, 247, 535
manganese/Mn oxides 16, 358, 380, 391, 393, 438–439, 440, 443, 497, 552, 624
Mn concretions 17
manure/manuring 117, 318
maps *see* soil maps/mapping
marble 179, 201, 227, 232, 234
Marbut, C. F. 5–6, 52, 108, 296, 337, 544
marine isotope stage (MIS) 345, 346, 630, 643, 644

marine sediment 191
marker bed 599, 642
marl 187, 188, 193, 225, 280, 600
marshes 190, 195
mass balance approach 174, 236, 238, 321, 323, 360, 442, 460–461, 583–587
mass data (mass-based data) 553, 571, 579
mass flux 460
mass movement/mass wasting 246, 291, 294, 306, 351, 437, 482, 525, 527
  effects on soils 169, 480, 546
  theoretical considerations 226, 325, 330, 456, 484, 538, 541
mass spectrometry 601
matric forces/tension 13, 84, 225, 367, 377, 433
  matric potential 83–85, 86, 430
maturation 351, 354
mature soil; *see also* zonal soils 108, 327, 336, 337, 470, 568
Matuyama Epoch 597
maximum limiting date *see* age
mean annual air temperature (MAAT) 143
mean annual soil temperature (MAST) 143, 144
mean residence time (MRT) dates 289, 290, 449, 604, 606, 608, 609, 611, 632
Mediterranean soils 201
Melanic Brunisol soils 147
melanization 47, 103, 168, 189, 197, 305, 323, 327, 330, 332, 352, 354, 356–357, 361, 362, 384, 398, 400, 459
melonhole 283
meltwater, glacial 193, 480
mesa and butte landscape 185
Mesisol soils 147
mesocombinations, soil 149
metagabbro 180, 182
metal oxyhydroxides 174
metals/metal cations 235, 320, 393, 402, 441, 444, 445–449
metapedogenesis 318
  metapedogenic factor 317
M horizons 259, 539, 541
micas 62, 67, 68–71, 79–639
mice 94
micrite/micritic carbonates 181, 349, 419, 420, 641
microagglomerates (silica) 422, 424
microaggregates *see* floccules
microbasins; *see also* Vertisols 282, 283, 289
microclimate *see* climate/macroclimate
microcline 63
microcombinations, soil 149
microfabric 244
microfauna/microorganisms 44, 354, 359, 380, 381, 405, 415, 419, 438, 439, 496, 611, 648
  microbial activity 44, 95, 382, 439, 624
  microbial gums 357
  microbial respiration 92
microfractures 231
microknolls; *see also* Vertisols 283, 289
micromorphology, soil 22–25, 26, 411, 548, 553, 569, 599, 623
  fabric 22, 24, 178
micronutrients 358
microprobe analysis 29
microrelief/microtopography, soil 149, 244, 252, 261, 268, 275, 283, 340, 430, 433, 443, 501–506
microscopy
  light 11, 23, 563, 650
  scanning electron (SEM) 23, 26–29, 188, 364, 376, 377, 420, 650
  transmission electron (TEM) 23, 30
microtopography *see* microrelief
Milankovitch cycles 342–343
mildews 98
millipedes 44, 97, 101, 102, 103–354
Mima mounds 256–257, 293
mineralization 47, 234, 268, 274, 313, 325, 338, 348, 352, 356–358, 388, 394, 442, 445, 450, 456, 610, 623, 624, 635
minerals/mineralogy 11, 54–81, 133, 168, 176, 177, 179, 191, 205, 206, 209, 210, 219, 226, 232, 238, 301, 312, 423, 526, 568, 578, 599, 633, 641
  clay 9, 176, 184, 236, 305, 352, 356, 362, 363, 365, 366, 370, 375, 376, 382, 386, 388, 435, 438, 452, 553, 568, 570, 579, 612, 623, 624, 625, 627, 628, 636
    as end products of weathering 226, 231–232, 233
    characteristic mineralogies 206
    utility in soil gemorphology 187, 201, 220, 222
    see also phyllosilicates/phyllosilicate minerals
  definition/concept 54
  effects on 456
  ferromagnesian 180, 184, 208, 442
  ferromagnetic 596
  heavy 220, 503, 522, 526, 581
  index 581
  in soils 82, 88, 226, 231, 619
  light 522, 526, 581
  magnetic 642
  mineralogical maturity 196, 393
  primary silicate 11, 61, 63, 209, 233, 616, 618, 640, 643
  primary 9, 54, 171, 178, 226, 331, 334, 354, 361, 424, 441, 447, 449, 625
  resistant/slowly weatherable 172, 174, 183, 202, 219, 220, 222, 225, 226, 323, 361, 393, 423, 452, 460, 535, 539, 580, 584, 637
  resistant/weatherable ratios 221–222, 526, 553, 580–583, 584
  secondary 9, 54, 61, 178, 226, 310, 375
  short-range-order 208, 209, 210
  weatherable/weak 37, 48, 168, 171, 177, 180, 197, 202, 226, 351, 354, 386, 392, 398, 452, 537, 570, 580, 584
minimum limiting date *see* age
mining/mines 318, 595
minnesotaite 66
Mitchell, P. B. 534
mites 43–45, 97, 101
mixed zones 623, 628
Mn : Fe ratio 439
mobile minerals/elements 553, 584, 641
modal profile 589
models/modeling
  computer-based 406
  conceptual/pedogenic 295, 298, 320, 323, 324, 342, 374, 378, 403, 411, 418, 419, 424, 437, 439, 444, 449, 525, 532, 534, 544
  process-based 297
moder humus 43, 44

moisture characteristic function/curve 85
moisture content, soil 87, 88, 122, 302, 305, 310
moisture/water flux 470, 472, 482
molds 98, 440, 602
mole rats 246
moles 247, 256
Mollisols 127, 388, 572, 608, 636
  pedogenesis 69, 126, 185, 244, 252, 291, 324, 335, 384, 460, 506, 573
  relationships to geomorphology 280, 454, 514
  taxonomy/classification 121
monazite 202
monogenetic soils 337, 338
monosiallitization 350
monosilicic acid 423, 424, 649
monsoons 199
montmorillonite 66, 67, 72, 202, 277
monzanite 584
mor humus 43, 354, 361, 442
moraines 301, 559, 569, 577, 612, 614
morphology, soil 9, 299
Mössbauer spectroscopy 80
mottles, soil 13, 17, 624
  high chroma (red) 68, 352, 381, 495
  low chroma (gray) 381, 491, 494, 495, 496, 501
  related to redox processes 177, 380, 382, 385, 483, 492, 493, 495, 499, 569
  see also redoximorphic features
mounds 242, 243, 247, 249–252, 253
  burial/ceremonial 318
  treethrow 245, 261, 262, 501, 502, 505
mucigel 96
mucilage 94
muck, 189, 351, 356
mudstone 180
mukkara 285, 286, 289, 290
mull humus 43–44, 354
muscovite 66, 67, 68–69, 79, 172, 202, 371
mushrooms 98, 99
mutualists 94, 97
mycorrhizae *see* fungi
myriapods 101

N cycle 101
N fixation 93
nacrite 65, 67
natric horizons 118, 122, 132, 349, 370, 427, 431, 511, 570, 624
  Natralbolls 431
  Natraqualfs 431
  Natrargids 460
  Natrudalfs 429
  Natrudolls 429
  Natrustalfs 429
  Natrustolls 426, 635
Natural Resources Conservation Service (NRCS) 52, 151
Nebraska Sand Hills 199
needle ice 265
nematodes 91, 94, 97, 101
neoformation, clay 186, 350, 356, 362, 393, 394, 398, 433, 504, 570, 636, 637
neon-21 612, 614, 615
neosilicates 61, 62, 63
neoskeletans 497
nepheline syenite 170
net precipitation 473
neutron attenuation 83
neutron probe 490
nickel 182, 183, 393, 461
nightcrawlers; *see also* earthworms 103–105
nine unit landsurface model 484–486
niobium 460
nitrates 380, 402, 424
nitrification 234
nitrites 402
nitrogen 252, 262, 310, 312, 313, 358, 482, 503
nodules 24, 174, 290, 381, 393, 412, 419, 422, 541, 543, 623
non-mounders (soil fauna) 249–252
Noncalcic Brown soils 109, 111, 124
nontronite 66, 67, 72, 391
normal soil 108, 336–338, 534
"not soil" 34, 159, 539, 578
numerical age dating 300, 549, 560, 596–618
nutrient content/fertility 252, 253, 259, 262, 323, 331, 387, 454, 537
  nutrient cycles/cycling 102, 105, 247, 252, 258, 330, 538

obliquity of the orbit 343
obsidian 182, 560
  obsidian hydration dating 552, 553, 560, 562
octahedral sheet 64, 65, 67, 382
octahedron/octahedral structure 55, 59
O horizons 37–46, 94, 354, 473–493
  definition/concept 36, 37
  fibric materials 38, 42, 128, 131, 356
  formation/genesis 42, 43, 44, 46, 48, 103, 240, 265, 274, 317, 355, 356, 452
  hemic O horizon material 38, 42, 128, 132, 356
  images 46
  sapric O horizon material 38, 42, 128, 488
  suffixes 38
  thickness 42, 45, 105, 190, 249, 442, 493, 495, 502, 503, 505, 593
olivine 61, 62, 68, 181, 182, 183, 202, 208
opal 422, 424
  opal, biogenic 62, 188, 359
  opal phytoliths 46, 206, 359, 423, 623, 649
optical density of the oxalate extract (ODOE) 308
orbital geometry 342
orders, soil map 148
orders, soil 108, 115
organic matter/carbon, soil
  change over time 319, 338, 384, 450, 590, 593
  composition 329, 610, 645
  composition re: organic acids *see* acids, organic
  content 10, 43, 252, 258, 262, 280, 310, 311, 315, 318, 338, 354, 381, 395, 479, 480, 483, 505, 538, 633, 635
    spatial variation in content 289, 305, 476, 481, 503, 509, 511
  content of ^{13}C (isotopes) 644
  context-related examples 373, 387, 399, 427, 429, 452, 455
  dating soil organic matter 602, 611, 618, 632
  distribution by depth 285, 289, 354, 386, 475, 623, 637
  gains/accumulation/production 210, 268, 321, 325, 335, 539
  losses of 458
  and paleosols 206, 624

organic matter/carbon, soil (*cont.*)
pedogenesis-related 173, 209, 239, 252, 253, 267, 271–274, 287, 324, 352, 384, 405, 411, 429, 440, 442, 443, 446
utility in soil geomorphology 219, 222, 569, 573, 642
organic soils/materials *see* Histosols
organo-metallic complexes *see* complexes, organo-metallic
organotrophs 97
ornithogenic soils 94–104
Orthels 274
orthoclase *see* feldspars, potassium feldspars
Orthods 453, 634
orthosilicates 61
ortstein 49, 443–444
osmotic potential 83, 84
outer layer 365, 366
outwash *see* glaciofluvial sediment
oven-drying 83
overbank deposits 191, 193
overland flow 471, 482, 484
oxalic acid 236
oxbow lakes 191
oxidation and reduction; *see also* redox processes 48, 59, 97, 129, 168, 173, 232, 271, 275, 298, 299, 302, 371, 380–382, 443, 492, 496, 510
oxidation/oxidizing conditions 32, 47, 51, 186, 276, 318, 325, 352, 379, 380, 381, 382, 392, 393, 443, 455, 456, 492, 495, 497, 499, 506, 610, 643
oxide clays 59, 121, 361, 363, 386, 388, 394, 395, 399, 579, 639
oxides
aluminum 56–58, 80–81
in soils 55–60
iron 57, 60, 68, 80–81, 92
manganese 60, 92
metal 55, 399, 579
oxidized zone 51
Oxisols 56, 75, 80, 127, 170, 173, 174, 176, 362, 385, 388, 390, 400, 401, 579, 634, 636
oxic horizons 120, 122, 386, 392, 393, 396, 570, 574, 624, 636
pedogenesis 126, 173, 400
relationships to geomorphology 396, 512, 513, 514
taxonomy/classification 121
oxygen isotopes 220, 642, 648
oxyhydroxides 187, 232, 393, 439

P waves 552, 559
paddies, rice 318, 382, 492, 498
paha 529, 530, 531, 532
Paleargids 334, 416
paleoclimate/paleoenvironments 317, 342, 345–346, 620, 630
soil/sediment properties and past environments 190, 359, 406, 415, 439, 627, 631, 632
theoretical considerations re pedogenesis 305, 465, 554, 568, 599, 621, 644
paleohumans 319
paleomagnetism 206, 346, 596–597, 618, 627, 630
paleosols 32, 215, 431, 480, 516, 517, 518, 534, 548, 590, 620
buried *see* buried soils/paleosols
definition/concept 52, 619
development of 573, 574, 631
exhumed 622, 626
impact on pedogenesis 204, 206, 427
paleopedology 22, 516, 522, 525, 578, 582, 605, 608, 619, 620
relict 523, 621
stacked 208, 643
utility in soil geomorphology 216, 217, 521, 597–599, 642, 646, 647, 649
see also Afton soil/paleosol; buried soils/paleosols; Farmdale paleosol; geosols; Sangamon, paleosol/surface; and Yarmouth-Sangamon paleosol
paleovegetation 646, 647, 652
Paleudults 163, 392, 499, 516
palimpsests 619, 620, 640, 641
pallid zone 178
Palouse Hills 306–307
paludification 351, 356
palygorskite 639
pans (various types, pedogenic) 390, 427, 454, 559
panspots *see* slickspots
parallelepipeds 283
parasites 94, 101
parent material 167–225, 314, 345, 421, 460
definition/concept 32, 50, 52, 167, 347, 348
impacts on pedogenesis and soil pattern 275, 298, 320, 373, 377, 396, 407, 411, 419, 444, 451, 453, 475, 633, 635, 640
isotropic 239
organic 189–191
residual 156, 170–191, 407
transported 191–215
types of 213, 276–277
uniform/uniformity 573, 579, 583, 584, 643
utility in soil geomorphology 150, 154, 466, 469, 470, 533, 553, 575, 636, 642
parna 197
particle size classes/distribution 10, 133, 197
patchiness 252, 262
pathogens 94, 97
pathways, pedogenic *see* pedogenesis/pedogenic pathways
patina 438
Paton, T.R. 534
patterned ground 245, 267–271, 273, 275, 276, 314
peat and peat bogs 43, 189, 267, 274–280, 351, 488, 600, 652
pebbles 12
peds
shape 20
size 19, 20
surfaces/faces 292, 363, 371, 497
Pedalfers 108
pediments/pedimentation 483, 519, 522, 525, 543, 544, 545, 546, 628
pedisediment 169, 477, 480, 525, 526, 543, 544, 546
pedisedimentation 543
pedo-weathering profile 52
Pedocals 108, 109, 403
pedoderm 620
pedodiagenesis 624
pedofeatures 24, 25
pedogenesis 167, 275, 325, 336, 347–461
as affected by parent material 169, 170, 192, 204, 212, 267, 533
associative 348, 353
definition 348
dissociative 348, 353

factors/controls on 245, 268, 292, 297, 312, 337, 479, 480, 502, 532, 537
impacts on parent material 216, 642
intensity of 586, 643
and paleosols 204, 624, 625, 635, 644
pedogeneticization 302, 456
pedogenic accessions 299, 329, 330, 335–336, 339, 465, 568, 589, 619
pedogenic compartments 447
pedogenic imprinting 174
pedogenic infiltration 473
pedogenic paradigms 297, 537
pedogenic pathways 126, 169, 215, 239, 267, 296, 320, 328, 331, 333–335, 338, 339, 360, 367, 370, 384, 450, 453, 459, 477, 588, 592, 594, 619, 633
pedogenic precipitation 473
pedogenic processes
  as the impact soil or parent material properties 114, 173
  as they impact taxonomy/classification 145
  definition/concept 7
  examples 368, 472
  theoretical considerations 169, 226, 279, 295, 296, 297, 320–322, 339, 341, 347, 353, 484, 486, 537–539, 595, 622, 633
  variation across space 276, 303, 313
pedogenic provinces 272, 276
pedogenic signature 209
pedogenic system 328, 471
and pedoturbation 239, 244, 267
progressive 296, 325–328, 329, 333, 336, 337–340, 347, 348, 502, 552, 588
rates of 300, 301, 302, 314, 336, 554, 587, 590, 592, 595, 599
regressive 239, 240, 326–328, 332, 333, 336, 338, 340, 347, 348, 568, 588, 630
relic processes/features 375
sequence of soil development 115
static 326, 327, 332, 583
steady state conditions 42, 86, 301, 325, 337, 338, 570, 588, 590, 593, 594, 644
and taxonomy/classification 114,145
theoretical considerations 295, 320, 322, 324, 337, 465, 534, 568, 606, 652
time$_{zero}$ considerations 196
top-down 324, 457, 459
upbuilding 457, 459
pedogeomorphic fabric 162
pedologic order 239, 240
pedology 4, 159, 320, 323, 337, 538, 582, 587
  geopedology 4
  pedologists 167, 312, 341, 492
pedoluvium 543
pedon 33–34, 35, 106, 107, 115, 148, 150, 160, 469, 476
pedoplasmation 173
pedo-stratigraphy 528
pedoturbation 145, 294, 328, 534, 535, 584
  definition/concept 239, 295, 326, 350, 353
  depth of, implications 174, 240
  effects on horizonation 48, 354, 356, 363, 367, 378, 384, 427, 442, 446, 460, 541, 579, 624
  effects on stratification 168, 219, 266, 546, 652
  aeroturbation 245, 246, 330, 351, 444
  anthroturbation 245, 246, 292–293, 330, 351
  aquaturbation 245–247, 330, 351
  arboturbation 261
  argilliturbation 72, 244, 245, 246, 277–291, 292, 329, 330, 333, 334, 335, 351, 362, 367, 435, 501
    rates of 289
  bioturbation 247–262, 537, 541, 616
    examples 615, 628
    impacts on pedogenesis/biomantle formation 173, 217, 242–243, 332, 398, 412–414, 442, 459, 534, 541–546, 606
    maximal depth 171, 253, 254, 535, 538, 541, 543
  bombturbation 294, 351
  congelliturbation 263
  cryoturbation 243, 245, 246, 263, 269, 273, 275, 277, 292, 314–315, 330, 351, 509, 588
    Cryepts 316, 635
    Cryids 636
    cryopedology/cryopedogenesis 263, 271
    Cryorthents 309
    Cryosols 272
    cryostatic pressure 275
    Cryumbrepts 509
    effects on classification/taxonomy 39, 129
    involutions 274
  crystalturbation 245, 246, 293, 329, 330, 351
  faunalturbation 244, 245, 246, 247–259, 329, 330, 332, 351, 442, 542
  floralturbation 245, 246, 259–262, 330, 351
  graviturbation 245, 246, 263, 291–292, 294, 330, 351
  impacturbation 245, 294, 330, 351
  proanisotropic 239–242, 244, 245, 247, 287, 290, 328, 329, 350
  progressive 242
  proisotropic 173, 239–242, 244, 247, 250, 262, 287, 328, 330, 350
  regressive 242, 324, 327
  seismiturbation 245, 246, 294, 330, 351
  utility in soil geomorphology 173, 561, 596, 606, 608, 613, 641
  various pedoturbation vectors 104, 263
penguins 104
perched water table *see* water table, perched
percolation/percolating soil water 85, 86
  impact on pedogenesis 85, 359, 366, 368, 373, 374, 375, 377, 378, 379, 394, 411–414, 429, 430, 628, 637
  theoretical 323, 363, 537
peridotite 68, 179, 182
periglacial activity/sediments/processes 212, 256, 264, 315, 527, 533
permafrost 264, 266–267, 269, 272, 373, 501
  discontinuous 266
  dry 266
  spatial extent and typical landscapes 263, 278
  table 266, 275, 293
  taxonomy/classification 39, 121

permafrost (*cont.*)
  utility in soil geomorphology/ paleoenvironmental reconstruction 212, 375, 527, 632, 636
permanent charge 209
permanent wilting point 84
permeability 18
  factors that affect permeability 13, 20, 87, 184, 231, 234, 263, 318, 419, 428
  impact on pedogenesis/weathering 171, 172, 227, 280, 323, 380, 382, 405, 425, 431, 489, 504, 507, 624
  permissivity 83
  saturated 18
  unsaturated 18
persistent/residual zones 534
pervection 265, 276, 350, 361
petrocalcic horizons 118, 122, 250, 329, 332, 335, 403, 404, 407, 412–414, 416, 417, 418, 422, 424, 568, 570, 611, 619, 624
  Petrocalcids 332, 416, 418
  petrosalic horizons 230
pH, soil 234
  as it affects weathering 189, 228, 232, 233, 385, 404, 405, 439, 455, 580
  depth trends 315, 455, 484
  effects on pedogenesis 189, 191, 209, 355, 362, 370, 394, 423, 424, 427, 441, 444, 451, 452
  factors that affect pH 184, 197, 234, 250, 253, 259, 280, 352, 359, 382, 395, 405, 447, 506
  index of 333
  of typical soils/settings 194, 206, 262, 289, 394, 428, 454, 503, 587
  utility in soil geomorphology 219, 222, 553, 569, 578, 624, 626
phases, soil 108
phenolic substances 43
Phillips, J. D. 339
phlogopite 66
phosphorous 209, 253, 262, 295, 319, 358, 386, 387, 569
photoautotrophs 97
photosynthesis/photosynthetic organisms 93, 645
phyllite 174
phyllosilicates/phyllosilicate minerals 9, 27, 62, 63–73, 232, 392
  1 : 1 layer silicates 55, 64, 65, 67–69, 121, 132, 182, 350, 382, 395, 398, 579, 639, 640
  2 : 1 layer silicates 55, 64, 65, 66, 68–73, 79, 208, 277, 350, 395, 398, 579
    dioctahedral 70
    trioctahedral 70
    2 : 1 : 1 layer silicates 73
  see also clay; minerals/mineralogy, clay
phylogenetic relationships 97
physical ripening 351, 375, 378, 461, 473
phytoburation 259
phytoliths *see* opal phytoliths
piezometer 85, 490
pimple mounds 256
Pinedale glacial advance 344, 571
pisoliths 390, 393, 413, 414, 417
pit, treethrow 261, 262, 324
pits and mounds *see* mounds
placic horizons 49, 120, 132, 443, 570, 624
  Placaquods 443
Plaggepts 635
Planosol soils 109, 113, 124, 384, 429
plants 93, 231, 234
plasma, soil 24, 25, 219, 225, 348, 353, 357, 361, 442, 476, 597
playas 194, 293, 409, 421, 422, 425, 433, 440, 486
Pleistocene Epoch 7, 206, 305, 315, 342, 406, 407, 410, 415, 486, 556, 568, 597, 611, 621
plinthite 39, 49, 132, 173, 177, 350, 391–392, 393, 396, 399, 501, 514, 535, 568, 570, 624, 636
  Plinthaquults 398
  Plinthohumults 398
  Plinthustults 398
Pliocene Epoch 342, 392, 407, 415, 518, 599
pluvial periods/conditions 345, 406, 409
pocket gophers 246, 249, 256
pocosins 488
podzolization 111, 197, 209, 276, 325, 328, 349, 352, 373, 440, 441, 442, 446, 641
  factors that promote/inhibit 104, 278, 305, 308, 318, 331, 451, 453
  Podzol soils 109, 111, 124
  Podzolic soils 147
  quantification
    POD Index 451, 578
    podzolization index 580, 581
  rates 453
  see also Spodosols
point bars 191
polar landscapes 267, 276, 441
polarity
  excursions 597
  normal 596
  reversed 596
  sub-chronozones 597
polarized light 31
pollen/palynology 190, 627, 651, 652
  pollen diagram 651, 652
polygenetic soils/polygenesis 167, 169, 209, 300, 335, 338–339, 386, 415, 520, 611, 621, 632, 640, 641, 647, 649
polyhedral model 55, 64
polymorphic soils 456, 458
polypedon 34, 35, 107, 146, 149, 152
pores in soils 27, 83, 85, 86, 292, 363, 367, 412, 414, 418, 509
  connectivity 17, 373, 430
  macropores 17–18, 103, 239, 252
  micropores, soil 17–18
  packing pores 17
  porosity in general 17–18, 20, 86, 101, 179, 209, 211, 227, 231, 234, 253, 264, 282, 302, 349, 357, 413, 427, 430, 461, 614
  space 13, 350, 425, 507
  size distribution 86, 173, 174, 368
  tortuosity 17
  vesicular 375, 434
positive feedback *see* feedback
post-depositional modifications (PDMs) 555
post-settlement alluvium 483
posts (on rocks, as a dating tool) 559
potassium 252, 253, 262, 358, 365, 484, 639
  potassium carbonate 233
  potassium hydroxide 233
potential energy (of soil water) 83
prairie dogs 243, 246
Prairie soils 109, 111, 124
precession of the equinoxes 343

precipitates 232
precipitation, gross 472, 473
precipitation, net 472
predators 101
preferential flow (of soil water) 87, 189, 414, 443
Pre-Illinoian drift/glacial advances 344, 516, 518, 521, 526
pressure plate 85
primary producers 93–96
prime symbol 37, 50
principle components analysis 301, 592
principle of ascendancy and descendancy 548
principle of cross-cutting relationships 548, 555
principle of original horizontality 548
principle of superposition 242, 548, 550, 551, 555
principle of uniformitarianism 7
priority factors: energy model 323
  priority factor: climate 323
  priority factor: relief 323
probe, push 150
process-systems model/approach (Simonson) 320–323, 324, 353
Profile Darkness Index (PDI) 573, 576
Profile Development Index; *see also* Harden Index 571, 575, 577
proglacial lakes 193
prokaya/prokaryotes 93
proto-imogolite theory/model 444, 449
protozoa 43, 91, 93, 94, 99, 101–103
proxy data (paleoclimate) 206
pseudogley model/conditions 381, 491
pseudopods 99
pseudoscorpions 46, 97, 101
Psammoturbels 273
puff 283
pumice 208
pygmy forest 454
pyrite 184, 186, 194, 352, 455
pyroclastic materials 195, 207, 208, 423, 424, 597
pyrophyllite 66, 67, 68
pyroxenes 61, 62, 68, 181, 521, 581, 583

quantification of soil development 215, 300, 461, 568
quartz 61, 174
  characteristics/properties 11, 79–197
  mineralogy/crystallography 62, 63
  pedogenesis/soil mineralogy 17, 172, 179, 182, 183, 184, 186, 189, 197, 208, 234, 359, 361, 371, 373, 386, 399, 423–424, 442, 492, 514, 537, 580
  quartz/feldspar ratios 454, 521, 583
  quartz monzanite 179
  Quartzipsamments 398
  taxonomy/classification 132
  utility in soil geomorphology/dating 188, 542, 614, 616, 618
  utility in soil quantification 323, 584
  veins/vein height 173, 176, 536, 539, 542, 552, 559, 566
  weathering-related 184, 228, 385, 423
quartzite 179, 227
Quaternary
  geology/geologists 52, 298
  Period (and sediment thereof) 342–345, 346, 599, 617, 620
  stratigraphy 516, 520, 522–529, 530, 620

radiation 87
radiocarbon dating 289, 290, 415, 433, 435, 438, 522, 529, 547, 549, 550, 599, 601, 604, 605, 606, 609, 612, 617, 618, 632, 641, 646, 651, 652
  bomb carbon 608
  calibration of ages/calibration curve 602, 603
  contamination of samples 601, 602, 608, 632
  uncertainty of ages 601
radius ratio 55
recarbonation 348, 359
red edge 499
red wigglers; *see also* earthworms 103
Red Desert soils 109, 110, 124
Red Podzolic soils 109, 112
Red-Yellow Podzolic soils 125
Reddish-Brown Lateritic soils 109, 112, 125
Reddish Brown soils 109, 110, 125
Reddish Chestnut soils 109, 110, 125
Reddish Prairie soils 109, 111, 125
Redness Index 401
Redness Rating 395, 400, 401, 573
redoximorphic features 59, 122, 131, 150, 152, 374, 381, 491, 492, 494, 496–501, 570, 624, 637
  redox concentrations 16, 381, 382, 391–392, 495, 498, 501
  redox conditions; *see also* oxidation and reduction 496
  redox depletions 371, 377–378, 381, 494, 498, 499, 501
  redox patterns 501, 624
  redox processes; *see also* oxidation and reduction 234, 278, 492
  reduced zone 177
  reduction/reducing conditions 380, 382, 384, 400, 455, 490, 492, 494, 495, 496–497, 506, 624, 642
reg 435, 436
regolith 32, 33, 170, 171, 213, 226, 236, 238, 291, 399, 614
  residual 32, 33, 170
  thickness of 237, 238
  transported
Regosolic soils
Regosol soils 125, 147
regression analysis 300, 302, 312, 451, 593, 594, 639
relative age/dating 300, 360, 438, 550, 554, 596, 597, 617
Relative Horizon Distinctness Index 219, 574
relative position (as a dating tool) 555
relief *see* topography/relief
remanent magnetism 596
removals/losses (from a soil or horizon) 321, 322, 326, 329, 330, 348, 349, 353, 382, 456, 460, 461, 470
Rendolls 225, 332, 399
Rendzina soils 109, 113, 125
reptiles 101, 244
residuum 32, 69, 156, 170, 171, 179, 184, 185, 187, 198, 213, 226, 301, 324, 393, 612, 627
  from limestone 170, 186–187, 189, 190, 201–202
resistant materials *see* minerals/mineralogy, soil
rhizobia 94, 97
rhizolith 419
rhizoplane 95
rhizosphere 95, 97, 100

Rhodoxeralfs 201
R horizon 36, 37
rhyolite 182, 208
ripening *see* phyical ripening
rock characteristics
  angularity (as a dating tool) 552, 556, 558
  pits (as a dating tool) 558
  staining (as a dating tool) 558
  structure 176, 398
  varnish 434, 438–439, 440, 444, 552, 556, 566, 636
rock salt *see also* halite 181, 227
rocks 226, 227, 231, 290, 435, 439, 543
  acidic 179, 227, 396, 418, 637
  basic 174, 179, 180–184, 396, 418, 637
  sedimentary 180, 184–189
  siliceous 179–180
  ultrabasic/ultramafic 174, 179, 182, 386
rodents 249
roots 94–96, 97
  excretions/exudates 95, 97, 227, 355, 357, 371, 445, 497
  impact on pedogenesis 93, 173, 242, 245, 259, 355, 357, 371, 375, 380, 404, 419, 420, 435, 458, 496–499, 611
  impact on weathering 227–231, 234
  respiration 92, 405, 647, 649
  rooting zone 34
  traces/casts 623, 624
  utility in paleoenvironmental reconstruction 602, 632, 647, 649
roundness 14, 222
Roxana silt 480, 521
rubification 187, 305, 352, 399–400, 401, 438, 553, 573
  rubefaction 168, 187
Ruhe, R. V. 516–529, 546, 580
Runge, E. C. A. 323
runoff 87, 127, 239, 306, 330, 331, 405, 440, 466, 472–476, 479, 480, 482, 486, 502, 509, 535, 612
runout zones 212, 217
rusts 98
rutile 202, 386, 542, 580, 581, 584

sabkhas 422
Sahara Desert 202, 207
saltate/saltation 195, 196
salts 349, 423, 425, 435, 437, 440
  salic horizons 119, 122, 128, 132, 329, 428, 636
  Salids 426, 428, 636
  saline soil water 127
  saline soils 84, 194, 436
  salinization 112, 276, 277, 278, 349, 425–427
  salt crusts 60, 230, 277
  salt cycle 426, 427
  salt weathering 229–230, 231, 236, 276
  salts in soils 85, 194, 236, 237, 285, 293, 321, 351, 402, 410, 421, 425, 436, 437, 507
  salty/saline soils 131, 146, 229, 318, 384, 487
  soluble salts 55, 78, 173, 252, 349, 424–433, 569, 624
sand bars 193
sand, eolian/dune *see* dunes; eolian materials/sediments, dune sand
sand/sandy soils 9, 86, 132, 168, 195, 203, 219, 253, 368, 384, 388, 429, 443, 446, 497, 499, 501, 537
sandstone 179, 180, 184, 185, 227, 393, 514
sand wedges 269, 274
Sangamon
  interglaciation 344, 520
  Late Sangamon paleosol/surface 520, 521–525, 542
  paleosol/surface 361, 431, 466, 480, 519, 520, 521–524, 548, 583, 604, 620, 625, 626, 627, 628, 631, 633, 642
sanidine 63
saponite 66, 67
saprolite 239, 467, 573
  Al-rich 170, 174
  characteristics/thickness 33, 171, 173, 174, 178, 387, 398, 400, 512, 515, 536
  definition/concept 39, 50
  impacts on pedogenesis 176, 393, 395, 399, 534, 535–541, 542
  massive 173
  saprolitization 147, 169, 174, 179, 394, 460, 461
  structured 172, 173
  zones 174–176, 514
saturation
  anthric saturation 492
  saturated conditions 84, 122, 212, 380, 392, 393, 461, 488, 489, 494, 497, 501, 643
  saturated flow 13, 86, 489
Sauer, C. 465
scatterplots 300, 302, 361, 481, 573, 580, 591, 594
schist 172, 179, 180, 227
Schmidt hammer 559
scoria 208
scorpions 97, 101
sea level 194, 195, 342–345
sea water 409, 455
secondary electrons 26
sediment flux 469, 471
sediment-limited systems/slopes 196
self-churning 282, 287
self-plowing 282
self-swallowing 282, 285
self-weight collapse 350, 375, 378, 461
sepiolite 639
sequence, pedogenic 299
sequum, soil 50, 373
sericite 70
serir 435
serpentines 62, 67–69, 70, 78, 80
  serpentine barrens 69
  serpentine soils 183
  serpentinite 179, 182, 187
sesquioxides 597
  and classification/taxonomy 37, 38, 39
  content 573
  as related to weathering 55, 174
  as related to pedogenesis 350, 364, 367, 375–376, 386, 387, 392–393, 424, 443, 445, 449, 450
  sesquioxidic nodules 174, 495, 499
Seventh Approximation 114
shadowing, topographic 303, 310, 475
shale 179, 180, 184, 185, 188, 227, 280, 421, 430, 635
Shaler, N. S. 5
Shantung Brown soils 109, 111
shattering 229
shear strength 212, 291, 331
shear stress 212
sheeting 227

shelf 283
shells 604
shredders 94, 101, 103
shrews 94
shrink–swell
  potential 19
  processes 72, 121, 173, 277, 280, 287, 288, 291, 335, 367, 378, 437
S horizons 259, 541, 542
Si/Al ratio 395
siallitization 350
Sibirtsev, N. M. 296
siderite 186
Sierozem soils 109, 110, 125
sieving 384
silica/silicon 38, 57, 174, 182, 232, 293, 350, 359, 382, 388, 390, 391, 392, 393, 394, 398, 422, 433, 579, 587, 639
  activity 637
  biogenic 206, 649
  precipitation in soils 624
  secondary/pedogenic 119, 347, 349, 350, 364, 371, 376–377, 379, 422, 424, 618
  silcrete 49, 291, 391, 396, 422
  siliceous duricrusts 422
  silicification 350, 422–424
  solubility 233
silica tetrahedron/tetrahedra 55, 61, 63
silicates 61–73, 209, 415, 422
  chain 61
  double-chain 61
  framework 61–63
  layer; *see also* phyllosilicates/phyllosilicate minerals 63, 184
  single-chain 61
silicic acid 233, 234, 382, 385
silimanite 202, 580, 581
siloxane surface 65
silt 9, 203, 350, 392, 429, 434, 437, 438, 584
  caps (on stones) 266, 272, 375
  silt : clay ratios 201
  translocation 265, 272, 292
siltstone 180, 227
silver 253
silverfish 101
similar series/soils 149, 151, 157
Simonson, R. W. 320, 323, 324
sinkholes 234
SiO_2/R_2O_3 ratio 391, 392, 395
skeletal grains/soil skeleton 9, 24, 25, 364, 376
  definition/concept 24, 353, 361
  impact on pedogenesis 354, 363, 371, 373, 376, 415, 447, 449, 461
  utility in soil geomorphology 221, 581, 623
skinks 104
slake/slaking 119, 229, 374, 422
slate 179, 227
slick spots 430–433, 480, 625
slickensides 39, 122, 244, 283–287, 289, 290, 512, 569, 624, 636
slides 212, 291, 480
slope/slopes
  aspect *see* aspect, topographic/slope
  asymmetry 309
  backwasting 525
  complex 478, 479
  compound 479
  concavity 154, 471, 482
  convergent/divergent 476
  convexity 154, 471, 480, 482
  curvature/shape 153, 310, 469, 472, 474, 482, 483, 489
  debris slope 480
  elements 475, 476, 477
    backslope 254, 470, 471–472, 476, 477, 479, 480, 482, 484, 489, 509, 511
    footslope 202, 330, 396, 468, 470, 477, 479, 482–484, 489, 510, 514
    shoulder 254, 430, 468–471, 475, 477, 479, 480–482, 484, 489, 499, 506
    summit 203, 298, 430, 468, 470, 471, 472, 477, 479, 480, 484, 486, 509, 511
    toeslope 215, 330, 470, 471, 475, 477, 479, 483, 484, 508, 509, 511, 514
  failure 212, 331
  flowlines/water pathways 469, 474, 476, 480, 482, 483, 489
  gradient/steepness 127, 153, 158, 163, 209, 212, 215–216, 267, 292, 298, 306, 310, 331, 405, 406, 471–476, 477, 502, 546, 555
  head 469, 470, 476
  inflection point(s) 475
  as landforms 298, 302, 306, 548, 549
  length 474, 475
  nose 310, 469, 470, 476, 525
  position 158, 476, 489
  processes 202, 263, 309, 471–479, 485, 486, 506–514, 532–535, 538–539
  side slopes 469
  simple 478
  stability/instability 215, 632
  variation 158
  waning/waxing 479
  wash 541
  slopewash 215, 460, 471, 475, 484, 486, 508, 514, 518, 525, 527, 535, 544, 548
sloughing zones 532, 534
slowly weatherable fraction *see* minerals resistant/slowly weatherable
slugs 103
slumps 212, 294, 482
S-matrix 24
smectite 62, 66, 67, 70, 71–72, 79–277
  conditions that favor its formation 179, 182, 210, 277, 280, 386, 504, 510, 514, 637, 638, 639
  impacts on pedogenesis/soils 82, 280, 289, 291, 330, 333, 335, 367
  what alters/weathers to it 68, 184, 334
  what it alters/weathers to 180, 233
Smith, G. D. 107, 114, 145, 146, 465, 625
smuts 98
snails 103, 206, 604
snakes 104
snow
  snowmelt 306, 316, 324, 366, 410, 415, 450, 491, 504, 509
  snowpack 87, 303, 306, 310, 312, 315, 450, 451, 502, 504, 505
soapstone 182
sodium 426–433
  exchangeable 20, 38, 262, 318, 365, 423, 428
  sodium carbonates 351
  sodium chloride 424
  sodium citrate-dithionite 447, 448, 593

sodium (*cont.*)
sodium pyrophosphate 447, 448
sodium salts/sodic soils 16, 293, 349, 428, 480, 487, 502
sodium sulfate 427
soil body *see* soil map unit
soil boundary *see* soil maps, boundaries
soil classification *see* classification, soil
soil climate 312, 406, 438, 508, 636, 639
soil continuum 107
soil cycles 532
soil (definition) 9, 106
soil densification 461
soil development
associated processes 238, 310, 467, 534, 546
definition/concept 348
geographic/spatial trends in 277
indices of 219, 222, 300, 572–587, 593
theroetical considerations 299, 321, 325, 327, 336, 466, 470, 568, 572
what impacts/affects soil development 206, 262, 302, 306, 310, 316, 408, 467, 474, 475, 484, 487, 501, 503, 504, 508, 562
soil evolution 108, 320, 337, 340, 348, 537, 538, 590, 619, 640
soil evolution model (Johnson and Watson-Stegner) 325–336, 338, 339
soil fertility 280
theoretical considerations/ position statements 325, 339, 465, 588
soil-forming interval 217, 547, 587, 625, 626, 630
soil forming factors *see* state factors
soil frost 450
concrete 264
granular 264
honeycomb 264
stalactite 264
soil gases/air/atmosphere 82, 91–92, 245, 247, 253, 380, 404, 405, 419, 434, 611, 647, 648
soil genesis (see "pedogenesis")
soil geography 3–6, 106, 160, 298, 322, 347, 537
soil hydrology 472–501
soil landscape analysis 158–163
soil maps/mapping 146
associations 149, 158
boundaries 107, 144, 150–157, 158, 160–163, 298
fuzziness/uncertainty 157
error and variability 156–158
generalization 146
legend 149
making/creating 150–156, 465, 525, 539
map units 34, 35, 146, 156, 158, 161, 475
consociations 149, 156, 163
contrast 149, 163
complexes 149, 163
inclusions 146, 149, 156
pattern 158
scale 146, 156
variability 150
see also soil survey/mapping
soil mixing *see* pedoturbation
soil moisture control section 127
soil moisture regimes 130
aquic 122, 128, 129, 130, 131, 490, 501, 637
aridic 128, 129, 130, 132, 400
peraquic 128, 129, 130, 131
perudic 130
torric *see* aridic soil moisture regime
udic 129, 130, 132, 374
ustic 129, 130, 132, 280
xeric 129, 130, 132, 399, 422, 638
soil organisms 93, 95, 349, 419
soil physical properties 82, 114
soil physics 82–92
soil productivity 245
soil profile 33–34, 35, 104, 106, 173, 206, 240, 291, 476, 617, 618
thickness 568, 571, 577
weighted values (statistically weighted by profile thickness) 570, 571, 644
soil quality 144
soil respiration 647, 649
soil science 4, 295, 320
soil separates 9
soil series 108, 114, 115, 150, 216
soil survey/mapping 114, 299, 469
National Cooperative Soil Survey 108, 115, 465
National Soil Survey Center 115
soil surveys (soil survey publications) 145
soil temperature regimes 133–143, 144, 357
cryic 128, 131, 144
frigid 144
hypergelic 144
hyperthermic 144
isofrigid 144
isohyperthermic 144
isomesic 144
isothermic 144
mesic 144
pergelic 144
subgelic 144
thermic 144, 450, 488
soil thickness *see* thickness, soil
soil type 108
soil water 82–84, 87, 331, 635
adsorption/adsorption 48, 82–83, 280, 290, 355, 382, 424, 439, 612
content/wetness 149, 163, 268, 469, 484
movement/flow 85–87, 231, 611
solution/water 18, 92, 356, 365, 446, 447
Soil Taxonomy see classification, soil
Solidi soils 429
solifluction 267, 274, 309, 314, 482, 525, 534
Solod soils 147, 426, 428, 429
solodization 112, 349, 384, 426, 427, 429–430
Solodized Solonetz soils 426, 428, 429
Solonchak soils 109, 112, 125, 426, 428
Solonetz soils 109, 112, 125, 147, 426, 428–429, 432, 433
solonization 112, 349, 426, 427–429
Soloth soils 109, 112, 125, 426, 429
Sols Bruns Acides 125
solubility 232
solum 50, 173
definition/concept 35, 36, 539
lower boundary (solum thickness) 50, 51, 159, 330, 338, 349, 524, 553, 554, 571, 595
factors that impact solum thickness; *see also* thickness, soil 305, 314, 316, 359, 482, 526, 568, 628

processes associated with the boundary 169
solution (chemical weathering process) 232
sombric horizon 120, 132
Sombric Brunisol soils 147
sorosilicates 61, 62, 63
sorted circles 267
sorting 17, 264
Southern Iowa Drift Plain 517–518, 526, 527, 528, 530, 545
sow bugs 101
space-filling model 55
"Spanish moss" 93
sparmicritisation 349, 418
spatial variability, soil 154, 156, 317, 321, 387, 425, 443, 589, 620
factors that affect 162, 287, 340–341, 506
soil patterns 159, 256, 262, 275, 320, 324, 339, 340
specific heat 133
specific heat capacity, soil 88–90
speleothems 603, 618
Sphagnum moss 132, 190
sphene 202
sphere-packing model 55
sphericity 14, 219, 222
spiders 97, 101
splays (delta and levee) 194
Spodosols 127, 210, 218, 262, 354, 447, 448, 460, 504, 578, 579, 636, 645
pedogenesis 126, 328, 356, 362, 452, 494, 578, 608, 636, 642
relationships to geomorphology 513, 514
spodic horizons 30, 119, 153, 440, 443, 450, 569, 591, 624, 633, 636
spodic materials 122, 636
taxonomy/classification 119, 121
spoil materials 320
sponge spicules 206, 652
springs/seeps 488, 625
springtails 97, 101
squirrels 240
stadials 345
stages of development 300
state-factor model/approach (Dokuchaev and Jenny) 5, 154, 296–300, 320, 321, 323, 327, 336, 341, 537, 587
state factors 6–7, 167, 197, 296–298, 299, 317, 320, 331–333, 337, 340, 347, 588, 619
age/time 296, 297–298, 300, 317, 323, 327, 342, 386, 388, 411, 555, 556, 568
anthropogenic 317
climate 296, 297, 312–317, 328, 337, 555, 561
organisms 296, 297, 310–312, 317, 323, 537
parent material 296, 297, 301–302, 310, 312, 323, 328
relief/topography 296, 297, 298, 302–310, 328, 469
utility in mapping 106, 153
see also functional-factorial models of soil development
staurolite 202, 583
stemflow 357, 472, 473
Steno, N. 548
stochastic complexity 340
stones/stony 12
stone circles 256, 632
stone lines/stone zones 174, 217, 256, 387, 396, 524, 525, 526, 527, 539, 542, 628
examples/images 176, 261, 516, 535, 541, 631
formation 224, 242–244, 254, 329, 363, 483, 506, 529, 535, 537, 538, 543–546
utility in soil geomorphology 188, 219, 222, 244
see also biomantles; lag concentrate/surface gravel
stone polygons 267
stone stripes 267
stony mantles 435
strain 349, 460
stratification/stratified materials/ layering 301
destruction of original stratification 264, 652
examples of stratified materials 191, 193, 209, 368
formal stratigraphic units 547
chronostratigraphic units 547, 548, 599
geochronometric unit 548
lithostratigraphic units 547, 620
pedostratigraphic unit 548, 620
Stratigraphic Code 547–548
stratigraphic column 217, 518, 642
tephrostratigraphic units 599
formation via pedogenesis/ pedoturbation 243, 544
impacts on pedogenesis 482
streptomycetes 97
structure, rock 49, 240
structure, soil/pedogenic 18–20, 21, 49, 82, 219, 222, 289, 388, 569, 623, 624
apedal/weak 434, 437
blocky 20
clods 18
columnar 20, 21, 118, 349, 429, 511, 512
crumb 20, 43, 146, 242
formation/preservation/destruction 14, 105, 173, 242, 252, 255, 275, 283, 398, 428, 429, 430, 434, 624
granular/crumb 20, 21, 116, 173, 178, 255, 275, 283, 357, 387, 453, 461, 512, 623
massive 19
peds 17, 18, 19, 247, 353, 357
platy 20, 21, 275, 375
prismatic/prisms 20, 21, 87, 119, 334, 374, 375, 378, 379, 422, 430, 437, 512
faces 371
single-grained 19
structureless 19
subsoil 296
wedge-shaped aggregates 283
subaerial exposure 466, 518, 548, 555, 619, 626, 634
subaqueous exposure 467
Subarctic Brown Forest soils 125
subgroups, taxonomic 115, 130–133, 134–139, 140, 492
suborders, taxonomic 115–122, 128–129
subsoil; *see also* B horizons 49, 285, 287, 378
sulfur 194, 411, 453, 455
Sulfaquents 455, 456, 635
Sulfaquepts 456
sulfates 60, 349, 380, 400, 402, 424, 455
sulfidation 352
sulfides 352
sulfidization 194, 352, 455
Sulfihemists 455

sulfur (*cont.*)
sulfites 402
sulfuric acid 55, 230, 352, 455
sulfuric horizon 132
sulfuricization 352, 455–456
superimposed soils 456, 625
surface area
and water retention 82–83
factors that affect surface area 13–14, 227, 230, 234, 280, 282
impact on pedogenesis/weathering 17, 197, 231, 302, 353–354, 356, 359, 381, 414, 438, 451, 501
surface boulder frequency (as a dating tool) 558
surface crust(s) 86, 405, 421, 425, 427, 433, 435, 436, 439
surfaces
age 144, 153, 298, 300–301, 407–408, 410, 411, 415, 422, 438, 523, 547, 561, 568, 579, 587, 591, 615
buried 549, 555, 619
constructional/aggradational/depositional; *see also* aggradation (additions to soil surface) 324, 467, 522, 526, 548, 616, 634
erosion/erosional 216, 467, 483, 521–525, 529, 531, 532, 543, 549, 550, 634
geomorphic 238, 341, 411, 466–468, 469, 517, 524, 525, 543, 547–551, 604–605, 614
instability/unstable surfaces 274
morphometry 468
roughness 555
stability/stable surfaces 169, 217, 324, 467, 532, 556, 568, 587, 616, 622, 628
surface exposure dating (SED) 437, 438, 554
surface tension (water) 82
surficial geology map 148, 167
suspension (transport mode) 195
swallowhole 235
swelling (of layers in clay minerals) 72, 86
swelling soils; *see also* Vertisols 82, 86, 285
symbiotic associations (symbionts) 93, 94
synchronous units/surfaces 547
Synthetic Alpine Slope (SAS) model 315, 316

taenolite 67, 70
talc 62, 66, 67, 68, 183
talf 468
talus 334
tectosilicates 61–63
temperature, soil 46, 86, 91
diurnal variation in soils 89, 91, 133, 143, 314, 439
tensiometer 85
tension, soil water 127
tephra 207, 208, 209, 210, 324, 422, 626
tephrochronology 208, 547, 597–599, 618
tephrostratigraphy 518
see also pyroclastic materials
termites 45, 74, 97, 100, 101–103, 236, 246, 247, 248, 249, 250, 253, 255–259, 388, 398, 501, 539, 544
mounds 102, 176, 239, 245, 251–541
casing 260
galleries 257
sheetings 257
terra rossa 187, 201–202, 506
Terra Rossa soils 109, 398
terraces 468
agricultural 318
marine 335, 454, 613
stream/fluvial 193, 345, 530, 577, 588, 590, 605, 606, 612
Tertiary Period 342
tessera 34
tetrahedral sheet 64, 65, 68
texture, soil 9–14, 19, 90, 156, 216, 247, 285, 301, 302, 331, 504, 579
and classification/taxonomy 151
and soil water 85, 86
classes 108
fine 183, 471
impact on pedogenesis/weathering 82, 168, 179, 378, 380, 406, 506, 633
textural triangle 10, 12, 202, 508
texture-contrast soils 217, 362, 534–537, 543
thallus/thalli 96, 236, 552, 562–566
thecamoebae 45
thermal properties, soil 88
thermal conductivity, soil and rock 87, 88, 90, 264, 268
thermal contraction 269, 271
thermal expansion 231
thermal fatigue 231
thickness, soil 189, 324, 326, 327, 456, 460
soil thickness model (Johnson) 324–325
thickness, soil (progressive/thickening) 325
thickness, soil (regressive/thinning) 325
thickness, soil (static) 325
see also solum, lower boundary
thin sections 30–31, 363, 388, 444, 569, 623
thinning, soil 328
Thorp, J. 114, 337
thresholds, general 331–335
intrinsic 291, 329, 330, 331, 333, 334, 335, 384, 419, 619
pedogenic 301, 332, 334, 339, 384, 568, 589, 590, 594
extrinsic 331
throughfall 357, 472, 473
throughflow 325, 443, 476, 480, 482, 484, 491, 510, 538, 541, 542
thrusting/thrust cones 289
tidal deposits 194
tidal marshes/backwaters 190, 195, 455
till, glacial 211, 216, 218, 324, 378, 431, 480, 506, 529, 621
ablation 211
basal 211, 375, 491
superglacial 211
time domain reflectometry (TDR) 83
time transgressive units/surfaces 204, 466, 467, 479, 521, 547, 548, 554, 589, 590, 622, 628
$time_{zero}$ 169–170, 300, 312, 340, 342, 547, 555, 573–574, 575, 579, 580, 584, 589–590, 595, 596, 606, 608, 621, 632
Tirs 278
titanium 174, 217, 222, 360, 391, 393, 407, 460, 579, 584–586, 587
titanomagetites 642
todorokite 60
tombstones 558, 562

tonalite 179, 184
tongues/tongueing *see* glossic tongues/tongueing
topaz 202, 542
topography/relief 122, 154, 156, 168, 267, 295, 298, 302, 306, 312, 314, 315, 410, 425, 469, 470, 475, 476, 480, 482, 486, 492, 589, 639
toposequences 300, 302, 443, 469
  topofunctions 299
Torrifluvents 418, 436
Torriorthents 225, 408, 460
tourmaline 62, 202, 323, 521, 542, 580, 581, 582, 583, 584
transformations 145, 186, 225, 321, 322, 329, 330, 334, 351, 353, 456, 461, 470
translocation (transfer) processes
  factors that affect translocation 225, 357, 371, 420, 437
  favorable settings/situations 400, 425–427, 497
  impact on weathering 173
  of clay; *see also* lessivage 50, 172, 173, 189, 225, 441
  of humus 354, 355, 610, 636
  of iron and aluminum; *see also* laterization; podzolization 50, 392, 440, 441–442, 452, 453
  theoretical considerations 312, 321, 329, 330, 348, 353, 356, 456, 470, 484, 584, 624, 641
transport processes on slopes; *see also* erosion; mass movement 171, 226, 388, 423
transport-limited systems/slopes 171, 196
travertine 181
treefall/treethrow; *see also* uprooting, tree 261, 351
tree rings *see* dendrochronology
tremolite 62
trioctahedral sheet 64, 65, 68
triple planation 538, 541
Tropaquods 514, 515
tropics/tropical landscapes/environments 143, 174, 177, 387, 392, 396–397, 399, 512, 514, 534, 539, 579, 643
Tropofibrists 514
tuco-tucos 256
tuff 207, 285, 423, 599
Tundra soils 109, 110, 125
tunnel structures 60
Turbels 274, 276–277, 309

U-Th (U-series) dating 566, 600, 618
Udalfs 453, 634, 636
Udepts 452, 453
Udipsamments 446
Udolls 356, 634, 635, 636
Udults 636
Ultisols 56, 80, 127, 218, 388, 392, 398, 399, 400, 401, 501, 514, 636
  pedogenesis 126, 173, 328, 579
  relationships to geomorphology 396, 454, 512, 513, 630
  taxonomy/classification 116, 121, 362
ultramafic/ultramafic rocks 68
ultraxerous soils 277
Umbrorthels 317
undercutting (of slopes) 212
underfit streams 193
uniformity value (UV) index 217, 221, 223, 224
unit cell 64
unit volume factor 461
units of pressure 83
unloading 227
unoxidized 32, 33
unsaturated flow 86, 225, 431, 433
unsorted 211
unsorted circles 267
unstratified 211
upbuilding, general 324, 325, 327, 348, 456
  developmental 169, 325, 327, 328, 330, 348, 456, 457, 521
  retardant 325, 327, 328, 330, 349, 456, 457
upfreezing 264–266, 267
uprooting, tree 239, 244, 264, 265, 266, 501, 502, 505
  impact on pedogenesis 217, 240, 243, 246, 501, 503, 504, 538, 541
  theoretical/conceptual 242, 259, 332
uranium-series dating 611
urban soils (earths) 293, 319
urbanthroturbation 293
"use and management", soil 145, 474
Ustalfs 569
Ustepts 569
Ustifluvents 332
Ustolls 430, 569, 609, 630

V 28–238 deep sea core 337, 345
van der Waals forces 365, 366
varves 193, 602
vegetation
  change 169, 310, 349, 373, 384, 450, 649
  cover 212, 275, 310, 314, 315, 331, 333, 340, 406, 450
ventifacts 196, 529
vermiculite 62, 66, 67, 68, 70, 71, 72, 79, 80, 179, 180, 182, 184, 189, 639
Vermudolls 241, 252, 357
Vermustolls 252, 357
vernadite 60
vertebrates 104
Vertisols 72, 116, 121, 185, 277, 282, 284, 286, 287, 288, 289, 290, 327, 329, 334, 335, 353, 388, 504, 636, 639
  "bowls" *see also* microbasins 285, 286, 287, 289
  "chimneys" *see also* "microknolls" 285, 286, 287, 289
  pedogenesis 126, 335, 362
  see also argilliturbation; Grumusol soils
vertization 351
vesicles 26
vetusols 621
viscosity (of soil water) 82
Vitrands 635
vizcacha 243
voids, soil 17, 24, 25, 349
  void, packing 26
volcanism/volcanic materials 119, 174, 195, 207, 209, 210, 350, 396, 422, 538, 560, 596, 597, 614
volume deformation 461
volumetric estimate/basis (statistical) 553, 571, 579, 587
vughs 25, 26, 29

water content, soil 82, 88, 503
  gravimetric water content 83
  volumetric water content 83
water flow in soils 82, 87
water holding capacity (of soil) 99, 101, 103
water retention (by soils) 82
water table 129, 302, 318, 486–492, 508, 633
  apparent; *see also* endosaturation 491
  depth to 302, 380, 476, 480, 489, 490, 492, 494, 502, 510
  effects on 122
  fluctuations/variability 247, 382, 454, 482, 490, 492
  impacts on pedogenesis 174, 190, 247, 250, 295, 310, 324, 327, 356, 366, 393, 400, 443, 445, 450, 453, 470, 472, 499, 502, 635, 639
  perched; *see also* episaturation 131, 489, 491–492
    above buried paleosols 204, 480, 625
    above fragipans 374, 378
    above permafrost 267, 274
    impacts on pedogenesis 182, 367, 370, 371, 379, 382, 424, 496, 542
    why it happens 87, 225, 299, 384, 413, 422, 428, 454, 479
water vapor transport (in soil) 90, 92
waves and currents 194, 195
waxes 442
weakly developed soils 121, 204
weathering 149, 168, 171, 184, 187, 208, 219, 226–238, 299, 321, 322, 325, 329, 337, 347, 353, 365, 375, 385, 445, 535, 539, 552, 557, 558, 642
  agents of 231, 445
  biotic 93, 99, 231, 234–236, 445, 447
  chemical *see also* decomposition 174, 227, 231–234, 407, 467
    definition/concept 231, 313, 351
    effects on pedogenesis/ mineralogy 178, 188, 370, 371, 379, 424, 426, 438, 441, 442, 444, 518, 636
    settings where it is common/ important 186, 238, 388, 479
  controls on 229, 236–238
  degree of/intensity/status 176, 222, 238, 276, 454, 458, 521, 553, 556, 558, 569, 579, 580, 628, 635
  depth of 171, 237
  extreme/intense 132, 350, 385–393, 399, 441, 444, 454, 520, 534, 639
  factors that promote/inhibit 209, 212, 274, 303–305, 316, 323, 356, 358, 366, 373, 431, 450, 499, 502, 506
  general definition 226
  index/indices 579, 632
  isovolumetric 172, 173, 460, 541
  losses 460, 587
  mechanical *see* weathering, physical
  of various minerals 70, 73, 206, 395, 623
  physical *see also* disintegration 70, 180, 184, 186, 227, 230, 236, 237, 238, 313, 314, 351
  pits 557
  products/by-products 231, 234, 236
W horizons 259, 541
  clay mineral examples 174, 178, 362, 370, 382, 386, 398, 444, 637
  fate/fluxes of 174, 234, 238, 538–539
  impacts on pedogenesis/weathering 48, 176, 226, 227, 237, 378, 379, 409, 452, 467, 556, 584
  in saprolite 171
  profile 32–33, 34, 51, 52, 175, 177–178, 236, 375, 396, 583
  rates of 82, 87, 197, 236, 237, 238, 456, 467, 637, 639
  ratios 522, 554, 623, 625
  regimes 637
  rinds 187, 551, 552, 553, 554, 555, 556–560, 561, 566
  resistance to 58, 226, 228
  sequence 641
  spheroidal 229, 230
  under acidic conditions 184, 202, 424
  under alkaline conditions 202, 280
  weatherable minerals see minerals/mineralogy, weatherable/weak
  weathering discontinuity/front 378
  weathering front 172, 173, 456, 467, 476, 538
  weathering surfaces 396
  weathered zone 518
  when clay minerals are the end product 67, 68, 71, 73, 208, 233
weathering-limited slopes/surfaces 171
weighted (statistically) soil properties 570–572
welding / soil welding 53, 456, 458, 480, 520, 522, 625, 626, 628
Wentworth scale 6
wet edge 499
wetlands 149, 318, 381, 486
wetness effect 204
wetting/drying 173, 185, 229, 231, 238, 292, 351, 357, 363, 375, 378, 382, 392, 422, 461, 499, 542
wetting fronts
  as impacted by texture 192, 302
  impact on pedgenesis 247, 291, 362, 363, 366–367, 368, 400–405, 425, 431, 434, 438
Whitney, M. 5, 107
Wiesenböden soils 109, 113
Wilde, S.A. 296
willemseite 66
Wisconsin ice sheet/glaciation 105, 211, 344, 480, 516, 518, 522, 528, 628
wombats 231, 247
Woodfordian glaciation 516, 521, 628
wood 603
woodlice 103

X-ray diffraction 71, 73–80
  diffractograms 76, 79
  diffractometer 76, 78
  differential X-ray diffraction 80
  diffraction peaks, 3rd order 77
  diffraction peaks, 2nd order 77
xenotime 584
Xeralfs 401, 636
Xerolls 306, 636
xerolysis 351, 352
Xerults 636

Yarmouth (and Yarmouth–Sangamon) surfaces 522, 525
Yarmouth interglaciation 344, 520
Yarmouth paleosol 520, 625, 640
Yarmouth–Sangamon paleosol 520, 521–529, 530, 542, 625, 630
easts 93, 98
Yellow Podzolic soils 109, 112
Yellowish-Brown Lateritic soils 109, 112
yttrium 584

zeolites 62, 63
zinc 253
zircon/zirconium 62
and quantitation of pedogenic and geomorphic processes 174, 323, 579, 583, 584–586
and weathering 202, 386, 393, 581
utility in soil geomorphology 173, 196, 217, 219, 222, 360, 521, 580, 582, 585
zoisite 202, 583
zonal soils *see also* normal soils 336, 337, 392
Zonal soils 5, 108, 109, 110–111, 112–113
zooflagellates 100

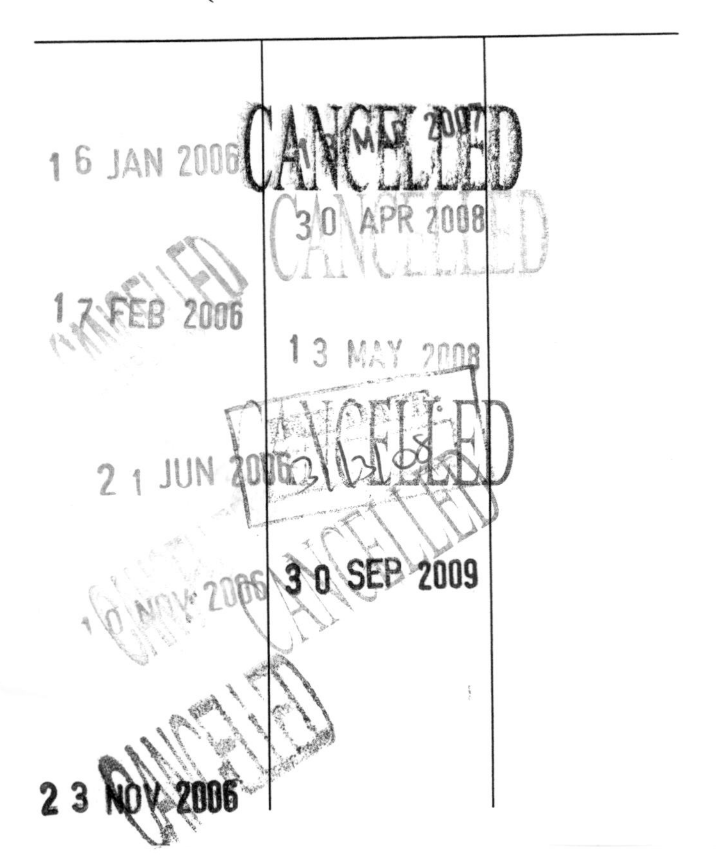
16 JAN 2006
17 FEB 2006
21 JUN 2006
10 NOV 2006
23 NOV 2006
30 APR 2008
13 MAY 2008
30 SEP 2009
CANCELLED